Lecture Notes in Computer Science 10624

More information about this series at http://www.springer.com/series/7410

Tsuyoshi Takagi · Thomas Peyrin (Eds.)

Advances in Cryptology – ASIACRYPT 2017

23rd International Conference on the Theory
and Applications of Cryptology and Information Security
Hong Kong, China, December 3–7, 2017
Proceedings, Part I

Editors
Tsuyoshi Takagi
The University of Tokyo
Tokyo
Japan

Thomas Peyrin
Nanyang Technological University
Singapore
Singapore

ISSN 0302-9743 ISSN 1611-3349 (electronic)
Lecture Notes in Computer Science
ISBN 978-3-319-70693-1 ISBN 978-3-319-70694-8 (eBook)
https://doi.org/10.1007/978-3-319-70694-8

Library of Congress Control Number: 2017957984

LNCS Sublibrary: SL4 – Security and Cryptology

Printed on acid-free paper

This Springer imprint is published by Springer Nature
The registered company is Springer International Publishing AG
The registered company address is: Gewerbestrasse 11, 6330 Cham, Switzerland

Preface

ASIACRYPT 2017, the 23rd Annual International Conference on Theory and Application of Cryptology and Information Security, was held in Hong Kong, SAR China, during December 3–7, 2017.

The conference focused on all technical aspects of cryptology, and was sponsored by the International Association for Cryptologic Research (IACR).

ASIACRYPT 2017 received 243 submissions from all over the world. The Program Committee selected 67 papers (from which two were merged) for publication in the proceedings of this conference. The review process was made by the usual double-blind peer review by the Program Committee consisting of 48 leading experts of the field. Each submission was reviewed by at least three reviewers, and five reviewers were assigned to submissions co-authored by Program Committee members. This year, the conference operated a two-round review system with rebuttal phase. In the first-round review the Program Committee selected the 146 submissions that were considered of value for proceeding to the second round. In the second-round review the Program Committee further reviewed the submissions by taking into account their rebuttal letter from the authors. All the selection process was assisted by 334 external reviewers. These three-volume proceedings contain the revised versions of the papers that were selected. The revised versions were not reviewed again and the authors are responsible for their contents.

The program of ASIACRYPT 2017 featured three excellent invited talks. Dustin Moody gave a talk entitled "The Ship Has Sailed: The NIST Post-Quantum Cryptography 'Competition'," Wang Huaxiong spoke on "Combinatorics in Information-Theoretic Cryptography," and Pascal Paillier gave a third talk. The conference also featured a traditional rump session that contained short presentations on the latest research results of the field. The Program Committee selected the work "Identification Protocols and Signature Schemes Based on Supersingular Isogeny Problems" by Steven D. Galbraith, Christophe Petit, and Javier Silva for the Best Paper Award of ASIACRYPT 2017. Two more papers, "Kummer for Genus One over Prime Order Fields" by Sabyasachi Karati and Palash Sarkar, and "A Subversion-Resistant SNARK" by Behzad Abdolmaleki, Karim Baghery, Helger Lipmaa, and Michał Zając were solicited to submit the full versions to the *Journal of Cryptology*. The program chairs selected Takahiro Matsuda and Bart Mennink for the Best PC Member Award.

Many people have contributed to the success of ASIACRYPT 2017. We would like to thank the authors for submitting their research results to the conference. We are very grateful to all of the Program Committee members as well as the external reviewers for their fruitful comments and discussions on their areas of expertise. We are greatly indebted to Duncan Wong and Siu Ming Yiu, the general co-chairs, for their efforts and overall organization. We would also like to thank Allen Au, Catherine Chan, Sherman S.M. Chow, Lucas Hui, Zoe Jiang, Xuan Wang, and Jun Zhang, the local

Organizing Committee, for their continuous supports. We thank Duncan Wong and Siu Ming Yiu for expertly organizing and chairing the rump session.

Finally, we thank Shai Halevi for letting us use his nice software for supporting all the paper submission and review process. We also thank Alfred Hofmann, Anna Kramer, and their colleagues for handling the editorial process of the proceedings published at Springer LNCS.

December 2017

Tsuyoshi Takagi
Thomas Peyrin

ASIACRYPT 2017

The 23rd Annual International Conference on Theory and Application of Cryptology and Information Security

Sponsored by the International Association for Cryptologic Research (IACR)

December 3–7, 2017, Hong Kong, SAR China

General Co-chairs

Name	Affiliation
Duncan Wong	CryptoBLK Limited
Siu Ming Yiu	The University of Hong Kong, SAR China

Program Co-chairs

Name	Affiliation
Tsuyoshi Takagi	University of Tokyo, Japan
Thomas Peyrin	Nanyang Technological University, Singapore

Program Committee

Name	Affiliation
Shweta Agrawal	IIT Madras, India
Céline Blondeau	Aalto University, Finland
Joppe W. Bos	NXP Semiconductors, Belgium
Chris Brzuska	TU Hamburg, Germany
Jie Chen	East China Normal University, China
Sherman S.M. Chow	The Chinese University of Hong Kong, SAR China
Kai-Min Chung	Academia Sinica, Taiwan
Nico Döttling	University of California, Berkeley, USA
Thomas Eisenbarth	Worcester Polytechnic Institute, USA
Dario Fiore	IMDEA Software Institute, Madrid, Spain
Georg Fuchsbauer	Inria and ENS, France
Steven Galbraith	Auckland University, New Zealand
Jian Guo	Nanyang Technological University, Singapore
Viet Tung Hoang	Florida State University, USA
Jérémy Jean	ANSSI, France
Jooyoung Lee	KAIST, South Korea
Dongdai Lin	Chinese Academy of Sciences, China
Feng-Hao Liu	Florida Atlantic University, USA
Stefan Mangard	Graz University of Technology, Austria
Takahiro Matsuda	AIST, Japan
Alexander May	Ruhr University Bochum, Germany
Bart Mennink	Radboud University, The Netherlands

Amir Moradi	Ruhr University Bochum, Germany
Pratyay Mukherjee	Visa Research, USA
Mridul Nandi	Indian Statistical Institute, India
Khoa Nguyen	Nanyang Technological University, Singapore
Miyako Ohkubo	NICT, Japan
Tatsuaki Okamoto	NTT Secure Platform Laboratories, Japan
Arpita Patra	Indian Institute of Science, India
Bart Preneel	KU Leuven, Belgium
Matthieu Rivain	CryptoExperts, France
Reihaneh Safavi-Naini	University of Calgary, Canada
Yu Sasaki	NTT Secure Platform Laboratories, Japan
Peter Schwabe	Radboud University, The Netherlands
Fang Song	Portland State University, USA
Francois-Xavier Standaert	UCL, Belgium
Damien Stehlé	ENS Lyon, France
Ron Steinfeld	Monash University, Australia
Rainer Steinwandt	Florida Atlantic University, USA
Mehdi Tibouchi	NTT Secure Platform Laboratories, Japan
Dominique Unruh	University of Tartu, Estonia
Gilles Van Assche	STMicroelectronics, Belgium
Serge Vaudenay	EPFL, Switzerland
Ingrid Verbauwhede	KU Leuven, Belgium
Ivan Visconti	University of Salerno, Italy
Lei Wang	Shanghai Jiaotong University, China
Meiqin Wang	Shandong University, China
Jiang Zhang	State Key Laboratory of Cryptology, China

Additional Reviewers

Masayuki Abe
Arash Afshar
Divesh Aggarwal
Shashank Agrawal
Ahmad Ahmadi
Mamun Akand
Gorjan Alagic
Joel Alwen
Abdelrahaman Aly
Miguel Ambrona
Elena Andreeva
Diego Aranha
Nuttapong Attrapadung
Sepideh Avizheh
Saikrishna Badrinarayanan
Shi Bai
Fatih Balli
Subhadeep Banik
Zhenzhen Bao
Hridam Basu
Alberto Batistello
Balthazar Bauer
Carsten Baum
Georg T. Becker
Christof Beierle
Sonia Beläd
Fabrice Benhamouda
Francesco Berti
Guido Bertoni
Sanjay Bhattacherjee
Jean-Francois Biasse
Begül Bilgin
Olivier Blazy
Johannes Bloemer
Sonia Mihaela Bogos
Sasha Boldyreva
Charlotte Bonte
Raphael Bost
Leif Both
Florian Bourse
Sébastien Canard
Brent Carmer
Wouter Castryck
Dario Catalano
Gizem Çetin
Avik Chakraborti
Nishanth Chandran

Julian Loss
Steve Lu
Xianhui Lu
Atul Luykx
Chang Lv
Vadim Lyubashevsky
Monosij Maitra
Mary Maller
Giorgia Azzurra Marson
Marco Martinoli
Daniel Masny
Sarah Meiklejohn
Peihan Miao
Michele Minelli
Takaaki Mizuki
Ahmad Moghimi
Payman Mohassel
Maria Chiara Molteni
Seyyed Amir Mortazavi
Fabrice Mouhartem
Köksal Mus
Michael Naehrig
Ryo Nishimaki
Anca Nitulescu
Luca Nizzardo
Koji Nuida
Kaisa Nyberg
Adam O'Neill
Tobias Oder
Olya Ohrimenko
Emmanuela Orsini
Elisabeth Oswald
Elena Pagnin
Pascal Paillier
Jiaxin Pan
Alain Passelègue
Sikhar Patranabis
Roel Peeters
Chris Peikert
Alice Pellet-Mary
Ludovic Perret
Peter Pessl
Thomas Peters
Christophe Petit
Duong Hieu Phan
Antigoni Polychroniadou
Romain Poussier
Ali Poustindouz
Emmanuel Prouff
Kexin Qiao
Baodong Qin
Sebastian Ramacher
Somindu C. Ramanna
Shahram Rasoolzadeh
Divya Ravi
Francesco Regazzoni
Jean-René Reinhard
Ling Ren
Joost Renes
Oscar Reparaz
Joost Rijneveld
Damien Robert
Jérémie Roland
Arnab Roy
Sujoy Sinha Roy
Vladimir Rozic
Joeri de Ruiter
Yusuke Sakai
Amin Sakzad
Simona Samardjiska
Olivier Sanders
Pascal Sasdrich
Alessandra Scafuro
John Schanck
Tobias Schneider
Jacob Schuldt
Gil Segev
Okan Seker
Binanda Sengupta
Sourav Sengupta
Jae Hong Seo
Masoumeh Shafienejad
Setareh Sharifian
Sina Shiehian
Kazumasa Shinagawa
Dave Singelée
Shashank Singh
Javier Silva
Luisa Siniscalchi
Daniel Slamanig
Benjamin Smith
Ling Song
Pratik Soni
Koutarou Suzuki
Alan Szepieniec
Björn Tackmann
Mostafa Taha
Raymond K.H. Tai
Katsuyuki Takashima
Atsushi Takayasu
Benjamin Hong Meng Tan
Qiang Tang
Yan Bo Ti
Yosuke Todo
Ni Trieu
Roberto Trifiletti
Thomas Unterluggauer
John van de Wetering
Muthuramakrishnan Venkitasubramaniam
Daniele Venturi
Dhinakaran Vinayagamurthy
Vanessa Vitse
Damian Vizár
Satyanarayana Vusirikala
Sebastian Wallat
Alexandre Wallet
Haoyang Wang
Minqian Wang
Wenhao Wang
Xiuhua Wang
Yuyu Wang
Felix Wegener
Puwen Wei
Weiqiang Wen
Mario Werner
Benjamin Wesolowski
Baofeng Wu
David Wu
Keita Xagawa
Zejun Xiang
Chengbo Xu
Shota Yamada
Kan Yang
Kang Yang
Kan Yasuda

Donggeon Yhee
Kazuki Yoneyama
Kisoon Yoon
Yu Yu
Zuoxia Yu
Henry Yuen
Aaram Yun
Mahdi Zamani
Greg Zaverucha
Cong Zhang
Jie Zhang
Kai Zhang
Ren Zhang
Wentao Zhang
Yongjun Zhao
Yuqing Zhu

Local Organizing Committee

Co-chairs

Duncan Wong	CryptoBLK Limited
Siu Ming Yiu	The University of Hong Kong, SAR China

Members

Lucas Hui (Chair)	The University of Hong Kong, SAR China
Catherine Chan (Manager)	The University of Hong Kong, SAR China
Jun Zhang	The University of Hong Kong, SAR China
Xuan Wang	Harbin Institute of Technology, Shenzhen, China
Zoe Jiang	Harbin Institute of Technology, Shenzhen, China
Allen Au	The Hong Kong Polytechnic University, SAR China
Sherman S.M. Chow	The Chinese University of Hong Kong, SAR China

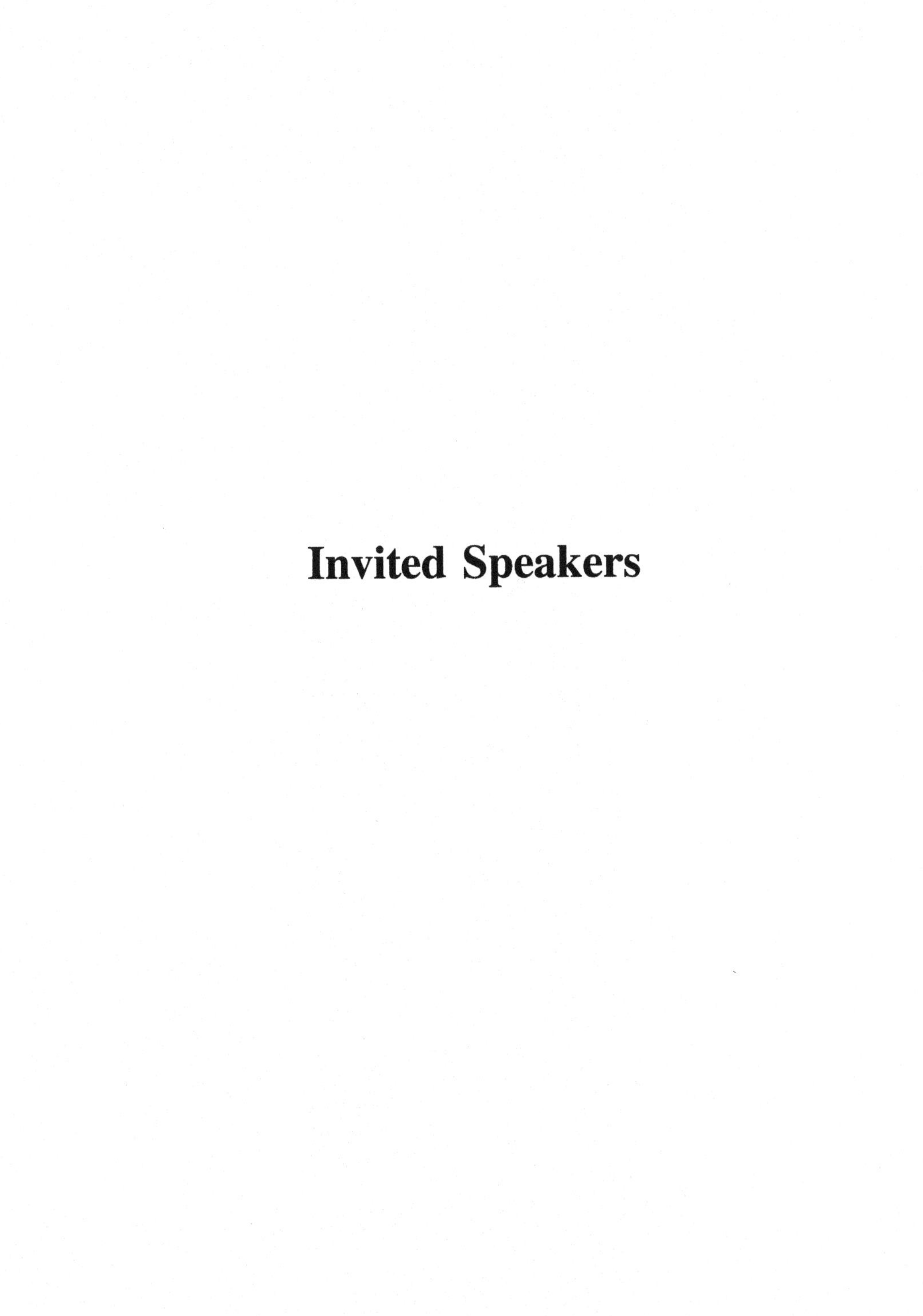

Invited Speakers

The Ship Has Sailed: the NIST Post-quantum Cryptography "Competition"

Dustin Moody

Computer Security Division, National Institute of Standards and Technology

Abstract. In recent years, there has been a substantial amount of research on quantum computers – machines that exploit quantum mechanical phenomena to solve mathematical problems that are difficult or intractable for conventional computers. If large-scale quantum computers are ever built, they will compromise the security of many commonly used cryptographic algorithms. In particular, quantum computers would completely break many public-key cryptosystems, including those standardized by NIST and other standards organizations.

Due to this concern, many researchers have begun to investigate post-quantum cryptography (also called quantum-resistant cryptography). The goal of this research is to develop cryptographic algorithms that would be secure against both quantum and classical computers, and can interoperate with existing communications protocols and networks. A significant effort will be required to develop, standardize, and deploy new post-quantum algorithms. In addition, this transition needs to take place well before any large-scale quantum computers are built, so that any information that is later compromised by quantum cryptanalysis is no longer sensitive when that compromise occurs.

NIST has taken several steps in response to this potential threat. In 2015, NIST held a public workshop and later published NISTIR 8105, Report on Post-Quantum Cryptography, which shares NIST's understanding of the status of quantum computing and post-quantum cryptography. NIST also decided to develop additional public-key cryptographic algorithms through a public standardization process, similar to the development processes for the hash function SHA-3 and the Advanced Encryption Standard (AES). To begin the process, NIST issued a detailed set of minimum acceptability requirements, submission requirements, and evaluation criteria for candidate algorithms, available at http://www.nist.gov/pqcrypto. The deadline for algorithms to be submitted was November 30, 2017.

In this talk, I will share the rationale on the major decisions NIST has made, such as excluding hybrid and (stateful) hash-based signature schemes. I will also talk about some open research questions and their potential impact on the standardization effort, in addition to some of the practical issues that arose while creating the API. Finally, I will give some preliminary information about the submitted algorithms, and discuss what we've learned during the first part of the standardization process.

Combinatorics in Information-Theoretic Cryptography

Huaxiong Wang

School of Physical and Mathematical Sciences,
Nanyang Technological University, Singapore
hxwang@ntu.edu.sg

Abstract. Information-theoretic cryptography is an area that studies cryptographic functionalities whose security does not rely on hardness assumptions from computational intractability of mathematical problems. It covers a wide range of cryptographic research topics such as one-time pad, authentication code, secret sharing schemes, secure multiparty computation, private information retrieval and post-quantum security etc., just to mention a few. Moreover, many areas in complexity-based cryptography are well known to benefit or stem from information-theoretic methods. On the other hand, combinatorics has been playing an active role in cryptography, for example, the hardness of Hamiltonian cycle existence in graph theory is used to design zero-knowledge proofs. In this talk, I will focus on the connections between combinatorics and information-theoretic cryptography. After a brief (incomplete) overview on their various connections, I will present a few concrete examples to illustrate how combinatorial objects and techniques are applied to the constructions and characterizations of information-theoretic schemes. Specifically, I will show

1. how perfect hash families and cover-free families lead to better performance in certain secret sharing schemes;
2. how graph colouring from planar graphs is used in constructing secure multiparty computation protocols over non-abelian groups;
3. how regular intersecting families are applied to the constructions of private information retrieval schemes.

Part of this research was funded by Singapore Ministry of Education under Research Grant MOE2016-T2-2-014(S).

Contents – Part I

Lattices

Homomorphic Encryptions

Access Control

Oblivious Protocols

Contents – Part II

Block Chains

Multi-party Protocols

Operating Modes Security Proofs

Contents – Part III

Asiacrypt 2017 Best Paper

Identification Protocols and Signature Schemes Based on Supersingular Isogeny Problems

Steven D. Galbraith[1(✉)], Christophe Petit[2], and Javier Silva[3]

[1] Mathematics Department, University of Auckland, Auckland, New Zealand
s.galbraith@auckland.ac.nz
[2] School of Computer Science, University of Birmingham, Birmingham, UK
christophe.f.petit@gmail.com
[3] Universitat Pompeu Fabra, Barcelona, Spain
javier.silva@upf.edu

Abstract. We provide a new identification protocol and new signature schemes based on isogeny problems. Our identification protocol relies on the hardness of the endomorphism ring computation problem, arguably the hardest of all problems in this area, whereas the only previous scheme based on isogenies (due to De Feo, Jao and Plût) relied on potentially easier problems. The protocol makes novel use of an algorithm of Kohel-Lauter-Petit-Tignol for the quaternion version of the ℓ-isogeny problem, for which we provide a more complete description and analysis. Our new signature schemes are derived from the identification protocols using the Fiat-Shamir (respectively, Unruh) transforms for classical (respectively, post-quantum) security. We study their efficiency, highlighting very small key sizes and reasonably efficient signing and verification algorithms.

1 Introduction

A recent research area is cryptosystems whose security is based on the difficulty of finding a path in the isogeny graph of supersingular elliptic curves [6,8,14,21,22]. Unlike other elliptic curve cryptosystems, the only known quantum algorithm for these problems, due to Biasse-Jao-Sankar [4], has exponential complexity. Hence, additional motivation for the study of these cryptosystems is that they are possibly suitable for post-quantum cryptography.

A large range of cryptographic primitives can now be based on isogeny assumptions. The work of Charles-Goren-Lauter [6] gave a collision-resistant hash function. Jao-De Feo [21] gave a key exchange protocol, De Feo-Jao-Plût [14] gave a public key encryption scheme and an interactive identification protocol, Jao-Soukharev [22] gave an undeniable signature, and Xi-Tian-Wang [41] gave a designated verifier signature. In this paper we focus on identification protocols and signature schemes.

A first identification protocol based on isogeny problems was proposed by De Feo-Jao-Plût [14], as an extension of the key exchange protocol of Jao-De Feo [21]. This scheme has the advantage of being simple to describe and easy to implement. On the other hand, it inherits the disadvantages of [21], in particular

T. Takagi and T. Peyrin (Eds.): ASIACRYPT 2017, Part I, LNCS 10624, pp. 3–33, 2017.
https://doi.org/10.1007/978-3-319-70694-8_1

it relies on a non-standard isogeny problem using small isogeny degrees, reveals auxiliary points, and uses special primes.

The fastest classical attack on this scheme has heuristic running time of $\tilde{O}(p^{1/4})$ bit operations, and the fastest quantum attack has running time of $\tilde{O}(p^{1/6})$. Several recent papers [17,19,29,32] have shown that revealing auxiliary points may be dangerous in certain contexts. It is therefore highly advisable to build cryptographic schemes on the most general, standard and potentially hardest isogeny problems.

Our main contribution in this paper is a new identification protocol with statistical zero-knowledge and computational soundness based on the endomorphism ring computation problem. The latter problem has been studied for some time in computational number theory, and is equivalent to computing an isogeny between two arbitrary given elliptic curves, without any restriction on the parameters and no extra information revealed. In contrast to the problem used in De Feo-Jao-Plût's protocol, this problem has heuristic classical complexity of $\tilde{O}(p^{1/2})$ bit operations, and quantum complexity $\tilde{O}(p^{1/4})$.

Our identification protocol is very similar to the standard sigma protocol for graph isomorphism.

The public key is a pair of elliptic curves (E_0, E_1) and the private key is an isogeny $\phi : E_0 \rightarrow E_1$. To interactively prove knowledge of ϕ one chooses a random isogeny $\psi : E_1 \rightarrow E_2$ and sends E_2 to the verifier. The verifier sends a bit b. If $b = 0$ the prover reveals ψ. If $b = 1$ the prover reveals an isogeny $\mu : E_0 \rightarrow E_2$. In either case, the verifier checks that the response is correct. The interaction is repeated a number of times until the verifier is convinced that the prover knows an isogeny from E_0 to E_1. However, the subtlety is that we cannot just set $\mu = \psi \circ \phi$, as then E_1 would appear on the path in the isogeny graph from E_0 to E_2 and so we would have leaked the private key. The crucial idea is to use the algorithm of Kohel-Lauter-Petit-Tignol [26] to produce an isogeny $\mu : E_0 \rightarrow E_2$ that is completely independent of ϕ. The mathematics behind the algorithm of Kohel-Lauter-Petit-Tignol goes beyond what usually arises in elliptic curve cryptography.

Our second contribution are secure digital signatures based on isogeny problems, which we construct using generic transforms from identification protocols. We use the well-known Fiat-Shamir transform [15] to obtain security against classical adversaries in the random oracle model. This is not known to be secure against quantum adversaries[1] so for post-quantum security we use another transform due to Unruh [33]. We provide a full description of the two resulting signature schemes. Our signatures have very small key sizes, and reasonably efficient signing and verification procedures. The full version of the paper also contains two signature schemes based on the De Feo-Jao-Plût ID-scheme.[2]

[1] To some extent, Fiat-Shamir signatures are in fact not believed to be secure against quantum adversaries [33], although they can be proven to be secure under certain conditions [34] which the schemes presented do not verify.

[2] These signatures schemes were independently proposed by Yoo et al. [42]; our versions have smaller signature sizes for the same security guarantees.

As an additional contribution, we carefully analyse the complexity of the quaternion isogeny algorithm of Kohel-Lauter-Petit-Tignol [26] for powersmooth norms, and we highlight a property of its output distribution (under a minor change) that had remained unnoticed until now. This contribution is of independent interest, and it might be useful for other schemes based on similar isogeny problems.

Outline. The paper is organized as follows. In Sect. 2 we give preliminaries on isogeny problems and random walks in isogeny graphs, as well as security definitions for identification protocols. In Sect. 3 we describe our new identification protocol based on the endomorphism ring computation problem. In Sect. 4 we present our signature schemes and summarize their main efficiency features. A full version of this paper is available on the IACR eprint server [18].

2 Preliminaries

2.1 Hard Problem Candidates Related to Isogenies

We summarize the required background on elliptic curves. For a more detailed exposition of the theory, see [31]. Let E, E' be two elliptic curves over a finite field $\mathbb{F}_q$. An *isogeny* $\varphi : E \to E'$ is a non-constant morphism from E to E' that maps the neutral element into the neutral element. The degree of an isogeny φ is the degree of φ as a morphism. An isogeny of degree ℓ is called an ℓ-isogeny. If φ is separable, then $\deg \varphi = \# \ker \varphi$. If there is a separable isogeny between two curves, we say that they are *isogenous*. Tate's theorem is that two curves E, E' over $\mathbb{F}_q$ are isogenous over $\mathbb{F}_q$ if and only if $\#E(\mathbb{F}_q) = \#E'(\mathbb{F}_q)$.

A separable isogeny can be identified with its kernel [40]. Given a subgroup G of E, we can use Vélu's formulae [39] to explicitly obtain an isogeny $\varphi : E \to E'$ with kernel G and such that $E' \cong E/G$. These formulas involve sums over points in G, so using them is efficient as long as $\#G$ is small. Kohel [25] and Dewaghe [12] have (independently) given formulae for the Vélu isogeny in terms of the coefficients of the polynomial defining the kernel, rather than in terms of the points in the kernel. Given a prime ℓ, the torsion group $E[\ell]$ contains exactly $\ell + 1$ cyclic subgroups of order ℓ, each one corresponding to a different isogeny.

A composition of n separable isogenies of degrees ℓ_i for $1 \leq i \leq n$ gives an isogeny of degree $N = \prod_i \ell_i$ with kernel a group G of order N. Conversely any isogeny whose kernel is a group of smooth order can be decomposed as a sequence of isogenies of small degree, hence can be computed efficiently. For any permutation σ on $\{1, \ldots, n\}$, by considering appropriate subgroups of G, one can write the isogeny as a composition of isogenies of degree $\ell_{\sigma(i)}$. Hence, there is no loss of generality in the protocols in our paper of considering chains of isogenies of increasing degree.

For each isogeny $\varphi : E \to E'$, there is a unique isogeny $\hat{\varphi} : E' \to E$, which is called the *dual isogeny* of φ, and which satisfies $\varphi\hat{\varphi} = \hat{\varphi}\varphi = [\deg \varphi]$. If we have two isogenies $\varphi : E \to E'$ and $\varphi' : E' \to E$ such that $\varphi\varphi'$ and $\varphi'\varphi$ are the identity in their respective curves, we say that φ, φ' are *isomorphisms*, and that E, E' are *isomorphic*. Isomorphism classes of elliptic curves over $\mathbb{F}_q$ can be labeled with their j-invariant [31, III.1.4(b)]. An isogeny $\varphi : E \to E'$ such that $E = E'$ is called an *endomorphism*. The set of endomorphisms of an elliptic curve, denoted by $\mathrm{End}(E)$, has a ring structure with the operations point-wise addition and function composition.

Elliptic curves can be classified according to their endomorphism ring. Over the algebraic closure of the field, $\mathrm{End}(E)$ is either an order in a quadratic imaginary field or a maximal order in a quaternion algebra. In the first case, we say that the curve is *ordinary*, whereas in the second case we say that the curve is *supersingular*. The endomorphism ring of a supersingular curve over a field of characteristic p is a maximal order $\mathcal{O}$ in the quaternion algebra $B_{p,\infty}$ ramified at p and ∞.

In the case of supersingular elliptic curves, there is always a curve in the isomorphism class defined over $\mathbb{F}_{p^2}$, and the j-invariant of the class is also an element of $\mathbb{F}_{p^2}$. A theorem by Deuring [11] gives an equivalence of categories between the j-invariants of supersingular elliptic curves over $\mathbb{F}_{p^2}$ up to Galois conjugacy in $\mathbb{F}_{p^2}$, and the maximal orders in the quaternion algebra $B_{p,\infty}$ up to the equivalence relation given by $\mathcal{O} \sim \mathcal{O}'$ if and only if $\mathcal{O} = \alpha^{-1}\mathcal{O}'\alpha$ for some $\alpha \in B_{p,\infty}^*$. Specifically, the equivalence of categories associates to every j-invariant a maximal order that is isomorphic to the endomorphism ring of any curve with that j-invariant. Furthermore, if E_0 is an elliptic curve with $\mathrm{End}(E_0) = \mathcal{O}_0$, there is a one-to-one correspondence (which we call the *Deuring correspondence*) between isogenies $\psi : E_0 \to E$ and left $\mathcal{O}_0$-ideals I. More details on the Deuring correspondence can be found in Chap. 41 of [37].

We now present some hard problem candidates related to supersingular elliptic curves, and discuss the related algebraic problems in the light of the Deuring correspondence.

Problem 1. *Let p, ℓ be distinct prime numbers. Let E, E' be two supersingular elliptic curves over $\mathbb{F}_{p^2}$ with $\#E(\mathbb{F}_{p^2}) = \#E'(\mathbb{F}_{p^2}) = (p+1)^2$, chosen uniformly at random. Find $k \in \mathbb{N}$ and an isogeny of degree ℓ^k from E to E'.*

Problem 2. *Let p, ℓ be distinct prime numbers. Let E be a supersingular elliptic curve over $\mathbb{F}_{p^2}$, chosen uniformly at random. Find $k_1, k_2 \in \mathbb{N}$, a supersingular elliptic curve E' over $\mathbb{F}_{p^2}$, and two distinct isogenies of degrees ℓ^{k_1} and ℓ^{k_2}, respectively, from E to E'.*

The hardness assumption of both problems has been used in [6] to prove preimage and collision-resistance of a proposed hash function. Variants of the first problem, in which some extra information is provided, were used by De Feo-Jao-Plût [14] to build an identification scheme, a key exchange protocol and a public-key encryption scheme. More precisely, the identification scheme in [14] relies on Problems 3 and 4 below (which De Feo, Jao and Plût call the *Computational Supersingular Isogeny* and *Decisional Supersingular Product* problems). In order to state them we need to introduce some notation. Let p be a prime of the form $\ell_1^{e_1}\ell_2^{e_2} \cdot f \pm 1$, and let E be a supersingular elliptic curve over $\mathbb{F}_{p^2}$. Let $\{R_1, S_1\}$ and $\{R_2, S_2\}$ be bases for $E[\ell_1^{e_1}]$ and $E[\ell_2^{e_2}]$, respectively.

Problem 3 (Computational Supersingular Isogeny). *Let $\phi_1 : E \to E'$ be an isogeny with kernel $\langle [m_1]R_1 + [n_1]S_1 \rangle$, where m_1, n_1 are chosen uniformly at random from $\mathbb{Z}/\ell_1^{e_1}\mathbb{Z}$, and not both divisible by ℓ_1. Given E' and the values $\phi_1(R_2), \phi_1(S_2)$, find a generator of $\langle [m_1]R_1 + [n_1]S_1 \rangle$.*

The fastest known algorithms for this problem use a meet-in-the-middle argument. The classical and quantum algorithm have heuristic running time respectively of $\tilde{O}(\ell_1^{e_1/2})$ and $\tilde{O}(\ell_1^{e_1/3})$ bit operations, which is respectively $\tilde{O}(p^{1/4})$ and $\tilde{O}(p^{1/6})$ in the context of De Feo-Jao-Plût [14].

Problem 4 (Decisional Supersingular Product). *Let E, E' be supersingular elliptic curves over $\mathbb{F}_{p^2}$ such that there exists an isogeny $\phi : E \to E'$ of degree $\ell_1^{e_1}$. Fix generators $R_2, S_2 \in E[\ell_2^{e_2}]$ and suppose $\phi(R_2)$ and $\phi(S_2)$ are given. Consider the two distributions of pairs (E_2, E_2') as follows:*

- *(E_2, E_2') such that there is a cyclic group $G \subseteq E[\ell_2^{e_2}]$ of order $\ell_2^{e_2}$ and $E_2 \cong E/G$ and $E_2' \cong E'/\phi(G)$.*
- *(E_2, E_2') where E_2 is chosen at random among the curves having the same cardinality as E_0, and $\phi' : E_2 \to E_2'$ is a random $\ell_1^{e_1}$-isogeny.*

The problem is, given (E, E') and the auxiliary points $(R_2, S_2, \phi(R_2), \phi(S_2))$, plus a pair (E_2, E_2'), to determine from which distribution the pair is sampled.

We stress that Problems 3 and 4 are potentially easier than Problems 1 and 2 because special primes are used and extra points are revealed. Furthermore, it is shown in Sect. 4 of [17] that if $\mathrm{End}(E)$ is known and one can find any isogeny from E to E' then one can compute the specific isogeny of degree $\ell_1^{e_1}$. The following problem, on the other hand, offers better foundations for cryptography based on supersingular isogeny problems.

Problem 5. *Let p be a prime number. Let E be a supersingular elliptic curve over $\mathbb{F}_{p^2}$, chosen uniformly at random. Determine the endomorphism ring of E.*

Note that it is essential that the curve is chosen randomly in this problem, as for special curves the endomorphism ring is easy to compute. Essentially, Problem 5 is the same as explicitly computing the forward direction of Deuring's correspondence. This problem was studied by Kohel in [25], in which an algorithm to solve it was obtained, but with expected running time $\tilde{O}(p)$. It was later improved by Galbraith to $\tilde{O}(p^{\frac{1}{2}})$, under heuristic assumptions [16]. Interestingly, the best *quantum* algorithm for this problem runs in time $\tilde{O}(p^{\frac{1}{4}})$, only providing a quadratic speedup over classical algorithms [4]. This has largely motivated the use of supersingular isogeny problems in cryptography.

Problem 6. *Let p be a prime number. Let E, E' be supersingular elliptic curves over $\mathbb{F}_{p^2}$, chosen uniformly at random.*[3] *Find an isogeny $E \to E'$.*

---

[3] The special case $E' = E$ occurs with negligible probability so it can be ignored.

Heuristically, if we can solve Problem 1 or Problem 6, then we can solve Problem 5. To compute an endomorphism of E, we take two random walks $\phi_1 : E \to E_1$ and $\phi_2 : E \to E_2$, and solve Problem 6 on the pair E_1, E_2, obtaining an isogeny $\psi : E_1 \to E_2$. Then the composition $\hat{\phi}_2\psi\phi_1$ is an endomorphism of E. Repeating the process, it is easy to find four endomorphisms that are linearly independent, thus generating a subring of $\mathrm{End}(E)$, and this subring is likely to be of small index so that the full ring can be recovered.

For the converse, suppose that we can compute the endomorphism rings of both E and E'. The strategy is to compute a module I that is a left ideal of $\mathrm{End}(E)$ and a right ideal of $\mathrm{End}(E')$ of appropriate norm, and to translate it back to the geometric setting to obtain an isogeny. This approach motivated the quaternion ℓ-isogeny algorithm of Kohel-Lauter-Petit-Tignol [26,28], which solves the following problem:

Problem 7. *Let p, ℓ be distinct prime numbers. Let $\mathcal{O}_0, \mathcal{O}_1$ be two maximal orders in $B_{p,\infty}$, chosen uniformly at random. Find $k \in \mathbb{N}$ and an ideal I of norm ℓ^k such that I is a left $\mathcal{O}_0$-ideal and its right order is isomorphic to $\mathcal{O}_1$.*

The algorithm can be adapted to produce ideals of B-powersmooth norm (meaning the norm is $\prod_i \ell_i^{e_i}$ where the ℓ_i are distinct primes and $\ell_i^{e_i} \le B$) for $B \approx \frac{7}{2}\log p$ and using $O(\log p)$ different primes, instead of ideals of norm a power of ℓ. We will use that version in our signature scheme.

For completeness we mention that ordinary curve versions of Problems 1 and 5 are not known to be equivalent, and in fact there is a subexponential algorithm for computing the endomorphism ring of ordinary curves [5], whereas the best classical algorithm known for computing isogenies is still exponential. There is, however, a subexponential quantum algorithm for computing an isogeny between ordinary curves [7], which is why the main interest in cryptography is the supersingular case.

2.2 Random Walks in Isogeny Graphs

Let $p \ge 5$ be a prime number. There are $N_p := \lfloor \frac{p}{12} \rfloor + \epsilon_p$ supersingular j-invariants in characteristic p, with $\epsilon_p = 0, 1, 1, 2$ when $p = 1, 5, 7, 11 \bmod 12$ respectively. For any prime $\ell \neq p$, one can construct a so-called isogeny graph, where each vertex is associated to a supersingular j-invariant, and an edge between two vertices is associated to a degree ℓ isogeny between the corresponding vertices.

Isogeny graphs are regular[4] with regularity degree $\ell + 1$; they are undirected since to any isogeny from j_1 to j_2 corresponds a dual isogeny from j_2 to j_1. Isogeny graphs are also very good *expander graphs* [20]; in fact they are optimal expander graphs in the following sense:

[4] One needs to pay close attention to the cases $j = 0$ and $j = 1728$ when counting isogenies, but this has no effect on our general schemes.

Definition 1 (Ramanujan graph). *Let G be a k-regular graph, and let $k, \lambda_2, \cdots, \lambda_r$ be the eigenvalues of the adjacency matrix sorted by decreasing order of the absolute value. Then G is a* Ramanujan graph *if*

$$\lambda_2 \leq 2\sqrt{k-1}.$$

This is optimal by the Alon-Boppana bound: given a family $\{G_N\}$ of k-regular graphs as above, and denoting by $\lambda_{2,N}$ the corresponding second eigenvalue of each graph G_N, we have $\liminf_{N\to\infty} \lambda_{2,N} \geq 2\sqrt{k-1}$. The Ramanujan property of isogeny graphs follows from the Weil conjectures proved by Deligne [10,30].

Let p and ℓ be as above, and let j be a supersingular invariant in characteristic p. We define a random step of degree ℓ from j as the process of randomly and uniformly choosing a neighbour of j in the ℓ-isogeny graph, and returning that vertex. For a composite degree $n = \prod_i \ell_i$, we define a random walk of degree n from j_0 as a sequence of j-invariants j_i such that j_i is a random step of degree ℓ_i from j_{i-1}. We do not require the primes ℓ_i to be distinct.

The output of random walks in expander graphs converge quickly to a uniform distribution. In our signature scheme we will be using random walks of *B-powersmooth* degree n, namely $n = \prod_i \ell_i^{e_i}$, with all prime powers $\ell_i^{e_i}$ smaller than some bound B, with B as small as possible. To analyse the ouptut distribution of these walks we will use the following generalization[5] of classical random walk theorems [20]:

Theorem 1 (Random walk theorem). *Let p be a prime number, and let j_0 be a supersingular invariant in characteristic p. Let j be the final j-invariant reached by a random walk of degree $n = \prod_i \ell_i^{e_i}$ from j_0. Then for every j-invariant $\tilde{j}$ we have*

$$\left|\Pr[j = \tilde{j}] - \frac{1}{N_p}\right| \leq \prod_i \left(\frac{2\sqrt{\ell_i}}{\ell_i + 1}\right)^{e_i}.$$

Proof: Let v_{tj} be the probability that the outcome of the first t random steps is a given vertex j, and let $v_t = (v_{tj})_j$ be vectors encoding these probabilities. Let v_0 correspond to an initial state of the walk at j_0 (so that $v_{0j_0} = 1$ and $v_{0j} = 0$ for all $i \neq j_0$). Let A_{ℓ_i} be the adjacency matrix of the ℓ_i-isogeny graph. Its largest eigenvalue is k. By the Ramanujan property the second largest eigenvalue is smaller than k in absolute value, so the eigenspace associated to $\lambda_1 = k$ is of dimension 1 and generated by the vector $u := (N_p^{-1})_j$ corresponding to the uniform distribution. Let λ_{2i} be the second largest eigenvalue of A_{ℓ_i} in absolute value.

If step t is of degree ℓ_i we have $v_t = \frac{1}{k} A_{\ell_i} v_{t-1}$. Moreover we have $||v_t - u||_2 \leq \frac{1}{k}\lambda_{2i}||v_{t-1} - u||_2$ since the eigenspace associated to k is of dimension 1. Iterating on all steps we deduce

$$||v_t - u||_2 \leq \prod_i |\tfrac{1}{k}\lambda_{2i}|^{e_i} ||v_0 - u||_2 \leq \prod_i |\tfrac{1}{k}\lambda_{2i}|^{e_i}$$

[5] Random walks theorems are usually stated for a single graph whereas our walks will switch from one graph to another, all with the same vertex set but different edges.

since $||v_0 - u||_2^2 = (1-\frac{1}{N_p})^2 + \frac{N_p-1}{N_p}(\frac{1}{N_p})^2 \leq 1 - \frac{2}{N_p} + \frac{2}{N_p^2} < 1$. Finally we have

$$\left|\Pr[j=\tilde{j}] - \frac{1}{N_p}\right| = ||v_t - u||_\infty \leq ||v_t - u||_2 \leq \prod_i |\tfrac{1}{k}\lambda_{2i}|^{e_i} \leq \prod_i \left(\frac{2\sqrt{\ell_i}}{\ell_i+1}\right)^{e_i},$$

where we have used the Ramanujan property to bound the eigenvalues. □

In our security proof we will want the right-hand term to be smaller than $(p^{1+\epsilon})^{-1}$ for an arbitrary positive constant ϵ, and at the same time we will want the powersmooth bound B to be as small as possible. The following lemma shows that taking $B \approx 2(1+\epsilon)\log p$ suffices asymptotically.

Lemma 1. *Let $\epsilon > 0$. There is a function $c_p = c(p)$ such that $\lim_{p\to\infty} c_p = 2(1+\epsilon)$, and, for each p,*

$$\prod_{\substack{\ell_i \text{ prime} \\ e_i := \max\{e | \ell_i^e < c_p \log p\}}} \left(\frac{\ell_i+1}{2\sqrt{\ell_i}}\right)^{e_i} > p^{1+\epsilon}.$$

Proof: Let B be an integer. We have

$$\prod_{\substack{\ell_i^{e_i} < B \\ \ell_i \text{ prime} \\ e_i \text{ maximal}}} \left(\frac{\ell_i+1}{2\sqrt{\ell_i}}\right)^{e_i} > \prod_{\substack{\ell_i < B \\ \ell_i \text{ prime}}} \left(\frac{\ell_i+1}{2\sqrt{\ell_i}}\right) > \prod_{\substack{\ell_i < B \\ \ell_i \text{ prime}}} \left(\frac{\sqrt{\ell_i}}{2}\right).$$

Taking logarithms, using the prime number theorem and replacing the sum by an integral we have

$$\log \prod_{\substack{\ell_i < B \\ \ell_i \text{ prime}}} \left(\frac{\sqrt{\ell_i}}{2}\right) = \sum_{\substack{\ell_i < B \\ \ell_i \text{ prime}}} \frac{1}{2}\log \ell_i - \sum_{\substack{\ell_i < B \\ \ell_i \text{ prime}}} \log 2 \approx \frac{1}{2}\int_1^B \log x \frac{1}{\log x} dx - \frac{B}{\log B} =$$

$$= \frac{1}{2}B - \frac{B}{\log B} \approx \frac{1}{2}B.$$

if B is large enough. Taking $B = c\log(p)$ where $c = 2(1+\epsilon)$ gives $\frac{1}{2}B = (1+\epsilon)\log p = \log(p^{1+\epsilon})$ which proves the lemma. □

2.3 Identification Schemes

In this section we recall the standard cryptographic notions of identification schemes. A good general reference is Chap. 8 of Katz [23]. A sigma-protocol is a three-move proof of knowledge of a relation. The notions of honest verifier zero-knowledge (HVZK) and 2-special soundness are standard and due to lack of space we do not recall them. In the special case of "hard relations" (see Definition 3 below), one can interpret a sigma-protocol as a public key identification scheme. Good general references are the lecture notes of Damgård [9] and Venturi [35]. All algorithms below are probabilistic polynomial-time (PPT) unless otherwise stated.

An identification scheme is an interactive protocol between two parties (a Prover and a Verifier), where the Prover aims to convince the Verifier that it knows some secret key without revealing anything about it. This is achieved by the Prover first committing to some value, then the Verifier sending a challenge, and finally the Prover providing some answer in accordance to the commitment, the challenge and the secret. We use the terminology and notation of Abdalla-An-Bellare-Namprempre [1] (also see Bellare-Poettering-Stebila [3]). We also introduce a notion of "recoverability" which is implicit in the Schnorr signature scheme and seems to be folklore in the field.

Definition 2. *A* canonical identification scheme *is* $\mathcal{ID} = (K, \mathcal{P}, \mathcal{V}, c)$ *where: K is a PPT algorithm (key generation) that on input a security parameter λ outputs a pair* (PK, SK)*; $\mathcal{P}$ is a PPT algorithm taking input* SK *and outputting a message; $c \geq 1$ is the (integer) bitlength of the challenge (a function of the security parameter λ); $\mathcal{V}$ is a deterministic polynomial-time verification algorithm that takes as input* PK *and a transcript and outputs* 0 *or* 1*. A* transcript of an honest execution *of the scheme $\mathcal{ID}$ is the sequence:* $\text{CMT} \leftarrow \mathcal{P}(\text{PK}, \text{SK})$, $\text{CH} \leftarrow \{0,1\}^c$, $\text{RSP} \leftarrow \mathcal{P}(\text{PK}, \text{SK}, \text{CMT}, \text{CH})$*. On an honest execution we require that* $\mathcal{V}(\text{PK}, \text{CMT}, \text{CH}, \text{RSP}) = 1$.

An impersonator *for $\mathcal{ID}$ is an algorithm $\mathcal{I}$ that plays the following game: $\mathcal{I}$ takes as input a public key* PK *and a set of transcripts of honest executions of the scheme $\mathcal{ID}$; $\mathcal{I}$ outputs* CMT*, receives* $\text{CH} \leftarrow \{0,1\}^c$ *and outputs* RSP*. We say that $\mathcal{I}$ wins if* $\mathcal{V}(\text{PK}, \text{CMT}, \text{CH}, \text{RSP}) = 1$*. The advantage of $\mathcal{I}$ is* $|\Pr(\mathcal{I} \text{ wins}) - \frac{1}{2^c}|$*. We say that $\mathcal{ID}$ is* secure against impersonation under passive attacks *if the advantage is negligible for all PPT adversaries.*

An ID-scheme $\mathcal{ID}$ is non-trivial *if $c \geq \lambda$.*

An ID-scheme is recoverable *if there is a deterministic polynomial-time algorithm Rec such that for any transcript* $(\text{CMT}, \text{CH}, \text{RSP})$ *of an honest execution we have*

$$Rec(\text{PK}, \text{CH}, \text{RSP}) = \text{CMT}.$$

One can transform any 2-special sound ID scheme into a non-trivial scheme by running t sessions in parallel, and this is secure for classical adversaries (see Sect. 8.3 of [23]). We will not need this result in the quantum case. One first generates $\text{CMT}_i \leftarrow \mathcal{P}(\text{PK}, \text{SK})$ for $1 \leq i \leq t$. One then samples $\text{CH} \leftarrow \{0,1\}^{ct}$ and parses it as $\text{CH}_i \in \{0,1\}^c$ for $1 \leq i \leq t$. Finally one computes $\text{RSP}_i \leftarrow P(\text{PK}, \text{SK}, \text{CMT}_i, \text{CH}_i)$. We define

$$\mathcal{V}(\text{PK}, \text{CMT}_1, \cdots, \text{CMT}_t, \text{CH}, \text{RSP}_1, \cdots, \text{RSP}_t) = 1$$

if and only if $\mathcal{V}(\text{PK}, \text{CMT}_i, \text{CH}_i, \text{RSP}_i) = 1$ for all $1 \leq i \leq t$. The successful cheating probability is then improved to $1/2^{ct}$, which is non-trivial when $t \geq \lambda/c$.

An ID-scheme is a special case of a sigma-protocol with respect to the relation defined by the instance generator K as $(\text{PK}, \text{SK}) \leftarrow K$, where we think of SK as a witness for PK. More generally, any sigma-protocol for a relation of a certain type can be turned into an identification scheme.

Definition 3. *(Definition 6 of [35]; Sect. 6 of [9]; Definition 15 of [33], where it is called "hard instance generator") A* hard relation *R on $Y \times X$ is one where there exists a PPT algorithm K that outputs pairs $(y, x) \in Y \times X$ such that $R(y, x) = 1$, but for all PPT adversaries $\mathcal{A}$*

$$\Pr[(y, x) \leftarrow K(1^\lambda); x' \leftarrow \mathcal{A}(y) : R(y, x') = 1] \leq \mathrm{negl}(\lambda).$$

The following result is essentially due to Feige, Fiat and Shamir [13] and has become folklore in this generality. For the proof see Theorem 5 of [35].

Theorem 2. *Let R be a hard relation with generator K and let $(\mathcal{P}, \mathcal{V})$ be the prover and verifier in a sigma-protocol for R with c-bit challenges for some integer $c \geq 1$. Suppose the sigma-protocol is complete, 2-special sound, and honest verifier zero-knowledge. Then $(K, \mathcal{P}, \mathcal{V}, c)$ is a canonical identification scheme that is secure against impersonation under (classical) passive attacks.*

There are standard constructions to construct signature schemes from identification protocols. Due to lack of space we refer to Abdalla-An-Bellare-Namprempre [1] (also see Bellare-Poettering-Stebila [3]). As discussed in the full version of the paper, our ID-schemes are recoverable, and this allows us to reduce the signature size compared with general constructions.

3 New Identification Protocol Based Endomorphism Ring Computation

We now present our main result. The main advantage of our identification protocol compared with De Feo-Jao-Plût's one is that security is based on the general problem of computing the endomorphism ring of a supersingular elliptic curve, or equivalently on computing an isogeny between two supersingular curves. In particular, the prime has no special property and no auxiliary points are revealed.

3.1 High Level Description

Our identification protocol is similar to the graph isomorphism zero-knowledge protocol, in which one reveals one of two graph isomorphisms, but never enough information to deduce the secret isomorphism.

As recalled in Sect. 2.1, although it is believed that computing endomorphism rings of supersingular elliptic curves is a hard computational problem in general, there are some particular curves for which it is easy. The following construction is explained in Lemma 2 of [26]. We choose $E_0 : y^2 = x^3 + Ax$ over a field $\mathbb{F}_{p^2}$ where $p \equiv 3 \pmod 4$ and $\#E_0(\mathbb{F}_{p^2}) = (p+1)^2$. We have $j(E_0) = 1728$. When $p = 3 \bmod 4$, the quaternion algebra $B_{p,\infty}$ ramified at p and ∞ can be canonically represented as $\mathbb{Q}\langle \mathbf{i}, \mathbf{j} \rangle = \mathbb{Q} + \mathbb{Q}\mathbf{i} + \mathbb{Q}\mathbf{j} + \mathbb{Q}\mathbf{k}$, where $\mathbf{i}^2 = -1$, $\mathbf{j}^2 = -p$ and $\mathbf{k} := \mathbf{ij} = -\mathbf{ji}$. The endomorphism ring of E_0 is isomorphic to the maximal order $\mathcal{O}_0$ with $\mathbb{Z}$-basis $\{1, \mathbf{i}, \frac{1+\mathbf{k}}{2}, \frac{\mathbf{i}+\mathbf{j}}{2}\}$. Indeed, there is an isomorphism of quaternion algebras

$\theta : B_{p,\infty} \to \text{End}(E_0) \otimes \mathbb{Q}$ sending $(1, \mathbf{i}, \mathbf{j}, \mathbf{k})$ to $(1, \phi, \pi, \pi\phi)$ where $\pi : (x, y) \to (x^p, y^p)$ is the Frobenius endomorphism, and $\phi : (x, y) \to (-x, \iota y)$ with $\iota^2 = -1$.

To generate the public and private keys, we take a random isogeny (walk in the graph) $\varphi : E_0 \to E_1$ and, using this knowledge, compute $\text{End}(E_1)$. The public information is E_1. The secret is $\text{End}(E_1)$, or equivalently a path from E_0 to E_1. Under the assumption that computing the endomorphism ring is hard, the secret key cannot be computed from the public key only.

Our scheme will require three algorithms, that are explained in detail in later sections.

Translate isogeny path to ideal: Given $E_0, \mathcal{O}_0 = \text{End}(E_0)$ and a chain of isogenies from E_0 to E_1, to compute $\mathcal{O}_1 = \text{End}(E_1)$ and a left $\mathcal{O}_0$-ideal I whose right order is $\mathcal{O}_1$.

Find new path: Given a left $\mathcal{O}_0$-ideal I corresponding to an isogeny $E_0 \to E_2$, to produce a new left $\mathcal{O}_0$-ideal J corresponding to an "independent" isogeny $E_0 \to E_2$ of powersmooth degree.

Translate ideal to isogeny path: Given $E_0, \mathcal{O}_0, E_2, I$ such that I is a left $\mathcal{O}_0$-ideal whose right order is isomorphic to $\text{End}(E_2)$, to compute a sequence of prime degree isogenies giving the path from E_0 to E_2.

Figure 1 gives the interaction between the prover and the verifier. We define L to be the product of all prime powers ℓ^e such that $\ell^e \leq B = 2(1+\epsilon)\log p$ for an arbitrary $\epsilon > 0$. In other words, let $\ell_1, \ldots, \ell_r$ be the list of all primes up to B and let $L = \prod_{i=1}^{r} \ell_i^{e_i}$ where $\ell_i^{e_i} \leq B < \ell_i^{e_i+1}$. Note that $r \approx B/\log(B)$ and so $L \approx p^{2(1+\epsilon)}$.

One can see that Fig. 1 gives a canonical, recoverable identification protocol, but it is not non-trivial as the challenge is only one bit.

---

1. The public key is a pair (E_0, E_1) and the private key is an isogeny $\varphi : E_0 \to E_1$.
2. The prover performs a random walk starting from E_1 of degree L in the graph, obtaining a curve E_2 and an isogeny $\psi : E_1 \to E_2$, and reveals E_2 to the verifier.
3. The verifier challenges the prover with a random bit $b \leftarrow \{0, 1\}$.
4. If $b = 0$, the prover sends ψ to the verifier.
   If $b = 1$, the prover does the following:
   - Compute $\text{End}(E_2)$ and translate the isogeny path between E_0 and E_2 into a corresponding ideal I giving the path in the quaternion algebra.
   - Use the **Find new path** algorithm to compute an "independent" path between $\text{End}(E_0)$ and $\text{End}(E_2)$ in the quaternion algebra, represented by an ideal J.
   - Translate the ideal J to an isogeny path η from E_0 to E_2.
   - Return η to the verifier.
5. The verifier accepts the proof if the answer to the challenge is indeed an isogeny between E_1 and E_2 or between E_0 and E_2, respectively.

Fig. 1. New identification scheme

The isogenies involved in this protocol are summarised in the following diagram:

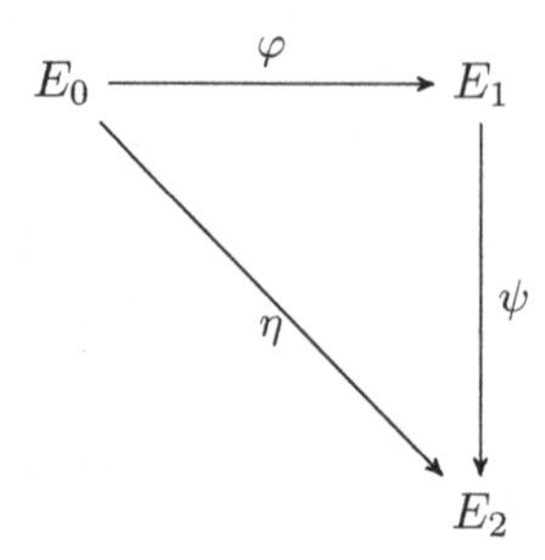

The two translation algorithms mentioned above in the $b = 1$ case will be described in Sect. 3.4. They rely on the fact that $\text{End}(E_0)$ is known. The algorithms are efficient when the degree of the random walk is powersmooth, and for this reason all isogenies in our protocols will be of powersmooth degree. The powersmooth version of the quaternion isogeny algorithm of Kohel-Lauter-Petit-Tignol will be described and analysed in Sect. 3.3. The random walks are taken of sufficiently large degree such that their output has close to uniform distribution, by Theorem 1 and Lemma 1.

We repeat the process to reduce the cheating probability. The computational hardness of Problem 5 remains essentially the same if the curves are chosen according to a distribution that is close to uniform. We can then prove:

Theorem 3. *Let λ be a security parameter and $t \geq \lambda$. If Problem 6 is computationally hard, then the identification scheme obtained from t parallel executions of the protocol in Fig. 1 is a non-trivial, recoverable canonical identification scheme that is secure against impersonation under (classical) passive attacks.*

The advantage of this protocol over De Feo-Jao-Plût's protocol is that it relies on a more standard and potentially harder computational problem. In the rest of this section we first give a proof of Theorem 3, then we provide details of the algorithms involved in our scheme.

3.2 Proof of Theorem 3

We shall prove that the sigma protocol in Fig. 1 is complete, 2-special sound and honest verifier zero-knowledge. It follows that t parallel executions of the protocol is non-trivial as well as being 2-special sound and HVZK. The theorem will then follow from Theorem 2 and Problem 6 (which implies that the relation being proved is a hard relation).

Completeness. Let φ be an isogeny between E_0 and E_1 of B-powersmooth degree, for $B = O(\log p)$. If the challenge received is $b = 0$, it is clear that the prover knows a valid isogeny $\psi : E_1 \to E_2$, so the verifier accepts the proof. If $b = 1$, the prover follows the procedure described above and the verifier accepts. In the next subsections we will show that this procedure is polynomial time.

2-special soundness. Let (E_0, E_1) be a public key for the scheme. Suppose we are given transcripts $(\text{CMT}, \{\text{CH}_1, \text{CH}_2\}, \{\text{RSP}_1, \text{RSP}_2\})$ for the single-bit scheme such that $\mathcal{V}(\text{PK}, \text{CMT}, \text{CH}_i, \text{RSP}_i) = 1$ for all $i \in \{1, 2\}$. Let $E_2 = \text{CMT}$. Since $\text{CH}_1 \neq \text{CH}_2$ the responses RSP_1 and RSP_2 therefore give two isogenies $\psi : E_1 \to E_2$, $\eta : E_0 \to E_2$.

Given these two valid answers an extraction algorithm can compute an isogeny $\phi : E_0 \to E_1$ as $\phi = \hat{\psi} \circ \eta$, where $\hat{\psi}$ is the dual isogeny of ψ. The extractor outputs ϕ, which is a solution to Problem 6. This is summarized in the following diagram.

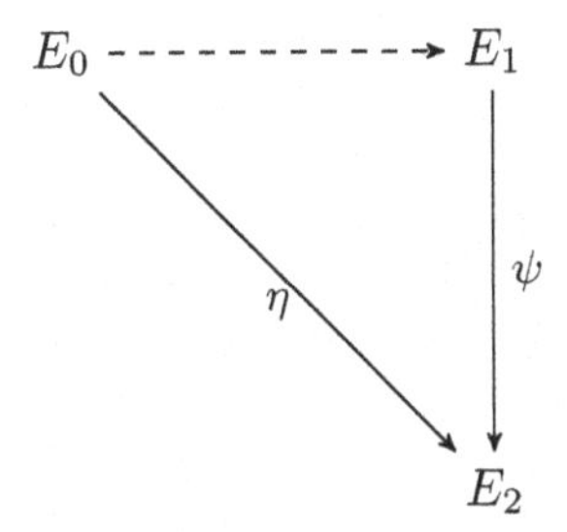

Honest verifier zero-knowledge. We shall prove that there exists a probabilistic polynomial time simulator $\mathcal{S}$ that outputs transcripts indistinguishable from transcripts of interactions with an honest verifier, in the sense that the two distributions are statistically close. Note that $\mathcal{O}_0 = \text{End}(E_0)$ is public information so is known to the simulator. The simulator starts by taking a random coin $b \leftarrow \{0, 1\}$.

- If $b = 0$, take a random walk from E_1 of powersmooth degree L, as in the real protocol, obtaining a curve E_2 and an isogeny $\psi : E_1 \to E_2$. The simulator outputs the transcript $(E_2, 0, \psi)$.

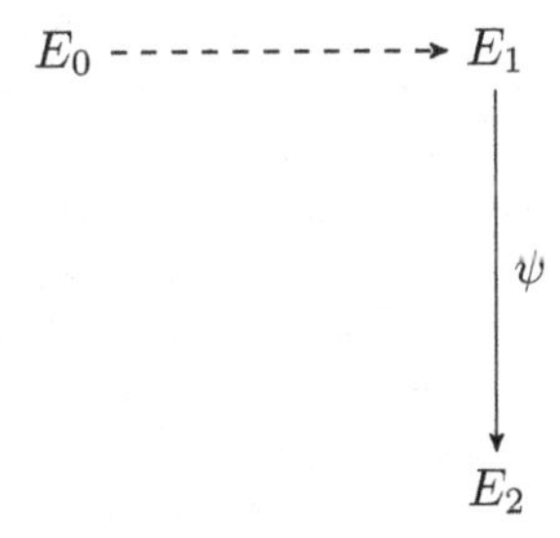

  In this case, it is clear that the distributions of every element in the transcript are the same as in the real interaction, as they are generated in the same way. This is possible because, when $b = 0$, the secret is not required for the prover to answer the challenge.
- If $b = 1$, take a random walk from E_0 of powersmooth degree L to obtain a curve E_2 and an isogeny $\mu : E_0 \to E_2$, then proceed as in Step 3 of Fig. 1 to produce another isogeny $\eta : E_0 \to E_2$. The simulator outputs the transcript $(E_2, 1, \eta)$.

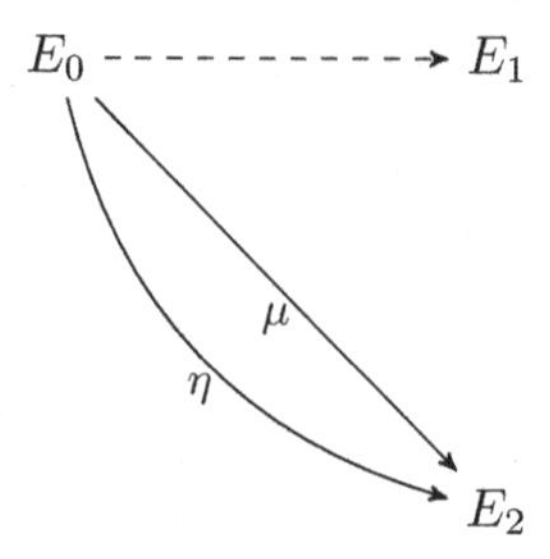

The reason to output η instead of μ is to ensure that the transcript distributions are indistinguishable from the distributions in the real scheme.

We first study the distribution of E_2 up to isomorphism. Let X_r be the output of the random walk from E_1 to produce $j(E_2)$ in the real interaction, and let X_s be the output of the random walk from E_0 to produce $j(E_2)$ in the simulation.

Let $\mathcal{G}$ be the set of all supersingular j-invariants, namely the vertex set of the isogeny graph. Note that $\#\mathcal{G} = N_p \approx p/12$. By Theorem 1 and Lemma 1, since the isogeny walks have degree L, we have, for any $j \in \mathcal{G}$

$$\left|\Pr(X_r = j) - \frac{1}{N_p}\right| \leq \frac{1}{p^{1+\epsilon}}, \qquad \left|\Pr(X_s = j) - \frac{1}{N_p}\right| \leq \frac{1}{p^{1+\epsilon}}.$$

Therefore

$$\begin{aligned}\sum_{j \in \mathcal{G}} |\Pr(X_r = j) - \Pr(X_s = j)| &\leq N_p \cdot \max_i |\Pr(X_r = j) - \Pr(X_s = j)| \\ &\leq N_p \cdot \left(\frac{1}{p^{1+\epsilon}} + \frac{1}{p^{1+\epsilon}}\right) \approx \frac{1}{6p^{\epsilon}}\end{aligned}$$

which is a negligible function of λ for any constant $\epsilon > 0$. In other words, the statistical distance, between the distribution of $j(E_2)$ in the real signing algorithm and the simulation, is negligible. Now, since η is produced in the same way from E_0 and E_2 in the simulation and in the real protocol execution, we have that the statistical distance between the distributions of η is also negligible. This follows from Lemma 2 in Sect. 3.3, which states that the output of the quaternion path algorithm does not depend on the input ideal, only on its ideal class.

3.3 Quaternion Isogeny Path Algorithm

In this section we sketch the quaternion isogeny algorithm from Kohel-Lauter-Petit-Tignol [26] and we evaluate its complexity when $p = 3 \bmod 4$. (The original paper does not give a precise complecity analysis; it is only claimed that the algorithm runs in heuristic probabilistic polynomial time.) This is the algorithm used for the **Find new path** procedure in the identification scheme.

The algorithm takes as input two maximal orders $\mathcal{O}, \mathcal{O}'$ in the quaternion algebra $B_{p,\infty}$, and it returns a sequence of left $\mathcal{O}$-ideals $I_0 = \mathcal{O} \supset I_1 \supset \ldots \supset I_e$ such that the right order of I_e is in the same equivalence class as $\mathcal{O}'$. In addition, the output is such that the index of I_{i+1} in I_i is a small prime for all i. The paper [26] focusses on the case where the norm of I_e is ℓ^e for some integer e, but it mentions that the algorithm can be extended to the case of powersmooth norms. We will only describe and use the powersmooth version. In our application there are some efficiency advantages from using isogenies whose degree is a product of small powers of distinct primes, rather than a large power of a small prime.

Note that the ideals returned by the quaternion isogeny path algorithm (or equivalently the right orders of these ideals) correspond to vertices of the path in the quaternion algebra graph, and to a sequence of j-invariants by Deuring's correspondence. In the next subsection we will describe how to make this correspondence explicit; here we focus on the quaternion algorithm itself.

An important feature of the algorithm is that paths between two arbitrary maximal orders $\mathcal{O}$ and $\mathcal{O}'$ are always constructed as a concatenation of two paths from each maximal order to a special maximal order. As mentioned above, in our protocol and the discussion below we fix $\mathcal{O}_0 = \langle 1, \mathbf{i}, \frac{1+\mathbf{k}}{2}, \frac{\mathbf{i}+\mathbf{j}}{2} \rangle$ where $\mathbf{i}^2 = -1$ and $\mathbf{j}^2 = -p$. General references for maximal orders and ideals in quaternion algebras are [36,37].

We focus on the case where $\mathcal{O} = \mathcal{O}_0$, and assume that instead of a second maximal $\mathcal{O}'$ we are given the corresponding left $\mathcal{O}_0$-ideal I as input (the two variants of the problem are equivalent). This will be sufficient for our use of the algorithm. We assume that I is given by a $\mathbb{Z}$ basis of elements in $\mathcal{O}_0$. Denote by $n(\alpha)$ and $n(I)$ the norm of an element or ideal respectively. The equivalence class of maximal orders defines an equivalence class of $\mathcal{O}_0$-ideals, where two ideals I and J are in the same class if and only if $I = Jq$ with $q \in B_{p,\infty}^*$. Therefore our goal is, given a left $\mathcal{O}_0$-ideal I, to compute another left $\mathcal{O}_0$-ideal J with powersmooth norm in the same ideal class. Further, in order to be able to later apply Algorithm 2, we require the norm of I to be odd (but the **Find new path** algorithm also allows to find even norm ideals if desired). Without loss of generality we assume there is no integer $s > 1$ such that $I \subset s\mathcal{O}_0$, and that $I \neq \mathcal{O}_0$. The algorithm proceeds as follows:

1. Compute an element $\delta \in I$ and an ideal $I' = I\bar{\delta}/n(I)$ of prime norm N.
2. Find $\beta \in I'$ with norm NS where S is powersmooth and odd.
3. Output $J = I'\bar{\beta}/N$.

Steps 1 and 3 of this algorithm rely on the following simple result [26, Lemma 5]: if I is a left $\mathcal{O}$-ideal of reduced norm N and α is an element of I, then $I\bar{\alpha}/N$ is a left $\mathcal{O}$-ideal of norm $n(\alpha)/N$. Clearly, I and J are in the same equivalence class.

To compute δ in Step 1, first a Minkowski-reduced basis $\{\alpha_1, \alpha_2, \alpha_3, \alpha_4\}$ of I is computed. To obtain Lemma 2 below we make sure that the Minkowski basis is uniformly randomly chosen among all such bases[6]. Then random elements $\delta = \sum_i x_i \alpha_i$ are generated with integers x_i in an interval $[-m, m]$, where m is determined later, until the norm of δ is equal to $n(I)$ times a prime. A probable prime suffices in this context (actually Step 1 is not strictly needed but aims

[6] In [26] an arbitrary Minkowski basis was chosen.

to simplify Step 2), so we can use the Miller-Rabin test to discard composite numbers with a large probability.

Step 2 is the core of the algorithm and actually consists of the following substeps:

2a. Find α such that $I' = \mathcal{O}_0 N + \mathcal{O}_0 \alpha$.
2b. Find $\beta_1 \in \mathcal{O}_0$ with odd powersmooth norm NS_1.
2c. Find $\beta_2 \in \mathbb{Z}\mathbf{j} + \mathbb{Z}\mathbf{k}$ such that $\alpha = \beta_1\beta_2 \bmod N\mathcal{O}_0$.
2d. Find $\beta_2' \in \mathcal{O}_0$ with odd powersmooth norm S_2 and $\lambda \in \mathbb{Z}_N^*$ such that $\beta_2' = \lambda\beta_2 \bmod N\mathcal{O}_0$.
2e. Set $\beta = \beta_1\beta_2'$.

In Step 2a we need $\alpha \in I'$ such that $\gcd(n(\alpha), N^2) = N$. This is easily achieved by taking α as a random small linear combination of a Minkowski basis, until the condition is met. Note that if $\alpha \in I'$ is such that $\gcd(n(\alpha), N^2) = N$ then $J := \mathcal{O}_0 N + \mathcal{O}_0\alpha \subseteq I'$ and $J \neq \mathcal{O}_0 N$. Since the norm of $\mathcal{O}_0 N$ is N^2 and N is prime it follows that the norm of J is N and so $J = I'$.

In Step 2b the algorithm actually searches for $\beta_1 = a + b\mathbf{i} + c\mathbf{j} + d\mathbf{k}$. A large enough powersmooth number S_1 is fixed a priori, then the algorithm generates small random values of c, d until the norm equation $a^2 + b^2 = S_1 - p(c^2 + d^2)$ can be solved efficiently using Cornacchia's algorithm (for example, until the right hand side is a prime equal to 1 modulo 4).

Step 2c is just linear algebra modulo N. As argued in [26] it has a negligible chance of failure, in which case one can just go back to Step 2b.

In Step 2d the algorithm a priori fixes S_2 large enough, then searches for integers a, b, c, d, λ with $\lambda \notin N\mathbb{Z}$ such that $N^2(a^2 + b^2) + p\left((\lambda C + cN)^2 + (\lambda D + dN)^2\right) = S_2$ where we have $\beta_2 = C\mathbf{j} + D\mathbf{k}$. If necessary S_2 is multiplied by a small prime such that $p(C^2 + D^2)S_2$ is a square modulo N, after which the equation is solved modulo N, leading to two solutions for λ. An arbitrary solution is chosen, and then looking at the equation modulo N^2 leads to a linear space of solutions for $(c, d) \in \mathbb{Z}_N$. The algorithm chooses random solutions until the equation

$$a^2 + b^2 = \left(S_2 - p^2\left((\lambda C + cN)^2 + (\lambda D + dN)^2\right)\right)/N^2$$

can be efficiently solved with Cornacchia's algorithm.

The overall algorithm is summarized in Algorithm 1. We now prove two lemmas on this algorithm. The first lemma shows that the output of this algorithm only depends on the ideal class of I but not on I itself. This is important in our identification protocol, as otherwise part of the secret isogeny φ could potentially be recovered from η. The second lemma gives a precise complexity analysis of the algorithm, where [26] only showed probabilistic polynomial time complexity. Both lemmas are of independent interest.

Lemma 2. *The output distribution of the quaternion isogeny path algorithm only depends on the equivalence class of its input. (In particular, the output distribution does not depend on the particular ideal class representative chosen for this input.)*

Proof: Let I_1 and I_2 be two left $\mathcal{O}_0$-ideals in the same equivalence class, namely there exists $q \in B_{p,\infty}^*$ such that $I_2 = I_1 q$. We show that the distribution of the ideal I' computed in Step 1 of the algorithm is identical for I_1 and I_2. As the inputs are not used anymore in the remaining of the algorithm this will prove the lemma.

In the first step the algorithm computes a Minkowski basis of its input, uniformly chosen among all possible Minkowski bases. Let $B_1 = \{\alpha_{11}, \alpha_{12}, \alpha_{13}, \alpha_{14}\}$ be a Minkowski basis of I_1. Then by multiplicativity of the norm we have that $B_2 = \{\alpha_{11}q, \alpha_{12}q, \alpha_{13}q, \alpha_{14}q\}$ is a Minkowski basis of I_2. The algorithm then computes random elements $\delta = \sum_i x_i \alpha_i$ for integers x_i in an interval $[-m, m]$. Clearly, for any element δ_1 computed when the input is I_1, there corresponds an element $\delta_2 = \delta_1 q$ computed when the input is I_2. This is repeated until the norm of δ is a prime times $n(I)$. As $n(I_2) = n(I_1)n(q)$ the stopping condition is equivalent for both. Finally, an ideal I of prime norm is computed as $I\bar{\delta}/n(I)$. Clearly when $\delta_2 = \delta_1 q$ we have $\frac{I_2 \bar{\delta}_2}{n(I_2)} = \frac{I_1 q \bar{q} \bar{\delta}_1}{n(q)n(I_1)} = \frac{I_1 \bar{\delta}_1}{n(I_1)}$. This shows that the prime norm ideal computed in Step 1 only depends on the equivalence class of the input. □

The expected running time given in the following lemma relies on several heuristics related to the factorization of numbers generated following certain distributions. Intuitively all these heuristics say that asymptotically those numbers behave in the same way as random numbers of the same size.

Lemma 3. *Let* $X := \max |c_{ij}|$ *where* $c_{ij} \in \mathbb{Z}$ *are integers such that* $c_{i1} + c_{i2}\mathbf{i} + c_{i3}\frac{1+\mathbf{k}}{2} + c_{i4}\frac{\mathbf{i}+\mathbf{j}}{2}$ *for* $1 \le i \le 4$ *forms a* $\mathbb{Z}$*-basis for* I*. If* $\log X = O(\log p)$ *then Algorithm 1 heuristically runs in time* $\tilde{O}(\log^3 p)$*, and produces an output of norm* S *with* $\log(S) \approx \frac{7}{2}\log(p)$ *which is* $(\frac{7}{2} + o(1))\log p$*-powersmooth.*

Proof: The Minkowski basis can be computed in $O(\log^2 X)$, for example using the algorithm of [27].

For generic ideals the reduced norms of all Minkowski basis elements[7] are in $O(\sqrt{p})$ (see [26, Sect. 3.1]). In the first loop we initially set $m = \lceil \log p \rceil$. Assuming heuristically that the numbers N generated behave like random numbers we expect the box to produce some prime number. The resulting N will be in $\tilde{O}(\sqrt{p})$. For some non generic ideals the Minkowski basis may contain two pairs of elements with norms respectively significantly smaller or larger than $O(\sqrt{p})$; in that case we can expect to finish the loop for smaller values of m by setting $x_3 = x_4 = 0$, and to obtain some N of a smaller size.

Rabin's pseudo-primality test performs a single modular exponentiation (modulo a number of size $\tilde{O}(\sqrt{p})$), and it is passed by composite numbers with a probability at most $1/4$. The test can be repeated r times to decrease this probability to $1/4^r$. Assuming heuristically that the numbers tested behave like random numbers the test will only be repeated a significant amount of times

[7] The reduced norm of an ideal element is the norm of this element divided by the norm of the ideal.

Algorithm 1. Find new path algorithm

Input: $\mathcal{O}_0 = \langle 1, \mathbf{i}, \frac{1+\mathbf{k}}{2}, \frac{\mathbf{i}+\mathbf{j}}{2} \rangle$, I a left $\mathcal{O}_0$-ideal. **Output:** J a left $\mathcal{O}_0$-ideal of powersmooth norm such that $I = Jq$ for some $q \in B_{p,\infty}$.

1: $\{\alpha_1, \alpha_2, \alpha_3, \alpha_4\}$ Minkowski-reduced basis of I.
2: $\alpha_i \leftarrow \{\pm\alpha_i\}$ for $i = 1, 2, 3, 4$.
3: **loop**
4: $\quad\{x_1, x_2, x_3, x_4\} \leftarrow [-m, m]^4$. Start with $m = \lceil \log p \rceil$ and do exhaustive search in the box, increasing m if necessary.
5: $\quad\delta := \sum_{i=1}^{4} x_i \alpha_i$
6: $\quad$**if** $N := n(\delta)/n(I)$ is prime **then return** $N, I' := I\bar{\delta}/n(I)$
7: Set an a priori powersmooth bound $s = \frac{7}{2} \log p$, and odd integers S_1, S_2 with $S_1 > p \log p$, $S_2 > p^3 \log p$ and s-powersmooth product $S_1 S_2$.
8: Choose $\alpha \in I$ such that $\gcd(n(\alpha), N^2) = N$, so that $I' = \mathcal{O}_0 N + \mathcal{O}_0 \alpha$.
9: **while** a, b are not found **do**
10: $\quad c, d \leftarrow [-m, m]^2$, for $m = \lfloor \sqrt{NS_1/2p} \rfloor$. Increase S_1 and s if necessary.
11: $\quad a, b \leftarrow$ Solution of $a^2 + b^2 = NS_1 - p(c^2 + d^2)$ (solve using Cornacchia's algorithm).
12: $\beta_1 = a + b\mathbf{i} + c\mathbf{j} + d\mathbf{k}$
13: Set β_2 as a solution of $\alpha = \beta_1\beta_2 \bmod N\mathcal{O}_0$ with $\beta_2 \in \mathbb{Z}\mathbf{j} + \mathbb{Z}\mathbf{k}$.
14: Write $\beta_2 = C\mathbf{j} + D\mathbf{k}$. Try small odd primes r in increasing order until we find one such that $\left(\frac{(C^2+D^2)S_2 r}{N}\right) = 1$, and set $S_2 = S_2 r$. Update s accordingly.
15: $\lambda \leftarrow$ Solution of $p\lambda^2(C^2 + D^2) = S_2 \bmod N$.
16: **while** a, b are not found **do**
17: $\quad c, d \leftarrow$ Solution of $p\lambda^2(C^2 + D^2) + 2p\lambda N(Cc + Dd) = S_2 \bmod N^2$.
18: $\quad a, b \leftarrow$ Solution of $a^2 + b^2 = \left(S_2 - p^2\left((\lambda C + cN)^2 + (\lambda D + dN)^2\right)\right)/N^2$ (solve using Cornacchia's algorithm). Increase S_2 and s if necessary.
19: $\beta_2' = a + b\mathbf{i} + c\mathbf{j} + d\mathbf{k}$
20: $J = I'\overline{\beta_1\beta_2'}/N$

on actual prime numbers, so in total it will be repeated $O(\log p)$ times. This leads to a total complexity of $\tilde{O}(\log^3 p)$ bit operations for the first loop using fast (quasi-linear) modular multiplication.

The other two loops involve solving equations of the form $x^2 + y^2 = M$. For such an equation to have solutions it is sufficient that M is a prime with $M = 1 \bmod 4$, a condition that is heuristically satisfied after $2 \log M$ random trials. Choosing S_1 and S_2 as in the algorithm ensures that the right-hand term of the equation is positive, and (assuming this term behaves like a random number of the same size) is of the desired form for some choices (c, d), at least heuristically. Cornacchia's algorithm runs in time $\tilde{O}(\log^2 M)$, which is also $\tilde{O}(\log^2 p)$ in the algorithm. The pseudo-primality tests will require $\tilde{O}(\log^3 p)$ operations in total, and their cost will dominate both loops.

Computing β_2 is just linear algebra modulo $N \approx \tilde{O}(\sqrt{p})$ and this cost can be neglected. The last two steps can similarly be neglected.

As a result, we get an overall cost of $\tilde{O}(\log^3 p)$ bit operations for the whole algorithm.

Let $s = \frac{7}{2}\log p$. We have $n(J) = n(I')n(\beta_1)n(\beta_2')/N^2$ so neglecting $\log\log$ factors $\log n(J) \approx \frac{1}{2}\log p + \log p + 3\log p - \log p = \frac{7}{2}\log p$. We make the heuristic assumption that $\log n(J) = (\frac{7}{2} + o(1))\log p$. Moreover heuristically $\prod_{p_i^{e_i} < s} p_i^{e_i} \approx (s)^{s/\log s} \approx p^{7/2+o(1)}$ so we can expect to find $S_1 S_2$ that is s-powersmooth and of the correct size. □

3.4 Step-by-Step Deuring Correspondence

We now discuss algorithms to convert isogeny paths into paths in the quaternion algebra, and vice versa. This will be necessary in our protocols as we are sending curves and isogenies, whereas the process uses the quaternion isogeny algorithm.

All the isogeny paths that we will need to translate in our signature scheme will start from the special j-invariant $j_0 = 1728$. We recall (see beginning of Sect. 3.1) that this corresponds to the curve E_0 with equation $y^2 = x^3 + x$ and endomorphism ring $\mathrm{End}(E_0) := \langle 1, \phi, \frac{1+\pi\phi}{2}, \frac{\pi+\phi}{2}\rangle$. Moreover there is an isomorphism of quaternion algebras sending $(1, \mathbf{i}, \mathbf{j}, \mathbf{k})$ to $(1, \phi, \pi, \pi\phi)$.

For any isogeny $\varphi : E_0 \to E_1$ of degree n, we can associate a left $\mathrm{End}(E_0)$-ideal $I = \mathrm{Hom}(E_1, E_0)\varphi$ of norm n, corresponding to a left $\mathcal{O}_0$-ideal with the same norm in the quaternion algebra $B_{p,\infty}$. Conversely every left $\mathcal{O}_0$-ideal arises in this way [25, Section 5.3]. In our protocol we will need to make this correspondence explicit, namely we will need to pair up each isogeny from E_0 with the correct $\mathcal{O}_0$-ideal. Moreover we need to do this for "large" degree isogenies to ensure a good distribution via our random walk theorem.

Translating an Ideal to an Isogeny Path. Let E_0 and $\mathcal{O}_0 = \mathrm{End}(E_0)$ be given, together with a left $\mathcal{O}_0$-ideal I corresponding to an isogeny of degree n. We assume I is given as a $\mathbb{Z}$-basis $\{\alpha_1, \ldots, \alpha_4\}$. The main idea to determine the corresponding isogeny explicitly is to determine its kernel [40].

Assume for the moment that n is a small prime. One can compute generators for all cyclic subgroups of $E_0[n]$, each one uniquely defining a degree n isogeny which can be computed with Vélu's formulae. A generator P then corresponds to the basis $\{\alpha_1, \ldots, \alpha_4\}$ if and only if $\alpha_j(P) = 0$ for all $1 \le j \le 4$. To evaluate $\alpha(P)$ with $\alpha \in I$ and $P \in E_0[n]$, we first write $\alpha = (u + v\mathbf{i} + w\mathbf{j} + x\mathbf{k})/2$, then we compute P' such that $[2]P' = P$ and finally we evaluate $[u]P' + [v]\phi(P') + [w]\pi(P') + [x]\pi(\phi(P'))$.

An alternative to trying all subgroups is to choose a pair $\{P_1, P_2\}$ of generators for $E_0[n]$ and, for some $\alpha \in I$, solve the discrete logarithm instance (if possible) $\alpha(P_2) = [x]\alpha(P_1)$. It follows that $\alpha(P_2 - [x]P_1) = 0$ and so we have determined a candidate point in the kernel of the isogeny. Both solutions are too expensive for large n.

When $n = \ell^e$ the degree n isogeny can be decomposed into a composition of e degree ℓ isogenies. If I is the corresponding left $\mathcal{O}_0$-ideal of norm ℓ^e, then $I_i := I \bmod \mathcal{O}_0\ell^i$ is a left $\mathcal{O}_0$-ideal of norm ℓ^i corresponding to the first i isogenies. Similarly if P is a generator for the kernel of the degree ℓ^e isogeny then $\ell^{e-i+1}P$ is the kernel of the degree ℓ^i isogeny corresponding to the first i steps. One can therefore

Algorithm 2. Translating ideal to isogeny path

Input: $\mathcal{O}_0 = \mathrm{End}(E_0) = \langle 1, \phi, \frac{1+\pi\phi}{2}, \frac{\pi+\phi}{2} \rangle$, $I = \langle \alpha_1, \alpha_2, \alpha_3, \alpha_4 \rangle, n = \prod_{i=1}^{r} \ell_i^{e_i}$ with $2 \nmid n$. **Output:** the isogeny corresponding to I through Deuring's correspondence.

for $i = 1, \ldots, r$ **do**
  Compute a basis $\{P_{i1}, P_{i2}\}$ for the $\ell_i^{e_i}$ torsion on E_0
  for $j = 1, 2$ **do**
    Compute P'_{ij} such that $P_{ij} = [2]P'_{ij}$
$\varphi_0 = [1]_{E_0}$
for $i = 1, \ldots, r$ **do**
  for $k = 1, 2, 3, 4$ **do**
    $\alpha_{ik} = \alpha_k$ with its coefficients reduced modulo $\ell_i^{e_i}$.
    Write $\alpha_{ik} = (u_{ik} + v_{ik}\mathbf{i} + w_{ik}\mathbf{j} + x_{ik}\mathbf{k})/2$.
    for $j = 1, 2$ **do**
      $P_{ijk} = [u_{ik}]P'_{ij} + [v_{ik}]\phi(P'_{ij}) + [w_{ik}]\pi(P'_{ij}) + [x_{ik}]\pi(\phi(P'_{ij}))$
  Solve ECDLP to compute Q_i of order $\ell_i^{e_i}$ such that $\alpha_{ik}(Q_i) = 0$ for all k
  Compute $\phi_i =$ isogeny with kernel $\langle \varphi_{i-1}(Q_i) \rangle$ (compute with Vélu's formulae).
  Set $\varphi_i = \phi_i \varphi_{i-1}$
Output $\varphi_0, \phi_1, \ldots, \phi_r$.

perform the matching of ideals with kernels step-by-step with successive approximations of I or P respectively. This algorithm is more efficient than the previous one, but it still requires to compute ℓ^e torsion points, which in general may be defined over a degree ℓ^e extension of $\mathbb{F}_{p^2}$. To ensure that the ℓ^e torsion is defined over $\mathbb{F}_{p^2}$ one can choose p such that $\ell^e \mid (p \pm 1)$ as in the De Feo-Jao-Plût protocols; however for general p this translation algorithm will still be too expensive.

We solve this efficiency issue by using powersmooth degree isogenies in our protocols. When $n = \prod_i \ell_i^{e_i}$ with distinct primes ℓ_i, one reduces to the prime power case as follows. For simplicity we assume that 2 does not divide n. The isogeny of degree n can be decomposed into a sequence of prime degree isogenies. For simplicity we assume the isogeny steps are always performed in increasing degree order; we can require that this is indeed the case in our protocols. Let $n_i := \prod_{j \leq i} \ell_j^{e_j}$. Using a Chinese Remainder Theorem-like representation, points in $E_0[n]$ can be represented as a sequence of points in $E_0[\ell_i^{e_i}]$. If I is a left $\mathcal{O}_0$-ideal of norm n and φ is the corresponding isogeny, then the kernel of $I \bmod \mathcal{O}_0\ell_i^{e_i}$ is the $\ell_i^{e_i}$ part of the kernel of φ, namely $\ker(I \bmod \mathcal{O}_0\ell_i^{e_i}) = [n/\ell_i^{e_i}] \ker \varphi$. Given a left $\mathcal{O}_0$-ideal I, Algorithm 2 progressively identifies the corresponding isogeny sequence.

In our protocols we will have $\ell_i^{e_i} = O(\log n) = O(\log p)$; moreover we will be using $O(\log p)$ different primes. The complexity of Algorithm 2 under these assumptions is given by the following lemma. Note that almost all primes ℓ_i are such that $\sqrt{B} < \ell_i \leq B$ and so $e_i = 1$, hence we ignore the obvious ℓ-adic speedups that can be obtained in the rare cases when ℓ_i is small.

Lemma 4. *Let $n = \prod \ell_i^{e_i}$ with $\log n = O(\log p)$ and $\ell_i^{e_i} = O(\log p)$. Then Algorithm 2 can be implemented to run in time $\tilde{O}(\log^6 p)$ bit operations for the first loop, and $\tilde{O}(\log^4 p)$ for the rest of the algorithm.*

Proof: Without any assumption on p the $\ell_i^{e_i}$ torsion points will generally be defined over $\ell_i^{e_i}$ degree extension fields, hence they will be of $O(\log^2 p)$ size. However the isogenies themselves will be rational, i.e. defined over $\mathbb{F}_{p^2}$. This means their kernel is defined by a polynomial over $\mathbb{F}_{p^2}$. Isogenies over $\mathbb{F}_{p^2}$ of degree d can be evaluated at any point in $\mathbb{F}_{p^2}$ using $O(d)$ field operations in $\mathbb{F}_{p^2}$.

Let $d = \ell_i^{e_i}$. To compute a basis of the d-torsion, we first factor the division polynomial over $\mathbb{F}_{p^2}$. This polynomial has degree $O(d^2) = O(\log(p)^2)$. Using the algorithm in [24] this can be done in $\tilde{O}(\log^4 p)$ bit operations. Since the isogenies are defined over $\mathbb{F}_{p^2}$, this will give factors of degree at most $(d-1)/2$, each one corresponding to a cyclic subgroup. We then randomly choose some factor with a probability proportional to its degree, and we factor it over its splitting field, until we have found a basis of the d-torsion. After $O(1)$ random choices we will have a basis of the d-torsion. Each factorization costs $\tilde{O}(\log^5 p)$ using the algorithm in [38], and verifying that two points generate the d-torsion can be done with $O(d)$ field operations. It then takes $O(d)$ field operations to compute generators for all kernels. As $r = O(\log p)$ we deduce that the first loop requires $\tilde{O}(\log^6 p)$ bit operations.

Computing P_{ijk} involves Frobenius operations and multiplications by scalars bounded by d (and so $O(\log\log p)$ bits). This requires $O(\log p)$ field operations, that is a total of $\tilde{O}(\log^3 p)$ bit operations. Any cyclic subgroup of order $\ell_i^{e_i}$ is generated by a point $Q_i = aP_{i1} + bP_{i2}$, and the image of this point by α_{ik} is $aP_{i1k} + bP_{i2k}$. One can determine the integers a, b by an ECDLP computation or by testing random choices. There are roughly $\ell_i^{e_i} = O(\log p)$ subgroups, and testing each of them requires at most $O(\log\log p)$ field operations, so finding Q_i requires $\tilde{O}(\log p)$ field operations. Evaluating $\varphi_{i-1}(Q_i)$ requires $O(\log^2 p)$ field operations. Computing the isogeny ϕ_i can be done in $O(\log p)$ field operations using Vélu's formulae. As $r = O(\log p)$ we deduce that the second loop requires $\tilde{O}(\log^4 p)$ bit operations. □

We stress that in our signature algorithm, Algorithm 2 will be run $O(\log p)$ times. However the torsion points are independent of both the messages and the keys, so they can be precomputed. Hence the "online" running time of Algorithm 2 is $\tilde{O}(\log^4 p)$ bit operations per execution.

Translating an Isogeny Path to an Ideal. Let $E_0, E_1, \ldots, E_r$ be an isogeny path and suppose $\varphi_i : E_0 \to E_i$ is of degree $n_i = \prod_{j\leq i} \ell_j^{e_j}$. We define $I_0 = \mathcal{O}_0$. Then for $i = 1, \ldots, r$ we compute an element $\alpha_i \in I_{i-1}$ and an ideal $I_i = I_{i-1}\ell_i^{e_i} + I_{i-1}\alpha_i$ that corresponds to the isogeny $\varphi_i = \phi_i \circ \ldots \circ \phi_1$. (We stress that the definition of I_i here differs from the previous subsection.) At step i, we use a basis of I_{i-1} to compute a quadratic form f_i that is the norm form of the ideal I_{i-1}. The roots of this quadratic form modulo $\ell_i^{e_i}$ correspond to candidates for α_i and hence I_i. Note that this correspondence is not injective: a priori there will be $O((\ell_i^{e_i})^3)$ roots but there are only $O(\ell_i^{e_i})$ corresponding ideals including the correct one. Our strategy is to pick random solutions to the quadratic form until the maps α_i and ϕ_i have the same kernels.

Algorithm 3. Translating isogeny path to ideal

Input: $E_0, E_1, \ldots, E_r$ isogeny path, $\phi_i : E_{i-1} \to E_i$ of degree $\ell_i^{e_i}$. **Output:** the ideal path $I_0, \ldots, I_r$ corresponding to the isogeny path.

1: Let $I_0 = \mathcal{O}_0$
2: **for** $i = 1, \ldots, r$ **do**
3: Find Q_i of order $\ell_i^{e_i}$ that generates the kernel of ϕ_i
4: Compute $[\beta](Q_i)$ for all $\beta \in \{1, \mathbf{i}, \frac{\mathbf{i}+\mathbf{j}}{2}, \frac{1+\mathbf{k}}{2}\}$
5: Let $\{\beta_1, \beta_2, \beta_3, \beta_4\}$ a basis of I_{i-1}
6: Let $f_i(w, x, y, z) = n(w\beta_1 + x\beta_2 + y\beta_3 + z\beta_4)$
7: **repeat**
8: Pick a random solution to $f_i(w, x, y, z) = 0 \bmod \ell_i^{e_i}$
9: Set $\alpha_i = w\beta_1 + x\beta_2 + y\beta_3 + z\beta_4$
10: **until** $[\alpha_i](Q_i) = 0$
11: Set $I_i = I_{i-1}\ell_i^{e_i} + I_{i-1}\alpha_i$
12: Perform basis reduction on I_i

In our protocols we will have $\ell_i^{e_i} = O(\log n) = O(\log p)$; moreover we will be using $O(\log p)$ different primes. The complexity of Algorithm 3 under these assumptions is given by the following lemma.

Lemma 5. *Let* $n = \prod_{i=1}^{r} \ell_i^{e_i}$ *with* $\log n = O(\log p)$ *and* $\ell_i^{e_i} = O(\log p)$, *and assume all the isogenies are defined over* $\mathbb{F}_{p^2}$. *Then Algorithm 3 can be implemented to run in expected time* $\tilde{O}(\log^4 p)$ *and the output is a* $\mathbb{Z}$*-basis with integers bounded by* X *such that* $\log X = O(\log p)$.

PROOF: We remind that without any assumption on p the $\ell_i^{e_i}$ torsion points will generally be defined over $\ell_i^{e_i}$ degree extension fields, hence they will be of $O(\log^2 p)$ size. Isogenies of degree d can be evaluated at any point using $O(d)$ field operations.

When the degree is odd the isogeny ϕ_i is naturally given by a polynomial ψ_i such that the roots of ψ_i correspond to the x-coordinates of affine points in $\ker \varphi_i$. To identify a generator Q_i we first factor ψ_i over $\mathbb{F}_{p^2}$. Using the algorithm in [38] this can be done with $\tilde{O}(\log^3 p)$ bit operations. We choose a random irreducible factor with a probability proportional to its degree, we use this polynomial to define a field extension of $\mathbb{F}_{p^2}$, and we check whether the corresponding point is of order $\ell_i^{e_i}$. If not we choose another irreducible factor and we repeat. We expect to only need to repeat this $O(1)$ times, and each step requires $\tilde{O}(\log p)$ bit operations. So the total cost for line 3 is $\tilde{O}(\log^3 p)$.

Step 4 requires $O(\log \log p)$ field operations to compute a point Q_i' such that $[2]Q_i' = Q_i$. After that it mostly requires $O(\log p)$ field operations to compute the Frobenius map. The total cost of this step is therefore $\tilde{O}(\log^3 p)$.

Basis elements for all the ideals I_i appearing in the algorithm can be reduced modulo $\mathcal{O}_0 n$, hence their coefficients are of size $\log n = O(\log p)$.

To compute a random solution to f_i modulo $\ell_i^{e_i}$, we choose uniformly random values for w, x, y, and when the resulting quadratic equation in z has solutions modulo $\ell_i^{e_i}$ we choose a random one. As $\ell_i^{e_i} = O(\log p)$ the cost of this step can be neglected. Computing $[\alpha_i](Q_i)$ requires $O(\log\log p)$ operations over a field of size $O(\log^2 p)$. On average we expect to repeat the loop $O(\ell_i^{e_i}) = O(\log p)$ times, resulting in a total cost of $\tilde{O}(\log^3 p)$. Computing each f_i costs $\tilde{O}(\log p)$ bit operations.

As $r = O(\log p)$ the total cost of the algorithm is $\tilde{O}(\log^4 p)$.

One can check that all integers in the algorithm are bounded in terms of n, and so coefficients are of size X where $\log X = O(\log n) = O(\log p)$. □

Recall that the condition $\log X = O(\log p)$ is needed in Lemma 3.

4 Classical and Post-Quantum Signature Schemes

Digital signatures are one of the most fundamental cryptographic primitives. It is well-known that they can be built from identification protocols using the Fiat-Shamir transform [15]. The resulting signatures are existentially unforgeable under adaptive chosen-message attacks (the standard security definition for signatures) with respect to classical adversaries, in the random oracle model. The transform is also secure against quantum adversaries under certain conditions [34], however these conditions are met by neither De Feo-Jao-Plût's protocol nor ours. In particular, soundness relies on computational assumptions in both protocols. However, one can replace the Fiat-Shamir transform with an alternative transform due to Unruh to achieve security against quantum adversaries [33].

This section explains the two signature schemes obtained from our new identification protocol. Due to lack of space we refer to Yoo et al. [42] and the full version of the paper [18] for the two signature schemes obtained from the De Feo-Jao-Plût ID-scheme.

4.1 Classical Signature Scheme Based on Endomorphism Ring Computation

In this section we fully specify the signature scheme resulting from applying a variant of the Fiat-Shamir transform to our new identification scheme based on the endomorphism ring computation problem, and we analyse its efficiency.

Key Generation Algorithm: On input a security parameter λ generate a prime p with 2λ bits, which is congruent to 3 modulo 4. Let $E_0 : y^2 = x^3 + Ax$ over $\mathbb{F}_p$ be supersingular, and let $\mathcal{O}_0 = \mathrm{End}(E_0)$. Fix B, S_1, S_2 as small as possible[8] such that $S_k := \prod_i \ell_{k,i}^{e_{k,i}}$, $\ell_{k,i}^{e_{k,i}} < B$, $\gcd(S_1, S_2) = 1$, and $\prod \left(\frac{2\sqrt{\ell_{k,i}}}{\ell_{k,i}+1}\right)^{e_{k,i}} < (p^{1+\epsilon})^{-1}$. Perform a random isogeny walk of degree S_1 from the curve E_0 with j-invariant $j_0 = 1728$ to a curve E_1 with j-invariant j_1. Compute $\mathcal{O}_1 = \mathrm{End}(E_1)$ and the ideal I corresponding to this isogeny. Choose a hash function H with $t = 2\lambda$ bits of output. The public key is $\textsc{pk} = (p, j_1, H)$ and the secret key is $\textsc{sk} = \mathcal{O}_1$, or equivalently I.

[8] The exact procedure is irrelevant here.

Signing Algorithm: On input a message m and keys (PK, SK), recover the parameters p and j_1. For $i = 1, \ldots, t$, generate a random isogeny walk w_i of degree S_2, ending at a j-invariant $j_{2,i}$. Compute $h := H(m, j_{2,1}, \ldots, j_{2,t})$ and parse the output as t challenge bits b_i. For $i = 1, \ldots, t$, if $b_i = 1$ use w_i and Algorithm 3 of Sect. 3.4 to compute the corresponding ideal I_i and hence its right order $\mathcal{O}_{2,i} = \text{End}(E_{2,i})$, then use the algorithm of Sect. 3.3 on input II_i to compute a "fresh" path between $\mathcal{O}_0$ and $\mathcal{O}_{2,i}$, and finally use Algorithm 2 to compute an isogeny path w_i' from j_0 to $j_{2,i}$. If $b_i = 0$ set $z_i := w_i$, otherwise set $z_i := w_i'$. Return the signature $\sigma = (h, z_1, \ldots, z_t)$.

Verification Algorithm: On input a message m, a signature σ and a public key PK, recover the parameters p and j_1. For each $1 \leq i \leq t$ one uses z_i to compute the image curve $E_{2,i}$ of the isogeny. Hence the verifier recovers the signature components $j_{2,i}$ for $1 \leq i \leq t$. The verifier then recomputes the hash $H(m, j_{2,1}, \ldots, j_{2,t})$ and checks that the value is equal to h, accepting the signature if this is the case and rejecting otherwise.

We now show that this scheme is a secure signature.

Theorem 4. *If Problem 6 is computationally hard then the signature scheme is secure in the random oracle model under a chosen message attack.*

PROOF: As shown in Sect. 3.2, if Problem 6 is computationally hard then the identification scheme (sigma protocol) has 2-special soundness and honest verifier zero-knowledge. Theorem 2 therefore implies that the identification scheme is secure against impersonation under passive attacks. It follows from Abdalla et al. [1] that the signature scheme is secure in the random oracle model. □

Efficiency: As the best classical algorithm for computing the endomorphism ring of a supersingular elliptic curve runs in time $\tilde{O}(\sqrt{p})$ one can take $\log p = 2\lambda$. By Theorem 1 and Lemma 1, taking $B \approx 2(1+\epsilon)\log p$ ensures that the outputs of random walks are distributed uniformly enough. Random walks then require $2(1+\epsilon)\log p$ bits to represent, so signatures are

$$t + \frac{t}{2}\left(2(1+\epsilon)\lceil\log p\rceil + \frac{7}{2}\lceil\log p\rceil\right)$$

bits on average, depending on the challenge bits. For λ bits of security, we choose $t = 2\lambda$, so the average signature length is approximately $2\lambda + (\lambda)(4(1+\epsilon)\lambda + 7\lambda) \approx (11 + 4\epsilon)\lambda^2 \approx 11\lambda^2$.

Private keys are $2(1+\epsilon)\log p \approx 4\lambda$ bits if a canonical representation of the kernel of the isogeny between E_0 and E_1 is stored. This can be reduced to 2λ bits for generic E_1: if I is the ideal corresponding to this isogeny, it is sufficient to store another ideal J in the same class, and for generic E_1 there exists one ideal of norm $n \approx \sqrt{p}$. To represent this ideal in the most efficient way, it is sufficient to give n and a second integer defining the localization of I at every prime factor ℓ of n, for canonical embeddings of $B_{p,\infty}$ into $M_2(\mathbb{Q}_\ell)$. This reduces storage costs to roughly 2λ bits. Public keys are $3\log p = 6\lambda$ bits. A signature mostly requires t calls to the Algorithms of Sects. 3.3 and 3.4, for a total cost of $\tilde{O}(\lambda^5)$. Verification requires to check $O(\lambda)$ isogeny walks, each one comprising $O(\lambda)$ steps with a cost $\tilde{O}(\lambda^3)$ each when modular polynomials are precomputed, hence a total cost of $\tilde{O}(\lambda^5)$ bit operations (under the same heuristic assumptions as in Lemma 3).

Optimization with Non Backtracking Walks: In our description of the signature scheme we have allowed isogeny paths to "backtrack". We made this choice to simplify the convergence analysis of random walks and because it does not affect the asymptotic complexity of our schemes significantly. However in practice at any concrete security parameter, it will be better to use non-backtracking random walks as they will converge more quickly to a uniform distribution [2].

4.2 Post-Quantum Signature Scheme Based on Endomorphism Ring Computation

We briefly describe the signature scheme arising from applying Unruh's transform to the identification protocol of Sect. 3.

Key Generation Algorithm: On input a security parameter λ generate a prime p with 4λ bits, which is congruent to 3 modulo 4. Let $E_0 : y^2 = x^3 + Ax$ over $\mathbb{F}_p$ be supersingular, and let $\mathcal{O}_0 = \text{End}(E_0)$. Set $t = 3\lambda$. Fix B, S_1, S_2 as in the key generation algorithm of Sect. 4.1. Perform a random isogeny walk of degree S_1 from the curve E_0 with j-invariant $j_0 = 1728$ to a curve E_1 with j-invariant j_1. Compute $\mathcal{O}_1 = \text{End}(E_1)$ and the ideal I corresponding to this isogeny.

Choose a hash function $H : \{0,1\}^* \to \{0,1\}^t$. Let $N \approx \frac{7}{2} \log p$ be an upper bound for the bitlength of the representation of any isogeny path in the algorithm. Let $G : \{0,1\}^N \to \{0,1\}^N$ be a hash function such that every element has polynomially many preimages. The public key is $\text{PK} = (p, j_1, H, G)$ and the secret key is $\text{SK} = \mathcal{O}_1$, or equivalently I.

Signing Algorithm: On input a message m and keys (PK, SK), recover the parameters p and j_1. For $i = 1, \ldots, t$ generate a random isogeny walk w_i of degree S_2, ending at a j-invariant $j_{2,i}$.

For $i = 1, \ldots, t$ apply Algorithm 3 of Sect. 3.4 to compute the ideal I_i corresponding to the isogeny path w_i, then use the algorithm of Sect. 3.3 on input II_i to compute a "fresh" ideal corresponding to a path between $\mathcal{O}_0$ and $\mathcal{O}_{2,i}$, and finally use Algorithm 2 to compute an isogeny path w_i' from j_0 to $j_{2,i}$.

Compute $g_{i,0} = G(w_i)$ and $g_{i,1} = G(w_i')$ for $1 \leq i \leq t$, where the bitstrings w_i and w_i' are padded with zeroes to become binary strings of length N. Compute $h := H(m, j_1, j_{2,1}, \ldots, j_{2,t}, g_{1,0}, g_{1,1}, \ldots, g_{t,0}, g_{t,1})$ and parse the output as t challenge bits h_i. For $i = 1, \ldots, t$, if $h_i = 0$ then set $\text{RSP}_i = w_i$ and if $h_i = 1$ then set $\text{RSP}_i = w_i'$. Return the signature $\sigma = (h, \text{RSP}_1, \ldots, \text{RSP}_t, g_{1,1-h_1}, \ldots, g_{t,1-h_t})$.

Verification Algorithm: On input a message m, a signature σ and a public key PK, recover the parameters p and j_1.

For each $1 \le i \le t$ one uses RSP_i to compute the image curve $E_{2,i}$ of the isogeny (if $h_i = 0$ then RSP_i is a path from E_1 and if $h_i = 1$ then it is a path from E_0). Hence the verifier recovers the j-invariants $j_{2,i}$ for $1 \le i \le t$.

The verifier then computes $g_{i,h_i} = G(\text{RSP}_i)$ for $1 \le i \le t$ (again padding to N bits using zeros). Finally the verifier computes the hash value

$$h' = H(m, j_1, j_{2,1}, \ldots, j_{2,t}, g_{1,0}, g_{1,1}, \ldots, g_{t,0}, g_{t,1}).$$

If $h' = h$ then the verifier accepts the signature and otherwise rejects.

We now show that this scheme is a secure signature.

Theorem 5. *If Problem 6 is computationally hard then the signature scheme is secure in the quantum random oracle model under a chosen message attack.*

PROOF: As shown in Sect. 3.2, if Problem 6 is computationally hard then the identification scheme (sigma protocol) has 2-special soundness and honest verifier zero-knowledge. A result of Unruh [33] then implies that the signature scheme is secure in the quantum random oracle model. □

Efficiency: For the same reasons as in the application of the Unruh transform applied to the De Feo-Jao-Plût scheme, this signature scheme is less efficient than its classical counterpart. Again, we only send half the values $g_{i,j}$, since the missing values can be recomputed by the signer.

The average signature size is $t + t((\log S_1 + N)/2) + tN$, on the basis that half the responses RSP_i can be represented using $\log S_1$ bits and half of them require N bits. For λ bits of security, we choose $\log p = 4\lambda$ and $t = 3\lambda$, so that $N = 14\lambda$ and $\log S_1 = (8 + \epsilon)\lambda$. Then the average signature size is approximately $75\lambda^2$.

4.3 Comparison

Tables 1 and 2 summarize the main efficiency features of four signature schemes based either on De Feo-Jao-Plût or on our new identification scheme, and on Fiat-Shamir or Unruh transforms. The numbers provided were obtained by optimizing signature sizes first, then signing and verification time and finally key sizes; other trade-offs are of course possible. The scheme based on the De Feo-Jao-Plût identification protocol and Unruh transform was discovered independently in [42]; the version we give incorporates optimizations that reduce the signature sizes for the same security guarantees[9]. Signatures based on De Feo-Jao-Plût identification protocol are simpler and somewhat more efficient than signatures based on our new identification protocol; however the latter have the advantage to rely on more standard and potentially harder computational problems. Schemes based on the Fiat-Shamir transform are more efficient than schemes based on Unruh's transform; however the latter provide security guarantees against quantum adversaries.

[9] Both signature sizes depend linearly on a parameter t which we fixed in a more conservative manner than Yoo et al. (see the full version of the paper for a discussion on this choice). With $t = 2\lambda$ their signatures are $69\lambda^2$ bits and ours are $48\lambda^2$ bits, and with $t = 3\lambda$ their signatures are $\lceil 103.5\lambda^2 \rceil$ bits and ours are $72\lambda^2$ bits. Tables 1 and 2 report values for $t = 3\lambda$.

Table 1. Asymptotic efficiency of four signature schemes using De Feo-Jao-Plût and our identification protocol, and Fiat-Shamir and Unruh transform, as a function of the security parameter λ. All sizes are in bits and computation costs are in bit operations.

	Private key size	Public key size	Signature size	Signing costs	Verification costs
DFJP + FS	2λ	28λ	$12\lambda^2$	$\tilde{O}(\lambda^3)$	$\tilde{O}(\lambda^3)$
Sec 3 + FS	2λ	6λ	$11\lambda^2$	$\tilde{O}(\lambda^5)$	$\tilde{O}(\lambda^5)$
DFJP + U	3λ	42λ	$72\lambda^2$	$\tilde{O}(\lambda^3)$	$\tilde{O}(\lambda^3)$
Sec 3 + U	4λ	12λ	$75\lambda^2$	$\tilde{O}(\lambda^5)$	$\tilde{O}(\lambda^5)$

Table 2. Concrete efficiency of our signature schemes at security levels of 128 and 256 bits. Security level provided are against classical or quantum adversaries for schemes based on the Fiat-Shamir or Unruh transforms respectively. All sizes are in bits.

	128 bit			256 bit		
	Private key	Public key	Signature	Private key	Public key	Signature
DFJP + FS	256	3584	196608	512	7168	786432
Sec 3 + FS	256	768	180224	512	1536	720896
DFJP + U	384	5376	1179648	768	10752	4718592
Sec 3 + U	512	1536	1228800	1024	3072	4915200

Table 1 and a quick comparison with RSA signatures suggest that isogeny-based signatures schemes may be efficiency enough for practical use. Indeed for RSA signatures, key sizes are cubic in the security parameter, and signing and verification times are respectively quasi-quadratic and quasi-linear in the key sizes (the latter assuming a small public key exponent is used), amounting to $\tilde{O}(\lambda^3)$ and $\tilde{O}(\lambda^6)$. As for concrete parameters, key sizes are much smaller for isogeny-based signatures than for RSA signatures and comparable to ECDSA signatures. Further work in this area should aim at decreasing signature sizes.

5 Conclusion

We provided both a new identification protocol and new signature schemes based on isogeny problems. While the only previous identification protocol based on isogeny problems relied on special and potentially easier variants of these problems [14], our protocol is based on what is arguably the hardest problem in this area, namely the endomorphism ring computation problem. A crucial ingredient for our protocol is the quaternion isogeny algorithm of Kohel-Lauter-Petit-Tignol [26] in the powersmooth case, for which we provide a more complete description and analysis. The signature schemes are derived using the Fiat-Shamir and Unruh transforms, respectively for classical and post-quantum security. We showed that they can have very small key sizes and reasonably efficient signing and verification algorithms compared to RSA signatures.

Isogeny problems are interesting in cryptography for their potential resistance to quantum algorithms, but they are also rather new in cryptography. Among all isogeny problems, the problem of computing the endomorphism ring of a supersingular elliptic curve is the most natural one to consider from an algorithmic number theory point of view, and it has in fact been studied since Kohel's PhD thesis in 1996. Yet, even this problem is far from having received the same scrutiny as more established cryptography problems like discrete logarithms or integer factoring. We hope that this paper will encourage the community to study its complexity.

Acknowledgement. We thank Dominique Unruh, David Pointcheval and Ali El Kaafarani for discussions related to this paper. Research from the second author was supported by a research grant from the UK government.

A Efficient Representations of Isogeny Paths and Other Data

Our schemes require representing/transmitting elliptic curves and isogenies. In this section we briefly explain how to represent certain mathematical objects appearing in our protocol as bitstrings in a canonical way so that minimal data needs to be sent and stored.

Let p be a prime number. Every supersingular j-invariant is defined over $\mathbb{F}_{p^2}$. A canonical representation of $\mathbb{F}_{p^2}$-elements is obtained via a canonical choice of degree 2 irreducible polynomial over $\mathbb{F}_p$. Canonical representations in any other extension fields are defined in a similar way. Although there are only about $p/12$ supersingular j-invariants in characteristic p, we are not aware of an efficient method to encode these invariants into $\log p$ bits, so we represent supersingular j-invariants with the $2\log p$ bits it takes to represent an arbitrary $\mathbb{F}_{p^2}$-element.

Elliptic curves are defined by their j-invariant up to isomorphism. Hence, rather than sending the coefficients of the elliptic curve equation, it suffices to send the j-invariant. For any invariant j there is a canonical elliptic curve equation $E_j : y^2 = x^3 + \frac{3j}{1728-j}x + \frac{2j}{1728-j}$ when $j \neq 0, 1728$, $y^2 = x^3 + 1$ when $j = 0$, and $y^2 = x^3 + x$ when $j = 1728$. The last one is used in our main signature scheme.

We now turn to representing chains $E_0, E_1, \ldots, E_n$ of isogenies $\phi_i : E_{i-1} \to E_i$ each of prime degree ℓ_i where $1 \leq i \leq n$. Here ℓ_i are always very small primes. A useful feature of our protocols is that isogeny chains can always be chosen such that the isogeny degrees are increasing $\ell_i \geq \ell_{i-1}$. First we need to discuss how to represent the sequence of isogeny degrees. If all degrees are equal to a fixed constant ℓ (e.g., $\ell = 2$) and if the length n is a fixed system parameter, then there is nothing to send. If the degrees are different then the most compact representation seems to be to compute and send

$$N = \prod_{i=1}^{n} \ell_i.$$

The receiver can recover the sequence of isogeny degrees from N by factoring using trial division and ordering the primes by size. This representation is possible due to our convention that the isogeny degrees are increasing and since the degrees are all small.

To represent the curves themselves in the chain of isogenies the simplest method is to send all the j-invariants $j_i = j(E_i) \in \mathbb{F}_{p^2}$ for $0 \leq i \leq n$. This requires $2(n+1)\log_2(p)$ bits. Note that the verifier is able to check the correctness of the isogeny chain by checking that $\Phi_{\ell_i}(j_{i-1}, j_i) = 0$ for all $1 \leq i \leq n$, where Φ_{ℓ_i} is the ℓ_i-th modular polynomial. The advantage of this method is that verification is relatively quick (just evaluating a polynomial that can be precomputed and stored).

The other naive method is to send the x-coordinate of a kernel point $P_i \in E_{j_i}$ on the canonical curve. This requires large bandwidth.

A refinement of the second method is used in our signature scheme based on the De Feo-Jao-Plût identification protocol, where ℓ is fixed and one can publish a point that defines the kernel of the entire isogeny chain. Precisely a curve E and points $R, S \in E[\ell^n]$ are fixed. Each integer $0 \leq \alpha < \ell^n$ defines a subgroup $\langle R + [\alpha]S \rangle$ and hence an ℓ^n isogeny. It suffices to send α, which requires $\log_2(\ell^n)$ bits. In the case $\ell = 2$ this is just n bits, which is smaller than all the other suggestions in this section.

The full version of the paper contains a more thorough discussion of optimisations and also an analysis of the complexity of computing isogeny chains.

References

1. Abdalla, M., An, J.H., Bellare, M., Namprempre, C.: From identification to signatures via the fiat-shamir transform: minimizing assumptions for security and forward-security. In: Knudsen, L.R. (ed.) EUROCRYPT 2002. LNCS, vol. 2332, pp. 418–433. Springer, Heidelberg (2002). https://doi.org/10.1007/3-540-46035-7_28
2. Alon, N., Benjamini, I., Lubetzky, E., Sodin, S.: Non-backtracking random walks mix faster. Commun. Contemp. Math. **9**(4), 585–603 (2007)
3. Bellare, M., Poettering, B., Stebila, D.: From identification to signatures, tightly: a framework and generic transforms. In: Cheon, J.H., Takagi, T. (eds.) ASIACRYPT 2016. LNCS, vol. 10032, pp. 435–464. Springer, Heidelberg (2016). https://doi.org/10.1007/978-3-662-53890-6_15
4. Biasse, J.-F., Jao, D., Sankar, A.: A quantum algorithm for computing isogenies between supersingular elliptic curves. In: Meier, W., Mukhopadhyay, D. (eds.) INDOCRYPT 2014. LNCS, vol. 8885, pp. 428–442. Springer, Cham (2014). https://doi.org/10.1007/978-3-319-13039-2_25
5. Bisson, G., Sutherland, A.V.: Computing the endomorphism ring of an ordinary elliptic curve over a finite field. J. Number Theory **131**(5), 815–831 (2011)
6. Charles, D.X., Lauter, K.E., Goren, E.Z.: Cryptographic hash functions from expander graphs. J. Cryptol. **22**(1), 93–113 (2009)
7. Childs, A.M., Jao, D., Soukharev, V.: Constructing elliptic curve isogenies in quantum subexponential time. J. Math. Cryptol. **8**(1), 1–29 (2014)
8. Costello, C., Longa, P., Naehrig, M.: Efficient algorithms for supersingular isogeny Diffie-Hellman. In: Robshaw, M., Katz, J. (eds.) CRYPTO 2016. LNCS, vol. 9814, pp. 572–601. Springer, Heidelberg (2016). https://doi.org/10.1007/978-3-662-53018-4_21

9. Damgård, I.: On σ-protocols. University of Aarhus, Department for Computer Science, Lecture Notes (2010)
10. Deligne, P.: La conjecture de Weil. I. Publications Mathématiques de l'Institut des Hautes Études Scientifiques **43**(1), 273–307 (1974)
11. Deuring, M.: Die Typen der Multiplikatorenringe elliptischer Funktionenkörper. Abhandlungen aus dem Mathematischen Seminar der Universität Hamburg **14**, 197–272 (1941). https://doi.org/10.1007/BF02940746
12. Dewaghe, L.: Isogénie entre courbes elliptiques. Util. Math. **55**, 123–127 (1999)
13. Feige, U., Fiat, A., Shamir, A.: Zero-knowledge proofs of identity. J. Cryptol. **1**(2), 77–94 (1988)
14. De Feo, L., Jao, D., Plût, J.: Towards quantum-resistant cryptosystems from supersingular elliptic curve isogenies. J. Math. Cryptol. **8**(3), 209–247 (2014)
15. Fiat, A., Shamir, A.: How to prove yourself: practical solutions to identification and signature problems. In: Odlyzko, A.M. (ed.) CRYPTO 1986. LNCS, vol. 263, pp. 186–194. Springer, Heidelberg (1987). https://doi.org/10.1007/3-540-47721-7_12
16. Galbraith, S.D.: Constructing isogenies between elliptic curves over finite fields. LMS J. Comput. Math **2**, 118–138 (1999)
17. Galbraith, S.D., Petit, C., Shani, B., Ti, Y.B.: On the security of supersingular isogeny cryptosystems. In: Cheon, J.H., Takagi, T. (eds.) ASIACRYPT 2016. LNCS, vol. 10031, pp. 63–91. Springer, Heidelberg (2016). https://doi.org/10.1007/978-3-662-53887-6_3
18. Galbraith, S.D., Petit, C., Silva, J.: Signature schemes based on supersingular isogeny problems. Cryptology ePrint Archive, Report 2016/1154 (2016). http://eprint.iacr.org/2016/1154
19. Gélin, A., Wesolowski, B.: Loop-abort faults on supersingular isogeny cryptosystems. In: Lange, T., Takagi, T. (eds.) PQCrypto 2017. LNCS, vol. 10346, pp. 93–106. Springer, Cham (2017). https://doi.org/10.1007/978-3-319-59879-6_6
20. Hoory, S., Linial, N., Wigderson, A.: Expander graphs and their applications. Bull. Amer. Math. Soc. **43**, 439–561 (2006)
21. Jao, D., De Feo, L.: Towards quantum-resistant cryptosystems from supersingular elliptic curve isogenies. In: Yang, B.-Y. (ed.) PQCrypto 2011. LNCS, vol. 7071, pp. 19–34. Springer, Heidelberg (2011). https://doi.org/10.1007/978-3-642-25405-5_2
22. Jao, D., Soukharev, V.: Isogeny-based quantum-resistant undeniable signatures. In: Mosca, M. (ed.) PQCrypto 2014. LNCS, vol. 8772, pp. 160–179. Springer, Cham (2014). https://doi.org/10.1007/978-3-319-11659-4_10
23. Katz, J.: Digital Signatures. Springer, Heidelberg (2010)
24. Kedlaya, K.S., Umans, C.: Fast polynomial factorization and modular composition. SIAM J. Comput. **40**(6), 1767–1802 (2011)
25. Kohel, D.: Endomorphism rings of elliptic curves over finite fields. Ph.D. thesis, University of California, Berkeley (1996)
26. Kohel, D., Lauter, K., Petit, C., Tignol, J.-P.: On the quaternion ℓ-isogeny path problem. LMS J. Comput. Math. **17A**, 418–432 (2014)
27. Nguyen, P.Q., Stehlé, D.: Low-dimensional lattice basis reduction revisited. ACM Trans. Algorithms **5**(4), 46 (2009)
28. Petit, C.: On the quaternion ℓ-isogeny problem. Presentation slides from a talk at the University of Neuchâtel, March 2015
29. Petit, C.: Faster algorithms for isogeny problems using torsion point images. In: ASIACRYPT 2017 (2017, to appear). http://eprint.iacr.org/2017/571
30. Pizer, A.K.: Ramanujan graphs and Hecke operators. Bull. Am. Math. Soc. **23**(1), 127–137 (1990)

31. Silverman, J.H.: The Arithmetic of Elliptic Curves. Springer, Heidelberg (1986)
32. Ti, Y.B.: Fault attack on supersingular isogeny cryptosystems. In: Lange, T., Takagi, T. (eds.) PQCrypto 2017. LNCS, vol. 10346, pp. 107–122. Springer, Cham (2017). https://doi.org/10.1007/978-3-319-59879-6_7
33. Unruh, D.: Non-interactive zero-knowledge proofs in the quantum random oracle model. In: Oswald, E., Fischlin, M. (eds.) EUROCRYPT 2015. LNCS, vol. 9057, pp. 755–784. Springer, Heidelberg (2015). https://doi.org/10.1007/978-3-662-46803-6_25
34. Unruh, D.: Post-quantum security of Fiat-Shamir. In: ASIACRYPT 2017 (2017, to appear). https://eprint.iacr.org/2017/398
35. Venturi, D.: Zero-knowledge proofs and applications. University of Rome, Lecture Notes (2015)
36. Vignéras, M.-F.: Arithmétique des algébres de quaternions. Springer, Heidelberg (1980)
37. Voight, J.: Quaternion algebras (2017). https://math.dartmouth.edu/~jvoight/quat-book.pdf
38. von zur Gathen, J., Shoup, V.: Computing Frobenius maps and factoring polynomials. Comput. Complex. **2**, 187–224 (1992)
39. Vélu, J.: Isogénies entre courbes elliptiques. Commun. de l'Académie royale des Sci. de Paris **273**, 238–241 (1971)
40. Waterhouse, W.C.: Abelian varieties over finite fields. Ann. scientifiques de l'ENS **2**, 521–560 (1969)
41. Xi, S., Tian, H., Wang, Y.: Toward quantum-resistant strong designated verifier signature from isogenies. Int. J. Grid Util. Comput. **5**(2), 292–296 (2012)
42. Yoo, Y., Azarderakhsh, R., Jalali, A., Jao, D., Soukharev, V.: A post-quantum digital signature scheme based on supersingular isogenies. In: Financial Crypto 2017 (2017)

Post-Quantum Cryptography

An Existential Unforgeable Signature Scheme Based on Multivariate Quadratic Equations

Kyung-Ah Shim(✉), Cheol-Min Park, and Namhun Koo

Division of Integrated Mathematics, National Institute for Mathematical Sciences, Daejeon, Republic of Korea
{kashim,mpcm,nhkoo}@nims.re.kr

Abstract. A multivariate quadratic public-key cryptography (MQ-PKC) is one of the most promising alternatives for classical PKC after the eventual coming of a quantum computer. We propose a new MQ-signature scheme, ELSA, based on a hidden layer of quadratic equations which is an important role in dramatically reducing the secret key size and computational complexity in signing. We prove existential unforgeability of our scheme against an adaptive chosen-message attack under the hardness of the MQ-problem induced by a public key of ELSA with a specific parameter set in the random oracle model. We analyze the security of ELSA against known attacks and derive a concrete parameter based on the security analysis. Performance of ELSA on a recent Intel processor is the fastest among state-of-the-art signature schemes including classical ones and Post-Quantum ones. It takes 6.3 μs and 13.39 μs for signing and verification, respectively. Compared to Rainbow, the secret size of the new scheme has reduced by a factor of 88% maintaining the same public key size.

Keywords: Isomorphism of polynomials problem · Direct attack · Existential unforgeability · Key recovery attack · Multivariate-quadratic problem

1 Introduction

Online banking, e-commerce, mobile communication, and cloud computing depend fundamentally on the security of the underlying cryptographic algorithms. Public-key cryptography (PKC) is particularly crucial since they provide digital signatures and establish secure communication without requiring in-person meetings. In 1996, Shor [49] proposed a quantum algorithm that solves the integer factorization problem and the discrete logarithm problem in finite fields and on elliptic curves in polynomial time. Thus, the existence of a sufficiently large quantum computer would be a real-world threat to break RSA, Diffie-Hellman key exchange, DSA and ECDSA the most widely used PKC in practice. There are four well-known classes of cryptographic primitives that are believed to remain secure in the presence of a quantum computer:

T. Takagi and T. Peyrin (Eds.): ASIACRYPT 2017, Part I, LNCS 10624, pp. 37–64, 2017.
https://doi.org/10.1007/978-3-319-70694-8_2

code-based cryptography (McEliece encryption [37]), lattice-based cryptography (NTRU [30]), hash-based cryptography (Merkle's hash-tree signatures [38]), and multivariate quadratic (MQ) cryptography (HFEv- [40], UOV [33]). These cryptographic primitives have been resist classical and quantum cryptanalysis which has inspired widespread confidence in their suitability as a post-quantum primitive.

MQ-PKC is based on the hardness of solving large systems of multivariate quadratic equations, called MQ-problem which is known to be NP-complete. To construct MQ-PKC, it needs a way to hide a trapdoor. In MQ-PKC, a public key is a system of multivariate quadratic polynomials and a trapdoor is hidden in secret affine layers using the ASA (affine-substitution-affine) structure. A long-standing challenge is to design PKC based on symmetric cipher components which are similar to those used in mainstream block ciphers such as AES. Solving this appealing but difficult challenge would not only increase the diversity in PKC, but might also help reducing the considerable performance gap between PKC and symmetric cryptography. One of the directions was to design public-key schemes from symmetric components. A typical symmetric cipher is built from layers of affine transformations (A) and S-boxes (S). This has been the mainstream of MQ-PKC. The security of the ASA structure relies on the hardness of the isomorphism-of-polynomials (IP) problem [40].

Several new ideas to build MQ-schemes from symmetric cipher components were recently introduced by Biryukov *et al.* [10] at Asiacrypt 2014. They used the so-called ASASA structure: combining two quadratic mappings S by interleaving random affine layers A. With quadratic S layers, the overall scheme has degree 4, so the polynomial description provided by the public key remains of reasonable size. This is very similar to the 2R scheme by Patarin and Goubin [43], which is broken by several attacks [8,18], including a powerful decomposition attack [25]. At Crypto 2015 and Asiacrypt 2015, Biryukov *et al.*'s two public-key encryption schemes are broken by key recovery attacks [27,39].

Since the first MQ-encryption scheme was proposed by Imai and Matsumoto [36], a number of MQ-schemes in this $\mathrm{MQ}+\mathrm{IP}$ paradigm have been proposed, i.e., these MQ-schemes are not solely based on the MQ-Problem, but also on some variants of the IP problem. Most of the MQ-schemes have been broken due to the uncertainty of the IP problem. There are only two exceptions from the MQ-IP paradigm: HFEv- variants [42,45] and Unbalanced Oil-and-Vinegar (UOV) variants [16,33] as signature schemes. MQ-schemes require simplicity of operations (matrices and vectors) and small fields avoid multiple-precision arithmetic. So, they require only modest computational resources, which makes them attractive for the use on low cost devices such as smart cards [11,12]. In particular, MQ-signature schemes in the $\mathrm{MQ}+\mathrm{IP}$ paradigm are superior to other competitors in terms of performance and signature size. Despite these advantages, MQ-schemes in the $\mathrm{MQ}+\mathrm{IP}$ paradigm has two main problems: (i) it has relatively large key sizes and (ii) all the schemes in the $\mathrm{MQ}+\mathrm{IP}$ paradigm have been proposed with actual parameters for practical, but they have no security reduction to the hardness of the MQ-problem. The reason for this is that they require a hidden structure which relies on the hardness of the IP problem. Moreover, cryptanalysis results of many MQ-schemes have shown that the IP-problem relies on the MinRank problem [14,24].

In the last years, a few researchers started designing provably secure MQ-schemes based on the hardness of random instances of the MQ-problem. At PKC 2012, Huang *et al.* [31] proposed a public-key encryption scheme with a security reduction to the hardness of solving a set of quadratic equations whose coefficients of highest degree are chosen according to a discrete Gaussian distributions. The other terms are chosen uniformly at random. Such a problem is a variant of the classical problem of solving a system of non-linear equations (PoSSo, PoSSo problem with degree 2 equations is the MQ-problem), which is known to be hard for random systems. They claimed that their variant is not easier than solving the PoSSo problem for random instances. At PKC 2014, Albrecht *et al.* [2] showed that Huang *et al.*'s new problem is reduced to an easy instance of the Learning With Errors problem. They concluded that one cannot find parameters for a secure and practical scheme: a public-key of at least 1.03 GB is required to achieve 80-bit security against the simplest of their attacks.

Another approach is to construction of an MQ-signature scheme from an identification scheme (IDS) based on the MQ-problem via the Fiat-Shamir transform. The resulting scheme [1] obtained from Sakumoto *et al.*'s IDS based on the MQ-problem [47] via the Fiat-Shamir transform is the first provably secure MQ-signature scheme, which solely relies on the MQ-problem. Recently, Chen *et al.* [13] implemented the resulting signature scheme, MQDSS in [1]. MQDSS solves the problem of large key sizes of MQ-PKC by removing the dependence of the IP-problem, but loses the most significant advantages of MQ-ones, fast performance and short signature size. Like this, the history of the design of public-key schemes show that the stronger security arguments the larger performance gap. Therefore, it still remains an open problem to design a practical MQ-signature scheme with a security reduction to the MQ-problem.

For most practical purposes, one still requires a signature scheme that is sufficiently fast and has a short signature size. There have been several attempts to design MQ-signature schemes with higher performance. Gligoroski *et al.* [28] proposed an MQ-signature scheme, MQQ-SIG, based on multivariate quadratic quasigroups (MQQ). MQQ-SIG is the shortest secret key among MQ-ones and the fastest in signing among known signature schemes, but it requires a huge public key which is about 5.7 times and 12,336 times larger than that of Rainbow and ECDSA, respectively. At PKC 2015, Faugère *et al.* [23] mounted polynomial-time key-recovery attacks on all known constructions based on MQQ. They broke an MQQ-SIG instance of an 80-bit security level in less than 2 days. An enhanced version of the Tame Triangular System scheme (enTTS) [15,52] uses very sparse polynomials which make enTTS very efficient in terms of secret key size and signing time, but its public key size is much bigger than other MQ-ones. In this paper, we provide a solution to the two problems of MQ-schemes in the MQ + IP paradigm by proposing a existential unforgeable MQ-signature scheme with a highly optimized practicability for both performance and signature size.

Our Contributions. We propose a new MQ-signature scheme, ELSA, with faster performance and shorter secret key.

- **A New Signature Scheme.** Our signature scheme is based on a hidden layer of quadratic equations. This method makes it possible to remove the use of the Gaussian elimination by reducing the complexity of signing from $\mathcal{O}(n^3)$ to $\mathcal{O}(n^2)$. It plays an important role in dramatically reducing the secret key size and computational cost in signing.
- **High Speed for Both Signing and Verification.** Our scheme is the fastest public-key signature scheme for both signing and verification among the state-of-the-art signature schemes including classical ones and Post-Quantum ones. We implement our scheme for a secure and optimal parameter at a 128-bit security level. Signing of ELSA is about 3.2 times and hundreds of times faster than that of Rainbow and MQDSS, respectively. Also, signing and verification of ELSA is about 17.2 times and 2.3 times faster than those of BLISS-BI, respectively, and signature size of BLISS-BI is about 8.9 times larger than that of ELSA, where BLISS-BI is currently the most efficient lattice-based signature scheme.
- **Shorter Secret Key Size.** Compared to Rainbow, the secret key size of ELSA has reduced by a factor of 88% maintaining the same public key size. Compared to enTTS, the public key size of ELSA have reduced by a factor of 40%.
- **Existential Unforgeability.** We prove existential unforgeability of ELSA against an adaptive chosen-message attack under the hardness of the MQ-problem induced by a public key of ELSA with a specific parameter set in the random oracle model.

Organization. The rest of the paper is organized as follows. In Sect. 2, we propose a new MQ-signature scheme, ELSA. In Sect. 3, we analyze the security of our scheme against all known attacks. In Sect. 4, we give a security proof of ELSA under the hardness of the MQ-problem in the random oracle model. We evaluate performance of our scheme for a secure and optimal parameter at the 128-bit security level and compare it to the state-of-the-art signature schemes in Sect. 5. We conclude in Sect. 6.

2 A New MQ-Signature Scheme

Here, we propose a new MQ-signature scheme based on a hidden layer of quadratic equations.

Let $\mathbb{F}_q$ be a finite field with elements q. A multivariate quadratic system $\mathcal{P} = (\mathcal{P}^{(1)}, \cdots, \mathcal{P}^{(m)})$ of m equations in n variables is defined by

$$\mathcal{P}^{(k)}(x_1, \cdots, x_n) = \sum_{i=1}^{n} \sum_{j=i}^{n} p_{ij}^{(k)} x_i x_j + \sum_{i=1}^{n} p_i^{(k)} x_i + p_0^{(k)},$$

for $k = 1, \cdots, m$, and $p_{ij}^{(k)}, p_i^{(k)}, p_0^{(k)} \in_R \mathbb{F}_q$. The main idea for the construction of MQ-signature schemes is to choose a system $\mathcal{F} : \mathbb{F}_q^n \rightarrow \mathbb{F}_q^m$ of m quadratic polynomials in n variables which can be easily inverted. We call $\mathcal{F}$ a central map. After that one chooses two affine or linear invertible maps S: $\mathbb{F}_q^m \to \mathbb{F}_q^m$ and T: $\mathbb{F}_q^n \to \mathbb{F}_q^n$ to hide the structure of the central map $\mathcal{F}$ in the public key. A public key is the composed quadratic map $\mathcal{P} = S \circ \mathcal{F} \circ T$ which is supposed to be hardly distinguishable from a random system and therefore be difficult to invert. The secret key consists of $(S, \mathcal{F}, T)$ which allows to invert $\mathcal{P}$.

2.1 Our Construction

To construct a new central map for an MQ-signature scheme, we need to define the following four index sets as

$$L = \{1, \cdots, l\},\ K = \{l+1, \cdots, l+k\},\ R = \{l+k+1, \cdots, l+k+r\},$$
$$U = \{l+k+r+1, \cdots, l+k+r+u\},$$

where $|L| = l, |K| = k, |R| = r$, and $|U| = u$. A central map is a multivariate quadratic system $\mathcal{F} = (\mathcal{F}^{(1)}, \cdots, \mathcal{F}^{(m)})$ of m equations and n variables defined by

$$\begin{cases} \mathcal{F}^{(1)}(\mathbf{x}) = L_1(\mathbf{x_{L+K+R}})R_{11}(\mathbf{x_{L+K}}) + \cdots + L_r(\mathbf{x_{L+K+R}})R_{1r}(\mathbf{x_{L+K}}) + \Phi_1(\mathbf{x_L}), \\ \quad\vdots \\ \mathcal{F}^{(k)}(\mathbf{x}) = L_1(\mathbf{x_{L+K+R}})R_{k1}(\mathbf{x_{L+K}}) + \cdots + L_r(\mathbf{x_{L+K+R}})R_{kr}(\mathbf{x_{L+K}}) + \Phi_k(\mathbf{x_L}), \end{cases}$$

$$\begin{cases} \mathcal{F}^{(k+1)}(\mathbf{x}) = L_1(\mathbf{x_{L+K+R}})R'_{11}(\mathbf{x}) + \cdots + L_r(\mathbf{x_{L+K+R}})R'_{1r}(\mathbf{x}) + \Psi_1(\mathbf{x_{L+K}}) + L'_1(\mathbf{x_{L+K+R}}), \\ \quad\vdots \\ \mathcal{F}^{(k+u)}(\mathbf{x}) = L_1(\mathbf{x_{L+K+R}})R'_{u1}(\mathbf{x}) + \cdots + L_r(\mathbf{x_{L+K+R}})R'_{ul}(\mathbf{x}) + \Psi_u(\mathbf{x_{L+K}}) + L'_u(\mathbf{x_{L+K+R}}), \end{cases}$$

where $\mathbf{x_L} = (x_1, \cdots, x_l)$, $\mathbf{x_{L+K}} = (x_1, \cdots, x_{l+k})$, $\mathbf{x_{L+K+R}} = (x_1, \cdots, x_{l+k+r})$, $\mathbf{x} = (x_1, \cdots, x_n)$, $m = k + u$ and $n = l + r + m$. We call $\mathcal{F}^{(i)}$ for $i = 1, \cdots, k$ and $\mathcal{F}^{(i)}$ for $i = k+1, \cdots, k+u$ polynomials in the first layer and the second layer, respectively.

How to Define L_i, R_{ij} and R'_{ij}.

- To define L_i, it needs to construct a hidden layer $\mathcal{L}$ of quadratic equations. L_i is a linear equation in variables $(x_1, \cdots, x_{l+k+r})$ for $i = 1, \cdots, r$. We define a system of r quadratic equations as

$$\mathcal{L} : \begin{cases} L(\mathbf{x_L})L_1(\mathbf{x_{L+K+R}}) = \xi_1, \\ \quad\vdots \\ L(\mathbf{x_L})L_r(\mathbf{x_{L+K+R}}) = \xi_r, \end{cases}$$

  where L is a linear equation in variables $(x_1, \cdots, x_l)$ and $\xi_i \in \mathbb{F}_q^*$. We choose random β_{ij} for $i = 1, \cdots, r$ and $j = 1, \cdots, l+k+r$ such that an $r \times r$

submatrix matrix $\Lambda_r = \begin{pmatrix} \beta_{1l+k+1} & \cdots\cdots & \beta_{1l+k+r} \\ \cdots & \cdots\cdots & \cdots \\ \beta_{rl+k+1} & \cdots\cdots & \beta_{rl+k+r} \end{pmatrix}$ of an $r \times (l+k+r)$ matrix Λ is invertible, where

$$\Lambda = \begin{pmatrix} \beta_{11} & \cdots\cdots & \beta_{1l+k+r} \\ \beta_{21} & \cdots\cdots & \beta_{2l+k+r} \\ \cdots & \cdots\cdots & \cdots \\ \beta_{r1} & \cdots\cdots & \beta_{rl+k+r} \end{pmatrix}$$

is a coefficient matrix of $(L_1, \cdots, L_r)$.

- Φ_i is a quadratic equation in variables $(x_1, \cdots, x_l)$ for $i = 1, \cdots, k$ defined by $\Phi_i = \sum_{j=1}^{l} \sum_{t=j}^{l} \varphi_{j,t}^i x_j x_t$, for $\varphi_{j,t}^i \in_R \mathbb{F}_q$.
- R_{ij} is a linear equation in variables $(x_1, \cdots, x_{l+k})$ for $i = 1, \cdots, k$ and $j = 1, \cdots, r$ such that a $k \times k$ submatrix matrix $\Theta_k = \begin{pmatrix} \alpha_{1l+1} & \cdots\cdots & \alpha_{1l+k} \\ \cdots & \cdots\cdots & \cdots \\ \alpha_{kl+1} & \cdots\cdots & \alpha_{kl+k} \end{pmatrix}$ of a $k \times (l+k)$ matrix Θ is invertible, where Θ is a coefficient matrix of $(L(\mathbf{x_L}) \cdot \mathcal{F}^{(1)} - L(\mathbf{x_L}) \cdot \Phi_1(\mathbf{x_L}), \cdots, L(\mathbf{x_L}) \cdot \mathcal{F}^{(k)} - L(\mathbf{x_L}) \cdot \Phi_k(\mathbf{x_L}))$ such that

$$\begin{pmatrix} L(\mathbf{x_L}) \cdot \mathcal{F}^{(1)} - L(\mathbf{x_L}) \cdot \Phi_1(\mathbf{x_L}) \\ L(\mathbf{x_L}) \cdot \mathcal{F}^{(2)} - L(\mathbf{x_L}) \cdot \Phi_2(\mathbf{x_L}) \\ \cdots \\ L(\mathbf{x_L}) \cdot \mathcal{F}^{(k)} - L(\mathbf{x_L}) \cdot \Phi_k(\mathbf{x_L}) \end{pmatrix} = \begin{pmatrix} \xi_1 R_{11}(\mathbf{x}) + \cdots + \xi_r R_{1r}(\mathbf{x_{L+K}}) \\ \xi_1 R_{21}(\mathbf{x}) + \cdots + \xi_r R_{2r}(\mathbf{x_{L+K}}) \\ \cdots \\ \xi_1 R_{k1}(\mathbf{x}) + \cdots + \xi_r R_{kr}(\mathbf{x_{L+K}}) \end{pmatrix}$$

$$= \begin{pmatrix} \alpha_{11} & \cdots\cdots & \alpha_{1l+k} \\ \alpha_{21} & \cdots\cdots & \alpha_{2l+k} \\ \cdots & \cdots\cdots & \cdots \\ \alpha_{k1} & \cdots\cdots & \alpha_{kl+k} \end{pmatrix} \cdot \begin{pmatrix} x_1 \\ x_2 \\ \cdots \\ x_{l+k} \end{pmatrix}.$$

- Ψ_i is a sparse polynomial in variables $(x_1, \cdots, x_{l+k})$ for $i = k+1, \cdots, k+u$ defined by

$$\Psi_i = \sum_{j=1}^{l+k} \psi_{i,j} x_j x_{(i+j-1) (\mathrm{mod}\ l+k)+1}$$

where $\psi_{i,j} \in_R \mathbb{F}_q$ so that the symmetric matrix of the quadratic part of each Ψ_i has rank $l+k$ and any crossterms in Ψ_i for all $i = k+1, \cdots, k+u$ don't overlap.

- L'_i is a linear equation in variables $(x_1, \cdots, x_{l+k+r})$ for $i = 1, \cdots, u$ defined by $L'_i = \sum_{j=1}^{l+k+r} \nu_j^i x_j$, where $\nu_j^i \in_R \mathbb{F}_q$.
- R'_{ij} is a linear equation in variables $(x_{l+k+r+1}, \cdots, x_n)$ for $i = 1, \cdots, u$ and $j = 1, \cdots, r$. We choose R'_{ij} such that a $u \times u$ submatrix $\Delta_u = \begin{pmatrix} \delta_{1l+k+r+1} & \cdots & \delta_{1n} \\ \cdots & & \cdots\cdots\cdots \\ \delta_{ul+k+r+1} & \cdots & \delta_{un} \end{pmatrix}$ of a $u \times n$ matrix $\Delta = \begin{pmatrix} \delta_{11} & \cdots & \delta_{1n} \\ \cdots\cdots & \cdots & \cdots \\ \delta_{u1} & \cdots & \delta_{un} \end{pmatrix}$ is invertible, where Δ is a coefficient matrix of $(L(\mathbf{x_L}) \cdot \mathcal{F}^{(k+1)} - L(\mathbf{x_L}) \cdot \Psi_1(\mathbf{x_{L+K}}) - L(\mathbf{x_L}) \cdot$

$L_1'(\mathbf{x_{L+K+R}}), \cdots, L(\mathbf{x_L})\cdot\mathcal{F}^{(k+u)} - L(\mathbf{x_L})\cdot\Psi_u(\mathbf{x_{L+K}}) - L(\mathbf{x_L})\cdot L_u'(\mathbf{x_{L+K+R}}))$ such that

$$\begin{pmatrix} L(\mathbf{x_L})\cdot\mathcal{F}^{(k+1)} - L(\mathbf{x_L})\cdot\Psi_1(\mathbf{x_{L+K}}) - L(\mathbf{x_L})\cdot L_1'(\mathbf{x_{L+K+R}}) \\ L(\mathbf{x_L})\cdot\mathcal{F}^{(k+2)} - L(\mathbf{x_L})\cdot\Psi_2(\mathbf{x_{L+K}}) - L(\mathbf{x_L})\cdot L_2'(\mathbf{x_{L+K+R}}) \\ \cdots \\ L(\mathbf{x_L})\cdot\mathcal{F}^{(k+u)} - L(\mathbf{x_L})\cdot\Psi_u(\mathbf{x_{L+K}}) - L(\mathbf{x_L})\cdot L_u'(\mathbf{x_{L+K+R}}) \end{pmatrix}$$

$$= \begin{pmatrix} \xi_1 R_{11}'(\mathbf{x}) + \cdots + \xi_r R_{1r}'(\mathbf{x}) \\ \xi_1 R_{21}'(\mathbf{x}) + \cdots + \xi_r R_{2r}'(\mathbf{x}) \\ \cdots \\ \xi_1 R_{u1}'(\mathbf{x}) + \cdots + \xi_r R_{ur}'(\mathbf{x}) \end{pmatrix} = \begin{pmatrix} \delta_{11} & \cdots & \delta_{1n} \\ \cdots & \cdots & \cdots \\ \delta_{u1} & \cdots & \delta_{un} \end{pmatrix} \cdot \begin{pmatrix} x_{l+k+r+1} \\ \cdots \\ x_n \end{pmatrix}.$$

- From this construction, we store only $(\mathbf{L}, \mathbf{L}', \Phi, \Psi^S, {\Theta_k}^{-1}, {\Lambda_r}^{-1}, {\Delta_u}^{-1})$ for $\mathcal{F}$ instead of all the coefficients of $\mathcal{F}$, where $\mathbf{L} = \{L, \xi_i\}_{i=1}^r$, $\mathbf{L}' = \{L_i'\}_{i=1}^{l+k+r}$, $\Phi = \{\Phi_i\}_{i=1}^k$ and $\Psi^S = \{\Psi_i\}_{i=1}^u$.

How to Invert the Central Map. Given $\gamma = (\gamma_1, \cdots, \gamma_m)$, to compute $\mathcal{F}^{-1}(\gamma) = \mathbf{s}$, i.e., to find $\mathbf{s}$ such that $\mathcal{F}(\mathbf{x}) = \gamma$, do the followings:

- In the first layer, compute $L(\mathbf{x_L})\cdot\mathcal{F}^{(i)} = L(\mathbf{x_L})\cdot\gamma_i$ for $i = 1, \cdots, k$ by getting a linear system of k equations with $l + k$ variables as

$$\begin{cases} \xi_1 R_{11}(\mathbf{x_{L+K}}) + \cdots + \xi_l R_{1r}(\mathbf{x_{L+K}}) = \gamma_1 \cdot L(\mathbf{x_L}) - \Phi_1(\mathbf{x_L})\cdot L(\mathbf{x_L}), \\ \vdots \\ \xi_1 R_{k1}(\mathbf{x_{L+K}}) + \cdots + \xi_r R_{kr}(\mathbf{x_{L+K}}) = \gamma_k \cdot L(\mathbf{x_L}) - \Phi_k(\mathbf{x_L})\cdot L(\mathbf{x_L}). \end{cases}$$

  - Choose a random Vinegar vector $\mathbf{s_L} = (s_1, \cdots, s_l) \in \mathbb{F}_q^l$. If $L(\mathbf{s}_L) = 0$ then choose another random Vinegar vector. Plug $\mathbf{s_L}$ into the above linear system by getting a new linear system of k equations with k variables.
  - Solve the linear system by computing

$$\begin{pmatrix} s_{l+1} \\ s_{l+2} \\ \cdots \\ s_{l+k} \end{pmatrix} = \Theta_k^{-1} \cdot \begin{pmatrix} \gamma_1 \cdot L(\mathbf{s_L}) - \Phi_1(\mathbf{s_L})\cdot L(\mathbf{s_L}) - c_1 \\ \gamma_2 \cdot L(\mathbf{s_L}) - \Phi_2(\mathbf{s_L})\cdot L(\mathbf{s_L}) - c_2 \\ \cdots \\ \gamma_k \cdot L(\mathbf{s_L}) - \Phi_k(\mathbf{s_L})\cdot L(\mathbf{s_L}) - c_k \end{pmatrix},$$

    where c_j is a constant derived from the linear equation $\xi_1 R_{j1}(\mathbf{x_{L+K}}) + \cdots + \xi_l R_{jr}(\mathbf{x_{L+K}})$ for $j = 1, \cdots, k$.
- In the hidden layer, plug $\mathbf{s_{L+K}} = (s_1, \ldots, s_{l+k})$ into a quadratic system $\mathcal{L}$ by getting a linear system of r equations with r variables as

$$\begin{cases} L_1(\mathbf{s_{L+K}}, x_{l+k+1}, \cdots, x_{l+k+r}) = L(\mathbf{s}_L)^{-1}\cdot\xi_1, \\ \vdots \\ L_r(\mathbf{s_{L+K}}, x_{l+k+1}, \cdots, x_{l+k+r}) = L(\mathbf{s}_L)^{-1}\cdot\xi_k, \end{cases}$$

where $L(\mathbf{s}_L) \neq 0$. Get a solution $(s_{l+k+1}, \cdots, s_{l+k+r})$ by computing

$$\begin{pmatrix} s_{l+k+1} \\ s_{l+k+2} \\ \cdots \\ s_{l+k+k} \end{pmatrix} = \Lambda_k^{-1} \cdot \begin{pmatrix} L(\mathbf{s}_L)^{-1} \cdot \xi_1 \\ L(\mathbf{s}_L)^{-1} \cdot \xi_2 \\ \cdots \\ L(\mathbf{s}_L)^{-1} \cdot \xi_k \end{pmatrix}.$$

– In the second layer, compute $L(\mathbf{x_L}) \cdot \mathcal{F}^{(i)} = L(\mathbf{x_L}) \cdot \gamma_i$ for $i = k+1, \cdots, k+u$ getting a linear system of u equations with $l+k+r+u$ variables as

$$\begin{cases} \xi_1 R'_{11}(\mathbf{x}) + \cdots + \xi_l R'_{1r}(\mathbf{x}) = \gamma_{k+1} \cdot L(\mathbf{x}_L) - \Psi_1(\mathbf{x}_L) \cdot L(\mathbf{x}_L) - L'_1(\mathbf{x_{L+K+R}}) \cdot L(\mathbf{x}_L), \\ \xi_1 R'_{21}(\mathbf{x}) + \cdots + \xi_l R'_{2r}(\mathbf{x}) = \gamma_{k+2} \cdot L(\mathbf{x}_L) - \Psi_2(\mathbf{x}_L) \cdot L(\mathbf{x}_L) - L'_2(\mathbf{x_{L+K+R}}) \cdot L(\mathbf{x}_L), \\ \cdots \\ \xi_1 R'_{u1}(\mathbf{x}) + \cdots + \xi_l R'_{ur}(\mathbf{x}) = \gamma_{k+u} \cdot L(\mathbf{x}_L) - \Psi_u(\mathbf{x}_L) \cdot L(\mathbf{x}_L) - L'_u(\mathbf{x_{L+K+R}}) \cdot L(\mathbf{x}_L), \end{cases}$$

and plug $\mathbf{s_{L+K+R}} = (s_1, \cdots, s_{l+k+r})$ into the linear system getting a linear system of u equations with u variables. Get a solution $(s_{l+k+r+1}, \cdots, s_{l+k+r+u})$ by computing

$$\begin{pmatrix} s_{l+k+r+1} \\ s_{l+k+r+2} \\ \cdots \\ s_{l+k+r+u} \end{pmatrix} = \Delta_u^{-1} \cdot \begin{pmatrix} \gamma_{k+1} \cdot L(\mathbf{s}_L) - \Psi_1(\mathbf{s}_L) \cdot L(\mathbf{s}_L) - c'_1 \\ \gamma_{k+2} \cdot L(\mathbf{s}_L) - \Psi_2(\mathbf{s}_L) \cdot L(\mathbf{s}_L) - c'_2 \\ \cdots \\ \gamma_{k+u} \cdot L(\mathbf{s}_L) - \Psi_u(\mathbf{s}_L) \cdot L(\mathbf{s}_L) - c'_u \end{pmatrix},$$

where c'_i is a constant of the linear equation $\xi_1 R'_{i1}(\mathbf{s_{L+K}}, \mathbf{x}_R) + \cdots + \xi_r R'_{ir}(\mathbf{s_{L+K}}, \mathbf{x}_R)$ for $i = 1, \cdots, u$.

– Finally, we get a solution $(s_1, \cdots, s_n)$ of $\mathcal{F}(\mathbf{x}) = \gamma$ by performing only three matrix multiplications and computation of quadratic terms without using the Gaussian elimination.

Now, we construct a new MQ-signature scheme based on this central map.

■ **ELSA** (**E**fficient **L**ayered **S**ign**a**ture Scheme).

- **KeyGen**(1^λ). For a security parameter λ, generate a public/secret key pair $<PK, SK> = <\mathcal{P}, (\widetilde{S}, \widetilde{T}, \mathcal{F} = (\mathbf{L}, \mathbf{L}', \Phi, \Psi^S, \widetilde{\Theta_r}, \widetilde{\Lambda_k}, \widetilde{\Delta_u}))>$ as
  - Choose randomly two affine maps $\widetilde{S}$ and $\widetilde{T}$. If neither $\widetilde{S}$ nor $\widetilde{T}$ is invertible then choose again, where $\widetilde{X} = X^{-1}$.
  - Choose randomly $\mathbf{L}, \Phi, \Psi^S, \widetilde{\Theta_r}, \widetilde{\Lambda_r}$ and $\widetilde{\Delta_u}$, where $\mathbf{L} = \{L, \xi_i\}_{i=1}^r$, $\mathbf{L}' = \{L'_i\}_{i=1}^u$, $\Phi = \{\Phi_i\}_{i=1}^k$ and $\Psi^S = \{\Psi_i\}_{i=1}^u$ satisfy all the conditions described above. If neither $\widetilde{\Theta_r}$, $\widetilde{\Lambda_k}$ nor $\widetilde{\Delta_u}$ is invertible then choose again. Compute $\mathcal{P}$ from $\mathcal{P} = S \circ \mathcal{F} \circ T$.
- **Sign**$(SK, \mathtt{m})$. Given a message $\mathtt{m}$,
  - Compute $h(\mathtt{m})$ and $\widetilde{S}(h(\mathtt{m})) = \gamma$, where $\gamma = (\gamma_1, \cdots, \gamma_m)$.
  - Compute $\mathbf{s}$ such that $\mathcal{F}^{-1}(\gamma) = \mathbf{s}$, i.e., $\mathcal{F}(\mathbf{s}) = \gamma$ as the above. Then $\mathbf{s} = (s_1, \cdots, s_n)$ is a solution of $F(\mathbf{x}) = \gamma$.
  - Compute $\widetilde{T}(\mathbf{s}) = \sigma$. Then σ is a signature of $\mathtt{m}$.

- **Verify**$(PK, \mathtt{m}, \sigma)$. Given a signature σ on $\mathtt{m}$ and a public key $\mathcal{P}$, check $\mathcal{P}(\sigma) = h(\mathtt{m})$. If it holds, accept σ, otherwise, reject it.

Remark 1. We now explain how the public key and secret key sizes of ELSA are calculated. The public key requires $\frac{m(n+1)(n+2)}{2}$ field elements as their coefficients. The secret maps S and T require $m(m+1)$ and $n(n+1)$ field elements, respectively. It requires $(l+r+l+1)$ field elements for $\mathbf{L}$, $u(l+k+r+1)$ field elements for $\mathbf{L}'$, $\frac{k(l+1)(l+2)}{2}$ field elements for Φ, $u(l+k)$ field elements for Ψ^S, k^2 field elements for $\widetilde{\Theta_k}$, r^2 field elements for $\widetilde{\Lambda_k}$ and u^2 field elements for $\widetilde{\Delta_u}$. Thus, the secret key requires $n(n+1) + m(m+1) + \frac{k(l+1)(l+2)}{2} + u(2l+2k+r+1) + (k^2+r^2+u^2) + (l+r+1)$ field elements.

2. UOV and Rainbow requires the use of Gaussian elimination for solving linear systems in signing. In these schemes, the majority of computational cost for signing count for that of the Gaussian elimination. In ELSA, only three matrix multiplications using ${\Theta_r}^{-1}, {\Lambda_k}^{-1}, {\Delta_u}^{-1}$ are required for solving the resulting linear systems in signing without using the Gaussian elimination. So, it achieves $\mathcal{O}(n^2)$ complexity in signing instead of $\mathcal{O}(n^3)$.

3 Security Analysis of ELSA

The security of all MQ-schemes in the MQ + IP paradigm is not only based on the MQ-Problem, but also on some variant of the Isomorphism of Polynomials (IP) problem. Furthermore, layered MQ-schemes require the hardness of the MinRank problem. These underlying problems are defined as follows:

- **Polynomial System Solving (PoSSo) Problem:** Given a system $\mathcal{P} = (\mathcal{P}^{(1)}, \cdots, \mathcal{P}^{(m)})$ of m nonlinear polynomials defined over $\mathbb{F}_q$ with degree of d in variables $(x_1, \cdots, x_n)$ and $\mathbf{y} = (y_1, \cdots, y_m) \in \mathbb{F}_q^m$, find $\mathbf{x}' = (x_1', \cdots, x_n') \in \mathbb{F}_q^n$ such that $\mathcal{P}(\mathbf{x}') = \mathbf{y}$, i.e., $\mathcal{P}^{(1)}(x_1', \cdots, x_n') = y_1, \cdots, \mathcal{P}^{(m)}(x_1', \cdots, x_n') = y_m$.
- **EIP (Extended Isomorphism of Polynomials) Problem:** Given a nonlinear multivariate system $\mathcal{P}$ such that $\mathcal{P} = S \circ \mathcal{F} \circ T$ for linear or affine maps S and T, and $\mathcal{F}$ belonging to a special class of nonlinear polynomial system $\mathcal{C}$, find a decomposition of $\mathcal{P}$ such that $\mathcal{P} = S' \circ \mathcal{F}' \circ T'$ for linear or affine maps S' and T', and $\mathcal{F}' \in \mathcal{C}$.
- **MinRank Problem:** Let $m, n, r, k \in \mathbb{N}$ and $r, m < n$. The MinRank(r) problem is, given $(M_1, \cdots, M_l) \in \mathbb{F}_q^{m \times n}$, find a non-zero k-tuple $(\lambda_1, \cdots, \lambda_k) \in \mathbb{F}_q^k$ such that $Rank(\sum_{i=1}^{k} \lambda_i M_i) \leq r$.

The PoSSo problem is proven to be NP-complete [26]. For efficiency, MQ-PKC restrict to quadratic polynomials. The PoSSo problem with all polynomials $(P^{(1)}, \cdots, P^{(m)})$ of degree 2 is called the MQ-Problem for multivariate quadratic. The IP problem was first described by Patarin at Eurocrypt'96 [40], there is not much known about the difficulty of the IP problem in contrast to the MQ-problem. The problem of finding a low rank linear combination of matrices was originally introduced in [48] as one of the natural questions in linear algebra, and the authors proved its NP-completeness.

A feature of MQ-PKC in the MQ + IP paradigm is that there exist a large number of different secret keys for a given public key [51]. Informally, suppose that $<\mathcal{P}, (S, \mathcal{F}, T)>$ is a public/secret key pair of an MQ-PKC, we call $(S', \mathcal{F}', T')$ is an equivalent key of $(S, \mathcal{F}, T)$ if $\mathcal{P} = S \circ \mathcal{F} \circ T = S' \circ \mathcal{F}' \circ T'$, where S' and T' are invertible affine maps, and $\mathcal{F}'$ preserves all zero coefficients of $\mathcal{F}$. The concept of equivalent keys plays a major role in the cryptanalysis of MQ-schemes. If an attacker finds any of the equivalent keys then he can forge a signature. Thus, the attacker wants to find an equivalent key with the simplest structure. Known attacks of MQ-schemes be divided into the following two classes:

- **Direct Attack.** Given a public key $\mathcal{P}$ and $\mathbf{y} \in \mathbb{F}_q^m$, find a solution $\mathbf{x} \in \mathbb{F}_q^n$ of $\mathcal{P}(\mathbf{x}) = \mathbf{y}$.
- **Key Recovery Attack (KRA).** Given $\mathcal{P} = S \circ \mathcal{F} \circ T$, find equivalent keys of $(S, \mathcal{F}, T)$:
  - KRAs using equivalent keys and good keys,
  - Rank-based KRAs to find linear combinations associated matrices at some given rank, to find nontrivial invariant subspaces of linear combinations associated matrices and so on: MinRank attack, HighRank attack, Kipnis-Shamir attack.

3.1 Direct Attacks

Direct attacks use equation solvers like XL and Gröbner basis algorithms such as Buchberger, F4 and F5 for solving the MQ-problem. Complexity of the MQ-Problem is determined by that of the HybridF5 (HF5) algorithm [7]. The basic idea is to guess some of the variables to create overdetermined systems before applying Faugère's F5 algorithm [22]. When doing so, one has to run the F5 algorithm several times to find a solution of the original system. When guessing k variables over $\mathbb{F}_q$, this number is given by q^k. The complexity of solving a semi-regular (random) system of m quadratic equations in n variables over $\mathbb{F}_q$ by HF5 can be estimated as

$$C_{HF5}(q, m, n) = min_{k \geq 0}\, q^k \cdot \mathcal{O}\left(\left(m \cdot \binom{n-k+d_{reg}-1}{d_{reg}}\right)^{\omega}\right),$$

where the degree of regularity d_{reg} is the index of the first non-positive coefficient in the $S_{m,n} = \dfrac{(1-z^2)^m}{(1-z)^n}$ and $2 \leq \omega \leq 3$ is the linear algebra constant of solving a linear system. The internal equations used by HF5 are very sparse and thus $\omega = 2$ can be used to obtain a lower bound on the complexity. If we really want to break a scheme, we either calculate the correct α or use $\omega = 2.8$ as an upper bound [50].

Using HF5 algorithm ($\omega = 2$), we summarize the lower bounds of the numbers of equations (m) for solving determined systems defined over $\mathbb{F}_{2^8}$ required to achieve given security levels in Table 1.

Table 1. Lower bounds of the numbers of quadratic equations for determined systems over $\mathbb{F}_{2^8}$ at each security level.

λ	80	96	128	192	256
m	26	31	43	68	93

3.2 Replacement Attacks

Our central map has a special feature for inverting: each central polynomial uses a linear combination of the products of two lines and additional quadratic terms. This feature and hidden quadratic systems make it possible to remove the use of the Gaussian elimination resulting in the reduction of signing cost and secret key size. In particular, L_i for $i = 1, \cdots, r$ are used in all the central polynomials $\mathcal{F}^{(i)}$ for $i = 1, \cdots, k+u$. Thus, one can replace L_i with a new variable via an appropriate changing of variables. More precisely, one can replace $L_i(\mathbf{x_{L+K+R}})$ with y_{l+k+i} for $i = 1, \cdots, r$ and x_j with y_j for $j = 1, \cdots, l+k, l+k+r+1, \cdots, l+k+r+u$. Then one gets a new central map, $\overline{\mathcal{F}} = (\overline{\mathcal{F}}^{(1)}, \cdots, \overline{\mathcal{F}}^{(m)})$ in the new variables $(y_1, \cdots, y_n)$ as

$$\begin{cases} \widehat{\mathcal{F}}^{(1)}(\mathbf{y}) = y_{l+k+1}R_{11}(\mathbf{y_{L+K}}) + \cdots + y_{l+k+r}R_{1r}(\mathbf{y_{L+K}}) + \Phi_1(\mathbf{y_L}), \\ \vdots \\ \widehat{\mathcal{F}}^{(k)}(\mathbf{y}) = y_{l+k+1}R_{k1}(\mathbf{y_{L+K}}) + \cdots + y_{l+k+r}R_{kr}(\mathbf{y_{L+K}}) + \Phi_k(\mathbf{y_L}), \end{cases}$$

$$\begin{cases} \widehat{\mathcal{F}}^{(k+1)}(\mathbf{y}) = y_{l+k+1}R'_{11}(\mathbf{y}) + \cdots + y_{l+k+r}R'_{1r}(\mathbf{y}) + \Psi_1(\mathbf{y_{L+K}}) + \overline{L_1}(\mathbf{y_{L+K+R}}), \\ \vdots \\ \widehat{\mathcal{F}}^{(k+u)}(\mathbf{y}) = y_{l+k+1}R'_{u1}(\mathbf{y}) + \cdots + y_{l+k+r}R'_{ur}(\mathbf{y}) + \Psi_u(\mathbf{y_{L+K}}) + \overline{L_u}(\mathbf{y_{L+K+R}}), \end{cases}$$

where $\mathbf{y_L} = (y_1, \cdots, y_l)$, $\mathbf{y_{L+K}} = (y_1, \cdots, y_{l+k})$ and $\mathbf{y_{L+K+R}} = (y_1, \cdots, y_{l+k+r})$. Then the public key can be written as

$$\mathcal{P} = S \circ (\mathcal{F} \circ T_R) \circ (T_R^{-1} \circ T) = S \circ \overline{\mathcal{F}} \circ \overline{T},$$

where $\overline{\mathcal{F}} = \mathcal{F} \circ T_R$, $\overline{T} = T_R^{-1} \circ T$ and T_R is an invertible map defined by

$$T_R(\mathbf{x^T}) = \begin{pmatrix} I_{L+K} & 0 & 0 \\ 0 & L_1 & 0 \\ 0 & L_2 & 0 \\ \cdots & \cdots & \cdots \\ 0 & L_r & 0 \\ 0 & 0 & I_U \end{pmatrix} \cdot \begin{pmatrix} x_1 \\ \cdots \\ x_n \end{pmatrix} = \begin{pmatrix} y_1 \\ \cdots \\ y_n \end{pmatrix},$$

where I_{L+K} and I_U are an $(l+k)\times(l+k)$-identity matrix and a $u\times u$-identity matrix, respectively. In this case, we can consider the public key $\mathcal{P} = S \circ \overline{\mathcal{F}} \circ \overline{T}$ with the secret key $(S, \overline{\mathcal{F}}, \overline{T})$ since $\overline{\mathcal{F}}$ is still invertible with the same way as in §2.1. We provide security analysis of ELSA against all attacks with respect to these two types of secret keys $(S, \mathcal{F}, T)$ and $(S, \overline{\mathcal{F}}, \overline{T})$ for the public key $\mathcal{P}$.

3.3 Key Recovery Attacks

In 2008, Ding *et al.* [17] presented Rainbow Band Separation (RBS) attacks on Rainbow. Later, Thomae [50] applied the attacks to other MQ-schemes using the concept of good keys which is a generalization of the RBS attacks. In this subsection, we analyze security of ELSA against the key recovery attacks (KRAs) using equivalent keys and good keys.

Let $F^{(i)}$ $(1 \le i \le m)$ be symmetric matrices associated to the homogeneous quadratic part of the i-th component of the central map $\mathcal{F}$. The matrices $F^{(i)}$ are depicted in Fig. 1, where white parts denote zero entries and gray parts denote arbitrary entries. The matrices are the same as those of Rainbow [44]. After mounting the replacement attack described in Sect. 3.2, we get symmetric matrices $\overline{F}^{(i)}$ $(1 \le i \le m)$ representing the quadratic part of the i-th component of $\overline{\mathcal{F}}^{(i)}$ which is depicted in Fig. 2.

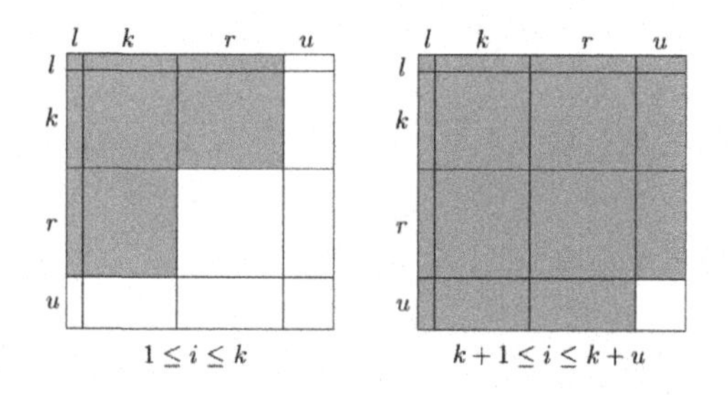

Fig. 1. Symmetric matrices for quadratic parts of $\mathcal{F}$.

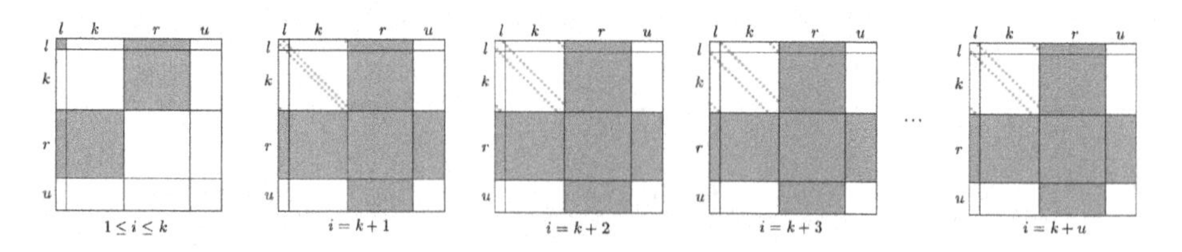

Fig. 2. Symmetric matrices for quadratic parts of $\overline{\mathcal{F}}$.

Analogously, we denote $P^{(i)}$ $(1 \le i \le m)$ be symmetric matrices representing the quadratic part of the i-th component of the public key $\mathcal{P}$. Due to the structure of $\mathcal{F}$, we know that certain coefficients in $\mathcal{F}^{(i)}$ are systematically zero. Since $\mathcal{P} = S \circ \mathcal{F} \circ T$, we obtain $\mathcal{F} = \widetilde{S} \circ \mathcal{P} \circ \widetilde{T}$, where $\widetilde{S} = S^{-1}$ and $\widetilde{T} = T^{-1}$. From this, we get the following equality:

$$\mathcal{F}^{(i)} = \widetilde{T}^{\mathsf{T}} \left(\sum_{j=1}^{m} \widetilde{s}_{ij} P^{(j)} \right) \widetilde{T}, \ \forall 1 \leq i \leq m.$$

The corresponding system of equations is:

$$f_{\beta\gamma}^{(i)} = \sum_{x=1}^{m} \sum_{y=1}^{n} \sum_{z=1}^{n} c_{yz}^{(x)} \widetilde{s}_{ix} \widetilde{t}_{y\beta} \widetilde{t}_{z\gamma} \tag{1}$$

for some coefficient $c_{yz}^{(x)}$, as we have already known that $f_{\beta\gamma}^{(i)} = 0$ for some i, β, γ by the construction of $\mathcal{F}$. Since the number equations obtained by (1) equals the number of zeros in all $\mathcal{F}^{(k)}$, we get $\dfrac{kr(r+1) + mu(u+1)}{2} + ku(n-u)$ cubic equations. The number of variables in $\widetilde{S}$ and $\widetilde{T}$ is $n^2 + m^2$. The number of equations for $\overline{\mathcal{F}}$ is

$$\frac{k(n-l)(n+l+1) + u(n-r)(n+r+1)}{2} - r[(l+k)k + (n-r)u] - u(l+k).$$

The complexity of solving such systems using HF5 is very large. To improve this complexity, we use the concept of equivalent keys [50,51]. Let $\mathbb{GL}_n(\mathbb{F}_q)$ be a general linear group of degree n over $\mathbb{F}_q$, for an integer n.

Definition 3.1 [Equivalent Key]. Let $S, S' \in \mathbb{GL}_m(\mathbb{F}_q)$ and $T, T' \in \mathbb{GL}_n(\mathbb{F}_q)$ and $\mathcal{F}, \mathcal{F}' \in \mathbb{F}_q[x_1, ..., x_n]^m$. We say that $(\mathcal{F}, S, T)$ is *equivalent* to $(\mathcal{F}', S', T')$ if and only if $S \circ \mathcal{F} \circ T = S' \circ \mathcal{F}' \circ T'$ and $\mathcal{F}|_I = \mathcal{F}'|_I$, that is, $\mathcal{F}$ and $\mathcal{F}'$ share the same structure when restricted to a fixed index set $I = \{I^{(1)}, \cdots I^{(m)}\}$.

If $S \circ \mathcal{F} \circ T = \mathcal{P} = S' \circ \mathcal{F}' \circ T'$, where $\mathcal{F}'$ preserves all systematic zero coefficients of $\mathcal{F}$ then we call S' and T' equivalent keys. Thus, an attacker who has any of equivalent keys can forge signatures on any messages. If we can find simpler equivalent keys, we can reduce the number of variables in S and T. If there are two invertible linear maps $\Sigma \in \mathbb{GL}_m(\mathbb{F}_q)$ and $\Omega \in \mathbb{GL}_n(\mathbb{F}_q)$ such that

$$\mathcal{P} = S \circ \Sigma^{-1} \circ (\Sigma \circ \mathcal{F} \circ \Omega) \circ \Omega^{-1} \circ T,$$

and $\mathcal{F}' (= \Sigma \circ \mathcal{F} \circ \Omega)$ and $\mathcal{F}$ have the same structure then $(\mathcal{F}', S', T')$ is an equivalent key.

For the original central map $\mathcal{F}$, its equivalent keys are the same as those of Rainbow since the matrices $(F^{(1)}, \cdots, F^{(m)})$ are the same as those of Rainbow [50]. Thus, the equivalent keys for $\mathcal{F}$ are of the form given in Fig. 3, in this case, $\mathcal{F}'^{(i)}$ also have same form as $\mathcal{F}'^{(i)}$ given in Fig. 1.

Next, we find equivalent keys for the central map $\overline{\mathcal{F}}$. To preserve the structure in second layer, we can find Ω and Σ of the form given in Fig. 4, so we

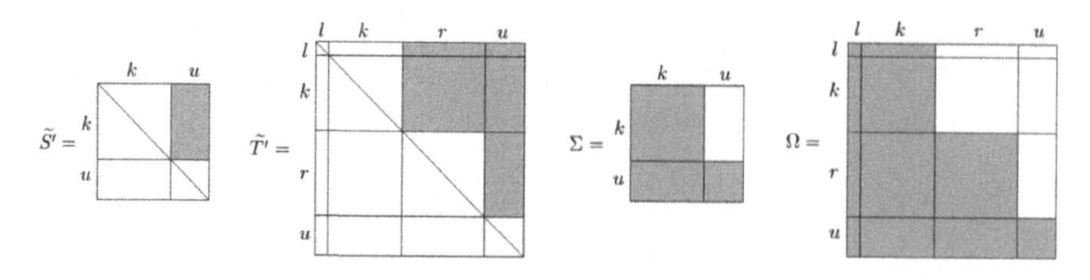

Fig. 3. Equivalent keys of ELSA w.r.t. $\mathcal{F}$.

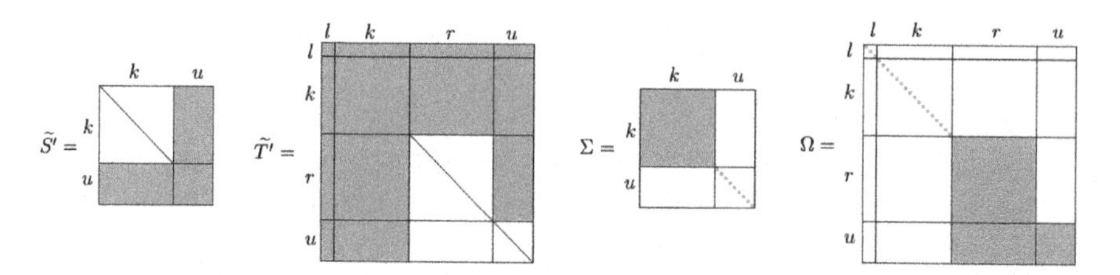

Fig. 4. Equivalent keys of ELSA w.r.t. $\overline{\mathcal{F}}$.

get equivalent keys of the form given in Fig. 4. However, we can find simpler equivalent keys than ones given in Fig. 4 to improve the complexity significantly by changing the preservation set, i.e., the set of indices for the quadratic terms with zero coefficients. For it, we consider the generalized version of $\overline{\mathcal{F}}$ denoted by $\widehat{\mathcal{F}}$ which is depicted in Fig. 5. So, we need to find equivalent keys $(\widehat{\mathcal{F}}', S', T')$ such that $\widehat{\mathcal{F}'}^{(i)}$ preserves the generalized version $\widehat{\mathcal{F}}^{(i)}$.

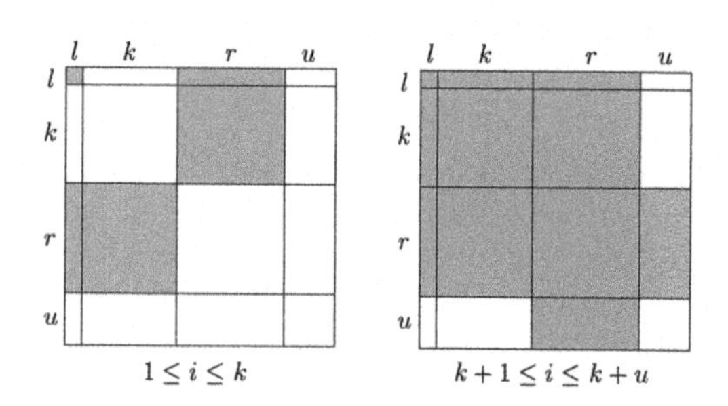

Fig. 5. $\widehat{\mathcal{F}}$: Generalized version of $\overline{\mathcal{F}}$.

Lemma 3.1. For the generalized central map $\widehat{\mathcal{F}}$ given in Fig. 5, we can find equivalent keys S' and T' of the form given in Fig. 6 with high probability, where gray parts denote arbitrary entries and white parts denote zero entries and there are ones at the diagonal.

Proof. As in [50], we can find Σ and Ω given in Fig. 6. With high probability, there exist equivalent keys (S', T') of the form given in Fig. 6. □

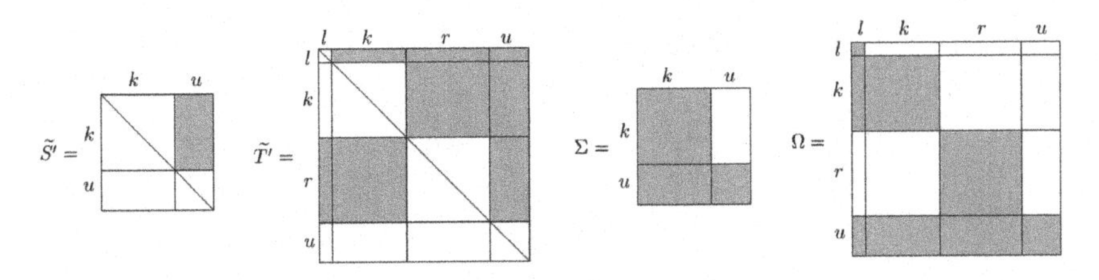

Fig. 6. Equivalent keys of ELSA w.r.t. $\widehat{\mathcal{F}}$.

After applying the transformations Σ and Ω in Lemma 3.1., we also get the central map $\widehat{\mathcal{F}}' = \Sigma \circ \widehat{\mathcal{F}} \circ \Omega$ as in Fig. 5. From the equivalent key given in Fig. 6, we get a system of $\frac{k(n-l)(n+l+1)}{2} - kr(l+k)$ cubic equations and $\frac{u^2(2l+2k+u+1)}{2}$ quadratic equations with $n(n-u)+k(u-k-r)-l^2-r^2$ variables. However, the complexity of solving such a system is still large: for ELSA with $(\mathbb{F}_q, l, k, r, u) = (\mathbb{F}_{2^8}, 6, 28, 30, 15)$, lower bound on the complexity of solving the system by HF5 is 2^{1696}. To further decrease this complexity, we use the notion of good keys which is a generalization of equivalent keys. Good keys don't preserve all the zero coefficients of $\mathcal{F}$, but just some of them. Hence, we can choose Σ and Ω more widely and further reduce the number of variables.

Definition 3.2 [Good Key]. Let $S, S'' \in \mathbb{GL}_n(\mathbb{F}_q)$ and $T, T'' \in \mathbb{GL}_m(\mathbb{F}_q)$ and $\mathcal{F}, \mathcal{F}'' \in \mathbb{F}_q[x_1, ..., x_n]^m$, and $J = \{J^{(1)}, \cdots, J^{(m)}\} \subset I = \{I^{(1)}, \cdots, I^{(m)}\}$ for all k with at least one $J^{(k)} \neq \phi$. We say that $(\mathcal{F}'', S'', T'')$ is a *good key* for $(\mathcal{F}, S, T)$ if and only if $S \circ \mathcal{F} \circ T = S'' \circ \mathcal{F}'' \circ T''$ and $\mathcal{F}|_J = \mathcal{F}''|_J$.

To find good keys, let $(\mathcal{F}', S', T')$ be an equivalent key for ELSA. If

$$\mathcal{P} = S' \circ \mathcal{F}' \circ T' = (S' \circ \Sigma'^{-1}) \circ (\Sigma' \circ \mathcal{F}' \circ \Omega') \circ (\Omega'^{-1} \circ T')$$

for some two linear maps $\Sigma' \in \mathbb{GL}_m(\mathbb{F}_q)$ and $\Omega' \in \mathbb{GL}_n(\mathbb{F}_q)$, and $\mathcal{F}'' = \Sigma' \circ \mathcal{F}' \circ \Omega'$ satisfies the condition in above definition, then

$$(\mathcal{F}'',\ S'',\ T'') = (\Sigma' \circ \mathcal{F}' \circ \Omega',\ S' \circ \Sigma'^{-1},\ \Omega'^{-1} \circ T'),$$

S'' and T'' are good keys. The following proposition shows the existence of good keys for ELSA.

Lemma 3.2. Let $(S', \widehat{\mathcal{F}}', T')$ be an equivalent key for ELSA given in Fig. 6. Then there are good keys $(S'', \widehat{\mathcal{F}}'', T'')$ of the form given in Fig. 7. Only the last column of $\widetilde{T''}$ contains arbitrary values in the first $l+k+r$ rows, which are equal to the corresponding values in $\widetilde{T'}$. Respectively, only u values of the k-th row of $\widetilde{S''}$ contain arbitrary values, which are equal to the corresponding values in $\widetilde{S'}$.

Proof. Using linear algebra, we can obtain unique Σ' and Ω' given in Fig. 7. It shows the existence of a good key (S'', T'') of the form given in Fig. 7. □

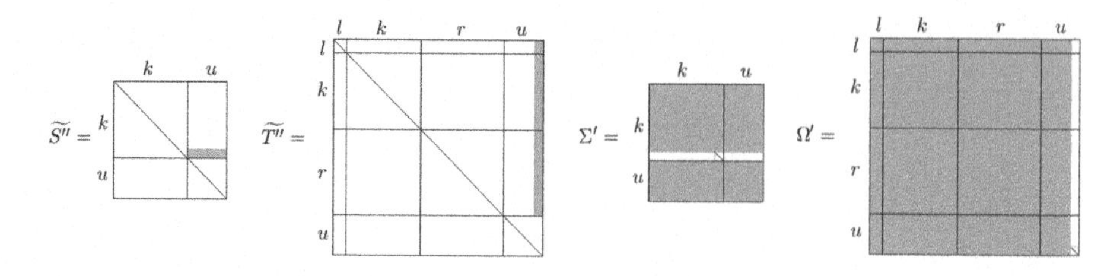

Fig. 7. Good keys of ELSA.

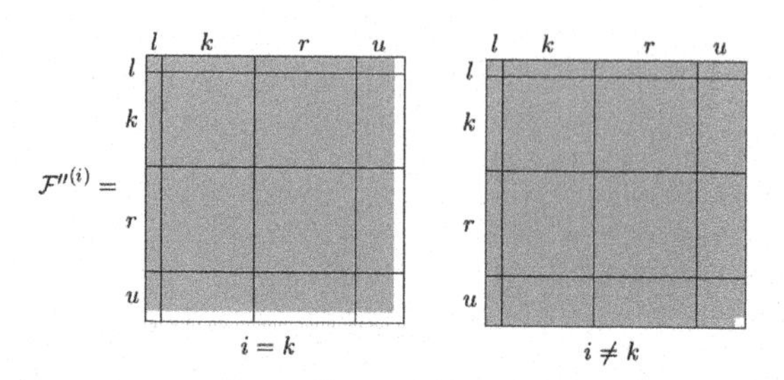

Fig. 8. Central map $\widehat{\mathcal{F}}''$ after applying Σ' and Ω' in Lemma 3.2.

Finally, after applying the transformations Σ' and Ω', we get the central map $\widehat{\mathcal{F}}'' = \Sigma' \circ \widehat{\mathcal{F}}' \circ \Omega'$ as given in Fig. 8. Finally, we obtain the following Theorem.

Theorem 3.1. The main complexity of the key recovery attack using good keys on ELSA is determined by solving $n-1$ bihomogeneous equations and m quadratic equations with n variables.

After obtaining one column of T' and one row of S', all the other parts of T' and S' are revealed by linear equations as in [50]. Consequently, we recover the equivalent keys T' and S'.

We find different equivalent keys for three types of central maps $\mathcal{F}$, $\overline{\mathcal{F}}$ and $\widehat{\mathcal{F}}$, where $\mathcal{F}$, $\overline{\mathcal{F}}$ and $\widehat{\mathcal{F}}$ are the original central map, the resulting central map after the replacement attack and the general version of $\overline{\mathcal{F}}$, respectively. The KRAs using equivalent keys for $\widehat{\mathcal{F}}$ are more effective than those for $\mathcal{F}$ and $\overline{\mathcal{F}}$. However, $\mathcal{F}$, $\overline{\mathcal{F}}$ and $\widehat{\mathcal{F}}$ have the same forms of good keys resulting in the same complexities given in Theorem 3.1. Table 2 shows improvements of lower bound ($\alpha = 2$) and upper bound ($\alpha = 2.8$) on the complexities of solving such a system by HF5 achieved by the KRAs using equivalent keys and good keys for ELSA with $(\mathbb{F}_q, l, k, r, u) = (\mathbb{F}_{2^8}, 6, 28, 30, 15)$.

Key Recovery Attacks using Linear Part of the Central Map. It is also known that some coefficients of linear terms in the central map are zero. This does not significantly affect the KRAs since the number of quadratic terms with zero coefficients is much larger than that of linear terms with zero coefficients. When we reduce the number of variables in good key recovery, we use Ω' where

Table 2. Lower-bounds/upper-bounds on the complexities of the KRAs using equivalent keys and good keys for ELSA under different forms of central maps, $\mathcal{F}$, $\overline{\mathcal{F}}$ and $\widehat{\mathcal{F}}$ with $(F_{2^8}, 6, 28, 30, 15)$

ELSA	# of Equ.	# of Vari.	d_{reg}	Comp. (low./upp.)
KRA ($\mathcal{F}$)	45,060(Cubic)	8,090	1017	$2^{9215}/2^{12901}$
KRA ($\overline{\mathcal{F}}$)	77,197(Cubic)	8,090	727	$2^{7263}/2^{10169}$
KRA ($\widehat{\mathcal{F}}$)	68,782(Cubic)	8,090	779	$2^{7632}/2^{10686}$
KRA Equi. ($\mathcal{F}$)	28,980(Cubic) + 16,080(Quad.)	2,400	53	$2^{760}/2^{1064}$
KRA Equi. ($\overline{\mathcal{F}}$)	77,197(Cubic)	5,731	425	$2^{4481}/2^{6274}$
KRA Equi. ($\widehat{\mathcal{F}}$)	59,332(Cubic) + 9,450(Quad.)	3,588	135	$2^{1696}/2^{2375}$
KRA Good	121(Quad.)	79	16	$2^{131}/2^{183}$

each coordinate function has at least $n-1$ linear terms (See Lemma 3.2). Even if $\mathcal{F}'^{(k)}$ has only one linear term for each k, $\mathcal{F}'^{(k)} \circ \Omega'$ has at least $n-1$ linear terms. Nevertheless, if there is no linear term in $\mathcal{F}$, we can get nm linear terms with zero coefficients of $\mathcal{F}' \circ \Omega'$ and n variables in the constant part of $\widetilde{T''}$ by choosing Ω and Ω' carefully satisfying Lemmas 3.1 and 3.2. Then we can set $\Sigma' = \left(\widetilde{S'}\right)^{-1} = S'$ so that the variables in $\widetilde{S''}$ are removed. Finally, we get a system of $m(n+1)$ quadratic equations with $2n-u$ variables. In this case, for ELSA with $(F_{2^8}, 6, 28, 30, 15)$, the complexity of solving this system by HF5 is 2^{71}.

3.4 Rank-Based Attacks

• **MinRank attack.** In MinRank attacks, one tries to find linear combinations $M = \sum_{i=1}^{m} \mu_i P^{(i)}$ of the matrices $P^{(i)}$, where M has a minimal rank. Underlying idea of an algorithm to solve this MinRank problem [48] is to search for a vector lying in the kernel of the desired linear combination M. Complexity of the MinRank attack is determined by that of finding the linear combination. Since the forms of symmetric matrices of ELSA w.r.t. $\mathcal{F}$ are the same as those of Rainbow, we can get that its complexity against the attack is q^{l+k+1} from [9,44]. Next, by using similar technique, we investigate the complexity of ELSA w.r.t. $\widehat{\mathcal{F}}$ against the attack in Proposition 3.1.

Proposition 3.1. The complexity of ELSA w.r.t. $\widehat{\mathcal{F}}$ against the MinRank attack is $min\{q^{l+2r-k+1}, q^{l+2r+1}, q^{2l+k+1}\}$.

Proof. In MinRank attacks, we must find a vector $v \in \mathbb{F}_q^n$ such that $v \in \ker P$, where P is a matrix with the minimal rank in Span$\{P^{(i)}\}$. The probability for finding such a vector is the same as that of finding $v' \in \mathbb{F}_q^n$ such that $v' \in \ker Q$, where Q is a matrix with the minimal rank in Span$\{\widehat{F}^{(i)}\}$ and $\widehat{F}^{(i)}$ is the matrix of the quadratic part of $\widehat{\mathcal{F}}$. More precisely, $\widehat{F}^{(i)}$ has of the form $\begin{pmatrix} * & 0 & * \\ 0 & 0 & * \\ * & * & 0 \end{pmatrix}$ in the first layer as given in Fig. 2. Then $\widehat{F}^{(i)} \cdot (0,*,0)^T = (0,0,*)^T$. Let $w_i = \widehat{F}^{(i)} \cdot (0,*,0)^T = (0,0,*)^T$. Then the probability that w_i is linearly dependent is

$$1 - \prod_{i=0}^{k-1}(1 - \frac{q^i}{q^r}) > 1/q^{r-k+1}.$$

Note that $\sum_{i=1}^{k} \lambda_i \widehat{F}^{(i)}$ has a minimal rank. Hence, the probability of $v' \in \ker(\sum_{i=1}^{k} \lambda_i \widehat{F}^{(i)})$ for a random vector v' and non-trivial λ_i is $1/q^{l+r} \cdot 1/q^{r-k+1} = 1/q^{l+2r-k+1}$, where the provability that the vector v' has of the form $(0,*,0)$ is $1/q^{l+r}$. Similarly, the probabilities for $\widehat{F}^{(i)} \cdot (*,0,0)$ and $\widehat{F}^{(i)} \cdot (0,0,*)$ are $1/q^{l+2r+1}$ and $1/q^{2l+k+1}$, respectively. □

Finally, the complexity of ELSA against the MinRank attack is $min\{q^{l+k+1}, q^{l+2r-k+1}, q^{l+2r+1}, q^{2l+k+1}\}$.

- **HighRank Attack.** In HighRank attacks, one tries to identify the variables appearing the lowest number of times in the central polynomials. The variables $x_{l+k+r+1}, \cdots, x_n$ appear only in the quadratic terms of the central polynomials $(\mathcal{F}^{(k+1)}, \cdots, \mathcal{F}^{(k+u)})$ of the second layer of ELSA. Thus, it is similar to that of Rainbow. As in [44], we get its complexity against the HighRank attacks is $q^u \cdot \frac{n^3}{6}$.
- **Kipnis-Shamir Attack (UOV Attack).** Kipnis-Shamir attack [34] was originally used to break the balanced Oil and Vinegar signature scheme [41]. We consider the generalization to the unbalanced case. We have already known that the complexity of ELSA w.r.t. $\mathcal{F}$ against the Kipnis-Shamir attack is $q^{n-2u-1} \cdot u^4$ as in [44] since the forms of symmetric matrices of ELSA w.r.t. $\mathcal{F}$ are the same as those of Rainbow.

Now, we give security analysis of ELSA with the central map $\widehat{\mathcal{F}}$ against the Kipnis-Shamir attacks. We first define the following four index sets as

$$D_1 = \{i|1 \le i \le l\},\ D_2 = \{i|l+1 \le i \le l+k\},$$
$$D_3 = \{i|l+k+1 \le i \le l+k+r\},\ D_4 = \{i|l+k+r+1 \le i \le n\}.$$

We define five meaningful subspaces of $\mathbb{F}_q^n$ for the attacks on ELSA as

$$V_{1000} = \{(x_1,\cdots,x_n)|x_i = 0, i \notin D_1\}, V_{0100} = \{(x_1,\cdots,x_n)|x_i = 0, i \notin D_2\},$$
$$V_{0010} = \{(x_1,\cdots,x_n)|x_i = 0, i \notin D_3\}, V_{0001} = \{(x_1,\cdots,x_n)|x_i = 0, i \notin D_4\},$$
$$V_{1110} = \{(x_1,\cdots,x_n)|x_i = 0, i \in D_4\}.$$

The goal of the attacks is to find the preimage of the above subspaces under an equivalent key T'. We use the following property: any linear combinations of the matrices $\widehat{F}^{(1)}, \cdots, \widehat{F}^{(m)}$ is of the form $\begin{pmatrix} * & * & * & 0 \\ * & * & * & 0 \\ * & * & * & * \\ 0 & 0 & * & 0 \end{pmatrix} \cdots (*)$ from Fig. 2. The following Theorems show why invariant subspaces exist with a certain probability.

Lemma 3.3. Let $\phi : \mathbb{F}_q^n \to \mathbb{F}_q^n$ be a linear transformation of the form $(*)$. Then we get that $\phi(V_{0001})$, $\phi(V_{1000})$, and $\phi(V_{0100})$ are subspaces of V_{0010}, V_{1110} and V_{1110}, respectively.

Note that the image of other subspaces except the three subspaces in Lemma 3.3 under the map ϕ is the full space $\mathbb{F}_q^n$.

Let $H = \sum_{i=1}^{m} \lambda_i \widehat{F}^{(i)}$ be a linear combination of the matrices $\widehat{F}^{(i)}$. Note that H has the form of $(*)$. Then we get the following Theorem as in [34].

Theorem 3.2. Assume that, for some k $(1 \le k \le m)$, the matrix $\widehat{F}^{(k)}$ is invertible. Then, the map $(\widehat{F}^{(k)})^{-1} \cdot H$ has nontrivial invariant subspace $\phi(V_{0001})$, $\phi(V_{1000})$ and $\phi(V_{0100})$ with probability not less than q^{-r+u}, q^{-k-r} and q^{-l-r}, respectively.

Proof. They are obtained from the following fact: $[(\widehat{F}^{(k)})^{-1} \cdot \widehat{F}^{(i)}](V_{0001}) \subset (\widehat{F}^{(k)})^{-1}(V_{0010})$ and $V_{0001} \subset (\widehat{F}^{(k)})^{-1}(V_{0010})$, let $\Phi = (\widehat{F}^{(k)})^{-1} \cdot \widehat{F}^{(i)}$, then as in [33], we have

$$Pr[\Phi(V_{0001}) \subset V_{0010}] \ge q^{-r+u},$$

where $u = dim(V_{0001})$ and $r = dim(V_{0010})$. Thus, we get a nontrivial invariant subspace V_{0001} with probability not less than q^{-r+u}. □

Theorem 3.3. Let $W = \sum_{i=1}^{m} \lambda_i P^{(i)}$ be a linear combination of the matrices $P^{(i)}$ and let $P^{(k)}$ (for some k, $1 \le k \le m$) be invertible. Then the map $(P^{(k)})^{-1} \cdot W$ has nontrivial invariant subspaces V_{0010}, V_{1110} and V_{1110} which are subspaces of $T^{-1}(V_{0010})$, $T^{-1}(V_{1000})$ and $T^{-1}(V_{0100})$ with probability not less than q^{-r+u}, q^{-k-r} and q^{-l-r}, respectively.

Proof. They are obtained from the Theorem 3.2 and the following:

$$(P^{(k)})^{-1} \cdot W = (P^{(k)})^{-1} \cdot \sum_{i=1}^{m} \lambda_i P^{(i)} = (T^T \cdot F^{(k)} \cdot T)^{-1} \cdot \sum_{i=1}^{m} \lambda_i \cdot (T^T \cdot \widehat{F}^{(i)} \cdot T)$$
$$= T^{-1} \cdot (\sum_{i=1}^{m} \lambda_i (\widehat{F}^{(k)})^{-1} \cdot \widehat{F}^{(i)}) \cdot T.$$

$$(P^{(k)})^{-1} \cdot W(T^{-1}(V_{0001})) = (T^{-1} \cdot (\sum_{i=1}^{m} \lambda_i (\widehat{F}^{(k)})^{-1} \cdot \widehat{F}^{(i)}) \cdot T)(T^{-1}(V_{0001}))$$
$$= T^{-1} \cdot (\sum_{i=1}^{m} \lambda_i (F^{(k)})^{-1} \cdot F^{(i)})(V_{0001}) \subset T^{-1}(V_{0010}).$$

Thus, we get a nontrivial invariant subspace V_{0001} with probability not less than q^{-r+u}. □

Consequently, the complexity of ELSA against the Kipnis-Shamir attack is $min\{q^{r-u}, q^{k+r}, q^{l+r}, q^{n-2u-1} \cdot u^4\}$.

Based on these security analysis, we can select secure parameter sets $(\mathbb{F}_q, l, k, r, u)$ that achieve given security levels.

4 Existential Unforgeability of ELSA

Here, we prove existential unforgeability of ELSA against an adaptive chosen-message attack under the hardness of the MQ-problem induced by a public key of ELSA.

4.1 Formal Security Model and Complexity Assumption

In this section, we describe formal security models of signature schemes. The most general security notion of signature schemes is existential unforgeability against an adaptive chosen-message attack. Its formal security model is defined as follows:

EXISTENTIAL UNFORGEABILITY AGAINST ADAPTIVE CHOSEN-MESSAGE ATTACKS (EUF-acma). An adversary $\mathcal{A}$'s advantage $Adv_{\mathcal{PKS},\mathcal{A}}$ is defined as its probability of success in the following game between a challenger $\mathcal{C}$ and $\mathcal{A}$:

- **Setup.** The challenger runs `Setup` algorithm and its resulting system parameters are given to $\mathcal{A}$.
- **Sign Queries.** $\mathcal{A}$ issues the following queries: adaptively, $\mathcal{A}$ requests a signature on a message $\mathtt{m}_i$, $\mathcal{C}$ returns a signature σ_i.
- **Output.** Eventually, $\mathcal{A}$ outputs σ^* on a message $\mathtt{m}^*$ and wins the game if
  (i) $\mathtt{Verify}(\mathtt{m}^*, \sigma^*) = 1$,
  (ii) $\mathtt{m}^*$ has never requested to the `Sign` oracle.

Definition 4.1. A forger $\mathcal{A}(t, g_H, q_S, \epsilon)$-breaks a signature scheme if $\mathcal{A}$ runs in time at most t, $\mathcal{A}$ makes at most q_H queries to the hash oracle, q_S queries to the signing oracle and $Adv_{\mathcal{PKS},\mathcal{A}}$ is at least ϵ. A signature scheme is (t, q_E, q_S, ϵ)-EUF-acma if no forger (t, q_H, q_S, ϵ)-breaks it in the above game.

Next, we need to define the following sets as:

- $\mathcal{MQ}_{ELSA}(\mathbb{F}_q, m, n)$: a set of all quadratic equations defined over $\mathbb{F}_q$ with m equations and n variables induced by all public keys of ELSA$(\mathbb{F}_q, l, k, r, u)$, where $m = k + u$ and $n = l + r + m$.
- $\mathcal{MQ}_R(\mathbb{F}_q, m, n)$: a set of all random quadratic equations defined over $\mathbb{F}_q$ of m equations and n variables.

Definition 4.2. We say that the MQ-problem in $\mathcal{MQ}_X(\mathbb{F}_q, m, n)$ is (t, ϵ)-hard if no t-time algorithm has advantage at least ϵ in solving the MQ-problem in $\mathcal{MQ}_X(\mathbb{F}_q, m, n)$.

To prove existential unforgeability of ELSA against an adaptive chosen-message attack, we want to find a reduction to the hardness of MQ-problem in $\mathcal{MQ}_{ELSA}(\mathbb{F}_q, m, n)$. The hardness of the MQ-problem for a system of m quadratic equations with n variables mainly depends on the selection of $\mathbb{F}_q$, m and n. However, the security of ELSA against the attacks presented in §3 depends on the selection of the specific parameter set $(\mathbb{F}_q, l, k, r, u)$ such that

$m = k + u$ and $n = l + r + m$. If the parameter set $(\mathbb{F}_q, l, k, r, u)$ is chosen be secure against the MinRank attack, HighRank attack and Kipnis-Shamir attack, then it remains only two attacks to consider: the direct attack and KRAs with good keys. In Theorem 3.1, we have shown that the security of KRAs with good keys for ELSA is still reduced to the intractability of the MQ-problem, i.e., the complexity of the KRAs using good keys on ELSA is determined by solving $n-1$ bihomogeneous equations and m quadratic equations with n variables.

4.2 Existential Unforgeability

Now, we prove existential unforgeability of ELSA against an adaptive chosen-message attack under the hardness of the MQ-problem induced by a public key of ELSA in the random oracle model.

Theorem 4.1. If the MQ-problem in $\mathcal{MQ}_{ELSA}(\mathbb{F}_q, m, n)$ is (t', ε')-hard, ELSA $(\mathbb{F}_q, l, k, r, u)$ is $(t, q_H, q_S, \varepsilon)$-EUF-acma, for any t and ε satisfying

$$\varepsilon \geq \mathsf{e} \cdot (q_S + 1) \cdot \varepsilon', \; t' \geq t + q_H \cdot c_V + q_S \cdot c_S,$$

where e is the base of the natural logarithm, and c_S and c_V are time for a signature generation and a signature verification, respectively, where $m = k + u$, and $n = l + r + m$ if the parameter set $(\mathbb{F}_q, l, k, r, u)$ is chosen to be secure against the MinRank attack, HighRank attack, Kipnis-Shamir attack and KRAs using good keys.

Proof. An instance $(\mathcal{P}, \eta)$ of the MQ-problem in $\mathcal{MQ}_{ELSA}(\mathbb{F}_q, m, n)$ is given, where $\mathcal{P}$ is a quadratic system of m equations and n variables. Suppose that $\mathcal{A}$ is a forger who breaks ELSA$(\mathbb{F}_q, l, k, r, u)$ with the target public key $\mathcal{P}$. We will construct an algorithm $\mathcal{B}$ which outputs a solution $\mathbf{x} \in \mathbb{F}_q^n$ such that $\mathcal{P}(\mathbf{x}) = \eta$ by using $\mathcal{A}$. Algorithm $\mathcal{B}$ performs the following simulation by interacting with $\mathcal{A}$.

Setup. Algorithm $\mathcal{B}$ sets $PK = \mathcal{P}$, which is a public key of ELSA$(\mathbb{F}_q, l, k, r, u)$.

At any time, $\mathcal{A}$ can query a random oracle H and `Sign` oracle. To answer these queries, $\mathcal{B}$ does the following:

H-Queries. For H-queries, $\mathcal{B}$ maintains a list of tuples $(\mathtt{m}_i, c_i, \tau_i)$ as explained below. We call this list H`-list`. When $\mathcal{A}$ queries H at $\mathtt{m}_i \in \{0, 1\}^*$,

1. If the query already appears on H`-list` in a tuple $(\mathtt{m}_i, c_i, \tau_i, \mathcal{P}(\tau_i))$ then $\mathcal{B}$ returns $H(\mathtt{m}_i) = \mathcal{P}(\tau_i)$.
2. Otherwise, $\mathcal{B}$ picks a random coin $c_i \in \{0, 1\}$ with $Pr[c_i = 0] = \frac{1}{q_S+1}$.
   - If $c_i = 1$ then $\mathcal{B}$ chooses a random $\tau_i \in \mathbb{F}_q^n$, adds a tuple $(\mathtt{m}_i, c_i, \tau_i, \mathcal{P}(\tau_i))$ to H`-list` and returns $H(\mathtt{m}_i) = \mathcal{P}(\tau_i)$.
   - If $c_i = 0$ then $\mathcal{B}$ adds $(\mathtt{m}_i, c_i, *, \eta)$ to H`-list` from the instance and returns $H(\mathtt{m}_i) = \eta$.

Sign Queries. When $\mathcal{A}$ makes a `Sign`-query on $\mathtt{m}_i$, $\mathcal{B}$ finds the corresponding tuple $(\mathtt{m}_i, c_i, \tau_i, \mathcal{P}(\tau_i))$ from H`-list`.

- If $c_i = 1$ then $\mathcal{B}$ responds with τ_i.
- If $c_i = 0$ then $\mathcal{B}$ reports failure and terminates.

All responses to Sign queries not aborted are valid. If $\mathcal{B}$ doesn't abort as a result of $\mathcal{A}$'s Sign query then $\mathcal{A}$'s view in the simulation is identical to its view in the real attack.

Output. Finally, $\mathcal{A}$ produces a signature τ^* on a message $\mathtt{m}^*$. If it is not valid then $\mathcal{B}$ reports failure and terminates. Otherwise, a query on m^* already appears on H-list in a tuple $(\mathtt{m}^*, c^*, \tau^*, \mathcal{P}(\tau^*))$: if $c_* = 1$ then reports failure and terminates. Otherwise, $c^* = 0$, i.e., $(c^*, \mathtt{m}^*, *, \eta)$, then $\mathcal{P}(\tau^*) = \eta$. Finally, $\mathcal{B}$ outputs τ^* is a solution of $\mathcal{P}$.

To show that $\mathcal{B}$ solves the given instance with probability at least ε', we analyze three events needed for $\mathcal{B}$ to succeed:

- E_1: $\mathcal{B}$ doesn't abort as a result of $\mathcal{A}$'s Sign query.
- E_2: $\mathcal{A}$ generates a valid and nontrivial signature forgery τ_i on $\mathtt{m}_i$.
- E_3: Event E_2 occurs, $c_i = 0$ for the tuple containing $\mathtt{m}_i$ in H-list.

Algorithm $\mathcal{B}$ succeeds if all of these events happen. The probability $Pr[E_1 \wedge E_3]$ is decomposed as

$$Pr[E_1 \wedge E_3] = Pr[E_1] \cdot Pr[E_2 \wedge E_1] \cdot Pr[E_3 | E_1 \wedge E_2] \quad \cdots \quad (**).$$

The probability that $\mathcal{B}$ doesn't abort as a result of $\mathcal{A}$'s Sign query is at least $(1 - \frac{1}{q_S+1})^{q_S}$ since $\mathcal{A}$ makes at most q_S queries to the Sign oracle. Thus, $Pr[E_1] \geq (1 - \frac{1}{q_S+1})^{q_S}$. If $\mathcal{B}$ doesn't abort as a result of $\mathcal{A}$'s Sign query then $\mathcal{A}$'s view is identical to its view in the real attack. Hence, $Pr[E_1 \wedge E_2] \geq \varepsilon$. Given that events E_1, E_2 and E_3 happened, $\mathcal{B}$ will abort if $\mathcal{A}$ generates a forgery with $c_i = 1$. Thus, all the remaining c_i are independent of $\mathcal{A}$'s view. Since $\mathcal{A}$ could not have issued a signature query for the output we know that c is independent of $\mathcal{A}$'s current view and therefore $Pr[c = 0 | E_1 \wedge E_2] = \frac{1}{q_S+1}$. Then we get $Pr[E_3 | E_1 \wedge E_2] \geq \frac{1}{q_S+1}$. From $(**)$, $\mathcal{B}$ produces the correct answer with probability at least

$$(1 - \frac{1}{q_S+1})^{q_S} \cdot \varepsilon \cdot \frac{1}{q_S+1} \geq \frac{1}{\mathsf{e}} \cdot \frac{\varepsilon}{(q_S+1)} \geq \varepsilon'.$$

Algorithm $\mathcal{B}$'s running time is the same as $\mathcal{A}$'s running time plus the time that takes to respond to q_H H-queries, and q_S Sign-queries. The H- and Sign-queries require a signature verification and a signature generation, respectively. We assume that a signature generation and a signature verification take time c_S and c_V, respectively. Thus, the total running time is at most $t' \geq t + q_H \cdot c_V + q_S \cdot c_S$. □

5 Selection of Parameter and Implementation

Here, we evaluate practical feasibility of ELSA targeting a recent Intel processor. We choose a secure and optimal parameter for ELSA and provide comparisons between ours, classical ones and Post-Quantum ones in terms of performance, key sizes and signature sizes.

5.1 Selection of Secure and Optimal Parameter

We want to select secure parameter set $(\mathbb{F}_q, l, k, r, u)$ for ELSA with the optimal secret key size at a 128-bit security level where $m = k + u$ and $n = l + r + m$. Based on our security analysis in Sect. 3, we choose $(\mathbb{F}_{2^8}, 6, 28, 30, 15)$ at the 128-bit security level. We summarize complexities of our parameter against the known attacks in Table 3. For computing of complexities against direct attacks and KRAs using good keys, we use HF5 with $\omega = 2$.

Table 3. Complexities of ELSA$(\mathbb{F}_{2^8}, 6, 28, 30, 15)$ against all the attacks.

$(\mathbb{F}_q, l, k, r, u)$	Direct	KRA (Good)	Kipnis-Shamir attack	MinRank	HighRank
$(\mathbb{F}_{2^8}, 6, 28, 30, 15)$	2^{131}	2^{131}	2^{136}	2^{280}	2^{143}

Table 4. Performance, key sizes and signature sizes of ours, classical-ones and post-quantum ones.

Scheme λ	Sig. size (bytes)	PK (bytes)	SK (bytes)	Sign (bytes)	Verify (bytes)	CPU
Classical ones						
RSA-3072[e] 128	361	384	3072	8,802,242	87,360	Intel Core i5-6600 3.3 GHz
ECDSA-256[e] 128	64	64	96	163,994	310,048	Intel Core i5-6600 3.3 GHz
ed25519[e] [4] 128	64	32	64	48,976	165,322	Intel Core i5-6600 3.3 GHz
Lattice-based						
TESLA-416[t] [3] 128	1,280	1,331,200	1,011,744	697,940	250,264	Intel Core i7-4770K(Haswell)
TESLA-768[t] [3] > 128	2,336	4,227,072	3,293,216	2,232,906	863,790	Intel Core i7-4770K(Haswell)
BLISS-BI [19,20] 128	700	875	250	358,400	102,000	Intel Core i7 3.4 GHz
ntrumls 439x[e] [29] 128	988	1,112	1,305	485,580	223,488	Intel Core i5-6600 3.3 GHz
Hash-based						
SPHINCS 256[s] [5] 256	41,000	1,056	1,088	51,636,372	1,451,004	Intel Xeon E3-1275 3.5 GHz
Code-based						
CFS [35] 80	75	20,968,300	4,194,300	4,200,000,000	–	Intel Xeon W3070 3.2 GHz
MQ-based						
MQDSS-31-64 [13] > 128	40,952	72	64	8,510,616	5,752,616	Intel Core i7-4770K 3.5 GHz
enTTS $(\mathbb{F}_{2^8}, 15, 60, 88)$ [15,52] 128	88	234,960	13,051	–	–	–
Rainbow[o] $(\mathbb{F}_{2^8}, 36, 21, 22)$ [6] 128	79	139,320	105,006	64,658	44,397	Intel Core i5-6600 3.3 GHz
ELSA $(\mathbb{F}_{2^8}, 6, 28, 30, 15)$ 128	**79**	**139,320**	**12,427**	**20,880**	**44,190**	Intel Core i5-6600 3.3 GHz

Sig. Size, PK and SK represent signature size, public key and secret key, respectively.
ed25519 is EdDSA signatures using Curve25519.
>128 means that the scheme achieves 2^{λ} security level, where $\lambda > 128$.
[t] The scheme has a tight security reduction to the underlying problem.
[s] The scheme is provably secure in the standard model.
[e] The result is given by the eBACS project [6].
[o] We implement Rainbow based on the code in [6] at the 128-bit security level on Intel Core i5-6600 3.3 GHz.

5.2 Result and Comparison

We implement ELSA($\mathbb{F}_{2^8}$, 6, 28, 30, 15) on an Intel Core i5-6600 3.3 GHz whose result is an average of 1,000 measurements for each function using the C++ programming language with g++ compiler. We follow the standard practice of disabling Turbo Boost and hyperthreading. For comparison, we also implement Rainbow($\mathbb{F}_{2^8}$, 36, 21, 22) on the same platform based on open source codes given by the eBACS project [6] since there is no record for Rainbow at the 128-bit security level. Table 4 gives benchmarking results of ELSA and compares the benchmarks to state-of-the-art results from the literatures or given by the eBACS project [6].

Our scheme is the fastest signature scheme in both signing and verification among classical ones and Post-Quantum ones. Compared to Rainbow, the secret key size of ELSA is reduced by a factor of 88% maintaining the same public key size. Compared to enTTS, the public key size of ELSA have reduced by a factor of 40%. Signing of ELSA is about 3.2 times faster than that of Rainbow. Signing and verification of ELSA is hundreds of times faster than those of MQDSS, respectively. Signing and verification of ELSA is about 17.2 times and 2.3 times faster than those of BLISS-BI, respectively. It takes 6 μs and 13.3 μs for signing and verification, respectively.

6 Conclusion

We have proposed a new MQ-signature scheme, ELSA, based on a hidden layer of quadratic equations. Our scheme is the fastest signature scheme in both signing and verification among classical ones as well as Post-Quantum ones. Compared to Rainbow, the secret key size in ELSA is reduced by a factor of 88% maintaining the same public key size. Signing of ELSA is about 3.2 times than that of Rainbow on Intel Core i5-6600. It takes 6.3 μs and 13.39 μs for signing and verification, respectively. There is still room for improvements in terms of performance. We believe that our scheme is a leading candidate for low-cost constrained devices. We have also shown that ELSA($\mathbb{F}_q, l, k, r, u$) is existential unforgeable against an adaptive chosen-message attack under the hardness of the MQ-problem in $\mathcal{MQ}_{ELSA}(\mathbb{F}_q, m, n)$ in the random oracle model. However, this reduction doesn't mean the reduction to the MQ-problem in $\mathcal{MQ}_R(\mathbb{F}_q, m, n)$ although it haven't been proved that the public key of the MQ-schemes could be distinguished from random one. It still remains an open problem to construct a high-speed MQ-signature schemes with a security reduction to the hardness of the MQ-problem in $\mathcal{MQ}_R(\mathbb{F}_q, m, n)$.

References

1. El Yousfi Alaoui, S.M., Dagdelen, Ö., Véron, P., Galindo, D., Cayrel, P.-L.: Extended security arguments for signature schemes. In: Mitrokotsa, A., Vaudenay, S. (eds.) AFRICACRYPT 2012. LNCS, vol. 7374, pp. 19–34. Springer, Heidelberg (2012). https://doi.org/10.1007/978-3-642-31410-0_2

2. Albrecht, M.R., Faugére, J.-C., Fitzpatrick, R., Perret, L., Todo, Y., Xagawa, K.: Practical cryptanalysis of a public-key encryption scheme based on new multivariate quadratic assumptions. In: Krawczyk, H. (ed.) PKC 2014. LNCS, vol. 8383, pp. 446–464. Springer, Heidelberg (2014). https://doi.org/10.1007/978-3-642-54631-0_26
3. Alkim, E., Bindel, N., Buchmann, J., Dagdelen, O., Schwabe, P.: TESLA: tightly-secure efficient signatures from standard lattices, Cryptology ePrint Archive: Report 2015/755 (2015)
4. Bernstein, D.J.: Curve25519: new diffie-hellman speed records. In: Yung, M., Dodis, Y., Kiayias, A., Malkin, T. (eds.) PKC 2006. LNCS, vol. 3958, pp. 207–228. Springer, Heidelberg (2006). https://doi.org/10.1007/11745853_14
5. Bernstein, D.J., et al.: SPHINCS: practical stateless hash-based signatures. In: Oswald, E., Fischlin, M. (eds.) EUROCRYPT 2015. LNCS, vol. 9056, pp. 368–397. Springer, Heidelberg (2015). https://doi.org/10.1007/978-3-662-46800-5_15
6. Bernstein, D.J., Lange, T.: eBACS: ECRYPT benchmarking of cryptographic systems. http://bench.cr.yp.to. Accessed 30 Sept 2016
7. Bettale, L., Faugére, J.-C., Perret, L.: Hybrid approach for solving multivariate systems over finite fields. J. Math. Cryptol. **3**, 177–197 (2009)
8. Biham, E.: Cryptanalysis of patarin's 2-round public key system with S boxes (2R). In: Preneel, B. (ed.) EUROCRYPT 2000. LNCS, vol. 1807, pp. 408–416. Springer, Heidelberg (2000). https://doi.org/10.1007/3-540-45539-6_28
9. Billet, O., Gilbert, H.: Cryptanalysis of rainbow. In: Prisco, R., Yung, M. (eds.) SCN 2006. LNCS, vol. 4116, pp. 336–347. Springer, Heidelberg (2006). https://doi.org/10.1007/11832072_23
10. Biryukov, A., Bouillaguet, C., Khovratovich, D.: Cryptographic schemes based on the ASASA structure: black-box, white-box, and public-key (extended abstract). In: Sarkar, P., Iwata, T. (eds.) ASIACRYPT 2014. LNCS, vol. 8873, pp. 63–84. Springer, Heidelberg (2014). https://doi.org/10.1007/978-3-662-45611-8_4
11. Bogdanov, A., Eisenbarth, T., Rupp, A., Wolf, C.: Time-area optimized public-key engines: $\mathcal{MQ}$-cryptosystems as replacement for elliptic curves? In: Oswald, E., Rohatgi, P. (eds.) CHES 2008. LNCS, vol. 5154, pp. 45–61. Springer, Heidelberg (2008). https://doi.org/10.1007/978-3-540-85053-3_4
12. Chen, A.I.-T., et al.: SSE implementation of multivariate PKCs on modern x86 CPUs. In: Clavier, C., Gaj, K. (eds.) CHES 2009. LNCS, vol. 5747, pp. 33–48. Springer, Heidelberg (2009). https://doi.org/10.1007/978-3-642-04138-9_3
13. Chen, M.-S., Hülsing, A., Rijneveld, J., Samardjiska, S., Schwabe, P.: From 5-pass $\mathcal{MQ}$-based identification to $\mathcal{MQ}$-based signatures. In: Cheon, J.H., Takagi, T. (eds.) ASIACRYPT 2016. LNCS, vol. 10032, pp. 135–165. Springer, Heidelberg (2016). https://doi.org/10.1007/978-3-662-53890-6_5
14. Courtois, N.T.: Efficient zero-knowledge authentication based on a linear algebra problem MinRank. In: Boyd, C. (ed.) ASIACRYPT 2001. LNCS, vol. 2248, pp. 402–421. Springer, Heidelberg (2001). https://doi.org/10.1007/3-540-45682-1_24
15. Czypek, P., Heyse, S., Thomae, E.: Efficient implementations of MQPKS on constrained devices. In: Prouff, E., Schaumont, P. (eds.) CHES 2012. LNCS, vol. 7428, pp. 374–389. Springer, Heidelberg (2012). https://doi.org/10.1007/978-3-642-33027-8_22
16. Ding, J., Schmidt, D.: Rainbow, a new multivariable polynomial signature scheme. In: Ioannidis, J., Keromytis, A., Yung, M. (eds.) ACNS 2005. LNCS, vol. 3531, pp. 164–175. Springer, Heidelberg (2005). https://doi.org/10.1007/11496137_12

17. Ding, J., Yang, B.-Y., Chen, C.-H.O., Chen, M.-S., Cheng, C.-M.: New differential-algebraic attacks and reparametrization of rainbow. In: Bellovin, S.M., Gennaro, R., Keromytis, A., Yung, M. (eds.) ACNS 2008. LNCS, vol. 5037, pp. 242–257. Springer, Heidelberg (2008). https://doi.org/10.1007/978-3-540-68914-0_15
18. Ding-Feng, Y., Kwok-Yan, L., Zong-Duo, D.: Cryptanalysis of "2R" schemes. In: Wiener, M. (ed.) CRYPTO 1999. LNCS, vol. 1666, pp. 315–325. Springer, Heidelberg (1999). https://doi.org/10.1007/3-540-48405-1_20
19. Ducas, L.: Accelerating bliss: the geometry of ternary polynomials, Cryptology ePrint Archive: Report 2014/874 (2014)
20. Ducas, L., Durmus, A., Lepoint, T., Lyubashevsky, V.: Lattice signatures and bimodal Gaussians. In: Canetti, R., Garay, J.A. (eds.) CRYPTO 2013. LNCS, vol. 8042, pp. 40–56. Springer, Heidelberg (2013). https://doi.org/10.1007/978-3-642-40041-4_3
21. Düll, M., Haase, B., Hinterwälder, G., Hutter, M., Paar, C., Sánchez, A.H., Schwabe, P.: High-speed curve25519 on 8-bit, 16-bit and 32-bit microcontrollers. Des. Codes Crypt. **77**(2–3), 493–514 (2015)
22. Faugére, J.-C.: A new efficient algorithm for computing Gröbner bases without reduction to zero (F5). In: ISSAC 2002, pp. 75–83 (2002)
23. Faugère, J.-C., Gligoroski, D., Perret, L., Samardjiska, S., Thomae, E.: A polynomial-time key-recovery attack on MQQ cryptosystems. In: Katz, J. (ed.) PKC 2015. LNCS, vol. 9020, pp. 150–174. Springer, Heidelberg (2015). https://doi.org/10.1007/978-3-662-46447-2_7
24. Faugère, J.-C., Levy-dit-Vehel, F., Perret, L.: Cryptanalysis of MinRank. In: Wagner, D. (ed.) CRYPTO 2008. LNCS, vol. 5157, pp. 280–296. Springer, Heidelberg (2008). https://doi.org/10.1007/978-3-540-85174-5_16
25. Faugére, J.-C., Perret, L.: High order derivatives and decomposition of multivariate polynomials. In: ACM International Symposium on Symbolic and Algebraic Computation, pp. 207–214 (2009)
26. Garey, M.R., Johnson, D.S.: Computers and Intractability: A Guide to the Theory of NP-Completeness. W.H. Freeman and Company, New York (1979)
27. Gilbert, H., Plût, J., Treger, J.: Key-recovery attack on the ASASA cryptosystem with expanding S-boxes. In: Gennaro, R., Robshaw, M. (eds.) CRYPTO 2015. LNCS, vol. 9215, pp. 475–490. Springer, Heidelberg (2015). https://doi.org/10.1007/978-3-662-47989-6_23
28. Gligoroski, D., Ødegård, R.S., Jensen, R.E., Perret, L., Faugère, J.-C., Knapskog, S.J., Markovski, S.: MQQ-SIG. In: Chen, L., Yung, M., Zhu, L. (eds.) INTRUST 2011. LNCS, vol. 7222, pp. 184–203. Springer, Heidelberg (2012). https://doi.org/10.1007/978-3-642-32298-3_13
29. Hoffstein, J., Pipher, J., Schanck, J.M., Silverman, J.H., Whyte, W.: Transcript secure signatures based on modular lattices. In: Mosca, M. (ed.) PQCrypto 2014. LNCS, vol. 8772, pp. 142–159. Springer, Cham (2014). https://doi.org/10.1007/978-3-319-11659-4_9
30. Hoffstein, J., Pipher, J., Silverman, J.H.: NTRU: a ring-based public key cryptosystem. In: Buhler, J.P. (ed.) ANTS 1998. LNCS, vol. 1423, pp. 267–288. Springer, Heidelberg (1998). https://doi.org/10.1007/BFb0054868
31. Huang, Y.-J., Liu, F.-H., Yang, B.-Y.: Public-key cryptography from new multivariate quadratic assumptions. In: Fischlin, M., Buchmann, J., Manulis, M. (eds.) PKC 2012. LNCS, vol. 7293, pp. 190–205. Springer, Heidelberg (2012). https://doi.org/10.1007/978-3-642-30057-8_12

32. Hülsing, A., Rijneveld, J., Song, F.: Mitigating multi-target attacks in hash-based signatures. In: Cheng, C.-M., Chung, K.-M., Persiano, G., Yang, B.-Y. (eds.) PKC 2016. LNCS, vol. 9614, pp. 387–416. Springer, Heidelberg (2016). https://doi.org/10.1007/978-3-662-49384-7_15
33. Kipnis, A., Patarin, J., Goubin, L.: Unbalanced oil and vinegar signature schemes. In: Stern, J. (ed.) EUROCRYPT 1999. LNCS, vol. 1592, pp. 206–222. Springer, Heidelberg (1999). https://doi.org/10.1007/3-540-48910-X_15
34. Kipnis, A., Shamir, A.: Cryptanalysis of the oil and vinegar signature scheme. In: Krawczyk, H. (ed.) CRYPTO 1998. LNCS, vol. 1462, pp. 257–266. Springer, Heidelberg (1998). https://doi.org/10.1007/BFb0055733
35. Landais, G., Sendrier, N.: Implementing CFS. In: Galbraith, S., Nandi, M. (eds.) INDOCRYPT 2012. LNCS, vol. 7668, pp. 474–488. Springer, Heidelberg (2012). https://doi.org/10.1007/978-3-642-34931-7_27
36. Matsumoto, T., Imai, H.: Public quadratic polynomial-tuples for efficient signature-verification and message-encryption. In: Barstow, D., Brauer, W., Brinch Hansen, P., Gries, D., Luckham, D., Moler, C., Pnueli, A., Seegmüller, G., Stoer, J., Wirth, N., Günther, C.G. (eds.) EUROCRYPT 1988. LNCS, vol. 330, pp. 419–453. Springer, Heidelberg (1988). https://doi.org/10.1007/3-540-45961-8_39
37. McEliece, R.: A public-key cryptosystem based on algebraic coding theory, DSN progress report 42–44. Jet Propulsion Laboratories, Pasadena (1978)
38. Merkle, R.C.: A digital signature based on a conventional encryption function. In: Pomerance, C. (ed.) CRYPTO 1987. LNCS, vol. 293, pp. 369–378. Springer, Heidelberg (1988). https://doi.org/10.1007/3-540-48184-2_32
39. Minaud, B., Derbez, P., Fouque, P.-A., Karpman, P.: Key-recovery attacks on ASASA. In: Iwata, T., Cheon, J.H. (eds.) ASIACRYPT 2015. LNCS, vol. 9453, pp. 3–27. Springer, Heidelberg (2015). https://doi.org/10.1007/978-3-662-48800-3_1
40. Patarin, J.: Hidden fields equations (HFE) and isomorphisms of polynomials (IP): two new families of asymmetric algorithms. In: Maurer, U. (ed.) EUROCRYPT 1996. LNCS, vol. 1070, pp. 33–48. Springer, Heidelberg (1996). https://doi.org/10.1007/3-540-68339-9_4
41. Patarin, J.: The oil and vinegar signature scheme. In: Dagstuhl Workshop on Cryptography, September 1997
42. Patarin, J., Courtois, N., Goubin, L.: QUARTZ, 128-bit long digital signatures. In: Naccache, D. (ed.) CT-RSA 2001. LNCS, vol. 2020, pp. 282–297. Springer, Heidelberg (2001). https://doi.org/10.1007/3-540-45353-9_21
43. Patarin, J., Goubin, L.: Asymmetric cryptography with S-boxes is it easier than expected to design efficient asymmetric cryptosystems? In: Han, Y., Okamoto, T., Qing, S. (eds.) ICICS 1997. LNCS, vol. 1334, pp. 369–380. Springer, Heidelberg (1997). https://doi.org/10.1007/BFb0028492
44. Petzoldt, A.: Selecting and reducing key sizes for multivariate cryptography, Ph.D. thesis (2013)
45. Petzoldt, A., Chen, M.-S., Yang, B.-Y., Tao, C., Ding, J.: Design principles for HFEv- based multivariate signature schemes. In: Iwata, T., Cheon, J.H. (eds.) ASIACRYPT 2015. LNCS, vol. 9452, pp. 311–334. Springer, Heidelberg (2015). https://doi.org/10.1007/978-3-662-48797-6_14
46. Pöppelmann, T., Oder, T., Güneysu, T.: High-performance ideal lattice-based cryptography on 8-bit ATxmega microcontrollers. In: Lauter, K., Rodríguez-Henríquez, F. (eds.) LATINCRYPT 2015. LNCS, vol. 9230, pp. 346–365. Springer, Cham (2015). https://doi.org/10.1007/978-3-319-22174-8_19

47. Sakumoto, K., Shirai, T., Hiwatari, H.: Public-key identification schemes based on multivariate quadratic polynomials. In: Rogaway, P. (ed.) CRYPTO 2011. LNCS, vol. 6841, pp. 706–723. Springer, Heidelberg (2011). https://doi.org/10.1007/978-3-642-22792-9_40
48. Shallit, J.O., Frandsen, G.S., Buss, J.F.: The computational complexity of some problems of linear algebra, BRICS series report, Aarhus, Denmark, RS-96-33. http://www.brics.dk/RS/96/33
49. Shor, P.W.: Polynomial-time algorithms for prime factorization and discrete logarithms on a quantum computer. SIAM J. Comput. **26**(5), 1484–1509 (1997)
50. Thomae, E.: About the security of multivariate quadratic public key schemes, Dissertation Thesis by Dipl. math. E. Thomae, RUB (2013)
51. Wolf, C., Preneel, B.: Large superfluous keys in $\mathcal{M}$ultivariate $\mathcal{Q}$uadratic asymmetric systems. In: Vaudenay, S. (ed.) PKC 2005. LNCS, vol. 3386, pp. 275–287. Springer, Heidelberg (2005). https://doi.org/10.1007/978-3-540-30580-4_19
52. Yang, B.-Y., Chen, J.-M.: Building secure tame-like multivariate public-key cryptosystems: the new TTS. In: Boyd, C., González Nieto, J.M. (eds.) ACISP 2005. LNCS, vol. 3574, pp. 518–531. Springer, Heidelberg (2005). https://doi.org/10.1007/11506157_43

Post-quantum Security of Fiat-Shamir

Dominique Unruh(✉)

University of Tartu, Tartu, Estonia
unruh@ut.ee

Abstract. The Fiat-Shamir construction (Crypto 1986) is an efficient transformation in the random oracle model for creating non-interactive proof systems and signatures from sigma-protocols. In classical cryptography, Fiat-Shamir is a zero-knowledge proof of knowledge assuming that the underlying sigma-protocol has the zero-knowledge and special soundness properties. Unfortunately, Ambainis, Rosmanis, and Unruh (FOCS 2014) ruled out non-relativizing proofs under those conditions in the quantum setting.

In this paper, we show under which strengthened conditions the Fiat-Shamir proof system is still post-quantum secure. Namely, we show that if we require the sigma-protocol to have computational zero-knowledge and *statistical* soundness, then Fiat-Shamir is a zero-knowledge simulation-sound proof system (but not a proof of knowledge!). Furthermore, we show that Fiat-Shamir leads to a post-quantum secure unforgeable signature scheme when additionally assuming a "dual-mode hard instance generator" for generating key pairs.

Keywords: Post-quantum security · Fiat-Shamir · Non-interactive proof systems · Signatures

1 Introduction

1.1 Background

Fiat-Shamir signatures. Signatures are (next to encryption) probably one of the most important constructs in modern cryptography. In search for efficient signature schemes, Fiat-Shamir [12] gave a construction for transforming many three-round identification schemes into signatures, using the random oracle. (The transformation was stated only for a specific case, but the general construction is an easy generalization. [12] also does not contain a complete security proof, but a proof was later provided by Pointcheval and Stern [20].) The Fiat-Shamir transform and variations thereof have since been used in a large number of constructions (signatures [21,23], group signatures [7], anonymous credentials [10], e-voting [1], anonymous attestation [9], etc.) The benefit of the Fiat-Shamir transform is that it combines efficiency with universality: The underlying identification scheme can be any so-called sigma-protocol (see below), this allows

T. Takagi and T. Peyrin (Eds.): ASIACRYPT 2017, Part I, LNCS 10624, pp. 65–95, 2017.
https://doi.org/10.1007/978-3-319-70694-8_3

for great flexibility in how public and secret key are related and enables the construction of more advanced signature schemes and related schemes such as group signatures, etc.

Non-interactive zero-knowledge proofs. At the first glance unrelated, but upon closer inspection intimately connected to signatures are non-interactive zero-knowledge proof of knowledge (NIZKPoK). In fact, Fiat-Shamir can also be seen as a highly efficient construction for NIZKPoKs in the random oracle model [11]. Basically, a NIZKPoK allows a prover to show his knowledge of a witness sk that stands in a given relation to a publicly known statement pk. From a NIZKPoK, we can derive a signature scheme: To sign a message m, the signer constructs a proof that he knows the secret key corresponding to the public key pk. (Of course, the message m needs to be included in the proof as well, we omit the details for now.) For this construction to work, the NIZKPoK needs to satisfy certain advanced security notions ("simulation-sound extractability");[1] Fiat-Shamir satisfies this notion in the classical setting [11]. Thus Fiat-Shamir doubles both as a signature scheme and as a NIZKPoK, leading to simple and highly efficient constructions of both.

The construction. In order to understand the rest of this introduction more easily, we sketch the construction of Fiat-Shamir (the precise definition is given in Definition 11). We will express it as a NIZKPoK since this makes the analysis more modular. (We study Fiat-Shamir as a signature scheme in Sect. 6.)

A sigma-protocol Σ is a three-message protocol: The prover (given a statement x and a corresponding valid witness w) sends a message com, called "commitment", to the verifier. The verifier (who knowns only the statement x) responds with a uniformly random "challenge" ch. Then the prover answers with his "response" $resp$, and the verifier checks whether $(com, ch, resp)$ is a valid interaction. If so, he accepts the proof of the statement x. In the following, we will assume that ch has superlogarithmic length, i.e., there are superpolynomially many different challenges. This can always be achieved by parallel-composing the sigma-protocol.

Given the sigma-protocol Σ, the Fiat-Shamir transform yields a non-interactive proof system: The prover P_{FS} internally executes the prover of the sigma-protocol to get the commitment com. Then he computes the challenge as $ch := H(x\|com)$ where H is a hash function, modeled as a random oracle. That is, instead of letting the verifier generate a random challenge, the prover produces it by hashing. This guarantees, at least on an intuitively level, that the prover does not have any control over the challenge, it is as if it was chosen randomly. Then the prover internally produces the response $resp$ corresponding to com and ch and sends the non-interactive proof $com\|resp$ to the verifier.

The Fiat-Shamir verifier V_{FS} computes $ch := H(x\|com)$ and checks whether $(com, ch, resp)$ is a valid interaction of the sigma-protocol.

---

[1] We do not know where this was first shown, a proof in the quantum case can be found in [26].

Note that numerous variants of the Fiat-Shamir are possible. For example, one could compute $ch := H(com)$ (omitting x). However, this variant of Fiat-Shamir is malleable, see [11].

Difficulties with Fiat-Shamir. The Fiat-Shamir transform is a deceptively simple construction, but proving its security turns out to be more involved that one would anticipate. To prove security (specifically, the unforgeability property in the signature setting, or the extractability in the NIZKPoK setting), we need simulate the interaction of the adversary with the random oracle, and then rerun the same interaction with slightly changed random oracle responses ("rewinding"). The first security proof by Fiat and Shamir [12] overlooked that issue.[2] Bellare and Rogaway [5, Sect. 5.2] also prove the security of the Fiat-Shamir transform (as a proof system) but simply claim the soundness without giving a proof (we assume that they also overlooked the difficulties involved).[3] The first complete security proof of the Fiat-Shamir as a signature scheme is by Pointcheval and Stern [20] who introduced the so-called "forking lemma", a central tool for analyzing the security of Fiat-Shamir (it allows us to analyze the rewinding used in the security proof). When considering Fiat-Shamir as a NIZKPoK, the first proof was given by Faust et al. [11]; they showed that Fiat-Shamir is zero-knowledge and simulation-sound extractable.[4] This short history of the security proofs indicates that Fiat-Shamir is more complicated than it may look at the first glance.

Further difficulties were noticed by Shoup and Gennaro [24] who point out that the fact that the Fiat-Shamir security proof uses rewinding can lead to considerable difficulties in the analysis of more complex security proofs (namely, it may lead to an exponential blowup in the running time of a simulator; Pointcheval and Stern [19] experienced similar problems). Fischlin [13] notes that the rewinding also leads to less tight reductions, which in turn may lead to longer key sizes etc. for protocols using Fiat-Shamir.

Another example of unexpected behavior: Assume Alice gets a n pairs of public keys (pk_{i0}, pk_{i1}), and then can ask for *one* of the secret keys for each pair (i.e., sk_{i0} or sk_{i1} is revealed, never both), and then Alice is supposed to prove using Fiat-Shamir that he knows *both* secret keys for *one* of the pairs. Intuitively, we expect Alice not to be able to do that (if Fiat Shamir is indeed

[2] The proof of [12, Lemma 6] claims without proof that a successful adversary cannot find a square root mod n of $\prod_{j=1}^{k} v_j^{c_j}$. In hindsight, this proof step would implicitly use the forking lemma [20] that was developed only nine years later. [12] also mentions a full version of their paper, but to the best of our knowledge no such full version has ever appeared.

[3] A "final paper" is also mentioned, but to the best of our knowledge never appeared.

[4] They only sketch the zero-knowledge property, though. Their proof sketch overlooks one required property of the sigma-protocol: unpredictable commitments (Definition 6). Without this (easy to achieve) property, at least the simulator constructed in [11] will not work correctly. Concurrently and independently, [6] also claims the same security properties, but the theorems are given without any proof or proof idea.

a proof of knowledge), but as we show in the full version [27], Fiat-Shamir does not guarantee that Alice cannot successfully produce a proof in this situation!

To circumvent all those problems, Fischlin [13] gave an alternative construction of NIZKPoKs and signature schemes in the random oracle model whose security proof does not use rewinding. However, their construction seems less efficient in terms of the computation performed by the prover (although this is not fully obvious if the tightness of the reduction is taken into account), and their construction requires an additional property (unique responses[5]) from the underlying sigma-protocol.

We do not claim that those difficulties in proving and using Fiat-Shamir necessarily speak against Fiat-Shamir. But they show one needs to carefully analyze which precise properties Fiat-Shamir provably has, and not rely on what Fiat-Shamir intuitively achieves.

Post-quantum security. In this paper we are interested in the post-quantum security of Fiat-Shamir. That is, under what conditions is Fiat-Shamir secure if the adversary has a quantum computer? In the post-quantum setting, the random oracle has to be modeled as a random function that can be queried in superposition[6] since a normal hash function can be evaluated in superposition as well (cf. [8]). Ambainis et al. [2] showed that in this model, Fiat-Shamir is insecure in general. More precisely, they showed that relative to certain oracles, there are sigma-protocols such that: The sigma-protocol satisfies the usual security properties. (Such as zero-knowledge and special soundness. These are sufficient for security in the classical case.) But when applying the Fiat-Shamir transform to it, the resulting NIZKPoK is not sound (and thus, as a signature, not unforgeable). Since this negative result is relative to specific oracles, it does not categorically rule out a security proof. However, it shows that no relativizing security proof exists, and indicates that it is unlikely that Fiat-Shamir can be shown post-quantum secure in general. Analogous negative results [2] hold for Fischlin's scheme [13].

Unruh [26] gave a construction of a NIZKPoK/signature scheme in the random oracle model that is avoids these problems and is post-quantum secure (simulation-sound extractable zero-knowledge/strongly unforgeable). However, Unruh's scheme requires multiple executions of the underlying sigma-protocol, leading to increased computational and communication complexity in comparison with Fiat-Shamir which needs only a single execution.[7] Furthermore,

---

[5] Unique responses: It is computationally infeasible to find two valid responses for the same commitment/challenge pair. See Definition 6 below.

[6] E.g., the adversary can produce states such as $\sum_x 2^{-|x|/2}|x\rangle \otimes |H(x)\rangle$.

[7] This assumes that the underlying sigma-protocol has a large challenge space. If the underlying sigma-protocol has a small challenge space (e.g., the challenge is a bit) then for Fiat-Shamir the sigma-protocol needs to be parallel composed first to increase its challenge space. In this case, the complexity of Fiat-Shamir and Unruh are more similar. (See, e.g., [14] that compares (optimizations of) Fiat-Shamir and Unruh for a specific sigma-protocol and concludes that Unruh has an overhead in communication complexity of merely 60% compared to Fiat-Shamir.).

Fiat-Shamir is simpler (in terms of the construction, if not the proof), and more established in the crypto community. In fact, a number of papers have used Fiat-Shamir to construct post-quantum secure signature schemes (e.g., [3,4,15–18]). The negative results by Ambainis et al. show that the post-quantum security of these schemes is hard to justify.[8] Thus the post-quantum security of Fiat-Shamir would be of great interest, both from a practical and theoretical point of view.

Is there a possibility to show the security of Fiat-Shamir notwithstanding the negative results from [2]? There are two options (besides non-relativizing proofs): (a) Unruh [25] introduced an additional condition for sigma-protocols, so-called "perfectly unique responses".[9] Unique responses means that for any commitment and challenge in a sigma-protocol, there exists at most one valid response. They showed that a sigma-protocol that additionally has perfect unique responses is a proof of knowledge while [2] showed that without unique responses, a sigma protocol will not in general be a proof of knowledge (relative to some oracle). Similarly, [2] does not exclude that Fiat-Shamir is post-quantum secure when the underlying sigma-protocol has perfectly unique responses.[10] (b) If we do not require extractability, but only require soundness (i.e., if we only want to prove that there exists a witness, not that we know it), then [2] does not exclude a proof that Fiat-Shamir is sound based on a sigma-protocol with perfect special soundness (but (computational) special soundness is not sufficient). In this paper, we mainly follow approach (b), but we also have some results related to research direction (a).

1.2 Our Contribution

Security of Fiat-Shamir as a proof system. We prove that Fiat-Shamir is post-quantum secure as a proof system. More precisely, we prove that it is zero-knowledge (using random-oracle programming techniques from [26]), and that it is sound (i.e., a proof of knowledge, using a reduction to quantum search). More precisely:

Theorem 1 (Post-quantum security of Fiat-Shamir – informal). *Assume that Σ has honest-verifier zero-knowledge and* statistical *soundness.*

Then the Fiat-Shamir proof system (P_{FS}, V_{FS}) is zero-knowledge and sound.[11]

[8] We stress that the *classical* security of these schemes is not in question. Also, not all these papers explicitly claim to have post-quantum security. However, they all give constructions that are based on supposedly quantum hard assumptions. Arguably, one of the main motivations for using such assumptions is post-quantum security. Thus the papers do not claim wrong results, but they would be considerably strengthened by a proof of the post-quantum security of Fiat-Shamir.

[9] It is called "strict soundness" in [25] but we use the term "unique responses" to match the language used elsewhere in the literature, e.g., [13].

[10] Interestingly, *computational* unique responses as in footnote 5 are shown not to be sufficient, even when we want only *computational* extractability/unforgeability.

[11] We stress: It is sound in the sense of a proof system, but not known to be a proof *of knowledge*.

The assumptions are the same as in the classical setting, except that instead of computational special soundness (as in the classical case), we need statistical soundness.[12] This is interesting, because it means that we need one of the properties of the sigma-protocol to hold unconditionally, even though we only want computational security in the end. However, [2] shows that this is necessary: when assuming only computational (special) soundness, they construct a counter-example to the soundness of Fiat-Shamir (relative to some oracle).

Simulation-soundness. In addition to the above, we also show that Fiat-Shamir has simulation-soundness. Simulation-soundness is a property that guarantees non-malleability, i.e., that an adversary cannot take a proof gotten from, say, an honest participant and transform it into a different proof (potentially for a different but related statement).[13] This is particularly important when using Fiat-Shamir to construct signatures (see below) because we would not want the adversary to transform one signature into a different signature. Our result is:

Theorem 2 (Simulation-soundness of Fiat-Shamir – informal). *Assume that Σ has honest-verifier zero-knowledge, statistical soundness, and unique responses.*

Then the Fiat-Shamir proof system (P_{FS}, V_{FS}) has simulation-soundness.

Note that unique responses are needed for this result even in the classical case. If we only require a slightly weaker form of simulation-soundness ("weak" simulation-soundness), then we can omit that requirement.

Signatures. Normally, the security of Fiat-Shamir signatures is shown by reducing it to the simulation-sound extractability of Fiat-Shamir (implicitly or explicitly). Unfortunately, we do not know whether Fiat-Shamir is extractable in the quantum setting. Thus, we need a new proof of the security of Fiat-Shamir signatures that only relies on simulation-soundness. We can do so by making additional assumptions about the way the key generator works: We call an algorithm G a "dual-mode hard instance generator" if G outputs a key pair (pk, sk) in such a way that pk is computationally indistinguishable from an invalid pk (i.e., a pk that has no corresponding sk). An example of such an instance generator would be: sk is chosen uniformly at random, and $pk := F(sk)$ for a pseudo-random generator F. Then we have:

Theorem 3 (Fiat-Shamir signatures – informal). *Assume that G is a dual-mode hard instance generator. Fix a sigma-protocol Σ (for showing that a given public key has a corresponding secret key). Assume that Σ has honest-verifier zero-knowledge, statistical soundness.*

Then the Fiat-Shamir signature scheme is unforgeable.

Note that classically, we only require that G is a hard instance generator. That is, given pk, it is hard to find sk. We leave it as an open problem whether this is sufficient in the post-quantum setting, too.

[12] That is, soundness has to hold against computationally unlimited adversaries.

[13] Formally, simulation-soundness is defined by requiring that soundness holds even when the adversary has access to a simulator that produces fake proofs.

Organization. In Sect. 2, we fix some simple notation. In Sect. 3, we discuss the (relatively standard) security notions for sigma-protocols used in this paper. In Sect. 4, we define security notions for non-interactive proof systems in the random oracle model. In Sect. 5 we give out main results, the security properties of Fiat-Shamir (zero-knowledge, soundness, simulation-soundness, ...). In Sect. 6, we show how to construct signature schemes from non-interactive zero-knowledge proof systems, in particular from Fiat-Shamir.

Readers who are interested solely in conditions under which Fiat-Shamir signatures are post-quantum secure but not in the security proofs may restrict their attention to Sects. 3 and 6 (in particular Corallary 23).

A full version with additional material on extractability appears online [27].

2 Preliminaries

$\mathrm{Fun}(n,m)$ is the set of all functions from $\{0,1\}^n$ to $\{0,1\}^m$. $a \oplus b$ denotes the bitwise XOR between bitstrings (of the same length).

If H is a function, we write $H(x := y)$ for the function H' with $H'(x) = y$ and $H'(x') = H(x')$ for $x' \neq x$. We call a list $ass = (x_1 := y_1, \dots, x_n := y_n)$ an *assignment-list*. We then write $H(ass)$ for $H(x_1 := y_1)(x_2 := y_2)\dots(x_n := y_n)$. (That is, H is updated to return y_i on input x_i, with assignments occurring later in ass taking precedence.)

We write $x \leftarrow A(\dots)$ to denote that the result of the algorithm/measurement A is assigned to x. We write $Q \leftarrow |\Psi\rangle$ or $Q \leftarrow \rho$ to denote that the quantum register Q is initialized with the quantum state $|\Psi\rangle$ or ρ, respectively. We write $x \xleftarrow{\$} M$ to denote that x is assigned a uniformly randomly chosen element of the set M.

If H is a classical function, then A^H means that A has oracle access to H in superposition (i.e., to the unitary $|x,y\rangle \to |x, y \oplus H(x)\rangle$).

Theorem 4 (Random oracle programming [26]**).** *Let $\ell^{in}, \ell^{out} \geq 1$ be a integers, and $H \xleftarrow{\$} \mathrm{Fun}(\ell^{in}_\eta, \ell^{out}_\eta)$. Let A_C be an algorithm, and A_0, A_2 be oracles algorithms, where A_0^H makes at most q_A queries to H, A_C is classical, and the output of A_C has collision-entropy at least k given A_C's initial state (which is classical). A_0, A_C, A_2 may share state. Then*

$$\Big| \Pr[b = 1 : A_0^H(), xcom \leftarrow A_C(), ch := H(xcom), b \leftarrow A_2^H(ch)] - \Pr[b = 1 : A_0^H(), xcom \leftarrow A_C(), ch \xleftarrow{\$} \{0,1\}^m, H(xcom) := ch, b \leftarrow A_2^H(ch)] \Big| \leq (4 + \sqrt{2})\sqrt{q_A}\, 2^{-k/4}.$$

Lemma 5 (Hardness of search [27]**).** *Let $H : \{0,1\}^n \to \{0,1\}^m$ be a uniformly random function. For any q-query algorithm A, it holds that* $\Pr[H(x) = 0 : x \leftarrow A^H()] \leq 32 \cdot 2^{-m} \cdot (q+1)^2$.

3 Sigma Protocols

In this paper, we will consider only proof systems for *fixed-length relations* . A fixed-length relation R_η is a family of relations on bitstrings such that:

For every η, there are values ℓ^x_η and ℓ^w_η such that $(x,w) \in R_\eta$ implies $|x| = \ell^x_\eta$ and $|w| = \ell^w_\eta$, and such that ℓ^x_η, ℓ^w_η can be computed in time polynomial in η. Given x, w, it can be decided in polynomial-time in η whether $(x,w) \in R_\eta$.

We now define sigma protocols and related concepts. The notions in this section are standard in the classical setting, and easy to adapt to the quantum setting. Note that the definitions are formulated without the random oracle, we only use the random oracle later for constructing non-interactive proofs out of sigma protocols.

A *sigma protocol* for a fixed-length relation R_η is a three-message proof system. It is described by the lengths $\ell^{com}_\eta, \ell^{ch}_\eta, \ell^{resp}_\eta$ of the "commitments", "challenges", and "responses" (those lengths may depend on η), by a quantum-polynomial-time[14] prover (P^1_Σ, P^2_Σ) and a deterministic polynomial-time verifier V_Σ. We will commonly denote statement and witness with x and w (with $(x,w) \in R$ in the honest case). The first message from the prover is $com \leftarrow P^1_\Sigma(1^\eta, x, w)$ and is called the *commitment* and satisfies $com \in \{0,1\}^{\ell^{com}}$, the uniformly random reply from the verifier is $ch \xleftarrow{\$} \{0,1\}^{\ell^{ch}}$ (called *challenge*), and the prover answers with a message $resp \leftarrow P^2_\Sigma(1^\eta, x, w, ch)$ (the *response*) that satisfies $resp \in \{0,1\}^{\ell^{resp}}$. We assume P^1_Σ, P^2_Σ to share classical or quantum state. Finally $V_\Sigma(1^\eta, x, com, ch, resp)$ outputs 1 if the verifier accepts, 0 otherwise.

Definition 6 (Properties of sigma protocols). *Let $(\ell^{com}_\eta, \ell^{ch}_\eta, \ell^{resp}_\eta, P^1_\Sigma, P^2_\Sigma, V_\Sigma)$ be a sigma protocol. We define:*

- ***Completeness:*** *For any quantum-polynomial-time algorithm A, there is a negligible μ such that for all η,*

$$\Pr[(x,w) \in R_\eta \ \wedge\ V_\Sigma(1^\eta, x, com, ch, resp) = 0 : (x,w) \leftarrow A(1^\eta), \\ com \leftarrow P^1_\Sigma(1^\eta, x, w),\ ch \xleftarrow{\$} \{0,1\}^{\ell^{ch}_\eta},\ resp \leftarrow P^2_\Sigma(1^\eta, x, w, ch)] \le \mu(\eta).$$

- ***Statistical soundness:*** *There is a negligible μ such that for any stateful classical (but not necessarily polynomial-time) algorithm A and all η, we have that*

$$\Pr[ok = 1 \wedge x \notin L_R : (x, com) \leftarrow A(1^\eta),\ ch \xleftarrow{\$} \{0,1\}^{\ell^{ch}}, \\ resp \leftarrow A(1^\eta, ch), ok \leftarrow V_\Sigma(1^\eta, x, com, ch, resp)] \le \mu(\eta).$$

- ***Honest-verifier zero-knowledge (HVZK):*** *There is a quantum-polynomial-time algorithm S_Σ (the simulator) such that for any stateful*

[14] Typically, P^1_Σ and P^2_Σ will be classical, but we do not require this since our results also hold for quantum P^1_Σ, P^2_Σ. But the inputs and outputs of P^1_Σ, P^2_Σ are classical.

quantum-polynomial-time algorithm A there is a negligible μ such that for all η and $(x,w) \in R_\eta$,

$$\begin{aligned}\big|\Pr[b=1 : (x,w) \leftarrow A(1^\eta), com \leftarrow P^1_\Sigma(1^\eta,x,w), ch \xleftarrow{\$} \{0,1\}^{\ell^{ch}_\eta},\\ resp \leftarrow P^2_\Sigma(1^\eta,x,w,ch), b \leftarrow A(1^\eta, com, ch, resp)]\\ -\Pr[b=1 : (x,w) \leftarrow A(1^\eta), (com,ch,resp) \leftarrow S(1^\eta,x),\\ b \leftarrow A(1^\eta, com, ch, resp)]\big| \le \mu(\eta).\end{aligned}$$

- ***Perfectly unique responses:*** *There exist no values $\eta, x, com, ch, resp, resp'$ with $resp \neq resp'$ and $V_\Sigma(1^\eta, x, com, ch, resp) = 1$ and $V_\Sigma(1^\eta, x, com, ch', resp') = 1$.*
- ***Unique responses:*** *For any quantum-polynomial-time A, the following is negligible:*

$$\Pr\big[resp \neq resp' \wedge V_\Sigma(1^\eta,x,com,ch,resp) = 1 \wedge V_\Sigma(1^\eta,x,com,ch',resp') = 1 : (x, com, ch, resp, resp') \leftarrow A(1^\eta)\big].$$

- ***Unpredictable commitments:*** *The commitment has superlogarithmic collision-entropy. In other words, there is a negligible μ such that for all η and $(x,w) \in R_\eta$,*

$$\Pr[com_1 = com_2 : com_1 \leftarrow P^1_\Sigma(1^\eta,x,w),\ com_2 \leftarrow P^1_\Sigma(1^\eta,x,w)] \le \mu(\eta).$$

Note: the "unpredictable commitments" property is non-standard, but satisfied by all sigma-protocols we are aware of. However, any sigma-protocol without unpredictable commitments can be transformed into one with unpredictable commitments by appending superlogarithmically many random bits to the commitment (that are then ignored by the verifier).

4 Non-interactive Proof Systems (Definitions)

In the following, let H always denote a function $\{0,1\}^{\ell^{in}_\eta} \to \{0,1\}^{\ell^{out}_\eta}$ where $\ell^{in}_\eta, \ell^{out}_\eta$ may depend on the security parameter η. Let $\mathrm{Fun}(\ell^{in}_\eta, \ell^{out}_\eta)$ denote the set of all such functions.

A non-interactive proof system (P,V) for a relation R_η consists of a quantum-polynomial-time algorithm P and a deterministic polynomial-time algorithm V, both taking an oracle $H \in \mathrm{Fun}(\ell^{in}_\eta, \ell^{out}_\eta)$. $\pi \leftarrow P^H(1^\eta, x, w)$ is expected to output a proof π for the statement x using witness w. We require that $|\pi| = \ell^\pi_\eta$ for some length ℓ^π_η. (i.e., the length of a proof π depends only on the security parameter.) And $ok \leftarrow V^H(1^\eta, x, \pi)$ is supposed to return $ok = 1$ if the proof π is valid for the statement x. Formally, we define:

Definition 7 (Completeness). *(P,V) has* completeness *for a fixed-length relation R_η iff for any polynomial-time oracle algorithm A there is a negligible μ such that for all η,*

$$\Pr[(x,w) \in R_\eta \ \wedge\ V^H(1^\eta,x,\pi) = 0 : H \xleftarrow{\$} \mathrm{Fun}(\ell^{in}_\eta, \ell^{out}_\eta), (x,w) \leftarrow A^H(1^\eta),\ \pi \leftarrow P^H(1^\eta,x,w)] \le \mu(\eta).$$

For the following definition, a *simulator* is a classical stateful algorithm S. Upon invocation, $S(1^\eta, x)$ returns a proof π. Additionally, S may reprogram the random oracle. That is, S may choose an assignment-list ass, and H will then be replaced by $H(ass)$.

Definition 8 (Zero-knowledge). *Given a simulator S, the oracle $S'(x, w)$ runs $S(1^\eta, x)$ and returns the latter's output. Given a prover P, the oracle $P'(x, w)$ runs $P(1^\eta, x, w)$ and returns the latter's output.*

A non-interactive proof system (P, V) is zero-knowledge *iff there is a quantum-polynomial-time simulator S such that for every quantum-polynomial-time oracle algorithm A there is a negligible μ such that for all η and all normalized density operators ρ,*

$$\Big| \Pr[b = 1 : H \xleftarrow{\$} \mathrm{Fun}(\ell_\eta^{in}, \ell_\eta^{out}),\ b \leftarrow A^{H,P'}(1^\eta, \rho)] \\ - \Pr[b = 1 : H \xleftarrow{\$} \mathrm{Fun}(\ell_\eta^{in}, \ell_\eta^{out}),\ b \leftarrow A^{H,S'}(1^\eta, \rho)] \Big| \leq \mu(\eta). \tag{1}$$

Here we quantify only over A that never query $(x, w) \notin R$ from the P' or S'-oracle.

Definition 9 (Soundness). *A non-interactive proof system (P, V) is* sound *iff for any quantum-polynomial-time oracle algorithm A, there is a negligible function μ, such that for all η and all normalized density operators ρ,*

$$\Pr[ok_V = 1 \wedge x \notin L_R : (x, \pi) \leftarrow A^H(1^\eta, \rho),\ ok_V \leftarrow V^H(1^\eta, x, \pi)] \leq \mu(\eta).$$

Here $L_R := \{x : \exists w.(x, w) \in R\}$.

In some applications, soundness as defined above is not sufficient. Namely, consider a security proof that goes along the following lines: We start with a game in which the adversary interacts with an honest prover. We replace the honest prover by a simulator. From the zero-knowledge property it follows that this leads to an indistinguishable game. And then we try to use soundness to show that the adversary in the new game cannot prove certain statements.

The last proof step will fail: soundness guarantees nothing when the adversary interacts with a simulator that constructs fake proofs. Namely, it could be that the adversary can take a fake proof for some statement and changes it into a fake proof for another statement of its choosing. (Technically, soundness cannot be used because the simulator programs the random oracle, and Definition 9 provides no guarantees if the random oracle is modified.)

An example where this problem occurs is the proof of Theorem 21 below (unforgeability of Fiat-Shamir signatures).

To avoid these problems, we adapt the definition of simulation-soundness [22] to the quantum setting. Roughly speaking, simulation-soundness requires that the adversary cannot produce wrong proofs π, even if it has access to a simulator that it can use to produce arbitrary fake proofs. (Of course, it does not count if the adversary simply outputs one of the fake proofs it got from the simulator. But we require that the adversary cannot produce any other wrong proofs.)

Definition 10 (Simulation-soundness). *A non-interactive proof system* (P, V) *is* simulation-sound *with respect to the simulator* S *iff for any quantum-polynomial-time oracle algorithm* A*, there is a negligible function* μ*, such that for all* η *and all normalized density operators* ρ*,*

$$\Pr[ok_V = 1 \land x \notin L_R \land (x, \pi) \notin \textsf{S-queries} : \\ (x, \pi) \leftarrow A^{H,S''}(1^\eta, \rho),\ ok_V \leftarrow V^{H_{final}}(1^\eta, x, \pi)] \leq \mu(\eta). \tag{2}$$

Here the oracle $S''(x)$ *invokes* $S(1^\eta, x)$*. And* H_{final} *refers to the value of the random oracle* H *at the end of the execution (recall that invocations of* S *may change* H*).* S-queries *is a list containing all queries made to* S'' *by* A*, as pairs of input/output. (Note that the input and output of* S'' *are classical, so such a list is well-defined.)*

We call (P, V) weakly simulation-sound *if the above holds with the following instead of* (2)*, where* S-queries *contains only the query inputs to* S''*:*

$$\Pr[ok_V = 1 \land x \notin L_R \land x \notin \textsf{S-queries} : \\ (x, \pi) \leftarrow A^{H,S''}(1^\eta, \rho),\ ok_V \leftarrow V^{H_{final}}(1^\eta, x, \pi)] \leq \mu(\eta). \tag{3}$$

When considering simulation-sound zero-knowledge proof systems, we will always implicitly assume that the same simulator is used for the simulation-soundness and for the zero-knowledge property.

5 Fiat-Shamir

For the rest of this paper, fix a sigma-protocol $\Sigma = (\ell_\eta^{com}, \ell_\eta^{ch}, \ell_\eta^{resp}, P_\Sigma^1, P_\Sigma^2, V_\Sigma)$ for a fixed-length relation R_η. Let $H : \{0,1\}^{\ell_\eta^x + \ell_\eta^{com}} \to \{0,1\}^{\ell_\eta^{ch}}$ be a random oracle.

Definition 11. *The* Fiat-Shamir proof system (P_{FS}, V_{FS}) *consists of the algorithms* P_{FS} *and* V_{FS} *defined in Fig. 1.*

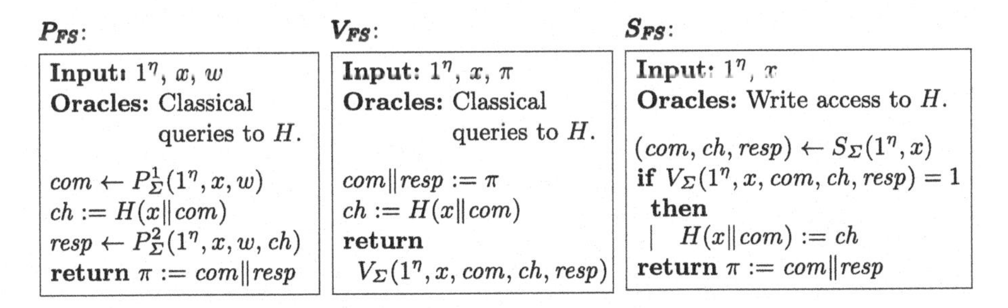

Fig. 1. Prover P_{FS} and verifier V_{FS} of the Fiat-Shamir proof system. S_{FS} is the simulator constructed in the proof of Theorem 14.

In the remainder of this section, we show the following result, which is an immediate combination of Theorems 14, 16, 17, and Lemma 13 below.

Theorem 12. *If Σ has completeness, unpredictable commitments, honest-verifier zero-knowledge, statistical soundness, then Fiat-Shamir (P_{FS}, V_{FS}) has completeness, zero-knowledge, and weak simulation-soundness.*

If Σ additionally has unique responses, then Fiat-Shamir has simulation-soundness.

5.1 Completeness

Lemma 13. *If Σ has completeness and unpredictable commitments, then Fiat-Shamir (P_{FS}, V_{FS}) has completeness.*

Interestingly, without unpredictable commitments, the lemma does not hold. Consider the following example sigma-protocol: Let $R_\eta := \{(x,w) : |x| = |w| = \eta\}$, $\ell^{com} := \ell^{ch} := \ell^{resp} := \eta$. Let $P^1_\Sigma(1^\eta, x, w)$ output $com := 0^\eta$. Let $P^2_\Sigma(1^\eta, x, w, ch)$ output $resp := ch$ if $ch \neq w$, and $resp := \overline{ch}$ else ($\overline{ch}$ is the bit-wise negation of ch). Let $V_\Sigma(1^\eta, x, com, ch, resp) = 1$ iff $|x| = \eta$ and $ch = resp$. This sigma-protocol has all the properties from Definition 6 except unpredictable commitments. Yet (P_{FS}, V_{FS}) does not have completeness: A can chose $x := 0^\eta$ and $w := H(0^\eta \| 0^\eta)$. For those choices of (x,w), $P_{FS}(x,w)$ will chose $com = 0^\eta$ and $ch = H(x\|com) = w$ and thus $resp = \overline{ch}$ and return $\pi = (com, \overline{ch})$. This proof will be rejected by V_{FS} with probability 1.

Proof of Lemma 13. Fix a polynomial-time oracle algorithm A. We need to show that $\Pr[win = 1 : \textit{Game 1}]$ *is negligible for the following game:*

Game 1 (Completeness). $H \xleftarrow{\$} \mathrm{Fun}(\ell^{in}_\eta, \ell^{out}_\eta)$, $(x,w) \leftarrow A^H(1^\eta)$, $\pi \leftarrow P^H_{FS}(1^\eta, x, w)$, $ok_V \leftarrow V^H_{FS}(1^\eta, x, \pi)$, $win := ((x,w) \in R_\eta \wedge ok_V = 0)$.

Let $P^{1,class}_\Sigma, P^{2,class}_\Sigma$ be classical implementations of P^1_Σ, P^2_Σ. (I.e., $P^{1,class}_\Sigma, P^{2,class}_\Sigma$ have the same output distribution but do not perform quantum computations or keep a quantum state. $P^{1,class}_\Sigma, P^{2,class}_\Sigma$ might not be polynomial-time, and the state they keep might not be polynomial space.)

We use Theorem 4 to transform Game 1. For a fixed η, let A^H_0 run $(x,w) \leftarrow A^H(1^\eta)$ (and return nothing). Let $A_C()$ run $com \leftarrow P^{1,class}_\Sigma(1^\eta, x, w)$ and return $x\|com$. Let $A^H_2(ch)$ run $resp \leftarrow P^{2,class}_\Sigma(1^\eta, x, w, ch)$ and $ok_V \leftarrow V_\Sigma(1^\eta, x, com, ch, resp)$ and return $b := win := ((x,w) \in R_\eta \wedge ok_V = 0)$. (Note: A_C and A^H_2 are not necessarily polynomial-time, we will only use that A^H_0 is polynomial-time.)

Let p_1, p_2 denote the first and second probability in Theorem 4, respectively. By construction, $p_1 = \Pr[win = 1 : \text{Game 1}]$.

Furthermore, $p_2 = \Pr[win = 1 : \text{Game 2}]$ for the following game:

Game 2. $H \xleftarrow{\$} \mathrm{Fun}(\ell^{in}_\eta, \ell^{out}_\eta)$, $(x,w) \leftarrow A^H(1^\eta)$, $com \leftarrow P^1_\Sigma(1^\eta, x, w)$, $ch \xleftarrow{\$} \{0,1\}^{\ell^{ch}}$, $resp \leftarrow P^2_\Sigma(1^\eta, x, w, ch)$, $ok_V \leftarrow V_\Sigma(1^\eta, x, com, ch, resp)$, $win := ((x,w) \in R_\eta \wedge ok_V = 0)$.

Then Theorem 4 implies that

$$\big|\Pr[win=1 : \text{Game 1}]-\Pr[win=1 : \text{Game 2}]\big| = |p_1-p_2| \leq (4+\sqrt{2})\sqrt{q_A}2^{-k/4} =: \mu \quad (4)$$

where q_A is the number of queries performed by A_0^H, and k the collision-entropy of $x\|com$. Since A is polynomial-time, q_A is polynomially bounded. And since Σ has unpredictable commitments, k is superlogarithmic. Thus μ is negligible.

Since Σ has completeness, $\Pr[win = 1 : \text{Game 2}]$ is negligible. From (4) it then follows that $\Pr[win = 1 : \text{Game 1}]$ is negligible. This shows that (P_{FS}, V_{FS}) has completeness. □

5.2 Zero-Knowledge

Theorem 14 (Fiat-Shamir is zero-knowledge). *Assume that Σ is honest-verifier zero-knowledge and has completeness and unpredictable commitments. Then the Fiat-Shamir proof system (P_{FS}, V_{FS}) is zero-knowledge.*

Proof. In this proof, we will in many places omit the security parameter η for readability. (E.g., we write $\{0,1\}^{\ell^{ch}}$ instead of $\{0,1\}^{\ell^{ch}_\eta}$ and $S_\Sigma(x)$ instead of $S_\Sigma(1^\eta, x)$.) It is to be understood that this is merely a syntactic omission, the variables and algorithms still depend on η.

To show that Fiat-Shamir is zero-knowledge, we first define a simulator S_{FS}, see Fig. 1. In the definition of S_{FS} we use the honest-verifier simulator S_Σ for Σ (see Definition 6) which exists since Σ is HVZK by assumption. Fix a quantum-polynomial-time adversary A, and a quantum state ρ (that may depend on η). Let q_H and q_P denote polynomial upper bounds on the number of queries performed by A to the random oracle H and the prover/simulator, respectively. We need to show that (1) is negligible (with $P := P_{FS}$ and $S := S_{FS}$). For this, we transform the lhs of (1) into the rhs of (1) using a sequences of games.

Game 1 (Real world). $b \leftarrow A^{H,P_{FS}}(\rho)$.

Game 2 (Programming H). $b \leftarrow A^{H,P^*}(\rho)$ *with the following oracle P^*:*

$P^*(x, w)$ runs $com \leftarrow P_\Sigma^1(x, w)$, $ch \xleftarrow{\$} \{0,1\}^{\ell^{ch}}$, $H(x\|com) := ch$, $resp \leftarrow P_\Sigma^2(x, w, ch)$. Then it returns $\pi := com\|resp$.

Notice that P^* reprograms the random oracle in a similar way as the simulator does. Thus, P^* is not a valid prover any more, but the game is well-defined nonetheless.

In order to relate Games 1 and 2, we define a hybrid game:

Game 3_i (Hybrid). $b \leftarrow A^{H,P'}(\rho)$ *where P' behaves as P_{FS} in the first i invocations, and as P^* (see Game 2) in all further invocations.*

Fix some $i \geq 0$ and some η. We will now bound $\big|\Pr[b = 1 : \text{Game } 3_i] - \Pr[b = 1 : \text{Game } 3_{i+1}]\big|$ by applying Theorem 4. Let $A_0^H()$ be an algorithm that executes

$A^{H,P'}(\rho)$ until just before the i-th query to P'.[15] Note that at that point, the query input x, w for the $(i+1)$-st P'-query are fixed. Let $P_\Sigma^{1,class}, P_\Sigma^{2,class}$ be classical implementations of P_Σ^1, P_Σ^2. (I.e., $P_\Sigma^{1,class}, P_\Sigma^{2,class}$ have the same output distribution but do not perform quantum computations or keep a quantum state. $P_\Sigma^{1,class}, P_\Sigma^{2,class}$ might not be polynomial-time.) Let $A_C()$ compute $com \leftarrow P_\Sigma^{1,class}(x,w)$ and return $x\|com$ if $(x,w) \in R$. (If $(x,w) \notin R$, $A_C()$ instead outputs a η uniformly random bits.) Let $A_2^H(ch)$ compute $resp \leftarrow P_\Sigma^{2,class}(x,w,ch)$, set $\pi := com\|resp$, and then finish the execution of A^H using π as the response of the $(i+1)$-st P'-query. A_2^H outputs the output of A^H. Note that in the execution of A_2^H, P' will actually behave like P^* and thus reprogram the random oracle H. A_2^H does not actually reprogram H (it only has readonly access to it), but instead maintains a list of all changes performed by P^* to simulate queries to H performed by A accordingly.

Since Σ has unpredictable commitments, the output of P_Σ^1 has collision-entropy $\geq k(\eta)$ for some superlogarithmic k, assuming $(x,w) \in R$. Hence the output of A_C has collision-entropy $\geq k' := \min\{\eta, k\}$.

Since A makes at most q_H queries to H, and at most q_P queries to the prover, and since P_{FS} and P^* make one and zero queries to H, respectively, A_0^H makes at most $q_A := q_H + q_P$ queries to H.

Let

$$\begin{aligned}P_{lhs} := \Pr[b = 1 : H \xleftarrow{\$} \mathrm{Fun}(\ell^x + \ell^{com}, \ell^{ch}), A_0^H(), x\|com \leftarrow A_C(),\\ ch := H(x\|com), b \leftarrow A_2^H(ch)],\\ P_{rhs} := \Pr[b = 1 : H \xleftarrow{\$} \mathrm{Fun}(\ell^x + \ell^{com}, \ell^{ch}), A_0^H(), x\|com \leftarrow A_C(),\\ ch \xleftarrow{\$} \{0,1\}^{\ell^{ch}}, H(x\|com) := ch, b \leftarrow A_2^H(ch)]\end{aligned}$$

Then, by Theorem 4,

$$\left|P_{lhs} - P_{rhs}\right| \leq (4 + \sqrt{2})\sqrt{q_A}2^{-k/4} =: \mu_1. \tag{5}$$

Since k is superlogarithmic, and $q_A = q_H + q_P$ is polynomially bounded, we have that μ_1 is negligible.

With those definitions, we have that

$$P_{lhs} = \Pr[b = 1 : \text{Game } 3_{i+1}] \tag{6}$$

because $x\|com \leftarrow A_C(), ch := H(x\|com)$ together with the steps $resp \leftarrow P_\Sigma^{2,class}(x,w,ch)$ and $\pi := com\|resp$ executed by A_2^H compute what P_{FS} would compute,[16] hence the $(i+1)$-st query is exactly what it would be in Game 3_{i+1}.

[15] Note that A_0^H has both ρ and the security parameter η hardcoded. This is no problem in the present case because Theorem 4 does not need A_0^H, A_C, A_2^H to be efficient.

[16] The case that $A_C()$ outputs η random bits when $(x,w) \notin R$ does not occur since A queries the prover only with $(x,w) \in R$ by Definition 8, and hence A_0^H only chooses x, w with $(x,w) \in R$.

And we have that

$$P_{rhs} = \Pr[b = 1 : \text{Game } 3_i] \tag{7}$$

because $x\|com \leftarrow A_C()$, $ch \xleftarrow{\$} \{0,1\}^{\ell^{ch}}$, $H(x\|com) := ch$, together with the steps $resp \leftarrow P_\Sigma^{2,class}(x, w, ch)$ and $\pi := com\|resp$ executed by A_2^H compute what P^* would compute, hence the i-st query is exactly what it would be in Game 3_i.

From (5)–(7), we have (for all i and η):

$$\big|\Pr[b = 1 : \text{Game } 3_{i+1}] - \Pr[b = 1 : \text{Game } 3_i]\big| \leq \mu_1 \tag{8}$$

Furthermore, we have that

$$\begin{aligned} &\Pr[b = 1 : \text{Game } 3_0] = \Pr[b = 1 : \text{Game } 2] \\ \text{and} \qquad &\Pr[b = 1 : \text{Game } 3_{q_P}] = \Pr[b = 1 : \text{Game } 1] \end{aligned} \tag{9}$$

by definition of the involved games. (For the second equality, we use that $A^{H,P'}$ makes at most q_P queries to P'.)

Thus we have

$$\begin{aligned} &\big|\Pr[b = 1 : \text{Game } 1] - \Pr[b = 1 : \text{Game } 2]\big| \\ &\overset{(9)}{=} \big|\Pr[b = 1 : \text{Game } 3_{q_P}] - \Pr[b = 1 : \text{Game } 3_0]\big| \\ &\leq \sum_{i=0}^{q_P - 1} \big|\Pr[b = 1 : \text{Game } 3_{i+1}] - \Pr[b = 1 : \text{Game } 3_i]\big| \\ &\overset{(8)}{\leq} \sum_{i=0}^{q_P - 1} \mu_1 = q_P \mu_1 =: \mu_2. \end{aligned} \tag{10}$$

Since μ_1 is negligible and q_P is polynomially bounded, μ_2 is negligible.

Game 4. *$b \leftarrow A^{H,P^{**}}(\rho)$ with the following oracle P^{**}:*

$P^{**}(x, w)$ runs: $com \leftarrow P_\Sigma^1(x, w)$, $ch \xleftarrow{\$} \{0,1\}^{\ell^{ch}}$, $resp \leftarrow P_\Sigma^2(x, w, ch)$, if $V_\Sigma(x, com, ch, resp) = 1$ then $H(x\|com) := ch$. Then it returns $\pi := com\|resp$.

By assumption, Σ has completeness. Furthermore, A never queries $(x, w) \notin R$ from P^{**} (see Definition 8). Thus with overwhelming probability, $V_\Sigma(x, com, ch, resp) = 1$ holds in each query to P^{**}. Thus with overwhelming probability, the condition $V_\Sigma(x, com, ch, resp) = 1$ in the if-statement is satisfied in each invocation of P^{**}, and P^{**} performs the same steps as P^*. Thus for some negligible μ_3 we have

$$\big|\Pr[b = 1 : \text{Game } 2] - \Pr[b = 1 : \text{Game } 4]\big| \leq \mu_3. \tag{11}$$

Let S_{FS} be as in Fig. 1.

Game 5. $b \leftarrow A^{H,S'_{FS}}$. *(Here $S'_{FS}(x, w)$ runs $S_{FS}(x)$, analogous to S' in Definition 8.)*

By definition, $P^{**}(x, w)$ performs the following steps:

- $com \leftarrow P^1_\Sigma(x, w)$, $ch \leftarrow \{0,1\}^{\ell^{ch}}$, $resp \leftarrow P^2_\Sigma(x, w, ch)$, if $V_\Sigma(x, com, ch, resp) = 1$ then $H(x\|com) := ch$.

In constract, S'_{FS} performs:

- $(com, ch, resp) \leftarrow S_\Sigma(x)$, if $V_\Sigma(x, com, ch, resp) = 1$ then $H(x\|com) := ch$.

By definition of honest-verifier zero-knowledge, $(com, ch, resp)$ as chosen in the first item is indistinguishable by a quantum-polynomial-time algorithm from $(com, ch, resp)$ as chosen second item, assuming $(x, w) \in R$. (And $(x, w) \in R$ is guaranteed since by Definition 8, A only queries $(x, w) \in R$ from the prover/simulator.) A standard hybrid argument then shows that no quantum-polynomial-time adversary can distinguish oracle access to P^{**} from oracle access to S'_{FS}. Hence

$$\left|\Pr[b = 1 : \text{Game 4}] - \Pr[b = 1 : \text{Game 5}]\right| \leq \mu_4 \tag{12}$$

for some negligible μ_4.

Altogether, we have

$$\left|\Pr[b = 1 : \text{Game 1}] - \Pr[b = 1 : \text{Game 5}]\right| \overset{(10)-(12)}{\leq} \mu_2 + \mu_3 + \mu_4.$$

Since μ_2, μ_3, and μ_4 are negligible, so is $\mu_2 + \mu_3 + \mu_4$. Thus (1) from Definition 8 is negligible. This shows that S_{FS} is a simulator as required by Definition 8, thus Fiat-Shamir is zero-knowledge. □

5.3 Soundness

Theorem 15. *Assume that Σ has statistical soundness. Then the Fiat-Shamir proof system (P_{FS}, V_{FS}) is sound.*

It may seem surprising that we need an information-theoretical property (statistical soundness of Σ) to get a computational property (soundness of (P_{FS}, V_{FS})). Might it not be sufficient to assume that Σ has computational soundness (or the somewhat stronger, computational special soundness)? Unfortunately, [2] shows that (relative to certain oracles), there is a sigma-protocol Σ with computational special soundness such that (P_{FS}, V_{FS}) is not sound. So, we cannot expect Theorem 15 to hold assuming only computational special soundness, at least not with a relativizing proof.[17]

The proof is based on the following observation: To produce a fake Fiat-Shamir proof, the adversary needs to find an input (x, com) to the random oracle H such that $ch := H(x\|com)$ is a challenge for which there exists a valid

[17] [2] leaves the possibility of a relativizing proof that Fiat-Shamir is secure if Σ has perfectly unique responses and computational special soundness, though. But then we have another information-theoretical assumption, namely perfectly unique responses.

response. We call such a challenge *promising.* (Additionally, the adversary needs to also find that response, but we do not make use of that fact.) So, to show that forging a proof is hard, we need to show that outputs of H that are promising are hard to find. Since the sigma-protocol has statistical soundness, there cannot be too many promising challenges (otherwise, an unlimited adversary would receive a promising challenge with non-negligible probability, compute the corresponding response, and break the statistical soundness of the sigma-protocol). By reduction to existing bounds on the quantum hardness of search in a random function, we then show that finding a promising challenge in H is hard.

Proof of Theorem 15. In this proof, we will in most places omit the security parameter η for readability. (E.g., we write ℓ^{ch} instead of ℓ^{ch}_η and $S_\Sigma(x)$ instead of $S_\Sigma(\eta, x)$.) It is to be understood that this is merely a syntactic omission, the variables and algorithms still depend on η.

Let $x \in \{0,1\}^{\ell^x}$, $com \in \{0,1\}^{com}$. We call a $ch \in \{0,1\}^{\ell^{ch}}$ promising for *(x, com) iff there exists a $resp \in \{0,1\}^{\ell^{resp}}$ such that $V_\Sigma(x, com, ch, resp) = 1$.*

Claim 1. *There is a negligible μ such that for any $x \in \{0,1\}^{\ell^x} \setminus L_R$ and any $com \in \{0,1\}^{\ell^{com}}$, there exist at most $\mu 2^{\ell^{ch}}$ promising ch.*

Since Σ has statistical soundness, by definition (Definition 6) there exists a negligible function μ such that for all $x \notin L_R$, all $com \in \{0,1\}^{\ell^{com}}$, and all A, we have:

$$\Pr[V_\Sigma(x, com, ch, resp) = 1 : ch \xleftarrow{\$} \{0,1\}^{\ell^{ch}}, resp \leftarrow A(x, com, ch)] \leq \mu. \quad (13)$$

Let A be the adversary that, given (x, com, ch) outputs some $resp$ with $V_\Sigma(x, com, ch, resp) = 1$ if it exists, and an arbitrary output otherwise. That is, whenever ch is promising for (x, com), A outputs $resp$ such that $V_\Sigma(x, com, ch, resp) = 1$. For any x, com, let $prom_{x,com}$ denote the number of promising ch. Then for all $x \notin L_R$ and all $com \in \{0,1\}^{\ell^{com}}$, we have

$$\begin{aligned} prom_{x,com} &= 2^{\ell^{ch}} \Pr[ch \text{ is promising for } (x, com) : ch \xleftarrow{\$} \{0,1\}^{\ell^{ch}}] \\ &\leq 2^{\ell^{ch}} \Pr[V_\Sigma(x, com, ch, resp) = 1 : ch \xleftarrow{\$} \{0,1\}^{\ell^{ch}}, \\ &\quad resp \leftarrow A(x, com, ch)] \overset{(13)}{\leq} 2^{\ell^{ch}} \mu. \end{aligned}$$

This shows the claim.

We now define an auxiliary distribution $\mathcal{D}$ on functions $f : \{0,1\}^{\ell^x + \ell^{com}} \rightarrow \{0,1\}^{\ell^{ch}}$ as follows: For each x, com, let $f(x\|com)$ be an independently chosen uniformly random promising ch. If no promising ch exists for (x, com), $f(x\|com) := 0^{\ell^{ch}}$.

Let A be a quantum-polynomial-time adversary that breaks the soundness of Fiat-Shamir given some initial state ρ. That is, δ is non-negligible where

$$\delta := \Pr[ok_V = 1 \land x \notin L_R : (x, com\|resp) \leftarrow A^H(\rho),\ ok_V \leftarrow V^H_{FS}(x, com\|resp)].$$

By definition of V_{FS}, we have that $ok_V = 1$ implies that $V_\Sigma(x, com, ch, resp) = 1$ where $ch := H(x\|com)$. In particular, $ch = H(x\|com)$ is promising for (x, com). Thus, if $ok_V = 1 \wedge x \notin L_R$ then $f(x\|com) = H(x\|com)$ with probability at least $1/(\mu 2^{\ell^{ch}})$ for $f \leftarrow \mathcal{D}$. Hence for uniformly random H,

$$\Pr[f(x\|com) = H(x\|com) : (x, com\|resp) \leftarrow A^H(\rho)] \geq \frac{\delta}{\mu 2^{\ell^{ch}}}. \tag{14}$$

Let $B^H(\rho)$ perform the following steps: It defines $H'(x\|com) := H(x\|com) \oplus f(x\|com)$. It invokes $(x, com\|resp) \leftarrow A^{H'}(\rho)$. It returns $x\|com$.

Let q be a polynomial upper bound for the number of queries performed by A. Although B may not be quantum-polynomial-time (f may not be efficiently computable), B performs only q queries since each query to H' can be implemented using one query to H.[18]

If H is uniformly random, then H' is uniformly random. Thus by (14), $H'(x\|com) = f(x\|com)$ with probability $\geq 2^{-\ell^{ch}}\delta/\mu$. Thus $H(x\|com) = 0^{\ell^{ch}}$ with probability $\geq 2^{-\ell^{ch}}\delta/\mu$. In other words, B finds a zero-preimage of H with probability $\geq 2^{-\ell^{ch}}\delta/\mu$. By Lemma 5, this implies that $2^{-\ell^{ch}}\delta/\mu \leq 32 \cdot 2^{-\ell^{ch}} \cdot (q+1)^2$. Hence $\delta \leq 32\mu \cdot (q+1)^2$. Since q is polynomially bounded (as A is quantum-polynomial-time) and μ is negligible, we have that δ is negligible.

Since this holds for all quantum-polynomial-time A, it follows that (P_{FS}, V_{FS}) is sound. □

5.4 Simulation-Soundness

We give two theorems on simulation-soundness, depending on whether the sigma-protocol has unique responses or not.

Theorem 16. (Fiat-Shamir is weakly simulation-sound). *Assume that Σ has statistical soundness.*

Then the Fiat-Shamir proof system (P_{FS}, V_{FS}) is weakly simulation-sound with respect to the simulator S_{FS} from Fig. 1.

Proof. In this proof, we will in most places omit the security parameter η for readability. (E.g., we write ℓ^{ch} instead of ℓ^{ch}_η and $S_\Sigma(x)$ instead of $S_\Sigma(\eta, x)$.) It is to be understood that this is merely a syntactic omission, the variables and algorithms still depend on η. For brevity, we will also omit the choosing of the random oracle H from all games. That is, every game implicitly starts with $H \xleftarrow{\$} \mathrm{Fun}(\ell^{in}, \ell^{out})$.

Fix a quantum-polynomial-time adversary A, and a density operator ρ. Let q_H and q_P denote polynomial upper bounds on the number of queries performed by A to the random oracle H and the prover/simulator, respectively. We need

[18] To implement the unitary $\mathbf{U}_{H'} : |a\|b\rangle \mapsto |a\|(b \oplus H'(a))\rangle$, B first invokes $\mathbf{U}_H : |a\|b\rangle \mapsto |a\|(b \oplus H(a))\rangle$ by using the oracle H, and then $\mathbf{U}_f : |a\|b\rangle \mapsto |a\|(b \oplus f(a))\rangle$ which B implements on its own.

to show that (3) *holds with* $V := V_{FS}$ *and* $S := S_{FS}$ *for some negligible* μ. *For this, we transform the game from* (3) *using a sequence of games until we reach a game where the adversary has a negligible success probability. The following game encodes the game from* (3)*: (We write* $com \| resp$ *instead of* π *to be able to explicitly refer to the two components of* π*.)*

Game 1 (Real world). $S_A \leftarrow \rho$. $x \| com \| resp \leftarrow A^{H,S_{FS}}(S_A)$. $ok_V \leftarrow V_{FS}^{H_{final}}(x, com \| resp)$. $win := \bigl(ok_V = 1 \wedge x \notin L_R \wedge x \notin \textsf{S-queries}\bigr)$.

Here we use H to refer to the initial value of the random oracle H, and H_{final} to the value of H after it has been reprogrammed by S_{FS}. (See Definition 10.)

We now show that in Game 1, we have

$$V_{FS}^{H_{final}}(x, com \| resp) = 1 \ \wedge \ x \notin \textsf{S-queries} \implies V_{FS}^{H}(x, com \| resp) = 1. \quad (15)$$

Assume for contradiction that (15) does not hold, i.e., that $V_{FS}^{H_{final}}(x, com \| resp) = 1$ and $x \notin \textsf{S-queries}$, but $V_{FS}^{H}(x, com \| resp) = 0$ in some execution of Game 1. Since V_{FS}^{H} queries H only for input $x \| com$, this implies that $H_{final}(x \| com) \neq H(x \| com)$. Since H is only reprogrammed by invocations of S_{FS}, $H(x \| com)$ must have been reprogrammed by S_{FS}. Consider the last query to S_{FS} that programmed $H(x \| com)$ (in case there are several). By construction of S_{FS}, that query had input x, in contradiction to $x \notin \textsf{S-queries}$. Thus our assumption that (15) does not hold was false. Thus (15) follows.

We now consider a variant of Game 1 where the verifier in the end gets access to H instead of H_{final}. (That is, we can think of H being reset to its original state without the simulator's changes.)

(In this and the following games, we will not need to refer to *com* and *resp* individually any more, so we just write π instead of $com \| resp$.)

Game 2 (Unchanged H). $S_A \leftarrow \rho$. $x \| \pi \leftarrow A^{H,S_{FS}}(S_A)$. $ok_V \leftarrow V_{FS}^{H}(x, \pi)$. $win := \bigl(ok_V = 1 \wedge x \notin L_R \wedge x \notin \textsf{S-queries}\bigr)$.

By (15), we get

$$\Pr[win : \text{Game } 2] \geq \Pr[win : \text{Game } 1]. \quad (16)$$

Furthermore, we have

$$\Pr[ok_V = 1 \wedge x \notin L_R : \text{Game } 2] \geq \Pr[win : \text{Game } 2].$$

We define an oracle algorithm B. When invoked as $B^H(S_A)$, it simulates an execution of $A^{H,S_{FS}}(S_A)$. Note that S_{FS} can program the random oracle H. In order to simulate this, B^H keeps track of the assignments ass_S made by S_{FS}, and then provides A with the oracle $H(ass_S)$ (i.e., H reprogrammed according to the assignment-list ass_S) instead of H. Then $B^H(S_A)$ will have the same distribution of outputs as $A^{H,S_{FS}}(S_A)$. (But of course, any reprogramming of H performed by the S_{FS} simulated by B will not have any effect beyond the execution of B. That is, the function H before and after the invocation of B^H will be the same.)

By construction of B (and because V_{FS} gets access to H and not H_{final} in (16)), we then have

$$\Pr[win : \text{Game 3}] = \Pr[ok_V = 1 \wedge x \notin L_R : \text{Game 2}].$$

Game 3 (Adversary B). $S_A \leftarrow \rho$. $x\|\pi \leftarrow B^H(S_A)$. $ok_V \leftarrow V_{FS}^H(x,\pi)$. $win := (ok_V = 1 \wedge x \notin L_R)$.

By Theorem 15, (P_{FS}, V_{FS}) is sound. Furthermore, since A and S_{FS} are quantum-polynomial-time, B is quantum-polynomial-time. Thus by definition of soundness (Definition 9), there is a negligible μ such that

$$\Pr[win : \text{Game 3}] \leq \mu.$$

Combining the inequalities from this proof, we get $\Pr[win : \text{Game 1}] \leq \mu + \mu'$. And $\mu + \mu'$ is negligible. Since Game 1 is the game from the definition of weak simulation soundness (Definition 10) for (P_{FS}, V_{FS}), and since A was an arbitrarily quantum-polynomial-time oracle algorithm, it follows that (P_{FS}, V_{FS}) is weakly simulation-sound. □

If we add another assumption about the sigma-protocol, we even can get (non-weak) simulation-soundness:

Theorem 17 (Fiat-Shamir is simulation-sound). *Assume that Σ has statistical soundness and unique responses.*

Then the Fiat-Shamir proof system (P_{FS}, V_{FS}) is simulation-sound with respect to the simulator S_{FS} from Fig. 1.

Unique responses are necessary in this theorem. As pointed out in [11], if Σ does not have unique responses, it cannot be simulation-sound, even in the classical case. Namely, if we do not require unique responses, it could be that whenever $(com, ch, resp\|0)$ is a valid proof in Σ, so is $(com, ch, resp\|1)$, and vice versa. Thus any valid Fiat-Shamir proof $com\|(resp\|0)$ could be efficiently transformed into another valid Fiat-Shamir proof $com\|(resp\|1)$ for the same statement. This would contradict the simulation-soundness of (P_{FS}, V_{FS}).

Proof. *In this proof, we will in most places omit the security parameter η for readability. (E.g., we write ℓ^{ch} instead of ℓ^{ch}_η and $S_\Sigma(x)$ instead of $S_\Sigma(\eta, x)$.) It is to be understood that this is merely a syntactic omission, the variables and algorithms still depend on η. For brevity, we will also omit the choosing of the random oracle H from all games. That is, every game implicitly starts with $H \xleftarrow{\$} \mathrm{Fun}(\ell^{in}, \ell^{out})$.*

Fix a quantum-polynomial-time adversary A, and a density operator ρ. Let q_H and q_P denote polynomial upper bounds on the number of queries performed by A to the random oracle H and the prover/simulator, respectively. We need to show that (2) holds with $V := V_{FS}$ and $S := S_{FS}$ for some negligible μ. For *this, we transform the game from* (2) *using a sequence of games until we reach a game where the adversary has a negligible success probability. The following game encodes the game from* (2)*: (We write* $com\|resp$ *instead of* π *to be able to explicitly refer to the two components of* π*.)*

Game 4 (Real world). $S_A \leftarrow \rho$. $x\|com\|resp \leftarrow A^{H,S_{FS}}(S_A)$. $ok_V \leftarrow V_{FS}^{H_{final}}(x, com\|resp)$. $win := \big(ok_V = 1 \land x \notin L_R \land (x, com\|resp) \notin \mathsf{S\text{-}queries}\big)$.

Here we use H to refer to the initial value of the random oracle H, and H_{final} to the value of H after it has been reprogrammed by S_{FS}. (See Definition 10.)

We define a variant of the random variable S-queries. Let S-queries* be the list of all S_{FS}-queries $(x', com'\|resp', ch')$ where x' was the input to S_{FS}, $com'\|resp'$ was the response of S_{FS}, and ch' was the value of $H(x'\|com')$ right after the query to S_{FS}. (Note that $H(x'\|com')$ may change later due to reprogramming.) Notice that the only difference between S-queries and S-queries* is that in the latter, we additionally track the values $ch' = H(x'\|com')$.

Let RespConflict denote the event that $V_\Sigma(x, com, H_{final}(x\|com), resp) = 1$ and that there is a query $(x', com'\|resp', ch') \in$ S-queries with $x' = x$, $com' = com$, $ch' = H_{final}(x\|com)$, and $resp' \neq resp$ and $V_\Sigma(x, com, ch', resp') = 1$.

Since Σ has unique responses, it follows that

$$\Pr[\mathsf{RespConflict} : \text{Game 4}] \leq \mu'$$

for some negligible μ'. (Otherwise, we could construct an adversary that simulates Game 4, and then searches for $(x, com\|resp', ch) \in$ S-queries with $V_\Sigma(x, com, ch, resp') = 1$ and $resp' \neq resp$.)

Thus

$$\Big|\Pr[win : \text{Game 4}] - \Pr[win \land \neg\mathsf{RespConflict} : \text{Game 4}]\Big| \leq \mu'.$$

We now show that in Game 4, we have

$$V_{FS}^{H_{final}}(x, com\|resp) = 1 \ \land\ (x, com\|resp) \notin \mathsf{S\text{-}queries} \ \land\ \neg\mathsf{RespConflict} \implies V_{FS}^{H}(x, com\|resp) = 1. \quad (17)$$

Assume for contradiction that (17) does not hold, i.e., that $V_{FS}^{H_{final}}(x, com\|resp) = 1$ and $(x, com\|resp) \notin$ S-queries and $\neg$RespConflict, but $V_{FS}^{H}(x, com\|resp) = 0$ in some execution of Game 4. Since V_{FS}^{H} queries H only for input $x\|com$, this implies that $H_{final}(x\|com) \neq H(x\|com)$. Since H is only reprogrammed by invocations of S_{FS}, $H(x\|com)$ must have been reprogrammed by S_{FS}. Consider the last query to S_{FS} that programmed $H(x\|com)$ (in case there are several). By construction of S_{FS}, that query had input x, and returns $(com, resp')$ for some $resp'$. In particular, $(x, com\|resp') \in$ S-queries. Let ch be the challenge chosen by S_{FS} in that query. Then $(x, com\|resp', ch) \in$ S-queries*. By construction of S_{FS}, we have $V_\Sigma(x, com, ch, resp') = 1$ (else H would not have been reprogrammed in that query) and $H_{final}(x\|com) = ch$ (because we are considering the last S_{FS}-query that programmed $H(x\|com)$). Since $(x, com\|resp) \notin \mathsf{S\text{-}queries}$ and $(x, com\|resp') \in \mathsf{S\text{-}queries}$, we have $resp \neq resp'$. Since $V_{FS}^{H_{final}}(x, com\|resp) = 1$ and $ch = H_{final}(x\|com)$, we have that $V_\Sigma(x, com, ch, resp) = 1$ by definition of V_{FS}. Summarizing, we have $V_\Sigma(x, com, ch, resp) = 1$ and $ch = H_{final}(x\|com)$ and $V_\Sigma(x, com, ch, resp') = 1$ and $(x, com\|resp', ch) \in \mathsf{S\text{-}queries}^*$ and $resp \neq resp'$. By definition of $\mathsf{RespConflict}$, this contradicts $\neg\mathsf{RespConflict}$. Thus our assumption that (17) does not hold was false. Thus (17) follows.

We now consider a variant of Game 4 where the verifier in the end gets access to H instead of H_{final}. (That is, we can think of H being reset to its original state without the simulator's changes.)

(In this and the following games, we will not need to refer to com and $resp$ individually any more, so we just write π instead of $com\|resp$.)

Game 5 (Unchanged H). $S_A \leftarrow \rho$. $x\|\pi \leftarrow A^{H,S_{FS}}(S_A)$. $ok_V \leftarrow V_{FS}^H(x,\pi)$. $win := \big(ok_V = 1 \wedge x \notin L_R \wedge (x,\pi) \notin \mathsf{S\text{-}queries}\big)$.

By (17), we get

$$\Pr[win : \text{Game 5}] \geq \Pr[win \wedge \neg\mathsf{RespConflict} : \text{Game 4}]. \tag{18}$$

Furthermore, we have

$$\Pr[ok_V = 1 \wedge x \notin L_R : \text{Game 5}] \geq \Pr[win : \text{Game 5}].$$

We define an oracle algorithm B. When invoked as $B^H(S_A)$, it simulates an execution of $A^{H,S_{FS}}(S_A)$. Note that S_{FS} can program the random oracle H. In order to simulate this, B^H keeps track of the assignments ass_S made by S_{FS}, and then provides A with the oracle $H(ass_S)$ (i.e., H reprogrammed according to the assignment-list ass_S) instead of H. Then $B^H(S_A)$ will have the same distribution of outputs as $A^{H,S_{FS}}(S_A)$. (But of course, any reprogramming of H performed by the S_{FS} simulated by B will not have any effect beyond the execution of B. That is, the function H before and after the invocation of B^H will be the same.)

By construction of B (and because V_{FS} gets access to H and not H_{final} in (18)), we then have

$$\Pr[win : \text{Game 6}] = \Pr[ok_V = 1 \wedge x \notin L_R : \text{Game 5}].$$

Game 6 (Adversary B). $S_A \leftarrow \rho$. $x\|\pi \leftarrow B^H(S_A)$. $ok_V \leftarrow V_{FS}^H(x,\pi)$. $win := \big(ok_V = 1 \wedge x \notin L_R\big)$.

By Theorem 15, (P_{FS}, V_{FS}) is sound. Furthermore, since A and S_{FS} are quantum-polynomial-time, B is quantum-polynomial-time. Thus by definition of soundness (Definition 9), there is a negligible μ such that

$$\Pr[win : \text{Game 6}] \leq \mu.$$

Combining the inequalities from this proof, we get $\Pr[win : \text{Game 4}] \leq \mu + \mu'$. And $\mu + \mu'$ is negligible. Since Game 4 is the game from Definition 10 for (P_{FS}, V_{FS}), and since A was an arbitrarily quantum-polynomial-time oracle algorithm, it follows that (P_{FS}, V_{FS}) is simulation-sound. $\square$

6 Signatures

Originally, Fiat-Shamir was constructed as a signature scheme [12]. Only later, [5] used the same idea to construct a non-interactive zero-knowledge proof. The fact that Fiat-Shamir gives rise to a secure signature scheme can be seen as a special case of its properties as a proof system. Namely, any non-interactive zero-knowledge proof system with simulation-sound extractability can be used as a signature scheme. In the quantum setting, [26] showed that their construction of simulation-sound extractable non-interactive proofs gives rise to a signature scheme in the same way. However, this approach does not show that Fiat-Shamir gives rise to a secure signature scheme because we are not able to prove that Fiat-Shamir is extractable. For analyzing Fiat-Shamir, we show under which conditions a simulation-sound zero-knowledge non-interactive proof system gives rise to a signature scheme. Combined with our results from Sect. 5, this implies security for Fiat-Shamir based signatures.

The basic idea of the construction of signatures from non-interactive proof systems (e.g., Fiat-Shamir) is the following: To sign a message m, one needs to show the knowledge of one's secret key. Thus, we need a relation R_η between public and secret keys, and we need an algorithm G to generate public/secret key pairs such that it is hard to guess the secret key (a "hard instance generator"). We formalize the definition below (Definition 20).

An example of a hard instance generator would be: $R_\eta := \{(x,w) : |w| = \eta \wedge x = f(w)\}$ for some quantum-one-way function f, and G picks w uniformly from $\{0,1\}^\eta$, sets $x := f(w)$, and returns (x,w).

Now a signature is just a proof of knowledge of the secret key. That is, the statement is the public key, and the witness is the secret key. However, a signature should be bound to a particular message. For this, we include the message m in the statement that is proven. That is, the statement that is proven consists of a public key and a message, but the message is ignored when determining whether a given statement has a witness or not. (In the definition below, this is formalized by considering an extended relation R'.) The simulation-soundness of the proof system will then guarantee that a proof/signature with respect to one message cannot be transformed into a proof/signature with respect to another message because this would mean changing the statement.

A signature scheme consists of three oracle algorithms: Keys are generated with $(pk, sk) \leftarrow KeyGen^H(1^\eta)$. The secret key sk is used to sign a message m using the signing algorithm $\sigma \leftarrow Sign^H(1^\eta, sk, m)$ to get a signature σ. And the signature is considered valid iff $Verify^H(1^\eta, pk, \sigma, m) = 1$.

An instance generator for a relation R_η is an algorithm G such that $G(1^\eta)$ outputs $(x,w) \in R_\eta$ with overwhelming probability.

We now describe how to use a simulation-sound zero-knowledge protocol (e.g., Fiat-Shamir) to construct a signature scheme:

Definition 18 (Signatures from non-interactive proofs). *Let G be an instance generator for a relation R_η. Fix a length ℓ_η^m. Let $R'_\eta := \{(x\|m, w) : |m| = \ell_\eta^m \wedge (x,w) \in R_\eta\}$. Let (P,V) be a non-interactive proof system for*

R'_η (in the random oracle model). Then we construct the signature scheme (KeyGen, Sign, Verify) with message space $\{0,1\}^{\ell^m_\eta}$ as follows:

- *$KeyGen^H(1^\eta)$: Pick $(x,w) \leftarrow G(1^\eta)$. Let $pk := x$, $sk := (x,w)$. Return (pk, sk).*
- *$Sign^H(1^\eta, sk, m)$ with $sk = (x,w)$: Run $\sigma \leftarrow P^H(1^\eta, x\|m, w)$. Return σ.*
- *$Verify^H(1^\eta, pk, \sigma, m)$ with $pk = x$: Run $ok \leftarrow V^H(1^\eta, x\|m, \sigma)$. Return ok.*

Note that we use a proof system for the relation R'_η instead of R_η. However, in most cases (including Fiat-Shamir) it is trivial to construct a proof system for R'_η given one for R_η. This is because any sigma-protocol for R_η is also a sigma-protocol for R'_η.[19] The only reason why we need to use R'_η is that we want to include the message m inside the statement (without logical significance), and R'_η allows us to do precisely that. (In the case of Fiat-Shamir, the overall effect will simply be to include m in the hash, see Definition 22.)

The security property we will prove is unforgeability. Unforgeability comes in two variants: weak unforgeability that ensures that the adversary cannot forge a signature for a message that has not been signed before, and strong unforgeability that additionally ensures that the adversary cannot even produce a different signature for a message that has been signed before. (Weak unforgeability is often just called unforgeability.) The definitions are standard, we include them here for completeness:

Definition 19 (Strong/weak unforgeability). *A signature scheme (KeyGen, Sign, Verify) is* strongly unforgeable *iff for all polynomial-time oracle algorithms A there exists a negligible μ such that for all η, we have*

$$\begin{aligned}&\Pr[ok = 1 \wedge (m^*, \sigma^*) \notin \mathbf{Sig}\text{-}\mathsf{queries}:\\&\quad H \leftarrow \mathrm{Fun}(\ell^{in}_\eta, \ell^{out}_\eta), (pk, sk) \leftarrow KeyGen^H(1^\eta),\\&\quad (\sigma^*, m^*) \leftarrow A^{H,\mathbf{Sig}}(1^\eta, pk), ok \leftarrow Verify^H(1^\eta, pk, \sigma^*, m^*)] \le \mu(\eta).\end{aligned} \tag{19}$$

Here **Sig** *is a* classical[20] *oracle that upon classical input m returns $Sign^H(1^\eta, sk, m)$. (But queries to H are quantum.) And* **Sig**-queries *is the list of all queries made to* **Sig**. *(I.e., when* **Sig** *is queried with m and σ, (m,σ) is added to the list* **Sig**-queries.*) And $\ell^{in}_\eta, \ell^{out}_\eta$ denote the input/output length of the random oracle used by the signature scheme.*

We call (KeyGen, Sign, Verify) weakly unforgeable *if the above holds with the following instead of (19), where* **Sig**-queries *contains only the query inputs made to* **Sig** *(i.e., m instead of (m,σ)):*

$$\begin{aligned}\Pr[ok = 1 \wedge m^* \notin \mathbf{Sig}\text{-}\mathsf{queries}: H \leftarrow \mathrm{Fun}(\ell^{in}_\eta, \ell^{out}_\eta), (pk, sk) \leftarrow KeyGen^H(1^\eta),\\ (\sigma^*, m^*) \leftarrow A^{H,\mathbf{Sig}}(1^\eta, pk), ok \leftarrow Verify^H(1^\eta, pk, \sigma^*, m^*)] \le \mu(\eta).\end{aligned}$$

[19] This is made formal by the construction of Σ' in the proof of Corollary 23.

[20] Formally, this means that **Sig** measures its input at the beginning of the each query.

In [26], a hard instance generator was defined as an algorithm that outputs a statement/witness pair such that it is hard on average to find a valid witness given only the statement. However, since we will do not assume a proof system with extractability, we need a stronger variant of this definition: A dual-mode hard instance generator requires more. While a hard instance generator requires that is it hard to find a witness for x, a dual-mode hard instance generator requires that it is hard to distinguish whether x even has a witness. In other words, we should not be able to distinguish x as returned by G from x^* as returned by an algorithm G^* that returns statements that do not have a witness (except with negligible probability). Formally:

Definition 20 (Dual-mode hard instance generator). *We call an algorithm G a* dual-mode hard instance generator *for a fixed-length relation R_η iff*

- *G is quantum-polynomial-time, and*
- *there is a negligible μ such that for every η, $\Pr[(x,w) \in R_\eta : (x,w) \leftarrow G(1^\eta)] \geq 1 - \mu(\eta)$, and*
- *for all quantum-polynomial-time algorithm A, there is a quantum-polynomial-time algorithm G^* and negligible μ_1, μ_2 such that for all η,*

$$\Big|\Pr[b=1 : (x,w) \leftarrow G(1^\eta),\ b \leftarrow A(1^\eta, x)] \\ - \Pr[b=1 : x \leftarrow G^*(1^\eta),\ b \leftarrow A(1^\eta, x)]\Big| \leq \mu_1(\eta).$$

and

$$\Pr[x \in L_R : x \leftarrow G^*(1^\eta)] \leq \mu_2(\eta).$$

Note that we allow G^* to depend on A. This is a slightly weaker requirement than requiring a universal G^*. We chose the weaker variant because it is sufficient for our proof below.

An example of a dual-mode hard instance generator is: Let $R_\eta := \{(x,w) : |w| = \eta \wedge x = F(w)\}$ for some quantum pseudorandom generator $F : \{0,1\}^\eta \to \{0,1\}^{2\eta}$, and G picks w uniformly from $\{0,1\}^\eta$, sets $x := F(w)$, and returns (x,w). The conditions from Definition 20 are satisfied for G^* which returns $x \xleftarrow{\$} \{0,1\}^{2\eta}$.

With this definition, we can state the main results of this section, namely the strong (weak) unforgeability of signatures constructed from non-interactive zero-knowledge proof systems that are (weakly) simulation-sound:

Theorem 21 (Unforgeability from simulation-soundness). *Fix a relation R_η. Let R'_η be defined as in Definition 18. If (P,V) is zero-knowledge and simulation-sound (weakly simulation-sound) for R'_η, and G is a dual-mode hard instance generator for R_η, then the signature scheme $(KeyGen, Sign, Verify)$ from Definition 18 is strongly unforgeable (weakly unforgeable).*

The proof is given in Sect. 6.1 below.

Fiat-Shamir. The two preceding theorems are formulated for generic simulation-sound zero-knowledge proof systems. By specializing Theorem 21 to the case that (P,V) is the Fiat-Shamir proof system, we get a signature scheme based on a dual-mode hard instance generator and a zero-knowledge sigma-protocol with statistical soundness. The resulting signature scheme is the following:

Definition 22 (Fiat-Shamir signatures). *Let G be an instance generator for a relation R_η. Fix a length ℓ_η^m. Then we construct the signature scheme $(KeyGen, Sign, Verify)$ with message space $\{0,1\}^{\ell_\eta^m}$ as follows:*

- $KeyGen^H(1^\eta)$*: Pick* $(x,w) \leftarrow G(1^\eta)$*. Let* $pk := x$*,* $sk := (x,w)$*. Return* (pk, sk)*.*
- $Sign^H(1^\eta, sk, m)$ *with* $sk = (x,w)$*:* $com \leftarrow P_\Sigma^1(1^\eta, x, w)$*.* $resp \leftarrow P_\Sigma^2(1^\eta, x, w, H(x\|m\|com))$*. Return* $\sigma := com\|resp$*.*
- $Verify^H(1^\eta, pk, \sigma, m)$ *with* $pk = x$ *and* $\sigma = com\|resp$*: Run* $ok \leftarrow V_\Sigma(1^\eta, x, com, H(x\|m\|com), resp)$*. Return* ok*.*

Corollary 23 (Fiat-Shamir signatures). *Assume that Σ is honest-verifier zero-knowledge, has completeness, has unpredictable commitments, and has statistical soundness for R_η, and that ℓ_η^{ch} is superlogarithmic. Assume that G is a dual-mode hard instance generator for R_η.*

Then the signature scheme $(KeyGen_{FS}, Sign_{FS}, Verify_{FS})$ from Definition 22 is weakly unforgeable.

If Σ additionally has unique responses, the signature scheme is strongly unforgeable.

Proof. *Let Σ' be the following sigma-protocol for R': The message lengths $\ell_\eta^{com}, \ell_\eta^{ch}, \ell_\eta^{resp}$ are the same as for Σ. For $x \in \{0,1\}^{\ell_\eta^x}$, $m \in \{0,1\}^{\ell_\eta^m}$, the prover $P_{\Sigma'}^1(1^\eta, (x\|m), w)$ runs $P_\Sigma^1(1^\eta, x, w)$, and $P_{\Sigma'}^2(1^\eta, (x\|m), w, ch)$ runs $P_\Sigma^2(1^\eta, x, w, ch)$. And $V_{\Sigma'}(1^\eta, x\|m, com, ch, resp)$ runs $V_\Sigma(1^\eta, x, com, ch, resp)$.*

It is easy to check that Σ' is honest-verifier zero-knowledge, has completeness, has unpredictable commitments, and has statistical soundness for R'_η. (Using the fact that Σ has these properties for R_η.) And ℓ^{ch} is superlogarithmic.

We apply the Fiat-Shamir construction (Definition 11) to Σ'. The resulting proof system (P_{FS}, V_{FS}) is zero-knowledge and weakly simulation-sound for R'_η by Theorems 14 and 16. Then we apply the construction of signatures (Definition 18) to (P_{FS}, V_{FS}) and G. By Theorem 21, the resulting signature scheme S is weakly unforgeable.

Finally, notice that this signature scheme S is the signature scheme from Definition 22. (By explicitly instantiating the constructions from Definition 11 and Definition 18 and the definition of Σ'.)

If Σ additionally has unique responses, then Σ' also has unique responses. Thus by Theroem 17, (P_{FS}, V_{FS}) is simulation-sound. Hence by Theorem 21, S is strongly unforgeable. □

6.1 Security Proof

Proof of Theorem 21. We prove the case of strong unforgeability (assuming simulation-soundness). The case of weak unforgeability is proven almost identically, we just have to replace all occurrences of $(m^*, \sigma^*) \notin$ **Sig**-queries *by* $m^* \notin$ **Sig**-queries *and* $(x^*, \pi^*) \notin$ S-queries *by* $x^* \notin$ S-queries.

In this proof, we will in many places omit the security parameter η *for readability. (E.g., we write* ℓ^m *instead of* ℓ^m_η *and* $Sign(sk, m)$ *instead of* $Sign(1^\eta, sk, m)$*.) It is to be understood that this is merely a syntactic omission, the variables and algorithms still depend on* η.

In the following, H *will always denote a uniformly random function from* $\mathrm{Fun}(\ell^{in}, \ell^{out})$*. That is, every game written below implicitly starts with* $H \xleftarrow{\$} \mathrm{Fun}(\ell^{in}, \ell^{out})$.

Fix a polynomial-time oracle algorithm A*. By definition of strong unforgeability (Definition 19), we need to show*

$$\Pr[win = 1 : Game\ 1] \leq \mu(\eta)$$

for some negligible μ *and the following game:*

Game 1 (Unforgeability). $(pk, sk) \leftarrow KeyGen^H()$, $(\sigma^*, m^*) \leftarrow A^{H,\mathbf{Sig}}(pk)$, $ok \leftarrow Verify^H(pk, \sigma^*, m^*)$. $win := (ok = 1 \ \wedge \ (m^*, \sigma^*) \notin \mathbf{Sig}\text{-}queries)$.

We will transform this game in several steps. First, we inline the definitions of **Sig** (Definition 19) and *KeyGen*, *Sign*, and *Verify* (Definition 18). This leads to the following game:

Game 2. $(x, w) \leftarrow G(1^\eta)$. $(x^*, \pi^*) \leftarrow B^{H,P^H}(x, w)$. $ok \leftarrow V^H(x^*, \pi^*)$. $win := (ok = 1 \ \wedge \ (x^*, \pi^*) \notin \text{S-queries})$.

Here B is a polynomial-time oracle algorithm that runs A with input $pk := x$, and that, whenever A queries **Sig** with input m, invokes P^H with input $(x \| m, w)$ instead. And when A returns some (m^*, σ^*), then B returns (x^*, π^*) with $x^* := x \| m^*$ and $\pi^* := \sigma^*$. And S-queries is the list of queries made to P^H. More precisely, when P^H is invoked with (x', w') and responds with π', then (x', π') is appended to S-queries.

We then have:

$$\Pr[win = 1 : \text{Game 1}] = \Pr[win = 1 : \text{Game 2}]$$

We now use the zero-knowledge property of (P, V). Let S be the simulator whose existence is guaranteed by Definition 8. Let S' be the oracle that on input $(x, w) \in R'$ runs $S(x)$ and returns the latter's output (as in Definition 8).

Then

$$\Big|\Pr[win = 1 : \text{Game 2}] - \Pr[win = 1 : \text{Game 3}]\Big| \leq \mu_1$$

for some negligible μ_1, and with the following game:

Game 3. $(x, w) \leftarrow G(1^\eta)$. $(x^*, \pi^*) \leftarrow B^{H,S'^H}(x, w)$. $ok \leftarrow V^{H_{final}}(x^*, \pi^*)$. $win := (ok = 1 \ \wedge \ (x^*, \pi^*) \notin \text{S-queries})$.

Here H_{final} is as in Definition 10, i.e., the value of the random oracle H after it has been reprogrammed by S.

By $x \le x^*$, we mean that x consists of the first ℓ^x bits of x^*. (I.e., $x^* = x\|m$ for some m.)

Game 4. $(x,w) \leftarrow G(1^\eta)$. $(x^*,\pi^*) \leftarrow B^{H,S'^H}(x,w)$. $ok \leftarrow V^{H_{final}}(x^*,\pi^*)$. $win := (ok = 1 \wedge x \le x^* \wedge (x^*,\pi^*) \notin \mathsf{S\text{-}queries})$.

Since B by construction always outputs $x^* = x\|m^*$, we have

$$\Pr[win = 1 : \text{Game 3}] = \Pr[win = 1 : \text{Game 4}].$$

Let $C^{H,S^H}(x)$ be a polynomial-time oracle algorithm that runs A with input $pk := x$, and that, whenever A queries **Sig** with input m, instead invokes S^H with input $x\|m$. And when A returns some (m^*,σ^*), then C returns (x^*,π^*) with $x^* := x\|m^*$ and $\pi^* := \sigma^*$.

Note that there are two differences between B^{H,S'^H} and C^{H,S^H}: First, C does not take w as input. Second, C invokes S^H instead of S'^H. Since $S'(x\|m,w)$ invokes $S(x\|m)$ whenever $(x\|m,w) \in R'$, B and C will differ only when $(x\|m,w) \notin R'$. By definition of R', this happens only when $(x,w) \notin R$. And this, in turn, happens with negligible probability since (x,w) are chosen by G, and G is a dual-mode hard instance generator. Thus there exists a negligible μ_2 such that

$$\Big|\Pr[win = 1 : \text{Game 4}] - \Pr[win = 1 : \text{Game 5}]\Big| \le \mu_2 \qquad \text{with}$$

Game 5. $(x,w) \leftarrow G(1^\eta)$. $(x^*,\pi^*) \leftarrow C^{H,S^H}(x)$. $ok \leftarrow V^{H_{final}}(x^*,\pi^*)$. $win := (ok = 1 \wedge x \le x^* \wedge (x^*,\pi^*) \notin \mathsf{S\text{-}queries})$.

Since G is a dual-mode hard instance generator, and since the computation in Game 5 after $(x,w) \leftarrow G(1^\eta)$ is quantum-polynomial-time[21] and does not use w, we have (by Definition 20) that there exists a quantum-polynomial-time G^* and a negligible μ_3 such that:

$$\Big|\Pr[win = 1 : \text{Game 5}] - \Pr[win = 1 : \text{Game 6}]\Big| \le \mu_3 \qquad \text{with}$$

Game 6. $x \leftarrow G^*(1^\eta)$. $(x^*,\pi^*) \leftarrow C^{H,S^H}(x)$. $ok \leftarrow V^{H_{final}}(x^*,\pi^*)$. $win := (ok = 1 \wedge x \le x^* \wedge (x^*,\pi^*) \notin \mathsf{S\text{-}queries})$.

Since G^* was chosen as in Definition 20, we have that $x \in L_R$ with some negligible probability μ_4 in Game 6. Thus

$$\Big|\Pr[win = 1 : \text{Game 6}] - \Pr[win = 1 : \text{Game 7}]\Big| \le \mu_4 \qquad \text{with}$$

[21] Note: to simulate the oracle H (which is a random function and thus has an exponentially large value-table), we use the fact from [28] that a $2q$-wise hash function cannot be distinguished from random by a q-query adversary. This allows us to simulate H using a $2q$-wise hash function for suitable polynomially-bounded q (that may depend on A).

Game 7. $x \leftarrow G^*(1^\eta)$. $(x^*,\pi^*) \leftarrow C^{H,S^H}(x)$. $ok \leftarrow V^{H_{final}}(x^*,\pi^*)$. $win := (ok = 1 \ \wedge x \notin L_R \ \wedge \ x \leq x^* \ \wedge \ (x^*,\pi^*) \notin \mathsf{S\text{-}queries})$.

By definition of R', we have that $x \notin L_R \wedge x \leq x^* \implies x^* \notin L_R$. Thus

$$\Pr[win : \text{Game 7}] \leq \Pr[ok = 1 \wedge x^* \notin L_R \wedge (x^*,\pi^*) \notin \mathsf{S\text{-}queries} : \text{Game 7}].$$

Since (P,V) is simulation-sound (Definition 10), and "$x \leftarrow G^*(1^\eta)$. $(x^*,\pi^*) \leftarrow C^{H,S^H}(x)$" can be executed by a quantum-polynomial-time oracle algorithm with oracle access to H and S^H, we have that there is a negligible μ_5 such that

$$\Pr[ok = 1 \wedge x^* \notin L_R \wedge (x^*,\pi^*) \notin \mathsf{S\text{-}queries} : \text{Game 7}] \leq \mu_5.$$

Combining all inequalities from this proof, we get that

$$\Pr[win : \text{Game 1}] \leq \mu_1 + \cdots + \mu_5 =: \mu.$$

The function μ is negligible since $\mu_1, \ldots, \mu_5$ are. Since A was arbitrary and quantum-polynomial-time, and Game 1 is the game from Definition 19, it follows that $(\mathit{KeyGen}, \mathit{Sign}, \mathit{Verify})$ is strongly unforgeable. □

Acknowledgments. I thank Andris Ambainis, and Ali El Kaafarani for valuable discussions, and Alexander Belov for breaking the Quantum Forking Conjecture upon which earlier versions of this work were based. This work was supported by institutional research funding IUT2-1 of the Estonian Ministry of Education and Research, the Estonian ICT program 2011–2015 (3.2.1201.13-0022), and by the Estonian Centre of Exellence in IT (EXCITE) funded by ERDF.

References

1. Adida, B.: Helios: web-based open-audit voting. In: USENIX Security Symposium 2008, pp. 335–348. USENIX (2008)
2. Ambainis, A., Rosmanis, A., Unruh, D.: Quantum attacks on classical proof systems (the hardness of quantum rewinding). In: FOCS 2014, pp. 474–483. IEEE (2014)
3. Bansarkhani, R.E., Kaafarani, A.E.: Post-quantum attribute-based signatures from lattice assumptions. IACR ePrint 2016/823 (2016)
4. Baum, C., Damgård, I., Oechsner, S., Peikert, C.: Efficient commitments and zero-knowledge protocols from ring-SIS with applications to lattice-based threshold cryptosystems. IACR ePrint 2016/997 (2016)
5. Bellare, M., Rogaway, P.: Random oracles are practical: a paradigm for designing efficient protocols. In: CCS 1993, pp. 62–73. ACM (1993)
6. Bernhard, D., Pereira, O., Warinschi, B.: How not to prove yourself: pitfalls of the fiat-shamir heuristic and applications to helios. In: Wang, X., Sako, K. (eds.) ASIACRYPT 2012. LNCS, vol. 7658, pp. 626–643. Springer, Heidelberg (2012). https://doi.org/10.1007/978-3-642-34961-4_38
7. Boneh, D., Boyen, X., Shacham, H.: Short group signatures. In: Franklin, M. (ed.) CRYPTO 2004. LNCS, vol. 3152, pp. 41–55. Springer, Heidelberg (2004). https://doi.org/10.1007/978-3-540-28628-8_3

8. Boneh, D., Dagdelen, Ö., Fischlin, M., Lehmann, A., Schaffner, C., Zhandry, M.: Random oracles in a quantum world. In: Lee, D.H., Wang, X. (eds.) ASIACRYPT 2011. LNCS, vol. 7073, pp. 41–69. Springer, Heidelberg (2011). https://doi.org/10.1007/978-3-642-25385-0_3
9. Brickell, E., Camenisch, J., Chen, L.: Direct anonymous attestation. In: ACM CCS 2004, pp. 132–145. ACM, New York (2004)
10. Camenisch, J., Lysyanskaya, A.: An efficient system for non-transferable anonymous credentials with optional anonymity revocation. In: Pfitzmann, B. (ed.) EUROCRYPT 2001. LNCS, vol. 2045, pp. 93–118. Springer, Heidelberg (2001). https://doi.org/10.1007/3-540-44987-6_7
11. Faust, S., Kohlweiss, M., Marson, G.A., Venturi, D.: On the non-malleability of the Fiat-Shamir transform. In: Galbraith, S., Nandi, M. (eds.) INDOCRYPT 2012. LNCS, vol. 7668, pp. 60–79. Springer, Heidelberg (2012). https://doi.org/10.1007/978-3-642-34931-7_5
12. Fiat, A., Shamir, A.: How to prove yourself: practical solutions to identification and signature problems. In: Odlyzko, A.M. (ed.) CRYPTO 1986. LNCS, vol. 263, pp. 186–194. Springer, Heidelberg (1987). https://doi.org/10.1007/3-540-47721-7_12
13. Fischlin, M.: Communication-efficient non-interactive proofs of knowledge with online extractors. In: Shoup, V. (ed.) CRYPTO 2005. LNCS, vol. 3621, pp. 152–168. Springer, Heidelberg (2005). https://doi.org/10.1007/11535218_10
14. Goldfeder, S., Chase, M., Zaverucha, G.: Efficient post-quantum zero-knowledge and signatures. IACR ePrint 2016/1110 (2016)
15. Gordon, S.D., Katz, J., Vaikuntanathan, V.: A group signature scheme from lattice assumptions. In: Abe, M. (ed.) ASIACRYPT 2010. LNCS, vol. 6477, pp. 395–412. Springer, Heidelberg (2010). https://doi.org/10.1007/978-3-642-17373-8_23
16. Libert, B., Ling, S., Mouhartem, F., Nguyen, K., Wang, H.: Signature schemes with efficient protocols and dynamic group signatures from lattice assumptions. In: Cheon, J.H., Takagi, T. (eds.) ASIACRYPT 2016. LNCS, vol. 10032, pp. 373–403. Springer, Heidelberg (2016). https://doi.org/10.1007/978-3-662-53890-6_13
17. Libert, B., Ling, S., Mouhartem, F., Nguyen, K., Wang, H.: Zero-knowledge arguments for matrix-vector relations and lattice-based group encryption. In: Cheon, J.H., Takagi, T. (eds.) ASIACRYPT 2016. LNCS, vol. 10032, pp. 101–131. Springer, Heidelberg (2016). https://doi.org/10.1007/978-3-662-53890-6_4
18. Ling, S., Nguyen, K., Wang, H.: Group signatures from lattices: simpler, tighter, shorter, ring-based. In: Katz, J. (ed.) PKC 2015. LNCS, vol. 9020, pp. 427–449. Springer, Heidelberg (2015). https://doi.org/10.1007/978-3-662-46447-2_19
19. Pointcheval, D., Stern, J.: Provably secure blind signature schemes. In: Kim, K., Matsumoto, T. (eds.) ASIACRYPT 1996. LNCS, vol. 1163, pp. 252–265. Springer, Heidelberg (1996). https://doi.org/10.1007/BFb0034852
20. Pointcheval, D., Stern, J.: Security proofs for signature schemes. In: Maurer, U. (ed.) EUROCRYPT 1996. LNCS, vol. 1070, pp. 387–398. Springer, Heidelberg (1996). https://doi.org/10.1007/3-540-68339-9_33
21. Pointcheval, D., Stern, J.: Security arguments for digital signatures and blind signatures. J. Cryptol. **13**(3), 361–396 (2000)
22. Sahai, A.: Non-malleable non-interactive zero knowledge and adaptive chosen-ciphertext security. In: FOCS 1999. IEEE (1999)
23. Schnorr, C.P.: Efficient signature generation by smart cards. J. Cryptol. **4**(3), 161–174 (1991)
24. Shoup, V., Gennaro, R.: Securing threshold cryptosystems against chosen ciphertext attack. J. Cryptol. **15**(2), 75–96 (2002)

25. Unruh, D.: Quantum proofs of knowledge. In: Pointcheval, D., Johansson, T. (eds.) EUROCRYPT 2012. LNCS, vol. 7237, pp. 135–152. Springer, Heidelberg (2012). https://doi.org/10.1007/978-3-642-29011-4_10
26. Unruh, D.: Non-interactive zero-knowledge proofs in the quantum random oracle model. In: Oswald, E., Fischlin, M. (eds.) EUROCRYPT 2015. LNCS, vol. 9057, pp. 755–784. Springer, Heidelberg (2015). https://doi.org/10.1007/978-3-662-46803-6_25
27. Unruh, D.: Post-quantum security of Fiat-Shamir. IACR ePrint 2017/398 (2017)
28. Zhandry, M.: Secure identity-based encryption in the quantum random oracle model. In: Safavi-Naini, R., Canetti, R. (eds.) CRYPTO 2012. LNCS, vol. 7417, pp. 758–775. Springer, Heidelberg (2012). https://doi.org/10.1007/978-3-642-32009-5_44

Symmetric Key Cryptanalysis

Improved Conditional Cube Attacks on Keccak Keyed Modes with MILP Method

Zheng Li[1], Wenquan Bi[1], Xiaoyang Dong[2](✉), and Xiaoyun Wang[1,2](✉)

[1] Key Laboratory of Cryptologic Technology and Information Security, Ministry of Education, Shandong University, Jinan, People's Republic of China
{lizhengcn,biwenquan}@mail.sdu.edu.cn

[2] Institute for Advanced Study, Tsinghua University, Beijing, People's Republic of China
{xiaoyangdong,xiaoyunwang}@tsinghua.edu.cn

Abstract. Conditional cube attack is an efficient key-recovery attack on Keccak keyed modes proposed by Huang *et al.* at EUROCRYPT 2017. By assigning bit conditions, the diffusion of a conditional cube variable is reduced. Then, using a greedy algorithm (Algorithm 4 in Huang *et al.*'s paper), Huang *et al.* find some ordinary cube variables, that do not multiply together in the $1st$ round and do not multiply with the conditional cube variable in the $2nd$ round. Then the key-recovery attack is launched. The key part of conditional cube attack is to find enough ordinary cube variables. Note that, the greedy algorithm given by Huang *et al.* adds ordinary cube variable without considering its bad effect, i.e. the new ordinary cube variable may result in that many other variables could not be selected as ordinary cube variable (they multiply with the new ordinary cube variable in the first round).

In this paper, we bring out a new MILP model to solve the above problem. We show how to model the CP-like-kernel and model the way that the ordinary cube variables do not multiply together in the $1st$ round as well as do not multiply with the conditional cube variable in the $2nd$ round. Based on these modeling strategies, a series of linear inequalities are given to restrict the way to add an ordinary cube variable. Then, by choosing the objective function of the maximal number of ordinary cube variables, we convert Huang *et al.*'s greedy algorithm into an MILP problem and the maximal ordinary cube variables are found.

Using this new MILP tool, we improve Huang *et al.*'s key-recovery attacks on reduced-round Keccak-MAC-384 and Keccak-MAC-512 by 1 round, get the first 7-round and 6-round key-recovery attacks, respectively. For Ketje Major, we conclude that when the nonce is no less than 11 lanes, a 7-round key-recovery attack could be achieved. In addition, for Ketje Minor, we use conditional cube variable with 6-6-6 pattern to launch 7-round key-recovery attack.

T. Takagi and T. Peyrin (Eds.): ASIACRYPT 2017, Part I, LNCS 10624, pp. 99–127, 2017.
https://doi.org/10.1007/978-3-319-70694-8_4

1 Introduction

Nowadays, the cryptanalysis progress of symmetric-key ciphers heavily depends on automated evaluation tools. Providing a reliable security evaluation is the key point for a cipher to be accepted by industry. Recently, cryptographic communities found that many classical cryptanalysis methods could be converted to mathematical optimization problems which aim to achieve the minimal or maximal value of an objective function under certain constraints. Mixed-integer Linear Programming (MILP) is the most widely studied technique to solve these optimization problems. One of the most successful applications of MILP is to search differential and linear trails. Mouha *et al.* [25] and Wu and Wang [30] first applied MILP method to count active Sboxes of word-based block ciphers. Then, at Asiacrypt 2014, by deriving some linear inequalities through the H-Representation of the convex hull of all differential patterns of Sbox, Sun *et al.* [29] extended this technique to search differential and linear trails. Another two important applications are to search integral distinguisher [31] and impossible differentials [8,27].

Keccak [3], designed by Bertoni *et al.*, has been selected as the new cryptographic hash function standard SHA-3. As one of the most important cryptographic standards, Keccak attracts lots of attention from the world wide researchers and engineers. Till now, many cryptanalysis results [7,10,11,18,19, 21,24] and evaluation tools [9,14,23] have been proposed, including the recent impressive collision attacks [26,28]. Since the robust design of Keccak, the cryptanalysis progress of Keccak is still limited. It must be pointed out that the automatic evaluation tools for Keccak are still needed to be enriched urgently.

At Eurocrypt 2015, Dinur *et al.* [12] for the first time considered the security of the Keccak keyed modes against cube-attack-like cryptanalysis and give some key recovery attacks on reduced-round Keccak-MAC and Keyak [5]. At CT-RSA 2015, Dobraunig *et al.* [15] evaluate the security of Ascon [16] against the cube-like cryptanalysis. Later, Dong *et al.* [17] applied the cube-like method to Ketje Sr [4] which adopts smaller state size of Keccak-p permutation. At Eurocrypt 2017, Huang *et al.* [20] introduced a new type of cube-like attack, called *conditional cube attack*, which takes advantage of the large state freedom of Keccak to find a so-called *conditional cube variable* that do not multiply with all the other *cube variables* (called *ordinary cube variables*) in the first round and second round of Keccak, meanwhile, all *ordinary cube variables* do not multiply with each other in the first round. Thus, the degree of output polynomial of reduced-round Keccak over the cube variables is reduced by 1 and a *conditional cube tester* is constructed. Then Li *et al.* [22] applied the *conditional cube attack* to reduced-round Ascon.

1.1 Our Contributions

For *conditional cube attack*, when the *conditional cube variable* is determined, the most important work is to find enough *ordinary cube variables* to launch the key recovery attack. In [20], Huang *et al.* gives the Algorithm 4 to search the *ordinary cube variables*. It is a greedy algorithm, it randomly selects a cube variable and adds to *ordinary cube variable* set, when the variable does not multiply with other *ordinary cube variables* in the set in the first round and does not multiply with *conditional cube variable* either in both the first and second round. The drawback is that it can hardly get the maximum number (optimal) of *ordinary cube variables*. Because, when a cube variable is added to *ordinary cube variable* set, many more variables which multiply with the new added cube variable in the first round will be discarded, which means that we add just one cube variable with the price that many variables lost the chance to be an *ordinary cube variable*. Actually, the search problem is an optimization problem. When the capacity of Keccak is large, the greedy algorithm is enough to find a proper *ordinary cube variable* set. However, when the capacity or the state size is small, the algorithm could hardly find enough *ordinary cube variables* and invalidate the *conditional cube attack*. In fact, for Keccak-MAC-512 and Keccak-MAC-384, only 5 round and 6 round attacks are achieved by Huang *et al.*'s algorithm. When the capacity is large or the internal state of Keccak sponge function is smaller than 1600-bit, e.g. 800-bit Ketje Minor, the number of *ordinary cube variables* is reduced significantly.

In this paper, we present a novel technique to search *ordinary cube variables* by using MILP method[1]. By modeling the relations between *ordinary cube variables* and *conditional cube variables* in the first and second round, modeling the so-called *CP-like-kernel* and *ordinary cube variables* chosen conditions, we construct a linear inequality system. The target object is the maximum number of *ordinary cube variables*. Based on this MILP tool, we improve Huang *et al.*'s attacks on Keccak-MAC and give some interesting results on Ketje Major and Minor, which are summarized in Table 1. In addition, we list our source code of the new MILP tool[2] in a public domain to enrich the automatic evaluation tools on Keccak and help the academic communities study Keccak much easier. The following are the main application results of the MILP tool.

1. It should be noted that, when the capacity reaches 768 or 1024, the cryptanalysis of Keccak becomes very hard. In fact, collision results on round-reduced Keccak-384 or Keccak-512 that are better than the birthday bound could respectively reach 4/3-round, while the preimage attacks [19,24] on

[1] Note that, in Huang *et al.*'s paper, a small MILP model is also introduced, however, it could only find some distinguisher attacks on Keccak hash function.

[2] https://github.com/lizhengcn/MILP_conditional_cube_attack.

the two versions could reach only 4 rounds. Based on our MILP tool, for Keccak-MAC-384, we find more than 63 *ordinary cube variables* and improve Huang *et al.*'s attack by 1 round, and get the very first 7-round key-recovery attack. For Keccak-MAC-512, we find more than 31 *ordinary cube variables* and improve Huang *et al.*'s attack by 1 round, and get the first 6-round key-recovery attack. These are the longest attacks that the cryptanalysis of Keccak with big capacity (768 or 1024) could reach.

2. For Ketje Major, we conclude that when the nonce is no less than 11 lanes, a 7-round conditional cube attack could work. In addition, we get the borderline length of the nonce for the 6-round key-recovery attack is 8 lanes.
3. For Ketje Minor, we use a new *conditional cube variable* and find 124 *ordinary cube variables*. Then a new 7-round key-recovery attack is proposed, which improved the previous best result by a factor of 2^{15}.

Table 1. Summary of key recovery attacks on Keccak keyed modes

Variant	Capacity	Attacked rounds	Time	Source
Keccak-MAC	768	6	2^{40}	[20]
		7	2^{75}	Sect. 5.1
	1024	5	2^{24}	[20]
		6	$2^{58.3}$	Sect. 5.2
Variant	Nonce	Attacked rounds	Time	Source
Ketje Major	Full	6	2^{64}	[17]
	Full	7	2^{96}	[17]
	$\geq$512	6	2^{41}	Sect. 6.1
	$\geq$704	7	2^{83}	Sect. 6.1
Ketje Minor	Full	6	2^{64}	[17]
	Full	7	2^{96}	[17]
	Full	6	2^{49}	Sect. 6.2
	Full	7	2^{81}	Sect. 6.2

Full: the attacks use maximum length of nonce.

1.2 Organization of the Paper

Section 2 gives some notations, and brief description on Keccak-permutations, Keccak-MAC, Ketje. Some related works are introduced in Sect. 3. Section 4 describes the MILP search model for conditional cube attack. Round-reduced key-recovery attacks on Keccak-MAC-384/512 are introduced in Sect. 5. Section 6 gives the applications to Ketje. Section 7 concludes this paper.

2 Preliminaries

2.1 Notations

S_i	the intermediate state after i-round of Keccak-p, for example $S_{0.5}$ means the intermediate state before χ in 1st round of Keccak-p,
A	used in tables: for Keccak-MAC, the initial state, for Ketje, the state after π^{-1} of Keccak-p^*,
$A[i][j]$	the 32/64-bit word indexed by $[i, j, *]$ of state A, $0 \leqslant i \leqslant 4$, $0 \leqslant j \leqslant 4$,
$A[i][j][k]$	the bit indexed by $[i, j, k]$ of state A,
v_i	the ith cube variable,
K	128-bit key, for Keccak-MAC, $K = k_0 \| k_1$, both k_0 and k_1 are 64-bit, for Ketje Major, $K = k_0 \| k_1 \| k_2$, k_0 is 56-bit, k_1 is 64-bit, k_2 is 8-bit, for Ketje Minor, $K = k_0 \| k_1 \| k_2 \| k_3 \| k_4$, k_0 is 24-bit, k_1,k_2 and k_3 are 32-bit, k_4 is 8-bit,
$k_i[j]$	the jth bit of k_i,
capacity	in Keccak-MAC, it is the all zero padding bits; in Ketje, it is the padding of nonce.

2.2 The Keccak-p permutations

The Keccak-p permutations are derived from the Keccak-f permutations [3] and have a tunable number of rounds. A Keccak-p permutation is defined by its width $b = 25 \times 2^l$, with $b \in \{25, 50, 100, 200, 400, 800, 1600\}$, and its number of rounds n_r, denoted as Keccak-$p[b]$. The round function R consists of five operations:

$$\mathtt{R} = \iota \circ \chi \circ \pi \circ \rho \circ \theta$$

Keccak-$p[b]$ works on a state A of size b, which can be represented as 5×5 $\frac{b}{25}$-bit lanes, as depicted in Fig. 1, $A[i][j]$ with i for the index of column and j for the index of row. In what follows, indexes of i and j are in set $\{0, 1, 2, 3, 4\}$ and they are working in modulo 5 without other specification.

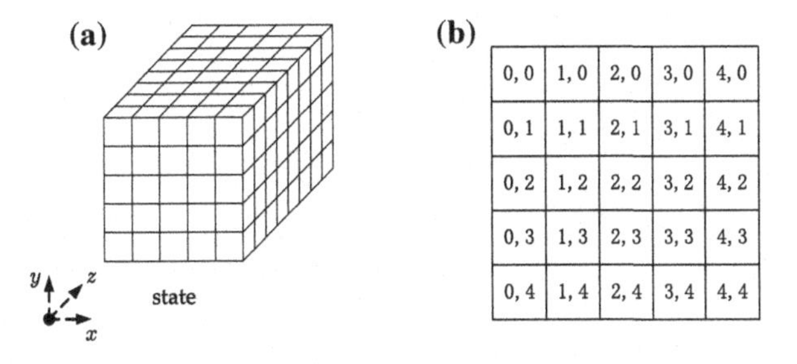

Fig. 1. (a) The Keccak State [3], (b) state A in 2-dimension

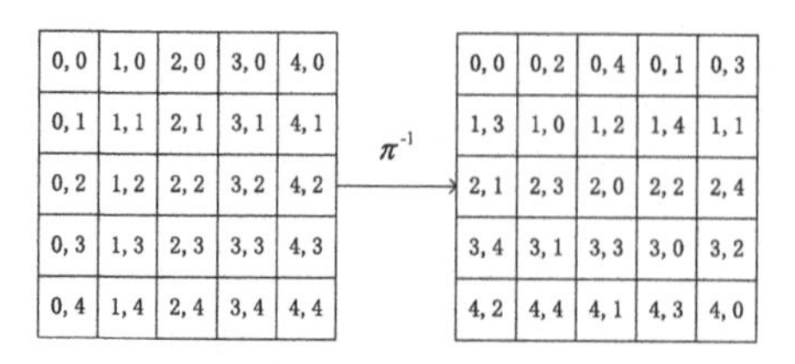

Fig. 2. π^{-1}

θ: $A[x][y] = A[x][y] \oplus \sum_{j=0}^{4} (A[x-1][j] \oplus (A[x+1][j] \lll 1))$.
ρ: $A[x][y] = A[x][y] \lll r[x, y]$.
π: $A[y][2x+3y] = A[x][y]$.
χ: $A[x][y] = A[x][y] \oplus ((\neg A[x+1][y]) \wedge A[x+2][y]$.
ι: $A[0][0] = A[0][0] \oplus RC$.

In Ketje v2, *the twisted permutations*, Keccak-$p^*[b] = \pi \circ$ Keccak-$p[b] \circ \pi^{-1}$, are introduced to effectively re-order the bits in the state. π^{-1} is the inverse of π, shown in Fig. 2.

$$\pi^{-1}: A[x+3y][x] = A[x][y].$$

2.3 Keccak-MAC

A MAC form of Keccak can be obtained by adding key as the prefix of message/nonce. As depicted in Fig. 3, the input of Keccak-MAC-n is concatenation of key and message and n is half of the capacity length.

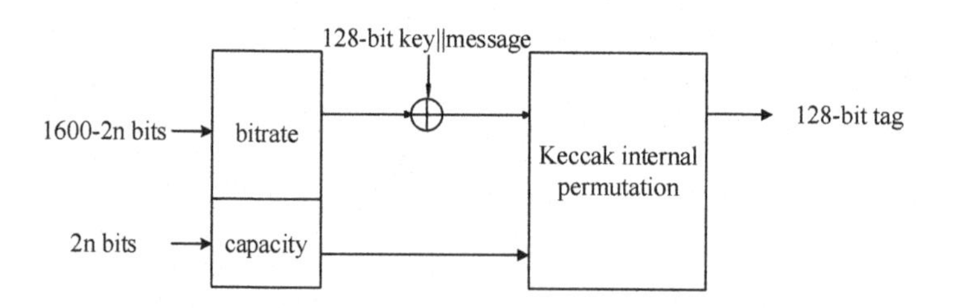

Fig. 3. Construction of Keccak-MAC-n

2.4 Ketje

Ketje [4] is a submission by Keccak team. It is a sponge-like construction. In Ketje v1, two instances are proposed, Ketje Sr and Jr with 400-bit and 200-bit state sizes, respectively. In the latest Ketje v2, another two instances Ketje Minor and Major are added to the family, with 800-bit and 1600-bit state sizes,

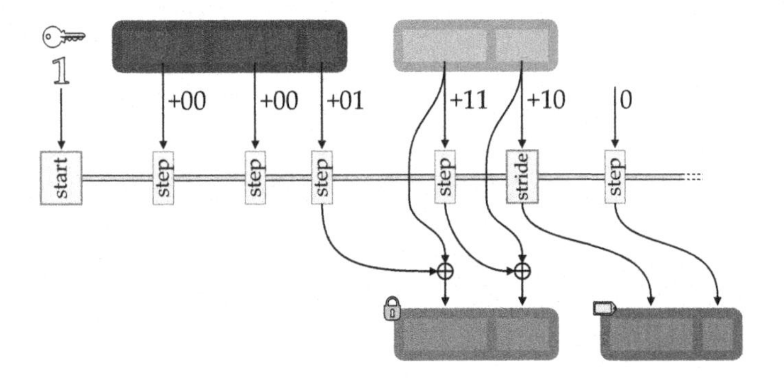

Fig. 4. Wrapping a header and a body with MonkeyWrap [4]

respectively. Ketje Sr is the primary recommendation. In the following, we give a brief overview about the Ketje v2. For a complete description, we refer to the design document [4].

The structure of Ketje is an authenticated encryption mode MonkeyWrap, shown Fig. 4, which is based on MonkeyDuplex [6]. It consists of four parts as follows:

1. **The initialization phase:** The initialization takes the secret key K, the public nonce N and some paddings as the initial state. Then $n_{start} = 12$ rounds Keccak-p^* is applied.
2. **Processing associated data:** ρ-bit blocks associated data are padded to $(\rho + 4)$-bit and absorbed by xoring them to the state, then $n_{step} = 1$ round Keccak-p^* is applied. If associated data is empty, this procedure is still needed to be applied which means an empty block is padded to $(\rho + 4)$-bit and then processed similarly.
3. **Processing the plaintext:** Plaintext is processed in ρ-bit blocks in a similar manner, with ciphertext blocks extracted from the state right after adding the plaintext.
4. **Finalization:** The finalization with $n_{stride} = 6$ rounds Keccak-p^* and a series of $n_{step} = 1$ round Keccak-p^*s are performed to get the required length of tag T.

In Ketje v2, four concrete instances are proposed, shown in Table 2. $n_{start} = 12$,$n_{step} = 1$ and $n_{stride} = 6$. For Ketje Minor and Major, the recommended key length is 128-bit, so the maximal length of nonce is $(800 - 128 - 18{=})654$ and $(1600 - 128 - 18{=})1454$ bits. This paper discusses the shortest length of nonce that a conditional cube attack could be applied.

Table 2. Four instances in Ketje v2

Name	f	ρ	Main use case
Ketje Jr	Keccak-p^*[200]	16	Lightweight
Ketje Sr	Keccak-p^*[400]	32	Lightweight
Ketje Minor	Keccak-p^*[800]	128	Lightweight
Ketje Major	Keccak-p^*[1600]	256	High performance

3 Related Work

3.1 Cube Attack

At EUROCRYPT 2009, Dinur and Shamir introduced the cube attack [13], in which the output bit of a symmetric cryptographic scheme can be regarded as a polynomial $f(k_0, \ldots, k_{n-1}, v_0, \ldots, v_{m-1})$ over $GF(2)$, $k_0, \ldots, k_{n-1}$ are the secret variables (the key bits), $v_0, \ldots, v_{m-1}$ are the public variables (e.g. IV or nonce bits).

Theorem 1 [13].

$$f(k_0, \ldots, k_{n-1}, v_0, \ldots, v_{m-1}) = t \cdot P + Q(k_0, \ldots, k_{n-1}, v_0, \ldots, v_{m-1}) \quad (1)$$

t is called maxterm and is a product of certain public variables, for example $(v_0, \ldots, v_{s-1})$, $1 \leq s \leq m$, denoted as cube C_t. None of the monomials in Q is divisible by t. P is called superpoly, which does not contain any variables of C_t. Then the sum of f over all values of the cube C_t (cube sum) is

$$\sum_{v'=(v_0,\ldots,v_{s-1})\in C_t} f(k_0, \ldots, k_{n-1}, v', v_s, \ldots, v_{m-1}) = P \quad (2)$$

where C_t contains all binary vectors of the length s, $v_s, \ldots, v_{m-1}$ are fixed to constant.

The basic idea is to find enough t whose P is linear and not a constant. This enables the key recovery through solving a system of linear equations.

3.2 Huang *et al.*'s Conditional Cube Attack

Conditional cube attack [20] was proposed by Huang *et al.* to attack Keccak keyed mode, including Keccak-MAC and Keyak. We quote some definitions and a theorem here.

Definition 1 [20]. *Cube variables that have propagation controlled in the first round and are not multiplied with each other after the second round of Keccak are called* ***conditional cube variables****. Cube variables that are not multiplied with each other after the first round and are not multiplied with any conditional cube variable after the second round are called* ***ordinary cube variables****.*

Theorem 2 [20]. *For $(n+2)$-round Keccak sponge function $(n > 0)$, if there are $p(0 \leq p < 2^n + 1)$ conditional cube variables $v_0, \ldots, v_{p-1}$, and $q = 2^{n+1} - 2p + 1$ ordinary cube variables, $u_0, \ldots, u_{q-1}$ (If $q = 0$, we set $p = 2^n + 1$), the term $v_0 v_1 \ldots v_{p-1} u_0 \ldots u_{q-1}$ will not appear in the output polynomials of $(n+2)$-round Keccak sponge function.*

Actually, we use the special case of the above theorem when $p = 1$. We describe it as a corollary for clearness.

Corollary 1. *For $(n + 2)$-round Keccak sponge function $(n > 0)$, if there is one conditional cube variable v_0, and $q = 2^{n+1} - 1$ ordinary cube variables, $u_0, \ldots, u_{q-1}$, the term $v_0 u_0 \ldots u_{q-1}$ will not appear in the output polynomials of $(n + 2)$-round Keccak sponge function.*

4 Modeling Search Strategy

Define $A[x][y][z] = 1$ when it is an *ordinary cube variable* or *conditional cube variable*, else $A[x][y][z] = 0$.

4.1 Modeling CP-like-kernel

In the Keccak submission document [3], the original concept is illustrated as following: if all columns in a state have even parity, θ is the identity, which is illustrated. The *conditional cube variable* used in this is set in CP-kernel to reach a reduced diffusion. At ASIACRYPT 2016, Guo *et al.* [19] assign $A[1][y]$, $y = 0, 1, 2, 3$, to be variables and $A[1][4] = \bigoplus_{i=0}^{3} A[1][y]$ so that variables in each column sum to 0. Then θ is the identity. In fact, when the parity of a column remains constant, the variables in the column do not propagate through θ operation. We denoted this property as a *CP-like-kernel*. In order to reduce the diffusion of *ordinary cube variables*, we set them as CP-like-kernel.

In CP-like-kernel, if certain column contain *ordinary cube variables*, then the number of the variables must be no less than two. If $n(n = 2, 3, 4, 5)$ bits in a column contain cube variables, we set the $n - 1$ bits to be independent *ordinary cube variables* and 1 bit variable to be the sum of the $n-1$ bits. So the constraints in modeling CP-like-kernel have the following two purposes:

1. Avoid the number of bits containing cube variable in each column from being one;
2. Record which column contains cube variables.

Given x, z, suppose $A[x][y][z], y = 0,1,2,3,4$ possibly contain *ordinary cube variables*. If there exists an *ordinary cube variable* in $A[x][y][z]$ for some y, then the dummy variable $d = 1$. Else $d = 0$. Then we get the following inequalities

$$\begin{cases} \sum_{y=0}^{4} A[x][y][z] \geq 2d \\ d \geq A[x][0][z] \\ d \geq A[x][1][z] \\ d \geq A[x][2][z] \\ d \geq A[x][3][z] \\ d \geq A[x][4][z] \end{cases} \tag{3}$$

For any x, z ($x = 0,1,\ldots,4, z = 0,1,\ldots,63$), denote the corresponding dummy variable as $d[x][z]$. $d[x][z]$ records whether the column $[x][z]$ contain cube variables as illustrated above. The $[x][z]$ column can provide $\sum_{y=0}^{4} A[x][y][z] - d[x][z]$ independent cube variables. The number of independent cube variables that the whole state can provide is to sum up the ones of all columns with $x = 0,1\ldots4, z = 0,1\ldots63$. Correspondingly, the objective function of the MILP model is set as

$$\sum_{x,y,z} A[x][y][z] - \sum_{x,z} d[x][z],$$

i.e. the number of cube variables in the whole state.

4.2 Modeling the First Round

We omit the θ operation in the first round, as it does not influence the distribution of cube variables according to the property of CP-like-kernel. With the help of SAGE [1], the Keccak round function can be operated in the form of algebraic symbols. So the internal the bits of state S_1 are describe as algebraic form functions about the bits of the initial state S_0. Using a easy search program, we know which two bits in state S_0 will be multiplied in S_1. Constraints are added according to the following two conditions:

1. (a) **Condition:** Any of the *ordinary cube variables* do not multiply with each other in the first round.
   (b) **Constraint:** If two bits $S_0[x_1][y_1][z_1]$ and $S_0[x_2][y_2][z_2]$ multiply, the constraint

   $$A[x_1][y_1][z_1] + A[x_2][y_2][z_2] \leq 1$$

   will be added to avoid their simultaneous selection as *ordinary cube variables*.

2. (a) **Condition:** The *conditional cube variable* does not multiply with any of the *ordinary cube variables* in the first round.
   (b) **Constraint:** If one bit $S_0[x][y][z]$ multiplies with the *conditional cube variable*, the constraint
   $$A[x][y][z] = 0$$
   will be added to avoid it from being selected as *ordinary cube variables.*

4.3 Modeling the Second Round

We list Property 1 for the conditions added to control the diffusion of the *conditional cube variable* v_0.

Property 1. In χ operation, denote the input and output state as X and Y respectively, one bit $X[x][y][z]$ only multiplies with two bits $X[x-1][y][z]+1$ and $X[x+1][y][z]$.

(1) If only one bit $X[x][y][z]$ contains variable v_0, conditions $X[x-1][y][z]+1=0$ and $X[x+1][y][z]=0$ can avoid v_0 from diffusing by χ.
(2) If only n bits $X[x_0][y_0][z_0], X[x_1][y_1][z_1] \ldots X[x_{n-1}][y_{n-1}][z_{n-1}]$ contain variable v_0, $2n$ conditions
$$X[x_0-1][y_0][z_0]+1=0, X[x_0+1][y_0][z_0]=0,$$
$$X[x_1-1][y_1][z_1]+1=0, X[x_1+1][y_1][z_1]=0,$$
$$\ldots$$
$$X[x_{n-1}-1][y_{n-1}][z_{n-1}]+1=0, X[x_{n-1}+1][y_{n-1}][z_{n-1}]=0$$
can avoid v_0 from diffusing by χ.

1. **Condition:** Under the above conditions added to the first round, the *conditional cube variable* does not multiply with any of the *ordinary cube variables* in the second round.
2. **Constraint:** If one bit $S_0[x][y][z]$ multiplies with the *conditional cube variable*, the constraint
$$A[x][y][z] = 0$$
will be added to avoid it from being selected as *ordinary cube variables.*

5 Applications to Round-Reduced Keccak-MAC

5.1 Attack on 7-Round Keccak-MAC-384

For Keccak-MAC-384 with 1600-bit state, rate occupies 832 bits, and capacity 768 bits. As Fig. 5 shows us, 128-bit key (k_0, k_1) locates at the first two yellow lanes, and *conditional cube variable* v_0 is set in CP-like-kernel as $S_0[2][0][0] = S_0[2][1][0] = v_0$ in blue, then the white bits represent nonce or message bits, all of which can be selected as *ordinary cube variables*, while the grey ones are initialized with all zero. Note that the lanes, which are possible to be *ordinary cube variables*, obey CP-like-kernel. List these lanes in a set $\mathbb{V}$:

$$\mathbb{V} = \{[0][1], [0][2], [1][1], [1][2], [2][0], [2][1], [2][2], [3][0], [3][1], [4][0], [4][1]\}$$

Additionally, the subset of $\mathbb{V}$, $\mathbb{V}_i, i = 0, 1 \ldots 4$ represents the set of lanes whose x-index equals 0, 1 ... 4 respectively.

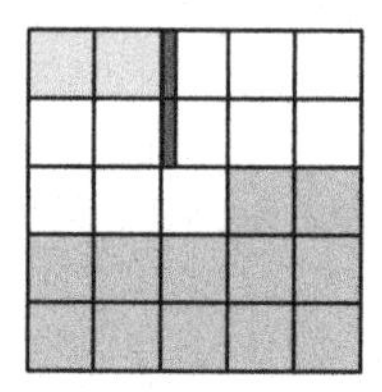

Fig. 5. The initial state of Keccak-MAC-384 (Color figure online)

According to the modeling search strategy illustrated in Sect. 4, we search for the maximal number of independent *ordinary cube variables*. The objective function is

$$\sum_{x,y \in \mathbb{V}, z \in \{0,1 \ldots 63\}} A[x][y][z] - \sum_{x \in \{0,1 \ldots 4\}, z \in \{0,1 \ldots 63\}} d[x][z],$$

To model the CP-like-kernel, constraints are in the following:

$$\begin{cases} \sum_{x,y \in \mathbb{V}_x, z} A[x][y][z] \geq 2d[x][z] \\ d[x][z] \geq A[x][y][z], y \in \mathbb{V}_x \end{cases} \quad for\ x = 0, 1 \ldots 4, z = 0, 1 \ldots 63 \tag{4}$$

The input state is initialized with key k, *conditional cube variable* v_0, possible *ordinary cube variables* v_i (placed in bit position $[x_i][y_i][z_i]$) and zero padding. After ρ, π, χ operation in the first round, the state is in the algebraic symbolic form of the initial state bits. If any $v_0 v_i$ exists, and the bit corresponding to i is $[x_i][y_i][z_i]$, constraint $A[x_i][y_i][z_i] = 0$ is added. If any $v_i v_j$ exists, the bit corresponding to i, j, constraint $A[x_i][y_i][z_i] + A[x_j][y_j][z_j] \leq 1$ is added. The above constraints are to avoid any multiplication in the first round among cube variables. Additionally, we add the four bit conditions around *conditional cube variable* v_0 before the first χ operation to reduce its diffusion. After θ, ρ, π, χ operation in the second round, similarly, if any $v_0 v_i$ exists, and the bit corresponding to i is $[x_i][y_i][z_i]$, constraint $A[x_i][y_i][z_i] = 0$ is added to avoid any *ordinary cube variables* from multiplying with v_0 in the second second. With the help of Gurobi [2], the objective function is optimized under all the above constraints, i.e. with all cube variables obeying CP-like-kernel, the maximum of cube variables is 65. Actually, 64 cube variables are enough to perform the 7-round attack on Keccak-MAC-384. Both the cube variables and conditions are listed in Table 3.

In 7-round attack on Keccak-MAC-384, $2^6 = 64$ cube variables are denoted by $v_0, v_1 \dots v_{63}$. Based on Corollary 1, v_0 is the conditional cube variable fixed in the beginning and $v_1, v_2 \dots v_{63}$ are ordinary cube variables found by MILP search strategy. We summarize the requirements as following:

(1) $v_0, v_1 \dots v_{63}$ do not multiply with each other in the first round;
(2) Under some conditions on key and nonce, v_0 does not multiply with any of $v_1, v_2 \dots v_{63}$ in the second round.

While all the nonce bits are constant, all the bit conditions are satisfied if and only if all the key bits are guessed correctly. Thus, zero sums over the 128-bit tag with cube variables set as Table 3 mean a correct key guess.

We analyze the time and data complexity of the attack: with the parameters set in Table 3, the 8 guessed key bits $k_0[5] + k_1[5]$, $k_0[60]$, $k_0[35]$, $k_0[54]$, $k_1[29]$, $k_0[7]$, $k_1[45]$, $k_0[18]$ can be recovered. The time complexity of one recovery is $2^8 * 2^{64}$. According to the property of permutation, it is totally symmetric in z-axis. Thus we can obtain corresponding parameters set with any rotation of i-bit $(0 \leq i < 64)$ in z-axis. Therefore, the guessed key bits rotated i-bit i.e. $k_0[i+5]$ $+ k_1[i+5]$, $k_0[i+60]$, $k_0[i+35]$, $k_0[i+54]$, $k_1[i+29]$, $k_0[i+7]$, $k_1[i+45]$, $k_0[i+18]$ can be recovered. Through simple count, for $0 \leq i < 8$, 70 independent key bits out of 128 key bits can be recovered, 8 iterations consumes $8 \times 2^8 \times 2^{64}$ and the remaining 58 key bits are left to exhaustive search consuming 2^{58}. Combine the two parts, the procedure consumes $8 \times 2^8 \times 2^{64} + 2^{58} = 2^{75}$ computations of 7-round of Keccak-MAC-384, correspondingly 2^{75} $(message, tag)$ pairs are needed. After the procedure above, all the 128 bits in k_0, k_1 can be recovered. Therefore, both time and data complexity of the attack are 2^{75}.

5.2 Attack on 6-Round Keccak-MAC-512

For Keccak-MAC-384 with 1600-bit state, rate occupies 832 bits, and capacity 768 bits. As Fig. 6 shows us, 128-bit key (k_0, k_1) locates at the first two yellow lanes, and *conditional cube variable* v_0 is set in CP-like-kernel as $S_0[2][0][0] = S_0[2][1][0] = v_0$ in blue, then the white bits represent nonce or message bits, all of which can be selected as *ordinary cube variables*, while the grey ones are initialized with all zero. Note that the lanes, which are possible to be *ordinary cube variables*, obey CP-like-kernel. List these lanes in a set $\mathbb{V}$:

$$\mathbb{V} = \{[2][0], [2][1], [3][0], [3][1]\}$$

Table 3. Parameters set for attack on 7-round Keccak-MAC-384

Ordinary Cube Variables
$A[2][0][1]{=}v_1$,$A[2][1][1]{=}v_2$,$A[2][2][1]{=}v_1 + v_2$,$A[3][0][3]{=}A[3][1][3]{=}v_3$, $A[2][0][5]{=}A[2][2][5]{=}v_4$,$A[1][1][7]{=}A[1][2][7]{=}v_5$,$A[2][0][8]{=}A[2][1][8]{=}v_6$, $A[3][0][9]{=}A[3][1][9]{=}v_7$,$A[4][0][10]{=}A[4][1][10]{=}v_8$,$A[2][1][11]{=}A[2][2][11]{=}v_9$, $A[2][0][12]{=}A[2][1][12]{=}v_{10}$,$A[4][0][12]{=}A[4][1][12]{=}v_{11}$,$A[3][0][13]{=}A[3][1][13]{=}v_{12}$, $A[2][1][14]{=}A[2][2][14]{=}v_{13}$,$A[4][0][14]{=}A[4][1][14]{=}v_{14}$,$A[0][1][15]{=}A[0][2][15]{=}v_{15}$, $A[1][1][15]{=}A[1][2][15]{=}v_{16}$,$A[2][1][15]{=}A[2][2][15]{=}v_{17}$,$A[2][1][18]{=}A[2][2][18]{=}v_{18}$, $A[2][1][19]{=}A[2][2][19]{=}v_{19}$,$A[2][0][20]{=}v_{20}$,$A[2][1][20]{=}v_{21}$,$A[2][2][20]{=}v_{20} + v_{21}$, $A[3][0][20]{=}A[3][1][20]{=}v_{22}$,$A[2][0][21]{=}A[2][2][21]{=}v_{23}$,$A[0][1][22]{=}A[0][2][22]{=}v_{24}$, $A[3][0][23]{=}A[3][1][23]{=}v_{25}$,$A[2][1][24]{=}A[2][2][24]{=}v_{26}$,$A[2][0][27]{=}A[2][2][27]{=}v_{27}$, $A[0][1][28]{=}A[0][2][28]{=}v_{28}$,$A[1][1][30]{=}A[1][2][30]{=}v_{29}$,$A[3][0][30]{=}A[3][1][30]{=}v_{30}$, $A[0][1][32]{=}A[0][2][32]{=}v_{31}$,$A[0][1][34]{=}A[0][2][34]{=}v_{32}$,$A[1][1][34]{=}A[1][2][34]{=}v_{33}$, $A[3][0][35]{=}A[3][1][35]{=}v_{34}$,$A[0][1][37]{=}A[0][2][37]{=}v_{35}$,$A[0][1][38]{=}A[0][2][38]{=}v_{36}$, $A[1][1][38]{=}A[1][2][38]{=}v_{37}$,$A[1][1][39]{=}A[1][2][39]{=}v_{38}$,$A[3][0][39]{=}A[3][1][39]{=}v_{39}$, $A[1][1][40]{=}A[1][2][40]{=}v_{40}$,$A[3][0][40]{=}A[3][1][40]{=}v_{41}$,$A[2][0][41]{=}A[2][1][41]{=}v_{42}$, $A[2][0][43]{=}A[2][1][43]{=}v_{43}$,$A[2][0][45]{=}A[2][1][45]{=}v_{44}$,$A[0][1][46]{=}A[0][2][46]{=}v_{45}$, $A[3][0][46]{=}A[3][1][46]{=}v_{46}$,$A[0][1][47]{=}A[0][2][47]{=}v_{47}$,$A[0][1][49]{=}A[0][2][49]{=}v_{48}$, $A[1][1][50]{=}A[1][2][50]{=}v_{49}$,$A[2][0][50]{=}A[2][1][50]{=}v_{50}$,$A[1][1][52]{=}A[1][2][52]{=}v_{51}$, $A[2][1][52]{=}A[2][2][52]{=}v_{52}$,$A[2][0][53]{=}A[2][1][53]{=}v_{53}$,$A[2][1][56]{=}A[2][2][56]{=}v_{54}$, $A[3][0][56]{=}A[3][1][56]{=}v_{55}$,$A[0][1][58]{=}A[0][2][58]{=}v_{56}$,$A[2][1][58]{=}A[2][2][58]{=}v_{57}$, $A[0][1][59]{=}A[0][2][59]{=}v_{58}$,$A[0][1][60]{=}A[0][2][60]{=}v_{59}$,$A[2][0][61]{=}v_{60}$,$A[2][1][61]{=}v_{61}$, $A[2][2][61]{=}v_{60} + v_{61}$,$A[2][0][62]{=}v_{62}$,$A[2][1][62]{=}v_{63}$, $A[2][2][62]{=}v_{62} + v_{63}$,$A[4][0][63]{=}A[4][1][63]{=}v_{64}$,
Conditional Cube Variable
$A[2][0][0]{=}A[2][1][0]{=}v_0$
Bit Conditions
$A[4][0][44] = A[4][1][44] + A[2][2][45]$, $A[2][0][4] = k_0[5] + k_1[5] + A[0][1][5] + A[2][1][4] + A[0][2][5] + A[2][2][4] + 1$, $A[2][0][59] = k_0[60] + A[2][1][59] + A[2][2][59] + 1$, $A[4][0][6] = A[2][0][7] + A[2][1][7] + A[4][1][6] + A[2][2][7] + A[3][1][7]$, $A[2][2][23]{=}A[2][0][23] + A[4][0][22] + A[2][1][23] + A[4][0][22]$, $A[2][2][46]{=}A[2][0][46] + A[4][0][45] + A[2][1][46] + A[4][1][45]$, $A[2][2][36]{=}A[2][0][36] + A[4][0][35] + A[2][1][36] + A[4][1][35]$, $A[2][2][63]{=}A[2][0][63] + A[4][0][62] + A[2][1][63] + A[4][1][62]$, $A[2][2][42]{=}A[2][0][42] + A[4][0][41] + A[2][1][42] + A[4][4][41]$, $A[0][2][35]{=}k_0[35] + A[3][0][36] + A[0][1][35] + A[3][1][36]$, $A[2][2][53]{=}k_0[54] + A[0][1][54] + A[0][2][54]$,$A[4][1][13]{=}A[2][0][14] + A[4][0][13]$, $A[1][2][29]{=}k_1[29] + A[3][0][28] + A[1][1][29] + A[3][1][28]$, $A[2][2][6]{=}k_0[7] + A[2][0][6] + A[0][1][7] + A[2][1][6] + A[0][2][7]$, $A[1][2][45]{=}k_1[45] + A[3][0][44] + A[1][1][45] + A[3][1][44]$,$A[4][1][19]{=}A[4][0][19]$, $A[2][2][31]{=}A[2][0][31] + A[4][0][30] + A[2][1][31] + A[4][1][30]$, $A[2][2][17]{=}k_0[18] + A[2][0][17] + A[0][1][18] + A[2][1][17] + A[0][2][18]$
Guessed Key Bits
$k_0[5] + k_1[5]$, $k_0[60]$,$k_0[35]$,$k_0[54]$, $k_1[29]$,$k_0[7]$,$k_1[45]$,$k_0[18]$

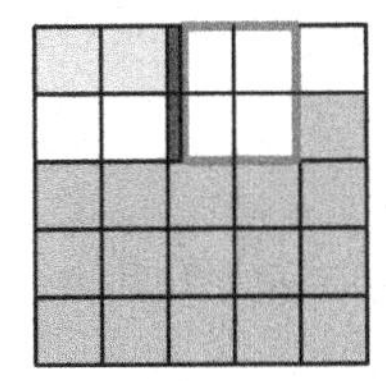

Fig. 6. The initial state of Keccak-MAC-512 (Color figure online)

Additionally, the subset of $\mathbb{V}$, $\mathbb{V}_i, i = 2, 3$ represents the set of lanes whose x-index equals 2, 3 respectively. According to the modeling search strategy illustrated in Sect. 4, we search the controllable bits for the most *ordinary cube variables.* The objective function is

$$\sum_{x,y \in \mathbb{V}, z \in \{0,1...63\}} A[x][y][z] - \sum_{x \in \{0,1...4\}, z \in \{0,1...63\}} d[x][z],$$

To model the CP-like-kernel, constraints are in the following according to Eq. 3:

$$\begin{cases} \sum_{x,y \in \mathbb{V}_x, z} A[x][y][z] \geq 2d[x][z] \\ d[x][z] \geq A[x][y][z], y \in \mathbb{V}_x \end{cases} for\ x = 2, 3, z = 0, 1 \dots 63 \quad (5)$$

The method of adding constraints to avoid multiplication is just the same as Keccak-MAC-384. With the help of Gurobi [2], the objective function is optimized under all the above constraints. The maximum of cube variables obeying CP-like-kernel is 26 (including a conditional cube variables). As the number of cube variables is not enough to perform the 6-round attack on Keccak-MAC-512, and many nonce bits are not utilized, we continue the search for appropriate *ordinary cube variables* among the single bits in lanes [0, 1], [1, 1], [4, 0].

Modeling the single bits

A single bit here means it is the only bit in its column that contains cube variable, exactly, it is set as a new *ordinary cube variable.* As the optimization according to CP-like-kernel above, most cube variables have been settled. Additionally, the state is so large as 1600-bit. Although a single bit diffuse to 11 bits after the first θ operation, it may not multiply with all the other cube variables in the first round, and not multiply with *conditional cube variable* v_0 in the second round. The objective function is the sum of all possible bits to be *ordinary cube variables.* Then, constraints are added to avoid the above two kinds of multiplication in the same way.

Another 6 single bits are found as 6 new *ordinary cube variables.* Totally, we find (6 + 26=)32 dimension cube and based on it a 6 round key-recovery attack on Keccak-MAC-512 is achieved. Both the cube variables and conditions are listed in Table 4.

Table 4. Parameters set for attack on 6-round KECCAK-MAC-512

Ordinary Cube Variables
$A[3][0][56]=A[3][1][56]=v_1$,$A[2][0][1]=A[2][1][1]=v_2$,$A[2][0][8]=A[2][1][8]=v_3$, $A[2][0][12]=A[2][1][12]=v_4$,$A[2][0][23]=A[2][1][23]=v_5$,$A[2][0][41]=A[2][1][41]=v_6$, $A[2][0][43]=A[2][1][43]=v_7$,$A[2][0][45]=A[2][1][45]=v_8$,$A[2][0][50]=A[2][1][50]=v_9$, $A[2][0][53]=A[2][1][53]=v_{10}$,$A[2][0][62]=A[2][1][62]=v_{11}$,$A[3][0][3]=A[3][1][3]=v_{12}$, $A[3][0][4]=A[3][1][4]=v_{13}$,$A[3][0][9]=A[3][1][9]=v_{14}$,$A[3][0][12]=A[3][1][12]=v_{15}$, $A[3][0][13]=A[3][1][13]=v_{16}$,$A[3][0][14]=A[3][1][14]=v_{17}$,$A[3][0][20]=A[3][1][20]=v_{18}$, $A[3][0][23]=A[3][1][23]=v_{19}$,$A[3][0][27]=A[3][1][27]=v_{20}$,$A[3][0][33]=A[3][1][33]=v_{21}$, $A[3][0][35]=A[3][1][35]=v_{22}$,$A[3][0][39]=A[3][1][39]=v_{23}$,$A[3][0][40]=A[3][1][40]=v_{24}$, $A[3][1][46]=A[3][0][46]=v_{25}$,$A[2][1][56]=v_{26}$,$A[4][0][12]=v_{27}$,$A[2][0][56]=v_{28}$, $A[0][1][33]=v_{29}$,$A[0][1][57]=v_{30}$,$A[4][0][60]=v_{31}$
Conditional Cube Variable
$A[2][0][0]=A[2][1][0]=v_0$
Bit Conditions
$A[4][0][44] = 0$,$A[2][0][59] = k_0[60] + A[2][1][59] + A[0][1][60] + 1$, $A[2][0][4] = k_0[5] + k_1[5] + A[0][1][5] + 1 + A[2][1][4]$, $A[4][0][6] = A[2][0][7] + A[2][1][7] + A[3][1][7]$, $A[2][0][46] = A[4][0][45] + A[2][1][46]$,$A[2][0][31] = A[4][0][30] + A[2][1][31]$, $A[4][0][3] = k_0[5] + k_1[5] + A[0][1][5] + 1$,$A[0][1][19] = k_0[19]$, $A[3][1][30] = k_0[29] + A[3][0][30] + A[0][1][29]$,$A[0][1][34] = k_0[34]$, $A[3][1][22] = k_0[21] + A[3][0][22] + A[0][1][21]$,$A[1][1][28] = k_1[28]$, $A[3][1][36] = k_0[35] + A[3][0][36] + A[0][1][35]$,$A[0][1][51] = 0$, $A[3][1][49] = k_0[48] + A[3][0][49] + A[0][1][48]$,$A[2][0][18] = A[2][1][18] + 1$, $A[3][1][41] = k_0[40] + A[3][0][41] + A[0][1][40]$,$A[4][0][8] = k_0[8] + k_1[7] + A[1][1][7]$, $A[3][0][63] = k_0[62] + A[0][1][60] + A[3][1][63]$,$A[4][0][51] = k_1[50] + A[1][1][50]$, $A[2][0][51] = k_0[52] + A[0][1][52] + A[2][1][51] + 1$, $A[2][0][63] = A[4][0][62] + A[2][1][63] + A[3][1][63]$, $A[2][1][58] = k_0[59] + A[2][0][58] + A[0][1][59]$, $A[4][0][13] = k_0[13] + k_1[12] + k_1[34] + A[1][1][12] + A[1][1][34] + 1$, $A[2][0][26] = A[3][0][26] + A[4][0][25] + A[2][1][26] + 1$, $A[3][0][16] = k_1[17] + A[1][1][17] + A[3][1][16]$, $A[1][1][24] = k_1[24] + A[4][0][25] + A[0][1][25] + 1$, $A[2][0][42] = A[4][0][41] + A[2][1][42] + A[3][1][42] + 1$, $A[4][0][40] = 1$, $A[2][0][17] = k_0[18] + A[0][1][18] + A[2][1][17]$, $A[3][0][15] = k_1[16] + A[1][1][16] + A[2][1][16] + A[3][1][15] + 1$, $A[3][0][6] = k_0[5] + A[0][1][5] + A[3][1][6] + 1$, $A[2][0][33] = A[2][1][33]$, $A[0][1][13] = k_0[13] + 1$, $A[3][0][59] = k_0[58] + A[0][1][58] + A[3][1][59] + 1$, $A[0][1][32] = k_0[32] + A[4][0][30] + A[1][1][32]$
Guessed Key Bits
$k_0[60]$, $k_0[5] + k_1[5]$, $k_0[19]$, $k_0[29]$, $k_0[34]$, $k_0[21]$, $k_0[35]$, $k_0[48]$, $k_0[40]$, $k_0[62]$, $k_0[52]$,$k_0[59]$, $k_0[13] + k_1[12] + k_1[34]$,$k_1[17]$,$k_1[28]$,$k_1[24]$, $k_0[8] + k_1[7]$, $k_0[18]$, $k_1[16]$,$k_0[5]$, $k_1[50]$, $k_0[13]$, $k_0[58]$, $k_0[32]$

In 6-round attack on Keccak-MAC-512, $2^5 = 32$ cube variables denoted by $v_0, v_1 \dots v_{31}$. Based on Corollary 1, v_0 is the conditional cube variable and $v_1, v_2 \dots v_{31}$ are ordinary cube variables. We summarize the requirements as following:

(1) $v_0, v_1 \dots v_{31}$ do not multiply with each other in the first round;
(2) Under some conditions on key and nonce, v_0 does not multiply with any of $v_1, v_2 \dots v_{31}$ in the second round.

All the bit conditions are satisfied if and only if all the key bits are guessed correctly. Thus, zero sums over the 128-bit tag with cube variables set as Table 4 suggest a correct key guess. Furthermore, the similar key recovery can be performed with any offset in z-axis.

We analyze the time and data complexity of the attack: 4 iterations in z-axis recover 72 key bits, and the remaining 56 key bits are recovered by exhaustive search, thus the procedure consumes $4 \times 2^{24} \times 2^{32} + 2^{56} = 2^{58.3}$ computations of 6-round initialization of Keccak-MAC-512, correspondingly $2^{58.3}$ $(message, tag)$ pairs are needed. After the procedure above, all the 128 bits in k_0, k_1 can be recovered. Therefore, both time and data complexity of the attack are $2^{58.3}$.

6 Attacks on Round-Reduced Initialization of Ketje

At 6 March 2017, the Keccak team announces the Ketje cryptanalysis prize to encourage the cryptanalysis.

6.1 Attacks on Round-Reduced Initialization of Ketje Major

Ketje Major operates on a 1600-bit state, the recommended key length is 128-bit, which is similar to Keccak-MAC. We focus on the instances with recommended 128-bit key. The number of nonce bits in Ketje Major is variable from 0 to 1454.

To explore the resistance against conditional cube attack of the different instances, we apply the MILP search strategy to search the possible cube variables in the instances with different lengths of nonce, and list the corresponding number of cube variables in Table 5. Similar to attacks on Keccak-MAC described in Sects. 5.1, 5.2, 32 cube variables are needed to perform 6-round attack, and 64 cube variables are needed to perform 7-round attack. Thus, Table 5 tells us that when the nonce is no less than 704 bits (11 lanes), cube variables are enough to perform 7-round attack on Ketje Major and 6-round attack on Ketje Major can be performed if the nonce is no less than 512 bits (8 lanes).

As instances with more nonce bits can directly use the parameters of instances with less nonce bits, we list the details of 6-round and 7-round attacks on Ketje Major with 512-bit and 704-bit nonce.

Table 5. The number of cube variables in CP-like-kernel in different nonces in Ketje Major

Nonce: bits(lanes)	Number of cube variables in CP-like-kernel
448(7)	21
512(8)	41
576(9)	50
640(10)	59
704(11)	75
832(13)	81

Table 6. Parameters set for attack on 6-round KETJE MAJOR

Ordinary Cube Variables
$A[4][1][2]=A[4][4][2]=v_1$, $A[4][1][4]=A[4][4][4]=v_2$, $A[4][1][10]=A[4][4][10]=v_3$, $A[4][1][11]=A[4][4][11]=v_4$,$A[3][0][14]=A[3][3][14]=v_5$,$A[3][0][17]=A[3][3][17]=v_6$, $A[4][1][19]=A[4][4][19]=v_7$,$A[4][1][20]=A[4][4][20]=v_8$,$A[4][1][27]=A[4][4][27]=v_9$, $A[3][0][28]=A[3][3][28]=v_{10}$,$A[4][1][28]=A[4][4][28]=v_{11}$,$A[3][0][33]=A[3][3][33]=v_{12}$, $A[3][0][36]=A[3][3][36]=v_{13}$,$A[3][0][37]=A[3][3][37]=v_{14}$,$A[4][1][38]=A[4][4][38]=v_{15}$, $A[3][0][45]=A[3][3][45]=v_{16}$,$A[4][1][59]=A[4][4][59]=v_{17}$,$A[4][1][60]=A[4][4][60]=v_{18}$, $A[2][2][18]=A[2][4][18]=v_{19}$,$A[2][2][19]=A[2][4][19]=v_{20}$,$A[2][2][51]=A[2][4][51]=v_{21}$, $A[2][2][27]=A[2][4][27]=v_{22}$,$A[2][2][28]=A[2][4][28]=v_{23}$,$A[2][2][52]=A[2][4][52]=v_{24}$, $A[2][2][53]=A[2][4][53]=v_{25}$,$A[2][2][36]=A[2][4][36]=v_{26}$,$A[2][2][37]=A[2][4][37]=v_{27}$, $A[2][2][39]=A[2][4][39]=v_{28}$,$A[2][2][55]=A[2][4][55]=v_{29}$,$A[2][2][60]=A[2][4][60]=v_{30}$, $A[2][2][62]=A[2][4][62]=v_{31}$
Conditional Cube Variable
$A[3][0][0]=A[3][3][0]=v_0$
Bit Conditions
$A[3][3][41]=k_1[42]+A[1][0][42]+A[3][0][41]+A[2][2][42]+A[1][3][42]+1$, $A[4][4][7]=A[3][0][7]+A[0][2][6]+A[3][3][7]$, $A[2][4][31]=k_1[31]+A[1][0][31]+A[3][0][30]+A[1][3][31]+A[3][3][30]+1$, $A[3][3][8]=A[3][0][8]+A[4][1][8]+A[0][2][7]$, $A[4][4][49]=A[2][1][50]+A[4][1][49]+A[2][2][50]+A[3][3][50]+A[2][4][50]$, $A[2][4][11]=A[2][1][11]+A[3][3][11]+1$, $A[2][4][61]=A[2][1][61]+A[2][2][61]+A[3][3][61]$, $A[0][2][38]=k_0[30]+k_1[38]+A[2][1][37]+1$, $A[4][4][12]=A[2][1][13]+A[4][1][12]+A[3][3][13]+A[2][4][13]$
Guessed Key Bits
$k_1[42]$,$k_1[31]$,$k_0[30]+k_1[38]$

Attack on 6-Round Initialization of Ketje Major. According to parameters set in Table 6, guess the 3 key bits listed, compute cube sums on variables $v_0, \ldots, v_{31}$, zero cube sums suggest a right key(i.e. 3 guessed key bits in Table 6).

It consumes $2^3 \times 2^{32} = 2^{35}$ computations of 6-round initialization of KETJE MAJOR. According to the property of permutation, it is totally symmetric in z-axis. Thus we can obtain corresponding parameters set with any rotation of i-bit ($0 \leq i < 64$) in z-axis. Therefore, 128 key bits can be recovered by 64 iterations for $0 \leq i < 64$, so the time complexity is $64 \times 2^3 \times 2^{32} = 2^{41}$.

Attack on 7-Round Initialization of Ketje Major. We use $A[1][0][0] = A[1][3][0] = v_0$ as condition cube variable. According to parameters set in Table 7, guess the 16 key bits listed, compute cube sums on variables $v_0, \ldots, v_{63}$, zero cube sums suggest a right key (i.e. 16 guessed key bits in Table 7). It consumes $2^{16} \times 2^{64} = 2^{80}$ computations of 7-round initialization of KETJE MAJOR. Similar to the case above, 46 key bits can be recovered by 4 iterations for $0 \leq i < 4$, and the remaining 82 key bits can be recovered by exhaustive search. The time complexity is $4 \times 2^{16} \times 2^{64} + 2^{82} = 2^{83}$.

6.2 Attacks on Round-Reduced Initialization of Ketje Minor

The state of Ketje Minor is 800-bit, which is the half of the state size of Keccak-MAC (1600-bit). As the upper part of Fig. 7 shows, in Huang *et al.*'s attack on Keccak-MAC, one *conditional cube variable* v_0 is chosen, placed in two black bits of S_0. After adding some conditions, the conditional cube variable v_0 is diffused to 22 bits shown in state $S_{1.5}$, we denote the diffusion pattern as 2-2-22. For the state of Ketje Minor is much smaller, the *conditional cube variable* in 2-2-22 pattern diffuses relatively much greater, there are only 26 *ordinary cube variables* in CP-like-kernel optimized with MILP search strategy, which is not enough for 7-round attack.

In order to solve the problem, we find a new *conditional cube variable.* As shown in the lower part of Fig. 7, after adding some conditions, the diffusion pattern is 6-6-6 and only 6 bits in $S_{1.5}$ contains the conditional cube variable. At last, we find enough *ordinary cube variables* with the MILP tool to launch the key-recovery attacks on 5/6/7-round reduced Ketje Minor.

In details, from S_0 to S_1, $\theta, \rho, \pi, \chi, \iota$ are operated in sequence. θ operation holds the distribution of v_0 according to CP-like-kernel. Operations ρ and π only permute the bit positions, while ι only adds a constant. Thus, we only need to control the diffusion of χ operation. We denote the state before χ operation in the first round as $S_{0.5}$. According to Property 1-(2), 12-bit conditions based on key and nonce are introduced to keep the 6 bits containing v_0 from diffusion. Then the diffusion of v_0 maintains the 6-6-6 pattern.

Attack on 5-Round Initialization of Ketje Minor. In 5-round attack, we choose $2^4 = 16$ cube variables denoted by $v_0, v_1 \ldots v_{15}$. Based on Corollary 1,

Table 7. Parameters set for attack on 7-round KETJE MAJOR

Ordinary Cube Variables
$A[3][2][0]=A[3][3][0]=v_1, A[1][0][1]=A[1][3][1]=v_2, A[4][1][4]=A[4][4][4]=v_3,$ $A[3][0][5]=v_4, A[3][2][5]=v_5, A[3][3][5]=v_4+v_5, A[1][0][7]=A[1][3][7]=v_6,$ $A[1][0][9]=A[1][3][9]=v_7, A[3][2][9]=A[3][3][9]=v_8, A[4][1][9]=A[4][4][9]=v_9,$ $A[3][0][10]=v_{10}, A[3][2][10]=v_{11}, A[3][3][10]=v_{10}+v_{11}, A[4][1][10]=A[4][4][10]=v_{12},$ $A[3][2][11]=A[3][3][11]=v_{13}, A[4][1][11]=A[4][4][11]=v_{14}, A[1][0][12]=A[1][3][12]=v_{15},$ $A[3][2][15]=A[3][3][15]=v_{16}, A[1][0][17]=A[1][3][17]=v_{17}, A[1][0][19]=A[1][3][19]=v_{18},$ $A[4][1][20]=A[4][4][20]=v_{19}, A[4][1][26]=A[4][4][26]=v_{20}, A[3][0][27]=A[3][2][27]=v_{21},$ $A[1][0][29]=A[1][3][29]=v_{22}, A[3][2][30]=A[3][3][30]=v_{23}, A[3][2][31]=A[3][3][31]=v_{24},$ $A[1][0][32]=A[1][3][32]=v_{25}, A[1][0][33]=A[1][3][33]=v_{26}, A[4][1][33]=A[4][4][33]=v_{27},$ $A[3][0][38]=A[3][2][38]=v_{28}, A[1][0][39]=A[1][3][39]=v_{29}, A[3][0][41]=A[3][3][41]=v_{30},$ $A[3][0][42]=A[3][2][42]=v_{31}, A[1][0][43]=A[1][3][43]=v_{32}, A[3][0][43]=A[3][3][43]=v_{33},$ $A[3][0][45]=A[3][2][45]=v_{34}, A[3][0][46]=v_{35}, A[3][2][46]=v_{36}, A[3][3][46]=v_{35}+v_{36},$ $A[3][0][47]=A[3][2][47]=v_{37}, A[3][0][48]=A[3][2][48]=v_{38}, A[3][0][49]=v_{39},$ $A[3][2][49]=v_{40}, A[3][3][49]=v_{39}+v_{40}, A[3][2][50]=A[3][3][50]=v_{41},$ $A[3][2][51]=A[3][3][51]=v_{42}, A[3][2][52]=A[3][3][52]=v_{43}, A[4][1][52]=A[4][4][52]=v_{44},$ $A[3][2][53]=A[3][3][53]=v_{45}, A[3][0][56]=v_{46}, A[3][2][56]=v_{47}, A[3][3][56]=v_{46}+v_{47},$ $A[3][2][60]=A[3][3][60]=v_{48}, A[4][1][61]=A[4][4][61]=v_{49}, A[1][0][62]=A[1][3][62]=v_{50},$ $A[3][2][63]=A[3][3][63]=v_{51}, A[2][2][20]=A[2][4][20]=v_{52}, A[2][1][26]=A[2][4][26]=v_{53},$ $A[1][0][4]=A[1][3][4]=v_{54}, A[2][2][33]=A[2][4][33]=v_{55}, A[2][1][35]=v_{56},$ $A[2][2][35]=v_{57}, A[2][4][35]=v_{56}+v_{57}, A[2][1][40]=A[2][2][40]=v_{58},$ $A[2][1][44]=A[2][2][44]=v_{59}, A[2][2][45]=A[2][4][45]=v_{60}, A[2][2][54]=A[2][4][54]=v_{61},$ $A[2][1][23]=A[2][2][23]=v_{62}, A[1][0][2]=A[1][3][2]=v_{63}$
Bit Conditions
$A[4][4][42]=k_1[41] + A[1][0][41] + A[4][1][42] + A[0][2][42] + A[1][3][41] + 1,$ $A[2][4][48]=k_0[38] + k_1[48] + A[1][0][48] + A[1][3][48] + A[0][2][46],$ $A[4][4][47]=k_1[46] + A[1][0][46] + A[4][1][47] + A[1][3][46]+ 1,$ $A[3][3][58]=k_1[59] + A[1][0][59] + A[3][0][58] + A[2][1][59] + A[3][2][58] + A[1][3][59],$ $A[3][3][17]=k_0[8] + A[3][0][17] + A[0][2][16]+ A[3][2][17],$ $A[3][3][26]=k_0[17] + A[3][0][26] + A[0][2][25] + A[3][2][26],$ $A[3][3][27]=k_0[18] + A[0][2][26],\ A[3][3][47]=k_0[38] + A[0][2][46],$ $A[3][3][7]=k_1[8] + A[1][0][8] + A[3][0][7] + A[3][2][7] + A[1][3][8],$ $A[3][3][48]=k_0[39] + A[0][2][47], A[4][4][44]=A[2][1][45] + A[4][1][44] + A[3][3][45],$ $A[3][3][55]=k_0[46] + A[3][0][55] + A[0][2][54] + A[3][2][55],$ $A[4][4][41]=A[2][0][42] + A[2][1][42] + A[4][1][41] + A[3][3][42] + A[2][4][42],$ $A[4][4][46]=k_1[45] + A[1][0][45] + A[4][1][46]+ A[0][2][46] + A[1][3][45] + 1,$ $A[2][4][52]=k_1[52] + A[1][0][52] + A[3][0][51] + A[1][3][52],$ $A[0][2][43]=k_0[35] + k_1[43] + A[2][0][42] + A[2][1][42] + A[2][4][42] + 1,$ $A[1][3][61]=k_1[61] + A[1][0][61] + A[3][0][60] + A[2][1][61],$ $A[0][2][44]=k_1[43] + A[2][1][45] + A[3][3][45] + 1$
Guessed Key Bits
$k_1[41], k_0[38] + k_1[48], k_1[46], k_1[59], k_0[8], k_0[17], k_0[18], k_0[38], k_1[8], k_0[39], k_0[46],$ $k_1[45], k_1[52], k_0[35] + k_1[43], k_1[61], k_1[43]$

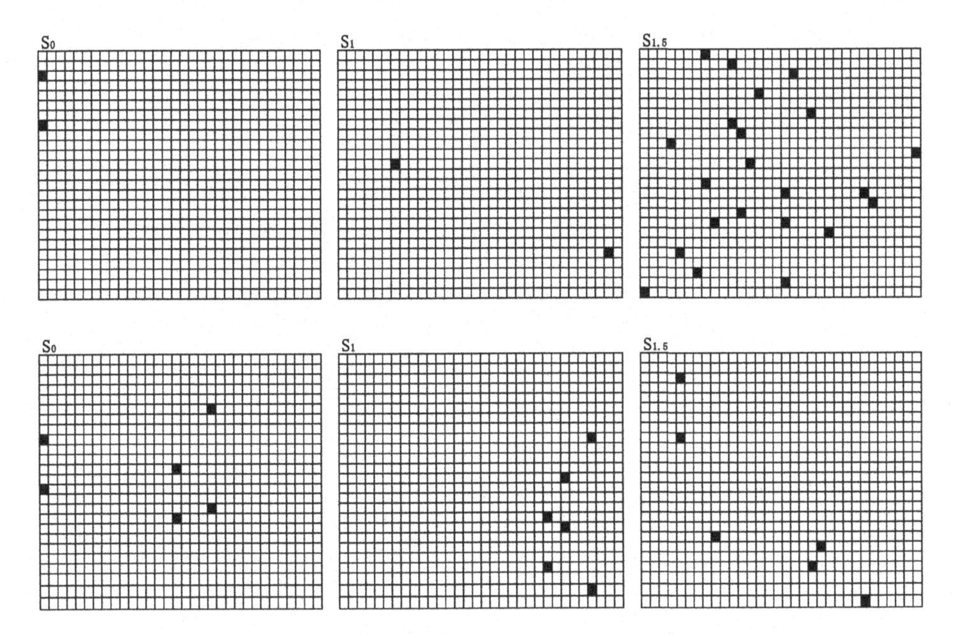

Fig. 7. Diffusions of the conditional cube variable in 2-2-22 and 6-6-6 Pattern in Ketje Minor

v_0 is the conditional cube variable and $v_1, v_2 \ldots v_{15}$ are ordinary cube variables. We summarize the requirements as following:

(1) $v_0, v_1 \ldots v_{15}$ do not multiply with each other in the first round;
(2) Under some conditions on key and nonce, v_0 does not multiply with any of $v_1, v_2 \ldots v_{15}$ in the second round.

Under (1), any of cube variables $v_0, v_1 \ldots v_{15}$ only exists as a one-degree term in the output of 1-round Ketje Minor, i.e. the degree of any bit in S_1 is no more than one. The degree of one round function is 2. When we say the degree of some state, we mean the highest degree among the cube variables in all terms of the state. If conditions in (2) are met, according to Corollary 1, the term $v_0 v_1 \ldots v_{15}$ will not appear in S_5, so the degree over cube variables $v_0, v_1 \ldots v_{15}$ is at most 15. Otherwise, the degree of S_5 is 16.

Thus, under given conditions on key and nonce, the cube sums of all bits in S_5 over $v_0, v_1 \ldots v_{15}$ are zero, otherwise the cube sums are random if those conditions are not met. Actually, $\rho = 128$ bits of S_5 are known in Ketje Minor. If the cube sum on each of the 128 bits is zero, we can determine that the corresponding conditions are satisfied.

Table 8. Parameters set for attack on the 5-round initialization of Ketje Minor

Conditional Cube Variable
$A[0][1][19]=A[0][3][19]=A[1][2][15]=A[1][3][15]=A[3][1][0]=A[3][2][0]=v_0$
Ordinary Cube Variables
$A[2][0][2]=A[2][1][2]=v_1, A[2][0][4]=A[2][1][4]=v_2, A[2][0][7]=A[2][1][7]=v_3,$ $A[2][0][11]=A[2][1][11]=v_4, A[2][0][12]=A[2][1][12]=v_5, A[2][0][20]=A[2][1][20]=v_6,$ $A[2][0][23]=A[2][1][23]=v_7, A[2][0][29]=A[2][1][29]=v_8, A[2][0][30]=A[2][1][30]=v_9,$ $A[3][0][3]=A[3][1][3]=v_{10}, A[3][0][6]=A[3][1][6]=v_{11}, A[3][0][12]=A[3][1][12]=v_{12},$ $A[3][0][13]=A[3][1][13]=v_{13}, A[3][0][17]=A[3][1][17]=v_{14}, A[3][0][21]=A[3][1][21]=v_{15}.$
Bit Conditions
$A[1][0][24]=k_1[24] + A[4][0][25] + A[4][1][25] + A[0][2][25] + A[1][2][24] + A[4][2][25] + A[1][3][24] + A[4][3][25] + A[1][4][24] + A[4][4][25] + 1$
$A[1][0][31]=k_1[31] + k_3[30] + A[3][0][30] + A[3][1][30] + A[1][2][31] + A[3][2][30] + A[1][3][31] + A[2][4][31] + A[3][4][30] + 1$
$A[1][0][19]=k_1[19] + k_3[18] + A[3][0][18] + A[2][1][19] + A[3][1][18] + A[1][2][19] + A[3][2][18] + A[1][3][19] + A[1][4][19] + A[3][4][18] + 1$
$A[0][1][16]=k_0[8] + k_3[17] + A[0][2][16] + A[3][2][17] + A[0][3][16] + A[4][3][17] + A[0][4][16] + A[3][4][17]$
$A[0][1][13]=k_0[5] + k_2[12] + A[0][2][13] + A[1][2][13] + A[0][3][13] + A[2][3][12] + A[0][4][13] + A[2][4][12]$
$A[1][0][20]=k_1[20] + A[4][0][21] + A[0][1][21] + A[4][1][21] + A[1][2][20] + A[4][2][21] + A[1][3][20] + A[4][3][21] + A[1][4][20] + A[4][4][21] + 1$
$A[1][0][10]=k_1[10] + k_3[9] + A[3][0][9] + A[3][1][9] + A[1][2][10] + A[3][2][9] + A[1][3][10] + A[2][3][10] + A[1][4][10] + A[3][4][9]$
$A[0][1][27]=k_0[19] + k_3[28] + A[3][0][28] + A[4][0][28] + A[3][1][28] + A[0][2][27] + A[3][2][28] + A[0][3][27] + A[0][4][27] + A[3][4][28] + 1$
$A[0][1][15]=k_0[7] + k_3[16] + A[3][0][16] + A[3][1][16] + A[0][2][15] + A[3][2][16] + A[4][2][16] + A[0][3][15] + A[0][4][15] + A[3][4][16]$
$A[0][1][20]=k_0[12] + k_3[21] + A[0][2][20] + A[3][2][21] + A[4][2][21] + A[0][3][20] + A[0][4][20] + A[3][4][21] + 1$
$A[0][1][26]=k_0[18] + k_2[25] + A[2][0][25] + A[2][1][25] + A[0][2][26] + A[0][3][26] + A[2][3][25] + A[0][4][26] + A[1][4][26] + A[2][4][25]$
$A[1][0][25]=k_1[25] + k_3[24] + A[2][0][25] + A[3][0][24] + A[3][1][24] + A[1][2][25] + A[3][2][24] + A[1][3][25] + A[1][4][25] + A[3][4][24] + 1$
Guessed Key Bits
$k_1[24], k_1[31] + k_3[30], k_1[19] + k_3[18], k_0[8] + k_3[17], k_0[5] + k_2[12],$ $k_1[20], k_1[10] + k_3[9], k_0[19] + k_3[28], k_0[7] + k_3[16], k_0[12] + k_3[21],$ $k_0[18] + k_2[25], k_1[25] + k_3[24]$

As Table 8 shows, the 12 bit conditions are related to key and nonce bits. We guess the 12 key bits with all the possible values. While all the nonce bits are constant, all the bit conditions are satisfied if and only if all the key bits are guessed correctly. Thus, zero sums over the 128 known bits of S_5 ($S_5[0][0], S_5[1][1], S_5[2][2], S_5[3][3]$[3]) with cube variables set as Table 8 mean a correct key guess. We give an example here for intuition, in which key is generated randomly and all the controllable nonce bits are set as zero.

128-bit key ($K = k_0||k_1||k_2||k_3||k_4$):

`1010000011010110011101001101110001110010000111011101110010110110`

`1111110010011101001011000101010100010111101000111100101100000101`

The correct value for the guessed key bits in Table 8 is `110111101010`.

guessed value: `000000000000`,

cube sums: `0xf0217c64, 0x8a61f7e1, 0x67f01330, 0xa9b1c06`

...

guessed value: `110111101010`,

cube sums: `0x0, 0x0, 0x0, 0x0`

...

guessed value: `000011010110`,

cube sums: `0xf4c1bc4, 0xea79d2a4, 0xc2880990, 0x8ae4140d`

...

guessed value: `111111111111`,

cube sums: `0x7b115312, 0xa9156874, 0x9cabc23, 0x6ecd5ef9`

Furthermore, we can perform the similar key recovery with any offset $0, 1 \ldots 31$ in z-axis. We analyze the time and data complexity of the attack: the procedure consumes $32 \times 2^{12} \times 2^{16} = 2^{33}$ computations of 5-round initialization of Ketje Minor, correspondingly 2^{33} $(nonce, plaintext, ciphertext, tag)$ pairs are needed. After the procedure above, all the 120 bits in k_0, k_1, k_2, k_3 can be recovered, and the remaining 8 bits of k_4 can be determined by brute search. Therefore, time complexity of the attack is 2^{33} computations of 5-round initialization of Ketje Minor, and data complexity is 2^{33} $(nonce, plaintext, ciphertext, tag)$ pairs.

Attack on 6-Round Initialization of Ketje Minor. In 6-round attack, similar to the 5-round attack, we choose $2^5 = 32$ cube variables denoted by $v_0, v_1 \ldots v_{31}$. Based on Corollary 1, v_0 is the conditional cube variable and $v_1, v_2 \ldots v_{31}$ are ordinary cube variables. We summarize the requirements as following:

(1) $v_0, v_1 \ldots v_{31}$ do not multiply with each other in the first round;
(2) Under some conditions on key and nonce, v_0 does not multiply with any of $v_1, v_2 \ldots v_{31}$ in the second round.

The recovery attack can be performed similarly to 5-round attack. While all the nonce bits are constant, all the bit conditions are satisfied if and only if all

[3] These four 32-bit words are the first four words after π which are the output bits of Keccak-p^*.

Table 9. Ordinary cube variables and bit conditions for attack on the 6-round initialization of Ketje Minor

Ordinary Cube Variables
$A[2][0][2]=A[2][1][2]=v_1, A[2][0][4]=A[2][1][4]=v_2, A[2][0][7]=A[2][1][7]=v_3,$ $A[2][0][11]=A[2][1][11]=v_4, A[2][0][12]=A[2][1][12]=v_5, A[2][0][20]=A[2][1][20]=v_6,$ $A[2][0][23]=A[2][1][23]=v_7, A[2][0][29]=A[2][1][29]=v_8, A[2][0][30]=A[2][1][30]=v_9,$ $A[3][0][3]=A[3][1][3]=v_{10}, A[3][0][6]=A[3][1][6]=v_{11}, A[3][0][12]=A[3][1][12]=v_{12},$ $A[3][0][13]=A[3][1][13]=v_{13}, A[3][0][17]=A[3][1][17]=v_{14}, A[3][0][21]=A[3][1][21]=v_{15},$ $A[3][0][22]=A[3][1][22]=v_{16}, A[3][0][26]=A[3][1][26]=v_{17}, A[3][0][31]=A[3][1][31]=v_{18},$ $A[4][0][0]=A[4][1][0]=v_{19}, A[4][0][5]=A[4][1][5]=v_{20}, A[4][0][8]=A[4][1][8]=v_{21},$ $A[4][0][15]=A[4][1][15]=v_{22}, A[4][0][18]=A[4][1][18]=v_{23}, A[4][0][22]=A[4][1][22]=v_{24},$ $A[4][0][24]=A[4][1][24]=v_{25}, A[0][2][1]=A[0][3][1]=v_{26}, A[0][2][5]=A[0][3][5]=v_{27},$ $A[0][2][10]=A[0][3][10]=v_{28}, A[0][2][15]=A[0][3][15]=v_{29}, A[0][2][31]=A[0][3][31]=v_{30},$ $A[1][2][4]=A[1][3][4]=v_{31}.$
Bit Conditions
$A[1][0][24]=k_1[24] + A[4][0][25] + A[4][1][25] + A[0][2][25] + A[1][2][24] + A[4][2][25] + A[1][3][24] + A[4][3][25] + A[1][4][24] + A[4][4][25] + 1$
$A[1][0][31]=k_1[31] + k_3[30] + A[3][0][30] + A[3][1][30] + A[1][2][31] + A[3][2][30] + A[1][3][31] + A[2][4][31] + A[3][4][30] + 1$
$A[1][0][19]=k_1[19] + k_3[18] + A[3][0][18] + A[2][1][19] + A[3][1][18] + A[1][2][19] + A[3][2][18] + A[1][3][19] + A[1][4][19] + A[3][4][18] + 1$
$A[0][1][16]=k_0[8] + k_3[17] + A[0][2][16] + A[3][2][17] + A[0][3][16] + A[4][3][17] + A[0][4][16] + A[3][4][17]$
$A[0][1][13]=k_0[5] + k_2[12] + A[0][2][13] + A[1][2][13] + A[0][3][13] + A[2][3][12] + A[0][4][13] + A[2][4][12]$
$A[1][0][20]=k_1[20] + A[4][0][21] + A[0][1][21] + A[4][1][21] + A[1][2][20] + A[4][2][21] + A[1][3][20] + A[4][3][21] + A[1][4][20] + A[4][4][21] + 1$
$A[1][0][10]=k_1[10] + k_3[9] + A[3][0][9] + A[3][1][9] + A[1][2][10] + A[3][2][9] + A[1][3][10] + A[2][3][10] + A[1][4][10] + A[3][4][9]$
$A[0][1][27]=k_0[19] + k_3[28] + A[3][0][28] + A[4][0][28] + A[3][1][28] + A[0][2][27] + A[3][2][28] + A[0][3][27] + A[0][4][27] + A[3][4][28] + 1$
$A[0][1][15]=k_0[7] + k_3[16] + A[3][0][16] + A[3][1][16] + A[3][2][16] + A[4][2][16] + A[0][4][15] + A[3][4][16]$
$A[0][1][20]=k_0[12] + k_3[21] + A[0][2][20] + A[3][2][21] + A[4][2][21] + A[0][3][20] + A[0][4][20] + A[3][4][21] + 1$
$A[0][1][26]=k_0[18] + k_2[25] + A[2][0][25] + A[2][1][25] + A[0][2][26] + A[0][3][26] + A[2][3][25] + A[0][4][26] + A[1][4][26] + A[2][4][25]$
$A[1][0][25]=k_1[25] + k_3[24] + A[2][0][25] + A[3][0][24] + A[3][1][24] + A[1][2][25] + A[3][2][24] + A[1][3][25] + A[1][4][25] + A[3][4][24] + 1$

the key bits are guessed correctly. Thus, zero sums over the 128 known bits of S_6 ($S_6[0][0], S_6[1][1], S_6[2][2], S_6[3][3]$) with conditional cube variable set as Table 8 and ordinary cube variables set as Table 9 mean a correct key guess. We give an example here for intuition, in which key is generated randomly and all the controllable nonce bits are set as zero.

128-bit key ($K = k_0||k_1||k_2||k_3||k_4$):
`1001011000001001100010100101011010101110110110011100100111011010`
`0011111110101101101001110111100101000101101110001110011101101101`
The correct value for the guessed key bits in Table 8 is `100001001100`.
guessed value: `000000000000`,
cube sums: `0x555b48a6, 0xcce8cd70, 0x9e41800d, 0x66b12d4f`
...
guessed value: `100001001100`,
cube sums: `0x0, 0x0, 0x0, 0x0`
...
guessed value: `010101101100`,
cube sums: `0xc61fa207, 0x24f02427, 0x3fed45e0, 0x36a8326d`
...
guessed value: `111111111111`,
cube sums: `0x834061d2, 0x14200817, 0xd56d2379, 0xc93e01f8`

We analyze the time and data complexity of the attack: the procedure consumes $32 \times 2^{12} \times 2^{32} = 2^{49}$ computations of 6-round initialization of Ketje Minor, correspondingly 2^{49} $(nonce, plaintext, ciphertext, tag)$ pairs are needed. After the procedure above, all the 120 bits in k_0, k_1, k_2, k_3 can be recovered, and the remaining 8 bits in k_4 can be determined by brute search. Therefore, both time and data complexity of the attack are 2^{49}.

Attack on 7-Round Initialization of Ketje Minor. In 7-round attack, similar to the 5/6-round attack, we choose $2^6 = 64$ cube variables denoted by $v_0, v_1 \dots v_{63}$. Based on Corollary 1, v_0 is the conditional cube variable and $v_1, v_2 \dots v_{63}$ are ordinary cube variables. We summarize the requirements as following:

(1) $v_0, v_1 \dots v_{63}$ do not multiply with each other in the first round;
(2) Under some conditions on key and nonce, v_0 does not multiply with any of $v_1, v_2 \dots v_{63}$ in the second round.

While all the nonce bits are constant, all the bit conditions are satisfied if and only if all the key bits are guessed correctly. Thus, zero sums over the 128 known bits of S_7 ($S_7[0][0], S_7[1][1], S_7[2][2], S_7[3][3]$) with conditional cube variable set as Table 8 and ordinary cube variables set as Table 10 mean a correct key guess.

We analyze the time and data complexity of the attack: the procedure consumes $32 \times 2^{12} \times 2^{64} = 2^{81}$ computations of 7-round initialization of Ketje Minor, correspondingly 2^{81} $(nonce, plaintext, ciphertext, tag)$ pairs are needed. After the procedure above, all the 120 bits in k_0, k_1, k_2, k_3 can be recovered, and the remaining 8 bits in k_4 can be determined by brute search. Therefore, both time and data complexity of the attack are 2^{81}.

Table 10. Ordinary cube variables for attack on the 7-round initialization of Ketje Minor

Ordinary Cube Variables
$A[0][1][0]=v_1$,$A[0][2][0]=v_2$,$A[0][3][0]=v_3$,$A[0][4][0]=v_1+v_2+v_3$,$A[0][2][1]=v_4$,
$A[0][3][1]=v_5$,$A[0][4][1]=v_4+v_5$,$A[0][1][2]=A[0][3][2]=v_6$,$A[0][2][3]=A[0][4][3]=v_7$,
$A[0][1][4]=v_8$,$A[0][3][4]=v_9$,$A[0][4][4]=v_8+v_9$,$A[0][1][5]=v_{10}$,$A[0][2][5]=v_{11}$,
$A[0][3][5]=v_{10}+v_{11}$,$A[0][1][6]=v_{12}$,$A[0][2][6]=v_{13}$,$A[0][4][6]=v_{12}+v_{13}$,
$A[0][1][8]=A[0][3][8]=v_{14}$,$A[0][1][9]=v_{15}$,$A[0][2][9]=v_{16}$,$A[0][3][9]=v_{17}$,
$A[0][4][9]=v_{15}+v_{16}+v_{17}$,$A[0][2][10]=v_{18}$,$A[0][3][10]=v_{19}$,$A[0][4][10]=v_{18}+v_{19}$,
$A[0][1][13]=v_{20}$,$A[0][3][13]=v_{21}$,$A[0][4][13]=v_{20}+v_{21}$,$A[0][1][14]=v_{22}$,$A[0][2][14]=v_{23}$,
$A[0][4][14]=v_{22}+v_{23}$,$A[0][1][15]=v_{24}$,$A[0][2][15]=v_{25}$,$A[0][3][15]=v_{24}+v_{25}$,
$A[0][1][16]=v_{26}$,$A[0][2][16]=v_{27}$,$A[0][4][16]=v_{26}+v_{27}$,$A[0][2][17]=A[0][4][17]=v_{28}$,
$A[0][2][19]=A[0][4][19]=v_{29}$,$A[0][3][21]=A[0][4][21]=v_{30}$,$A[0][1][22]=v_{31}$,
$A[0][2][22]=v_{32}$,$A[0][3][22]=v_{33}$,$A[0][4][22]=v_{31}+v_{32}+v_{33}$,$A[0][2][23]=A[0][4][23]=v_{34}$,
$A[0][1][24]=v_{35}$,$A[0][3][24]=v_{36}$,$A[0][4][24]=v_{35}+v_{36}$,$A[0][1][25]=v_{37}$,$A[0][3][25]=v_{38}$,
$A[0][4][25]=v_{37}+v_{38}$,$A[0][1][27]=A[0][4][27]=v_{39}$,$A[0][1][30]=A[0][2][30]=v_{40}$,
$A[0][1][31]=v_{41}$,$A[0][2][31]=v_{42}$,$A[0][3][31]=v_{43}$,$A[0][4][31]=v_{41}+v_{42}+v_{43}$,
$A[2][0][1]=A[2][4][1]=v_{44}$,$A[2][0][2]=v_{45}$,$A[2][1][2]=v_{46}$,$A[2][3][2]=v_{47}$,
$A[2][4][2]=v_{45}+v_{46}+v_{47}$,$A[2][1][3]=v_{48}$,$A[2][3][3]=v_{49}$,$A[2][4][3]=v_{48}+v_{49}$,
$A[2][0][4]=v_{50}$,$A[2][1][4]=v_{51}$,$A[2][3][4]=v_{50}+v_{51}$,$A[2][0][5]=A[2][4][5]=v_{52}$,
$A[2][0][7]=v_{53}$,$A[2][1][7]=v_{54}$,$A[2][3][7]=v_{53}+v_{54}$,$A[2][0][9]=A[2][3][9]=v_{55}$,
$A[2][0][10]=A[2][4][10]=v_{56}$,$A[2][0][11]=v_{57}$,$A[2][1][11]=v_{58}$,$A[2][4][11]=v_{57}+v_{58}$,
$A[2][0][12]=v_{59}$,$A[2][1][12]=v_{60}$,$A[2][4][12]=v_{59}+v_{60}$,$A[2][1][13]=v_{61}$,$A[2][3][13]=v_{62}$,
$A[2][4][13]=v_{61}+v_{62}$,$A[2][1][14]=A[2][4][14]=v_{63}$.

7 Conclusion

In this paper, we comprehensively study the conditional cube attack against Keccak keyed modes. In order to find enough ordinary cube variables in low degrees of freedom of Keccak keyed modes, we introduce a new MILP model. We show how to model the CP-like-kernel and model the way that the ordinary cube variables do not multiply together in the first round as well as do not multiply with the conditional cube variable in the second round. Then, a series of linear inequality system are brought out, which accurately restrict the way to add an ordinary cube variable. Then, by choosing the objective function of the maximal number of ordinary cube variables, we convert Huang *et al.*'s greedy algorithm into an MILP problem and maximal number of ordinary cube variables is determined. Based on this method, we extend the best previous attacks on round-reduced Keccak-MAC-384 and Keccak-MAC-512 by 1 round, and achieve the first 7-round and 6-round key-recovery attacks, respectively. In addition, we give some results on Ketje Major and Minor and get the best results on these two ciphers (Tables 7, 10 and 11).

Currently, the cryptanalysis progress of symmetric-key ciphers heavily depends on automated evaluation tools. For many reasons, the cryptanalysis

Table 11. Bit conditions for attack on the 7-round initialization of Ketje Minor

Bit Conditions
$A[1][0][24]=k_1[24] + A[4][0][25] + A[4][1][25] + A[0][2][25] + A[1][2][24] + A[4][2][25] + A[1][3][24] + A[4][3][25] + A[1][4][24] + A[4][4][25] + 1$
$A[1][0][31]=k_1[31] + k_3[30] + A[3][0][30] + A[3][1][30] + A[1][2][31] + A[3][2][30] + A[1][3][31] + A[2][4][31] + A[3][4][30] + 1$
$A[1][0][19]=k_1[19] + k_3[18] + A[3][0][18] + A[2][1][19] + A[3][1][18] + A[1][2][19] + A[3][2][18] + A[1][3][19] + A[1][4][19] + A[3][4][18] + 1$
$A[3][0][17]=k_0[8] + k_3[17] + A[3][1][17] + A[3][2][17] + A[0][3][16] + A[4][3][17] + A[3][4][17]$
$A[1][2][13]=k_0[5] + k_2[12] + A[0][2][13] + A[2][3][12]$
$A[1][0][20]=k_1[20] + A[4][0][21] + A[0][1][21] + A[4][1][21] + A[1][2][20] + A[4][2][21] + A[1][3][20] + A[4][3][21] + A[1][0][4][20] + A[4][4][21] + 1$
$A[1][0][10]=k_1[10] + k_3[9] + A[3][0][9] + A[3][1][9] + A[1][2][10] + A[3][2][9] + A[1][3][10] + A[2][3][10] + A[1][4][10] + A[3][4][9]$
$A[3][0][28]=k_0[19] + k_3[28] + A[4][0][28] + A[3][1][28] + A[0][2][27] + A[3][2][28] + A[0][3][27] + A[3][4][28] + 1$
$A[3][0][16]=k_0[7] + k_3[16] + A[3][1][16] + A[3][2][16] + A[4][2][16] + A[0][4][15] + A[3][4][16]$
$A[3][0][21]=k_0[12] + k_3[21] + A[0][1][20] + A[3][1][21] + A[0][2][20] + A[3][2][21] + A[4][2][21] + A[0][3][20] + A[0][4][20] + A[3][4][21] + 1$
$A[1][4][26]=k_0[18] + k_2[25] + A[2][0][25] + A[0][1][26] + A[2][1][25] + A[0][2][26] + A[0][3][26] + A[2][3][25] + A[0][4][26] + A[2][4][25]$
$A[1][0][25]=k_1[25] + k_3[24] + A[2][0][25] + A[3][0][24] + A[3][1][24] + A[1][2][25] + A[3][2][24] + A[1][3][25] + A[1][4][25] + A[3][4][24] + 1$

of the new SHA-3 standard Keccak is very hard and limited, more evaluation tools on Keccak are urgently needed. The MILP method introduced in this paper enriches the Keccak tools, and helps academic communities study Keccak much easier.

Acknowledgments. We would like to thank the anonymous reviewers of Asiacrypt 2017 who helped us to improve this paper. This work is supported by China's 973 Program (No. 2013CB834205), the National Key Research and Development Program of China (No. 2017YFA0303903), the National Natural Science Foundation of China (Nos. 61672019, 61402256), the Fundamental Research Funds of Shandong University (No. 2016JC029), National Cryptography Development Fund (No. MMJJ20170121), Zhejiang Province Key R&D Project (No. 2017C01062).

References

1. http://www.sagemath.org/
2. http://www.gurobi.com/
3. Berton, G., Daemen, J., Peeters, M., Assche, G.V.: The KECCAK sponge function family. http://keccak.noekeon.org/

4. Berton, G., Daemen, J., Peeters, M., Assche, G.V., Keer, R.V.: CAESAR submission: KETJE v2 (2016). http://competitions.cr.yp.to/round3/ketjev2.pdf
5. Berton, G., Daemen, J., Peeters, M., Assche, G.V., Keer, R.V.: CAESAR submission: Keyak v2 (2016). http://competitions.cr.yp.to/round3/keyakv22.pdf
6. Bertoni, G., Daemen, J., Peeters, M., Assche, G.: Duplexing the sponge: single-pass authenticated encryption and other applications. In: Miri, A., Vaudenay, S. (eds.) SAC 2011. LNCS, vol. 7118, pp. 320–337. Springer, Heidelberg (2012). https://doi.org/10.1007/978-3-642-28496-0_19
7. Boura, C., Canteaut, A., De Cannière, C.: Higher-order differential properties of KECCAK and *Luffa*. In: Joux, A. (ed.) FSE 2011. LNCS, vol. 6733, pp. 252–269. Springer, Heidelberg (2011). https://doi.org/10.1007/978-3-642-21702-9_15
8. Cui, T., Jia, K., Fu, K., Chen, S., Wang, M.: New automatic search tool for impossible differentials and zero-correlation linear approximations. IACR Cryptology ePrint Archive 2016/689 (2016). http://eprint.iacr.org/2016/689
9. Daemen, J., Van Assche, G.: Differential propagation analysis of Keccak. In: Canteaut, A. (ed.) FSE 2012. LNCS, vol. 7549, pp. 422–441. Springer, Heidelberg (2012). https://doi.org/10.1007/978-3-642-34047-5_24
10. Dinur, I., Dunkelman, O., Shamir, A.: New attacks on Keccak-224 and Keccak-256. In: Canteaut, A. (ed.) FSE 2012. LNCS, vol. 7549, pp. 442–461. Springer, Heidelberg (2012). https://doi.org/10.1007/978-3-642-34047-5_25
11. Dinur, I., Dunkelman, O., Shamir, A.: Collision attacks on up to 5 rounds of SHA-3 using generalized internal differentials. In: Moriai, S. (ed.) FSE 2013. LNCS, vol. 8424, pp. 219–240. Springer, Heidelberg (2014). https://doi.org/10.1007/978-3-662-43933-3_12
12. Dinur, I., Morawiecki, P., Pieprzyk, J., Srebrny, M., Straus, M.: Cube attacks and cube-attack-like cryptanalysis on the round-reduced Keccak sponge function. In: Oswald, E., Fischlin, M. (eds.) EUROCRYPT 2015. LNCS, vol. 9056, pp. 733–761. Springer, Heidelberg (2015). https://doi.org/10.1007/978-3-662-46800-5_28
13. Dinur, I., Shamir, A.: Cube attacks on tweakable black box polynomials. In: Joux, A. (ed.) EUROCRYPT 2009. LNCS, vol. 5479, pp. 278–299. Springer, Heidelberg (2009). https://doi.org/10.1007/978-3-642-01001-9_16
14. Dobraunig, C., Eichlseder, M., Mendel, F.: Heuristic tool for linear cryptanalysis with applications to CAESAR candidates. In: Iwata, T., Cheon, J.H. (eds.) ASIACRYPT 2015. LNCS, vol. 9453, pp. 490–509. Springer, Heidelberg (2015). https://doi.org/10.1007/978-3-662-48800-3_20
15. Dobraunig, C., Eichlseder, M., Mendel, F., Schläffer, M.: Cryptanalysis of ASCON. In: Nyberg, K. (ed.) CT-RSA 2015. LNCS, vol. 9048, pp. 371–387. Springer, Cham (2015). https://doi.org/10.1007/978-3-319-16715-2_20
16. Dobraunig, C., Eichlseder, M., Mendel, F., Schläffer, M.: Ascon v1.2. Submission to the CAESAR Competition (2016)
17. Dong, X., Li, Z., Wang, X., Qin, L.: Cube-like attack on round-reduced initialization of Ketje Sr. IACR Trans. Symmetric Cryptol. **2017**(1), 259–280 (2017). http://tosc.iacr.org/index.php/ToSC/article/view/594
18. Duc, A., Guo, J., Peyrin, T., Wei, L.: Unaligned rebound attack: application to Keccak. In: Canteaut, A. (ed.) FSE 2012. LNCS, vol. 7549, pp. 402–421. Springer, Heidelberg (2012). https://doi.org/10.1007/978-3-642-34047-5_23

19. Guo, J., Liu, M., Song, L.: Linear structures: applications to cryptanalysis of round-reduced KECCAK. In: Cheon, J.H., Takagi, T. (eds.) ASIACRYPT 2016. LNCS, vol. 10031, pp. 249–274. Springer, Heidelberg (2016). https://doi.org/10.1007/978-3-662-53887-6_9
20. Huang, S., Wang, X., Xu, G., Wang, M., Zhao, J.: Conditional cube attack on reduced-round Keccak sponge function. In: Coron, J.-S., Nielsen, J.B. (eds.) EUROCRYPT 2017. LNCS, vol. 10211, pp. 259–288. Springer, Cham (2017). https://doi.org/10.1007/978-3-319-56614-6_9
21. Jean, J., Nikolić, I.: Internal differential boomerangs: practical analysis of the round-reduced Keccak-f permutation. In: Leander, G. (ed.) FSE 2015. LNCS, vol. 9054, pp. 537–556. Springer, Heidelberg (2015). https://doi.org/10.1007/978-3-662-48116-5_26
22. Li, Z., Dong, X., Wang, X.: Conditional cube attack on round-reduced ASCON. IACR Trans. Symmetric Cryptol. **2017**(1), 175–202 (2017)
23. Mella, S., Daemen, J., Assche, G.V.: New techniques for trail bounds and application to differential trails in KECCAK. IACR Trans. Symmetric Cryptol. **2017**(1), 329–357 (2017). http://tosc.iacr.org/index.php/ToSC/article/view/597
24. Morawiecki, P., Pieprzyk, J., Srebrny, M.: Rotational cryptanalysis of round-reduced KECCAK. In: Moriai, S. (ed.) FSE 2013. LNCS, vol. 8424, pp. 241–262. Springer, Heidelberg (2014). https://doi.org/10.1007/978-3-662-43933-3_13
25. Mouha, N., Wang, Q., Gu, D., Preneel, B.: Differential and linear cryptanalysis using mixed-integer linear programming. In: Wu, C.-K., Yung, M., Lin, D. (eds.) Inscrypt 2011. LNCS, vol. 7537, pp. 57–76. Springer, Heidelberg (2012). https://doi.org/10.1007/978-3-642-34704-7_5
26. Qiao, K., Song, L., Liu, M., Guo, J.: New collision attacks on round-reduced keccak. In: Coron, J.-S., Nielsen, J.B. (eds.) EUROCRYPT 2017. LNCS, vol. 10212, pp. 216–243. Springer, Cham (2017). https://doi.org/10.1007/978-3-319-56617-7_8
27. Sasaki, Y., Todo, Y.: New impossible differential search tool from design and cryptanalysis aspects. In: Coron, J.-S., Nielsen, J.B. (eds.) EUROCRYPT 2017. LNCS, vol. 10212, pp. 185–215. Springer, Cham (2017). https://doi.org/10.1007/978-3-319-56617-7_7
28. Song, L., Liao, G., Guo, J.: Non-full sbox linearization: applications to collision attacks on round-reduced KECCAK. In: Katz, J., Shacham, H. (eds.) CRYPTO 2017. LNCS, vol. 10402, pp. 428–451. Springer, Cham (2017). https://doi.org/10.1007/978-3-319-63715-0_15
29. Sun, S., Hu, L., Wang, P., Qiao, K., Ma, X., Song, L.: Automatic security evaluation and (related-key) differential characteristic search: application to SIMON, PRESENT, LBlock, DES(L) and other bit-oriented block ciphers. In: Sarkar, P., Iwata, T. (eds.) ASIACRYPT 2014. LNCS, vol. 8873, pp. 158–178. Springer, Heidelberg (2014). https://doi.org/10.1007/978-3-662-45611-8_9
30. Wu, S., Wang, M.: Security evaluation against differential cryptanalysis for block cipher structures. IACR Cryptology ePrint Archive 2011/551 (2011)
31. Xiang, Z., Zhang, W., Bao, Z., Lin, D.: Applying MILP method to searching integral distinguishers based on division property for 6 lightweight block ciphers. In: Cheon, J.H., Takagi, T. (eds.) ASIACRYPT 2016. LNCS, vol. 10031, pp. 648–678. Springer, Heidelberg (2016). https://doi.org/10.1007/978-3-662-53887-6_24

Automatic Search of Bit-Based Division Property for ARX Ciphers and Word-Based Division Property

Ling Sun[1,2], Wei Wang[1], and Meiqin Wang[1,2,3](✉)

[1] Key Laboratory of Cryptologic Technology and Information Security, Ministry of Education, Shandong University, Jinan 250100, China
lingsun@mail.sdu.edu.cn, {weiwangsdu,mqwang}@sdu.edu.cn
[2] Science and Technology on Communication Security Laboratory, Chengdu 610041, China
[3] State Key Laboratory of Cryptology, P.O. Box 5159, Beijing 100878, China

Abstract. Division property is a generalized integral property proposed by Todo at Eurocrypt 2015. Previous tools for automatic searching are mainly based on the Mixed Integer Linear Programming (MILP) method and trace the division property propagation at the bit level. In this paper, we propose automatic tools to detect ARX ciphers' division property at the bit level and some specific ciphers' division property at the word level.

For ARX ciphers, we construct the automatic searching tool relying on Boolean Satisfiability Problem (SAT) instead of MILP, since SAT method is more suitable in the search of ARX ciphers' differential/linear characteristics. The propagation of division property is translated into a system of logical equations in Conjunctive Normal Form (CNF). Some logical equations can be dynamically adjusted according to different initial division properties and stopping rule, while the others corresponding to r-round propagations remain the same. Moreover, our approach can efficiently identify some optimized distinguishers with lower data complexity. As a result, we obtain a 17-round distinguisher for SHACAL-2, which gains four more rounds than previous work, and an 8-round distinguisher for LEA, which covers one more round than the former one.

For word-based division property, we develop the automatic search based on Satisfiability Modulo Theories (SMT), which is a generalization of SAT. We model division property propagations of basic operations and S-boxes by logical formulas, and turn the searching problem into an SMT problem. With some available solvers, we achieve some new distinguishers. For CLEFIA, 10-round distinguishers are obtained, which cover one more round than the previous work. For the internal block cipher of Whirlpool, the data complexities of 4/5-round distinguishers are improved. For Rijndael-192 and Rijndael-256, 6-round distinguishers are presented, which attain two more rounds than the published ones. Besides, the integral attacks for CLEFIA are improved by one round with the newly obtained distinguishers.

Keywords: Automatic search · Division property · ARX · SAT/SMT

T. Takagi and T. Peyrin (Eds.): ASIACRYPT 2017, Part I, LNCS 10624, pp. 128–157, 2017.
https://doi.org/10.1007/978-3-319-70694-8_5

1 Introduction

Automatic tools for cryptanalysis play a more and more important role in the design and cryptanalysis of symmetric ciphers. One common direction to construct automatic tools is to transform the searching problems into some mathematical problems, so that some existing solvers can be invoked. The involved mathematical problems can be roughly divided into three categories, which are Boolean Satisfiability Problem (SAT)/Satisfiability Modulo Theories (SMT) problem [7,16,24,32], Mixed Integer Linear Programming (MILP) problem [10,25,39,45], and Constraint Programming (CP) problem [12,38]. At the very start, the researches on automatic search of distinguishers concentrated on detecting differential and linear characteristics, since differential [4] and linear [20] cryptanalysis are two of the most powerful techniques in cryptanalysis of symmetric-key primitives. Recently, with the advent of division property [41], which is a generalized integral property, some researches about automatic searching for division property arose.

Division property was proposed by Todo [41] at Eurocrypt 2015, which was originally used to search integral distinguishers of block cipher structures. Due to the newly identified division property, at Crypto 2015, MISTY1 [21] was broken by Todo for the first time. Later, Todo and Morii [42] introduced the bit-based division property at FSE 2016, which propagates each bit independently, and a 14-round integral distinguisher for SIMON32 [3] was detected. Depending on the partition of the internal state, the methods behind the obtained distinguishers can be divided into three categories. (1) *state-based* division property: evaluate the division properties of some generalized structures. Todo [41] finished the extensive research for 2-branch Feistel structure and SPN on the whole state. Related works were provided in [5]. (2) *word-based* division property: evaluate the division properties of some specific ciphers at the word level. Todo [41] implemented the search for a variety of AES-like ciphers with 4-bit S-boxes, and the 6-round integral distinguisher [40] for MISTY1 was obtained based on this method. Some works on this topic were introduced in [34,35,46]. (3) *bit-based* division property: evaluate the propagation of division property at the bit level. Note that it is more likely to obtain better distinguishers under a more subtle partition since more information can be taken into account. All published automatic tools of integral distinguishers based on division property focused on the bit level. At Asiacrypt 2016, Xiang et al. [45] applied MILP method to search integral distinguishers with bit-based division property. Soon after, the automatic search of integral distinguishers based on MILP method for ARX ciphers was proposed in [36]. Many other automatic tools relying on MILP and CP can be found in [37,38,47].

ARX ciphers constitute a broad class of symmetric-key cryptographic algorithms, and are composed of a small set of simple operations such as modular addition, bit rotation, bit shift and XOR. To claim the security of ARX ciphers, one way is to prove the security bounds just as Dinu et al. showed in [9], where a long trail design strategy for ARX ciphers with provable bounds was proposed. The other is to estimate the maximum number of rounds of the detectable distinguishers which heavily relied on automatic tools, and the searching of distinguishers is converted into an SAT/SMT problem or MILP problem. The results show that SAT/SMT based methods [18,24,32] outperform MILP based methods [10] in the search of differential/linear characteristics for ARX ciphers. Hence, for bit-based division property, it is worth exploring whether automatic tools based on SAT/SMT method can be constructed and provide better performance for ARX primitives.

Although the search of bit-based division property can take advantage of more details, it is infeasible to trace the division property propagation at the bit level for some ciphers with large state and complicated operations, such as Rijndael [8] with 256-bit block size. In order to get the tradeoff between accuracy and practicability as we detect the division property, we also consider building automatic tool to search integral distinguishers on account of word-based division property.

Our Contributions. For the integral cryptanalysis, we construct automatic searching tools of bit-based division property for ARX ciphers and word-based division property for some specific ciphers. The key point is to translate the propagation of division property into an SAT/SMT problem and control the function calls. Specifically, the contributions can be summarized as follows:

- For ARX ciphers, we propose automatic tools to search integral distinguishers using bit-based division property. First, we model the division property propagations of the three basic operations, i.e., **Copy**, `AND`, and `XOR`, and present formulas in Conjunctive Normal Form (CNF) for them. Then, the concrete equations for the modular addition operation to depict bit-based division property propagation can be achieved. The initial division property and stopping rule are transformed to logical equations, too. At last, the propagation of division property for ARX cipher is described by a system of logical equations in CNF, where some logical formulas can be dynamically adjusted according to different initial division properties of the input multi-set and final division properties of the output multi-set, and the others corresponding to r-round propagations remain the same.
- For integral cryptanalysis, it is better to adopt distinguishers with less data requirements, and our approach can efficiently identify some optimal[1] distinguishers which require less chosen plaintexts among the distinguishers with the same length. Our searching approach is composed of two algorithms. The first one restricts the search scope of initial division property and determines the maximum number of rounds of distinguishers achieved in our model. The second one optimizes the distinguishers based on the first algorithm's output.
- For word-based division property, we construct automatic tool based on SMT method. We first study how to model division property propagations of basic operations by logical formulas. Moreover, by exclusion method, we construct formulas to depict the possible propagations calculated by the Substitution

[1] The integral distinguishers are optimal under the search strategies defined in this paper.

rule. With some available solvers, we can efficiently search integral distinguishers by setting initial division property and stopping rule rationally. Finally, the problem of searching division property can be transformed into an SMT problem.
- New integral distinguishers are detected for some ARX ciphers, such as SHACAL-2 [13], LEA [14], and HIGHT [15]. With the two algorithms mentioned above, the number of initial division properties required to be evaluated for SHACAL-2 is reduced from $2^{79.24}$ to 410, so that we can easily obtain a 17-round integral distinguisher with data complexity 2^{241} chosen plaintexts, which achieves four more rounds than previous work. For LEA, an 8-round distinguisher is identified, which covers one more round than the one found by MILP method [36]. For HIGHT, although the lengths and data requirements of the newly obtained distinguishers are not improved, some of them have more zero-sum bits than those proposed in [36].
- New word-based division properties are presented for some specific ciphers. For CLEFIA [31], we discover 10-round distinguishers, which attain one more round than the one proposed in [19]. With the newly obtained distinguishers for CLEFIA, we can improve the previous integral attacks by one round. The

Table 1. Summary of integral distinguishers.

Cipher	Block size	Key size	Round†	Length‡	$\log_2$(Data)	Balanced bits	Reference
SHACAL-2	256	128–512	64	12	1	32	[43]
				13	32	1	[30]
				17	241	7	Section 5.1
LEA	128	128/192/256	24/28/32	6	32	1	[14]
				6	32	2	[36]
				7	96	1	[36]
				8	118	1	Section 5.1
CLEFIA	128	128/192/256	18/22/26	6	32	32	[31]
				8	96	32	[31]
				9	112	32	[19]
				9	105	24	[29]
				10	127	64	Section 5.2
Rijndael	192	128/192/256	12/12/14	4	24	160	[23]
				4	176	192	[41]
				6	160	64	Section 5.2
	256	128/192/256	14	4	24	64	[11,23]
				4	232	256	[41]
				6	160	64	Section 5.2
Whirlpool	512	-	10	4	64	512	[22]
				5	488	512	[41]
				5	384	512	Section 5.2

†The number of encryption rounds.
‡The number of rounds covered by the distinguisher.

data requirements of 4/5-round integral distinguishers for the internal block cipher of Whirlpool [1] are reduced. As to Rijndael-192 and Rijndael-256 [8], 6-round distinguishers are proposed, which cover two more rounds than the previous work.

Our main results and the comparisons are listed in Tables 1 and 3.

The rest of the paper is organized as follows. In Sect. 2, some notations and background are introduced. Section 3 focuses on the automatic search of integral distinguishers with bit-based division property for ARX ciphers. The automatic method relying on SMT to search integral distinguishers in accordance with word-based division property is provided in Sect. 4. Section 5 presents some applications of the developed automatic tools. We conclude the paper in Sect. 6.

2 Preliminary

2.1 Notations

For any $a \in \mathbb{F}_2^n$, its i-th element is denoted as $a[i]$, where the bit positions are labeled in big-endian, and the Hamming weight $w(a)$ is calculated by $w(a) = \sum_{i=0}^{n-1} a[i]$. For any $\boldsymbol{a} = (a_0, a_1, \ldots, a_{m-1}) \in \mathbb{F}_2^{\ell_0} \times \mathbb{F}_2^{\ell_1} \times \cdots \times \mathbb{F}_2^{\ell_{m-1}}$, the vectorial Hamming weight of $\boldsymbol{a}$ is defined as $W(\boldsymbol{a}) = (w(a_0), w(a_1), \ldots, w(a_{m-1})) \in \mathbb{Z}^m$. For any $\boldsymbol{k} \in \mathbb{Z}^m$ and $\boldsymbol{k'} \in \mathbb{Z}^m$, we define $\boldsymbol{k} \succeq \boldsymbol{k'}$ if $k_i \geq k'_i$ for all i. Otherwise, $\boldsymbol{k} \not\succeq \boldsymbol{k'}$.

For any set $\mathbb{K}$, $|\mathbb{K}|$ denotes the number of elements in $\mathbb{K}$. $\emptyset$ stands for an empty set. Denote $\mathbb{Z}_m$ the set $\{0, 1, \ldots, m\}$.

Definition 1 (Bit Product Function). *Assume $u \in \mathbb{F}_2^n$ and $x \in \mathbb{F}_2^n$. The Bit Product Function π_u is defined as*

$$\pi_u(x) = \prod_{i=0}^{n-1} x[i]^{u[i]}.$$

For $\boldsymbol{u} = (u_0, u_1, \ldots, u_{m-1}) \in \mathbb{F}_2^{\ell_0} \times \mathbb{F}_2^{\ell_1} \times \cdots \times \mathbb{F}_2^{\ell_{m-1}}$, let $\boldsymbol{x} = (x_0, x_1, \ldots, x_{m-1}) \in \mathbb{F}_2^{\ell_0} \times \mathbb{F}_2^{\ell_1} \times \cdots \times \mathbb{F}_2^{\ell_{m-1}}$ be the input, the Bit Product Function $\pi_{\boldsymbol{u}}$ is defined as

$$\pi_{\boldsymbol{u}}(\boldsymbol{x}) = \prod_{i=0}^{m-1} \pi_{u_i}(x_i).$$

2.2 Division Property

The original integral distinguishers mainly focus on the propagation of ALL and BALANCE properties [17]. While, the division property, proposed by Todo at Eurocrypt 2015 [41], is a generalized integral property, which traces the implicit properties between traditional ALL and BALANCE properties. First, a set of plaintexts, whose division property follows initial division property, is chosen. Then, the division property of the set of texts encrypted over one round is deduced from the propagation rules. And so on, we can exploit the division property over several rounds, and determine the existence of the integral distinguishers. In the following, we briefly recall the definition of division property, and propagation rules for basic operations involved in the encryption process.

Definition 2 (Division Property [41]**).** *Let $\mathbb{X}$ be a multi-set whose elements take values from $\mathbb{F}_2^{\ell_0} \times \mathbb{F}_2^{\ell_1} \times \cdots \times \mathbb{F}_2^{\ell_{m-1}}$. When the multi-set $\mathbb{X}$ has the division property $\mathcal{D}_{\mathbb{K}}^{\ell_0,\ell_1,\ldots,\ell_{m-1}}$, where $\mathbb{K}$ denotes a set of m-dimensional vectors whose i-th element takes a value between 0 and ℓ_i, it fulfills the following conditions:*

$$\bigoplus_{\boldsymbol{x}\in\mathbb{X}} \pi_{\boldsymbol{u}}(\boldsymbol{x}) = \begin{cases} \textit{unknown} & \textit{if there is } \boldsymbol{k} \in \mathbb{K} \textit{ s.t. } W(\boldsymbol{u}) \succeq \boldsymbol{k}, \\ 0 & \textit{otherwise.} \end{cases}$$

Remark 1. Note that ℓ_0, ℓ_1, ..., ℓ_{m-1} are restricted to 1 when we consider **bit-based division property**.

Propagation Rules for Division Property.

Rule 1 (Substitution [41]**).** *Let F be a function that consists of m S-boxes, where the bit length and the algebraic degree of the i-th S-box is ℓ_i and d_i bits, respectively. The input and the output take values from $\mathbb{F}_2^{\ell_0} \times \mathbb{F}_2^{\ell_1} \times \cdots \times \mathbb{F}_2^{\ell_{m-1}}$, and $\mathbb{X}$ and $\mathbb{Y}$ denote the input and output multi-sets, respectively. Assuming that $\mathbb{X}$ has division property $\mathcal{D}_{\mathbb{K}}^{\ell_0,\ell_1,\ldots,\ell_{m-1}}$, where $\mathbb{K}$ denotes a set of m-dimensional vectors whose i-th element takes a value between 0 and ℓ_i, the division property of $\mathbb{Y}$ is $\mathcal{D}_{\mathbb{K}'}^{\ell_0,\ell_1,\ldots,\ell_{m-1}}$, where*[2]

$$\mathbb{K}' = \left\{ \left(\left\lceil \frac{k_0}{d_0} \right\rceil, \left\lceil \frac{k_1}{d_1} \right\rceil, \cdots, \left\lceil \frac{k_{m-1}}{d_{m-1}} \right\rceil \right) \middle| \boldsymbol{k} = (k_0, k_1, \ldots, k_{m-1}) \in \mathbb{K} \right\}.$$

Rule 2 (Copy [41]**).** *Let F be a copy function, where the input x takes value from $\mathbb{F}_2^n$ and the output is calculated as $(y_0, y_1) = (x, x)$. Let $\mathbb{X}$ and $\mathbb{Y}$ be the input and output multi-sets, respectively. Assuming that $\mathbb{X}$ has the division property $\mathcal{D}_{\{k\}}^n$, the division property of $\mathbb{Y}$ is $\mathcal{D}_{\mathbb{K}'}^{n,n}$, where*

$$\mathbb{K}' = \{(k-i, i) | 0 \leqslant i \leqslant k\}.$$

Rule 3 (`XOR` [41]**).** *Let F be an* `XOR` *function, where the input (x_0, x_1) takes value from $\mathbb{F}_2^n \times \mathbb{F}_2^n$ and the output is calculated as $y = x_0 \oplus x_1$. Let $\mathbb{X}$ and $\mathbb{Y}$ be the input and output multi-sets, respectively. Assuming that $\mathbb{X}$ has division property $\mathcal{D}_{\mathbb{K}}^{n,n}$, the division property of $\mathbb{Y}$ is $\mathcal{D}_{\{k'\}}^n$, where*

$$k' = \min\{k_0 + k_1 | (k_0, k_1) \in \mathbb{K}\}.$$

Here, if k' is larger than n, the propagation characteristic of division property is aborted. Namely, a value of $\bigoplus_{y\in\mathbb{Y}} \pi_v(y)$ is 0 for all $v \in \mathbb{F}_2^n$.

[2] The same vector is not inserted twice, similarly hereinafter.

Rule 4 (Split [41]). *Let F be a split function, where the input x is an element belonging to $\mathbb{F}_2^n$ and the output is calculated as $y_0 \| y_1 = x$, where (y_0, y_1) takes value from $\mathbb{F}_2^{n_0} \times \mathbb{F}_2^{n-n_0}$. Let $\mathbb{X}$ and $\mathbb{Y}$ be the input and output multi-sets, respectively. Assuming that $\mathbb{X}$ has the division property $\mathcal{D}_{\{k\}}^n$, the division property of $\mathbb{Y}$ is $\mathcal{D}_{\mathbb{K}'}^{n_0, n-n_0}$, where*

$$\mathbb{K}' = \{(k-i, i) | 0 \leqslant i \leqslant k, k-i \leqslant n_0, i \leqslant n-n_0\}.$$

Rule 5 (Concatenation [41]). *Let F be a concatenation operation, where the input (x_0, x_1) takes value from $\mathbb{F}_2^{n_0} \times \mathbb{F}_2^{n_1}$ and the output is calculated as $y = x_0 \| x_1$. Let $\mathbb{X}$ and $\mathbb{Y}$ be the input and output multi-sets, respectively. Assuming that $\mathbb{X}$ has the division property $\mathcal{D}_{\mathbb{K}}^{n_0, n_1}$, the division property of $\mathbb{Y}$ is $\mathcal{D}_{\{k'\}}^{n_0+n_1}$, where*

$$k' = \min\{k_0 + k_1 | (k_0, k_1) \in \mathbb{K}\}.$$

The above rules are defined at the word level, while, when it comes to bit-based division property, **Copy** and `XOR` rules can be applied, naturally. Another important propagation rule under bit-based division property is `AND`, which is stated in the following.

Rule 6 (Bit-based `AND` [42]). *Let F be an `AND` function, where the input (x_0, x_1) takes value from $\mathbb{F}_2 \times \mathbb{F}_2$, and the output is calculated as $y = x_0 \wedge x_1$. Let $\mathbb{X}$ and $\mathbb{Y}$ be the input and output multi-sets, respectively. Assuming that $\mathbb{X}$ has division property $\mathcal{D}_{\mathbb{K}}^{1,1}$, the division property of $\mathbb{Y}$ is $\mathcal{D}_{\mathbb{K}'}^1$, where*

$$\mathbb{K}' = \left\{ \left\lceil \frac{k_0 + k_1}{2} \right\rceil \middle| \boldsymbol{k} = (k_0, k_1) \in \mathbb{K} \right\}.$$

Similar to differential/linear characteristic in differential/linear cryptanalysis, the concatenation of r division properties of the internal states constitutes an r-round division trail, which is formally defined in the following.

Definition 3 (Division Trail [45]). *Let f be the round function of an iterated block cipher. Assume that the input multi-set has division property $\mathcal{D}_{\{\boldsymbol{k}\}}^{\ell_0, \ell_1, \cdots, \ell_{m-1}}$, and the internal state after i rounds has division property $\mathcal{D}_{\mathbb{K}_i}^{\ell_0, \ell_1, \cdots, \ell_{m-1}}$. Thus we have the following chain of division property propagations:*

$$\{\boldsymbol{k}\} \triangleq \mathbb{K}_0 \xrightarrow{f} \mathbb{K}_1 \xrightarrow{f} \mathbb{K}_2 \xrightarrow{f} \cdots \xrightarrow{f} \mathbb{K}_r.$$

Moreover, for any vector $\boldsymbol{k}_i^* \in \mathbb{K}_i$ $(i \geqslant 1)$, there must exist a vector $\boldsymbol{k}_{i-1}^* \in \mathbb{K}_{i-1}$ such that $\boldsymbol{k}_{i-1}^*$ can propagate to $\boldsymbol{k}_i^*$ by propagation rules. Furthermore, for $(\boldsymbol{k}_0, \boldsymbol{k}_1, \ldots, \boldsymbol{k}_r) \in \mathbb{K}_0 \times \mathbb{K}_1 \times \cdots \times \mathbb{K}_r$, if $\boldsymbol{k}_{i-1}$ can propagate to $\boldsymbol{k}_i$ for all $i \in \{1, 2, \ldots, r\}$, we call $(\boldsymbol{k}_0, \boldsymbol{k}_1, \ldots, \boldsymbol{k}_r)$ an r-round division trail.

The propagation of division property round by round will eventually lead to a multi-set without integral property. The following proposition can be used to detect whether a set has integral property or not, which helps us to decide when to stop propagating.

Proposition 1 (Set without Integral Property [45]). *Assume* $\mathbb{X}$ *is a multiset satisfying division property* $\mathcal{D}_{\mathbb{K}}^{\ell_0,\ell_1,\cdots,\ell_{m-1}}$*, then* $\mathbb{X}$ *does not have integral property if and only if* $\mathbb{K}$ *contains all vectors with vectorial Hamming weight* 1.

Distinguishing Attacks with Division Property.
Suppose the output division property of an integral distinguisher has balanced property on b bits. Once the sum for each of the b bits is zero, the distinguisher $\mathcal{D}$ outputs '1'; otherwise, outputs '0'. The success rate of the distinguishing attack p is composed of two cases: one is $\mathcal{D}$ outputs '1' when the oracle $\mathcal{O}$ is a concrete cipher $\mathcal{F}$ actually, the other is $\mathcal{D}$ outputs '0' when $\mathcal{O}$ is a random permutation $\mathcal{RP}$. For $\mathcal{F}$, the balanced property holds with probability 1, while for $\mathcal{RP}$ is 2^{-b}. Assuming that the probability of whether the oracle is $\mathcal{F}$ or $\mathcal{RP}$ is 0.5, it is clear that $p = 0.5 \cdot 1 + 0.5 \cdot (1 - 2^{-b}) = 1 - 2^{-b-1}$, which is 0.75 for $b = 1$, and is count for distinguishing attack.

In order to increase the success rate, we can repeat the distinguishing attack with different chosen-plaintext structures. For an n-bit cipher, suppose that the input division property requires that t bits need to be traversed. Then, the number of times the distinguishers can be replayed is at most 2^{n-t}. The data complexity of the distinguishing attack need to be discussed accordingly.

2.3 SAT and SMT Problems

In computer science, the Boolean Satisfiability Problem (SAT) [6] is the problem of determining if there exists an interpretation that satisfies a given Boolean formula. In other words, it discusses whether the variables involved in a given Boolean formula can be consistently replaced by the value `True` or `False` so that the formula is evaluated to be `True`. If this is the case, the formula is called *satisfiable*.

The Satisfiability Modulo Theories (SMT) [2] problem is a decision problem for logical formulas expressed in classical first-order logic with equality. An SMT instance is a generalization of SAT instance in which various sets of variables are replaced by predicates from a variety of underlying theories. SMT formulas provide a much richer modeling language than is possible with SAT formulas.

To solve SAT and SMT problems, there are many openly available solvers, and we use CryptoMiniSat[3] and STP[4], respectively. In order to search integral distinguishers efficiently, we adopt the C++ interface of CryptoMiniSat and the Python interface of STP.

3 Automatic Search of Bit-Based Division Property for ARX Ciphers

For ARX ciphers, since SAT/SMT method [18,24,32] is more suitable to search for differential/linear characteristics than MILP method [10], we construct the

[3] https://github.com/msoos/cryptominisat.
[4] http://stp.github.io/.

automatic searching tool relying on SAT instead of MILP. First, we model the division property propagations of three basic operations, i.e., **Copy**, `AND`, and `XOR`, and construct formulas in Conjunctive Normal Form (CNF) for them. Then, the model used to describe bit-based division property propagation for the modular addition operation is constructed based on the three basic models. By setting initial division property and stopping rule appropriately, the problem of searching integral distinguishers using bit-based division property for ARX ciphers can be converted into an SAT problem, and settled efficiently.

3.1 Models of Basic Operations at the Bit Level

We consider the division property propagations of the three basic operations (**Copy**, `AND` and `XOR`) at the bit level, and the input and output are composed of bit variables which take a value of 0 or 1. Then the division trails of each operation correspond to vectors formed by the input and output variables. To depict the propagations of these operations, we translate the rules in Sect. 2.2 into formulas in CNF, of which the solutions correspond to all the possible division trails. More specifically, we first determine all the vectors corresponding to division trails, and then exclude those impossible vector values by logical formulas. We call this idea the *exclusion method*. By analyzing all the possible division trails of bit-based **Copy**, `AND` and `XOR` operations, we construct models to describe bit-based division property propagations for them.

Model 1 (Bit-based Copy). *Denote* $(a) \xrightarrow{Copy} (b_0, b_1)$ *a division trail of* **Copy** *operation, the following logical equations are sufficient to depict the propagation of bit-based division property,*

$$\begin{cases} \overline{b_0} \vee \overline{b_1} = 1 \\ a \vee b_0 \vee \overline{b_1} = 1 \\ a \vee \overline{b_0} \vee b_1 = 1 \\ \overline{a} \vee b_0 \vee b_1 = 1 \end{cases}.$$

Proof: Let (a, b_0, b_1) be the 3-bit vector composed of the input and output division properties. For an arbitrary 3-bit vector, it has eight possible values, which are

$$(\mathbf{0}, \mathbf{0}, \mathbf{0}), (0, 0, 1), (0, 1, 0), (0, 1, 1), (1, 0, 0), (\mathbf{1}, \mathbf{0}, \mathbf{1}), (\mathbf{1}, \mathbf{1}, \mathbf{0}), (1, 1, 1).$$

When restricting to **Copy** operation, there are three division trails corresponding to the values in bold above. Thus, $(*, 1, 1)$, $(0, 0, 1)$, $(0, 1, 0)$, and $(1, 0, 0)$ are impossible cases required to be excluded, where $*$ can take 0 or 1.

In order to eliminate $(*, 1, 1)$, we assert $\overline{b_0} \vee \overline{b_1} = 1$. With this assertion, (a, b_0, b_1) cannot take values of the form $(*, 1, 1)$. Then, after eliminating all impossible cases in a similar way, we obtain the set of formulas in CNF to describe bit-based division property propagation of **Copy** operation. □

When it comes to bit-based AND operation, similar to the procedure for **Copy** operation, we consider all the possible division trails. Denote (a_0, a_1) the bit variables representing the input division property of AND operation, and let b be the bit variable standing for the output division property. Obviously, there are four division trails for AND operation, which are $(0,0) \rightarrow (0)$, $(1,0) \rightarrow (1)$, $(0,1) \rightarrow (1)$, and $(1,1) \rightarrow (1)$. Therefore, the set of logical equations have four solutions corresponding to (a_0, a_1, b), i.e., $(0,0,0)$, $(0,1,1)$, $(1,0,1)$, and $(1,1,1)$. Thus, we need to delete the impossible ones as follows.

Model 2 (Bit-based AND). *Denote* $(a_0, a_1) \xrightarrow{AND} (b)$ *a division trail of AND function, the following logical equations are sufficient to describe bit-based division property propagation of AND operation,*

$$\begin{cases} \overline{a_1} \vee b = 1 \\ a_0 \vee a_1 \vee \overline{b} = 1 \\ \overline{a_0} \vee b = 1 \end{cases}.$$

For bitwise XOR operation, only three division trails are possible, which are $(0,0,0)$, $(0,1,1)$, $(1,0,1)$, and the model can be constructed in a similar way.

Model 3 (Bit-based XOR). *Denote* $(a_0, a_1) \xrightarrow{XOR} (b)$ *a division trail of XOR function, the following logical equations are sufficient to evaluate the bit-based division property through XOR operation,*

$$\begin{cases} \overline{a_0} \vee \overline{a_1} = 1 \\ a_0 \vee a_1 \vee \overline{b} = 1 \\ a_0 \vee \overline{a_1} \vee b = 1 \\ \overline{a_0} \vee a_1 \vee b = 1 \end{cases}.$$

For specific ciphers, such as HIGHT [15], TEA [44], and XTEA [26], we also encounter cases where the number of output branches for **Copy** operation or the number of input branches for XOR operation is more than 2. The exclusion method can be generalized accordingly, and we omit it for space limitation.

Let $\boldsymbol{x} = (x_0, x_1, \ldots, x_{n-1})$, $\boldsymbol{y} = (y_0, y_1, \ldots, y_{n-1})$, and $\boldsymbol{z} = (z_0, z_1, \ldots, z_{n-1})$, which is the modular addition of $\boldsymbol{x}$ and $\boldsymbol{y}$, be n-bit vectors. Then the Boolean function of z_i can be iteratively expressed as follows[5].

$$\begin{aligned} &z_{n-1} = x_{n-1} \oplus y_{n-1} \oplus c_{n-1},\ c_{n-1} = 0, \\ &z_i = x_i \oplus y_i \oplus c_i,\ c_i = x_{i+1} \cdot y_{i+1} \oplus (x_{i+1} \oplus y_{i+1}) \cdot c_{i+1}, \\ &i = n-2, n-3, \ldots, 0. \end{aligned} \tag{1}$$

In this way, the modular addition can be decomposed into **Copy**, AND, and XOR operations, and the model to depict its propagation is summarized as follows.

[5] Note that the bit positions are labeled in big-endian.

Model 4 (Modular Addition). *Let* $(a_0, \ldots, a_{n-1}, b_0, \ldots, b_{n-1}, d_0, \ldots, d_{n-1})$ *be a division trail of n-bit modular addition operation, to describe the division property propagation, the* ***Copy****, AND, and XOR models should be applied in the order specified as follows,*

$$\begin{cases}
(a_{n-1}) \xrightarrow{Copy} (a_{n-1,0}, a_{n-1,1}) \\
(b_{n-1}) \xrightarrow{Copy} (b_{n-1,0}, b_{n-1,1}) \\
(a_{n-1,0}, b_{n-1,0}) \xrightarrow{XOR} (d_{n-1}) \\
(a_{n-1,1}, b_{n-1,1}) \xrightarrow{AND} (v_0) \\
(v_0) \xrightarrow{Copy} (g_0, r_0) \\
(a_{n-2}) \xrightarrow{Copy} (a_{n-2,0}, a_{n-2,1}, a_{n-2,2}) \\
(b_{n-2}) \xrightarrow{Copy} (b_{n-2,0}, b_{n-2,1}, b_{n-2,2}) \\
\left.\begin{array}{l}
(a_{n-i,0}, b_{n-i,0}, g_{i-2}) \xrightarrow{XOR} (d_{n-i}) \\
(a_{n-i,1}, b_{n-i,1}) \xrightarrow{AND} (v_{i-1}) \\
(a_{n-i,2}, b_{n-i,2}) \xrightarrow{XOR} (m_{i-2}) \\
(m_{i-2}, r_{i-2}) \xrightarrow{AND} (q_{i-2}) \\
(v_{i-1}, q_{i-2}) \xrightarrow{XOR} (w_{i-2}) \\
(w_{i-2}) \xrightarrow{Copy} (g_{i-1}, r_{i-1}) \\
(a_{n-i-1}) \xrightarrow{Copy} (a_{n-i-1,0}, a_{n-i-1,1}, a_{n-i-1,2}) \\
(b_{n-i-1}) \xrightarrow{Copy} (b_{n-i-1,0}, b_{n-i-1,1}, b_{n-i-1,2})
\end{array}\right\} \text{iterated for } i = 2, \ldots, n-2, \\
(a_{1,0}, b_{1,0}, g_{n-3}) \xrightarrow{XOR} (d_1) \\
(a_{1,1}, b_{1,1}) \xrightarrow{AND} (v_{n-2}) \\
(a_{1,2}, b_{1,2}) \xrightarrow{XOR} (m_{n-3}) \\
(m_{n-3}, r_{n-3}) \xrightarrow{AND} (q_{n-3}) \\
(v_{n-2}, q_{n-3}) \xrightarrow{XOR} (w_{n-3}) \\
(a_0, b_0, w_{n-3}) \xrightarrow{XOR} (d_0)
\end{cases}$$

where $a_{i,j}$, $b_{i,j}$, v_i, m_i, g_i, r_i, q_i, *and* w_i *are intermediate variables, and their usage is illustrated in Table 2. In this model,* $(12n - 19)$ *intermediate variables are introduced in total, which include* $(3n-4)$ $a_{i,j}$*'s,* $(3n-4)$ $b_{i,j}$*'s,* $(n-1)$ v_i*'s,* $(n-2)$ m_i*'s,* $(n-2)$ g_i*'s,* $(n-2)$ r_i*'s,* $(n-2)$ q_i*'s, and* $(n-2)$ w_i*'s.*

Model 4 deals with the case where the two input branches of the modular addition operation are variables. When it comes to the modular addition of a variable and an unknown constant (subkey), the corresponding propagation models can be deduced similarly as discussed in [36], and we omit it due to space limitation.

To sum up, the bit-based division property propagations through all kinds of basic operations in ARX ciphers are converted into sets of logical equations. We first construct SAT model which characterizes one round bit-based division property propagation, then an SAT problem depicting r-round division trails can be obtained by repeating this procedure for r times.

Table 2. Illustration of Intermediate Variables for Modular Addition Operation.

Distribution of intermediate variables	
$z_{n-1} = \underbrace{x_{n-1}}_{a_{n-1,0}} \oplus \underbrace{y_{n-1}}_{b_{n-1,0}}$	
$z_{n-2} = \underbrace{x_{n-2}}_{a_{n-2,0}} \oplus \underbrace{y_{n-2}}_{b_{n-2,0}} \oplus \underbrace{c_{n-2}}_{g_0}$	$\underbrace{c_{n-2}}_{v_0} = \underbrace{x_{n-1}}_{a_{n-1,1}} \underbrace{y_{n-1}}_{b_{n-1,1}}$
$z_{n-3} = \underbrace{x_{n-3}}_{a_{n-3,0}} \oplus \underbrace{y_{n-3}}_{b_{n-3,0}} \oplus \underbrace{c_{n-3}}_{g_1}$	$\underbrace{c_{n-3}}_{w_0} = \overbrace{\underbrace{x_{n-2}}_{a_{n-2,1}} \underbrace{y_{n-2}}_{b_{n-2,1}}}^{v_1} \oplus \overbrace{\overbrace{(\underbrace{x_{n-2}}_{a_{n-2,2}} \oplus \underbrace{y_{n-2}}_{b_{n-2,2}})}^{m_0} \overbrace{c_{n-2}}^{r_0}}^{q_0}$
$z_{n-4} = \underbrace{x_{n-4}}_{a_{n-4,0}} \oplus \underbrace{y_{n-4}}_{b_{n-4,0}} \oplus \underbrace{c_{n-4}}_{g_1}$	$\underbrace{c_{n-4}}_{w_1} = \overbrace{\underbrace{x_{n-3}}_{a_{n-3,1}} \underbrace{y_{n-3}}_{b_{n-3,1}}}^{v_2} \oplus \overbrace{\overbrace{(\underbrace{x_{n-3}}_{a_{n-3,2}} \oplus \underbrace{y_{n-3}}_{a_{n-3,2}})}^{m_1} \overbrace{c_{n-3}}^{r_1}}^{q_1}$
...	...
$z_1 = \underbrace{x_1}_{a_{1,0}} \oplus \underbrace{y_1}_{b_{1,0}} \oplus \underbrace{c_1}_{g_{n-3}}$	$\underbrace{c_1}_{w_{n-4}} = \overbrace{\underbrace{x_2}_{a_{2,1}} \underbrace{y_2}_{b_{2,1}}}^{v_{n-3}} \oplus \overbrace{\overbrace{(\underbrace{x_2}_{a_{2,2}} \oplus \underbrace{y_2}_{b_{2,2}})}^{m_{n-4}} \overbrace{c_2}^{r_{n-4}}}^{q_{n-4}}$
$z_0 = x_0 \oplus y_0 \oplus c_0$	$\underbrace{c_0}_{w_{n-3}} = \overbrace{\underbrace{x_1}_{a_{1,1}} \underbrace{y_1}_{b_{1,1}}}^{v_{n-2}} \oplus \overbrace{\overbrace{(\underbrace{x_1}_{a_{1,2}} \oplus \underbrace{y_1}_{b_{1,2}})}^{m_{n-3}} \overbrace{c_1}^{r_{n-3}}}^{q_{n-3}}$

3.2 Initial Division Property and Stopping Rule

We propose a 'dynamic' searching, which can set the initial division property and stopping rule more efficiently. In the C++ interface of CryptoMiniSat, there is a function called `solver()`, which takes 'assumptions' as parameter, so that we can adjust the 'assumptions', instead of the original model, and invoke `solver()` calls to search for integral distinguishers under different initial division properties and output division properties automatically. In our model, 'assumptions' are composed of two parts of logical equations: one is determined by the initial division property, and another is deduced from the stopping rule.

Initial Division Property. Denote $(a_0, a_1, \dots, a_{n-1})$ the variables representing bit-based division property of the input multi-set. For example, suppose that the initial division property is $\boldsymbol{k}_0 = (0, \underbrace{1, \dots, 1}_{n-1})$. To evaluate the propagation under $\boldsymbol{k}_0$, we set the first part of the assumptions by logical equations, i.e., $a_0 = 0$, $a_1 = 1$, ..., $a_{n-1} = 1$. If we want to test division property under another initial division property, only logical equations involved in the assumptions need to be changed.

Stopping Rule. The stopping rule is formulated according to Proposition 1. When it comes to the bit-based division property, a multi-set $\mathbb{X}$, whose elements take values from $\mathbb{F}_2^n$, does not have integral property if and only if its division property contains all the n unit vectors. Hence, we need to check all the n unit vectors one by one. Denote $(b_0, b_1, \ldots, b_{n-1})$ the variables representing bit-based division property of the output multi-set after r rounds. For each $i \in \{0, 1, \ldots, n-1\}$, we set the second part of the assumptions by $b_i = 1$ and $b_j = 0$ $(j \neq i)$. Together with the initial division property, the two parts of parameters are determined for the `solver()` function, and the searching algorithm can be transformed into an SAT problem. If it is 'satisfiable' for the i-th unit vector, it means that the output division property contains the i-th unit vector. Once it is satisfiable for each unit vector, the output division property contains all unit vectors, and the corresponding multi-set, i.e., the outputs of the r-th round, does not have any integral property, and the propagation should stop and an $(r-1)$-round distinguisher is obtained. Only if there is at least one index j, such that the problem is not satisfiable for the j-th unit vector, we proceed to the $(r+1)$-th round and evaluate the division property in a similar way.

3.3 Algorithms to Find Optimal Distinguishers

According to the discussion of the above subsections, the propagation of division property for ARX cipher is depicted by a system of logical equations in CNF. Some logical formulas can be dynamically adjusted according to different initial division properties of the input set and final division properties of the output set, while the others corresponding to r-round propagations remain the same. To obtain an optimal integral distinguisher, many candidates of initial division properties need to be tested. However, we could not afford such computations for too many candidates in practice.

In order to break through the difficulty, we put forward an efficient searching approach, which is composed of two algorithms. The first one restricts the search scope of initial division property and detects the number of rounds of the optimal distinguisher achieved under our model. For the instance of SHACAL-2, the search scope is significantly reduced from 256 bits to 17 bits. The second one detects the concrete optimal distinguishers efficiently based on the first algorithm's output. With these two algorithms, we drastically reduce the number of initial division properties required to be evaluated. For example, for the 17-round distinguisher with data complexity 2^{241} chosen plaintexts for SHACAL-2, which is provided in Sect. 5.1, the direct search requires us to test $\sum_{i=1}^{256-241} \binom{256}{i} \approx 2^{79.24}$ initial division properties. While in our algorithms, only 410 initial division properties are tested, and the distinguisher is identified.

The design of the two algorithms is based on the *embedded property* below. For different initial division properties $\boldsymbol{k}_0$ and $\boldsymbol{k}_1$ *s.t.*, $\boldsymbol{k}_0 \succeq \boldsymbol{k}_1$, there in no need to test $\boldsymbol{k}_1$, if the output multi-set under $\boldsymbol{k}_0$ does not have integral property, likewise, it is not necessary to test $\boldsymbol{k}_0$, if the output multi-set under $\boldsymbol{k}_1$ has integral property.

Proposition 2 (Embedded Property). *Let E_r be an r-round iterated encryption algorithm, f be the round function, which only composes of* ***Substitution, Copy, XOR, Split,*** *and* ***Concatenation*** *operations. Suppose that the input and the output take values from $\mathbb{F}_2^n = \mathbb{F}_2^{\ell_0} \times \mathbb{F}_2^{\ell_1} \times \cdots \times \mathbb{F}_2^{\ell_{m-1}}$, $\boldsymbol{k}_0$ and $\boldsymbol{k}_1$ are two initial division properties with $W(\boldsymbol{k}_0) \succeq W(\boldsymbol{k}_1)$. If the output multi-set under $\boldsymbol{k}_0$ does not have integral property, then the output multi-set under $\boldsymbol{k}_1$ has no integral property.*

Proof: Define

$$\mathbb{S}_{\boldsymbol{k}}^n = \{\boldsymbol{a} = (a_0, a_1, \ldots, a_{m-1}) | W(\boldsymbol{a}) \succeq W(\boldsymbol{k})\},$$

and

$$\mathbb{S}_{\mathbb{K}}^n = \bigcup_{\boldsymbol{k} \in \mathbb{K}} \mathbb{S}_{\boldsymbol{k}}^n.$$

Suppose that there are two sets $\mathbb{K}_0$ and $\mathbb{K}_1$ belonging to $\mathbb{Z}_{\ell_0} \times \mathbb{Z}_{\ell_1} \times \cdots \times \mathbb{Z}_{\ell_{m-1}}$, with $\mathbb{S}_{\mathbb{K}_0}^n \subseteq \mathbb{S}_{\mathbb{K}_1}^n$. $\mathcal{D}_{\mathbb{K}_0}^n \xrightarrow{f} \mathcal{D}_{\mathbb{K}_0'}^n$ and $\mathcal{D}_{\mathbb{K}_1}^n \xrightarrow{f} \mathcal{D}_{\mathbb{K}_1'}^n$ stand for the division property propagations through one round. By the definition of division property, it is sufficient to prove that $\mathbb{S}_{\mathbb{K}_0'}^n \subseteq \mathbb{S}_{\mathbb{K}_1'}^n$, which can be accomplished by separately proving for every basic operation. We take the substitution operation as an example, and the other operations can be proved similarly.

Now, denote $\mathcal{D}_{\mathbb{K}_0}^n \xrightarrow{\text{S}} \mathcal{D}_{\mathbb{K}_0'}^n$ and $\mathcal{D}_{\mathbb{K}_1}^n \xrightarrow{\text{S}} \mathcal{D}_{\mathbb{K}_1'}^n$ the division property propagations through substitution layer, where $\mathbb{S}_{\mathbb{K}_0}^n \subseteq \mathbb{S}_{\mathbb{K}_1}^n$. For every $\boldsymbol{k}_0' \in \mathbb{K}_0'$, there exists $\boldsymbol{k}_0 \in \mathbb{K}_0$, such that $(\boldsymbol{k}_0, \boldsymbol{k}_0')$ constitutes a division trail of the substitution operation. Since $\mathbb{S}_{\mathbb{K}_0}^n \subseteq \mathbb{S}_{\mathbb{K}_1}^n$, there will be a $\boldsymbol{k}_1 \in \mathbb{K}_1$ with $W(\boldsymbol{k}_0) \succeq W(\boldsymbol{k}_1)$. By Rule 1, we have $W(\boldsymbol{k}_0') \succeq W(\boldsymbol{k}_1')$, which implies that $\mathbb{S}_{\boldsymbol{k}_0'}^n \subseteq \mathbb{S}_{\boldsymbol{k}_1'}^n$. Thus,

$$\mathbb{S}_{\mathbb{K}_0'}^n = \bigcup_{\boldsymbol{k}_0' \in \mathbb{K}_0'} \mathbb{S}_{\boldsymbol{k}_0'}^n \subseteq \bigcup_{\boldsymbol{k}_1' \in \mathbb{K}_1'} \mathbb{S}_{\boldsymbol{k}_1'}^n = \mathbb{S}_{\mathbb{K}_1'}^n.$$

□

Algorithm 1: Detecting the Maximum Number of Rounds and Restricting the Search Scope. Denote the n vectors with Hamming weight $n-1$ as $\boldsymbol{in}_i = (\underbrace{1, \ldots, 1}_{i}, 0, \underbrace{1, \ldots, 1}_{n-i-1})$, $0 \leqslant i \leqslant n-1$. Let $\boldsymbol{out}_j = (\underbrace{0, \ldots, 0}_{j}, 1, \underbrace{0, \ldots, 0}_{n-j-1})$, $0 \leqslant j \leqslant n-1$, be the n unit vectors. For $0 \leqslant i \leqslant n-1$, we evaluate the bit-based division property propagation under the initial division property $\boldsymbol{in}_i$, and check whether the output division property of the r-th round contains all n unit vectors, i.e., the problem is satisfiable for each $\boldsymbol{out}_j$ $(0 \leqslant j \leqslant n-1)$ under the fixed $\boldsymbol{in}_i$. If for all $\boldsymbol{in}_i$ $(0 \leqslant i \leqslant n-1)$ and $\boldsymbol{out}_j$ $(0 \leqslant j \leqslant n-1)$, the problem is satisfiable, we conclude that $(r-1)$ is the maximum number of rounds based on our model. Otherwise, we proceed to the $(r+1)$-th round and evaluate the division property in a similar way. When the maximum number of rounds r_m is determined, the index i of the corresponding $\boldsymbol{in}_i$ leading to the longest distinguisher is stored in a set $\mathbb{S}$. The output of Algorithm 1 is the maximum number of round r_m and an index set $\mathbb{S}$.

Algorithm 1. Detecting the Maximum Number of Rounds & Restricting the Search Scope

```
Input: Objective algorithm E
Output: The maximum number of rounds r_m of integral distinguisher, the
        index set S
1  r = 0, S = ∅, flag = 1, r_m = 0;
2  while flag==1 do
3      r = r + 1;
4      flag = 0;
5      for i = 0; i < n do
6          let the initial division property be in_i;
7          for j = 0; j < n do
8              let the output division property be out_j;
9              solve the r-round SAT problem under the assumptions;
10             if the problem is not satisfiable then
11                 flag=1;
12                 break;
13         if flag == 1 then
14             break;
15 r = r − 1, r_m = r;
16 if r_m == 0 then
17     return r_m, S;
18 for i = 0; i < n do
19     let the initial division property be in_i;
20     evaluate its division property after r_m-round propagation;
21     if there is zero-sum bit then
22         S = S ∪ {i};
23         continue;
24 return r_m, S;
```

Although we have detected r_m-round distinguishers, the data requirement to implement the integral cryptanalysis is 2^{n-1}. And the distinguisher with lower data complexity is more interesting, so we proceed Algorithm 2 to optimize the distinguishers obtained in Algorithm 1.

Algorithm 2: Detecting the Optimal Distinguisher. Let the index set $\mathbb{S} = \{j_0, j_1, \ldots, j_{|\mathbb{S}|-1}\}$ be the output of Algorithm 1. With Proposition 2, we claim that the elements in the complementary set $\overline{\mathbb{S}} = \{0, 1, \ldots, n-1\} \backslash \mathbb{S}$ of $\mathbb{S}$ refer to the 'necessary' bit indexes to obtain an r_m-round integral distinguisher. In other words, if any bit whose index belongs to $\overline{\mathbb{S}}$ is set to '0' in the initial division property, the division property after r_m-round propagation will have no integral property. In this sense, we call $\overline{\mathbb{S}}$ the *necessary set*, whose elements are called *necessary indexes*, and the corresponding bit must be fixed to '1', while, $\mathbb{S}$ is called the *sufficient set*, and the elements in $\mathbb{S}$ are called *sufficient indexes*.

Algorithm 2. Detecting the Optimal Distinguisher

Input: Objective algorithm E, the maximum number of rounds r_m, the sufficient set $\mathbb{S}$

Output: A list *List* representing Optimal integral distinguishers

```
1  flag = 0, List = ∅, k0 = (0, 0, ..., 0);
2  k_0 = 0, k_1 = 0, ..., k_{n-1} = 0;
3  for i = 0; i < n do
4      if i ∉ S then
5          k_i = 1;
6  t = 0;
7  while flag == 0 do
8      for every t-tuple (i_0, i_1, ..., i_{t-1}) of S do
9          for i ∈ S do
10             if i ∈ {i_0, i_1, ..., i_{t-1}} then
11                 k_i = 1;
12             else if i ∈ S\{i_0, i_1, ..., i_{t-1}} then
13                 k_i = 0;
14         let the initial division property be k0 = (k_0, k_1, ..., k_{n-1});
15         evaluate its bit-based division property after r_m-round propagation;
16         if there is zero-sum bit then
17             flag = 1;
18             break;
19     t = t + 1;
20 t = t − 1;
21 for every t-tuple (i_0, i_1, ..., i_{t-1}) of S do
22     InActive = S\{i_0, i_1, ..., i_{t-1}};
23     for i ∈ S do
24         if i ∈ {i_0, i_1, ..., i_{t-1}} then
25             k_i = 1;
26         else if i ∈ S\{i_0, i_1, ..., i_{t-1}} then
27             k_i = 0;
28     let the initial division property be k0 = (k_0, k_1, ..., k_{n-1});
29     evaluate its bit-based division property after r_m-round propagation;
30     ZeroSum = ∅;
31     for i = 0; i < n do
32         if the i-th output bit satisfies zero-sum property then
33             ZeroSum = ZeroSum ∪ {i};
34     if ZeroSum ≠ ∅ then
35         List = List ∪ {InActive, ZeroSum};
36 return List;
```

To reduce the data complexity, we need to analyze whether the bits with sufficient indexes can be set to '0'. The possibility of reducing data complexity lies in the size of $\mathbb{S}$. If $|\mathbb{S}| = 1$, there is no margin to further reduce the data complexity, and we obtain integral distinguishers with data complexity 2^{n-1} chosen plaintexts. In case of $|\mathbb{S}| > 1$, we firstly set all bits corresponding to $\overline{\mathbb{S}}$ in initial division property to '1' while the other bits are set to '0', and check whether there is zero-sum bit after r_m-round propagation. If it is indeed the case, we get an integral distinguisher with data complexity $2^{n-|\mathbb{S}|}$ chosen plaintexts. Otherwise, we gradually increase the number of '1's in the positions indicated by the sufficient indexes, and check whether zero-sum bit exists or not. The concrete description of this procedure can be found in Algorithm 2. After executing this algorithm, the return value will be the optimal distinguishers under our model.

Remark 2. Note that Step 8 in Algorithm 2 requests us to check out $\frac{|\mathbb{S}|!}{(|\mathbb{S}|-t)!\cdot t!}$ different initial division properties. When $|\mathbb{S}|$ is very large, the time taken to perform this **for** loop gradually increases with t growing. But, for all the ciphers analyzed in this paper, $|\mathbb{S}|$ is not very large and the runtime is acceptable.

4 Automatic Search of Word-Based Division Property

When the state of the cipher is very large, such as 256-bit, and the involved operations are very complicated, it is hard to trace the division property propagation at the bit level. In this section, we concentrate on automatic search of word-based division property efficiently. First, we study how to model division property propagations of basic operations by logical formulas at the word level. Secondly, by exclusion method, we construct formulas to depict the possible propagations calculated by Substitution rule. By setting initial division property and stopping rule rationally, the problem of searching division property can be transformed into an SMT problem, which is a generalization of SAT and can be efficiently settled with some openly available solvers.

4.1 Models of Basic Operations at the Word Level

We study the division property propagations of the basic operations at the word level. Different from Sect. 3, the input and output are variables in $\mathbb{F}_2^n$, and more kinds of formulas, such as inequalities, can be handled by SMT, so that the translation from the rules introduced in Sect. 2.2 to constraints are more flexible. We just list the models as follows.

Model 5 (Word-based Copy). *Denote* $(a) \xrightarrow{\textit{Copy}} (b_0, b_1)$ *a division trail of an n-bit* ***Copy*** *function, the following constraints are sufficient to describe the division property propagation of* ***Copy*** *operation,*

$$\begin{cases} a \leqslant n \\ b_0 \leqslant n \\ b_1 \leqslant n \\ a = b_0 + b_1 \end{cases}.$$

Model 6 (Word-based XOR). *Denote* $(a_0, a_1) \xrightarrow{XOR} (b)$ *a division trail of n-bit XOR operation, the following constraints are sufficient to depict the division property propagation of XOR operation,*

$$\begin{cases} a_0 \leqslant n \\ a_1 \leqslant n \\ b \leqslant n \\ a_0 + a_1 = b \end{cases}.$$

Model 7 (Split). *Let F be the split function in Rule 4. Denote* $(a) \xrightarrow{F} (b_0, b_1)$ *a division trail of F, the following constraints are sufficient to describe the division property propagation of* ***Split*** *operation,*

$$\begin{cases} a \leqslant n \\ b_0 \leqslant n_0 \\ b_1 \leqslant n - n_0 \\ a = b_0 + b_1 \end{cases}.$$

Model 8 (Concatenation). *Let F be the concatenation function in Rule 5. Denote* $(a_0, a_1) \xrightarrow{F} (b)$ *a division trail of F, the following constraints are sufficient to depict the division property propagation of* ***Concatenation*** *operation,*

$$\begin{cases} a_0 \leqslant n_0 \\ a_1 \leqslant n_1 \\ b \leqslant n_0 + n_1 \\ a_0 + a_1 = b \end{cases}.$$

Many ciphers take Maximum Distance Separable (MDS) matrices over finite field as linear mappings, such as the MixColumn operation for AES [28]. Todo [41] proposed a dedicated function called Partition to handle the division property propagation through MixColumn operation. We generalize it into SMT model in order to deal with some ciphers involving MDS matrices.

Model 9 (Partition/MixColumn). *Let* $F(x) = M \cdot x$, *where M is an MDS matrix over* $(\mathbb{F}_2^m)^s$. *Denote* $(a_0, a_1, \ldots, a_{s-1}) \xrightarrow{F} (b_0, b_1, \ldots, b_{s-1})$ *a division trail, the following constraints are sufficient to propagate the division property,*

$$\begin{cases} a_i \leqslant m, i = 0, 1, \ldots, s-1 \\ b_j \leqslant m, j = 0, 1, \ldots, s-1 \\ a_0 + a_1 + \cdots + a_{s-1} = b_0 + b_1 + \cdots + b_{s-1} \end{cases}.$$

4.2 Modelling S-Box

Since conventional division property is propagated at the word level, we do not need to precisely depict S-box, and use Rule 1 instead. By Rule 1, we find that the output multi-set follows $\mathcal{D}^m_{\lceil \frac{k}{d} \rceil}$ if the input multi-set satisfies $\mathcal{D}^m_k$ for an m-bit S-box with degree d. Accordingly, we deduce possible propagations for S-box, which are converted into SMT model by exclusion method mentioned in Sect. 3.

Model 10 (4-bit S-box with Degree 3). *Denote* $(x) \xrightarrow{S_{(4)}} (y)$ *a division trail of 4-bit S-box* $S_{(4)}$*, whose algebraic degree is 3, where* $x = (x[0], x[1], x[2])$ *and* $y = (y[0], y[1], y[2])$ *are supposed to be 3-bit vectors. Then, the following constraints are sufficient to describe the propagation of division property,*

$$\begin{cases} x \leqslant 4 \\ y \leqslant 4 \\ x[0] \vee \overline{y[0]} = 1 \\ \overline{x[0]} \vee x[1] \vee x[2] \vee y[0] = 1 \\ \overline{y[1]} = 1 \\ x[0] \vee \overline{x[1]} \vee \overline{y[0]} \vee y[1] \vee y[2] = 1 \\ x[0] \vee x[1] \vee \overline{x[2]} \vee y[0] \vee y[1] \vee \overline{y[2]} = 1 \\ x[0] \vee x[1] \vee x[2] \vee y[0] \vee y[1] \vee \overline{y[2]} = 1 \end{cases}.$$

Proof: Note that for a 4-bit S-box with algebraic degree 3, the possible propagations are $(0) \xrightarrow{S_{(4)}} (0)$, $(1) \xrightarrow{S_{(4)}} (1)$, $(2) \xrightarrow{S_{(4)}} (1)$, $(3) \xrightarrow{S_{(4)}} (1)$, and $(4) \xrightarrow{S_{(4)}} (4)$, and the natural constraints deduced from Rule 1 are $x \leqslant 4$ and $y \leqslant 4$. After adding these two natural constraints, the number of possible combinations of $(x[0], x[1], x[2], y[0], y[1], y[2])$ reduces to 25, which are

$$\begin{array}{l} (\mathbf{0,0,0,0,0,0}), (0,0,0,0,0,1), (0,0,0,0,1,0), (0,0,0,0,1,1), (0,0,0,1,0,0), \\ (0,0,1,0,0,0), (\mathbf{0,0,1,0,0,1}), (0,0,1,0,1,0), (0,0,1,0,1,1), (0,0,1,1,0,0), \\ (0,1,0,0,0,0), (\mathbf{0,1,0,0,0,1}), (0,1,0,0,1,0), (0,1,0,0,1,1), (0,1,0,1,0,0), \\ (0,1,1,0,0,0), (\mathbf{0,1,1,0,0,1}), (0,1,1,0,1,0), (0,1,1,0,1,1), (0,1,1,1,0,0), \\ (1,0,0,0,0,0), (1,0,0,0,0,1), (1,0,0,0,1,0), (1,0,0,0,1,1), (\mathbf{1,0,0,1,0,0}). \end{array}$$

The five vectors in bold are what we expect. After observation, $(0,*,*,1,*,*)$, $(1,0,0,0,*,*)$, $(*,*,*,*,1,*)$, $(0,1,*,0,0,0)$, $(0,0,1,0,0,0)$ and $(0,0,0,0,0,1)$ are impossible cases, where $*$ takes 0 or 1.

In order to eliminate $(0,*,*,1,*,*)$, we assert $x[0] \vee \overline{y[0]} = 1$. With this assertion, $(x[0], x[1], x[2], y[0], y[1], y[2])$ cannot take values of the form $(0,*,*,1,*,*)$. After eliminating all impossible cases one by one, we obtain the set of logical formulas to describe division property propagation of $S_{(4)}$. □

For 8-bit S-box with degree 7, possible propagations are $(0) \rightarrow (0)$, $(1) \rightarrow (1)$, $(2) \rightarrow (1)$, $(3) \rightarrow (1)$, $(4) \rightarrow (1)$, $(5) \rightarrow (1)$, $(6) \rightarrow (1)$, $(7) \rightarrow (1)$, and $(8) \rightarrow (8)$, and the model can be constructed in a similar way.

Model 11 (8-bit S-box with Degree 7). *Denote* $(x) \xrightarrow{S_{(8)}} (y)$ *a division trail of 8-bit S-box* $S_{(8)}$*, whose algebraic degree is 7, where* $x = (x[0], x[1], x[2], x[3])$ *and* $y = (y[0], y[1], y[2], y[3])$ *are supposed to be 4-bit vectors. Then, the following* *constraints are sufficient to describe the possible propagations,*

$$\begin{cases} x \leqslant 8 \\ y \leqslant 8 \\ \overline{x[0]} \vee y[0] = 1 \\ x[0] \vee \overline{y[0]} = 1 \\ y[1] = 0 \\ y[2] = 0 \\ \overline{x[3]} \vee y[0] \vee y[1] \vee y[2] \vee y[3] = 1 \\ \overline{x[2]} \vee y[0] \vee y[1] \vee y[2] \vee y[3] = 1 \\ \overline{x[1]} \vee y[0] \vee y[1] \vee y[2] \vee y[3] = 1 \\ x[0] \vee x[1] \vee x[2] \vee x[3] \vee y[0] \vee y[1] \vee y[2] \vee \overline{y[3]} = 1 \end{cases}.$$

For other types of S-boxes, exclusion method can be applied and constraints to depict division property propagations can be constructed similarly.

4.3 Initial Division Property and Stopping Rule

Just as in Sect. 3, to make the searching algorithm dynamic, the initial division property and stopping rule are inserted into assumptions. In the Python interface of STP, the function, which accepts 'assumptions' as parameter, is called `check()`.

Denote $(a_0, a_1, \ldots, a_{m-1})$ the variables representing division property of the input multi-set. For example, suppose that the initial division property is $\boldsymbol{k} = (k_0, k_1, \ldots, k_{m-1})$. To propagate division property under $\boldsymbol{k}$, we set the first part of the assumptions by logical formulas, i.e., $a_0 = k_0$, $a_1 = k_1$, …, and $a_{m-1} = k_{m-1}$. Only logical formulas involved in the assumptions are required to be replaced if we want to test division property under another initial division property.

Restricted to conventional division property, Proposition 1 claims that a multi-set $\mathbb{X} \in \mathbb{F}_2^n = \mathbb{F}_2^{\ell_0} \times \mathbb{F}_2^{\ell_1} \times \cdots \times \mathbb{F}_2^{\ell_{m-1}}$ does not have integral property if and only if its division property contains all vectors with vectorial Hamming weight being 1. In order to determine whether r-round integral property exists or not under a fixed initial division property, we make m `check()` calls to test m vectors with vectorial Hamming weight 1. If all the corresponding SMT problems are satisfiable, the r-round output set has no integral property and an $(r-1)$-round distinguisher is obtained. Otherwise, we go on to the $(r+1)$-th round and evaluate the division property in a similar way.

5 Applications

In this section, we provide some new distinguishers based on the searching methods proposed in Sects. 3 and 4. We first present results for some ARX ciphers, whose integral distinguishers are obtained by evaluating bit-based division property, and then turn to the word-based division property of some specific ciphers.

5.1 Bit-Based Division Properties for ARX Ciphers

Application to SHACAL-2. SHACAL-2 [13] is a 256-bit block cipher and has been selected as one of the four block ciphers by NESSIE. Its round function is based on the compression function of the hash function SHA-2 [27], and is iterated for 64 times. SHACAL-2 supports variable key lengths up to 512 bits, yet it should not be used with a key shorter than 128 bits. An illustration of the round function can be found in Fig. 1, where K^r and W^r are round key and round constant, Maj, Ch, $\sum_0$, and $\sum_1$ are defined as follows,

$$\begin{aligned}
Maj(X,Y,Z) &= (X \cdot Y) \oplus (X \cdot Z) \oplus (Y \cdot Z),\\
Ch(X,Y,Z) &= (X \cdot Y) \oplus (\overline{X} \cdot Z),\\
\sum\nolimits_0(X) &= (X \ggg 2) \oplus (X \ggg 13) \oplus (X \ggg 22),\\
\sum\nolimits_1(X) &= (X \ggg 6) \oplus (X \ggg 11) \oplus (X \ggg 25).
\end{aligned}$$

Since the values of K^r and W^r do not influence the bit-based division property propagation, and we will not introduce them here. For more information, please refer to [13].

Firstly, Algorithm 1 in Sect. 3.3 is implemented and we find that the longest distinguisher under our model can achieve 17 rounds. At the same time, we obtain the sufficient set $\mathbb{S} = \{22-31, 153-159\}$. Then, for $r = 17$ and $\mathbb{S}$, Algorithm 2 is performed. Finally, we obtain a 17-round integral distinguisher with data complexity 2^{241} chosen plaintexts, which is

$$\text{Inactive Bits: } \{23-31, 154-159\} \xrightarrow{\text{17 Rounds}} \text{Zero-sum Bits: } \{249-255\},$$

where the bit indexes for the input and output are labeled as 0, 1, ..., 255 from left to right, and the bit indexes are labeled in a similar way in the remaining of this subsection. In order to identify this distinguisher, we try 256 initial division properties when implementing Algorithm 1, and $1 + \binom{17}{1} + \binom{17}{2} = 154$ initial division properties are evaluated when performing Algorithm 2. In total, with 410 tests under different initial division properties, we obtain the optimal distinguisher, while $\sum_{i=1}^{256-241} \binom{256}{i} \approx 2^{79.24}$ initial division properties are required to be tested for the direct search instead of using Algorithms 1 and 2.

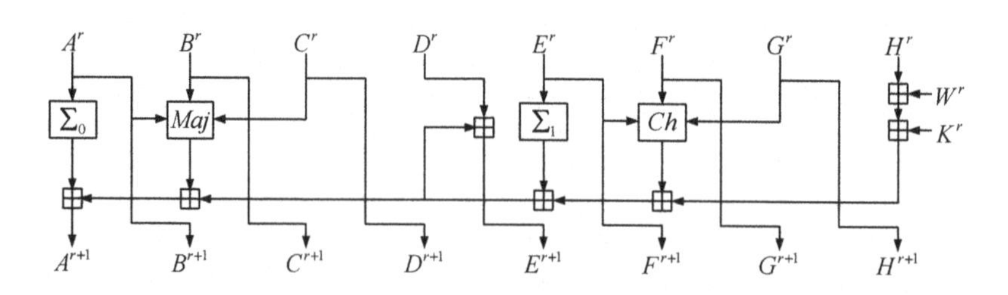

Fig. 1. The round function of SHACAL-2.

As far as we know, the best integral distinguisher in the literature is the 13-round one proposed in [30], and the newly obtained one covers four more rounds.

Applications to Other ARX Ciphers. Besides SHACAL-2, many ARX ciphers are analyzed, including LEA [14], HIGHT [15], and SPECK family of block ciphers [3], and we only list the results for space limitation.

For LEA, we obtain an 8-round integral distinguisher with data complexity 2^{118} chosen plaintexts, which is

$$\text{Inactive Bits: } \{27-31, 59-63\} \xrightarrow{8 \text{ Rounds}} \text{Zero-sum Bits: } \{36\}.$$

Comparing to the 7-round distinguishers based on MILP method provided in [36], we gain one more round.

Six integral distinguishers with data complexity 2^{63} chosen plaintexts are detected for HIGHT, which are

$$\text{Inactive Bits: } \{14\} \xrightarrow{18 \text{ Rounds}} \text{Zero-sum Bits: } \{6,7\},$$
$$\text{Inactive Bits: } \{15\} \xrightarrow{18 \text{ Rounds}} \text{Zero-sum Bits: } \{6,7\},$$
$$\text{Inactive Bits: } \{31\} \xrightarrow{18 \text{ Rounds}} \text{Zero-sum Bits: } \{7\},$$
$$\text{Inactive Bits: } \{46\} \xrightarrow{18 \text{ Rounds}} \text{Zero-sum Bits: } \{38,39\},$$
$$\text{Inactive Bits: } \{47\} \xrightarrow{18 \text{ Rounds}} \text{Zero-sum Bits: } \{38,39\},$$
$$\text{Inactive Bits: } \{63\} \xrightarrow{18 \text{ Rounds}} \text{Zero-sum Bits: } \{39\}.$$

Note that the third one and the last one are same to the 18-round distinguishers in [36], which are obtained under MILP method. And the other four distinguishers we identified have more zero-sum bits under the same data requirement.

For all versions of SPECK family of block ciphers, we obtain 6-round integral distinguishers. The data requirements are 2^{31} for SPECK32, 2^{45} for SPECK48, 2^{61} for SPECK64, 2^{93} for SPECK96, and 2^{125} for SPECK128.

All of the experiments are conducted on a server, and we use at most four 2.30 GHz Intel® Xeon® CPU E5-2670 v3 processors. All the SAT based experiments are implemented by the C++ interface of CryptoMiniSat5, using at most 4 threads. The runtimes to obtain the optimal distinguishers for SHACAL-2, LEA, and HIGHT are 6 h, 30 min, and 15 min, respectively, and the runtimes for all variants of SPECK take less than 6 min.

5.2 Word-Based Division Property for Some Specific Ciphers

Application to CLEFIA. CLEFIA [31] is a 128-bit block cipher supporting key lengths of 128, 192, and 256 bits, and it has been adopted as one of the ISO/IEC international standards in lightweight cryptography. The number of rounds, are 18, 22 and 26 for 128-bit, 192-bit and 256-bit keys, respectively. The round function follows a 4-branch Type-2 Generalized Feistel Network [48] with two parallel F functions (F_0, F_1). The 128-bit state value can be regarded as concatenation of four 32-bit words, and the input of the r-th round is denoted by $(X^r[0], X^r[1], X^r[2], X^r[3])$. One round of encryption is illustrated in Fig. 2, where $RK^r[0]$ and $RK^r[1]$ denote round keys.

Aiming at searching integral distinguishers for CLEFIA as long as possible, we first evaluate the division property under 16 initial division properties $\boldsymbol{in}_i$, $0 \leqslant i \leqslant 15$, whose i-th element is set to 7, and the others are set to 8. Then, we obtain eight 10-round integral distinguishers with data complexity 2^{127} chosen plaintexts. We also evaluate the division property under another 16 initial division properties $\boldsymbol{in}'_i$, $0 \leqslant i \leqslant 15$, whose i-th element is set to 6, and the others are set to 8. However, there is no integral property after 10-round propagation under $\boldsymbol{in}'_i$. Besides, 120 initial division properties with two elements being 7 and the others being 8 are also considered, and no integral property is detected. Thus, the 10-round integral distinguishers with data complexity 2^{127} chosen plaintexts probably are the best integral distinguishers using word-based division property. The initial division properties of these 10-round distinguishers are listed as follows.

$$\begin{aligned}
&(7,8,8,8,\ 8,8,8,8,\ 8,8,8,8,\ 8,8,8,8), (8,7,8,8,\ 8,8,8,8,\ 8,8,8,8,\ 8,8,8,8),\\
&(8,8,7,8,\ 8,8,8,8,\ 8,8,8,8,\ 8,8,8,8), (8,8,8,7,\ 8,8,8,8,\ 8,8,8,8,\ 8,8,8,8),\\
&(8,8,8,8,\ 8,8,8,8,\ 7,8,8,8,\ 8,8,8,8), (8,8,8,8,\ 8,8,8,8,\ 8,7,8,8,\ 8,8,8,8),\\
&(8,8,8,8,\ 8,8,8,8,\ 8,8,7,8,\ 8,8,8,8), (8,8,8,8,\ 8,8,8,8,\ 8,8,8,7,\ 8,8,8,8).
\end{aligned}$$

After 10-round propagation, all the 10-round distinguishers have eight zero-sum bytes, which are labeled as $\{4-7, 12-15\}$, and the bytes are labeled as 0, 1, ..., 15 from left to right.

To our knowledge, the longest integral distinguishers for CLEFIA cover 9 rounds [19, 29], and these newly found distinguishers achieve one more round. With the 10-round distinguishers, we can recover the key of 13-round CLEFIA-128 with one more round than [19], where the precomputation, partial sum technique and exhaustive search can be adopted similarly. The data, time and memory complexities are 2^{127} chosen plaintexts, 2^{120} encryptions and 2^{100} bytes, respectively. The integral attacks for CLEFIA-192 and CLEFIA-256 can be improved by one round, too.

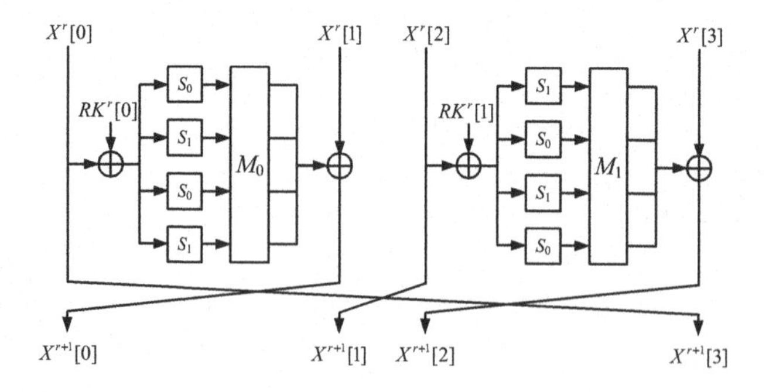

Fig. 2. Round function of CLEFIA.

Table 3. Data requirements to construct r-round integral distinguishers.

Cipher	$\log_2(Data)$				Reference
	$r=3$	$r=4$	$r=5$	$r=6$	
Rijndael-192	8	24	64	160	Section 5.2
	56	176	-	-	[41]
Rijndael-256	8	16	32	160	Section 5.2
	56	232	-	-	[41]
Whirlpool	8	56	384	-	Section 5.2
	56	344	488	-	[41]

Applications to Other Ciphers. We also implement the method in Sect. 4 to search integral distinguishers for many other ciphers.

For the internal block cipher of Whirlpool [1], comparing to the results given by Todo [41], we improve the data complexities of integral distinguishers for different rounds, which can be found in Table 3. For Rijndael-192 and Rijndael-256 [8], we extend the length of distinguishers comparing to the best results proposed by Todo [41], and the experimental results can be found in Table 3. The integral distinguishers for Whirlpool, Rijndael-192, and Rijndael-256 are provided in Appendix A.

We also implement our automatic tool to search integral distinguishers for MISTY1, MISTY2 [21], and KASUMI [33]. For MISTY1, we obtain the same distinguisher found by Todo [40]. As to MISTY2, a 7-round integral distinguisher with data complexity 2^{32} chosen plaintexts is found, which is same to the best one proposed in [34]. A 5-round integral distinguisher starting from the second round with data complexity 2^{48} chosen plaintexts is obtained for KASUMI. Comparing to the best 5-round one proposed in [35] with data complexity 2^{53} chosen plaintexts by using division property, our newly found distinguisher requires less data.

All the SMT based tests are implemented in the Python interface of STP2.0, using single thread. The runtimes for all the ciphers analyzed in this section only take few minutes.

6 Conclusion

In this paper, we propose the automatic searching tools for the integral distinguishers based on bit-based division property for ARX ciphers and word-based division property. For ARX ciphers, the automatic searching tool relying on SAT instead of MILP is constructed, since SAT method is more suitable in the search of ARX ciphers' differential/linear characteristics. First, the models, which are composed of logical formulas in CNF, to describe bit-based division property propagations for three basic operations, i.e., **Copy**, `AND`, and `XOR`, are provided by exclusion method. Then, we give the model of the modular addition based on the three basic models. After setting initial division property and stopping rule appropriately, the problem of searching integral distinguishers using bit-based division property for ARX ciphers can be converted into an SAT problem.

Besides, to get the optimal distinguisher, two algorithms are proposed. The first one restricts the search scope of initial division property and detects the round of optimal distinguisher achieved under our model. The second one detects the concrete optimal distinguishers efficiently based on the first algorithm's output.

We realize the automatic search of word-based division property with SMT method. We first show how to model division property propagations of basic operations by logical formulas. Moreover, by exclusion method, we construct formulas to depict the possible propagations calculated by Substitution rule. By setting initial division property and stopping rule rationally, the problem of searching division property can be transformed into an SMT problem, and we can efficiently search integral distinguishers with some openly available solvers.

As a result, we improve the previous integral distinguishers for SHACAL-2, LEA, CLEFIA, Rijndael-192, and Rijndael-256 according to the number of rounds. Moreover, the integral attacks for CLEFIA are improved by one round with the newly obtained distinguishers.

Discussion on the superiority to MILP method. We think it is hard to give a comprehensive comparison between MILP and SAT, and try to reflect the efficiency of SAT for ARX ciphers by recording the time spent on the search for the same distinguisher with a fixed initial division property under the same computation resource. The experimental results show that SAT model performs better than MILP model. As an illustration, for the optimal distinguisher of SHACAL-2, CryptoMiniSat returns the result after about 24 s, while MILP optimizer (Gurobi 7.0.2) takes about 44000 s, which is almost 1650 times as long as the SAT solver. Thus, it seems that SAT model is more suitable to search division properties for ARX ciphers.

Discussion on the optimality and completeness of the search. We confirm that the integral distinguishers are optimal under the search strategies defined in this paper. However, we cannot guarantee the completeness. If a more dedicated model for the modular addition is proposed, better integral distinguishers for ARX ciphers may be detected, which will be a future work.

Acknowledgements. The authors would like to thank the anonymous reviewers of Asiacrypt 2017 for their helpful comments. This work was supported by the 973 Program (No. 2013CB834205), NSFC Projects (No. 61572293), Science and Technology on Communication Security Laboratory of China (No. 9140c110207150c11050), as well as Chinese Major Program of National Cryptography Development Foundation (No. MMJJ20170102).

A Integral Distinguishers of Whirlpool and Rijndael

A.1 Integral Distinguishers of Whirlpool

Note the intermediate state of the internal block cipher of Whirlpool can be represented by an 8×8 matrix of bytes, and the indexes of the involved bytes are illustrated in Fig. 3.

0	1	2	3	4	5	6	7
8	9	10	11	12	13	14	15
16	17	18	19	20	21	22	23
24	25	26	27	28	29	30	31
32	33	34	35	36	37	38	39
40	41	42	43	44	45	46	47
48	49	50	51	52	53	54	55
56	57	58	59	60	61	62	63

Fig. 3. Indexes for whirlpool.

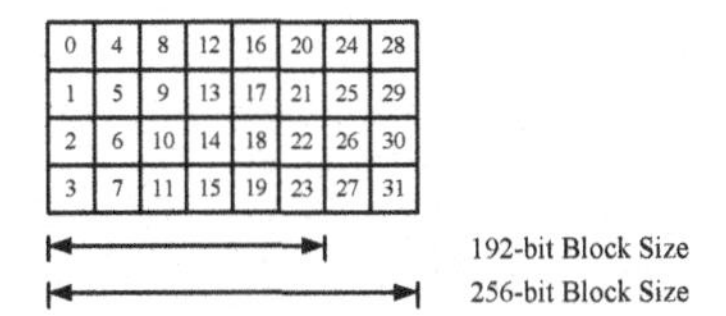

0	4	8	12	16	20	24	28
1	5	9	13	17	21	25	29
2	6	10	14	18	22	26	30
3	7	11	15	19	23	27	31

Fig. 4. Indexes for Rijndael-192 and Rijndael-256.

The integral distinguishers obtained in the paper are illustrated as follows.

$$\text{Active Bytes: } \{0\} \xrightarrow{\text{3 Rounds}} \text{Zero-sum Bytes: } \{0-63\},$$
$$\text{Active Bytes: } \{0, 22, 29, 36, 43, 50, 57\} \xrightarrow{\text{4 Rounds}} \text{Zero-sum Bytes: } \{0-63\},$$
$$\text{Active Bytes: } \{0-5, 8-12, 15-19, 22-26, 29-33,$$
$$36-40, 43-47, 50-55, 57-62\}$$
$$\xrightarrow{\text{5 Rounds}} \text{Zero-sum Bytes: } \{0-63\}.$$

A.2 Integral Distinguishers of Rijndael

For Rijndael family of block ciphers, the internal state can be treated as a $4 \times N_b$ matrix of bytes, where N_b is the number of 32-bit words in the block. The indexes for the matrix is shown in Fig. 4.

The integral distinguishers for Rijndael-192 mentioned in the paper are listed as follows:

$$\text{Active Bytes: } \{0\} \xrightarrow{\text{3 Rounds}} \text{Zero-sum Bytes: } \{0-23\},$$
$$\text{Active Bytes: } \{0, 5, 10\} \xrightarrow{\text{4 Rounds}} \text{Zero-sum Bytes: } \{16-23\},$$
$$\text{Active Bytes: } \{0, 4, 5, 9, 10, 14, 15, 19\} \xrightarrow{\text{5 Rounds}} \text{Zero-sum Bytes: } \{12-19\},$$
$$\text{Active Bytes: } \{0-7, 9-12, 14-17, 19-22\}$$
$$\xrightarrow{\text{6 Rounds}} \text{Zero-sum Bytes: } \{0-7\}.$$

And the distinguishers we found for Rijndael-256 are presented below:

$$\text{Active Bytes: } \{0\} \xrightarrow{\text{3 Rounds}} \text{Zero-sum Bytes: } \{0-31\},$$
$$\text{Active Bytes: } \{0, 5\} \xrightarrow{\text{4 Rounds}} \text{Zero-sum Bytes: } \{8-11, 20-31\},$$
$$\text{Active Bytes: } \{0, 5, 14, 19\} \xrightarrow{\text{5 Rounds}} \text{Zero-sum Bytes: } \{8-11, 24-27\},$$
$$\text{Active Bytes: } \{0, 3-5, 8, 9, 12-14, 16-19, 21-23, 26, 27, 30, 31\}$$
$$\xrightarrow{\text{6 Rounds}} \text{Zero-sum Bytes: } \{8-11, 24-27\}.$$

References

1. Barreto, P.S., Rijmen, V.: The Whirlpool hashing function. In: First Open NESSIE Workshop, Leuven, Belgium, vol. 13, p. 14 (2000)
2. Barrett, C.W., Sebastiani, R., Seshia, S.A., Tinelli, C.: Satisfiability modulo theories. Handb. Satisf. **185**, 825–885 (2009)
3. Beaulieu, R., Shors, D., Smith, J., Treatman-Clark, S., Weeks, B., Wingers, L.: The SIMON and SPECK lightweight block ciphers. In: Proceedings of the 52nd Annual Design Automation Conference, San Francisco, CA, USA, 7–11 June 2015, pp. 175:1–175:6 (2015)
4. Biham, E., Shamir, A.: Differential cryptanalysis of DES-like cryptosystems. In: Menezes, A.J., Vanstone, S.A. (eds.) CRYPTO 1990. LNCS, vol. 537, pp. 2–21. Springer, Heidelberg (1991). https://doi.org/10.1007/3-540-38424-3_1
5. Biryukov, A., Khovratovich, D., Perrin, L.: Multiset-algebraic cryptanalysis of reduced Kuznyechik, Khazad, and secret SPNs. IACR Trans. Symmetric Cryptol. **2016**(2), 226–247 (2016)
6. Cook, S.A.: The complexity of theorem-proving procedures. In: Proceedings of the Third Annual ACM Symposium on Theory of Computing, pp. 151–158. ACM (1971)
7. Courtois, N.T., Bard, G.V.: Algebraic cryptanalysis of the data encryption standard. In: Galbraith, S.D. (ed.) Cryptography and Coding 2007. LNCS, vol. 4887, pp. 152–169. Springer, Heidelberg (2007). https://doi.org/10.1007/978-3-540-77272-9_10
8. Daemen, J., Rijmen, V.: The Design of Rijndael: AES - The Advanced Encryption Standard. Information Security and Cryptography. Springer, Heidelberg (2002)
9. Dinu, D., Perrin, L., Udovenko, A., Velichkov, V., Großschädl, J., Biryukov, A.: Design strategies for ARX with provable bounds: SPARX and LAX. In: Cheon, J.H., Takagi, T. (eds.) ASIACRYPT 2016. LNCS, vol. 10031, pp. 484–513. Springer, Heidelberg (2016). https://doi.org/10.1007/978-3-662-53887-6_18
10. Fu, K., Wang, M., Guo, Y., Sun, S., Hu, L.: MILP-based automatic search algorithms for differential and linear trails for speck. In: Peyrin, T. (ed.) FSE 2016. LNCS, vol. 9783, pp. 268–288. Springer, Heidelberg (2016). https://doi.org/10.1007/978-3-662-52993-5_14
11. Galice, S., Minier, M.: Improving integral attacks against Rijndael-256 up to 9 rounds. In: Vaudenay, S. (ed.) AFRICACRYPT 2008. LNCS, vol. 5023, pp. 1–15. Springer, Heidelberg (2008). https://doi.org/10.1007/978-3-540-68164-9_1
12. Gerault, D., Minier, M., Solnon, C.: Constraint programming models for chosen key differential cryptanalysis. In: Rueher, M. (ed.) CP 2016. LNCS, vol. 9892, pp. 584–601. Springer, Cham (2016). https://doi.org/10.1007/978-3-319-44953-1_37
13. Handschuh, H., Naccache, D.: SHACAL: a family of block ciphers. Submission to the NESSIE project (2002)
14. Hong, D., Lee, J.-K., Kim, D.-C., Kwon, D., Ryu, K.H., Lee, D.-G.: LEA: a 128-bit block cipher for fast encryption on common processors. In: Kim, Y., Lee, H., Perrig, A. (eds.) WISA 2013. LNCS, vol. 8267, pp. 3–27. Springer, Cham (2014). https://doi.org/10.1007/978-3-319-05149-9_1
15. Hong, D., et al.: HIGHT: a new block cipher suitable for low-resource device. In: Goubin, L., Matsui, M. (eds.) CHES 2006. LNCS, vol. 4249, pp. 46–59. Springer, Heidelberg (2006). https://doi.org/10.1007/11894063_4

16. Kamal, A.A., Youssef, A.M.: Applications of SAT solvers to AES key recovery from decayed key schedule images. In: Fourth International Conference on Emerging Security Information Systems and Technologies, SECURWARE 2010, Venice, Italy, 18–25 July 2010, pp. 216–220 (2010)
17. Knudsen, L., Wagner, D.: Integral cryptanalysis. In: Daemen, J., Rijmen, V. (eds.) FSE 2002. LNCS, vol. 2365, pp. 112–127. Springer, Heidelberg (2002). https://doi.org/10.1007/3-540-45661-9_9
18. Kölbl, S., Leander, G., Tiessen, T.: Observations on the SIMON block cipher family. In: Gennaro, R., Robshaw, M. (eds.) CRYPTO 2015. LNCS, vol. 9215, pp. 161–185. Springer, Heidelberg (2015). https://doi.org/10.1007/978-3-662-47989-6_8
19. Li, Y., Wu, W., Zhang, L.: Improved integral attacks on reduced-round CLEFIA block cipher. In: Jung, S., Yung, M. (eds.) WISA 2011. LNCS, vol. 7115, pp. 28–39. Springer, Heidelberg (2012). https://doi.org/10.1007/978-3-642-27890-7_3
20. Matsui, M.: Linear cryptanalysis method for DES cipher. In: Helleseth, T. (ed.) EUROCRYPT 1993. LNCS, vol. 765, pp. 386–397. Springer, Heidelberg (1994). https://doi.org/10.1007/3-540-48285-7_33
21. Matsui, M.: New block encryption algorithm MISTY. In: Biham, E. (ed.) FSE 1997. LNCS, vol. 1267, pp. 54–68. Springer, Heidelberg (1997). https://doi.org/10.1007/BFb0052334
22. Mendel, F., Rechberger, C., Schläffer, M., Thomsen, S.S.: The rebound attack: cryptanalysis of reduced whirlpool and Grøstl. In: Dunkelman, O. (ed.) FSE 2009. LNCS, vol. 5665, pp. 260–276. Springer, Heidelberg (2009). https://doi.org/10.1007/978-3-642-03317-9_16
23. Minier, M., Phan, R.C.-W., Pousse, B.: Distinguishers for ciphers and known key attack against Rijndael with large blocks. In: Preneel, B. (ed.) AFRICACRYPT 2009. LNCS, vol. 5580, pp. 60–76. Springer, Heidelberg (2009). https://doi.org/10.1007/978-3-642-02384-2_5
24. Mouha, N., Preneel, B.: Towards finding optimal differential characteristics for ARX: application to Salsa20. Technical report, Cryptology ePrint Archive, Report 2013/328 (2013)
25. Mouha, N., Wang, Q., Gu, D., Preneel, B.: Differential and linear cryptanalysis using mixed-integer linear programming. In: Wu, C.-K., Yung, M., Lin, D. (eds.) Inscrypt 2011. LNCS, vol. 7537, pp. 57–76. Springer, Heidelberg (2012). https://doi.org/10.1007/978-3-642-34704-7_5
26. Needham, R.M., Wheeler, D.J.: TEA extensions. Report, Cambridge University, Cambridge, UK, October 1997
27. PUB. FIPS 180–2: Secure hash standard (SHS). US Department of Commerce, National Institute of Standards and Technology (NIST) (2012)
28. Rijmen, V., Daemen, J.: Advanced encryption standard. In: Proceedings of Federal Information Processing Standards Publications, National Institute of Standards and Technology, pp. 19–22 (2001)
29. Shibayama, N., Kaneko, T.: A new higher order differential of CLEFIA. IEICE Trans. **97–A**(1), 118–126 (2014)
30. Shin, Y., Kim, J., Kim, G., Hong, S., Lee, S.: Differential-linear type attacks on reduced rounds of SHACAL-2. In: Wang, H., Pieprzyk, J., Varadharajan, V. (eds.) ACISP 2004. LNCS, vol. 3108, pp. 110–122. Springer, Heidelberg (2004). https://doi.org/10.1007/978-3-540-27800-9_10
31. Shirai, T., Shibutani, K., Akishita, T., Moriai, S., Iwata, T.: The 128-bit blockcipher CLEFIA (extended abstract). In: Biryukov, A. (ed.) FSE 2007. LNCS, vol. 4593, pp. 181–195. Springer, Heidelberg (2007). https://doi.org/10.1007/978-3-540-74619-5_12

32. Song, L., Huang, Z., Yang, Q.: Automatic differential analysis of ARX block ciphers with application to SPECK and LEA. In: Liu, J.K., Steinfeld, R. (eds.) ACISP 2016. LNCS, vol. 9723, pp. 379–394. Springer, Cham (2016). https://doi.org/10.1007/978-3-319-40367-0_24
33. KASUMI Specification: Specification of the 3GPP confidentiality and integrity algorithms. Version, vol. 1, pp. 8–17
34. Sugio, N., Igarashi, Y., Kaneko, T.: Integral characteristics of MISTY2 derived by division property. In: 2016 International Symposium on Information Theory and Its Applications, ISITA 2016, Monterey, CA, USA, 30 October–2 November 2016, pp. 151–155 (2016)
35. Sugio, N., Igarashi, Y., Kaneko, T., Higuchi, K.: New integral characteristics of KASUMI derived by division property. In: Choi, D., Guilley, S. (eds.) WISA 2016. LNCS, vol. 10144, pp. 267–279. Springer, Cham (2017). https://doi.org/10.1007/978-3-319-56549-1_23
36. Sun, L., Wang, W., Liu, R., Wang, M.: MILP-aided bit-based division property for ARX-based block cipher. IACR Cryptology ePrint Archive, 2016:1101 (2016)
37. Sun, L., Wang, W., Wang, M.: MILP-aided bit-based division property for primitives with non-bit-permutation linear layers. IACR Cryptology ePrint Archive, 2016:811 (2016)
38. Sun, S., Gerault, D., Lafourcade, P., Yang, Q., Todo, Y., Qiao, K., Hu, L.: Analysis of AES, SKINNY, and others with constraint programming. IACR Trans. Symmetric Cryptol. **2017**(1), 281–306 (2017)
39. Sun, S., Hu, L., Wang, P., Qiao, K., Ma, X., Song, L.: Automatic security evaluation and (related-key) differential characteristic search: application to SIMON, PRESENT, LBlock, DES(L) and other bit-oriented block ciphers. In: Sarkar, P., Iwata, T. (eds.) ASIACRYPT 2014. LNCS, vol. 8873, pp. 158–178. Springer, Heidelberg (2014). https://doi.org/10.1007/978-3-662-45611-8_9
40. Todo, Y.: Integral cryptanalysis on full MISTY1. In: Gennaro, R., Robshaw, M. (eds.) CRYPTO 2015. LNCS, vol. 9215, pp. 413–432. Springer, Heidelberg (2015). https://doi.org/10.1007/978-3-662-47989-6_20
41. Todo, Y.: Structural evaluation by generalized integral property. In: Oswald, E., Fischlin, M. (eds.) EUROCRYPT 2015. LNCS, vol. 9056, pp. 287–314. Springer, Heidelberg (2015). https://doi.org/10.1007/978-3-662-46800-5_12
42. Todo, Y., Morii, M.: Bit-based division property and application to SIMON family. In: Peyrin, T. (ed.) FSE 2016. LNCS, vol. 9783, pp. 357–377. Springer, Heidelberg (2016). https://doi.org/10.1007/978-3-662-52993-5_18
43. Wen, L., Wang, M.: Integral zero-correlation distinguisher for ARX block cipher, with application to SHACAL-2. In: Susilo, W., Mu, Y. (eds.) ACISP 2014. LNCS, vol. 8544, pp. 454–461. Springer, Cham (2014). https://doi.org/10.1007/978-3-319-08344-5_32
44. Wheeler, D.J., Needham, R.M.: TEA, a tiny encryption algorithm. In: Preneel, B. (ed.) FSE 1994. LNCS, vol. 1008, pp. 363–366. Springer, Heidelberg (1995). https://doi.org/10.1007/3-540-60590-8_29
45. Xiang, Z., Zhang, W., Bao, Z., Lin, D.: Applying MILP method to searching integral distinguishers based on division property for 6 lightweight block ciphers. In: Cheon, J.H., Takagi, T. (eds.) ASIACRYPT 2016. LNCS, vol. 10031, pp. 648–678. Springer, Heidelberg (2016). https://doi.org/10.1007/978-3-662-53887-6_24
46. Zhang, H., Wu, W.: Structural evaluation for generalized feistel structures and applications to LBlock and TWINE. In: Biryukov, A., Goyal, V. (eds.) INDOCRYPT 2015. LNCS, vol. 9462, pp. 218–237. Springer, Cham (2015). https://doi.org/10.1007/978-3-319-26617-6_12

47. Zhang, W., Rijmen, V.: Division cryptanalysis of block ciphers with a binary diffusion layer. IACR Cryptology ePrint Archive, 2017:188 (2017)
48. Zheng, Y., Matsumoto, T., Imai, H.: On the construction of block ciphers provably secure and not relying on any unproved hypotheses. In: Brassard, G. (ed.) CRYPTO 1989. LNCS, vol. 435, pp. 461–480. Springer, New York (1990). https://doi.org/10.1007/0-387-34805-0_42

Collisions and Semi-Free-Start Collisions for Round-Reduced RIPEMD-160

Fukang Liu[1], Florian Mendel[2], and Gaoli Wang[1(✉)]

[1] Shanghai Key Laboratory of Trustworthy Computing, School of Computer Science and Software Engineering, East China Normal University, Shanghai, China
liufukangs@163.com, glwang@sei.ecnu.edu.cn
[2] Graz University of Technology, Graz, Austria
florian.mendel@iaik.tugraz.at

Abstract. In this paper, we propose an improved cryptanalysis of the double-branch hash function RIPEMD-160 standardized by ISO/IEC. Firstly, we show how to theoretically calculate the step differential probability of RIPEMD-160, which was stated as an open problem by Mendel *et al.* at ASIACRYPT 2013. Secondly, based on the method proposed by Mendel *et al.* to automatically find a differential path of RIPEMD-160, we construct a 30-step differential path where the left branch is sparse and the right branch is controlled as sparse as possible. To ensure the message modification techniques can be applied to RIPEMD-160, some extra bit conditions should be pre-deduced and well controlled. These extra bit conditions are used to ensure that the modular difference can be correctly propagated. This way, we can find a collision of 30-step RIPEMD-160 with complexity 2^{67}. This is the first collision attack on round-reduced RIPEMD-160. Moreover, by a different choice of the message words to merge two branches and adding some conditions to the starting point, the semi-free-start collision attack on the first 36-step RIPEMD-160 from ASIACRYPT 2013 can be improved. However, the previous way to pre-compute the equation $T^{\lll S_0} \boxplus C_0 = (T \boxplus C_1)^{\lll S_1}$ costs too much. To overcome this obstacle, we are inspired by Daum's *et al.* work on MD5 and describe a method to reduce the time complexity and memory complexity to pre-compute that equation. Combining all these techniques, the time complexity of the semi-free-start collision attack on the first 36-step RIPEMD-160 can be reduced by a factor of $2^{15.3}$ to $2^{55.1}$.

Keywords: RIPEMD-160 · Semi-free-start collision · Collision · Hash function · Compression function

1 Introduction

A cryptographic hash function is a function which takes arbitrary long messages as input and output a fixed-length hash value of size n bits. There are three

T. Takagi and T. Peyrin (Eds.): ASIACRYPT 2017, Part I, LNCS 10624, pp. 158–186, 2017.
https://doi.org/10.1007/978-3-319-70694-8_6

basic requirements for a hash function, which are preimage resistance, second-preimage resistance and collision resistance. Most standardized hash functions are based on the Merkle-Damgård paradigm [2,12] and iterate a compression function H with fixed-size input to compress arbitrarily long messages. Therefore, the compression function itself should satisfy equivalent security requirements so that the hash function can inherit from it. There are two attack models on the compression function. One is called free-start collision attack, the other is semi-free-start collision attack. The free-start collision attack is to find two different pairs of message and chaining value (CV, M), (CV', M') which satisfy $\mathrm{H}(CV, M) = \mathrm{H}(CV', M')$. The semi-free-start collision attack works in the same way apart from an additional condition that $CV = CV'$. The last decade has witnessed the fall of a series of hash functions such as MD4, MD5, SHA-0 and SHA-1 since many break-through results on hash functions cryptanalysis [15,20–23] were obtained. All of these hash functions belong to the MD-SHA family, whose design strategy is based on the utilization of additions, rotations, xor and boolean functions in an unbalanced Feistel network.

RIPEMD family can be considered as a subfamily of the MD-SHA-family since RIPEMD-0 [1] is the first representative and consists of two MD4-like functions computed in parallel with totally 48 steps. The security of RIPEMD-0 was first put into question by Dobbertin [4] and a practical collision attack on it was proposed by Wang *et al.* [20]. In order to reinforce the security of RIPEMD-0, Dobbertin, Bosselaers and Preneel [3] proposed two strengthened versions of RIPEMD-0 in 1996, which are RIPEMD-128 and RIPEMD-160 with 128/160 bits output and 64/80 steps, respectively. In order to make both computation branches more distinct from each other, not only different constants, but also different rotation values, message insertion schedules and boolean functions are used for RIPEMD-128 and RIPEMD-160 in their both branches.

For RIPEMD-128, there has been a series of analysis on it [5,8,16–18], threatening its security. As for RIPEMD-160, Mendel *et al.* [11] proposed an improved method to automatically find the differential path of RIPEMD-160 at ASIACRYPT 2013. With their method, they found a 48-step differential path and a 36-step differential path. Based on the two differential paths, Mendel *et al.* [11] mounted a semi-free-start collision attack on 42-step RIPEMD-160 and a semi-free-start collision attack on the first 36-step RIPEMD-160. Additionally, they also proposed an open problem to theoretically calculate the step differential probability. Besides, there are also some other analytical results on RIPEMD-160, such as a preimage attack [13] on 31-step RIPEMD-160, a distinguisher on up to 51 steps of the compression function [14], a practical semi-free-start collision attack on 36 steps of the compression function [9] (not starting from the first step), and a semi-free-start collision attack on 48-step RIPEMD-160 [19]. However, RIPEMD-160 is yet unbroken and is widely used in the implementations of security protocols as a ISO/IEC standard.

In 2005, Daum investigated the probability computation of T-functions (a function for which the i-th output bit depends only on the i first lower bits of all input words) in his PhD thesis [6]. More specifically, he proposed a method to calculate the probability that T satisfies the equation $(T \boxplus C_0)^{\lll S} = T^{\lll S} \boxplus C_1$

where C_0 and C_1 are constants. According to our analysis of the open problem to calculate the step differential probability of RIPEMD-160, we find that calculating such a probability is equivalent to calculating the probability that the modular difference of the internal states is correctly propagated and the bit conditions on the internal states hold. Although Daum's work can be used to calculate the probability that the modular difference is correctly propagated, it can't solve the open problem completely since the probability that one bit condition on the internal state holds is not 1/2 any more. However, by considering the calculation of the probability that T satisfies the equation $(T \boxplus C_0)^{\lll S} = T^{\lll S} \boxplus C_1$ from a different perspective, we can deduce some useful characteristics of T which can be used to calculate the probability that the bit conditions hold. In this way, we can solve the open problem completely.

This paper is organized as follows. In Sect. 2, we briefly describe the algorithm of RIPEMD-160. In Sect. 3, we describe our method to calculate the step differential probability. In Sect. 4, we describe our improved way to pre-compute the equation $T^{\lll S_0} \boxplus C_0 = (T \boxplus C_1)^{\lll S_1}$. In Sect. 5, we describe the collision attack on the first 30-step RIPEMD-160. In Sect. 6, we describe the improved semi-free-start collision attack on the first 36-step RIPEMD-160. Finally, we conclude the paper in Sect. 7.

Our Contributions

1. Our method to theoretically calculate the step differential probability consists of two steps. At first, we consider the probability that the modular difference of the internal states holds, which will help obtain some characteristics of Q_i (Q_i is referred to Sect. 2.2). Then, for each characteristics of Q_i, the probability that the bit conditions on the internal states hold under the condition that this characteristic of Q_i holds can be calculated. In this way, the theoretical calculation of the step differential probability of RIPEMD-160 becomes feasible.
2. We deduce a useful property from the PhD thesis of Daum [6]. Based on it, we can convert solving the equation $T^{\lll S_0} \boxplus C_0 = (T \boxplus C_1)^{\lll S_1}$ into solving the equation $T^{\lll S_0} \boxplus C_2 = T^{\lll S_1}$. By analyzing the expectation of the number of the solutions to the equation if given many pairs of $(C_0,\ C_1)$, we can claim that our new method to obtain the solutions at the phase of merging only costs 4 times of checking the equation $T^{\lll S_0} \boxplus C_0 = (T \boxplus C_1)^{\lll S_1}$ on average, thus having a negligible influence on the efficiency compared with the previous method [5,11]. Moreover, both the time complexity and memory complexity of our new method to pre-compute the equation is 2^{32}, which is much smaller than the strategy by constructing a table of size 2^{64} to store the solutions.
3. By using the technique described in [11] to automatically find a differential path for RIPEMD-160, we can construct a 30-step differential path where the left branch is sparse and the right branch is controlled as sparse as possible. For the left branch, we leave it holding probabilistically. For the right branch, we apply the message modification techniques [20] to it. However, according to our analysis of the open problem to theoretically calculate the step

differential probability of RIPEMD-160, the differential path of RIPEMD-160 holds only when both the bit conditions and the modular difference of the internal states hold. That's different from MD4 since the differential path of MD4 holds only when the bit conditions on the internal states hold. Since the message modification can only be used to ensure the bit conditions hold, the difficulty is how to have the modular difference of the internal states hold when applying it to RIPEMD-160. Fortunately, we discover that we can add some extra bit conditions on the internal states to have the modular difference hold. Therefore, before applying the message modification, we have to pre-deduce these extra bit conditions on the internal states by considering the characteristics of Q_i. After obtaining the newly added extra bit conditions, by adjusting the message modification techniques so that it can be applied to RIPEMD-160, we can mount a 30-step collision attack on RIPEMD-160 with probability 2^{-67}.

4. Based on the 36-step differential path, by a different choice of message words to merge both branches, we can improve the time complexity of the merging phase. Moreover, based on the characteristics of Q_{15}, we can add some extra bit conditions on Y_{11} at the phase of finding a starting point to further improve our attack. The improved semi-free-start collision attack on the first 36-step RIPEMD-160 is $2^{55.1}$, which is much smaller than the previous best known result (Table 1).

Table 1. Summary of preimage and collision attack on RIPEMD-160.

Target	Attack type	Steps	Complexity	Ref
Comp. function	Preimage	31	2^{148}	[13]
Hash function	Preimage	31	2^{155}	[13]
Comp. function	Semi-free-start collision	36[a]	low	[9]
Comp. function	Semi-free-start collision	36	$2^{70.4}$	[11]
Comp. function	Semi-free-start collision	36	$2^{55.1}$	New
Comp. function	Semi-free-start collision	42[a]	$2^{75.5}$	[11]
Comp. function	Semi-free-start collision	48[a]	$2^{76.4}$	[19]
Hash function	Collision	30	2^{67}	New

[a]An attack starts at an intermediate step.

2 Description of RIPEMD-160

RIPEMD-160 is a 160-bit hash function that uses the Merkle-Damgård construction as domain extension algorithm: the hash function is built by iterating a 160-bit compression function H which takes as input a 512-bit message block M_i and a 160-bit chaining variables CV_i:

$$CV_{i+1} = H(CV_i, M_i)$$

where a message M to hash is padded beforehand to a multiple of 512 bits and the first chaining variable is set to the predetermined initial value IV, that is $CV_0 = IV$. We refer to [3] for a detailed description of RIPEMD-160.

2.1 Notations

For a better understanding of this paper, we introduce the following notations.

1. $\lll$, $\ggg$, $\oplus$, $\vee$, $\wedge$ and $\neg$ represent respectively the logic operation: *rotate left*, *rotate right*, *exclusive or*, *or*, *and*, *negate*.
2. $\boxplus$ and $\boxminus$ represent respectively the modular addition and modular substraction on 32 bits.
3. $M = (m_0, m_1, \ldots, m_{15})$ and $M' = (m'_0, m'_1, \ldots, m'_{15})$ represent two 512-bit message blocks.
4. $\Delta m_i = m'_i - m_i$ represents the modular difference between two message words m_i and m'_i.
5. K^l_j and K^r_j represent the constant used at the left and right branch for round j.
6. Φ^l_j and Φ^r_j represent respectively the 32-bit boolean function at the left and right branch for round j.
7. X_i, Y_i represent respectively the 32-bit internal state of the left and right branch updated during step i for compressing M.
8. X'_i, Y'_i represent respectively the 32-bit internal state of the left and right branch updated during step i for compressing M'.
9. $X_{i,j}$, $Y_{i,j}$ represent respectively the j-th bit of X_i and Y_i, where the least significant bit is the 0th bit and the most significant bit is the 31st bit.
10. Q_i represents the 32-bit temporary state of the right branch updated during step i for compressing M.
11. s^l_i and s^r_i represent respectively the rotation constant used at the left and right branch during step i.
12. $\pi_1(i)$ and $\pi_2(i)$ represent the index of the message word used at the left and right branch during step i.
13. $[Z]_i$ represents the i-th bit of the 32-bit Z.
14. $[Z]_{j\sim i}$ $(0 \leq i < j \leq 31)$ represents the i-th bit to the j-th bit of the 32-bit word Z.
15. $x_i[j]$, $x_i[-j]$ (x can be X and Y) is the resulting value by only changing the j-th bit of x_i. $x_i[j]$ is obtained by changing the j-th bit of x_i from 0 to 1. $x_i[-j]$ is obtained by changing the j-th bit of x_i from 1 to 0.
16. $x_i[\pm j_1, \pm j_2, \ldots, \pm j_l]$ (x can be X and Y) is the value by changing the j_1-th, j_2-th, j_l-th bit of x_i. The "+" sign means the bit is changed from 0 to 1, and the "−" sign means the bit is changed from 1 to 0.
17. P(A) is the probability of the event A.

2.2 RIPEMD-160 Compression Function

The RIPEMD-160 compression function is a wider version of RIPEMD-128, which is based on MD4, but with the particularity that it consists of two different and almost independent parallel instances of it. We differentiate the two computation branches by left and right branch. The compression function consists of 80 steps divided into 5 rounds of 16 steps each in both branches.

Table 2. Boolean functions and round constants in RIPEMD-160

Round j	ϕ_j^l	ϕ_j^r	K_j^l	K_j^r	Function	Expression
0	XOR	ONX	0x00000000	0x50a28be6	XOR(x, y, z)	$x \oplus y \oplus z$
1	IFX	IFZ	0x5a827999	0x5c4dd124	IFX(x, y, z)	$(x \wedge y) \oplus (\neg x \wedge z)$
2	ONZ	ONZ	0x6ed9eba1	0x6d703ef3	IFZ(x, y, z)	$(x \wedge z) \oplus (y \wedge \neg z)$
3	IFZ	IFX	0x8f1bbcdc	0x7a6d76e9	ONX(x, y, z)	$x \oplus (y \vee \neg z)$
4	ONX	XOR	0xa953fd4e	0x00000000	ONZ(x, y, z)	$(x \vee \neg y) \oplus z$

Initialization. The 160-bit input chaining variable CV_i is divided into five 32-bit words h_i (i=0, 1, 2, 3, 4), initializing the left and right branch 160-bit internal state in the following way:

$$X_{-4} = h_0^{\ggg 10}, \quad X_{-3} = h_4^{\ggg 10}, \quad X_{-2} = h_3^{\ggg 10}, \quad X_{-1} = h_2, \quad X_0 = h_1.$$
$$Y_{-4} = h_0^{\ggg 10}, \quad Y_{-3} = h_4^{\ggg 10}, \quad Y_{-2} = h_3^{\ggg 10}, \quad Y_{-1} = h_2, \quad Y_0 = h_1.$$

Particularly, CV_0 corresponds to the following five 32-bit words:

$$X_{-4} = Y_{-4} = \texttt{0xc059d148}, X_{-3} = Y_{-3} = \texttt{0x7c30f4b8}, X_{-2} = Y_{-2} = \texttt{0x1d840c95},$$
$$X_{-1} = Y_{-1} = \texttt{0x98badcfe}, X_0 = Y_0 = \texttt{0xefcdab89}.$$

The Message Expansion. The 512-bit input message block is divided into 16 message words m_i of size 32 bits. Each message word m_i will be used once in every round in a permuted order π for both branches.

The Step Function. At round j, the internal state is updated in the following way.

$$X_i = X_{i-4}^{\lll 10} \boxplus (X_{i-5}^{\lll 10} \boxplus \Phi_j^l(X_{i-1}, X_{i-2}, X_{i-3}^{\lll 10}) \boxplus m_{\pi_1(i)} \boxplus K_j^l)^{\lll s_i^l},$$
$$Y_i = Y_{i-4}^{\lll 10} \boxplus (Y_{i-5}^{\lll 10} \boxplus \Phi_j^r(Y_{i-1}, Y_{i-2}, Y_{i-3}^{\lll 10}) \boxplus m_{\pi_2(i)} \boxplus K_j^r)^{\lll s_i^r},$$
$$Q_i = Y_{i-5}^{\lll 10} \boxplus \Phi_j^r(Y_{i-1}, Y_{i-2}, Y_{i-3}^{\lll 10}) \boxplus m_{\pi_2(i)} \boxplus K_j^r,$$

where i = (1, 2, 3, ..., 80) and j = (0, 1, 2, 3, 4). The details of the boolean functions and round constants for RIPEMD-160 are displayed in Table 2. As for other parameters, you can refer to [3].

The Finalization. A finalization and a feed-forward is applied when all 80 steps have been computed in both branches. The five 32-bit words $h_i^{'}$ composing the output chaining variable are computed in the following way.

$$h_0^{'} = h_1 \boxplus X_{79} \boxplus Y_{78})^{\lll 10},$$
$$h_1^{'} = h_2 \boxplus X_{78}^{\lll 10} \boxplus Y_{77}^{\lll 10},$$
$$h_2^{'} = h_3 \boxplus X_{77}^{\lll 10} \boxplus Y_{76}^{\lll 10},$$
$$h_3^{'} = h_4 \boxplus X_{76}^{\lll 10} \boxplus Y_{80},$$
$$h_4^{'} = h_0 \boxplus X_{80} \boxplus Y_{79}.$$

3 Calculate the Step Differential Probability

In [11], Mendel *et al.* pointed out that it is not as easy to calculate the differential probability for each step of a given differential path of RIPEMD-160 as that of RIPEMD-128. The main reason is that the step function in RIPEMD-160 is no longer a T-function. Therefore, the accurate calculation of the differential probability becomes very hard. However, we can divide the calculation of the step differential probability into two steps. Define as μ the event that all bit conditions on the internal state hold, as ν the event that the modular difference of the internal state holds. Although Daum has proposed a method [6] to calculate $P(\nu)$, we will use a different method to calculate it, since our goal is not only to calculate $P(\nu)$ but also to obtain some useful characteristics of Q_i. Then, we can leverage the deduced characteristics and the bit conditions on the internal states to calculate $P(\mu\nu)$. In this way, the step differential probability $P(\mu\nu)$ can be obtained. We use the step function of the right branch as an example and give its description below. We will show how to deduce the useful characteristics of Q_i and calculate $P(\mu\nu)$.

3.1 Description of the Open Problem

Since the step function of RIPEMD-160 at both branches has the same form, we take the right branch as an example to describe the open problem.

$$Y_i = Y_{i-4}^{\lll 10} \boxplus (Y_{i-5}^{\lll 10} \boxplus \Phi_j^l(Y_{i-1}, Y_{i-2}, Y_{i-3}^{\lll 10}) \boxplus m_{\pi_2(i)} \boxplus K_j^r)^{\lll s_i^r}.$$

To ensure the given differential path holds, we need to impose conditions on some bits of Y_i and control the modular difference of Y_i. The open problem is how to calculate the probability that both the bit conditions on Y_i and the modular difference of Y_i are satisfied under the condition that all conditions on $Y_{i-1}, Y_{i-2}, Y_{i-3}, Y_{i-4}, Y_{i-5}$ are satisfied. For example, according to the differential path displayed in Table 16, we know that:

$$\begin{aligned}
&Y'_{15} = Y_{15}[-5,-20,-26],\ Y'_{14} = Y_{14}[5,11,22],\ Y'_{13} = Y_{13}[-9,-24,26,-30],\\
&Y'_{12} = Y_{12}[0,-15,21],\ Y'_{11} = Y_{11}[1,10,12,15,-16,24,26,-28],\\
&Y'_{10} = Y_{10}[-3,21,22,23,24,25,26,-28],\ \Delta m_3 = 0.
\end{aligned}$$

Firstly, we use Y_{10}, Y_{11}, Y_{12}, Y_{13}, Y_{14}, m_3 to calculate Y_{15}. Then, we use $Y'_{10}, Y'_{11}, Y'_{12}, Y'_{13}, Y'_{14}, m'_3$ to calculate Y'_{15}. Then, the differential probability for step 15 is equal to the probability that $Y'_{15} = Y_{15} \boxminus 2^5 \boxminus 2^{20} \boxminus 2^{26}$ and that all bit conditions on Y_{15} are satisfied.

3.2 The Probability of $(T \boxplus C_0)^{\lll S} = T^{\lll S} \boxplus C_1$

Given two constants C_0 and C_1, Daum has described a method [6] to calculate the probability that T satisfies $(T \boxplus C_0)^{\lll S} = T^{\lll S} \boxplus C_1$ $(1 \leq S \leq 31)$. However, we consider the problem from a different perspective by considering the characteristics of T which satisfies such an equation. In this way, we can not only calculate the probability of this equation, but also can obtain the characteristics of T for further use to theoretically calculate the step differential probability.

Let $R_0||R_1 = T \boxplus C_0$, where R_0 is an S-bit variable representing the higher S bits of $T \boxplus C_0$ and R_1 is a (32-S)-bit variable representing the lower (32-S) bits of $T \boxplus C_0$. Let $R_1'||R_0' = T^{\lll S} \boxplus C_1$, where R_1' is a (32-S)-bit variable representing the higher (32-S) bits of $T^{\lll S} \boxplus C_1$ and R_0' is an S-bit variable representing the lower S bits of $T^{\lll S} \boxplus C_1$. Then, the probability of $(T \boxplus C_0)^{\lll S} = T^{\lll S} \boxplus C_1$ $(1 \leq S \leq 31)$ is equal to $\mathrm{P}(R_0 = R_0'$ and $R_1 = R_1')$. Since

$$
\begin{aligned}
R_0 &\equiv [T]_{31\sim(32-S)} + [C_0]_{31\sim(32-S)} + carry_0 \; mod \; (2^S),\\
R_0' &\equiv [T]_{31\sim(32-S)} + [C_1]_{(S-1)\sim 0} \; mod \; (2^S),\\
R_1 &\equiv [T]_{(31-S)\sim 0} + [C_0]_{(31-S)\sim 0} \; mod \; (2^{32-S}),\\
R_1' &\equiv [T]_{(31-S)\sim 0} + [C_1]_{31\sim S} + carry_1 \; mod \; (2^{32-S}),
\end{aligned}
$$

where $carry_0$ represents the carry from the (31-S)-th bit to the (32-S)-th when calculating $T \boxplus C_0$, and $carry_1$ represents the carry from the (S-1)-th bit to the S-th bit when calculating $T^{\lll S} \boxplus C_1$. For simplicity, we define as κ the event that $carry_0 = 0$ and as ω the event that $carry_1 = 0$. Therefore,

$$
\begin{aligned}
&P(R_0 = R_0') = P(\kappa \; and \; [C_0]_{31\sim(32-S)} =\\
&[C_1]_{(S-1)\sim 0}) + P(\overline{\kappa} \; and \; [C_0]_{31\sim(32-S)} + 1 \equiv [C_1]_{(S-1)\sim 0} \; mod \; (2^S)),\\
&P(R_1 = R_1') = P(\omega \; and \; [C_0]_{(31-S)\sim 0} =\\
&[C_1]_{31\sim S}) + P(\overline{\omega} \; and \; [C_0]_{(31-S)\sim 0} \equiv [C_1]_{31\sim S} + 1 \; mod \; (2^{32-S})).
\end{aligned}
$$

We denote the positions of the bits of $[C_0]_{(31-S)\sim 0}$ equal to 1 by $k_0, k_1, \ldots, k_n$ and denote the positions of the bits of $[C_1]_{(S-1)\sim 0}$ equal to 1 by $r_0, r_1, \ldots, r_m$. Then, the value of $P(\kappa)$ and $P(\omega)$ can be directly deduced as below:

1. If $[C_0]_{(31-S)\sim 0} = 0$, then $P(\kappa) = 1$. Otherwise, $P(\kappa) = 1 - \sum_{i=0}^{n} 2^{-(32-S-k_i)}$.
2. If $[C_1]_{(S-1)\sim 0} = 0$, then $P(\omega) = 1$. Otherwise, $P(\omega) = 1 - \sum_{i=0}^{m} 2^{-(S-r_i)}$.

Thus, we can compute $P(R_0 = R_0' \; and \; R_1 = R_1')$ in this way:

1. If $[C_0]_{(31-S)\sim 0} = [C_1]_{31\sim S}$ and $[C_0]_{31\sim(32-S)} = [C_1]_{(S-1)\sim 0}$, then $P(R_0 = R_0' \; and \; R_1 = R_1') = P(\kappa) \times P(\omega)$.
2. If $[C_0]_{(31-S)\sim 0} = [C_1]_{31\sim S}$ and $[C_0]_{31\sim(32-S)} + 1 \equiv [C_1]_{(S-1)\sim 0} \; mod \; (2^S)$, then $P(R_0 = R_0' \; and \; R_1 = R_1') = P(\overline{\kappa}) \times P(\omega)$.
3. If $[C_0]_{(31-S)\sim 0} \equiv [C_1]_{31\sim S} + 1 \; mod \; (2^{32-S})$ and $[C_0]_{31\sim(32-S)} = [C_1]_{(S-1)\sim 0}$, then $P(R_0 = R_0' \; and \; R_1 = R_1') = P(\kappa) \times P(\overline{\omega})$.
4. If $[C_0]_{(31-S)\sim 0} \equiv [C_1]_{31\sim S} + 1 \; mod \; (2^{32-S}))$ and $[C_0]_{31\sim(32-S)} + 1 \equiv [C_1]_{(S-1)\sim 0} \; mod \; (2^S))$, then $P(R_0 = R_0' \; and \; R_1 = R_1') = P(\overline{\kappa}) \times P(\overline{\omega})$.
5. If C_0 and C_1 doesn't belong to any of the above four cases, then $P(R_0 = R_0' \; and \; R_1 = R_1') = 0$.

According to the above method to calculate $P(R_0 = R'_0 \text{ and } R_1 = R'_1)$, the following property can be directly deduced. (In fact, we can also deduce it by using the Corollary 4.14 in [6].)

Property 1. Given random constants C_0 and C_1 of 32 bits each, there exists a T of 32 bits which satisfies $(T \boxplus C_0)^{\lll S} = T^{\lll S} \boxplus C_1$ if and only if (C_0, C_1) satisfies one of the following equations:

1. $C_1 = (C_0 \boxminus 1)^{\lll S}$, and $[C_1]_{(S-1)\sim 0} \neq 0$.
2. $C_1 = (C_0 \boxplus 2^{32-S})^{\lll S}$, and $[C_0]_{(31-S)\sim 0} \neq 0$.
3. $C_1 = (C_0 \boxplus 2^{32-S} \boxminus 1)^{\lll S}$, and $[C_1]_{(S-1)\sim 0} \neq 0, [C_0]_{(31-S)\sim 0} \neq 0$.
4. $C_1 = C_0^{\lll S}$.

Example. In the following, we give an example how to calculate the probability of $(T\boxplus \texttt{0x80bfd9ff})^{\lll 12} = T^{\lll 12}\boxplus \texttt{0xfd9ff80c}$. To have a better understanding of our method to calculate the probability, we explain it by Table 3.

Table 3. Calculation of the probability

T	31	30	29	28	27	26	25	24	23	22	21	20	19	18	17	16	15	14	13	12	11	10	9	8	7	6	5	4	3	2	1	0
T																																
C_0	1	0	0	0	0	0	0	0	1	0	1	1	1	1	1	1	1	1	0	1	1	0	0	1	1	1	1	1	1	1	1	1
	R_0												R_1																			
$T^{\lll 12}$	19	18	17	16	15	14	13	12	11	10	9	8	7	6	5	4	3	2	1	0	31	30	29	28	27	26	25	24	23	22	21	20
$T^{\lll 12}$																																
C_1	1	1	1	1	1	1	0	1	1	0	0	1	1	1	1	1	1	1	1	1	1	0	0	0	0	0	0	0	1	1	0	0
	R'_1																				R'_0											

According to Table 3, we can find the following relationship between C_0 and C_1:

$$[C_0]_{19\sim 0} = [C_1]_{31\sim 12},\ [C_0]_{31\sim 20} + 1 \equiv [C_1]_{11\sim 0} \bmod (2^{12}).$$

Therefore, we can get $P((T\boxplus \texttt{0x80bfd9ff})^{\lll 12} = T^{\lll 12}\boxplus \texttt{0xfd9ff80c}) = P(\overline{\kappa}) \times P(\omega)$. By considering the characteristics of T, $P(\overline{\kappa})$ and $P(\omega)$ can be calculated as below:

$$\begin{aligned}
P(\overline{\kappa}) &= P([T]_{19} = 1) + P([T]_{19\sim 18} = 01) + P([T]_{19\sim 17} = 001) \\
&+ P([T]_{19\sim 16} = 0001) + P([T]_{19\sim 15} = 00001) + P([T]_{19\sim 14} = 000001) \\
&+ P([T]_{19\sim 12} = 00000011) + P([T]_{19\sim 11} = 000000101) \\
&+ P([T]_{19\sim 8} = 000000100111) + P([T]_{19\sim 7} = 0000001001101) \\
&+ P([T]_{19\sim 6} = 00000010011001) + P([T]_{19\sim 5} = 000000100110001) \\
&+ P([T]_{19\sim 4} = 0000001001100001) + P([T]_{19\sim 3} = 00000010011000001) \\
&+ P([T]_{19\sim 2} = 000000100110000001) + P([T]_{19\sim 1} = 0000001001100000001) \\
&+ P([T]_{19\sim 0} = 00000010011000000001) \\
&= \Sigma_{i=1}^{6} 2^{-i} + 2^{-8} + 2^{-9} + \Sigma_{i=12}^{20} 2^{-i}.
\end{aligned}$$

$$\begin{aligned}
P(\omega) &= 1 - P([T]_{31} = 1) - P([T]_{31-23} = 011111111) - P([T]_{31-22} = 0111111101) \\
&= 1 - (2^{-1} + 2^{-9} + 2^{-10}).
\end{aligned}$$

Thus, $P((T \boxplus \texttt{0x80bfd9ff})^{\lll 12} = T^{\lll 12} \boxplus \texttt{0xfd9ff80c}) \approx 2^{-1}$. In this example, we call $[T]_{19} = 1$ one possible characteristic of T, and we call $[T]_{31} = 1$ one impossible characteristic of T. Totally, there are 17 possible characteristics of T and 3 impossible characteristics of T.

3.3 Calculating the Step Differential Probability

We use the step function of the right branch to explain our method to calculate the step differential probability. Let $\Delta = Y_i' \boxminus Y_i$, $\Delta_{i-5} = Y_{i-5}'^{\lll 10} \boxminus Y_{i-5}^{\lll 10}$, $\Delta_{i-4} = Y_{i-4}'^{\lll 10} \boxminus Y_{i-4}^{\lll 10}$, $\Delta F = \Phi_j^l(Y_{i-1}', Y_{i-2}', Y_{i-3}'^{\lll 10}) \boxminus \Phi_j^l(Y_{i-1}, Y_{i-2}, Y_{i-3}^{\lll 10})$, then $P(\nu) = P(\Delta = \Delta_{i-4} \boxplus (\Delta_{i-5} \boxplus \Delta F \boxplus \Delta m_{\pi_2(i)} \boxplus Q_i)^{\lll s_i^r} \boxminus Q_i^{\lll s_i^r})$. Given the differential path and the bit conditions to control the differential propagation, Δ, Δ_{i-5}, Δ_{i-4}, ΔF and $\Delta m_{\pi_2(i)}$ are all fixed. Let $C_0 = \Delta_{i-5} \boxplus \Delta F \boxplus \Delta m_{\pi_2(i)}$ and $C_1 = \Delta \boxminus \Delta_{i-4}$, we can obtain that $P(\nu) = P((Q_i \boxplus C_0)^{\lll s_i^r} = Q_i^{\lll s_i^r} \boxplus C_1)$, which can be quickly calculated as described in Sect. 3.2.

Observe that when calculating Y_i, there are conditions on some bits of Y_{i-4} and Y_i, i.e., some bits of Y_{i-4} and Y_i are fixed. In addition, in order to make the modular difference of Y_i satisfied, there are some constraints on Q_i. By analyzing the constraints carefully, the characteristics of Q_i can be discovered, which will make the theoretical calculation of $P(\mu\nu)$ feasible. By the following example, we will introduce how to leverage the characteristics of Q_i and the bit conditions on Y_{i-4} and Y_i to calculate $P(\mu\nu)$. The general case can be handled in the same way.

Example. For the given differential path in Table 16, we know that

$$\begin{aligned}
\Delta F &= ONX(Y_{14}', Y_{13}', (Y_{12}')^{\lll 10}) \boxminus ONX(Y_{14}, Y_{13}, Y_{12}^{\lll 10}) = \texttt{0xbffa20},\\
Y_{11}'^{\lll 10} &= Y_{11}^{\lll 10}[-26, 25, 22, 20, 11, -6, 4, 2],\ \Delta_{11} = Y_{11}'^{\lll 10} \boxminus Y_{11}^{\lll 10} = \texttt{0xfe5007d4},\\
Y_{10}'^{\lll 10} &= Y_{10}^{\lll 10}[31, -13, -6, 4, 3, 2, 1, 0],\ \Delta_{10} = Y_{10}'^{\lll 10} \boxminus Y_{10}^{\lll 10} = \texttt{0x7fffdfdf},\\
\Delta &= Y_{15}' - Y_{15} = \texttt{0xfbefffe0},\ \Delta m_3 = 0.
\end{aligned}$$

Therefore, Q_{15} has to satisfy the equation $(Q_{15} \boxplus \texttt{0x80bfd9ff})^{\lll 12} = Q_{15}^{\lll 12} \boxplus \texttt{0xfd9ff80c}$. According to the example in Sect. 3.2, the characteristics of Q_{15} which satisfies such an equation can be deduced and we display it in Table 4.

Let $a = Q_{15}^{\lll 12}$, $b = Y_{11}^{\lll 10}$, $d = Y_{15}$, since $Y_{15} = Y_{11}^{\lll 10} \boxplus Q_{15}^{\lll 12}$, we can obtain that $d = a \boxplus b$. In addition, we denote by c_i the carry from the(i-1)-th bit to the i-th bit when calculating $a \boxplus b$. Thus,

$$[d]_i = [a]_i \oplus [b]_i \oplus c_i,\ (c_0 = 0,\ 0 \le i \le 31).$$

Define as A_i the event that $[a]_i = 0$, as B_i the event that $[b]_i = 0$, as λ_i the event that $c_i = 0$, as D_i the event that $Y_{15,i} = 0$, as ν_{15} the event that $Y_{15}' - Y_{15} = \texttt{0xfbefffe0}$, as η_{15} the event that all the 7 conditions on Y_{15} hold. For a better understanding of our method, we display the calculation of Y_{15} in Table 5. Then, $P(\eta_{15}\nu_{15})$ can be calculated as follows:

Table 4. The characteristics of Q_{15}

i	χ_i (Characteristic)	Type	i	χ_i (Characteristic)	Type
1	$[Q_{15}]_{31} = 1$	Impossible	11	$[Q_{15}]_{19\sim 11} = 000000101$	Possible
2	$[Q_{15}]_{31\sim 23} = 011111111$	Impossible	12	$[Q_{15}]_{19\sim 8} = 000000100111$	Possible
3	$[Q_{15}]_{31\sim 22} = 0111111101$	Impossible	13	$[Q_{15}]_{19\sim 7} = 0000001001101$	Possible
4	$[Q_{15}]_{19} = 1$	Possible	14	$[Q_{15}]_{19\sim 6} = 00000010011001$	Possible
5	$[Q_{15}]_{19\sim 18} = 01$	Possible	15	$[Q_{15}]_{19\sim 5} = 000000100110001$	Possible
6	$[Q_{15}]_{19\sim 17} = 001$	Possible	16	$[Q_{15}]_{19\sim 4} = 0000001001100001$	Possible
7	$[Q_{15}]_{19\sim 16} = 0001$	Possible	17	$[Q_{15}]_{19\sim 3} = 00000010011000001$	Possible
8	$[Q_{15}]_{19\sim 15} = 00001$	Possible	18	$[Q_{15}]_{19\sim 2} = 000000100110000001$	Possible
9	$[Q_{15}]_{19\sim 14} = 000001$	Possible	19	$[Q_{15}]_{19\sim 1} = 0000001001100000001$	Possible
10	$[Q_{15}]_{19\sim 12} = 00000011$	Possible	20	$[Q_{15}]_{19\sim 0} = 00000010011000000001$	Possible

Table 5. Calculation of $Y_{15} = Y_{11}^{\lll 10} \boxplus Q_{15}^{\lll 12}$

$Q_{15}^{\lll 12}$	19	18	17	16	15	14	13	12	11	10	9	8	7	6	5	4
$Y_{11}^{\lll 10}$	21	20	19	18	17	16	15	14	13	12	11	10	9	8	7	6
Y_{15}	31	30	29	28	27	26	25	24	23	22	21	20	19	18	17	16
$Q_{15}^{\lll 12}(a)$																
$Y_{11}^{\lll 10}(b)$	-	1	-	-	1	1	0	-	-	0	-	0	-	-	1	0
$Y_{15}(d)$	-	-	-	-	-	1	-	-	-	1	-	1	0	-	-	-

$Q_{15}^{\lll 12}$	3	2	1	0	31	30	29	28	27	26	25	24	23	22	21	20
$Y_{11}^{\lll 10}$	5	4	3	2	1	0	31	30	29	28	27	26	25	24	23	22
Y_{15}	15	14	13	12	11	10	9	8	7	6	5	4	3	2	1	0
$Q_{15}^{\lll 12}(a)$																
$Y_{11}^{\lll 10}(b)$	0	0	0	1	0	-	0	-	-	1	-	0	1	0	-	-
$Y_{15}(d)$	-	-	-	-	1	-	-	-	-	-	1	-	-	1	-	-

$$
\begin{aligned}
P(\eta_{15}\nu_{15}) &= P(\lambda_{12}\eta_{15}\nu_{15}) + P(\overline{\lambda_{12}}\eta_{15}\nu_{15}), \\
P(\lambda_{12}\eta_{15}\nu_{15}) &= \Sigma_{i=4}^{20} P(\overline{D_{26}D_{22}D_{20}}D_{19}\chi_i \mid \lambda_{12}) \times \{P(\overline{D_{11}D_5D_2}\lambda_{12}) \\
&\quad - \Sigma_{i=1}^{3}[P(\overline{D_{11}D_5D_2}\lambda_{12} \mid \chi_i) \times P(\chi_i)]\}, \\
P(\overline{\lambda_{12}}\eta_{15}\nu_{15}) &= \Sigma_{i=4}^{20} P(\overline{D_{26}D_{22}D_{20}}D_{19}\chi_i \mid \overline{\lambda_{12}}) \times \{P(\overline{D_{11}D_5D_2\lambda_{12}}) \\
&\quad - \Sigma_{i=1}^{3}[P(\overline{D_{11}D_5D_2\lambda_{12}} \mid \chi_i) \times P(\chi_i)]\}.
\end{aligned}
$$

However, according to the characteristics of Q_{15}, we know that $[Q_{15}]_{31}$ is always 0 if $Y'_{15} \boxminus Y_{15}$ = `0xfbefffe0`, which implies that $P(\overline{\lambda_{12}} \mid \nu_{15}) = 0$ and $P(\lambda_{12} \mid \nu_{15}) = 1$. Therefore, calculating $P(\lambda_{12}\eta_{15}\nu_{15})$ is enough. Take the calculation of $P(\overline{D_{11}D_5D_2}\lambda_{12}) - \Sigma_{i=1}^{3}[P(\overline{D_{11}D_5D_2}\lambda_{12} \mid \chi_i) \times P(\chi_i)]$ as an example. Firstly, we calculate $P(\overline{D_{11}D_5D_2}\lambda_{12} \mid \chi_3)$. As Table 6 shows, the calculation is detailed as below.

Table 6. Calculation of $P(\overline{D_{11}D_5D_2}\lambda_{12} \mid \chi_3)$

$Q_{15}^{\lll 12}$	31	30	29	28	27	26	25	24	23	22	21	20
$Y_{11}^{\lll 10}$	1	0	31	30	29	28	27	26	25	24	23	22
Y_{15}	11	10	9	8	7	6	5	4	3	2	1	0
$Q_{15}^{\lll 12}(a)$	0	1	1	1	1	1	1	1	0	1	-	-
$Y_{11}^{\lll 10}(b)$	0	-	0	-	-	1	-	0	1	0	-	-
$Y_{15}(d)$	1	-	-	-	-	-	1	-	-	1	-	-

$P(\overline{D_{11}D_5D_2}\lambda_{12} \mid \chi_3) = P(\overline{D_{11}D_5D_2} \mid \chi_3)$.

$P(\overline{D_{11}D_5D_2} \mid \chi_3) = P(\overline{\lambda_{11}D_5D_2} \mid \chi_3) = P(\overline{B_{10}}) \times P(\overline{D_5D_2} \mid \chi_3) + P(B_{10}) \times P(\overline{\lambda_{10}D_5D_2} \mid \chi_3)$

$= \frac{1}{2} \times P(\overline{D_5D_2} \mid \chi_3) + \frac{1}{2} \times P(\overline{\lambda_{10}D_5D_2} \mid \chi_3)$.

$P(\overline{D_5D_2} \mid \chi_3) = P(B_5) \times P(\lambda_5\overline{D_2} \mid \chi_3) + P(\overline{B_5}) \times P(\overline{\lambda_5D_2} \mid \chi_3)$

$= \frac{1}{2} \times P(\lambda_5\overline{D_2} \mid \chi_3) + \frac{1}{2} \times P(\overline{\lambda_5D_2} \mid \chi_3) = \frac{1}{2} \times P(\overline{D_2} \mid \chi_3)$.

$P(\overline{D_2} \mid \chi_3) = P(\lambda_2)$.

$P(\lambda_2) = P(A_1B_1) + [P(\overline{A_1}B_1) + P(A_1\overline{B_1})] \times P(\lambda_1) = \frac{1}{4} + \frac{1}{2} \times P(\lambda_1)$.

$P(\lambda_1) = P(A_1B_1) + P(\overline{A_1}B_1) + P(A_1\overline{B_1}) = \frac{3}{4}$.

$P(\overline{\lambda_{10}D_5D_2} \mid \chi_3) = P(\overline{\lambda_9D_5D_2} \mid \chi_3) = P(\overline{B_8}) \times P(\overline{D_5D_2} \mid \chi_3) + P(B_8) \times P(\overline{\lambda_8D_5D_2} \mid \chi_3)$

$= \frac{1}{2} \times P(\overline{D_5D_2} \mid \chi_3) + \frac{1}{2} \times P(\overline{\lambda_8D_5D_2} \mid \chi_3)$.

$P(\overline{\lambda_8D_5D_2} \mid \chi_3) = P(\overline{B_7}) \times P(\overline{D_5D_2} \mid \chi_3) + P(B_7) \times P(\overline{\lambda_7D_5D_2} \mid \chi_3)$

$= \frac{1}{2} \times P(\overline{D_5D_2} \mid \chi_3) + \frac{1}{2} \times P(\overline{\lambda_7D_5D_2} \mid \chi_3)$.

$P(\overline{\lambda_7D_5D_2} \mid \chi_3) = P(\overline{D_5D_2} \mid \chi_3)$.

Therefore, $P(\overline{D_{11}D_5D_2}\lambda_{12} \mid \chi_3) = \frac{5}{16}$. In the same way, we can obtain that $P(\overline{D_{11}D_5D_2}\lambda_{12} \mid \chi_2) = \frac{1}{4}$, $P(\overline{D_{11}D_5D_2}\lambda_{12} \mid \chi_1) = \frac{159}{1024}$ and $P(\overline{D_{11}D_5D_2}\lambda_{12}) = P(\overline{D_{11}D_5D_2}) = \frac{1}{8}$. Hence,

$$P(\overline{D_{11}D_5D_2}\lambda_{12}) - \Sigma_{i=1}^{3}[P(\overline{D_{11}D_5D_2}\lambda_{12} \mid \chi_i) \times P(\chi_i)]$$
$$= \frac{1}{8} - \frac{1}{2} \times \frac{159}{1024} - \frac{1}{2^9} \times \frac{1}{4} - \frac{1}{2^{10}} \times \frac{5}{16} \approx \frac{1}{16} = 2^{-4}.$$

Since

$$\Sigma_{i=4}^{20}P(\overline{D_{26}D_{22}D_{20}}D_{19}\chi_i \mid \lambda_{12}) = \Sigma_{i=4}^{20}[P(\overline{D_{26}D_{22}D_{20}}D_{19} \mid \chi_i\lambda_{12}) \times P(\chi_i \mid \lambda_{12})]$$
$$= \Sigma_{i=4}^{20}[P(\overline{D_{26}D_{22}D_{20}}D_{19} \mid \chi_i\lambda_{12}) \times P(\chi_i)],$$

and $P(\overline{D_{26}D_{22}D_{20}}D_{19} \mid \chi_i\lambda_{12})$ $(4 \leq i \leq 20)$ can be calculated in the same way as above, the value of $\Sigma_{i=4}^{20}P(\overline{D_{26}D_{22}D_{20}}D_{19}\chi_i \mid \lambda_{12})$ can be obtained. Thus, the probability of the step function can be calculated.

In summary, in order to theoretically calculate the step differential probability for step i, we should firstly deduce the characteristics of Q_i so that the modular difference can be correctly propagated. Then, for each characteristics of Q_i, the calculation of the probability that the bit conditions hold is changed to calculating the probability that $A+B=C$ where only part bits of A and B are fixed and some bits of C are restricted to fixed values. When all characteristics of Q_i are considered, the step differential probability can be obtained.

4 Solving the Equation $T^{\lll S_0} \boxplus C_0 = (T \boxplus C_1)^{\lll S_1}$

When using the method proposed by Landelle and Peyrin to analyze RIPEMD-128 and RIPEMD-160 [5], an equation like $T^{\lll S_0} \boxplus C_0 = (T \boxplus C_1)^{\lll S_1}$ is always constructed. In order to reduce the time complexity of the merging phase, pre-computing the equation becomes a feasible way. However, in the previous analysis [5,11], the method of pre-computing the equation costs too much time and memory. In this section, we propose a method to reduce the time complexity and memory complexity. Based on *Property 1*, given a constant C_1, if there exists a solution to the equation $(T \boxplus C_1)^{\lll S_1} = T^{\lll S_1} \boxplus C_2$, then C_2 can only take the following four possible values:

1. $C_2 = (C_1 \boxminus 1)^{\lll S_1}$, and $[C_2]_{(S_1-1)\sim 0} \neq 0$.
2. $C_2 = (C_1 \boxplus 2^{32-S_1})^{\lll S_1}$, and $[C_1]_{(31-S_1)\sim 0} \neq 0$.
3. $C_2 = (C_1 \boxplus 2^{32-S_1} \boxminus 1)^{\lll S_1}$, and $[C_2]_{(S_1-1)\sim 0} \neq 0, [C_1]_{(31-S_1)\sim 0} \neq 0$.
4. $C_2 = C_1^{\lll S_1}$.

Therefore, given a constant C_1, we can compute and store the four possible values of C_2 based on the relationship between C_1 and C_2 as above. Then, for each value of C_2, we need to solve the equation $T^{\lll S_0} \boxplus C_0 = T^{\lll S_1} \boxplus C_2$. Let $C_3 = C_0 \boxminus C_2$, the equation becomes $T^{\lll S_0} \boxplus C_3 = T^{\lll S_1}$. Therefore, we only need to pre-compute the equation $T^{\lll S_0} \boxplus C_3 = T^{\lll S_1}$. Then, in order to obtain the solutions to the equation $T^{\lll S_0} \boxplus C_0 = (T \boxplus C_1)^{\lll S_1}$, we only need to guess four possible values of C_2. For each guessed value of C_2, the solutions to the equation $T^{\lll S_0} \boxplus C_3 = T^{\lll S_1}$ can be quickly obtained. For the obtained solution T, we have to verify whether it satisfies the equation $(T \boxplus C_1)^{\lll S_1} = T^{\lll S_1} \boxplus C_2$ since T satisfies it with probability. Pre-computing the equation $T^{\lll S_0} \boxplus C_3 = T^{\lll S_1}$ only costs 2^{32} time and 2^{32} memory, which is much smaller.

The expectation of the number of the solution to $T^{\lll S_0} \boxplus C_0 = (T \boxplus C_1)^{\lll S_1}$ also has an influence on the time complexity of the merging phase. Since it is not mentioned in the previous analysis, it is necessary to give a theoretical value. Consider the equation $T^{\lll S_0} \boxplus C_0 = (T \boxplus C_1)^{\lll S_1}$. Once we fix one constant, supposing that is C_0, and then exhaust all the 2^{32} possible values of T, the corresponding C_1 can be obtained. Since more than one value of T might correspond to the same C_1, one value of C_1 will correspond to more than one value of T if C_0 is fixed. We show it in Table 7.

Table 7. Number of the solutions

T	0	...	i	...	j	...	0xffffffff
C_1	x	...	x	...	x	...	y

C_1	0	...	i	...	j	...	k	...	0xffffffff
T		...	$NULL$	...	$T_{i_2}, T_{i_3}, T_{i_4}$	...	$T_{i_5}, T_{i_6}, T_{i_7}, T_{i_8}$	...	

When C_0 is fixed and C_1 is random, we denote by ε the number of the solutions, and denote by p_i the probability of that there are i solutions to the equation. In addition, we denote by N_i the number of C_1 which corresponds to i solutions to the equation. Suppose there are at most n solutions to the equation. Then, we can deduce that

$$N_1 + 2N_2 + \ldots + nN_n = 2^{32},$$
$$p_i = \frac{N_i}{2^{32}},$$
$$E(\varepsilon) = p_1 + 2p_2 + \ldots + np_n = \frac{N_1 + 2N_2 + \ldots + nN_n}{2^{32}} = 1.$$

Therefore, the number of expected solutions to $T^{\lll S_0} \boxplus C_0 = (T \boxplus C_1)^{\lll S_1}$ is 1. In the same way, we can obtain that the number of expected solutions to $T^{\lll S_0} \boxplus C_3 = T^{\lll S_1}$ is also 1.

In conclusion, given many pairs of (C_0, C_1), we can calculate the four corresponding possible values of C_2 at first. Since the number of expected solutions to $T^{\lll S_0} \boxplus C_3 = T^{\lll S_1}$ is 1, we will obtain four possible solutions to $T^{\lll S_0} \boxplus C_3 = T^{\lll S_1}$ on average for the four values of C_2. However, we need to further check whether the four solutions T satisfy $T^{\lll S_0} \boxplus C_0 = (T \boxplus C_1)^{\lll S_1}$. Since the expectation of the number of the solution to $T^{\lll S_0} \boxplus C_0 = (T \boxplus C_1)^{\lll S_1}$ is 1, we will obtain one solution to $T^{\lll S_0} \boxplus C_0 = (T \boxplus C_1)^{\lll S_1}$ on average. Therefore, when solving the equation $T^{\lll S_0} \boxplus C_0 = (T \boxplus C_1)^{\lll S_1}$, only four times of check is enough on average, which is very quick. Therefore, the time complexity of solving the equation is 2^2.

5 Collision Attack on the First 30-Step RIPEMD-160

By constructing a 30-step differential path, where the left branch is sparse and the right branch is controlled as sparse as possible, applying the message modification techniques proposed by Wang [20] to the right branch while the left branch remains probabilistic, it is possible to mount a collision attack on 30-step RIPEMD-160 with probability 2^{-67}. The 30-step differential path is shown in Table 8. Using the single-step modification and multi-step modification in [20], the bit conditions on the internal states can be satisfied. As mentioned before, the differential path holds only when both the modular difference of the internal states and the bit conditions hold, which is different from MD4. However, the message modification techniques can't be directly used to ensure the modular difference of the internal states holds. Moreover, the probability that the modular difference of the internal states holds has a great effect on the phase of the message modification, the reason for this will be discussed later. Therefore, how to have the modular difference of the internal states hold when using the message modification becomes an urgent problem to be solved. According to the previous part to calculate the step differential probability, we can change such a problem into how to ensure Q_i satisfies its corresponding equation so that ΔY_i holds when using the message modification.

Table 8. 30-step differential path, where $m'_{15} = m_{15} \boxplus 2^{24}$, and $\Delta m_i = 0$ $(0 \leqslant i \leqslant 14)$. Note that the symbol n represents that a bit changes to 1 from 0, u represents that a bit changes to 0 from 1, and - represents that the bit value is free.

X_i		$\pi_1(i)$	Y_i		$\pi_2(i)$
-4	--------------------------------		-4	--------------------------------	
-3	--------------------------------		-3	--------------------------------	
-2	--------------------------------		-2	--------------------------------	
-1	--------------------------------		-1	--------------------------------	
00	--------------------------------	00	00	--------------------------------	05
01	--------------------------------	01	01	--------------------------------	14
02	--------------------------------	02	02	--------------------------------	07
03	--------------------------------	03	03	--------------------------------	00
04	--------------------------------	04	04	--------------------------------	09
05	--------------------------------	05	05	--------------------------------	02
06	--------------------------------	06	06	--------------------------------	11
07	--------------------------------	07	07	--------------------------------	04
08	--------------------------------	08	08	--------------------------------	13
09	--------------------------------	09	09	-----1-1-1----------------------	06
10	--------------------------------	10	10	----000000-1--1---0000--1-001010	15
11	--------------------------------	11	11	-0--0---000011011001000000 0nuuuu	08
12	--------------------------------	12	12	nuuuuuuuuuuuuuuuu0n0n00----01100	01
13	--------------------------------	13	13	0unn1uu-111-1-1--nuunn11011011un	10
14	--------------------------------	14	14	-100001111----1-10nu10101-nu1-11	03
15	--------------------------------	15	15	00---01111-0u-u-101000-u----0-01	12
16	-------------------------------n	07	16	111-n1uu000n1n--0001n----nuuuuuu	06
17	-------------------------------0	04	17	1u1-1--un--0111-00u10unnn-nnn01-	11
18	---------------------1---------1	13	18	01------0n-011--1n0000----0-00-1	03
19	---------------------0----------	01	19	1u------1--100--010----------1-1	07
20	---------------------n----------	10	20	-0--------1---------0nu11---11-0	00
21	---------------------0----------	06	21	-1-----1011-----11111-101-------	13
22	-----------1---------1----------	15	22	u-----001-u-----------1u------00	05
23	n----------0--------------------	03	23	1--------------0-----01-------n-	10
24	0----------n--------------------	12	24	1--------------1----0-1-------00	14
25	1----------0----------1---------	00	25	1----n-----0--------1---------01	15
26	-1---------1----------0---------	09	26	-----------0--------unn---------	08
27	-0--------------------n---------	05	27	-u------------------------------	12
28	-n--------------------0---------	02	28	--------------------------------	04
29	-0----------1-------------------	14	29	--------------------------------	09
30	--------------------------------	11	30	--------------------------------	01

Other Conditions
$Y_{11,31} \bigvee \neg Y_{10,21} = 1$, $Y_{11,29} \bigvee \neg Y_{10,19} = 1$, $Y_{11,28} \bigvee \neg Y_{10,18} = 1$, $Y_{11,26} \bigvee \neg Y_{10,16} = 1$, $Y_{11,25} \bigvee \neg Y_{10,15} = 1$, $Y_{11,24} \bigvee \neg Y_{10,14} = 1$.
$Y_{14,21} = 1$, $Y_{14,20} = 1$, $Y_{14,9} = 1$ (We use the three conditions); Or $Y_{15,21} = 1$, $Y_{14,21} = 0$, $Y_{14,20} = 0$, $Y_{14,19} = 0$.
$Y_{15,6} = 1$, $Y_{14,6} = 0$, $Y_{15,5} = 1$; Or $Y_{14,6} = 1$, $Y_{15,5} = 0$ (We use the two conditions).
$Y_{15,29} = 0$, $Y_{15,28} = 0$, $Y_{15,27} = 1$.
$Y_{18,28} = Y_{17,28}$, $Y_{18,21} = Y_{17,21}$, $Y_{18,16} = Y_{17,16}$.
$Y_{19,17} = Y_{18,17}$, $Y_{19,8} = Y_{18,8}$, $Y_{19,1} = Y_{18,1}$.
$Y_{20,24} = Y_{19,24}$.
$Y_{22,19} = Y_{21,19}$, $Y_{22,20} = Y_{21,20}$.
$Y_{28,19} = Y_{27,19}$, $Y_{28,20} = Y_{27,20}$, $Y_{28,21} = Y_{27,21}$.
$X_{15,0} = X_{14,22}$.
$X_{22,31} = X_{21,21}$.

5.1 Deducing Extra Bit Conditions to Control the Characteristics of Q_i

Given a differential path, both the bit conditions on the internal states and the equations that all Q_i have to satisfy are fixed. The differential path holds only when all these bit conditions hold and all Q_i satisfy their corresponding equations. Although the message modification techniques proposed by Wang can be used to ensure the bit conditions on the internal states hold, it can't be directly used to ensure Q_i satisfies its corresponding equation. However, if we can add some extra bit conditions on Y_i and Y_{i-4} to ensure Q_i always satisfies its corresponding equation, the influence of Q_i can be eliminated. Then, the message modification ensures that both the bit conditions and the modular differences of the internal state hold at the same time. Taking Q_{13} as an example, we show how to deduce the extra bit conditions on Y_{13} and Y_9.

Based on the 30-step differential path in Table 8, we can obtain that Q_{13} has to satisfy the equation $(Q_{13} \boxplus \texttt{0x6ffba800})^{\lll 14} = Q_{13}^{\lll 14} \boxplus \texttt{0xea001bff}$ so that the modular difference ΔY_{13} holds, from which we can deduce the characteristics of Q_{13} as described before. We only choose two possible characteristics of Q_{13}, which are $[Q_{13}]_{31} = 0$ and $[Q_{13}]_{17} = 1$. By applying the single-step message modification, all the bit conditions on Y_{13} and Y_9 can be satisfied, which means that some bits of them are fixed. Considering the relationship between Y_{13} and Y_9:

$$Q_{13}^{\lll 14} = Y_{13} \boxminus Y_9^{\lll 10},$$

our goal is to ensure the two bit conditions on Q_{13} are satisfied under the condition that some bits of Y_{13} and Y_9 are already fixed. We show the calculation of $Q_{13}^{\lll 14} = Y_{13} \boxminus Y_9^{\lll 10}$ in Table 9, which will help understand how to accurately deduce the extra bit conditions.

Table 9. The calculation of $Q_{13}^{\lll 14} = Y_{13} \boxminus Y_9^{\lll 10}$

Y_{13}	0100	1uu-	111-	1-1-	-nuu	nn11	0110	11un
$Y_9^{\lll 10}$	10--	----	----	----	--10	----	---1	-1-1
$Q_{13}^{\lll 14}$	1---	----	----	----	--0-	----	----	----

If we impose four bit conditions on Y_9, which are $Y_{9,2} = 0$, $Y_{9,3} = 1$, $Y_{9,20} = 0$, $Y_{9,21} = 1$, the two bit conditions on Q_{13} will hold with probability 1. In other words, if all the bit conditions (including the extra conditions) on Y_9 and Y_{13} hold, the equation $(Q_{13} \boxplus \texttt{0x6ffba800})^{\lll 14} = Q_{13}^{\lll 14} \boxplus \texttt{0xea001bff}$ will always hold. Therefore, by adding four extra conditions on Y_9, the message modification can ensure both the bit conditions on Y_{13} and the modular difference ΔY_{13} hold.

Sometimes, however, adding many extra conditions costs too much. Therefore, for some special cases, we use a dynamic way to add fewer conditions to ensure that Q_i satisfies its corresponding equation with probability 1 or close to 1. For example, in order to ensure that the modular difference ΔY_{23} holds, Q_{23} has to satisfy the equation $(Q_{23} \boxplus \texttt{0x81000001})^{\lll 9} = Q_{23}^{\lll 9} \boxplus \texttt{0x102}$, from which we can deduce the characteristics of Q_{23}. Then, we choose one possible characteristic, which is $[Q_{23}]_{31} = 1$. In this way, Q_{23} satisfies its corresponding equation with probability $1 - 2^{-23} \approx 1$. By considering the calculation of $Q_{23}^{\lll 9} = Y_{23} \boxminus Y_{19}^{\lll 10}$ as shown in Table 10, we describe how to dynamically determine the bit conditions on Y_{23}.

Table 10. The calculation of $Q_{23}^{\lll 9} = Y_{23} \boxminus Y_{19}^{\lll 10}$

Y_{23}	1---	----	----	---0	----	-01-	----	--n-
$Y_{19}^{\lll 10}$	1u--	----	1--1	00--	010-	----	----	-1-1
$Q_{23}^{\lll 9}$	----	----	----	----	----	---1	----	----

According to the multi-step message modification [20], we should deal from lower bits to higher bits to correct Y_{23}. Therefore, we compare $[Y_{23}]_{7\sim 0}$ with $[Y_{19}^{\lll 10}]_{7\sim 0}$ only when $Y_{23,1}$ has been corrected. For different relationships between them, we should determine the bit conditions differently. By dynamically determine the conditions on Y_{23} in this way, we can ensure Q_{23} satisfies its corresponding equation with probability close to 1 by applying the message modification to correct $Y_{23,8}$.

1. If $[Y_{23}]_{7\sim 0} \geq [Y_{19}^{\lll 10}]_{7\sim 0}$, we add a condition $Y_{23,8} \bigoplus Y_{19,30} = 1$.
2. If $[Y_{23}]_{7\sim 0} < [Y_{19}^{\lll 10}]_{7\sim 0}$, we add a condition $Y_{23,8} \bigoplus Y_{19,30} = 0$.

As described above, we can deduce many extra bit conditions on the internal states, and they are displayed in Table 11. Then we can take these newly added bit conditions into consideration when applying the message modification techniques. In this way, both the bit conditions and the modular difference of the internal states can be satisfied at the same time.

Table 11. Equations of Q_i for the 30-step differential path and extra conditions to control the equations

Equation: $(Q_i \boxplus in)^{\lll shift} = Q_i^{\lll shift} \boxplus out$				
i	shift	in	out	Extra conditions
11	8	0x1000000	0x1	$Y_{7,24} = 1$
12	11	0x15	0xa800	$Y_{8,21} = 0$, $Y_{8,19} = 0$
13	14	0x6ffba800	0xea001bff	$Y_{9,3} = 1$, $Y_{9,2} = 0$, $Y_{9,21} = 1$, $Y_{9,20} = 0$
14	14	0x40400001	0x1010	$Y_{10,31} = 0$
15	12	0xafffff5f	0xfff5fb00	$Y_{15,9} = 0$, $Y_{11,31} = 1$
16	6	0x9d020	0x2740800	
17	9	0x85f87f2	0xbf0fe410	$Y_{13,20} = 1$, $Y_{13,18} = 0$, $Y_{17,28} = 0$, $Y_{17,26} = 1$, $Y_{13,16} = 0$
18	7	0x0	0x0	
19	15	0xffffd008	0xe8040000	$Y_{15,21} = 0$
20	7	0xd75fbffc	0xafdffdec	
21	12	0x10200813	0x812102	$Y_{21,6} = 1$, $Y_{17,28} = 0$, $Y_{21,10} = Y_{17,0}$
22	8	0xff7edffe	0x7edffeff	$Y_{22,30} = 1$, $Y_{18,21} = 1$, $Y_{22,2} = Y_{18,24}$, $Y_{22,3} = Y_{18,25}$, $Y_{22,4} = Y_{18,26}$, $Y_{22,5} = Y_{18,27}$, $Y_{22,6} = Y_{18,28}$, $Y_{22,7} = Y_{18,29}$
23	9	0x81000001	0x102	If $[Y_{23}]_{7\sim 0} \geq [Y_{19}^{\lll 10}]_{7\sim 0}$, then $Y_{23,8} \bigoplus Y_{19,30} = 1$ If $[Y_{23}]_{7\sim 0} < [Y_{19}^{\lll 10}]_{7\sim 0}$, then $Y_{23,8} \bigoplus Y_{19,30} = 0$
24	11	0xffffff00	0xfff80000	
25	7	0x80000	0x4000000	
26	7	0x1000800	0x80040000	
27	12	0x7ffc0000	0xbffff800	
28	7	0x0	0x0	
29	6	0xc0000000	0xfffffff0	
30	15	0x10	0x80000	

5.2 Multi-step Modification for RIPEMD-160

After obtaining the newly added bit conditions, we need to apply the message modification techniques to correct the bits of the internal states. Since the single-step modification is relatively simple, we refer the interested readers to [20] for more details. The following is an example to correct the three bit conditions on Y_1 by single-step modification. For the first round, we can correct the bit conditions on the internal states in this similar way.

$$Y_1 \longleftarrow Y_1 \oplus (\overline{Y_{1,3}} \lll 3) \oplus (\overline{Y_{1,14}} \lll 14) \oplus (\overline{Y_{1,29}} \lll 29).$$
$$m_5 \longleftarrow (Y_1 \boxminus Y_{-3}^{\lll 10})^{\ggg 8} \boxminus (Y_{-4}^{\lll 10} \boxplus ONX(Y_0, Y_{-1}, Y_{-2}^{\lll 10}) \boxplus K_0^r).$$

For the internal states after the first round, the multi-step modification should be applied. However, the step function of RIPEMD-160 is no longer a T-function. Therefore, the multi-step modification for RIPEMD-160 is slightly different from that for MD4 [20]. We take correcting $Y_{17,4}$, $Y_{17,3}$ and $Y_{23,16}$ as three examples to show three types of multi-step modification for RIPEMD-160.

Table 12. Message modification for correcting $Y_{17,4}$

			Modify m	New internal state	Q
Y_9	m_{13}	7	$m_{13} \longleftarrow m_{13} \boxplus (Q_9^{\lll 7} \boxplus 2^{27})^{\ggg 7} \boxminus Q_9)$	$Y_9^{new} = Y_9[27]$	Q_9 is changed.
Y_{10}	m_6	7	$m_6 \longleftarrow (Y_{10} \boxminus Y_6^{\lll 10})^{\ggg 7} \boxminus Y_5^{\lll 10} \boxminus ONX(Y_9^{new}, Y_8, Y_7^{\lll 10}) \boxminus K_0^r$	Y_{10}	Q_{10} stays the same.
Y_{11}	m_{15}	8	$m_{15} \longleftarrow (Y_{11} \boxminus Y_7^{\lll 10})^{\ggg 8} \boxminus Y_6^{\lll 10} \boxminus ONX(Y_{10}, Y_9^{new}, Y_8^{\lll 10}) \boxminus K_0^r$	Y_{11}	Q_{11} stays the same.
Y_{12}	m_8	11	$m_8 \longleftarrow (Y_{12} \boxminus Y_8^{\lll 10})^{\ggg 11} \boxminus Y_7^{\lll 10} \boxminus ONX(Y_{11}, Y_{10}, Y_9^{new \lll 10}) \boxminus K_0^r$	Y_{12}	Q_{12} stays the same.
Y_{13}	m_1	14	$m_1 \longleftarrow (Y_{13} \boxminus Y_9^{new \lll 10})^{\ggg 14} \boxminus Y_8^{\lll 10} \boxminus ONX(Y_{12}, Y_{11}, Y_{10}^{\lll 10}) \boxminus K_0^r$	Y_{13}	Q_{13} is changed.
Y_{14}	m_{10}	14	$m_{10} \longleftarrow (Y_{14} \boxminus Y_{10}^{\lll 10})^{\ggg 14} \boxminus Y_9^{new \lll 10} \boxminus ONX(Y_{13}, Y_{12}, Y_{11}^{\lll 10}) \boxminus K_0^r$	Y_{14}	Q_{14} stays the same.

In order to correct $Y_{17,4}$, we can change the 27th bit of m_6. Therefore, we can change the 27th bit of Y_9 by changing the value of m_{13}. Then, modify m_6, m_{15}, m_8, m_1, m_{10} to have Y_i $(10 \leq i \leq 14)$ remaining the same. In this way, $Y_{17,4}$ can be corrected. According to Table 12, we can find that Q_9 and Q_{13} are changed during the phase of message modification. Since there is no constraints on Q_9, it doesn't matter if Q_9 is changed. However, Q_{13} has to satisfy the equation $(Q_{13} \boxplus \mathtt{0x6ffba800})^{\lll 14} = Q_{13}^{\lll 14} \boxplus \mathtt{0xea001bff}$ so that the modular difference ΔY_{13} holds. Thus, we have to consider the influence of its change. As introduced in the previous part, we have added some extra conditions on Y_9 to ensure Q_{13} will always satisfy this equation under the condition that all bit conditions on Y_9 and Y_{13} hold. Although Y_9 is changed when correcting $Y_{17,4}$, it won't have an influence on the conditions added to control the characteristics of Q_{13}, which means that Q_{13} still satisfies its corresponding equation even though it is changed. The main reason is that we have controlled the characteristics of Q_{13} by the newly added bit conditions and such a Q_{13} will always satisfy its corresponding equation. If we don't pre-deduce the extra bit conditions to control the characteristics of Q_{13}, the equation $(Q_{13} \boxplus \mathtt{0x6ffba800})^{\lll 14} = Q_{13}^{\lll 14} \boxplus \mathtt{0xea001bff}$ may not hold any more since Q_{13} has been changed. In other word, $Y_{17,4}$ may be probabilistically corrected. And the probability is equal to the probability that the equation

Table 13. Message modification for correcting $Y_{23,16}$

			Modify m	New internal state	Extra Conditions
Y_1	m_5	8	$m_5 \longleftarrow m_5 \boxplus 2^7$	$Y_1^{new} = Y_1[15]$	$(Q_1 \boxplus 2^7)^{\lll 8} = Q_1^{\lll 8} \boxplus 2^{15}$.
Y_2	m_{14}	9	$m_{14} = (Y_2 \boxminus Y_{-2}^{\lll 10})^{\ggg 9} \boxminus Y_{-3}^{\lll 10} \boxminus ONX(Y_1^{new}, Y_0, Y_{-1}^{\lll 10}) \boxminus K_0^r$	Y_2	
Y_3	m_7	9		Y_3	$Y_{0,5} = 0.$
Y_4	m_0	11		Y_4	$Y_{2,25} = 1.$
Y_5	m_9	13	$m_9 \longleftarrow (Y_5 \boxminus Y_1^{new \lll 10})^{\ggg 13} \boxminus Y_0^{\lll 10} \boxminus ONX(Y_4, Y_3, Y_2^{\lll 10}) \boxminus K_0^r$	Y_5	
Y_6	m_2	15	$m_2 \longleftarrow (Y_6 \boxminus Y_2^{\lll 10})^{\ggg 15} \boxminus Y_1^{new \lll 10} \boxminus ONX(Y_5, Y_4, Y_3^{\lll 10}) \boxminus K_0^r$	Y_6	

Table 14. Message modification for correcting $Y_{17,3}$

			Modify m	New internal state	Q
Y_6	m_2	15	$Y_6^{new} \longleftarrow [Y_{10} \boxminus (Y_5^{\lll 10} \boxplus ONX(Y_9, Y_8, Y_7^{\lll 10}) \boxplus m_6 \boxplus 2^{26} \boxplus K_0^r)^{\lll 7}]^{\ggg 10}$ $m_2 \longleftarrow (Y_6^{new} \boxminus Y_2^{\lll 10})^{\ggg 15} \boxminus Y_1^{\lll 10} \boxminus ONX(Y_5, Y_4, Y_3^{\lll 10}) \boxminus K_0^r$	Y_6^{new}	Q_6 is changed.
Y_7	m_{11}	15	$m_{11} \longleftarrow (Y_7 \boxminus Y_3^{\lll 10})^{\ggg 15} \boxminus Y_2^{\lll 10} \boxminus ONX(Y_6^{new}, Y_5, Y_4^{\lll 10}) \boxminus K_0^r$	Y_7	Q_7 stays the same.
Y_8	m_4	5	$m_4 \longleftarrow (Y_8 \boxminus Y_4^{\lll 10})^{\ggg 5} \boxminus Y_3^{\lll 10} \boxminus ONX(Y_7, Y_6^{new}, Y_5^{\lll 10}) \boxminus K_0^r$	Y_8	Q_8 stays the same.
Y_9	m_{13}	7	$m_{13} \longleftarrow (Y_9 \boxminus Y_5^{\lll 10})^{\ggg 7} \boxminus Y_4^{\lll 10} \boxminus ONX(Y_8, Y_7, Y_6^{new \lll 10}) \boxminus K_0^r$	Y_9	Q_9 stays the same.
Y_{10}	m_6	7	$m_6 \longleftarrow m_6 \boxplus 2^{26}$	Y_{10}	Q_{10} is changed.
Y_{11}	m_{15}	8	$m_{15} \longleftarrow (Y_{11} \boxminus Y_7^{\lll 10})^{\ggg 8} \boxminus Y_6^{new \lll 10} \boxminus ONX(Y_{10}, Y_9, Y_8^{\lll 10}) \boxminus K_0^r$	Y_{11}	Q_{11} stays the same.

$(Q_{13} \boxplus \texttt{0x6ffba800})^{\lll 14} = Q_{13}^{\lll 14} \boxplus \texttt{0xea001bff}$ holds, which is about $2^{-0.5}$. Moreover, if we correct n bits of Y_{17} by using the strategy as Table 12 displays and don't pre-deduce the extra bit conditions, the probability that they are right corrected is about $2^{-0.5n}$, which will have a great effect on the probability to mount the collision attack on 30-step RIPEMD-160. Therefore, it is significant to pre-deduce the extra bit conditions to control the characteristics of Q_i, which will improve the time complexity of the message modification.

In order to correct $Y_{23,16}$, we can change the 7th bit of m_5. As displayed in Table 13, by modifying some message words and adding some extra conditions on the internal states, $Y_{23,16}$ can be corrected. For the strategy in Table 13, $Y_{23,16}$ can be corrected with probability that the equation $(Q_1 \boxplus 2^7)^{\lll 8} = Q_1^{\lll 8} \boxplus 2^{15}$ holds, which is $1 - 2^{-17} \approx 1$. Therefore, we can ignore the influence of this equation. Sometimes, however, such an equation holds with a low probability, which is bad for the correcting. In order to eliminate the influence, we can use the same idea in Sect. 5.1 to pre-deduce some extra bit conditions to control the characteristics of Q_i so that Q_i will satisfy such an equation.

In order to correct $Y_{17,3}$, we can change the 26th bit of m_6. Firstly, we compute a new value of Y_6 so that Y_{10} can stay the same only by adding 2^{26} to m_6. Then, a new value of m_2 can be obtained. To have Y_i $(7 \leq i \leq 11)$ remaining the same, m_{11}, m_4, m_{13}, m_6, m_{15} should be accordingly modified. As for strategy displayed in Table 14, it is because there is no condition on Y_6 that we can choose such a method to correct $Y_{17,3}$. Since there is no condition on Y_3 either, $Y_{18,31}$ can be corrected by using the similar strategy.

The multi-step message modification is summarized in Table 15. In this table, we also display some extra bit conditions to control the characteristics of Q_1 and Q_4 so that the newly added bit conditions on them for message modification can be satisfied. Although some of the equations of Q_1 and Q_4 remain uncontrolled, they will hold with probability close to 1.

Table 15. Summarization of the multi-step modification for Y_i $(17 \leq i \leq 23)$

Chaining variables	Bits to be corrected (i)	Chaining variables used	Extra Conditions
Y_{17}	1,2,12,13,14,15,23,24,30,31,21	Y_5, Y_6, Y_7, Y_8, Y_9, Y_{10}	$Y_5[i-19]$.
Y_{17}	4,5,7,8,9,10,17,18,19,20,26,27,28	Y_9, Y_{10}, Y_{11}, Y_{12}, Y_{13}, Y_{14}	$Y_9[i-9]$.
Y_{17}	11,29	Y_8, Y_9, Y_{10}, Y_{11}, Y_{12}, Y_{13}	$Y_8[i-9]$, $Y_{7,i-19}=1$.
Y_{17}	3	Y_6, Y_7, Y_8, Y_9, Y_{10}, Y_{11}	
Y_{18}	2,3,5,11,12,13,14,15,18,19,20,28,30	Y_2, Y_3, Y_4, Y_5, Y_6, Y_7	$Y_2[i-23]$.
Y_{18}	0,10,16,21,22,23	Y_4, Y_5, Y_6, Y_7, Y_8, Y_9	$Y_4[i-23]$, $Y_{5,i-13}=0$.
Y_{18}	31	Y_3, Y_4, Y_5, Y_6, Y_7, Y_8	
Y_{19}	19	Y_{15}, Y_{16}, Y_{17}, Y_{18}	$Y_{15}[16]$, $Y_{14,6}=1$, $Y_{16,26}=Y_{17,26}$.
Y_{20}	0,2,3,7,8,9,10,11,21,24,30	Y_1, Y_2, Y_3, Y_4, Y_5, Y_6	$Y_1[i-7]$, $Y_{0,i-17}=1$.
Y_{21}	7,8,9,13,15,22,23,24,30	Y_4, Y_5, Y_6, Y_7, Y_8, Y_9	$Y_4[i-1]$, $Y_{4,28}=1$, $Y_{4,27}=1$, $Y_{4,26}=1$, $Y_{0,19}=0$, $Y_{0,16}=0$,$Y_{4,5}=1$, $Y_{0,28}=0$, $Y_{0,27}=0$.
Y_{21}	6,10,11,12,14,21	Y_1, Y_2, Y_3, Y_4, Y_5, Y_6	$Y_1[i-22]$, $Y_{0,i}=0$, $Y_{2,i-12}=0$.
Y_{22}	0,1,2,3,4,5,6,7,8,9, 19,20,21,23,24,25,30,31	Y_8, Y_9, Y_{10}, Y_{11}, Y_{12}, Y_{13}	$Y_8[i-8]$, $Y_{7,i-18}=0$.
Y_{23}	8,9,10,16,31	Y_1, Y_2, Y_3, Y_4, Y_5, Y_6	$Y_1[i-1]$, $Y_{0,i-11}=0$, $Y_{2,i+9}=1$, $Y_{1,29}=1$, $Y_{-3,20}=0$, $Y_{-3,19}=0$. $Y_{1,6}=0$, $Y_{-3,29}=1$, $Y_{-3,28}=1$.
$Y_{4,28}=1$, $Y_{4,27}=1$, $Y_{4,26}=1$, $Y_{0,19}=0$, $Y_{0,16}=0$ are used to control: $(Q_4 \boxplus 2^{18})^{\lll 11} = Q_4^{\lll 11} \boxplus 2^{29}$.			
$Y_{4,5}=1$, $Y_{0,28}=0$, $Y_{0,27}=0$ are used to control: $(Q_4 \boxplus 2^{28})^{\lll 11} = Q_4^{\lll 11} \boxplus 2^{7}$ and $(Q_4 \boxplus 2^{29})^{\lll 11} = Q_4^{\lll 11} \boxplus 2^{8}$.			
$Y_{1,29}=1$, $Y_{-3,20}=0$, $Y_{-3,19}=0$ are used to control: $(Q_1 \boxplus 2^{22})^{\lll 8} = Q_1^{\lll 8} \boxplus 2^{30}$.			
$Y_{1,6}=0$, $Y_{-3,29}=1$, $Y_{-3,28}=1$ are used to control: $(Q_1 \boxplus 2^{31})^{\lll 8} = Q_1^{\lll 8} \boxplus 2^{7}$.			

5.3 Complexity Evaluation

For the left branch, we don't apply any message modification techniques to it. By randomly generating message words, we test the probability that the left branch holds. According to our experiments, the probability is about 2^{-29}.

For the right branch, we can use the message modification techniques to correct the bits of Y_i $(17 \leq i \leq 23)$. However, we can't find a way to correct all the bits of them, thus leaving 14 bit conditions remaining uncontrolled, i.e., 13 bits of Y_{19} and 1 bit of Y_{23}. Besides, to ensure Q_{20} can satisfy its corresponding equation with probability 1, some extra bit conditions on Y_{20} and Y_{16} should be added. However, it is difficult to have all these newly added bit conditions hold by using the message modification techniques, which will cause a lower probability. Therefore, we leave Q_{20} holding with probability about 2^{-1}. For Q_i $(11 \leq i \leq 23,\ i \neq 20)$, by correcting the newly added extra bit conditions, they will satisfy their corresponding equations with probability 1 or close to 1.

For Y_i $(24 \leq i \leq 30)$, since it is difficult to correct the 20 bit conditions on them, we leave them holding probabilistically. In addition, Q_i $(24 \leq i \leq 30)$ satisfy their corresponding equations with probability about 2^{-3}. Therefore, the right branch holds with probability about $2^{-14-1-20-3} = 2^{-38}$.

When applying the message modification techniques, we add 26 bit conditions on Y_0 and 4 bit conditions on Y_{-3}. Therefore, we need to use two message blocks (M_1, M_2) to mount the 30-step collision attack. M_1 is used to generate such a hash value that the bit conditions on Y_0 and Y_{-3} have been satisfied when compressing M_2, which costs $2^{26+4} = 2^{30}$ time. In conclusion, the 30-step collision attack succeeds with probability of about $2^{-29-38} = 2^{-67}$, and the time complexity is about $2^{67} + 2^{30} \approx 2^{67}$. The implementation of this attack is available at https://github.com/Crypt-CNS/RIPEMD160-30Step.git.

6 Improved Semi-Free-Start Collision Attack

6.1 36-Step Semi-Free-Start Collision Path

Mendel *et al.* [11] improved the techniques in [7,10], and used the improved algorithm to find two differential paths of RIPEMD-160. One is a 48-step semi-free-start collision path, the other is a 36-step semi-free-start collision path. Since we focus on the semi-free-start collision attack on the first 36-step RIPEMD-160, we only introduce the 36-step semi-free-start collision path. The differential path is displayed in Table 16. In order to have a full understanding of our improvement, it is necessary to briefly introduce the method proposed by Landelle and Peyrin [5].

The main idea of the method can be divided into three steps. Firstly, the attacker chooses the internal states in both branches and fixes some message words to ensure the non-linear parts. This step is called *find a starting point.*

Table 16. 36-step differential path, where $m_7' = m_7 \boxplus 2^4 \boxplus 2^{15} \boxplus 2^{30}$, and $\Delta m_i = 0$ $(i \neq 7, 0 \leqslant i \leqslant 15)$. Note that the symbol n represents that a bit changes to 1 from 0, u represents that a bit changes to 0 from 1, and - represents that the bit value is free.

i	X_i	$\pi_1(i)$	i	Y_i	$\pi_2(i)$
-4	--------------------------------		-4	--------------------------------	
-3	--------------------------------		-3	--------------------------------	
-2	--------------------------------		-2	--------------------------------	
-1	--------------------------------		-1	--------------------------------	
00	--------------------------------	00	00	--------------------------------	05
01	--------------------------------	01	01	--1--------------1----------1---	14
02	--------------------------------	02	02	--0---10---------10-----0---0---	07
03	--------------------------------	03	03	------1n---------1n----0n---0---	00
04	--------------------------------	04	04	100----n0-11-n0--------110--n0--	09
05	--------------------------------	05	05	n1-1---00----00n0-1---000---1101	02
06	--------------------------------	06	06	n--1000111110un---uuuuuuuu11-1u-	11
07	--------------------------------	07	07	uuu00un-n1u011nn000001101-01-n00	04
08	------n-uuuuuuuuuuu-nuuu-uuuuuuu	08	08	11u0uu--1-u1u0n1nn-nu-100---000u	13
09	--uun-nn-n---nnunnuu-------n--nn	09	09	0-uu011--01-u00011000n-nn01--111	06
10	-----n--unuun-u-u-----nn----u---	10	10	--1u1nnnnnn0110010-0-0-00100u1-0	15
11	--n---nuuu--nu--un-n------------	11	11	0--u-n1n---1--1un--n-n--100001n-	08
12	-----u----n-nnnnnnnnnnnnnnn-----	12	12	10-110----n0---0u1--1----0-111-n	01
13	--------------------------------	13	13	1u--0n-u-----------1-1u--10---1-	10
14	--------------------------------	14	14	----1----n-00--0----n111--n1-0--	03
15	--------------------------------	15	15	-----u---1-u0-------1-----u--1--	12
16	--------------------------------	07	16	---------------------------0----	06
17	--------------------------------	04	17	---------------------------1----	11
18	--------------------------------	13	18	--------------------------------	03
19	--------------------------------	01	19	--------------------------------	07
20	--------------------------------	10	20	--------------------------------	00
21	--------------------------------	06	21	--------------------------------	13
22	--------------------------------	15	22	--------------------------------	05
23	--------------------------------	03	23	--------------------------------	10
24	--------------------------------	12	24	--------------------------------	14
25	--------------------------------	00	25	--------------------------------	15
26	--------------------------------	09	26	--------------------------------	08
27	--------------------------------	05	27	--------------------------------	12
28	--------------------------------	02	28	--------------------------------	04
29	--------------------------------	14	29	--------------------------------	09
30	--------------------------------	11	30	--------------------------------	01
31	--------------------------------	08	31	--------------------------------	02
32	--------------------------------	03	32	--------------------------------	15
33	--------------------------------	10	33	--------------------------------	05
34	--------------------------------	14	34	--------------------------------	01
35	--------------------------------	04	35	--------------------------------	03
36	--------------------------------		36	--------------------------------	

Secondly, the attacker uses the remaining free message words to merge both branches to ensure that the chaining variables in both branches are the same by computing backward from the middle. At last, the rest of the differential path in both branches are verified probabilistically by computing forward from the middle.

6.2 Finding a Starting Point

Different from the choice of the message words for merging in [11], we set m_3 free at the phase of finding a starting point and use it at the phase of merging.

Table 17. The starting point, where $m'_7 = m_7 \boxplus 2^4 \boxplus 2^{15} \boxplus 2^{30}$,and $\Delta m_i = 0.(i \neq 7, 0 \leqslant i \leqslant 15)$. Note that the word messages marked in green are all fixed. Those marked in black are all free while the one marked in red is to be inserted difference in.

```
Xi                                     π1(i) Yi                                     π2(i)
-4 -------- -------- -------- --------       -4 -------- -------- -------- --------
-3 -------- -------- -------- --------       -3 -------- -------- -------- --------
-2 -------- -------- -------- --------       -2 -------- -------- -------- --------
-1 -------- -------- -------- --------       -1 -------- -------- -------- --------
00 -------- -------- -------- --------  00   00 -------- -------- -------- --------  05
01 -------- -------- -------- --------  01   01 --1----- -------- -1------ ----1---  14
02 -------- -------- -------- --------  02   02 10000010 11111100 01000110 00110000  07
03 -------- -------- -------- --------  03   03 0111101n 11000000 01n10010 n0100001  00
04 -------- -------- -------- --------  04   04 1001110n 00110n01 01100011 1001n000  09
05 -------- -------- -------- --------  05   05 n1010110 0111000n 00110000 01111101  02
06 11011100 10101101 01110010 01011001  06   06 n0010001 11110un1 00uuuuuu uu1111u1  11
07 01001101 01110100 11010011 11101011  07   07 uuu00un1 n1u011nn 00000110 11011n00  04
08 001100n0 uuuuuuuu uuu1nuuu 1uuuuuuu  08   08 11u0uu11 10u1u0n1 nn0nu110 0010000u  13
09 00uun0nn 1n110nnu nnuu1011 001n10nn  09   09 01uu0111 0011u000 11000n0n n0100111  06
10 10110n11 unuun0u0 u00100nn 1100u011  10   10 011u1nnn nnn01100 10000000 0100u110  15
11 10n101nu uu11nu10 un1n0100 01011100  11   11 011u1n1n 1011011u n00n1n11 100001n1  08
12 00011u00 11n1nnnn nnnnnnnn nnn01101  12   12 10011010 11n00110 u1011000 0001111n  01
13 11111000 01111111 01000011 00010100  13   13 1u100n0u 11110100 000111u0 01010111  10
14 10010011 00110110 11101010 00010010  14   14 01001000 1n000110 0000n111 11n10000  03
15 -------- -------- -------- --------  15   15 -----u-- -1-u0--- ----1--- --u--1--  12
16 -------- -------- -------- --------  07   16 -------- -------- -------- ---0----  06
17 -------- -------- -------- --------  04   17 -------- -------- -------- ---1----  11
18 -------- -------- -------- --------  13   18 -------- -------- -------- --------  03
19 -------- -------- -------- --------  01   19 -------- -------- -------- --------  07
20 -------- -------- -------- --------  10   20 -------- -------- -------- --------  00
21 -------- -------- -------- --------  06   21 -------- -------- -------- --------  13
22 -------- -------- -------- --------  15   22 -------- -------- -------- --------  05
23 -------- -------- -------- --------  03   23 -------- -------- -------- --------  10
24 -------- -------- -------- --------  12   24 -------- -------- -------- --------  14
25 -------- -------- -------- --------  00   25 -------- -------- -------- --------  15
26 -------- -------- -------- --------  09   26 -------- -------- -------- --------  08
27 -------- -------- -------- --------  05   27 -------- -------- -------- --------  12
28 -------- -------- -------- --------  02   28 -------- -------- -------- --------  04
29 -------- -------- -------- --------  14   29 -------- -------- -------- --------  09
30 -------- -------- -------- --------  11   30 -------- -------- -------- --------  01
31 -------- -------- -------- --------  08   31 -------- -------- -------- --------  02
32 -------- -------- -------- --------  03   32 -------- -------- -------- --------  15
33 -------- -------- -------- --------  10   33 -------- -------- -------- --------  05
34 -------- -------- -------- --------  14   34 -------- -------- -------- --------  01
35 -------- -------- -------- --------  04   35 -------- -------- -------- --------  03
36 -------- -------- -------- --------       36 -------- -------- -------- --------
```

Message Words	m_0	m_1	m_2	m_3	m_4	m_5	m_6	m_7
Value	*	0x67dbd0a9	*	*	0x5cd30b65	*	0x651c397d	*

Message Words	m_8	m_9	m_{10}	m_{11}	m_{12}	m_{13}	m_{14}	m_{15}
Value	0x050ff865	*	0xa9f94c09	0x509bf856	0x0588c327	0x86671566	*	0xc3349b51

In this way, we can improve the successful probability of merging. However, the right branch is not fully satisfied any more, thus resulting in an uncontrolled probability in the right branch.

According to the characteristics of Q_{15} displayed in Table 4, we observe that $[Q_{15}]_{31} = 0$. According to Table 5, we can find that if $Y_{11,0} = 1$, $Y_{11,29} = 1$, $Y_{11,30} = 1$ are satisfied at the phase of finding a starting point, $Y_{15,11} = 1$ will hold with a much higher probability, thus improving the uncontrolled probability in the right branch.

By adding three more bit conditions on $Y_{11,0}$, $Y_{11,29}$, $Y_{11,30}$ and setting m_3 free, using the technique for finding a starting point in [11], we obtain a new starting point displayed in Table 17.

6.3 Probability Neglected While Computing Backward

Based on the differential path in Table 16, we know that $\Delta X_5 = 0$, $\Delta X_4 = 0$, $\Delta X_3 = 0$, $\Delta Y_1 = 0$, $\Delta Y_0 = 0$, $\Delta Y_{-1} = 0$, $\Delta Y_{-2} = 0$ while $\Delta X_8 \neq 0$, $\Delta X_9 \neq 0$, $\Delta Y_3 \neq 0$, $\Delta Y_4 \neq 0$, $\Delta Y_5 \neq 0$. At the phase of finding a starting point, $\Delta X_5 = 0$ and $\Delta Y_1 = 0$ have been satisfied. However, for the original algorithm [11] to merge both branches, the conditions that $\Delta X_4 = 0$, $\Delta X_3 = 0$, $\Delta Y_0 = 0$, $\Delta Y_{-1} = 0$, $\Delta Y_{-2} = 0$ have been neglected. We define the probability of these conditions as *neglected probability*.

According to the conditions $\Delta X_4 = 0$, $\Delta X_3 = 0$, $\Delta Y_0 = 0$, $\Delta Y_{-1} = 0$ and $\Delta Y_{-2} = 0$, we can get the following equations:

$$\begin{aligned}
0 &= (X_9' \boxminus X_5^{\lll 10})^{\ggg 11} \boxminus (X_9 \boxminus X_5^{\lll 10})^{\ggg 11} \boxminus (XOR(X_8', X_7', X_6'^{\lll 10}) \\
&\quad \boxminus XOR(X_8, X_7, X_6^{\lll 10})), \\
0 &= (X_8' \boxminus X_4^{\lll 10})^{\ggg 9} \boxminus (X_8 \boxminus X_4^{\lll 10})^{\ggg 9} \boxminus (m_7' \boxminus m_7), \\
0 &= (Y_5' \boxminus Y_1^{\lll 10})^{\ggg 13} \boxminus (Y_5 \boxminus Y_1^{\lll 10})^{\ggg 13} \boxminus (ONX(Y_4', Y_3', Y_2^{\lll 10}) \\
&\quad \boxminus ONX(Y_4, Y_3, Y_2^{\lll 10})), \\
0 &= (Y_4' \boxminus Y_0^{\lll 10})^{\ggg 11} \boxminus (Y_4 \boxminus Y_0^{\lll 10})^{\ggg 11} \boxminus (ONX(Y_3', Y_2, Y_1^{\lll 10}) \\
&\quad \boxminus ONX(Y_3, Y_2, Y_1^{\lll 10})), \\
0 &= (Y_3' \boxminus Y_{-1}^{\lll 10})^{\ggg 9} \boxminus (Y_3 \boxminus Y_{-1}^{\lll 10})^{\ggg 9} \boxminus (m_7' \boxminus m_7).
\end{aligned}$$

Observing the five equations above, it is easy to find that there are some similarities between them. Therefore, we can change the problem of calculating the probability that the five equations hold into calculating the probability that T satisfies $(T \boxplus C_0)^{\ggg S} = T^{\ggg S} \boxplus C_1$. Let $T' = T^{\ggg S}$, the equation becomes $T'^{\lll S} \boxplus C_0 = (T' \boxplus C_1)^{\lll S}$, whose probability can be calculated as introduced before.

For equation $(X_9' \boxminus X_5^{\lll 10})^{\ggg 11} \boxminus (X_9 \boxminus X_5^{\lll 10})^{\ggg 11} \boxminus (XOR(X_8', X_7', X_6'^{\lll 10}) \boxminus XOR(X_8, X_7, X_6^{\lll 10})) = 0$, $X_9' \boxminus X_9 = \texttt{0xdb459013}$, $XOR(X_8', X_7', X_6'^{\lll 10}) \boxminus XOR(X_8, X_7, X_6^{\lll 10}) = \texttt{0x25b68b3}$, $C_0 = \texttt{0xdb459013} \boxminus 0 = \texttt{0xdb459013}$, $C_1 = \texttt{0x25b68b3}$. Therefore, $P(\Delta X_4 = 0) = P(T^{\lll 11} \boxplus \texttt{0xdb459013} = (T \boxplus \texttt{0x25b68b3})^{\lll 11}) \approx 2^{-11.7}$. In the same way, we can obtain that

$$P(\Delta X_3 = 0) = P(T^{\lll 9} \boxplus \texttt{0x1002081} = (T \boxplus \texttt{0x40008010})^{\lll 9}) \approx 2^{-8.4},$$
$$P(\Delta Y_0 = 0) = P(T^{\lll 13} \boxplus \texttt{0x80010000} = (T \boxplus \texttt{0xfffc0008})^{\lll 13}) =\approx 2^{-1},$$
$$P(\Delta Y_{-1} = 0) = P(T^{\lll 11} \boxplus \texttt{0x1040008} = (T \boxplus \texttt{0x1002080})^{\lll 11}) \approx 1,$$
$$P(\Delta Y_{-2} = 0) = P(T^{\lll 9} \boxplus \texttt{0x1002080} = (T \boxplus \texttt{0x40008010})^{\lll 9}) \approx 2^{-0.4}.$$

Therefore, the *negelected probability* is $2^{-11.7-8.4-1-0.4} = 2^{-21.5}$. In order to eliminate the influence of the *negelected probability* at the phase of merging, for a given starting point, we can pre-compute the valid m_9 that makes $\Delta X_4 = 0$ and $\Delta X_3 = 0$ satisfied, which costs 2^{32} time and about $2^{32} \times P(\Delta X_4 = 0) \times P(\Delta X_3 = 0) = 2^{32-11.7-8.4} = 2^{12.9}$ memory. Then, at the phase of merging, given one valid m_9, we can firstly compute and store the valid m_2 that makes $Y_{1,3} = 1$, $Y_{1,14} = 1$, $Y_{1,29} = 1$, $\Delta Y_0 = 0$ and $\Delta Y_{-1} = 0$ satisfied, which costs 2^{29} time and about $2^{29} \times P(\Delta Y_0 = 0) \times P(\Delta Y_1 = 0) = 2^{28}$ memory. After choosing the valid m_9 and m_2, only the condition $\Delta Y_{-2} = 0$ has an influence on the merging, whose probability is $P(\Delta Y_{-2} = 0) \approx 2^{-0.4}$.

6.4 Merging both Branches with $m_0, m_2, m_3, m_5, m_7, m_9, m_{14}$

At the merging phase, our target is to use the remaining free message words to obtain a perfect match on the values of the five initial chaining variables of both branches. Our procedure of merging is detailed as below.

Step 1: Choose a valid value of m_9, then compute until X_4 in the left branch. Fix $Y_{1,3} = 1$, $Y_{1,14} = 1$, $Y_{1,29} = 1$ and exhaust all the 2^{29} possible values of Y_1. Then compute and store the valid m_2 that makes $\Delta Y_0 = 0$, $\Delta Y_1 = 0$ satisfied. We denote the valid number of m_2 by $VNUM$ and define the array that stores the valid m_2 as $VALIDM2[]$.

Step 2: Set random values to m_7, then compute until X_2 in the left branch.

Step 3: Set $m_2 = VALIDM2[index]$ (initialize $index$ as 0), Y_1 and Y_0 can be computed based on the following equation. If $index$ becomes $VNUM$ again, goto Step 2.

$$Y_1^{\lll 10} = (Y_6 \boxminus Y_2^{\lll 10})^{\ggg 15} \boxminus (ONX(Y_5, Y_4, Y_3^{\lll 10}) \boxplus m_2 \boxplus K_0^r),$$
$$Y_0^{\lll 10} = (Y_5 \boxminus Y_1^{\lll 10})^{\ggg 13} \boxminus (XOR(Y_4, Y_3, Y_2^{\lll 10}) \boxplus m_9 \boxplus K_0^r).$$

Step 4: Since $X_0 = Y_0$ and we have obtained the value of Y_0 at Step 3, we can compute X_0, X_1 and m_5 as follows. $X_0 = Y_0$, $X_1^{\lll 10} = X_5 \boxminus (X_0^{\lll 10} \boxplus ONX(X_4, X_3, X_2^{\lll 10}) \boxplus m_4 \boxplus K_0^l)^{\lll 5}$, $m_5 = (X_6 \boxminus X_2^{\lll 10})^{\ggg 8} \boxminus (X_1^{\lll 10} \boxplus ONX(X_5, X_4, X_3^{\lll 10}) \boxplus K_0^l)$.

Step 5: We can use the conditions $X_{-1} = Y_{-1}$ and $X_{-2} = Y_{-2}$ to construct an equation system of m_0 and m_3. Observe the step functions:

$$X_{-1}^{\lll 10} = (X_4 \boxminus X_0^{\lll 10})^{\ggg 12} \boxminus (XOR(X_3, X_2, X_1^{\lll 10}) \boxplus m_3 \boxplus K_0^l),$$
$$X_{-2}^{\lll 10} = (X_3 \boxminus X_{-1}^{\lll 10})^{\ggg 15} \boxminus (XOR(X_2, X_1, X_0^{\lll 10}) \boxplus m_2 \boxplus K_0^l),$$
$$Y_{-1}^{\lll 10} = (Y_4 \boxminus Y_0^{\lll 10})^{\ggg 11} \boxminus (ONX(Y_3, Y_2, Y_1^{\lll 10}) \boxplus m_0 \boxplus K_0^r),$$
$$Y_{-2}^{\lll 10} = (Y_3 \boxminus Y_{-1}^{\lll 10})^{\ggg 9} \boxminus (ONX(Y_2, Y_1, Y_0^{\lll 10}) \boxplus m_7 \boxplus K_0^r).$$

Let $A = (X_4 \boxminus X_0^{\lll 10})^{\ggg 12} \boxminus (XOR(X_3, X_2, X_1^{\lll 10}) \boxplus K_0^l)$, $B = (Y_4 \boxminus Y_0^{\lll 10})^{\ggg 11} \boxminus (ONX(Y_3, Y_2, Y_1^{\lll 10}) \boxplus K_0^r)$, $C = XOR(X_2, X_1, X_0^{\lll 10}) \boxplus m_2 \boxplus K_0^l$, $D = ONX(Y_2, Y_1, Y_0^{\lll 10}) \boxplus m_7 \boxplus K_0^r$, $T' = X_3 \boxminus A \boxplus m_3$, $T = T'^{\ggg 15}$, $C_0 = Y_3 \boxminus X_3$, $C_1 = D \boxminus C$. According to the condition $X_{-1} = Y_{-1}$, we can obtain one equation: $A \boxminus m_3 = B \boxminus m_0$. According to the condition $X_{-2} = Y_{-2}$, we can obtain another equation: $T^{\lll 15} \boxplus C_0 = (T \boxplus C_1)^{\lll 9}$. As introduced before, we can obtain its solutions by 2^2 computations on average. If there is no solution, goto Step 3. It is essential that all solutions should be taken into consideration since there may be more than one solution to the equation $T^{\lll 15} \boxplus C_0 = (T \boxplus C_1)^{\lll 9}$.

Step 6: Compute X_{-1} and Y_{-1} by m_3. Since $\Delta Y_{-2} = 0$ holds with probability, we have to check whether Y_{-1} satisfies the equation $0 = (Y_3' \boxminus Y_{-1}^{\lll 10})^{\ggg 9} \boxminus (Y_3 \boxminus Y_{-1}^{\lll 10})^{\ggg 9} \boxminus (m_7' \boxminus m_7)$. If this equation doesn't hold for all pairs of (m_0, m_3), goto Step 3.

Step 7: Compute X_{-2}, Y_{-2}, X_{-3}, Y_{-3} and m_{14}.

Step 8: This is the uncontrolled part of merging. At this point, all freedom degree have been used and the last condition $X_{-4} = Y_{-4}$ will hold with probability 2^{-32}.

Verification. We have verified the merging phase by implementation. Based on the starting point in Table 17, we choose a valid value of $m_9 = \texttt{0x471fba32}$, and the number of the corresponding valid m_2 is $\texttt{0xfcf2100}$. The following is an instance obtained by carrying out the merging phase.

$$m_0 = \texttt{0x678c8c36}, m_2 = \texttt{0x5293b823}, m_3 = \texttt{0xd90c1aa9}, m_5 = \texttt{0x13d3dff6},$$
$$m_7 = \texttt{0x794a60c6}, m_{14} = \texttt{0xee8e443e}, Y_{-4} = \texttt{0xd055ce6}, Y_{-3} = \texttt{0xdf979ac7},$$
$$Y_{-2} = \texttt{0xae4836b3}, Y_{-1} = \texttt{0x57b6f5fb}, Y_0 = \texttt{0x6b9ec934}.$$

6.5 Uncontrolled Probability

Firstly, we give the theoretical calculation of the uncontrolled probability of the left branch.

$$\begin{aligned}
P(\Delta X_{15} = 0) &= P(T^{\lll 9} \boxplus \texttt{0xf0bfff7f} = (T \boxplus \texttt{0x7f785fff})^{\lll 9}) \\
&= (2^{-1} + 2^{-2} + 2^{-3} + 2^{-4} + 2^{-9} + \Sigma_{i=11}^{23} 2^{-i}) \times (1 - 2^{-1} - \Sigma_{i=3}^{9} 2^{-i}) \\
&= \frac{\texttt{0x3ca85f7f}}{2^{32}} \approx 2^{-2.1}, \\
P(\Delta X_{16} = 0) &= P(T^{\lll 8} \boxplus \texttt{0x40008010} = (T \boxplus \texttt{0xf400081})^{\lll 8}) \\
&= 2^{-4} \times (2^{-2} + 2^{-17} + 2^{-24}) \\
&= \frac{\texttt{0x4000810}}{2^{32}} \approx 2^{-6}.
\end{aligned}$$

Therefore, the theoretical value of the uncontrolled probability of the left branch is about $2^{-2.1-6} = 2^{-8.1}$.

Secondly, we use our method to evaluate the uncontrolled probability of the right branch. Since we add three bit conditions on Y_{11} at the phase of finding a starting point, it is necessary to fix the values of the three bits before calculation. Then, we obtain that the probability that the modular difference of Y_{15} and the seven bit conditions (as showed in Table 16) on Y_{15} hold is about $2^{-7.6}$. The probability that the modular difference of Y_{16} and the one bit condition (as showed in Table 16) on Y_{16} hold is about 2^{-2}. The probability that the modular difference of Y_{17} and the one bit condition (as showed in Table 16) on Y_{17} hold is about 2^{-2}. The probability that the modular difference of Y_{18} holds is about 1. The probability that the modular difference of Y_{19} holds is about $2^{-0.4}$. Besides, there are five more bit conditions on Y_{15}, Y_{16} and Y_{17}, which are $Y_{15,0} = Y_{16,0}$, $Y_{15,15} = Y_{16,15}$, $Y_{15,21} = Y_{16,21}$, $Y_{16,15} = Y_{17,15}$ and $Y_{16,30} = Y_{17,30}$. Therefore, with our method to calculate the step differential probability, the uncontrolled probability of the right branch is about $2^{-7.6-2-2-0.4-5} \approx 2^{-17}$.

Then, we consider the uncontrolled probability of both branches for a specific starting point in Table 17. We can calculate the uncontrolled probability of the left branch in this way: exhaust all 2^{32} possible values of m_{14} and count the number of m_{14} which makes $\Delta X_{15} = 0$ and $\Delta X_{16} = 0$ satisfied. According to the experiment, the valid number of m_{14} is `0x1020000` and thus the uncontrolled probability of the left branch is about 2^{-8}. For the uncontrolled probability of the right branch, we can exhaust all 2^{32} possible values of m_3 and count the number of m_3 which makes the conditions on Y_{15}, Y_{16}, Y_{17}, Y_{18}, Y_{19} satisfied. According to the experiment, the valid number of m_3 is `0x9f64` and thus the uncontrolled probability of the right branch is about $2^{-16.68}$. We have to stress this is the uncontrolled probability of both branches for a specific starting point. Comparing this result with the theoretical value, we observe that they are almost the same, which implies that our method to theoretically calculate the step differential probability is reliable.

Moreover, during the merging phase, we can not control the value matching on the first IV word, and it adds another factor 2^{-32}. Since the expected value of the number of the solution to $T^{\lll S_0} \boxplus C_0 = (T \boxplus C_1)^{\lll S_1}$ is 1, its influence on the probability can be ignored. What's more, $Y'_{-2} = Y_{-2}$ holds with probability $2^{-0.4}$. Therefore, the total uncontrolled probability is $2^{-32-8.1-17-0.4} = 2^{-57.5}$, which is much higher than the original one $2^{-72.6}$. Given a starting point, the degree of freedom left is 32+28+12=72 since m_7, m_2, m_9 can take 2^{32}, 2^{28}, 2^{12} possible values respectively. Besides, we can generate many staring points to mount the semi-free-start collision attack on the first 36-step RIPEMD-160. Therefore, the degree of freedom is enough.

6.6 Complexity Evaluation

Firstly, we consider the complexity of the merging phase. Based on the fact that $X_{-4} = Y_{-4}$ holds with probability 2^{-32}, $Y'_{-2} = Y_{-2}$ holds with probability $2^{-0.4}$, and the expectation of the number of the solution to $T^{\lll S_0} \boxplus C_0 = (T \boxplus C_1)^{\lll S_1}$ is 1, we can give an estimation of the running times of each step at the merging phase. We estimate that Step 7 to Step 8 will run for 2^{32} times, Step 6 will run for $2^{32+0.4} = 2^{32.4}$ times, Step 3 to Step 5 will run for $2^{32+0.4} = 2^{32.4}$ times, Step 2 will run for $2^{32.4-28} = 2^{4.4}$ times, Step 1 will run for only one time. Since Step 2 contains about 2-step computation of the step function, Step 3 to

Step 5 contains about $(8+2^2=12)$-step computation of the step function, Step 6 contains about 2-step computation of the step function, and Step 7-8 contains 5-step computations of the step function, we estimate the complexity of the merging phase as $2^{4.4} \times 2/72 + 2^{32.4} \times 12/72 + 2^{32.4} \times 2/72 + 2^{32} \times 5/72 \approx 2^{30}$. Taking the uncontrolled probability of both branches into consideration, the complexity becomes $2^{30+17+8.1} = 2^{55.1}$.

Next, we consider the memory complexity of the merging phase. Given a valid m_9, computing the valid values of m_2 and storing the results costs 2^{29} time and 2^{28} memory. At the pre-computing phase, pre-computing the valid values of m_9 and storing the results costs 2^{32} time and $2^{12.9}$ memory. In addition, pre-computing the equation $T^{\lll 15} \boxplus C_0 = T^{\lll 9}$ costs 2^{32} time and 2^{32} memory. Since the probability of the 36-semi-free-start collision attack is $2^{-57.5}$, one valid m_9 is enough for the improved attack. Therefore, at the merging phase, the memory complexity is $2^{32} + 2^{28}$. Since the time complexity of computing valid m_2, m_9 and pre-computing the equation is much smaller than $2^{55.1}$, it can be ignored. In summary, the time complexity of the semi-free-start collision attack on RIPEMD-160 reduced to 36 steps is $2^{55.1}$ and the memory requirements are $2^{32} + 2^{28} + 2^{12.9} \approx 2^{32}$. The implementation of this attack is available at https://github.com/Crypt-CNS/RIPEMD160-36Step.git.

7 Conclusion

In this paper, we propose a feasible method to theoretically calculate the step differential probability, which was stated as an open problem at ASIACRYPT 2013. Besides, we propose a method to reduce the time complexity and memory complexity to pre-compute the equation $T^{\lll S_0} \boxplus C_0 = (T \boxplus C_1)^{\lll S_1}$. Based on our analysis of the expectation of the number of the solutions to this equation, we conclude that our new way to obtain the solutions only costs four times of checking. In addition, we construct a differential path where the left branch is sparse and the right branch is controlled as sparse as possible. Using the message modification techniques and deducing some extra bit conditions based on the equation that Q_i has to satisfy, it is possible to mount a 30-step collision attack on RIPEMD-160 with probability about 2^{-67}. What's more, based on the 36-step differential path found by Mendel *et al.*, we take a different strategy to choose the message words for merging. In this way, we improve the time complexity of the semi-free-start attack on the first 36-step RIPEMD-160. Compared with the best analytical result of this attack on RIPEMD-160, we reduce the time complexity from $2^{70.4}$ to $2^{55.1}$. Moreover, our improvement also brings us some insights into the choice of message words for merging. Therefore, the message words for merging should be determined with care, which will make a difference.

Acknowledgements. The authors would like to thank the anonymous reviewers for their helpful comments and suggestions. Fukang Liu and Gaoli Wang are supported by the National Natural Science Foundation of China (Nos. 61572125, 61632012, 61373142), and Shanghai High-Tech Field Project (No. 16511101400). Florian Mendel has been supported by the Austrian Science Fund (FWF) under grant P26494-N15.

References

1. Bosselaers, A., Preneel, B. (eds.): RIPE 1992. LNCS, vol. 1007. Springer, Heidelberg (1995). https://doi.org/10.1007/3-540-60640-8
2. Damgård, I.B.: A design principle for hash functions. In: Brassard, G. (ed.) CRYPTO 1989. LNCS, vol. 435, pp. 416–427. Springer, New York (1990). https://doi.org/10.1007/0-387-34805-0_39
3. Dobbertin, H., Bosselaers, A., Preneel, B.: RIPEMD-160: a strengthened version of RIPEMD. In: Gollmann, D. (ed.) FSE 1996. LNCS, vol. 1039, pp. 71–82. Springer, Heidelberg (1996). https://doi.org/10.1007/3-540-60865-6_44
4. Dobbertin, H.: RIPEMD with two-round compress function is not collision-free. J. Cryptol. **10**(1), 51–69 (1997)
5. Landelle, F., Peyrin, T.: Cryptanalysis of full RIPEMD-128. In: Johansson, T., Nguyen, P.Q. (eds.) EUROCRYPT 2013. LNCS, vol. 7881, pp. 228–244. Springer, Heidelberg (2013). https://doi.org/10.1007/978-3-642-38348-9_14
6. Daum, M.: Cryptanalysis of hash functions of the MD4-family (2005). http://www-brs.ub.ruhr-uni-bochum.de/netahtml/HSS/Diss/DaumMagnus/diss.pdf
7. Mendel, F., Nad, T., Schläffer, M.: Finding SHA-2 characteristics: searching through a minefield of contradictions. In: Lee, D.H., Wang, X. (eds.) ASIACRYPT 2011. LNCS, vol. 7073, pp. 288–307. Springer, Heidelberg (2011). https://doi.org/10.1007/978-3-642-25385-0_16
8. Mendel, F., Nad, T., Schläffer, M.: Collision attacks on the reduced dual-stream hash function RIPEMD-128. In: Canteaut, A. (ed.) FSE 2012. LNCS, vol. 7549, pp. 226–243. Springer, Heidelberg (2012). https://doi.org/10.1007/978-3-642-34047-5_14
9. Mendel, F., Nad, T., Scherz, S., Schläffer, M.: Differential attacks on reduced RIPEMD-160. In: Gollmann, D., Freiling, F.C. (eds.) ISC 2012. LNCS, vol. 7483, pp. 23–38. Springer, Heidelberg (2012). https://doi.org/10.1007/978-3-642-33383-5_2
10. Mendel, F., Nad, T., Schläffer, M.: Improving local collisions: new attacks on reduced SHA-256. In: Johansson, T., Nguyen, P.Q. (eds.) EUROCRYPT 2013. LNCS, vol. 7881, pp. 262–278. Springer, Heidelberg (2013). https://doi.org/10.1007/978-3-642-38348-9_16
11. Mendel, F., Peyrin, T., Schläffer, M., Wang, L., Wu, S.: Improved cryptanalysis of reduced RIPEMD-160. In: Sako, K., Sarkar, P. (eds.) ASIACRYPT 2013. LNCS, vol. 8270, pp. 484–503. Springer, Heidelberg (2013). https://doi.org/10.1007/978-3-642-42045-0_25
12. Merkle, R.C.: One way hash functions and DES. In: Brassard, G. (ed.) CRYPTO 1989. LNCS, vol. 435, pp. 428–446. Springer, New York (1990). https://doi.org/10.1007/0-387-34805-0_40
13. Ohtahara, C., Sasaki, Y., Shimoyama, T.: Preimage attacks on step-reduced RIPEMD-128 and RIPEMD-160. In: Lai, X., Yung, M., Lin, D. (eds.) Inscrypt 2010. LNCS, vol. 6584, pp. 169–186. Springer, Heidelberg (2011). https://doi.org/10.1007/978-3-642-21518-6_13
14. Sasaki, Y., Wang, L.: Distinguishers beyond three rounds of the RIPEMD-128/-160 compression functions. In: Bao, F., Samarati, P., Zhou, J. (eds.) ACNS 2012. LNCS, vol. 7341, pp. 275–292. Springer, Heidelberg (2012). https://doi.org/10.1007/978-3-642-31284-7_17

15. Stevens, M., Bursztein, E., Karpman, P., Albertini, A., Markov, Y.: The first collision for full SHA-1. In: Katz, J., Shacham, H. (eds.) CRYPTO 2017. LNCS, vol. 10401, pp. 570–596. Springer, Cham (2017). https://doi.org/10.1007/978-3-319-63688-7_19
16. Wang, G., Wang, M.: Cryptanalysis of reduced RIPEMD-128. J. Softw. **19**(9), 2442–2448 (2008)
17. Wang, G.: Practical collision attack on 40-step RIPEMD-128. In: Benaloh, J. (ed.) CT-RSA 2014. LNCS, vol. 8366, pp. 444–460. Springer, Cham (2014). https://doi.org/10.1007/978-3-319-04852-9_23
18. Wang, G., Yu, H.: Improved cryptanalysis on RIPEMD-128. IET Inf. Secur. **9**(6), 354–364 (2015)
19. Wang, G., Shen, Y., Liu, F.: Cryptanalysis of 48-step RIPEMD-160. IACR Trans. Symmetric Cryptol. **2017**(2), 177–202 (2017)
20. Wang, X., Lai, X., Feng, D., Chen, H., Yu, X.: Cryptanalysis of the hash functions MD4 and RIPEMD. In: Cramer, R. (ed.) EUROCRYPT 2005. LNCS, vol. 3494, pp. 1–18. Springer, Heidelberg (2005). https://doi.org/10.1007/11426639_1
21. Wang, X., Yu, H.: How to break MD5 and other hash functions. In: Cramer, R. (ed.) EUROCRYPT 2005. LNCS, vol. 3494, pp. 19–35. Springer, Heidelberg (2005). https://doi.org/10.1007/11426639_2
22. Wang, X., Yu, H., Yin, Y.L.: Efficient collision search attacks on SHA-0. In: Shoup, V. (ed.) CRYPTO 2005. LNCS, vol. 3621, pp. 1–16. Springer, Heidelberg (2005). https://doi.org/10.1007/11535218_1
23. Wang, X., Yin, Y.L., Yu, H.: Finding collisions in the full SHA-1. In: Shoup, V. (ed.) CRYPTO 2005. LNCS, vol. 3621, pp. 17–36. Springer, Heidelberg (2005). https://doi.org/10.1007/11535218_2

Linear Cryptanalysis of DES with Asymmetries

Andrey Bogdanov and Philip S. Vejre(✉)

Technical University of Denmark, Kongens Lyngby, Denmark
{anbog,psve}@dtu.dk

Abstract. Linear cryptanalysis of DES, proposed by Matsui in 1993, has had a seminal impact on symmetric-key cryptography, having seen massive research efforts over the past two decades. It has spawned many variants, including multidimensional and zero-correlation linear cryptanalysis. These variants can claim best attacks on several ciphers, including PRESENT, Serpent, and CLEFIA. For DES, none of these variants have improved upon Matsui's original linear cryptanalysis, which has been the best known-plaintext key-recovery attack on the cipher ever since. In a revisit, Junod concluded that when using 2^{43} known plaintexts, this attack has a complexity of 2^{41} DES evaluations. His analysis relies on the standard assumptions of right-key equivalence and wrong-key randomisation.

In this paper, we first investigate the validity of these fundamental assumptions when applied to DES. For the right key, we observe that strong linear approximations of DES have more than just one dominant trail and, thus, that the right keys are in fact *inequivalent* with respect to linear correlation. We therefore develop a new right-key model using Gaussian mixtures for approximations with several dominant trails. For the wrong key, we observe that the correlation of a strong approximation after the partial decryption with a wrong key still shows much *non-randomness*. To remedy this, we propose a novel wrong-key model that expresses the wrong-key linear correlation using a version of DES with more rounds. We extend the two models to the general case of multiple approximations, propose a likelihood-ratio classifier based on this generalisation, and show that it performs better than the classical Bayesian classifier.

On the practical side, we find that the distributions of right-key correlations for multiple linear approximations of DES exhibit exploitable *asymmetries*. In particular, not all sign combinations in the correlation values are possible. This results in our improved multiple linear attack on DES using 4 linear approximations at a time. The lowest computational complexity of $2^{38.86}$ DES evaluations is achieved when using $2^{42.78}$ known plaintexts. Alternatively, using 2^{41} plaintexts results in a computational complexity of $2^{49.75}$ DES evaluations. We perform practical experiments to confirm our model. To our knowledge, this is the best attack on DES.

Keywords: Linear cryptanalysis · DES · Mixture models · Right-key equivalence · Wrong-key randomisation · Linear hull · Multiple linear

T. Takagi and T. Peyrin (Eds.): ASIACRYPT 2017, Part I, LNCS 10624, pp. 187–216, 2017.
https://doi.org/10.1007/978-3-319-70694-8_7

1 Introduction

Accepted as a standard in 1976 by the National Bureau of Standards (later NIST), DES can now celebrate its fortieth birthday. Being a highly influential cipher, it has inspired much cryptanalysis. Triple-DES is still massively deployed in conservative industries such as banking. Moreover, it is used to secure about 3% of Internet traffic [1].

The first attack on the full DES came in 1992, where Biham and Shamir demonstrated that *differential cryptanalysis* enabled a key recovery using 2^{47} *chosen plaintexts* in time 2^{37} [2]. The year after, in 1993, Matsui introduced a new cryptanalytic technique, *linear cryptanalysis*, which DES proved especially susceptible to. While the first iteration of the attack required 2^{47} *known plaintexts* [21], Matsui soon improved his attack to only require 2^{43} known texts, taking 2^{43} time to recover the key. This complexity estimate was lowered to 2^{41} by Junod in [17]. In [18], Knudsen and Mathiassen lower the complexity to 2^{42} plaintexts, however this attack uses *chosen plaintexts.*

In this paper we present the first successful attack on full DES using multiple linear approximations. By developing new models for the correlation distributions, and by exploiting asymmetries in the right-key distribution, we obtain an improved key-recovery attack. Using $2^{42.78}$ known plaintexts, the attack recovers the key in time equal to $2^{38.86}$ DES encryptions.

1.1 Previous Work and Problems

Linear cryptanalysis has proven to be widely applicable, and has spawned many variants and generalisations. Amongst them are differential-linear cryptanalysis [19], multiple linear cryptanalysis [3,16], multidimensional linear cryptanalysis [14,15], zero-correlation linear cryptanalysis [5,6], multivariate linear cryptanalysis [8], etc. These techniques have successfully been applied to a wide range of ciphers, including Serpent [14,23], PRESENT [8,9], Camellia and CLEFIA [4], and CAST-256 [27].

Matsui first introduced the concept of a *linear approximation* of a block cipher in [21]. If we denote the encryption of a plaintext $\mathcal{P}$ using key K by $\mathcal{C} = E_K(\mathcal{P})$, then a linear approximation of this cipher is a pair of masks, (α, β), which indicate some bits of the plaintext and ciphertext. The idea is to find α and β such that the sum of plaintext bits indicated by α is strongly correlated to the sum of ciphertext bits indicated by β. A measure of the strength of a linear approximation is the *linear correlation*, defined by

$$C_K(\alpha, \beta) = 2 \cdot \Pr(\langle \alpha, x \rangle \oplus \langle \beta, E_K(x) \rangle = 0) - 1,$$

where $\langle \cdot, \cdot \rangle$ is the canonical inner product. Matsui showed how an approximation with linear correlation that deviates significantly from zero can be used to attack the cipher, and found such approximations for DES. The attack procedure was formalised as Algorithm 2, in which an attacker obtains plaintext-ciphertext pairs over r rounds of a cipher. The attacker then guesses the outer round keys in order to encrypt/decrypt the outer rounds, and compute the correlation over $r - 2$ rounds.

Standard assumptions for linear cryptanalysis on DES. In [17] Junod revisited Matsui's attack, and concluded that Matsui's original complexity was slightly overestimated. Junod instead estimated that the attack could be performed in time 2^{41} using the same number of known plaintexts. Central to both Matui's and Junod's analysis are two assumptions.

Assumption A (Right-Key Equivalence). *For a linear approximation* (α, β)*, the magnitude of the correlation,* $|C_K(\alpha, \beta)|$*, does not deviate significantly from its expected value over all keys, that is,* $|C_K(\alpha, \beta)| = \mathrm{E}(|C_K(\alpha, \beta)|)$.

Problem 1: Insufficient Right-Key Distribution: The assumption of right-key equivalence is usually the result of assuming that the magnitude of the linear correlation is determined by a single dominant trail. This further implies that the linear correlation only takes on two values over the key space. However, in [24], Nyberg first introduced the concept of a *linear hull*, i.e. the collection of all trails of a linear approximation, and showed that Assumption A is not true in general. In [7], Bogdanov and Tischhauser gave a refined version of Assumption A, which takes a larger part of the hull into account. However, to the best of our knowledge, no thorough exploration of the right-key distribution for DES has been conducted, and it is unclear how accurate Assumption A is in this context.

Assumption B (Wrong-Key Randomisation). *In the context of Algorithm 2, the correlation of a linear approximation* (α, β) *is equal to 0 for all wrong guesses of the outer round keys.*

Problem 2: Unrealistic Wrong-Key Distribution: The assumption of wrong-key randomisation implies that if an attacker guesses the wrong outer round keys in Algorithm 2, the resulting texts pairs behave in a completely random way, i.e. the linear correlation will be equal to zero. A refined version of this assumption was given by Bogdanov and Tischhauser in [7], where the wrong-key distribution was given as the Gaussian distribution $\mathcal{N}(0, 2^{-n})$, where n is the block size. This distribution matches that of an ideal permutation. Neither of these assumptions have been verified for DES. Indeed, DES exhibits very strong linear approximations, and it is not clear if a wrong key guess is sufficient to make the linear correlation close to that of an ideal permutation.

Linear cryptanalysis of DES with multiple approximations. While several models for using multiple approximations for linear cryptanalysis have been proposed, see e.g. [3,8,14–16,26], the application to DES has been very limited. In [16], Kaliski and Robshaw specifically note that their approach is limited when applied to DES. In [26], Semaev presents an alternative approach, but does not obtain better results than Matsui's original attack.

The most promising attempt was given in [3] by Biryukov et al. Under Assumption A, when using M approximations, the key space can be partitioned into at most 2^M key classes based on the signs of the M linear correlations. This allowed Biyukov et al. to describe the correlation of each key class as an

M-variate normal distribution $\mathcal{N}_M(\boldsymbol{\mu}_i, 1/N \cdot \mathbf{I})$, where $\mathbf{I}$ is an $M \times M$ identity matrix, and the mean vector is given by

$$\boldsymbol{\mu}_i = (s_{i,1}|C_K(T_1)|, \ldots, s_{i,M}|C_K(T_M)|)^\top,$$

where $s_{i,j} \in \{-1, 1\}$ describes the sign combination of the i'th key class. Based on this, they developed a Bayesian classifier, in order to decide between a correct or incorrect guess of the round keys in Algorithm 2.

Problem 3: Applying Multiple Linear Cryptanalysis to DES: While Biryukov et al. demonstrate that their method of using multiple approximations can potentially reduce the complexity of Matsui's attack, they also note that the structure of DES makes it difficult to arbitrarily use a large number of approximations. As such, they did not present a new attack on DES. Similar observations were made by Kaliski and Robshaw in [16]. To the best of our knowledge, no other variants of linear cryptanalysis which uses multiple approximations have been able to outperform Matsui's original attack.

1.2 Our Contributions

More Accurate Right-Key Model for DES. In Sect. 3 we consider Problem 1, i.e. the fundamental problem of the DES right-key distribution. We enumerated over 1000 trails for the linear approximation used by Matsui, and calculated the resulting correlation distribution for 1 million keys. We demonstrate in Sect. 3.2 that while this distribution does have two modes symmetric around zero, each mode does not consist of a single value, as predicted by Assumption A. Indeed, it is not even the case that each mode takes on a simple Gaussian distribution. As such, one cannot consider different keys to have equivalent behaviour.

We therefore develop a new model for the right-key distribution in Sect. 3.3. This model is given below, and expresses the distribution as a mixture of Gaussian components. An example of this model applied to DES is shown in Fig. 1.

Model A (Right-Key Equivalence for One Approximation). *Consider a linear approximation (α, β) of r rounds of DES. The distribution of the linear correlation $C_K(\alpha, \beta)$ over the key space is approximately given by a Gaussian mixture for some weights λ_i and components $\mathcal{N}(\mu_i, \sigma_i^2)$, $i = 1, \ldots, \ell$.*

Applying this model to the approximations used by Matsui, we show that it is able to accurately describe the observed distribution. Moreover, it is interesting to note that the component associated with the dominant trail *only accounts for 30% of the correlation, contrasting Assumption* A. We furthermore apply the mixture model to describe the full correlation distribution observed during an attack. We note that when the number of texts used in the attack is small, the right-key distribution originally given by Matsui is a good approximation. However, we stress that the cryptanalyst should carefully examine the right-key distribution when this is not the case.

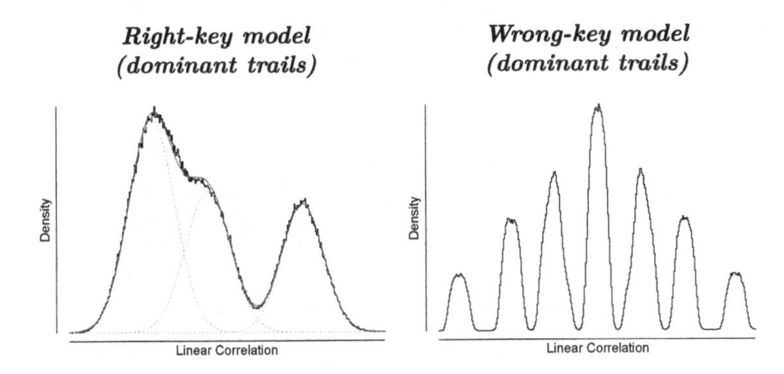

Fig. 1. Our new models for the distributions of linear correlation over the key space for DES. The distributions are expressed as Gaussian mixtures. The model shows a deviation from the standard assumptions of right-key equivalence and wrong-key randomisation.

New Wrong-Key Model for DES. In Sect. 4 we consider Problem 2. In order to obtain a wrong-key model that more accurately describes the case of a wrong key guess in Algorithm 2, we propose the following new approach.

Model B (Non-random Wrong-Key Distribution). *Consider an Algorithm 2 style attack on r rounds of DES using a linear approximation (α, β) over $r-2$ rounds. Let R_K be the keyed round function of DES, and let $E_K^{\star}$ denote the r-round encryption function. For a wrong guess of the outer round keys, the correlation will be distributed as for the cipher*

$$E'_K(x) = R_{K_a}^{-1}(E_K^{\star}(R_{K_b}^{-1}(x))), \tag{1}$$

where K_a and K_b are chosen uniformly at random.

This model accurately matches the situation of guessing the wrong outer round keys in an Algorithm 2 attack. We enumerated over 900 trails for the linear approximation used by Matsui for the cipher E', and calculated the resulting correlation distribution for 1 million keys. The result is shown in Fig. 1. While the distribution has mean zero, the shape of the distribution does not match Assumption B, nor that of the revised version by Bogdanov and Tischhauser, as its variance is much larger than 2^{-n}. As is the case for the right-key distribution, the wrong-key distribution is also not a simple Gaussian, but rather some Gaussian mixture. Again, for low data complexities, we demonstrate that a Gaussian model is sufficient to describe the wrong-key distribution observed during an attack, but advise caution when the data complexity is close to full codebook.

Multiple Linear Cryptanalysis with Asymmetries. In Sects. 5 and 6 we remedy Problem 3. We develop a classifier for M approximations based on the likelihood-ratio of the right-key and wrong-key distributions developed in Sects. 3 and 4. This classifier is given by

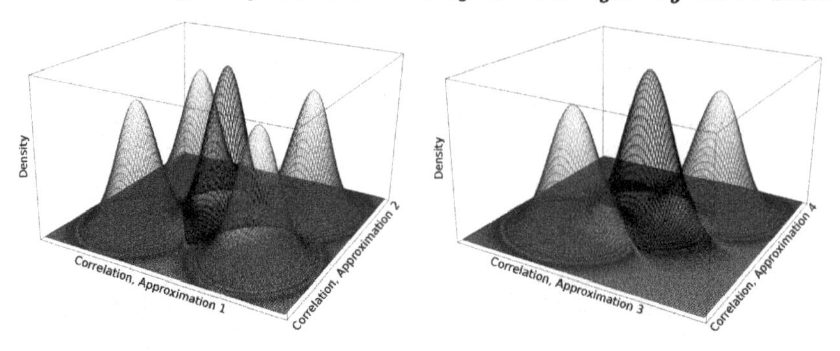

Fig. 2. An illustration of the difference between a symmetric and an asymmetric joint distribution of linear correlation for two approximations over the key space. The right-key distribution is blue, while the wrong-key distribution is red. (Color figure online)

$$\Lambda(\mathbf{x}) = \frac{\sum_{i=1}^{\ell} \lambda_i \phi_M(\mathbf{x}; \boldsymbol{\mu}_i, \boldsymbol{\Sigma}_i + (2^{-n} + 1/N)\mathbf{I})}{\phi_M(\mathbf{x}; \mathbf{0}, \boldsymbol{\Sigma}_W + (2^{-n} + 1/N)\mathbf{I})},$$

where ϕ is the probability density function (PDF) of the Gaussian distribution. The wrong-key distribution is a simple M-variate Gaussian. The right-key distribution is a mixture of at most 2^M, M-variate components based on the signs of the M correlations. In contracts to the work in [3], we do not partition the key space, but express the correlation distribution over the entire key space. Also in contrast to this work, our classifier directly takes the wrong-key distribution into account. We demonstrate how this improves the classifier.

We make the interesting observation that if the right-key distribution is asymmetric, that is, if the number of components is less than 2^M, we obtain a stronger classifier. This situation is demonstrated in Fig. 2. From this example, one can get an intuitive understanding of how an asymmetric distribution makes it easier to distinguish between right-key and wrong-key. We therefore propose the term *symmetry factor*, namely the ratio between number of components and 2^M, and conjecture that a lower symmetry factor will result in a stronger attack.

First Successful Multiple Linear Cryptanalysis of DES. By using the asymmetric classifier in Sect. 6, we give the first attack on full DES using multiple linear approximations which improves Matsui's original attack. We use two sets of four linear approximations. Using $2^{42.78}$ known plaintexts, the attack recovers the key in time equal to $2^{38.86}$ encryptions, with a success probability of 85%. This is 4.4 times faster than Junod's estimate of Matsui's attack, and uses $2^{40.2}$ fewer texts. We confirm these results by measuring the actual correlation distributions using this number of texts for 1300 random keys, and computing the resulting advantage of our classifier. We find that the model fits our practical results very well. Alternatively, we can lower the data complexity to 2^{41}, and recover the key in time $2^{49.76}$, with a success probability of 80%. Our attack is compared to previous attacks on full DES in Table 1.

Table 1. Comparison of key-recovery attacks on full DES.

Technique	Data complexity	Time complexity	Success probability	Attack scenario	Source
Differential	$2^{47.00}$	$2^{37.00}$	58%	Chosen plaintext	[2]
Linear	$2^{43.00}$	$2^{43.00}$	85%	Known plaintext	[22]
Linear	$2^{43.00}$	$2^{41.00}$	85%	Known plaintext	[17]
Multiple linear	$2^{42.78}$	$2^{38.86}$	85%	Known plaintext	Sect. 6
Multiple linear	$2^{41.00}$	$2^{49.76}$	80%	Known plaintext	Sect. 6

2 Linear Cryptanalysis of DES

In 1993, Matsui introduced the concept of linear cryptanalysis and applied it to derive a key-recovery attack on the full 16-round DES [21,22]. In this section, we briefly outline the attack. We then give an overview of the assumptions Matsui made in his analysis, and show the resulting complexity of the attack. Moreover, we show a variant of linear cryptanalysis due to Biryukov et al. [3], which will be important for the remaining part of this work.

2.1 Basics of Linear Cryptanalysis

We consider a block cipher with block length n and key length κ. We denote the encryption of plaintext $\mathcal{P} \in \mathbb{F}_2^n$ under key $K \in \mathbb{F}_2^\kappa$ by $E_K(\mathcal{P})$. The idea of linear cryptanalysis is to find a *linear approximation* $(\alpha, \beta) \in \mathbb{F}_2^n \times \mathbb{F}_2^n$ such that the magnitude of its *linear correlation*, defined by

$$C_K(\alpha, \beta) = 2 \cdot \Pr(\langle \alpha, x \rangle \oplus \langle \beta, E_K(x) \rangle = 0) - 1,$$

is large. Here, $\langle \cdot, \cdot \rangle$ denotes the canonical inner product on $\mathbb{F}_2^n$. Thus, the correlation is a measure of how often the parity bit $\langle \alpha, \mathcal{P} \rangle$ of the plaintext is equal to the parity bit $\langle \beta, \mathcal{C} \rangle$ of the ciphertext. We expect a strong cipher to only have approximations with linear correlation close to 0, and hence a correlation value that deviates significantly from 0 indicates a weakness of the cipher.

For Feistel ciphers, such as DES, the linear correlation of an approximation (α, β) can be calculated by considering so called *linear trails* of the cipher. We define a single-round linear trail of DES as the triple $(u, t, v) \in \mathbb{F}_2^n \times \mathbb{F}_2^m \times \mathbb{F}_2^n$, where m is the size of a single round key. The linear correlation of this single-round trail is then defined as

$$C_{K_r}(u, t, v) = 2 \cdot \Pr(\langle u, x \rangle \oplus \langle v, R_{K_r}(x) \rangle = \langle t, K_r \rangle) - 1,$$

where R_{K_r} is the DES round-function using the r'th round key K_r. We now define a linear trail T over r rounds as a collection of single-round trails (u_i, t_i, u_{i+1}), $i = 0, \ldots, r-1$, as well as the *correlation contribution* of the trail T as [11,12] $C_K(T) = \prod_{i=0}^{r-1} C_{K_i}(u_i, t_i, u_{i+1})$. We will also make use of the concept of an associated *key trail* $\bar{T}$ of a trail T. The key trail is defined as the concatenation of the t_i, $i = 0, \ldots, r-1$.

Daemen and Rijmen demonstrated that the correlation contribution of a trail can be written as [11,12]

$$C_K(T) = (-1)^{s_T \oplus \langle \bar{T}, \bar{K} \rangle} |C_K(T)|, \tag{2}$$

where s_T is a sign bit specific to the trail T, and $\bar{K}$ denotes the concatenation of the round keys K_i. Moreover, under the assumption of independent round keys, $|C_K(T)|$ is independent of the key. Thus, the correlation contribution of a trail T has a fixed magnitude for all keys, but the sign is determined by the round key bits indicated by the key trail $\bar{T}$. Finally, Daemen and Rijmen give the correlation over all r rounds for some approximation (α, β) as [11,12]

$$C_K(\alpha, \beta) = \sum_{u_0 = \alpha, u_r = \beta} C_K(T) = \sum_{u_0 = \alpha, u_r = \beta} (-1)^{s_T \oplus \langle \bar{T}, \bar{K} \rangle} |C_K(T)|, \tag{3}$$

i.e. the sum of the correlation contributions of all trails from α to β.

2.2 Matsui's Approach

Matsui's key observation was that DES exhibits linear trails where the correlation contribution deviates significantly from zero. Consider the full 16-round DES, let $\mathcal{P}$ be the plaintext, and let $\mathcal{C}$ be the ciphertext. Let $[i_0, \ldots, i_\ell]$ denote an element in $\mathbb{F}_2^n$ whose i_j'th components are 1, $j = 0, \ldots, \ell$, while all other components are 0. Then, over 14 rounds of DES, the approximations

$$\gamma_1 = ([7, 18, 24], [7, 18, 24, 29, 47]) \quad \text{and} \quad \delta_3 = ([15, 39, 50, 56, 61], [39, 50, 56]),$$

both have trails with correlation contribution $C_K(T) = \pm 2^{-19.75}$ [22]. From (2) we can determine one bit of information if we know the sign of $C_K(T)$, namely the parity $\langle \bar{T}, \bar{K} \rangle$ of the round key bits indicated by the key trail $\bar{T}$. Let k_f denote the key-bits of round key K_0 required to partially encrypt a plaintext $\mathcal{P}$ one round and calculate $\langle \alpha, R_{K_0}(\mathcal{P}) \rangle$, and let k_b denote the key-bits of round key K_{r-1} required to partially decrypt the ciphertext $\mathcal{C}$ one round and calculate $\langle \beta, R_{K_{r-1}}^{-1}(\mathcal{C}) \rangle$. Matsui developed the following general approach in order to determine $|k_f| + |k_b| + 1$ key bits, formalised as Algorithm 2.

Algorithm 2

1. Obtain N plaintext-ciphertext pairs.
2. For each guess of the key-bits (k_f, k_b), partially encrypt/decrypt each plaintext-ciphertext pair $(\mathcal{P}, \mathcal{C})$ and calculate the number of times L_i the input parity $\langle \alpha, R_{R_0}(\mathcal{P}) \rangle$ is equal to the output partiy $\langle \beta, R_{R_{r-1}}^{-1}(\mathcal{C}) \rangle$ for the i'th guess, $i = 1, \ldots, 2^{|k_f| + |k_b|}$.

3. For each counter L_i, if $L_i > N/2$, guess that the sign bit $\langle \bar{T}, \bar{K} \rangle = s_T$, otherwise guess that $\langle \bar{T}, \bar{K} \rangle = s_T \oplus 1$.
4. For any counter L_i with $|T_i - N/2| > \Gamma$, for a predetermined value Γ, guess the remaining $\kappa - (|k_f| + |k_b| + 1)$ bits of the master key K, and determine the correct value of K through trial encryption.

For his attack on DES, Matsui performed Algorithm 2 once for γ_1 and once for δ_3, determining 26 bits before guessing the remaining 30 bits of K. In his analysis of the success rate and complexity of the attack, Matsui assumed that the linear correlation of the approximations γ_1 and δ_3 were only determined by a single trail T. The idea is that the correlation contribution of T is much larger than that of all other trails – a so called *dominant trail.* We will call the associated key trail $\bar{T}$ of such a trail a *dominant key trail.* In the presence of such a dominant trail, $C_K(\alpha, \beta)$ only takes on two values over the key space. This can be seen from Eq. (3), as the case of a dominant trail implies that this sum only has one term. Under this assumption, Matsui concluded that when using 2^{43} texts, there is an 85% probability of recovering the key at a time complexity of 2^{43} DES encryptions. In a later analysis of Matsui's attack [17], Junod concluded that the actual computational complexity is closer to 2^{41} DES encryptions.

2.3 Biryukov et al. – Multiple Approximations

A natural extension of Matsui's linear cryptanalysis is to attempt to use multiple linear approximations simultaneously. The first attempt at developing such a framework was by Kaliski and Robshaw in [16]. This work has the limitation that all linear approximations must have the same dominant key trail, and the approximations were assumed to be statistically independent. Moreover, as Kaliski and Robshaw note, the application of this method to DES is very limited.

Another approach was undertaken by Biryukov et al. in [3]. Here, the approximations can in principle be picked arbitrarily, but the framework still requires the assumption of one dominant trail for each approximation, and independence between approximations. Due to these restrictions, the foundations of multidimensional linear cryptanalysis was developed in e.g. [14,15]. While this approach has been applied with great success to a large range of ciphers, no results have been shown on DES. Thus, Matsui's single linear cryptanalysis still provides the best results on this cipher.

Let us briefly reconsider the method by Biryukov et al., assuming the use of M linear approximations. The idea is to partition the key space into at most 2^M classes based on the parity of the $\langle \bar{T}_i, \bar{K} \rangle$, where $\bar{T}_i$ is the dominant key trail of the i'th approximation. An Algorithm 2 type attack is then performed: For each guess of the key-bits (k_f, k_b), the vector $(L_{i,1}, \ldots, L_{i,M})$ is calculated, and the likelihood of that vector belonging to each of the key classes is computed. The right guess of (k_f, k_b) should yield one class with high likelihood, and the class then indicates at most M parity bits, $\langle \bar{T}_i, \bar{K} \rangle$. Central to the analysis of [3] are the following two assumptions:

Assumption 1 (Right-Key Equivalence). *For a linear approximation (α, β), the magnitude of the correlation, $|C_K(\alpha, \beta)|$, does not deviate significantly from its expected value over all keys, that is, $|C_K(\alpha, \beta)| = \mathrm{E}(|C_K(\alpha, \beta)|)$.*

Assumption 2 (Wrong-Key Randomisation). *For Algorithm 2, the correlation of a linear approximation (α, β) is 0 for all wrong guesses of (k_f, k_b).*

The assumption of right-key equivalence implies that the linear approximation has one dominant trail, say T, and consequently the distribution of the correlation over the key space only takes on two values, namely $\pm|C_K(T)|$. Thus, the natural partitioning of the key space for M approximations is the partitioning induced by the sign of the correlations, i.e. the vector $((-1)^{\langle \bar{T}_1, \bar{K} \rangle}, \ldots, (-1)^{\langle \bar{T}_M, \bar{K} \rangle})$. In practice however, the correlations are calculated from the counters $L_{i,j}$. The joint distribution of the resulting measured correlations, for some specific key class, is given in [3] as an M-variate normal distribution, described in the following model.

Model 1 (Right-Key Partitioning for Multiple Approximations [3]). *Consider a set of linear approximations $(\alpha_1, \beta_1), \ldots, (\alpha_M, \beta_M)$ of r rounds of DES. Then, the key space can be partitioned into at most 2^M key classes based on the signs of the correlations. The undersampled distribution of the linear correlation vector, using N texts and restricted to the i'th key class, denoted by $C_i^N(\boldsymbol{\alpha}, \boldsymbol{\beta})$, is an M-variate normal distribution*

$$C_i^N(\boldsymbol{\alpha}, \boldsymbol{\beta}) \sim \mathcal{N}_M(\boldsymbol{\mu}_i, 1/N \cdot \mathbf{I}).$$

The mean vector of the i'th key class is given by $\boldsymbol{\mu}_i[j] = s_{i,j}|C_K(T_i)|$, where $s_{i,j} \in \{-1, 1\}$ describes the sign combination of the i'th key class, $j = 1, \ldots, M$.

Based on this model, a Bayesian classifier is constructed. We refer to Sect. 5 for the details. While the approach presented by Biryukov et al. seems promising, it has yet to result in an improved attack on DES. To the best of our knowledge, no other variants of linear cryptanalysis which uses multiple approximations have been able to outperform Matsui's original attack. Moreover, while updated versions of Assumptions 1 and 2 have been applied to other ciphers, no such work exists for DES. In the following, we address these concerns. We consider the right-key distribution in Sect. 3, and the wrong-key distribution in Sect. 4. Using the results obtained in these sections, we develop an improved linear attack on DES in Sects. 5 and 6.

3 Right-Key Correlation for DES: Key Inequivalence

In this section, we consider the correlation distribution of DES approximations over the key space. In Sect. 3.1, we consider current models for this distribution, as well as the undersampled distribution. In Sect. 3.2, we enumerate a large number of trails for DES, and show that, contrary to Assumption 1, the absolute value of the correlation does vary significantly as the key changes. In fact, the correlation distribution has a complicated structure. In Sect. 3.3, we develop a new model for this correlation based on Gaussian mixtures, which is able to accurately describe this structure. Moreover, we extend the model to describe the full undersampled correlation distribution over keys for multiple approximations.

3.1 The Correlation Distribution of a Single Approximation

As mentioned, most linear cryptanalysis of DES assumes that each linear approximation has one dominant trail, determining the magnitude of the absolute correlation. This idea is effectively expressed by Assumption 1. Consider, for example, one of the approximations used by Matsui, γ_1. This approximation has a primary trail T_A over 14 rounds of DES with correlation contribution $C_K(T_A) = \pm 2^{-19.75}$. In [24], Nyberg first introduced the concept of a linear hull, i.e. the collection of all trails of a linear approximation, and showed that Assumption 1 is not true in general. For γ_1, the trail with second largest correlation contribution, T', has contribution $C_K(T') = \pm 2^{-25.86}$. While the contribution from this trail is not large enough to change the sign of the linear correlation $C_K(\gamma_1)$, or increase/decrease the magnitude of the correlation much, it does not match the model given in Assumption 1. When including the second trail, the correlation distribution does not take on only two distinct values, but four.

Signal/noise decomposition. In order to refine Assumption 1, Bogdanov and Tischhauser considered a *signal/noise decomposition* of the hull in [7]. Consider a situation in which d dominant trails of an approximation (α, β) are known. We call this collection of trails the *signal*, and define the *signal correlation* as the sum of their correlation contributions

$$C'_K(\alpha, \beta) = \sum_{i=1}^{d} (-1)^{s_{T_i} \oplus \langle \bar{T}_i, \bar{K} \rangle} |C_K(T_i)|.$$

The remaining part of the hull is unknown, and is modelled as *noise*, with the distribution $\mathcal{N}(0, 2^{-n})$. Then, the refined right-key equivalence assumption of [7] states that the correlation of (α, β) is given by the sum of the signal correlation and the noise:

$$C_K(\alpha, \beta) = C'_K(\alpha, \beta) + \mathcal{N}(0, 2^{-n}).$$

Since the approximations we will typically consider in the context of DES have quite high correlation, the addition of the noise term will not make a significant difference. However, we include it for completeness.

Undersampling. The cryptanalyst is most often not interested in having to obtain the full codebook to exactly measure the linear correlation $C_K(\alpha, \beta)$. Therefore, the undersampled distribution is of great interest. Let

$$C_K^N(\alpha, \beta) = \frac{2}{N} \#\{x_i, i = 1, \ldots, N | \langle \alpha, x_i \rangle \oplus \langle \beta, E_K(x_i) \rangle = 0\} - 1$$

be the empirical value of $C_K(\alpha, \beta)$ measured using N text pairs. Here, we assume that x_i is drawn uniformly at random with replacement from $\mathbb{F}_2^n$. Matsui first considered the distribution of $C_K^N(\alpha, \beta)$ over the key space under Assumption 1. In this case, Matsui used the Gaussian distribution $C_K^N(\alpha, \beta) \sim \mathcal{N}(C_K(\alpha, \beta), 1/N)$. While no proof is given in [21], one can show this result via a Gaussian approximation to the binomial distribution, assuming that $|C_K(\alpha, \beta)|$ is small.

3.2 Exploring the Signal Distribution of DES

On the basis of the signal/noise model, we now turn our attention to the signal distribution of DES approximations. By computing the signal correlation C_K' for a large number of trails, we are able to get a good idea of the actual distribution of the correlation C_K. We first describe how the signal trails were enumerated.

Our trail enumeration algorithm. We implemented a bounded breadth-first search in order to enumerate trails of DES approximations over 14 rounds. The algorithm consists of two search phases and a matching phase. Consider an approximation (α, β). The first search phase searches for trails in the forward direction, from round one to round seven. The search starts with α as an input mask to the first round, and then finds t and v such that the single round trails (α, t, v) has non-zero correlation. This process is then repeated for each trail with v as input mask to the second round, etc. The second search phase is similar, but searches backwards from β.

The searches are bounded in two ways. First, we only consider trails that activate at most three S-Boxes in each round. Second, we limit the number of trails which are kept in each round. This is done in such a way that only the trails with largest absolute correlation contribution are kept. This ensures a locally optimal choice, although no such guarantee can be made globally. The number of trails kept is determined by the *branching factor* B, such that in the i'th round of the search, $i \cdot B$ trails are kept.

After the two search phases, each trail found in the forward direction is matched to any trail in the backwards direction which shares the same mask in the middle. In this way, we obtain a number of trails of (α, β) over 14 rounds. Globally optimal trails will have a good chance of being enumerated if the branching factor B is chosen sufficiently large. In the following, we set $B = 1$ million, which means that we can find at most 7 million trails in each search direction. Note that the number of trails eventually discovered by the algorithm highly depends on the number of rounds and the approximation under consideration. We performed the enumeration for the eight approximations given in Table 2 using 20 Intel Xeon Processor E5-2680 cores. The enumeration took about 8 CPU hours.

Computing the Signal Distribution. Using the algorithm described above, we enumerated 1126 trails of the approximation γ_1 over 14 rounds, and calculated the signal correlation

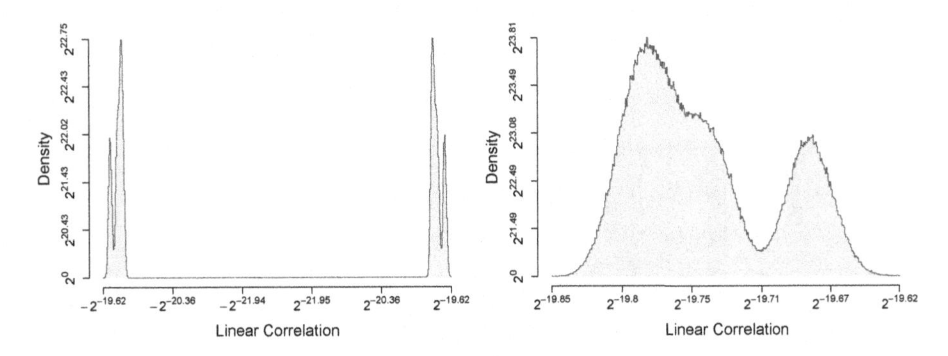

Fig. 3. The signal distribution of linear correlation for the approximation γ_1 over 14 rounds of DES. The signal correlation was calculated using 1126 trails and 1 million randomly drawn keys. The trails had an absolute correlation contribution between $2^{-43.61}$ and $2^{-19.75}$. The left plot shows the two main modes, symmetric around zero. The right plot shows only the positive half of the distribution.

$$C'_K(\gamma_1) = \sum_{i=1}^{1126} (-1)^{s_{T_i} \oplus \langle \bar{T}_i, \bar{K} \rangle} |C_K(T_i)|,$$

for 1 million randomly drawn keys. The trails we found have an absolute correlation contribution between $2^{-43.61}$ and $2^{-19.75}$, and include the dominant trail used by Matsui in [22]. The resulting distribution can be seen in Fig. 3.

The left part of the figure shows the full distribution over the key space. At this scale, the distribution resembles the one described in Sect. 2; there are two very prominent modes symmetric around zero, with peaks around $\pm 2^{-19.75}$, corresponding to the correlation contribution of the dominant trail. However, the right part of the plot, showing the positive half of the distribution, largely contradicts Assumption 1 of key equivalence. While the mean of the distribution is $2^{-19.75}$, it also has a non-negligible standard deviation of $2^{-24.71}$. Moreover, the distribution is not Gaussian. The correlations cluster around three values, namely $2^{-19.79}$, $2^{-19.75}$, and $2^{-19.68}$. Interestingly, the probability density is larger around the cluster with the lowest correlation value.

Under the signal/noise model, adding the noise distribution $\mathcal{N}(0, 2^{-n})$ gives us a good estimate of the actual distribution of the correlation $C_K(\gamma_1)$. However, due to the large variance of the signal distribution, the effect of the noise term is negligible in this case. Thus, the distribution in Fig. 3 should be quite close to the actual distribution. This poses a fundamental problem, as none of the analysis of linear cryptanalysis applied to DES accounts for this type of distribution. Indeed, it is not clear how the distribution of the undersampled correlation, C_K^N, looks, which is essential to know when determining the complexity of linear attacks.

3.3 A New Mixture Model for Single and Multiple Approximations

To relieve the problems discussed in Sect. 3.2, we now propose a model for the correlation distribution based on *Gaussian mixtures*. Consider a distribution in

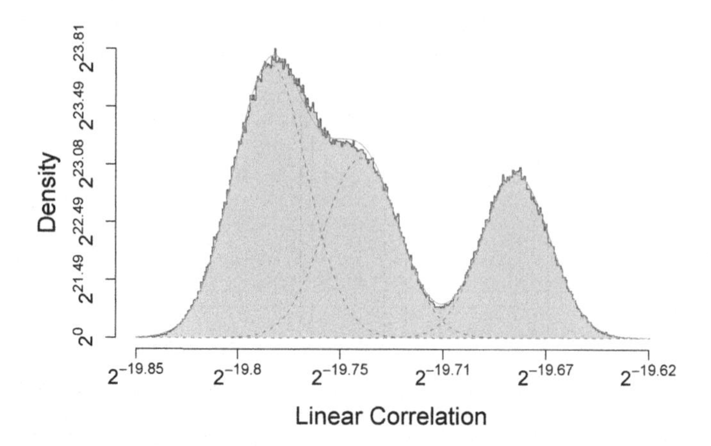

Fig. 4. A Gaussian mixture fitted to the correlation distribution of the linear approximation γ_1 over 14 rounds of DES. The individual components are shown in red, the mixture density is shown in green, and the measured distribution is shown in blue. Under this model, only 30% of the distribution is attributed to the Gaussian component associated with the dominant trail. (Color figure online)

which each sample is drawn from one of ℓ Gaussian distributions. Each Gaussian is called a *component*. The probability of the sample being drawn from the i'th component is λ_i, usually called the *weights*, with $\sum \lambda_i = 1$. The probability density function (PDF) of such a distribution is given by

$$f(x) = \sum_{i=1}^{\ell} \lambda_i \phi(x; \mu_i, \sigma_i^2),$$

where $\phi(x; \mu_i, \sigma_i^2)$ is the PDF of the i'th Gaussian distribution, having mean μ_i and variance σ_i^2 [20]. We will denote the distribution itself by $\mathcal{M}(\lambda_i, \mu_i, \sigma_i^2, \ell)$. We then propose the following model.

Model 2 (Right-Key Inequivalence for One Approximation). *Consider a linear approximation (α, β) of r rounds of DES. The distribution of the linear correlation $C_K(\alpha, \beta)$ over the key space is approximately given by a Gaussian mixture for some weights λ_i and components $\mathcal{N}(\mu_i, \sigma_i^2)$, $i = 1, \ldots, \ell$. That is,*

$$C_K(\alpha, \beta) \sim \mathcal{M}(\lambda_i, \mu_i, \sigma_i^2, \ell).$$

We note that the signal/noise decomposition easily applies to this model. If we determine that the signal correlation follows a Gaussian mixture, i.e. $C'_K(\alpha, \beta) \sim \mathcal{M}(\lambda'_i, \mu'_i, \sigma_i^{2\prime}, \ell')$ for some appropriate parameters, then we can approximate the actual correlation distribution by adding the noise distribution:

$$C_K(\alpha, \beta) \sim \mathcal{M}(\lambda'_i, \mu'_i, \sigma_i^{2\prime}, \ell') + \mathcal{N}(0, 2^{-n}).$$

We apply Model 2 to the distribution obtained in Sect. 3.2. The result of fitting a Gaussian mixture model with three components to the positive part of the signal distribution is shown in Fig. 4. We first note that the mixture model fits the measured signal distribution quite well. The parameters are

$$\begin{aligned}\lambda_1 &= 0.45, \quad \mu_1 = 2^{-19.79}, \quad \sigma_1^2 = 2^{-52.40},\\ \lambda_2 &= 0.30, \quad \mu_2 = 2^{-19.75}, \quad \sigma_2^2 = 2^{-52.37},\\ \lambda_3 &= 0.25, \quad \mu_3 = 2^{-19.68}, \quad \sigma_3^2 = 2^{-52.68}.\end{aligned}$$

The second mixture component has mean equal to the correlation contribution of the dominant trail, but this component only contributes to 30% of the full distribution. In fact, the main part of the contribution, 45%, can be attributed to the first component, which has a slightly lower mean. This demonstrates that considering only the contribution of the dominant trail can be misleading, even when the remaining trails have a far lower correlation contribution. In general, one should consider as large a part of the hull as possible. Nevertheless, for attacks with relatively low data complexity, the actual distribution can easily be hidden, as we shall see next.

The undersampled mixture. In Sect. 3.2, we recalled that under the assumption of a dominant trail, the distribution of the undersampled correlation C_K^N is given by the Gaussian $\mathcal{N}(C_K, 1/N)$. We state the following equivalent result in the setting of Model 2 and give an outline of the proof.

Theorem 1 (Undersampled distribution). *Assuming Model 2, the undersampled correlation distribution of an approximation (α, β) obtained using N random text pairs is given by*

$$C_K^N(\alpha, \beta) \sim \mathcal{M}(\lambda_i, \mu_i, \sigma_i^2, \ell) + \mathcal{N}(0, 1/N).$$

Proof. For any fixed key k, C_k^N is distributed as $\mathrm{Bin}(N, C_k)$ over the random text sample, which can be approximated by $\mathcal{N}(C_k, 1/N)$ if C_k is small. That is, $C_K^N \mid K = k \sim \mathcal{N}(C_k, 1/N)$. The PDF of the compound distribution C_K^N, i.e. without the conditioning on K, is given by

$$p_{C_K^N}(y) = \int \phi(y; x, 1/N) \cdot \sum_{i=1}^{\ell} \lambda_i \phi(x; \mu_i, \sigma_i^2) dx,$$

which can be shown to be equal to

$$p_{C_K^N}(y) = \sum_{i=1}^{\ell} \lambda_i \phi(y; \mu_i, \sigma_i^2 + 1/N).$$

This is a Gaussian mixture where each component can be written as $\mathcal{N}(\mu_i, \sigma_i^2) + \mathcal{N}(0, 1/N)$. But since we add the second distribution with probability one, the same distribution can be obtained by first drawing from the original mixture, and then adding the distribution $\mathcal{N}(0, 1/N)$, finishing the proof. □

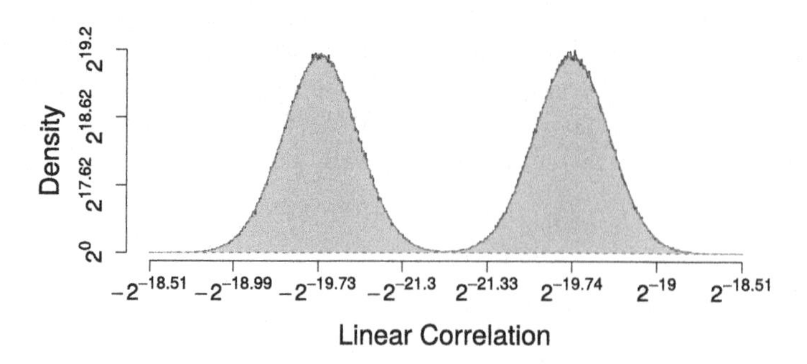

Fig. 5. The distribution of the undersampled linear correlation of γ_1, $C'_K + \mathcal{N}(0, 2^{-n}) + \mathcal{N}(0, 1/N)$, over 14 rounds of DES, with $N = 2^{43}$. C'_K was measured using 1126 trails over 1 million randomly drawn keys. A Gaussian mixture with two components have been fitted to the distribution. The components are shown in red, while the full distribution is shown in green. (Color figure online)

If the number of texts N is relatively large, the model can be somewhat simplified. If we wanted to apply Model 2 and Theorem 1 directly to the case of γ_1, we would model the measured correlation as

$$C_K^N(\gamma_1) = \mathcal{M}(\lambda_i, \mu_i, \sigma_i^2, 6) + \mathcal{N}(0, 2^{-n}) + \mathcal{N}(0, 1/N), \tag{4}$$

using six components for the Gaussian mixture. However, the details of the mixture are easily lost at high levels of undersampling, as can be seen in Fig. 5. Here, we have shown the distribution

$$C'_K(\gamma_1) + \mathcal{N}(0, 2^{-n}) + \mathcal{N}(0, 1/N),$$

where $N = 2^{43}$. The resulting distribution can be described as a Gaussian mixture with two components, instead of six. Each component has variance roughly equal to $1/N$, and the means are $\pm 2^{-19.75}$, i.e. the correlation contribution of the dominant trail. This agrees with the models used by e.g. Matsui and Biryukov, et al., but we stress that this is only true when N is relatively small compared to the linear correlation. In particular, for ciphers with strong dominant trails, $1/N$ needs to be larger than the variance of the positive/negative part of the distributions. For values of N close to the full codebook, this is not true (unless the approximation is extremely weak), and the distribution of C_K cannot be ignored. However, this simplification will help greatly when we consider the joint distribution of multiple approximations in the next subsection.

The Gaussian mixture of multiple approximations. Model 2 and the results of Sect. 3.3 can be generalised to consider the case of multiple linear approximations. Let $C_K(\boldsymbol{\alpha}, \boldsymbol{\beta})$ denote the vector of correlations of M linear approximations, $(C_K(\alpha_1, \beta_1), \ldots, C_K(\alpha_M, \beta_M))^\top$. In the following, we will restrict ourselves to the case where the signal distributions, $C'_K(\alpha_i, \beta_i)$, each have two distinct modes: one positive and one negative. This allows us to split the joint signal distribution, $C'_K(\boldsymbol{\alpha}, \boldsymbol{\beta})$, into at most 2^M components determined by the signs of $C'_K(\boldsymbol{\alpha}, \boldsymbol{\beta})$. In the case of relatively low values of N, we propose the following model.

Model 3 (Right-Key Mixture for Multiple Approximations). *Consider a set of linear approximations $(\alpha_1, \beta_1), \ldots, (\alpha_M, \beta_M)$ of r rounds of DES. The undersampled distribution of the linear correlation vector over the key space, $C_K^N(\boldsymbol{\alpha}, \boldsymbol{\beta})$, is approximately given by an M-variate Gaussian mixture, namely*

$$C_K^N(\boldsymbol{\alpha}, \boldsymbol{\beta}) \sim \mathcal{M}_M(1/\ell, \boldsymbol{\mu}_i, \boldsymbol{\Sigma}_i + 1/N \cdot \mathbf{I}, \ell),$$

where $\ell \leq 2^M$. Moreover, the parameters of the mixture components are given by

$$\begin{aligned}\boldsymbol{\mu}_i &= \mathrm{E}(C_K(\boldsymbol{\alpha}, \boldsymbol{\beta}) | s_{i,j} \cdot C_K(\alpha_i, \beta_i) > 0, j = 1, \ldots, M),\\ \boldsymbol{\Sigma}_i &= \mathrm{Cov}(C_K(\boldsymbol{\alpha}, \boldsymbol{\beta}) | s_{i,j} \cdot C_K(\alpha_i, \beta_i) > 0, j = 1, \ldots, M),\end{aligned}$$

where $s_{i,j} \in \{-1, 1\}$ describes the sign combination of the i'th component.

As for the case of a single approximation, the signal/noise decomposition applies to this model, resulting in an undersampled distribution of the form

$$C_K^N(\boldsymbol{\alpha}, \boldsymbol{\beta}) \sim \mathcal{M}_M(1/\ell, \boldsymbol{\mu}'_i, \boldsymbol{\Sigma}'_i + (2^{-n} + 1/N)\mathbf{I}, \ell).$$

The signal parameters, $\boldsymbol{\mu}'_i$ and $\boldsymbol{\Sigma}'_i$, can be estimated by enumerating an appropriate number of trails and then calculating $C'_K(\boldsymbol{\alpha}, \boldsymbol{\beta})$ for a large number of keys.

This model bears some resemblance to the one given by Biryukov et al. in [3]. While both models use the signs of the correlation vector to split the distribution into several Gaussians, our model captures the entire key space in one distribution, whereas the model in [3] partitions the key space into at most 2^M parts which are considered separately. Additionally, we do not make any assumption about the independence of the linear approximations. As such, $\boldsymbol{\Sigma}_i$ need not be diagonal matrices, and not all 2^M sign combinations need to be present. While the possibility of $\ell < 2^M$ is briefly mentioned in [3], all experiments were done such that $\ell = 2^M$. As we shall see in Sect. 5, the case of $\ell < 2^M$ allows for stronger attacks. Moreover, an improved attack on full DES was not presented in [3]. We apply our model to obtain a key-recovery attack on full DES in Sect. 6. First, however, we turn our attention to the wrong-key distribution.

4 Wrong-Key Correlation for DES: Non-random Behaviour

In this section, we consider the correlation distribution of DES approximations in the case of a wrong key guess in Algorithm 2. This distribution is essential, as the effectiveness of the algorithm is determined by how different the right-key and wrong-key distributions are. In Sect. 4.1, we consider the current models for the wrong-key distribution. In Sect. 4.2, we develop a new model for the wrong-key distribution of DES, and show that the distribution obtained under this model deviates significantly from that considered in Sect. 4.1. Nevertheless, as for the right-key in Sect. 3, we show that the deviation has little impact when the number of texts used in the attack is relatively small.

4.1 The Current Ideal Wrong-Key Distribution

The assumption of wrong-key randomisation, Assumption 2, used by Matsui in [22] and by Biryukov et al. in [3], predicts that a wrong guess of the outer round keys in Algorithm 2 should result in an approximation with correlation zero. This is motivated by the idea that if we encrypt/decrypt using the wrong key, we are doing something equivalent to encrypting two extra rounds. This should result in a linear correlation much closer to zero, as we are essentially considering the correlation over $r+4$ rounds instead of r rounds. However, as shown by Daemen and Rijmen in [13], even a linear approximation of an ideal permutation will approximately have the correlation distribution

$$C_K(\alpha, \beta) \sim \mathcal{N}(0, 2^{-n}),$$

where n is the blocksize. Since we intuitively cannot do "worse" than an ideal cipher, the correlation of a wrong guess should follow this distribution. This consideration led Bogdanov and Tischhauser to present an updated wrong-key randomisation hypothesis in [7], in which the wrong key correlation follows this ideal Gaussian distribution. However, if we consider the case of DES where, even over 14 rounds, strong linear approximations exist, the wrong-key correlation might not be close to the ideal distribution. We consider this problem next.

4.2 A New Non-random Wrong-Key Distribution

Consider the scenario in which an attacker obtains a plaintext-ciphertext pair computed over r rounds of a cipher, and attempts to encrypt the plaintext one round, and decrypt the ciphertext one round, in order to calculate the correlation of an approximation over $r-2$ rounds. If the attacker uses the wrong round keys for the encryption/decryption, she essentially obtains a plaintext/ciphertext pair of some related cipher with $r+2$ rounds. Motivated by this, we propose the following wrong-key model for linear cryptanalysis on DES.

Model 4 (Non-random Wrong-Key Distribution). *Consider an Algorithm 2 style attack on r rounds of DES using a linear approximation (α, β) over $r-2$ rounds. Let R_K be the keyed round function of DES, and let $E_K^\star$ denote the r-round encryption function. For a wrong guess of the outer round keys, the correlation will be distributed as for the cipher*

$$E'_K(x) = R_{K_a}^{-1}(E_K^\star(R_{K_b}^{-1}(x))), \tag{5}$$

where K_a and K_b are chosen uniformly at random.

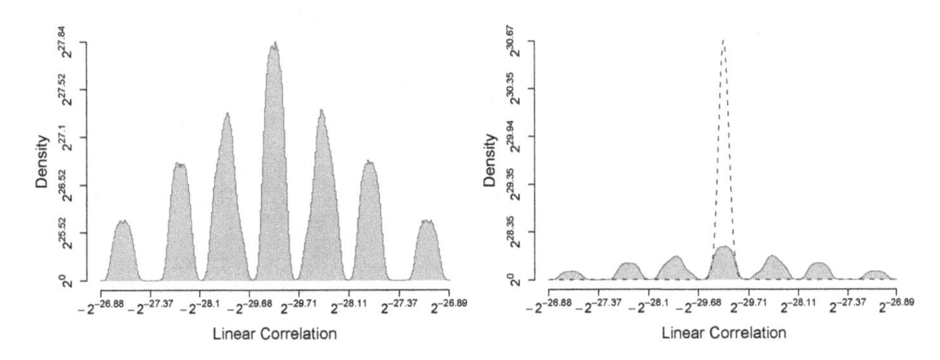

Fig. 6. The distribution of linear correlation for the approximation γ_1 over 18 rounds of DES with randomly chosen outer round keys. The correlation was calculated using 954 trails and 1 million randomly drawn keys. The distribution is close to zero, but the variance is $2^{-56.08}$. To the right, the distribution is compared to that of an ideal permutation, i.e. the Gaussian $\mathcal{N}(0, 2^{-64})$.

For DES, where encryption and decryption are similar, this can reasonably be simplified to $E'_K(x) = E^{r+2}_K$, where the outer round keys are randomly chosen.

In light of this, we considered the approximation γ_1 over 18 rounds of DES, with randomly chosen outer round keys. Using the algorithm described in Sect. 3.2, with $B = 1$ million, we enumerated 954 trails of this approximation. Using 20 Intel Xeon Processor E5-2680 cores, the enumeration took about 15 CPU hours. We then calculated the resulting signal correlation for 1 million keys. The trails had an absolute correlation contribution between $2^{-45.84}$ and $2^{-28.75}$. The distribution is shown in Fig. 6. We note that the result is similar for the other approximations given in Table 2.

As was the case for the right-key distribution, this wrong-key distribution appears to be approximately a Gaussian mixture. More importantly, while the distribution is symmetric around zero, the variance is much larger than that of an ideal permutation: $2^{-56.08}$ compared to 2^{-64}. This shows that, while the added four rounds make the correlation weaker, the assumption of a resulting ideal distribution is optimistic. For attacks that use a data complexity close to the full codebook, this assumption could result in a overestimate of success probability or an underestimate of attack complexity. Moreover, if the cryptanalyst only appends/prepends one round to the approximation, this effect could be significant.

The undersampled distribution. While the distribution in Fig. 6 is far from ideal, the actual distribution of the correlation matters little if the level of undersampling is significant. If we apply signal/noise decomposition and Theorem 1 to our estimate of the wrong-key distribution, with the number of texts $N = 2^{43}$, we obtain the result shown in Fig. 7. We see here that it is sufficient to use a single Gaussian distribution to approximate the undersampled wrong-key correlation distribution. If this distribution is similar for other approximations, it will

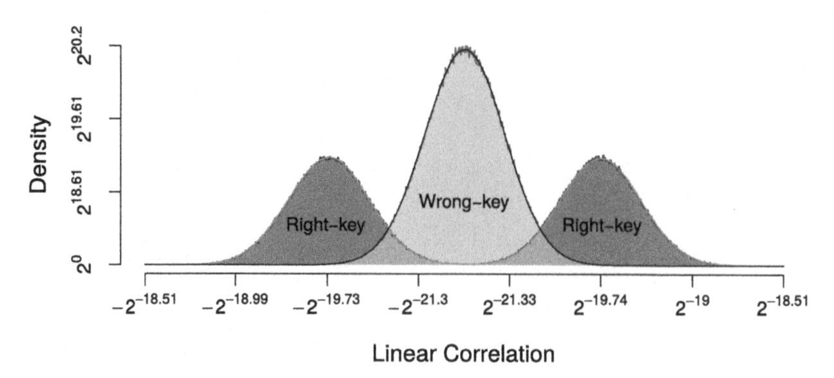

Fig. 7. Undersampled right-key (blue) and wrong-key (red) distributions for the approximation γ_1 with $N = 2^{43}$. The signal distributions were measured using 1 million randomly drawn keys. A Gaussian mixture has been fitted to the right-key distribution (green), while a single Gaussian distribution was fitted to the wrong-key distribution (black). (Color figure online)

be sufficient to model the joint wrong-key correlation distribution of M approximations as an M-variate Gaussian distribution. Thus, if $\mathbf{\Sigma}_W$ is the covariance matrix of the signal correlation of the M approximations over E'_K, then the undersampled wrong-key distribution will approximately be given by

$$C_K^N(\boldsymbol{\alpha}, \boldsymbol{\beta}) \sim \mathcal{N}(\mathbf{0}, \mathbf{\Sigma}_W + (2^{-n} + 1/N)\mathbf{I}),$$

if $1/N$ is sufficiently large.

Using Model 3 for the right-key and Model 4 for the wrong-key distribution, we develop a classifier that uses both these distributions in the following section.

5 Classifying Keys Using Asymmetric Distributions

In Sect. 3, we developed a model for the linear correlation distribution of a correct key-guess in Algorithm 2, namely a multivariate Gaussian mixture model. In Sect. 4, we similarly developed a simple multivariate Gaussian model for the linear correlation distribution of a wrong key-guess. Using these two distributions, we now develop a classifier based on the likelihood-ratio, which can be used in Algorithm 2 to decide between potential right and wrong key guesses. We first present the classifier given in [3] in Sect. 5.1. We then introduce our new classifier in Sect. 5.2, and compare the performance of the two in Sect. 5.3.

In the following, we will consider the two sets of four linear approximations over 14 rounds of DES given in Table 2. While it is difficult to visualise the joint distribution of more than three approximations, Fig. 8 shows the pairwise joint distributions of the approximations γ_1, γ_2, γ_3, and γ_4, as well as the marginal distributions, for $N = 2^{43}$. Note that the joint distributions of γ_1 and γ_3, as well as that of γ_2 and γ_4, only have two components. We will explore this phenomenon in Sect. 5.4, and show that such distributions can improve our classifier.

Table 2. The top table specifies two sets of four linear approximations over 14 rounds of DES, and gives the correlation contribution of their dominant trail, as well as the sign bit of that trail. The bottom table specifies the set of non-zero bits of the associated dominant key trails, where t_i^j is the j'th bit of t_i.

Linear approximation	Dominant key trail	$\lvert C_K(T_\cdot)\rvert$	$s_{T_\cdot}$
$\gamma_1 = ([7,18,24],[7,18,24,29,47])$	$\bar{T}_A$	$2^{-19.75}$	1
$\gamma_2 = ([7,18,24],[7,18,24,29,44,48])$	$\bar{T}_B$	$2^{-20.48}$	1
$\gamma_3 = ([7,18,24,29],[7,18,24,47])$	$\bar{T}_A$	$2^{-20.75}$	0
$\gamma_4 = ([7,18,24,29],[7,18,24,44,48])$	$\bar{T}_B$	$2^{-20.07}$	1
$\delta_1 = ([15,39,50,56],[39,50,56,61])$	$\bar{T}_C$	$2^{-20.75}$	0
$\delta_2 = ([12,16,39,50,56],[39,50,56,61])$	$\bar{T}_D$	$2^{-20.07}$	1
$\delta_3 = ([15,39,50,56,61],[39,50,56])$	$\bar{T}_C$	$2^{-19.75}$	1
$\delta_4 = ([12,16,39,50,56,61],[39,50,56])$	$\bar{T}_D$	$2^{-20.48}$	1

Key trail	Non-zero key mask bits	Key trail	Non-zero key mask bits
$\bar{T}_A$	$\{t_1^{22}, t_2^{44}, t_3^{22}, t_5^{22}, t_6^{44}, t_7^{22}, t_9^{22}, t_{10}^{44}, t_{11}^{22}, t_{13}^{22}\}$	$\bar{T}_B$	$\bar{T}_A \backslash t_{13}^{22} \cup \{t_{13}^{19}, t_{13}^{23}\}$
$\bar{T}_C$	$\{t_0^{22}, t_2^{22}, t_3^{44}, t_4^{22}, t_6^{22}, t_7^{44}, t_8^{22}, t_{10}^{22}, t_{11}^{44}, t_{12}^{22}\}$	$\bar{T}_D$	$\bar{T}_C \backslash t_0^{22} \cup \{t_0^{19}, t_0^{23}\}$

5.1 The Bayesian Classifier of Biryukov et al.

Consider an Algorithm 2 style attack using M linear approximations. Let $\mathcal{K}_R$ denote the space of correct guesses of the key-bits (k_f, k_b), and let $\mathcal{K}_W$ denote the space of wrong guesses. We have to classify each key-guess as either an incorrect guess or a potential correct guess, based on the measured linear correlation vector $\mathbf{x}$. Let $f_R(\mathbf{x}) = \Pr(\mathbf{x} \mid (k_f, k_b) \in \mathcal{K}_R)$ be the PDF of the right-key correlation distribution. We define the Bayesian classifier, BC, as the following decision rule

$$BC(\mathbf{x}) = \begin{cases} \text{If } B(\mathbf{x}) > \Gamma, \text{ decide that } (k_f, k_b) \in \mathcal{K}_R, \\ \text{otherwise, decide that } (k_f, k_b) \in \mathcal{K}_W, \end{cases}$$

where $B(\mathbf{x}) = f_R(\mathbf{x})$. Under Model 3, $B(\mathbf{x})$ is given as the Gaussian mixture

$$B(\mathbf{x}) - \sum_{i=1}^{\ell} \lambda_i \phi_M(\mathbf{x}; \boldsymbol{\mu}_i, \boldsymbol{\Sigma}_i + (2^{-n} + 1/N)\mathbf{I}).$$

This exact classifier is not described in [3], but it is essentially identical to the one developed there. The difference is that in [3], each component of f_R is considered separately, and so ℓ scores are produced for each key guess. The classifier BC should be functionally equivalent to this approach, but this representation allows for easy comparison to the likelihood-ratio classifier we propose next.

5.2 Our Likelihood Classifier

We now propose a new classifier based in the likelihood-ratio. As opposed to the Bayesian classifier, the likelihood classifier directly takes the wrong-key distribution into account. To this end, let $f_W(\mathbf{x}) = \Pr(\mathbf{x} \mid (k_f, k_b) \in \mathcal{K}_R)$ be the PDF

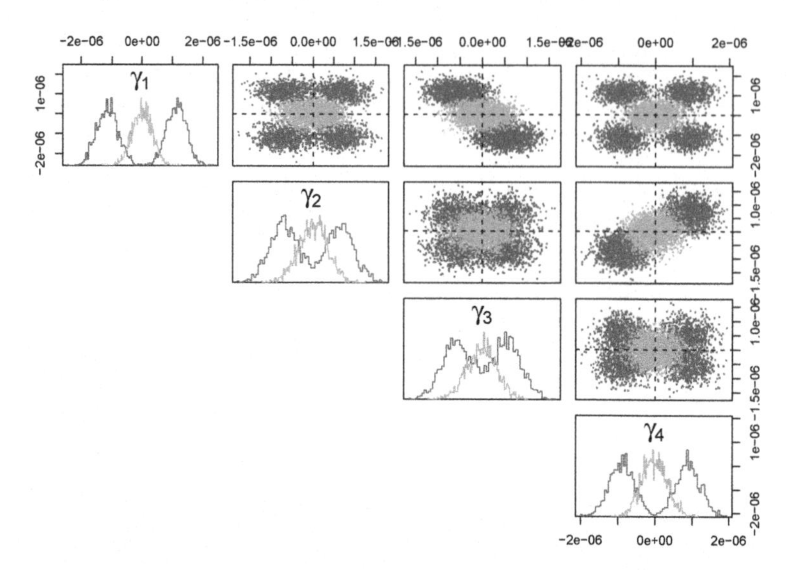

Fig. 8. Histograms and pairwise distributions of the undersampled correlations of approximations $\gamma_1, \ldots, \gamma_4$ given in Table 2. The right-key distributions are shown in blue, the wrong-key distributions are shown in red. The number of texts is $N = 2^{43}$. Note that since γ_1 and γ_3 have the same dominant key trail, their joint distribution only has two components. Likewise for γ_2 and γ_4. (Color figure online)

of the wrong-key correlation distribution. Then the *likelihood-ratio* is defined as $\Lambda(\mathbf{x}) = f_R(\mathbf{x})/f_W(\mathbf{x})$. For the right-key and wrong-key distributions described in Sects. 3 and 4, this is equal to

$$\Lambda(\mathbf{x}) = \frac{\sum_{i=1}^{\ell} \lambda_i \phi_M(\mathbf{x}; \boldsymbol{\mu}_i, \boldsymbol{\Sigma}_i + (2^{-n} + 1/N)\mathbf{I})}{\phi_M(\mathbf{x}; \mathbf{0}, \boldsymbol{\Sigma}_W + (2^{-n} + 1/N)\mathbf{I})},$$

where $\mathbf{x}$ is an observed value of correlations for M approximations. A large value of $\Lambda(\mathbf{x})$ will then indicate a likely correct key guess, while a low value will indicate a wrong key guess. Thus, we define the *likelihood classifier LC* as the following decision rule

$$LC(\mathbf{x}) = \begin{cases} \text{If } \Lambda(\mathbf{x}) > \Gamma, \text{decide that } (k_f, k_b) \in \mathcal{K}_R, \\ \text{otherwise, decide that } (k_f, k_b) \in \mathcal{K}_W. \end{cases}$$

In light of this definition, two important concepts are the success probability and advantage of the classifier. Formally, we define the success probability and advantage, respectively, as

$$P_S = 1 - \Pr(\Lambda(\mathbf{x}) < \Gamma \mid (k_f, k_b) \in \mathcal{K}_R), \quad (6)$$

$$a = -\log_2(\Pr(\Lambda(\mathbf{x}) \geq \Gamma \mid (k_f, k_b) \in \mathcal{K}_W)), \quad (7)$$

in accordance with the usual definition [25]. We usually choose Γ such that we achieve a certain success probability. Under our proposed model, the involved probabilities cannot be explicitly stated. Thus, we must rely on simulations to calculate these values. Since simulating values from a Gaussian distribution is easy, this is not a problem. Using this approach, we now compare the performance of the likelihood classifier and the Bayesian classifier.

5.3 Decision Boundaries

The likelihood classifier LC divides the M-dimensional cube $[-1, 1]^M$ into two regions separated by the *decision boundary*, namely where $\Lambda(\mathbf{x}) = \Gamma$. On one side of the decision boundary, observations are classified as belonging to the right-key distribution, while observations from the other side are classified as belonging to the wrong-key distribution. By visualising this decision boundary, we can get a better understanding of the classifier.

In the following, we consider the eight approximations given in Table 2, over 14 rounds of DES. We enumerated between 1100 and 1400 trails for each approximation and calculated the signal correlations for 1 million random keys, in order to estimate $\boldsymbol{\mu}_i$ and $\boldsymbol{\Sigma}_i$. The same was done over E'_K, where between 950 and 1100 trails were enumerated, in order to estimate $\boldsymbol{\Sigma}_W$. For each data point, we added noise drawn from $\mathcal{N}_M(\mathbf{0}, (2^{-n} + 1/N)\mathbf{I})$, according to the signal/noise decomposition and Theorem 1. This allows us to simulate $\Lambda(\mathbf{x})$ and $B(\mathbf{x})$ for varying values of N and calculate the resulting decision boundary and advantage.

Consider the pair of approximations γ_1 and δ_1 and let $N = 2^{43}$. We simulate $\Lambda(\mathbf{x})$ and $B(\mathbf{x})$ for each data point as described above, and then fix a threshold value for each classifier such that $P_S = 0.90$, cf. Eq. (6). The resulting decision boundaries, as well as the related probability distributions, are shown in Fig. 9. In this case, the likelihood classifier obtains an advantage of 5.5 bits, while the Bayesian classifier only has an advantage of 3.1 bits. By considering the decision boundary, it is clear why this is the case. Since the Bayesian classifier only uses information about the right-key distribution, it simply creates a decision boundary around each component of the mixture which is large enough to obtain the desired success probability. In view of the information that is available to the classifier, this makes sense, since observations close to the mean of component have a larger chance of being a correct key guess. Because of this, the parts of the right-key distribution which is farthest away from the wrong-key distribution is also discarded as unlikely candidates. This in turn requires the decision boundary to be wider than actually needed, and the advantage is therefore quite low due to an increased number of false positives.

The likelihood classifier on the other hand does use information about the wrong-key distribution. The decision boundary is created such that there is a good boundary between each component and the wrong-key distribution. Any observation that is sufficiently far away from the wrong-key distribution is deemed a likely correct key guess, no matter how extreme the observation is in the right-key distribution. Thus, extreme points in the right-key distribution are not "wasted", allowing for a tight decision boundary around the wrong-key distribution, yielding a larger advantage.

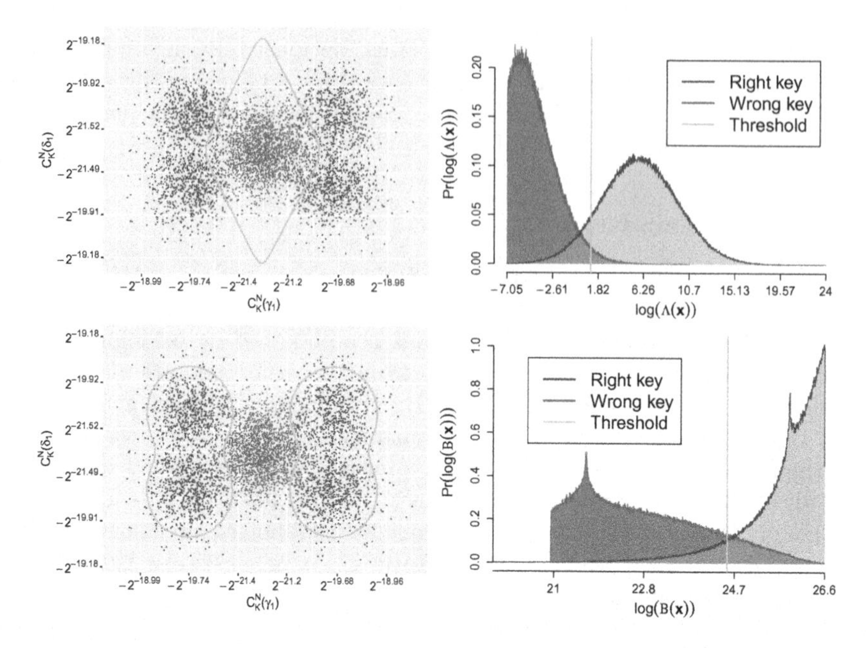

Fig. 9. Left: The joint distribution of $C_K^N(\gamma_1)$ and $C_K^N(\delta_1)$, with $N = 2^{43}$, are shown for both a right key guess (blue) and a wrong key guess (red). The decision boundaries for a success probability of 90% are drawn for the likelihood-ratio classifier (top) and the Bayesian classifier (bottom). Right: The corresponding probability distributions of $\Lambda(\mathbf{x})$ (top) and $B(\mathbf{x})$ (bottom) as well as the threshold value. The likelihood ratio classifier obtains an advantage of 5.5 bits, while the Bayesian classifier obtains an advantage of 3.1 bits. (Color figure online)

For the approximations used here, all sign combinations of the correlation vector are possible. In terms of the mixture model, the number of components is $\ell = 2^M$. We now turn our attention to the case where $\ell < 2^M$.

5.4 Observations on the Asymmetric Distribution

As shown in Sect. 3.2, the sign of the signal correlation $C'_K(\gamma_1)$ for a given key is determined by the parity $\langle \bar{T}_A, \bar{K} \rangle$, where $\bar{T}_A$ is the dominant key trail. Consider the two approximations γ_1 and γ_3 given in Table 2. Both approximations have the same dominant key trail, and since their sign bits s_T are different, the sign of their correlation will therefore always be opposite. In the terminology of Sect. 3.3, the number of components ℓ of the Gaussian mixture is strictly less than 2^M. We will call such a distribution *asymmetric*. On the other hand, the two approximations γ_1 and δ_1 have different dominant key-trails, and therefore all four sign combinations of their correlations are possible. In this case, $\ell = 2^M$, and we call such a distribution *symmetric*.

For γ_1 and δ_1, the decision boundary for the likelihood classifier was shown in Fig. 9. For γ_1 and γ_3, the decision boundary is shown in Fig. 10. Here, the "missing" components in the first and third quadrant are clearly shown, while

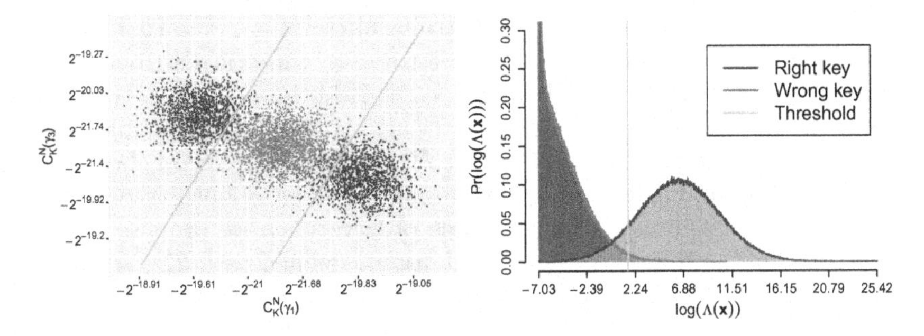

Fig. 10. Left: The joint distribution of $C_K^N(\gamma_1)$ and $C_K^N(\gamma_3)$, with $N = 2^{43}$, are shown for a right key guess (blue) and a wrong key guess (red). The decision boundaries for a success probability of 90% are drawn for the likelihood-ratio classifier. Right: The probability distributions of $\Lambda(\mathbf{x})$ as well as the threshold value. The classifier obtains an advantage of 6.2 bits. (Color figure online)

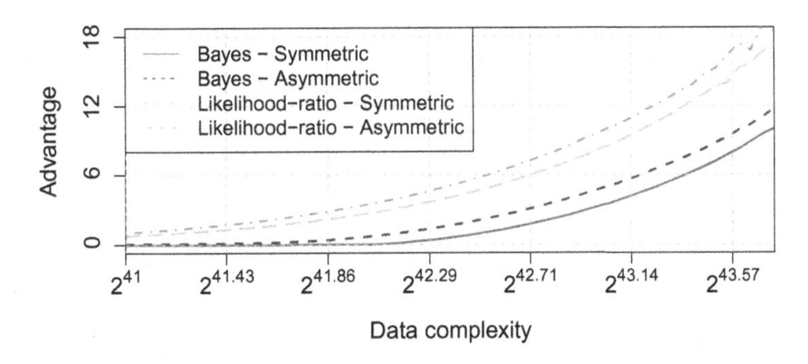

Fig. 11. A comparison of the advantage obtained by using the Bayesian classifier and the likelihood ratio classifier on both symmetric and asymmetric correlation distributions. The symmetric distribution uses the set of approximations $\{\gamma_1, \gamma_2, \delta_1, \delta_2\}$ while the asymmetric distribution uses the set $\{\gamma_1, \gamma_2, \gamma_3, \gamma_4\}$.

the wrong-key distribution is still symmetric around zero. We note that, all else being equal, the classifier on the asymmetric distribution achieves an increased advantage of 0.7 bits. Moreover, the comparison here is fair, since the strength of δ_1 is the same as that of γ_3. The reason for this increase is apparent when we compare the two decision boundaries. For the asymmetric distribution, the decision boundary is such that even extreme points in the wrong-key distribution towards the first and third quadrant are easily classified as wrong key guesses. This decreases the number of false positives, increasing the advantage.

This improvement in the classifier for asymmetric distributions generally extends to higher dimensions, where the effect can be even more pronounced. Indeed, for larger M, ℓ can be much smaller than 2^M. In the example above, we had $\ell = 2$ while $2^M = 4$. Consider now the set of approximations $\{\gamma_1, \gamma_2, \gamma_3, \gamma_4\}$. A shown in Table 2, these approximations only have two distinct dominant key trails, implying that the set has an asymmetric distribution with $\ell = 4 < 2^M = 16$. Figure 11 compares the advantage of this set of approximations to the set

$\{\gamma_1, \gamma_2, \delta_1, \delta_2\}$, which has a symmetric distribution, i.e. $\ell = 2^M = 16$. In general, we observe that the classifiers are stronger for the asymmetric distribution, with an increase in advantage of 1.4 bits for $N = 2^{43}$. Additionally, the better performance of the likelihood classifier is quite clear, consistently obtaining a larger advantage over the Bayesian classifier. For $N = 2^{43}$, the likelihood classifier has an advantage 4.9 bits higher than the Bayesian classifier on both the symmetric and asymmetric distribution. Due to these observations, we propose the term *symmetry factor* for these types of distributions, defined as $\ell/2^M$. A distribution with symmetry factor one is a symmetric distribution, while a symmetry factor less than one indicates an asymmetric distribution. We conjecture that, all else being equal, a lower symmetry factor will result in a stronger classifier.

6 Improved Attack on DES

Using the results from the previous sections, we now mount a key-recovery attack on DES using eight linear approximations. We will use two sets of four linear approximations, $\{\gamma_1, \gamma_2, \gamma_3, \gamma_4\}$ and $\{\delta_1, \delta_2, \delta_3, \delta_4\}$ over 14 rounds, as given in Table 2. The attack is mostly identical to Matsui's Algorithm 2. As such, we obtain N plaintext-ciphertext pairs over 16 rounds, guess the key-bits required to partially encrypt/decrypt the texts and compute the linear correlations, and then use the likelihood classifier to categorise each guess as a likely wrong or right key guess. For each guess, we further gain some parity bits of the key based on the signs of the correlations.

6.1 Attack Description

Table 3 shows the key- and text-bits relevant to the attack. For both sets of approximations, we need to know 29 bits of the plaintext/ciphertext, designated $t_{f,\cdot}/t_{b,\cdot}$, and we will guess 24 bits of the first/last round key, designated $k_{f,\cdot}/k_{b,\cdot}$. Moreover, the signs of $C_K^N(\gamma_1)$, $C_K^N(\gamma_4)$, $C_K^N(\delta_3)$, and $C_K^N(\delta_2)$, will allow us to deduce the parity bits p_A, p_B, p_C, and p_D. Thus, the attacker will learn a total of 52 bits of the master key, and will have to guess the remaining 4 bits. In the following, we assume that the distribution parameters $\boldsymbol{\mu}_{i,\cdot}$, $\boldsymbol{\Sigma}_{i,\cdot}$, and $\boldsymbol{\Sigma}_{W,\cdot}$ have been determined before the attack, as described in Sect. 3.3. Moreover, we assume that $\lambda_i = 1/\ell$ for all i. The attack is then given as follows:

- **Distillation**
  1. Obtain N plaintext-ciphertext pairs.
  2. Create two vectors $\mathbf{t}_\gamma$ and $\mathbf{t}_\delta$ of size 2^{29} each. $\mathbf{t}_\gamma[i]$ (similarly $\mathbf{t}_\delta$) is equal to the number of text pairs such that the bits $(t_{f,\gamma}, t_{b,\gamma})$ are equal to i.
- **Analysis**
  1. For each guess of $(k_{f,\gamma}, k_{b,\gamma})$, calculate the vector

$$\mathbf{c}_\gamma = (C_K^N(\gamma_1), C_K^N(\gamma_2), C_K^N(\gamma_3), C_K^N(\gamma_4))^\top,$$

     by partially encrypting/decrypting the data in $\mathbf{t}_\gamma$. Do similarly for the δ-approximations to calculate $\mathbf{c}_\delta$.

Table 3. This table specifies the key/text bits involved in the attack, as well as the parity key bits derived. X^i denotes the i'th bit of X.

Forward key bits guessed				#bits
$k_{f,\gamma}$	$\{K_0^{18}, \ldots, K_0^{23}\}$	$k_{f,\delta}$	$\{K_0^{24}, \ldots, K_0^{35}, K_0^{42}, \ldots, K_0^{47}\}$	6+18
Backward key bits guessed				#bits
$k_{b,\gamma}$	$\{K_{15}^{24}, \ldots, K_{15}^{35}, K_{15}^{42}, \ldots, K_{15}^{47}\}$	$k_{b,\delta}$	$\{K_{15}^{18}, \ldots, K_{15}^{23}\}$	18+6
Plaintext bits stored				#bits
$t_{f,\gamma}$	$\{\mathcal{P}^{11}, \ldots, \mathcal{P}^{16}, \mathcal{P}^{39}, \mathcal{P}^{50}, \mathcal{P}^{56}\}$			9
$t_{f,\delta}$	$\{\mathcal{P}^{0}, \mathcal{P}^{7}, \mathcal{P}^{15}, \ldots, \mathcal{P}^{24}, \mathcal{P}^{27}, \ldots, \mathcal{P}^{31}, \mathcal{P}^{44}, \mathcal{P}^{47}, \mathcal{P}^{48}\}$			20
Ciphertext bits stored				#bits
$t_{b,\gamma}$	$\{\mathcal{C}^{0}, \mathcal{C}^{7}, \mathcal{C}^{15}, \ldots, \mathcal{C}^{24}, \mathcal{C}^{27}, \ldots, \mathcal{C}^{31}, \mathcal{C}^{44}, \mathcal{C}^{47}, \mathcal{C}^{48}\}$			20
$t_{b,\delta}$	$\{\mathcal{C}^{11}, \ldots, \mathcal{C}^{16}, \mathcal{C}^{39}, \mathcal{C}^{50}, \mathcal{C}^{56}\}$			9
Parity bits obtained from signs				
p_A	$K_1^{22} \oplus K_2^{44} \oplus K_3^{22} \oplus K_5^{22} \oplus K_6^{44} \oplus K_7^{22} \oplus K_9^{22} \oplus K_{10}^{44} \oplus K_{11}^{22} \oplus K_{13}^{22}$			
p_B	$p_A \oplus K_{13}^{22} \oplus K_{13}^{19} \oplus K_{13}^{23}$			
p_C	$K_0^{22} \oplus K_2^{22} \oplus K_3^{44} \oplus K_4^{22} \oplus K_6^{22} \oplus K_7^{44} \oplus K_8^{22} \oplus K_{10}^{22} \oplus K_{11}^{44} \oplus K_{12}^{22}$			
p_D	$p_C \oplus K_0^{22} \oplus K_0^{19} \oplus K_0^{23}$			

2. Calculate

$$\Lambda(\mathbf{c}_\gamma) = \frac{\frac{1}{4}\sum_{i=1}^{4} \phi_M(\mathbf{c}_\gamma; \boldsymbol{\mu}_{i,\gamma}, \boldsymbol{\Sigma}_{i,\gamma} + (2^{-n} + 1/N)\mathbf{I})}{\phi_M(\mathbf{c}_\gamma; \mathbf{0}, \boldsymbol{\Sigma}_{W,\gamma} + (2^{-n} + 1/N)\mathbf{I})},$$

   for each guess of $(k_{f,\gamma}, k_{b,\gamma})$. If $\Lambda(\mathbf{c}_\gamma) \leq \Gamma_\gamma$, discard the key guess. Likewise, calculate $\Lambda(\mathbf{c}_\delta)$ for each guess of $(k_{f,\delta}, k_{b,\delta})$. If $\Lambda(\mathbf{c}_\delta) \leq \Gamma_\delta$, discard the key guess.
3. For each surviving key guess, determine the four bits p_A, p_B, p_C, p_D based on the signs of $\mathbf{c}_\gamma$ and $\mathbf{c}_\delta$.

- **Search**
  1. For each remaining guess of $(k_{f,\gamma}, k_{b_\gamma}, k_{f,\delta}, k_{b,\delta})$, guess the last 4 bits of the master key, and verify the guess by trial encryption.

6.2 Attack Complexity

In the following, we assume that one computational unit is the time it takes to perform one round of DES. The computational complexity of the distillation phase is $\mathcal{O}(N)$, while the memory complexity is $\mathcal{O}(2 \cdot 2^{29})$. For the analysis phase, each C_K^N can be calculated for all key guesses in time $\mathcal{O}((|k_{f,\cdot}| + |k_{b,\cdot}|)2^{|k_{f,\cdot}|+|k_{b,\cdot}|+1.6})$ using the FFT method presented in [10]. In total, step 1 of the analysis phase can be completed in time $\mathcal{O}(2 \cdot 4 \cdot 24 \cdot 2^{25.6}) \approx \mathcal{O}(2^{33.18})$. Step 2 requires the calculation of $\ell + 1$ terms for each key-guess of the type $(\mathbf{x} - \boldsymbol{\mu})^\top \boldsymbol{\Sigma}^{-1}(\mathbf{x} - \boldsymbol{\mu})$, to calculate the normal probabilities. Each term can be computed in time $\mathcal{O}(2M^3)$. Thus, step 2 takes a total of $\mathcal{O}(2 \cdot 2^{24} \cdot 5 \cdot 4^3) \approx \mathcal{O}(2^{33.32})$ time. Step 3 takes $\mathcal{O}(2 \cdot 2^{24-a_\gamma} + 2 \cdot 2^{24-a_\delta})$ time, where a_γ and a_δ is the advantage of the classifiers in step 2. The analysis step requires $\mathcal{O}(2^{24-a_\gamma} + 2^{24-a_\delta})$ memory to store the surviving key guesses. The search phase

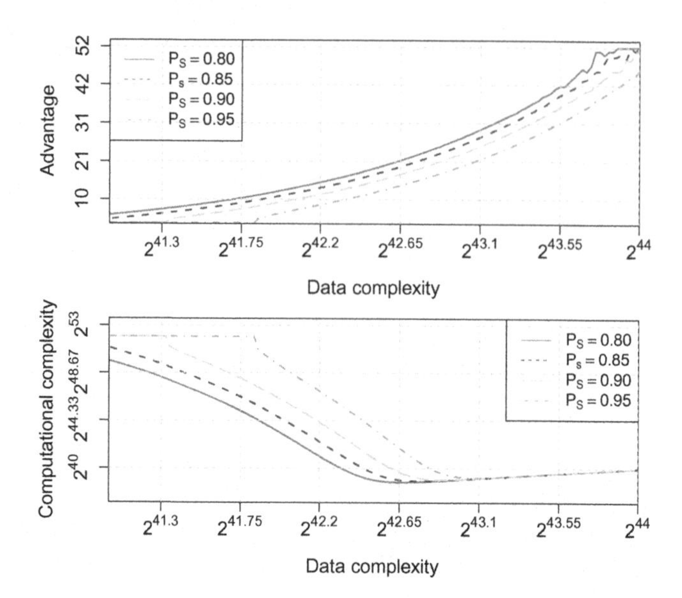

Fig. 12. Top: Combined advantage of the two likelihood classifiers using approximations in Table 2. The success probabilities include the probability of guessing the four parity bits correctly. Bottom: The computational complexity of our key-recovery attack on DES. Each curve has a clear minimum where the trade-off between the data complexity and the strength of the classifiers is optimal.

requires $\mathcal{O}(16 \cdot 2^{48-(a_\gamma+a_\delta)} \cdot 2^{56-52}) = \mathcal{O}(16 \cdot 2^{56-(a_\gamma+a_\delta+4)})$ time and negligible memory. Dividing everything by 16 to get the total number of full DES encryptions, the computational complexity is approximately

$$\mathcal{O}(N \cdot 2^{-4} + 2^{29.18} + 2^{29.32} + 2^{21-a_\gamma} + 2^{21-a_\delta} + 2^{52-(a_\gamma+a_\delta)}).$$

Thus, the attack complexity depends on the advantage of the two classifiers, which in turn depends on the choice of Γ_γ and Γ_δ. Note that step 3 of the analysis phase is not guaranteed to succeed, so the threshold values must be chosen such that the overall success probability of the attack is P_S. Namely, if P_γ and P_δ is the success probabilities of the two classifiers, and Q_γ and Q_δ is the success probabilities of determining the parity bits, then we fix Γ_γ and Γ_δ such that $P_\gamma \cdot P_\delta \cdot Q_\gamma \cdot Q_\delta = P_S$. Using the data obtained in Sect. 5.3, we calculated the total advantage $a_\gamma + a_\delta + 4$ for different N and different values of the success probability P_S. The results are shown in Fig. 12, along with the corresponding attack complexities. For low data complexities, the search phase is dominant, and so the $2^{52-(a_\gamma+a_\delta)}$ term determines the time complexity. For high data complexities, however, the $N \cdot 2^{-4}$ term is dominant. This gives each complexity curve a clear minimum. In a comparison to Matsui's attack, we see that for $P_S = 85\%$, the minimum is achieved at $N = 2^{42.775}$ where the computational complexity is $2^{38.86}$ DES encryptions. This is 17.6 times faster than Matsui's attack estimate (or 4.4 times faster than Junod's estimate of the attack in [17]) using $2^{40.2}$ fewer texts.

6.3 Experimental Verification

While it would be possible to carry out the attack in practice, we would need to do this for many keys to get an idea of the actual advantage, making the experiment infeasible. Instead, we measured the actual values of $\mathbf{c}_\gamma$ and $\mathbf{c}_\delta$ over 14 and 18 rounds of DES (the right key and wrong key, respectively) with $N = 2^{42.78}$ for randomly chosen keys. This can be done in a bitsliced manner, and is therefore faster than performing the actual attack, while giving us all the information we need to verify our model. Using several months of CPU time, we collected 1300 data points for the right key and wrong key distributions. We first note that the observed distributions closely match those predicted by the model in e.g. Figure 8. Moreover, we obtain the advantages $a_\gamma = 6.72$ and $a_\delta = 10.31$, which would give us a complexity of $2^{38.88}$ – very close to that predicted by our model.

References

1. TLS stats from 1.6 billion connections to mozilla.org. https://jve.linuxwall.info/blog/index.php?post/2016/08/04/TLS-stats-from-1.6-billion-connections-to-mozilla.org. Accessed 07 Sep 2017
2. Biham, E., Shamir, A.: Differential cryptanalysis of the full 16-round DES. In: Brickell, E.F. (ed.) CRYPTO 1992. LNCS, vol. 740, pp. 487–496. Springer, Heidelberg (1993). https://doi.org/10.1007/3-540-48071-4_34
3. Biryukov, A., Cannière, C.D., Quisquater, M.: On multiple linear approximations. In: Franklin, M. (ed.) CRYPTO 2004. LNCS, vol. 3152, pp. 1–22. Springer, Heidelberg (2004). https://doi.org/10.1007/978-3-540-28628-8_1
4. Bogdanov, A., Geng, H., Wang, M., Wen, L., Collard, B.: Zero-correlation linear cryptanalysis with FFT and improved attacks on ISO standards Camellia and CLEFIA. In: Lange, T., Lauter, K., Lisoněk, P. (eds.) SAC 2013. LNCS, vol. 8282, pp. 306–323. Springer, Heidelberg (2014). https://doi.org/10.1007/978-3-662-43414-7_16
5. Bogdanov, A., Rijmen, V.: Zero-correlation linear cryptanalysis of block ciphers. IACR Cryptology ePrint Archive 2011, 123 (2011)
6. Bogdanov, A., Rijmen, V.: Linear hulls with correlation zero and linear cryptanalysis of block ciphers. Des. Codes Cryptogr. **70**(3), 369–383 (2014)
7. Bogdanov, A., Tischhauser, E.: On the wrong key randomisation and key equivalence hypotheses in Matsui's algorithm 2. In: Moriai, S. (ed.) FSE 2013. LNCS, vol. 8424, pp. 19–38. Springer, Heidelberg (2014). https://doi.org/10.1007/978-3-662-43933-3_2
8. Bogdanov, A., Tischhauser, E., Vejre, P.S.: Multivariate linear cryptanalysis: the past and future of PRESENT. IACR Cryptology ePrint Archive 2016, 667 (2016)
9. Cho, J.Y.: Linear cryptanalysis of reduced-round PRESENT. In: Pieprzyk, J. (ed.) CT-RSA 2010. LNCS, vol. 5985, pp. 302–317. Springer, Heidelberg (2010). https://doi.org/10.1007/978-3-642-11925-5_21
10. Collard, B., Standaert, F.-X., Quisquater, J.-J.: Improving the time complexity of Matsui's linear cryptanalysis. In: Nam, K.-H., Rhee, G. (eds.) ICISC 2007. LNCS, vol. 4817, pp. 77–88. Springer, Heidelberg (2007). https://doi.org/10.1007/978-3-540-76788-6_7
11. Daemen, J.: Cipher and hash function design strategies based on linear and differential cryptanalysis. Ph.D. thesis, KU Leuven (1995)

12. Daemen, J., Rijmen, V.: The Design of Rijndael: AES - The Advanced Encryption Standard. Information Security and Cryptography. Springer, Heidelberg (2002). https://doi.org/10.1007/978-3-662-04722-4
13. Daemen, J., Rijmen, V.: Probability distributions of correlation and differentials in block ciphers. J. Math. Cryptol. **1**(3), 221–242 (2007)
14. Hermelin, M., Cho, J.Y., Nyberg, K.: Multidimensional linear cryptanalysis of reduced round serpent. In: Mu, Y., Susilo, W., Seberry, J. (eds.) ACISP 2008. LNCS, vol. 5107, pp. 203–215. Springer, Heidelberg (2008). https://doi.org/10.1007/978-3-540-70500-0_15
15. Hermelin, M., Cho, J.Y., Nyberg, K.: Multidimensional extension of Matsui's algorithm 2. In: Dunkelman, O. (ed.) FSE 2009. LNCS, vol. 5665, pp. 209–227. Springer, Heidelberg (2009). https://doi.org/10.1007/978-3-642-03317-9_13
16. Kaliski Jr., B.S., Robshaw, M.J.B.: Linear cryptanalysis using multiple approximations. In: Desmedt, Y.G. (ed.) CRYPTO 1994. LNCS, vol. 839, pp. 26–39. Springer, Heidelberg (1994). https://doi.org/10.1007/3-540-48658-5_4
17. Junod, P.: On the complexity of Matsui's attack. In: Vaudenay, S., Youssef, A.M. (eds.) SAC 2001. LNCS, vol. 2259, pp. 199–211. Springer, Heidelberg (2001). https://doi.org/10.1007/3-540-45537-X_16
18. Knudsen, L.R., Mathiassen, J.E.: A chosen-plaintext linear attack on DES. In: Goos, G., Hartmanis, J., Leeuwen, J., Schneier, B. (eds.) FSE 2000. LNCS, vol. 1978, pp. 262–272. Springer, Heidelberg (2001). https://doi.org/10.1007/3-540-44706-7_18
19. Langford, S.K., Hellman, M.E.: Differential-linear cryptanalysis. In: Desmedt, Y.G. (ed.) CRYPTO 1994. LNCS, vol. 839, pp. 17–25. Springer, Heidelberg (1994). https://doi.org/10.1007/3-540-48658-5_3
20. Lindsay, B.G.: Mixture models: theory, geometry and applications. In: NSF-CBMS Regional Conference Series in Probability and Statistics, pp. i–163. JSTOR (1995)
21. Matsui, M.: Linear cryptanalysis method for DES Cipher. In: Helleseth, T. (ed.) EUROCRYPT 1993. LNCS, vol. 765, pp. 386–397. Springer, Heidelberg (1994). https://doi.org/10.1007/3-540-48285-7_33
22. Matsui, M.: The first experimental cryptanalysis of the data encryption standard. In: Desmedt, Y.G. (ed.) CRYPTO 1994. LNCS, vol. 839, pp. 1–11. Springer, Heidelberg (1994). https://doi.org/10.1007/3-540-48658-5_1
23. Nguyen, P.H., Wu, H., Wang, H.: Improving the algorithm 2 in multidimensional linear cryptanalysis. In: Parampalli, U., Hawkes, P. (eds.) ACISP 2011. LNCS, vol. 6812, pp. 61–74. Springer, Heidelberg (2011). https://doi.org/10.1007/978-3-642-22497-3_5
24. Nyberg, K.: Linear approximation of block ciphers. In: Santis, A. (ed.) EUROCRYPT 1994. LNCS, vol. 950, pp. 439–444. Springer, Heidelberg (1995). https://doi.org/10.1007/BFb0053460
25. Selçuk, A.A.: On probability of success in linear and differential cryptanalysis. J. Cryptol. **21**(1), 131–147 (2008)
26. Semaev, I.A.: New results in the linear cryptanalysis of DES. IACR Cryptology ePrint Archive 2014, 361 (2014). http://eprint.iacr.org/2014/361
27. Zhao, J., Wang, M., Wen, L.: Improved linear cryptanalysis of CAST-256. J. Comput. Sci. Technol. **29**(6), 1134–1139 (2014)

Yoyo Tricks with AES

Sondre Rønjom[1,2](✉), Navid Ghaedi Bardeh[2], and Tor Helleseth[2,3]

[1] Nasjonal sikkerhetsmyndighet, Oslo, Norway
[2] Department of Informatics, University of Bergen, Bergen, Norway
{sondre.ronjom,navid.bardeh,tor.helleseth}@uib.no
[3] Forsvarets Forskningsinstitutt (FFI), Kjeller, Norway

Abstract. In this paper we present new fundamental properties of SPNs. These properties turn out to be particularly useful in the adaptive chosen ciphertext/plaintext setting and we show this by introducing for the first time key-independent *yoyo-distinguishers* for 3- to 5-rounds of AES. All of our distinguishers beat previous records and require respectively $3, 4$ and $2^{25.8}$ data and essentially zero computation except for observing differences. In addition, we present the first key-independent distinguisher for 6-rounds AES based on yoyos that preserve impossible zero differences in plaintexts and ciphertexts. This distinguisher requires an impractical amount of $2^{122.83}$ plaintext/ciphertext pairs and essentially no computation apart from observing the corresponding differences. We then present a very favorable key-recovery attack on 5-rounds of AES that requires only $2^{11.3}$ data complexity and 2^{31} computational complexity, which as far as we know is also a new record. All our attacks are in the adaptively chosen plaintext/ciphertext scenario.

Our distinguishers for AES stem from new and fundamental properties of generic SPNs, including generic SAS and SASAS, that can be used to preserve zero differences under the action of exchanging values between existing ciphertext and plaintext pairs. We provide a simple distinguisher for 2 generic SP-rounds that requires only 4 adaptively chosen ciphertexts and no computation on the adversaries side. We then describe a generic and deterministic yoyo-game for 3 generic SP-rounds which preserves zero differences in the middle but which we are not capable of exploiting in the generic setting.

Keywords: SPN · AES · Zero-Differences · Secret-key distinguisher · Impossible Differences · Key-recovery

1 Introduction

Block ciphers are typically designed by iterating an efficiently implementable round function many times in the hope that the resulting composition behaves like a randomly drawn permutation. The designer is typically constrained by various practical criterion, e.g. security target, implementation boundaries, and specialised applications, that might lead the designer to introduce symmetries and structures in the round function as a compromise between efficiency and security.

T. Takagi and T. Peyrin (Eds.): ASIACRYPT 2017, Part I, LNCS 10624, pp. 217–243, 2017.
https://doi.org/10.1007/978-3-319-70694-8_8

In the compromise, a round function is iterated enough times to make sure that any symmetries and structural properties that might exist in the round function vanish. Thus, a round function is typically designed to increasingly de-correlate with structure and symmetries after several rounds.

Yoyo game cryptanalysis was introduced by Biham et al. in [1] for cryptanalysis of 16 rounds of SKIPJACK. The yoyo game, similarly to Boomerang attacks [2], is based on adaptively making new pairs of plaintexts and ciphertexts that preserve a certain property inherited from the original pair. The ciphertext and/or plaintext space can in this case typically be partitioned into subsets of plaintexts or pairs of plaintexts closed under exchange operations where all pairs in a set satisfy the same property. A typical situation is that a pair of plaintexts and/or ciphertexts satisfy a certain zero difference after a few rounds and where new pairs of ciphertexts and plaintexts that satisfy the same zero difference can be formed simply by swapping words or bytes between the corresponding ciphertexts or plaintexts that preserve the same property. This type of cryptanalysis is typically structural and has previously been particularly successful on Feistel designs. Recently, Biryukov et al. [3] used the yoyo game to provide generic distinguishers against Feistel Networks with secret round functions up to 7 rounds. Boomerang attacks are quite similar to yoyos and in [4], Biryukov describes 5- and 6-round Boomerang key-recovery attacks on AES. Other types of structural attacks relevant in our setting include *invariant subspace attacks* [5,6] and *subspace trail cryptanalysis* [7,8]. Moreover, one may also note the paper by Ferguson et al. [9] and its implied result on 5-round AES.

Low data- and computational-complexity distinguishers and key-recovery attacks on round-reduced block ciphers have recently gained renewed interest in the literature. There are several reasons for this. In one direction cryptanalysis of block ciphers have focused on maximising the number of rounds that can be broken without exhausting the full codebook and key space. This often leads to attacks marginally close to that of pure brute force. These are attacks that typically have been improved over time based on many years of cryptanalysis. The most successful attacks often become de-facto standard methods of cryptanalysis for a particular block cipher and might discourage anyone from pursuing new directions in cryptanalysis that do not reach the same number of rounds. This in itself might hinder new breakthroughs, thus it can be important to investigate new promising ideas that might not have reached their full potential yet. New methods of cryptanalysis that break or distinguish fewer rounds faster and with lower complexity than established cryptanalysis are therefore interesting in this process. Many constructions employ reduced round AES as part of their design. Reduced versions of AES have nice and well-studied properties that can be favorably as components of larger designs (see for instance Simpira [10]). The state of the art analysis of low-complexity cryptanalysis of AES was presented in Bouillaguet et al. [11] and further improved by Derbez et al. [12] using automatic search tools. When it comes to distinguishers, the best attacks on 3-round AES are either based on truncated differentials [13] or integral cryptanalysis [14]. While the integral attack needs 2^8 chosen plaintexts, a truncated differential needs $2^{4.3}$ texts at the expense of a little more computation.

When it comes to 4 rounds of AES, an impossible differential attack requires $2^{16.25}$ chosen plaintexts and roughly the same amount of computation. Then at Crypto 2016, Sun et al. [15] presented the first 5-round key-dependent distinguisher for AES that was later improved to $2^{98.2}$ chosen plaintexts with 2^{107} computations. Later, at Eurocrypt 2017, Grassi et al. [8] proposed a 5-round *key-independent* distinguisher for 5-rounds AES that requires 2^{32} chosen texts and computational cost of $2^{35.6}$ look-ups into memory of size 2^{36} bytes with a 99% probability success rate.

1.1 Our Contribution

We present for the first time applications of cryptanalysis based on the yoyo game introduced in [1] to generic Substitution Permutation Networks(SPNs) that iterate a generic round function $A \circ S$ where S is a non-linear layer consisting of at least two concatenated s-boxes and A is a generic affine transformation. The s-boxes and affine layers can be all different. This way it resembles the setting of SAS and SASAS [16] cryptanalysis.

First we provide a generic framework for the yoyo game on generic SP-networks. Then we show that distinguishing two generic SP rounds with a yoyo game requires encryption and decryption of in total one pair of plaintexts and one pair of ciphertexts respectively and no computational effort on the adversaries part; the distinguisher is purely structural and immediate. We then provide a generic framework for analysing 3 rounds generic SPN which seems to be the maximum possible number of generic rounds achievable with a deterministic yoyo game and a generic SPN. We then apply our generic results to the most well-studied block cipher in use today, AES. Since an even number of AES-rounds can be viewed as a generic SPN with half the number of rounds, our 2- and 3-round generic SPN distinguishers apply directly to 4- and 6-rounds of AES. We extend the generic distinguishers to cover 3- and 5-rounds AES in a natural way, including a new secret key recovery for 5 rounds. All of our secret key distinguishers improve on previously published results both in terms of time and data complexity.

1.2 Overview of this Paper and Main Results

In Sect. 2 we analyse generic SPNs formed by iterating a round function consisting of concatenated layers of s-boxes and generic affine linear transformations. In Sect. 2.1 we describe a simple yoyo distinguisher for two non-secret but generic SPN rounds that is purely structural and requires only one chosen pair of plaintexts and one (adaptively) chosen pair of ciphertexts. The distinguisher involves no computation. In Sect. 2.2 we describe generic zero differential properties for 3-round SPN that are preserved in a yoyo game. If the difference of a pair of plaintexts (or ciphertexts) is zero in particular words in the middle rounds, then the yoyo game preserves this property and can be used to generate "infinitely" many new pairs with the exact same zero difference pattern in the middle rounds with probability 1. The current drawback in the generic setting is that the adversary needs a way to distinguish this condition.

Table 1. Comparison of key-recovery on 5 rounds of AES

Attack	Rounds	Data	Computation	Memory	Ref.
MitM	5	8 CP	2^{64}	2^{56}	[17]
Imp. polyt	5	15 CP	2^{70}	2^{41}	[18]
Integral	5	2^{11} CP	$2^{45.7}$	small	[19]
Imp. diff	5	$2^{31.5}$ CP	2^{33}	2^{38}	[20]
Boomerang	5	2^{39} ACC	2^{39}	2^{33}	[4]
Yoyo	5	$2^{11.3}$ ACC	2^{31}	small	Sect. 3.5

Table 2. Secret-key distinguishers for AES

Property	Rounds	Data	Cost	Key-independent	Ref.
Trun. Diff	3	$2^{4.3}$ CP	$2^{11.5}$ XOR	✓	[7,21]
Integral	3	2^{8} CP	2^{8} XOR	✓	[14]
Yoyo	3	3 ACC	2 XOR	✓	Sect. 3.1
Imp. Diff	4	$2^{16.25}$ CP	$2^{22.3}$ M	✓	[20]
Integral	4	2^{32} CP	2^{32} XOR	✓	[14]
Yoyo	4	4 ACC	2 XOR	✓	Sect. 3.2
Struct. Diff	5	2^{33}	$2^{36.6}$ M	✓	[8]
Imp. Diff	5	$2^{98.2}$ CP	2^{107} M		[7]
Integral	5	2^{128} CC	2^{128} XOR		[15]
Yoyo	5	$2^{25.8}$ ACC	$2^{24.8}$ XOR	✓	Sect. 3.3
Yoyo	6	$2^{122.83}$ ACC	$2^{121.83}$ XOR	✓	Sect. 3.4

In Sect. 3 we apply our generic results to AES. Two rounds of AES, essentially corresponding to four parallel superboxes and a large linear super-mixlayer, can essentially be viewed as one generic SP-round and thus our generic method can be applied directly to it. We begin Sect. 3.1 by presenting a simple distinguisher for 3 rounds of AES. It requires one chosen pair of plaintexts and one chosen ciphertext and no computation. Then in Sect. 3.2 we directly apply the generic yoyo distinguisher presented in Sect. 2.1 to 4 rounds of AES. It requires one chosen pair of plaintexts and one chosen pair of ciphertexts, and no computation. In Sect. 3.3 we extend the 4-round yoyo distinguisher to 5 rounds by testing for pairs derived by the yoyo game that obey unusual byte collisions in the plaintext (or ciphertext) differences. Then, in Sect. 3.4 we apply the theory on 3 generic SP-rounds presented in Sect. 2.2 directly to form the first 6-round AES secret key distinguisher using $2^{122.83}$ texts derived from multiple yoyo-games.

The current best key-recovery attacks for 5-rounds of AES are found in Table 1, while a list of the current best secret-key distinguishers for 3 to 6 rounds is given in Table 2. We have adopted that the data complexity is measured in minimum number of chosen plaintexts/ciphertexts CP/CC or/and adaptive chosen plaintexts/ciphertexts ACP/ACC. Time complexity is measured in equivalent encryptions (E), memory accesses (M) or XOR operations (XOR) - adopting that 20 M $\approx$ 1 Round of Encryption.

2 Yoyo Analysis of Generic SPNs

We start off by analysing permutations on $\mathbb{F}_q^n$ for $q = 2^k$ of the form

$$F(x) = S \circ L \circ S \circ L \circ S \tag{1}$$

where S is a large s-box formed by concatenating n smaller s-boxes s over $\mathbb{F}_q$ and where L is a linear transformation acting on elements of $\mathbb{F}_q^n$. *Notice that our results apply directly to* $S \circ A \circ S \circ A \circ S$ *where* A *are affine transformations, however restricting to linear transformations* L *slightly simplifies our presentation.* We call an element of $\mathbb{F}_q$ a *word* and a vector of words $\alpha = (\alpha_0, \alpha_1, \ldots, \alpha_{n-1}) \in \mathbb{F}_q^n$ is called a *state* to emphasize that we are thinking of states in a block cipher. A vector of words in $\mathbb{F}_q^n$ can also be viewed as a list of k-bit vectors of length n. The Hamming weight of a vector $x = (x_0, x_1, \ldots, x_{n-1})$ is defined as the number of nonzero components in the vector and is denoted by $wt(x)$.

We refer to the small s-boxes operating on words in $\mathbb{F}_q$ as *component* s-boxes to not confuse them with the large s-box S they are part of and that operates on $\mathbb{F}_q^n$.

Two vectors (or a pair of differences of two vectors) that are different can be zero in the same positions, and we will be interested in comparing two differences according to their zero positions so it is useful to define the following.

Definition 1 (The zero difference pattern). *Let* $\alpha \in \mathbb{F}_q^n$ *and define the* zero difference pattern

$$\nu(\alpha) = (z_0, z_1, \ldots, z_{n-1})$$

that returns a binary vector in $\mathbb{F}_2^n$ *where* $z_i = 1$ *indicates that* α_i *is zero or* $z_i = 0$ *otherwise.*

The complement of a zero difference pattern is often called *activity pattern* in literature. For a linear transformation L on $\mathbb{F}_q^n$ we generally do not have that the differences $\alpha \oplus \beta$ and $L(\alpha) \oplus L(\beta)$ are zero in the same positions for random values of α and β. However, permutations S do and the zero difference pattern is thus an invariant of S.

Lemma 1. *For two states* $\alpha, \beta \in \mathbb{F}_q^n$*, the zero pattern of their difference is preserved through* S*, hence*

$$\nu(\alpha \oplus \beta) = \nu(S(\alpha) \oplus S(\beta)).$$

Proof. Since S is a permutation s-box this leads to $\alpha_i \oplus \beta_i = 0$ if and only if $s(\alpha_i) \oplus s(\beta_i) = 0$, thus the pattern vector of the difference is preserved through S.

Although this is rather trivial, we leave it as a lemma for later reference. The following definition is central to the paper.

Definition 2. *For a vector* $v \in \mathbb{F}_2^n$ *and a pair of states* $\alpha, \beta \in \mathbb{F}_q^n$ *define a new state* $\rho^v(\alpha, \beta) \in \mathbb{F}_q^n$ *such that the i'th component is defined by*

$$\rho^v(\alpha, \beta)_i = \alpha_i v_i \oplus \beta_i(v_i \oplus 1).$$

This is equivalent to

$$\rho^v(\alpha, \beta)_i = \begin{cases} \alpha_i & \text{if } v_i = 1, \\ \beta_i & \text{if } v_i = 0. \end{cases}$$

Notice that $(\alpha', \beta') = (\rho^v(\alpha, \beta), \rho^v(\beta, \alpha))$ is a new pair of states formed by swapping individual words between α and β according to the binary coefficients of v. From the definition it can be seen that

$$\rho^v(\alpha, \beta) \oplus \rho^v(\beta, \alpha) = \alpha \oplus \beta. \tag{2}$$

Let $\overline{v}$ be the complement of v. Note that $\rho^{\overline{v}}(\alpha, \beta) = \rho^v(\beta, \alpha)$ and therefore $\{\rho^v(\alpha, \beta), \rho^v(\beta, \alpha))\} = \{\rho^{\overline{v}}(\alpha, \beta), \rho^{\overline{v}}(\beta, \alpha))\}$, implying that v and $\overline{v}$ result in the same pair. The maximum number of possible unique pairs (α', β') generated this way is 2^{n-1} (including the original pair). The maximum is only attainable if $\alpha_i \neq \beta_i$ for all $0 \leq i < n$. Assume this is the case. If we restrict v to the 2^{n-1} binary vectors in $\mathbb{F}_2^n$ with the last coefficient set to 0 we span exactly 2^{n-1} unique pairs, including (α, β). If $v = (0, 0, 0, \ldots, 0)$ is avoided, we generate only new pairs $(\rho^v(\alpha, \beta), \rho^v(\beta, \alpha))$ all unequal to (α, β).

The function ρ^v has some interesting properties. We leave the following in as a lemma for further reference.

Lemma 2. *Let* $\alpha, \beta \in \mathbb{F}_q^n$ *and* $v \in \mathbb{F}_2^n$. *Then we have that* ρ *commutes with the s-box layer,*

$$\rho^v(S(\alpha), S(\beta)) = S(\rho^v(\alpha, \beta))$$

and thus

$$S(\alpha) \oplus S(\beta) = S(\rho^v(\alpha, \beta)) \oplus S(\rho^v(\beta, \alpha)).$$

Proof. S operates independently on individual words and so the result follows trivially from the definition of ρ^v.

Lemma 3. *For a linear transformation* $L(x) = L(x_0, x_1, x_2, \ldots, x_{n-1})$ *acting on n words we have that*

$$L(\alpha) \oplus L(\beta) = L(\rho^v(\alpha, \beta)) \oplus L(\rho^v(\beta, \alpha))$$

for any $v \in \mathbb{F}_2^n$.

Proof. Due to the linearity of L it follows that $L(x) \oplus L(y) = L(x \oplus y)$. Moreover, due to relation (2), $\rho^v(\alpha, \beta) \oplus \rho^v(\beta, \alpha) = \alpha \oplus \beta$ and thus $L(\rho^v(\alpha, \beta)) \oplus L(\rho^v(\beta, \alpha)) = L(\alpha) \oplus L(\beta)$.

So from Lemma 3 we have that $L(\alpha) \oplus L(\beta) = L(\rho^v(\alpha,\beta)) \oplus L(\rho^v(\beta,\alpha))$ and from Lemma 2 we showed that $S(\alpha) \oplus S(\beta) = S(\rho^v(\alpha,\beta)) \oplus S(\rho^v(\beta,\alpha))$. This implies that

$$L(S(\alpha)) \oplus L(S(\beta)) = L(S(\rho^v(\alpha,\beta))) \oplus L(S(\rho^v(\beta,\alpha))) \tag{3}$$

however it does not generally hold that

$$S(L(\alpha)) \oplus S(L(\beta)) = S(L(\rho^v(\alpha,\beta))) \oplus S(L(\rho^v(\beta,\alpha))). \tag{4}$$

However it is easy to see that the zero difference pattern does not change when we apply L or S to any pair $\alpha' = \rho^v(\alpha,\beta)$ and $\beta' = \rho^v(\beta,\alpha)$. As described in Lemma 3 we clearly have that

$$\nu(L(\alpha) \oplus L(\beta)) = \nu(L(\rho^v(\alpha,\beta)) \oplus L(\rho^v(\beta,\alpha)))$$

and the differences are zero in exactly the same positions. For the function S we notice that if the difference between two words α_i and β_i before a component s-box is zero, then it is also zero after the component s-box. In other words,

$$\nu(S(\alpha) \oplus S(\beta)) = \nu(S(\rho^v(\alpha,\beta)) \oplus S(\rho^v(\beta,\alpha))).$$

Thus it follows that, although (4) does not hold, we do have that

$$\nu(S(L(\alpha)) \oplus S(L(\beta))) = \nu(S(L(\rho^v(\alpha,\beta))) \oplus S(L(\rho^v(\beta,\alpha)))) \tag{5}$$

always holds, i.e. words are zero in exactly the same positions in the difference for any pair $\alpha' = \rho^v(\alpha,\beta)$ and $\beta' = \rho^v(\beta,\alpha)$ through L and S.

The yoyo attack is heavily based on the simple result in the lemma below that summarises the properties above.

Theorem 1. *Let $\alpha,\beta \in \mathbb{F}_q^n$ and $\alpha' = \rho^v(\alpha,\beta)$, $\beta' = \rho^v(\beta,\alpha)$ then*

$$\nu(S \circ L \circ S(\alpha) \oplus S \circ L \circ S(\beta)) = \nu(S \circ L \circ S(\alpha') \oplus S \circ L \circ S(\beta'))$$

Proof. The proof follows from the three steps below:

- Lemma 2 implies that $S(\alpha) \oplus S(\beta) = S(\alpha') \oplus S(\beta')$.
- The linearity of L gives $L(S(\alpha)) \oplus L(S(\beta)) = L(S(\alpha')) \oplus L(S(\beta'))$.
- Using Lemma 1 it follows, since S is a permutation and preserves the zero difference pattern, that
  $\nu(S(L(S(\alpha))) \oplus S(L(S(\beta)))) = \nu(S(L(S(\alpha'))) \oplus S(L(S(\beta'))))$.

2.1 Yoyo Distinguisher for Two Generic SP-Rounds

Two full generic rounds are equal to $G_2' = L \circ S \circ L \circ S$. However, to simplify the presentation, we remove the final linear layer and restrict our attention to

$$G_2 = S \circ L \circ S. \tag{6}$$

We show that G_2 can be distinguished, with probability 1, using two plaintexts and two ciphertexts. The distinguisher is general, in the sense that each of the S and L transformations can be different and does not require any computation on the adversaries part.

The idea is simple. If we fix a pair of plaintexts $p^0, p^1 \in \mathbb{F}_q^n$ with a particular zero difference pattern $\nu(p^0 \oplus p^1)$, then from the corresponding ciphertexts $c^0 = G_2(p^0)$ and $c^1 = G_2(p^1)$ we can construct a pair of new ciphertexts c'^0 and c'^1 that decrypt to a pair of new plaintexts p'^0, p'^1 whose difference has exactly the same zero difference pattern. Moreover this is deterministic and holds with probability 1.

Theorem 2. *(Generic yoyo game for 2-rounds).*
Let $p^0 \oplus p^1 \in \mathbb{F}_q^n$, $c^0 = G_2(p^0)$ *and* $c^1 = G_2(p^1)$. *Then for any* $v \in \mathbb{F}_2^n$ *let* $c'^0 = \rho^v(c^0, c^1)$ *and let* $c'^1 = \rho^v(c^1, c^0)$. *Then*

$$\begin{aligned}\nu(G_2^{-1}(c'^0) \oplus G_2^{-1}(c'^1)) &= \nu(p'^0 \oplus p'^1)\\ &= \nu(p^0 \oplus p^1).\end{aligned}$$

Proof. This follows directly from Eq. 3 and we have that

$$L^{-1}(S^{-1}(c^0)) \oplus L^{-1}(S^{-1}(c^1)) = L^{-1}(S^{-1}(\rho^v(c^0, c^1))) \oplus L^{-1}(S^{-1}(\rho^v(c^1, c^0))).$$

Since the differences are equal the r.h.s. difference and the l.h.s. differences are zero in exactly the same words. Thus we must have that $\nu(G_2^{-1}(c'^0) \oplus G_2^{-1}(c'^1)) = \nu(p^0 \oplus p^1)$.

By symmetry, the exact same property obviously also holds in the decryption direction.

What Theorem 2 states is that if we pick a pair of plaintexts p^0 and p^1 with a zero difference $\nu(p^0 \oplus p^1)$, we can encrypt the pair to a pair of ciphertexts c^0 and c^1, construct a new set of ciphertexts $c'^0 = \rho^v(c^0, c^1)$ and $c'^1 = \rho^v(c^1, c^0)$ (simply interchanging words between the original pair) then decrypt to a pair of new plaintexts with the exact same zero difference pattern. Thus this leaves us with a straight-forward distinguisher that requires two plaintexts and two adaptively chosen ciphertexts. There is no need for any computation on the adversaries part as the result is immediate.

By symmetry we could of course instead have started with a pair of ciphertexts with a given zero difference pattern and instead adaptively picked a new pair of plaintexts that would decrypt to a new pair of ciphertexts whose zero difference pattern corresponds to the first ciphertext pair.

2.2 Analysis of Three Generic SP-Rounds

In this section we show that there is a powerful deterministic difference symmetry in 3 generic SPN rounds, where we cut away the final linear L-layer, and analyse

$$G_3 = S \circ L \circ S \circ L \circ S \tag{7}$$

where we also omit numbering the transformations to indicate that they can be all different.

The three round property follows from the two round property. We have already shown in Theorem 2 that for two states α and β it follows that

$$\nu(G_2^{-1}(\rho^v(G_2(\alpha), G_2(\beta))) \oplus G_2^{-1}(\rho^v(G_2(\beta), G_2(\alpha)))) = \nu(\alpha \oplus \beta).$$

Since G_2 and G_2^{-1} have identical forms, it also follows that

$$\nu(G_2(\rho^v(G_2^{-1}(\alpha), G_2^{-1}(\beta))) \oplus G_2(\rho^v(G_2^{-1}(\beta), G_2^{-1}(\alpha)))) = \nu(\alpha \oplus \beta)$$

Also, from Lemma 1 we know that zero difference patterns are preserved through s-box layers S, that is

$$\nu(\alpha \oplus \beta) = \nu(S(\alpha) \oplus S(\beta)).$$

Thus, assuming a particular zero difference pattern in front of the middle S-layer in (7) is equivalent to assuming the same zero difference pattern after it. Hence, the following Theorem follows straightforwardly.

Theorem 3. *(Generic yoyo game for 3-rounds).*
Let $G_3 = S \circ L \circ S \circ L \circ S$. If $p^0, p^1 \in \mathbb{F}_q^n$ and $c^0 = G_3(p^0)$ and $c^1 = G_3(p^1)$. Then

$$\nu(G_2(\rho^{v_1}(p^0, p^1)) \oplus G_2(\rho^{v_1}(p^1, p^0))) = \nu(G_2^{-1}(\rho^{v_2}(c^0, c^1)) \oplus G_2^{-1}(\rho^{v_2}(c^1, c^0)))$$

for any $v_1, v_2 \in \mathbb{F}_2^n$. Moreover, for any $z \in \mathbb{F}_2^n$, let $R_P(z)$ denote the pairs of plaintexts where $\nu(G_2(p^0) \oplus G_2(p^1)) = z$ and $R_C(z)$ the pairs of ciphertexts where $\nu(G_2^{-1}(c^0) \oplus G_2^{-1}(c^1)) = z$. Then it follows that

$$(G_3(\rho^v(p^0, p^1)), G_3(\rho^v(p^1, p^0))) \in R_C(z)$$

for any $(p^0, p^1) \in R_P(z)$ while

$$(G_3^{-1}(\rho^v(c^0, c^1)), G_3^{-1}(\rho^v(c^1, c^0))) \in R_P(z)$$

for any $(c^0, c^1) \in R_C(z)$.

Thus, from a single pair of plaintexts p_1, p_2, we can continuously generate new elements that with probability 1 belong to $R_C(z)$ and $R_P(z)$, which contain exactly the pairs of plaintexts and ciphertexts that have difference pattern z in the middle.

A distinguisher for this requires first to test a number of pairs until there is one that has a particular Hamming weight of the zero difference pattern, then try to distinguish this case when it happens. The probability that a random pair of plaintexts has a sum with zero difference pattern containing exactly m zeros (the difference is non-zero in exactly m words) in the middle is $\binom{n}{m}\frac{(q-1)^{-m}}{q^n}$ where $q = 2^k$. Thus we need to test approximately the inverse of that number of pairs to expect to find one correct pair, and thus require that the complexity of any distinguisher which succeeds in detecting the right condition times the number of pairs to be checked is substantially less than brute force.

We leave this case with a small observation. Assume that we have in fact found a pair of plaintexts that belongs to $R_P(z)$ for a particular zero difference pattern of Hamming weight $n - m$. In this case we will continuously generate new pairs that are guaranteed to be contained in $R_P(z)$ and $R_C(z)$. However we need a way to distinguish that this is in fact happening. Let A be the affine layer in an SASAS construction. Assume that $S^{-1}(c_1) = x \oplus z$ while $S^{-1}(c_2) = y \oplus z$ where $A^{-1}(x)$ and $A^{-1}(y)$ are non-zero only in the positions where z is zero, while $A^{-1}(z)$ is non-zero only in the positions where z is zero. It follows that x and y belong to a linear subspace U of dimension $m - n$ while z must belong to the complementary subspace V of dimension m such that $U \oplus V = \mathbb{F}_q^n$. Hence, a problem for further study is to investigate whether $c_1 \oplus c_2 = S(x \oplus z) \oplus S(y \oplus z)$ for $x, y \in U$ and $z \in V$ has particular generic distinguishing properties. A distinguisher for this would of course apply equally well to the plaintext side, i.e. $p_1 \oplus p_2 = S^{-1}(x' \oplus z') \oplus S^{-1}(y' \oplus z')$ for $x', y' \in U'$ and $z' \in V'$.

3 Applications to AES

The round function in AES [19] is most often presented in terms of operations on 4×4 matrices over $\mathbb{F}_q$ where $q = 2^8$. One round of AES consists of SubBytes (SB), ShiftRows (SR), MixColumns (MC) and AddKey (AK). SubBytes applies a fixed 8-bit s-box to each byte of the state, ShiftRows rotates each row a number of times, while MixColumns applies a fixed linear transformation to each column. Four-round AES can be described with the *superbox* representation [22–24] operating on four parallel 32-bit words (or elements of $\mathbb{F}_{2^8}^4$) instead of bytes where leading and trailing linear layers are omitted for sake of clarity. A similar description of AES is given in [25]. The superbox representation of AES now consists of iterating four parallel keyed 32-bit sboxes and a large linear "super"-layer. Thus while one round is a composition $AK \circ MC \circ SR \circ SB$, two rounds of AES is equal to the composition

$$(AK \circ MC \circ SR \circ SB) \circ (AK \circ MC \circ SR \circ SB). \tag{8}$$

Since we are only considering differences, we can leave out AddKey(AK) for sake of clarity. Now since SR commutes with SB we can rewrite (8) as

$$R^{2\prime} = MC \circ SR \circ (SB \circ MC \circ SB) \circ SR. \tag{9}$$

The initial SR has no effect in our analysis, thus we leave it out and only consider

$$R^2 = MC \circ SR \circ (SB \circ MC \circ SB) = MC \circ SR \circ S$$

where $S = SB \circ MC \circ SB$ constitutes four parallel super-boxes acting independently on 32-bits of the state. In terms of the generic SPN-analysis in the

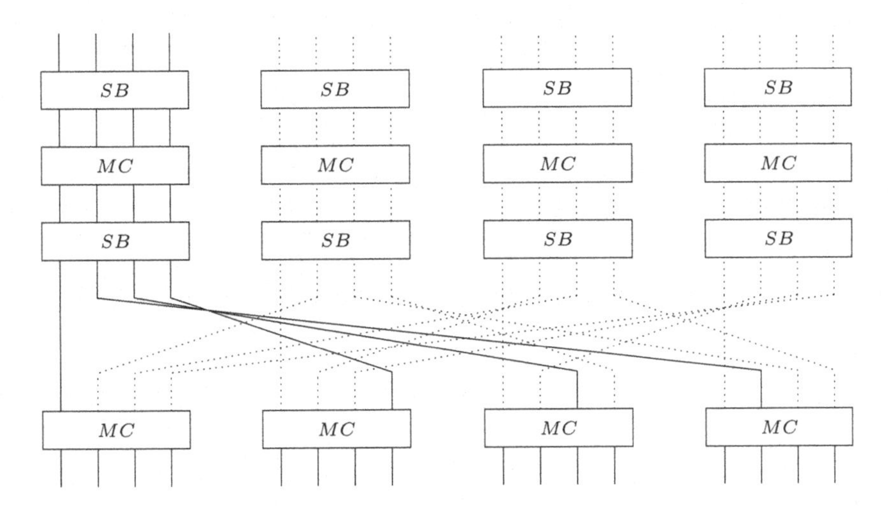

Fig. 1. Two rounds AES in the super-box representation

previous section, the state of AES consists of four words where each word contains four bytes. This is equivalent to the superbox [23] representation shown in Fig. 1 where the initial ShiftRows has been omitted. Thus, let $S = SB \circ MC \circ SB$ and $L = SR \circ MC \circ SR$. Four rounds of AES can be viewed in a similar way simply by repeating the composition of (9)

$$R^{4'} = MC \circ SR \circ S \circ L \circ S \circ SR$$

and we end up with

$$R^4 = S \circ L \circ S \tag{10}$$

if we omit the linear layers before the first and after the last s-box layer. The designers of AES used this representation to provide an elegant proof that the lower bound on the number of active s-boxes over 4 rounds is $5 \cdot 5 = 25$. The number of active super boxes in Fig. 3 due to the linear layer L is at least 5, while the minimum number of active s-boxes inside a super-box is also 5 due to the MixColumns matrix, thus the total number of active s-boxes is at least 25.

Similarly, 6 rounds of AES is equal to

$$R^{6'} = MC \circ SR \circ S \circ L \circ S \circ L \circ S \circ SR \tag{11}$$

which, when the leading and trailing linear layers are removed, becomes

$$R^6 = S \circ L \circ S \circ L \circ S. \tag{12}$$

Thus two rounds of AES can be viewed as one generic SPN-round consisting of a state-vector of 4 words from $\mathbb{F}_{2^8}^4$ consisting of one s-box layer of 4 parallel concatenated superboxes and one large linear layer. Therefore can any even number of rounds be seen as half the number of generic SP rounds. It follows that our generic analysis presented in the previous section applies directly to 4 and 6 rounds of AES.

Since two rounds of AES correspond to one generic SPN-round, our generic analysis on SP-rounds does not cover odd rounds of AES. However, we extend the 2- and 4-round distinguishers by one round in a natural way by exploiting properties of one AES round. The following observation is used to extend our distinguishers to 3 and 5 rounds. First, 3 rounds of AES can be written as $Q \circ S$ where $Q = SB \circ MC \circ SR$ by adding a round at the end and 5 rounds of AES can be written as $S \circ L \circ S \circ Q'$ where $Q' = SR \circ MC \circ SB$ where a round is added at the beginning of 4 AES rounds. We have again omitted the trailing and leading linear layers. Both our 3-rounds distinguisher and our 5-round distinguishers exploit properties of one AES-round, and in particular the effect of MixColumns in Q and Q'.

Definition 3. *Let $Q = SB \circ MC \circ SR$.*

For a binary vector $z \in \mathbb{F}_2^4$ of weight t let V_z denote the subspace of $q^{4 \cdot (4-t)}$ states $x = (x_0, x_1, x_2, x_3)$ formed by setting x_i to all possible values of $\mathbb{F}_q^4$ if $z_i = 0$ or to zero otherwise. Then, for any state $a = (a_0, a_1, a_2, a_3)$, let

$$T_{z,a} = \{Q(a \oplus x) \,|x \in V_z\}.$$

It is important to note that the sets $T_{z,a}$ in Definition 3 in practice depend on variable round keys xored in between layers which we assume are implicit from the context in which we use them.

Let H_i denote the image of the i'th word in $SR(a \oplus x)$ when x is in V_z. Observe that $|H_i| = q^{4-t}$. Then define

$$T_i^{z,a} = SB \circ MC(H_i).$$

Since SB and MC operate on each word individually then $T_{z,a}$ has $T_i^{z,a}$ as its i'th word.

Lemma 4. *The set $T_{z,a}$ satisfies*

$$T_{z,a} = T_0^{z,a} \times T_1^{z,a} \times T_2^{z,a} \times T_3^{z,a}$$

where $|T_i^{z,a}| = q^{4-wt(z)}$.

Proof. In Fig. 2 it is easy to see that each word contributes one independent byte to each word after SR. Thus if $4 - t$ words are nonzero, and each word contributes a single independent byte to each word after applying SR, it follows that each word after SR can take on exactly q^{4-t} independent values. Since H_i denotes the set of q^{4-t} possible values that word i can have after SR and MC and SB operate independently and in parallel on words, it follows that $T_i^{z,a} = SB \circ MC(H_i)$.

It is not hard to see that the inverse of Q', Q'^{-1}, enjoys the same property.

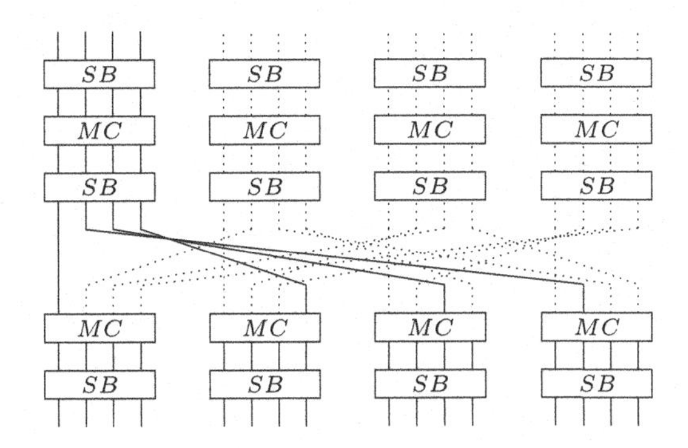

Fig. 2. Three rounds $SB \circ MC \circ SR \circ S = Q \circ S$

We ask the reader to keep one thing in mind. In our analysis, the last $MC \circ SR$ layers and first SR layers are omitted for simplicity of presentation. Thus, when we say that we swap ciphertexts c^0 and c^1 to form $c'^0 = \rho^v(c^0, c^1)$ and $c'^1 = \rho^v(c^1, c^0)$, then for it to work for the real full round AES, we instead apply the transformation

$$c'^0 = MC \circ SR(\rho^v(SR^{-1} \circ MC^{-1}(c^0), SR^{-1} \circ MC^{-1}(c^1)))$$

and

$$c'^1 = MC \circ SR(\rho^v(SR^{-1} \circ MC^{-1}(c^1), SR^{-1} \circ MC^{-1}(c^0))).$$

Similarly, when we swap words in the plaintexts we need to account for the extra SR-layer in the full rounds, i.e.

$$p'^0 = SR^{-1}(\rho^v(SR(p^0), SR(p^1)))$$

and

$$p'^1 = SR^{-1}(\rho^v(SR(p^1), SR(p^0))).$$

All our results, except for the 6-round distinguisher, have been, and are easy, to verify experimentally on a laptop and require only to implement very simple operations. In the following sections we present our results on AES.

For the pseudo-codes we use Algorithm 1 to simplify the presentation. If the input pairs are distinct in at least two words, which happens with probability $(1 - 2^{-94})$, then this algorithm always returns a new text. If a pair in fact is equal in exactly three words, then the pair is simply discarded. Since we use a simplified swap operation, we have to go a few more rounds back and forth in the yoyo game to construct pairs, instead of returning all possible swap-pairs at once for a ciphertext pair or plaintext pair.[1]

[1] C-code for our attacks can be found at https://github.com/sondrer/YoyoTricksAES.

Algorithm 1. Swaps the first word where texts are different and returns one text

```
function SIMPLESWAP(x^0, x^1)                                    //x^0 ≠ x^1
    x'^0 ← x^1
    for i from 0 to 3 do
        if x_i^0 ≠ x_i^1 then
            x'_i^0 ← x_i^0
            return x'^0
        end if
    end for
end function
```

3.1 Yoyo Distinguisher for Three Rounds of AES

Two rounds of AES correspond to one generic SPN round and can be distinguished trivially. If we consider three full AES rounds minus the linear layer before the first and after the last s-box layer, we have that

$$\begin{aligned} R^3 &= SB \circ MC \circ SR \circ S \\ &= Q \circ S \end{aligned}$$

which is depicted in Fig. 2. The basis for our distinguisher is implicit in Lemma 4. Let p^0 and p^1 denote two plaintexts with $z = \nu(p^0 \oplus p^1)$ and $wt(z) = t$, i.e. the difference between the plaintexts is zero in t of the words. Due to Lemma 1 we have that $\nu(S(p^0) \oplus S(p^1)) = \nu(p^0 \oplus p^1)$ thus the zero difference pattern is preserved through the superbox layer S. Then, since $S(p^0)$ and $S(p^1)$ can be regarded as two states that differ only in $4-t$ words, it follows from Lemma 4 that both $Q(S(p^0)) = c^0$ and $Q(S(p^1)) = c^1$ belong to the same set $T_{z,a}$, which are generally unknown due to the addition of secret keys. However, the ciphertexts themselves span subsets of $T_{z,a}$. Since both c^0 and c^1 are in the same set $T_{z,a}$, it follows that each word c_i^0 and c_i^1 of $c^0 = (c_0^0, c_1^0, c_2^0, c_3^0)$ and $c^1 = (c_0^1, c_1^1, c_2^1, c_3^1)$ are drawn from the same subsets $T_i^{z,a} \subset \mathbb{F}_q^4$ of size q^{4-t}. In particular, the set

$$T'_{z,a} = \{c_0^0, c_0^1\} \times \{c_1^0, c_1^1\} \times \{c_2^0, c_2^1\} \times \{c_3^0, c_3^1\}$$

must be a subset of $T_{z,a}$ of at most size 2^4, where $\{c_i^0, c_i^1\}$ is a subset of $T_i^{z,a}$ as shown in Lemma 4. In other words, if we pick any ciphertext $c' \neq c^0, c^1$ from $T'_{z,a}$ it follows that $\nu(Q^{-1}(c') \oplus S(p^0)) = \nu(Q^{-1}(c') \oplus S(p^1)) = \nu(S(p^0) \oplus S(p^1))$ and in particular, $\nu(R^{-3}(c') \oplus p^0) = \nu(R^{-3}(c') \oplus p^1) = \nu(p^0 \oplus p^1)$. This implies a straightforward distinguisher for 3 rounds of AES that requires two chosen plaintexts and one adaptively chosen ciphertext. To simplify, the adversary picks two plaintexts p^0 and p^1 that differ in only one word such that $t = wt(\nu(p^0 \oplus p^1)) = 3$. The corresponding ciphertexts c^0 and c^1 specify a subset $T'_{z,a} \subset T_{z,a}$ of size 2^4 including the original ciphertexts. Thus if the adversary picks any ciphertext $c' = \rho^v(c^0, c^1) \in T'_{z,a}$ not equal to c^0 or c^1 and asks for the decryption of it p', then with probability 1 we have that $\nu(p' \oplus p^0) = \nu(p' \oplus p^1) = \nu(p^0 \oplus p^1)$.

Algorithm 2. Distinguisher for 3 rounds of AES

Input: A pair of plaintext with $wt(\nu(p^0 \oplus p^1)) = 3$
Output: 1 for an AES, -1 otherwise.

```
// r-round AES enc/dec without first SR and last SR ∘ MC
c^0 ← enc_k(p^0, 3), c^1 ← enc_k(p^1, 3)        //encrypt pair
c' ← SIMPLESWAP(c^0, c^1)
p' ← dec_k(c', 3)                               //decrypt c'
if ν(p' ⊕ p^1) = ν(p^0 ⊕ p^1) then
    return 1.
else
    return −1
end if
```

Thus the difference of p' and any of the initial plaintexts p^i is zero in exactly the same words as the initially chosen difference $p^0 \oplus p^1$. This can only happen at random with probability 2^{-96}.

3.2 Yoyo Distinguisher for Four Rounds of AES

In this section we present a remarkably simple and efficient distinguisher for 4-rounds AES. For 4 rounds of AES we simply apply the distinguisher for 2 rounds generic SPN in Sect. 2.1. Four rounds of AES is equal to

$$R'^4 = MC \circ SR \circ S \circ L \circ S \circ SR$$

where $S \circ L \circ S$ consists of the "super-layers" in AES. To simplify the notation, we omit the last layer of $MC \circ SR$ together with the initial SR-layer and apply the distinguisher directly to

$$R^4 = S \circ L \circ S.$$

Following Sect. 2.1, the adversary picks a pair of plaintexts p^0 and p^1 whose difference is zero in exactly t out of four words. The adversary then asks for the encryption of the plaintexts and receives the corresponding ciphertexts c^0 and c^1 and picks a new pair of ciphertexts $c'^0 = \rho^v(c^0, c^1)$ and $c'^1 = c'^0 \oplus c^0 \oplus c^1$ for any $v \in \mathbb{F}_2^4$. That is, he makes a new pair of ciphertexts simply by exchanging any words between the c^0 and c^1. The new pair of ciphertexts now decrypts to a pair of new plaintexts p'^0 and p'^1 that has exactly the same zero difference pattern as p^0 and p^1. By symmetry, the same distinguisher works in the decryption direction.

3.3 Yoyo Distinguisher for Five Rounds of AES

We can extend the 4-round distinguisher to a 5-round distinguisher by combining the 4-round yoyo distinguisher together with the observation used in the 3-rounds

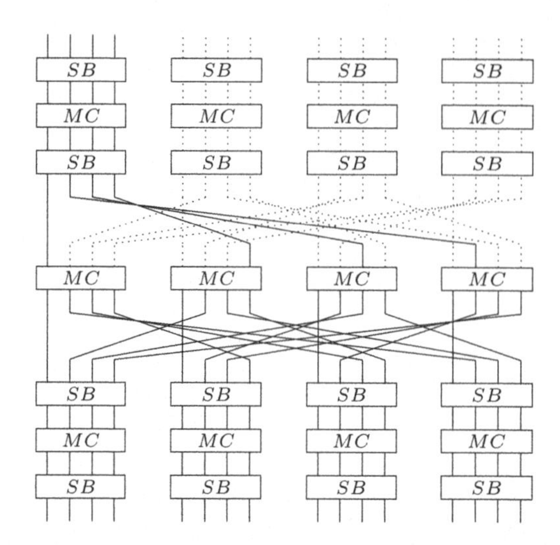

Fig. 3. The structure of $S \circ L \circ S$

Algorithm 3. Distinguisher for 4 rounds of AES

Input: A pair of plaintexts with $wt(\nu(p^0 \oplus p^1)) = 3$
Output: 1 for an AES, -1 otherwise.
// r-round AES enc/dec without first SR and last $SR \circ MC$
$c^0 \leftarrow enc_k(p^0, 4)$, $c^1 \leftarrow enc_k(p^1, 4)$ *//encrypt pair*
$c'^0 \leftarrow$ SimpleSWAP(c^0, c^1), $c'^1 \leftarrow$ SimpleSWAP(c^1, c^0)
$p'^0 \leftarrow dec_k(c'^0, 4)$, $p'^1 \leftarrow dec_k(c'^1, 4)$ *//decrypt pair*
if $\nu(p'^0 \oplus p'^1) = \nu(p^0 \oplus p^1)$ **then**
  return 1.
else
  return -1
end if

distinguisher and described in Lemma 4. We add a round $MC \circ SB$, shifting out the SR-layer in that round, at the beginning of 4 rounds $S \circ L \circ S \circ SR$ and get

$$\begin{aligned} R^5 &= S \circ L \circ S \circ SR \circ MC \circ SB \\ &= S \circ L \circ S \circ Q' \\ &= R^4 \circ Q' \end{aligned}$$

as depicted in Fig. 4 where $Q' = SR \circ MC \circ SB$. Notice that $Q'^{-1} = SB^{-1} \circ MC^{-1} \circ SR^{-1}$ enjoys the same property as Q in Lemma 4, though with inverse components. The main idea of our distinguisher is that if the difference between two plaintexts after the first round (essentially after Q') is zero in t out of 4 words, we apply the yoyo game and return new plaintext pairs that are zero in exactly the same words after one round. Then, due to Lemma 4, the plaintexts must reside in the same sets and this is a property we will exploit in our distinguisher. In particular, assume that we have two plaintexts p^0 and p^1 where

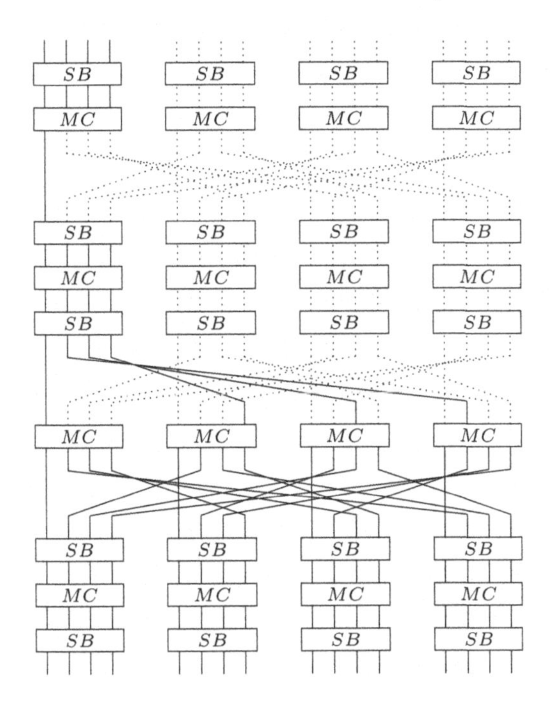

Fig. 4. Five Rounds $R^4 \circ Q'$

$Q'(p^0) \oplus Q'(p^1) = a^0 \oplus a^1$ is zero in 3 out of 4 words. Then since each byte of a word is mapped to distinct words through SR^{-1}, it follows from Lemma 4 that p^0 and p^1 belongs to the same set $T_{z,a} = T_0^{z,a} \times T_1^{z,a} \times T_2^{z,a} \times T_3^{z,a}$ for $wt(z) = 3$ and where each set $T_i^{z,a} \subset \mathbb{F}_q^4$ has size exactly $q = q^{4-wt(z)}$. In other words, if a pair of plaintexts p^0 and p^1 encrypt one round (through Q') to a pair of intermediate states whose difference is zero in 3 out of 4 words, then p^0 and p^1 have probability q^{-1} of having the same value in a particular word. We do not actually know the sets $T_i^{z,a}$ since they are key-dependent, but we know their size. However, due to the MixColumns matrix M we can add an even more fine grained property and we now explain the last property that we use in our distinguisher. The 4×4 MixColumns matrix M satisfy $wt(x) + wt(xM) \geq 5$ for any non-zero $x \in \mathbb{F}_{2^8}^4$. In particular, if x has $t > 0$ non-zero bytes, then $x \cdot M$ has at least $5 - t$ non-zero bytes. In other words, if x has $4 - t$ zeros, then $x \cdot M$ can not contain t or more zeros. This follows because the total number of non-zero bytes before and after M can not be less than 5, and this therefore means that the total number of zeros before and after M can not be more than $8 - 5 = 3$. The same property holds for the inverse M^{-1} of M. We add it as a Lemma for reference.

Lemma 5. *Let M denote a 4×4 MixColumns matrix and $x \in \mathbb{F}_q^4$. If t bytes in x are zero, then $x \cdot M$ or $x \cdot M^{-1}$ can not contain $4 - t$ or more zeros.*

Algorithm 4. Distinguisher for 5 rounds of AES

Output: 1 for an AES, -1 otherwise.
// r-round AES enc/dec without first SR and last $SR \circ MC$
$cnt1 \leftarrow 0$
while $cnt1 < 2^{13.4}$ **do**
  $cnt1 \leftarrow cnt1 + 1$
  $p^0, p^1 \leftarrow$ generate random pair with $wt(\nu(p^0 \oplus p^1)) = 3$
  $cnt2 \leftarrow 0$, $WrongPair \leftarrow False$
  while $cnt2 < 2^{11.4}$ & $WrongPair = False$ **do**
    $cnt2 \leftarrow cnt2 + 1$
    $c^0 \leftarrow enc_k(p^0, 5)$, $c^1 \leftarrow enc_k(p^1, 5)$
    $c'^0 \leftarrow$ SIMPLESWAP(c^0, c^1), $c'^1 \leftarrow$ SIMPLESWAP(c^1, c^0)
    $p'^0 \leftarrow dec_k(c'^0, 5)$, $p'^1 \leftarrow dec_k(c'^1, 5)$
    $p \leftarrow (p'^0 \oplus p'^1)$
    for i from 0 to 3 **do**
      if $wt(\nu(p_i)) \geq 2$ **then**
        $WrongPair \leftarrow True$
      end if
    end for
    $p^0 \leftarrow$ SIMPLESWAP(p'^0, p'^1), $p^1 \leftarrow$ SIMPLESWAP(p'^1, p'^0)
  end while
  if $WrongPair = False$ **then**
    return 1. *//Did not find difference with two or more zeros*
  end if
end while
return -1.

Proof. Follows directly from the well-known properties of M.

Thus, if a pair of plaintexts encrypt one round to a pair of states $Q'(a)$ and $Q'(b)$ that has a zero difference pattern of weight t (only $4-t$ out of four words are active), we have the following, where $Q' = SR \circ MC \circ SB$.

Theorem 4. *Let a and b denote two states where the zero difference pattern $\nu(Q'(a) \oplus Q'(b))$ has weight t. Then the probability that any $4-t$ bytes are simultaneously zero in a word in the difference $a \oplus b$ is q^{t-4}. When this happens, all bytes in the difference are zero.*

Proof. The proof follows from the explanation above. First of all, it follows from Lemma 4 that two words in the same position of a and b are drawn from a same set $T_i^{z,a}$ of size q^{4-t} where $wt(z) = t$. Thus words in the same positions of a and b are equal with probability $q^{-(4-t)} = q^{t-4}$. Since t out of 4 words are zero in $Q'(a) \oplus Q'(b)$, we have that t bytes are zero in each word in the difference $SR^{-1}(a) \oplus SR^{-1}(b)$ at the input to $SB^{-1} \circ MC^{-1}$. Due to Lemma 5, it follows that $4-t$ bytes can not be zero in each word in the difference after MC^{-1}. This is preserved in the difference through SB^{-1} and xor with constants.

We now have the machinery to build a distinguisher for 5 rounds of AES. First the adversary picks enough pairs of plaintexts so that he can expect to find one that the difference has exactly t zero words after one round. Let $B = MC \circ s^4$ where s^4 denotes the concatenation of 4 copies of the AES s-box. Then $Q' = SR \circ MC \circ SB$ can be seen as four parallel applications of B, one on each word, composed with ShiftRows. If two words are equal on input to B, they are equal at the output and thus their difference is zero. So the adversary picks pairs whose difference is nonzero in exactly one word. Then he tries enough pairs (p^0, p^1) until the difference of the output word of the active B contains t zero bytes. When this difference is passed through the SR-layer, it means that t words are zero in the state difference $Q'(p^0) \oplus Q'(p^1)$. If he then plays the yoyo game on the next four rounds, the yoyo returns with at most 7 new pairs of plaintexts (p'^0, p'^1) that satisfy the exact same zero difference pattern after one round. Hence, if the initial pair (p^0, p^1) satisfy $z = \nu(Q'(p^0) \oplus Q'(p^1))$, then each of the new pairs returned by the yoyo obey the same property. In particular, each returned pair of plaintexts obey Theorem 4 which can then be used to distinguish on probability of collisions in bytes of words.

The distinguisher is now straightforward. The probability that a pair p^0 and p^1, with zero difference pattern of weight 3, is encrypted through Q' (essentially one round) to a pair of states, with zero difference pattern of weight t, can be well approximated by

$$p_b(t) = \binom{4}{t} q^{-t}$$

where $q = 2^8$. Thus, in a set of $p_b(t)^{-1}$ pairs $\mathcal{P}$ we expect to find one satisfying pair. Now, for each pair in $\mathcal{P}$ we need a way to distinguish the case when we hit a correct pair. For a random pair of plaintexts, the probability that $4 - t$ bytes are zero simultaneously in any of the 4 words is roughly

$$4p_b(4-t) = 4 \cdot \binom{4}{t} \cdot q^{t-4}$$

while, for a correct case, it is $4 \cdot q^{t-4}$. Hence, for each pair of initial plaintexts in $\mathcal{P}$, we need to generate roughly $p_b(4-t)^{-1}/4$ pairs with the yoyo game to distinguish wrong pairs from the right pair. Thus, in total, the adversary needs to test a set $\mathcal{P}$ containing at least $p_b(t)^{-1}$ pairs, and for each pair he needs to generate roughly $p_b(4-t)^{-1}/4$ new plaintext pairs using the yoyo game. Thus, the total data complexity is

$$2 \cdot (p_b(t)^{-1} \cdot (4 \cdot p_b(4-t))^{-1}) = \frac{p_b(t)^{-1} \cdot p_b(4-t)^{-1}}{2}.$$

For $t \in \{1, 3\}$ we get a data complexity of 2^{27} while for $t = 2$ we get away with roughly $2^{25.8}$. Since the yoyo returns at most 7 new pairs per plaintext pair, we have to repeat the yoyo on new received plaintext pairs by applying the swap-technique in Definition 2 to new plaintexts back and forth over 5 rounds until enough pairs are gathered. Thus, the adversary can always continue the test a few times on a right pair to ensure that the condition is met, but this does not contribute to the overall data complexity (Fig. 5).

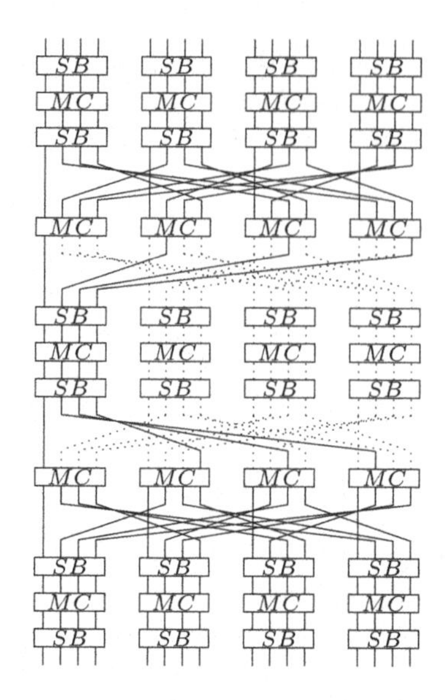

Fig. 5. Six Rounds $S \circ L \circ S \circ L \circ S$

3.4 Impossible Differential Yoyo Distinguisher for 6 Rounds

In this section we present for the first time a secret key distinguisher for 6 rounds of AES that requires $2^{122.83}$ adaptively chosen plaintexts and ciphertexts. We apply the 3-rounds yoyo game described in Sect. 2.2 for generic SPNs directly to 6 rounds of AES in the superbox form $R^6 = S \circ L \circ S \circ L \circ S$ where S is a superbox and $L = SR \circ MC \circ SR$ is the linear superlayer. In Sect. 2.2, we were unable to come up with a generic distinguisher for 3 generic SP-rounds. However, for AES we can exploit impossible zero difference conditions induced by the L layer. Moreover, we use the fact that the total number of active super-boxes over four full AES rounds is at least 5 (see for instance [23]). This also means that the minimal number of active words over the first $L \circ S$ layer is 5. Moreover, if we pick a pair of plaintexts (p^0, p^1) whose difference is zero in exactly two words after $L \circ S$, it follows that the plaintexts themselves must differ in at least three words since the total number of active words is at least 5. In other words, a pair of plaintexts with a difference containing two zero-words can not partially decrypt through the inverse of $L \circ S$ to a pair whose difference is zero in two or more words also. Now, if a pair of plaintexts (p^0, p^1) partially encrypt through $L \circ S$ to a pair of states that are zero in two words with zero difference pattern z, then using the yoyo game on new ciphertext and plaintext pairs back and forth generates "infinitely" many pairs that have the exact same zero difference pattern after $L \circ S$. But since $\nu(L \circ S(p^0) \oplus L \circ S(p^1)) = \nu(S \circ L \circ S(p^0) \oplus S \circ L \circ S(p^1)) = z$, it follows that the difference between pairs of ciphertexts can not contain two or more zero-words either. Hence, if the adversary has one pair of plaintexts (p^0, p^1) that satisfy $\nu(L \circ S(p^0) \oplus L \circ S(p^1)) = 2$, he can generate as many new pairs of plaintext and ciphertext pairs as he wants using the yoyo, and all of

Algorithm 5. Distinguisher for 6 rounds of AES

```
Input: Set P contains 2^61.4 plaintext pairs (p^0, p^1)
Output: 1 for an AES, -1 otherwise.
  // r-round AES enc/dec without first SR and last SR ∘ MC
  for (p^0, p^1) ∈ P do
    WrongPair ← False, counter ← 0
    while counter < 2^60.4 & WrongPair = False do
      if wt(ν(p^0 ⊕ p^1)) ≥ 2 then
        WrongPair ← True                    //Too many zeros in difference
      end if
      c^0 ← enc_k(p^0, 6), c^1 ← enc_k(p^1, 6)
      if wt(ν(c^0 ⊕ c^1)) ≥ 2 then
        WrongPair ← True                    //Too many zeros in difference
      end if
      c'^0 ← SIMPLESWAP(c^0, c^1), c'^1 ← SIMPLESWAP(c^1, c^0)
      p'^0 ← dec_k(c'^0, 6), p' ← dec_k(c'^1, 6)
      p^0 ← SIMPLESWAP(p'^0, p'^1), p^1 ← SIMPLESWAP(p'^1, p'^0)
      counter ← counter + 1
    end while
    if WrongPair = False then
      return 1.
    end if
  end for
  return −1.
```

these will have the exact same zero difference in the middle. Moreover, none of these plaintext and ciphertext pairs can ever collide in two or more words. This suggests a simple, though impractical, distinguisher for 6 rounds of AES.

First we have that $wt(\nu(L \circ S(p^0) \oplus L \circ S(p^1))) = t$ with a probability $\binom{4}{t}(2^{32} - 1)^{4-t}/2^{32 \cdot 4}$ that can be well approximated by

$$p_w(t) = \binom{4}{t} 2^{-t \cdot 32}$$

for a random pair of plaintexts. If the pair is correct, there is generally a

$$p_w(4-t) = \binom{4}{4-t} 2^{(4-t) \cdot (-32)}$$

probability of a collision in $4 - t$ words at the same time in either the ciphertext or the plaintext. Thus by testing both plaintext and ciphertext pairs, the probability becomes $2 \cdot p_w(4-t)$ that the difference of random pairs of plaintexts or ciphertexts are zero in $4 - t$ words. But this is impossible in AES for right pairs, since the total number of zero words in the differences $p^0 \oplus p^1$ and $L \circ S(p^0) \oplus L \circ S(p^1)$ can not be more than 3 due to the L-layer. Thus a straightforward distinguisher is as follows. We prepare $p_w(t)^{-1}$ pairs in a set $\mathcal{P}$. For each pair, we run the yoyo game back and forth until we have generated $\frac{p_w(4-t)^{-1}}{2}$

pairs. If there is a collision in $4-t$ words we discard the pair from $\mathcal{P}$ and continue with a new pair until there are no such collisions. In that case, we conclude a success for the distinguisher. The total data complexity of the 6-rounds distinguisher is

$$D(t) = p_w(t)^{-1} \cdot p_w(4-t)^{-1}$$

where the lowest data-complexity is found by setting $t = 2$ such that $D(t) = 2^{122.83}$.

This is as far as we know the first 6-round secret key distinguisher for AES that does not use the full codebook. Philosophically, one could argue that $2 \cdot r$ AES-rounds should exhibit the same properties as r generic SP-round consisting of four s-boxes and one large linear layer. However, to conclude that 6 rounds of AES is weaker than the 3 rounds generic SP, requires a deeper study of the 3-rounds yoyo game in Sect. 2.1.

3.5 A 5-Round Key Recovery Yoyo on AES

The 5-round key-recovery is formed by adding a round $Q' = SR \circ MC \circ SB$ in front of $S \circ L \circ S$, just like in Sect. 3.3, and aim at finding the first round-key xored in front of $R^5 = S \circ L \circ S \circ Q$. The MixColumns matrix M in AES is defined by the circular matrix

$$\begin{bmatrix} \alpha & \alpha \oplus 1 & 1 & 1 \\ 1 & \alpha & \alpha \oplus 1 & 1 \\ 1 & 1 & \alpha & \alpha \oplus 1 \\ \alpha \oplus 1 & 1 & 1 & \alpha \end{bmatrix}.$$

The function $MC \circ SB$ works on each word of the state independently, thus assume we pick two pairs of plaintexts p^0 and p^1 where the first words are given by $p_0^0 = (0, i, 0, 0)$ and $p_0^1 = (z, z \oplus i, 0, 0)$ where z is a random non-zero element of $\mathbb{F}_q$. The three other words are equal for the two plaintexts. Let $k_0 = (k_{0,0}, k_{0,1}, k_{0,2}, k_{0,3})$ denote the key-bytes XORed with the first word of the plaintext. Then the difference between the first words after the partial encryption of the two plaintexts through $MC \circ SB \circ AK$ becomes

$$\begin{aligned} \alpha b_0 \oplus (\alpha \oplus 1) b_1 &= y_0 \\ b_0 \oplus \alpha b_1 &= y_1 \\ b_0 \oplus b_1 &= y_2 \\ (\alpha \oplus 1) b_0 \oplus b_1 &= y_3. \end{aligned}$$

where $b_0 = s(k_{0,0}) \oplus s(z \oplus k_{0,0})$ and $b_1 = s(k_{0,1} \oplus z \oplus i) \oplus s(k_{0,1} \oplus i)$, where $s(x)$ is the AES-sbox. Since the plaintexts are equal in the last two bytes, this part cancels out in the difference. In particular, if we look at the third equation,

$$s(k_{0,0}) \oplus s(z \oplus k_{0,0}) \oplus s(k_{0,1} \oplus z \oplus i) \oplus s(k_{0,1} \oplus i) = y_2$$

it is not hard to see that it is zero for $i \in \{k_{0,0} \oplus k_{0,1}, z \oplus k_{0,0} \oplus k_{0,1}\}$. Thus, if we let i run through all values of $\mathbb{F}_q$, we are guaranteed that there are at least two values for which the third equation is zero.

We prepare a set $\mathcal{P}$ of plaintext pairs as follows. For each i, generate a pair of plaintexts p^0 and p^1 where the first word of p^0 is $p_0^0 = (0, i, 0, 0)$ while the first word in the second text is $p_0^1 = (z, z \oplus i, 0, 0)$. Then encrypt this pair with five rounds to ciphertexts c^0 and c^1. Then pick 5 new ciphertext pairs $c'^0, c'^1 = (\rho^v(c^0, c^1), \rho^v(c^1, c^0))$ and return the corresponding plaintexts p'^0 and p'^1. If a pair is of the correct form, the first words of p'^0 and p'^1 will satisfy

$$M \circ s^4(p_0'^0 \oplus k_0) \oplus M \circ s^4(p_0'^1 \oplus k_0) = (z_0, z_1, 0, z_3) \tag{13}$$

where M is the MC matrix and s^4 is the concatenation of 4 parallel s-boxes. The adversary can now test each of the 2^{24} remaining candidate keys and determines whether the third coordinate in (13) is zero for all 5 pairs of plaintexts returned by the yoyo, where we already known that $k_{0,0} \oplus k_{0,1} \in \{i, i \oplus z\}$ for known values i and z. This equation holds for all 5 pairs at random with probability $2^{-8 \cdot 5} = 2^{-40}$ thus a false positive might occur with probability 2^{-16} when testing 2^{24} keys. In practice, the adversary can always remove uncertainty by generating a few more pairs when the test succeeds on the first five pairs, since this happens rarely and does not affect the total data complexity per attack. Thus, the total number of adaptively chosen plaintexts needed for finding a correct pair is $2^8 \cdot 5$ pairs, which corresponds to

$$D = 2 \cdot 2^8 \cdot 5 \approx 2^{11}$$

ciphertexts and plaintexts.

For the total computational complexity, testing key guesses for one set should be 2^{24} instead of $2 \cdot 2^{24}$, because it is sufficient to test $k_{0,1} = k_{0,0} \oplus i$ and there is no need to test $k_{0,1} = k_{0,0} \oplus i \oplus z$, considering i will run over all 2^8 possible values and $k_{0,0}$ also runs over all 2^8 possible values. This corresponds to $2^{29.3}$ s-box look-ups because to check that the third component in Eq. (13) is zero for each key on 5 pairs one needs $2 \cdot 4$ s-box look-ups for each of the 5 pairs. This has to be done 2^8 times, one run for each pair in $\mathcal{P}$. Thus the total number of s-box look-ups corresponds to $2^{24} \cdot 2 \cdot 4 \cdot 5 \cdot 2^8 = 2^{29.3+8} = 2^{37.3}$. This roughly corresponds to 2^{31} 5-rounds of AES where we assume that one round costs 16 s-box lookups. Notice that when we have found one of the correct subkeys, it is trivial (and extremely efficient) to determine the rest of the subkeys.

Since the adversary now knows the first subkey k_0, he can make a pair of words $a_0', b_0' \in \mathbb{F}_{2^8}^4$ that differ only in their first byte. He then makes a new pair by first applying the inverse MixColumns matrix M^{-1} and then applying the inverse s-box to each byte of the pair of words. Finally he XORs the first subkey k_0 to each word which results in a pair of words a_0 and b_0. He can now make a pair of full plaintexts $p^0 = (a_0, 0, 0, 0)$ and $p^1 = (b_0, 0, 0, 0)$ (they must be equal in the last three words) whose difference after $SR \circ MC \circ SB \circ AK$ is guaranteed to be non-zero in only the first word. However, this first initial pair of plaintexts p^0 and p^1 is useless for recovering the remaining subkeys since they

Algorithm 6. Key recovery for 5 rounds of AES

Input: Set $\mathcal{P}$ contains 2^8 plaintext pairs (p^0, p^1) where $p_0^0 = (0, i, 0, 0)$ and $p_0^1 = (1, 1 \oplus i, 0, 0)$ for $i = 0, ..., 2^8 - 1$ ($p_j^0 = p_j^1 = 0$ for $j = 1, 2, 3$)
Output: Secret key k_0

```
// r-round AES enc/dec without first SR and last SR ∘ MC
for i from 0 to 2^8 − 1 do
    p^0 ← 0, p^1 ← 0                                  //initialize to all zero state
    p_0^0 ← (0, i, 0, 0), p_0^1 ← (1, 1 ⊕ i, 0, 0)
    S ← {(p^0, p^1)}
    while len(S) < 5 do                               //generate 4 new plaintext pairs
        c^0 ← enc_k(p^0, 5), c^1 ← enc_k(p^1, 5)
        c'^0 ← SimpleSWAP(c^0, c^1), c'^1 ← SimpleSWAP(c^1, c^0)
        p'^0 ← dec_k(c'^0, 5), p'^1 ← dec_k(c'^1, 5)
        p^0 ← SimpleSWAP(p'^0, p'^1), p^1 ← SimpleSWAP(p'^1, p'^0)
        S ← S ∪ {(p^0, p^1)}
    end while
    for all 2^24 remaining key candidates k_0 do
        for all (p^0, p^1) ∈ S do
            //Check if third equation l_3(a ⊕ b) = 0 in (13) holds
            if l_3(s^4(p_0^0 ⊕ k_0) ⊕ s^4(p_0^1 ⊕ k_0)) ≠ 0 then
                break and jump to next key
            end if
        end for
        return k_0;                    //Eq. 13 is zero for all values and k_0 is correct
    end for
end for
```

are equal in the three last words. But the adversary can now use this initial pair to generate m new plaintext pairs p'^0 and p'^1 using the yoyo that are with high probability different in their three last words and satisfy the same condition after $SR \circ MC \circ SB \circ AK$. In particular, $MC \circ SB \circ AK(p'^0) \oplus MC \circ SB \circ AK(p'^1)$ has exactly one active byte in each word which yields simple relations that can be used for recovering the remaining subkeys. If we continue with attacking the second subkey k_1, it follows that each of the m pairs returned by the yoyo now satisfy a relation

$$M \circ s^4(p_1'^0 \oplus k_1) \oplus M \circ s^4(p_1'^1 \oplus k_1) = (0, w, 0, 0) \tag{14}$$

for an unknown plaintext pair dependent variable w and fixed k_1. Since (14) holds, we must also have that the relation

$$M^{-1} \cdot (0, w, 0, 0) = w \cdot (\alpha_0, \alpha_1, \alpha_2, \alpha_3) = s^4(p_1'^0 \oplus k_1) \oplus s^4(p_1'^1 \oplus k_1) \tag{15}$$

holds for an unknown variable w and known values $(\alpha_0, \alpha_1, \alpha_2, \alpha_3)$ determined by the second column in M^{-1}. Thus, when one keybyte of k_1 is guessed, the remaining keybytes are determined by simple relations between byte values in Eq. 15 and leaves out at most spending $4 \cdot 2^8$ guesses to find the correct key.

The remaining subkeys are found by solving similar equations determined by the cyclically equivalent columns of M^{-1}.

The adversary needs at least 2 pairs of plaintexts to recover a unique solution for one of the remaining subkeys. However, since he is recovering 3 subkeys at once, and wants all of them to be correct, he should test the solutions against 4 pairs to leave a comfortable margin against false positives in repeated attacks. Thus, since the first initial pair is useless, the adversary typically uses 5 pairs to recover the full key with a comfortable margin. However, even if the attacker brute-forced each 32-bit subkey individually against the relations in (15) above it would not affect the total complexity. Thus, guessing the remaining subkeys does not add to the total time and data complexity, and so recovering the full round key is dominated by guessing the first subkey which costs roughly 2^{31} 5-round AES encryptions and $2^{11.32}$ adaptively chosen ciphertexts and plaintexts (corresponding to $2^{10.32}$ pairs).

4 Conclusion

In this paper we describe new and fundamental properties of SPNs. Our new analysis show that AES is particularly susceptible to adaptive cryptanalysis for up to 6 rounds. We emphasize this by providing new key-independent secret key distinguishers for 3- to 5-rounds AES that breaks all previous records in the literature, in addition to the first key-independent secret-key distinguisher for 6-rounds AES. In addition, we have described a 5-round secret key recovery that requires only $2^{11.3}$ plaintexts/ciphertexts and 2^{31} computations. Our results apply directly to similar designs and opens up the way for new and interesting applications in cryptanalysis.

Acknowledgements. We thank the anonymous reviewers for their valuable comments and suggestions. This Research was supported by the Norwegian Research Council.

References

1. Biham, E., Biryukov, A., Dunkelman, O., Richardson, E., Shamir, A.: Initial observations on Skipjack: cryptanalysis of Skipjack-3XOR. In: Tavares, S., Meijer, H. (eds.) SAC 1998. LNCS, vol. 1556, pp. 362–375. Springer, Heidelberg (1999). https://doi.org/10.1007/3-540-48892-8_27
2. Wagner, D.: The boomerang attack. In: Knudsen, L. (ed.) FSE 1999. LNCS, vol. 1636, pp. 156–170. Springer, Heidelberg (1999). https://doi.org/10.1007/3-540-48519-8_12
3. Biryukov, A., Leurent, G., Perrin, L.: cryptanalysis of Feistel networks with secret round functions. In: Dunkelman, O., Keliher, L. (eds.) SAC 2015. LNCS, vol. 9566, pp. 102–121. Springer, Cham (2016). https://doi.org/10.1007/978-3-319-31301-6_6
4. Biryukov, A.: The boomerang attack on 5 and 6-round reduced AES. In: Dobbertin, H., Rijmen, V., Sowa, A. (eds.) AES 2004. LNCS, vol. 3373, pp. 11–15. Springer, Heidelberg (2005). https://doi.org/10.1007/11506447_2

5. Leander, G., Abdelraheem, M.A., AlKhzaimi, H., Zenner, E.: A cryptanalysis of PRINTCIPHER: the invariant subspace attack. In: Rogaway, P. (ed.) CRYPTO 2011. LNCS, vol. 6841, pp. 206–221. Springer, Heidelberg (2011). https://doi.org/10.1007/978-3-642-22792-9_12
6. Leander, G., Minaud, B., Rønjom, S.: A generic approach to invariant subspace attacks: cryptanalysis of Robin, iSCREAM and Zorro. In: Oswald, E., Fischlin, M. (eds.) EUROCRYPT 2015. LNCS, vol. 9056, pp. 254–283. Springer, Heidelberg (2015). https://doi.org/10.1007/978-3-662-46800-5_11
7. Grassi, L., Rechberger, C., Rønjom, S.: Subspace trail cryptanalysis and its applications to AES. IACR Trans. Symmetric Cryptol. **2016**(2), 192–225 (2016)
8. Grassi, L., Rechberger, C., Rønjom, S.: A new structural-differential property of 5-Round AES. In: Coron, J.-S., Nielsen, J.B. (eds.) EUROCRYPT 2017. LNCS, vol. 10211, pp. 289–317. Springer, Cham (2017). https://doi.org/10.1007/978-3-319-56614-6_10
9. Ferguson, N., Kelsey, J., Lučks, S., Schneier, B., Stay, M., Wagner, D., Whiting, D.: Improved cryptanalysis of Rijndael. In: Goos, G., Hartmanis, J., Leeuwen, J., Schneier, B. (eds.) FSE 2000. LNCS, vol. 1978, pp. 213–230. Springer, Heidelberg (2001). https://doi.org/10.1007/3-540-44706-7_15
10. Gueron, S., Mouha, N.: Simpira v2: a family of efficient permutations using the AES round function. In: Cheon, J.H., Takagi, T. (eds.) ASIACRYPT 2016. LNCS, vol. 10031, pp. 95–125. Springer, Heidelberg (2016). https://doi.org/10.1007/978-3-662-53887-6_4
11. Bouillaguet, C., Derbez, P., Dunkelman, O., Fouque, P.A., Keller, N., Rijmen, V.: Low-data complexity attacks on AES. IEEE Trans. Inf. Theory **58**(11), 7002–7017 (2012)
12. Derbez, P., Fouque, P.-A.: Automatic search of meet-in-the-middle and impossible differential attacks. In: Robshaw, M., Katz, J. (eds.) CRYPTO 2016. LNCS, vol. 9815, pp. 157–184. Springer, Heidelberg (2016). https://doi.org/10.1007/978-3-662-53008-5_6
13. Knudsen, L.R.: Truncated and higher order differentials. In: Preneel, B. (ed.) FSE 1994. LNCS, vol. 1008, pp. 196–211. Springer, Heidelberg (1995). https://doi.org/10.1007/3-540-60590-8_16
14. Daemen, J., Knudsen, L., Rijmen, V.: The block cipher Square. In: Biham, E. (ed.) FSE. LNCS, vol. 1267, pp. 149–165. Springer, Heidelberg (1997). https://doi.org/10.1007/BFb0052343
15. Sun, B., Liu, M., Guo, J., Qu, L., Rijmen, V.: New insights on AES-Like SPN ciphers. In: Robshaw, M., Katz, J. (eds.) CRYPTO 2016. LNCS, vol. 9814, pp. 605–624. Springer, Heidelberg (2016). https://doi.org/10.1007/978-3-662-53018-4_22
16. Biryukov, A., Shamir, A.: Structural cryptanalysis of SASAS. In: Pfitzmann, B. (ed.) EUROCRYPT 2001. LNCS, vol. 2045, pp. 395–405. Springer, Heidelberg (2001). https://doi.org/10.1007/3-540-44987-6_24
17. Derbez, P.: Meet-in-the-middle on AES. In: Ph.D. thesis. Ecole normale supieure de Paris - ENS Paris (2013)
18. Tiessen, T.: Polytopic cryptanalysis. In: Fischlin, M., Coron, J.-S. (eds.) EUROCRYPT 2016. LNCS, vol. 9665, pp. 214–239. Springer, Heidelberg (2016). https://doi.org/10.1007/978-3-662-49890-3_9
19. Daemen, J., Rijmen, V.: The Design of Rijndael: AES - The Advanced Encryption Standard. Springer, Heidelberg (2002). https://doi.org/10.1007/978-3-662-04722-4
20. Biham, E., Keller, N.: Cryptanalysis of reduced variants of Rijndael. In: 3rd AES Conference, vol. 230 (2000)

21. Biryukov, A., Khovratovich, D.: Two new techniques of side-channel cryptanalysis. In: Paillier, P., Verbauwhede, I. (eds.) CHES 2007. LNCS, vol. 4727, pp. 195–208. Springer, Heidelberg (2007). https://doi.org/10.1007/978-3-540-74735-2_14
22. Rijmen, V.: Cryptanalysis and design of iterated block ciphers. Doctoral Dissertation, K.U. Leuven (1997)
23. Daemen, J., Rijmen, V.: Plateau characteristics. IET Inf. Secur. **1**(1), 11–17 (2007)
24. Daemen, J., Rijmen, V.: Understanding two-round differentials in AES. In: Prisco, R., Yung, M. (eds.) SCN 2006. LNCS, vol. 4116, pp. 78–94. Springer, Heidelberg (2006). https://doi.org/10.1007/11832072_6
25. Gilbert, H.: A simplified representation of AES. In: Sarkar, P., Iwata, T. (eds.) ASIACRYPT 2014. LNCS, vol. 8873, pp. 200–222. Springer, Heidelberg (2014). https://doi.org/10.1007/978-3-662-45611-8_11

New Key Recovery Attacks on Minimal Two-Round Even-Mansour Ciphers

Takanori Isobe[1(✉)] and Kyoji Shibutani[2]

[1] University of Hyogo, Kobe, Japan
takanori.isobe@ai.u-hyogo.ac.jp
[2] Nagoya University, Nagoya, Japan
kyoji.shibutani@nagoya-u.jp

Abstract. We propose new key recovery attacks on the two minimal two-round n-bit Even-Mansour ciphers that are secure up to $2^{2n/3}$ queries against distinguishing attacks proved by Chen et al. Our attacks are based on the meet-in-the-middle technique which can significantly reduce the data complexity. In particular, we introduce novel matching techniques which enable us to compute one of the two permutations without knowing a part of the key information. Moreover, we present two improvements of the proposed attack: one significantly reduces the data complexity and the other reduces the time complexity. Compared with the previously known attacks, our attack first breaks the birthday barrier on the data complexity although it requires chosen plaintexts. When the block size is 64 bits, our attack reduces the required data from 2^{45} known plaintexts to 2^{26} chosen plaintexts with keeping the time complexity required by the previous attacks. Furthermore, by increasing the time complexity up to 2^{62}, the required data is further reduced to 2^8, and $DT = 2^{70}$, where DT is the product of data and time complexities. We show that our low-data attack on the minimal n-bit two-round Even-Mansour ciphers requires $DT = 2^{n+6}$ in general cases. Since the proved lower bound on the required DT for the one-round n-bit Even-Mansour ciphers is 2^n, our results imply that adding one round to the one-round Even-Mansour ciphers does not sufficiently improve the security against key recovery attacks.

Keywords: Block cipher · Even-Mansour ciphers · Meet-in-the-middle attack · Key recovery · Partial invariable pair · Matching with the input-restricted public permutation

1 Introduction

1.1 Even-Mansour Cipher

The *Even-Mansour cipher* consisting of two direct key XORs separated by one public permutation was proposed in 1991 [9,10]. Since then, it has been considered as one of the simplest block cipher design. Indeed, its description is rather simple:

$$E_{K_0,K_1}(x) = P(x \oplus K_0) \oplus K_1,$$

T. Takagi and T. Peyrin (Eds.): ASIACRYPT 2017, Part I, LNCS 10624, pp. 244–263, 2017.
https://doi.org/10.1007/978-3-319-70694-8_9

where P is an n-bit fixed and public permutation with two n-bit secret keys K_0 and K_1.

Bogdanov et al. generalized it as the multiple-round Even-Mansour constructions, and presented the first security bounds against distinguishing attacks for them [1]. As opposed to the original Even-Mansour cipher, the multiple-round Even-Mansour construction can comprise t independent public permutations on n-bit words separated by n-bit independent key additions:

$$E^{(t)}(x) = E_{K_0,\ldots,K_t}(x) = P_t(\cdots(P_2(P_1(x \oplus K_0) \oplus K_1) \oplus K_2)\cdots) \oplus K_t.$$

There has been a series of results towards the provable security of the iterated Even-Mansour ciphers with independently and randomly drawn permutations since then. The aforementioned work [1] proves that at least $2^{\frac{2n}{3}}$ queries are required to distinguish $E^{(t)}$ with $t \geq 2$ from a random permutation and conjectures that the bound is roughly $2^{\frac{t}{t+1}n}$. Steinberger [16] improves this result by proving that the bound of $2^{\frac{3}{4}n}$ holds for $t \geq 3$. Lampe et al. [14] prove a security of $2^{\frac{t}{t+2}n}$ for all even values of t, which is slightly lower than conjectured. Chen and Steinberger [4] have managed to prove the conjectured $2^{\frac{t}{t+1}n}$ bound on the number of queries required for a distinguishing attack, and then Hoang and Tessaro proved the exact bound of it [12].

1.2 Minimal Construction

The original Even-Mansour cipher, which only consists of a single permutation surrounded by key XORs, ensures security up to $2^{n/2}$ queries of the adversary who has access to the encryption function E_K and the internal permutation P [9,10]. Even and Mansour proved an information-theoretic bound that any attack on the scheme must satisfy the equation of $DT = \Omega(2^n)$, where D and T are the data and time complexities, i.e. the number of queries to the encryption function E_K and the permutation P, respectively. The case of $(D, T) = (2^{n/2}, 2^{n/2})$ satisfies the bound of $2^{n/2}$ queries. Shortly after the introduction of the scheme, Daemen [5] presented a key recovery attack matching the bound $DT = O(2^n)$ in the chosen-plaintext model. Dunkelman et al. [8] proposed the slidex attack and its application to close the gap between the upper and lower bounds on the security of the Even-Mansour scheme for a variety of tradeoff points. Moreover, they specifically consider the minimalistic single-key Even-Mansour, with $K_0 = K_1$, which provides exactly the same security. As pointed out by Dunkelman et al. [8], this construction is *minimal* in the sense that if one removes any component, i.e. either the addition of one of the keys, or the permutation P, the construction becomes trivially breakable.

Chen et al. [3] proved that two variants of two-round Even-Mansour ciphers are secure up to $2^{2n/3}$ queries against distinguishing attacks, while the one-round Even-Mansour cipher guarantees security up to birthday bound, namely $2^{n/2}$. One consists of two independent n-bit permutations P_1 and P_2, and a single n-bit key K:

$$\textbf{(2EM-1)} \qquad E_K^{(2)}(x) = P_2(P_1(x \oplus K) \oplus K) \oplus K.$$

The other consists of a single n-bit permutation P, and a single n-bit key K with a simple key scheduling function π,

$$\textbf{(2EM-2)} \qquad E'^{(2)}_K(x) = P(P(x \oplus K) \oplus \pi(K)) \oplus K,$$

where π is any linear orthomorphism of $\mathbb{F}_2^n$. Hereafter we refer to $E^{(2)}_K$ and $E'^{(2)}_K$ as 2EM-1 and 2EM-2, respectively. These constructions can be considered as *minimal* two-round Even-Mansour ciphers delivering security beyond the birthday bound, since they have no redundant component for the security. The proved lower bounds of 2EM-1 and 2EM-2 for distinguishing attacks by Chen et al. [3] are captured in Fig. 1. Regarding tightness of their security bounds for distinguishing attacks, Gazi proposed a generic distinguishing attack with the time complexity of $2^{n-1/2 \log_2 D}$ for any D [11], i.e. $DT^2 = 2^{2n}$. The attack matches the proved bound only in the specific case $(D, T) = (2^{2n/3}, 2^{2n/3})$.

Along with the distinguishing attacks, several key recovery attacks on 2EM-1 construction have been presented [7,15]. Unlike the one-round Even-Mansour construction, for the two-round Even-Mansour ciphers, a dedicated information-theoretic bound on D and T for any attack including *key recovery attacks* has not been known. At least, D and T required for key recovery attacks on the two-round constructions must satisfy $DT = \Omega(2^n)$ which is the bound for the one-round construction. Moreover, since a distinguishing attack is directly derived from a key recovery attack, D and T for the key recovery attacks must follow the lower bounds for distinguishing attacks on 2EM-1 and 2EM-2 given by Chen et al. [3]. For $n = 64$, Nikolić et al. proposed the first key recovery attacks on 2EM-1 requiring the time complexity of 2^{61} with 2^{59} known plaintexts [15]. Dinur et al. generalized it and reduced the data requirements to 2^{45}, while keeping the time complexity [7]. Therefore, the published best upper bound on DT is estimated as 2^{105} for $n = 64$. Since it is much larger than the lower bound for the one-round Even-Mansour ($DT = 2^{64}$), the two-round Even-Mansour cipher seems more secure against key recovery attacks than the one-round Even-Mansour cipher. However, due to the gap between the proved lower bound and the presented upper bound, the accurate security of the two-round construction is still unknown and it is an important open problem in the field of symmetric cryptography.

1.3 Our Contributions

In this paper, we propose new key recovery attacks on the two *minimal* two-round Even-Mansour ciphers 2EM-1 and 2EM-2. First, we present a basic attack on 2EM-1 by using the advanced meet-in-the-middle technique which potentially reduces the data complexity. In particular, we introduce novel matching techniques called *partial invariable pair* and *matching with input-restricted public permutation*, which enable us to compute one of the two permutations without knowing a part of the key information. Then, we improve the basic attack: one significantly reduces data complexity (low-data attack) and the other reduces

Table 1. Summary of our results for 2-round Even-Mansour ciphers.

Time T	Data D	DT	Condition of the parameters of a, b, and d
Basic attack (Sect. 3)			
2^{n-a-1}	2^{n-a} CP	$2^{2(n-a-1)}$	$a \cdot 2^a - 1 \leq n - a$
Low-data attack (Sect. 4)			
2^{n-a}	$2^{n-(a+d)}$ CP	$2^{2(n-a)-d}$	$a \cdot 2^a + d \leq n - a$
Time-optimized attack (Sect. 5)			
2^{n-b} †1	2^{n-a} CP	2^{2n-a-b}	$b \cdot 2^a + (b-a) \leq n - b$

†1: The attack includes 2^{n-a} memory access. CP: Chosen Plaintext

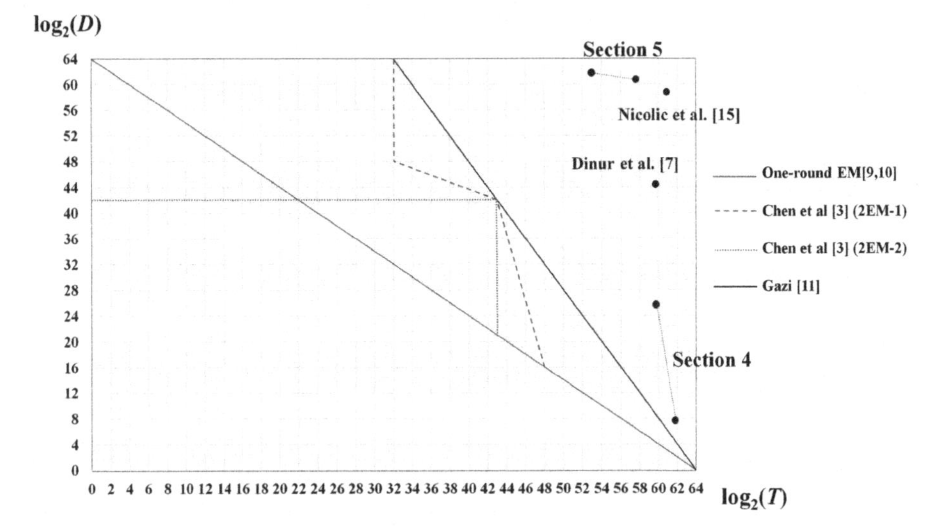

Fig. 1. Comparison of the previous results and our results when $n = 64$. The blue solid line is the lower bound of the one-round Even-Mansour cipher, i.e. $DT = \Omega(2^n)$ [9,10]. The blue dashed and dot lines are the lower bound for distinguishing attacks on 2EM-1 and 2EM-2 by Chen et al. [3], respectively. The black solid line is the upper bound for distinguishing attacks on 2EM-1 and 2EM-2 by Gazi [11] (Color figure online).

time complexity (time-optimized attack) by dynamically finding partial invariable pairs. Our results are summarized in Table 1. In our attacks, there are some tradeoff points of data and time complexities by choosing the parameters of a, b and d under the conditions. We emphasize that all of our attacks do not contain any operation over 2^n, and can be regarded as exponential-advantage attacks as with the previous attack [7,15].

Figure 1 and Table 2 show the comparative results with the previous attacks when $n = 64$. Our attacks can drastically reduce the required data from 2^{45} to 2^{26} with keeping the time complexity of the previous attacks [7,15], although our attacks require chosen plaintexts. By increasing time complexity up to 2^{62}, the required data is further reduced to 2^8. Since the previous attacks are based

Table 2. Comparison of the previous results and our results for key recovery attacks on 2-round Even-Mansour ciphers when $n = 64$.

Time T	Data D	DT	Reference
One-round EM: $E_K^{(1)}(x) = P(x \oplus K) \oplus K)$ [9,10]			
2^{64-x}	$2^x CP$	2^{64}	[5]
2^{64-x}	$2^x KP$	2^{64}	[8]
Two-round EM1: $E_K^{(2)}(x) = P_2(P_1(x \oplus K) \oplus K) \oplus K$			
2^{61}	$2^{59} KP$	2^{120}	[15]
$2^{60.1}$	$2^{45} KP$	$2^{105.1}$	[7]
2^{60}	$2^{26} CP$	2^{86}	Sect. 4
2^{62}	2^8 CP	2^{70}	Sect. 4
2^{58} †2	2^{61} CP	2^{119}	Sect. 5
2^{53} †3	2^{62} CP	2^{115}	Sect. 5
Two-round EM2: $E_K'^{(2)}(x) = P(P(x \oplus K) \oplus \pi(K)) \oplus K$			
2^{60}	$2^{26} CP$	2^{86}	Sect. 6
2^{62}	2^8 CP	2^{70}	Sect. 6
2^{58} †2	2^{61} CP	2^{119}	Sect. 6
2^{53} †3	2^{62} CP	2^{115}	Sect. 6

†2: The attack includes 2^{61} memory accesses.
†3: The attack includes 2^{62} memory accesses.
KP: Known Plaintext, CP: Chosen Plaintext

on multi collisions of the n-bit state, they cannot break the birthday barrier of data and time complexity. On the other hand, our attacks essentially exploit multi collisions of one part of the state, which we call partial invariable pairs in this paper. The required time and data complexity for finding such invariable pairs are much less than those required for finding multi collisions of the whole state. Therefore, our attacks are feasible even if the required data is restricted to be less than $2^{n/2}$. In the time-optimized attacks, we can reduce the computation cost of the internal permutation to 2^{53}, but it requires 2^{62} memory accesses. Basically, it is hard to fairly compare the costs of one encryption and one memory access, because these costs strongly depend on the execution environments, the size of the table, and the underlying permutation. Thus, we do not claim that our time-optimized attacks sufficiently reduce the time complexity required for the previously known key recovery attacks. However, obviously the cost of encryptions is non trivially reduced. We believe that it is an interesting tradeoff to know the concrete security of the minimal two-round Even-Mansour construction. Finally, we show that all of our attacks on 2EM-1 can be applied to the other minimal variant 2EM-2.

The minimum value of DT for $n = 64$ is estimated as 2^{70}, which is close to the proved lower bound for the single Even-Mansour cipher $DT = 2^{64}$ and Chen et al.'s lower bounds for distinguishing attacks on the two-round

Even-Mansour ciphers as shown in Fig. 1. Table 1 shows that our low-data attack requires $DT = 2^{2(n-a)-d}$ and $DT^2 = 2^{3(n-a)-d}$ for any n as long as $a \cdot 2^a + d \leq n - a$. When choosing $a = 2$, the maximum d is $d = n - 10$, and thus DT and DT^2 are estimated as $DT = 2^{n+6}$ and $DT^2 = 2^{2n+4}$ for any n, respectively. These results reveal that adding one round does not sufficiently improve the key recovery security with respect to the product of D and T, while there has not been attacks with time complexity less than the birthday bound unlike the single Even-Mansour cipher.

1.4 Outline of the Paper

The remainder of the paper is organized as follows. In Sect. 2, we introduce the specification of Even-Mansour ciphers and review previous work. In Sect. 3, we explain a basic attack on 2EM-1. Then Sects. 4 and 5 present the improved attacks on 2EM-1 with respect to data and time complexities, respectively. In Sect. 6, we show that our attacks on 2EM-1 are applicable to 2EM-2. Section 7 is the conclusion.

2 Even-Mansour Ciphers

In this section, we introduce the two minimal two-round Even-Mansour ciphers we focus on this paper, and review the previous results on the ciphers.

2.1 Two-Round Even-Mansour Ciphers

Let $P_1, \ldots, P_t$: $\{0,1\}^n \rightarrow \{0,1\}^n$ be independent public permutations and let $K_0, \ldots, K_t \in \{0,1\}^n$ be round keys. The t-round Even-Mansour cipher $E^{(t)}$: $\{0,1\}^n \times \{0,1\}^{n(t+1)} \rightarrow \{0,1\}^n$ consists of t public permutations and $(t+1)$ key injections is defined as follows [1]:

$$E^{(t)}(x) = E_{K_0,\ldots,K_t}(x) = P_t(\cdots(P_2(P_1(x \oplus K_0) \oplus K_1) \oplus K_2)\cdots) \oplus K_t,$$

where x is an n-bit input of $E^{(t)}$.

In this paper, we focus on the following two variants of two-round Even-Mansour ciphers ($t = 2$), which are provably secure up to $2^{2n/3}$ queries of the encryption function and the internal permutation(s) [3]:

$$\textbf{(2EM-1)} \qquad E_K^{(2)}(x) = P_2(P_1(x \oplus K) \oplus K) \oplus K,$$

$$\textbf{(2EM-2)} \qquad E_K'^{(2)}(x) = P(P(x \oplus K) \oplus \pi(K)) \oplus K,$$

where P_1, P_2 and P are independent n-bit permutations, K is n-bit key, and π is any linear orthomorphism of $\mathbb{F}_2^n$. As examples of orthomorphism, a simple rotation, Feistel-like construction (e.g. π: $(x, y) \mapsto (y, x \oplus y)$), field multiplication (e.g. π: $(x) \mapsto (x \times c)$, where $c \neq 0, 1$) are well known. Figure 2 illustrates these constructions. These constructions are regarded as *minimal* Even-Mansour ciphers delivering security beyond the birthday bound, since removing any component causes security to drop back to $O(2^{n/2})$ queries.

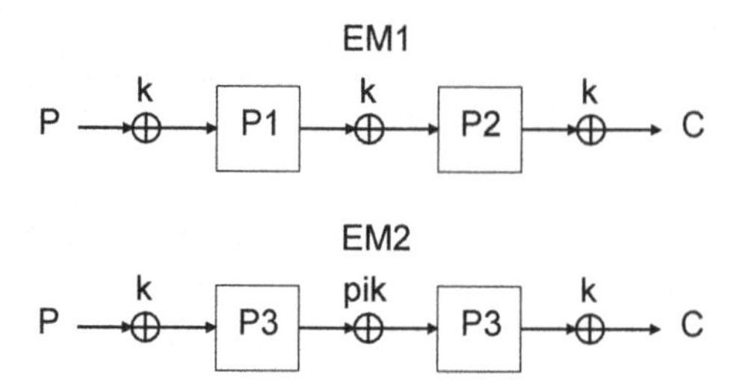

Fig. 2. Minimal two-round Even-Mansour ciphers, 2EM-1 and 2EM-2.

2.2 Previous Key Recovery Attacks on 2EM-1

Along with results on provable security of 2EM-1 [1,3,4,11,14,16], several key recovery attacks on the construction have been published [7,15].

Nikolić et al. proposed the first key recovery attacks on 2EM-1 [15]. They considered the graph of the function $P'(x) = x \oplus P(x)$ and showed that vertices with a large in-degree in this graph can be exploited to bypass an additional round of 2EM-1. Specifically, they define a keyed function inside of $E_K^{(2)}(x)$ as $Q(K, x) = K \oplus P(x \oplus K)$. Since the same key K is XORed before and after the first permutation P, the relation of $x \oplus Q(K, x) = (x \oplus K) \oplus P(x \oplus K)$ holds for any key K. If some output values of P' appear more than the average, then we can predict the value of $Q(K, x)$ with a higher probability than expected even when K is unknown. Then, K can be recovered by using the relation of $K = Q(K, x) \oplus E_K^{(2)}(x)$. In this attack, they exploit t-way multi collisions on the value $P'(x) = x \oplus P(x)$, namely $x_1, x_2, \ldots, x_t$ such that $x_1 \oplus P(x_1) = x_2 \oplus P(x_2) =, \ldots, = x_t \oplus P(x_t) = v$ for some value of v. Using it, $Q(K, x)$ can be guessed with a probability which is t times higher than the expected 2^{-n} without knowing K. For $n = 64$, their attack can recover the key of $E_K^{(2)}(x)$ with time complexity of 2^{61} and 2^{59} known plaintexts [15]. After that, Dinur et al. generalized their attack using concepts from graph theory [7]. In particular, they estimated the highest expected in-degree in the bipartite graph of $P'(x) = x \oplus P(x)$ depending on the number of input size. By considering all the vertices with an in-degree of at least 8, they reduced the data requirements to 2^{45}, while keeping the time complexity. Therefore, the published upper bound of DT is estimated as $2^{105} (= 2^{60} \times 2^{45})$ for $n = 64$. Since it is significantly larger than the bound of one-round Even-Mansour ($DT = 2^{64}$), two-round Even-Mansour cipher seems to sufficiently improve the key recovery security of the one-round Even-Mansour cipher. However, due to the gap between the proved lower bound [3,9,10] and the presented upper bound [7,15], the accurate security of the two-round construction is still unknown and it is an important open problem.

3 Basic Attacks on 2EM-1

This section presents a basic attack on 2EM-1, $E_K^{(2)}$ consisting of two public permutations P_1 and P_2 interleaved with three identical key injections by K.

Our attack is based on the meet-in-the-middle (MitM) framework [2,13], i.e., two functions f and g from $E_K^{(2)}$ are independently computable while the previous attacks [6] use a multi-collision technique.

In our attack, we introduce a novel matching technique called *matching with input-restricted public permutation*, which enables us to compute one of two permutations without knowing a part of the key information. Our new matching technique is based on *partial invariable pairs*, which is used for constructing an input restricted table for any permutation.

3.1 Definitions of Invariable Pair

Let f be an n-bit keyed function using a k-bit key, namely f_K: $\{0,1\}^n \to \{0,1\}^n$, where $K \in \{0,1\}^k$. We use the following two notations for an input-output pair.

Definition 1 (Invariable Pair [13]). *If there exists an input-output pair* (x, y) *of* f *such that* $f_K(x) = y$ *for any* K*, such an input-output pair* (x, y) *is defined by an invariable input-output pair of* f.

Definition 2 ((Target) Partial Invariable Pair). *If there exists a pair of a fixed input and a* b*-bit partial output* (x, y') *of* f *such that* $tr_b(f_K(x)) = y'$ *for any* K*, such a pair* (x, y') *is defined by a partial* b*-bit invariable input-output pair of* f*, where* $y' \in \{0,1\}^b$ $(b \leq n)$ *and* $tr_b(y)$ *represents a* b*-bit truncation of an* n*-bit output* y.

If the value of b *bits to be fixed is predetermined, it is called a target partial* b*-bit invariable input-output pair.*

3.2 How to Find Partial Invariable Pair

Assuming that an a-bit key is involved in f, the procedure for finding a b-bit partial invariable input-output pair is given as follows:

Step 1: Set an n-bit input x randomly.
Step 2: Compute $y_1 = f_K(x)$ with a key K from the set of 2^a keys.
Step 3: Store b bits of y_1 $(b \leq n)$ as $y'(= tr_b(y_1))$.
Step 4: Compute $y_2 = f_{K'}(x)$ with another key K' from the set of 2^a keys, where $K' \neq K$.
Step 5: Check whether b bits of y_2 are equal to y' at the same position. If so, repeat Steps 4 and 5. Then, if all possible K' are checked, output (x, y') as a b-bit partial invariable input-output pair of f. Otherwise, go back to Step 1 and restart with a different x.

The probability of the matching in Step 5 is 2^{-b} assuming that f is a sufficiently random function. Thus, the complexity of finding a b-bit partial invariable pair is estimated as $1/(2^{-b})^{2^a-1}$. If b bits of y_1 are predetermined, which is called target partial b-bit invariable input-output pair, the required complexity of finding such a pair is estimated as $1/(2^{-b})^{2^a}$.

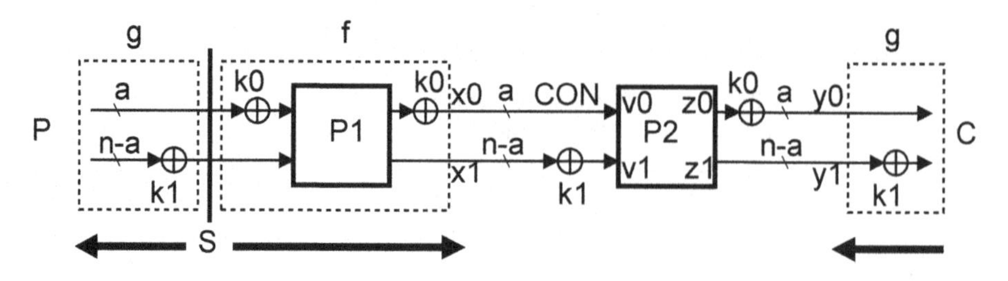

Fig. 3. Attack overview of 2EM-1.

3.3 Attack Overview

As illustrated in Fig. 3, we first divide the n-bit key K into an a-bit k_0 and the remaining $(n-a)$-bit k_1. Then, we introduce a function f that consists of P_1 and two key injections by k_0, and a function g that consists of an initial key injection by k_1 and a final key injection by k_1. Note that f and g are independently computed with k_0 and k_1, respectively. An output of f and an input of g are represented as $(x_0|x_1)$ and $(y_0|y_1)$, also an input and an output of P_2 are denoted as $(v_0|v_1)$ and $(z_0|z_1)$, respectively, where $x_0, v_0, k_0, z_0, y_0 \in \{0,1\}^a$, $x_1, v_1, k_1, z_1, y_1 \in \{0,1\}^{n-a}$.

At first glance, it seems to be difficult to do the matching between f and g, because f and g need k_1 and k_0 to compute the matching state around P_2, respectively. Thus, if the underlying permutation has sufficiently good diffusion property such as AES-128 with a fixed-key, it seems infeasible to construct the matching. To overcome this problem, we introduce a novel matching technique called matching with input-restricted public permutation.

Matching with Input-Restricted Public Permutation. The idea behind our new technique is to construct the input-restricted table of P_2 to find the corresponding n-bit value $(v_0|v_1)$ from only an $(n-a)$-bit value $y_1(=z_1)$ without knowing k_0 while computing g. In a straightforward way, given a value of $y_1(=z_1)$, $2^a(=2^n/2^{n-a})$ candidates of $(v_0|v_1)$ are found with 2^a P_2 computations. Since all k_0 values are tested in the function g, it totally requires $2^n(=2^{n-a} \times 2^a)$ P_2 computations. Thus, two functions f and g from $E_K^{(2)}$ are not independently computed.

In order to get rid of this problem, b bits of inputs $(v_0|v_1)$ are fixed, then the precomputation table of P_2, indexed by values of $(n-a)$-bit y_1, is constructed with less than 2^n P_2 computations, namely 2^{n-b}. Given an $(n-a)$-bit y_1, $2^{n-b}/2^{n-a} = 2^{a-b}$ candidates of $(v_0|v_1)$ are found with only one memory access of the precomputation table without the knowledge of k_0. If $a=b$, it is expected that one candidate is left. If $a \leq b$, it is expected that less than one candidate is left.

Since a *partial invariable pair* in function f allows us to fix b bits of inputs $(v_0|v_1)$ during the MitM procedure, the combination use of two techniques, the partial invariable pair and the matching with precomputation, enables us to mount a MitM attack on 2EM-1.

In summary, we mount the attack on $E_K^{(2)}$ by using *partial invariable pair* to the permutation P_1 in conjunction with *the matching with the input-restricted public permutation* technique to the permutation P_2.

3.4 Attack Procedure

The attack consists of an offline and an online phase. In the offline phase, a b-bit partial invariable pair of f is found, then an input-restricted precomputation table of P_2 is constructed. In the online phase, the MitM attack is mounted by using the precomputation table and querying the encryption oracle $E_K^{(2)}$.

In this attack, more than a bits of input of P_2 are not fixed by k_0, because an $(n-a)$-bit k_1 is used between f and P_2. Thus we consider the case where a is equal to b, which is optimal with respect to the time complexity.

Offline Phase

Step 1: Find an a-bit partial invariable pair of f, (S, x_0) such that $tr_a(f_{k_0}(S)) = x_0$ for any a-bit k_0.
Step 2: For all a-bit k_0, compute the remaining data of the invariable pair, and make a table of $(k_0^{(i)}, x_1^{(i)})$ such that $f_{k_0^{(i)}}(S) = (x_0|x_1^{(i)})$, where $1 \le i \le 2^a$.
Step 3: For all $(n-a)$-bit v_1, compute $(n-a)$-bit value z_1 by P_2, then make a table of $(v_1^{(j)}, z_1^{(j)})$, where $v_0 = x_0$, $P_2(v_0|v_1^{(j)}) = (z_0^{(j)}|z_1^{(j)})$ and $1 \le j \le 2^{n-a}$.

Online Phase

Step 1: Guess an $(n-a)$-bit k_1 and compute the corresponding plaintext P from the start state S and k_1.
Step 2: Send P to the encryption oracle $E_K^{(2)}$, then obtain the corresponding ciphertext C.
Step 3: Compute an $(n-a)$-bit y_1 from C and k_1.
Step 4: Look for an index d in the table of $(v_1^{(j)}, z_1^{(j)})$ such that $z_1^{(d)} = y_1$. If there is no such index, go back to Step 1.
Step 5: Compute $x_1' = v_1^{(d)} \oplus k_1$, and check if there exists an index e in the table of $(k_0^{(i)}, x_1^{(i)})$ satisfying $x_1' = x_1^{(e)}$. If there is no such index, go back to Step 1.
Step 6: Check if $P_2(v_0|v_1^{(d)}) = (z_0'|z_1^{(d)})$ holds, where z_0' is computed from C and $k_0^{(e)}$. If so, $K' = (k_0^{(e)}|k_1)$ is regarded as the correct key. Otherwise, go back to Step 1.

3.5 Evaluation

Here, we evaluate the cost required in each phase.

Offline Phase. In Steps 1 and 2, the time complexity required for finding an a-bit partial invariable pair is $(2^a)^{2^a-1}$ P_1 computations and the required

Table 3. Summary of computational costs for the basic attack on 2EM-1.

	Offline		Online		
	Time	Memory	Time	Data	Memory
	$((2^a)^{2^a-1} + 2^{n-a})\ \mathcal{P}$	2^a and 2^{n-a}	(2^{n-a}) MA	2^{n-a}	2^a and 2^{n-a}
$n = 32, a = 3$	2^{28} enc.	2^{29}	2^{29} MA	2^{29}	2^{29}
$n = 64, a = 4$	2^{60} enc.	2^{60}	2^{60} MA	2^{60}	2^{60}
$n = 128, a = 4$	2^{123} enc.	2^{124}	2^{124} MA	2^{124}	2^{124}
$n = 256, a = 5$	2^{250} enc.	2^{251}	2^{251} MA	2^{251}	2^{251}

$\mathcal{P}$: Internal permutation call, MA: Memory Access.
enc.: encryption function call (= 2 permutation calls).

memory is 2^a blocks. In Step 3, the required time complexity is 2^{n-a} P_2 computations and the required memory is 2^{n-a} blocks.

-Time complexity: $((2^a)^{2^a-1} + 2^{n-a})\ \mathcal{P}$ computations, where $\mathcal{P}$ denotes P_1 or P_2,
-Memory: 2^a and 2^{n-a} blocks.

For simplicity, hereafter computation costs for P_1 and P_2 are assumed to be the same and it denotes $\mathcal{P}$ computations. In addition, the cost of one encryption call is approximately estimated as two $\mathcal{P}$ computations.

Online Phase. Steps 1 to 4 are performed 2^{n-a} times. These steps include two XOR operations in Steps 1 and 3 and one memory access in Step 4. Note that, in Step 4, about one candidate is expected to be found due to the relation of $a = b$, if P_2 is a sufficiently good permutation. Step 5 is performed 2^{n-a} times with one XOR operation and one memory access. It is expected that $2^a/2^{n-a} = 2^{-n+2a}$ candidates will survive in Step 5.

We assume that the cost of one memory access in step 4 is sufficiently larger than one XOR operation and memory access in step 5, because the size of table in step 4 for the matching with input-restricted public permutation is much larger than one in step 5. Then, the time complexity of Steps 1 to 5 is approximately estimated as (2^{n-a}) memory accesses (MA). Step 6 is mounted only $2^{-n+2a} \times 2^{n-a} = 2^a$ times with $\mathcal{P}$ computations. Step 2 requires 2^{n-a} data, since a bits of state S are fixed when computing the function g with each k_1.

-Time complexity: 2^a $\mathcal{P}$ computations + (2^{n-a}) MA,
-Data complexity: 2^{n-a} chosen plaintexts.

Summary. The computational costs of offline and online phases for the basic attacks on the 2EM-1 are estimated as Table 3, where we choose a so that time complexity is minimized. Specifically, we freely choose a as long as it holds the condition of $(2^a)^{2^a-1} \leq 2^n$. If $(2^a)^{2^a-1}$ is less than 2^{n-a}, time complexity is estimated as 2^{n-a} $\mathcal{P}$ computations (2^{n-a-1} encryptions) in the offline phase and 2^{n-a} memory accesses in the online phase. Thus, maximizing a is optimal with

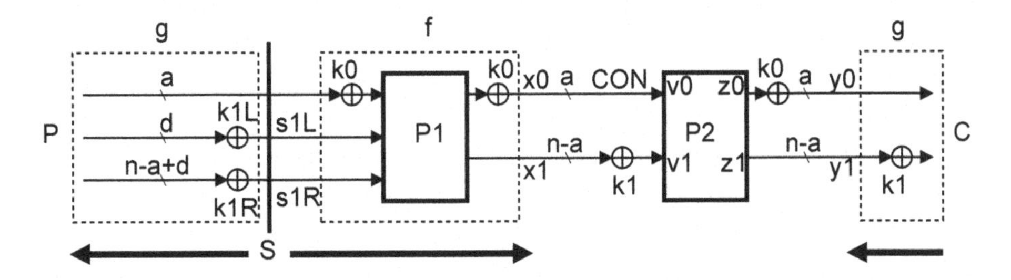

Fig. 4. Attack overview of low-data attacks on 2EM-1.

respect to time complexity. Assuming the cost of memory access is sufficiently smaller than that of the encryption[1], DT is expressed as $DT = 2^{n-a}/2 \times 2^{n-a} = 2^{2(n-a-1)}$ under the condition of $(2^a)^{2^a-1} \leq 2^{n-a}$.

4 Low-Data Attacks on 2EM-1

In this section, we introduce low-data attacks on 2EM-1 based on the attack in Sect. 3. The low-data attacks aim to reduce data requirement (i.e. access to encryption oracle $E_K^{(2)}$) by fixing parts of plaintexts while keeping lower time complexity than that of the brute force attack. In our attacks, the $(n-a)$-bit k_1 is further divided into a d-bit k_{1L} and an $(n-(a+d))$-bit k_{1R}. A start state S and a plaintext P are represented as $S = (s_0|s_{1L}|s_{1R})$ and $P = (p_0|p_{1L}|p_{1R})$, respectively, where $s_0, p_0 \in \{0,1\}^a$, $s_{1L}, p_{1L} \in \{0,1\}^d$ and $s_{1R}, p_{1R} \in \{0,1\}^{n-(a+d)}$.

The main idea is to control s_{1L} depending on k_{1L} so that the d bits of $(s_{1L} \oplus k_{1L})$ are always fixed. If s_0 is also fixed, $(a+d)$ bits of a plaintext are always fixed, i.e., the required data is reduced to $2^{n-(a+d)}$. To be more specific, given a value k_{1L}, a b-bit *target* partial invariable pair of f, (S, x_0) is dynamically found in the online phase, where s_0 is fixed and s_{1L} is chosen such that $(s_{1L} \oplus k_{1L})$ is fixed (Fig. 4).

4.1 Attack Procedure

In the offline phase, a b-bit input-restricted precomputation table of P_2 is constructed, where b is assumed to be equal to a. In the online phase, the MitM attack is mounted by dynamically finding a b-bit target partial invariable pair of f by the precomputation table.

[1] For example, if the underlying permutation is AES-128 with a fixed-key, one P computation requires about 160 memory accesses to compute 160 S-boxes. However, since the comparison of these costs heavily depends on the execution environments, the size of the table, and underlying permutation, we just assume that the cost of memory access is sufficiently smaller than that of the encryption.

Offline Phase

Step 1: Fix a-bit v_0 and $(a+d)$-bit of plaintext to $p_0|p_{1L}$.
Step 2: For all $(n-a)$-bit v_1, compute an $(n-a)$-bit value z_1 by P_2, then make a table of $(v_1^{(j)}, z_1^{(j)})$, where $P_2(v_0|v_1^{(j)}) = (z_0^{(j)}|z_1^{(j)})$ and $1 \leq j \leq 2^{n-b}$.

Online Phase

Step 1: Guess a d-bit k_{1L}, and compute the c-bit s_{1L} as $s_{1L} = k_{1L} \oplus p_{1L}$.
Step 2: Find a a-bit target partial invariable pair of f, $tr_b(f_{k_0}(S)) = (x_0)$ where s_0 and s_{1L} are fixed. Then make a table of (k_0^i, x_1^i) for all a-bit k_0, where $1 \leq i \leq 2^a$.
Step 3: Guess k_{1R} and compute the corresponding plaintext P from the start state S and k_1.
Step 4: Send P to the encryption oracle E_K^2, then obtain the corresponding ciphertext C.
Step 5: Compute $(n-a)$-bit y_1 from k_1 and C.
Step 6: Look for an index d in the table of $(v_1^{(j)}, z_1^{(j)})$ such that $z_1^{(d)} = y_1$. If there is no such index, go back to Step 3.
Step 7: Compute $x_1' = v_1^{(j)} \oplus k_1$, and check if there exists an index e in the table of $(k_0^{(i)}, x_1^{(i)})$ satisfying $x_1' = x_1^{(e)}$. If there is no such index, go back to Step 3.
Step 8: Check if $P_2(v_0|v_1^{(d)}) = (z_0'|z_1^{(d)})$ holds, where z_0' is computed from C and k_1^e. If so, $K' = (k_0^{(e)}|k_1)$ is regarded as the correct key. Otherwise go back to Step 3.
Step 9: For all d-bit k_{1L}, repeat Steps 1 to 8.

4.2 Evaluation

This section gives estimations of the cost required for our low-data attacks on 2EM-1.

Offline Phase. Step 2 in the offline phase requires 2^{n-a} $\mathcal{P}$ computations and 2^{n-a} blocks memory.

-Time complexity: (2^{n-a}) $\mathcal{P}$ computations,
-Memory: 2^{n-a} blocks.

Online Phase. Step 2 requires $(2^a)^{2^a}$ $\mathcal{P}$ computations to find a-bit target partial invariable pair of f by changing $n-(a+d)$-bit s_{1R}. Thus, it should hold the equation of $(2^a)^{2^a} < 2^{n-(a+d)}$, namely $a \times 2^a < n-(a+d)$. Step 2 is performed 2^d times with $(2^a)^{2^a}$ $\mathcal{P}$ computations and 2^a memory. Steps 3 to 6 are performed 2^{n-a} times. These steps include two XOR operations in Steps 3 and Step 5, one memory access in Step 6. Note that Step 4 is performed under the chosen-plaintext setting. Step 7 is performed $2^{a-a} \times 2^{n-a} = 2^{n-a}$ times with one XOR and one memory access. The required time complexity of Steps 3 to 7

Table 4. Summary of computational costs for the low-data attack on the 2-round Even-Mansour cipher.

	Offline		Online		
	Time	Memory	Time	Data	Memory
	$(2^{n-a})\ \mathcal{P}$	2^a and 2^{n-a}	$((2^a)^{2^a} \cdot 2^d + 2^a)\ \mathcal{P} + (2^{n-a})$ MA	$2^{n-(a+d)}$	2^{n-a}
$n = 32$					
$a = 3, d = 9$	2^{28} enc.	2^{29}	2^{30} enc. + 2^{29} MA	2^{20}	2^{29}
$a = 2, d = 22$	2^{29} enc.	2^{30}	2^{30} enc. + 2^{30} MA	2^8	2^{30}
$n = 64$					
$a = 3, d = 35$	2^{60} enc.	2^{61}	2^{58} enc. + 2^{61} MA	2^{26}	2^{61}
$a = 2, d = 54$	2^{61} enc.	2^{62}	2^{61} enc. + 2^{62} MA	2^8	2^{62}
$n = 128$					
$a = 4, d = 60$	2^{123} enc.	2^{124}	2^{123} enc. + 2^{124} MA	2^{64}	2^{124}
$a = 3, d = 101$	2^{124} enc.	2^{125}	2^{124} enc. + 2^{125} MA	2^{24}	2^{125}
$a = 2, d = 118$	2^{125} enc.	2^{126}	2^{125} enc. + 2^{126} MA	2^8	2^{126}
$n = 256$					
$a = 5, d = 160$	2^{250} enc.	2^{251}	2^{250} enc. + 2^{251} MA	2^{160}	2^{251}
$a = 2, d = 118$	2^{253} enc.	2^{254}	2^{253} enc. + 2^{254} MA	2^8	2^{254}

is approximately estimated as (2^{n-a}) memory accesses because the size of table in step 6 for the matching with input-restricted public permutation is assumed to be much larger than one in step 7. Step 8 is mounted only $2^{-n+2a} \times 2^{n-a} = 2^a$ times with $\mathcal{P}$ computations. The $2^{n-(a+d)}$ data is required in Step 4.

-Time complexity: $((2^a)^{2^a} \cdot 2^d + 2^a)\ \mathcal{P}$ computations + (2^{n-a}) MA,
-Memory: 2^a blocks,
-Data: $2^{n-(a+d)}$ chosen plaintexts.

Summary. The computational costs for the low-data attack on 2EM-1 are estimated as Table 4. For $n = 64$, 128, 256, data complexity is drastically reduced compared to the basic attack and previous attacks [7,15] while keeping the time complexity of basic attacks. Moreover, by increasing time complexity, i.e. choosing small a, the required data can be reduced to 2^8, where it does not include any 2^n operations. Assuming the cost of memory access is sufficiently smaller than that of the encryption, DT is expressed as $DT = 2^{n-a} \times 2^{n-(a+d)} = 2^{2(n-a)-d}$ under the condition of $(2^a)^{2^a} \cdot 2^d \leq 2^{n-a}$. Once n and a are determined, the maximal d is easily obtained from the condition. The minimal value of DT of $n = 64, 128$, and 256 are $2^{70} (= 2^{62} \times 2^8)$, $2^{134} (= 2^{126} \times 2^8)$, and $2^{262} (= 2^{254} \times 2^8)$, respectively. These are very close to the bound for single Even-Mansour cipher, i.e. 2^{64}, 2^{128}, and 2^{256}.

The bounds by low-data attacks can be generalized for any n as follows: when choosing $a = 2$, the maximum d is $d = n - a - 8$, and then DT is estimated as $DT = 2^{n+6}$ for any value of n, respectively.

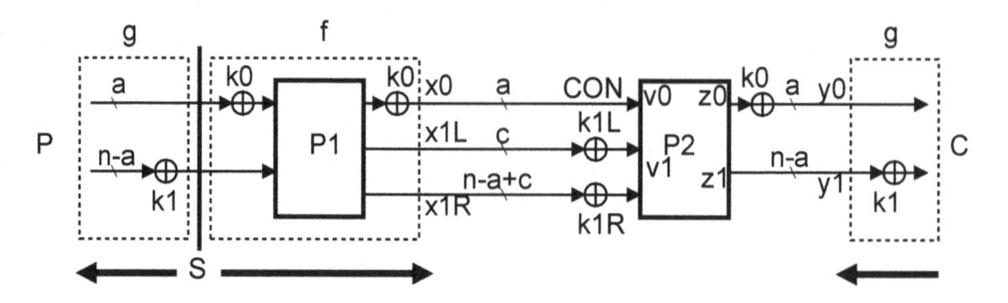

Fig. 5. Attack overview of time-optimized attacks on 2EM-1.

5 Time-Optimized Attacks on 2EM-1

In this section, we try to reduce the cost of $\mathcal{P}$ computations (i.e. access to the internal permutation oracle) of the basic attacks presented in Sect. 3. In this attack, an $(n-a)$-bit k_1 is further divided into a c-bit k_{1L} and an $(n-(a+c))$-bit k_{1R}. Similarly, x_1 and v_1 are represented as $x_1 = (x_{1L}|x_{1R})$ and $v_1 = (v_{1L}|v_{1R})$, respectively, where $x_{1L}, v_{1L} \in \{0,1\}^c$ and $x_{1R}, v_{1R} \in \{0,1\}^{n-(a+c)}$ (see Fig. 5).

The cost for $\mathcal{P}$ computations is dominated by the cost for constructing an input-restricted public permutation table in the offline phase, whose cost is estimated as 2^{n-a} $\mathcal{P}$ computations and 2^{n-a} memory. If additional c bits of the input of P_2 can also be fixed, it is reduced to $2^{n-(a+c)}$ $\mathcal{P}$ computations and memory. However, the additional c bits are not fixed in the online phase even if a c-bit x_{1L} is fixed in f, since such a c-bit input of $\mathcal{P}$ depends on k_{1L} between f and P_2, and all values of k_{1L} are tested during the MitM procedure. To solve this problem, we control x_{1L} depending on k_{1L} so that the c bits of $x_{1L} \oplus k_{1L}$ are always fixed in order to reduce the computational cost. In particular, given a value k_{1L} in the online phase, a $b(= a+c)$-bit target partial invariable pair of f is dynamically found.

5.1 Attack Procedure

In the offline phase, a $b(= a+c)$-bit input-restricted precomputation table of P_2 is constructed. In the online phase, the MitM attack is mounted by dynamically finding a b-bit target partial invariable pair of f by using the precomputation table.

Offline Phase

Step 1: Fix an a-bit v_0 and a c-bit v_{1L}.
Step 2: For all $(n-b)$-bit v_{1R}, compute $(n-a)$-bit values z_1 by P_2, then make a table of $(v_{1R}^{(j)}, z_1^{(j)})$, where $P_2(v_0|v_{1L}|v_{1R}^{(j)}) = (z_0^{(j)}|z_1^{(j)})$ and $1 \leq j \leq 2^{n-b}$

Online Phase

Step 1: Guess a c-bit k_{1L}, and choose a c-bit x_{1L} as $x_{1L} = k_{1L} \oplus v_{1L}$.
Step 2: Find a b-bit target partial invariable pair of f, $tr_b(f_{k_0}(S)) = (x_0|x_{1L})$ where b-bit $x_0|x_{1L}$ is fixed as $v_0|(k_{1L} \oplus v_{1L})$, and make a table of (k_0^i, x_{1R}^i) for all a-bit k_0, where $1 \le i \le 2^a$.
Step 3: Guess k_{1R} and compute the corresponding plaintext P from the start state S and k_1.
Step 4: Send P to the encryption oracle E_K^2, then obtain the corresponding ciphertext C.
Step 5: Compute an $(n-a)$-bit y_1 from k_1 and C.
Step 6: Look for an index d in the table of $(v_{1R}^{(j)}, z_1^{(j)})$ such that $z_1^{(d)} = y_1$. If there is no such index, go back to Step 3.
Step 7: Compute $x'_{1R} = v_{1R}^{(j)} \oplus k_{1R}$, and check if there exists an index e in the table of $(k_0^{(i)}, x_{1R}^{(i)})$ satisfying $x'_{1R} = x_{1R}^{(e)}$. If there is no such index, go back to Step 3.
Step 8: Check if $P_2(v_0|v_{1L}|v_{1R}^{(d)}) = (z'_0|z_1^{(d)})$ holds, where z'_0 is computed from C and $k_{1R}^{(e)}$. If so, $K' = (k_0^{(e)}|k_1)$ is regarded as the correct key. Otherwise go back to Step 3.
Step 9: For all c-bit k_{1L}, repeat Steps 1 to 8.

5.2 Evaluation

We evaluate each cost of our time-optimized attack on 2EM-1.

Offline Phase. Step 2 requires 2^{n-b} $\mathcal{P}$ computations and 2^{n-b} blocks memory.

-Time complexity: (2^{n-b}) $\mathcal{P}$ computations,
-Memory: 2^{n-b} blocks.

Online Phase. Step 2 is performed 2^c times with $(2^b)^{2^a}$ $\mathcal{P}$ computations and 2^a memory. Steps 3 to 6 are performed 2^{n-a} times. These steps include two XOR operations in Steps 3 and 5 and one memory access in Step 6. Note that, in Step 6, it is expected that there exist $2^{a-b}(= 2^{n-b}/2^{n-a})$ desired pairs, if P_2 is a sufficiently good permutation. Step 7 is performed $2^{a-b} \times 2^{n-a} = 2^{n-b}$ times with one XOR operation and one memory access. The required time complexity of Steps 3 to 7 is approximately estimated as (2^{n-a}) memory accesses, assuming 2^{n-a} is sufficiently larger than 2^{n-b}. Step 8 is mounted only $2^{-n+2a} \times 2^{n-a} = 2^a$ times with $\mathcal{P}$ computations. Step 4 requires 2^{n-a} data, since a bits of state S are fixed when computing the g function with each k_1.

-Time complexity: $((2^b)^{2^a} \cdot 2^c + 2^a)$ $\mathcal{P}$ computations + (2^{n-a}) MA,
-Memory: 2^a blocks,
-Data: 2^{n-a} chosen plaintexts.

Table 5. Summary of computational costs for the time-optimized attack on the 2-round Even-Mansour cipher.

	Offline		Online		
	Time	Memory	Time	Data	Memory
	$(2^{n-b})\ \mathcal{P}$	2^{n-b}	$((2^b)^{2^a} \cdot 2^{b-a} + 2^a)\ \mathcal{P} + (2^{n-a}\ )$ MA	2^{n-a}	2^{n-b}
			$n = 32$		
$a = 2, b = 6$	2^{25} enc	2^{26}	2^{27} enc. + 2^{30} MA	2^{30}	2^{26}
			$n = 64$		
$a = 3, b = 6$	2^{57} enc	2^{58}	2^{60} enc. + 2^{61} MA	2^{61}	2^{58}
$a = 2, b = 11$	2^{52} enc	2^{52}	2^{52} enc. + 2^{62} MA	2^{62}	2^{53}
			$n = 128$		
$a = 4, b = 7$	2^{120} enc	2^{121}	2^{115} enc. + 2^{124} MA	2^{124}	2^{121}
$a = 2, b = 20$	2^{107} enc	2^{103}	2^{98} enc. + 2^{126} MA	2^{126}	2^{103}
			$n = 256$		
$a = 5, b = 7$	2^{248} enc	2^{249}	2^{226} enc. + 2^{251} MA	2^{251}	2^{249}
$a = 2, b = 42$	2^{213} enc	2^{206}	2^{208} enc. + 2^{254} MA	2^{254}	2^{205}

Summary. The computational costs for the time-optimized attack on the 2EM-1 are estimated as Table 5. For n = 64, 128, 256, time complexity is reduced by properly choosing the values of b, although number of memory access is unchanged. Basically, it is very hard to compare the cost of encryption and memory access because it strongly depend on the execution environments, the size of table and the underlying permutation. Thus, we do not claim that time complexity is sufficiently improved by this algorithm. However, obviously the cost of encryptions are significantly reduced. We believe that it is an interesting tradeoff.

6 Application to 2EM-2

Our key recovery attacks on 2EM-1 are applicable to the other minimized construction 2EM-2:

$$E_K^{\prime(2)}(x) = P(P(x \oplus K) \oplus \pi(K)) \oplus K,$$

where π is any linear orthomorphism of $\mathbb{F}_2^n$. In this section, we consider the 2EM-2 whose π is Feistel-like construction (π: $(x, y) \mapsto (x \oplus y, x)$) as an example. The same idea is naturally applied to another candidate of π.

Recall that the point of our attacks is to find (target) partial invariable pairs to mount the matching with input-restricted public permutations in the line of the meet-in-the-middle attack. To take care of the key scheduling function π, we further divide the $(n-a)$-bit k_1 into an $(n/2-a)$-bit k_{1L} and an $(n/2)$-bit k_{1R} in the basic attack on 2EM-1. Then, $\pi(K)$ is expressed as $(k_{1R} \oplus (k_{1L}||k_0))||(k_{1L}||k_0)$, and 2EM-2 is illustrated as shown in Fig. 6. Here, a-bit k_0, which is for partial invariable pairs, is used twice after the first

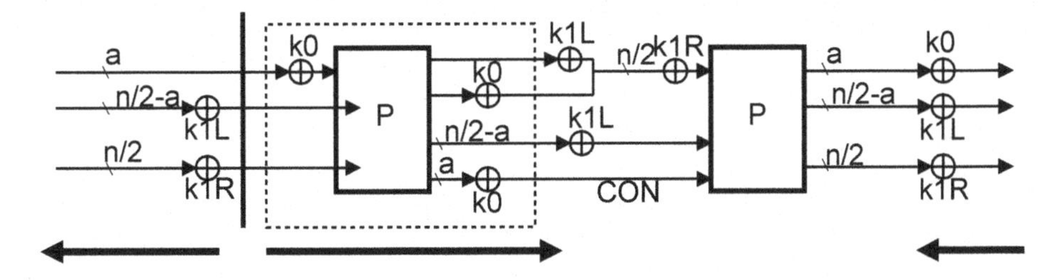

Fig. 6. Application to 2EM-2.

permutation P. Since the bottom a-bit input of the second permutation P is affected by only the value of k_0 after the first permutation, i.e. k_{1L} and k_{1R} does not affect the a-bit input of the second permutation, we can use partial invariable pairs of a-bit k_0 to fix a bits of inputs of second permutation. Using it, we can mount basic attacks on 2EM-2 in the same manner of 2EM-1. For low-data attacks or time-optimized attacks, by dynamically finding invariable pairs of k_0 in the online phase to fix the part of the plaintext or inputs of the second permutation, we can mount the same attacks of 2EM-1 to 2EM-2.

In the case of different linear orthomorphism functions as a key scheduling function, our attacks are feasible as long as we can find partial invariable pairs that fix the part of inputs of the second permutation. In the other examples such as a simple rotation and a field multiplication (e.g. π: $(x) \mapsto (x \times c)$, where $c \neq 0, 1$), there exist such invariable pairs of a-bit k_0 that is able to fix a-bit of inputs of the second permutation, because these orthomorphism functions are not full diffusion function in which an input bit affect any bit of the output.

7 Conclusion

In this paper, we proposed new key recovery attacks on the two minimal two-round Even-Mansour ciphers. Our attacks are based on the advanced meet-in-the-middle technique combined with our novel matching approach called *partial invariable pair* and *the matching with the input-restricted public permutation*. We presented the first attack that the data complexity is less than the birthday barrier, i.e. $2^{n/2}$, on the minimal two-round n-bit Even-Mansour ciphers, although in the chosen-plaintext setting. Then, by dynamically finding partial invariable pairs, the further improvements on the attacks that require the less data or the less time complexity were shown. We emphasize that our low-data attack on the two-round 64-bit Even-Mansour ciphers requires only 2^8 chosen plaintexts. In this case, the minimum value of the product of time and data complexity is 2^{70} which is close to the proved lower bound on the product of time and data complexity for the one-round Even-Mansour ciphers (2^{64}). Our results revealed that adding one round to the one-round Even-Mansour ciphers does not sufficiently improve the security against the key recovery attacks.

Acknowledgments. This work was supported in part by Grant-in-Aid for Young Scientist (B) (KAKENHI 17K12698) for Japan Society for the Promotion of Science.

References

1. Bogdanov, A., Knudsen, L.R., Leander, G., Standaert, F.-X., Steinberger, J., Tischhauser, E.: Key-alternating ciphers in a provable setting: encryption using a small number of public permutations. In: Pointcheval, D., Johansson, T. (eds.) EUROCRYPT 2012. LNCS, vol. 7237, pp. 45–62. Springer, Heidelberg (2012). https://doi.org/10.1007/978-3-642-29011-4_5
2. Bogdanov, A., Rechberger, C.: A 3-subset meet-in-the-middle attack: cryptanalysis of the lightweight block cipher KTANTAN. In: Biryukov, A., Gong, G., Stinson, D.R. (eds.) SAC 2010. LNCS, vol. 6544, pp. 229–240. Springer, Heidelberg (2011). https://doi.org/10.1007/978-3-642-19574-7_16
3. Chen, S., Lampe, R., Lee, J., Seurin, Y., Steinberger, J.: Minimizing the two-round Even-Mansour cipher. In: Garay, J.A., Gennaro, R. (eds.) CRYPTO 2014. LNCS, vol. 8616, pp. 39–56. Springer, Heidelberg (2014). https://doi.org/10.1007/978-3-662-44371-2_3
4. Chen, S., Steinberger, J.: Tight security bounds for key-alternating ciphers. In: Nguyen, P.Q., Oswald, E. (eds.) EUROCRYPT 2014. LNCS, vol. 8441, pp. 327–350. Springer, Heidelberg (2014). https://doi.org/10.1007/978-3-642-55220-5_19
5. Daemen, J.: Limitations of the Even-Mansour construction. In: Imai, H., Rivest, R.L., Matsumoto, T. (eds.) ASIACRYPT 1991. LNCS, vol. 739, pp. 495–498. Springer, Heidelberg (1993). https://doi.org/10.1007/3-540-57332-1_46
6. Dinur, I., Dunkelman, O., Keller, N., Shamir, A.: Key recovery attacks on 3-round Even-Mansour, 8-step LED-128, and full AES^2. In: Sako, K., Sarkar, P. (eds.) ASIACRYPT 2013. LNCS, vol. 8269, pp. 337–356. Springer, Heidelberg (2013). https://doi.org/10.1007/978-3-642-42033-7_18
7. Dinur, I., Dunkelman, O., Keller, N., Shamir, A.: Key recovery attacks on iterated Even-Mansour encryption schemes. J. Cryptol. **29**(4), 697–728 (2016)
8. Dunkelman, O., Keller, N., Shamir, A.: Minimalism in cryptography: the Even-Mansour scheme revisited. In: Pointcheval, D., Johansson, T. (eds.) EUROCRYPT 2012. LNCS, vol. 7237, pp. 336–354. Springer, Heidelberg (2012). https://doi.org/10.1007/978-3-642-29011-4_21
9. Even, S., Mansour, Y.: A construction of a cipher from a single pseudorandom permutation. In: Imai, H., Rivest, R.L., Matsumoto, T. (eds.) ASIACRYPT 1991. LNCS, vol. 739, pp. 210–224. Springer, Heidelberg (1993). https://doi.org/10.1007/3-540-57332-1_17
10. Even, S., Mansour, Y.: A construction of a cipher from a single pseudorandom permutation. J. Cryptol. **10**(3), 151–162 (1997)
11. Gaži, P.: Plain versus randomized cascading-based key-length extension for block ciphers. In: Canetti, R., Garay, J.A. (eds.) CRYPTO 2013. LNCS, vol. 8042, pp. 551–570. Springer, Heidelberg (2013). https://doi.org/10.1007/978-3-642-40041-4_30
12. Hoang, V.T., Tessaro, S.: Key-alternating ciphers and key-length extension: exact bounds and multi-user security. In: Robshaw, M., Katz, J. (eds.) CRYPTO 2016. LNCS, vol. 9814, pp. 3–32. Springer, Heidelberg (2016). https://doi.org/10.1007/978-3-662-53018-4_1
13. Isobe, T.: A single-key attack on the full GOST block cipher. J. Cryptol. **26**(1), 172–189 (2013)

14. Lampe, R., Patarin, J., Seurin, Y.: An asymptotically tight security analysis of the iterated Even-Mansour cipher. In: Wang, X., Sako, K. (eds.) ASIACRYPT 2012. LNCS, vol. 7658, pp. 278–295. Springer, Heidelberg (2012). https://doi.org/10.1007/978-3-642-34961-4_18
15. Nikolić, I., Wang, L., Wu, S.: Cryptanalysis of round-reduced LED. In: Moriai, S. (ed.) FSE 2013. LNCS, vol. 8424, pp. 112–129. Springer, Heidelberg (2014). https://doi.org/10.1007/978-3-662-43933-3_7
16. Steinberger, J.P.: Improved security bounds for key-alternating ciphers via Hellinger distance. IACR Cryptology ePrint Archive 2012/481 (2012)

Lattices

Large Modulus Ring-LWE $\geq$ Module-LWE

Martin R. Albrecht(✉) and Amit Deo(✉)

Information Security Group, Royal Holloway, University of London, London, UK
martin.albrecht@royalholloway.ac.uk, amit.deo.2015@rhul.ac.uk

Abstract. We present a reduction from the module learning with errors problem (MLWE) in dimension d and with modulus q to the ring learning with errors problem (RLWE) with modulus q^d. Our reduction increases the LWE error rate α by a quadratic factor in the ring dimension n and a square root in the module rank d for power-of-two cyclotomics. Since, on the other hand, MLWE is at least as hard as RLWE, we conclude that the two problems are polynomial-time equivalent. As a corollary, we obtain that the RLWE instance described above is equivalent to solving lattice problems on *module* lattices. We also present a self reduction for RLWE in power-of-two cyclotomic rings that halves the dimension and squares the modulus while increasing the error rate by a similar factor as our MLWE to RLWE reduction. Our results suggest that when discussing hardness to drop the RLWE/MLWE distinction in favour of distinguishing problems by the module rank required to solve them.

Keywords: Security reduction · Learning with errors · Lattice-based cryptography

1 Introduction

Lattice-based cryptography has emerged as a central area of research in the pursuit of designing quantum-safe primitives and advanced cryptographic constructions. For example, lattice-based schemes have been proposed for public-key encryption [Reg09, LP11], key exchange protocols [LP11, ADPS16, BCD+16], digital signatures [BG14, DDLL13], identity-based encryption [GPV08, DLP14] and fully homomorphic encryption schemes [Gen09, BGV12, GSW13].

A fundamental problem in lattice-based cryptography is the Learning with Errors problem (LWE) [Reg05]. For a given dimension n, modulus q and error distribution χ, samples of the LWE *distribution* in normal-form are constructed as $(\mathbf{a}, b = \frac{1}{q}\langle \mathbf{a}, \mathbf{s}\rangle + e \bmod 1)$, where $\mathbf{a} \in \mathbb{Z}_q^n$ is chosen uniformly at random and all components of the secret $\mathbf{s} \in \mathbb{Z}_q^n$ and e are drawn from the distribution χ.

The research of Albrecht was supported by EPSRC grant "Bit Security of Learning with Errors for Post-Quantum Cryptography and Fully Homomorphic Encryption" (EP/P009417/1). The research of Deo was supported by the EPSRC and the UK government as part of the Centre for Doctoral Training in Cyber Security at Royal Holloway, University of London (EP/K035584/1).

T. Takagi and T. Peyrin (Eds.): ASIACRYPT 2017, Part I, LNCS 10624, pp. 267–296, 2017.
https://doi.org/10.1007/978-3-319-70694-8_10

Distinguishing the LWE distribution from uniform is known as the decision LWE problem, whereas finding the secret $\mathbf{s}$ is known as the search LWE problem.

The seminal work of Regev [Reg05] establishes reductions from standard problems such as finding short vectors in general lattices to LWE, suggesting that LWE is indeed a difficult problem to solve. In particular, the ability to solve LWE in dimension n implies an efficient algorithm to find somewhat short vectors in *any* n-dimensional lattice. The concrete and asymptotic hardness of LWE has recently been surveyed in [APS15, HKM17]. Although LWE has proven to be a versatile ingredient for cryptography, it suffers from large key sizes (quadratic in the dimension) which motivated the development of more efficient LWE variants.

The Ring Learning with Errors problem (RLWE) was introduced in [LPR10]. RLWE can be seen as a specialisation of LWE where n-dimensional vectors are replaced by polynomials of degree smaller than n. Informally, for RLWE we first choose a ring R of dimension n, modulus q and error distribution χ over a related space of dimension n denoted $K_{\mathbb{R}}$. Then, to sample the RLWE distribution, we sample $a \in R/qR$ uniformly, a secret polynomial s in a suitable space and error e according to χ. We then output $(a, b = \frac{1}{q} a \cdot s + e \bmod R^{\vee})$ as the RLWE sample where $R^{\vee}$ denotes the dual of the ring R. A complete and more precise definition is given in Sect. 2.3. Similar to the case of plain LWE, the decision problem is to distinguish the RLWE distribution from uniform and the search problem is to find the secret s. As alluded to above, the RLWE problem generally offers an increase in efficiency over plain LWE. Intuitively, this can be seen by considering each RLWE sample as a structured set of n LWE samples.

It has been shown that RLWE is at least as hard as standard lattice problems on *ideal* lattices [LPR10, PRSD17]. However, these ideal lattice problems have received much less attention than their analogues on general lattices. Furthermore, some problems that are presumed hard on general lattices such as GapSVP are actually easy on ideal lattices and a recent series of works [CGS14, CDPR16, CDW17] showed that finding short vectors in ideal lattices is potentially easier on a quantum computer than in the general case. More precisely, the length of the short vectors found in quantum polynomial time are a sub-exponential multiple of the length of the shortest vector in the lattice. Currently, it is not known how to efficiently find such vectors in general lattices efficiently. However, the vectors that can be found in quantum polynomial time are mainly of theoretical interest since they are still too long to affect current RLWE-based cryptography. Another important caveat to note is that if there was a way to find even shorter vectors in ideal lattices, RLWE could still prove to be a difficult problem. This is due to the fact that RLWE has not been proven to be *equivalent* to finding short vectors in ideal lattices, i.e. the problem might be *strictly* harder.

It is worth noting that the reductions from lattice problems to LWE resp. RLWE [Reg05, LPR10, PRSD17] mentioned above have no dependency on q apart from the requirement that q must exceed some lower bound that depends on the dimension and error distribution. In these reductions, the class of lattices is simply defined by the dimension in plain LWE and the ring in the case of RLWE. Similarly, the approximation factors defining the lattice problems are also independent of q.

This interpretation of known hardness results is inconsistent with the current state-of-the-art cryptanalytic techniques for solving LWE. The cost of all known strategies scales with q [HKM17].

Indeed, for LWE it is well-known [BLP+13] that we can trade the size of the fundamental security parameter n and the modulus q without affecting security, as long as $n \log q$ remains constant. Furthermore, in the case of plain LWE we can choose n freely, reducing our dependence on large q to increase security. However, in the case of RLWE the analogue reduction to [BLP+13] is not known and the choice of ring R — and hence the dimension n — can lead to practical implementation advantages and a simpler interpretation of formally defined RLWE (see Sect. 3.1). Typically, a power-of-two cyclotomic ring is used, i.e. a ring isomorphic to $\mathbb{Z}[X]/\langle X^n + 1\rangle$ with $n = 2^k$. In addition to its simplicity, this choice also improves performance due to its amenability to FFT-based algorithms. In fact, power-of-two cyclotomic rings have proven extremely popular in the literature and dominate the design space, e.g. [LMPR08, Gen10, BGV12, DDLL13, BCNS15, ADPS16]. However, as stressed in [LPR13], "powers of two are sparsely distributed, and the desired concrete security level for an application may call for a ring dimension much smaller than the next-largest power of two. So restricting to powers of two could lead to key sizes and runtimes that are at least twice as large as necessary." Alternatively, if an implementation wishes to support intermediate field sizes, a new implementation of multiplication in the intermediate ring is required to achieve comparable performance.

The Module Learning with Errors problem (MLWE) [BGV12, LS15] was proposed to address shortcomings in both LWE and RLWE by interpolating between the two. It will be defined formally in Sect. 2. For now, one way to informally view the MLWE problem is to take the RLWE problem and replace the single ring elements (a and s) with module elements over the same ring. Using this intuition, RLWE can be seen as MLWE with module rank 1.

As expected, MLWE comes with hardness guarantees given by lattice problems based on a certain class of lattices. In this case, the lattices are generated by modules as opposed to ideals in the RLWE case and in contrast to RLWE, it has been shown that MLWE is *equivalent* to natural hard problems over these lattices. Indeed, solving the approximate shortest vector problem on module lattices for polynomial approximation factors would permit solving MLWE (and thus RLWE) efficiently. We note that this reduction, too, only has a mild dependency on q. Furthermore, MLWE has been suggested as an interesting option to hedge against potential attacks exploiting the algebraic structure of RLWE [CDW17]. Thus, MLWE might be able to offer a better level of security than RLWE, while still offering performance advantages over plain LWE.

An example illustrating the flexibility of MLWE is given by the CRYSTALS suite [BDK+17, DLL+17], where MLWE is used to build both key encapsulation and signature schemes. The advantage of using modules when implementing such systems is that the concrete-security/efficiency trade-off is highly tunable. Remembering that working in power-of-two dimensional rings enables efficient implementations, we can fix our ring and then change the rank of the module as desired. For example, suppose we were working in a module over a ring of dimension $n = 256$, then we can increase the effective dimension from 1024 to 1280 by simply increasing the rank of the module. This effective dimension would not be attainable using power-of-two dimensional rings in RLWE. Thus, MLWE promises to adjust the security level with much greater granularity than efficient RLWE instantiations and implementations for one security level can easily be extended to other security levels.

Contributions. After some preliminaries in Sect. 2, our main contribution is a reduction from MLWE in dimension d over some general ring R/qR to RLWE in R/q^dR. This was posed as an open problem in [LS15]. Our solution is given in Theorem 1 and Corollary 1. In Sect. 3.1, we carry out a tighter analysis of the reduction for power-of-two cyclotomic rings. It turns out that for the decision variants, we cannot obtain satisfactory bounds for our reduction to preserve non-negligible advantage unless we allow for super polynomial q and *absolute* noise in addition to negligible *noise rate*. We address this problem in Sect. 4 by considering the search variants. An instantiation of Corollary 3 for *power-of-two cyclotomic rings* is the following:

Corollary. *There exists an efficient reduction from search MLWE in modulus q, rank d and error rate α to search RLWE in modulus q^d and error rate $\alpha \cdot n^2\sqrt{d}$.*

In essence, this says that RLWE with modulus q^d is at least as hard as MLWE with modulus q and module rank d in the same ring. More generally, Corollary 3 shows that there is a freedom to trade between the rank of module and the modulus as long as we hold $d \log q = d' \log q'$ fixed for cyclotomic power-of-two rings. This means that for any decrease in d, we can always balance this off by increasing q exponentially without loss of security.

Our reduction is an application of the main result of Brakerski et al. [BLP+13] in the context of MLWE. In its simplest form, the reduction proceeds from the observation that for $\mathbf{a}, \mathbf{s} \in \mathbb{Z}_q^d$ with $\mathbf{s}$ small it holds that

$$q^{d-1} \cdot \langle \mathbf{a}, \mathbf{s} \rangle \approx \left(\sum_{i=0}^{d-1} q^i \cdot a_i \right) \cdot \left(\sum_{i=0}^{d-1} q^{d-i-1} \cdot s_i \right) \bmod q^d = \tilde{a} \cdot \tilde{s} \bmod q^d.$$

It should be noted that we incur an extra factor of $n^{3/2}\, d^{1/2}$ in error rate expansion when comparing our results to those in [BLP+13]. The extra factor of $n^{3/2}$ arises since we need to drown an (unknown) discrete Gaussian over an (unknown) lattice determined by the secret of the input MLWE instance. Naturally, the factor of d accounts for summing Gaussians when compressing the MLWE sample in rank d into a RLWE sample.

The error distribution of the output in our reduction is an ellipsoidal Gaussian (with bounded widths) as opposed to a spherical one. This type of error distribution appears in the standard hardness result for RLWE [LPR10] and should not be considered unusual. However, we also describe how to perform a reduction from search MLWE to spherical error search RLWE using Rényi divergence arguments (see Sect. 4.1). This is a tool that has recently received attention in lattice-based cryptography because it allows to tighten security reductions for search (and some decisional) problems [LSS14,BLL+15,BGM+16,LLM+16].

In Sect. 5, we present self-reductions from power-of-two RLWE in dimension n and modulus q to RLWE in dimension $n/2$ and modulus q^2 following the same strategy. Here, the error rate typically expands from α to $\tilde{\mathcal{O}}(\alpha \cdot n^{9/4})$ if we have access to $\mathcal{O}(1)$ samples and wish to preserve a non-negligible success probability.

Finally, in Appendix A, we show how to achieve the same flexibility as MLWE-based constructions for public-key encryption by *explicitly* only considering RLWE elements but relying on a MLWE/large modulus RLWE assumption resp. relying on the leftover hash lemma.

Interpretation. Our reduction along with the standard hardness results for MLWE [LS15] implies that RLWE with modulus q^d and error rate α is at least as hard as solving the approximate lattice problem Module-SIVP over power-of-two cyclotomic rings. The approximation factor in this case is $\gamma = \tilde{\mathcal{O}}(n^{5/2}\, d^{1/2})$. As there are also converse reductions from RLWE to Module-SIVP e.g. the dual attack [MR09] which requires finding short vectors in a module lattice, these observations imply RLWE is equivalent to Module-SIVP. Previous hardness results only stated that RLWE is at least as hard as Ideal-SIVP [LPR10].[1] We note, though, that it is not known if Module-SIVP is strictly harder than Ideal-SIVP.

Our results suggest that the distinction between MLWE and RLWE does not yield a hardness hierarchy. There are two different interpretations of this implication. The first and perhaps suspicious conclusion is that MLWE should not be used to hedge against powerful algorithms solving RLWE for *any* modulus. However, such an algorithm would essentially solve RLWE over any power-of-two cyclotomic field by our reduction in Sect. 5. Furthermore, as already mentioned in [BLP+13], an adversary solving our output RLWE instance with modulus q^d and any dimension n implies an adversary that can solve the standard LWE problem in dimension d and modulus q given n samples. While such an adversary cannot be ruled out in principle, it cannot be enabled by the algebraic structure of RLWE or ideal lattices. However, we note that this line of argument is less powerful when restricting to small constant d.

On the other hand, assuming that such a powerful adversary does not exist, an alternative interpretation is that our results suggest that the difficulty of solving RLWE increases with the size of the modulus when keeping dimension n and noise rate α (roughly) constant. This interpretation is consistent with cryptanalytic results as the best, known algorithms for solving LWE depend on q [APS15,HKM17] and the analogous result for LWE in [BLP+13]. Indeed, our output RLWE instance in modulus q^d has noise of size at least $q^{d/2}$. Thus, our

[1] Except for RLWE instances with modulus q^n which are known to be as hard as LWE in dimension n and modulus q [BLP+13].

RLWE output instances *cannot* be solved by finding short vectors in lattices of module rank 2 using standard primal or dual attacks in contrast to typical RLWE instances used in the literature. This augments standard reductions from RLWE resp. MLWE to Ideal-SIVP resp. Module-SIVP [Reg05, LPR10, LS15] which do not by themselves suggest that the problem becomes harder with increasing q.

2 Preliminaries

An n-dimensional lattice is a discrete subgroup of $\mathbb{R}^n$. Any lattice Λ can be seen as the set of all integer linear combinations of a set of basis vectors $\{\mathbf{b}_1, \ldots, \mathbf{b}_j\}$. That is, $\Lambda := \left\{\sum_{i=1}^{j} z_i \mathbf{b}_i : z_i \in \mathbb{Z}^n \text{ for } i = 1, \ldots, j\right\}$. The lattices we will be considering will have full rank i.e. $j = n$. We use the matrix $\mathbf{B} = [\mathbf{b}_1, \ldots, \mathbf{b}_n]$ to denote a basis. $\tilde{\mathbf{B}}$ is used to denote the Gram-Schmidt orthogonalisation of columns in $\mathbf{B}$ (from left to right) and $\|\mathbf{B}\|$ is the length of the longest vector (in Euclidean norm) of the basis given by $\mathbf{B}$. Additionally, for any $\mathbf{x} \in \mathbb{R}^n$, we write $\|\mathbf{x}\|$ to denote the standard Euclidean norm of $\mathbf{x}$. The dual of a lattice Λ is defined as $\Lambda^* = \{\mathbf{x} \in \text{span}(\Lambda) : \forall\, \mathbf{y} \in \Lambda, \langle \mathbf{x}, \mathbf{y} \rangle \in \mathbb{Z}\}$ where $\langle \cdot, \cdot \rangle$ is an inner product.

Given a matrix $\mathbf{M} \in \mathbb{C}^{m \times n}$, the singular values of $\mathbf{M}$ are defined to be the positive square roots of the eigenvalues of $\mathbf{M}^\dagger \mathbf{M}$ where $\mathbf{M}^\dagger$ denotes the conjugate transpose of $\mathbf{M}$. The matrix $\mathbf{M}^\dagger \mathbf{M}$ takes a diagonal form in some orthonormal basis of $\mathbb{R}^n$ since it is self-adjoint. We write $\sigma_i(\mathbf{M})$ for the ith singular value of $\mathbf{M}$ where $\sigma_1(\mathbf{M}) \geq \cdots \geq \sigma_n(\mathbf{M})$. We also denote the identity matrix in n dimensions using $\mathbb{I}_n$. In addition to the conjugate transpose denoted by $(\cdot)^\dagger$, the transpose of a matrix or vector will be denoted by $(\cdot)^T$. The complex conjugate of $z \in \mathbb{C}$ will be written as $\bar{z}$.

The uniform probability distribution over some finite set $\mathcal{S}$ will be denoted $U(\mathcal{S})$. If s is sampled from a distribution D, we write $s \leftarrow_\$ D$. Also, we let $\mathbf{s} = (s_0, \ldots, s_{d-1}) \leftarrow_\$ D^d$ denote the act of sampling each component s_i according to D independently. We also write $\text{Supp}(D)$ to mean the support of the distribution D. Note that we use standard big-$\mathcal{O}$ notation where $\tilde{\mathcal{O}}$ hides logarithmic factors.

For any algebraic number field K, an element $x \in K$ is said to be integral if it is a root of some monic polymonial with integer coefficients. The set of all integral elements forms the ring of integers of K denoted by $\mathcal{O}_K$. We also denote isomorphisms via the symbol $\simeq$.

2.1 Coefficient Embeddings

Let $K := \mathbb{Q}(\zeta)$ be an algebraic number field of degree n where $\zeta \in \mathbb{C}$ is an algebraic number. Then for any $s \in K$, we can write $s = \sum_{i=0}^{n-1} s_i \cdot \zeta^i$ where $s_i \in \mathbb{Q}$. We can embed this field element into $\mathbb{R}^n$ by associating it with its vector of coefficients s_{vec}. Therefore, for any $s \in K$ we have $s_{vec} = (s_0, \ldots, s_{n-1})^T$.

We can also represent multiplication by $s \in K$ in this coefficient embedding using matrices. The appropriate matrix will be denoted by $\text{rot}(s) \in \mathbb{R}^{n \times n}$.

In particular, for $r, s, t \in K$ with $r = st$, we have that $r_{vec} = \text{rot}(s) \cdot t_{vec}$. Note that the matrix $\text{rot}(s)$ must be invertible with inverse $\text{rot}(s^{-1})$ for $s \neq 0$. The explicit form of $\text{rot}(s)$ depends on the particular field K. In the case where K is a cyclotomic power-of-two field, i.e. $K = \mathbb{Q}[X]/\langle X^n + 1\rangle$ for power-of-two n, we have

$$\text{rot}(s) = \begin{bmatrix} s_0 & -s_{n-1} & -s_{n-2} & \cdots & \cdots & -s_1 \\ s_1 & s_0 & -s_{n-1} & \ddots & \ddots & -s_2 \\ \vdots & \vdots & \ddots & \ddots & \ddots & \vdots \\ \vdots & \vdots & \ddots & \ddots & \ddots & \vdots \\ s_{n-1} & s_{n-2} & \cdots & \cdots & \cdots & s_0 \end{bmatrix}. \tag{1}$$

2.2 Canonical Embeddings

We will often use canonical embeddings to endow field elements with a geometry. A number field $K(\zeta)$ has $n = r_1 + 2r_2$ field homomorphisms $\sigma_i : K \to \mathbb{C}$ fixing each element of $\mathbb{Q}$. Let $\sigma_1, \dots, \sigma_{r_1}$ be the real embeddings and $\sigma_{r_1+1}, \dots, \sigma_{r_1+2r_2}$ be complex. The complex embeddings come in conjugate pairs, so we have $\sigma_i = \overline{\sigma_{i+r_2}}$ for $i = r_1 + 1, \dots, r_1 + r_2$ if we use an appropriate ordering of the embeddings. Define

$$H := \{\mathbf{x} \in \mathbb{R}^{r_1} \times \mathbb{C}^{2r_2} : x_i = \overline{x_{i+r_2}}, i = r_1 + 1, \dots, r_1 + r_2\}.$$

and let $(\mathbf{e}_i)_{i=1}^n$ be the (orthonormal) basis assumed in the above definition of H. We can easily change to the basis $(\mathbf{h}_i)_{i=1}^n$ defined by

- $\mathbf{h}_i = \mathbf{e}_i$ for $i = 1, \dots, r_1$
- $\mathbf{h}_i = \frac{1}{\sqrt{2}}(\mathbf{e}_i + \mathbf{e}_{i+r_2})$ for $i = r_1 + 1, \dots, r_1 + r_2$
- $\mathbf{h}_i = \frac{\sqrt{-1}}{2}(\mathbf{e}_i - \mathbf{e}_{i+r_2})$ for $i = r_1 + r_2 + 1, \dots, r_1 + 2r_2$

to see that $H \simeq \mathbb{R}^n$ as an inner product space. The *canonical embedding* is defined as $\sigma_C : K \to \mathbb{R}^{r_1} \times \mathbb{C}^{2r_2}$ where

$$\sigma_C(x) := (\sigma_1(x), \dots, \sigma_n(x)).$$

The image of any field element under the canonical embedding lies in the space H, so we can always represent $\sigma_C(x)$ via the real vector $\sigma_H(x) \in \mathbb{R}^n$ through the change of basis described above. So for any $x \in K$, $\sigma_H(x) = U_H^\dagger \cdot \sigma_C(x)$ where the unitary matrix is given by

$$U_H = \begin{bmatrix} \mathbb{I}_{r_1} & 0 & 0 \\ 0 & \frac{1}{\sqrt{2}}\mathbb{I}_{r_2} & \frac{i}{\sqrt{2}}\mathbb{I}_{r_2} \\ 0 & \frac{1}{\sqrt{2}}\mathbb{I}_{r_2} & \frac{-i}{\sqrt{2}}\mathbb{I}_{r_2} \end{bmatrix} \in \mathbb{C}^{n \times n}. \tag{2}$$

Addition and multiplication of field elements is carried out component-wise in the canonical embedding, i.e. for any $x, y \in K$, $\sigma_C(xy)_i = \sigma_C(x)_i \cdot \sigma_C(y)_i$ and $\sigma_C(x + y) = \sigma_C(x) + \sigma_C(y)$. Multiplication is not component-wise for σ_H.

Specifically, in the basis $(\mathbf{e}_i)_{i=1}^n$, we have that multiplication by $x \in K$ can be written as left multiplication by the matrix $X_{ij} = \sigma_i(x)\delta_{ij}$ where δ_{ij} is the Kronecker delta. Therefore, in the basis $(\mathbf{h}_i)_{i=1}^n$, the corresponding matrix is $X_H = U_H^\dagger X U_H \in \mathbb{R}^{n\times n}$ which is not diagonal in general. However, for any X_H, we have $X_H \cdot X_H^T = X_H \cdot X_H^\dagger = U_H^\dagger X X^\dagger U_H$. Explicitly, $(X_H \cdot X_H^T)_{ij} = |\sigma_i(x)|^2\delta_{ij}$ i.e. $X_H \cdot X_H^T$ is a diagonal matrix. Likewise for $X_H^T \cdot X_H$. Therefore, the singular values of X_H are precisely given by $|\sigma_i(x)|$ for $i = 1, \ldots, n$.

Remark 1. We use $\sigma_i(\cdot)$ to denote both singular values and embeddings of field elements. If the argument is a matrix, it should be assumed that we are referring to singular values. Otherwise, $\sigma_i(\cdot)$ denotes a field embedding.

For a ring R contained in field K, we define the canonical embedding of the module R^d into the space H^d in the obvious way, i.e. by embedding each component of R^d into H separately. Furthermore, if we have a matrix of ring elements $\mathbf{G} \in R^{d'\times d}$ for integers d and d', we denote the action of $\mathbf{G}$ on R^d in canonical space H^d as $\mathbf{G}_H \in \mathbb{R}^{nd'\times nd}$. It is well-known that the dimension of $\mathcal{O}_K$ as a $\mathbb{Z}$-module is equal to the degree of K over $\mathbb{Q}$, meaning that the lattice $\sigma_H(R)$ is of *full rank.*

2.3 Ring-LWE and Module-LWE

Let R be some ring with field of fractions K and dual $R^\vee := \{x \in K : \mathrm{Tr}(xR) \subseteq \mathbb{Z}\}$. Also let $K_\mathbb{R} = K \otimes_\mathbb{Q} \mathbb{R}$ and define $\mathbb{T}_{R^\vee} := K_\mathbb{R}/R^\vee$. Note that distributions over $K_\mathbb{R}$ are sampled by choosing an element of the space H (as defined in Sect. 2.2) according to the distribution and mapping back to $K_\mathbb{R}$ via the isomorphism $H \simeq K_\mathbb{R}$. For example, sampling the Gaussian distribution D_α over $K_\mathbb{R}$ is done by sampling D_α over $H \simeq \mathbb{R}^n$ and then mapping back to $K_\mathbb{R}$. In all definitions below, let Ψ be a *family* of distributions over $K_\mathbb{R}$ and D be a distribution over $R_q^\vee$ where $R_q^\vee := R^\vee/(qR^\vee)$ and $R_q := R/(qR)$.

Definition 1 (RLWE Distribution). *For $s \in R_q^\vee$ and error distribution ψ over $K_\mathbb{R}$, we sample the ring learning with errors (RLWE) distribution $A_{q,s,\psi}^{(R)}$ over $R_q \times \mathbb{T}_{R^\vee}$ by outputting $(a, \frac{1}{q}(a \cdot s) + e \bmod R^\vee)$, where $a \leftarrow_\$ U(R_q)$ and $e \leftarrow_\$ \psi$.*

Definition 2 (Decision/Search RLWE problem). *The* decision *ring learning with errors problem $RLWE_{m,q,\Psi}^{(R)}(D)$ entails distinguishing m samples of $U(R_q \times \mathbb{T}_{R^\vee})$ from $A_{q,s,\psi}^{(R)}$ where $s \leftarrow_\$ D$ and ψ is an arbitrary distribution in Ψ.*

The search *variant s-$RLWE_{m,q,\Psi}^{(R)}(D)$ entails obtaining the secret $s \leftarrow_\$ D$.*

Definition 3 (MLWE Distribution). *Let $M := R^d$. For $\boldsymbol{s} \in (R_q^\vee)^d$ and an error distribution ψ over $K_\mathbb{R}$, we sample the module learning with error distribution $A_{d,q,\boldsymbol{s},\psi}^{(M)}$ over $(R_q)^d \times \mathbb{T}_{R^\vee}$ by outputting $(\boldsymbol{a}, \frac{1}{q}\langle \boldsymbol{a}, \boldsymbol{s}\rangle + e \bmod R^\vee)$ where $\boldsymbol{a} \leftarrow_\$ U((R_q)^d)$ and $e \leftarrow_\$ \psi$.*

Definition 4 (Decision/Search MLWE problem). *Let $M = R^d$. The* decision *module learning with errors problem $MLWE^{(M)}_{m,q,\Psi}(D)$ entails distinguishing m samples of $U((R_q)^d \times \mathbb{T}_{R^\vee})$ from $A^{(M)}_{q,\boldsymbol{s},\psi}$ where $\boldsymbol{s} \leftarrow_{\$} D^d$ and ψ is an arbitrary distribution in Ψ.*

The search *variant $s\text{-}MLWE^{(M)}_{m,q,\Psi}(D)$ entails obtaining the secret element $\boldsymbol{s} \leftarrow_{\$} D^d$.*

When $\Psi = \{\psi\}$, we replace Ψ by ψ in all of the definitions above. It can be shown that the *normal form* of the above problems where the secret distribution is a discretized version of the error distribution is at least as hard as the case where the secret is uniformly distributed. Therefore, it is customary to assume the normal form when discussing hardness.

2.4 Statistical Distance and Rényi Divergence

Definition 5 (Statistical Distance). *Let P and Q be distributions over some discrete domain X. The statistical distance between P and Q is defined as $\Delta(P, Q) := \sum_{i \in X} |P(i) - Q(i)|/2$. For continuous distributions, replace the sum by an appropriate integral.*

Claim. If P and Q are two probability distributions such that $P(i) \geq (1-\epsilon)Q(i)$ for all i, then $\Delta(P, Q) \leq \epsilon$.

We will also make use of the Rényi divergence as an alternative to the statistical distance to measure the similarity between two distributions.

Definition 6. *(Rényi Divergence) For any distributions P and Q such that $\mathrm{Supp}(P) \subseteq \mathrm{Supp}(Q)$, the Rényi divergence of P and Q of order $a \in [1, \infty]$ is given by*

$$R_a(P||Q) = \begin{cases} \exp\left(\sum_{x \in Supp(P)} P(x) \log \frac{P(x)}{Q(x)}\right) & \text{for } a = 1, \\ \left(\sum_{x \in Supp(P)} \frac{P(x)^a}{Q(x)^{a-1}}\right)^{\frac{1}{a-1}} & \text{for } a \in (1, \infty), \\ \max_{x \in Supp(P)} \frac{P(x)}{Q(x)} & \text{for } a = \infty. \end{cases}$$

For the case where P and Q are continuous distributions, we replace the sums by integrals and let $P(x)$ and $Q(x)$ denote probability densities. We also give a collection of well-known results on the Rényi divergence (cf. [LSS14]), many of which can be seen as multiplicative analogues of standard results for statistical distance. The proof of this lemma is given in [vEH14, LSS14].

Lemma 1 (Useful facts on Rényi divergence). *Let $a \in [1, +\infty]$. Also let P and Q be distributions such that $Supp(P) \subseteq Supp(Q)$. Then we have:*

- ***Increasing Function of the Order:*** *The function $a \mapsto R_a(P||Q)$ is non-decreasing, continuous and tends to $R_\infty(P||Q)$ as $a \to \infty$.*
- ***Log Positivity:*** *$R_a(P||Q) \geq R_a(P||P) = 1$.*

- ***Data Processing Inequality:*** *$R_a\left(P^f||Q^f\right) \leq R_a\left(P||Q\right)$ for any function f where P^f and Q^f denote the distributions induced by performing the function f on a sample from P and Q respectively.*
- ***Multiplicativity:*** *Let P and Q be distributions on a pair of random variables (Y_1, Y_2). Let $P_{2|1}(\cdot|y_1)$ and $Q_{2|1}(\cdot|y_1)$ denote the distributions of Y_2 under P and Q respectively given that $Y_1 = y_1$. Also, for $i \in \{1, 2\}$ denote the marginal distribution of Y_i under P resp. Q as P_i resp. Q_i. Then*
  - $R_a\left(P||Q\right) = R_a\left(P_1||Q_1\right) \cdot R_a\left(P_2||Q_2\right)$.
  - $R_a\left(P||Q\right) = R_\infty\left(P_1||Q_1\right) \cdot \max_{y_1 \in Supp(P_1)} R_a\left(P_{2|1}(\cdot|y_1)||Q_{2|1}(\cdot|y_1)\right)$.
- ***Probability Preservation:*** *Let $E \subseteq Supp(Q)$ be an arbitrary event. If $a \in (1, \infty)$, then $Q(E) \geq P(E)^{\frac{a}{a-1}}/R_a\left(P||Q\right)$. Furthermore, we have $Q(E) \geq P(E)/R_\infty\left(P||Q\right)$.*
- ***Weak Triangle Inequality:*** *Let P_1, P_2 and P_3 be three probability distributions such that $Supp(P_1) \subseteq Supp(P_2) \subseteq Supp(P_3)$. Then*

$$R_a\left(P_1||P_3\right) \leq \begin{cases} R_a\left(P_1||P_2\right) \cdot R_\infty\left(P_2||P_3\right), \\ R_\infty\left(P_1||P_2\right)^{\frac{a}{a-1}} \cdot R_a\left(P_2||P_3\right) \text{ if } a \in (1, +\infty). \end{cases}$$

2.5 Gaussian Measures

Definition 7 (Continuous Gaussian distribution). *The Gaussian function of parameter r and centre c is defined as*

$$\rho_{r,c}(x) = \exp\left(-\pi(x-c)^2/r^2\right)$$

and the Gaussian distribution $D_{r,c}$ is the probability distribution whose probability density function is given by $\frac{1}{r}\rho_{r,c}$.

Definition 8 (Multivariate Gaussian distribution). *Let $\Sigma = S^T S$ for some rank-n matrix $S \in \mathbb{R}^{m \times n}$. The multivariate Gaussian function with covariance matrix Σ centred on $\mathbf{c} \in \mathbb{R}^n$ is defined as*

$$\rho_{S,\mathbf{c}}(\boldsymbol{x}) = \exp\left(-\pi(\boldsymbol{x}-\mathbf{c})^T\left(S^T S\right)^{-1}(\boldsymbol{x}-\mathbf{c})\right)$$

and the corresponding multivariate Gaussian distribution denoted $D_{S,\mathbf{c}}$ is defined by the density function $\frac{1}{\sqrt{\det(\Sigma)}}\rho_{S,\mathbf{c}}$.

Note that if the centre c is omitted, it should be assumed that $c = 0$. If the covariance matrix is diagonal, we describe it using the vector of its diagonal entries. For example, suppose that $\left(S^T S\right)_{ij} = (s_i)^2\delta_{ij}$ and let $\mathbf{s} = (s_1, \ldots s_n)^T$. Then we would write $D_{\mathbf{s}}$ to denote the centred Gaussian distribution D_S.

We are often interested in families of Gaussian distributions. For $\alpha > 0$, we write $\Psi_{\leq \alpha}$ to denote the set of Gaussian distributions with diagonal covariance matrix of parameter $\mathbf{r}$ satisfying $r_i \leq \alpha$ for all i.

We also have discrete Gaussian distributions i.e. normalised distributions defined over some discrete set (typically lattices or lattice cosets). The notation for a discrete Gaussian over some n-dimensional lattice Λ and coset vector $\mathbf{u} \in \mathbb{R}^n$ with parameter r is $D_{\Lambda+\mathbf{u},r}$. This distribution has probability mass function $\frac{1}{\rho_r(\Lambda+\mathbf{u})}\rho_r$ where $\rho_r(\Lambda+\mathbf{u}) = \sum_{\mathbf{x}\in\Lambda+\mathbf{u}} \rho_r(\mathbf{x})$. It was shown in [GPV08] that we can efficiently sample from a (not too narrow) discrete Gaussian over a lattice to within negligible statistical distance. It was further shown that we can actually sample the discrete Gaussian precisely in [BLP+13]. This result is given below as Lemma 2.

Lemma 2 (Lemma 2.3 in [BLP+13], Sampling discrete Gaussians). *There is a probabilistic polynomial-time algorithm that, given a basis $\boldsymbol{B}$ of an n-dimensional lattice $\Lambda = \mathcal{L}(\boldsymbol{B})$, $\mathbf{c} \in \mathbb{R}^n$ and parameter $r \geq \|\tilde{\boldsymbol{B}}\| \cdot \sqrt{\ln(2n+4)/\pi}$ outputs a sample distributed according to $D_{\Lambda+\mathbf{c},r}$.*

Next we define the smoothing parameter of a lattice followed by a collection of lemmas that we will make use of.

Definition 9 (Smoothing parameter). *For a lattice Λ and any $\epsilon > 0$, the* smoothing parameter *$\eta_\epsilon(\Lambda)$ is defined as the smallest $s>0$ s.t. $\rho_{1/s}(\Lambda^* \backslash \{\boldsymbol{0}\}) \leq \epsilon$.*

Lemma 3 (Lemma 3.1 in [GPV08], Upper bound on smoothing parameter). *For any $\epsilon > 0$ and n-dimensional lattice Λ with basis $\boldsymbol{B}$,*

$$\eta_\epsilon(\Lambda) \leq \|\tilde{\boldsymbol{B}}\| \sqrt{\ln(2n(1+1/\epsilon))/\pi}.$$

Lemma 4 (Claim 3.8 in [Reg09], Sums of Gaussians over cosets). *For any n-dimensional lattice Λ, $\epsilon > 0$, $r \geq \eta_\epsilon(\Lambda)$ and $\mathbf{c} \in \mathbb{R}^n$, we have*

$$\rho_r(\Lambda + \mathbf{c}) \in \left[\frac{1-\epsilon}{1+\epsilon}, 1\right] \cdot \rho_r(\Lambda).$$

The claim $R_\infty(D_{\mathbf{t}}||Y) \leq \frac{1+\epsilon}{1-\epsilon}$ in the lemma below follows immediately from the proof given in [LS15].

Lemma 5 (Adapted from Lemma 7 in [LS15], Drowning ellipsoidal discrete Gaussians). *Let Λ be an n-dimensional lattice, $\boldsymbol{u} \in \mathbb{R}^n$, $\boldsymbol{r} \in (R^+)^n$, $\sigma > 0$ and $t_i = \sqrt{r_i^2 + \sigma^2}$ for all i. Assume that $\min_i r_i\sigma/t_i \geq \eta_\epsilon(\Lambda)$ for some $\epsilon \in (0,1/2)$. Consider the continuous distribution Y on $\mathbb{R}^n$ obtained by sampling from $D_{\Lambda+\mathbf{u},\mathbf{r}}$ and then adding a vector from D_σ. Then we have $\Delta(Y, D_{\mathbf{t}}) \leq 4\epsilon$ and $R_\infty(D_{\mathbf{t}}||Y) \leq \frac{1+\epsilon}{1-\epsilon}$.*

In the lemma below, ring elements are sampled in the *coefficient* embedding.

Lemma 6 (Adapted from Lemma 4.1 in [SS13], Upper bound on least singular value). *Let n be a power of two and $R = \mathbb{Z}[X]/\langle X^n+1\rangle$. Then for any $\delta \in (0,1)$, $t \geq \sqrt{2\pi}$ and $\sigma \geq \frac{t}{\sqrt{2\pi}} \cdot \eta_\delta(\mathbb{Z}^n)$, we have*

$$\Pr_{b \leftarrow\$ D_{\mathbb{Z}^n,\sigma}}\left[\frac{1}{\sigma_n(\mathrm{rot}(b))} \geq \frac{t\sqrt{2}}{\sigma\sqrt{n}}\right] \leq \frac{1+\delta}{1-\delta} \cdot \frac{n\sqrt{2\pi e}}{t}.$$

3 Reduction for General Rings

In this section, we show how to reduce an MLWE instance in module rank d and modulus q to an MLWE instance in rank d' and modulus q'. The particular case where $d' = 1$ yields a reduction from MLWE to RLWE. We start by describing the high-level intuition behind the reduction for the case $d' = 1$ and where the modulus goes from q to q^d. In this case, our strategy is to map $(\mathbf{a}, \mathbf{s}) \in (R_q)^d \times (R_q^\vee)^d$ to $(\tilde{a}, \tilde{s}) \in R_q \times R_{q'}^\vee$ aiming to satisfy the approximate equation

$$\frac{1}{q}\langle \mathbf{a}, \mathbf{s}\rangle \approx \frac{1}{q^d}(\tilde{a} \cdot \tilde{s}) \bmod R^\vee. \tag{3}$$

We then map from b to $\tilde{b} \approx b \bmod R^\vee$. For $q = \Omega(\text{poly}(n))$, if we take $\tilde{s} = (q^{d-1}, \ldots, 1)^T \cdot \mathbf{s}$ and $\tilde{a} = (1, \ldots, q^{d-1})^T \cdot \mathbf{a}$, we obtain

$$\begin{aligned} \frac{1}{q^d}(\tilde{a} \cdot \tilde{s}) &= \frac{1}{q}\langle \mathbf{a}, \mathbf{s}\rangle + \frac{1}{q^2}(\ldots) + \frac{1}{q^3}(\ldots) + \ldots \bmod R \\ &\approx \frac{1}{q}\langle \mathbf{a}, \mathbf{s}\rangle \bmod R. \end{aligned} \tag{4}$$

This mapping satisfies the requirement but leads to a narrow, yet non-standard error distribution. The reduction in Theorem 1 is a generalisation of the above idea. Specifically, take $\mathbf{G} \in (R)^{d' \times d}$ and $\tilde{\mathbf{s}} = \mathbf{G} \cdot \mathbf{s} \bmod (q'R)^{d'}$. Then we simply require that

$$\frac{1}{q'}\sum_{i=1}^{d'}\sum_{j=1}^{d} \tilde{a}_i g_{ij} s_j \approx \frac{1}{q}\sum_{j=1}^{d} a_j s_j \bmod R^\vee. \tag{5}$$

This requirement can be satisfied if we choose $\tilde{\mathbf{a}}$ such that

$$\frac{1}{q'}\sum_{i=1}^{d'} \tilde{a}_i g_{ij} \approx \frac{1}{q} a_j \bmod R \tag{6}$$

for $j = 1, \ldots, d$. To carry out this strategy, we will sample $\tilde{\mathbf{a}}$ over an appropriate lattice defined by $\mathbf{G}$ in the canonical embedding. The main challenge in applying this strategy is that we want the error in the new MLWE sample to follow a standard error distribution, i.e. a *continuous* Gaussian.

Theorem 1. *Let R be the ring of integers of some algebraic number field K of degree n, let d, d', q, q' be integers, $\epsilon \in (0, 1/2)$, and $\boldsymbol{G} \in R^{d' \times d}$. Also, fix $\boldsymbol{s} = (s_1, \ldots, s_d) \in (R_q^\vee)^d$. Further, let $\boldsymbol{B}_\Lambda$ be some known basis of the lattice $\Lambda = \frac{1}{q'}\boldsymbol{G}_H^T R^{d'} + R^d$ (in the canonical embedding), $\boldsymbol{B}_R$ be some known basis of R in H and*

$$r \geq \max\begin{cases} \|\tilde{\boldsymbol{B}}_\Lambda\| \cdot \sqrt{2\ln(2nd(1+1/\epsilon))/\pi} \\ \frac{1}{q}\|\tilde{\boldsymbol{B}}_R\| \cdot \sqrt{2\ln(2nd(1+1/\epsilon))/\pi} \\ \frac{1}{q}\max_i \|\tilde{\boldsymbol{B}}_{s_i R}\| \cdot \frac{1}{\min_k |\sigma_k(s_i)|} \cdot \sqrt{2\ln(2n(1+1/\epsilon))/\pi} \end{cases}$$

where $\boldsymbol{B}_{s_i R}$ *is a basis of* $s_i R$ *in the canonical embedding. There exists an efficient probabilistic mapping* $\mathcal{F} : (R_q)^d \times \mathbb{T}_{R^\vee} \longrightarrow (R_{q'})^{d'} \times \mathbb{T}_{R^\vee}$ *such that:*

1. *The output distribution given uniform input* $\mathcal{F}(U((R_q)^d \times \mathbb{T}_{R^\vee}))$ *is within statistical distance* 4ϵ *of the uniform distribution over* $(R_{q'})^{d'} \times \mathbb{T}_{R^\vee}$*.*
2. *Let* $M = R^d$*,* $M' = R^{d'}$ *and define* $B := \max_{i,j} |\sigma_i(s_j)|$*. The distribution of* $\mathcal{F}(A^{(M)}_{q,\mathbf{s},D_\alpha})$ *is within statistical distance* $(4d+6)\epsilon$ *of* $A^{(M')}_{q',G\mathbf{s},D_{\alpha'}}$ *where* $(\boldsymbol{\alpha}')_i^2 = \alpha^2 + r^2(\beta^2 + \sum_{j=1}^{d} |\sigma_i(s_j)|^2)$ *and* β *satisfies* $\beta^2 \geq B^2 d$*.*

Proof. We use the canonical embedding on each component of R^d individually, e.g. $\mathbf{a}_H = (\sigma_H(a_1), \ldots, \sigma_H(a_d)) \in H^d \simeq \mathbb{R}^{nd}$ and similarly for other module elements. We will also refer to the canonical embedding of R as simply R to ease notation. Suppose we are given $(\mathbf{a}, b) \in (R_q)^d \times \mathbb{T}_{R^\vee}$. The mapping $\mathcal{F}$ is performed as follows:

1. Sample $\mathbf{f} \leftarrow D_{\Lambda - \frac{1}{q}\mathbf{a}_H, r}$. Note that the parameter r is large enough so that we can sample the discrete Gaussian efficiently by Lemma 2.
2. Let $\mathbf{v} = \frac{1}{q}\mathbf{a}_H + \mathbf{f} \in \Lambda / R^d$ and set $\mathbf{x} \in (R_{q'})^{d'}$ to be a random solution of $\frac{1}{q'}\mathbf{G}_H^T \mathbf{x} = \mathbf{v} \bmod R^d$. Then set $\tilde{\mathbf{a}} \in M'$ to be the unique element of M' such that $\tilde{\mathbf{a}}_H = \mathbf{x}$.
3. Sample $\tilde{e}$ from the distribution $D_{r\beta}$ over $K_\mathbb{R} \simeq H$ for some $\beta > B\sqrt{d}$ and set $\tilde{b} = b + \tilde{e}$.
4. Finally, output $(\tilde{\mathbf{a}}, \tilde{b}) \in (R_{q'})^{d'} \times \mathbb{T}_{R^\vee}$.

Distribution of $\tilde{a}$. Suppose that $\mathbf{a} \in (R_q)^d$ was drawn uniformly at random. Step 2 of the reduction can be performed by adding a random element of the basis of solutions to $\frac{1}{q'}\mathbf{G}_H^T \mathbf{y} = 0 \bmod R^d$ to a particular solution of $\frac{1}{q'}\mathbf{G}_H^T \mathbf{x} = \mathbf{v} \bmod R^d$. In order to show that $\tilde{\mathbf{a}}$ is *nearly* uniform random, we will show that the vector $\mathbf{x}$ is *nearly* uniform random over the set $(R_{q'})^{d'}$. Note that every $\mathbf{x} \in (R_{q'})^{d'}$ is a solution to $\frac{1}{q'}\mathbf{G}_H^T \mathbf{x} = \mathbf{v} \bmod R^d$ for some $\mathbf{v}$ and the number of solutions to this equation in $(R_{q'})^{d'}$ for each $\mathbf{v}$ is the same. Thus, proving that $\mathbf{v}$ is *almost* uniform suffices. Observe that $r \geq \eta_\epsilon(\Lambda)$. Therefore, Lemma 4 tells us that for any particular $\bar{\mathbf{a}} \in (R_q)^d$ and $\bar{\mathbf{f}} \in \Lambda - \frac{1}{q}\bar{\mathbf{a}}_H$, we have

$$\begin{aligned} \Pr[\mathbf{a} = \bar{\mathbf{a}} \wedge \mathbf{f} = \bar{\mathbf{f}}] &= q^{-nd} \cdot \rho_r(\bar{\mathbf{f}}) / \rho_r(\Lambda - \frac{1}{q}\bar{\mathbf{a}}_H) \\ &= \frac{q^{-nd}}{\rho_r(\Lambda)} \cdot \frac{\rho_r(\Lambda)}{\rho_r(\Lambda - \frac{1}{q}\bar{\mathbf{a}}_H)} \cdot \rho_r(\bar{\mathbf{f}}) \\ &\in C \cdot \left[1, \frac{1+\epsilon}{1-\epsilon}\right] \cdot \rho_r(\bar{\mathbf{f}}) \end{aligned} \tag{7}$$

where $C := q^{-nd}/\rho_r(\Lambda)$ is a constant. By summing this equation over appropriate values of $\bar{\mathbf{a}}$ and $\bar{\mathbf{f}}$, Lemma 4 tells us that for any coset $\bar{\mathbf{v}} \in \Lambda / R^d$,

$$\begin{aligned}\Pr[\mathbf{v} = \bar{\mathbf{v}}] &\in C \cdot \left[1, \frac{1+\epsilon}{1-\epsilon}\right] \cdot \rho_r(q^{-1}R^d + \bar{\mathbf{v}}) \\ &\in C \cdot \rho_r(q^{-1}R^d) \cdot \left[1, \frac{1+\epsilon}{1-\epsilon}\right] \cdot \frac{\rho_r(q^{-1}R^d + \bar{\mathbf{v}})}{\rho_r(q^{-1}R^d)} \\ &\in C' \cdot \left[\frac{1-\epsilon}{1+\epsilon}, \frac{1+\epsilon}{1-\epsilon}\right]\end{aligned} \tag{8}$$

where $C' := C\rho_r(q^{-1}R^d)$. Note that we may apply Lemma 4 here since we know that $r \geq \eta_\epsilon((q)^{-1}R^d)$ by Lemma 3. This allows us to conclude that the distribution of $\mathbf{v}$ is within statistical distance $1 - [(1-\epsilon)/(1+\epsilon)]^2 \leq 4\epsilon$ of the uniform distribution. This means that $\mathbf{x}$ is uniformly random over $(R_{q'})^{d'}$ to within statistical distance 4ϵ implying that $\tilde{\mathbf{a}}$ is uniform random over $(R_{q'})^{d'}$ to within statistical distance 4ϵ. It is also clear that $\tilde{b}$ is exactly uniform random given that b is uniform random. This proves the first claim (uniform-to-uniform).

Distribution of $-\boldsymbol{f}$. In our analysis of the resulting error, it will be useful to understand the distribution of the vector $-\mathbf{f}$ for fixed $\tilde{\mathbf{a}}$ (and thus fixed $\mathbf{v} = \bar{\mathbf{v}}$). Note that fixing a value $\mathbf{f} = \bar{\mathbf{f}}$ fixes $\frac{1}{q}\mathbf{a} = \bar{\mathbf{v}} - \bar{\mathbf{f}} \bmod R^d$. By summing over all appropriate values of $\mathbf{f}$ in Eq. 7, one can show that the distribution of $-\mathbf{f}$ for any fixed $\tilde{\mathbf{a}}$ is within statistical distance $1 - (1-\epsilon)(1+\epsilon) \leq 2\epsilon$ of $D_{\frac{1}{q}R^d - \bar{\mathbf{v}}, r}$.

Distribution of the error. Suppose we are given the MLWE sample $(\mathbf{a}, b = \frac{1}{q}\langle\mathbf{a}, \mathbf{s}\rangle + e) \in (R_q)^d \times \mathbb{T}_{R^\vee}$ where $e \in K_\mathbb{R}$ is drawn from D_α. We have already shown that our map outputs $\tilde{\mathbf{a}} \in (R_{q'})^{d'}$ that is *almost* uniformly random. Now we condition on a fixed $\tilde{\mathbf{a}} = \bar{\tilde{\mathbf{a}}}$ and analyse the distribution of

$$(\tilde{b} - \frac{1}{q'}\langle\bar{\tilde{\mathbf{a}}} \cdot \tilde{\mathbf{s}}\rangle) \bmod R^\vee. \tag{9}$$

Let $\mathbf{f}_i \in \mathbb{R}^n$ be the vector consisting of the ith block of n entries of $\mathbf{f} \in \mathbb{R}^{nd}$ for $i = 1, \ldots, d$. Using the fact that $\tilde{\mathbf{s}} = \mathbf{G}\mathbf{s}$ and that $R^\vee$ is closed under multiplication by elements of R, we can rewrite this as

$$(\tilde{b} - \frac{1}{q'}\langle\bar{\tilde{\mathbf{a}}} \cdot \tilde{\mathbf{s}}\rangle) = \sum_{i=1}^{d} s_i \cdot \sigma_H^{-1}(-\mathbf{f}_i) + \tilde{e} + e \bmod R^\vee. \tag{10}$$

In fact, we want to analyse the RHS of the above equation in canonically embedded space. To do so, define the invertible matrix $S_{i,H} := U_H S_i U_H^\dagger \in \mathbb{R}^{n\times n}$ where U_H is given in Eq. (2) and S_i is the diagonal matrix with the field embeddings of s_i along the diagonal i.e. $[S_i]_{jk} = \sigma_j(s_i)\delta_{jk}$. Note that $S_{i,H}$ is the matrix representing multiplication by s in the basis $(\mathbf{h}_i)_{i=1}^n$ of H. Therefore, in canonical space, the error is given by

$$\sum_{i=1}^{d} S_{i,H} \cdot (-\mathbf{f}_i) + \sigma_H(\tilde{e}) + \sigma_H(e) \bmod R^\vee \tag{11}$$

where $\sigma_H(\tilde{e})$ and $\sigma_H(e)$ are distributed as $D_{r\beta}$ and D_α respectively. Also, letting $\bar{\mathbf{v}}_i$ denote the ith block of n coordinates of $\bar{\mathbf{v}}$, we know that $-\mathbf{f}_i$ is *almost* distributed as $D_{\frac{1}{q}R-\bar{\mathbf{v}}_i,r}$. It then follows that $S_{i,H} \cdot (-\mathbf{f}_i)$ is close in distribution to $D_{\frac{1}{q}S_{i,H}\cdot R - S_{i,H}\cdot\bar{\mathbf{v}}_i, r(S_{i,H})^T}$ i.e. an ellipsoidal discrete Gaussian. In fact the covariance matrix $r^2 S_{i,H} S_{i,H}^T$ is diagonal with respect to our basis $(\mathbf{h}_i)_{i=1}^n$ of $\mathbb{R}^n$ (see Sect. 2.2) with eigenvalues given by $r^2|\sigma_j(s_i)|^2$ for $j = 1, \ldots, n$. Note that we can conceptualise $\sigma_H(\tilde{e})$ as $\sum_{i=1}^d \tilde{e}^{(i)}$ where each $\tilde{e}^{(i)}$ is distributed as a continuous spherical Gaussian in $\mathbb{R}^n$ with parameter $\gamma_i \geq rB$. Recalling that $-\mathbf{f}$ is distributed as $D_{\frac{1}{q}R^d-\bar{\mathbf{v}},r}$ to within statistical distance 2ϵ, we can now apply Lemma 5 d times to conclude that

$$\sum_{i=1}^d S_{i,H} \cdot (-\mathbf{f}_i) + \sigma_H(\tilde{e}) = \sum_{i=1}^d S_{i,H} \cdot (-\mathbf{f}_i) + \tilde{e}^{(i)} \tag{12}$$

is distributed as the continuous Gaussian with a diagonal covariance matrix to within statistical distance $2\epsilon + 4d\epsilon$. In particular, the diagonal entries of the convariance matrix are given by $r^2\left(\beta^2 + \sum_{j=1}^d |\sigma_i(s_j)|^2\right)$ for $i = 1, \ldots, n$. Considering the original error term $\sigma_H(e)$ that follows the distribution D_α completes the proof. □

Remark 2. It is permissible to take $B := \min_{i,j} |\sigma_j(s_i)|$ in the above theorem. However, this will not save us any asymptotic factors in the output error distribution so we use $B := \max_{i,j} |\sigma_j(s_i)|$ to allow for cleaner looking bounds.

The following corollary specialises to a map from MLWE in module rank d to d/k and from modulus q to q^k for general rings. Taking $k = d$ constitutes a reduction from MLWE to RLWE. Note that the new secret distribution is non-standard in general, but we can always use the usual re-randomizing process to obtain a uniform secret. We also highlight the fact that the lower bound on r is not particularly tight due to a loose upper bound on the quantities $\|\tilde{\mathbf{B}}_{s_i R}\|$. This issue is addressed for power-of-two cyclotomics in Sect. 3.1. In fact, for a general cyclotomic ring R, it holds that $\|\mathbf{B}_{s_i R}\| = \|\sigma_H(s_i)\|$.

Corollary 1. *Let R be a ring with basis $\boldsymbol{B}_R$ in the canonical embedding and χ be a distribution satisfying*

$$\Pr_{s \leftarrow\!\!\$\, \chi}\left[\max_i |\sigma_i(s)| > B\right] \leq \delta \text{ and } \Pr_{s \leftarrow\!\!\$\, \chi}\left[\max_{i,j} \frac{|\sigma_i(s)|}{|\sigma_j(s)|} > B'\right] \leq \delta'$$

for some (B, δ) and (B', δ'). Also let $\alpha > 0$ and take any $\epsilon \in (0, 1/2)$. For any $k > 1$ that divides d and

$$r \geq \max\begin{cases} \frac{1}{q}\ \|\tilde{\boldsymbol{B}}_R\| \cdot \sqrt{2\ln(2nd(1+1/\epsilon))/\pi} \\ \frac{1}{q}\ B'\ \|\tilde{\boldsymbol{B}}_R\| \cdot \sqrt{2\ln(2nd(1+1/\epsilon))/\pi} \end{cases},$$

there is an efficient reduction from $MLWE^{(R^d)}_{m,q,\Psi_{\leq\alpha}}(\chi^d)$ *to* $MLWE^{(R^{d/k})}_{m,q^k,\Psi_{\leq\alpha'}}$ $(\mathbf{G}\cdot\chi^d)$ *for* $\mathbf{G} = \mathbb{I}_{d/k} \otimes (1, q, \ldots, q^{k-1}) \in R^{d/k \times d}$ *and*

$$(\alpha')^2 \geq \alpha^2 + 2r^2B^2d.$$

Moreover, this reduction reduces the advantage by at most $[1 - (1 - \delta - \delta')^d] + (4d + 10)\epsilon m$.

Proof. We run the reduction from Theorem 1, taking $q' = q^k$, $\beta^2 \geq B^2d$ and $\mathbf{G} \in R^{d/k\times d}$ as in the corollary statement. First, note that $\|\tilde{\mathbf{B}}_{s_iR}\| \leq \max_j |\sigma_j(s_i)| \cdot \|\tilde{\mathbf{B}}_R\|$ by considering multiplication in the canonical embedding and Lemma 2 from [ABB10]. In the *coefficient* embedding, we have that $\mathbf{G} = \mathbb{I}_{d/k} \otimes (1, q, \ldots, q^{k-1}) \otimes \mathbb{I}_n$ and the lattice of interest is $\frac{1}{q^k}\mathbf{G}^T\mathbb{Z}^{nd/k} + \mathbb{Z}^{nd}$ with basis $\mathbf{B} = \mathbb{I}_{d/k} \otimes \mathbf{Q} \otimes \mathbb{I}_n$ where

$$\mathbf{Q} = \begin{bmatrix} q^{-1} & q^{-2} & \cdots & q^{-k} \\ & q^{-1} & \cdots & q^{1-k} \\ & & \ddots & \vdots \\ & & & q^{-1} \end{bmatrix}.$$

To move from the coefficient embedding to the canonical embedding, we simply multiply by the matrix $\mathbf{B}_{R^d} := \mathbb{I}_d \otimes \mathbf{B}_R$. Therefore, in the canonical embedding, the basis is given by $\mathbf{B}_\Lambda = \mathbb{I}_{d/k} \otimes \mathbf{Q} \otimes \mathbf{B}_R$. Orthogonalising from left to right, we can see that $\|\tilde{\mathbf{B}}_\Lambda\|$ is precisely $\frac{1}{q}\|\tilde{\mathbf{B}}_R\|$.

Let E be the event that $\max_i |\sigma_i(s)| \leq B$ and F be the event $\max_{i,j} \frac{|\sigma_i(s)|}{|\sigma_j(s)|} \leq B'$ where $s \leftarrow_\$ \chi$. The fact that $P(E \cap F) = P(E) + P(F) - P(E \cup F) \geq P(E) + P(F) - 1 \geq 1 - \delta - \delta'$ implies the result. □

3.1 Power-of-Two Cyclotomic Rings

We now give a more specific account of Theorem 1 in the case where R for power-of-two is a cyclotomic ring, i.e. $R = \mathbb{Z}[X]/\langle X^n + 1\rangle$ for power-of-two n. We will also be considering discrete Gaussian secret distributions and normal form MLWE. The corollary given in this section is almost identical to Corollary 1 apart from the definition of the pairs (B, δ) and (B', δ'). This change makes the corollary amenable to known results for discrete Gaussian secret distributions.

It can be shown that the map taking the canonical embedding to the coefficient embedding is a scaled isometry with scaling factor $1/\sqrt{n}$. In particular, the canonical to coefficient embedding map sends a spherical Gaussian r to $r/\sqrt{n}$. Furthermore, the dual ring is given by $R^\vee := \frac{1}{n} \cdot R$ and takes the simple form of $\frac{1}{n}\mathbb{Z}^n$ in the coefficient embedding.

Let $\tau > 0$. We will be considering the case where the secret s is drawn from $D_{R^\vee,\tau}$ (and then reduced modulo $qR^\vee$). In the coefficient embedding, this is equivalent to drawing the secret from the distribution $D_{\frac{1}{n}\mathbb{Z}^n,\tau/\sqrt{n}}$.

Let S_H be the matrix of multiplication by s in the canonical embedding. For cyclotomic power-of-two rings, there is a simple relationship between components of the canonical embedding $\sigma_i(s)$ and the singular values of the matrix $\text{rot}(s)$. Let $\mathbf{B}_R = \sqrt{n} \cdot U$ denote the scaled isometry mapping from coefficient space to canonical space where U is unitary. Then we have ${S_H}^T S_H = U^{-1} \cdot \text{rot}(s)^T \text{rot}(s) \cdot U$. Since ${S_H}^T S_H$ is diagonal with elements given by $|\sigma_i(s)|^2$, the eigenvalues of $\text{rot}(s)^T \text{rot}(s)$ are exactly these diagonal elements. This implies $|\sigma_i(s)|$ are exactly the singular values of $\text{rot}(s)$. We will use this fact in the next claim.

Lemma 7. *Let $R = \mathbb{Z}[X]/\langle X^n + 1\rangle$ for some power-of-two n. Then for any $\delta \in (0,1)$, $t \geq \sqrt{2\pi}$ and $\tau \geq \frac{t}{\sqrt{2\pi n}} \cdot \eta_\delta(\mathbb{Z}^n)$, we have*

$$\Pr_{s \leftarrow\!\$\, D_{R^\vee,\tau}}\left[\frac{1}{\min_i |\sigma_i(s)|} \geq \frac{t\sqrt{2}}{\tau}\right] \leq \frac{1+\delta}{1-\delta} \cdot \frac{n\sqrt{2\pi e}}{t}.$$

Proof. Let $b = ns$. The distribution of b is $D_{\mathbb{Z}^n,\tau\sqrt{n}}$. Let $\sigma_n(\text{rot}(b))$ denote the least singular value of $\text{rot}(b)$. Now we can write

$$\begin{aligned}\Pr_{s \leftarrow\!\$\, D_{R^\vee,\tau}}\left[\frac{1}{\min_i |\sigma_i(s)|} \geq \frac{t\sqrt{2}}{\tau}\right] &= \Pr_{s \leftarrow\!\$\, D_{\frac{1}{n}\mathbb{Z}^n,\frac{\tau}{\sqrt{n}}}}\left[\frac{1}{\sigma_n(\text{rot}(s))} \geq \frac{t\sqrt{2}}{\tau}\right] \\ &= \Pr_{b \leftarrow\!\$\, D_{\mathbb{Z}^n,\tau\sqrt{n}}}\left[\frac{1}{\sigma_n(\text{rot}(b))} \geq \frac{t\sqrt{2}}{(\tau\sqrt{n}) \cdot \sqrt{n}}\right] \\ &\leq \frac{1+\delta}{1-\delta} \cdot \frac{n\sqrt{2\pi e}}{t}\end{aligned}$$

where the inequality comes from Lemma 6. □

In the proof of the following lemma, we will say that a distribution D over $\mathbb{Z}^n$ is (B,δ)-bounded for real numbers $B, \delta > 0$ if $\Pr_{\mathbf{x} \leftarrow\!\$\, D}[\|\mathbf{x}\| > B] \leq \delta$.

Lemma 8. *Let $R = \mathbb{Z}[X]/\langle X^n + 1\rangle$ for some power-of-two n. Then for any $\delta \in (0,1)$ and $\tau \geq 0$,*

$$\Pr_{s \leftarrow\!\$\, D_{R^\vee,\tau}}\left[\|\sigma_H(s)\| > C\tau\sqrt{n\log(n/\delta)}\right] \leq \delta$$

for some universal constant $C > 0$. We also have that

$$\Pr_{s \leftarrow\!\$\, D_{R^\vee,\tau}}\left[\|\sigma_H(s)\| > \tau\sqrt{n}\right] \leq 2^{-n}.$$

Proof. Take $B > 0$ and let $b = ns$. We have

$$\begin{aligned}\Pr_{s \leftarrow\!\$\, D_{R^\vee,\tau}}\left[\|\sigma_H(s)\| > B\right] &= \Pr_{s \leftarrow\!\$\, D_{\frac{1}{n}\mathbb{Z}^n,\frac{\tau}{\sqrt{n}}}}\left[\|s_{vec}\| > B/\sqrt{n}\right] \\ &= \Pr_{b \leftarrow\!\$\, D_{\mathbb{Z}^n,\tau\sqrt{n}}}\left[\|b\| > B\sqrt{n}\right].\end{aligned}$$

As mentioned in [BLP+13], we know that $D_{\mathbb{Z}^n,r}$ is $(Cr\sqrt{n\log(n/\delta)}, \delta)$-bounded for some universal constant $C > 0$ by taking a union bound over the n coordinates. Furthermore, an application of Lemma 1.5 in [Ban93] implies that $D_{\mathbb{Z}^n,r}$ is $(r\sqrt{n}, 2^{-n})$-bounded. Applying these results completes the proof. □

Corollary 2. *Let $R = \mathbb{Z}[X]/\langle X^n + 1\rangle$ for power-of-two n and χ be a distribution over $R^\vee$ satisfying*

$$\Pr_{s \leftarrow\$ \chi}[\|\sigma_H(s)\| > B_1] \leq \delta_1 \text{ and } \Pr_{s \leftarrow\$ \chi}\left[\max_j \frac{1}{|\sigma_j(s)|} \geq B_2\right] \leq \delta_2$$

for some (B_1, δ_1) and (B_2, δ_2). Also let $\alpha > 0$ and take any $\epsilon \in (0, 1/2)$. For any $k > 1$ that divides d,

$$r \geq \left(\frac{\max\{\sqrt{n}, B_1 B_2\}}{q}\right) \cdot \sqrt{2\ln(2nd(1+1/\epsilon))/\pi},$$

there is an efficient reduction from $MLWE^{(R^d)}_{m,q,\Psi_{\leq\alpha}}(\chi^d)$ to $MLWE^{(R^{d/k})}_{m,q^k,\Psi_{\leq\alpha'}}(\boldsymbol{G}\cdot\chi^d)$ for $\boldsymbol{G} = \mathbb{I}_{d/k} \otimes (1, q, \ldots, q^{k-1}) \in R^{d/k \times d}$ and

$$(\alpha')^2 \geq \alpha^2 + 2r^2 B_1^2 d.$$

Moreover, this reduction reduces the advantage by at most $[1 - (1 - \delta_1 - \delta_2)^d] + (4d + 10)\epsilon m$.

Proof. We apply Theorem 1 taking $\beta^2 \geq B_1^2 d$. For power-of-two cyclotomic rings, $\|\mathbf{B}_{s_i R}\| = \|\sigma_H(s)\|$. Furthermore, if $B_1 \geq \|\sigma_H(s)\|$, then it is guaranteed that $B_1 \geq \max_i |\sigma_i(s)|$. The rest of the proof is the same as in Corollary 1. □

To put the above corollary into context, we now discuss the pairs (B_1, δ_1) and (B_2, δ_2) when the secret distribution χ is $D_{R^\vee,\tau}$. From Lemma 8, for any $\delta_1 \in (0, 1)$, we have $B_1 = \mathcal{O}(\tau\sqrt{n\log(n/\delta_1)})$. Next, for any $\delta_2 \in (0, 1)$, we fix the parameter δ from Lemma 7 (e.g. $\delta = 1/2$) and take t from Lemma 7 proportional to n/δ_2. Then, as long as $\tau \geq \mathcal{O}(\sqrt{n\log(n)}/\delta_2)$, we can take $B_2 = \mathcal{O}(n/(\tau\delta_2))$. To summarize, we may take:

- $B_1 = \mathcal{O}(\tau\sqrt{n\log(n/\delta_1)})$ for arbitrary $\tau > 0$ and $\delta_1 \in (0, 1)$
- $B_2 = \tilde{\mathcal{O}}\left(\frac{n}{\tau\delta_2}\right)$ for $\tau \geq \mathcal{O}(\sqrt{n\log(n)}/\delta_2)$ and any $\delta_2 \in (0, 1)$
- $B_1 B_2 = \tilde{\mathcal{O}}\left(\frac{n\sqrt{n\log(n/\delta_1)}}{\delta_2}\right)$ for $\tau \geq \mathcal{O}(\sqrt{n\log(n)}/\delta_2)$ and any $\delta_1, \delta_2 \in (0, 1)$.

In an ideal setting, we would like to conclude that a probabilistic polynomial-time (PPT) algorithm that solves RLWE with non-negligible advantage implies a PPT algorithm capable of solving MLWE with non-negligible advantage. In order to achieve this, it is necessary that the loss in advantage incurred by any reduction should be negligible in the security parameter λ. Therefore, we would require that δ_1, δ_2 and ϵ all be negligible in the corollaries above. The requirement that δ_2 be negligible is particularly troublesome since this implies that B_1 and B_2 are super-polynomial in λ if we want to use the results above. This would mean that the resulting error in our reduction would also be super-polynomial. In particular, the case of normal form MLWE where $\tau = \alpha q$ ($= \mathsf{poly}(n)$) is not covered by the analysis given in the case that δ_2 is negligible. This issue will be addressed in Sect. 4 where we show that taking $\delta_2 = \mathcal{O}(1/d)$ suffices when considering *search* variants.

Yet, the analysis given so far remains relevant for sufficiently good algorithms for solving RLWE. For example, given access to an algorithm solving decision RLWE with advantage $1/\mathsf{poly}(\lambda)$, it would be adequate to consider δ_1, δ_2 and ϵ as $1/\mathsf{poly}(\lambda)$. These choices lead to a reduction from MLWE to RLWE (with polynomial noise) with $1/\mathsf{poly}(\lambda)$ loss in advantage which is acceptable given a sufficiently effective algorithm for solving RLWE.

4 Search Reductions Using Rényi Divergence

Given our analysis of the reduction explicited in Theorem 1, it is fairly straightforward to obtain analogous results based on Rényi divergence. We will show that our reduction can be used to solve search MLWE with non-negligible probability given an algorithm for solving search RLWE with non-negligible success probability. Note that this result could potentially be derived from statistical distance arguments, but we choose to use the Rényi divergence because it later allows us to reduce to a strictly spherical error distribution while increasing the width of the resulting error distribution only by small powers of n. In contrast, statistical distance arguments require the drowning noise to increase by super-polynomial factors. This is because we require negligible statistical distances to target distributions whereas we only require that Rényi divergences are $\mathcal{O}(1)$ to obtain meaningful results.

Theorem 2. *Let R be the ring of integers of some algebraic number field K of degree n, d, d', q, q' be integers, $\epsilon \in (0, 1/2)$, and $\boldsymbol{G} \in R^{d' \times d}$. Also, fix $\boldsymbol{s} = (s_1, \ldots, s_d) \in (R_q^\vee)^d$. Further, let $\boldsymbol{B}_\Lambda$ be some known basis of the lattice $\Lambda = \frac{1}{q'} \boldsymbol{G}_H^T R^{d'} + R^d$ (in the canonical embedding), $\boldsymbol{B}_R$ be some known basis of R in H and*

$$r \geq \max \begin{cases} \|\tilde{\boldsymbol{B}}_\Lambda\| \cdot \sqrt{2\ln(2nd(1+1/\epsilon))/\pi} \\ \frac{1}{q} \|\tilde{\boldsymbol{B}}_R\| \cdot \sqrt{2\ln(2nd(1+1/\epsilon))/\pi} \\ \frac{1}{q} \|\tilde{\boldsymbol{B}}_{s_i R}\| \cdot \frac{1}{\min_k |\sigma_k(s_i)|} \cdot \sqrt{2\ln(2n(1+1/\epsilon))/\pi} \end{cases}$$

where $\boldsymbol{B}_{s_i R}$ is a basis of $s_i R$ in the canonical embedding. Let $M = R^d$, $M' = R^{d'}$ and define $B := \max_{i,j} |\sigma_i(s_j)|$. There exists an efficient probabilistic mapping $\mathcal{F} : (R_q)^d \times \mathbb{T}_{R^\vee} \longrightarrow (R_{q'})^{d'} \times \mathbb{T}_{R^\vee}$ such that

$$R_\infty \left(A^{(M')}_{q', \boldsymbol{G}\boldsymbol{s}, D_{\boldsymbol{\alpha}'}} || \mathcal{F}(A^{(M)}_{q, \boldsymbol{s}, D_\alpha}) \right) \leq \left(\frac{1+\epsilon}{1-\epsilon} \right)^{d+3}$$

where $(\boldsymbol{\alpha}')_i^2 = \alpha^2 + r^2(\beta^2 + \sum_{j=1}^d |\sigma_i(s_j)|^2)$ and β satisfies $\beta^2 \geq B^2 d$.

Proof. We take the mapping $\mathcal{F}$ described in the proof of Theorem 1 and adopt the same notation. Recall that $(\tilde{\mathbf{a}}, \tilde{b})$ denotes the output of $\mathcal{F}$. Denote the distribution of interest $\mathcal{F}(A^{(M)}_{q,\mathbf{s},D_\alpha})$ as $\tilde{A}^{(M')}_{q',\mathbf{Gs},\tilde{D}}$ i.e. the distribution of $(\tilde{\mathbf{a}},\tilde{b})$ given that $(\mathbf{a},b)$ follows the distribution $A^{(M)}_{q,\mathbf{s},D_\alpha}$.

Distribution of $\tilde{\mathbf{a}}$. Let K_{sol} denote the number of solutions to the equation $\frac{1}{q'}\mathbf{G}_H^T\mathbf{x} = \mathbf{v} \bmod R^d$ and K_v the number of possible vectors $\mathbf{v}$. Recall that K_{sol} is constant in $\mathbf{v}$. For any $\bar{\tilde{\mathbf{a}}} \in R^{d'}_{q'}$, we have (from Eq. (8)) that

$$\Pr[\tilde{\mathbf{a}} = \bar{\tilde{\mathbf{a}}}] = \sum_{\bar{\mathbf{v}}\in\Lambda/\mathbb{Z}^{nd}} \Pr[\tilde{\mathbf{a}} = \bar{\tilde{\mathbf{a}}}|\mathbf{v} = \bar{\mathbf{v}}] \cdot \Pr[\mathbf{v} = \bar{\mathbf{v}}]$$

$$\geq C' \cdot \left(\frac{1-\epsilon}{1+\epsilon}\right)\frac{1}{K_{sol}} \geq \left(\frac{1-\epsilon}{1+\epsilon}\right)^2 \cdot \frac{1}{K_{sol}K_v}.$$

Note that picking $\tilde{\mathbf{a}}$ at random is identical to choosing $\mathbf{v}$ at random followed by picking a uniformly random solution to $\frac{1}{q'}\mathbf{G}_H^T\mathbf{x} = \mathbf{v} \bmod R^d$. Therefore, the distribution of $\tilde{\mathbf{a}}$ which we denote by $D^{(\tilde{\mathbf{a}})}$ satisfies

$$R_\infty\left(U(R^{d'}_{q'})||D^{(\tilde{\mathbf{a}})}\right) \leq \left(\frac{1+\epsilon}{1-\epsilon}\right)^2. \tag{13}$$

Distribution of $-\boldsymbol{f}$. Previously, we concluded that the distribution of $-\mathbf{f}$ was close in statistical distance to $D_{\frac{1}{q}R^d-\bar{\mathbf{v}},r}$ conditioned on some fixed $\tilde{\mathbf{a}}$. Once again, summing over appropriate values of $\mathbf{f}$ in Eq. (7) tells us that

$$\Pr[-\mathbf{f} = \bar{\mathbf{f}}|\tilde{\mathbf{a}} = \bar{\tilde{\mathbf{a}}}] \geq C \cdot \rho_r(\bar{\mathbf{f}}) \geq \frac{1-\epsilon}{1+\epsilon} \cdot \frac{\rho_r(\bar{\mathbf{f}})}{\rho_r(\frac{1}{q}R^d - \bar{\mathbf{v}})}.$$

Therefore, writing $D^{(-\mathbf{f})}$ as the distribution of $-\mathbf{f}$, we see that

$$R_\infty\left(D_{\frac{1}{q}R^d-\bar{\mathbf{v}},r}||D^{(-\mathbf{f})}\right) \leq \frac{1+\epsilon}{1-\epsilon}.$$

Distribution of the error term. We now analyse the distribution of the error term given in Eq. (10). Let $\mathbf{f}_i$ denote the ith block of n consecutive coordinates of $\mathbf{f} \in \mathbb{R}^{nd}$ Once again, we split the RHS of this error term and analyse it as $\sum_{i=1}^d \left(S_{i,H}^T \cdot (-\mathbf{f}_i) + \tilde{e}^{(i)}\right) + e$ where each $\tilde{e}^{(i)}$ is sampled independently from a continuous Gaussian on $\mathbb{R}^n$ with parameter $\gamma_i \geq rB$. Let $D^{(i)}$ denote the distribution of $\left(S_{i,H}^T \cdot (-\mathbf{f}_i) + \tilde{e}^{(i)}\right)$. We now use the data-processing inequality with the function $(-\mathbf{f}, \tilde{e}^{(1)}, \ldots, \tilde{e}^{(d)}) \longmapsto (S_{1,H}^T \cdot (-\mathbf{f}_1) + \tilde{e}^{(1)}, \ldots, S_{d,H}^T \cdot (-\mathbf{f}_d) + \tilde{e}^{(d)})$. For $i = 1, \ldots, d$, define $Y^{(i)}$ as the distribution obtained by sampling from $D_{\frac{1}{q}S_{i,H}R+S_{i,H}\cdot\bar{\mathbf{v}}_i,r(S_{i,H}^T)}$ and then adding a vector sampled from D_{γ_i}. Note that $Y^{(i)}$ is the distribution of $\mathbf{S}_i^T \cdot (-\mathbf{f}_i) + \tilde{e}^{(i)}$ in the case that the distribution of $-\mathbf{f}$ is *exactly* $D_{\frac{1}{q}R^d-\bar{\mathbf{v}},r}$. Let $D_\gamma = D_{\gamma_1} \times \cdots \times D_{\gamma_d}$. The data-processing inequality for Rényi divergence implies that

$$R_\infty\left(Y^{(1)} \times \cdots \times Y^{(d)}||D^{(1)} \times \cdots \times D^{(d)}\right) \leq R_\infty\left(D_{\frac{1}{q}R^d-\bar{\mathbf{v}},r} \times D_\gamma||D^{(-\mathbf{f})} \times D_\gamma\right)$$

$$\leq \frac{1+\epsilon}{1-\epsilon}.$$

Now we apply Lemma 5 by recalling that the covariance matrix $S_{i,H}^T S_{i,H}$ is diagonal with elements $|\sigma_j(s_i)|$ for $j = 1, \ldots n$. This allows us to conclude that for $i = 1, \ldots, d$,

$$R_\infty \left(D_{(\gamma_i^2 + r^2 S_{i,H}^T S_{i,H})^{1/2}} || Y^{(i)} \right) \leq \frac{1+\epsilon}{1-\epsilon}.$$

By first applying the data-processing inequality to the function that sums the samples and then considering the triangle inequality and independence, the above equation implies that

$$\begin{aligned} R_\infty \left(D_{(\alpha^2 + r^2\beta^2 + r^2 \sum_{i=1}^d S_{i,H}^T S_{i,H})^{1/2}} || \tilde{D} \right) &\leq \frac{1+\epsilon}{1-\epsilon} \cdot \prod_{i=1}^d R_\infty \left(D_{(\gamma^2 + r^2 S_{i,H}^T S_{i,H})^{1/2}} || Y^{(i)} \right) \\ &\leq \left(\frac{1+\epsilon}{1-\epsilon} \right)^{d+1} \end{aligned} \tag{14}$$

where $\tilde{D}$ is the distribution of the RHS of Eq. (10) (i.e. the sum of the distributions $D^{(i)}$).

Distribution of the reduction's output. We now complete the proof by combining the results above.

$$\begin{aligned} R_\infty \left(A_{q', \mathbf{Gs}, D_{\alpha'}}^{(M')} || \tilde{A}_{q', \mathbf{Gs}, \tilde{D}}^{(M')} \right) &\leq \left(\frac{1+\epsilon}{1-\epsilon} \right)^2 \cdot R_\infty \left(D_{\alpha'} || \tilde{D} \right) \\ &\leq \left(\frac{1+\epsilon}{1-\epsilon} \right)^2 \cdot \left(\frac{1+\epsilon}{1-\epsilon} \right)^{d+1} \end{aligned}$$

where the first inequality comes from the multiplicative property of Rényi divergence along with the inequality in (13) and the second comes from the weak triangle inequality along with (14). □

Corollary 3. *For power-of-two n, let $R = \mathbb{Z}[X]/\langle X^n + 1 \rangle$, m be a positive integer and χ be a distribution over $R^\vee$ satisfying*

$$\Pr_{s \leftarrow\$ \chi} [||\sigma_H(s)|| > B_1] \leq \delta_1 \text{ and } \Pr_{s \leftarrow\$ \chi} \left[\max_j \frac{1}{|\sigma_j(s)|} \geq B_2 \right] \leq \delta_2$$

for some (B_1, δ_1) and (B_2, δ_2). Also let $\alpha > 0$. For any $k > 1$ that divides $d > 1$ and

$$r \geq \left(\frac{\max\{\sqrt{n}, B_1 B_2\}}{q} \right) \cdot \sqrt{2 \ln(2nd(1 + m(d+3)))/\pi},$$

there exists an efficient reduction from search $MLWE_{m,q,\Psi_{\leq \alpha}}^{(R^d)}(\chi^d)$ *to* search $MLWE_{m,q^k,\Psi_{\leq \alpha'}}^{(R^{d/k})}(U(R_q^\vee))$ *for*

$$(\alpha')^2 \geq \alpha^2 + 2r^2 B_1^2 d.$$

In particular, if there is an algorithm solving search $MLWE^{(R^{d/k})}_{m,q^k,\Psi_{\leq \alpha'}}(U(R_q^\vee))$ *with success probability* p*, then for* search $MLWE^{(R^d)}_{m,q,\Psi_{\leq \alpha}}(\chi^d)$ *an algorithm exists which succeeds with probability at least* $[1-(\delta_1+\delta_2)]^d \cdot p/8$.

Proof. We use the reduction and analysis from Theorem 2 with $\beta^2 \geq B_1^2 d$ and $\mathbf{G} = I_{d/k} \otimes (1, q, \ldots, q^{k-1}) \in R^{d/k \times d}$ followed by a standard re-randomization of the resulting secret. Since we sample d such ring elements, we are in the realm of Theorem 2 with probability at least $(1-(\delta_1+\delta_2))^d$. Since we have m samples, we must raise the Rényi divergence in Theorem 2 to the mth power. Taking $\epsilon = \frac{1}{m(d+3)}$ ensures that $\left(\frac{1+\epsilon}{1-\epsilon}\right)^{(d+3)m} \leq 8$. The result now follows from the probability preservation property of the Rényi divergence and the fact that we can reverse the mapping between secrets. □

The results of this section are far more satisfying than the analysis given in the previous section when analysing a secret distribution of the form $D_{R^\vee,\tau}$. Let us assume that the probability of success p of an algorithm for solving RLWE is non-negligible. Then all we require is that $\delta_1, \delta_2 = O(1/d)$ in order to solve the search MLWE with non-negligible success probability. Therefore, we may take $B_1 = \tilde{\mathcal{O}}(\tau\sqrt{n})$ and $B_2 = \mathcal{O}(dn/\tau)$ for this secret distribution as long as $\tau \geq \tilde{\mathcal{O}}(d\sqrt{n})$. In this case, we have $\alpha' = \tilde{\mathcal{O}}(\tau n^2 \sqrt{d}/q)$. This simplifies to $\alpha' = \tilde{\mathcal{O}}(\alpha n^2 \sqrt{d})$ when considering the normal form of MLWE where $\tau = \alpha q$. Therefore, we see that even for typical error and secret distributions with polynomial standard deviations, search MLWE is not qualitatively harder than search RLWE with larger modulus, i.e. an efficient algorithm for the latter implies an efficient algorithm for the former.

4.1 Strictly Spherical Error Distributions

We will now present a lemma that allows us to reduce from MLWE to RLWE with a *spherical* error distribution.

Lemma 9. *For integers* m, n, *let* $\boldsymbol{M} \in \mathbb{R}^{m \times n}$ *be a matrix with non-zero singular values* σ_i *for* $i = 1, \ldots, n$ *and take* $\beta^2 \geq \sigma_1^2$. *Then*

- $R_2\left(D_{r\beta} || D_{r(\beta^2\mathbb{I}+\boldsymbol{M}^T\boldsymbol{M})^{1/2}}\right) \leq \left(1+\frac{\sigma_1^4}{\beta^4}\right)^{n/2}$,
- $R_\infty\left(D_{r\beta} || D_{r(\beta^2\mathbb{I}+\boldsymbol{M}^T\boldsymbol{M})^{1/2}}\right) \leq \left(1+\frac{\sigma_1^2}{\beta^2}\right)^{n/2}$.

We can now extend Theorem 2 to get a spherical output error distribution by applying the above Lemma to the final result along with the triangle inequality. In particular, the Rényi divergences given in Theorem 2 increase by factors of $\left(1+\frac{d^4 \max_{i,j}|\sigma_j(s_i)|^4}{\beta^4}\right)^{n/2}$ and $\left(1+\frac{d^2 \max_{i,j}|\sigma_j(s_i)|^2}{\beta^2}\right)^{n/2}$ for orders 2 and ∞ respectively. Therefore, when applying the theorem to m MLWE samples, we require that β increase by factors of $(mn)^{1/4}$ for order 2 and $(mn)^{1/2}$ for infinite order to ensure $\mathcal{O}(1)$ Rényi divergences. These ideas will be concretised in the proof of Theorem 3 in the next section.

5 Reducing RLWE in (n, q) to $(n/2, q^2)$

Throughout this entire section, we assume that n is a power of two. The reduction strategy is to represent polynomial multiplications in ring dimension n using $n \times n$ matrices by working in the *coefficient* embedding. The reduction follows the same blueprint as in Sect. 3 apart from the fact that we are no longer working exclusively in the canonical embedding. Since we are considering power-of-two cyclotomic rings, polynomial multiplication is always represented by a matrix of the form given in Eq. (1). Going from ring dimension n to $n/2$ just halves the dimension of these matrices. For clarity, we adopt the notation $R_{n,q} = \mathbb{Z}_q[X]/\langle X^n + 1\rangle$ and $R_n = \mathbb{Z}[X]/\langle X^n + 1\rangle$.

Our aim is to reduce RLWE in dimension and modulus (n, q) to RLWE in $(n/2, q^2)$ via some mapping: $a \in R_{n,q} \longmapsto \tilde{a} \in R_{n/2,q^2}$, $b \in \mathbb{T}_{R_n{}^\vee} \longmapsto \tilde{b} \in \mathbb{T}_{R_{n/2}{}^\vee}$, $s \in R_{n,q}{}^\vee \longmapsto \tilde{s} \in R_{n/2,q^2}{}^\vee$. We can start by defining a relationship between $\text{rot}(s)$ and $\text{rot}(\tilde{s})$. In order to make clear the distinction between the two rings, we denote $n \times n$ matrices associated with multiplications in $R_{n,q}$ by writing the subscript n, q. Given $\mathbf{G}, \mathbf{H} \in \mathbb{Z}^{n/2 \times n}$, the linear relationship will be defined via the equation

$$\text{rot}(\tilde{s})_{n/2,q^2} = 2 \cdot \mathbf{H} \cdot \text{rot}(s)_{n,q} \cdot \mathbf{G}^T. \tag{15}$$

Note that the factor of 2 is present to account for the fact that the new secret should be in the dual ring $R_{n/2,q^2}{}^\vee = \frac{2}{n}R$ and the matrix $\mathbf{H}$ ensures that we end up with a square matrix $\text{rot}(\tilde{s})_{n/2,q^2}$. We also need to be careful that $\mathbf{G}$ and $\mathbf{H}$ are chosen so the matrix $\text{rot}(\tilde{s})_{n/2,q^2}$ has the correct form. Define the map between b and $\tilde{b}$ (up to some Gaussian error) as

$$\tilde{b}_{vec} \approx 2\mathbf{H} \cdot b_{vec}.$$

In order for the reduction to work, we require that $\tilde{b} \approx \tilde{a} \cdot \tilde{s}/q^2 \bmod R_{n/2}{}^\vee$ i.e.

$$2 \cdot \mathbf{H} \cdot \text{rot}(s)_{n,q} \cdot \frac{1}{q} a_{vec} \approx 2 \cdot \mathbf{H} \cdot \text{rot}(s)_{n,q} \cdot \mathbf{G}^T \cdot \frac{1}{q^2} \tilde{a}_{vec} \bmod 2/n.$$

It is easy to see that we can satisfy this requirement by choosing $\tilde{a}$ such that

$$\frac{1}{q} a_{vec}^T = \frac{1}{q^2} \mathbf{G}^T \cdot \tilde{a}_{vec}^T \bmod 1.$$

Explicit forms for our choice of $\mathbf{G}$ and $\mathbf{H}$ are

$$\mathbf{G} = \mathbb{I}_{n/2} \otimes (1, q) \in \mathbb{Z}^{n/2 \times n}, \tag{16}$$

$$\mathbf{H} = \mathbb{I}_{n/2} \otimes (1, 0) \in \mathbb{Z}^{n/2 \times n}. \tag{17}$$

Claim. Take $\mathbf{G}$ and $\mathbf{H}$ as above. Then $\text{rot}(\tilde{s})_{n/2,q^2}$ is of the correct form (i.e. represents multiplication by some polynomial in $(R_{n/2,q^2})$).

Proof. We can write simple explicit forms $(\mathbf{G}^T)_{kl} = \delta_{k,2l-1} + q\delta_{k,2l}$ and $(\mathbf{H})_{ij} = \delta_{2i-1,j}$. Then the matrix multiplication $\mathbf{H} \cdot \text{rot}(s)_{n,q} \cdot \mathbf{G}^T$ yields $(\text{rot}(\tilde{s})_{n/2,q^2})_{il} = (\text{rot}(s)_{n,q})_{2i-1,2l-1} + (q\text{rot}(s)_{n,q})_{2i-1,2l}$ which is of the correct form. $\square$

Note that the mapping between secrets is

$$s = \sum_{i=0}^{n-1} s_i \cdot X^i \longmapsto \tilde{s} = (s_0 - q s_{n-1}) + \sum_{i=1}^{n/2-1} (s_{2i} + q s_{2i-1}) \cdot X^i. \quad (18)$$

Now the proof of correctness for this reduction is essentially the same as Theorem 2 with a few alterations. One of the more important changes is that we use Lemma 9 and target a spherical error. We do this to ensure that multiplication by $\mathbf{H}$ leads to a Gaussian with parameters that we can easily bound.

Theorem 3. *Let n be a power of two, q be an integer, fix $s \in R_{n,q}{}^\vee$ and*

$$r \geq \frac{1}{q} \cdot \max\left\{1, \frac{\|s_{vec}\|}{\sigma_n(\mathrm{rot}(s))}\right\} \cdot \sqrt{2\ln(2n(1+1/\epsilon))/\pi}.$$

Further, let $\sigma_1 := \sigma_1(\mathrm{rot}(s))$ and $\beta \geq 2\sigma_1\sqrt{n}$.

For any $\alpha > 0$, there exists an efficient mapping $\mathcal{F} : R_{n,q} \times \mathbb{T}_{R_{n,q}{}^\vee} \to R_{n/2,q} \times \mathbb{T}_{R_{n/2,q^2}{}^\vee}$ such that

- $R_2\left(A^{R_{n/2}}_{q^2,\tilde{s},D_{\alpha'}} \| \mathcal{F}(A^{R_n}_{q,s,D_\alpha})\right) \leq \left(\frac{1+\epsilon}{1-\epsilon}\right)^4 \cdot \left(1 + \frac{16n^2\sigma_1^4}{\beta^4}\right)^{n/2}$,
- $R_\infty\left(A^{R_{n/2}}_{q^2,\tilde{s},D_{\alpha'}} \| \mathcal{F}(A^{R_n}_{q,s,D_\alpha})\right) \leq \left(\frac{1+\epsilon}{1-\epsilon}\right)^4 \cdot \left(1 + \frac{4n\sigma_1^2}{\beta^2}\right)^{n/2}$

where $\tilde{s}$ is given in Eq. (18) and $(\alpha')^2 = 4\alpha^2 + r^2\beta^2$.

Proof. Suppose we are given $(a, b) \in R_{n,q} \times \mathbb{T}_{R_{n,q}{}^\vee}$ and take $\mathbf{G}, \mathbf{H} \in \mathbb{Z}^{n/2 \times n}$ as in Eqs. (16) and (17) respectively. The mapping $\mathcal{F}$ is performed as follows:

1. Sample $\mathbf{f} \leftarrow D_{\Lambda - \frac{1}{q}a_{vec}, r}$ over the lattice $\Lambda = \frac{1}{q^2}\mathbf{G}^T\mathbb{Z}^{n/2} + \mathbb{Z}^n$. Note that the parameter r is large enough so we can sample the discrete Gaussian efficiently by Lemma 2 since $\|\tilde{\mathbf{B}}_\Lambda\| = q^{-1}$.
2. Let $\mathbf{v} = \frac{1}{q}a_{vec} + \mathbf{f} \in \Lambda/\mathbb{Z}^n$ and set $\mathbf{x}$ to be a random solution of $\frac{1}{q^2}\mathbf{G}^T\mathbf{x} = \mathbf{v} \bmod 1$. Then set $\tilde{a} \in R_{n/2,q^2}$ to be the unique polynomial such that $\tilde{a}_{vec} = \mathbf{x}$.
3. Sample $\tilde{e}$ from the distribution $D_{r\beta}$ over $K_\mathbb{R} \simeq H \simeq \mathbb{R}^{n/2}$ and set $\tilde{b} = 2\mathbf{H} \cdot b + \tilde{e} \in \mathbb{T}_{R_{n/2,q^2}{}^\vee}$.
4. Finally, output $(\tilde{a}, \tilde{b}) \in (R_{n/2,q^2}) \times \mathbb{T}_{R_{n/2,q^2}{}^\vee}$.

Distribution of $\tilde{a}$: We can precisely repeat the argument given in the proof of Theorem 2 after noting that $r \geq \eta_\epsilon(\Lambda)$ and $r \geq \eta_\epsilon(q^{-1}\mathbb{Z}^n)$. The only conceptual difference is that we are now working in the coefficient embedding. Denoting the distribution of $\tilde{a}$ given uniform a by $D^{(\tilde{a})}$, we find that

$$R_\infty\left(U(R_{n/2,q^2}) \| D^{(\tilde{a})}\right) \leq \left(\frac{1+\epsilon}{1-\epsilon}\right)^2. \quad (19)$$

Distribution of the error: We now condition on fixed $\tilde{a} = \bar{\tilde{a}}$ and set $\bar{\mathbf{v}} = \mathbf{G}^T \bar{\tilde{a}}_{vec}$. Denoting the distribution of $-\mathbf{f}$ as $D^{(-\mathbf{f})}$ we also have that

$$R_\infty \left(D_{\frac{1}{q}\mathbb{Z}^n - \bar{\mathbf{v}}, r} || D^{(-\mathbf{f})} \right) \leq \left(\frac{1+\epsilon}{1-\epsilon} \right).$$

All that remains is to analyse the distribution of

$$\left(\tilde{b} - \frac{1}{q^2} \tilde{a} \cdot \tilde{s} \right)_{vec} = 2\mathbf{H} \cdot \mathrm{rot}(s) \cdot (-\mathbf{f}) + 2\mathbf{H} \cdot e_{vec} + \tilde{e}_{vec} \bmod 2/n \tag{20}$$

$$= 2\mathbf{H} \cdot (\mathrm{rot}(s) \cdot (-\mathbf{f}) + e_{vec} + \tilde{e}^*_{vec}) \bmod 2/n \tag{21}$$

where $\tilde{e}^*_{vec}$ (resp. e_{vec}) is drawn from the spherical distribution $D_{r\beta/(2\sqrt{n})}$ (resp. $D_{\alpha/\sqrt{n}}$). Note that the $\sqrt{n}$ factors take into account that we are working in the coefficient embedding.

The distribution of $\mathrm{rot}(s) \cdot D_{\frac{1}{q}\mathbb{Z}^n - \bar{\mathbf{v}}, r}$ is $D_{\frac{1}{q}\mathrm{rot}(s)\mathbb{Z}^n - \mathrm{rot}(s)\bar{\mathbf{v}}, r \cdot \mathrm{rot}(s)^T}$. By working in the orthogonal basis where the covariance matrix $\mathrm{rot}(s)^T \mathrm{rot}(s)$ is diagonal, we can apply Lemma 5. We also apply the data-processing inequality on $(-\mathbf{f}, \tilde{e}^*_{vec}) \longmapsto -\mathrm{rot}(s) \cdot \mathbf{f} + \tilde{e}^*_{vec}$ along with the triangle inequality to obtain

$$R_\infty \left(D_{err} || D^{(-\mathrm{rot}(s) \cdot \mathbf{f} + \tilde{e}^*_{vec})} \right) \leq \left(\frac{1+\epsilon}{1-\epsilon} \right) \cdot \left(\frac{1+\epsilon}{1-\epsilon} \right), \tag{22}$$

where D_{err} is a continuous Gaussian distribution with covariance $\Sigma = r^2(\frac{\beta^2}{4n}\mathbb{I} + \mathrm{rot}(s)^T \mathrm{rot}(s))$ and $D^{(-\mathrm{rot}(s) \cdot \mathbf{f} + \tilde{e}^*_{vec})}$ is the *exact* distribution of $-\mathrm{rot}(s) \cdot \mathbf{f} + \tilde{e}^*_{vec}$.

Distance to spherical error: We now apply Lemma 9 to find that

$$R_2 \left(D_{r\beta/(2\sqrt{n})} || D_{err} \right) \leq \left(1 + \frac{16 n^2 \sigma_1^4}{\beta^4} \right)^{n/2},$$

$$R_\infty \left(D_{r\beta/(2\sqrt{n})} || D_{err} \right) \leq \left(1 + \frac{4 n \sigma_1^2}{\beta^2} \right)^{n/2}.$$

Finally, using the weak triangle inequality with intermediate distribution $2\mathbf{H} \cdot D_{err}$ and the data-processing inequality, we obtain

$$R_2 \left(2\mathbf{H} \cdot D_{((r\beta)^2/(4n) + \alpha^2/n)^{1/2}} || D^{(RHS)} \right) \leq \left(\frac{1+\epsilon}{1-\epsilon} \right)^2 \cdot \left(1 + \frac{16 n^2 \sigma_1^4}{\beta^4} \right)^{n/2},$$

$$R_\infty \left(2\mathbf{H} \cdot D_{((r\beta)^2/(4n) + \alpha^2/n)^{1/2}} || D^{(RHS)} \right) \leq \left(\frac{1+\epsilon}{1-\epsilon} \right)^2 \cdot \left(1 + \frac{4 n \sigma_1^2}{\beta^2} \right)^{n/2}$$

where $D^{(RHS)}$ is the distribution of the RHS in Eq. (20).

Distribution of the reduction output: We conclude by combining the above results in the same way as in the proof of Theorem 2. We must also scale up by a factor of $\sqrt{n}$ to account for the fact that we have been working in the coefficient embedding. □

Corollary 4. *Let n be a power of two and χ be a distribution over $R_n^\vee$ satisfying*

$$\Pr_{s \leftarrow \$ \chi}[\|\sigma_H(s)\| > B_1] \leq \delta_1 \text{ and } \Pr_{s \leftarrow \$ \chi}\left[\max_j \frac{1}{|\sigma_j(s)|} \geq B_2\right] \leq \delta_2$$

for some (B_1, δ_1) and (B_2, δ_2). Also, let $\alpha > 0$ and $\epsilon \in (0, 1/2)$. For any

$$r \geq \frac{1}{q} \cdot \max\{1, B_1 B_2\} \cdot \sqrt{2\ln(2n(4m+1))/\pi},$$

let $(\alpha_c')^2 = 4\alpha^2 + 4r^2 B_1^2 (mn)^{2c}$. Suppose there exists an algorithm solving search $RLWE_{m,q^2,D_{\alpha_c'}}^{(R_{n/2})}(U(R_{n/2,q^2}^\vee))$ for $c = 1/4$ (resp. $c = 1/2$) with success probability $p_{1/4}$ (resp. $p_{1/2}$). Then there exists algorithms solving $RLWE_{m,q,D_\alpha}^{(R_n)}(\chi)$ with success probabilities at least $(1 - (\delta_1 + \delta_2))\frac{p_{1/4}^2}{8e^{1/2}}$ and $(1 - (\delta_1 + \delta_2))\frac{p_{1/2}}{8e^{1/2}}$.

Proof. We will be applying the reduction in Theorem 3 with $\epsilon = 1/(4m)$ along with a re-randomizing of the secret. We take $\beta = 2B_1(mn)^c$ in the theorem. Recall that for power-of-two cyclotomic rings, we have $\|\sigma_H(s)\| = \sqrt{n}\|s_{vec}\|$, $\min_j |\sigma_j(s)| = \sigma_n(\text{rot}(s))$ and $\max_j |\sigma_j(s)| = \sigma_1(\text{rot}(s))$. This means that we are able to apply the reduction and analysis of Theorem 3 with probability at least $1 - (\delta_1 + \delta_2)$. Since we have m samples, we need to raise the Rényi divergences to the mth power. Therefore, in the case that $c = 1/4$ (resp. $c = 1/2$), we have that the Rényi divergence of order 2 (resp. order ∞) is upper bounded by $8 \cdot e^{1/2}$. Note that the reduction defines a reversible map between the secrets. Therefore, the result is obtained by running the reduction, re-randomizing the secret, solving the resulting search RLWE instance and then mapping back to the original secret. □

Typically, we would have access to $m = \mathcal{O}(1)$ RLWE samples. Considering the normal form of RLWE with secret distribution $D_{R^\vee,\alpha q}$, we can take the parameters B_1 and B_2 to be $\tilde{\mathcal{O}}(\alpha q\sqrt{n})$ and $\tilde{\mathcal{O}}(n/(\alpha q))$ respectively. Therefore, the above corollary says that if we can solve RLWE in dimension $n/2$, modulus q^2 and error rate $\alpha \cdot n^{9/4}$ with non-negligible probability in polynomial time, then we can also solve RLWE with dimension n, modulus q and error rate α is polynomial time with non-negligible probability.

Acknowledgements. We thank Adeline Roux-Langlois and Léo Ducas for helpful discussions on an earlier draft of this work. We also thank our anonymous referees for their feedback.

A Design Space for RLWE Public-Key Encryption

Recall the simple public-key encryption scheme from [LPR10] which serves as the blueprint for many subsequent constructions. The scheme publishes a public-key $(a, b = a \cdot s + e)$, where both s and e are small elements from the ring of integers of a power-of-two cyclotomic field. Encryption of some polynomial m with $\{0, 1\}$ coefficients is then performed by sampling short r, e_1, e_2 and outputting:

$$(u,\, v) = \big(a \cdot r + e_1, \quad b \cdot r + e_2 + \lfloor q/2 \rfloor \cdot m \bmod q\big).$$

The decryption algorithm computes

$$u \cdot s - v = (a \cdot r + e_1) \cdot s - (a \cdot s + e) \cdot r - e_2 - \lfloor q/2 \rfloor \cdot m.$$

Let σ be the norm of s, e, r, e_1, e_2. Clearly, the final message will have noise of norm $\geq \sigma^2$. Thus to ensure correct decryption, q has a quadratic dependency on σ. As a consequence, in this construction, increasing σ and q can only reduce security by increasing the gap between noise and modulus.

However, this issue can be avoided and is avoided in MLWE-based constructions by picking some $\sigma' < \sigma$ at the cost of publishing more samples in the public key. For example, if $d = 2$ the public key becomes

$$((a', b'),\, (a'', b'')) = ((a', a' \cdot s + e'),\, (a'', a'' \cdot s + e'')),$$

where $s, e'e,''$ have norm σ. Encryption of some $\{0, 1\}$ polynomial m is then performed by sampling short $r', r'', e_1, \, e_2$ with norm σ' and outputting

$$(u,\, v) = (a' \cdot r' + a'' \cdot r'' + e_1, \quad b' \cdot r' + b'' \cdot r'' + e_2 + \lfloor q/2 \rfloor \cdot m \bmod q)\,.$$

The decryption algorithm computes

$$u \cdot s - v = (a' \cdot r' + a'' \cdot r'' + e_1) \cdot s - (a' \cdot s + e') \cdot r' - (a'' \cdot s + e'') \cdot r'' - e_2 - \lfloor q/2 \rfloor \cdot m.$$

The security of the public key reduces to the hardness of RLWE in dimension n with modulus q and noise size σ as before. The security of encryptions reduces to the hardness of MLWE in dimension $d = 2$ over ring dimension n, modulus q and noise size σ', i.e. the level of security is maintained for $\sigma' < \sigma$ by increasing the dimension. While we still require $q > \sigma \cdot \sigma'$, the size of σ' can be reduced at the cost of increasing d resp. by relying on RLWE with modulus q^d. Finally, note that we may think of Regev's original encryption scheme [Reg09] as one extreme corner of this design space (for LWE) with $d = 2\, n \log q$, where r', r'' are binary and where $e_1, e_2 = 0, 0$. That is, in the construction above, we can replace the Module-LWE assumption by the leftover hash lemma if d is sufficiently big.

References

[ABB10] Agrawal, S., Boneh, D., Boyen, X.: Efficient lattice (H)IBE in the standard model. In: Gilbert [Gil10], pp. 553–572

[ADPS16] Alkim, E., Ducas, L., Pöppelmann, T., Schwabe, P.: Post-quantum key exchange - a new hope. In: Holz, T., Savage, S. (eds.) 25th USENIX Security Symposium, USENIX Security 16, pp. 327–343. USENIX Association (2016)

[APS15] Albrecht, M.R., Player, R., Scott, S.: On the concrete hardness of learning with errors. J. Math. Cryptol. **9**(3), 169–203 (2015)

[Ban93] Banaszczyk, W.: New bounds in some transference theorems in the geometry of numbers. Math. Ann. **296**(1), 625–635 (1993)

[BCD+16] Bos, J.W., Costello, C., Ducas, L., Mironov, I., Naehrig, M., Nikolaenko, V., Raghunathan, A., Stebila, D.: Frodo: take off the ring! Practical, quantum-secure key exchange from LWE. In: Weippl, E.R., Katzenbeisser, S., Kruegel, C., Myers, A.C., Halevi, S. (eds.) ACM CCS 16: 23rd Conference on Computer and Communications Security, pp. 1006–1018. ACM Press, October 2016

[BCNS15] Bos, J.W., Costello, C., Naehrig, M., Stebila, D.: Post-quantum key exchange for the TLS protocol from the ring learning with errors problem. In: 2015 IEEE Symposium on Security and Privacy, pp. 553–570. IEEE Computer Society Press, May 2015

[BDK+17] Bos, J., Ducas, L., Kiltz, E., Lepoint, T., Lyubashevsky, V., Schanck, J.M., Schwabe, P., Stehlé, D.: CRYSTALS - kyber: a CCA-secure module-lattice-based KEM. Cryptology ePrint Archive, Report 2017/634 (2017). http://eprint.iacr.org/2017/634

[BG14] Bai, S., Galbraith, S.D.: An improved compression technique for signatures based on learning with errors. In: Benaloh, J. (ed.) CT-RSA 2014. LNCS, vol. 8366, pp. 28–47. Springer, Cham (2014). https://doi.org/10.1007/978-3-319-04852-9_2

[BGM+16] Bogdanov, A., Guo, S., Masny, D., Richelson, S., Rosen, A.: On the hardness of learning with rounding over small modulus. In: Kushilevitz, E., Malkin, T. (eds.) TCC 2016. LNCS, vol. 9562, pp. 209–224. Springer, Heidelberg (2016). https://doi.org/10.1007/978-3-662-49096-9_9

[BGV12] Brakerski, Z., Gentry, C., Vaikuntanathan, V.: (Leveled) fully homomorphic encryption without bootstrapping. In: Goldwasser, S. (ed.) ITCS 2012: 3rd Innovations in Theoretical Computer Science, pp. 309–325. Association for Computing Machinery, January 2012

[BLL+15] Bai, S., Langlois, A., Lepoint, T., Stehlé, D., Steinfeld, R.: Improved security proofs in lattice-based cryptography: using the Rényi divergence rather than the statistical distance. In: Iwata, T., Cheon, J.H. (eds.) ASIACRYPT 2015. LNCS, vol. 9452, pp. 3–24. Springer, Heidelberg (2015). https://doi.org/10.1007/978-3-662-48797-6_1

[BLP+13] Brakerski, Z., Langlois, A., Peikert, C., Regev, O., Stehlé, D.: Classical hardness of learning with errors. In: Boneh, D., Roughgarden, T., Feigenbaum, J. (eds.) 45th Annual ACM Symposium on Theory of Computing, pp. 575–584. ACM Press, June 2013

[CDPR16] Cramer, R., Ducas, L., Peikert, C., Regev, O.: Recovering short generators of principal ideals in cyclotomic rings. In: Fischlin, M., Coron, J.-S. (eds.) EUROCRYPT 2016. LNCS, vol. 9666, pp. 559–585. Springer, Heidelberg (2016). https://doi.org/10.1007/978-3-662-49896-5_20

[CDW17] Cramer, R., Ducas, L., Wesolowski, B.: Short stickelberger class relations and application to ideal-SVP. In: Coron, J.-S., Nielsen, J.B. (eds.) EUROCRYPT 2017. LNCS, vol. 10210, pp. 324–348. Springer, Cham (2017). https://doi.org/10.1007/978-3-319-56620-7_12

[CG13] Canetti, R., Garay, J.A. (eds.): CRYPTO 2013. LNCS, vol. 8042. Springer, Heidelberg (2013). https://doi.org/10.1007/978-3-642-40041-4

[CGS14] Campbell, P., Groves, M., Shepherd, D.: Soliloquy: a cautionary tale. In: ETSI 2nd Quantum-Safe Crypto Workshop, pp. 1–9 (2014)

[DDLL13] Ducas, L., Durmus, A., Lepoint, T., Lyubashevsky, V.: Lattice signatures and bimodal Gaussians. In: Canetti and Garay [CG13], pp. 40–56

[DLL+17] Ducas, L., Lepoint, T., Lyubashevsky, V., Schwabe, P., Seiler, G., Stehle, D.: CRYSTALS - Dilithium: digital signatures from module lattices. Cryptology ePrint Archive, Report 2017/633 (2017). http://eprint.iacr.org/2017/633

[DLP14] Ducas, L., Lyubashevsky, V., Prest, T.: Efficient identity-based encryption over NTRU lattices. In: Sarkar, P., Iwata, T. (eds.) ASIACRYPT 2014. LNCS, vol. 8874, pp. 22–41. Springer, Heidelberg (2014). https://doi.org/10.1007/978-3-662-45608-8_2

[Gen09] Gentry, C.: Fully homomorphic encryption using ideal lattices. In: Mitzenmacher, M. (ed.) 41st Annual ACM Symposium on Theory of Computing, pp. 169-178. ACM Press, May/June 2009

[Gen10] Gentry, C.: Toward basing fully homomorphic encryption on worst-case hardness. In: Rabin, T. (ed.) CRYPTO 2010. LNCS, vol. 6223, pp. 116–137. Springer, Heidelberg (2010). https://doi.org/10.1007/978-3-642-14623-7_7

[Gil10] Gilbert, H. (ed.): EUROCRYPT 2010. LNCS, vol. 6110. Springer, Heidelberg (2010). https://doi.org/10.1007/978-3-642-13190-5

[GPV08] Gentry, C., Peikert, C., Vaikuntanathan, V.: Trapdoors for hard lattices and new cryptographic constructions. In: Ladner, R.E., Dwork, C. (eds.) 40th Annual ACM Symposium on Theory of Computing, pp. 197–206. ACM Press, May 2008

[GSW13] Gentry, C., Sahai, A., Waters, B.: Homomorphic encryption from learning with errors: conceptually-simpler, asymptotically-faster, attribute-based. In: Canetti and Garay [CG13], pp. 75–92

[HKM17] Herold, G., Kirshanova, E., May, A.: On the asymptotic complexity of solving LWE. Des. Codes Crypt. 1–29 (2017). https://link.springer.com/journal/10623/onlineFirst/page/6

[LLM+16] Libert, B., Ling, S., Mouhartem, F., Nguyen, K., Wang, H.: Signature schemes with efficient protocols and dynamic group signatures from lattice assumptions. In: Cheon, J.H., Takagi, T. (eds.) ASIACRYPT 2016. LNCS, vol. 10032, pp. 373–403. Springer, Heidelberg (2016). https://doi.org/10.1007/978-3-662-53890-6_13

[LMPR08] Lyubashevsky, V., Micciancio, D., Peikert, C., Rosen, A.: SWIFFT: a modest proposal for FFT hashing. In: Nyberg, K. (ed.) FSE 2008. LNCS, vol. 5086, pp. 54–72. Springer, Heidelberg (2008). https://doi.org/10.1007/978-3-540-71039-4_4

[LP11] Lindner, R., Peikert, C.: Better key sizes (and attacks) for LWE-based encryption. In: Kiayias, A. (ed.) CT-RSA 2011. LNCS, vol. 6558, pp. 319–339. Springer, Heidelberg (2011). https://doi.org/10.1007/978-3-642-19074-2_21

[LPR10] Lyubashevsky, V., Peikert, C., Regev, O.: On ideal lattices and learning with errors over rings. In: Gilbert [Gil10], pp. 1–23

[LPR13] Lyubashevsky, V., Peikert, C., Regev, O.: A toolkit for ring-LWE cryptography. In: Johansson, T., Nguyen, P.Q. (eds.) EUROCRYPT 2013.

LNCS, vol. 7881, pp. 35–54. Springer, Heidelberg (2013). https://doi.org/10.1007/978-3-642-38348-9_3

[LS15] Langlois, A., Stehlé, D.: Worst-case to average-case reductions for module lattices. Des. Codes Crypt. **75**(3), 565–599 (2015)

[LSS14] Langlois, A., Stehlé, D., Steinfeld, R.: GGHLite: more efficient multilinear maps from ideal lattices. In: Nguyen, P.Q., Oswald, E. (eds.) EUROCRYPT 2014. LNCS, vol. 8441, pp. 239–256. Springer, Heidelberg (2014). https://doi.org/10.1007/978-3-642-55220-5_14

[MR09] Micciancio, D., Regev, O.: Lattice-based cryptography. In: Bernstein, D.J., Buchmann, J., Dahmen, E. (eds.) Post-Quantum Cryptography, pp. 147–191. Springer, Heidelberg (2009). https://doi.org/10.1007/978-3-540-88702-7_5

[PRSD17] Peikert, C., Regev, O., Stephens-Davidowitz, N.: Pseudorandomness of ring-LWE for any ring and modulus. In: STOC 2017 (2017)

[Reg05] Regev, O.: On lattices, learning with errors, random linear codes, and cryptography. In: Gabow, H.N., Fagin, R. (eds.) 37th Annual ACM Symposium on Theory of Computing, pp. 84–93. ACM Press, May 2005

[Reg09] Regev, O.: On lattices, learning with errors, random linear codes, and cryptography. J. ACM **56**(6), 34:1–34:40 (2009). Article no. 34

[SS13] Stehlé, D., Steinfeld, R.: Making NTRUEncrypt and NTRUSign as secure as standard worst-case problems over ideal lattices. Cryptology ePrint Archive, Report 2013/004 (2013). http://eprint.iacr.org/2013/004

[vEH14] van Erven, T., Harremos, P.: Rényi divergence and kullback-leibler divergence. IEEE Trans. Inf. Theory **60**(7), 3797–3820 (2014)

Revisiting the Expected Cost of Solving uSVP and Applications to LWE

Martin R. Albrecht[1], Florian Göpfert[2,3], Fernando Virdia[1(✉)], and Thomas Wunderer[3(✉)]

[1] Information Security Group, Royal Holloway, University of London, Egham, UK
martin.albrecht@royalholloway.ac.uk, fernando.virdia.2016@rhul.ac.uk
[2] rockenstein AG, Würzburg, Germany
fgoepfert@cdc.informatik.tu-darmstadt.de
[3] TU Darmstadt, Darmstadt, Germany
twunderer@cdc.informatik.tu-darmstadt.de

Abstract. Reducing the Learning with Errors problem (LWE) to the Unique-SVP problem and then applying lattice reduction is a commonly relied-upon strategy for estimating the cost of solving LWE-based constructions. In the literature, two different conditions are formulated under which this strategy is successful. One, widely used, going back to Gama & Nguyen's work on predicting lattice reduction (Eurocrypt 2008) and the other recently outlined by Alkim et al. (USENIX 2016). Since these two estimates predict significantly different costs for solving LWE parameter sets from the literature, we revisit the Unique-SVP strategy. We present empirical evidence from lattice-reduction experiments exhibiting a behaviour in line with the latter estimate. However, we also observe that in some situations lattice-reduction behaves somewhat better than expected from Alkim et al.'s work and explain this behaviour under standard assumptions. Finally, we show that the security estimates of some LWE-based constructions from the literature need to be revised and give refined expected solving costs.

Keywords: Cryptanalysis · Lattice-based cryptography · Learning with Errors · Lattice reduction

1 Introduction

The *Learning with Errors* problem (LWE) has attained a central role in cryptography as a key hard problem for building cryptographic constructions,

The research of Albrecht was supported by EPSRC grant "Bit Security of Learning with Errors for Post-Quantum Cryptography and Fully Homomorphic Encryption" (EP/P009417/1) and EPSRC grant "Multilinear Maps in Cryptography" (EP/L018543/1). The research of Göpfert and Wunderer was supported by the DFG as part of project P1 within the CRC 1119 CROSSING. The research of Virdia was supported by the EPSRC and the UK government as part of the Centre for Doctoral Training in Cyber Security at Royal Holloway, University of London (EP/K035584/1).

T. Takagi and T. Peyrin (Eds.): ASIACRYPT 2017, Part I, LNCS 10624, pp. 297–322, 2017.
https://doi.org/10.1007/978-3-319-70694-8_11

e.g. quantum-safe public-key encryption/key exchange and signatures schemes [Reg09, LP11, ADPS16, BG14a], fully homomorphic encryption [BV11, GSW13] and obfuscation of some families of circuits [BVWW16].

Informally, LWE asks to recover a secret vector $\mathbf{s} \in \mathbb{Z}_q^n$, given a matrix $\mathbf{A} \in \mathbb{Z}_q^{m\times n}$ and a vector $\mathbf{c} \in \mathbb{Z}_q^m$ such that $\mathbf{As} + \mathbf{e} = \mathbf{c} \bmod q$ for a short error vector $\mathbf{e} \in \mathbb{Z}_q^m$ sampled coordinate-wise from an error distribution χ. The decision variant of LWE asks to distinguish between an LWE instance $(\mathbf{A}, \mathbf{c})$ and uniformly random $(\mathbf{A}, \mathbf{c}) \in \mathbb{Z}_q^{m\times n} \times \mathbb{Z}_q^m$. To assess the security provided by a given set of parameters n, χ, q, two strategies are typically considered: the *dual* strategy finds short vectors in the lattice

$$q\Lambda^* = \left\{\mathbf{x} \in \mathbb{Z}_q^m \mid \mathbf{x} \cdot \mathbf{A} \equiv 0 \bmod q\right\},$$

i.e. it solves the *Short Integer Solutions* problem (SIS). Given such a short vector $\mathbf{v}$, we can decide if an instance is LWE by computing $\langle \mathbf{v}, \mathbf{c}\rangle = \langle \mathbf{v}, \mathbf{e}\rangle$ which is short whenever $\mathbf{v}$ and $\mathbf{e}$ are sufficiently short [MR09]. This strategy was recently revisited for small, sparse secret instances of LWE [Alb17]. The *primal* strategy finds the closest vector to $\mathbf{c}$ in the integral span of columns of $\mathbf{A}$ mod q [LP11], i.e. it solves the corresponding *Bounded Distance Decoding* problem (BDD) directly. Writing $[\mathbf{I}_n|\mathbf{A}']$ for the reduced row echelon form of $\mathbf{A}^T \in \mathbb{Z}_q^{n\times m}$ (with high probability and after appropriate permutation of columns), this task can be reformulated as solving the *unique Shortest Vector Problem* (uSVP) in the $m+1$ dimensional q-ary lattice

$$\Lambda = \mathbb{Z}^{m+1} \cdot \begin{pmatrix} \mathbf{I}_n & \mathbf{A}' & 0 \\ \mathbf{0} & q\,\mathbf{I}_{m-n} & 0 \\ \mathbf{c}^T & & t \end{pmatrix} \tag{1}$$

by Kannan's embedding [Kan87] with embedding factor t.[1] Indeed, BDD and uSVP are polynomial-time equivalent for small approximation factors up to $\sqrt{n/\log n}$ [LM09]. The lattice Λ has volume $t \cdot q^{m-n}$ and contains a vector of norm $\sqrt{\|\mathbf{e}\|^2 + t^2}$ which is unusually short, i.e. the gap between the first and second Minkowski minimum $\lambda_2(\Lambda)/\lambda_1(\Lambda)$ is large.

Alternatively, if the secret vector $\mathbf{s}$ is also short, there is a second established embedding reducing LWE to uSVP (cf. Eq. (4)). When the LWE instance under consideration is in *normal form*, i.e. the secret $\mathbf{s}$ follows the noise distribution, the geometries of the lattices in (1) and (4) are the same, which is why without loss of generality we only consider (1) in this work save for Sect. 5.

To find short vectors, lattice reduction [LLL82, Sch87, GN08a, HPS11, CN11, MW16] can be applied. Thus, to establish the cost of solving an LWE instance, we may consider the cost of lattice reduction for solving uSVP.

Two conflicting estimates for the success of lattice reduction in solving uSVP are available in the literature. The first is going back to [GN08b] and was developed in [AFG14, APS15, Gö16, HKM17] for LWE. This estimate is commonly

[1] Alternatively, we can perform lattice reduction on the q-ary lattice spanned by $\mathbf{A}^T$, i.e. the lattice spanned by the first m rows of (1), followed by an enumeration to find the closest (projected) lattice point to (the projection of) $\mathbf{c}$ [LP11, LN13].

relied upon by designers in the literature, e.g. [BG14a, CHK+17, CKLS16a, CLP17, ABB+17]. The second estimate was recently outlined in [ADPS16] and is relied upon in [BCD+16, BDK+17]. We will use the shorthand *2008 estimate* for the former and *2016 estimate* for the latter. As illustrated in Fig. 1, the predicted costs under these two estimates differ greatly. For example, considering $n = 1024$, $q \approx 2^{15}$ and χ a discrete Gaussian with standard deviation $\sigma = 3.2$, the former predicts a cost of $\approx 2^{355}$ operations, whereas the latter predicts a cost of $\approx 2^{287}$ operations in the same cost model for lattice reduction.[2]

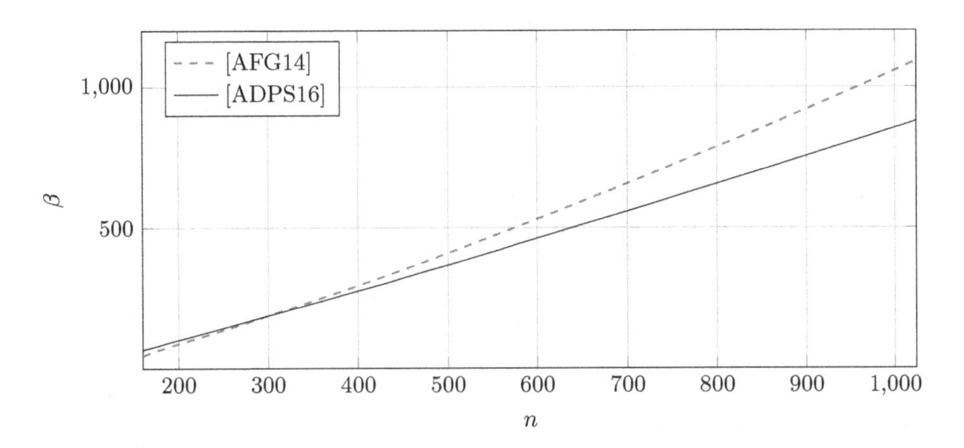

Fig. 1. Required block size β according to the estimates given in [AFG14, ADPS16] for modulus $q = 2^{15}$, standard deviation $\sigma = 3.2$ and increasing n; for [AFG14] we set $\tau = 0.3$ and $t = 1$. Lattice-reduction runs in time $2^{\Omega(\beta)}$.

Our Contribution. Relying on recent progress in publicly available lattice-reduction libraries [FPL17, FPY17], we revisit the embedding approach for solving LWE resp. BDD under some reasonable assumptions about the LWE error distribution. After some preliminaries in Sect. 2, we recall the two competing estimates from the literature in Sect. 3. Then, in Sect. 4, we expand on the exposition from [ADPS16] followed by presenting the results of running 23,000 core hours worth of lattice-reduction experiments in medium to larger block sizes β. Our results confirm that lattice-reduction largely follows the behaviour expected from the 2016 estimate [ADPS16]. However, we also find that in our experiments the attack behaves somewhat better than expected.[3] In Sect. 4.3, we then explain the observed behaviour of the BKZ algorithm under the *Geometric Series Assumption* (GSA, see below) and under the assumption that the unique

[2] Assuming that an SVP oracle call in dimension β costs $2^{0.292\,\beta+16.4}$ [BDGL16, APS15], where $+16.4$ takes the place of $o(\beta)$ from the asymptotic formula and is based on experiments in [Laa14].

[3] We note that this deviation from the expectation has a negligible impact on security estimates for cryptographic parameters.

shortest vector is distributed in a random direction relative to the rest of the basis. Finally, using the 2016 estimate, we show that some proposed parameters from the literature need to be updated to maintain the currently claimed level of security in Sect. 5. In particular, we give reduced costs for solving the LWE instances underlying TESLA [ABB+17] and the somewhat homomorphic encryption scheme in [BCIV17]. We also show that under the revised, corrected estimate, the primal attack performs about as well on SEAL v2.1 parameter sets as the dual attack from [Alb17].

2 Preliminaries

We write vectors in lower-case bold, e.g. $\mathbf{a}$, and matrices in upper-case bold, e.g. $\mathbf{A}$. We write $\langle\cdot,\cdot\rangle$ for the inner products and $\cdot$ for matrix-vector products. By abuse of notation we consider vectors to be row resp. column vectors depending on context, such that $\mathbf{v}\cdot\mathbf{A}$ and $\mathbf{A}\cdot\mathbf{v}$ are meaningful. We write $\mathbf{I}_m$ for the $m \times m$ identity matrix over whichever base ring is implied from context. We write $\mathbf{0}_{m\times n}$ for the $m \times n$ all zero matrix. If the dimensions are clear from the context, we may omit the subscripts.

2.1 Learning with Errors

The Learning with Errors (LWE) problem is defined as follows.

Definition 1 (LWE [Reg09]). *Let n, q be positive integers, χ be a probability distribution on $\mathbb{Z}$ and $\mathbf{s}$ be a secret vector in $\mathbb{Z}_q^n$. We denote by $L_{\mathbf{s},\chi}$ the probability distribution on $\mathbb{Z}_q^n \times \mathbb{Z}_q$ obtained by choosing $\mathbf{a} \in \mathbb{Z}_q^n$ uniformly at random, choosing $e \in \mathbb{Z}$ according to χ and considering it in $\mathbb{Z}_q$, and returning $(\mathbf{a}, c) = (\mathbf{a}, \langle\mathbf{a},\mathbf{s}\rangle + e) \in \mathbb{Z}_q^n \times \mathbb{Z}_q$.*

Decision-LWE *is the problem of deciding whether pairs $(\mathbf{a}, c) \in \mathbb{Z}_q^n \times \mathbb{Z}_q$ are sampled according to $L_{\mathbf{s},\chi}$ or the uniform distribution on $\mathbb{Z}_q^n \times \mathbb{Z}_q$.*

Search-LWE *is the problem of recovering $\mathbf{s}$ from $(\mathbf{a}, c) = (\mathbf{a}, \langle\mathbf{a},\mathbf{s}\rangle + e) \in \mathbb{Z}_q^n \times \mathbb{Z}_q$ sampled according to $L_{\mathbf{s},\chi}$.*

We may write LWE instances in matrix form $(\mathbf{A}, \mathbf{c})$, where rows correspond to samples $(\mathbf{a}_i, c_i)$. In many instantiations, χ is a discrete Gaussian distribution with standard deviation σ. Throughout, we denote the number of LWE samples considered as m. Writing $\mathbf{e}$ for the vector of error terms, we expect $\|\mathbf{e}\| \approx \sqrt{m}\sigma$.

2.2 Lattices

A lattice is a discrete subgroup of $\mathbb{R}^d$. Throughout, d denotes the dimension of the lattice under consideration and we only consider full rank lattices, i.e., lattices $\Lambda \subset \mathbb{R}^d$ such that $\mathsf{span}_{\mathbb{R}}(\Lambda) = \mathbb{R}^d$. A lattice $\Lambda \subset \mathbb{R}^d$ can be represented by a basis $\mathbf{B} = \{\mathbf{b}_1, \ldots, \mathbf{b}_k\}$, i.e., $\mathbf{B}$ is linearly independent and $\Lambda = \mathbb{Z}\mathbf{b}_1 + \cdots + \mathbb{Z}\mathbf{b}_k$. We write $\mathbf{b}_i$ for basis vectors and $\mathbf{b}_i^*$ for the corresponding Gram-Schmidt vectors.

We write $\Lambda(\mathbf{B})$ for the lattice generated by the rows of the matrix $\mathbf{B}$, i.e. all integer-linear combinations of the rows of $\mathbf{B}$. The volume of a lattice $\text{Vol}(\Lambda)$ is the absolute value of the determinant of any basis and it holds that $\text{Vol}(\Lambda) = \prod_{i=1}^{d} \|\mathbf{b}_i^*\|$. We write $\lambda_i(\Lambda)$ for *Minkowski's successive minima*, i.e. the radius of the smallest ball centred around zero containing i linearly independent lattice vectors. The *Gaussian Heuristic* predicts

$$\lambda_1(\Lambda) \approx \sqrt{\frac{d}{2\pi e}} \text{Vol}(\Lambda)^{1/d}.$$

For a lattice basis $\mathbf{B} = \{\mathbf{b}_1, \ldots, \mathbf{b}_d\}$ and for $i \in \{1, \ldots, d\}$ let $\pi_{\mathbf{B},i}(\mathbf{v})$ denote the orthogonal projection of $\mathbf{v}$ onto $\{\mathbf{b}_1, \ldots, \mathbf{b}_{i-1}\}$, where $\pi_{\mathbf{B},1}$ is the identity. We extend the notation to sets of vectors in the natural way. Since usually the basis $\mathbf{B}$ is clear from the context, we omit it in the notation and simply write π_i instead of $\pi_{\mathbf{B},i}$. Since Sect. 4.3 relies heavily on size reduction, we recall its definition and reproduce the algorithm in Algorithm 1.

Definition 2. *Let* $\mathbf{B}$ *be a basis,* $\mathbf{b}_i^*$ *its Gram-Schmidt vectors and*

$$\mu_{i,j} = \langle \mathbf{b}_i, \mathbf{b}_j^* \rangle / \langle \mathbf{b}_j^*, \mathbf{b}_j^* \rangle,$$

then $\mathbf{B}$ *basis is* size reduced *if* $|\mu_{i,j}| \leq 1/2$ *for* $1 \leq j \leq i \leq n$.

Data: lattice basis $\mathbf{B}$
Data: top index i
Data: start index $1 \leq s < i$
1 **for** j *from* $i-1$ *to* s **do**
2 $\quad \mu_{ij} \leftarrow \langle \mathbf{b}_i, \mathbf{b}_j^* \rangle / \langle \mathbf{b}_j^*, \mathbf{b}_j^* \rangle$;
3 $\quad \mathbf{b}_i \leftarrow \mathbf{b}_i - \lfloor \mu_{ij} \rceil \mathbf{b}_j$;
4 **end**

Algorithm 1: Size reduction

2.3 Lattice Reduction

Informally, lattice reduction is the process of improving the quality of a lattice basis. To express the output quality of a lattice reduction, we may relate the shortest vector in the output basis to the volume of the lattice in the *Hermite-factor regime* or to the shortest vector in the lattice, in the *approximation-factor regime*. Note that any algorithm finding a vector with approximation-factor α can be used to solve Unique-SVP with a gap $\lambda_2(\Lambda)/\lambda_1(\Lambda) < \alpha$.

The best known theoretical bound for lattice reduction is attained by Slide reduction [GN08a]. In this work, however, we consider the BKZ algorithm (more precisely: BKZ 2.0 [Che13], cf. Sect. 4.2) which performs better in practice. The BKZ-β algorithm repeatedly calls an SVP oracle for finding (approximate) shortest vectors in dimension or *block size* β. It has been shown that after polynomially many calls to the SVP oracle, the basis does not change much more [HPS11]. After BKZ-β reduction, we call the basis *BKZ-β reduced* and in the Hermite-factor regime assume [Che13] that this basis contains a vector of length $\|\mathbf{b}_1\| = \delta_0^d \cdot \mathrm{Vol}(L)^{1/d}$ where

$$\delta_0 = (((\pi\beta)^{1/\beta}\beta)/(2\pi e))^{1/(2(\beta-1))}.$$

Furthermore, we generally assume that for a BKZ-β reduced basis of $\Lambda(\mathbf{B})$ the Geometric Series Assumption holds.

Definition 3 (Geometric Series Assumption [Sch03]**).** *The norms of the Gram-Schmidt vectors after lattice reduction satisfy*

$$\|\mathbf{b}_i^*\| = \alpha^{i-1} \cdot \|\mathbf{b}_1\| \text{ for some } 0 < \alpha < 1.$$

Combining the GSA with the root-Hermite factor $\|\mathbf{b}_1\| = \delta_0^d \cdot \mathrm{Vol}(\Lambda)^{1/d}$ and $\mathrm{Vol}(\Lambda) = \prod_{i=1}^d \|\mathbf{b}_i^*\|$, we get $\alpha = \delta_0^{-2d/(d-1)} \approx \delta_0^{-2}$ for the GSA.

3 Estimates

As highlighted above, two competing estimates exist in the literature for when block-wise lattice reduction will succeed in solving uSVP instances such as (1).

3.1 2008 Estimate

A first systematic experimental investigation into the behavior of lattice reduction algorithms LLL, DEEP and BKZ was provided in [GN08b]. In particular, [GN08b] investigates the behavior of these algorithms for solving Hermite-SVP, Approx-SVP and Unique-SVP for families of lattices used in cryptography.

For Unique-SVP, the authors performed experiments in small block sizes on two classes of semi-orthogonal lattices and on Lagarias-Odlyzko lattices [LO83], which permit to estimate the gap $\lambda_2(\Lambda)/\lambda_1(\Lambda)$ between the first and second minimum of the lattice. For all three families, [GN08b] observed that LLL and BKZ seem to recover a unique shortest vector with high probability whenever $\lambda_2(\Lambda)/\lambda_1(\Lambda) \geq \tau\delta_0^d$, where $\tau < 1$ is an empirically determined constant that depends on the lattice family and algorithm used.

In [AFG14] an experimental analysis of solving LWE based on the same estimate was carried out for lattices of the form (1). As mentioned above, this lattice contains an unusually short vector $\mathbf{v} = (\mathbf{e} \mid t)$ of squared norm $\lambda_1(\Lambda)^2 = \|\mathbf{v}\|^2 = \|\mathbf{e}\|^2 + t^2$. Thus, when $t = \|\mathbf{e}\|$ resp. $t = 1$ this implies $\lambda_1(\Lambda) \approx \sqrt{2m}\sigma$ resp. $\lambda_1(\Lambda) \approx \sqrt{m}\sigma$, with σ the standard deviation of $\mathbf{e}_i \leftarrow_{\$} \chi$. The second minimum $\lambda_2(\Lambda)$ is assumed to correspond to the Gaussian Heuristic for the lattice. Experiments in [AFG14] using LLL and BKZ (with block sizes 5 and 10) confirmed the 2008 estimate, providing constant values for τ for lattices of the form (1), depending on the chosen algorithm, for a 10% success rate. Overall, τ was found to lie between 0.3 and 0.4 when using BKZ.

Still focusing on LWE, in [APS15] a closed formula for δ_0 is given in function of n, σ, q, τ, which implicitly assumes $t = \|\mathbf{e}\|$. In [Gö16] a bound for δ_0 in the [GN08b] model for the case of $t = 1$, which is always used in practice, is given. In [HKM17], a related closed formula is given, directly expressing the asymptotic running time for solving LWE using this approach.

3.2 2016 Estimate

In [ADPS16] an alternative estimate is outlined. The estimate predicts that $\mathbf{e}$ can be found if[4]

$$\sqrt{\beta/d}\,\|(\mathbf{e} \mid 1)\| \approx \sqrt{\beta}\sigma \le \delta_0^{2\beta-d}\,\mathrm{Vol}(\Lambda(\mathbf{B}))^{1/d}, \tag{2}$$

under the assumption that the Geometric Series Assumption holds (until a projection of the unusually short vector is found). The brief justification for this estimate given in [ADPS16] notes that this condition ensures that the projection of $\mathbf{e}$ orthogonally to the first $d - \beta$ (Gram-Schmidt) vectors is shorter than the expectation for $\mathbf{b}^*_{d-\beta+1}$ under the GSA and thus would be found by the SVP oracle when called on the last block of size β. Hence, for any β satisfying (2), the actual behaviour would deviate from that predicted by the GSA. Finally, the argument can be completed by appealing to the intuition that a deviation from expected behaviour on random instances—such as the GSA—leads to a revelation of the underlying structural, secret information.[5]

4 Solving uSVP

Given the significant differences in expected solving time under the two estimates, cf. Fig. 1, and recent progress in publicly available lattice-reduction libraries enabling experiments in larger block sizes [FPL17, FPY17], we conduct a more detailed examination of BKZ's behaviour on uSVP instances. For this, we first explicate the outline from [ADPS16] to establish the expected behaviour, which we then experimentally investigate in Sect. 4.2. Overall, our experiments confirm the expectation. However, the algorithm behaves somewhat better than expected, which we then explain in Sect. 4.3.

For the rest of this section, let $\mathbf{v}$ be a unique shortest vector in some lattice $\Lambda \subset \mathbb{R}^d$, i.e. in case of (1) we have $\mathbf{v} = (\mathbf{e} \mid t)$ where we pick $t = 1$.

4.1 Prediction

Projected norm. In what follows, we assume the unique shortest vector $\mathbf{v}$ is drawn from a spherical distribution or is at least "not too skewed" with respect

[4] [ADPS16] has $2\beta - d - 1$ in the exponent, which seems to be an error.

[5] We note that observing such a deviation implies solving Decision-LWE.

to the current basis. As a consequence, following [ADPS16], we assume that all orthogonal projections of $\mathbf{v}$ onto a k-dimensional subspace of $\mathbb{R}^d$ have expected norm $(\sqrt{k}/\sqrt{d})\,\|\mathbf{v}\|$. Note that this assumption can be dropped by adapting (2) to $\|\mathbf{v}\| \leq \delta_0^{2\beta-d}\,\mathrm{Vol}(\Lambda)^{\frac{1}{d}}$ since $\|\pi_{d-\beta+1}(\mathbf{v})\| \leq \|\mathbf{v}\|$.

Finding a projection of the short vector. Assume that β is chosen minimally such that (2) holds. When running BKZ the length of the Gram-Schmidt basis vectors of the current basis converge to the lengths predicted by the GSA. Therefore, at some point BKZ will find a basis $\mathbf{B} = \{\mathbf{b}_1, \ldots, \mathbf{b}_d\}$ of Λ for which we can assume that the GSA holds with root Hermite factor δ_0. Now, consider the stage of BKZ where the SVP oracle is called on the last full projected block of size β with respect to $\mathbf{B}$. Note that the projection $\pi_{d-\beta+1}(\mathbf{v})$ of the shortest vector is contained in the lattice

$$\Lambda_{d-\beta+1} := \Lambda\left(\pi_{d-\beta+1}(\mathbf{b}_{d-\beta+1}), \ldots, \pi_{d-\beta+1}(\mathbf{b}_d)\right),$$

since

$$\pi_{d-\beta+1}(\mathbf{v}) = \sum_{i=d-\beta+1}^{d} \nu_i \pi_{d-\beta+1}(\mathbf{b}_i) \in \Lambda_{d-\beta+1}, \text{ where } \nu_i \in \mathbb{Z} \text{ with } \mathbf{v} = \sum_{i=1}^{d} \nu_i \mathbf{b}_i.$$

By (2), the projection $\pi_{d-\beta+1}(\mathbf{v})$ is in fact expected to be the shortest non-zero vector in $\Lambda_{d-\beta+1}$, since it is shorter than the GSA's estimate for $\lambda_1(\Lambda_{d-\beta+1})$, i.e.

$$\|\pi_{d-\beta+1}(\mathbf{v})\| \approx \frac{\sqrt{\beta}}{\sqrt{d}}\,\|\mathbf{v}\| \leq \delta_0^{-2(d-\beta)+d}\mathrm{Vol}(\Lambda)^{\frac{1}{d}}.$$

Hence the SVP oracle will find $\pm\pi_{d-\beta+1}(\mathbf{v})$ and BKZ inserts

$$\mathbf{b}^{\mathsf{new}}_{d-\beta+1} = \pm \sum_{i=d-\beta+1}^{d} \nu_i \mathbf{b}_i$$

into the basis $\mathbf{B}$ at position $d - \beta + 1$, as already outlined in [ADPS16]. In other words, by finding $\pm\pi_{d-\beta+1}(\mathbf{v})$, BKZ recovers the last β coefficients $\nu_{d-\beta+1}, \ldots, \nu_d$ of $\mathbf{v}$ with respect to the basis $\mathbf{B}$.

Finding the short vector. The above argument can be extended to an argument for the full recovery of $\mathbf{v}$. Consider the case that in some tour of BKZ-β, a projection of $\mathbf{v}$ was found at index $d - \beta + 1$. Then in the following tour, by arguments analogous to the ones above, a projection of $\mathbf{v}$ will likely be found at index $d - 2\beta + 2$, since now it holds that

$$\pi_{d-2\beta+2}(\mathbf{v}) \in \Lambda_{d-2\beta+2} := \Lambda\left(\pi_{d-2\beta+2}(\mathbf{b}_{d-2\beta+2}), \ldots, \pi_{d-2\beta+2}(\mathbf{b}^{\mathsf{new}}_{d-\beta+1})\right).$$

Repeating this argument for smaller indices shows that after a few tours $\mathbf{v}$ will be recovered. Furthermore, noting that BKZ calls LLL which in turn calls size reduction, i.e. Babai's nearest plane [Bab86], at some index $i > 1$ size reduction will recover $\mathbf{v}$ from $\pi_i(\mathbf{v})$. In particular, it is well-known that size reduction (Algorithm 1) will succeed in recovering $\mathbf{v}$ whenever

$$\mathbf{v} \in \mathbf{b}_{d-\beta+1}^{\mathsf{new}} + \left\{ \sum_{i=1}^{d-\beta} c_i \cdot \mathbf{b}_i^* : c_i \in \left[-\frac{1}{2}, \frac{1}{2}\right] \right\}. \quad (3)$$

4.2 Observation

The above discussion naturally suggests a strategy to verify the expected behaviour. We have to verify that the projected norms $\|\pi_i(\mathbf{v})\| = \|\pi_i(\mathbf{e} \mid 1)\|$ do indeed behave as expected and that $\pi_{d-\beta+1}(\mathbf{v})$ is recovered by BKZ-β for the minimal $\beta \in \mathbb{N}$ satisfying (2). Finally, we have to measure when and how $\mathbf{v} = (\mathbf{e} \mid 1)$ is eventually recovered.

Thus, we ran lattice-reduction on many lattices constructed from LWE instances using Kannan's embedding. In particular, we picked the entries of $\mathbf{s}$ and $\mathbf{A}$ uniformly at random from $\mathbb{Z}_q$, the entries of $\mathbf{e}$ from a discrete Gaussian distribution with standard deviation $\sigma = 8/\sqrt{2\pi}$, and we constructed our basis as in (1) with embedding factor $t = 1$. For parameters (n, q, σ), we then estimated the minimal pair (in lexicographical order) (β, m) to satisfy (2).

Implementation. To perform our experiments, we used SageMath 7.5.1 [S+17] in combination with the `fplll` 5.1.0 [FPL17] and `fpylll` 0.2.4dev [FPY17] libraries. All experiments were run on a machine with Intel(R) Xeon(R) CPU E5-2667 v2 @ 3.30GHz cores ("strombenzin") resp. Intel(R) Xeon(R) CPU E5-2690 v4 @ 2.60GHz ("atomkohle"). Each instance was reduced on a single core, with no parallelisation.

Our BKZ implementation inherits from the implementation in `fplll` and `fpylll` of BKZ 2.0 [Che13] algorithm. As in BKZ 2.0, we restricted the enumeration radius to be approximately the size of the Gaussian Heuristic for the projected sublattice, apply recursive BKZ-β' preprocessing with a block size $\beta' < \beta$, make use of extreme pruning [GNR10] and terminate the algorithm when it stops making significant progress. We give simplified pseudocode of our implementation in Algorithm 2. We ran BKZ for at most 20 tours using `fplll`'s default pruning and preprocessing strategies and, using `fplll`'s default auto abort strategy, terminated the algorithm whenever the slope of the Gram Schmidt vectors did not improve for five consecutive tours. Additionally, we aborted if a vector of length $\approx \|\mathbf{v}\|$ was found in the basis (in line 15 of Algorithm 2).

Implementations of block-wise lattice reduction algorithms such as BKZ make heavy use of LLL [LLL82] and size reduction. This is to remove linear dependencies introduced during the algorithm, to avoid numerical stability issues and to improve the performance of the algorithm by moving short vectors to the front earlier. The main modification in our implementation is that calls to LLL during preprocessing and postprocessing are restricted to the current block, not

```
    Data: LLL-reduced lattice basis B
    Data: block size β, preprocessing block size β'
 1  repeat                                                        // tour
 2      for κ ← 1 to d do                                         // step_κ
 3          size reduction from index 1 to κ (inclusive);
 4          ℓ ← ||b*_κ||;
            // extreme pruning + recursive preprocessing
 5          repeat until termination condition met
 6              rerandomise π_κ(b_{κ+1}, ..., b_{κ+β−1});
 7              LLL on π_κ(b_κ, ..., b_{κ+β−1});
 8              BKZ-β' on π_κ(b_κ, ..., b_{κ+β−1});
 9              v ← SVP on π_κ(b_κ, ..., b_{κ+β−1});
10              if v ≠ ⊥ then
11                  extend B by inserting v into B at index κ + β;
12                  LLL on π_κ(b_κ, ..., b_{κ+β}) to remove linear dependencies;
13                  drop row with all zero entries;
14              end
15          size reduction from index 1 to κ (inclusive);
16          if ℓ = ||b*_κ|| then
17              yield ⊤;
18          else
19              yield ⊥;
20          end
21      end
22      if ⊤ for all κ then
23          return;
24      end
```

Algorithm 2: Simplified BKZ 2.0 Algorithm

touching any other vector, to aid analysis. That is, in Algorithm 2, LLL is called in lines 7 and 12 and we modified these LLL calls not to touch any row with index smaller than κ, not even to perform size reduction.

As a consequence, we only make use of vectors with index smaller than κ in lines 3 and 15. Following the implementations in [FPL17,FPY17], we call size reduction from index 1 to κ before (line 3) and after (line 15) the innermost loop with calls to the SVP oracle. These calls do not appear in the original description of BKZ. However, since the innermost loop re-randomises the basis when using extreme pruning, the success condition of the original BKZ algorithm needs to be altered. That is, the algorithm cannot break the outer loop once it makes no more changes as originally specified. Instead, the algorithm terminates if it does not find a shorter vector at any index κ. Now, the calls to size reduction ensure that the comparison at the beginning and end of each step κ is meaningful even when the Gram-Schmidt vectors are only updated lazily in the underlying implementation. That is, the call to size reduction triggers an internal update of the underlying Gram-Schmidt vectors and are hence implementation artefacts. The reader may think of these size reduction calls as explicating calls otherwise hidden behind calls to LLL and we stress that our analysis applies to BKZ as commonly implemented, our changes merely enable us to more easily predict and experimentally verify the behaviour.

We note that the break condition for the innermost loop at line 5 depends on the pruning parameters chosen, which control the success probability of enumeration. Since it does not play a material role in our analysis, we simply state that some condition will lead to a termination of the innermost loop.

Finally, we recorded the following information. At the end of each step κ during lattice reduction, we recorded the minimal index i such that $\pi_i(\mathbf{v})$ is in $\mathrm{span}(\mathbf{b}_1, \ldots, \mathbf{b}_i)$ and whether $\pm\mathbf{v}$ itself is in the basis. In particular, to find the index i in the basis $\mathbf{B}$ of $\pi_i(\mathbf{v})$ given $\mathbf{v}$, we compute the coefficients of $\mathbf{v}$ in basis $\mathbf{B}$ (at the current step) and pick the first index i such that all coefficients with larger indices are zero. Then, we have $\pi_i(\mathbf{b}_i) = c \cdot \pi_i(\mathbf{v})$ for some $c \in \mathbb{R}$. From the algorithm, we expect to have found $\pm\pi_i(\mathbf{b}_i) = \pi_i(\mathbf{v})$ and call i the index of the projection of $\mathbf{v}$.

Results. In Fig. 2, we plot the average norms of $\pi_i(\mathbf{v})$ against the expectation $\sqrt{d-i+1}\,\sigma \approx \sqrt{\frac{d-i+1}{d}}\sqrt{m \cdot \sigma^2 + 1}$, indicating that $\sqrt{d-i+1}\,\sigma$ is a close approximation of the expected lengths except perhaps for the last few indices.

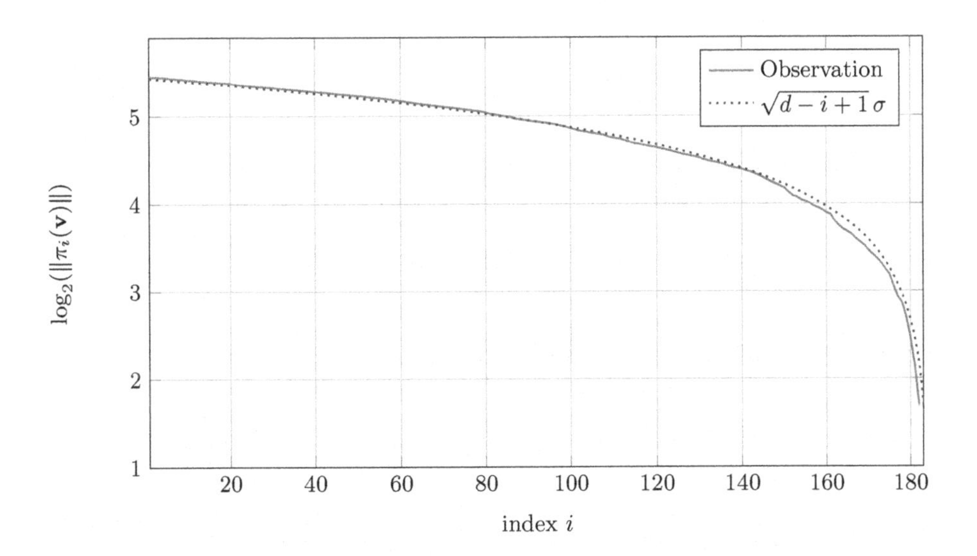

Fig. 2. Expected and average observed norms $\|\pi_i(\mathbf{v})\|$ for 16 bases (LLL-reduced) and vectors $\mathbf{v}$ of dimension $d = m+1$ and volume q^{m-n} with LWE parameters $n = 65, m = 182, q = 521$ and standard deviation $\sigma = 8/\sqrt{2\pi}$.

Recall that, as illustrated in Fig. 3, we expect to find the projection $\pi_{d-\beta+1}(\mathbf{v})$ when (β, d) satisfy (2), eventually leading to a recovery of $\mathbf{v}$, say, by an extension of the argument for the recovery of $\pi_{d-\beta+1}(\mathbf{v})$. Our experiments, summarised in

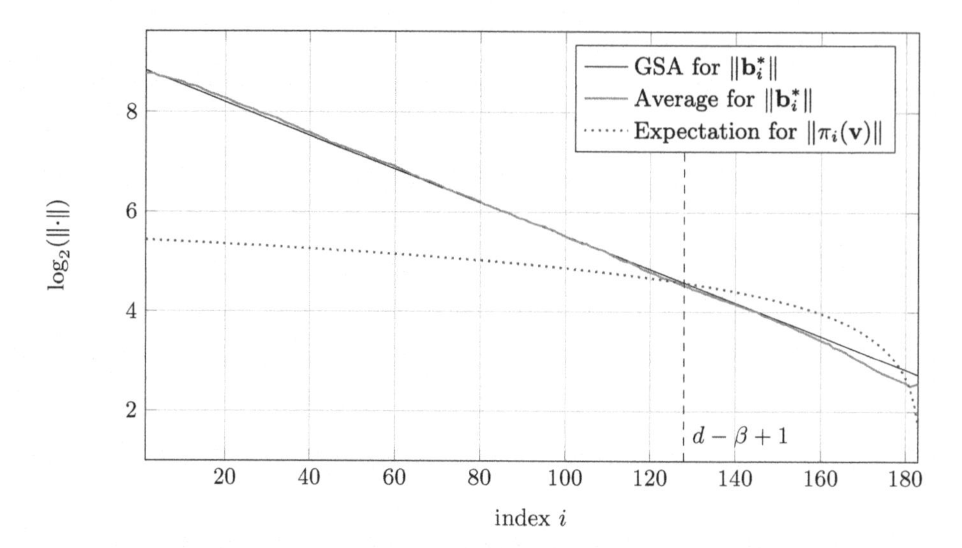

Fig. 3. Expected and observed norms for lattices of dimension $d = m + 1 = 183$ and volume q^{m-n} after BKZ-β reduction for LWE parameters $n = 65, m = 182, q = 521$ and standard deviation $\sigma = 8/\sqrt{2\pi}$ and $\beta = 56$ (minimal (β, m) such that (2) holds). Average of Gram-Schmidt lengths is taken over 16 BKZ-β reduced bases of random q-ary lattices, i.e. *without* an unusually short vector.

Table 1, show a related, albeit not identical behaviour. Defining a cut-off index $c = d - 0.9\beta + 1$ and considering $\pi_\kappa(\mathbf{v})$ for $\kappa < c$, we observe that the BKZ algorithm typically first recovers $\pi_\kappa(\mathbf{v})$ which is immediately followed by the recovery of $\mathbf{v}$ in the same step. In more detail, in Fig. 4 we show the measured probability distribution of the index κ such that $\mathbf{v}$ is recovered from $\pi_\kappa(\mathbf{v})$ in the same step. Note that the mean of this distribution is smaller than $d - \beta + 1$. We explain this bias in Sect. 4.3.

The recovery of $\mathbf{v}$ from $\pi_\kappa(\mathbf{v})$ can be effected by one of three subroutines: either by a call to LLL, by a call to size reduction, or by a call to enumeration that recovers $\mathbf{v}$ directly. Since LLL itself contains many calls to size reduction, and enumeration being lucky is rather unlikely, size reduction is a good place to start the investigation. Indeed, restricting the LLL calls in Algorithm 2 as outlined in Sect. 2.3, identifies that size reduction suffices. That is, to measure the success rate of size reduction recovering $\mathbf{v}$ from $\pi_\kappa(\mathbf{v})$, we observe size reduction acting on $\pi_\kappa(\mathbf{v})$. Here, we consider size reduction to fail in recovering $\mathbf{v}$ if it does not recover $\mathbf{v}$ given $\pi_\kappa(\mathbf{v})$ for $\kappa < c$ with $c = d - 0.9\beta + 1$, regardless of whether $\mathbf{v}$ is finally recovered at a later point either by size reduction on a new projection, or by some other call in the algorithm such as an SVP oracle call at a smaller index. As shown in Table 1, size reduction's success rate is close to 1. Note that the cut-off index c serves to limit underestimating the success rate: intuitively we do not expect size reduction to succeed when starting from a projection with larger index, such as $\pi_{d-\gamma+1}(\mathbf{v})$ with $\gamma < 10$. We discuss this in Sect. 4.3.

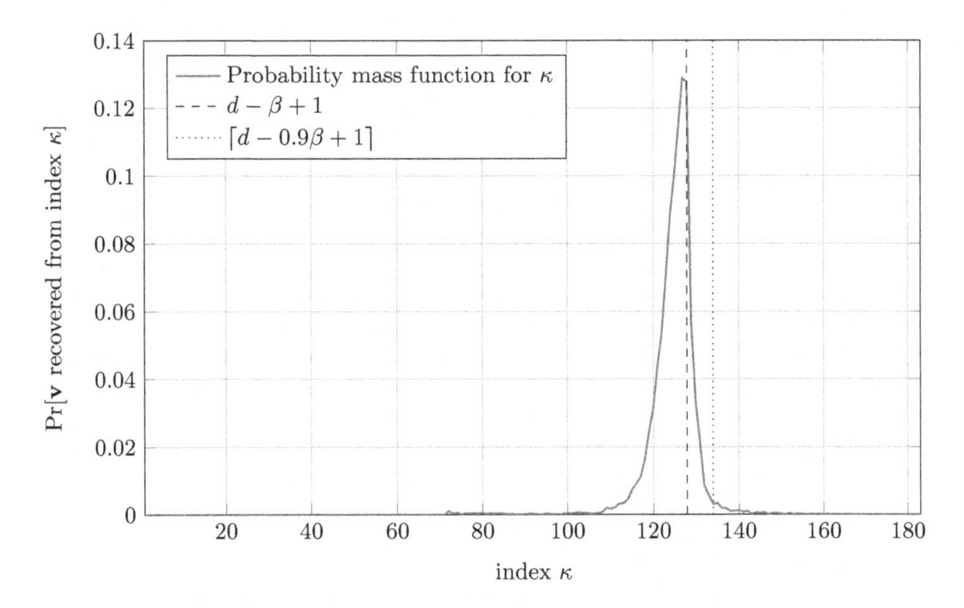

Fig. 4. Probability mass function of the index κ from which size reduction recovers $\mathbf{v}$, calculated over 10,000 lattice instances with LWE parameters $n = 65, m = 182, q = 521$ and standard deviation $\sigma = 8/\sqrt{2\pi}$, reduced using $\beta = 56$. The mean of the distribution is $\approx$124.76 while $d - \beta + 1 = 128$.

Overall, Table 1 confirms the prediction from [ADPS16]: picking $\beta = \beta_{2016}$ to be the block size predicted by the 2016 estimate leads to a successful recovery of $\mathbf{v}$ with high probability.

4.3 Explaining Observation

As noted above, our experiments indicate that the algorithm behaves better than expected by (2). Firstly, the BKZ algorithm does not necessarily recover a projection of $\mathbf{v}$ at index $d - \beta + 1$. Instead, the index κ at which we recover a projection $\pi_\kappa(\mathbf{v})$ follows a distribution with a centre below $d - \beta + 1$, cf. Fig. 4. Secondly, size reduction usually immediately recovers $\mathbf{v}$ from $\pi_\kappa(\mathbf{v})$. This is somewhat unexpected, since we do not have the guarantee that $|c_i| \leq 1/2$ as required in the success condition of size reduction given in (3).

Finding the projection. To explain the bias towards a recovery of $\pi_\kappa(\mathbf{v})$ for some $\kappa < d - \beta + 1$, note that if (2) holds then for the parameter sets *in our experiments* the lines for $\|\pi_i(\mathbf{v})\|$ and $\|\mathbf{b}_i^*\|$ intersect twice (cf. Fig. 3). Let $d - \gamma + 1$ be the index of the second intersection. Thus, there is a good chance that $\|\pi_{d-\gamma+1}(\mathbf{v})\|$ is a shortest vector in the lattice spanned by the last projected block of some small rank γ and will be placed at index $d-\gamma+1$. As a consequence,

Table 1. Overall success rate ("$\mathbf{v}$") and success rate of size reduction ("same step") for solving LWE instances characterised by n, σ, q with m samples, standard deviation $\sigma = 8/\sqrt{2\pi}$, minimal (β_{2016}, m_{2016}) such that $\sqrt{b_{2016}}\,\sigma \leq \delta_0^{2\beta_{2016}-(m_{2016}+1)} q^{(m_{2016}-n)/(m_{2016}+1)}$ with δ_0 in function of β_{2016}. The column "β" gives the actual block size used in experiments. The "same step" rate is calculated over all successful instances where $\mathbf{v}$ is found before the cut-off point c and for the instances where exactly $\pi_{d-b+1}(\mathbf{v})$ is found (if no such instance is found, we do not report a value). In the second case, the sample size is smaller, since not all instances recover $\mathbf{v}$ from exactly $\kappa = d - \beta + 1$. The column "time" lists average solving CPU time for one instance, in seconds. Note that our changes to the algorithm and our extensive record keeping lead to an increased running time of the BKZ algorithm compared to [FPL17,FPY17]. Furthermore, the occasional longer running time for smaller block sizes is explained by the absence of early termination when $\mathbf{v}$ is found.

n	q	β_{2016}	m_{2016}	β	#	$\mathbf{v}$	Same step		Time
							$\kappa < c$	$\kappa = d - \beta + 1$	
65	521	56	182	56	10000	93.3%	99.7%	99.7%	1,131.4
				51		52.8%	98.8%	97.3%	1,359.3
				46		4.8%	96.4%	85.7%	1,541.2
80	1031	60	204	60	1000	94.2%	99.6%	100.0%	2,929.0
				55		60.6%	99.3%	96.5%	2,458.5
				50		8.9%	97.6%	100.0%	1,955.0
				45		0.2%	100.0%	—	1,568.1
100	2053	67	243	67	500	88.8%	99.8%	100.0%	28,803.7
				62		39.6%	99.5%	100.0%	19,341.9
				57		5.8%	100.0%	100.0%	7,882.2
				52		0.2%	0.0%	—	3,227.0
108	2053	77	261	77	5	100.0%	100.0%	100.0%	351,094.2
110	2053	78	272	78	5	100.0%	100.0%	100.0%	1,012,634.8

all projections $\pi_i(\mathbf{v})$ with $i > d - \gamma + 1$ will be zero and $\pi_{d-\beta-\gamma+1}(\mathbf{v})$ will be contained in the β-dimensional lattice

$$\Lambda_{d-\beta-\gamma+1} := \Lambda\left(\pi_{d-\beta-\gamma+1}(\mathbf{b}_{d-\beta-\gamma+1}), \ldots, \pi_{d-\beta-\gamma+1}(\mathbf{b}_{d-\gamma+1})\right),$$

enabling it to be recovered by BKZ-β at an index $d - \beta - \gamma + 1 < d - \beta + 1$. Thus, BKZ in our experiments behaves better than predicted by (2). We note that another effect of this second intersection is that, for very few instances, it directly leads to a recovery of $\mathbf{v}$ from $\pi_{d-\beta-\gamma+1}(\mathbf{v})$.

Giving a closed formula incorporating this effect akin to (2) would entail to predict the index γ and then replace β with $\beta+\gamma$ in (2). However, as illustrated in Fig. 3, neither does the GSA hold for the last 50 or so indices of the basis [Che13] nor does the prediction $\sqrt{d-i+1}\,\sigma$ for $\|\pi_{d-1+1}(\mathbf{v})\|$.

We stress that while the second intersection often occurs for parameter sets within reach of practical experiments, it does not always occur for all parameter sets. That is, for many large parameter sets (n, α, q), e.g. those in [ADPS16], a choice of β satisfy (2) does *not* lead to a predicted second intersection at some larger index. Thus, this effect may highlight the pitfalls of extrapolating experimental lattice-reduction data from small instances to large instances.

Finding the short vector. In what follows, we assume that the projected norm $\|\pi_{d-k}(\mathbf{v})\|$ is indeed equal to this expected norm (cf. Fig. 2). We further assume that $\pi_i(\mathbf{v})$ is distributed in a random direction with respect to the rest of the basis. This assumption holds for LWE where the vector $\mathbf{e}$ is sampled from a (near) spherical distribution. We also note that we can rerandomise the basis and thus the relative directions. Under this assumption, we show that size reduction recovers the short vector $\mathbf{v}$ with high probability. More precisely, we show:

Claim 1. *Let $\mathbf{v} \in \Lambda \subset \mathbb{R}^d$ be a unique shortest vector and $\beta \in \mathbb{N}$. Assume that (2) holds, the current basis is $\mathbf{B} = \{\mathbf{b}_1, \ldots, \mathbf{b}_d\}$ such that $\mathbf{b}_\kappa^* = \pi_\kappa(\mathbf{v})$ for $\kappa = d - \beta + 1$ and*

$$\mathbf{v} = \mathbf{b}_k + \sum_{i=1}^{k-1} \nu_i \mathbf{b}_i$$

for some $\nu_i \in \mathbb{Z}$, and the GSA holds for $\mathbf{B}$ until index κ. If the size reduction step of BKZ-β is called on $\mathbf{b}_\kappa$, it recovers $\mathbf{v}$ with high probability over the randomness of the basis.

Note that if BKZ has just found a projection of $\mathbf{v}$ at index κ, the current basis is as required by Claim 1. Now, let $\nu_i \in \mathbb{Z}$ denote the coefficients of $\mathbf{v}$ with respect to the basis $\mathbf{B}$, i.e.

$$\mathbf{v} = \mathbf{b}_{d-\beta+1} + \sum_{i=1}^{d-\beta} \nu_i \mathbf{b}_i.$$

Let $\mathbf{b}_{d-\beta+1}^{(d-\beta+1)} = \mathbf{b}_{d-\beta+1}$, where the superscript denotes a step during size reduction. For $i = d - \beta, d - \beta - 1, \ldots, 1$ size-reduction successively finds $\mu_i \in \mathbb{Z}$ such that

$$\mathbf{w}_i = \mu_i \pi_i(\mathbf{b}_i) + \pi_i(\mathbf{b}_{d-\beta+1}^{(i+1)}) = \mu_i \mathbf{b}_i^* + \pi_i(\mathbf{b}_{d-\beta+1}^{(i+1)})$$

is the shortest element in the coset

$$L_i := \{\mu \mathbf{b}_i^* + \pi_i(\mathbf{b}_{d-\beta+1}^{(i+1)}) | \mu \in \mathbb{Z}\}$$

and sets

$$\mathbf{b}_{d-\beta+1}^{(i)} := \mu_i \mathbf{b}_i + \mathbf{b}_{d-\beta+1}^{(i+1)}.$$

Note that if $\mathbf{b}_{d-\beta+1}^{(i+1)} = \mathbf{b}_{d-\beta+1} + \sum_{j=i+1}^{d-\beta} \nu_j \mathbf{b}_j$, as in the first step $i = d - \beta$, then we have that

$$\pi_i(\mathbf{v}) = \nu_i \mathbf{b}_i^* + \pi_i(\mathbf{b}_{d-\beta+1}^{(i+1)}) \in L_i$$

is contained in L_i and hence

$$L_i = \pi_i(\mathbf{v}) + \mathbb{Z}\mathbf{b}_i^*.$$

If the projection $\pi_i(\mathbf{v})$ is in fact the shortest element in L_i, for the newly defined vector $\mathbf{b}_{d-\beta+1}^{(i)}$ it also holds that

$$\mathbf{b}_{d-\beta+1}^{(i)} = \nu_i \mathbf{b}_i + \mathbf{b}_{d-\beta+1}^{(i+1)} = \mathbf{b}_{d-\beta+1} + \sum_{j=i}^{d-\beta} \nu_j \mathbf{b}_j.$$

Hence, if $\pi_i(\mathbf{v})$ is the shortest element in L_i for all i, size reduction finds the shortest vector

$$\mathbf{v} = \mathbf{b}_{d-\beta+1}^{(1)}$$

and inserts it into the basis at position $d - \beta + 1$, replacing $\mathbf{b}_{d-\beta+1}$.

It remains to argue that with high probability p for every i we have that the projection $\pi_i(\mathbf{v})$ is the shortest element in L_i. The success probability p is given by

$$p = \prod_{i=1}^{d-\beta} p_i,$$

where the probabilities p_i are defined as

$$p_i = \Pr\left[\pi_i(\mathbf{v}) \text{ is the shortest element in } \pi_i(\mathbf{v}) + \mathbb{Z}\mathbf{b}_i^*\right].$$

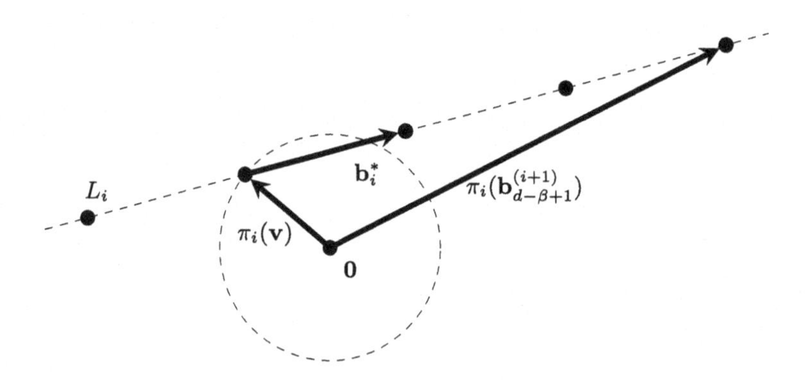

Fig. 5. Illustration of a case such that $\pi_i(\mathbf{v})$ is the shortest element on L_i.

For each i the probability p_i is equal to the probability that

$$\|\pi_i(\mathbf{v})\| < \min\{\|\pi_i(\mathbf{v}) + \mathbf{b}_i^*\|, \|\pi_i(\mathbf{v}) - \mathbf{b}_i^*\|\}$$

as illustrated in Fig. 5. To approximate the probabilities p_i, we model them as follows. By assumption, we have

$$r_i := \|\pi_i(\mathbf{v})\| = (\sqrt{d-i+1}/\sqrt{d})\,\|\mathbf{v}\| \text{ and } R_i := \|\mathbf{b}_i^*\| = \delta_0^{-2(i-1)+d}\mathrm{Vol}(\Lambda)^{\frac{1}{d}},$$

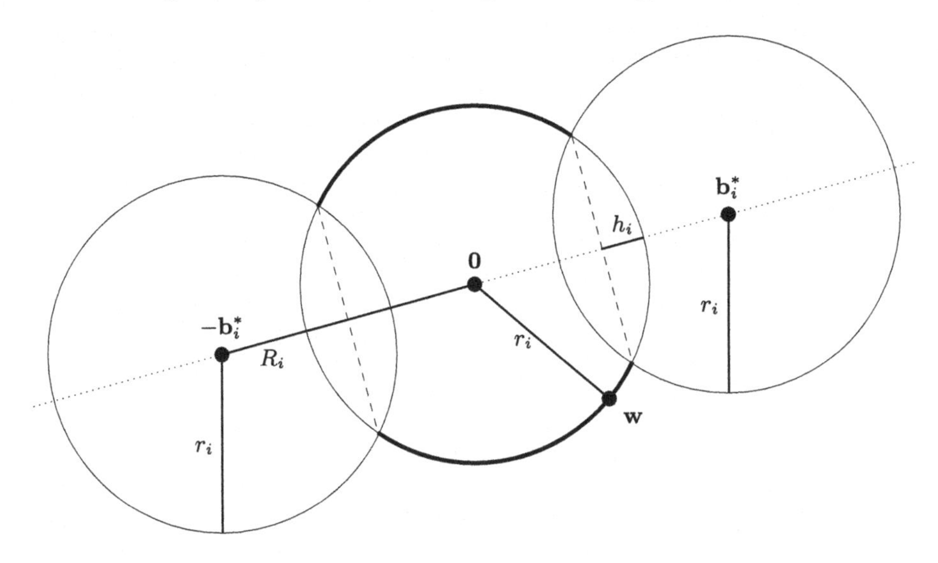

Fig. 6. Illustration of the success probability p_i in $\mathbb{R}^2$. If $\mathbf{w}$ is on the thick part of the circle, step i of size reduction is successful.

and that $\pi_i(\mathbf{v})$ is uniformly distributed with norm r_i. We can therefore model p_i as described in the following and illustrated in Fig. 6.

Pick a point $\mathbf{w}$ with norm r_i uniformly at random. Then the probability p_i is approximately the probability that $\mathbf{w}$ is closer to $\mathbf{0}$ than it is to $\mathbf{b}_i^*$ and to $-\mathbf{b}_i^*$, i.e.

$$r_i < \min\{\|\mathbf{w} - \mathbf{b}_i^*\|, \|\mathbf{w} + \mathbf{b}_i^*\|\}.$$

Calculating this probability leads to the following approximation of p_i

$$p_i \approx \begin{cases} 1 - \frac{2A_{d-i+1}(r_i,h_i)}{A_{d-i+1}(r_i)} & \text{if } R_i < 2r_i \\ 1 & \text{if } R_i \geq 2r_i \end{cases},$$

where $A_{d-i+1}(r_i)$ is the surface area of the sphere in $\mathbb{R}^{d-i+1}$ with radius r_i and $A_{d-i+1}(r_i, h_i)$ is the surface area of the hyperspherical cap of the sphere in $\mathbb{R}^{d-i+1}$ with radius r_i of height h_i with $h_i = r_i - R_i/2$. Using the formulas provided in [Li11], an easy calculation leads to

$$p_i \approx \begin{cases} 1 - \frac{\int_0^{2\frac{h_i}{r_i} - \left(\frac{h_i}{r_i}\right)^2} t^{((d-i)/2)-1}(1-t)^{-1/2}dt}{B(\frac{d-i}{2}, \frac{1}{2})} & \text{if } R_i < 2r_i \\ 1 & \text{if } R_i \geq 2r_i \end{cases},$$

where $B(\cdot, \cdot)$ denotes the Euler beta function. Note that $R_i \geq 2r_i$ corresponds to (3).

Estimated success probabilities p for different block sizes β are plotted in Fig. 7. Note that if we assume equality holds in (2), the success probability p only depends on the block size β and not on the specific lattice dimension, volume of the lattice, or the length of the unique short vector, since then the ratios between the predicted norms $\|\pi_{d-\beta+1-k}(\mathbf{v})\|$ and $\left\|\mathbf{b}^*_{d-\beta+1-k}\right\|$ only depend on β for all $k = 1, 2, \ldots$, since

$$\frac{\|\pi_{d-\beta+1-k}(\mathbf{v})\|}{\left\|\mathbf{b}^*_{d-\beta+1-k}\right\|} = \frac{\frac{\sqrt{\beta}\sqrt{\beta+k}}{\sqrt{\beta}\sqrt{d}}\|\mathbf{v}\|}{\delta_0^{2(\beta+k)-d}\operatorname{Vol}(\Lambda)^{\frac{1}{d}}} = \frac{\frac{\sqrt{\beta+k}}{\sqrt{\beta}}\delta_0^{2\beta-d}\operatorname{Vol}(\Lambda)^{\frac{1}{d}}}{\delta_0^{2(\beta+k)-d}\operatorname{Vol}(\Lambda)^{\frac{1}{d}}} = \frac{\sqrt{\beta+k}}{\sqrt{\beta}}\delta_0^{-2k}$$

and the estimated success probability only depends on these ratios.

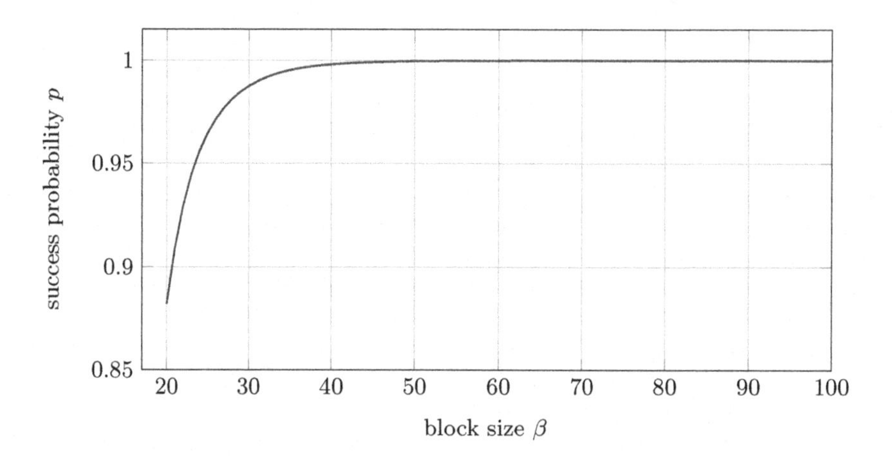

Fig. 7. Estimated success probability p for varying block sizes β, assuming β is chosen minimal such that (2) holds.

The prediction given in Fig. 7 is in line with the measured probability of finding $\mathbf{v}$ in the same step when its projection $\pi_{d-\beta+1}(\mathbf{v})$ is found as reported in Table 1 for $\beta = \beta_{2016}$ and $m = m_{2016}$. Finally, note that by the above analysis we do not expect to recover $\mathbf{v}$ from a projection $\pi_{d-\gamma+1}(\mathbf{v})$ for some small $\gamma \ll \beta$ except with small probability.

5 Applications

Section 4 indicates that (2) is a reliable indicator for when lattice-reduction will succeed in recovering an unusually short vector. Furthermore, as illustrated in Fig. 1, applying (2) lowers the required block sizes compared to the 2008 model which is heavily relied upon in the literature. Thus, in this section we evaluate the impact of applying the revised estimates to various parameter sets from the literature. Indeed, for many schemes we find that their parameters need to be adapted to maintain the currently claimed level of security.

Many of the schemes considered below feature an unusually short secret $\mathbf{s}$ where $s_i \leftarrow_{\$} \{-B, \ldots, B\}$ for some small $B \in \mathbb{Z}_q$. Furthermore, some schemes pick the secret to also be sparse such that most components of $\mathbf{s}$ are zero. Thus, before we apply the revised 2016 estimate, we briefly recall the alternative embedding due to Bai and Galbraith [BG14b] which takes these small (and sparse) secrets into account.

5.1 Bai and Galbraith's Embedding

Consider an LWE instance in matrix form $(\mathbf{A}, \mathbf{c}) \equiv (\mathbf{A}, \mathbf{A} \cdot \mathbf{s} + \mathbf{e}) \in \mathbb{Z}_q^{m \times n} \times \mathbb{Z}_q^m$. By inspection, it can be seen that the vector $(\nu \mathbf{s} \mid \mathbf{e} \mid 1)$, for some $\nu \neq 0$, is contained in the lattice Λ

$$\Lambda = \left\{ \mathbf{x} \in (\nu\mathbb{Z})^n \times \mathbb{Z}^{m+1} \;\middle|\; \mathbf{x} \cdot \left(\frac{1}{\nu}\mathbf{A} \mid \mathbf{I}_m \mid -\mathbf{c} \right)^\top \equiv 0 \bmod q \right\}, \tag{4}$$

where ν allows to balance the size of the secret and the noise. An $(n+m+1) \times (n+m+1)$ basis $\mathbf{M}$ for Λ can be constructed as

$$\mathbf{M} = \begin{pmatrix} \nu\mathbf{I}_n & -\mathbf{A}^\top & \mathbf{0} \\ \mathbf{0} & q\mathbf{I}_m & \mathbf{0} \\ \mathbf{0} & \mathbf{c} & 1 \end{pmatrix}.$$

Indeed, $\mathbf{M}$ is full rank, $\det(\mathbf{M}) = \mathrm{Vol}(\Lambda)$, and the integer span of $\mathbf{M} \subseteq \Lambda$, as we can see by noting that

$$\begin{pmatrix} \nu\mathbf{I}_n & -\mathbf{A}^\top & \mathbf{0} \\ \mathbf{0} & q\mathbf{I}_m & \mathbf{0} \\ \mathbf{0} & \mathbf{c} & 1 \end{pmatrix} \left(\frac{1}{\nu}\mathbf{A} \mid \mathbf{I}_m \mid -\mathbf{c} \right)^\top = (\mathbf{A} - \mathbf{A} \mid q\mathbf{I}_m \mid \mathbf{c} - \mathbf{c})^\top \equiv \mathbf{0} \bmod q.$$

Finally, note that $(\mathbf{s} \mid * \mid 1) \cdot \mathbf{M} = (\nu \mathbf{s} \mid \mathbf{e} \mid 1)$ for suitable values of $*$. If $\mathbf{s}$ is small and/or sparse, choosing $\nu = 1$, the vector $(\mathbf{s} \mid \mathbf{e} \mid 1)$ is unbalanced, i.e. $\frac{\|\mathbf{s}\|}{\sqrt{n}} \ll \frac{\|\mathbf{e}\|}{\sqrt{m}} \approx \sigma$, where σ is the standard deviation of the LWE error distribution. We may then want to rebalance it by choosing an appropriate value of ν such that $\|(\nu \mathbf{s} \mid \mathbf{e} \mid 1)\| \approx \sigma\sqrt{n+m}$. Rebalancing preserves $(\nu \mathbf{s} \mid \mathbf{e} \mid 1)$ as the unique shortest vector in the lattice, while at the same time increasing the volume of the lattice being reduced, reducing the block size required by (2).

If $\mathbf{s} \xleftarrow{\$} \{-1, 0, 1\}^n$ we expect $\|\nu \mathbf{s}\|^2 \approx \frac{2}{3}\nu^2 n$. Therefore, we can chose $\nu = \sqrt{\frac{3}{2}}\sigma$ to obtain $\|\nu \mathbf{s}\| \approx \sigma\sqrt{n}$, so that $\|(\mathbf{s} \mid \mathbf{e} \mid 1)\| \approx \sigma\sqrt{n+m}$. Similarly, if $w < n$ entries of $\mathbf{s}$ are non-zero from $\{-1, 1\}$, we have $\|\nu \mathbf{s}\|^2 = w\,\nu^2$. Choosing $\nu = \sqrt{\frac{n}{w}}\sigma$, we obtain a vector $\nu \mathbf{s}$ of length $\sigma\sqrt{n}$.

In the case of sparse secrets, combinatorial techniques can also be applied [HG07, BGPW16, Alb17]. Given a secret $\mathbf{s}$ with at most $w < n$ non-zero entries, we guess k entries of $\mathbf{s}$ to be 0, therefore decreasing the dimension of the lattice to consider. For each guess, we then apply lattice reduction to recover the remaining components of the vector $(\mathbf{s} \mid \mathbf{e} \mid 1)$. Therefore, when estimating the overall complexity for solving such instances, we find $\min_k\{1/p_k \cdot C(n-k)\}$ where $C(n)$ is the cost of running BKZ on a lattice of dimension n and p_k is the probability of guessing correctly.

5.2 Estimates

In what follows, we assume that the geometry of (4) is sufficiently close to that of (1) so that we transfer the analysis as is. Furthermore, we will denote applying (2) from [ADPS16] for Kannan's embedding as "Kannan" and applying (2) for Bai and Galbraith's embedding [BG14b] as "Bai-Gal". Unless stated otherwise, we will assume that calling BKZ with block size β in dimension d costs $8\,d\,2^{0.292\,\beta+16.4}$ operations [BDGL16, Alb17].

Lizard [CKLS16b, CKLS16a] is a PKE scheme based on the Learning With Rounding problem, using a small, sparse secret. The authors provide a reduction to LWE, and security parameters against classic and quantum adversaries, following their analysis. In particular, they cost BKZ by a single call to sieving on a block of size β. They estimate this call to cost $\beta\,2^{c\,\beta}$ operations where $c = 0.292$ for classical adversaries, $c = 0.265$ for quantum ones and $c = 0.2075$ as a lower bound for sieving ("paranoid"). Applying the revised 2016 cost estimate for the primal attack to the parameters suggested in [CKLS16b] (using their sieving cost model as described above) reduces the expected costs, as shown in Table 2. We note that in the meantime the authors of Lizard have updated their parameters in [CKLS16a].

Table 2. Bit complexity estimates λ for solving Lizard PKE [CKLS16b] as given in [CKLS16b] and using Kannan's resp. Bai and Galbraith's embedding under the 2016 estimate. The dimension of the LWE secret is n. In all cases, BKZ-β is estimated to cost $\beta\,2^{c\,\beta}$ operations.

	Classical			Quantum			Paranoid		
n, $\log_2 q$, σ	386, 11, 2.04			414, 11, 2.09			504, 12, 4.20		
Cost	β	d	λ	β	d	λ	β	d	λ
[CKLS16b]	418	—	130.8	456	—	129.7	590	—	131.6
Kannan	372	805	117.2	400	873	114.6	567	1120	126.8
Bai-Gal	270	646	88.5	297	692	86.9	372	833	85.9

HElib [GHS12a, GHS12b] is an FHE library implementing the BGV scheme [BGH13]. A recent work [Alb17] provides revised security estimates for HELib by employing a dual attack exploiting the small and sparse secret, using the same cost estimate for BKZ as given at the beginning of this section. In Table 3 we provide costs for a primal attack using Kannan's and Bai and Galbraith's embeddings. Primal attacks perform worse than the algorithm described [Alb17], but, as expected, under the 2016 estimate the gap narrows.

Table 3. Solving costs for LWE instances underlying HELib as given in [Alb17] and using Kannan's resp. Bai and Galbraith's embedding under the 2016 estimate. The dimension of the LWE secret is n. In all cases, BKZ-β is estimated to cost $8d\,2^{0.292\,\beta+16.4}$ operations.

80 bit security															
n	1024			2048			4096			8192			16384		
$\log_2 q$, σ	47, 3.2			87, 3.2			167, 3.2			326, 3.2			638, 3.2		
Cost	β	d	λ	β	d	λ	β	d	λ	β	d	λ	β	d	λ
[Alb17] SILKE$_{\text{sparse}}$	105	—	61.3	111	—	65.0	112	—	67.0	123	—	70.2	134	—	73.1
Kannan	156	2096	76.0	166	4003	79.8	171	7960	82.3	176	15606	84.7	180	31847	86.9
Bai-Gal	137	1944	70.3	152	3906	75.9	163	7753	79.9	169	16053	82.9	173	32003	85.9
128 bit security															
n	1024			2048			4096			8192			16384		
$\log_2 q$, σ	38, 3.2			70, 3.2			134, 3.2			261, 3.2			511, 3.2		
Cost	β	d	λ	β	d	λ	β	d	λ	β	d	λ	β	d	λ
[Alb17] SILKE$_{\text{sparse}}$	138	—	73.2	145	—	77.4	151	—	81.2	163	—	84.0	149	—	86.4
Kannan	225	2076	96.1	238	4050	100.9	245	8011	103.9	250	16017	106.4	257	31635	109.4
Bai-Gal	189	1901	86.6	211	3830	94.4	204	7348	99.3	185	13543	102.8	204	28236	105.9

SEAL [CLP17] is an FHE library by Microsoft, based on the FV scheme [FV12]. Up to date parameters are given in [CLP17], using the same cost model for BKZ as mentioned at the beginning of this section. In Table 4, we provide complexity estimates for Kannan's and Bai and Galbraith's embeddings under the 2016 estimate. Note that the gap in solving time between the dual and primal attack reported in [Alb17] is closed for SEAL v2.1 parameters.

Table 4. Solving costs for parameter choices in SEAL v2.1 as given in [CLP17], using [Alb17] as implemented in the current [APS15] estimator commit `84014b6` ("[Alb17]+"), and using Kannan's resp. Bai and Galbraith's embedding under the 2016 estimate. In all cases, BKZ-β is estimated to cost $8d\,2^{0.292\,\beta+16.4}$ operations.

n, $\log_2 q$, σ	1024, 35, 3.19			2048, 60, 3.19			4096, 116, 3.19			8192, 226, 3.19			16384, 435, 3.19		
Cost	β	d	λ	β	d	λ	β	d	λ	β	d	λ	β	d	λ
[CLP17]	230	—	97.6	282	—	115.1	297	—	119.1	307	—	123.1	329	—	130.5
[Alb17]+	255	—	104.9	298	—	118.4	304	—	121.2	310	—	124.0	328	—	130.2
Kannan	257	2085	105.5	304	4041	120.2	307	8047	122.0	312	15876	124.5	328	31599	130.1
Bai-Gal	237	1984	99.6	288	4011	115.5	299	8048	119.7	309	15729	123.6	326	31322	129.5

TESLA [BG14a, ABBD15] is a signature scheme based on LWE. Post-quantum secure parameters in the quantum random oracle model were recently proposed in [ABB+17]. In Table 5, we show that these parameters need to be increased to maintain the currently claimed level of security under the 2016 estimate. Note that [ABB+17] maintains a gap of $\approx \log_2 n$ bits of security between the best known attack on LWE and claimed security to account for a loss of security in the reduction.

Table 5. Bit complexity estimates for solving TESLA parameter sets [ABB+17]. The entry "[ABB+17]+" refers to reproducing the estimates from [ABB+17] using a current copy of the estimator from [APS15] which uses $t = 1$ instead of $t = \|\mathbf{e}\|$, as a consequence the values in the respective rows are slightly lower than in [ABB+17]. We compare with Kannan's embedding under the 2016 estimate. Classically, BKZ-β is estimated to cost $8d\,2^{0.292\,\beta+16.4}$ operations; quantumly BKZ-β is estimated to cost $8d\,\sqrt{\beta^{0.0225\,\beta} \cdot 2^{0.4574\,\beta}/2^{\beta/4}}$ operations in [ABB+17].

	TESLA-0			TESLA-1			TESLA-2		
n, $\log_2 q$, σ	644, 31, 55			804, 31, 57			1300, 35, 73		
Cost	β	d	λ	β	d	λ	β	d	λ
Classical									
[ABB+17]	—	—	110.0	—	—	142.0	—	—	204.0
[ABB+17]+	255	—	110.0	358	—	140.4	563	—	200.9
Kannan	248	1514	102.4	339	1954	129.3	525	3014	184.3
Post-Quantum									
[ABB+17]	—	—	71.0	—	—	94.0	—	—	142.0
[ABB+17]+	255	—	68.5	358	—	90.7	563	—	136.4
Kannan	248	1415	61.5	339	1954	81.1	525	3014	122.4

BCIV17 [BCIV17] is a somewhat homomorphic encryption scheme obtained as a simplification of the FV scheme [FV12] and proposed as a candidate for enabling privacy friendly energy consumption forecast computation in smart grid settings. The authors propose parameters for obtaining 80 bits of security, derived using the estimator from [APS15] available at the time of publication. As a consequence of applying (2), we observe a moderate loss of security, as reported in Table 6.

Table 6. Solving costs for proposed Ring-LWE parameters in [BCIV17] using Kannan's resp. Bai and Galbraith's embedding under the 2016 estimate. In both cases, BKZ-β is estimated to cost $8d\,2^{0.292\,\beta+16.4}$ operations.

80 bit security							
$n = 4096$, $\log_2 q = 186$, $\sigma = 102$							
Embedding	β	d	λ	Embedding	β	d	λ
Kannan	156	8105	77.9	Bai-Gal	147	7818	75.3

Acknowledgements. We thank Léo Ducas and Rachel Player for helpful discussions.

References

[ABB+17] Alkim, E., Bindel, N., Buchmann, J., Dagdelen, Ö., Eaton, E., Gutoski, G., Krämer, J., Pawlega, F.: Revisiting TESLA in the quantum random oracle model. In: Lange, T., Takagi, T. (eds.) PQCrypto 2017. LNCS, vol. 10346, pp. 143–162. Springer, Cham (2017). https://doi.org/10.1007/978-3-319-59879-6_9

[ABBD15] Alkim, E., Bindel, N., Buchmann, J., Dagdelen, Ö.: TESLA: tightly-secure efficient signatures from standard lattices. Cryptology ePrint Archive, Report 2015/755 (2015). http://eprint.iacr.org/2015/755

[ADPS16] Alkim, E., Ducas, L., Pöppelmann, T., Schwabe, P.: Post-quantum key exchange - a new hope. In: Holz, T., Savage, S. (eds.) 25th USENIX Security Symposium, USENIX Security 16, pp. 327–343. USENIX Association (2016)

[AFG14] Albrecht, M.R., Fitzpatrick, R., Göpfert, F.: On the efficacy of solving LWE by reduction to unique-SVP. In: Lee, H.-S., Han, D.-G. (eds.) ICISC 2013. LNCS, vol. 8565, pp. 293–310. Springer, Cham (2014). https://doi.org/10.1007/978-3-319-12160-4_18

[Alb17] Albrecht, M.R.: On dual lattice attacks against small-secret LWE and parameter choices in HElib and SEAL. In: Coron, J.-S., Nielsen, J.B. (eds.) EUROCRYPT 2017. LNCS, vol. 10211, pp. 103–129. Springer, Cham (2017). https://doi.org/10.1007/978-3-319-56614-6_4

[APS15] Albrecht, M.R., Player, R., Scott, S.: On the concrete hardness of learning with errors. J. Math. Cryptol. **9**(3), 169–203 (2015)

[Bab86] Babai, L.: On lovász' lattice reduction and the nearest lattice point problem. Combinatorica **6**(1), 1–13 (1986)

[BCD+16] Bos, J.W., Costello, C., Ducas, L., Mironov, I., Naehrig, M., Nikolaenko, V., Raghunathan, A., Stebila, D.: Frodo: take off the ring! Practical, quantum-secure key exchange from LWE. In: Weippl, E.R., Katzenbeisser, S., Kruegel, C., Myers, A.C., Halevi, S. (eds.) ACM CCS 16, pp. 1006–1018. ACM Press, October 2016

[BCIV17] Bos, J.W., Castryck, W., Iliashenko, I., Vercauteren, F.: Privacy-friendly forecasting for the smart grid using homomorphic encryption and the group method of data handling. In: Joye, M., Nitaj, A. (eds.) AFRICACRYPT 2017. LNCS, vol. 10239, pp. 184–201. Springer, Cham (2017). https://doi.org/10.1007/978-3-319-57339-7_11

[BDGL16] Becker, A., Ducas, L., Gama, N., Laarhoven, T.: New directions in nearest neighbor searching with applications to lattice sieving. In: Krauthgamer, R. (ed.) 27th SODA, pp. 10–24. ACM-SIAM, January 2016

[BDK+17] Bos, J., Ducas, L., Kiltz, E., Lepoint, T., Lyubashevsky, V., Schanck, J.M., Schwabe, P., Stehlé, D.: CRYSTALS - kyber: a CCA-secure module-lattice-based KEM. Cryptology ePrint Archive, Report 2017/634 (2017). http://eprint.iacr.org/2017/634

[BG14a] Bai, S., Galbraith, S.D.: An improved compression technique for signatures based on learning with errors. In: Benaloh, J. (ed.) CT-RSA 2014. LNCS, vol. 8366, pp. 28–47. Springer, Cham (2014). https://doi.org/10.1007/978-3-319-04852-9_2

[BG14b] Bai, S., Galbraith, S.D.: Lattice decoding attacks on binary LWE. In: Susilo, W., Mu, Y. (eds.) ACISP 2014. LNCS, vol. 8544, pp. 322–337. Springer, Cham (2014). https://doi.org/10.1007/978-3-319-08344-5_21

[BGH13] Brakerski, Z., Gentry, C., Halevi, S.: Packed ciphertexts in LWE-based homomorphic encryption. In: Kurosawa, K., Hanaoka, G. (eds.) PKC 2013. LNCS, vol. 7778, pp. 1–13. Springer, Heidelberg (2013). https://doi.org/10.1007/978-3-642-36362-7_1

[BGPW16] Buchmann, J., Göpfert, F., Player, R., Wunderer, T.: On the hardness of LWE with binary error: revisiting the hybrid lattice-reduction and meet-in-the-middle attack. In: Pointcheval, D., Nitaj, A., Rachidi, T. (eds.) AFRICACRYPT 2016. LNCS, vol. 9646, pp. 24–43. Springer, Cham (2016). https://doi.org/10.1007/978-3-319-31517-1_2

[BV11] Brakerski, Z., Vaikuntanathan, V.: Efficient fully homomorphic encryption from (standard) LWE. In: Ostrovsky, R. (ed.) 52nd FOCS, pp. 97–106. IEEE Computer Society Press, October 2011

[BVWW16] Brakerski, Z., Vaikuntanathan, V., Wee, H., Wichs, D.: Obfuscating conjunctions under entropic ring LWE. In: Sudan, M. (ed.) ITCS 2016, pp. 147–156. ACM, January 2016

[Che13] Chen, Y.: Réduction de réseau et sécurité concrète du chiffrement complètement homomorphe. Ph.D. thesis, Paris 7 (2013)

[CHK+17] Cheon, J.H., Han, K., Kim, J., Lee, C., Son, Y.: A practical post-quantum public-key cryptosystem based on spLWE. In: Hong, S., Park, J.H. (eds.) ICISC 2016. LNCS, vol. 10157, pp. 51–74. Springer, Cham (2017). https://doi.org/10.1007/978-3-319-53177-9_3

[CKLS16a] Cheon, J.H., Kim, D., Lee, J., Song, Y.: Lizard: cut off the tail! Practical post-quantum public-key encryption from LWE and LWR. Cryptology ePrint Archive, Report 2016/1126 (2016). http://eprint.iacr.org/2016/1126

[CKLS16b] Cheon, J.H., Kim, D., Lee, J., Song, Y.: Lizard: cut off the tail! Practical post-quantum public-key encryption from LWE and LWR. Cryptology ePrint Archive, Report 2016/1126 (20161222:071525) (2016). http://eprint.iacr.org/2016/1126/20161222:071525

[CLP17] Chen, H., Laine, K., Player, R.: Simple encrypted arithmetic library - SEAL v2.1. Cryptology ePrint Archive, Report 2017/224 (2017). http://eprint.iacr.org/2017/224

[CN11] Chen, Y., Nguyen, P.Q.: BKZ 2.0: better lattice security estimates. In: Lee, D.H., Wang, X. (eds.) ASIACRYPT 2011. LNCS, vol. 7073, pp. 1–20. Springer, Heidelberg (2011). https://doi.org/10.1007/978-3-642-25385-0_1

[FPL17] The FPLLL development team: FPLLL, a lattice reduction library (2017). https://github.com/fplll/fplll

[FPY17] The FPYLLL development team: FPYLLL, a Python (2 and 3) wrapper for FPLLL (2017). https://github.com/fplll/fpylll

[FV12] Fan, J., Vercauteren, F.: Somewhat practical fully homomorphic encryption. Cryptology ePrint Archive, Report 2012/144 (2012). http://eprint.iacr.org/2012/144

[GHS12a] Gentry, C., Halevi, S., Smart, N.P.: Homomorphic evaluation of the AES circuit. Cryptology ePrint Archive, Report 2012/099 (2012). http://eprint.iacr.org/2012/099

[GHS12b] Gentry, C., Halevi, S., Smart, N.P.: Homomorphic evaluation of the AES circuit. In: Safavi-Naini and Canetti [SNC12], pp. 850–867

[GN08a] Gama, N., Nguyen, P.Q.: Finding short lattice vectors within Mordell's inequality. In: Ladner, R.E., Dwork, C. (ed.) 40th ACM STOC, pp. 207–216. ACM Press, May 2008

[GN08b] Gama, N., Nguyen, P.Q.: Predicting lattice reduction. In: Smart, N. (ed.) EUROCRYPT 2008. LNCS, vol. 4965, pp. 31–51. Springer, Heidelberg (2008). https://doi.org/10.1007/978-3-540-78967-3_3

[GNR10] Gama, N., Nguyen, P.Q., Regev, O.: Lattice enumeration using extreme pruning. In: Gilbert, H. (ed.) EUROCRYPT 2010. LNCS, vol. 6110, pp. 257–278. Springer, Heidelberg (2010). https://doi.org/10.1007/978-3-642-13190-5_13

[GSW13] Gentry, C., Sahai, A., Waters, B.: Homomorphic encryption from learning with errors: conceptually-simpler, asymptotically-faster, attribute-based. In: Canetti, R., Garay, J.A. (eds.) CRYPTO 2013. LNCS, vol. 8042, pp. 75–92. Springer, Heidelberg (2013). https://doi.org/10.1007/978-3-642-40041-4_5

[Gö16] Göpfert, F.: Securely instantiating cryptographic schemes based on the learning with errors assumption. Ph.D. thesis, Technische Universität Darmstadt (2016). http://tuprints.ulb.tu-darmstadt.de/5850/

[HG07] Howgrave-Graham, N.: A hybrid lattice-reduction and meet-in-the-middle attack against NTRU. In: Menezes, A. (ed.) CRYPTO 2007. LNCS, vol. 4622, pp. 150–169. Springer, Heidelberg (2007). https://doi.org/10.1007/978-3-540-74143-5_9

[HKM17] Herold, G., Kirshanova, E., May, A.: On the asymptotic complexity of solving LWE. Des. Codes Crypt. (2017). https://link.springer.com/article/10.1007/s10623-016-0326-0

[HPS11] Hanrot, G., Pujol, X., Stehlé, D.: Analyzing blockwise lattice algorithms using dynamical systems. In: Rogaway, P. (ed.) CRYPTO 2011. LNCS, vol. 6841, pp. 447–464. Springer, Heidelberg (2011). https://doi.org/10.1007/978-3-642-22792-9_25

[Kan87] Kannan, R.: Minkowski's convex body theorem and integer programming. Math. Oper. Res. **12**(3), 415–440 (1987)

[Laa14] Laarhoven, T.: Sieving for shortest vectors in lattices using angular locality-sensitive hashing. Cryptology ePrint Archive, Report 2014/744 (2014). http://eprint.iacr.org/2014/744

[Li11] Li, S.: Concise formulas for the area and volume of a hyperspherical cap. Asian J. Math. Stat. **4**(1), 66–70 (2011)

[LLL82] Lenstra, A.K., Lenstra Jr., H.W., Lovász, L.: Factoring polynomials with rational coefficients. Math. Ann. **261**(4), 515–534 (1982)

[LM09] Lyubashevsky, V., Micciancio, D.: On bounded distance decoding, unique shortest vectors, and the minimum distance problem. In: Halevi, S. (ed.) CRYPTO 2009. LNCS, vol. 5677, pp. 577–594. Springer, Heidelberg (2009). https://doi.org/10.1007/978-3-642-03356-8_34

[LN13] Liu, M., Nguyen, P.Q.: Solving BDD by enumeration: an update. In: Dawson, E. (ed.) CT-RSA 2013. LNCS, vol. 7779, pp. 293–309. Springer, Heidelberg (2013). https://doi.org/10.1007/978-3-642-36095-4_19

[LO83] Lagarias, J.C., Odlyzko, A.M.: Solving low-density subset sum problems. In: 24th FOCS, pp. 1–10. IEEE Computer Society Press, November 1983

[LP11] Lindner, R., Peikert, C.: Better key sizes (and attacks) for LWE-based encryption. In: Kiayias, A. (ed.) CT-RSA 2011. LNCS, vol. 6558, pp. 319–339. Springer, Heidelberg (2011). https://doi.org/10.1007/978-3-642-19074-2_21

[MR09] Micciancio, D., Regev, O.: Lattice-based cryptography. In: Bernstein, D.J., Buchmann, J., Dahmen, E. (eds.) Post-Quantum Cryptography, pp. 147–191. Springer, Heidelberg (2009). https://doi.org/10.1007/978-3-540-88702-7_5

[MW16] Micciancio, D., Walter, M.: Practical, predictable lattice basis reduction. In: Fischlin, M., Coron, J.-S. (eds.) EUROCRYPT 2016. LNCS, vol. 9665, pp. 820–849. Springer, Heidelberg (2016). https://doi.org/10.1007/978-3-662-49890-3_31

[Reg09] Regev, O.: On lattices, learning with errors, random linear codes, and cryptography. J. ACM **56**(6), 1–40 (2009)

[S+17] Stein, W., et al.: Sage Mathematics Software Version 7.5.1. The Sage Development Team (2017). http://www.sagemath.org

[Sch87] Schnorr, C.-P.: A hierarchy of polynomial time lattice basis reduction algorithms. Theoret. Comput. Sci. **53**, 201–224 (1987)

[Sch03] Schnorr, C.P.: Lattice reduction by random sampling and birthday methods. In: Alt, H., Habib, M. (eds.) STACS 2003. LNCS, vol. 2607, pp. 145–156. Springer, Heidelberg (2003). https://doi.org/10.1007/3-540-36494-3_14

[SNC12] Safavi-Naini, R., Canetti, R. (eds.): CRYPTO 2012. LNCS, vol. 7417. Springer, Heidelberg (2012). https://doi.org/10.1007/978-3-642-32009-5

Coded-BKW with Sieving

Qian Guo[1,2(✉)], Thomas Johansson[1], Erik Mårtensson[1], and Paul Stankovski[1]

[1] Department of Electrical and Information Technology, Lund University, Lund, Sweden
{qian.guo,thomas.johansson,erik.martensson,paul.stankovski}@eit.lth.se
[2] ICTEAM/ELEN/Crypto Group, Université Catholique de Louvain, Ottignies-Louvain-la-Neuve, Belgium

Abstract. The Learning with Errors problem (LWE) has become a central topic in recent cryptographic research. In this paper, we present a new solving algorithm combining important ideas from previous work on improving the BKW algorithm and ideas from sieving in lattices. The new algorithm is analyzed and demonstrates an improved asymptotic performance. For Regev parameters $q = n^2$ and noise level $\sigma = n^{1.5}/(\sqrt{2\pi}\log_2^2 n)$, the asymptotic complexity is $2^{0.895n}$ in the standard setting, improving on the previously best known complexity of roughly $2^{0.930n}$. Also for concrete parameter instances, improved performance is indicated.

Keywords: LWE · BKW · Coded-BKW · Lattice codes · Lattice sieving

1 Introduction

Post-quantum crypto, the area of cryptography in the presence of quantum computers, is currently a major topic in the cryptographic community. Cryptosystems based on hard problems related to lattices are currently intensively investigated, due to their possible resistance against quantum computers. The major problem in this area, upon which cryptographic primitives can be built, is the *Learning with Errors* (LWE) problem.

LWE is an important, efficient and versatile problem. Some famous applications of LWE are to construct Fully Homomorphic encryption schemes [14–16,21]. A major motivation for using LWE is its connections to lattice problems, linking the difficulty of solving LWE (on average) to the difficulty of solving instances of some (worst-case) famous lattice problems. Let us state the LWE problem.

Definition 1. *Let n be a positive integer, q a prime, and let $\mathcal{X}$ be an error distribution selected as the discrete Gaussian distribution on $\mathbb{Z}_q$. Fix $\mathbf{s}$ to be a*

Q. Guo et al.—Supported by the Swedish Research Council (Grants No. 2015-04528).
Q. Guo—Supported in part by the European Commission and by the Belgian Fund for Scientific Research through the SECODE project (CHIST-ERA Program).

T. Takagi and T. Peyrin (Eds.): ASIACRYPT 2017, Part I, LNCS 10624, pp. 323–346, 2017.
https://doi.org/10.1007/978-3-319-70694-8_12

secret vector in $\mathbb{Z}_q^n$*, chosen according to a uniform distribution. Denote by* $L_{\mathbf{s},\mathcal{X}}$ *the probability distribution on* $\mathbb{Z}_q^n \times \mathbb{Z}_q$ *obtained by choosing* $\mathbf{a} \in \mathbb{Z}_q^n$ *uniformly at random, choosing an error* $e \in \mathbb{Z}_q$ *according to* $\mathcal{X}$ *and returning*

$$(\mathbf{a}, z) = (\mathbf{a}, \langle \mathbf{a}, \mathbf{s} \rangle + e)$$

in $\mathbb{Z}_q^n \times \mathbb{Z}_q$*. The (search)* LWE *problem is to find the secret vector* $\mathbf{s}$ *given a fixed number of samples from* $L_{\mathbf{s},\mathcal{X}}$*.*

The definition above gives the *search* LWE problem, as the problem description asks for the recovery of the secret vector $\mathbf{s}$. Another variant is the *decision* LWE problem. In this case the problem is to distinguish between samples drawn from $L_{\mathbf{s},\mathcal{X}}$ and a uniform distribution on $\mathbb{Z}_q^n \times \mathbb{Z}_q$. Typically, we are then interested in distinguishers with non-negligible advantage.

For the analysis of algorithms solving the LWE problem in previous work, there are essentially two different approaches. One being the approach of calculating the specific number of operations needed to solve a certain instance for a particular algorithm, and comparing specific complexity numbers. The other approach is asymptotic analysis. Solvers for the LWE problem with suitable parameters are expected to have fully exponential complexity, say bounded by 2^{cn} as n tends to infinity. Comparisons between algorithms are made by deriving the coefficient c in the asymptotic complexity expression.

1.1 Related Work

We list the three main approaches for solving the LWE problem in what follows. A good survey with concrete complexity considerations is [6] and for asymptotic comparisons, see [25].

The first class is the algebraic approach, which was initialized by Arora-Ge [8]. This work was further improved by Albrecht et al., using Gröbner bases [2]. Here we point out that this type of attack is mainly, asymptotically, of interest when the noise is very small. For extremely small noise the complexity can be polynomial.

The second and most commonly used approach is to rewrite the LWE problem as a lattice problem, and therefore lattice reduction algorithms [17,43], sieving and enumeration can be applied. There are several possibilities when it comes to reducing the LWE problem to some hard lattice problem. One is a direct approach, writing up a lattice from the samples and then to treat the search LWE problem as a Bounded Distance Decoding (BDD) problem [34,35]. One can also reduce the BDD problem to an UNIQUE-SVP problem [5]. Another variant is to consider the distinguishing problem in the dual lattice [37]. Lattice-based algorithms have the advantage of not using an exponential number of samples.

The third approach is the BKW-type algorithms.

BKW Variants. The BKW algorithm was originally proposed by Blum et al. [12] for solving the Learning Parity with Noise (LPN) problem (LWE for $q = 2$). It resembles Wagner's generalized birthday approach [44].

For the LPN case, there have been a number of improvements to the basic BKW approach. In [33], transform techniques were introduced to speed up the search part. Further improvements came in work by Kirchner [28], Bernstein and Lange [11], Guo et al. [22], Zhang et al. [46], Bogos and Vaudenay [13].

Albrecht et al. were first to apply BKW to the LWE problem [3], which they followed up with Lazy Modulus Switching (LMS) [4], which was further improved by Duc et al. in [20]. The basic BKW approach for LWE was improved in [23] and [29], resulting in an asymptotic improvement. These works improved by reducing a variable number of positions in each step of the BKW procedure as well as introducing a coding approach. Although the two algorithms were slightly different, they perform asymptotically the same and we refer to the approach as coded-BKW. It was proved in [29] that the asymptotic complexity for Regev parameters (public-key cryptography parameter) $q = n^2$ and noise level $\sigma = n^{1.5}/(\sqrt{2\pi}\log_2^2 n)$ is $2^{0.930n}$, the currently best known asymptotic performance for such parameters.

Sieving Algorithms. A key part of the algorithm to be proposed is the use of sieving in lattices. The first sieving algorithm for solving the shortest vector problem was proposed by Ajtai et al. in [1], showing that SVP can be solved in time and memory $2^{\Theta(n)}$. Subsequently, we have seen the NV-sieve [40], List-sieve [38], and provable improvement of the sieving complexity using the birthday paradox [24,41].

With heuristic analysis, [40] started to derive a complexity of $2^{0.415n}$, followed by GaussSieve [38], 2-level sieve [45], 3-level sieve [47] and overlattice-sieve [10]. Laarhoven started to improve the lattice sieving algorithms employing algorithmic breakthroughs in solving the nearest neighbor problem, angular LSH [30], and spherical LSH [32]. The asymptotically most efficient approach when it comes to time complexity is Locality Sensitive Filtering (LSF) [9] with both a space and time complexity of $2^{0.292n+o(1)n}$. Using quantum computers, the complexity can be reduced to $2^{0.265n+o(1)n}$ (see [31]) by applying Grover's quantum search algorithm.

1.2 Contributions

We propose a new algorithm for solving the LWE problem combining previous combinatorial methods with an important algorithmic idea – using a sieving approach. Whereas BKW combines vectors to reduce positions to zero, the previously best improvements of BKW, like coded-BKW, reduce more positions but at the price of leaving a small but in general nonzero value in reduced positions. These values are considered as additional noise. As these values increase in magnitude for each step, because we add them together, they have to be very small in the initial steps. This is the reason why in coded-BKW the number of positions reduced in a step is increasing with the step index. We have to start with a small number of reduced positions, in order to not obtain a noise that is too large.

The proposed algorithm tries to solve the problem of the growing noise from the coding part (or LMS) by using a sieving step to make sure that the noise from treated positions does not grow, but stays approximately of the same size. The basic form of the new algorithm then contains two parts in each iterative step. The first part reduces the magnitude of some particular positions by finding pairs of vectors that can be combined. The second part performs a sieving step covering all positions from all previous steps, making sure that the magnitude of the resulting vector components is roughly as in the already size-reduced part of the incoming vectors.

We analyze the new algorithm from an asymptotic perspective, proving a new improved asymptotic performance. Again, for asymptotic Regev parameters $q = n^2$ and noise level $\sigma = n^{1.5}$, the result is a time and space complexity of $2^{0.895n}$, which is a significant asymptotic improvement. We also get a first quantum acceleration for Regev's parameter setting by using the performance of sieving in the quantum setting. We additionally sketch on an algorithmic description for concrete instances. Here we can use additional non-asymptotic improvements and derive approximate actual complexities for specific instances, demonstrating its improved performance also for concrete instances.

1.3 Organization

The remaining parts of the paper are organized as followed. We start with some preliminaries in Sect. 2, including more basics on LWE, discrete Guassians and sieving in lattices. In Sect. 3 we review the details of the BKW algorithm and some recent improvements. Section 4 just contains a simple reformulation. In Sect. 5 we give the new algorithm in its basic form and in Sect. 6 we derive the optimal parameter selection and perform the asymptotic analysis. In Sect. 7 we briefly overview how the algorithm could be extended in the non-asymptotic case, including additional non-asymptotic improvements. We compute some rough estimates on expected number of operations to solve specific instances, indicating improved performance in comparison with previous complexity results. We conclude the paper in Sect. 8.

2 Background

2.1 Notations

Throughout the paper, the following notations are used.

- We write $\log(\cdot)$ for the base 2 logarithm and $\ln(\cdot)$ for the natural logarithm.

- In an n-dimensional Euclidean space $\mathbb{R}^n$, by the norm of a vector $\mathbf{x} = (x_1, x_2, \ldots, x_n)$ we refer to its L_2-norm, defined as

$$||\mathbf{x}|| = \sqrt{x_1^2 + \cdots + x_n^2}.$$

We then define the Euclidean distance between two vectors $\mathbf{x}$ and $\mathbf{y}$ in $\mathbb{R}^n$ as $||\mathbf{x} - \mathbf{y}||$.
- For an $[N, K]$ linear code, N denotes the code length and K denotes the dimension.

2.2 LWE Problem Description

Rather than giving a more formal definition of the decision version of LWE, we instead reformulate the search LWE problem, because our main purpose is to investigate its solving complexity. Assume that m samples

$$(\mathbf{a}_1, z_1), (\mathbf{a}_2, z_2), \ldots, (\mathbf{a}_m, z_m),$$

are drawn from the LWE distribution $L_{\mathbf{s},\mathcal{X}}$, where $\mathbf{a}_i \in \mathbb{Z}_q^n, z_i \in \mathbb{Z}_q$. Let $\mathbf{z} = (z_1, z_2, \ldots, z_m)$ and $\mathbf{y} = (y_1, y_2, \ldots, y_m) = \mathbf{sA}$. We can then write

$$\mathbf{z} = \mathbf{sA} + \mathbf{e},$$

where $\mathbf{A} = \begin{bmatrix}\mathbf{a}_1^{\mathrm{T}} \ \mathbf{a}_2^{\mathrm{T}} \cdots \mathbf{a}_n^{\mathrm{T}}\end{bmatrix}$, $z_i = y_i + e_i = \langle \mathbf{s}, \mathbf{a}_i \rangle + e_i$ and $e_i \stackrel{\$}{\leftarrow} \mathcal{X}$. Therefore, we have reformulated the search LWE problem as a decoding problem, in which the matrix $\mathbf{A}$ serves as the generator matrix for a linear code over $\mathbb{Z}_q$ and $\mathbf{z}$ is the received word. We see that the problem of searching for the secret vector $\mathbf{s}$ is equivalent to that of finding the codeword $\mathbf{y} = \mathbf{sA}$ such that the Euclidean distance $||\mathbf{y} - \mathbf{z}||$ is minimal.

The Secret-Noise Transformation. An important transformation [7,28] can be applied to ensure that the secret vector follows the same distribution $\mathcal{X}$ as the noise. The procedure works as follows. We first write $\mathbf{A}$ in systematic form via Gaussian elimination. Assume that the first n columns are linearly independent and form the matrix $\mathbf{A_0}$. We then define $\mathbf{D} = \mathbf{A_0}^{-1}$ and write $\hat{\mathbf{s}} = \mathbf{sD}^{-1} - (z_1, z_2, \ldots, z_n)$. Hence, we can derive an equivalent problem described by $\hat{\mathbf{A}} = (\mathbf{I}, \hat{\mathbf{a}}_{n+1}^{\mathrm{T}}, \hat{\mathbf{a}}_{n+2}^{\mathrm{T}}, \cdots, \hat{\mathbf{a}}_m^{\mathrm{T}})$, where $\hat{\mathbf{A}} = \mathbf{DA}$. We compute

$$\hat{\mathbf{z}} = \mathbf{z} - (z_1, z_2, \ldots, z_n)\hat{\mathbf{A}} = (\mathbf{0}, \hat{z}_{n+1}, \hat{z}_{n+2}, \ldots, \hat{z}_m).$$

Using this transformation, one can assume that each entry in the secret vector is now distributed according to $\mathcal{X}$.

The noise distribution $\mathcal{X}$ is usually chosen as the discrete Gaussian distribution, which will be briefly discussed in Sect. 2.3.

2.3 Discrete Gaussian Distribution

We start by defining the discrete Gaussian distribution over $\mathbb{Z}$ with mean 0 and variance σ^2, denoted $D_{\mathbb{Z},\sigma}$. That is, the probability distribution obtained by assigning a probability proportional to $\exp(-x^2/2\sigma^2)$ to each $x \in \mathbb{Z}$. Then the discrete Gaussian distribution $\mathcal{X}$ over $\mathbb{Z}_q$ with variance σ^2 (also denoted $\mathcal{X}_\sigma$) can be defined by folding $D_{\mathbb{Z},\sigma}$ and accumulating the value of the probability mass function over all integers in each residue class modulo q.

Following the path of previous work [3], we assume that in our discussed instances, the discrete Gaussian distribution can be approximated by the continuous counterpart. For instance, if X is drawn from $\mathcal{X}_{\sigma_1}$ and Y is drawn from $\mathcal{X}_{\sigma_2}$, then $X+Y$ is regarded as being drawn from $\mathcal{X}_{\sqrt{\sigma_1^2+\sigma_2^2}}$. This approximation is widely adopted in literature.

The sample complexity for distinguishing. To estimate the solving complexity, we need to determine the number of required samples to distinguish between the uniform distribution on $\mathbb{Z}_q$ and $\mathcal{X}_\sigma$. Relying on standard theory from statistics, using either previous work [34] or Bleichenbacher's definition of bias [39], we can conclude that the required number of samples is

$$\mathcal{O}\left(e^{2\pi\left(\frac{\sigma\sqrt{2\pi}}{q}\right)^2}\right).$$

2.4 Sieving in Lattices

We here give a brief introduction to the sieving idea and its application in lattices for solving the shortest vector problem (SVP). For an introduction to lattices, the SVP problem, and sieving algorithms, see e.g. [9].

In sieving, we start with a list $\mathcal{L}$ of relatively short lattice vectors. If the list size is large enough, we will obtain many pairs of $\mathbf{v}, \mathbf{w} \in \mathcal{L}$, such that $||\mathbf{v} \pm \mathbf{w}|| \leq \max\{||\mathbf{v}||, ||\mathbf{w}||\}$. After reducing the size of these lattice vectors a polynomial number of times, one can expect to find the shortest vector.

The core of sieving is thus to find a close enough neighbor $\mathbf{v} \in \mathcal{L}$ efficiently, for a vector $\mathbf{w} \in \mathcal{L}$, thereby reducing the size by further operations like addition or subtraction. This is also true for our newly proposed algorithm in a later section, since by sieving we solely desire to control the size of the added/subtracted vectors. For this specific purpose, many famous probabilistic algorithms have been proposed, e.g., Locality Sensitive Hashing (LSH) [27], Bucketing coding [19], May-Ozerov's algorithm [36] in the Hamming metric with important applications to decoding binary linear codes.

In the Euclidean metric, the state-of-the-art algorithm in the asymptotic sense is Locality Sensitive Filtering (LSF) [9], which requires $2^{0.2075n+o(n)}$ samples. In the classic setting, the time and memory requirements are both in the order of $2^{0.292n+o(n)}$. The constant hidden in the running time exponent can be reduced to 0.265 in the scenario of quantum computing. In the remaining part of the paper, we choose the LSF algorithm for the best asymptotic performance when we need to instantiate the sieving method.

3 The BKW Algorithm

The BKW algorithm is the first sub-exponential algorithm for solving the LPN problem, originally proposed by Blum et al. [12]. It can also be trivially adopted to the LWE problem, with single-exponential complexity.

3.1 Plain BKW

The algorithm consists of two phases: the reduction phase and the solving phase. The essential improvement comes from the first phase, whose underlying fundamental idea is the same as Wagner's generalized birthday algorithm [44]. That is, using an iterative collision procedure on the columns in the matrix $\mathbf{A}$, one can reduce its row dimension step by step, and finally reach a new LWE instance with a much smaller dimension. The solving phase can then be applied to recover the secret vector. We describe the core procedure of the reduction phase, called a plain BKW step, as follows. Let us start with $\mathbf{A}_0 = \mathbf{A}$.

Dimension reduction: In the i-th iteration, we look for combinations of two columns in $\mathbf{A}_{i-1}$ that add (or subtract) to zero in the last b entries. Suppose that one finds two columns $\mathbf{a}_{j_1,i-1}^{\mathrm{T}}, \mathbf{a}_{j_2,i-1}^{\mathrm{T}}$ such that

$$\mathbf{a}_{j_1,i-1} \pm \mathbf{a}_{j_2,i-1} = [* * \cdots * \underbrace{0\, 0 \cdots 0}_{b \text{ symbols}}],$$

where $*$ means any value. We then generate a new vector $\mathbf{a}_{j,i} = \mathbf{a}_{j_1,i-1} \pm \mathbf{a}_{j_2,i-1}$. We obtain a new generator matrix $\mathbf{A}_i$ for the next iteration, with its dimension reduced by b, if we remove the last b all-zero positions with no impact on the output of the inner product operation. We also derive a new "observed symbol" as $z_{j,i} = z_{j_1,i-1} \pm z_{j_2,i-1}$.
A trade-off: After one step of this procedure, we can see that the new noise variable $e_{j,i} = e_{j_1,i-1} \pm e_{j_2,i-1}$. If the noise variables $e_{j_1,i-1}$ and $e_{j_2,i-1}$ both follow the Gaussian distribution with variance σ_{i-1}^2, then the new noise variable $e_{j,i}$ is considered Gaussian distributed with variance $\sigma_i^2 = 2\sigma_{i-1}^2$.

After t_0 iterations, we have reduced the dimension of the problem to $n - t_0 b$. The final noise variable is thus a summation of 2^{t_0} noise variables generated from the LWE oracle. We therefore know that the noise connected to each column is of the form

$$e = \sum_{j=1}^{2^{t_0}} e_{i_j},$$

and the total noise is approximately Gaussian with variance $2^{t_0} \cdot \sigma^2$.

The remaining solving phase is to solve this transformed LWE instance. This phase does not affect its asymptotic complexity but has significant impact on its actual running time for concrete instances.

Similar to the original proposal [12] for solving LPN, which recovers 1 bit in the secret vector via majority voting, Albrecht et al. [3] exhaust one secret entry using a distinguisher. The complexity is further reduced by Duc et al. [20] using Fast Fourier Transform (FFT) to recover several secret entries simultaneously.

3.2 Coded-BKW

As described above, in each BKW step, we try to collide a large number of vectors $\mathbf{a}_i$ in a set of positions denoted by an index set I. We denote this sub-vector of a vector $\mathbf{a}$ as a $\mathbf{a}_I$. We set the size of the collision set to be $(\frac{q^b-1}{2})$, a very important parameter indicating the final complexity of the algorithm.

In this part we describe another idea that, instead of zeroing out the vector $\mathbf{a}_I$ by collisions, we try to collide vectors to make $\mathbf{a}_I$ small. The advantage of this idea is that one can handle more positions in one step for the same size of the collision set.

This idea was first formulated by Albrecht et al. in PKC 2014 [4], aiming for solving the LWE problem with a small secret. They proposed a new technique called Lazy Modulus Switching (LMS). Then, in CRYPTO 2015, two new algorithms with similar underlying algorithmic ideas were proposed independently in [23] and [29], highly enhancing the performance in the sense of both asymptotic and concrete complexity. Using the secret-noise transformation, these new algorithms can be used to solve the standard LWE problem.

In this part we use the notation from [23] to describe the BKW variant called coded-BKW, as it has the best concrete performance, i.e., it can reduce the magnitude of the noise by a constant factor compared with its counterpart technique LMS.

The core step – the coded-BKW step – can be described as follows.

Considering step i in the reduction phase, we choose a q-ary $[n_i, b]$ linear code, denoted $\mathcal{C}_i$, that can be employed to construct a lattice code, e.g., using Construction A (see [18] for details). The sub-vector $\mathbf{a}_I$ can then be written in terms of its two constituents, the codeword part $\mathbf{c}_I \in \mathcal{C}_i$ and an error part $\mathbf{e}_I \in \mathbb{Z}_q^{N_i}$. That is,

$$\mathbf{a}_I = \mathbf{c}_I + \mathbf{e}_I. \tag{1}$$

We rewrite the inner product $\langle \mathbf{s}_I, \mathbf{a}_I \rangle$ as

$$\langle \mathbf{s}_I, \mathbf{a}_I \rangle = \langle \mathbf{s}_I, \mathbf{c}_I \rangle + \langle \mathbf{s}_I, \mathbf{e}_I \rangle .$$

We can cancel out the part $\langle \mathbf{s}_I, \mathbf{c}_I \rangle$ by subtracting two vectors mapped to the same codeword, and the remaining difference is the noise. Due to symmetry, the size of the collision set can be $\frac{q^b-1}{2}$, as in the plain BKW step.

If we remove n_i positions in the i-th step, then we have removed $\sum_{i=1}^{t} n_i$ positions ($n_i \geq b$) in total. Thus, after guessing the remaining secret symbols in the solving phase, we need to distinguish between the uniform distribution and the distribution representing a sum of noise variables, i.e.,

$$z = \sum_{j=1}^{2^t} e_{i_j} + \sum_{i=1}^{n} s_i (E_i^{(1)} + E_i^{(2)} + \cdots + E_i^{(t)}), \tag{2}$$

where $E_i^{(h)} = \sum_{j=1}^{2^{t-h+1}} \hat{e}_{i_j}^{(h)}$ and $\hat{e}_{i_j}^{(h)}$ is the noise introduced in the h-th coded-BKW step. Here at most one error term $E_i^{(h)}$ is non-zero for one position in the index set, and the overall noise can be estimated according to Eq. (2).

The remaining problem is to analyze the noise level introduced by coding. In [23], it is assumed that every $E_i^{(h)}$ is close to a Gaussian distribution, which is tested in implementation. Based on known results on lattice codes, the standard deviation σ introduced by employing a q-ary $[N, k]$ linear code is estimated by

$$\sigma \approx q^{1-k/N} \cdot \sqrt{G(\Lambda_{N,k})}, \tag{3}$$

where $G(\Lambda_{N,k})$ is a code-related parameter satisfying

$$\frac{1}{2\pi e} < G(\Lambda_{N,k}) \leq \frac{1}{12}.$$

In [23], the chosen codes are with varying rates to ensure that the noise contribution of each position is equal. This is principally similar to the operation of changing the modulus size in each reduction step in [29].

4 A Reformulation

Let us reformulate the LWE problem and the steps in the different algorithms in a matrix form. Recall that we have the LWE samples in the form $\mathbf{z} = \mathbf{sA} + \mathbf{e}$. We write this as

$$(\mathbf{s}, \mathbf{e}) \begin{pmatrix} \mathbf{A} \\ \mathbf{I} \end{pmatrix} = \mathbf{z}. \tag{4}$$

The entries in the unknown left-hand side vector $(\mathbf{s}, \mathbf{e})$ are all i.i.d. The matrix above is denoted as $\mathbf{H_0} = \begin{pmatrix} \mathbf{A} \\ \mathbf{I} \end{pmatrix}$ and it is a known quantity, as well as $\mathbf{z}$.

By multiplying Eq. (4) from the right with special matrices $\mathbf{P_i}$ we are going to reduce the size of columns in the matrix. Starting with

$$(\mathbf{s}, \mathbf{e})\mathbf{H_0} = \mathbf{z},$$

we find a matrix $\mathbf{P_0}$ and form $\mathbf{H_1} = \mathbf{H_0 P_0}$, $\mathbf{z_1} = \mathbf{z P_0}$, resulting in

$$(\mathbf{s}, \mathbf{e})\mathbf{H_1} = \mathbf{z_1}.$$

Continuing this process for t steps, we have formed $\mathbf{H_t} = \mathbf{H_0 P_0} \cdots \mathbf{P_{t-1}}$, $\mathbf{z}_t = \mathbf{z P_0} \cdots \mathbf{P_{t-1}}$.

Plain BKW can be described as each $\mathbf{P_i}$ having columns with only two nonzero entries, both from the set $\{-1, 1\}$. The BKW procedure subsequently cancels rows in the $\mathbf{H_i}$ matrices in a way such that $\mathbf{H_t} = \begin{pmatrix} \mathbf{0} \\ \mathbf{H'_t} \end{pmatrix}$, where columns of $\mathbf{H'_t}$ have 2^t non-zero entries[1]. The goal is to minimize the magnitude of the column entries in $\mathbf{H_t}$. The smaller magnitude, the larger advantage in the corresponding samples.

The improved techniques like LMS and coded-BKW reduce the $\mathbf{H_t}$ similar to the BKW, but improves by using the fact that the top rows of $\mathbf{H'_t}$ do not have to be canceled to $\mathbf{0}$. Instead, entries are allowed to be of the same norm as in the $\mathbf{H'_t}$ matrix.

5 A BKW-Sieving Algorithm for the LWE Problem

The algorithm we propose uses a similar structure as the coded-BKW algorithm. The new idea involves changing the BKW step to also include a sieving step. In this section we give the algorithm in a simple form, allowing for some asymptotic analysis. We exclude some steps that give non-asymptotic improvements. These are considered in a later section. We assume that each entry in the secret vector $\mathbf{s}$ is distributed according to $\mathcal{X}$.

5.1 Initial Guessing Step

We select a few entries of $\mathbf{s}$ and guess these values (according to $\mathcal{X}$). We run through all likely values and for each of them we do the steps below. Based on a particular guess, the sample equations need to be rewritten accordingly.

For simplicity, the remaining unknown values are still denoted $\mathbf{s}$ after this guessing step and the length of $\mathbf{s}$ is still denoted n.

5.2 Transformation Steps

We start with some simplifying notation. The n positions in columns in $\mathbf{A}$ (first n positions in columns of $\mathbf{H}$) are considered as a concatenation of smaller vectors. We assume that these vectors have lengths which are $n_1, n_2, n_3, \ldots, n_t$, in such a way that $\sum_{i=1}^{t} n_i = n$. Also, let $N_j = \sum_{i=1}^{j} n_i$, for $j = 1, 2, \ldots, t$.

Before explaining the algorithmic steps, we introduce two notions that will be used later.

Notation CodeMap$(\mathbf{h}, i)$: We assume, following the idea of Coded-BKW, that we have fixed a lattice code $\mathcal{C}_i$ of length n_i. The vector $\mathbf{h}$ fed as input to CodeMap is first considered only restricted to the positions $N_{i-1}+1$ to N_i, i.e., as a vector of length n_i. This vector, denoted $\mathbf{h}_{[N_{i-1}+1, N_i]}$ is then mapped to the closest codeword in $\mathcal{C}_i$. This closest codeword is denoted CodeMap$(\mathbf{h}, i)$.

[1] Sometimes we get a little fewer than 2^t entries since 1 s can overlap. However, this probability is low and does not change the analysis.

The code $\mathcal{C}_i$ needs to have an associated procedure of quickly finding the closest codeword for any given vector. One could then use a simple code or a more advanced code. From an asymptotic viewpoint, it does not matter, but in a practical implementation there can be a difference. We are going to select the parameters in such a way that the distance to the closest codeword is expected to be no more than $\sqrt{n_i} \cdot B$, where B is a constant.

Notation Sieve($\mathcal{L}_\Delta, i, \sqrt{N_i} \cdot B$): The input $\mathcal{L}_\Delta$ contains a list of vectors. We are only considering them restricted to the first N_i positions. This procedure will find all differences between any two vectors such that the norm of the difference restricted to the first N_i positions is less than $\sqrt{N_i} \cdot B$. All such differences are put in a list $\mathcal{S}_\Delta$ which is the output of the procedure.

We assume that the vectors in the list $\mathcal{L}_\Delta$ restricted to the first N_i positions, all have a norm of about $\sqrt{N_i} \cdot B$. Then the problem is solved by algorithms for sieving in lattices, for example using Locality-Sensitive Hashing/Filtering.

For the description of the main algorithm, recall that

$$(\mathbf{s}, \mathbf{e})\mathbf{H_0} = \mathbf{z},$$

where $\mathbf{H_0} = \begin{pmatrix} \mathbf{A} \\ \mathbf{I} \end{pmatrix}$. We are going to perform t steps to transform $\mathbf{H_0}$ into $\mathbf{H_t}$ such that the columns in $\mathbf{H_t}$ are "small". Again, we look at the first n positions in a column corresponding to the $\mathbf{A}$ matrix. Since we are only adding or subtracting columns using coefficients in $\{-1, 1\}$, the remaining positions in the column are assumed to contain 2^i nonzero positions either containing a -1 or a 1, after i steps.

5.3 A BKW-Sieving Step

We are now going to fix an average level of "smallness" for a position, which is a constant denoted B, as above. The idea of the algorithm is to keep the norm of considered vectors of some length n' below $\sqrt{n'} \cdot B$.

A column $\mathbf{h} \in \mathbf{H_0}$ will now be processed by first computing $\Delta =$ CodeMap($\mathbf{h}$, 1). Then we place $\mathbf{h}$ in the list $\mathcal{L}_\Delta$. After running through all columns $\mathbf{h} \in \mathbf{H_0}$ they have been sorted into K lists $\mathcal{L}_\Delta$.

We then run through the lists, each containing roughly m/K columns. We perform a sieving step, according to $\mathcal{S}_\Delta = Sieve(\mathcal{L}_\Delta, \sqrt{N_1} \cdot B)$. The result is a list of vectors, where the norm of each vector restricted to the first N_1 positions is less than $\sqrt{N_1} \cdot B$. The indices of any two columns, i_j, i_k are kept in such a way that we can compute a new received symbol $z = z_{i_j} - z_{i_k}$. All vectors in all lists $\mathcal{S}_\Delta$ are now put as columns in $\mathbf{H_1}$. We now have a matrix $\mathbf{H_1}$ where the norm of each column restricted to the first n_1 positions is less than $\sqrt{N_1} \cdot B$. This is the end of the first step.

Next, we repeat roughly the same procedure another $t - 1$ times. A column $\mathbf{h} \in \mathbf{H_{i-1}}$ will now be processed by first computing $\Delta =$ CodeMap($\mathbf{h}, i$). We place $\mathbf{h}$ in the list $\mathcal{L}_\Delta$. After running through all columns $\mathbf{h} \in \mathbf{H_{i-1}}$ they have been sorted in K lists $\mathcal{L}_\Delta$.

We run through all lists, where each list contains roughly m/K columns. We perform a sieving step, according to $\mathcal{S}_\Delta = Sieve(\mathcal{L}_\Delta, i, \sqrt{N_i} \cdot B)$. The result is a list of vectors where the norm of each vector restricted to the first N_i positions is less than $\sqrt{N_i} \cdot B$. A new received symbol is computed. All vectors in all lists $\mathcal{S}_\Delta$ are now put as columns in $\mathbf{H_i}$. We get a matrix $\mathbf{H_i}$ where the norm of each column restricted to the first N_i positions is less than $\sqrt{N_i} \cdot B$. This is repeated for $i = 2, \ldots, t$. We assume that the parameters have been chosen in such a way that each matrix $\mathbf{H_i}$ can have m columns.

After performing these t steps we end up with a matrix $\mathbf{H_t}$ such that the norm of columns restricted to the first n positions is bounded by $\sqrt{n} \cdot B$ and the norm of the last m positions is roughly $2^{t/2}$. Altogether, this should result in samples generated as

$$\mathbf{z} = (\mathbf{s}, \mathbf{e})\mathbf{H_t}.$$

The values in the $\mathbf{z}$ vector are then roughly Gaussian distributed, with variance $\sigma^2 \cdot (nB^2 + 2^t)$. By running a distinguisher on the created samples $\mathbf{z}$ we can verify whether our initial guess is correct or not. After restoring some secret value, the whole procedure can be repeated, but for a smaller dimension.

5.4 Algorithm Summary

Algorithm 1. Coded-BKW with Sieving (main steps)

Input: Matrix $\mathbf{A}$ with n rows and m columns, received vector $\mathbf{z}$ of length m and algorithm parameters $t, n_i, 1 \leq i \leq t$, B

change the distribution of the secret vector (Gaussian elimination)
for i from 1 to t **do**:
  for all columns $\mathbf{h} \in \mathbf{H_{i-1}}$ **do**:
    $\Delta =$ CodeMap$(\mathbf{h}, i)$
    put $\mathbf{h}$ in list $\mathcal{L}_\Delta$
  for all lists $\mathcal{L}_\Delta$ **do**:
    $\mathcal{S}_\Delta =$ Sieve$(\mathcal{L}_\Delta, i, \sqrt{N_i} \cdot B)$
    put all $\mathcal{S}_\Delta$ as columns in $\mathbf{H_i}$
exhaustively guess the $\mathbf{s}_n$ entry using hypothesis testing

A summary of coded-BKW with sieving is contained in Algorithm 1.

Note that one may also use some advanced distinguisher, e.g., the FFT distinguisher, which is important to the concrete complexity but not the asymptotic performance.

5.5 High-Level Comparison with Previous BKW Versions

A high-level comparison between the behaviors of plain BKW, coded-BKW and coded-BKW with sieving is shown in Fig. 1.

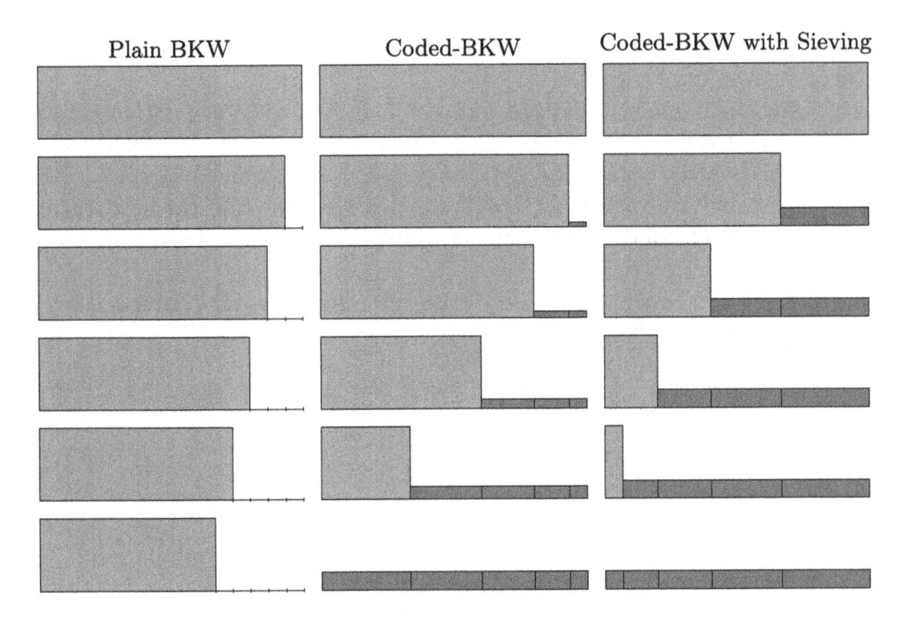

Fig. 1. A high level explanation of how the different versions of the BKW algorithm work.

Initially the average norm of all elements in a sample vector **a** is around $q/2$, represented by the first row in the figure. Plain BKW then gradually works towards a zero vector by adding/subtracting vectors in each step such that a fixed number of positions get canceled out to 0.

The idea of coded-BKW is to not cancel out the positions completely, and thereby allowing for longer steps. The positions that are not canceled out increase in magnitude by a factor of $\sqrt{2}$ in each step. To end up with an evenly distributed noise vector in the end we can let the noise in the new almost canceled positions increase by a factor of $\sqrt{2}$ in each step. Thus we can gradually increase the step size.

When reducing positions in coded-BKW, the previously reduced positions increase in magnitude by a factor of $\sqrt{2}$. However, the sieving step in coded-BKW with sieving makes sure that the previously reduced positions do not increase in magnitude. Thus, initially, we do not have to reduce the positions as much as in coded-BKW. However, the sieving process gets more expensive the more positions we work with, and we must therefore gradually decrease the step size.

6 Parameter Selection and Asymptotic Analysis

After each step, positions that already have been treated should remain at some given magnitude B. That is, the average (absolute) value of a treated position should be very close to B. This property is maintained by the way in which we apply the sieving part at each reduction step. After t steps we have therefore produced vectors of average norm $\sqrt{n} \cdot B$.

Assigning the number of samples to be $m = 2^k$, where 2^k is a parameter that will decide the total complexity of the algorithm, we will end up with roughly $m = 2^k$ samples after t steps. As already stated, these received samples will be roughly Gaussian with variance $\sigma^2 \cdot (nB^2 + 2^t)$. We assume that the best strategy is to keep the magnitudes of the two different contributions of the same order, so we choose $nB^2 \approx 2^t$.

Furthermore, in order to be able to recover a single secret position using m samples, we need

$$m = \mathcal{O}\left(e^{4\pi^2 \cdot \frac{\sigma^2 \cdot (nB^2 + 2^t)}{q^2}}\right).$$

Thus, we have

$$\ln 2 \cdot k = 4\pi^2 \cdot \frac{\sigma^2 \cdot (nB^2 + 2^t)}{q^2},$$

or

$$B = \sqrt{\frac{\ln 2}{8\pi^2} \cdot k/n} \cdot \frac{q}{\sigma}. \tag{5}$$

The expression for t is then

$$2^t = \frac{\ln 2}{8\pi^2} \cdot k \cdot q^2/\sigma^2. \tag{6}$$

Each of the t steps should deliver $m = 2^k$ vectors of the form described before.

Since we have two parts in each reduction step, we need to analyze these parts separately. First, consider performing the first part of reduction step number i using coded-BKW with an $[n_i, d_i]$ linear code, where the parameters n_i and d_i at each step are chosen for optimal (global) performance. We sort the 2^k vectors into 2^{d_i} different lists. Here the coded-BKW step guarantees that all the vectors in a list, restricted to the n_i considered positions, have an average norm less than $\sqrt{n_i} \cdot B$ if the codeword is subtracted from the vector. So the number of lists (2^{d_i}) has to be chosen so that this norm restriction is true. Then, after the coded-BKW step, the sieving step should leave the average norm over the N_i positions unchanged, i.e., less than $\sqrt{N_i} \cdot B$.

Since all vectors in a list can be considered to have norm $\sqrt{N_i} \cdot B$ in these N_i positions, the sieving step needs to find any pair that leaves a difference between two vectors of norm at most $\sqrt{N_i} \cdot B$. From the theory of sieving in lattices, we know that heuristics imply that a single list should contain at least $2^{0.208N_i}$ vectors to be able to produce the same number of vectors. The time and space complexity is $2^{0.292N_i}$ if LSF is employed.

Let us adopt some further notation. As we expect the number of vectors to be exponential we write $k = c_0 n$ for some c_0. Also, we adopt $q = n^{c_q}$ and $\sigma = n^{c_s}$. Then $B = C \cdot n^{c_q - c_s}$ and $t = \log_2 D + (2(c_q - c_s) + 1)\log_2 n$, for some constants C and D.

6.1 Asymptotics of Coded-BKW with Sieving

We assume exponential overall complexity and write it as 2^{cn} for some coefficient c to be determined. Each step is additive with respect to complexity, so we assume that we can use 2^{cn} operations in each step.

In the t steps we are choosing $n_1, n_2, \ldots$ positions for each step.

The number of buckets needed for the first step of coded-BKW is $(C' \cdot n^{c_s})^{n_1}$, where C' is another constant. In each bucket the dominant part in the time complexity is the sieving cost $2^{\lambda n_1}$, for a constant λ. The overall complexity, the product of these expressions, should match the bound 2^{cn}, and thus we choose n_1 such that $(C' \cdot n^{c_s})^{n_1} \approx 2^{cn} \cdot 2^{-\lambda n_1}$.

Since $B = C \cdot n^{c_q - c_s}$ we can in the first step use n_1 positions in such a way that $(C' \cdot n^{c_s})^{n_1} \approx 2^{cn} \cdot 2^{-\lambda n_1}$, where C' is another constant and λ is the constant hidden in the exponent of the sieving complexity.

Taking the log, $c_s \log n \cdot n_1 + \log C' n_1 = cn - \lambda n_1$. Therefore, we obtain

$$n_1 = \frac{cn}{c_s \log n + \lambda + \log C'}.$$

To simplify expressions, we use the notation $W = c_s \log n + \lambda + \log C'$.

For the next step, we get $W \cdot n_2 = cn - \lambda n_1$, which simplifies in asymptotic sense to

$$n_2 = \frac{cn}{W}\left(1 - \frac{\lambda}{W}\right).$$

Continuing in this way, we have $W \cdot n_i = cn - \lambda \sum_{j=1}^{i-1} n_j$ and we can obtain an asymptotic expression for n_i as

$$n_i = \frac{cn}{W}\left(1 - \frac{\lambda}{W}\right)^{i-1}.$$

After t steps we have $\sum_{i=1}^{t} n_i = n$, so we observe that

$$\sum_{i=1}^{t} n_i = \frac{cn}{W} \sum_{i=1}^{t} \left(1 - \frac{\lambda}{W}\right)^{i-1},$$

which simplifies to

$$n = \sum_{i=1}^{t} n_i = \frac{cn}{\lambda}\left(1 - \left(1 - \frac{\lambda}{W}\right)^{t}\right).$$

Now, we know that

$$c = \lambda \left(1 - \left(1 - \frac{\lambda}{W}\right)^{t}\right)^{-1}.$$

Since t and W are both of order $\Theta(\log n)$ that tend to infinity as n tends to infinity, we have that

$$c = \lambda \left(1 - \left(1 - \frac{\lambda}{W}\right)^{\frac{W}{\lambda} \cdot \frac{t\lambda}{W}}\right)^{-1} \to \lambda \left(1 - e^{-\frac{t\lambda}{W}}\right)^{-1},$$

when $n \to \infty$.

Since $t/W \to (1 + 2(c_q - c_s))/c_s$ when $n \to \infty$ this finally gives us

$$c = \frac{\lambda}{1 - e^{-\lambda(1+2(c_q-c_s))/c_s}}.$$

Theorem 1. *The time and space complexity of the proposed algorithm is* $2^{(c+o(1))n}$, *where*

$$c = \frac{\lambda}{1 - e^{-\lambda(1+2(c_q-c_s))/c_s}},$$

and $\lambda = 0.292$ *for classic computers and* 0.265 *for quantum computers.*

Proof. Since $c > \lambda$, there are exponential samples left for the distinguishing process. One can adjust the constants in (5) and (6) to ensure the success probability of hypothesis testing close to 1. □

6.2 Asymptotics When Using Plain BKW Pre-processing

In this section we show that Theorem 1 can be improved for certain LWE parameters. Let $t = t_0 + t_1$. We first derive the following lemma.

Lemma 1. *It is asymptotically beneficial to perform* t_0 *plain BKW steps, where* t_0 *is of order* $(2(c_q - c_s) + 1 - c_s/\lambda \cdot \ln(c_q/c_s)) \log n$, *if*

$$\frac{c_s}{\lambda} \ln \frac{c_q}{c_s} < 2(c_q - c_s) + 1.$$

Proof. Suppose in each plain BKW steps, we zero-out b positions. Therefore, we have that

$$q^b = \tilde{O}(2^{cn}),$$

and it follows that asymptotically b is of order $cn/(c_q \log n)$.

Because the operated positions in each step will decrease using coded-BKW with sieving, it is beneficial to replace a step of coded-BKW with sieving by a pre-processing step of plain BKW, if the allowed number of steps is large. We compute t_1 such that for $i \geq t_1$, $n_i \leq b$. That is,

$$\frac{cn}{W}\left(1 - \frac{\lambda}{W}\right)^{t_1 - 1} = \frac{cn}{c_q \log n}.$$

Thus, we derive that t_1 is of order $c_s/\lambda \cdot \ln(c_q/c_s) \cdot \log n$. □

If we choose $t_0 = t - t_1$ plain BKW steps, where t_1 is of order $c_s/\lambda \cdot \ln(c_q/c_s) \cdot \log n$ as in Lemma 1, then

$$n - t_0 b = \sum_{i=1}^{t_1} n_i = \frac{cn}{\lambda}\left(1 - \left(1 - \frac{\lambda}{W}\right)^{t_1}\right).$$

Thus

$$1 - \frac{c}{c_q}\left(2\left(c_q - c_s\right) + 1 - \frac{c_s}{\lambda}\ln\left(\frac{c_q}{c_s}\right)\right) = \frac{c}{\lambda}\left(1 - \frac{c_s}{c_q}\right).$$

Finally, we have the following theorem for characterizing its asymptotic complexity.

Theorem 2. *If $c > \lambda$ and $\frac{c_s}{\lambda}\ln\frac{c_q}{c_s} < 2(c_q - c_s) + 1$, then the time and space complexity of the proposed algorithm with plain BKW pre-processing is $2^{(c+o(1))n}$, where*

$$c = \left(\lambda^{-1}\left(1 - \frac{c_s}{c_q}\right) + c_q^{-1}\left(2\left(c_q - c_s\right) + 1 - \frac{c_s}{\lambda}\ln\left(\frac{c_q}{c_s}\right)\right)\right)^{-1},$$

and $\lambda = 0.292$ for classic computer and 0.265 for quantum computers.

Proof. The proof is similar to that of Theorem 1. □

6.3 CaseStudy: Asymptotic Complexity of Regev's Instances

In this part we present a case-study on the asymptotic complexity of Regev's parameter sets, a family of LWE instances with significance in public key cryptography.

Regev's instances: We pick that $q \approx n^2$ and $\sigma = n^{1.5}/(\sqrt{2\pi}\log_2^2 n)$ as suggested in [42].

Table 1. Asymptotic complexity of Regev's LWE parameter setting

Algorithm	Complexity exponent (c)
QS-BKW(w/ p)	0.8856
S-BKW(w/ p)	0.8951
S-BKW(w/o p)	0.9054
BKW2	0.9299
ENUM/DUAL	4.6720
DUAL-EXPSamples	1.1680

The asymptotic complexity of Regev's LWE instances is shown in Table 1. For this parameter set, we have $c_q = 2$ and $c_s = 1.5$, and the previously best algorithms in the asymptotic sense are the coded-BKW variants [23,29] (denoted BKW2 in this table) with time complexity $2^{0.9299n+o(n)}$. The item ENUM/DUAL represents the run time exponent of lattice reduction approaches using polynomial samples and exponential memory, while DUAL-EXPSamples represents the run time exponent of lattice reduction approaches using exponential samples and memory. Both values are computed according to formulas from [25], i.e., $2c_{BKZ} \cdot c_q/(c_q - c_s)^2$ and $2c_{BKZ} \cdot c_q/(c_q - c_s + 1/2)^2$, respectively. Here c_{BKZ} is chosen to be 0.292, the best constant that can be achieved heuristically [9].

We see from the table that the newly proposed algorithm coded-BKW with sieving outperforms the previous best algorithms asymptotically. For instance, the simplest strategy without plain BKW pre-processing, denoted S-BKW(w/o p), costs $2^{0.9054n+o(n)}$ operations, with pre-processing, the time complexity, denoted S-BKW(w/ p) is $2^{0.8951n+o(n)}$. Using quantum computers, the constant hidden in the exponent can be further reduced to 0.8856, shown in the table as QS-BKW(w/ p). Note that the exponent of the algebraic approach is much higher than that of the BKW variants for Regev's LWE instances.

6.4 A Comparison with the Asymptotic Complexity of Other Algorithms

A comparison between the asymptotic time complexity of coded-BKW with sieving and the previous best single-exponent algorithms is shown in Fig. 2, similar to the comparison made in [26]. We use pre-processing with standard BKW steps (see theorem 2), since that reduces the complexity of the coded-BKW with sieving algorithm for the entire plotted area.

Use of exponential space is assumed. Access to an exponential number of samples is also assumed. In the full version of this paper we will add a comparison between coded-BKW with sieving and the other algorithms, when only having access to a polynomial number of samples.

First of all we notice that coded-BKW with sieving behaves best for most of the area where BKW2 used to be the best algorithm. It also outperforms the dual algorithm with exponential number of samples on some areas where that algorithm used to be the best. It is also worth mentioning that the Regev instance is well within the area where coded-BKW with sieving performs best.

The area where BKW2 outperforms coded-BKW with sieving is not that cryptographically interesting, since $c_s < 0.5$, and thus Regev's reduction proof does not apply. By also allowing pre-processing with coded-BKW steps our new algorithm is guaranteed to be the best BKW algorithm. We will add an asymptotic expression for this version of the algorithm and update the plot in the full version of our paper.

It should be noted that the Arora-Ge algorithm has the best time complexity of all algorithms for $c_s < 0.5$.

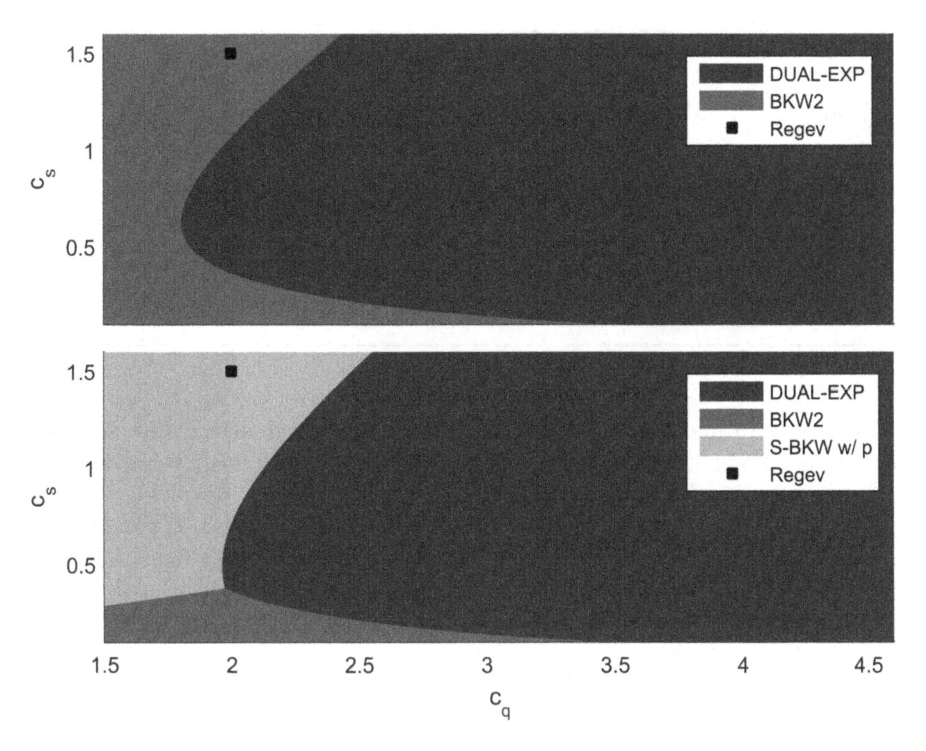

Fig. 2. A comparison between the asymptotic behavior of the best single-exponent algorithms for solving the LWE problem for different values of c_q and c_s. The different colored areas correspond to where the corresponding algorithm beats the other algorithms in that subplot.

7 Instantiations for Better Concrete Complexity

Until now we have described the new algorithm and analyzed its asymptotic performance. In this section, the practical instantiations for a better concrete solving complexity will be further discussed. We will include techniques and heuristics that may be negligible in the asymptotic sense, but has significant impact in practice.

The instantiation of Algorithm 1 for a good concrete complexity is shown in Algorithm 2, consisting of four main steps other than the pre-processing of changing the secret distribution. We include some coded-BKW steps which can be more efficient than sieving when the dimension of the elimination part of the vectors is small. Similar to [23], we also change the final step of guessing and performing hypothesis tests to a step of subspace hypothesis testing using the Fast Fourier Transformation (FFT). The required number of samples can thus be estimated as

$$8 \cdot l \cdot \log_2(q) \cdot e^{4\pi^2 \cdot \frac{\sigma_{final}^2}{q^2}},$$

Algorithm 2. Concrete instantiation of Algorithm 1

Input: Matrix $\mathbf{A}$ with n rows and m columns, received vector $\mathbf{z}$ of length m and parameters $t = t_0 + t_{1a} + t_{1b}$, n_i values and B.

change the distribution of the secret vector (Gaussian elimination)
perform t_0 plain BKW steps
perform t_{1a} coded-BKW steps
perform t_{1b} steps of coded-BKW with sieving
perform subspace hypothesis testing using FFT

where l is the dimension of the lattice codes employed in the last subspace hypothesis testing steps, and σ^2_{final} is the variance of final noise variable.

In the computation, we also assume that the complexity for solving the SVP problem with dimension k using heuristic sieving algorithms is $2^{0.292k+16.4}$ for $k > 90$; otherwise, for a smaller dimension k, the constant hidden in the run time exponent is chosen to be 0.387. This estimation was used by Albrecht et al. in their LWE-estimator [6]. They in turn derived the constant 16.4 from the experimental result in [30].

Table 2. Concrete time complexity for solving Regev's LWE instances.

n	q	σ	Complexity ($\log_2 \#\mathbb{Z}_2$)	
			This paper	†Coded-BKW [23]
256	65,537	25.53	134	149
512	262,147	57.06	252	292

†This estimation is also adopted in Albrecht's LWE-estimator [6].

A conservative estimate of the concrete complexity for solving Regev's LWE instances with dimension 256 and 512 is shown in Table 2. The complexity numbers are much smaller than those in [23], and the improvement is more significant when $n = 512$, with a factor of about 2^{40}. Therefore, other than its asymptotic importance, it can be argued that the algorithm coded-BKW with sieving is also efficient in practice.

We compare with the complexity numbers shown in [23] since their estimation also excludes the unnatural selection heuristic. Actually, we could have smaller complexity numbers by taking the influence of this heuristic into consideration, similar to what was done in the eprint version of [29].

The unnatural selection technique is a very useful heuristic to reduce the variance of noise in the vectorial part, as reported in the simulation section of [23]. On the other hand, the analysis of this heuristic is a bit tricky, and should be treated carefully.

We have implemented the new algorithm and tested it on small LWE instances. For ease of implementation, we chose to simulate using the LMS technique together with Laarhoven's angular LSH strategy [30]. While the simulation results bear no implications for the asymptotics, the performance for these small instances was as expected.

8 Conclusions and Future Work

In the paper we have presented a new BKW-type algorithm for solving the LWE problem. This algorithm, named coded-BKW with sieving, combines important ideas from two recent algorithmic improvements in lattice-based cryptography, i.e., coded-BKW and heuristic sieving for SVP, and outperforms the previously known approaches for important parameter sets in public-key cryptography.

For instance, considering Regev's parameter set, we have demonstrated an exponential asymptotic improvement, reducing the time and space complexity from $2^{0.930n}$ to $2^{0.895n}$ in the classic setting. We also obtain the first quantum acceleration for this parameter set, further reducing the complexity to $2^{0.886n}$ if quantum computers are provided.

This algorithm also has significant non-asymptotic improvements for some concrete parameters, compared with the previously best BKW variants. But one should further investigate the analysis when heuristics like unnatural selection are taken into consideration, in order to fully exploit its power on suggesting accurate security parameters for real cryptosystems.

The newly proposed algorithm definitely also has importance in solving many LWE variants with specific structures, e.g., the BINARY-LWE problem and the RING-LWE problem. An interesting research direction is to search for more applications, e.g., solving hard lattice problems, as in [29].

References

1. Ajtai, M., Kumar, R., Sivakumar, D.: A sieve algorithm for the shortest lattice vector problem. In: Proceedings of The Thirty-third Annual ACM Symposium on Theory of Computing, pp. 601–610. ACM (2001)
2. Albrecht, M., Cid, C., Faugere, J.C., Robert, F., Perret, L.: Algebraic algorithms for LWE problems. Cryptology ePrint Archive, report 2014/1018 (2014)
3. Albrecht, M.R., Cid, C., Faugere, J.C., Fitzpatrick, R., Perret, L.: On the complexity of the BKW algorithm on LWE. Des. Codes Crypt. **74**(2), 325–354 (2015)
4. Albrecht, M.R., Faugère, J.-C., Fitzpatrick, R., Perret, L.: Lazy modulus switching for the BKW algorithm on LWE. In: Krawczyk, H. (ed.) PKC 2014. LNCS, vol. 8383, pp. 429–445. Springer, Heidelberg (2014). https://doi.org/10.1007/978-3-642-54631-0_25
5. Albrecht, M.R., Fitzpatrick, R., Göpfert, F.: On the efficacy of solving LWE by reduction to unique-SVP. In: Lee, H.-S., Han, D.-G. (eds.) ICISC 2013. LNCS, vol. 8565, pp. 293–310. Springer, Cham (2014). https://doi.org/10.1007/978-3-319-12160-4_18

6. Albrecht, M.R., Player, R., Scott, S.: On the concrete hardness of learning with errors. J. Math. Cryptol. **9**(3), 169–203 (2015)
7. Applebaum, B., Cash, D., Peikert, C., Sahai, A.: Fast cryptographic primitives and circular-secure encryption based on hard learning problems. In: Halevi, S. (ed.) CRYPTO 2009. LNCS, vol. 5677, pp. 595–618. Springer, Heidelberg (2009). https://doi.org/10.1007/978-3-642-03356-8_35
8. Arora, S., Ge, R.: New algorithms for learning in presence of errors. In: Aceto, L., Henzinger, M., Sgall, J. (eds.) ICALP 2011. LNCS, vol. 6755, pp. 403–415. Springer, Heidelberg (2011). https://doi.org/10.1007/978-3-642-22006-7_34
9. Becker, A., Ducas, L., Gama, N., Laarhoven, T.: New directions in nearest neighbor searching with applications to lattice sieving. In: Proceedings of the Twenty-Seventh Annual ACM-SIAM Symposium on Discrete Algorithms, pp. 10–24. Society for Industrial and Applied Mathematics (2016)
10. Becker, A., Gama, N., Joux, A.: A sieve algorithm based on overlattices. LMS J. Comput. Math. **17**(A), 49–70 (2014)
11. Bernstein, D.J., Lange, T.: Never trust a bunny. In: Hoepman, J.-H., Verbauwhede, I. (eds.) RFIDSec 2012. LNCS, vol. 7739, pp. 137–148. Springer, Heidelberg (2013). https://doi.org/10.1007/978-3-642-36140-1_10
12. Blum, A., Kalai, A., Wasserman, H.: Noise-tolerant learning, the parity problem, and the statistical query model. J. ACM **50**(4), 506–519 (2003)
13. Bogos, S., Vaudenay, S.: Optimization of LPN solving algorithms. In: Cheon, J.H., Takagi, T. (eds.) ASIACRYPT 2016. LNCS, vol. 10031, pp. 703–728. Springer, Heidelberg (2016). https://doi.org/10.1007/978-3-662-53887-6_26
14. Brakerski, Z.: Fully homomorphic encryption without modulus switching from classical gapSVP. In: Safavi-Naini, R., Canetti, R. (eds.) CRYPTO 2012. LNCS, vol. 7417, pp. 868–886. Springer, Heidelberg (2012). https://doi.org/10.1007/978-3-642-32009-5_50
15. Brakerski, Z., Vaikuntanathan, V.: Efficient fully homomorphic encryption from (standard) LWE. In: Proceedings of the 2011 IEEE 52nd Annual Symposium on Foundations of Computer Science, pp. 97–106. IEEE Computer Society (2011)
16. Brakerski, Z., Vaikuntanathan, V.: Fully homomorphic encryption from ring-LWE and security for key dependent messages. In: Rogaway, P. (ed.) CRYPTO 2011. LNCS, vol. 6841, pp. 505–524. Springer, Heidelberg (2011). https://doi.org/10.1007/978-3-642-22792-9_29
17. Chen, Y., Nguyen, P.Q.: BKZ 2.0: better lattice security estimates. In: Lee, D.H., Wang, X. (eds.) ASIACRYPT 2011. LNCS, vol. 7073, pp. 1–20. Springer, Heidelberg (2011). https://doi.org/10.1007/978-3-642-25385-0_1
18. Conway, J.H., Sloane, N.J.A.: Sphere Packings, Lattices and Groups, vol. 290. Springer, Heidelberg (2013)
19. Dubiner, M.: Bucketing coding and information theory for the statistical high-dimensional nearest-neighbor problem. IEEE Trans. Inf. Theory **56**(8), 4166–4179 (2010)
20. Duc, A., Tramèr, F., Vaudenay, S.: Better algorithms for LWE and LWR. In: Oswald, E., Fischlin, M. (eds.) EUROCRYPT 2015. LNCS, vol. 9056, pp. 173–202. Springer, Heidelberg (2015). https://doi.org/10.1007/978-3-662-46800-5_8
21. Gentry, C., Sahai, A., Waters, B.: Homomorphic encryption from learning with errors: conceptually-simpler, asymptotically-faster, attribute-based. In: Canetti, R., Garay, J.A. (eds.) CRYPTO 2013. LNCS, vol. 8042, pp. 75–92. Springer, Heidelberg (2013). https://doi.org/10.1007/978-3-642-40041-4_5

22. Guo, Q., Johansson, T., Löndahl, C.: Solving LPN using covering codes. In: Sarkar, P., Iwata, T. (eds.) ASIACRYPT 2014. LNCS, vol. 8873, pp. 1–20. Springer, Heidelberg (2014). https://doi.org/10.1007/978-3-662-45611-8_1
23. Guo, Q., Johansson, T., Stankovski, P.: Coded-BKW: solving LWE using lattice codes. In: Gennaro, R., Robshaw, M. (eds.) CRYPTO 2015. LNCS, vol. 9215, pp. 23–42. Springer, Heidelberg (2015). https://doi.org/10.1007/978-3-662-47989-6_2
24. Hanrot, G., Pujol, X., Stehlé, D.: Algorithms for the shortest and closest lattice vector problems. In: Chee, Y.M., Guo, Z., Ling, S., Shao, F., Tang, Y., Wang, H., Xing, C. (eds.) IWCC 2011. LNCS, vol. 6639, pp. 159–190. Springer, Heidelberg (2011). https://doi.org/10.1007/978-3-642-20901-7_10
25. Herold, G., Kirshanova, E., May, A.: On the asymptotic complexity of solving LWE. IACR Cryptology ePrint Archive 2015, 1222 (2015). http://eprint.iacr.org/2015/1222
26. Herold, G., Kirshanova, E., May, A.: On the asymptotic complexity of solving LWE. J. Des. Codes Crypt. 1–29 (2017). https://doi.org/10.1007/s10623-016-0326-0
27. Indyk, P., Motwani, R.: Approximate nearest neighbors: towards removing the curse of dimensionality. In: Proceedings of the Thirtieth Annual ACM Symposium on Theory of Computing, pp. 604–613. ACM (1998)
28. Kirchner, P.: Improved generalized birthday attack. Cryptology ePrint Archive, report 2011/377 (2011). http://eprint.iacr.org/
29. Kirchner, P., Fouque, P.-A.: An improved BKW algorithm for LWE with applications to cryptography and lattices. In: Gennaro, R., Robshaw, M. (eds.) CRYPTO 2015. LNCS, vol. 9215, pp. 43–62. Springer, Heidelberg (2015). https://doi.org/10.1007/978-3-662-47989-6_3
30. Laarhoven, T.: Sieving for shortest vectors in lattices using angular locality-sensitive hashing. In: Gennaro, R., Robshaw, M. (eds.) CRYPTO 2015. LNCS, vol. 9215, pp. 3–22. Springer, Heidelberg (2015). https://doi.org/10.1007/978-3-662-47989-6_1
31. Laarhoven, T., Mosca, M., Van De Pol, J.: Finding shortest lattice vectors faster using quantum search. Des. Codes Crypt. **77**(2–3), 375–400 (2015)
32. Laarhoven, T., de Weger, B.: Faster sieving for shortest lattice vectors using spherical locality-sensitive hashing. In: Lauter, K., Rodríguez-Henríquez, F. (eds.) LATINCRYPT 2015. LNCS, vol. 9230, pp. 101–118. Springer, Cham (2015). https://doi.org/10.1007/978-3-319-22174-8_6
33. Levieil, É., Fouque, P.-A.: An improved LPN algorithm. In: Prisco, R., Yung, M. (eds.) SCN 2006. LNCS, vol. 4116, pp. 348–359. Springer, Heidelberg (2006). https://doi.org/10.1007/11832072_24
34. Lindner, R., Peikert, C.: Better key sizes (and attacks) for LWE-based encryption. In: Kiayias, A. (ed.) CT-RSA 2011. LNCS, vol. 6558, pp. 319–339. Springer, Heidelberg (2011). https://doi.org/10.1007/978-3-642-19074-2_21
35. Liu, M., Nguyen, P.Q.: Solving BDD by enumeration: an update. In: Dawson, E. (ed.) CT-RSA 2013. LNCS, vol. 7779, pp. 293–309. Springer, Heidelberg (2013). https://doi.org/10.1007/978-3-642-36095-4_19
36. May, A., Ozerov, I.: On computing nearest neighbors with applications to decoding of binary linear codes. In: Oswald, E., Fischlin, M. (eds.) EUROCRYPT 2015. LNCS, vol. 9056, pp. 203–228. Springer, Heidelberg (2015). https://doi.org/10.1007/978-3-662-46800-5_9
37. Micciancio, D., Regev, O.: Lattice-based cryptography. In: Bernstein, D.J., Buchmann, J., Dahmen, E. (eds.) Post-Quantum Cryptography, pp. 147–191. Springer, Berlin Heidelberg (2009)

38. Micciancio, D., Voulgaris, P.: Faster exponential time algorithms for the shortest vector problem. In: Proceedings of the Twenty-first Annual ACM-SIAM Symposium on Discrete Algorithms, pp. 1468–1480. SIAM (2010)
39. Mulder, E.D., Hutter, M., Marson, M.E., Pearson, P.: Using Bleichenbacher's solution to the hidden number problem to attack nonce leaks in 384-bit ECDSA: extended version. J. Crypt. Eng. **4**(1), 33–45 (2014). http://dx.doi.org/10.1007/s13389-014-0072-z
40. Nguyen, P.Q., Vidick, T.: Sieve algorithms for the shortest vector problem are practical. J. Math. Cryptol. **2**(2), 181–207 (2008). http://dx.doi.org/10.1515/JMC.2008.009
41. Pujol, X., Stehlé, D.: Solving the shortest lattice vector problem in time $2^{2.465n}$. IACR Cryptology ePrint Archive 2009, 605 (2009). http://eprint.iacr.org/2009/605
42. Regev, O.: On lattices, learning with errors, random linear codes, and cryptography. J. ACM **56**(6), 34:1–34:40 (2009). http://doi.acm.org/10.1145/1568318.1568324
43. Schnorr, C.P., Euchner, M.: Lattice basis reduction: improved practical algorithms and solving subset sum problems. Math. Program. **66**(1–3), 181–199 (1994)
44. Wagner, D.: A generalized birthday problem. In: Yung, M. (ed.) CRYPTO 2002. LNCS, vol. 2442, pp. 288–304. Springer, Heidelberg (2002). https://doi.org/10.1007/3-540-45708-9_19
45. Wang, X., Liu, M., Tian, C., Bi, J.: Improved Nguyen-Vidick heuristic sieve algorithm for shortest vector problem. In: Proceedings of the 6th ACM Symposium on Information, Computer and Communications Security, pp. 1–9. ACM (2011)
46. Zhang, B., Jiao, L., Wang, M.: Faster algorithms for solving LPN. In: Fischlin, M., Coron, J.-S. (eds.) EUROCRYPT 2016. LNCS, vol. 9665, pp. 168–195. Springer, Heidelberg (2016). https://doi.org/10.1007/978-3-662-49890-3_7
47. Zhang, F., Pan, Y., Hu, G.: A three-level sieve algorithm for the shortest vector problem. In: Lange, T., Lauter, K., Lisoněk, P. (eds.) SAC 2013. LNCS, vol. 8282, pp. 29–47. Springer, Heidelberg (2014). https://doi.org/10.1007/978-3-662-43414-7_2

Sharper Bounds in Lattice-Based Cryptography Using the Rényi Divergence

Thomas Prest(✉)

Paris, France
tprest@protonmail.com
https://tprest.github.io/

Abstract. The Rényi divergence is a measure of divergence between distributions. It has recently found several applications in lattice-based cryptography. The contribution of this paper is twofold.

First, we give theoretic results which renders it more efficient and easier to use. This is done by providing two lemmas, which give tight bounds in very common situations – for distributions that are tailcut or have a bounded relative error. We then connect the Rényi divergence to the max-log distance. This allows the Rényi divergence to indirectly benefit from all the advantages of a distance.

Second, we apply our new results to five practical usecases. It allows us to claim 256 bits of security for a floating-point precision of 53 bits, in cases that until now either required more than 150 bits of precision or were limited to 100 bits of security: rejection sampling, trapdoor sampling (61 bits in this case) and a new sampler by Micciancio and Walter. We also propose a new and compact approach for table-based sampling, and squeeze the standard deviation of trapdoor samplers by a factor that provides a gain of 30 bits of security in practice.

Keywords: Rényi divergence · Security proofs · Lattice-based cryptography · Gaussian sampling

1 Introduction

An essential tool in cryptography is the use of divergence measures to prove the security of cryptographic schemes. As an introductory example, we consider the statistical distance Δ. It verifies a probability preservation property, which states that for any two distributions $\mathcal{P}, \mathcal{Q}$ and any measureable event E over the support of $\mathcal{P}$ and $\mathcal{Q}$, we have

$$\mathcal{Q}(E) \geq \mathcal{P}(E) - \Delta(\mathcal{P}, \mathcal{Q}). \tag{1}$$

In a cryptographic context, a useful abstraction is to modelize a cryptographic scheme as relying on some ideal distribution $\mathcal{Q}$ and the success of an attacker against this scheme as an event E. If $\Delta(\mathcal{P}, \mathcal{Q})$ is negligible, the Eq. 1 will allow to say that a scheme secure with $\mathcal{Q}$ will stay secure if one replaces $\mathcal{Q}$ by an "imperfect" distribution $\mathcal{P}$. Many other measures can be used to provide security arguments in cryptography (see e.g. [Cac97]).

T. Takagi and T. Peyrin (Eds.): ASIACRYPT 2017, Part I, LNCS 10624, pp. 347–374, 2017.
https://doi.org/10.1007/978-3-319-70694-8_13

The Rényi divergence. In the subfield of lattice-based crytography, the Rényi divergence [R61] has been used for cryptographic proofs in several recent works. Noted R_a, it is somewhat trickier to use than the statistical distance. First, it is parameterized by a value $a \in [0, +\infty]$, and has different properties depending on a. It is not a distance, as it is asymmetric and does not verify the triangle inequality; the lack of these two properties can be problematic in security proofs. Interestingly, it also verifies a probability preservation property. For any event $E \subseteq \mathrm{Supp}(\mathcal{Q})$ and $a \in (1, +\infty)$, we have

$$\mathcal{Q}(E) \geq \mathcal{P}(E)^{a/(a-1)} / R_a(\mathcal{P} \| \mathcal{Q}). \tag{2}$$

The Eq. 2 is not additive like Eq. 1, but rather multiplicative. We will later see that in the context of search problems, it allows to give tighter bounds in practice.

1.1 Floating-Point in Lattice-Based Cryptography

Lattice-based cryptography has proven to be a serious candidate for post-quantum cryptography. It is efficient and allows to instantiate a wide range of cryptographic primitives. Some lattice-based schemes [DDLL13, ADPS16] have even already been deployed in large-scale projects.[1]

A notable characteristic of lattice-based cryptography is that it often makes extensive use of floating-point arithmetic, for several reasons.

Gaussians. The first vector for the use of floating-point arithmetic in lattice-based cryptography is the widespread need to sample from discrete Gaussian distributions. When done by standard approaches like precomputed tables, [Pei10] the required precision is rather high and renders the use of these tables cumbersome if not impractical.

On the other hand, bitwise approaches [DDLL13] have been developed to circumvent these floating-point issues, but they can be somewhat tricky to implement.

Rejection sampling. In the early lattice-based signature schemes GGH [GGH97] and NTRUSign [HHGP+03], there existed a correlation between the secret key and the distribution of the signatures. This subsequently led to several key-recovery attacks [GJSS01, GS02, NR06, Wan10, DN12b] which broke the signature schemes and their evolutions.

A provably secure countermeasure was introduced by Lyubashevsky [Lyu09]. The idea is to use rejection sampling as a final step, in order to "factor out" the correlation between the key and the distribution of the signatures.

This paradigm was instantiated in [Lyu12, GLP12, DDLL13, PDG14, POG15]. Now, in the existing implementations [DDLL13], this step is *not* done in floating-point. Because of precision concerns, another approach based on combining Bernoulli samples was chosen. We will see in Sect. 4.3 that this approach also has several drawbacks.

[1] [Str14] and https://www.imperialviolet.org/2016/11/28/cecpq1.html.

Trapdoor sampling. In lattice-based cryptography, the tool that makes the most intensive use of floating-point arithmetic is arguably trapdoor sampling. Introduced by Gentry et al. [GPV08], it is a cornerstone of lattice-based cryptography, as it has numerous applications such as hash-and-sign and identity-based encryption in the random oracle model [GPV08], signatures in the standard model [CHKP10, Boy10], hierarchical IBE [CHKP10, ABB10a, ABB10b], attribute-based encryption [Boy13, BGG+14], and much more.

The existing algorithms [Kle00, GPV08, Pei10, MP12] heavily rely on floating-point arithmetic and they perform between $O(n \log n)$ and $O(n^2)$ floating-point operations. However, the best available estimations require 150 bits of precision for a security of 256 bits, which is completely impractical.

As we can see, floating-point arithmetic can be found everywhere in lattice-based cryptography. However, if often comes with high precision, which makes it impractical as it stands.

1.2 Our Contributions

Theory. We provide theoretic tools related to the use of the Rényi divergence in cryptographic proofs. They make it not only simpler to use, but also very efficient in some easily-identifiable situations.

1. We establish two lemmas that bound the Rényi divergence of related distributions in two very common situations in lattice-based cryptography. The first lemma concerns tailcut distributions, and for this reason we call it the tailcut lemma. The second one involves distributions which relative error is bounded, so we call it the relative error lemma. The second lemma is particularly powerful in the sense that it often allows to take very aggressive parameters.
2. We show that taking $a = 2\lambda$ allows to have tight and efficient Rényi divergence-based security arguments for cryptographic schemes based on search problems. We also derive simple and explicit conditions on distributions that allow to easily replace a distribution by another in this context.
3. A simple and versatile distance of divergence was recently introduced by Micciancio and Walter [MW17], the max-log distance. We establish a "reverse Pinsker" inequality between it and the Rényi divergence. An immediate consequence is that we may benefit from the best of both worlds: the versatility of the max-log distance, and the power of the Rényi divergence.

Practice. Our results are not purely theoretic. In Sect. 4, we present five applications of them in lattice-based cryptography.

1. We start by the study of a sampler recently introduced by Micciancio and Walter [MW17]. We show that for this sampler, the security analysis provided by [MW17] can be improved and we can claim a full security of 256 bits instead of the 100 bits claimed in [MW17].

2. We revisit the table-based approach (see e.g. [Pei10]) for sampling distributions such as discrete Gaussians. By a Rényi divergence-based analysis combined to a little tweak on the precomputed table, we reduce the storage size by an order of magnitude, both in theory and in practice (where we gain a factor 9). Our improvement seems highly composable with other techniques related to precomputed tables.
3. We analyze the rejection sampling step of BLISS [DDLL13]. We show that it can be done simply and efficiently in floating-point, simultaneously eliminating the issues – code complexity, side-channel attacks, table storage, etc. – that plagued the only previously existing approach.
4. We then study trapdoor samplers [Kle00, GPV08, Pei10]. We improve the usual bounds on the standard deviation σ by obtaining a new bound which is both smaller and essentially independent of the security level λ. In practice, we gain about 30 bits of security compared to a statistical distance-based analysis.
5. The last contribution is also related to trapdoor samplers. We show that a precision of 64 bits allows 256 bits of security, whereas previous estimations [LP15, Pre15] required a precision of 150 bits.

A word on the security parameter and number of queries. In order to make our results as simple as possible and to derive explicit bounds, we consider in this paper that the security level λ and the number of queries q_s verify $\lambda \leq 256$ and $q_s \leq 2^{64}$. The first choice is arguably standard.

For the bound on q_s, we consider that making more than 2^{64} signature queries would be extremely costly and, unlike queries to e.g. a hash function, require the presence of the target to attack. In addition, it would be easily detectable by the target and so we believe it to be impractical.

Finally, a more pragmatic reason comes from NIST's current call for proposals for post-quantum cryptography,[2] which explicitly assumes that an attacker can make no more than 2^{64} signatures queries (resp. decryption queries).

However, if one decides to take $q_s > 2^{64}$, our results could be easily adapted, but their efficiency would be impacted.

1.3 Related Works

In the context of lattice-based cryptography, Stehlé, Steinfeld and their coauthors [LSS14, LPSS14, BLL+15] have used the Rényi divergence to derive better parameters for cryptographic schemes. The Rényi divergence has also been used by [BGM+16] to improve security proofs, and in [TT15], which aims to improve the proofs from [BLL+15].

A few papers [PDG14, DLP14] used a third metric, the Kullback-Leibler divergence – actually the Rényi divergence of order 1 –, but the Rényi divergence has since then given better results [BLL+15, this work].

[2] http://csrc.nist.gov/groups/ST/post-quantum-crypto/.

Precision issues have been tackled by [DN12a], which resorted to lazy Gaussian sampling but still didn't eliminate high-precision. A precision analysis of trapdoor samplers by Prest [Pre15] gave about 120 bits of precision for $\lambda = 192$ – which we extrapolate to 150 for $\lambda = 256$. A recent work by Saarinen [Saa15] has also claimed that using p-bit fixed point approximation achieves $2p$ bits of security, but this was proven to be incorrect by [MW17], which also introduced the max-log distance.

Finally, recent works [BS16, Mir17] have studied the usefulness of the Rényi divergence in the context of differential privacy and have independently come up with results similar to our relative error lemma.

1.4 Roadmap

Section 2 introduces the notations and tools that we will use throughout the paper, including the Rényi divergence.

Section 3 is dedicated to our theoretic results. We first present the tailcut and relative error lemmas, as well as typical usecases for their applications. We give a framework for using them in cryptographic proofs, along with explicit bounds. Finally, we establish a connection between the Rényi divergence and the max-log distance.

Section 4 presents five applications of our theoretic results. We first give a tighter analysis of a sampler from [MW17], then we revisit the standard table-based approach for sampling Discrete distributions. We then show that rejection sampling in BLISS can be done simply in floating-point arithmetic. To conclude, we study trapdoor samplers and provide improved bounds on the standard deviation and precision with which they can be used.

Section 5 concludes this article and presents related open problems.

2 Preliminaries

2.1 Notations

Cryptographic parameters. When clear from context, let λ be the security level of a scheme and q_s the number of public queries that an attacker can make. In this article, we consider that $\lambda \leq 256$ and $q_s \leq 2^{64}$.

Probabilities. For any distribution $\mathcal{D}$, we denote its support by $\mathrm{Supp}(\mathcal{D})$. We may abbreviate the statistical distance and Kullback-Leibler divergence by SD and KLD. As a mnemonic device, we will often refer to $\mathcal{D}$ as some perfect distribution, and to $\mathcal{D}_\delta$ as a distribution close to $\mathcal{D}$ in a sense parameterized by δ.

Matrices and vectors. Matrices will usually be in bold uppercase (e.g. $\mathbf{B}$), vectors in bold lowercase (e.g. $\mathbf{v}$) and scalars in italic (e.g. s). Vectors are represented as rows. The p-norm of an vector $\mathbf{v}$ is denoted by $\|\mathbf{v}\|_p$, and by convention $\|\mathbf{v}\| = \|\mathbf{v}\|_2$. Let $\|\mathbf{B}\|_2 = \max_{\mathbf{x}\neq 0} \|\mathbf{xB}\|_2/\|\mathbf{x}\|_2$ be the spectral norm of a matrix, it is also the maximum of its singular values and is sometimes denoted by $s_1(\mathbf{B})$. For $\mathbf{B} = (b_{ij})_{i,j}$, we define the max norm of $\mathbf{B}$ as $\|\mathbf{B}\|_{\max} = \max_{i,j} |b_{ij}|$.

Gram-Schmidt orthogonalization. An important tool in lattice-based cryptography is the Gram-Schmidt orthogonalization of a full-rank matrix $\mathbf{B}$, which is the unique factorization $\mathbf{B} = \mathbf{L} \cdot \tilde{\mathbf{B}}$ such that $\mathbf{L}$ is lower triangular with 1's on the diagonal, and $\tilde{\mathbf{B}}$ is orthogonal. Noting $\tilde{\mathbf{B}} = (\tilde{\mathbf{b}}_i)_i$, it allows to define the Gram-Schmidt norm, defined as $\|\mathbf{B}\|_{\text{GS}} = \max_i \|\tilde{\mathbf{b}}_i\|$.

Lattices and Gaussians. A lattice will be denoted by Λ. For a matrix $\mathbf{B} \in \mathbb{R}^{n \times m}$, let $\Lambda(\mathbf{B})$ be the lattice generated by $\mathbf{B}$: $\Lambda(\mathbf{B}) = \mathbb{Z}^n \cdot \mathbf{B}$. We define the Gaussian function $\rho_{\sigma,\mathbf{c}}$ as $\rho_{\sigma,\mathbf{c}}(\mathbf{x}) = \exp(-\|\mathbf{x} - \mathbf{c}\|^2/2\sigma^2)$, and the Gaussian distribution $D_{\Lambda,\sigma,\mathbf{c}}$ over a lattice as

$$D_{\Lambda,\sigma,\mathbf{c}}(\mathbf{x}) = \frac{\rho_{\sigma,\mathbf{c}}(\mathbf{x})}{\sum_{\mathbf{z}\in\Lambda} \rho_{\sigma,\mathbf{c}}(\mathbf{z})}$$

The parameter $\mathbf{c}$ may be omitted when it is equal to zero.

Smoothing parameter. For $\epsilon > 0$, we define the smoothing parameter $\eta_\epsilon(\Lambda)$ of a lattice as the smallest value $\sigma > 0$ such that $\rho_{1/\sigma}(\Lambda^\star \backslash \mathbf{0}) \leq \epsilon$. We carefully note that in the existing literature, some definitions take the smoothing parameter to be our definition multiplied by a factor $\sqrt{2\pi}$. A useful bound on the smoothing parameter is given by [MR07]:

$$\eta_\epsilon(\mathbb{Z}^n) \leq \frac{1}{\pi}\sqrt{\frac{1}{2}\log\left(2n\left(1 + \frac{1}{\epsilon}\right)\right)}. \tag{3}$$

2.2 The Rényi Divergence

We define the Rényi divergence in the same way as [BLL+15].

Definition 1 (Rényi divergence). *Let $\mathcal{P}, \mathcal{Q}$ be two distributions such that $Supp(\mathcal{P}) \subseteq Supp(\mathcal{Q})$. For $a \in (1, +\infty)$, we define the Rényi divergence of order a by*

$$R_a(\mathcal{P}\|\mathcal{Q}) = \left(\sum_{x\in Supp(\mathcal{P})} \frac{\mathcal{P}(x)^a}{\mathcal{Q}(x)^{a-1}}\right)^{\frac{1}{a-1}}.$$

In addition, we define the Rényi divergence of order $+\infty$ by

$$R_\infty(\mathcal{P}\|\mathcal{Q}) = \max_{x\in Supp(\mathcal{P})} \frac{\mathcal{P}(x)}{\mathcal{Q}(x)}.$$

Again, this definition is slightly different from some other existing definitions, which take the log of ours. However, it is more convenient for our purposes. Generic (resp. cryptographic) properties of the Rényi divergence can be found in [vEH14] (resp. [BLL+15]). We recall the most important ones.

Lemma 1 [BLL+15, **Lemma 2.9**]. *For two distributions $\mathcal{P}, \mathcal{Q}$ and two families of distributions $(\mathcal{P}_i)_i, (\mathcal{Q}_i)_i$, the Rényi divergence verifies the following properties:*

- ***Data processing inequality.*** *For any function f, $R_a(\mathcal{P}^f \| \mathcal{Q}^f) \leq R_a(\mathcal{P} \| \mathcal{Q})$.*
- ***Multiplicativity.*** *$R_a(\prod_i \mathcal{P}_i \| \prod_i \mathcal{Q}_i) = \prod_i R_a(\mathcal{P}_i \| \mathcal{Q}_i)$.*
- ***Probability preservation.*** *For any event $E \subseteq Supp(\mathcal{Q})$ and $a \in (1, +\infty)$,*

$$\mathcal{Q}(E) \geq \mathcal{P}(E)^{a/(a-1)} / R_a(\mathcal{P} \| \mathcal{Q}),$$
$$\mathcal{Q}(E) \geq \mathcal{P}(E) / R_\infty(\mathcal{P} \| \mathcal{Q}).$$

However, we note that the Rényi divergence is not a distance. In Sect. 3.4, we circumvent this issue by linking the Rényi divergence to the max-log distance.

3 Main Results

In this section, we present our theoretic results: the tailcut lemma and relative error lemma for bounding the Rényi divergence between distributions, a generic framework for using these lemmas and a "reverse Pinsker" inequality that connects the Rényi divergence to the max-log distance.

3.1 The Tailcut Lemma

This first lemma may arguably be considered as folklore; it is already briefly mentioned in e.g. [BLL+15]. Here we make it explicit, as applications of it arise naturally in lattice-based cryptography, especially whenever Gaussians distributions are used.

Lemma 2 (Tailcut). *Let $\mathcal{D}, \mathcal{D}_\delta$ be two distributions such that:*

- *$\exists \delta > 0$ such that $\frac{\mathcal{D}_\delta}{\mathcal{D}} \leq 1 + \delta$ over $Supp(\mathcal{D}_\delta)$*

Then for $a \in (1, +\infty]$:

$$R_a(\mathcal{D}_\delta \| \mathcal{D}) \leq 1 + \delta$$

Proof. We note $S = \text{Supp}(\mathcal{D}_\delta)$. If $a \neq +\infty$:

$$R_a(\mathcal{D}_\delta \| \mathcal{D})^{a-1} = \sum_{x \in S} \frac{\mathcal{D}_\delta(x)^a}{\mathcal{D}(x)^{a-1}} \leq (1+\delta)^{a-1} \sum_{x \in S} \mathcal{D}_\delta(x) \leq (1+\delta)^{a-1},$$

which yields the result. If $a = +\infty$, the result is immediate. □

We may also refer to Lemma 2 as the tailcut lemma. For the rest of the paper, $\mathcal{D}$ will typically refer to a "perfect" distribution, and $\mathcal{D}_\delta$ to a distribution which is close to $\mathcal{D}$ in a sense parameterized by δ.

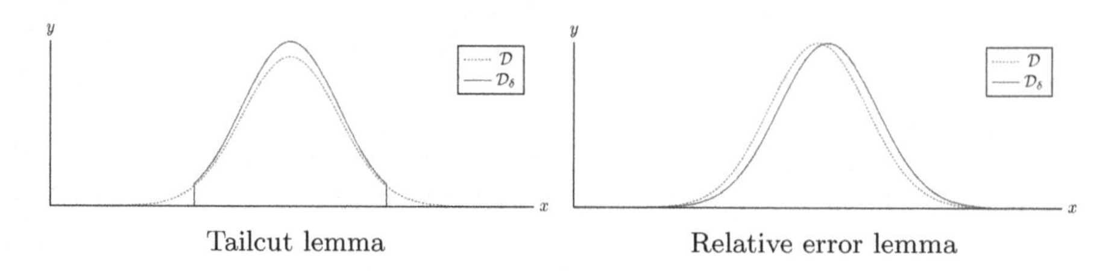

Fig. 1. Typical usecases for the tailcut lemma and the relative error lemma

Usecases. As its name implies, the tailcut lemma is adapted to situations where $\mathcal{D}_\delta$ is a "tailcut" of $\mathcal{D}$: we discard a set $T \subseteq \text{Supp}(\mathcal{D})$ such that $\mathcal{D}(T) \leq \delta$. In order to still have a true measure of probability, the remaining probabilities are scaled by a factor $\frac{1}{1-\mathcal{D}(T)} \approx 1 + \mathcal{D}(T) \leq 1 + \delta$, and we note $\mathcal{D}_\delta$ the new distribution. Lemma 2 gives a relation of closeness between $\mathcal{D}$ and $\mathcal{D}_\delta$ in this case, which is illustrated by the Fig. 1.

3.2 The Relative Error Lemma

In our second lemma, the conditions are slightly stricter than for the tailcut lemma, but as a compensation the result is a much stronger closeness relation. It is somewhat similar to the [PDG14, Lemma 2] for the KLD, but allows tighter security arguments.

Lemma 3 (Relative error). *Let $\mathcal{D}, \mathcal{D}_\delta$ be two distributions such that:*

- *$Supp(\mathcal{D}_\delta) = Supp(\mathcal{D})$*
- *$\exists \delta > 0$ such that $1 - \delta \leq \frac{\mathcal{D}_\delta}{\mathcal{D}} \leq 1 + \delta$ over $Supp(\mathcal{D}_\delta)$*

Then, for $a \in (1, +\infty)$:

$$R_a(\mathcal{D}_\delta || \mathcal{D}) \leq \left(1 + \frac{a(a-1)\delta^2}{2(1-\delta)^{a+1}}\right)^{\frac{1}{a-1}} \underset{\delta \to 0}{\sim} 1 + \frac{a\delta^2}{2}$$

Proof. Let $f_a : (x, y) \mapsto \frac{y^a}{(x+y)^{a-1}}$. We compute values of f_a and its derivatives around $(0, y)$:

$$\begin{aligned} f_a(x,y) &= y & &\text{for } x = 0 \\ \frac{\partial f_a}{\partial x}(x,y) &= 1 - a & &\text{for } x = 0 \\ \frac{\partial^2 f_a}{\partial x^2}(x,y) &= a(a-1)y^a(x+y)^{-a-1} & & \\ &\leq \frac{a(a-1)}{(1-\delta)^{a+1}y} & &\text{for } |x| \leq \delta \cdot y \end{aligned}$$

We now use partial Taylor bounds. If $|x| \leq \delta \cdot y$, then:

$$f_a(x,y) \leq f_a(0,y) + \frac{\partial f_a}{\partial x}(0,y) \cdot x + \frac{a(a-1)\delta^2}{2(1-\delta)^{a+1}} \cdot y$$

Let $S = \mathrm{Supp}(\mathcal{D}_\delta)$. Taking $y = \mathcal{D}_\delta(i)$, $x = \mathcal{D}(i) - \mathcal{D}_\delta(i)$, then summing i all over S and using the fact that $\sum_{i \in S} \mathcal{D}_\delta(i) = \sum_{i \in S} \mathcal{D}(i) = 1$ yields the result:

$$R_a(\mathcal{D}_\delta \| \mathcal{D}) = \sum_{i \in S} \frac{\mathcal{D}_\delta(i)^a}{\mathcal{D}(i)^{a-1}} \leq 1 + \frac{a(a-1)\delta^2}{2(1-\delta)^{a+1}}$$

□

We may also refer to Lemma 3 as the relative error lemma.

Usecases. The relative error lemma can be used when the relative error between $\mathcal{D}_\delta$ and $\mathcal{D}$ is bounded. This may typically happen when the probabilities of $\mathcal{D}$ are stored in floating-point with a precision $\log_2 \delta$ – though we will see that it is not limited to this situation. Again, this is illustrated by Fig. 1.

3.3 Security Arguments Using the Rényi Divergence

We consider a cryptographic scheme making q_s queries to either a perfect distribution $\mathcal{D}$ or an imperfect distribution $\mathcal{D}_\delta$. Let E be an event breaking the scheme by solving a search problem, and ε (resp. ε_δ) the probability that this event occurs under the use of $\mathcal{D}$ (resp. $\mathcal{D}_\delta$). We suppose that $\varepsilon_\delta \geq 2^{-\lambda}$. By the data processing and probability preservation inequalities:

$$\begin{aligned} \varepsilon &\geq \varepsilon_\delta^{a/(a-1)} / R_a(\mathcal{D}_\delta^{q_s} \| \mathcal{D}^{q_s}) \\ &\geq \varepsilon_\delta^{a/(a-1)} / R_a(\mathcal{D}_\delta \| \mathcal{D})^{q_s} \end{aligned}$$

We can choose any value in $(1, +\infty)$ for a, but small values for a impact the tightness of the reduction and large values impact its efficiency. Setting $a = 2\lambda$ seems to be a good compromise. Indeed, we then have $\varepsilon_\delta^{a/(a-1)} \geq \varepsilon_\delta / \sqrt{2}$, so we lose at most half a bit of security in the process.

Our goal is now to have $R_a(\mathcal{D}_\delta \| \mathcal{D})^{q_s} = \Omega(1)$, so that we have an almost tight security reduction. In this regard, having $R_a(\mathcal{D}_\delta \| \mathcal{D}) \leq 1 + \frac{1}{4q_s}$ is enough, since it yields $R_a(\mathcal{D}_\delta \| \mathcal{D})^{q_s} \leq e^{1/4} \leq \sqrt{2}$ by a classic inequality.[3]

This yields $\epsilon > 2^{-\lambda-1}$. By contraposition, a $(\lambda+1)$-bit secure scheme with $\mathcal{D}$ will be at least λ-bit secure when replacing $\mathcal{D}$ by $\mathcal{D}_\delta$ if the following condition is met:

$$R_a(\mathcal{D}_\delta \| \mathcal{D}) \leq 1 + \frac{1}{4q_s} \qquad \text{for } a = 2\lambda \tag{4}$$

We make two important remarks: first, this analysis is valid only for cryptographic schemes relying on *search problems*. This is the case for all the applications we consider in this paper, but for cryptographic schemes relying on decision problems, one may rather rely on SD-based, KLD-based analyses, or on specific Rényi divergence-based analyses as in [BLL+15, Sect. 4].

Second, the savings provided by our analysis heavily rely on the fact that the number of queries is limited. This was already observed in [BLL+15].

[3] $(1 + x/n)^n \leq e^x$ for $x, n > 0$.

Practical Implications. We consider a cryptographic scheme with $\lambda+1 \leq 257$ bits of security making $q_s \leq 2^{64}$ queries to a distribution $\mathcal{D}$. Replacing $\mathcal{D}$ by another distribution $\mathcal{D}_\delta$ will make the scheme lose at most one bit of security, provided that one of these conditions is verified:

$$\frac{\mathcal{D}_\delta}{\mathcal{D}} \leq 1+\delta \text{ for } \delta = \frac{1}{4q_s} \tag{5}$$

$$\text{Supp}(\mathcal{D}_\delta) = \text{Supp}(\mathcal{D}), \text{ and } 1-\delta \leq \frac{\mathcal{D}_\delta}{\mathcal{D}} \leq 1+\delta \text{ for } \frac{\delta^2}{(1-\delta)^{d+1}} \leq \frac{1}{4\lambda q_s} \tag{6}$$

Equation 5 comes from the tailcut lemma with Eq. 4, and Eq. 6 from the relative error lemma with Eq. 4. For $\lambda \leq 256$ and $q_s \leq 2^{64}$:

- the condition 5 translates to $\delta \leq 2^{-66}$,
- the condition 6 translates to $\delta \leq 2^{-37}$.

3.4 Relation to the Max-Log Distance

In [MW17], Micciancio and Walter introduced a new metric, the max-log distance. They argue that this metric is both easy to use and allows to have sharp bounds in cryptographic proofs.

In Lemma 4, we show that the log of the Rényi divergence is bounded (up to a constant) by the square of the max-log distance. It can be seen as a "reverse" analogue of Pinsker inequality for the SD and KLD, so we call it the reverse Pinsker inequality.

Definition 2 (max-log distance [MW17]). *The max-log distance between two distributions $\mathcal{P}$ and $\mathcal{Q}$ over the same support S is*

$$\Delta_{\text{ML}}(\mathcal{P}, \mathcal{Q}) = \max_{x \in S} |\log \mathcal{P}(x) - \log \mathcal{Q}(x)|$$

Lemma 4 (Reverse Pinsker inequality). *For two distributions $\mathcal{P}, \mathcal{Q}$ of common support, we have:*

$$R_a(\mathcal{P}\|\mathcal{Q}) \leq \left(1 + \frac{a(a-1)(e^{\Delta_{\text{ML}}(\mathcal{P},\mathcal{Q})}-1)^2}{2(2-e^{\Delta_{\text{ML}}(\mathcal{P},\mathcal{Q})})^{a+1}}\right)^{\frac{1}{a-1}} \underset{\Delta_{\text{ML}} \to 0}{\sim} 1 + \frac{a\Delta_{\text{ML}}(\mathcal{P},\mathcal{Q})^2}{2}$$

Proof. We note $\Delta_{\text{ML}}(\mathcal{P}, \mathcal{Q}) = \delta$ for some $\delta \geq 0$. We have:

$$\begin{aligned} \Delta_{\text{ML}}(\mathcal{P}, \mathcal{Q}) = \delta &\Rightarrow \forall x \in S, |\log \mathcal{P}(x) - \log \mathcal{Q}(x)| \leq \delta \\ &\Rightarrow \forall x \in S, e^{-\delta} \leq \tfrac{\mathcal{P}(x)}{\mathcal{Q}(x)} \leq e^{\delta} \\ &\Rightarrow R_a(\mathcal{P}\|\mathcal{Q}) \leq \left(1 + \tfrac{a(a-1)(e^\delta-1)^2}{2(2-e^\delta)^{a+1}}\right)^{\frac{1}{a-1}} \end{aligned}$$

The first implication applies the definition of the max-log distance, the second one passes to the exponential, the third one applies the relative error lemma. □

There are two implications from Lemma 4. First, we can add the max-log distance to our tools. Unlike the Rényi divergence, it is actually a distance, which is often useful when performing security analyses.

Second, the Lemma 4 provides evidence that the Rényi divergence gives sharper bounds than the max-log distance, as the log of the former is essentially bounded by the square of the second.

In addition, we point out that the max-log distance is defined only for distributions with a common support. For example, it cannot be applied to tailcut distributions. It is nevertheless a useful measure. One may for example use it if a true distance is needed, and then fall back to the Rényi divergence using Lemma 4.

4 Applications

In this section we provide five applications of our results. In all the cases studied, we manage to claim 256 bits of security while lowering the precision requirements to be less than 53 bits (or 61 bits for the last application). All the concrete bounds are obtained for $\lambda \leq 256$ and $q_s \leq 2^{64}$.

This bound of 53 bits is important. Floating-point with 53 bits of precision corresponds to the double precision type in the IEEE 754 standard, and is very often available in software – see e.g. the type `double` in C. In many cases, it can also be simulated using fixed-point numbers of 64 bits of precision, which can be done easily and efficiently, in particular over 64-bit architectures.

4.1 Tighter Analysis of the Micciancio-Walter Sampler

The first application of our results is also arguably the simplest. A new Gaussian sampler over $\mathbb{Z}$ was recently introduced by Micciancio and Walter [MW17]. They provide a security analysis using the max-log distance [MW17, Lemma 5.5].

Later, at the end of [MW17, Sect. 5.3], this lemma is used to argue that for a given set of parameters, if we note $\mathcal{Q}$ a perfect Gaussian distribution and $\mathcal{P}$ the output of the new sampler, we have $\Delta_{\mathrm{ML}}(\mathcal{P}\|\mathcal{Q}) \leq 2^{-52}$. This in turn allows them to claim about 100 bits of security.

A tighter analysis. We now prove that a Rényi divergence-based analysis gives tighter bounds than the max-log distance-based analysis from [MW17]. This analysis is done completely in black box, as we do not need to know anything about the sampler, except the fact that $\Delta_{\mathrm{ML}}(\mathcal{P}\|\mathcal{Q}) \leq 2^{-52}$. Applying the reverse Pinsker inequality (Lemma 4) yields $R_a(\mathcal{P}\|\mathcal{Q}) \leq 1 + 2^{-96}$ for any $a \leq 512$.

Following the security argument of Sect. 3.3 and in particular Eqs. 4 and 6, this allows us to claim that the use of this sampler is secure for 256 bits of security and $q_s = 2^{64}$ queries. This remains the case even if we ask up to 2^{94} queries, which we believe is more than enough for any practical application.

4.2 Revisiting the Table Approach

We now study a more generic problem, namely sampling distributions over $\mathbb{Z}$. We consider situations where the use of precomputed tables is practical: this includes but is not limited to (pseudo-)Gaussians with parameters known in advance.

We revisit the table-based approach. First, we show that the standard approach based on the cumulative distribution function (see e.g. [Pei10]) suffers from precision issues for a large class of distributions: light-tailed distributions. Informally, these are distributions which tails have a negligible weight (like Gaussians). They also happen to be widespread in lattice-based cryptography.

We then introduce a new approach based on the conditional density function. We show that for light-tailed distributions, it behaves in a much nicer way. To conclude, we take a real-life example and show that in terms of space, the new approach allows to gain an order of magnitude compared to the standard approach.

Definition 3. *For a distribution $\mathcal{D}$ over $S \subseteq \mathbb{Z}$, we call cumulative distribution function of $\mathcal{D}$ and note* $\mathrm{CDF}_{\mathcal{D}}$ *the function defined over S by*

$$\mathrm{CDF}_{\mathcal{D}}(z) = \sum_{i \le z} \mathcal{D}(z)$$

Classical CDF sampling. To sample from $\mathcal{D}$, a standard approach is to store a precomputed table of $\mathrm{CDF}_{\mathcal{D}}$, draw a uniform deviate $u \leftarrow [0,1]$ and output $z = \min\{i \in S | \mathrm{CDF}_{\mathcal{D}}(i) \ge u\}$. In practice, we will not store the complete CDF table. If $\mathcal{D} = D_{\mathbb{Z},c,\sigma}$ is a discrete Gaussian, then we store the values for $z \in (c - k_0\sigma, c + k_0\sigma) \cap \mathbb{Z}$ with a given precision p_0; here, k_0 is a "tailcut bound" which we can fix by either a SD or Rényi divergence argument. We now estimate the requirements in the context of λ bits of security and $m \cdot q_s$ queries.[4]

SD-based analysis. Using [GPV08, Lemma 4.2], we have $k_0 = \sqrt{2(\lambda + \log_2 m)}$. Each $\mathcal{D}(z) = \mathrm{CDF}_{\mathcal{D}}(z) - \mathrm{CDF}_{\mathcal{D}}(z-1)$ should be known with absolute precision $\lambda + \log_2 m$, so we may take $p_0 = \lambda + \log_2 m$.

Rényi divergence-based analysis. From the tailcut lemma (see also Eq. 5), it is sufficient to take $k_0 = \sqrt{2\log_2(4mq_s)}$. From the relative error lemma, each $\mathcal{D}(z)$ should be known with relative precision $\log_2 \delta$ verifying Eq. 6. For our choices of λ and q_s, this yields $k_0 \le \sqrt{2(66 + \log_2 m)}$ and $\log_2 \delta \le 37 + \log_2 m$.

For $\lambda = 256$, we divide the number of precomputed elements by about 1.87. A naive interpretation of the analyses above may also lead us to divide the precision p_0 by $(\lambda + \log_2 m)/(37 + \log_2 m) \approx 6.9$. However, the next paragraph will expose why we cannot simply do that.

[4] The call to a sampler over $\mathbb{Z}$ is often done several times per query. In the context of signatures, we typically have $m =$ the lattice dimension. Here we take $m = 2^{10}$.

Precision issues in the case of light-tailed distributions. In the previous paragraph, there is a slight but important difference between the SD and Rényi divergence analyses. The precision is given absolutely in the first case, and relatively in the second case. It is actually this relativity that allows us to use the relative error lemma in the second case, but it comes at a price: it is not efficient anymore to use the CDF table.

We present here an example explaining why this is the case: let $\mathcal{D}_2$ be the distribution defined over $\mathbb{N}^*$ by $\mathcal{D}_2(k) = 2^{-k}$. One can show that $\mathrm{CDF}_{\mathcal{D}_2}(k) = 1 - 2^{-k}$, so from a machine perspective, $\mathrm{CDF}_{\mathcal{D}_2}(k)$ will be rounded to 1 as soon as $k > p_0$. As a consequence, the probability output of the CDF table-based algorithm will be 0 for any $k > p_0 + 1$ and we will not be able to use the relative error lemma at all.

This problem is common to light-tailed distributions, including Gaussian-like distributions. As the CDF converges very fast to 1, we have to store it in high precision in order for it to be meaningful. This is not satisfactory from a practical viewpoint.

Conditional density sampling. A simple way around the aforementioned problem is to use the *conditional density function* instead of the CDF. First, we give its definition.

Definition 4. *For a distribution $\mathcal{D}$ over $\mathbb{N}$, we call conditional density function of $\mathcal{D}$ and note $\mathrm{CoDF}_{\mathcal{D}}$ the function defined by $\mathrm{CoDF}(z) = \mathcal{D}(z)/(\sum_{i \geq z} \mathcal{D}(i))$.*

In other words, $\mathrm{CoDF}(z)$ is the probability that a random variable X of distribution $\mathcal{D}$ takes the value z, *conditioned* to the fact that X is bigger or equal to z.[5] A way to use the CoDF to sample from $\mathcal{D}$ is given by Algorithm 1, a variation of the CDF sampler.

Algorithm 1. CoDF sampler

Require: A precomputed table of $\mathrm{CoDF}_{\mathcal{D}}$
Ensure: $z \leftarrow \mathcal{D}$
1: $z \leftarrow 0$
2: $u \leftarrow [0, 1]$ uniformly
3: **while** $u \geq \mathrm{CoDF}_{\mathcal{D}}(z)$ **do**
4: $\quad z \leftarrow z + 1$
5: $\quad u \leftarrow [0, 1]$ uniformly
6: **Return** z

It is easy to show that the expected number of loops in Algorithm 1 is the mean of $\mathcal{D}$. It outputs z with probability $\prod_{i<z} [1 - \mathrm{CoDF}_{\mathcal{D}}(i)] \cdot \mathrm{CoDF}_{\mathcal{D}}(z)$, which by a telescopic product is equal to

$$\frac{\sum_{i>0} \mathcal{D}(i)}{\sum_{i\geq 0} \mathcal{D}(i)} \times \frac{\sum_{i>1} \mathcal{D}(i)}{\sum_{i\geq 1} \mathcal{D}(i)} \times \cdots \times \frac{\sum_{i>z-1} \mathcal{D}(i)}{\sum_{i\geq z-1} \mathcal{D}(i)} \times \frac{\mathcal{D}(z)}{\sum_{i\geq z} \mathcal{D}(i)} = \mathcal{D}(z) \quad (7)$$

[5] We note that the support is now $S \subseteq \mathbb{N}$ instead of $S \subseteq \mathbb{Z}$, but switching between the two cases is algorithmically easy.

and therefore, Algorithm 1 is correct. However, in practice Algorithm 1 will be used with precomputed values which are only correct up to a given precision. Lemma 5 provides an analysis of the algorithm is this case.

Lemma 5. *For a distribution $\mathcal{D}$ of support $S \subseteq \mathbb{N}$, let $f = \textsc{CoDF}_\mathcal{D}$ be the* CoDF *of $\mathcal{D}$, and f_δ be an approximation of f such that, over S:*

$$\begin{array}{l} 1-\delta \leq \frac{f_\delta}{f} \leq 1+\delta \\ 1-\delta \leq \frac{1-f_\delta}{1-f} \leq 1+\delta \end{array} \tag{8}$$

Let $\mathcal{D}_\delta$ be the output distribution of the Algorithm 1 using a precomputed table of f_δ instead of f. Then, for any $z \in S$:

$$1-\delta z \underset{0 \leftarrow \delta}{\sim} (1-\delta)^z \leq \frac{\mathcal{D}_\delta(z)}{\mathcal{D}(z)} \leq (1+\delta)^z \underset{\delta \to 0}{\sim} 1+\delta z$$

Proof. We have

$$\begin{aligned} \mathcal{D}_\delta(z) &= \prod_{i<z} [1 - f_\delta(i)] \cdot f_\delta(z) \\ \Rightarrow (1-\delta)^z \prod_{i<z} [1-f(i)] \cdot f(z) \leq \mathcal{D}_\delta(z) &\leq (1+\delta)^z \prod_{i<z} [1-f(i)] \cdot f(z) \\ \Rightarrow (1-\delta)^z \cdot \mathcal{D}(z) \qquad\qquad \leq \mathcal{D}_\delta(z) &\leq (1+\delta)^z \cdot \mathcal{D}(z) \end{aligned}$$

The first implication comes from Eq. 8, the second one from Eq. 7. □

Provided that the CoDF is stored with enough precision, Lemma 5 gives us an inequality that allows to use the relative error lemma. Now, the interesting part is that for light-tailed distributions, the CoDF does not converge to 1 as fast as the CDF, which is important if we want the lower part of Eq. 8 to be true. For example, if $\mathcal{D} = D_{\mathbb{Z},1}$, we have $\text{CDF}_\mathcal{D}(z) - \text{CDF}_\mathcal{D}(z-1) = O(e^{-z^2/2})$, whereas $1 - \textsc{CoDF}_\mathcal{D}(z) = O(e^{-z})$. This allows to store $\textsc{CoDF}_\mathcal{D}$ in small precision and still remain able to use Lemma 5.

Of course, one may argue that z can be arbitrarily big. However, in practice we will not sample from a distribution $\mathcal{D}$ of infinite support directly but rather from a tailcut distribution of $\mathcal{D}$, in the bounds provided by the tailcut lemma, so z will not take too large values and we will be able to store $\textsc{CoDF}_\mathcal{D}$ efficiently.

Solving the precision issues. Going back to the example of the distribution $\mathcal{D}_2$, the Table 1 shows how $\text{CDF}_{\mathcal{D}_2}(k)$ and $\textsc{CoDF}_{\mathcal{D}_2}(k)$ are stored in machine precision, and how it impacts the associated sampler.

For the CDF-based sampler, due to precision issues, it samples from a distribution $\mathcal{D}_2'$ which has a probability 0 for elements in the tail of $\mathcal{D}_2$. In contrast, the CoDF-based sampler approximates $\mathcal{D}_2$ correctly even for elements in the tail of $\mathcal{D}_2$.

Application: sampling over $D^+_{\mathbb{Z},\sigma_2}$ in BLISS. An important step of the signature scheme BLISS consists of sampling $z \leftarrow D^+_{\mathbb{Z},\sigma_2}$, where $\sigma_2 \approx 0.85$.

Table 1. Precomputed values of CDF and CoDF of $\mathcal{D}_2$ as stored in 53 bits precision. The stored value of $\mathrm{CDF}_{\mathcal{D}_2}(k)$ quickly becomes 1, leading to the associated algorithm sampling from some incorrect distribution $\mathcal{D}_2'$ instead of $\mathcal{D}_2$.

k	1	2	3	...	54	55	...
$\mathrm{CDF}_{\mathcal{D}_2}(k)$	1/2	3/4	7/8	...	1	1	...
$\mathcal{D}_2'(k)$	1/2	1/4	1/8	...	0	0	...
$\mathrm{CoDF}_{\mathcal{D}_2}(k)$	1/2	1/2	1/2	...	1/2	1/2	...
$\mathcal{D}_2(k)$	1/2	1/4	1/8	...	2^{-54}	2^{-55}	...

In BLISS, this is done in a bitwise rejection sampling fashion [DDLL13, Algorithm 10], which is very efficient in hardware but not so much in software. In addition, the structure of the Algorithm 10 from [DDLL13] exposes it to side-channel attacks in the lines of [EFGT17] (see also Sect. 4.3). Instead, one can sample efficiently from $D^+_{\mathbb{Z},\sigma_2}$ using a precomputed table T:

- With a CDF+SD approach, T must have 20 elements of 266 bits each, which amounts to about 5 300 bits.
- With a CoDF+Rényi divergence approach and using Lemma 5, T must have 11 elements of about 53 bits each, which amounts to about 600 bits.[6]

Here, the CoDF+Rényi divergence approach makes us gain an order of magnitude in storage requirements. Another notable advantage is that it is particularly fit to a fixed-point implementation, which might make it easier to implement in hardware. In addition, it is generic in the sense that it can be applied to a large class of distributions over $\mathbb{N}$ (or $\mathbb{Z}$).

An open question is how to make Algorithm 1 constant-time and protected against side-channel attacks. The trivial way to make it constant-time is to always read the whole table, but this may incur a significant overhead.

4.3 Simpler and More Secure Rejection Sampling in BLISS

We recall that the context and motivation of doing rejection sampling in lattice-based cryptography is exposed in Sect. 1.1. We now focus our attention on the signature scheme BLISS [DDLL13]. In BLISS, the final step of the signature consists of this step:

$$\text{Accept with probability } p = 1/\left(M \exp(-\frac{\|\mathbf{Sc}\|^2}{2\sigma^2}) \cosh(\frac{\langle \mathbf{z}, \mathbf{Sc}\rangle}{\sigma^2})\right) \quad (9)$$

where $\mathbf{S}$ is the secret key, σ, M are public parameters and $\mathbf{c}, \mathbf{z}$ are part of the signature. In the original scheme and all the implementations that we are aware

[6] Actually, storing the 11 elements as 64-bit integers yields better relative precision and is easier to handle in practice.

of [LD13,Pop14,Str14], this step is implemented by the means of combining several Bernoulli distributions dependent of the bits of $\|\mathbf{Sc}\|^2$ and $\langle \mathbf{z}, \mathbf{Sc} \rangle$.

There are two drawbacks from this approach. First, the algorithm described in [DDLL13] for performing this step is rather sophisticated, and as a result it takes a significant portion of the coding effort in [LD13,Pop14,Str14].

The second drawback is that this algorithm is actually vulnerable to side-channel attacks: Espitau et al. [EFGT17] have shown that a side-channel analysis of the signature traces can recover both $\|\mathbf{Sc}\|^2$ and $\langle \mathbf{z}, \mathbf{Sc} \rangle$, and from it the secret key. Interestingly, it might be possible to extend this attack to a timing attack, in which case the implementation of Strongswan [Str14], deployed on Windows, Linux, Mac OS, Android and iOS platforms, could also suffer from it.

Simple Rejection Sampling. We observe that the step 9 doesn't need to be made exactly. We can simply compute a value p_δ such that $1 - \delta \leq \frac{p_\delta}{p} \leq 1 + \delta$, sample $u \leftarrow [0, 1]$ uniformly and accept if and only if $p_\delta \geq u$. By Eq. 6, it is sufficient that p is computed with a relative error 2^{-37}. This can be done easily:

1. **In software,** one may simply resort to a standard implementation of the exp function, such as the one provided `math.h` for the C language. As long as the relative precision provided is more than 37 bits of precision, we can use Eq. 6. We note that implementations of $\exp(\cdot)$ usually provide at least 53 bits of precision, which is more than enough for our purposes.
2. **In hardware,** an implementation of the exp function may not always be available. There are many ways around this issue, we present two of them:
   - One may use Padé approximants as an efficient way to compute exp. Padé approximants are generalizations of Taylor series: they approximate a function f by a polynomial fraction $\frac{P_n}{Q_m}$ instead of a polynomial P_n. They usually converge extremely fast, and in the case of the exp function, the relative error between $\exp(z)$ and its Padé approximant is less than 2^{-37} for an approximation of order 4 and $|z| < 1/2$.[7] A more detailed analysis is provided in appendix, Sect. A.1.
   - Another solution is to precompute the values $\exp(\frac{2^i}{2\sigma^2})$ for a small number of values $i \in \mathbb{N}$. This then allows to compute $\exp(\frac{z}{2\sigma^2})$ for any $z = \sum_i z_i 2^i$, since $\exp(\frac{z}{2\sigma^2}) = \prod_{z_i=1} \exp(\frac{2^i}{2\sigma^2})$.[8] For the parameters given by [DDLL13], $\|\mathbf{Sc}\|^2$ and $\langle \mathbf{z}, \mathbf{Sc} \rangle$ are integers and are less than 37 bits, which means that we would need to store at most 37 precomputed values.

   For the two proposed solutions, a very pessimistic analysis estimates that we perform less than 80 elementary floating-point operations to compute p. While it might seem a lot for 3 exponentials, it is negligible compared to the total cost of a signature, which is around $O(n \log n)$ for $n = 512$ in the BLISS scheme. In addition, all the techniques we propose are easy to protect against side-channel attacks.

---

[7] It is easy reduce any input z to the case $|z| < 1/2$ by taking $z' \leftarrow z \bmod (\ln 2)$ and observing that $e^{\ln 2} = 2$. The precision loss is negligible.

[8] For negative values, exp may be computed by inversion, or if it is not available, by also precomputing $\exp(-\frac{2^i}{2\sigma^2})$.

We note that our software solution and our hardware solution based on Padé approximants do not require to store any precomputed table.

In BLISS, explicitely computing the rejection bound as we did was discarded because of precision concerns. We note that all the security analysis in BLISS was performed using the SD, with only subsequent work [PDG14, BLL+15] using more adequate measures of divergence. Using the SD in our case would have required us compute transcendental functions with a precision 2^{λ}, which is impractical. The relative error lemma is the key which allows to argue that a floating-point approach is secure.

4.4 Squeezing the Standard Deviation of Trapdoor Samplers

Context. The two last sections are related to the most generic and powerful type of Gaussian sampling: trapdoor sampling. Algorithms for performing trapdoor sampling [Kle00, GPV08, Pei10, MP12] are essentially randomized variants of Babai's round-off and nearest plane algorithms [Bab85, Bab86]. For suitable parameters, they are statistically indistinguishable from a perfect Gaussian $D_{\Lambda,\sigma,\mathbf{c}}$.

For a cryptographic use, we want σ to be as small as possible in order to have the highest security guarantees. However, σ cannot be too small: if it is, then the trapdoor samplers will not behave anymore like perfect Gaussian oracles.[9] At the extreme case $\sigma = 0$, the samplers become deterministic and leak the shape of the basis used for sampling, exposing the associated schemes to key-recovery attacks described earlier. To avoid that, samplers usually come with lower bounds on σ for using it securely (see e.g. Theorem 1 for Klein's sampler [Kle00, GPV08]).

Roadmap. Before continuing, we establish the roadmap for this section and the next one. In this section, we show that, if σ is large enough, a Gaussian sampler with infinite precision is as secure as an ideal Gaussian. In the next one, we show that a Gaussian sampler with finite precision is as secure as one with infinite precision. Of course, such analyses are already known. Our contribution here is to use the Rényi divergence to have more aggressive parameters for σ and the precision of the sampler (Fig. 2).

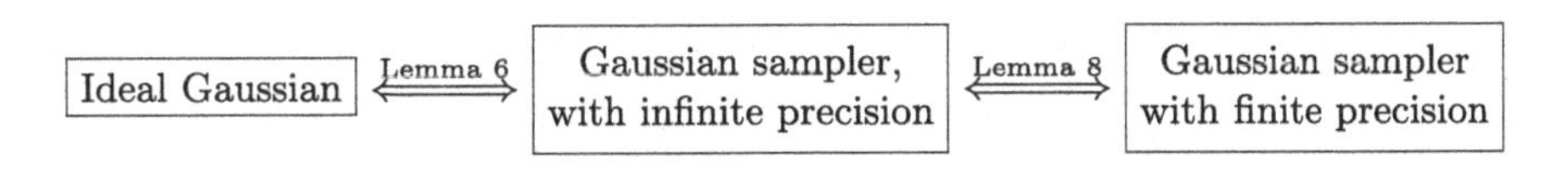

Fig. 2. Roadmap for asserting the security of a practical Gaussian sampler

Klein's sampler. We cannot analyse all the existing samplers in this article, so we now focus our attention on Klein's sampler [Kle00, GPV08]. It is described in Algorithm 2.

[9] If they did behave like perfect Gaussians when $\sigma \to 0$, then they would effectively solve the closest vector problem, which is a NP-hard problem.

Algorithm 2. $\text{KLEIN}_{\mathbf{L},\sigma}(\mathbf{t})$

Require: $\sigma \geq \eta_\epsilon(\mathbb{Z}^n) \cdot \|\mathbf{B}\|_{GS}$, the Gram-Schmidt orthogonalization $\mathbf{B} = \mathbf{L} \cdot \tilde{\mathbf{B}}$ and the values $\sigma_j = \sigma / \|\tilde{\mathbf{b}}_j\|$ for $j \in \{1, \dots, n\}$
Ensure: A vector $\mathbf{z}$ such that $\mathbf{zB} \leftarrow D_{\Lambda(\mathbf{B}),\sigma,\mathbf{tB}}$
1: **for** $j = n, \dots, 1$ **do**
2: $\quad c_j \leftarrow t_j + \sum_{i>j}(t_j - z_j)L_{ij}$
3: $\quad z_j \leftarrow D_{\mathbb{Z},\sigma_j,c_j}$
4: **return z**

An associated lower bound on σ for using Algorithm 2 is given in Theorem 1.

Theorem 1 ([DN12a, Theorem 1], concrete version of [GPV08, Theorem 4.1]). *Let* $\epsilon = 2^{-\lambda}$. *If* $\sigma \geq \eta_\epsilon(\mathbb{Z}^n) \cdot \|\mathbf{B}\|_{GS}$, *then the SD between* $\text{KLEIN}_{\mathbf{L},\sigma}(\mathbf{t}) \cdot \mathbf{B}$ *and the perfect discrete Gaussian* $D_{\Lambda(\mathbf{B}),\sigma,\mathbf{tB}}$ *is upper bounded by* $2^{-\lambda}$.

Combined to a standard SD-based argument, Theorem 1 establishes that σ must be proportional to $\sqrt{\lambda}$ in order to claim λ bits of security when using Algorithm 2. A better bound was established in [DLP14] but it remains proportional to $\sqrt{\lambda}$. In Lemma 6, we establish a bound that is both (almost) independent of λ and smaller.

Lemma 6 (Rényi divergence of Klein's sampler). *For any* $\epsilon \in (0, 1/4)$, *if* $\sigma \geqslant \eta_\epsilon(\mathbb{Z}^n) \cdot \|\mathbf{B}\|_{GS}$ *then the Rényi divergence between* $\mathcal{D} = D_{\Lambda(\mathbf{B}),\sigma,\mathbf{tB}}$ *and the output distribution* $\mathcal{D}_\epsilon$ *of* $\text{KLEIN}_{\mathbf{L},\sigma}(\mathbf{t}) \cdot \mathbf{B}$ *verifies*

$$R_a(\mathcal{D}_\epsilon \| \mathcal{D}) \leq \left(1 + \frac{a(a-1)\delta^2}{2(1-\delta)^{a+1}}\right)^{\frac{1}{a-1}} \underset{\delta \to 0}{\sim} 1 + \frac{a\delta^2}{2},$$

where $\delta = \left(\frac{1+\epsilon/n}{1-\epsilon/n}\right)^n - 1 \approx 2\epsilon$.

Proof. We note $\mathbf{v} = \mathbf{zB}$ and $\mathbf{c} = \mathbf{tB}$. As detailed in [GPV08], the probability that $\text{KLEIN}_{\mathbf{L},\sigma}(\mathbf{t})$ outputs a given $\mathbf{z}$ is proportional to

$$\prod_{i=1}^{n} \frac{1}{\rho_{\sigma_j,c_j}(\mathbb{Z})} \cdot \rho_{\sigma,\mathbf{c}}(\mathbf{v})$$

for $\sigma_j = \sigma / \|c_j\|$ and some $c_j \in \mathbb{R}$ that depends on $\mathbf{t}$ and $\mathbf{B}$. By assumption, $\sigma_j \geq \eta_\epsilon(\mathbb{Z}^n) \geq \eta_{\epsilon/n}(\mathbb{Z})$, therefore $\rho_{\sigma_j,c_j}(\mathbb{Z}) \in [\frac{1-\epsilon/n}{1+\epsilon/n}, 1] \cdot \rho_{\sigma_j}(\mathbb{Z})$ by [MR04, Lemma 4.4]. Since $\mathcal{D}(\mathbf{v})$ is proportional to $\rho_{\sigma,\mathbf{c}}(\mathbf{v})$ and $\mathcal{D}, \mathcal{D}_\epsilon$ both sum up to one, we have

$$\left(\frac{1-\epsilon/n}{1+\epsilon/n}\right)^n \leq \frac{\mathcal{D}_\epsilon}{\mathcal{D}} \leq \left(\frac{1+\epsilon/n}{1-\epsilon/n}\right)^n,$$

from which we may conclude by using the relative error lemma. $\square$

Plugging this result with the relative error lemma, we may use Klein's sampler with $\delta \approx 2\epsilon$ verifying Eq. 6, instead of $\epsilon \leq 2^{-\lambda}$ with the SD and $\epsilon \leq 2^{-\lambda/2}$ with the KLD [DLP14]. Compared to a SD-based analysis, this allows to squeeze σ by a factor $\sqrt{\lambda/38}$ that can be as large as ≈ 2.60 for $\lambda = 256$.

While it might seem a small gain, the security of trapdoor samplers is very sensitive to standard deviations variations. We estimate that this factor 2.60 allows to gain up to 30 bits of security (this claim is supported by e.g. [Pre15, Table 6.1]). A similar analysis for Peikert's sampler [Pei10] yields a similar gain.

4.5 Trapdoor Sampling in Standard Precision

For our last application of the Rényi divergence, we conclude our analysis of Klein's sampler (Algorithm 2), by performing its precision analysis. This section shows that it can be used safely in small precision.

First, we give a lemma that bounds the ratio of two Gaussian sums in $\mathbb{Z}$ with slightly different centers and standard deviations.

Lemma 7 (Ratio of Gaussian Sums in $\mathbb{Z}$). *Let two arbitrary centers $t, \bar{t} \in \mathbb{R}$ and standard deviations $\sigma, \bar{\sigma} > 0$. Let the Gaussian functions $\rho(z) = \rho_{\sigma,t}(z)$, $\bar{\rho}(z) = \rho_{\bar{\sigma},\bar{t}}(z)$ and the distributions $\mathcal{D}(z) = \rho(z)/\rho(\mathbb{Z})$, $\bar{\mathcal{D}}(z) = \bar{\rho}(z)/\bar{\rho}(\mathbb{Z})$. Let $u(z) = \frac{(z-\bar{t})^2}{2\bar{\sigma}^2} - \frac{(z-t)^2}{2\sigma^2}$. Then*

$$e^{-\mathbb{E}_{z \leftarrow \mathcal{D}}[u]} \leq \frac{\bar{\rho}(\mathbb{Z})}{\rho(\mathbb{Z})} \leq e^{-\mathbb{E}_{z \leftarrow \bar{\mathcal{D}}}[u]}$$

Proof. We first prove the left inequality. We have

$$\begin{aligned} \bar{\rho}(z) &= e^{-u(z)}\rho(z) \\ \Rightarrow \tfrac{\bar{\rho}(z)}{\rho(\mathbb{Z})} &= e^{-u(z)}\mathcal{D}(z) \\ \Rightarrow \tfrac{\bar{\rho}(\mathbb{Z})}{\rho(\mathbb{Z})} &= \mathbb{E}_{z \leftarrow \mathcal{D}}[e^{-u(z)}] \\ \Rightarrow \tfrac{\bar{\rho}(\mathbb{Z})}{\rho(\mathbb{Z})} &\geq e^{-\mathbb{E}_{z \leftarrow \mathcal{D}}[u(z)]} \end{aligned}$$

where the last inequality comes from Jensen's inequality: since e is convex, $\mathbb{E}[e^{-u}] \geq e^{\mathbb{E}[-u]}$. Following the same reasoning, one gets

$$\left(\bar{\mathcal{D}}(z)e^{u(z)} = \frac{\rho(z)}{\bar{\rho}(\mathbb{Z})}\right) \Rightarrow \left(\mathbb{E}_{z \leftarrow \bar{\mathcal{D}}}[e^u] = \frac{\rho(\mathbb{Z})}{\bar{\rho}(\mathbb{Z})}\right) \Rightarrow \left(\frac{\bar{\rho}(\mathbb{Z})}{\rho(\mathbb{Z})} \leq e^{-\mathbb{E}_{z \leftarrow \bar{\mathcal{D}}}[u]}\right)$$

□

This lemma is useful in the sense that it provides a relative error bound, which will be used in the next lemma in order use the relative error lemma. We now give a bound on the required precision for using safely Klein's sampler.

Lemma 8. *Let $\mathcal{D}$ (resp. $\bar{\mathcal{D}}$) be the output distribution of Algorithm 2 over the input $\mathbf{t}$ (resp. $\bar{\mathbf{t}}$), using precomputed values $(\mathbf{L}, (\sigma_j)_j)$ (resp. $(\bar{\mathbf{L}}, (\bar{\sigma}_j)_j)$). Let $\delta, \epsilon \in (0, .01)$. We note:*

- $T = n\|\mathbf{L}\|_{\max}(1.1 + \sigma\sqrt{2\pi} \cdot \|\mathbf{B}^{-1}\|_2)$
- $C = 1.3n\delta(\frac{T\sqrt{2\pi}}{\eta_\epsilon(\mathbb{Z}^n)} + 2\pi + 1)$

If we have the following (error) bounds on the input of Algorithm 2:

- $\mathbf{t} \in [-.5, .5]^n$
- $\|\bar{\mathbf{t}} - \mathbf{t}\|_\infty \leq \delta$
- $|\bar{\sigma_j} - \sigma_j| \leq \delta\sigma_j$ *for all* j
- $\|\bar{\mathbf{L}} - \mathbf{L}\|_{\max} \leq \delta\|\mathbf{L}\|_{\max}$

Then we have this inequality:

$$e^{-C} \leq \frac{\bar{\mathcal{D}}}{\mathcal{D}} \leq e^C.$$

The Lemma 8 covers – but is not limited to – the case where $\mathbf{L}$ and the $(\sigma_j)_j$'s are known up to a relative error, and $\mathbf{t}$ up to an absolute error. For any $\mathbf{z} \in \mathbb{Z}^n$, $D_{\mathbb{Z}^n,\sigma,\mathbf{z}+\mathbf{t}} = \mathbf{z} + D_{\mathbb{Z}^n,\sigma,\mathbf{t}}$, so it is perfectly reasonable to suppose $\mathbf{t} \in$ [-.5, .5]^n.

Proof. This proof is rather long, so we explain its outline first. In ①, we establish a bound $A \leq \frac{\mathcal{D}(\mathbf{z})}{\bar{\mathcal{D}}(\mathbf{z})} \leq B$, for some expressions A, B. In ②, we establish $|A| \leq C$ and ③, we establish $|B| \leq C$. We conclude in ④.
① Let $\mathbf{z} = \sum_j \hat{z}_j \in \mathbb{Z}^n$ be a possible output of both samplers. We note $\mathbf{v} = \mathbf{zB}$ and $\mathbf{c} = \mathbf{tB}$. There exist a unique n-tuple $(c_j)_j$ (resp. $(\bar{c_j})_j$) such that at each step j, $\mathcal{E}$ (resp. $\bar{\mathcal{E}}$) samples a discrete Gaussian in $\mathbb{Z}$ around c_j (resp. $\bar{c_j}$).

The probability that $\mathbf{z}$ is output by $\mathcal{E}$ is $\mathcal{D}(\mathbf{z}) = \prod_j \mathcal{D}_j(\hat{z}_j) = \prod_j \frac{\rho_j(\hat{z}_j)}{\rho_j(\mathbb{Z})}$, where $\rho_j = \rho_{\mathbb{Z},\sigma_j,c_j}$ is uniquely defined by $\mathbf{z}$. Similarly, $\bar{\mathcal{D}}(\mathbf{z}) = \prod_j \frac{\bar{\rho}_j(\hat{z}_j)}{\bar{\rho}_j(\mathbb{Z})}$, where $\bar{\rho}_j = \rho_{\mathbb{Z},\bar{\sigma}_j,\bar{c_j}}$. We have

$$\frac{\mathcal{D}(\mathbf{z})}{\bar{\mathcal{D}}(\mathbf{z})} = \prod_j \frac{\rho_j(\hat{z}_j)}{\rho_j(\mathbb{Z})} \frac{\bar{\rho}_j(\mathbb{Z})}{\bar{\rho}_j(\hat{z}_j)} = \prod_j \frac{\rho_j(\hat{z}_j)}{\bar{\rho}_j(\hat{z}_j)} \frac{\bar{\rho}_j(\mathbb{Z})}{\rho_j(\mathbb{Z})}$$

For each j, let $u_j(z) = \frac{(z-\bar{c_j})^2}{2\bar{\sigma}_j^2} - \frac{(z-c_j)^2}{2\sigma_j^2}$. Lemma 7 yields:

$$e^{-\mathbb{E}_{z\leftarrow\mathcal{D}_j}[u_j]} \leq \frac{\bar{\rho}_j(\mathbb{Z})}{\rho_j(\mathbb{Z})} \leq e^{-\mathbb{E}_{z\leftarrow\bar{\mathcal{D}}_j}[u_j]}$$

So that we have:

$$\sum_j \left[u_j(\hat{z}_j) - \mathbb{E}_{z\leftarrow\mathcal{D}_j}[u_j]\right] \leq \log\left(\frac{\mathcal{D}(\mathbf{z})}{\bar{\mathcal{D}}(\mathbf{z})}\right) \leq \sum_j \left[u_j(\hat{z}_j) - \mathbb{E}_{z\leftarrow\bar{\mathcal{D}}_j}[u_j]\right] \quad (10)$$

Let A and B be the left and right terms of the Eq. 10. If we can bound A and B, then we will be able to conclude by the relative error lemma.
② Now, we bound A. We write $\bar{\sigma}_j = (1 + \delta_{\sigma_j})\sigma_j$, where each $|\delta_{\sigma_j}| \leq \delta$ by hypothesis. Developing u_j yields:

$$u_j(z_j) = \frac{1}{2(1+\delta_{\sigma_j})^2\sigma_j^2}\left[(c_j-\bar{c_j})^2 + 2(c_j-\bar{c_j})(z_j-c_j) - (2\delta_{\sigma_j}+\delta_{\sigma_j}^2)(z_j-c_j)^2\right] \quad (11)$$

In order to bound $c_j - \bar{c_j}$, we note that numerically, c_j is exactly $t_j + \langle \mathbf{t} - \mathbf{z}, \mathbf{l}_j \rangle$, where $\mathbf{l}_j$ is the j-th row of $(\mathbf{L}^t - I_n)$. Noting $\bar{\mathbf{t}} = \mathbf{t} + \delta_{\mathbf{t}}$, $\bar{\mathbf{l}}_j = \mathbf{l}_j + \delta_{\mathbf{l}_j}$ and $L = \|\mathbf{L}\|_{\max}$, we have:

$$\bar{c_j} = c_j + \delta_{\mathbf{t},j} + \langle \delta_{\mathbf{t}}, \mathbf{l}_j \rangle + \langle \mathbf{t} - \mathbf{z}, \delta_{\mathbf{l}_j} \rangle + \langle \delta_{\mathbf{t}}, \delta_{\mathbf{l}_j} \rangle$$

Thus

$$\begin{aligned} |\bar{c_j} - c_j| &\leq \delta_{\mathbf{t},j} + \|\delta_{\mathbf{t}}\| \|\mathbf{l}_j\| + \|\delta_{\mathbf{l}_j}\| \|\mathbf{t} - \mathbf{z}\| + \|\delta_{\mathbf{t}}\| \|\delta_{\mathbf{l}_j}\| \\ &\leq \delta(nL+1) + \delta n L \sigma \sqrt{2\pi} \cdot \|\mathbf{B}^{-1}\|_2 + \delta^2 n L \\ &\leq \delta \cdot T \end{aligned} \tag{12}$$

In Eq. 12, we used the fact that:

- $\|\delta_{\mathbf{t}}\| \leq \delta \sqrt{n}$
- $\|\delta_{\mathbf{l}_j}\| \leq \delta \|\mathbf{l}_j\| \leq \delta \sqrt{n} L$
- $\|\mathbf{t} - \mathbf{z}\| \leq \|\mathbf{c} - \mathbf{v}\| \cdot \|\mathbf{B}^{-1}\|_2 \leq \sigma \sqrt{2\pi n} \cdot \|\mathbf{B}^{-1}\|_2$, with the last inequality coming from [MR07, Lemma 4.4] (see Lemma 10 in the appendix)

We have:

$$\begin{aligned} A &= \sum_j \frac{1}{2(1+\delta_{\sigma_j})^2 \sigma_j^2} \left[2(c_j - \bar{c_j})(\hat{z}_j - c_j - \mathbb{E}_{z_j \leftarrow \mathcal{D}_j}[z_j - c_j]) - (2\delta_{\sigma_j} + \delta_{\sigma_j}^2)[(\hat{z}_j - c_j)^2 - \mathbb{E}_{z_j \leftarrow \mathcal{D}_j}[(z_j - c_j)^2]\right] \\ |A| &\leq \sum_j \frac{1.1}{2\sigma_j^2} \left[2|c_j - \bar{c_j}|(|\hat{z}_j - c_j| + \sqrt{2\pi}\epsilon\sigma_j) + 2\delta[(\hat{z}_j - c_j)^2 + \sigma_j^2 + 2\pi\epsilon\sigma_j^2]\right] \\ &\leq \frac{1.1}{\sigma^2} \sum_j \left[\delta T(\|\tilde{\mathbf{b}}_j\|^2 \cdot |\hat{z}_j - c_j| + \|\tilde{\mathbf{b}}_j\| \sqrt{2\pi}\epsilon\sigma) + \delta[\|\tilde{\mathbf{b}}_j\|^2 (\hat{z}_j - c_j)^2 + \sigma^2 + 2\pi\epsilon\sigma^2]\right] \\ &\leq \frac{1.1\delta}{\sigma^2} \left[T \max_j \|\tilde{\mathbf{b}}_j\| (\|\mathbf{v} - \mathbf{c}\|_1 + \sqrt{2\pi}\epsilon\sigma n) + [\|\mathbf{v} - \mathbf{c}\|^2 + n\sigma^2 + 2\pi n \epsilon \sigma^2]\right] \\ &\leq 1.1\delta \left[T(n\sqrt{2\pi} + \sqrt{2\pi}\epsilon n)/\eta_\epsilon(\mathbb{Z}^n) + [2\pi n + n + 2\pi n \epsilon]\right] \\ &\leq 1.2\delta n \left[T\sqrt{2\pi}/\eta_\epsilon(\mathbb{Z}^n) + 2\pi + 1\right] \end{aligned} \tag{13}$$

In Eq. 13, the first line develops the formula for A by using Eq. 11. For the second line, we use [MR07, Lemma 4.2] (see Lemma 9 in the appendix) to bound the two expected values and the term 1.1 to absorb parasitic terms in δ_{σ_j} and ϵ.

The third line replaces σ_j by $\sigma / \|\tilde{\mathbf{b}}_j\|$ and $|c_j - \bar{c_j}|$ by the bound $\delta \cdot T$ from Eq. 12. For the fourth line, we notice that $\sum_j \|\tilde{\mathbf{b}}_j\| \cdot |\hat{z}_j - c_j| = \|\mathbf{v} - \mathbf{c}\|_1$ and $\sum_j \|\tilde{\mathbf{b}}_j\|^2 \cdot (\hat{z}_j - c_j)^2 = \|\mathbf{v} - \mathbf{c}\|_2^2$ (both equalities follow directly from the Lemma 4.4 of [GPV08]).

In the fifth line, we use the bounds $\|\mathbf{v} - \mathbf{c}\|_2 \leq \sigma\sqrt{2\pi n}$, and $\|\mathbf{v} - \mathbf{c}\|_1 \leq \sigma n \sqrt{2\pi}$: the first one comes from [MR07, Lemma 4.4], and the second one follows from the fact that there exists a vector $\mathbf{u}$ with coefficients being only ± 1 such that $\|\mathbf{v} - \mathbf{c}\|_1 = |\langle \mathbf{v} - \mathbf{c}, \mathbf{u} \rangle|$. Applying the Cauchy-Schwartz theorem yields the bound. The last line simplifies as much as possible the expression.

③ We now bound B, the right part of Eq. 10. We can write u_j as follows:

$$u_j(z_j) = \frac{1}{\bar{\sigma}_j^2} \left[-(1+\delta_{\sigma_j})^2 (c_j - \bar{c_j})^2 + 2(1+\delta_{\sigma_j})^2 (c_j - \bar{c_j})(z_j - c_j) - (2\delta_{\sigma_j} + \delta_{\sigma_j}^2)(z_j - \bar{c_j})^2\right] \tag{14}$$

To bound B, we replace the u_j in each $u_j(\hat{z}_j)$ by the expression in Eq. 11, and the u_j in each $\mathbb{E}_{z \leftarrow \bar{\mathcal{D}}_j}[u_j]$ by the expression of Eq. 14. This yields:

$$\begin{aligned} |B| \leq \sum_j \frac{1.1(c_j - \bar{c_j})^2}{\bar{\sigma}_j^2} &+ \sum_j \frac{1.1}{2\sigma_j^2}\left[2|c_j - \bar{c_j}| \cdot |\hat{z}_j - c_j| + 2|\delta_{\sigma_j}| \cdot |\hat{z}_j - c_j|^2\right] \\ &+ \sum_j \frac{1}{2\bar{\sigma}_j^2}\left[2|c_j - \bar{c_j}| \cdot |\mathbb{E}_{z \leftarrow \bar{\mathcal{D}}_j}[z_j - \bar{c_j}]| + 2|\delta_{\sigma_j}| \cdot \mathbb{E}_{z \leftarrow \bar{\mathcal{D}}_j}[(z_j - \bar{c_j})^2]\right] \\ \leq \quad \frac{1.1n(\delta \cdot T)^2}{\eta_\epsilon(\mathbb{Z}^n)^2} &+ 1.1n\delta[T\sqrt{2\pi}/\eta_\epsilon(\mathbb{Z}^n) + 2\pi + 1] \\ &+ 1.1\delta\epsilon[T\sqrt{2\pi}/\eta_\epsilon(\mathbb{Z}^n) + 2\pi], \end{aligned}$$

where the bound over $|B|$ is obtained using the same techniques as for $|A|$. Overall, we see that $|A|, |B| \leq C$.

④ To conclude, we have $-C \leq \log(\frac{\mathcal{D}(\mathbf{z})}{\bar{\mathcal{D}}(\mathbf{z})}) \leq C$, so $e^{-C} \leq \frac{\mathcal{D}(\mathbf{z})}{\bar{\mathcal{D}}(\mathbf{z})} \leq e^C$. □

Practical implications of Lemma 8. We can now easily – given a few simplifications – apply the relative error lemma. Even though in theory we have $\|\mathbf{M}\|_2 \leq n\|\mathbf{M}\|_{\mathrm{GS}}$, this is a worst-case bound [Pei10, Lemma 5.1]. In practice, it is reasonable to assume $\|\mathbf{B}\|_2 = O(\sqrt{\log n}) \cdot \|\mathbf{B}\|_{\mathrm{GS}}$, with a small constant factor in the big O [Pre15, Sect. 6.5.2].[10]

In addition, we make the simplification $\|\mathbf{B}^{-1}\|_{\mathrm{GS}} \approx \|\mathbf{B}\|_{\mathrm{GS}}^{-1}$,[11] which gives $\sigma\|\mathbf{B}^{-1}\|_2 \approx \sqrt{\log n} \cdot \eta_\epsilon(\mathbb{Z}^n)$. It is also easy to make $\|\mathbf{L}\|_{\max} = 1$, so we consider that this is the case. Removing terms which are clearly negligible, and since $e^C \underset{C \to 0}{\sim} 1 + C$, we have

$$1 - C' \leq \frac{\bar{\mathcal{D}}}{\mathcal{D}} \leq 1 + C', \qquad \text{with } C' \approx 8 \cdot n^2 \sqrt{\log n} \cdot \delta. \tag{15}$$

For typical values of n (say, $n = 1024$), we can take $\delta = 2^{-37}/C' \approx 2^{-61}$, which is secure as per the argument of Sect. 3.3. Therefore, precision 61 is sufficient to securely use Klein's sampler.

5 Conclusion and Open Problems

To conclude, we expose a few perspectives and open problems that we have encountered. Most of them are related to implementing the techniques we have introduced, but in our opinion extending our techniques to decision problems is probably the most challenging question.

The revisited table approach. It remains to see how the CoDF-based algorithm we proposed in Sect. 4.2 can be efficiently implemented and protected against side-channel attacks. Our approach also seems highly composable with existing techniques, and it would be interesting to find combinations that achieve

[10] Or alternatively, $\|\mathbf{B}\|_2 = O(\sqrt{\log q}) \cdot \|\mathbf{B}\|_{\mathrm{GS}}$ (see e.g. [Pei10, Lemma 5.2]).

[11] As an example, for NTRU matrices, this is true up to a factor 1.17^2 [DLP14].

better overall efficiency.[12] For example, a natural question would be to see how to combine it with Knuth-Yao trees (see e.g. [DG14]).

Rejection sampling in practice. The techniques that we described in Sect. 4.3 remain to be implemented, to assess their efficiency and whether they can easily be made impervious against side-channel attacks.

Precision analysis of trapdoor samplers. It would be interesting to apply the precision analysis of Sect. 4.5 to other samplers, such as the one of [Pei10]. A promising candidate would be a randomized variant of Ducas and Prest's fast Fourier nearest plane [DP16]. The fast Fourier transform is known to be very stable numerically, and since this algorithm has the same structure, it seems likely that it will inherit this stability and require less than 53 bits of precision.

Decision problems. All the applications that we give are in the context of search problems. We would like to achieve the same efficiency for decision problems: as of today, one can use decision-to-search tricks in the random oracle model as in e.g. [DLP14, Sect. 4] or the results from [BLL+15, Sect. 4]. However, none of these solutions is fully satisfying and having efficient and generic Rényi security arguments for decision problems remain open.

Acknowledgements. I would like to thank Fabrice Mouhartem, Damien Stehlé and Michael Walter for useful discussions. I am also grateful to Ange Martinelli, Daniele Micciancio, Thomas Ricosset and the anonymous reviewers of ASIACRYPT 2017 for their insightful comments which helped to improve the quality of this paper.

This work has been supported in part by the BPI-funded project "RISQ".

A Appendix

A.1 Padé Approximants

In this section, we give a very succinct explanation of Padé approximants in the context that interests us. A more detailed introduction can be found in e.g. [Ass06]. Informally, Padé approximants can be described as generalizations of Taylor series, as the latter approximate $(n+1)$-differentiable functions as

$$f(x) = P_n(x) + O(z^{n+1}),$$

with P_n a polynomial of degree n, whereas Padé approximants provide an approximation of the form

$$Q_m(x)f(x) = P_n(x) + O(z^{n+m+1}),$$

with P_n and Q_m being polynomials of degree n and m.

[12] In a sense, this is what we did at the end of Sect. 4.2, as the algorithm 10 from [DDLL13] is meant to be used in conjunction with two other Algorithms (11 and 12).

While Padé approximants are in general much trickier to compute than their Taylor series counterparts, such approximants are well known for the exponential function. Let $m = n$ and

$$P_n(x) = Q_n(-x) = \sum_{k=0}^{n} \frac{(2n-k)!n!x^k}{(2n)!(n-k)!k!}. \tag{16}$$

Then we have [Pad92]:

$$\left| \frac{P_n(x)}{Q_n(x)} - e^x \right| = \frac{(n!)^2 x^{2n+1} e^x}{(2n)!(2n+1)!}(1 + o(1)) \tag{17}$$

Since our goal is to have a relative error less than 2^{-37}, taking $(m, |x|) \leq (4, .5)$ or $(m, |x|) \leq (5, 1)$ is sufficient.

A.2 Classical Lemmas

Lemma 9 [MR07, **Lemma 4.2**]. *Let Λ be a n-dimensional lattice, $\mathbf{c} \in \mathbb{R}^n$, $\mathbf{u} \in \mathbb{R}^n$ a vector of norm 1 and reals $\epsilon \in (0,1)$, $\sigma \geq 2\eta_\epsilon(\Lambda)$. The following inequalities hold:*

$$\left| \mathbb{E}_{\mathbf{x} \leftarrow D_{\Lambda,\sigma,\mathbf{c}}}[\langle \mathbf{x} - \mathbf{c}, \mathbf{u} \rangle] \right| \leq \frac{\sqrt{2\pi}\epsilon\sigma}{1-\epsilon}$$

$$\left| \mathbb{E}_{\mathbf{x} \leftarrow D_{\Lambda,\sigma,\mathbf{c}}}[\langle \mathbf{x} - \mathbf{c}, \mathbf{u} \rangle^2] - \sigma^2 \right| \leq \frac{2\pi\epsilon\sigma^2}{1-\epsilon}$$

Lemma 10 [MR07, **Lemma 4.4**]. *Let Λ be a n-dimensional lattice, $\mathbf{c} \in \mathbb{R}^n$, and reals $\epsilon \in (0,1)$, $\sigma \geq \eta_\epsilon(\Lambda)$. We have:*

$$\mathbb{P}_{\mathbf{x} \leftarrow D_{\Lambda,\sigma,\mathbf{c}}}[\|\mathbf{x} - \mathbf{c}\| \geq \sigma\sqrt{2\pi n}] \leq \frac{1+\epsilon}{1-\epsilon} 2^{-n}$$

References

[ABB10a] Agrawal, S., Boneh, D., Boyen, X.: Efficient lattice (H)IBE in the standard model. In: Gilbert [Gil10], pp. 553–572

[ABB10b] Agrawal, S., Boneh, D., Boyen, X.: Lattice basis delegation in fixed dimension and shorter-ciphertext hierarchical IBE. In: Rabin [Rab10], pp. 98–115

[ADPS16] Alkim, E., Ducas, L., Pöppelmann, T., Schwabe, P.: Post-quantum key exchange - a new hope. In: Holz, T., Savage, S. (eds.) 25th USENIX Security Symposium, USENIX Security 16, Austin, TX, USA, 10–12 August 2016, pp. 327–343. USENIX Association (2016)

[Ass06] Van Assche, W.: Padé and hermite-padé approximation and orthogonality (2006)

[Bab85] Babai, L.: On Lovász' lattice reduction and the nearest lattice point problem. In: Mehlhorn, K. (ed.) STACS 1985. LNCS, vol. 182, pp. 13–20. Springer, Heidelberg (1985). https://doi.org/10.1007/BFb0023990

[Bab86] Babai, L.: On lovasz' lattice reduction and the nearest lattice point problem. Combinatorica **6**(1), 1–13 (1986)

[BGG+14] Boneh, D., Gentry, C., Gorbunov, S., Halevi, S., Nikolaenko, V., Segev, G., Vaikuntanathan, V., Vinayagamurthy, D.: Fully key-homomorphic encryption, arithmetic circuit ABE and compact garbled circuits. In: Nguyen and Oswald [NO14], pp. 533–556

[BGM+16] Bogdanov, A., Guo, S., Masny, D., Richelson, S., Rosen, A.: On the hardness of learning with rounding over small modulus. In: Kushilevitz, E., Malkin, T. (eds.) TCC 2016. LNCS, vol. 9562, pp. 209–224. Springer, Heidelberg (2016). https://doi.org/10.1007/978-3-662-49096-9_9

[BLL+15] Bai, S., Langlois, A., Lepoint, T., Stehlé, D., Steinfeld, R.: Improved security proofs in lattice-based cryptography: using the Rényi divergence rather than the statistical distance. In: Iwata, T., Cheon, J.H. (eds.) ASIACRYPT 2015. LNCS, vol. 9452, pp. 3–24. Springer, Heidelberg (2015). https://doi.org/10.1007/978-3-662-48797-6_1

[Boy10] Boyen, X.: Lattice mixing and vanishing trapdoors: a framework for fully secure short signatures and more. In: Nguyen, P.Q., Pointcheval, D. (eds.) PKC 2010. LNCS, vol. 6056, pp. 499–517. Springer, Heidelberg (2010). https://doi.org/10.1007/978-3-642-13013-7_29

[Boy13] Boyen, X.: Attribute-based functional encryption on lattices. In: Sahai, A. (ed.) TCC 2013. LNCS, vol. 7785, pp. 122–142. Springer, Heidelberg (2013). https://doi.org/10.1007/978-3-642-36594-2_8

[BS16] Bun, M., Steinke, T.: Concentrated differential privacy: simplifications, extensions, and lower bounds. In: Hirt, M., Smith, A. (eds.) TCC 2016. LNCS, vol. 9985, pp. 635–658. Springer, Heidelberg (2016). https://doi.org/10.1007/978-3-662-53641-4_24

[Cac97] Cachin, C.: Entropy measures and unconditional security in cryptography. Ph.D. thesis (1997)

[CHKP10] Cash, D., Hofheinz, D., Kiltz, E., Peikert, C.: Bonsai trees, or how to delegate a lattice basis. In: Gilbert [Gil10], pp. 523–552

[DDLL13] Ducas, L., Durmus, A., Lepoint, T., Lyubashevsky, V.: Lattice signatures and bimodal Gaussians. In: Canetti, R., Garay, J.A. (eds.) CRYPTO 2013. LNCS, vol. 8042, pp. 40–56. Springer, Heidelberg (2013). https://doi.org/10.1007/978-3-642-40041-4_3

[DG14] Dwarakanath, N.C., Galbraith, S.D.: Sampling from discrete gaussians for lattice-based cryptography on a constrained device. Appl. Algebra Eng. Commun. Comput. **25**(3), 159–180 (2014)

[DLP14] Ducas, L., Lyubashevsky, V., Prest, T.: Efficient identity-based encryption over NTRU lattices. In: Sarkar, P., Iwata, T. (eds.) ASIACRYPT 2014. LNCS, vol. 8874, pp. 22–41. Springer, Heidelberg (2014). https://doi.org/10.1007/978-3-662-45608-8_2

[DN12a] Ducas, L., Nguyen, P.Q.: Faster Gaussian lattice sampling using lazy floating-point arithmetic. In: Wang and Sako [WS12], pp. 415–432

[DN12b] Ducas, L., Nguyen, P.Q.: Learning a Zonotope and more: cryptanalysis of NTRUSign countermeasures. In: Wang and Sako [WS12], pp. 433–450

[DP16] Ducas, L., Prest, T.: Fast fourier orthogonalization. In: Abramov, S.A., Zima, E.V., Gao, X.-S. (eds.) Proceedings of the ACM on International Symposium on Symbolic and Algebraic Computation, ISSAC 2016, Waterloo, ON, Canada, 19–22 July 2016, pp. 191–198. ACM (2016)

[EFGT17] Espitau, T., Fouque, P.-A., Gérard, B., Tibouchi, M.: Generalized Howgrave-Graham-Szydlo and side-channel attacks against BLISS. Publication status unknown (2017). https://almasty.lip6.fr/~espitau/bin/SCBliss

[GGH97] Goldreich, O., Goldwasser, S., Halevi, S.: Public-key cryptosystems from lattice reduction problems. In: Kaliski, B.S. (ed.) CRYPTO 1997. LNCS, vol. 1294, pp. 112–131. Springer, Heidelberg (1997). https://doi.org/10.1007/BFb0052231

[Gil10] Gilbert, H. (ed.): EUROCRYPT 2010. LNCS, vol. 6110. Springer, Heidelberg (2010). https://doi.org/10.1007/978-3-642-13190-5

[GJSS01] Gentry, C., Jonsson, J., Stern, J., Szydlo, M.: Cryptanalysis of the NTRU signature scheme (NSS) from Eurocrypt 2001. In: Boyd, C. (ed.) ASIACRYPT 2001. LNCS, vol. 2248, pp. 1–20. Springer, Heidelberg (2001). https://doi.org/10.1007/3-540-45682-1_1

[GLP12] Güneysu, T., Lyubashevsky, V., Pöppelmann, T.: Practical lattice-based cryptography: a signature scheme for embedded systems. In: Prouff, E., Schaumont, P. (eds.) CHES 2012. LNCS, vol. 7428, pp. 530–547. Springer, Heidelberg (2012). https://doi.org/10.1007/978-3-642-33027-8_31

[GPV08] Gentry, C., Peikert, C., Vaikuntanathan, V.: Trapdoors for hard lattices and new cryptographic constructions. In: Ladner, R.E., Dwork, C. (eds.) 40th ACM STOC, Victoria, British Columbia, Canada, 17–20 May 2008, pp. 197–206. ACM Press (2008)

[GS02] Gentry, C., Szydlo, M.: Cryptanalysis of the revised NTRU signature scheme. In: Knudsen, L.R. (ed.) EUROCRYPT 2002. LNCS, vol. 2332, pp. 299–320. Springer, Heidelberg (2002). https://doi.org/10.1007/3-540-46035-7_20

[HHGP+03] Hoffstein, J., Howgrave-Graham, N., Pipher, J., Silverman, J.H., Whyte, W.: NTRUSign: digital signatures using the NTRU lattice. In: Joye, M. (ed.) CT-RSA 2003. LNCS, vol. 2612, pp. 122–140. Springer, Heidelberg (2003). https://doi.org/10.1007/3-540-36563-X_9

[Kle00] Klein, P.N.: Finding the closest lattice vector when it's unusually close. In: SODA (2000)

[LD13] Lepoint, T., Ducas, L.: Proof-of-concept software implementation of BLISS (2013). http://bliss.di.ens.fr/bliss-06-13-2013.zip

[LP15] Lyubashevsky, V., Prest, T.: Quadratic time, linear space algorithms for Gram-Schmidt orthogonalization and Gaussian sampling in structured lattices. In: Oswald, E., Fischlin, M. (eds.) EUROCRYPT 2015. LNCS, vol. 9056, pp. 789–815. Springer, Heidelberg (2015). https://doi.org/10.1007/978-3-662-46800-5_30

[LPSS14] Ling, S., Phan, D.H., Stehlé, D., Steinfeld, R.: Hardness of k-LWE and applications in traitor tracing. In: Garay, J.A., Gennaro, R. (eds.) CRYPTO 2014. LNCS, vol. 8616, pp. 315–334. Springer, Heidelberg (2014). https://doi.org/10.1007/978-3-662-44371-2_18

[LSS14] Langlois, A., Stehlé, D., Steinfeld, R.: GGHLite: more efficient multilinear maps from ideal lattices. In: Nguyen and Oswald [NO14], pp. 239–256

[Lyu09] Lyubashevsky, V.: Fiat-Shamir with aborts: applications to lattice and factoring-based signatures. In: Matsui, M. (ed.) ASIACRYPT 2009. LNCS, vol. 5912, pp. 598–616. Springer, Heidelberg (2009). https://doi.org/10.1007/978-3-642-10366-7_35

[Lyu12] Lyubashevsky, V.: Lattice signatures without trapdoors. In: Pointcheval and Johansson [PJ12], pp. 738–755

[Mir17] Mironov, I.: Renyi differential privacy. In: Proceedings of 30th IEEE Computer Security Foundations Symposium (2017). http://arxiv.org/abs/1702.07476

[MP12] Micciancio, D., Peikert, C.: Trapdoors for lattices: simpler, tighter, faster, smaller. In: Pointcheval and Johansson [PJ12], pp. 700–718

[MR04] Micciancio, D., Regev, O.: Worst-case to average-case reductions based on Gaussian measures. In: 45th FOCS, Rome, Italy, 17–19 October 2004, pp. 372–381. IEEE Computer Society Press (2004)

[MR07] Micciancio, D., Regev, O.: Worst-case to average-case reductions based on Gaussian measures. SIAM J. Comput. **37**, 267–302 (2007)

[MW17] Micciancio, D., Walter, M.: Gaussian sampling over the integers: efficient, generic, constant-time. In: Katz, J., Shacham, H. (eds.) CRYPTO 2017. LNCS, vol. 10402, pp. 455–485. Springer, Cham (2017). https://doi.org/10.1007/978-3-319-63715-0_16

[NO14] Nguyen, P.Q., Oswald, E. (eds.): EUROCRYPT 2014. LNCS, vol. 8441. Springer, Heidelberg (2014). https://doi.org/10.1007/978-3-642-55220-5

[NR06] Nguyen, P.Q., Regev, O.: Learning a parallelepiped: cryptanalysis of GGH and NTRU signatures. In: Vaudenay, S. (ed.) EUROCRYPT 2006. LNCS, vol. 4004, pp. 271–288. Springer, Heidelberg (2006). https://doi.org/10.1007/11761679_17

[Pad92] Padé, H.: Sur la représentation approchée d'une fonction par des fractions rationnelles. Ph.D. thesis (1892)

[PDG14] Pöppelmann, T., Ducas, L., Güneysu, T.: Enhanced lattice-based signatures on reconfigurable hardware. In: Batina, L., Robshaw, M. (eds.) CHES 2014. LNCS, vol. 8731, pp. 353–370. Springer, Heidelberg (2014). https://doi.org/10.1007/978-3-662-44709-3_20

[Pei10] Peikert, C.: An efficient and parallel Gaussian sampler for lattices. In: Rabin [Rab10], pp. 80–97

[PJ12] Pointcheval, D., Johansson, T. (eds.): EUROCRYPT 2012. LNCS, vol. 7237. Springer, Heidelberg (2012). https://doi.org/10.1007/978-3-642-29011-4

[POG15] Pöppelmann, T., Oder, T., Güneysu, T.: High-performance ideal lattice-based cryptography on 8-bit ATxmega microcontrollers. In: Lauter, K., Rodríguez-Henríquez, F. (eds.) LATINCRYPT 2015. LNCS, vol. 9230, pp. 346–365. Springer, Cham (2015). https://doi.org/10.1007/978-3-319-22174-8_19

[Pop14] Poppelmann, T.: Proof-of-concept hardware implementation of BLISS (2014). https://www.sha.rub.de/media/attachments/files/2014/09/lattice_processor_final_publication.zip

[Pre15] Prest, T.: Gaussian sampling in lattice-based cryptography. Theses, École Normale Supérieure December 2015

[Rab10] Rabin, T. (ed.): CRYPTO 2010. LNCS, vol. 6223. Springer, Heidelberg (2010). https://doi.org/10.1007/978-3-642-14623-7

[R61] Rnyi, A.: On measures of entropy and information. In: Proceedings of the Fourth Berkeley Symposium on Mathematical Statistics and Probability, Volume 1: Contributions to the Theory of Statistics, Berkeley, California, pp. 547–561. University of California Press (1961)

[Saa15] Saarinen, M.-J.O.: Gaussian sampling precision in lattice cryptography. Cryptology ePrint Archive, Report 2015/953 (2015). http://eprint.iacr.org/2015/953

[Str14] Swan, S.: Bimodal lattice signature scheme (BLISS) (2014). https://wiki.strongswan.org/projects/strongswan/wiki/BLISS

[TT15] Takashima, K., Takayasu, A.: Tighter security for efficient lattice cryptography via the Rényi divergence of optimized orders. In: Au, M.-H., Miyaji, A. (eds.) ProvSec 2015. LNCS, vol. 9451, pp. 412–431. Springer, Cham (2015). https://doi.org/10.1007/978-3-319-26059-4_23

[vEH14] van Erven, T., Harremoës, P.: IEEE Trans. Inf. Theory **60**(7), 3797–3820 (2014)

[Wan10] Wan, A.: Learning, cryptography, and the average case. Ph.D. thesis, Columbia University (2010)

[WS12] Wang, X., Sako, K. (eds.): ASIACRYPT 2012. LNCS, vol. 7658. Springer, Heidelberg (2012). https://doi.org/10.1007/978-3-642-34961-4

Homomorphic Encryptions

Faster Packed Homomorphic Operations and Efficient Circuit Bootstrapping for TFHE

Ilaria Chillotti[2(✉)], Nicolas Gama[1,2], Mariya Georgieva[3], and Malika Izabachène[4]

[1] Inpher, Lausanne, Switzerland
[2] Laboratoire de Mathématiques de Versailles, UVSQ, CNRS, Université Paris-Saclay, 78035 Versailles, France
ilaria.chillotti@uvsq.fr
[3] Gemalto, 6 rue de la Verrerie, 92190 Meudon, France
[4] CEA LIST, Point Courrier 172, 91191 Gif-sur-Yvette Cedex, France

Abstract. In this paper, we present several methods to improve the evaluation of homomorphic functions in TFHE, both for fully and for leveled homomorphic encryption. We propose two methods to manipulate packed data, in order to decrease the ciphertext expansion and optimize the evaluation of look-up tables and arbitrary functions in RingGSW based homomorphic schemes. We also extend the automata logic, introduced in [12,19], to the efficient leveled evaluation of weighted automata, and present a new homomorphic counter called TBSR, that supports all the elementary operations that occur in a multiplication. These improvements speed-up the evaluation of most arithmetic functions in a packed leveled mode, with a noise overhead that remains additive. We finally present a new circuit bootstrapping that converts LWE into low-noise RingGSW ciphertexts in just 137 ms, which makes the leveled mode of TFHE composable, and which is fast enough to speed-up arithmetic functions, compared to the gate-by-gate bootstrapping given in [12]. Finally, we propose concrete parameter sets and timing comparison for all our constructions.

Keywords: FHE · Leveled · Bootstrapping · LWE · GSW · Packing · Weighted automata · Arithmetic

1 Introduction

Fully homomorphic encryption (FHE) allows arbitrary computations over encrypted data, without decrypting them. The first construction was proposed in 2009 by Gentry [20], which introduced a new technique called bootstrapping to handle the noise propagation in ciphertexts. Although many efforts have been done since this first proposal to improve FHE, it remains too slow for real world applications. The most promising constructions are [5,21,30]. We focus on constructions based on the LWE problem, introduced by Regev in

T. Takagi and T. Peyrin (Eds.): ASIACRYPT 2017, Part I, LNCS 10624, pp. 377–408, 2017.
https://doi.org/10.1007/978-3-319-70694-8_14

2005 [28], and its ring variants [25]. Some public implementations are available, namely Helib [22,23], FV-NFlib [24] and SEAL [29], based on BGV [5,18], and FHEW [17] and TFHE [14], based on GSW [12,17,21].

BGV-based schemes use in general slow operations, but they can treat a lot of bits at the same time, so they can pack and batch many operations in a SIMD manner, like in GPUs. Further, the set of operations that are efficient with BGV are very constrained by the parameter set. Some parameters allow very fast vectorial sums and products modulo a fixed modulus (as in AES). But with these parameters, a comparison, a classical addition, extracting one bit or more complicated bit operations (as in SHA-256) are very slow.

On the other hand, recent developments have shown that GSW operations can evaluate very fast independent elementary operations on bits, like in a CPU. In the TFHE scheme (presented in [12] and based on GSW [21] and its ring variant [17]) the elementary operations are all the binary gates. Therefore, it is easy to represent any function that has few gates, and the running time is simply proportional to their number. A few methods have been proposed to perform multibit or packed/batched operations with GSW-based schemes. For instance, [4] extends the bootstrapping of FHEW [17] to evaluate non-linear functions with a few input bits. Unfortunately, the parameter sizes must increase exponentially with the number of bits in the plaintext space. Until this work, it was not clear how to perform efficient evaluations on packed data or batch operations, as it is in BGV-based schemes.

Homomorphic encryption falls in two families: leveved (LHE) and fully (FHE) homomorphic encryption. Informally, in LHE, for each function, there exist parameters that can homomorphically evaluate it[1]. The structure of the function to be evaluated (multiplicative depth in BGV or depth of compositions of branching algorithms for GSW) translates into a noise overhead, and the parameters must be chosen large enough to support this noise bound. This concept is represented in the paper by the notion of parameter levels. In FHE, a single parameter set allows to evaluate any function. This generalized definition implies that FHE is a particular case of LHE.

In many FHE schemes, the elementary operations consist in leveled gates with a symmetric noise propagation formula, and where non-linear gates cost more than linear ones. The papers [3,26] improve the efficiency of fully homomorphic implementations by optimizing the placement of bootstrapping between the gates throughout the circuit. This strategy does not really apply to GSW schemes that strongly rely on the asymmetric noise propagation formula, in which all circuits are expressed as deterministic automata or branching algorithms, because the depth of the circuit has a very small impact on the noise.

The TFHE construction of [12] proposes two modes of operations: a FHE mode composed of bootstrapped binary gates, and a LHE mode which can evaluate a deterministic automata or a branching algorithm and which supports

[1] To simplify, we include the key size and the noise rate.

very large depth of transitions.[2] Note however that in the LHE mode of [12], inputs and outputs have different types, which makes it non-composable. In this paper we optimize both FHE/LHE modes, and we solve the non-composability constraint.

Our Contribution. In this paper, we improve the TFHE construction of [12] for both FHE and LHE modes.

We first propose a *blind rotation* algorithm that we describe in Sect. 2. In FHE mode, this algorithm contributes to the acceleration of the gate bootstrapping of [12], and its implementation is now included in the core of the TFHE library [14]. This algorithm is also one of the building block we use to improve the LHE mode of TFHE.

Because of the asymmetric noise propagation, operating over packed ciphertexts in GSW-based schemes is harder than in BGV-based schemes. We describe two different techniques, that we call *horizontal packing* and *vertical packing*, that can be used to improve the evaluation of leveled circuits. An arbitrary function from $\{0,1\}^n \rightarrow \mathbb{T}^p$ can be represented as a truth table with p columns and 2^s rows. By packing these coefficients horizontally, the homomorphic evaluation of the function can be batched, and the p outputs can be produced in parallel. This technique is classical, but is only efficient for very large p. We propose another technique, called *vertical packing*, which packs the coefficients column-wise, and which achieves its maximal speed-up also when p is equal to 1.

We also extend the deterministic finite automata framework proposed in [12, 19] by working with *deterministic weighted finite automata*. For most multibit arithmetic functions, such as addition, multiplication and maximum value, these latter allow to compute the whole output in a running time that would have previously produced only a single bit.

Indeed, when an arithmetic operation is evaluated by a deterministic automata, the only bit of information that is retained is whether or not the destination state is accepting, and the rest of the path (that contains a lot of information on the result) is forgotten. Thus, we need to evaluate one automata for each bit of the result. Instead, by assigning a vector of weights on each transition, we are able to retain enough information along the path to get all the bits of the result at once, in a single pass of the automata. This decreases the complexity of these operations by at least one order of magnitude. Furthermore, we propose a new homomorphic counter (called *TBSR*) that supports homomorphically all the basic operations related to the multiplication: incrementation, division by 2 and extraction of bits. This technique gives another speed-up by a factor equal to the bit-size of the input. We show how to use it to represent the $O(d^2)$ (with d equal to the size of the input) schoolbook multi-addition or multiplication circuits, without increasing the homomorphic depth and with very low noise overhead.

[2] The TFHE construction is implemented and publicly available [14]. The actual running timings are 13 ms for each bootstrapped binary gate in FHE mode, and 34 µs per transition in LHE mode. The implementation also includes optimizations described in Sect. 2.

Our last contribution solves the main problem of the leveled mode of TFHE, which is the non-composability, due to the fact that inputs and outputs are of different types. The inputs are in fact RingGSW ciphertexts, while the outputs are LWE ciphertexts. We introduce a new bootstrapping, called *circuit bootstrapping*, that allows to transform LWE ciphertexts back to RingGSW ciphertexts that can be reused as inputs in leveled circuits. The implementation of this circuit bootstrapping is publicly available [14] and it runs in 137 ms, improving all previous techniques. The introduction of the circuit bootstrapping closes the loop and allows all the new techniques previously described to be applied also in a FHE mode. To show how these techniques improve homomorphic evaluations, we propose several examples with concrete parameters and running time. For instance, we show that we can evaluate a 10 bits to 1 bit ($\{0,1\}^{10} \to \{0,1\}$) look-up table in 340 µs and we can bootstrap the output in just 137 ms.

Paper organization. We first review mathematical definitions for the continuous LWE and RingGSW encryption over the torus and review the algorithmic procedures for the homomomorphic evaluation of gates. In particular, we extend the keyswitching algorithm to evaluate public or private $\mathbb{Z}$-module morphisms, and explain how it can be used to pack, unpack and move data across slots of a ciphertext. In Sect. 3, we show various techniques to speed-up operations on packed data: horizontal and vertical packing, our method to evaluate arithmetic functions via weighted automata and our TBSR counter technique. In Sect. 4, we introduce our *circuit bootstrapping* algorithm which makes it possible to connect gates of either RingGSW or LWE types and give the practical execution timings we have obtained. Section 5 depicts all our complexity results for different parameters set.

2 Preliminaries

This section introduces and revisits some basic concepts to understand the rest of the paper. The homomorphic constructions we present are based on the LWE problem, presented by Regev in 2005 [28], and on the GSW construction, proposed by Gentry-Sahai-Waters in 2013 [21]. We use the generalized definitions of TLWE and TGSW (the T stands for the torus representation) proposed in [12], and extend some of their results.

2.1 Background on TFHE

We denote by λ the security parameter. The set $\{0,1\}$ is written as $\mathbb{B}$. The real torus $\mathbb{R}/\mathbb{Z} = \mathbb{R} \mod 1$ of real numbers mod 1 is denoted by $\mathbb{T}$. $\mathfrak{R}$ is the ring $\mathbb{Z}[X]/(X^N+1)$ of integer polynomials modulo X^N+1, and $\mathbb{T}_N[X]$ is the module $\mathbb{R}[X]/(X^N+1) \mod 1$ of torus polynomials, where N is a power of 2. $\mathbb{B}_N[X]$ denotes the subset of $\mathfrak{R}$ of polynomials with binary coefficients. Note that $\mathbb{T}$ is a $\mathbb{Z}$-module and that $\mathbb{T}_N[X]$ is a $\mathfrak{R}$-module. The set of vectors of size n in E is denoted by E^n, and the set of $n \times m$ matrices with entries in E is noted $\mathcal{M}_{n,m}(E)$. As before, $\mathbb{T}^n$ (resp. $\mathbb{T}_N[X]^n$) and $\mathcal{M}_{n,m}(\mathbb{T})$ (resp. $\mathcal{M}_{n,m}(\mathbb{T}_N[X])$) are $\mathbb{Z}$-modules (resp. $\mathfrak{R}$-modules).

Distance, Lipschitzian functions, Norms. We use the standard ℓ_p-distance over $\mathbb{T}$, and use the (more convenient but improper) $\|x\|_p$ notation to denote the distance between 0 and x. Note that it satisfies $\forall m \in \mathbb{Z}$, $\|m \cdot \boldsymbol{x}\|_p \leq |m| \, \|\boldsymbol{x}\|_p$. For an integer or torus polynomial a modulo $X^N + 1$, we write $\|a\|_p$ the norm of its unique representative coefficients of degree $\leq N - 1$. The notion of lipschitzian function always refers to the ℓ_∞ distance: a function f is R-lipschitzian iff. $\|f(x) - f(y)\|_\infty \leq R \, \|x - y\|_\infty$ for all inputs x, y.

TLWE. TLWE is a generalized and scale invariant version of the LWE problem, proposed by Regev in 2005 [28], over the Torus $\mathbb{T}$.

Given a small linear lipshtitzian function φ_s from $\mathbb{T}_N[X]^{k+1}$ to $\mathbb{T}_N[X]$ (that depends on the secret key) and which we'll call the *phase* function, the TLWE encryption of $\mu \in \mathbb{T}_N[X]$ simply consists in picking a ciphertext c which is a Gaussian approximation of a preimage $\varphi_s^{-1}(\mu)$. If the Gaussian noise is small enough, the distribution of $\varphi_s(c)$ (over the probability space Ω of all possible choices of Gaussian noise) remains *concentrated* around the message μ, *i.e.* included in a ball of radius $< \frac{1}{4}$ around μ. Because this distribution is concentrated, it allows to properly define the intuitive notions of *expectation* and *variance*, which would in general not exist over the Torus: in this case, the expectation of $\varphi_s(c)$ is the original message μ, and its variance is equal to the variance of the Gaussian noise that was added during encryption. We refer to [12] for a general definition of Ω-space, concentrated distribution, expectation, variance, Gaussian and sub-Gaussian distributions over the Torus.

More precisely, a TLWE secret key $\boldsymbol{s} \in \mathbb{B}_N[X]^k$ is a vector of k binary polynomials of degree N. We assume that each coefficient of the secret key is chosen uniformly, so the key has $n = kN$ bits of entropy.

Definition 2.1 (TLWE, phase). TLWE *ciphertexts or samples are* $\mathbf{c} = (\boldsymbol{a}, b) \in \mathbb{T}_N[X]^{k+1}$ *that fall in one of the three cases:*

- *Noiseless Trivial of* μ*:* $\boldsymbol{a} = \mathbf{0}$ *and* $b = \mu$*. Note that this sample is independent of the secret key.*
- *Fresh* TLWE *sample of* μ *of standard deviation* α*:* $\boldsymbol{a}$ *is uniformly in* $\mathbb{T}_N[X]^k$ *and* b *follows a continuous Gaussian of standard deviation* α *centered in* $\mu + \boldsymbol{s} \cdot \boldsymbol{a}$*, where the variance is* α^2*. In the following, we will write* $(\boldsymbol{a}, b) \in \mathsf{TLWE}_{\boldsymbol{s},\alpha}(\mu)$.
- *Combination of* TLWE *samples:* $\mathbf{c} = \sum_{j=1}^{p} r_j \cdot \mathbf{c}_j$ *is a* TLWE *sample, where* $\mathbf{c}_1, \ldots, \mathbf{c}_p$ *are* TLWE *sample under the same key and* $r_1, \ldots, r_p$ *in* $\mathbb{Z}$ *or* $\mathfrak{R}$.

The phase of a sample $\mathbf{c}$ *is defined as* $\varphi_{\boldsymbol{s}}(\mathbf{c}) = b - \boldsymbol{s} \cdot \boldsymbol{a}$.

Like in [12], we say that a TLWE sample $\boldsymbol{c}$ is *valid* iff there exists a key $\boldsymbol{s} \in \mathbb{B}_N[X]^k$ such that the distribution of the phase $\varphi_{\boldsymbol{s}}(\boldsymbol{c})$ is concentrated. The message of a sample $\boldsymbol{c}$, written $\mathsf{msg}(\boldsymbol{c})$ is defined as the expectation of its phase over the Ω-probability space. We will write $\boldsymbol{c} \in \mathsf{TLWE}_{\boldsymbol{s}}(\mu)$ iff $\mathsf{msg}(\boldsymbol{c}) = \mu$. The error of a TLWE sample $\boldsymbol{c}$, $\mathsf{Err}(\boldsymbol{c})$ is then computed as $\varphi(\boldsymbol{c}) - \mathsf{msg}(\boldsymbol{c})$. The variance of the error will be denoted $\mathsf{Var}(\mathsf{Err}(\boldsymbol{c}))$ and its maximal amplitude $\|\mathsf{Err}(\boldsymbol{c})\|_\infty$.

The message of a fresh sample in $\mathsf{TLWE}_{\boldsymbol{s},\alpha}(\mu)$ is μ and its variance is α^2. The message function is linear: $\mathsf{msg}(\sum_{j=1}^{p} r_j \cdot \boldsymbol{c_j})$, where $c_j \in \mathsf{TLWE}_{\boldsymbol{s}}(r_j\mu_j)$ is equal to $\sum_j^p r_j\mu_j$ provided that the variance $\mathsf{Var}(\mathsf{Err}(\boldsymbol{c})) \leq \sum_{j=1}^{p} \|r_j\|_2^2 \cdot \mathsf{Var}(\mathsf{Err}(\boldsymbol{c_j}))$ and the maximal amplitude $\|\mathsf{Err}(\boldsymbol{c})\|_\infty \leq \sum_{j=1}^{p} \|r_j\|_1 \cdot \|\mathsf{Err}(\boldsymbol{c_j})\|_\infty$ remains small.

This definition of message has the great advantages to be linear, continuous, and that it works with infinite precision even over the continuous torus. In the practical case where the message is known to belong to a discrete subset $\mathcal{M}$ of $\mathbb{T}_N[X]$ and that the noise amplitude of $\mathbf{c}$ is smaller than the packing radius of $\mathcal{M}$, then the decryption algorithm can retrieve the message in practice by rounding the phase of the sample to its nearest element in $\mathcal{M}$. For example with $\mathcal{M} = \{(0, 1/2)\}[X]$, the packing is $1/4$ and thus the samples of variance smaller than $(1/2^{10})$ are decryptable with overwhelming probability.

Distinguishing TLWE encryptions of $\mathbf{0}$ from random samples in $\mathbb{T}_N[X]^k \times \mathbb{T}_N[X]$ is equivalent to the LWE problem initially defined by Regev [28] and its ring [25] and Scale invariant [6,11,13] variants.

The main parameters of TLWE are the noise rate α and the key entropy n, and the security parameter is a function of those parameters, as specified in [12, Sect. 6]. By choosing $N = 1$ and k large, TLWE-problem is the (Scalar) binary-TLWE-problem. When N large and $k = 1$, TLWE is binary-$\mathsf{RingLWE}$.

TGSW. In the same line as TLWE, TGSW generalizes the GSW encryption scheme, proposed by Gentry, Sahai and Waters in 2013 [21]. The gadget matrix H is defined with respect to a base $B_g \in \mathbb{N}$ as the $((k+1)\ell) \times (k+1)$ matrix with ℓ repeated super-decreasing $\mathbb{T}$-polynomials $(1/B_g, \ldots, 1/B_g^\ell)$ as:

$$H = \left(\begin{array}{c|c|c} 1/B_g & \ldots & 0 \\ \vdots & \ddots & \vdots \\ 1/B_g^\ell & \ldots & 0 \\ \hline \vdots & \ddots & \vdots \\ \hline 0 & \ldots & 1/B_g \\ \vdots & \ddots & \vdots \\ 0 & \ldots & 1/B_g^\ell \end{array}\right) \in \mathcal{M}_{(k+1)\ell,k+1}(\mathbb{T}_N[X]). \tag{1}$$

With this choice of gadget, it is possible to efficiently decompose elements of $\mathbb{T}_N[X]^{k+1}$ as a small linear combination of rows of H. As in [12], we use approximate decomposition. For a quality parameter $\beta \in \mathbb{R}_{>0}$ and a precision $\epsilon \in \mathbb{R}_{>0}$, we call $Dec_{H,\beta,\epsilon}(\boldsymbol{v})$ the (possibly randomized) algorithm that outputs a small vector $\boldsymbol{u} \in \mathfrak{R}^{(k+1)\ell}$, such that $\|\boldsymbol{u}\|_\infty \leq \beta$ and $\|\boldsymbol{u} \cdot H - \boldsymbol{v}\|_\infty \leq \epsilon$. In this paper we will always use this gadget H with the decomposition in base B_g, so we have $\beta = B_g/2$ and $\epsilon = 1/2B_g^\ell$.

TGSW *samples.* Let $\boldsymbol{s} \in \mathbb{B}_N[X]^k$ be a TLWE secret key and $H \in \mathcal{M}_{(k+1)\ell,k+1}(\mathbb{T}_N[X])$ the gadget previously defined. A TGSW sample C of a message $\mu \in \mathfrak{R}$ is equal to the sum $C = Z + \mu \cdot H \in \mathcal{M}_{(k+1)\ell,k+1}(\mathbb{T}_N[X])$ where $Z \in \mathcal{M}_{(k+1)\ell,k+1}(\mathbb{T}_N[X])$ is a matrix such that each line is a random TLWE sample of 0 under the same key.

A sample $C \in \mathcal{M}_{(k+1)\ell,k+1}(\mathbb{T}_N[X])$ is a valid TGSW sample iff there exists a unique polynomial $\mu \in \mathfrak{R}/H^{\perp}$ and a unique key s such that each row of $C - \mu \bullet H$ is a valid TLWE sample of 0 w.r.t. the key s. We denote $\mathsf{msg}(C)$ the message μ of C. By extension, we can define the phase of a TGSW sample C as the list of the $(k+1)\ell$ TLWE phases of each line of C, and the error as the list of the $(k+1)\ell$ TLWE errors of each line of C.

In addition, if we linearly combine TGSW samples $C_1, \ldots, C_p$ of messages $\mu_1, \ldots, \mu_p$ with the same keys and independent errors, s.t. $C = \sum_{i=1} e_i \cdot C_i$ is a sample of message $\sum_{i=1}^{p} e_i \cdot \mu_i$. The variance $\mathsf{Var}(C) = \sum_{i=1}^{p} \|e_i\|_2^2 \cdot \mathsf{Var}(C_i)$ and noise infinity norm $\|\mathsf{Err}(C)\|_\infty = \sum_{i=1}^{p} \|e_i\|_1 \cdot \|\mathsf{Err}(C)\|_\infty$. And, the lipschitzian property of the phase is preserved, i.e. $\|\varphi_s(A)\|_\infty \leq (Nk+1)\|A\|_\infty$.

Homomorphic Properties. As GSW, TGSW inherits homomorphic properties. We can define the internal product between two TGSW samples and the external $\boxdot$ product already defined and used in [7,12]. The external product is almost the GSW product [21], except that only one vector needs to be decomposed.

Definition 2.2 (External product). *We define the product $\boxdot$ as*

$$\begin{aligned} \boxdot : \mathsf{TGSW} \times \mathsf{TLWE} &\longrightarrow \mathsf{TLWE} \\ (A, \boldsymbol{b}) &\longmapsto A \boxdot \boldsymbol{b} = Dec_{\boldsymbol{h},\beta,\epsilon}(\boldsymbol{b}) \cdot A. \end{aligned}$$

The following theorem on the noise propagation of the external product was shown in [12, Sect. 3.2]:

Theorem 2.3 (External Product). *If A is a valid* TGSW *sample of message μ_A and $\boldsymbol{b}$ is a valid* TLWE *sample of message $\mu_{\boldsymbol{b}}$, then $A \boxdot \boldsymbol{b}$ is a* TLWE *sample of message $\mu_A \cdot \mu_{\boldsymbol{b}}$ and $\|Err(A \boxdot \boldsymbol{b})\|_\infty \leq (k+1)\ell N\beta \|Err(A)\|_\infty + \|\mu_A\|_1 (1 + kN)\epsilon + \|\mu_A\|_1 \|Err(\boldsymbol{b})\|_\infty$ (worst case), where β and ϵ are the parameters used in the decomposition algorithm. If $\|Err(A \boxdot \boldsymbol{b})\|_\infty \leq 1/4$ then $A \boxdot \boldsymbol{b}$ is a valid* TRLWE *sample. And assuming the heuristic 2.4, we have that $Var(Err(A \boxdot \boldsymbol{b})) \leq (k+1)\ell N\beta^2 Var(Err(A)) + (1+kN)\|\mu_A\|_2^2 \epsilon^2 + \|\mu_A\|_2^2 Var(Err(\boldsymbol{b}))$.*

There also exists an internal product between two TGSW samples, already presented in [1,12,17,19,21], and which consists in $(k+1)\ell$ independent $\boxdot$ products, and maps to the product of integer polynomials on plaintexts, and turns TGSW encryption into a ring homomorphism. Since we do not use this internal product in our constructions, so we won't detail it.

Independence heuristic. All our average-case bounds on noise variances rely on the independence heuristic below. They usually corresponds to the square-root of the worst-case bounds which don't need this heuristic. As already noticed in [17], this assumption matches experimental results.

Assumption 2.4 (Independence Heuristic). We assume that all the error coefficients of TLWE or TGSW samples of the linear combinations we consider are independent and concentrated. In particular, we assume that they are σ-subgaussian where σ is the square-root of their variance.

Notations. In the rest of the paper, the notation TLWE is used to denote the (scalar) binary TLWE problem, while for the ring mode, we use the notation TRLWE. TGSW is only used in ring mode with notation TRGSW, to keep uniformity with the TRLWE notation.

Sum-up of elementary homomorphic operations. Table 1 summarizes the possible operations on plaintexts that we can perform in LHE mode, and their correspondence over the ciphertexts. All these operations are expressed on the continuous message space $\mathbb{T}$ for TLWE and $\mathbb{T}_N[X]$ for TRLWE. As previously mentionned, all samples contain noise, the user is free to discretize the message space accordingly to allow practical exact decryption. All these algorithms will be described in the next sections.

Table 1. TFHE elementary operations - In this table, all μ_i's denote plaintexts in $\mathbb{T}_N[X]$ and $\mathbf{c}_i$ the corresponding TRLWE ciphertext. The m_i's are plaintexts in $\mathbb{T}$ and $\mathfrak{c}$ their TLWE ciphertext. The b_i's are bit messages and C_i their TRGSW ciphertext. The ϑ_i's are the noise variances of the respective ciphertexts. In the translation, w is in $\mathbb{T}_N[X]$. In the rotation, the u_i's are integer coefficients. In the $\mathbb{Z}[X]$-linear combination, the v_i's are integer polynomials in $\mathbb{Z}[X]$.

Operation	Plaintext	Ciphertext	Variance
Translation	$\mu + w$	$\mathbf{c} + (0, w)$	ϑ
Rotation	$X^{u_i}\mu$	$X^{u_i}\mathbf{c}$	ϑ
$\mathbb{Z}[X]$-linear	$\sum v_i \mu_i$	$\sum v_i \mathbf{c}_i$	$\sum \lVert v_i \rVert_2^2 \vartheta_i$
SampleExtract	$\sum \mu_i X^i \rightarrow \mu_p$	SampleExtract (Sect. 2.2)	ϑ
$\mathbb{Z}$-linear	$f(m_1, \ldots, m_p)$	$\mathsf{PubKS}_{\mathsf{KS}}(f, \mathfrak{c}_1, \ldots, \mathfrak{c}_p)$ (Algorithm 1)	$R^2\vartheta + n \log\left(\frac{1}{\alpha}\right) \mathrm{Cst}_{KS}$
	R-lipschitzian	$\mathsf{PrivKS}_{KS(f)}(\mathfrak{c}_1, \ldots, \mathfrak{c}_p)$ (Algorithm 2)	$R^2\vartheta + np \log\left(\frac{1}{\alpha}\right) \mathrm{Cst}_{KS}$
Ext. product	$b_1 \cdot \mu_2$	$C_1 \boxdot \mathbf{c_2}$ (Theorem 2.3)	$b_1\vartheta_2 + \mathrm{Cst}_{\mathsf{TRGSW}}\vartheta_1$
CMux	$b_1 ? \mu_2 : \mu_3$	$\mathtt{CMux}(C_1, \mathbf{c_2}, \mathbf{c_3})$ (Lemma 2.7)	$\max(\vartheta_2, \vartheta_3) + \mathrm{Cst}_{\mathsf{TRGSW}}\vartheta_1$
$\mathbb{T}$-non-linear	$X^{-\varphi(c_1)}\mu_2$	BlindRotate (Algorithm 3)	$\vartheta + n\mathrm{Cst}_{\mathsf{TRGSW}}$
Bootstrapping	decrypt$(c)?m : 0$	Gate Bootstrapping (Algorithm 4)	Cst

2.2 Key Switching Revisited

In the following, we instantiate TRLWE and TRGSW with different parameter sets and we keep the same name for the variables $n, N, \alpha, \ell, B_g, \ldots$, but we alternate between bar over and bar under variables to differentiate input and output parameters. In order to switch between keys in different parameter sets, but also to switch between the scalar and polynomial message spaces $\mathbb{T}$ and $\mathbb{T}_N[X]$, we use slightly generalized notions of sample extraction and keyswitching. Namely, we give to keyswitching algorithms the ability to homomorphically evaluate linear morphisms f from any $\mathbb{Z}$-module $\mathbb{T}^p$ to $\mathbb{T}_N[X]$. We define two flavors, one for a publicly known f, and one for a secret f encoded in the keyswitching key. In the following, we denote $\mathsf{PubKS}(f, \mathsf{KS}, \mathfrak{c})$ and $\mathsf{PrivKS}(\mathsf{KS}^{(f)}, \mathfrak{c})$ the output of Algorithms 1 and 2 on input the functional keyswitching keys KS and $\mathsf{KS}^{(f)}$ respectively and ciphertext $\mathfrak{c}$.

Algorithm 1. TLWE-to-TRLWE Public Functional Keyswitch

Input: p LWE ciphertexts $\mathfrak{c}^{(z)} = (\mathfrak{a}^{(z)}, \mathfrak{b}^{(z)}) \in \mathsf{TLWE}_{\mathfrak{K}}(\mu_z)$ for $z = 1, \dots, p$, a public R-lipschitzian morphism f from $\mathbb{T}^p$ to $\mathbb{T}_N[X]$, and $KS_{i,j} \in \mathsf{TRLWE}_{\underline{K}}(\frac{\mathfrak{K}_i}{2^j})$.
Output: A TRLWE sample $c \in \mathsf{TRLWE}_{\underline{K}}(f(\mu_1, \dots, \mu_p))$
1: **for** $i \in [\![1, n]\!]$ **do**
2: Let $a_i = f(\mathfrak{a}_i^{(1)}, \dots, \mathfrak{a}_i^{(p)})$
3: let $\tilde{a}_i$ be the closest multiple of $\frac{1}{2^t}$ to a_i, thus $\|\tilde{a}_i - a_i\|_\infty < 2^{-(t+1)}$
4: Binary decompose each $\tilde{a}_i = \sum_{j=1}^{t} \tilde{a}_{i,j} \cdot 2^{-j}$ where $\tilde{a}_{i,j} \in \mathbb{B}_N[X]$
5: **end for**
6: **return** $(0, f(\mathfrak{b}^{(1)}, \dots, \mathfrak{b}^{(p)})) - \sum_{i=1}^{n} \sum_{j=1}^{t} \tilde{a}_{i,j} \times \mathsf{KS}_{i,j}$

Theorem 2.5. *(Public KeySwitch) Given p LWE ciphertexts* $\mathfrak{c}^{(z)} \in \mathsf{TLWE}_{\mathfrak{K}}(\mu_z)$ *and a public R-lipschitzian morphism f of* $\mathbb{Z}$*-modules, from* $\mathbb{T}^p$ *to* $\mathbb{T}_N[X]$*, and* $\mathsf{KS}_{i,j} \in \mathsf{TRLWE}_{\underline{K},\underline{\gamma}}(\frac{\mathfrak{K}_i}{2^j})$ *with standard deviation* $\underline{\gamma}$*, Algorithm 1 outputs a* TRLWE *sample* $c \in \mathsf{TRLWE}_{\underline{K}}(f(\mu_1, \dots, \mu_p))$ *where:*

- $\|Err(c)\|_\infty \leq R\,\|Err(\mathfrak{c})\|_\infty + ntN\mathcal{A}_{\mathsf{KS}} + n2^{-(t+1)}$ *(worst case),*
- $Var(Err(c)) \leq R^2 Var(Err(\mathfrak{c})) + ntN\vartheta_{\mathsf{KS}} + n2^{-2(t+1)}$ *(average case), where* $\mathcal{A}_{\mathsf{KS}}$ *and* $\vartheta_{\mathsf{KS}} = \gamma^2$ *are respectively the amplitude and the variance of the error of* KS.

We have a similar result when the function is private. In this algorithm, we extend the input secret key $\mathfrak{K}$ by adding a $(n+1)$-th coefficient equal to -1, so that $\varphi_{\mathfrak{K}}(\mathfrak{c}) = -\mathfrak{K} \cdot \mathfrak{c}$. A detailed proof for both the private and the public keyswitching is given in the full version.

Theorem 2.6. *(Private KeySwitch) Given p* TLWE *ciphertexts* $\mathfrak{c}^{(z)} \in \mathsf{TLWE}_{\mathfrak{K}}(\mu_z)$*,* $\mathsf{KS}_{i,j} \in \mathsf{TRLWE}_{\underline{K},\underline{\gamma}}(f(0, \dots, \frac{\mathfrak{K}_i}{2^j}, \dots, 0))$ *where f is a private R-lipschitzian morphism of* $\mathbb{Z}$*-modules, from* $\mathbb{T}^p$ *to* $\mathbb{T}_N[X]$*, Algorithm 2 outputs a* TRLWE *sample* $c \in \mathsf{TRLWE}_{\underline{K}}(f(\mu_1, \dots, \mu_p))$ *where*

- $\|Err(c)\|_\infty \leq R\,\|Err(\mathfrak{c})\|_\infty + (n+1)R2^{-(n+1)} + p(n+1)\mathcal{A}_{\mathsf{KS}}$ *(worst-case),*
- $Var(Err(c)) \leq R^2 Var(Err(\mathfrak{c})) + (n+1)R^2 2^{-2(n+1)} + p(n+1)\vartheta_{\mathsf{KS}}$ *(average case), where* $\mathcal{A}_{\mathsf{KS}}$ *and* $\vartheta_{\mathsf{KS}} = \gamma^2$ *are respectively the amplitude and the variance of the error of* KS.

Algorithm 2. TLWE-to-TRLWE Private Functional Keyswitch

Input: p TLWE ciphertexts $\mathfrak{c}^{(z)} \in \mathsf{TLWE}_{\mathfrak{K}}(\mu_z)$, $\mathsf{KS}_{z,i,j} \in \mathsf{TRLWE}_{\underline{K}}(f(0, \dots, 0, \frac{\mathfrak{K}_i}{2^j}, 0, \dots, 0))$ where f is a secret R-lipschitzian morphism from $\mathbb{T}^p$ to $\mathbb{T}_N[X]$ and $\frac{\mathfrak{K}_i}{2^j}$ is at position z (also, $\mathfrak{K}_{n+1} = -1$ by convention).
Output: A TRLWE sample $c \in \mathsf{TRLWE}_{\underline{K}}(f(\mu_1, \dots, \mu_p))$.
1: **for** $i \in [\![1, n+1]\!]$, $z \in [\![1, p]\!]$ **do**
2: Let $\tilde{c}_i^{(z)}$ be the closest multiple of $\frac{1}{2^t}$ to $\mathfrak{c}_i^{(z)}$, thus $|\tilde{c}_i^{(z)} - \mathfrak{c}_i^{(z)}| < 2^{-(t+1)}$
3: Binary decompose each $\tilde{c}_i^{(z)} = \sum_{j=1}^{t} \tilde{c}_{i,j}^{(z)} \cdot 2^{-j}$ where $\tilde{c}_{i,j}^{(z)} \in \{0, 1\}$
4: **end for**
5: **return** $-\sum_{z=1}^{p} \sum_{i=1}^{n+1} \sum_{j=1}^{t} \tilde{c}_{i,j}^{(z)} \cdot \mathsf{KS}_{z,i,j}$

Sample Packing and Sample Extraction. A TRLWE message is a polynomial with N coefficients, which can be viewed as N slots over $\mathbb{T}$. It is easy to homomorphically extract a coefficient as a scalar TLWE sample. To that end, we will use the following convention in the rest of the paper: for all $n = kN$, a binary vector $\mathfrak{K} \in \mathbb{B}^n$ can be interpreted as a TLWE key, or alternatively as a TRLWE key $K \in \mathbb{B}_N[X]^k$ having the same sequence of coefficients. Namely, K_i is the polynomial $\sum_{j=0}^{N-1} \mathfrak{K}_{N(i-1)+j+1} X^j$. In this case, we say that K is the TRLWE interpretation of $\mathfrak{K}$, and $\mathfrak{K}$ is the TLWE interpretation of K.

Given a TRLWE sample $\boldsymbol{c} = (\boldsymbol{a}, b) \in \mathsf{TRLWE}_K(\mu)$ and a position $p \in [0, N-1]$, we call $\mathsf{SampleExtract}_p(\boldsymbol{c})$ the TLWE sample $(\mathfrak{a}, \mathfrak{b})$ where $\mathfrak{b} = b_p$ and $\mathfrak{a}_{N(i-1)+j+1}$ is the $(p-j)$-th coefficient of a_i (using the N-antiperiodic indexes). This extracted sample encodes the p-th coefficient μ_p with at most the same noise variance or amplitude as $\boldsymbol{c}$. In the rest of the paper, we will simply write $\mathsf{SampleExtract}(\boldsymbol{c})$ when $p = 0$. In the next Section, we will show how the KeySwitching and the SampleExtract are used to efficiently pack, unpack and move data across the slots, and how it differs from usual packing techniques.

2.3 Gate Bootstrapping Overview

This lemma on the evaluation of the CMux-gate extends Theorems 5.1 and 5.2 in [12] from the message space $\{0, 1/2\}$ to $\mathbb{T}_N[X]$:

Lemma 2.7 (CMux Gate). *Let $\boldsymbol{d_1}, \boldsymbol{d_0}$ be TRLWE samples and let $C \in \mathsf{TRGSW}_{\boldsymbol{s}}(\{0,1\})$. Then, $msg(\mathtt{CMux}(C, \boldsymbol{d_1}, \boldsymbol{d_0})) = msg(C)?msg(\boldsymbol{d_1}):msg(\boldsymbol{d_0})$. And we have $\|Err(\mathtt{CMux}(C, \boldsymbol{d_1}, \boldsymbol{d_0}))\|_\infty \leq \max(\|Err(\boldsymbol{d_0})\|_\infty, \|Err(\boldsymbol{d_1})\|_\infty) + \eta(C)$, where $\eta(C) = (k+1)\ell N\beta \|Err(C)\|_\infty + (kN+1)\epsilon$. Furthermore, under Assumption 2.4, we have: $Var(Err(\mathtt{CMux}(C, \boldsymbol{d_1}, \boldsymbol{d_0}))) \leq \max(Var(Err(\boldsymbol{d_0})), Var(Err(\boldsymbol{d_1}))) + \vartheta(C)$, where $\vartheta(C) = (k+1)\ell N\beta^2 Var(Err(C)) + (kN+1)\epsilon^2$.*

The proof is the same as for Theorems 5.1 and 5.2 in [12] because the noise of the output does not depend on the value of the TRLWE message.

Blind rotate. In the following, we give faster sub-routine for the main loop of Algorithm 3 in [12]. The improvement consists in a new CMux formula in the for loop of the Algorithm 3 instead of the formula in Algorithm 3 of [12]. The BlindRotate algorithm multiplies the polynomial encrypted in the input TRLWE ciphertext by an encrypted power of X. Theorem 2.8 follows from the fact that Algorithm 3 calls p times the CMux evaluation from Lemma 2.7.

Theorem 2.8. *Let $H \in \mathcal{M}_{(k+1)\ell,k+1}(\mathbb{T}_N[X])$ the gadget matrix and $Dec_{H,\beta,\epsilon}$ its efficient approximate gadget decomposition algorithm with quality β and precision ϵ defining TRLWE and TRGSW parameters. Let $\alpha \in \mathbb{R}_{\geq 0}$ be a standard deviation, $\mathfrak{K} \in \mathbb{B}^n$ be a TLWE secret key and $K \in \mathbb{B}_N[X]^k$ be its TRLWE interpretation. Given one sample $\boldsymbol{c} \in \mathsf{TRLWE}_K(\boldsymbol{v})$ with $\boldsymbol{v} \in \mathbb{T}_N[X]$, $p+1$ integers $a_1, \ldots, a_p$ and $b \in \mathbb{Z}/2N\mathbb{Z}$, and p TRGSW ciphertexts $C_1, \ldots, C_p$, where each $C_i \in \mathsf{TRGSW}_{K,\alpha}(s_i)$ for $s_i \in \mathbb{B}$. Algorithm 3 outputs a sample $\mathsf{ACC} \in \mathsf{TRLWE}_K(X^{-\rho} \cdot \boldsymbol{v})$ where $\rho = b - \sum_{i=1}^{p} s_i a_i$ such that*

- $\|\mathsf{Err}(\mathsf{ACC})\|_\infty \leq \|\mathsf{Err}(\mathbf{c})\|_\infty + p(k+1)\ell N\beta\mathcal{A}_C + p(1+kN)\epsilon$ *(worst case)*,
- $\mathsf{Var}(\mathsf{Err}(\mathsf{ACC})) \leq \mathsf{Var}(\mathsf{Err}(\mathbf{c})) + p(k+1)\ell N\beta^2\vartheta_C + p(1+kN)\epsilon^2$ *(average case)*, *where* $\vartheta_C = \alpha^2$ *and* $\mathcal{A}_C$ *are the variance and amplitudes of* $\mathsf{Err}(C_i)$.

Algorithm 3. BlindRotate

Input: A TRLWE sample $\mathbf{c}$ of $v \in \mathbb{T}_N[X]$ with key K.
1: $p+1$ int. coefficients $a_1, \dots, a_p, b \in \mathbb{Z}/2N\mathbb{Z}$
2: p TRGSW samples $C_1, \dots, C_p$ of $s_1, \dots, s_p \in \mathbb{B}$ with key K
Output: A TRLWE sample of $X^{-\rho}.v$ where $\rho = b - \sum_{i=1}^{p} s_i.a_i \mod 2N$ with key K
3: $\mathsf{ACC} \leftarrow X^{-b} \bullet \mathbf{c}$
4: **for** $i = 1$ **to** p
5: $\quad \mathsf{ACC} \leftarrow \mathsf{CMux}(C_i, X^{a_i} \cdot \mathsf{ACC}, \mathsf{ACC})$
6: **return** ACC

We define $\mathsf{BlindRotate}(\mathbf{c}, (a_1, \dots, a_p, b), C)$, the procedure described in Algorithm 3 that outputs the TLWE sample ACC as in Theorem 2.8.

Algorithm 4. Gate Bootstrapping TLWE-to-TLWE (calling Algorithm 3)

Input: A constant $\mu_1 \in \mathbb{T}$, a TLWE sample $\underline{\mathfrak{c}} = (\underline{\mathfrak{a}}, \underline{\mathfrak{b}}) \in \mathsf{TLWE}_{\underline{\mathfrak{K}},\underline{\eta}}(x \cdot \frac{1}{2})$, with $x \in \mathbb{B}$
a bootstrapping key $\mathrm{BK}_{\underline{\mathfrak{K}} \to \bar{\mathfrak{K}}, \bar{\alpha}} = (\mathrm{BK}_i)_{i \in [\![1,\underline{n}]\!]}$,
Output: A TLWE sample $\bar{\mathfrak{c}} = (\bar{\mathfrak{a}}, \bar{\mathfrak{b}}) \in \mathsf{TLWE}_{\bar{\mathfrak{K}},\bar{\eta}}(x \cdot \mu_1)$
1: Let $\mu = \frac{1}{2}\mu_1 \in \mathbb{T}$ (Pick one of the two possible values)
2: Let $\tilde{b} = \lfloor 2\bar{N}\underline{\mathfrak{b}} \rceil$ and $\tilde{a}_i = \lfloor 2\bar{N}\underline{\mathfrak{a}}_i \rceil \in \mathbb{Z}$ **for each** $i \in [\![1, \underline{n}]\!]$
3: Let $v := (1 + X + \dots + X^{\bar{N}-1}) \cdot X^{\frac{\bar{N}}{2}} \cdot \mu \in \mathbb{T}_{\bar{N}}[X]$
4: $\mathsf{ACC} \leftarrow \mathsf{BlindRotate}((\mathbf{0}, v), (a_1, \dots, a_n, b), (\mathrm{BK}_1, \dots, \mathrm{BK}_n))$
5: Return $(\mathbf{0}, \mu) + \mathsf{SampleExtract}(\mathsf{ACC})$

Gate Bootstrapping (TLWE-to-TLWE)

Theorem 2.9 (Gate Bootstrapping (TLWE-to-TLWE)). *Let* $\bar{H} \in \mathcal{M}_{(\bar{k}+1)\bar{\ell},\bar{k}+1}(\mathbb{T}_{\bar{N}}[X])$ *the gadget matrix and* $Dec_{\bar{H},\bar{\beta},\bar{\epsilon}}$ *its efficient approximate gadget decomposition algorithm, with quality* $\bar{\beta}$ *and precision* $\bar{\epsilon}$ *defining* TRLWE *and* TRGSW *parameters. Let* $\underline{\mathfrak{K}} \in \mathbb{B}^{\underline{n}}$ *and* $\bar{\mathfrak{K}} \in \mathbb{B}^{\bar{n}}$ *be two* TLWE *secret keys, and* $\bar{K} \in \mathbb{B}_{\bar{N}}[X]^{\bar{k}}$ *be the* TRLWE *interpretation of the key* $\bar{\mathfrak{K}}$, *and let* $\bar{\alpha} \in \mathbb{R}_{\geq 0}$ *be a standard deviation. Let* $BK_{\underline{\mathfrak{K}} \to \bar{\mathfrak{K}},\bar{\alpha}}$ *be a bootstrapping key, composed by the* $\underline{n}$ TRGSW *encryptions* $BK_i \in \mathsf{TRGSW}_{\bar{K},\bar{\alpha}}(\underline{\mathfrak{K}}_i)$ *for* $i \in [\![1, \underline{n}]\!]$. *Given one constant* $\mu_1 \in \mathbb{T}$, *and one sample* $\underline{\mathfrak{c}} \in \mathbb{T}^{\underline{n}+1}$ *whose coefficients are all multiples of* $\frac{1}{2\bar{N}}$, *Algorithm 4 outputs a* TLWE *sample* $\bar{\mathfrak{c}} \in \mathsf{TLWE}_{\bar{\mathfrak{K}}}(\mu)$ *where* $\mu = 0$ *iff.* $|\varphi_{\underline{\mathfrak{K}}}(\underline{\mathfrak{c}})| < \frac{1}{4}$, $\mu = \mu_1$ *otherwise and such that:*

- $\|\mathsf{Err}(\bar{\mathfrak{c}})\|_\infty \leq \underline{n}(\bar{k}+1)\bar{\ell}\bar{N}\bar{\beta}\bar{\mathcal{A}}_{BK} + \underline{n}(1+\bar{k}\bar{N})\bar{\epsilon}$ *(worst case)*,
- $\mathsf{Var}(\mathsf{Err}(\bar{\mathfrak{c}})) \leq \underline{n}(\bar{k}+1)\bar{\ell}\bar{N}\bar{\beta}^2\bar{\vartheta}_{BK} + \underline{n}(1+\bar{k}\bar{N})\bar{\epsilon}^2$ *(average case)*, *where* $\bar{\mathcal{A}}_{BK}$ *is the amplitude of* BK *and* $\bar{\vartheta}_{BK}$ *its variance s.t.* $\mathsf{Var}(\mathsf{Err}(BK_{\underline{\mathfrak{K}} \to \bar{K},\bar{\alpha}})) = \bar{\alpha}^2$.

Sketch of Proof. Algorithm 4 is almost the same as Algorithm 3 in [12] except that the main loop has been put in a separate algorithm (Algorithm 3) at line 2. In addition, the final KeySwitching has been removed which suppresses two terms in the norm inequality of the error. Note that the output is encrypted with the same key as the bootstrapping key. Another syntactic difference is that the input sample is a multiple of $1/2N$ (which can be achieved by rounding all its coefficients). Also, a small difference in the way we associate CMux operations removes a factor 2 in the noise compared to the previous gate bootstrapping procedure, and it is also faster.

Homomorphic operations (revisited) ***via*** **Gate Bootstrapping.** The fast bootstrapping of [17] and improved in [4,12] is presented for Nand gates. They evaluate a single Nand operation and they refresh the result to make it usable for the next operations. Other elementary gates are presented: the And, Or, Xor (and trivially Nor, Xnor, AndNot, etc. since NOT is cheap and noiseless). The term *gate bootstrapping* refers to the fact that this fast bootstrapping is performed after every gate evaluation[3].

The ternary Mux gate ($\texttt{Mux}(c, d_0, d_1) = c?d_1 : d_0 = (c \wedge d_1) \oplus ((1-c) \wedge d_0)$, for $c, d_0, d_1 \in \mathbb{B}$) is generally expressed as a combination of 3 binary gates. As already mentioned in [17], we can improve the Mux evaluation by performing the middle $\oplus$ as a regular addition before the final KeySwitching. Indeed, this xor has at most one operand which is true, and at this location, it only affects a negligible amount of the final noise, and is compensated by the fact that we save a factor 2 in the gate bootstrapping in the blind rotation from Algorithm 3. Overall, the ternary Mux gate can be evaluated in FHE mode by evaluating only two gate bootstrappings and one public keyswitch. We call this procedure *native* MUX, which computes:

- $c \wedge d_1$ via a gate bootstrapping (Algorithm 4) of $(\mathbf{0}, -\frac{1}{8}) + \boldsymbol{c} + \boldsymbol{d_1}$;
- $(1-c) \wedge d_0$ via a gate bootstrapping (Algorithm 4) of $(\mathbf{0}, \frac{1}{8}) - \boldsymbol{c} + \boldsymbol{d_0}$;
- a final keyswitch on the sum (Algorithm 1) which dominates the noise.

This native Mux is therefore bootstrappable with the same parameters as in [12]. More details are given in the full version. In the rest of the paper, when we compare different homomorphic techniques, we refer to the gate-bootstrapping mode as the technique consisting in evaluating small circuits expressed with any binary gates and/or the native Mux, and we use the following experimental timings (see Sect. 5):

Gate bootstrapping mode	
Pre-bootstrap 1 bit	$t_{GB} = 13\,\text{ms}$
Time per any binary gate (And, Or, Xor, ...)	$t_{GB} = 13\,\text{ms}$
Time per MUX	$2t_{GB} = 26\,\text{ms}$

[3] Actually, the gate bootstrapping technique can be used even if we do not need to evaluate a specific gate, but just to refresh noisy ciphertexts.

3 Leveled Homomorphic Circuits

Various packing techniques have already been proposed for homomorphic encryption, for instance the Lagrange embedding in Helib [22,23], the diagonal matrices encoding in [27] or the CRT encoding in [2]. The message space is often a finite ring (e.g. $\mathbb{Z}/p\mathbb{Z}$), and the packing function is in general chosen as a ring isomorphism that preserves the structure of $\mathbb{Z}/p\mathbb{Z}^N$. This way, elementary additions or products can be performed simultaneously on N independent slots, and thus, packing is in general associated to the concept of batching a single operation on multiple datasets. These techniques has some limitations, especially if in the whole program, each function is only run on a single dataset, and most of the slots are unused. This is particularly true in the context of GSW evaluations, where functions are split into many branching algorithms or automata, that are each executed only once.

In this paper, packing refers to the canonical coefficients embedding function, that maps N Scalar-TLWE messages $\mu_0, \ldots, \mu_{N-1} \in \mathbb{T}$ into a single TRLWE message $\mu(X) = \sum_{i=0}^{N-1} \mu_i X^i$. This function is a $\mathbb{Z}$-module isomorphism. Messages can be homomorphically unpacked from any slot using the (noiseless) SampleExtract procedure. Reciprocally, we can repack, move data across the slots, or clear some slots by using our public functional key switching from Algorithm 1 to evaluate respectively the canonical coefficient embedding function (i.e. the identity), a permutation, or a projection. Since these functions are 1-lipschitzian, by Theorem 2.5, these keyswitch operations only induce a linear noise overhead. It is arguably more straightforward than the permutation network technique used in Helib. But as in [2,10,15], our technique relies on a circular security assumption, even in the leveled mode since our keyswitching key encrypts its own key bits[4].

We now analyse how packing can speed-up TGSW leveled computations, first for lookup tables or arbitrary functions, and then for most arithmetic functions.

3.1 Arbitrary Functions and Look-Up Tables

The first class of functions that we analyse are arbitrary functions $f : \mathbb{B}^d \rightarrow \mathbb{T}^s$. Such functions can be expressed with a Look-Up Table (LUT), containing a list of 2^d input values (each one composed by d bits) and corresponding LUT values for the s sub-functions (1 element in $\mathbb{T}$ per sub-function f_j).

In order to compute $f(x)$, where $x = \sum_{i=0}^{d-1} x_i 2^i$ is a d-bit integer, the classical evaluation of such function, as proposed in [8,12] consists in evaluating the s subfunctions separately, and each of them is a binary decision tree composed by $2^d - 1$ `CMux` gates. The total complexity of the classical evaluation requires therefore to execute about $s \cdot 2^d$ `CMux` gates. Let's call $o_j = f_j(x) \in \mathbb{T}$ the j-th output of $f(x)$, for $j = 0, \ldots, s-1$. Figure 1 summarizes the idea of the computation of o_j.

[4] Circular security assumption could still be avoided in leveled mode if we accept to work with many keys.

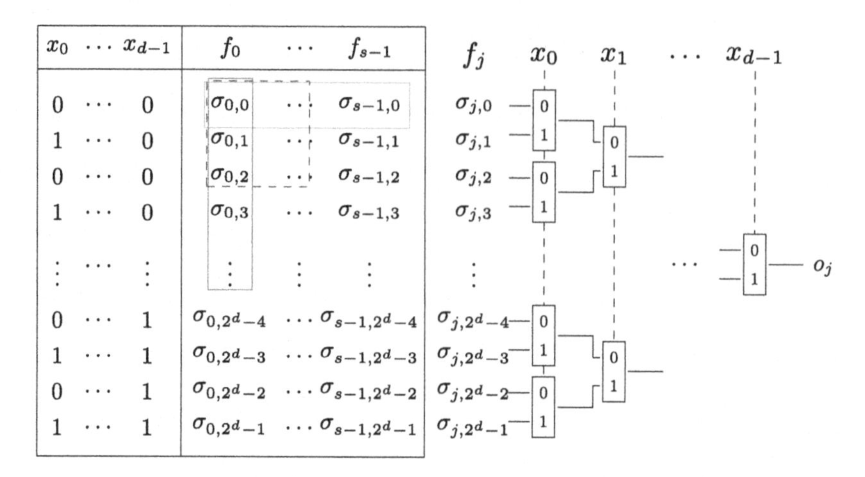

Fig. 1. LUT with CMux tree - Intuitively, the horizontal rectangle encircles the bits packed in the horizontal packing, while the vertical rectangle encircles the bits packed in the vertical packing. The dashed square represents the packing in the case where the two techniques are mixed. The right part of the figure represents the evaluation of the sub-function f_j on $x = \sum_{i=0}^{d-1} x_i 2^i$ via a CMux binary decision tree.

In this section we present two techniques, that we call horizontal and vertical packing, that can be used to improve the evaluation of a LUT.

Horizontal packing corresponds exactly to batching. In fact, it exploits the fact that the s subfunctions evaluate the same CMux tree with the same inputs on different data, which are the s truth tables. For each of the 2^d possible input values, we pack the LUT values of the s sub-functions in the first s slots of a single TRLWE ciphertext (the remaining $N - s$ are unused). By using a single 2^d size CMux tree to select the right ciphertext and obtain the s slots all at once, which is overall s times faster than the classical evaluation. Our *vertical packing* is very different from the batching technique. The basic idea is to pack several LUT values of a single sub-function in the same ciphertext, and to use both CMux and blind rotations to extract the desired value. Unlike batching, this can also speed up functions that have only a single bit of output. In the following we detail these two techniques.

In order to evaluate $f(x)$, the total amount of homomorphic CMux gates to be evaluated is $s(2^d - 1)$. If the function f is public, trivial samples of the LUT values $\sigma_{j,0}, \ldots, \sigma_{j,N-1}$ are used as inputs in the CMux gates. If f is private, the LUT values $\sigma_{j,0}, \ldots, \sigma_{j,N-1}$ are given encrypted. An analysis of the noise propagation in the binary decision CMux tree was already given in [12,19].

Horizontal Packing. The idea of the *Horizontal Packing* is to evaluate all the outputs of the function f together, instead of evaluating all the f_j separately. This is possible by using TRLWE samples as the message space is $\mathbb{T}_N[X]$. In fact, we could encrypt up to N LUT values $\sigma_{j,h}$ (for a fixed $h \in [\![0, 2^d - 1]\!]$) per TRLWE sample and evaluate the binary decision tree as described before. The number of CMux gates to evaluate is $\lceil \frac{s}{N} \rceil (2^d - 1)$. This technique is optimal if the size s of the output is a multiple of N. Unfortunately, s is in general $\leq N$, the number of gates to evaluate remains $2^d - 1$, which is only s times smaller than the non-packed approach, and is not advantageous if s is small. Lemma 3.1 specifies the noise propagation and it follows immediately from Lemma 2.7 and from the construction of the binary decision CMux tree, which has depth d.

Lemma 3.1 (Horizontal Packing). *Let $\boldsymbol{d}_0, \ldots, \boldsymbol{d}_{2^d-1}$ be* TRLWE *samples*[5] *such that $\boldsymbol{d}_h \in \mathsf{TRLWE}_K(\sum_{j=0}^{s} \sigma_{j,h} X^j)$ for $h \in [\![0, 2^d - 1]\!]$. Here the $\sigma_{j,h}$ are the LUT values relative to an arbitrary function $f : \mathbb{B}^d \to \mathbb{T}^s$. Let $C_0, \ldots, C_{d-1}$ be* TRGSW *samples, such that $C_i \in \mathsf{TRGSW}_K(x_i)$ with $x_i \in \mathbb{B}$ (for $i \in [\![0, d-1]\!]$), and $x = \sum_{i=0}^{d-1} x_i 2^i$. Let $\boldsymbol{d}$ be the* TRLWE *sample output by the f evaluation of the binary decision* CMux *tree for the LUT (described in Fig. 1). Then, using the same notations as in Lemma 2.7 and setting $\mathsf{msg}(\boldsymbol{d}) = f(x)$:*

- $\|Err(\boldsymbol{d})\|_\infty \leq \mathcal{A}_{\mathsf{TRLWE}} + d \cdot ((k+1)\ell N \beta \mathcal{A}_{\mathsf{TRGSW}} + (kN+1)\epsilon)$ *(worst case),*
- $Var(Err(\boldsymbol{d})) \leq \vartheta_{\mathsf{TRLWE}} + d \cdot ((k+1)\ell N \beta^2 \vartheta_{\mathsf{TRGSW}} + (kN+1)\epsilon^2)$ *(average case), where $\mathcal{A}_{\mathsf{TRLWE}}$ and $\mathcal{A}_{\mathsf{TRGSW}}$ are upper bounds of the infinite norm of the errors of the* TRLWE *samples ant the* TRGSW *samples respectively and $\vartheta_{\mathsf{TRLWE}}$ and $\vartheta_{\mathsf{TRGSW}}$ are upper bounds of their variances.*

Vertical Packing. In order to improve the evaluation of the LUT, we propose a second optimization called *Vertical Packing.* As for the horizontal packing we use the TRLWE encryption to encode N values at the same time. But now, instead of packing the LUT values $\sigma_{j,h}$ with respect to a fixed $h \in [\![0, 2^d - 1]\!]$ i.e. "horizontally", we pack N values $\sigma_{j,h}$ "vertically", with respect to a fixed $j \in [\![0, s-1]\!]$. Then, instead of just evaluating a full CMux tree, we use a different approach. If the LUT values are packed in boxes, our technique first uses a packed CMux tree to select the right box, and then, a blind rotation (Algorithm 3) to find the element inside the box.

We now explain how to evaluate the function f, or just one of its sub-functions f_j, on a fixed input $x = \sum_{i=0}^{d-1} x_i 2^i$. We assume we know the LUT associated to f_j as in Fig. 1. For retrieving the output of $f_j(x)$, we just have to return the LUT value $\sigma_{j,x}$ in position x.

Let $\delta = \log_2(N)$. We analyse the general case where 2^d is a multiple of $N = 2^\delta$. The LUT of f_j, which is a column of 2^d values, is now packed as $2^d/N$ TRLWE ciphertexts $\boldsymbol{d}_0, \ldots, \boldsymbol{d}_{2^{d-\delta}-1}$, where each $\boldsymbol{d}_k$ encodes N consecutive LUT values $\sigma_{j,kN+0}, \ldots, \sigma_{j,kN+N-1}$. To retrieve $f_j(x)$, we first need to select the block that contains $\sigma_{j,x}$. This block has the index $p = \lfloor x/N \rfloor$, whose bits are the $d - \delta$ most significant bits of x. Since the TRGSW encryption of these bits are among our inputs, one can use a CMux tree to select this block $\boldsymbol{d_p}$. Then, $\sigma_{j,x}$ is the ρ-th coefficient of the message of $\boldsymbol{d_p}$ where $\rho = x \mod N = \sum_{i=0}^{\delta-1} x_i 2^i$. The

[5] The TRLWE samples can be trivial samples, in the case where the function f and its LUT are public.

Algorithm 5. Vertical Packing LUT of $f_j : \mathbb{B}^d \to \mathbb{T}$ (calling Algorithm 3)

Input: A list of $\frac{2^d}{N}$ TRLWE samples $\boldsymbol{d}_p \in \mathsf{TRLWE}_K(\sum_{i=0}^{N-1} \sigma_{j,pN+i} X^i)$ for $p \in [\![0, \frac{2^d}{N} - 1]\!]$, a list of d TRGSW samples $C_i \in \mathsf{TRGSW}_K(x_i)$, with $x_i \in \mathbb{B}$ and $i \in [\![0, d-1]\!]$,
Output: A TLWE sample $\mathfrak{c} \in \mathsf{TLWE}_{\mathfrak{K}}(o_j = f_j(x))$, with $x = \sum_{i=0}^{d-1} x_i 2^i$
1: Evaluate the binary decision CMux tree of depth $d - \delta$, with TRLWE inputs $\boldsymbol{d}_0, \ldots, \boldsymbol{d}_{\frac{2^d}{N}-1}$ and TRGSW inputs $C_\delta, \ldots, C_{d-1}$, and output a TRLWE sample $\boldsymbol{d}$
2: $\boldsymbol{d} \leftarrow \mathsf{BlindRotate}(\boldsymbol{d}, (2^0, \ldots, 2^{\delta-1}, 0), (C_0, \ldots, C_{\delta-1}))$
3: Return $\mathfrak{c} = \mathsf{SampleExtract}(\boldsymbol{d})$

bits of ρ are the δ least significant bits of x, which are also available as TRGSW ciphertexts in our inputs. We can therefore use a blind rotation (Algorithm 3) to homomorphically multiply $\boldsymbol{d_p}$ by $X^{-\rho}$, which brings the coefficient $\sigma_{j,x}$ in position 0, and finally, we extract it with a SampleExtract. Algorithm 5 details the evaluation of $f_j(x)$.

The entire cost of the evaluation of $f_j(x)$ with Algorithm 5 consists in $\frac{2^d}{N} - 1$ CMux gates and a single blind rotation, which corresponds to δ CMux gates. Overall, we get a speed-up by a factor N on the evaluation of each partial function, so a factor N in total.

Lemma 3.2 (Vertical Packing LUT of f_j). *Let $f_j : \mathbb{B}^d \to \mathbb{T}$ be a sub-function of the arbitrary function f, with LUT values $\sigma_{j,0}, \ldots, \sigma_{j,2^d-1}$. Let $\boldsymbol{d}_0, \ldots, \boldsymbol{d}_{\frac{2^d}{N}-1}$ be* TRLWE *samples, such that $\boldsymbol{d}_p \in \mathsf{TRLWE}_K(\sum_{i=0}^{N-1} \sigma_{j,pN+i} X^i)$ for $p \in [\![0, \frac{2^d}{N} - 1]\!]$[6]. Let $C_0, \ldots, C_{d-1}$ be* TRGSW *samples, such that $C_i \in \mathsf{TRGSW}_K(x_i)$, with $x_i \in \mathbb{B}$ and $i \in [\![0, d-1]\!]$.*

Then Algorithm 5 outputs a TLWE *sample $\mathfrak{c}$ such that $\mathsf{msg}(\mathfrak{c}) = f_j(x) = o_j$ where $x = \sum_{i=0}^{d-1} x_i 2^i$ and using the same notations as in Lemma 2.7 and Theorem 2.8, we have:*

- $\|\mathsf{Err}(\boldsymbol{d})\|_\infty \leq \mathcal{A}_{\mathsf{TRLWE}} + d \cdot ((k+1)\ell N \beta \mathcal{A}_{\mathsf{TRGSW}} + (1+kN)\epsilon)$ *(worst case),*
- $\mathsf{Var}(\mathsf{Err}(\boldsymbol{d})) \leq \vartheta_{\mathsf{TRLWE}} + d \cdot ((k+1)\ell N \beta^2 \vartheta_{\mathsf{TRGSW}} + (1+kN)\epsilon^2)$ *(average case), where $\mathcal{A}_{\mathsf{TRLWE}}$ and $\mathcal{A}_{\mathsf{TRGSW}}$ are upper bounds of the infinite norm of the errors in the* TRLWE *samples ant the* TRGSW *samples respectively, while $\vartheta_{\mathsf{TRLWE}}$ and $\vartheta_{\mathsf{TRGSW}}$ are upper bounds of the variances.*

Proof (Sketch). The proof follows immediately from the results of Lemma 2.7 and Theorem 2.8, and from the construction of the binary decision CMux tree. In particular, the first CMux tree has depth $(d - \delta)$ and the blind rotation evaluates δ CMux gates, which brings a total factor d in the depth. As the CMux depth is the same as in horizontal packing, the noise propagation matches too.

[6] If the sub-function f_j and its LUT are public, the LUT values $\sigma_{j,0}, \ldots, \sigma_{j,2^d-1}$ can be given in clear. This means that the TRLWE samples $\boldsymbol{d}_p$, for $p \in [\![0, \frac{2^d}{N} - 1]\!]$ are given as trivial TRLWE samples $\boldsymbol{d}_p \leftarrow (\boldsymbol{0}, \sum_{i=0}^{N-1} \sigma_{j,pN+i} X^i)$ in input to Algorithm 5.

Remark 1 As previously mentioned, the horizontal and vertical packing techniques can be mixed together to improve the evaluation of f in the case where s and d are both small, i.e. the previous two methodology cannot be applied separately but we have $2^d \cdot s > N$. In particular, if we pack $s = x$ coefficients horizontally and $y = N/x$ coefficients vertically, we need $\lceil 2^d/y \rceil - 1$ CMux gates plus one vertical packing LUT evaluation in order to evaluate f, which is equivalent to $\log_2(y)$ CMux evaluations. The result is composed of the first x TLWE samples extracted.

3.2 Arithmetic Operations via Weighted Automata

In [12], the arithmetic operations were evaluated via deterministic finite automata using CMux gates. It was made possible thanks to the fact that the messages were binary. In this paper, the samples on which we perform the arithmetic operations pack several torus values together. A more powerful tool is thus needed to manage the evaluations in an efficient way. Deteministic weighted finite automata (det-WFA) are deterministic finite automata where each transition contains an additional weight information. By reading a word, the outcome of a det-WFA is the sum of all weights encountered along the path (here, we work with an additive group), whereas the outcome of a deterministic finite automata (DFA) is just a boolean that states whether the destination state is accepting. The weights of a det-WFA can be seen as a memory that stores the bits of the partial result, all along the evaluation path. Let's take for example the evaluation of the MAX circuit, that takes in input two d-bit integers and returns the maximal value between them. With DFA, to retrieve all the d bits of the result we need d different automata, for a total of $O(d^2)$ transitions. By introducing the weights, all the bits of the result are given in one pass after only $O(d)$ transitions. To our knowledge, our paper is the first one introducing this tool on the FHE context. In this section, we detail the use of det-WFA to evaluate some arithmetic functions largely used in applications, such as addition (and multi-addition), multiplication, squaring, comparison and max, etc. We refer to [9,16] for further details on the theory of weighted automata.

Definition 3.3 (Deterministic weighted finite automata (det-WFA)). *A deterministic weighted finite automata (det-WFA) over a group $(S, \oplus)$ is a tuple $\mathfrak{A} = (Q, i, \Sigma, \mathcal{T}, F)$, where Q is a finite set of states, i is the initial state, Σ is the alphabet, $\mathcal{T} \subseteq Q \times \Sigma \times S \times Q$ is the set of transitions and $F \subseteq Q$ is the set of final states. Every transition itself is a tuple $t = q \xrightarrow{\sigma,\nu} q'$ from the state q to the state q' by reading the letter σ with weight $w(t)$ equal to ν, and there is at most one transition per every pair (q, σ).*

Let $P = (t_1, \ldots, t_d)$ be a path, with $t_j = q_{j-1} \xrightarrow{\sigma_j,\nu_j} q_j$. The word $\boldsymbol{\sigma} = \sigma_1 \ldots \sigma_d \in \Sigma^d$ induced by P is accepted by the det-WFA $\mathfrak{A}$ if $q_0 = i$ and $q_d \in F$. The weight $w(\boldsymbol{\sigma})$ of a word $\boldsymbol{\sigma}$ is equal to $\bigoplus_{j=1}^{d} w(t_j)$, where the $w(t_j)$ are all the weights of the transitions in P: $\boldsymbol{\sigma}$ is called the label of P. Note that every label induces a single path (i.e. there is only one possible path per word).

Remark 2. In our applications, we fix the alphabet $\Sigma = \mathbb{B}$. Definition 3.3 restraints the WFA to the deterministic (the non-deterministic case is not supported), complete and universally accepting case (i.e. all the words are accepted). In the general case, the additive group would be replaced by a semiring $(S, \oplus, \otimes, 0, 1)$. In the rest of the paper we set $(S, \oplus)$ as $(\mathbb{T}_N[X], +)$.

Theorem 3.4 (Evaluation of det-WFA). *Let $\mathfrak{A} = (Q, i, \mathbb{B}, \mathcal{T}, F)$ be a det-WFA with weights in $(\mathbb{T}_N[X], +)$, and let $|Q|$ denote the total number of states. Let $C_0, \ldots, C_{d-1}$ be d valid TRGSW_K samples of the bits of a word $\boldsymbol{\sigma} = \sigma_0 \ldots \sigma_{d-1}$. By evaluating at most $d \cdot |Q|$ $\mathtt{CMux}$ gates, we output a TRLWE sample $\mathbf{d}$ that encrypts the weight $w(\boldsymbol{\sigma})$, such that (using the same notations as in Lemma 2.7)*

- $\|\mathsf{Err}(\mathbf{d})\|_\infty \leq d \cdot ((k+1)\ell N \beta \mathcal{A}_{\mathsf{TRGSW}} + (kN+1)\epsilon)$ *(worst case),*
- $\mathsf{Var}(\mathsf{Err}(\mathbf{d})) \leq d \cdot ((k+1)\ell N \beta^2 \vartheta_{\mathsf{TRGSW}} + (kN+1)\epsilon^2)$ *(average case),* *where $\mathcal{A}_{\mathsf{TRGSW}}$ is an upper bound on the infinite norm of the error in the TRGSW samples and $\vartheta_{\mathsf{TRGSW}}$ is an upper bound of their variance. Moreover, if all the words connecting the initial state to a fixed state $q \in Q$ have the same length, then the upper bound on the number of $\mathtt{CMux}$ to evaluate decreases to $|Q|$.*

Proof (Sketch). This theorem generalizes Theorem 5.4 of [12] for det-WFA. The automaton is still evaluated from the last letter σ_{d-1} to the first one σ_0, using one TRLWE ciphertext $\mathbf{c}_{j,q}$ per position $j \in [\![0, d-1]\!]$ in the word and per state $q \in Q$. Before reading a letter, all the TRLWE samples $\mathbf{c}_{d,q}$, for $q \in Q$, are initialized to zero. When processing the j-th letter σ_j, each pair of transitions $q \xrightarrow{0,\nu_0} q_0$ and $q \xrightarrow{1,\nu_1} q_1$ is evaluated as $\mathbf{c}_{j,q} = \mathtt{CMux}(C_j, \mathbf{c}_{j+1,q_1} + (\mathbf{0}, \nu_1), \mathbf{c}_{j+1,q_0} + (\mathbf{0}, \nu_0))$. The final result is $\mathbf{c}_{0,i}$, which encodes $w(\boldsymbol{\sigma})$ by induction on the $\mathtt{CMux}$ graph. Since translations are noiseless, the output noise corresponds to a depth-d of $\mathtt{CMux}$. Like in [12], the last condition implies that only $|Q|$ of the $d|Q|$ $\mathtt{CMux}$ are accessible and need to be evaluated. □

MAX. In order to evaluate the MAX circuit of two d-bit integers, $x = \sum_{i=0}^{d-1} x_i 2^i$ and $y = \sum_{i=0}^{d-1} y_i 2^i$, we construct a det-WFA that takes in input all the bits $x_{d-1}, \ldots, x_0$ of x and $y_{d-1}, \ldots, y_0$ of y, and outputs the maximal value between them. The idea is to enumerate the x_i and y_i, starting from the most significant bits down to the least significant ones. The det-WFA described in Fig. 2 has 3 principal states (A, B, E) and 4 intermediary states ((A), (B), $(E, 1)$, $(E, 0)$), which keeps track of which number is the maximum, and in case of equality what is the last value of x_i. A weight $+\frac{1}{2}X^i$ is added on all the transitions that reads the digit 1 from the maximum. Overall, the next lemma, which is a direct consequence of Theorem 3.4, shows that the Max can be computed by evaluating only $5d$ $\mathtt{CMux}$ gates, instead of $\Theta(d^2)$ with classical deterministic automata.

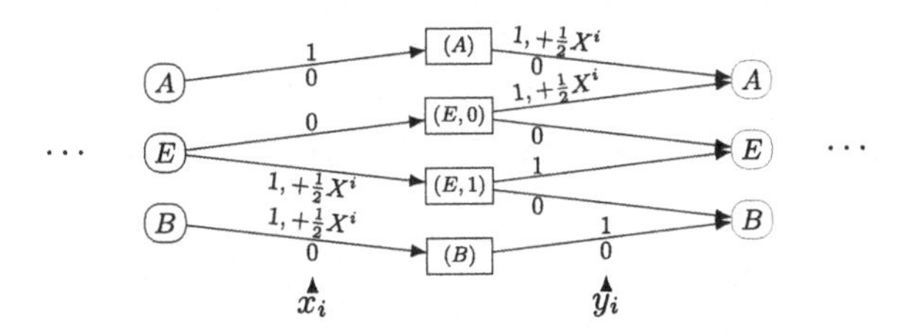

Fig. 2. Max: det-WFA - The states A and (A) mean that y is the maximal value, the states B and (B) mean that x is the maximal value, and finally, the states E, $(E, 1)$ and $(E, 0)$ mean that x and y are equals on the most significant bits. If the current state is A or B, the following state will stay the same. The initial state is E. If the current state is E, after reading x_i there are two possible intermediate states: $(E, 1)$ if $x_i = 1$ and $(E, 0)$ if $x_i = 0$. After reading the value of y_i, the 3 possible states A, B and E are possible. The det-WFA is repeated as many times as the bit length of the integers evaluated and the weights are given in clear.

Remark 3. In practice, to evaluate the MAX function, we convert the det-WFA in a circuit that counts $5d$ CMux gates. Roughly speaking, we have to read the automata in the reverse. We initialize 5 states A, B, E_0, E_1, E as null TRLWE samples. Then, for i from $d - 1$ to 0, we update the states as follows:

$$\begin{cases} E_0 := \texttt{CMux}(C_i^y, A + (\mathbf{0}, \frac{1}{2}X^i), E); \\ E_1 := \texttt{CMux}(C_i^y, E, B); \\ A := \texttt{CMux}(C_i^y, A + (\mathbf{0}, \frac{1}{2}X^i), A); \\ E := \texttt{CMux}(C_i^x, E_1 + (\mathbf{0}, \frac{1}{2}X^i), E_0); \\ B := \texttt{CMux}(C_i^x, B + (\mathbf{0}, \frac{1}{2}X^i), B). \end{cases}$$

Here the C_i^x and C_i^y are TRGSW encryptions of the bits x_i and y_i respectively, and they are the inputs. The output of the evaluation is the TRLWE sample E, which contains the maximal value.

Lemma 3.5 (Evaluation of Max det-WFA). *Let $\mathfrak{A}$ be the det-WFA of the Max, described in Fig. 2. Let $C_0^x, \ldots, C_{d-1}^x, C_0^y, \ldots, C_{d-1}^y$ be TRGSW_K samples of the bits of x and y respectively. By evaluating $5d$ CMux gates (depth $2d$), the Max det-WFA outputs a TRLWE sample $\mathbf{d}$ encrypting the maximal value between x and y and (with same notations as in Lemma 2.7)*

- $\|Err(\mathbf{d})\|_\infty \leq 2d \cdot ((k+1)\ell N\beta \mathcal{A}_{\mathsf{TRGSW}} + (kN+1)\epsilon)$ *(worst case);*
- $Var(Err(\mathbf{d})) \leq 2d \cdot ((k+1)\ell N\beta^2 \vartheta_{\mathsf{TRGSW}} + (kN+1)\epsilon^2)$ *(average case).*
  Here $\mathcal{A}_{\mathsf{TRGSW}}$ and $\vartheta_{\mathsf{TRGSW}}$ are upper bounds of the amplitude and of the variance of the errors in the TRGSW samples.

Multiplication. For the multiplication we use the same approach and we construct a det-WFA which maps the schoolbook multiplication. We illustrate the construction on the example of the multiplication between two 2-bits integers $x = x_1x_0$ and $y = y_1y_0$. After an initial step of bit by bit multiplication, a multi-addition (shifted of one place on the left for every line) is performed. The bits of the final result are computed as the sum of each column with carry.

The det-WFA computes the multiplication by keeping track of the partial sum of each column in the states, and by using the transitions to update these sums. For the multiplication of 2-bits integers, the automaton (described in Fig. 3) has 6 main states (i, c_0, c_{10}, c_{11}, c_{20}, c_{21}), plus 14 intermediary states that store the last bit read (noted with capital letters and parenthesis). The value of the i-th output bit is put in a weight on the last transition of each column.

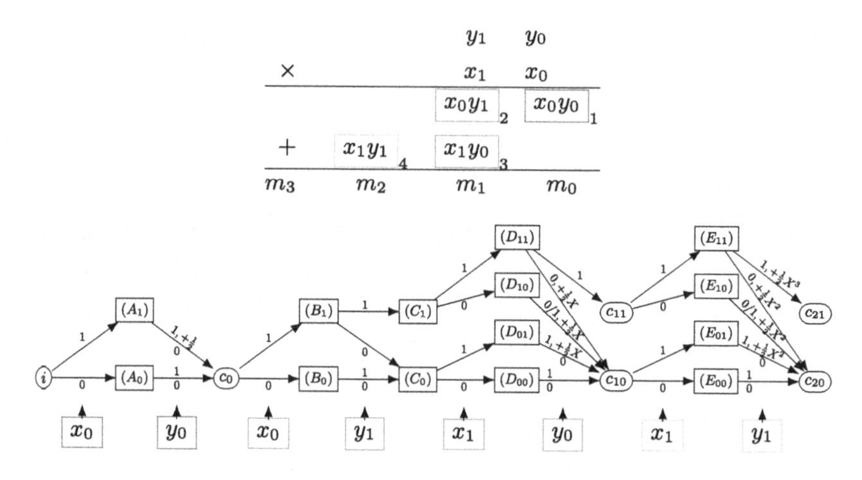

Fig. 3. Schoolbook 2-bits multiplication and corresponding det-WFA

For the generic multiplication of two d-bits integers, we can upper bound the number of states by $4d^3$, instead of $\Theta(d^4)$ with one classical automata per output bit. For a more precise number of states we wrote a C++ program to eliminate unreachable states and refine the leading coefficient. The depth is $2d^2$ and the noise evaluation can be easily deducted by previous results. The same principle can be used to construct the **multi-addition**, and its det-WFA is slightly simpler (one transition per bit in the sum instead of two).

3.3 TBSR Counter Techniques

We now present another design which is specific to the multi-addition (or its derivatives), but which is faster than the generic construction with weighted automata. The idea is to build an homomorphic scheme that can represent small integers, say between 0 and $N = 2^p$, and which is dedicated to only the three elementary operations used in the multi addition algorithm, namely:

1. Extract any of the bits of the value as a TLWE sample;
2. Increment the value by 1 and
3. Integer division of the value by 2.

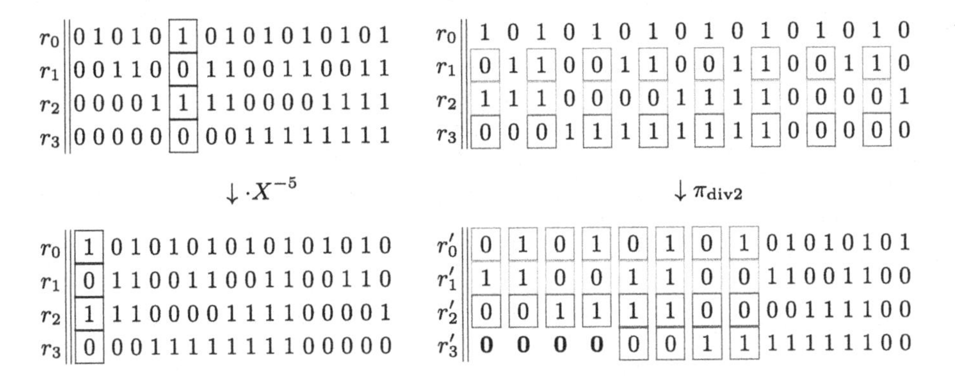

Fig. 4. TBSR - example of addition +5 and division by 2.

We will now explain the basic idea, and then, we will show how to implement it efficiently on TRLWE ciphertexts.

For $j \in [0, p = \log_2(N)]$ and $k, l \in \mathbb{Z}$, we call $B_{j,k}^{(l)}$ the j-th bit of $k + l$ in the little endian signed binary representation. The latter form very simple binary sequence: $B_0^{(0)} = (0, 1, 0, 1, \ldots)$ is 2-periodic, $B_1^{(0)} = (0, 0, 1, 1, 0, 0, 1, 1 \ldots)$ is 4-periodic, more generally, for all $j \in [0, p]$ and $l \in \mathbb{Z}$, $B_j^{(l)}$ is 2^j-antiperiodic, and is the left shift of $B_j^{(0)}$ by l positions. Therefore, it suffices to have $2^j \leq N$ consecutive values of the sequence to (blindly) deduce all the remaining bits. And most importantly, for each integer $k \in \mathbb{Z}$, $(B_{0,k}^{(l)}, B_{1,k}^{(l)}, \ldots, B_{p,k}^{(l)})$ is the (little endian signed) binary representation of $l + k \mod 2N$. We now suppose that an integer l in $[0, N-1]$ is represented by its Bit Sequence Representation, defined as $BSR(l) = [B_0^{(l)}, \ldots, B_p^{(l)}]$. And we see how to compute $BSR(l+1)$ and $BSR(\lfloor l/2 \rfloor)$ using only copy and negations operations on bits at a fixed position which does not depend on l (blind computation). Then, we will see how to represent these operations homomorphically on TRLWE ciphertexts (Fig. 4).

Increment: Let $U = [u_0, \ldots, u_p]$ be the BSR of some unknown number $l \in [0, N-1]$. Our goal is to compute $V = [v_0, \ldots, v_p]$ which is the BSR of $l + 1$. Again, we recall that it suffices to define the sequence v_i on N consecutive values, the rest is deduced by periodicity. To map the increment operation, all we need to do is shifting the sequences by 1 position: $v_{j,k} := u_{j,k+1}$ for all $k \in [0, N-1]$. Indeed, this operation transforms each $B_{j,k}^{(l)}$ into $B_{j,k+1}^{(l)} = B_{j,k}^{(l+1)}$, and the output V is the BSR of $l + 1$.

Integer division by two: Let $U = [u_0, \ldots, u_p]$ be the BSR of some unknown number $l \in [0, N-1]$. Our goal is to compute $V = [v_0, \ldots, v_p]$ which is the BSR of $\lfloor \frac{l}{2} \rfloor$. First, we note that the integer division by 2 corresponds to a right shift over the bits. Thus for $j \in [0, p-1]$ and $k \in \mathbb{N}$, we can set $v_{j,k} = u_{j+1,2k}$. Indeed, $u_{j+1,2k}$ is the $j+1$-th bit of $l + 2k$ is the j-th bit of its half $\lfloor l/2 \rfloor + k$, which is our desired $v_{j,k} = B_{j,k}^{(\lfloor l/2 \rfloor)}$. This is unfortunately not enough to reconstruct the last sequence v_p, since we have no information on the $p+1$-th bits in U. However, in our case, we can reconstruct this last sequence directly. First, the numbers $\lfloor\frac{l}{2}\rfloor + k$ for $k \in [0, N/2-1]$ are all $< N$, so we can blindly set the corresponding $v_{p,k} = 0$. Then, we just need to note that $(u_{p,0}, \ldots, u_{p,N-1})$ is $N-l$ times 0 followed by l times 1, and our target $(v_{p,N/2}, \ldots, v_{p,N-1})$ must consist $N/2-l$ times 0 followed by $\lfloor l/2 \rfloor$ times 1. Therefore, our target can be filled with the even positions $(u_{p,0}, u_{p,2}, \ldots, u_{p,N-2})$. To summarize, division by 2 corresponds to the following blind transformation:

$$\begin{cases} v_{j,k} &= u_{j+1,2k} \text{ for } j \in [0, p-1], k \in [0, N-1] \\ v_{p,k} &= 0 \text{ for } k \in [0, \frac{N}{2}-1] \\ v_{p,N/2+k} &= u_{p,2k} \text{ for } k \in [0, \frac{N}{2}-1] \end{cases}$$

We now explain how we can encode these BSR sequences on TRLWE ciphertexts, considering that all the coefficients need to be in the torus rather than in $\mathbb{B}$, and that we need to encode sequences that are either N-periodic or N-antiperiodic. Furthermore, since the cyclic shift of coefficients is heavily used in the increment operation, we would like to make it correspond to the multiplication by X, which has a similar behaviour on coefficients of torus polynomials. Therefore, this is our basic encoding of the BSR sequences: Let $U = [u_0, \ldots, u_p]$ be the BSR of some unknown number $l \in [0, N-1]$, For $j \in [0, p-1]$, we represent u_j with the polynomial $\mu_i = \sum_{k=0}^{N-1} \frac{1}{2} u_{j,k} X^k$, and we represent the last u_p with the polynomial $\mu_p = \sum_{k=0}^{N-1} (\frac{1}{2} u_{p,k} - \frac{1}{4}) X^k$. This simple rescaling between the bit representation U and the torus representation $M = [\mu_0, \ldots, \mu_p]$ is bijective. Using this encoding, the integer division transformation presented above immediately rewrites into this affine function, which transforms the coefficients $(\mu_{j,k})_{j \in [1,p], k \in [0,2,\ldots,2N-2]} \in \mathbb{T}^{pN}$ into $(\mu'_0, \ldots, \mu'_p)$ as follow:

$$\pi_{\text{div2}}: \begin{cases} \mu'_{j,k} &= \mu_{j+1,2k} \text{ for } j \in [0, p-2], k \in [0, N-1] \\ \mu'_{p-1,k} &= \mu_{p,2k} + \frac{1}{4} \text{ for } k \in [0, N-1] \\ \mu'_{p,k} &= -\frac{1}{4} \text{ for } k \in [0, \frac{N}{2}-1] \\ \mu'_{p,N/2+k} &= \mu_{p,2k} \text{ for } k \in [0, \frac{N}{2}-1] \end{cases}$$

Finally, we call TBSR ciphertext of an unknown integer $l \in [0, N-1]$ a vector $C = [c_0, \ldots, c_p]$ of TRLWE ciphertexts of message $[\mu_0, \ldots, \mu_p]$.

Definition 3.6 (TBSR encryption).

- *Params and keys:* TRLWE *parameters N with secret key $K \in \mathbb{B}_N[X]$, and a circular-secure keyswitching key $KS_{K \to K,\gamma}$ from K to itself, noted just KS.*
- TBSRSet(l): *return a vector of trivial* TRLWE *ciphertexts encoding the torus representation of $[B_0^{(l)}, \ldots, B_p^{(l)}]$.*
- TBSRBitExtract$_j(C)$: *Return* SampleExtract$_0(c_j)$ *when $j < p$.*[7]
- TBSRIncrement(C): *Return $X^{-1}.C$.*

[7] For the p-th bit, one would return SampleExtract$(c_p) + (\mathbf{0}, \frac{1}{4})$, but it is always 0 if $l \in [0, N-1]$.

- $\mathsf{TBSRDiv2}(C)$: *Use KS to evaluate π_{div2} homomorphically on C. Since it is a 1-lipschitzian affine function, this means: apply the public functional* KeySswitch *to KS, the linear part of π_{div2} and C, and then, translate the result by the constant part of π_{div2}.*

Theorem 3.7 (TBSR operations). *Let N,K, and KS be TBSR parameters..., and C a TBSR ciphertext of l with noise amplitude η (or noise variance ϑ). Then for $j \in [0, p-1]$, $\mathsf{TBSRBitExtract}_j(C)$ is a $LWE_{\mathfrak{K}}$ ciphertext of the j-th bit of l, over the message space $\{0, \frac{1}{2}\}$, with noise amplitude (resp. variance) $\leq \eta$ (resp. $\leq \vartheta$). If $l \leq N-2$, $\mathsf{TBSRIncrement}(C)$ is a TBSR ciphertext of $l+1$ with noise amplitude (resp. variance) $\leq \eta$ (resp. $\leq \vartheta$). $C' = \mathsf{TBSRDiv2}(C)$ is a TBSR ciphertext of $\lfloor l/2 \rfloor$ such that:*

- $\|\mathit{Err}(C')\|_\infty \leq \mathcal{A} + N^2 t \mathcal{A}_{\mathsf{KS}} + N 2^{-(t+1)}$ *(worst-case);*
- $\mathit{Var}(\mathit{Err}(C')) \leq \vartheta + N^2 t \vartheta_{\mathsf{KS}} + N 2^{-2(t+1)}$ *(average case).*

Proof (sketch). Correctness has already been discussed, the noise corresponds to the application of a public keyswitch on the same key: with $n = N$.

Using the TBSR counter for a multi-addition or a multiplication.

The TBSR counter allows to perform a multi-addition or multiplication using the school-book elementary algorithms. This leads to a leveled multiplication circuit with KeySwitching which is quadratic instead of cubic with weighted automata.

Lemma 3.8 *Let N,B_g,ℓ and* KS *be TBSR and* TRGSW *parameters with the same key K, We suppose that each TBSR ciphertext has $p \leq 1 + \log(N)$* TRLWE *ciphertexts. And let (A_i) and (B_i) for $i \in [0, d-1]$ be* TRGSW*-encryptions of the bits of two d-bits integers (little endian), with the same noise amplitude $\mathcal{A}_A$ (resp. variance ϑ_A).*

Then, there exists an algorithm (see the full version for more details) that computes all the bits of the product within $2d^2p$ CMux and $(2d-2)p$ public keyswitch, and the output ciphertexts satisfy:

- $\|\mathit{Err}(\mathit{Out})\|_\infty \leq 2d^2((k+1)\ell N \beta \mathcal{A}_A + (kN+1)\epsilon) + (2d-2)(N^2 t \mathcal{A}_{\mathsf{KS}} + N 2^{-(t+1)})$;
- $\mathit{Var}(\mathit{Err}(\mathit{Out})) \leq 2d^2((k+1)\ell N \beta^2 \vartheta_A + (kN+1)\epsilon^2) + (2d-2)(N^2 t \vartheta_{\mathsf{KS}} + N 2^{-2(t+1)})$.

4 Combining Leveled with Bootstrapping

In the previous sections, we presented efficient leveled algorithms for some arithmetic operations, but the input and output have different types (e.g. TLWE/TRGSW) and we can't compose these operations, like in a usual algorithm. In fully homomorphic mode, connecting the two becomes possible if we have an efficient bootstrapping between TLWE and TRGSW ciphertexts. Fast bootstrapping procedures have been proposed in [12,17], and the external product Theorem 2.3 from [7,12] has contributed to accelerate leveled operations. Unfortunately, these bootstrapping cannot output GSW ciphertexts. Previous solutions proposed in [1,19,21] based on the internal product are not practical. In this section, we propose an efficient technique to convert back TLWE ciphertexts to TRGSW, that runs in 137 ms. We call it *circuit bootstrapping.*

Our goal is to convert a TLWE sample with large noise amplitude over some binary message space (e.g. amplitude $\frac{1}{4}$ over $\{0, \frac{1}{2}\}$), into a TRGSW sample with a low noise amplitude $<2^{-20}$ over the integer message space $\{0, 1\}$.

In all previous constructions, the TLWE decryption consists in a circuit, which is then evaluated using the internal addition and multiplication laws over TRGSW ciphertexts. The target TRGSW ciphertext is thus the result of an arithmetic expression over TRGSW ciphertexts. Instead, we propose a more efficient technique, which reconstructs the target directly from its very sparse internal structure. Namely, a TRGSW ciphertext of a message $\mu \in \{0, 1\}$ is a vector of $(k+1)\ell$ TRLWE ciphertexts. Each of these TRLWE ciphertexts encrypts the same message as μh_i, where h_i is the corresponding line of the gadget matrix H. Depending on the position of the row (which can be indexed by $u \in [1, k+1]$ and $j \in [1, \ell]$), this message is $\mu - K_u \cdot Bg^{-j}$ where K_u is the u-th polynomial of the secret key and $K_{k+1} = -1$. So we can use ℓ times the TLWE-to-TLWE bootstrapping of [12] to obtain a TLWE sample of each message in $\{\mu Bg^{-1}, \ldots, \mu B_g^{-\ell}\}$. Then we use the private key-switching technique to "multiply" these ciphertexts by the secret $-K_u$, to reconstruct the correct message.

4.1 Circuit Bootstrapping (TLWE-to-TRGSW)

Our circuit bootstrapping, detailed in Algorithm 6, crosses 3 levels of noise and encryption (Fig. 5). Each level has its own key and parameters set. In order to distinguish the different levels, we use an intuitive notation with bars. The upper bar will be used for level 2 variables, the under bar for the level 0 variables and level 1 variables will remain without any bar. The main difference between the three levels of encryption is the amount of noise supported. Indeed, the higher the level is, the smaller is the noise. Level 0 corresponds to ciphertexts with very large noise (typically, $\underline{\alpha} \geq 2^{-11}$). Level 0 parameters are very small, computations are almost instantaneous, but only a very limited amount of linear operations are tolerated. Level 1 corresponds to medium noise (typically, $\alpha \geq 2^{-30}$). Ciphertexts in level 1 have medium size parameters, which allows for relatively fast operations, and for instance a leveled homomorphic evaluation of a relatively large automata, with transition timings described in Sect. 5 of [12]. Level 2 corresponds to ciphertexts with small noise (typically, $\bar{\alpha} \geq 2^{-50}$). This level corresponds to the limit of what can be mapped over native 64-bit operations. Practical values and details are given in Sect. 5.

Our circuit bootstrapping consists in three parts:

- TLWE-to-TLWE **Pre-keyswitch.** The input of the algorithm is a TLWE sample with a large noise amplitude over the message space $\{0, \frac{1}{2}\}$. Without loss of generality, it can be keyswitched to a level 0 TLWE ciphertext

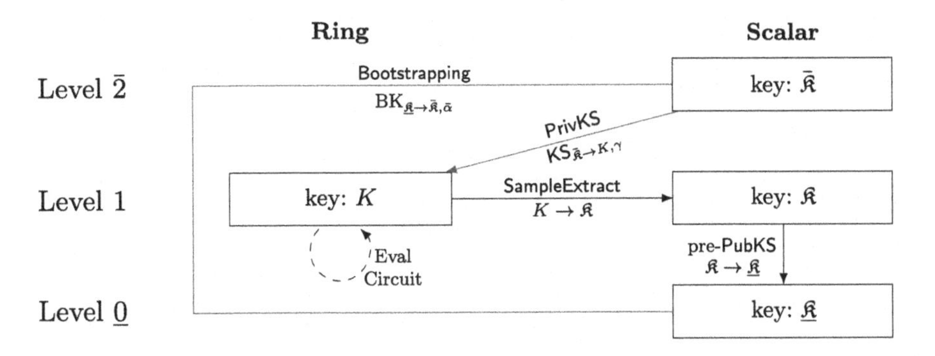

Fig. 5. The figure represents the three levels of encryption on which our construction shifts. The arrows show the operations that can be performed inside each level or how to move from a level to another. In order to distinguish the objects with respect to their level, we adopted the intuitive notations "superior bar" for level 2, "no bar" for level 1 and "under bar" for level 0. We highlight in blue the different stages of the circuit bootstrapping (whose detailed description is given below).

$\underline{\mathfrak{c}} = (\underline{\mathfrak{a}}, \underline{\mathfrak{b}}) \in \mathsf{TLWE}_{\underline{\mathfrak{K}},\underline{\eta}}(\mu \cdot \frac{1}{2})$, of a message $\mu \in \mathbb{B}$ with respect to the small secret key $\underline{\mathfrak{K}} \in \mathbb{B}^{\underline{n}}$ and a large standard deviation $\underline{\eta} \in \mathbb{R}$ (typically, $\underline{\eta} \leq 2^{-5}$ to guaranty correct decryption with overwhelming probability). This step is standard.

- TLWE-to-TLWE **Bootstrapping** (Algorithm 4): Given a level 2 bootstrapping key $\mathrm{BK}_{\underline{\mathfrak{K}}\to\bar{\mathfrak{K}},\bar{\alpha}} = (\mathrm{BK}_i)_{i\in[\![1,\underline{n}]\!]}$ where $\mathrm{BK}_i \in \mathsf{TRGSW}_{\bar{K},\bar{\alpha}}(\underline{\mathfrak{K}}_i))$, we use ℓ times the TLWE-to-TLWE Bootstrapping algorithm (Algorithm 4) on $\underline{\mathfrak{c}}$, to obtain ℓ TLWE ciphertexts $\bar{\mathfrak{c}}^{(1)}, \ldots, \bar{\mathfrak{c}}^{(\ell)}$ where $\bar{\mathfrak{c}}^{(w)} \in \mathsf{TLWE}_{\bar{\mathfrak{K}},\bar{\eta}}(\mu \cdot \frac{1}{B_g{}^w})$, with respect to the same level 2 secret key $\bar{\mathfrak{K}} \in \mathbb{B}^{\bar{n}}$, and with a fixed noise parameter $\bar{\eta} \in \mathbb{R}$ which does not depend on the input noise. If the bootstrapping key has a level 2 noise $\bar{\alpha}$, we expect the output noise $\bar{\eta}$ to remain smaller than level 1 value.
- TLWE-to-TRLWE **private key-switching** (Algorithm 2): Finally, to reconstruct the final TRGSW ciphertext of μ, we simply need to craft a TRLWE ciphertext which has the same phase as $\mu \cdot \boldsymbol{h_i}$, for each row of the gadget matrix H. Since $\boldsymbol{h_i}$ contains only a single non-zero constant polynomial in position $u \in [1, k+1]$ whose value is $\frac{1}{B_g^w}$ where $w \in [1, \ell]$, the phase of $\mu \cdot \boldsymbol{h_i}$ is $\mu K_u \cdot \frac{1}{B_g^w}$ where K_u is the u-th term of the key K. If we call f_u the (secret) morphism from $\mathbb{T}$ to $\mathbb{T}_N[X]$ defined by $f_u(x) = K_u \cdot x$, we just need to apply f_u homomorphically to the TLWE sample $\bar{\mathfrak{c}}^{(w)}$ to get the desired TRLWE sample. Since f_u is 1-lipschitzian (for the infinity norm), this operation be done with additive noise overhead via the private functional keyswitch (Algorithm 2).

Theorem 4.1 (Circuit Bootstrapping Theorem). *Let $n, \alpha, N, k, B_g, \ell, H, \epsilon$ denote* TRLWE/TRGSW *level 1 parameters, and the same variables names with underbars/upperbars for level 0 and 2 parameters. Let $\underline{\mathfrak{K}} \in \mathbb{B}^{\underline{n}}$, $\mathfrak{K} \in \mathbb{B}^n$ and $\bar{\mathfrak{K}} \in$*

Algorithm 6. Circuit Bootstrapping (calling Algorithms 4 and 2)

Input: A level 0 TLWE sample $\underline{\mathfrak{c}} = (\underline{\mathfrak{a}}, \underline{\mathfrak{b}}) \in \mathsf{TLWE}_{\underline{\mathfrak{K}},\underline{\eta}}(\mu \cdot \frac{1}{2})$, with $\mu \in \mathbb{B}$, a bootstrapping key $\mathrm{BK}_{\underline{\mathfrak{K}} \to \bar{\mathfrak{K}}, \bar{\alpha}} = (\mathrm{BK}_i \in \mathsf{TRGSW}_{\bar{K},\bar{\alpha}}(\underline{\mathfrak{K}}_i))_{i \in [\![1,\underline{n}]\!]}$, $k+1$ private keyswitch keys $\mathsf{KS}^{(f_u)}_{\bar{\mathfrak{K}} \to K,\gamma}$ corresponding to the functions $f_u(x) = -K_u \cdot x$ when $u \le k$, and $f_{k+1}(x) = 1 \cdot x$.

Output: A level 1 TRGSW sample $C \in \mathsf{TRGSW}_{K,\eta}(\mu)$

1: **for** $w = 1$ **to** ℓ

2: $\quad \bar{\mathfrak{c}}^{(w)} \leftarrow \mathsf{Bootstrapping}_{\mathrm{BK}, \frac{1}{B_g^w}}(\underline{\mathfrak{c}})$

3: $\quad$ **for** $u = 1$ **to** $k + 1$

4: $\quad\quad \mathbf{c}^{(u,w)} = \mathsf{PrivKS}(\mathsf{KS}^{(f_u)}, \bar{\mathfrak{c}}^{(w)})$

5: Return $C = (\mathbf{c}^{(u,w)})_{1 \le u \le k+1, 1 \le w \le \ell}$

$\mathbb{B}^{\bar{n}}$, *be a level 0, 1 and 2* TLWE *secret keys, and* $\underline{K}, K, \bar{K}$ *their respective* TRLWE *interpretation. Let* $BK_{\underline{\mathfrak{K}} \to \bar{\mathfrak{K}}, \bar{\alpha}}$ *be a bootstrapping key, composed by the* $\underline{n}$ TRGSW *encryptions* $BK_i \in \mathsf{TRGSW}_{\bar{K},\bar{\alpha}}(\underline{\mathfrak{K}}_i)$ *for* $i \in [\![1, \underline{n}]\!]$. *For each* $u \in [\![1, k+1]\!]$, *let* f_u *be the morphism from* $\mathbb{T}$ *to* $\mathbb{T}_N[X]$ *defined by* $f_u(x) = K_u \cdot x$, *and* $\mathsf{KS}^{f_u}_{\bar{\mathfrak{K}} \to K,\gamma} = (\mathsf{KS}^{(u)}_{i,j} \in \mathsf{TRLWE}_{K,\gamma}((\bar{\mathfrak{K}}_i K_u \cdot 2^{-j})))_{i \in [\![1,\bar{n}]\!], j \in [\![1,t]\!]}$ *be the corresponding private-key-switching key. Given a level 0* TLWE *sample* $\underline{\mathfrak{c}} = (\underline{\mathfrak{a}}, \underline{\mathfrak{b}}) \in \mathsf{TLWE}_{\underline{\mathfrak{K}}}(\mu \cdot \frac{1}{2})$, *with* $\mu \in \mathbb{B}$, *the Algorithm 6 outputs a level 1* TRGSW *sample* $C \in \mathsf{TRGSW}_K(\mu)$ *such that*

- $\|\mathsf{Err}(C)\|_\infty \le \underline{n}(\bar{k}+1)\bar{\ell}\bar{N}\bar{\beta}\mathcal{A}_{BK} + \underline{n}(1 + \bar{k}\bar{N})\bar{\epsilon} + \bar{n}2^{-(t+1)} + \bar{n}t\mathcal{A}_{\mathsf{KS}}$ *(worst);*
- $\mathsf{Var}(\mathsf{Err}(C)) \le \underline{n}(\bar{k}+1)\bar{\ell}\bar{N}\bar{\beta}^2\bar{\vartheta}_{BK} + \underline{n}(1+\bar{k}\bar{N})\bar{\epsilon}^2 + \bar{n}2^{-2(t+1)} + \bar{n}t\vartheta_{\mathsf{KS}}$ *(average).*

Here $\bar{\vartheta}_{BK} = \bar{\alpha}^2$ *and* $\mathcal{A}_{BK}$ *is the variance and amplitude of* $\mathsf{Err}(BK_{\underline{\mathfrak{K}} \to \bar{K},\bar{\alpha}})$, *and* $\vartheta_{\mathsf{KS}} = \gamma^2$ *and* $\mathcal{A}_{KS}$ *are the variance and amplitude of* $\mathsf{Err}(\mathsf{KS}_{\bar{\mathfrak{K}} \to K,\gamma})$.

Proof (sketch). The output TRGSW ciphertext is correct, because by construction, the i-th TRLWE component $\mathbf{c}^{(u,w)}$ has the correct message $\mathsf{msg}(\mu \cdot H_i) = \mu K_u / B_g^w$. $\mathbf{c}^{(u,w)}$ is obtained by chaining one TLWE-to-TLWE bootstrapping (Algorithm 4) with one private key switchings, as in Algorithm 2. The values of maximal amplitude and variance of $\mathsf{Err}(C)$ are directly obtained from the partial results of Lemma 2.9 and Theorem 2.6. In total, Algorithm 6 performs exactly ℓ bootstrappings (Algorithm 4), and $\ell(k+1)$ private key switchings (Algorithm 2). □

Comparison with previous bootstrappings for TGSW. The circuit bootstrapping we just described evaluates a quasilinear number of level-2 external products, and a quasilinear number of level 1 products in the private keyswitchings. With the parameters proposed in the next section, it runs in 0.137 s for a 110-bit security parameter, level 2 operations take 70% of the running time, and the private keyswitch the remaining 30%.

Our circuit bootstrapping is not the first bootstrapping algorithm that outputs a TRGSW ciphertext. Many constructions have previously been proposed and achieve valid asymptotical complexities, but very few concrete parameters are proposed. Most of these constructions are recalled in the last section of [19]. In all of them, the bootstrapped ciphertext is obtained as an arithmetic expression on TRGSW ciphertexts involving linear combinations and internal products. First, all the schemes based on scalar variants of TRGSW suffer from a slowdown of a factor at least quadratic in the security parameter, because the products of small matrices with polynomial coefficients (via FFT) are replaced with large dense matrix products. Thus, bootstrapping on TGSW variants would require days of computations, instead of the 0.137 s we propose. Now, assuming that all the bootstrapping uses (Ring) instantiations of TRGSW, the design in [8] based on the expansion of the decryption circuit via Barrington theorem, as well as the expression as a minimal deterministic automata of the same function in [19] require a quadratic number of internal level 2 TRGSW products, which is much slower than what we propose. Finally, the CRT variant in [1,19] uses only a quasi-linear number of products, but since it uses composition between automata, these products need to run in level 3 instead of level 2, which induces a huge slowdown (a factor 240 in our benchs), because elements cannot be represented on 64-bits native numbers.

5 Comparison and Practical Parameters

We now explicit the practical parameters for our scheme, and we give the running time comparison for the evaluation of the homomorphic circuits described before in LHE and FHE mode (with or without the new optimization techniques).

In [12] the timing for the gate bootstrapping was 52 ms. We improved it to 13 ms: a speed up of a factor 2 is due to the dedicated assembly FFT for $X^N + 1$ in double precision. An additional speed ups (by a factor 1.5) is due to a new choice of parameters, for the same security level (in particular the ℓ TRGSW parameter is reduced to 2 instead of 3). Finally, we replaced the core of the gate bootstrapping with the simpler CMux and blindRotate (Algorithm 3) described in Sect. 2, which gives the last 1.33x speed-up. For the same reason, the external product is now executed in 34 µs. We added these optimizations to the public repository of the TFHE library [14]. A experimental measurement of the noise confirmed the average case bounds on the variance, predicted under the independance assumption.

As a consequence, all binary gates are executed in 13 ms, and the native bootstrapped MUX (also described in Sect. 2) gate takes 26 ms on a 64-bit single core (i7-4910MQ) at 2.90 GHz. Starting from all these considerations, we implemented our circuit bootstrapping as a proof of concept. The code is available in the experimental repository of TFHE [14]. We perform a Circuit Bootstrapping in 0.137 s. One of the main constraints to obtain this performance is to ensure that all the computations are feasible and correct under 53 bits of floating point precision, in order to use the fast FFT. This requires to refine the parameters of the scheme. We verified the accuracy of the FFT with a slower but exact Karatsuba implementation of the polynomial product.

Concrete Parameters. In our three levels, we used the following TRLWE and TRGSW parameter sets, which have at least 110-bits of security, according to the security analysis in [12].

Level	Minimal noise α	n	B_g	ℓ
0	$\underline{\alpha} = 2^{-15.33}$	$\underline{n} = 500$	N.A	N.A.
1	$\alpha = 2^{-32.33}$	$n = 1024$	$B_g = 2^8$	$\ell = 2$
2	$\bar{\alpha} = 2^{-45.33}$	$\bar{n} = 2048$	$\bar{B}_g = 2^9$	$\bar{\ell} = 4$

Since we assume circular security, we will use only one key per level, and the following keyswitch parameters (in the leveled setting, the reader is free to increase the number of keys if he does not wish to assume circularity).

Level	t	γ	Usage
$1 \to 0$	$\underline{t} = 12$	$\underline{\gamma} = 2^{-14}$	Circuit Bootstap, Pre-KS
$2 \to 1$	$\bar{t} = 30$	$\bar{\gamma} = 2^{-31}$	Circuit Bootstap, Step 4 in Algorithm 6
$1 \to 1$	$t = 24$	$\gamma = 2^{-24}$	TBSR

Thus, we get these noise variances in input or in output

Output TLWE	Fresh TRGSW in LHE	TRGSW Output of CB	Bootst. key
$\vartheta \leq 2^{-10,651}$	$\vartheta = 2^{-60}$	$\vartheta \leq 2^{-47.03}$	$\vartheta_{BK} = 2^{-88}$

And finally, this table summarizes the timings (Core i7-4910MQ laptop), noises overhead, and maximal depth of all our primitives.

	CPU Time	Var Noise add	Max depth
Circuit bootstrap	$t_{CB} = 137\,\text{ms}$	N.A	N.A
Fresh CMux	$t_{XP} = 34\,\mu\text{s}$	$2^{-23.99}$	16384
CB CMux	$t_{XP} = 34\,\mu\text{s}$	$2^{-20.86}$	3466
PubKS_{TBSR}	$t_{KS} = 180\,\text{ms}$	$2^{-23.42}$	16384

More details on these parameter choices are provided in the full version.

Time Comparison. With these parameters, we analyse the (single-core) execution timings for the evaluation of the LUT, MAX and Multiplication in LHE and FHE mode.

In the LHE mode (left hand side of Fig. 6), all inputs are fresh ciphertexts (either TRLWE or TRGSW) and we compare the previous versions [12] (without packing/batching or gate bootstapping) with the new optimizations i.e. horizontal/vertical packing; with weighted automata or with TBSR techniques. In the

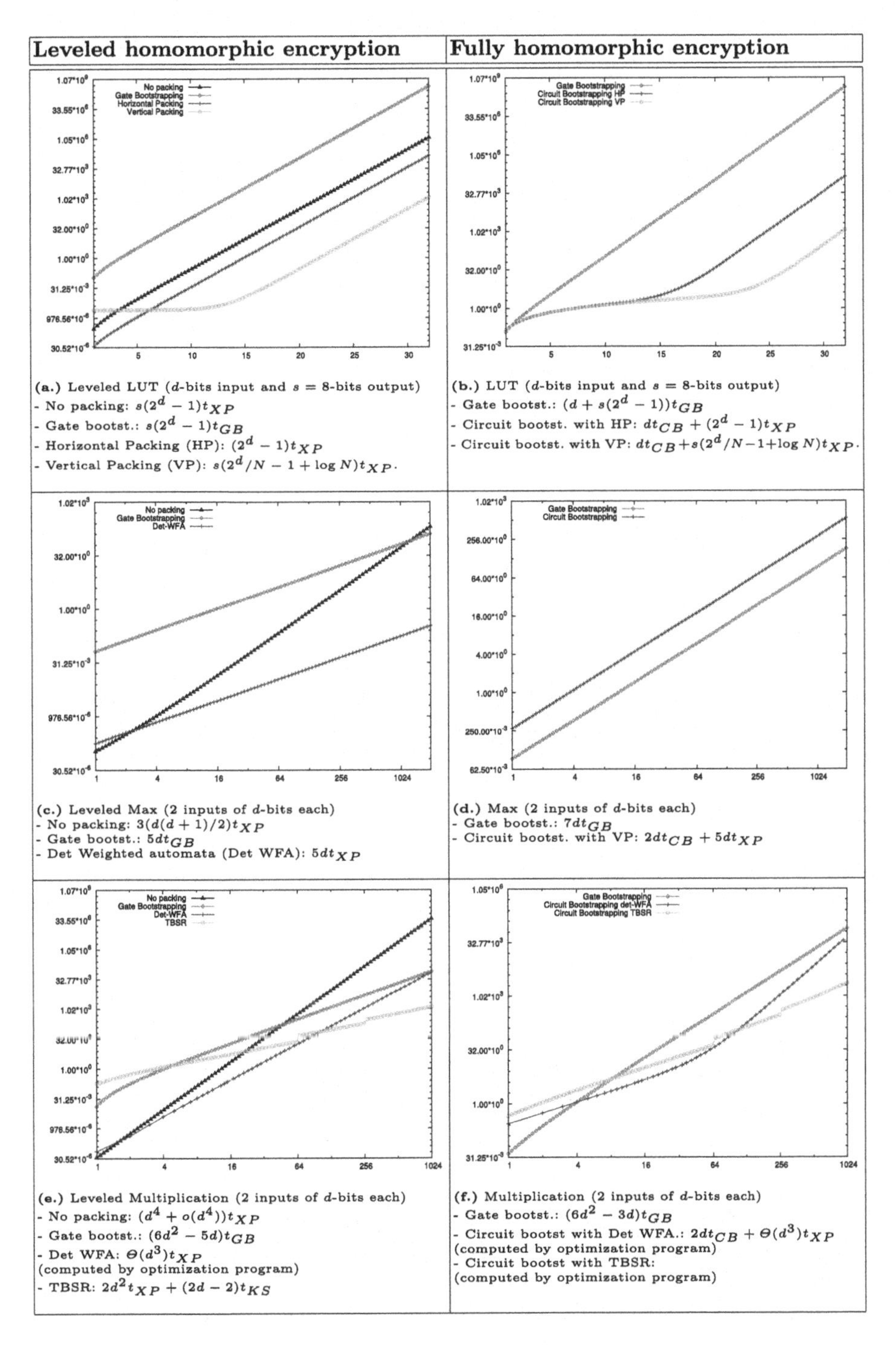

Fig. 6. The y coordinate represents the running time in seconds (in logscale), the x coordinate represents the number of bits in the input (in logscale for c–f).

FHE mode (right hand side of Fig. 6), all inputs and outputs are TLWE samples on the $\{0, \frac{1}{2}\}$ message space with noise amplitude $\frac{1}{4}$. Each operation starts by bootstrapping its inputs. We compare the gate-by-gate bootstapping strategy with the mixed version where we use leveled encryption with circuit bootsrapping. Our goal is to identify which method is better for each of the 6 cases. We observe that compared to the gate bootstrapping, we obtain a huge speed-up for the homomorphic evaluation of arbitrary function in both LHE and FHE mode, in particular, we can evaluate a 8 bits to 1 bit lookup table and bootstrap the output in just 137 ms, or evaluate an arbitrary 8 bits to 8 bits function in 1.096 s, and an arbitrary 16 bits to 8 bits function in 2.192 s in FHE mode. For the multiplication in LHE mode, it is better to use the weighted automata technique when the number is less than 128 bits, and the TBSR counter after that. In the FHE mode, the weighted automata becomes faster than gate-bootstrapping after 4 bits of inputs, then the TBSR optimization becomes faster for >64 bits inputs.

6 Conclusion

In this paper we improved the efficiency of TFHE, by proposing some new packing techniques. For the first time we use det-WFA in the context of homomorphic encryption to optimize the evaluation of arithmetic circuits, and we introduced the TBSR counter. By combining these optimizations, we obtained a significant timing speed-up and decrease the ciphertext overhead for TLWE and TRGSW based encryption. We also solved the problem of non universal composability of TFHE leveled gates, by proposing the efficient circuit bootstrapping that runs in 134 ms; we implemented it in the TFHE project [14].

Acknowledgements. This work has been supported in part by the CRYPTOCOMP project.

References

1. Alperin-Sheriff, J., Peikert, C.: Faster bootstrapping with polynomial error. In: Garay, J.A., Gennaro, R. (eds.) CRYPTO 2014. LNCS, vol. 8616, pp. 297–314. Springer, Heidelberg (2014). https://doi.org/10.1007/978-3-662-44371-2_17
2. Benarroch, D., Brakerski, Z., Lepoint, T.: FHE over the integers: decomposed and batched in the post-quantum regime. Cryptology ePrint Archive, 2017/065
3. Benhamouda, F., Lepoint, T., Mathieu, C., Zhou, H.: Optimization of bootstrapping in circuits. In: ACM-SIAM, pp. 2423–2433 (2017)
4. Biasse, J.-F., Ruiz, L.: FHEW with efficient multibit bootstrapping. In: Lauter, K., Rodríguez-Henríquez, F. (eds.) LATINCRYPT 2015. LNCS, vol. 9230, pp. 119–135. Springer, Cham (2015). https://doi.org/10.1007/978-3-319-22174-8_7
5. Brakerski, Z., Gentry, C., Vaikuntanathan, V.: (leveled) fully homomorphic encryption without bootstrapping. In: ITCS, pp. 309–325 (2012)
6. Brakerski, Z., Langlois, A., Peikert, C., Regev, O., Stehlé, D.: Classical hardness of learning with errors. In: Proceedings of 45th STOC, pp. 575–584. ACM (2013)

7. Brakerski, Z., Perlman, R.: Lattice-based fully dynamic multi-key FHE with short ciphertexts. In: Robshaw, M., Katz, J. (eds.) CRYPTO 2016. LNCS, vol. 9814, pp. 190–213. Springer, Heidelberg (2016). https://doi.org/10.1007/978-3-662-53018-4_8
8. Brakerski, Z., Vaikuntanathan, V.: Lattice-based FHE as secure as PKE. In: ITCS, pp. 1–12 (2014)
9. Buchsbaum, A.L., Giancarlo, R., Westbrook, J.R.: On the determinization of weighted finite automata. SIAM J. Comput. **30**(5), 1502–1531 (2000)
10. Cheon, J.H., Coron, J.-S., Kim, J., Lee, M.S., Lepoint, T., Tibouchi, M., Yun, A.: Batch fully homomorphic encryption over the integers. In: Johansson, T., Nguyen, P.Q. (eds.) EUROCRYPT 2013. LNCS, vol. 7881, pp. 315–335. Springer, Heidelberg (2013). https://doi.org/10.1007/978-3-642-38348-9_20
11. Cheon, J.H., Stehlé, D.: Fully homomophic encryption over the integers revisited. In: Oswald, E., Fischlin, M. (eds.) EUROCRYPT 2015. LNCS, vol. 9056, pp. 513–536. Springer, Heidelberg (2015). https://doi.org/10.1007/978-3-662-46800-5_20
12. Chillotti, I., Gama, N., Georgieva, M., Izabachène, M.: Faster fully homomorphic encryption: bootstrapping in less than 0.1 seconds. In: Cheon, J.H., Takagi, T. (eds.) ASIACRYPT 2016. LNCS, vol. 10031, pp. 3–33. Springer, Heidelberg (2016). https://doi.org/10.1007/978-3-662-53887-6_1
13. Chillotti, I., Gama, N., Georgieva, M., Izabachène, M.: A homomorphic LWE based e-voting scheme. In: Takagi, T. (ed.) PQCrypto 2016. LNCS, vol. 9606, pp. 245–265. Springer, Cham (2016). https://doi.org/10.1007/978-3-319-29360-8_16
14. Chillotti, I., Gama, N., Georgieva, M., Izabachène, M.: TFHE: fast fully homomorphic encryption library, August 2016. https://tfhe.github.io/tfhe/
15. Coron, J.-S., Lepoint, T., Tibouchi, M.: Scale-invariant fully homomorphic encryption over the integers. In: Krawczyk, H. (ed.) PKC 2014. LNCS, vol. 8383, pp. 311–328. Springer, Heidelberg (2014). https://doi.org/10.1007/978-3-642-54631-0_18
16. Droste, M., Gastin, P.: Weighted automata and weighted logics. In: Droste, M., Kuich, W., Vogler, H. (eds.) Handbook of Weighted Automata. Monographs in Theoretical Computer Science. An EATCS Series, pp. 175–211. Springer, Heidelberg (2009). https://doi.org/10.1007/978-3-642-01492-5_5
17. Ducas, L., Micciancio, D.: FHEW: bootstrapping homomorphic encryption in less than a second. In: Oswald, E., Fischlin, M. (eds.) EUROCRYPT 2015. LNCS, vol. 9056, pp. 617–640. Springer, Heidelberg (2015). https://doi.org/10.1007/978-3-662-46800-5_24
18. Fan, J., Vercauteren, F.: Somewhat practical fully homomorphic encryption (2012). https://eprint.iacr.org/2012/144
19. Gama, N., Izabachène, M., Nguyen, P.Q., Xie, X.: Structural lattice reduction: generalized worst-case to average-case reductions. In: EUROCRYPT 2016, ePrint Archive, 2014/283
20. Gentry, C.: Fully homomorphic encryption using ideal lattices. In: STOC (2009)
21. Gentry, C., Sahai, A., Waters, B.: Homomorphic encryption from learning with errors: conceptually-simpler, asymptotically-faster, attribute-based. In: Canetti, R., Garay, J.A. (eds.) CRYPTO 2013. LNCS, vol. 8042, pp. 75–92. Springer, Heidelberg (2013). https://doi.org/10.1007/978-3-642-40041-4_5
22. Halevi, S., Shoup, I.V.: HElib - an implementation of homomorphic encryption, September 2014. https://github.com/shaih/HElib/
23. Halevi, S., Shoup, V.: Algorithms in HElib. In: Garay, J.A., Gennaro, R. (eds.) CRYPTO 2014. LNCS, vol. 8616, pp. 554–571. Springer, Heidelberg (2014). https://doi.org/10.1007/978-3-662-44371-2_31

24. Lepoint, T.: FV-NFLlib: library implementing the Fan-Vercauteren homomorphic encryption scheme, May 2016. https://github.com/CryptoExperts/FV-NFLlib
25. Lyubashevsky, V., Peikert, C., Regev, O.: On ideal lattices and learning with errors over rings. In: Gilbert, H. (ed.) EUROCRYPT 2010. LNCS, vol. 6110, pp. 1–23. Springer, Heidelberg (2010). https://doi.org/10.1007/978-3-642-13190-5_1
26. Paindavoine, M., Vialla, B.: Minimizing the number of bootstrappings in fully homomorphic encryption. In: Dunkelman, O., Keliher, L. (eds.) SAC 2015. LNCS, vol. 9566, pp. 25–43. Springer, Cham (2016). https://doi.org/10.1007/978-3-319-31301-6_2
27. Hiromasa, R., Abe, M., Okamoto, T.: Packing messages and optimizing bootstrapping in GSW-FHE. In: Katz, J. (ed.) PKC 2015. LNCS, vol. 9020, pp. 699–715. Springer, Heidelberg (2015). https://doi.org/10.1007/978-3-662-46447-2_31
28. Regev, O.: On lattices, learning with errors, random linear codes, and cryptography. In: STOC, pp. 84–93 (2005)
29. SEAL. Simple encrypted arithmetic library. https://sealcrypto.codeplex.com/
30. van Dijk, M., Gentry, C., Halevi, S., Vaikuntanathan, V.: Fully homomorphic encryption over the integers. In: Gilbert, H. (ed.) EUROCRYPT 2010. LNCS, vol. 6110, pp. 24–43. Springer, Heidelberg (2010). https://doi.org/10.1007/978-3-642-13190-5_2

Homomorphic Encryption for Arithmetic of Approximate Numbers

Jung Hee Cheon[1(✉)], Andrey Kim[1], Miran Kim[2], and Yongsoo Song[1]

[1] Seoul National University, Seoul, Republic of Korea
{jhcheon,kimandrik,lucius05}@snu.ac.kr
[2] University of California, San Diego, USA
mrkim@ucsd.edu

Abstract. We suggest a method to construct a homomorphic encryption scheme for approximate arithmetic. It supports an approximate addition and multiplication of encrypted messages, together with a new *rescaling* procedure for managing the magnitude of plaintext. This procedure truncates a ciphertext into a smaller modulus, which leads to rounding of plaintext. The main idea is to add a noise following significant figures which contain a main message. This noise is originally added to the plaintext for security, but considered to be a part of error occurring during approximate computations that is reduced along with plaintext by rescaling. As a result, our decryption structure outputs an approximate value of plaintext with a predetermined precision.

We also propose a new batching technique for a RLWE-based construction. A plaintext polynomial is an element of a cyclotomic ring of characteristic zero and it is mapped to a message vector of complex numbers via complex canonical embedding map, which is an isometric ring homomorphism. This transformation does not blow up the size of errors, therefore enables us to preserve the precision of plaintext after encoding. In our construction, the bit size of ciphertext modulus grows linearly with the depth of the circuit being evaluated due to rescaling procedure, while all the previous works either require an exponentially large size of modulus or expensive computations such as bootstrapping or bit extraction. One important feature of our method is that the precision loss during evaluation is bounded by the depth of a circuit and it exceeds at most one more bit compared to unencrypted approximate arithmetic such as floating-point operations. In addition to the basic approximate circuits, we show that our scheme can be applied to the efficient evaluation of transcendental functions such as multiplicative inverse, exponential function, logistic function and discrete Fourier transform.

Keywords: Homomorphic encryption · Approximate arithmetic

1 Introduction

Homomorphic encryption (HE) is a cryptographic scheme that enables homomorphic operations on encrypted data without decryption. Many of HE schemes

T. Takagi and T. Peyrin (Eds.): ASIACRYPT 2017, Part I, LNCS 10624, pp. 409–437, 2017.
https://doi.org/10.1007/978-3-319-70694-8_15

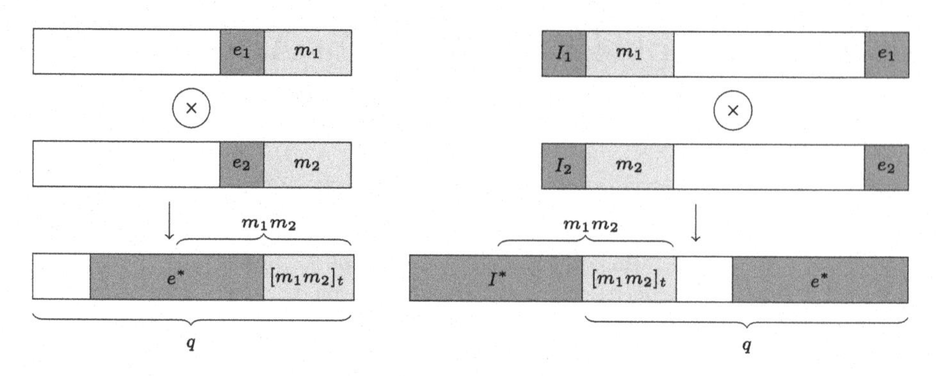

Fig. 1. Homomorphic multiplications of BGV-type HE schemes (left) and FV-type HE schemes (right)

(e.g. [2,4–7,12,13,18,19,21,25,26,33]) have been suggested following Gentry's blueprint [23]. HE can be applied to the evaluation of various algorithms on encrypted financial, medical, or genomic data [11,29,31,36,41].

Most of real-world data contain some errors from their true values. For instance, a measured value of quantity has an observational error from its true value and sampling error can be made as only a sample of the whole population is being observed in statistics. In practice, data should be discretized (quantized) to an approximate value such as floating-point number, in order to be represented by a finite number of bits in computer systems. In this case, an approximate value may substitute the original data and a small rounding error does not have too much effect on computation result. For the efficiency of approximate arithmetic, we store a few numbers of significant digits (e.g. most significant bits, MSBs) and carry out arithmetic operations between them. The resulting value should be rounded again by removing some inaccurate least significant bits (LSBs) to maintain the bit size of significant (mantissa).

Unfortunately this rounding operation has been considered difficult to perform on HE since it is not simply represented as a small-degree polynomial. Previous approaches to approximate arithmetic require similar multiplicative depth and complexity to the case of bootstrapping for extraction of MSBs [1,27]. Other methods based on exact integer operations [16,20] require an exponentially large bit size of ciphertext modulus with the depth of the circuit to ensure correctness.

We point out that the decryption structures of existing HE schemes are not appropriate for arithmetic of indiscreet spaces. For a plaintext modulus t and a ciphertext modulus q, BGV-type HE schemes [5,19,25,33] have a decryption structure of the form $\langle \mathbf{c}_i, sk\rangle = m_i + te_i \pmod q$. Therefore, the MSBs of $m_1 + m_2$ and m_1m_2 are destroyed by inserted errors e_i during homomorphic operations. On the other hand, the decryption structure of FV-type HE schemes [2,4,22] is $\langle \mathbf{c}_i, sk\rangle = qI_i + (q/t)m_i + e_i$ for some I_i and e_i. Multiplication of two ciphertexts satisfies $\langle \mathbf{c}^*, sk\rangle = qI^* + (q/t)m_1m_2 + e^*$ for $I^* = tI_1I_2 + I_1m_2 + I_2m_1$ and $e^* \approx t(I_1e_2 + I_2e_1)$, so the MSBs of resulting message are also destroyed (see Fig. 1 for an illustration). HE schemes with matrix ciphertexts [21,26] support homomorphic operations over the integers

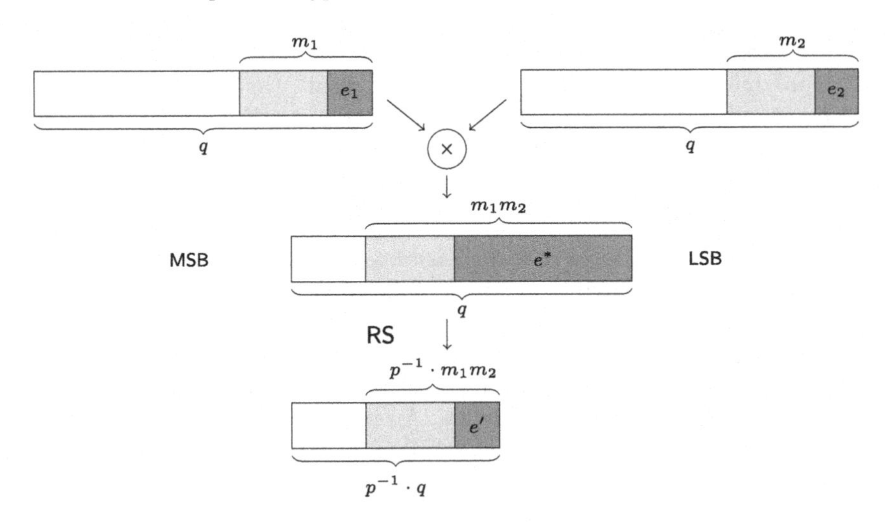

Fig. 2. Homomorphic multiplication and rescaling for approximate arithmetic

(or integral polynomials) but the error growth depends on the size of plaintexts. As a result, previous HE schemes are required to have an exponentially large ciphertext modulus with the depth of a circuit for approximate arithmetic.

Homomorphic Encryption for Approximate Arithmetic. The purpose of this paper is to present a method for efficient approximate computation on HE. The main idea is to treat an encryption noise as part of error occurring during approximate computations. That is, an encryption $\boldsymbol{c}$ of message m by the secret key sk will have a decryption structure of the form $\langle \boldsymbol{c}, sk \rangle = m + e \pmod{q}$ where e is a small error inserted to guarantee the security of hardness assumptions such as the learning with errors (LWE), the ring- LWE (RLWE) and the NTRU problems. If e is small enough compared to the message, this noise is not likely to destroy the significant figures of m and the whole value $m' = m + e$ can replace the original message in approximate arithmetic. One may multiply a scale factor to the message before encryption to reduce the precision loss from encryption noise.

For homomorphic operations, we always maintain our decryption structure small enough compared to the ciphertext modulus so that computation result is still smaller than q. However, we still have a problem that the bit size of message increases exponentially with the depth of a circuit without rounding. To address this problem, we suggest a new technique - called *rescaling* - that manipulates the message of ciphertext. Technically it seems similar to the modulus-switching method suggested by Brakerski and Vaikuntanatan [6], but it plays a completely different role in our construction. For an encryption $\boldsymbol{c}$ of m such that $\langle \boldsymbol{c}, sk \rangle = m + e \pmod{q}$, the rescaling procedure outputs a ciphertext $\lfloor p^{-1} \cdot \boldsymbol{c} \rceil \pmod{q/p}$, which is a valid encryption of m/p with noise about e/p. It reduces the size of ciphertext modulus and consequently removes the error located in the LSBs of messages, similar to the rounding step of fixed/floating-point arithmetic, while almost preserving the precision of plaintexts.

The composition of homomorphic operation and rescaling mimics the ordinary approximate arithmetic (see Fig. 2). As a result, the bit size of a required ciphertext modulus grows linearly with the depth of a circuit rather than exponentially. We also prove that this scheme is almost optimal in the sense of precision: precision loss of a resulting message is at most one bit more compared to unencrypted floating-point arithmetic.

Encoding Technique for Packing Messages. It is inevitable to encrypt a vector of multiple plaintexts in a single ciphertext for efficient homomorphic computation. The plaintext space of previous RLWE-based HE schemes is a cyclotomic polynomial ring $\mathbb{Z}_t[X]/(\Phi_M(X))$ of a finite characteristic. A plaintext polynomial could be decoded as a vector of plaintext values into a product of finite fields by a ring isomorphism [38,39]. An inserted error is placed separately from the plaintext space so it may be removed by using plaintext characteristic after carrying out homomorphic operations.

On the other hand, a plaintext of our scheme is an element of a cyclotomic ring of characteristic zero and it embraces a small error which is inserted from encryption to ensure the security or occurs during approximate arithmetic. Hence we adapt an *isometric ring homomorphism* - the complex canonical embedding map. It preserves the size of polynomials so that a small error in a plaintext polynomial is not blow up during encoding/decoding procedures.

Let $\mathbb{H} = \{(z_j)_{j\in\mathbb{Z}_M^*} : z_{-j} = \overline{z_j}, \forall j \in \mathbb{Z}_M^*\} \subseteq \mathbb{C}^{\Phi(M)}$ and let T be a subgroup of the multiplicative group $\mathbb{Z}_M^*$ satisfying $\mathbb{Z}_M^*/T = \{\pm 1\}$. The native plaintext space of our scheme is the set of polynomials in the cyclotomic ring $\mathcal{R} = \mathbb{Z}[X]/(\Phi_M(X))$ with magnitude bounded by ciphertext modulus. The decoding procedure first transforms a plaintext polynomial $m(X) \in \mathcal{R}$ into a complex vector $(z_j)_{j\in\mathbb{Z}_M^*} \in \mathbb{H}$ by the canonical embedding map σ and then sends it to a vector $(z_j)_{j\in T}$ using the natural projection π: $\mathbb{H} \to \mathbb{C}^{\phi(M)/2}$. The encoding method is almost the inverse of the decoding procedure, but a round-off algorithm is required for discretization so that the output becomes an integral polynomial. In short, our encoding function is given by

$$\begin{array}{ccccccc}
\mathbb{C}^{\phi(M)/2} & \xrightarrow{\pi^{-1}} & \mathbb{H} & \xrightarrow{\lfloor\cdot\rceil_{\sigma(\mathcal{R})}} & \sigma(\mathcal{R}) & \xrightarrow{\sigma^{-1}} & \mathcal{R} \\
\boldsymbol{z} = (z_i)_{i\in T} & \longmapsto & \pi^{-1}(\boldsymbol{z}) & \longmapsto & \left\lfloor \pi^{-1}(\boldsymbol{z}) \right\rceil_{\sigma(\mathcal{R})} & \longmapsto & \sigma^{-1}\left(\left\lfloor \pi^{-1}(\boldsymbol{z}) \right\rceil_{\sigma(\mathcal{R})} \right)
\end{array}$$

where $\lfloor\cdot\rceil_{\sigma(\mathcal{R})}$ denotes the rounding to a close element in $\sigma(\mathcal{R})$.

Homomorphic Evaluation of Approximate Arithmetic. One important feature of our method is that the precision loss during homomorphic evaluation is bounded by depth of a circuit and it is at most one more bit compared to unencrypted approximate arithmetic. Given encryptions of d messages with η bits of precision, our HE scheme of depth $\lceil \log d \rceil$ computes their product with $(\eta - \log d - 1)$ bits of precision in d multiplications while unencrypted approximate arithmetic such as floating-point multiplication can compute a significant with $(\eta - \log d)$ bits of precision. On the other hand, the previous methods require $\Omega(\eta^2 d)$ homomorphic computations by using bitwise encryption or need a large plaintext space of bit size $\Omega(\eta d)$ unless relying on expensive computations such as bootstrapping or bit extraction.

In our scheme, the required bit size of the largest ciphertext modulus can be reduced down to $O(\eta \log d)$ by performing the rescaling procedure after multiplication of ciphertexts. The parameters are smaller than for the previous works and this advantage enables us to efficiently perform the approximate evaluation of *transcendental* functions such as the exponential, logarithm and trigonometric functions by the evaluation of their Taylor series expansion. In particular, we suggest a specific algorithm for computing the multiplicative inverse with reduced complexity, which enables the efficient evaluation of rational functions.

We verify our algorithms by implementation on a machine with an Intel Core i5 running at 2.9 GHz processor using a parameter set with 80-bit security level. It takes about 0.45 s for multiplicative inverse of ciphertext with 14 bits of precision, yielding an amortized rate of 0.11 ms per slot. We can also evaluate the exponential function using its Taylor expansion and it results in an amortized time per slots of 0.16 ms.

In a cloud-computing environment, a large amount of data is being generated and one needs to handle these huge data collections. Our scheme could be a practical solution for data analysis as it allows the encryption of much information in a single ciphertext so we can parallelize both space and computation together. For example, we improved the homomorphic evaluation of logistic function using a batching technique, which can be used in a disease prediction analysis. Our implementation homomorphically evaluated the degree seven Taylor polynomial of logistic function in about 0.13 ms per slot (and less than 0.54 s total) compared to 30 s and 1.8 s of evaluation time of [3,9] without parallelization, respectively.

Another example is evaluating discrete Fourier transform homomorphically using a *fast Fourier transform*(FFT) algorithm. We follow the encoding method of [15] for roots of unity in polynomial ring so that it does not consume ciphertext level during evaluation. We also apply our rescaling procedure for operations to Hadamard space and a batching technique, which results in a much smaller parameter and amortized evaluation time, respectively. We could process the standard processing (FFT-Hadamard product of two vectors-inverse FFT) of dimension 2^{13} in 22 min (0.34 s per slot) on a machine with four cores compared to 17 min of previous work [16] with six processors with no batching technique. Based on evaluation of discrete Fourier transform, we can securely compute the exact multiplication of integral polynomials by removing the fractional part of an approximate result. Likewise, our HE for approximate arithmetic can be applied to exact computation when the result has a specific format or property.

Follow-up. We provide an open-source implementation of our HE library (`HEAAN`) and algorithms in the C++ language. The source code is available at `github` [10]. We introduced `HEAAN` at a workshop for the standardization of HE hosted by Microsoft Research.[1]

[1] https://www.microsoft.com/en-us/research/event/homomorphic-encryption-standardization-workshop/.

There are some follow-up works on application of this paper to a secure control of cyber-physical system [28] and a gradient descent algorithm for privacy-preserving logistic regression of biomedical data [30].

Related Works. A substantial number of studies have concerned about the processing of real numbers over encryption. Jäschke and Armknecht [27] observed that a rational number can be approximated to an integer by multiplying with a power of two and rounding. An integer is encoded in a binary fashion, so that each bit is encrypted separately. The one performing homomorphic multiplication can bring the product to the required precision by simply discarding the ciphertexts which corresponds to the last LSBs. However, bitwise encryption causes a huge number of computation of ciphertexts for a single rounding operation. The other method is to scale them to integers, but a plaintext modulus is exponential in the length of message. For example, Arita and Nakasato [1] scale the fixed point numbers by a power of two and then represent them as scalars in a polynomial ring with an enlarged plaintext modulus. In order to realize homomorphic multiplication of encrypted fixed point numbers, it needs a right shift by a number equal to the precision. However, it requires a considerable amount of computations including a bit extraction operation.

On the other hand, Dowlin et al. [20] present an efficient method to represent fixed-point numbers, which are encoded as integral polynomials with coefficients in the range $(-\frac{1}{2}B, \frac{1}{2}B)$ using its base-B representation for an odd integer $B \geq 3$. Costache et al. [16] analyze the representations of [20] and compute the lower bound of plaintext modulus. However, exact arithmetic of fixed point numbers causes required the size of plaintext modulus to grow exponentially with the depth of a circuit.

Road-map. Section 2 briefly introduces notations and some preliminaries about algebras and the RLWE problem. Section 3 presents a homomorphic encryption scheme for approximate arithmetic and analyzes the noise growth during basic homomorphic operations. In Sect. 4, we suggest some algorithms to homomorphically evaluate typical approximate circuits, multiplicative inverse, exponential function, logistic function and discrete Fourier transform. We also compute the theoretical precision of the outputs. In Sect. 5, we perform the implementation of our scheme for the evaluations of circuits described in Sect. 4.

2 Preliminaries

2.1 Basic Notation

All logarithms are base 2 unless otherwise indicated. We denote vectors in bold, e.g. $\mathbf{a}$, and every vector in this paper will be a column vector. We denote by $\langle \cdot, \cdot \rangle$ the usual dot product of two vectors. For a real number r, $\lfloor r \rceil$ denotes the nearest integer to r, rounding upwards in case of a tie. For an integer q, we identify $\mathbb{Z} \cap (-q/2, q/2]$ as a representative of $\mathbb{Z}_q$ and use $[z]_q$ to denote the reduction of the integer z modulo q into that interval. We use $x \leftarrow D$ to denote the sampling x according to a distribution D. It denotes the sampling from the uniform distribution over D when D is a finite set. We let λ denote the security parameter throughout the paper: all known valid attacks against the cryptographic scheme under scope should take $\Omega(2^\lambda)$ bit operations.

2.2 The Cyclotomic Ring and Canonical Embedding

For a positive integer M, let $\Phi_M(X)$ be the M-th cyclotomic polynomial of degree $N = \phi(M)$. Let $\mathcal{R} = \mathbb{Z}[X]/(\Phi_M(X))$ be the ring of integers of a number field $\mathbb{Q}[X]/(\Phi_M(X))$. We write $\mathcal{R}_q = \mathcal{R}/q\mathcal{R}$ for the residue ring of $\mathcal{R}$ modulo an integer q. An arbitrary element of the cyclotomic ring $\mathcal{S} = \mathbb{R}[X]/(\Phi_M(X))$ of *real* polynomials will be represented as a polynomial $a(X) = \sum_{j=0}^{N-1} a_j X^j$ of degree less than N and identified with its coefficient vector $(a_0, \ldots, a_{N-1}) \in \mathbb{R}^N$. We define the relevant norms on the coefficient vector of a such as $\|a\|_\infty$ and $\|a\|_1$.

We write $\mathbb{Z}_M^* = \{x \in \mathbb{Z}_M : \gcd(x, M) = 1\}$ for the multiplicative group of units in $\mathbb{Z}_M$. Recall that the canonical embedding of $a \in \mathbb{Q}[X]/(\Phi_M(X))$ into $\mathbb{C}^N$ is the vector of evaluation values of a at the roots of $\Phi_M(X)$. We naturally extend it to the set of real polynomials $\mathcal{S}$ so $\sigma(a)$ will be defined as $(a(\zeta_M^j))_{j \in \mathbb{Z}_M^*} \in \mathbb{C}^N$ for any $a \in \mathcal{S}$ where $\zeta_M = \exp(-2\pi i/M)$ denotes a primitive M-th roots of unity. The ℓ_∞-norm of $\sigma(a)$ is called the *canonical embedding norm* of a, denoted by $\|a\|_\infty^{\mathsf{can}} = \|\sigma(a)\|_\infty$. This measurement will be used to analyze the size of polynomials throughout this paper. The canonical embedding norm $\|\cdot\|_\infty^{\mathsf{can}}$ satisfies the following properties:

- For all $a, b \in \mathcal{S}$, we have $\|a \cdot b\|_\infty^{\mathsf{can}} \leq \|a\|_\infty^{\mathsf{can}} \cdot \|b\|_\infty^{\mathsf{can}}$.
- For all $a \in \mathcal{S}$, we have $\|a\|_\infty^{\mathsf{can}} \leq \|a\|_1$.
- There is a ring constant c_M depending only on M such that $\|a\|_\infty \leq c_M \cdot \|a\|_\infty^{\mathsf{can}}$ for all $a \in \mathcal{S}$.

The ring constant is obtained by $c_M = \|\mathsf{CRT}_M^{-1}\|_\infty$ where CRT_M is the CRT matrix for M, i.e., the Vandermonde matrix over the complex primitive M-th roots of unity, and the norm for a matrix $U = (u_{ij})_{0 \leq i,j < N}$ is defined by $\|U\|_\infty = \max_{0 \leq i < N} \left\{ \sum_{j=0}^{N-1} |u_{ij}| \right\}$. Refer [17] for a discussion of c_M.

2.3 Gaussian Distributions and RLWE Problem

We first define the space

$$\mathbb{H} = \{\boldsymbol{z} = (z_j)_{j \in \mathbb{Z}_M^*} \in \mathbb{C}^N : z_j = \overline{z_{-j}}, \forall j \in \mathbb{Z}_M^*\},$$

which is isomorphic to $\mathbb{R}^N$ as an inner product space via the unitary basis matrix

$$U = \begin{pmatrix} \frac{1}{\sqrt{2}} I & \frac{i}{\sqrt{2}} J \\ \frac{1}{\sqrt{2}} J & \frac{-i}{\sqrt{2}} I \end{pmatrix}$$

where I is the identity matrix of size $N/2$ and J is its reversal matrix.

For $r > 0$, we define the Gaussian function $\rho_r\colon \mathbb{H} \to (0, 1]$ as $\rho_r(\boldsymbol{z}) = \exp(-\pi\|\boldsymbol{z}\|_2^2/r^2)$. Denote by Γ_r the continuous Gaussian probability distribution whose density is given by $r^{-N} \cdot \rho_r(\boldsymbol{z})$. Now one can extend this to an elliptical Gaussian distribution $\Gamma_{\mathbf{r}}$ on $\mathbb{H}$ as follows: let $\mathbf{r} = (r_1, \ldots, r_N) \in (\mathbb{R}^+)^N$ be a vector of positive real numbers, then a sample from $\Gamma_{\mathbf{r}}$ is given by $U \cdot \boldsymbol{z}$ where each entry of $\boldsymbol{z} = (z_i)$ is chosen independently from the (one-dimensional) Gaussian distribution Γ_{r_i} on $\mathbb{R}$. This also gives a distribution $\Psi_{\mathbf{r}}$ on $\mathbb{Q}[X]/(\Phi_M(X)) \otimes \mathbb{R}$. That is, $\mathsf{CRT}_M^{-1} \cdot U \cdot \boldsymbol{z}$ gives us the coordinates with respect to the polynomial basis $1, X, X^2, \ldots, X^{N-1}$.

In practice, one can discritize the continuous Gaussian distribution $\Psi_{\mathbf{r}}$ by taking a valid rounding $\lfloor \Psi_{\mathbf{r}} \rceil_{\mathcal{R}^\vee}$. Refer [34,35] for explaining the methods in more details. We use this discrete distribution as the RLWE error distribution.

Here we define the RLWE distribution and decisional problem associated with it. Let $\mathcal{R}^\vee$ be the dual fractional ideal of $\mathcal{R}$ and write $\mathcal{R}_q^\vee = \mathcal{R}^\vee/q\mathcal{R}^\vee$. For a positive integer modulus $q \geq 2$, $s \in \mathcal{R}_q^\vee$, $\mathbf{r} \in (\mathbb{R}^+)^N$ and an error distribution $\chi := \lfloor \Psi_{\mathbf{r}} \rceil_{\mathcal{R}^\vee}$, we define $A_{N,q,\chi}(s)$ as the RLWE distribution obtained by sampling $a \leftarrow \mathcal{R}_q$ uniformly at random, $e \leftarrow \chi$ and returning $(a, a \cdot s + e) \in \mathcal{R}_q \times \mathcal{R}_q^\vee$.

The (decision) ring learning with errors, denoted by $\mathsf{RLWE}_{N,q,\chi}(\mathcal{D})$, is a problem to distinguish arbitrarily many independent samples chosen according to $A_{N,q,\chi}(s)$ for a random choice of s sampled from the distribution $\mathcal{D}$ over $\mathcal{R}^\vee$ from the same number of uniformly random and independent samples from $\mathcal{R}_q \times \mathcal{R}_q^\vee$.

3 Homomorphic Encryption for Approximate Arithmetic

In this section, we describe a method to construct a HE scheme for approximate arithmetic on encrypted data. Given encryptions of m_1 and m_2, this scheme allows us to securely compute encryptions of approximate values of $m_1 + m_2$ and $m_1 m_2$ with a predetermined precision. The main idea of our construction is to treat an inserted noise of RLWE problem as part of am error occurring during approximate computation. The most important feature of our scheme is the *rounding* operation of plaintexts. Just like the ordinary approximate computations using floating-point numbers, the rounding operation removes some LSBs of message and makes a trade-off between size of numbers and precision loss.

Our concrete construction is based on the BGV scheme [5] with a multiplication method by raising the ciphertext modulus [25], but our methodology can be applied to most of existing HE schemes. Appendix A shows a description of LWE-based HE scheme for approximate arithmetic.

3.1 Decryption Structure of Homomorphic Encryption for Approximate Arithmetic

Most of existing HE schemes perform operations on a modulo space such as $\mathbb{Z}_t$ and $\mathbb{Z}_t[X]/(\Phi_M(X))$. In other words, they aim to compute a ciphertext which encrypts some LSBs of a resulting message after homomorphic computation. For example, in the case of BGV-type schemes [5,25,33], plaintexts are placed in the lowest bits of ciphertext modulus, that is, an encryption $\boldsymbol{c}$ of a message m with respect to a secret sk has a decryption structure of the form $\langle \boldsymbol{c}, sk \rangle = m + te \pmod{q}$. A multiplication of encryptions of m_1, m_2 preserves some LSBs of $m_1 m_2$ (i.e., $[m_1 m_2]_t$), while its MSBs (i.e., $\lfloor m_1 m_2 / t \rceil$) are destroyed by errors. On the other hand, FV-type schemes [2,4,22] put messages in the left-most bits of ciphertext modulus, so that their decryption structures satisfy $\langle \boldsymbol{c}, sk \rangle = \lfloor q/t \rceil \cdot m + e \pmod{q}$. However, the MSBs of the resulting message are also destroyed during homomorphic multiplication between $\langle \boldsymbol{c}_i, sk \rangle = q \cdot I_i + \lfloor q/t \rceil \cdot m_i + e_i$, each of which contains an additional error I_i in the left position of message.

Our goal is to carry out approximate arithmetic over encrypted data, or equivalently, compute the MSBs of a resulting message after homomorphic operations. The main idea is to add an encryption noise following significant figures of an input message. More precisely, our scheme has a decryption structure of the form $\langle \boldsymbol{c}, sk \rangle = m + e \pmod{q}$ for some small error e. We insert this encryption error to guarantee the security of scheme, but it will be considered as an error that arises during approximate computations. That is, the output of decryption algorithm will be treated as an approximate value of the original message with a high precision. The size of a plaintext will be small enough compared to the ciphertext modulus for homomorphic operations so that the result of an arithmetic computation is still smaller than the ciphertext modulus.

There are some issues that we need to consider more carefully. In unencrypted approximate computations, small errors may blow up when applying operations in succession, so it is valuable to consider the proximity of a calculated result to the exact value of an algorithm. Similarly, encrypted plaintexts in our scheme will contain some errors and they might be increased during homomorphic evaluations. Thus we compute an upper bound of errors and predict the precision of resulting values.

The management of the size of messages is another issue. If we compute a circuit of multiplicative depth L without rounding of messages, then the bit size of an output value will exponentially grow with L. This naive method is inappropriate for practical usage because it causes a huge ciphertext modulus. To resolve this problem, we suggest a new technique which divides intermediate values by a base. It allows us to discard some inaccurate LSBs of a message while an error is still kept relatively small compared to the message. This method leads to maintain the size of messages almost same and make the required ciphertext modulus linear in the depth L.

3.2 Plaintext Encoding for Packing

The batching technique in HE system allows us to encrypt multiple messages in a single ciphertext and enables a parallel processing in SIMD manner. In practice, we take its advantage to parallelize computations and reduce the memory and complexity. A ring of finite characteristic has been used as a plaintext space in the previous RLWE-based HE schemes. A small error, which is located in a separated place in a ciphertext modulus, is inserted to ensure security and it may be removed after carrying out homomorphic operations. Then an output polynomial is decoded into a message vector with respect to the CRT-based encoding technique [38,39]. Meanwhile, a plaintext of our scheme is a polynomial contained in a ring of characteristic zero and it embraces am error for security, so an inserted error cannot be removed after decryption.

Intuition. A native plaintext space of our RLWE-based construction can be understood as the set of polynomials $m(X) \in \mathcal{S}$ such that $\|m\|_\infty^{\mathsf{can}} \ll q$. The roots of a cyclotomic polynomial $\Phi_M(X)$ are the complex primitive roots of unity in the extension field $\mathbb{C}$. We evaluate a plaintext polynomial at these roots in order to transform it into a vector of complex numbers, so the (extended) canonical embedding map σ: $\mathcal{S} \to \mathbb{C}^N$ plays a role of decoding algorithm.

For technical details, we first point out that the image of canonical embedding map is the subring $\mathbb{H} = \{(z_j)_{j\in\mathbb{Z}_M^*} : z_j = \overline{z_{-j}}\}$ of $\mathbb{C}^N$. Let T be a multiplicative subgroup of $\mathbb{Z}_M^*$ satisfying $\mathbb{Z}_M^*/T = \{\pm 1\}$. Then $\mathbb{H}$ can be identified with $\mathbb{C}^{N/2}$ via the natural projection π, defined by $(z_j)_{j\in\mathbb{Z}_M^*} \mapsto (z_j)_{j\in T}$. Then our decoding algorithm is to transform an arbitrary polynomial $m(X) \in \mathcal{R}$ into a complex vector $\boldsymbol{z}$ such that $\boldsymbol{z} = \pi \circ \sigma(m) \in \mathbb{C}^{N/2}$.

The encoding algorithm is defined as the inverse of decoding procedure. Specifically, it encodes an input vector $\boldsymbol{z} = (z_i)_{i\in T}$ in a polynomial $m(X) = \sigma^{-1} \circ \pi^{-1}(\boldsymbol{z})$ where $\pi^{-1}(\boldsymbol{z})[j]$ is z_j if $j \in T$, and $\overline{z_{-j}}$ otherwise. Note that the encoding/decoding algorithms are isometric ring isomorphisms between $(\mathcal{S}, \|\cdot\|_\infty^{\mathsf{can}})$ and $(\mathbb{C}^{N/2}, \|\cdot\|_\infty)$, so the size of plaintexts and errors are preserved via these transformations.

Since $\pi^{-1}(\boldsymbol{z})$ might not be contained in the image of canonical embedding map, we need to discritize $\pi^{-1}(\boldsymbol{z})$ to an element of $\sigma(\mathcal{R})$. Recall that $\mathcal{R}$ has a $\mathbb{Z}$-basis $\{1, X, \ldots, X^{N-1}\}$ and it yields a rank-N ideal lattice $\sigma(\mathcal{R})$ having basis $\{\sigma(1), \sigma(X), \ldots, \sigma(X^{N-1})\}$. The goal of rounding process is to find a vector, denoted by $\lfloor \pi^{-1}(\boldsymbol{z}) \rceil_{\sigma(\mathcal{R})}$, with a rounding error $\|\pi^{-1}(\boldsymbol{z}) - \lfloor \pi^{-1}(\boldsymbol{z}) \rceil_{\sigma(\mathcal{R})}\|_\infty$. There are several round-off algorithms including the coordinate-wise randomized rounding. See [35] for details.

A rounding error may destroy the significant figures of a message during encoding procedure. Hence we recommend to multiply a scaling factor $\Delta \geq 1$ to a plaintext before rounding in order to preserve its precision. Our encoding/decoding algorithms are explicitly given as follows:

- $\mathsf{Ecd}(\boldsymbol{z}; \Delta)$. For a $(N/2)$-dimensional vector $\boldsymbol{z} = (z_i)_{i\in T}$ of complex numbers, the encoding procedure first expands it into the vector $\pi^{-1}(\boldsymbol{z}) \in \mathbb{H}$ and computes its discretization to $\sigma(\mathcal{R})$ after multiplying a scaling factor Δ. Return the corresponding integral polynomial $m(X) = \sigma^{-1}(\lfloor \Delta \cdot \pi^{-1}(\boldsymbol{z}) \rceil_{\sigma(\mathcal{R})}) \in \mathcal{R}$.
- $\mathsf{Dcd}(m; \Delta)$. For an input polynomial $m \in \mathcal{R}$, output the vector $\boldsymbol{z} = \pi \circ \sigma(\Delta^{-1} \cdot m)$, i.e., the entry of $\boldsymbol{z}$ of index $j \in T$ is $z_j = \Delta^{-1} \cdot m(\zeta_M^j)$.

As a toy example, let $M = 8$ (i.e., $\Phi_8(X) = X^4 + 1$) and $\Delta = 64$. Let $T = \{\zeta_8, \zeta_8^3\}$ for the root of unity $\zeta_8 = \exp(-2\pi i/8)$. For a given vector $\boldsymbol{z} = (3+4i, 2-i)$, the corresponding real polynomial $\frac{1}{4}(10+4\sqrt{2}X+10X^2+2\sqrt{2}X^3)$ has evaluation values $3+4i$ and $2-i$ at ζ_8 and ζ_8^3, respectively. Then the output of encoding algorithm is $m(X) = 160+91X+160X^2+45X^3 \leftarrow \mathsf{Ecd}(\boldsymbol{z};\Delta)$, which is the closest integral polynomial to $64 \cdot \frac{1}{4}(10+4\sqrt{2}X+10X^2+2\sqrt{2}X^3)$. Note that $64^{-1} \cdot (m(\zeta_8), m(\zeta_8^3)) \approx (3.0082+4.0026i, 1.9918-0.9974i)$ is approximate to the input vector $\boldsymbol{z}$ with a high precision.

3.3 Leveled Homomorphic Encryption Scheme for Approximate Arithmetic

The purpose of this subsection is to construct a leveled HE scheme for approximate arithmetic. For convenience, we fix a base $p > 0$ and a modulus q_0, and let $q_\ell = p^\ell \cdot q_0$ for $0 < \ell \le L$. The integer p will be used as a base for scaling in approximate computation. For a security parameter λ, we also choose a parameter $M = M(\lambda, q_L)$ for cyclotomic polynomial. For a level $0 \le \ell \le L$, a ciphertext of level ℓ is a vector in $\mathcal{R}_{q_\ell}^k$ for a fixed integer k. Our scheme consists of five algorithm ($\mathsf{KeyGen}, \mathsf{Enc}, \mathsf{Dec}, \mathsf{Add}, \mathsf{Mult}$) with constants B_{clean} and $B_{\mathsf{mult}}(\ell)$ for noise estimation. For convenience, we will describe a HE scheme over the polynomial ring $\mathcal{R} = \mathbb{Z}[X]/(\Phi_M(X))$.

- $\mathsf{KeyGen}(1^\lambda)$. Generate a secret value sk, a public information pk for encryption, and a evaluation key evk.
- $\mathsf{Enc}_{pk}(m)$. For a given polynomial $m \in \mathcal{R}$, output a ciphertext $\boldsymbol{c} \in \mathcal{R}_{q_L}^k$. An encryption $\boldsymbol{c}$ of m will satisfy $\langle \boldsymbol{c}, sk \rangle = m + e \pmod{q_L}$ for some small e. The constant B_{clean} denotes an encryption bound, i.e., error polynomial of a fresh ciphertext satisfies $\|e\|_\infty^{\mathsf{can}} \le B_{\mathsf{clean}}$ with an overwhelming probability.
- $\mathsf{Dec}_{sk}(\boldsymbol{c})$. For a ciphertext $\boldsymbol{c}$ at level ℓ, output a polynomial $m' \leftarrow \langle \boldsymbol{c}, sk \rangle \pmod{q_\ell}$ for the secret key sk.

Unlike the most of existing schemes, our scheme does not have a separate plaintext space from an inserted error. An output $m' = m + e$ of decryption algorithm is slightly different from the original message m, but it can be considered to be an approximate value for approximate computations when $\|e\|_\infty^{\mathsf{can}}$ is small enough compared to $\|m\|_\infty^{\mathsf{can}}$. The intuition of approximate encryption has been partially used previously, for example, a switching key for homomorphic multiplication in [4–6,12] or an evaluation key for the squashed decryption circuit in [13,18] are encrypted in a similar way.

The algorithms for homomorphic operations are required to satisfy the following properties.

- $\mathsf{Add}(\boldsymbol{c}_1, \boldsymbol{c}_2)$. For given encrypts of m_1 and m_2, output an encryption of $m_1 + m_2$. An error of output ciphertext is bounded by sum of two errors in input ciphertexts.
- $\mathsf{Mult}_{evk}(\boldsymbol{c}_1, \boldsymbol{c}_2)$. For a pair of ciphertexts $(\boldsymbol{c}_1, \boldsymbol{c}_2)$, output a ciphertext $\boldsymbol{c}_{\mathsf{mult}} \in \mathcal{R}_{q_\ell}^k$ which satisfies $\langle \boldsymbol{c}_{\mathsf{mult}}, sk \rangle = \langle \boldsymbol{c}_1, sk \rangle \cdot \langle \boldsymbol{c}_2, sk \rangle + e_{\mathsf{mult}} \pmod{q_\ell}$ for some polynomial $e_{\mathsf{mult}} \in \mathcal{R}$ with $\|e_{\mathsf{mult}}\|_\infty^{\mathsf{can}} \le B_{\mathsf{mult}}(\ell)$.

We may adapt the techniques of existing HE schemes over the ring $\mathcal{R}$ to construct a HE scheme for approximate arithmetic. For example, the ring-based

BGV scheme [5], its variant with multiplication by raising ciphertext modulus [25] ($k = 2$), or the NTRU scheme [33] ($k = 1$) can be used as a base scheme. Our scheme has its own distinct and unique characteristic represented by the following *rescaling* procedure.

- $\mathsf{RS}_{\ell\to\ell'}(\mathbf{c})$. For a ciphertext $\mathbf{c} \in \mathcal{R}_{q_\ell}^k$ at level ℓ and a lower level $\ell' < \ell$, output the ciphertext $\mathbf{c}' \leftarrow \left\lfloor \frac{q_{\ell'}}{q_\ell} \mathbf{c} \right\rceil$ in $\mathcal{R}_{q_{\ell'}}^k$, i.e., $\mathbf{c}'$ is obtained by scaling $\frac{q_{\ell'}}{q_\ell}$ to the entries of $\mathbf{c}$ and rounding the coefficients to the closest integers. We will omit the subscript $\ell \to \ell'$ when $\ell' = \ell - 1$.

For an input ciphertext $\mathbf{c}$ of a message m such that $\langle \mathbf{c}, sk\rangle = m + e \pmod{q_\ell}$, the output ciphertext $\mathbf{c}'$ of rescaling procedure satisfies $\langle \mathbf{c}', sk\rangle = \frac{q_{\ell'}}{q_\ell} m + (\frac{q_{\ell'}}{q_\ell} e + e_{\mathsf{scale}}) \pmod{q_{\ell'}}$. Let $\boldsymbol{\tau} = \frac{q_{\ell'}}{q_\ell}\mathbf{c} - \mathbf{c}'$ and assume that an error polynomial $e_{\mathsf{scale}} = \langle \boldsymbol{\tau}, sk\rangle$ is bounded by some constant B_{scale}. Then the output ciphertext becomes an encryption of $\frac{q_{\ell'}}{q_\ell} m$ with a noise bounded by $\frac{q_{\ell'}}{q_\ell}\|e\|_\infty^{\mathsf{can}} + B_{\mathsf{scale}}$.

Technically this procedure is similar to the modulus-switching algorithm [5], but it has a completely different role in our construction. The rescaling algorithm divides a plaintext by an integer to remove some inaccurate LSBs as a rounding step in usual approximate computations using floating-point numbers or scientific notation. The magnitude of messages can be maintained almost the same during homomorphic evaluation, and thus the required size of the largest ciphertext modulus grows linearly with the depth of the circuit being evaluated.

Tagged Informations. A homomorphic operation has an effect on the size of plaintext and the growth of message and noise. Each ciphertext will be tagged with bounds of a message and an error in order to dynamically manage their magnitudes. Hence, a full ciphertext will be of the form $(\mathbf{c}, \ell, \nu, B)$ for a ciphertext vector $\mathbf{c} \in \mathcal{R}_{q_\ell}^k$, a level $0 \le \ell \le L$, an upper bound $\nu \in \mathbb{R}$ of message and an upper bound $B \in \mathbb{R}$ of noise. Table 1 shows the full description of our scheme and homomorphic operations for ciphertexts with tagged information.

Table 1. Description of our scheme

$\mathsf{Enc}_{pk}:$	$m \mapsto (\mathbf{c}, L, \nu, B_{\mathsf{clean}})$ for some $\nu \ge \|m\|_\infty^{\mathsf{can}}$
$\mathsf{Dec}_{sk}:$	$(\mathbf{c}, \ell, \nu, B) \mapsto (\langle \mathbf{c}, sk\rangle \pmod{q_\ell}, B)$
$\mathsf{RS}_{\ell\to\ell'}:$	$(\mathbf{c}', \ell, \nu, B) \mapsto (\mathbf{c}, \ell', p^{\ell'-\ell}\cdot\nu, p^{\ell'-\ell}\cdot B + B_{\mathsf{scale}})$
$\mathsf{Add}:$	$((\mathbf{c}_1, \ell, \nu_1, B_1), (\mathbf{c}_2, \ell, \nu_2, B_2)) \mapsto (\mathbf{c}_{\mathsf{add}}, \ell, \nu_1 + \nu_2, B_1 + B_2)$
$\mathsf{Mult}_{evk}:$	$((\mathbf{c}_1, \ell, \nu_1, B_1), (\mathbf{c}_2, \ell, \nu_2, B_2))$
	$\mapsto (\mathbf{c}_{\mathsf{mult}}, \ell, \nu_1\nu_2, \nu_1 B_2 + \nu_2 B_1 + B_1 B_2 + B_{\mathsf{mult}})$

Homomorphic Operations of Ciphertexts at Different Levels. When given encryptions $\mathbf{c}, \mathbf{c}'$ of m, m' belong to the different levels ℓ and $\ell' < \ell$, we should bring a ciphertext $\mathbf{c}$ at a larger level ℓ to the smaller level ℓ' before homomorphic operation. There are two candidates: simple modular reduction and the RS procedure. It should be chosen very carefully by considering the scale of messages because the simple modular reduction $\boldsymbol{c} \mapsto \boldsymbol{c} \pmod{q_{\ell'}}$ preserves the plaintext while RS procedure changes the plaintext from m to $\frac{q_{\ell'}}{q_\ell} m$ as in Fig. 3. Throughout this paper, we perform simple modulus reduction to the smaller modulus before computation on ciphertexts at different levels unless stated otherwise.

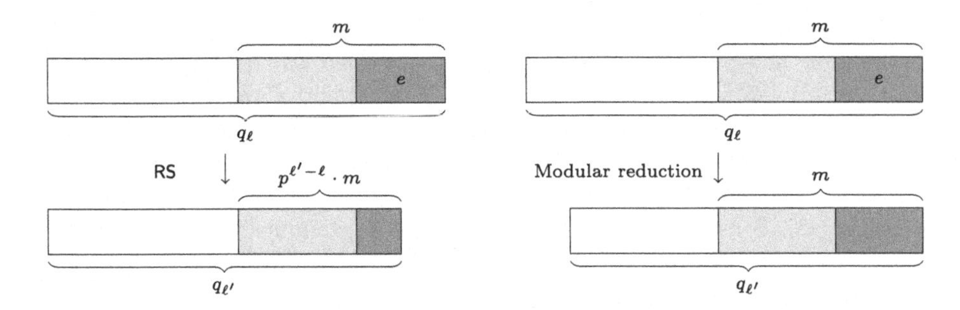

Fig. 3. Rescaling and simple modular reduction

3.4 Concrete Construction of RLWE-based HE Scheme

The performance of our construction and the noise growth depend on the base HE scheme. Moreover, a more accurate noise estimation can be done if we choose a specific one. We take the BGV scheme [5] with multiplication method by raising ciphertext modulus [25] as the underlying scheme of our concrete construction and implementation. From Costache and Smart's comparison [14], it seems to be the most efficient among the existing RLWE-based schemes.

For security and simplicity, we will use *power-of-two* degree cyclotomic rings. In this case, the dual ideal $\mathcal{R}^\vee = N^{-1} \cdot \mathcal{R}$ of $\mathcal{R} = \mathbb{Z}[X]/(X^N + 1)$ is simply a scaling of the ring. The RLWE problem is informally described by transforming samples $(a, b = a \cdot s' + e') \in \mathcal{R}_q \times \mathcal{R}_q^\vee$ into $(a, b = a \cdot s + e) \in \mathcal{R}_q \times \mathcal{R}_q$ where $s = s' \cdot N \in \mathcal{R}$ and $e = e' \cdot N \in \mathcal{R}$, so that the coefficients of e can be sampled independently from the discrete Gaussian distribution.

We will also choose the ring of Gaussian integers $\mathbb{Z}[i]$ as a discrete subspace of $\mathbb{C}$ for implementation. Another advantage of power-of-two degree cyclotomic rings is the efficient rounding operation $\lfloor \cdot \rceil_{\mathcal{R}^\vee}$ in dual fractional ideal $\mathcal{R}^\vee$. Since the columns of matrix CRT_M defined in Sect. 2.2 are mutually orthogonal, the encoding of plaintext can be efficiently done by rounding coefficients to the nearest integers after multiplication with the matrix CRT_M^{-1}.

We adopt the notation of some distributions on from [25]. For a real $\sigma > 0$, $\mathcal{DG}(\sigma^2)$ samples a vector in $\mathbb{Z}^N$ by drawing its coefficient independently from the discrete Gaussian distribution of variance σ^2. For an positive integer h, $\mathcal{HWT}(h)$ is the set of signed binary vectors in $\{0, \pm 1\}^N$ whose Hamming weight is exactly h. For a real $0 \le \rho \le 1$, the distribution $\mathcal{ZO}(\rho)$ draws each entry in the vector from $\{0, \pm 1\}^N$, with probability $\rho/2$ for each of -1 and $+1$, and probability being zero $1 - \rho$.

- $\mathsf{KeyGen}(1^\lambda)$.
  - Given the security parameter λ, choose a power-of-two $M = M(\lambda, q_L)$, an integer $h = h(\lambda, q_L)$, an integer $P = P(\lambda, q_L)$ and a real value $\sigma = \sigma(\lambda, q_L)$.
  - Sample $s \leftarrow \mathcal{HWT}(h)$, $a \leftarrow \mathcal{R}_{q_L}$ and $e \leftarrow \mathcal{DG}(\sigma^2)$. Set the secret key as $sk \leftarrow (1, s)$ and the public key as $pk \leftarrow (b, a) \in \mathcal{R}_{q_L}^2$ where $b \leftarrow -as + e \pmod{q_L}$.
  - Sample $a' \leftarrow \mathcal{R}_{P \cdot q_L}$ and $e' \leftarrow \mathcal{DG}(\sigma^2)$. Set the evaluation key as $evk \leftarrow (b', a') \in \mathcal{R}_{P \cdot q_L}^2$ where $b' \leftarrow -a's + e' + Ps^2 \pmod{P \cdot q_L}$.
- $\mathsf{Ecd}(\boldsymbol{z}; \Delta)$. For a $(N/2)$-dimensional vector $\boldsymbol{z} = (z_j)_{j \in T} \in \mathbb{Z}[i]^{N/2}$ of Gaussian integers, compute the vector $\lfloor \Delta \cdot \pi^{-1}(\boldsymbol{z}) \rceil_{\sigma(\mathcal{R})}$. Return its inverse with respect to canonical embedding map.
- $\mathsf{Dcd}(m; \Delta)$. For an input polynomial $m(X) \in \mathcal{R}$, compute the corresponding vector $\pi \circ \sigma(m)$. Return the closest vector of Gaussian integers $\boldsymbol{z} = (z_j)_{j \in T} \in \mathbb{Z}[i]^{N/2}$ after scaling, i.e., $z_j = \left\lfloor \Delta^{-1} \cdot m(\zeta_M^j) \right\rceil$ for $j \in T$.
- $\mathsf{Enc}_{pk}(m)$. Sample $v \leftarrow \mathcal{ZO}(0.5)$ and $e_0, e_1 \leftarrow \mathcal{DG}(\sigma^2)$. Output $v \cdot pk + (m + e_0, e_1) \pmod{q_L}$.
- $\mathsf{Dec}_{sk}(\boldsymbol{c})$. For $\boldsymbol{c} = (b, a)$, output $b + a \cdot s \pmod{q_\ell}$.
- $\mathsf{Add}(\boldsymbol{c}_1, \boldsymbol{c}_2)$. For $\boldsymbol{c}_1, \boldsymbol{c}_2 \in \mathcal{R}_{q_\ell}^2$, output $\boldsymbol{c}_{\mathsf{add}} \leftarrow \boldsymbol{c}_1 + \boldsymbol{c}_2 \pmod{q_\ell}$.
- $\mathsf{Mult}_{evk}(\boldsymbol{c}_1, \boldsymbol{c}_2)$. For $\boldsymbol{c}_1 = (b_1, a_1), \boldsymbol{c}_2 = (b_2, a_2) \in \mathcal{R}_{q_\ell}^2$, let $(d_0, d_1, d_2) = (b_1 b_2,\ a_1 b_2 + a_2 b_1, a_1 a_2) \pmod{q_\ell}$. Output $\boldsymbol{c}_{\mathsf{mult}} \leftarrow (d_0, d_1) + \lfloor P^{-1} \cdot d_2 \cdot evk \rceil \pmod{q_\ell}$.
- $\mathsf{RS}_{\ell \to \ell'}(\boldsymbol{c})$. For $\boldsymbol{c} \in \mathcal{R}_{q_\ell}^2$, output $\boldsymbol{c}' \leftarrow \left\lfloor \frac{q_{\ell'}}{q_\ell} \boldsymbol{c} \right\rceil \in \pmod{q_\ell'}$.

Throughout this paper, we use non-integral polynomial as plaintext for convenience of analysis, so that a ciphertext $(\boldsymbol{c} \in \mathcal{R}_{q_\ell}^2, \ell, \nu, B)$ will be called a valid encryption of $m \in \mathcal{S}$ if $\|m\|_\infty^{\mathsf{can}} \leq \nu$ and $\langle \boldsymbol{c}, sk \rangle = m + e \pmod{q_\ell}$ for some polynomial $e \in \mathcal{S}$ with $\|e\|_\infty^{\mathsf{can}} \leq B$. The following lemmas give upper bounds on noise growth after encryption, rescaling and homomorphic operations. See Appendix B for proofs.

Lemma 1 (Encoding and Encryption). *Encryption noise is bounded by* $B_{\mathsf{clean}} = 8\sqrt{2}\sigma N + 6\sigma\sqrt{N} + 16\sigma\sqrt{hN}$. *If* $\boldsymbol{c} \leftarrow \mathsf{Enc}_{pk}(m)$ *and* $m \leftarrow \mathsf{Ecd}(\boldsymbol{z}; \Delta)$ *for some* $\boldsymbol{z} \in \mathbb{Z}[i]^{N/2}$ *and* $\Delta > N + 2B_{\mathsf{clean}}$, *then* $\mathsf{Dcd}(\mathsf{Dec}_{sk}(\boldsymbol{c})) = \boldsymbol{z}$.

Lemma 2 (Rescaling). *Let* $(\boldsymbol{c}, \ell, \nu, B)$ *be an encryption of* $m \in \mathcal{S}$. *Then* $(\boldsymbol{c}', \ell', p^{\ell'-\ell} \cdot \nu, p^{\ell'-\ell} \cdot B + B_{\mathsf{scale}})$ *is a valid encryption of* $p^{\ell'-\ell} \cdot m$ *for* $\boldsymbol{c}' \leftarrow \mathsf{RS}_{\ell \to \ell'}(\boldsymbol{c})$ *and* $B_{\mathsf{scale}} = \sqrt{N/3} \cdot (3 + 8\sqrt{h})$.

Lemma 3 (Addition/Multiplication). *Let* $(\boldsymbol{c}_i, \ell, \nu_i, B_i)$ *be encryptions of* $m_i \in \mathcal{S}$ *for* $i = 1, 2$, *and let* $\boldsymbol{c}_{\mathsf{add}} \leftarrow \mathsf{Add}(\boldsymbol{c}_1, \boldsymbol{c}_2)$ *and* $\boldsymbol{c}_{\mathsf{mult}} \leftarrow \mathsf{Mult}_{evk}(\boldsymbol{c}_1, \boldsymbol{c}_2)$. *Then* $(\boldsymbol{c}_{\mathsf{add}}, \ell, \nu_1 + \nu_2, B_1 + B_2)$ *and* $(\boldsymbol{c}_{\mathsf{mult}}, \ell, \nu_1 \nu_2, \nu_1 B_2 + \nu_2 B_1 + B_1 B_2 + B_{\mathsf{mult}}(\ell))$ *are valid encryptions of* $m_1 + m_2$ *and* $m_1 m_2$, *respectively, where* $B_{\mathsf{ks}} = 8\sigma N/\sqrt{3}$ *and* $B_{\mathsf{mult}}(\ell) = P^{-1} \cdot q_\ell \cdot B_{\mathsf{ks}} + B_{\mathsf{scale}}$.

Permutations over the Plaintext Slots. It is known that the Galois group $\mathcal{G}\mathsf{al} = \mathcal{G}\mathsf{al}(\mathbb{Q}(\zeta_M)/\mathbb{Q})$ consists of the mappings $\kappa_k \colon m(X) \mapsto m(X^k) \pmod{\Phi_M(X)}$ for a polynomial $m(X) \in \mathcal{R}$ and all k co-prime with M, and that it is isomorphic to $\mathbb{Z}_M^*$. As describe in [24], applying the transformation κ_k to the polynomials is very useful for the permutation on a vector of plaintext values.

For example, a plaintext polynomial $m(X)$ is decoded into a vector of evaluations at the specific points, i.e., $(m(\zeta_M^j))_{j\in T}$ for a subgroup T of $\mathbb{Z}_M^*$ satisfying $\mathbb{Z}_M^*/T = \{\pm 1\}$. For any $i, j \in T$, there is an element $\kappa_k \in \mathcal{G}\mathsf{al}$ which sends an element in the slot of index i to an element in the slot of index j. That is, for a vector of plaintext values $\boldsymbol{z} = (z_j)_{j\in T} \in \mathbb{C}^{N/2}$ with the corresponding polynomial $m(X) = \sigma^{-1} \circ \pi^{-1}(\boldsymbol{z})$, if $k = j^{-1} \cdot i \pmod{M}$ and $m' = \kappa_k(m)$, then we have $z'_j = m'(\zeta_M^j) = m(\zeta_M^{jk}) = m(\zeta_M^i) = z_i$. Hence the element in the slot of index j of m' is the same as that in the slot of index i of m.

Given an encryption $\boldsymbol{c}$ of a message $m \in \mathcal{R}$ with a secret key $sk = (1, s)$, we denote $\kappa_k(\boldsymbol{c})$ the vector obtained by applying κ_k to the entries of ciphertext $\boldsymbol{c}$. It follows from [24] that $\kappa_k(\boldsymbol{c})$ is a valid encryption of $\kappa_k(m)$ with respect to the secret $\kappa_k(s)$. In addition, the key-switching technique can be applied to the ciphertext $\kappa_k(\boldsymbol{c})$ in order to get an encryption of the same message with respect to the original secret s.

Relative Error. The decryption result of a ciphertext is an approximate value of plaintext, so the noise growth from homomorphic operations may cause some negative effect such as loss of significance. Hence it needs to dynamically manage the bound of noise of ciphertexts for a correct understanding of the outputs. A full ciphertext $(\boldsymbol{c}, \ell, \nu, B)$ contains upper bounds of plaintext and noise, but sometimes it is convenient to consider the relative error defined by $\beta = B/\nu$.

For example, it is easy to see that the addition of ciphertexts with relative errors $\beta_i = B_i/\nu_i$ produces a ciphertext with a relative error bounded by $\max_i\{\beta_i\}$. In other case, if we multiply two ciphertexts $(\boldsymbol{c}_1, \ell, \nu_1, B_1), (\boldsymbol{c}_2, \ell, \nu_2, B_2)$ and scale down to a lower level ℓ' (as floating-point multiplication does), it produces a ciphertext at level ℓ' with a relative error

$$\beta' = \beta_1 + \beta_2 + \beta_1\beta_2 + \frac{B_{\mathsf{mult}}(\ell) + p^{\ell-\ell'} \cdot B_{\mathsf{scale}}}{\nu_1\nu_2}$$

from Lemmas 2 and 3. This relative error is very close to $\beta_1 + \beta_2$ similar to the case of unencrypted floating-point multiplication under an appropriate choice of parameter and level.

4 Homomorphic Evaluation of Approximate Arithmetic

In this section, we describe some algorithms for evaluating some circuits commonly used in practical applications and analyze error growth of an output ciphertext based on our concrete construction. We start with the homomorphic evaluations of typical circuits such as addition and multiplication by constants, monomial, and polynomial. These can be extended to approximate series for analytic functions such as multiplicative inverse and exponential function. The required parameters and precision of results will be also analyzed together.

For the convenience of analysis, we will assume that the term $\beta_1\beta_2 + (B_{\mathsf{mult}}(\ell) + p^{\ell-\ell'} \cdot B_{\mathsf{scale}})/(\nu_1\nu_2)$ is always bounded by a fixed constant β_*, so the relative error of ciphertext $\mathbf{c}' \leftarrow \mathsf{RS}_{\ell\to\ell'}(\mathsf{Mult}(\mathbf{c}_1, \mathbf{c}_2))$ satisfies the inequality $\beta' \le \beta_1 + \beta_2 + \beta_*$. We will discuss about the choice of β_* and check the validity of this assumption at the end of Sect. 4.1.

4.1 Polynomial Functions

The goal of this subsection is to suggest an algorithm for evaluating an arbitrary polynomial, and analyze its complexity and precision of output ciphertext. We start with the constant addition and multiplication functions $f(x) = x + a$ and $f(x) = ax$ for a constant $a \in \mathcal{R}$.

Lemma 4 (Addition/Multiplication by Constant). *Let* $(\mathbf{c}, \ell, \nu, B)$ *be an encryption of* $m \in \mathcal{S}$. *For a constant* $a \in \mathcal{R}$, *let* $\mathbf{c}_{\mathsf{a}} \leftarrow \mathbf{c} + (a, 0) \pmod{q_\ell}$ *and* $\mathbf{c}_{\mathsf{m}} \leftarrow a \cdot \mathbf{c} \pmod{q_\ell}$. *Then* $(\mathbf{c}_{\mathsf{a}}, \ell, \nu + \|a\|_\infty^{\mathsf{can}}, B)$ *and* $(\mathbf{c}_{\mathsf{m}}, \ell, \|a\|_\infty^{\mathsf{can}} \cdot \nu, \|a\|_\infty^{\mathsf{can}} \cdot B)$ *are valid encryptions of* $m + a$ *and* am, *respectively.*

Proof. There is a polynomial $e \in \mathcal{R}$ such that $\langle \mathbf{c}, sk \rangle = m + e \pmod{q_\ell}$ and $\|e\|_\infty^{\mathsf{can}} \le B$. It is obvious that $\langle \mathbf{c}_{\mathsf{a}}, sk \rangle = a + \langle \mathbf{c}, sk \rangle = (a + m) + e \pmod{q_\ell}$. We also have $\langle \mathbf{c}_{\mathsf{m}}, sk \rangle = a \cdot (m + e) = am + ae \pmod{q_\ell}$ and $\|a \cdot e\|_\infty^{\mathsf{can}} \le \|a\|_\infty^{\mathsf{can}} \cdot B$. □

Now we describe an algorithm to evaluate the power polynomial x^d for a power of two integer d. For simplicity, we assume that the bound ν of message m is equal to the base p.

For an input polynomial $m \in \mathcal{R}$ of size $\|m\|_\infty^{\mathsf{can}} \le p$, Algorithm 1 repeatedly performs the rescaling procedure after each of squaring step to maintain the size of message, thus the output of Algorithm 1 is an encryption of the scaled value $p \cdot f(m/p) = m^d/p^{d-1}$. The following lemma explains the correctness of Algorithm 1 and gives the relative error of the output ciphertext.

Algorithm 1. Power polynomial $f(x) = x^d$ of power-of-two degree

---

1: **procedure** POWER($\mathbf{c} \in \mathcal{R}_{q_\ell}^2, d = 2^r$)
2: $\quad$ $\mathbf{c}_0 \leftarrow \mathbf{c}$
3: $\quad$ **for** $j = 1$ to r **do**
4: $\quad\quad$ $\mathbf{c}_j \leftarrow \mathsf{RS}(\mathsf{Mult}(\mathbf{c}_{j-1}, \mathbf{c}_{j-1}))$
5: $\quad$ **end for**
6: $\quad$ **return** $\mathbf{c}_r$
7: **end procedure**

Lemma 5. *Let $(\mathbf{c}, \ell, p, \beta_0 \cdot p)$ be an encryption of $m \in \mathcal{S}$ and d be a power-of-two integer. Then Algorithm 1 outputs a valid encryption $(\mathbf{c}_r, \ell - r, p, \beta_d \cdot p)$ of m^d/p^{d-1} for some real number $\beta_d \leq d \cdot \beta_0 + (d-1) \cdot \beta_*$.*

Proof. We argue by induction on j. It is easy to see that $(\mathbf{c}_j, \ell - j, p, \beta_{2^j} \cdot p)$ is an encryption of m^{2^j}/p^{2^j-1} for some real number $\beta_{2^j} \leq 2 \cdot \beta_{2^{j-1}} + \beta_*$. After r iterations, it produces an encryption $(\mathbf{c}_r, \ell - r, p, \beta_{2^r} \cdot p)$ of m^{2^r}/p^{2^r-1} for some β_{2^r} such that $\beta_{2^r} \leq 2^r \cdot \beta_0 + (2^r - 1) \cdot \beta_*$. □

Algorithm 1 can be extended to an algorithm which evaluates an arbitrary polynomial. Similar to the previous case, this extended algorithm outputs an encryption of the scaled value $p \cdot f(m/p) = m^d/p^{d-1}$.

Lemma 6. *Let $(\mathbf{c}, \ell, p, B)$ be an encryption of $m \in \mathcal{S}$ and let d be a positive integer. Then one can compute a valid encryption $(\mathbf{c}', \ell - \lceil \log d \rceil, p, \beta_d \cdot p)$ of m^d/p^{d-1} for some real number $\beta_d \leq d \cdot \beta_0 + (d-1) \cdot \beta_*$.*

Lemma 7 (Polynomial). *Let $f(x) = \sum_{j=0}^{d} a_j x^j$ be a nonzero polynomial of coefficients a_j in $\mathcal{R}$ and of degree d. Let $(\mathbf{c}, \ell, p, \beta_0 \cdot p)$ be an encryption of $m\mathcal{S}$. Then one can compute a valid encryption $(\mathbf{c}', \ell - \lceil \log d \rceil, M_f, \beta_d \cdot M_f)$ of $p \cdot f(m/p)$ for $M_f = p \cdot \sum_{j=0}^{d} \|a_j\|_\infty^{\mathsf{can}}$ and for some real number $\beta_d \leq d \cdot \beta_0 + (d-1) \cdot \beta_*$.*

If the relative error of input ciphertext satisfies $\beta_0 \leq \beta_*$, the relative error of the resulting ciphertext is bounded by $\beta_d \leq d \cdot \beta_0 + (d-1) \cdot \beta_* \leq 2d \cdot \beta_0$. Hence, the precision loss is bounded by $(\log d + 1)$ bits, which is comparable to loss of significance occurring in unencrypted numerical computations. The evaluation of polynomial of degree d can be done in d homomorphic multiplications between ciphertext of depth $r = \lceil \log d \rceil$ by computing the encryptions of $m, m^2/p, \ldots, m^d/p^{d-1}$ simultaneously. We may apply the Paterson-Stockmeyer algorithm [37] to the evaluation procedure. Then a degree d polynomial can be evaluated using $O(\sqrt{d})$ multiplications, which gives a similar upper bound on relative error as the naive approach.

Let us return to the assumption $\beta_1\beta_2 + (B_{\mathsf{mult}}(\ell) + p^{\ell-\ell'} \cdot B_{\mathsf{scale}})/(\nu_1\nu_2) \leq \beta_*$. We will choose β_* as an upper bound of relative errors of fresh ciphertexts in our scheme. After evaluation of circuits of depth less than $(L-1)$, the resulting ciphertext will have a relative error less than $2^L \cdot \beta_*$. It means that the first term $\beta_1\beta_2$ will be bounded by $2^{L+1} \cdot \beta_*^2$ after evaluation. The condition $2^{L+1} \cdot \beta_*^2 \leq \frac{1}{2}\beta_*$, or equivalently $\beta_* \leq 2^{-L-2}$, seems to be natural; otherwise the relative error becomes $2^{L+1} \cdot \beta_* \geq 2^{-1}$ after evaluation, so the decryption result will have almost no information. Thus we have $\beta_1 + \beta_2 \leq \frac{1}{2}\beta_*$. The second term is equal to $(p^{\ell'-\ell} \cdot B_{\mathsf{mult}}(\ell) + B_{\mathsf{scale}})/\nu'$ where $\nu' = p^{\ell'-\ell} \cdot \nu_1\nu_2$ is the message bound of new ciphertext obtained by rescaling after multiplication. The numerator is asymptotically bounded by $p^{\ell'-\ell} \cdot B_{\mathsf{mult}}(\ell) + B_{\mathsf{scale}} = O(N)$. If the message bound always satisfies $\nu' \geq p$ as in our algorithms, the second term is $(B_{\mathsf{mult}}(\ell) + p^{\ell-\ell'} \cdot B_{\mathsf{scale}})/(\nu_1\nu_2) = O(p^{-1} \cdot N)$ which is smaller than a half of relative error of fresh ciphertext because $\beta_* \geq p^{-1} \cdot B_{\mathsf{clean}} = \Omega(p^{-1} \cdot \sigma N)$.

4.2 Approximate Polynomials and Multiplicative Inverse

We now homomorphically evaluate complex analytic functions $f(x)$ using their Taylor decomposition $f(x) = T_d(x) + R_d(x)$ for $T_d(x) = \sum_{j=0}^{d} \frac{f^{(j)}(0)}{j!} x^j$ and $R_d(x) = f(x) - T_d(x)$. Lemma 7 can be utilized to evaluate the rounded polynomial of scaled Taylor expansion $\lfloor p^u \cdot T_d \rceil (x)$ of $f(x)$ for some non-negative integers u and d, which outputs an approximate value of $p^{u+1} \cdot f(m/p)$. The bound of error is obtained by aggregating the error occurring during evaluation, the rounding error and the error of the remainder term $p^{u+1} \cdot R_d(m/p)$. In the case of RLWE-based constructions, we should consider the corresponding plaintext vector $\pi \circ \sigma(m) = (z_j)_{j \in T}$ and convergence of series in each slot.

As an example, the exponential function $f(x) = \exp(x)$ has the Taylor polynomial $T_d(x) = \sum_{j=0}^{d} \frac{1}{j!} x^j$ and the remaining term is bounded by $|R_d(x)| \leq \frac{e}{(d+1)!}$ when $|x| \leq 1$. Assume that we are given an encryption $(\mathbf{c}, \ell, p, \beta_0 \cdot p)$ of m. With the input ciphertext $\mathbf{c}$ and the polynomial $\lfloor p^u \cdot T_d \rceil (x)$, one can compute an encryption of $p^{u+1} \cdot T_d(m/p)$. We see that an error of the resulting ciphertext is bounded by

$$dp + p^{u+1} \cdot \sum_{j=1}^{d} \frac{1}{j!}(j \cdot \beta_0 + (j-1)\beta_*) \leq dp + p^{u+1} \cdot (e\beta_0 + \beta_*).$$

If we write $\exp(m/p) := \sigma^{-1} \circ \pi^{-1}((\exp(z_j/p))_{j \in T})$, the output ciphertext can be also viewed as an encryption of $p^{u+1} \cdot \exp(m/p)$ of the form $(\mathbf{c}', \ell - \lceil \log d \rceil, \nu', B')$ for $\nu' = p^{u+1} \cdot e$ and $B' = dp + p^{u+1} \cdot (e\beta_0 + \beta_* + \frac{e}{(d+1)!})$, and its relative error is bounded by $\beta' \leq (\beta_0 + \beta_* \cdot e^{-1}) + (p^{-u} \cdot d \cdot e^{-1} + \frac{1}{(d+1)!})$. If $\beta_0 \geq \beta_*$, then we may take integers d and u satisfying $(d+1)! \geq 4\beta_0^{-1}$ and $p^u \geq 2\beta_0^{-1} \cdot d$ to make the relative error less than $2\beta_0$. In this case, the precision loss during evaluation of exponential function is less than one bit.

In the case of multiplicative inverse, we adopt an algorithm described in [8] to get a better complexity. Assuming that a complex number x satisfies $|\hat{x}| \leq 1/2$ for $\hat{x} = 1 - x$, we get

$$x(1+\hat{x})(1+\hat{x}^2)(1+\hat{x}^{2^2})\cdots(1+\hat{x}^{2^{r-1}}) = 1 - \hat{x}^{2^r}. \tag{1}$$

Note that $|\hat{x}^{2^r}| \leq 2^{-2^r}$, and it converges to one as r goes to infinity. Hence, $\prod_{j=0}^{r-1}(1+\hat{x}^{2^j}) = x^{-1}(1-\hat{x}^{2^r})$ can be considered as an approximate multiplicative inverse of x with 2^r bits of precision.

For homomorphic evaluation, we change a scale and assume that a complex number z_j satisfies $|\hat{z_j}| \leq p/2$ for $\hat{z_j} = p - z_j$. The standard approach starts by normalizing those numbers to be in the unit interval by setting $x = z_j/p$. Since we cannot multiply fractions over encrypted data, the precision point should move to the left for each term of (1). That is, we multiply both sides of the Eq. (1) by p^{2^r} and then it yields

$$z_j(p+\hat{z_j})(p^{2^1}+\hat{z_j}^{2^1})(p^{2^2}+\hat{z_j}^{2^2})\cdots(p^{2^{r-1}}+\hat{z_j}^{2^{r-1}}) = p^{2^r} - \hat{z_j}^{2^r}.$$

Therefore, the product $p^{-2^r} \cdot \prod_{i=0}^{r-1}(p^{2^i} + \hat{z}_j^{\,2^i})$ can be seen as the approximate inverse of z_j with 2^r bits of precision. Let $\hat{\boldsymbol{z}} = (\hat{z}_j)_{j\in T}$ and $\boldsymbol{z}^{-1} = (z_j^{-1})_{j\in T}$. Algorithm 2 takes an encryption of $\hat{m} = \sigma^{-1}\circ\pi^{-1}(\hat{\boldsymbol{z}})$ as an input and outputs an encryption of its scaled multiplicative inverse $p^2 \cdot (\sigma^{-1}\circ\pi^{-1}(\boldsymbol{z}^{-1}))$ by evaluating the polynomial $\prod_{j=0}^{r-1}(p^{2^j} + \hat{m}^{2^j})$. The precision of the resulting ciphertext and the optimal iterations number r will be analyzed in the following lemma.

Algorithm 2. Inverse function $f(x) = x^{-1}$

```
1: procedure INVERSE(c ∈ R²_{q_ℓ}, r)
2:     p ← (p, 0)
3:     c_0 ← c
4:     v_1 ← p + c_0 (mod q_{ℓ−1})
5:     for j = 1 to r − 1 do
6:         c_j ← RS(Mult(c_{j−1}, c_{j−1}))
7:         v_{j+1} ← RS(Mult(v_j, p + c_j))
8:     end for
9:     return v_r
10: end procedure
```

Lemma 8 (Multiplicative Inverse). *Let $(\boldsymbol{c}, \ell, p/2, B_0 = \beta_0 \cdot p/2)$ be an encryption of $\hat{m} \in \mathcal{S}$ and let $m = p - \hat{m}$. Then Algorithm 2 outputs a valid encryption $(\boldsymbol{v}_r, \ell - r, 2p, \beta \cdot 2p)$ of $m' = p \cdot \prod_{i=0}^{r-1}(1 + (\hat{m}/p)^{2^i})$ for some $\beta \le \beta_0 + r\beta_*$.*

Proof. From Lemma 4, $(\boldsymbol{v}_1, \ell-1, 3p/2, B_0)$ is a valid encryption of $p+\hat{m}$ and its relative error is $\beta_1' = \beta_0/3$. It also follows from Lemma 5 that $(\boldsymbol{c}_j, \ell-j, 2^{-2^j}\cdot p, \beta_j \cdot 2^{-2^j}\cdot p)$ is a valid encryption of $\hat{m}^{2^j}/p^{2^j-1}$ for some real number $\beta_j \le 2^j\cdot(\beta_0+\beta_*)$, and so $(\boldsymbol{p}+\boldsymbol{c}_j, \ell-j, (1+2^{-2^j})p, \beta_j\cdot 2^{-2^j}\cdot p)$ is a valid encryption of $p+\hat{m}^{2^j}/p^{2^j-1} = (p^{2^j}+\hat{m}^{2^j})/p^{2^j-1}$ with a relative error $\beta_j' \le \beta_j/(2^{2^j}+1) \le 2^j\cdot(\beta_0+\beta_*)/(2^{2^j}+1)$, respectively.

Using the induction on j, we can show that

$$\left(\boldsymbol{v}_j, \ell-j, p\cdot\prod_{i=0}^{j-1}(1+2^{-2^i}), \beta_j''\cdot p\cdot\prod_{i=0}^{j-1}(1+2^{-2^i})\right)$$

is a valid encryption of $\prod_{i=0}^{j-1}(p^{2^i}+\hat{m}^{2^i})/p^{2^j-2} = p\cdot\prod_{i=0}^{j-1}(1+(\hat{m}/p)^{2^i})$ with a relative error $\beta_j'' \le \sum_{i=0}^{j-1}\beta_i' + (j-1)\cdot\beta_*$. Note that the message is bounded by $p\cdot\prod_{i=0}^{j-1}(1+2^{-2^i}) = (2p)\cdot(1-2^{-2^j}) < 2p$ and the relative error satisfies

$$\beta_j'' \le \left(\sum_{i=0}^{j-1}\frac{2^i}{2^{2^i}+1}\right)\cdot(\beta_0+\beta_*) + (j-1)\cdot\beta_* \le \beta_0 + j\cdot\beta_*$$

from the fact that $\sum_{i=0}^{\infty} \frac{2^i}{2^{2^i}+1} = 1$. Therefore, the output $\boldsymbol{v}_r$ of Algorithm 2 represents a valid encryption $(\boldsymbol{v}_r, \ell - r, 2p, \beta \cdot 2p)$ of $m' = p \cdot \prod_{i=0}^{r-1}(1 + (\hat{m}/p)^{2^i})$ for some $\beta \leq \beta_0 + r \cdot \beta_*$. □

Let $m^{-1}(X) := \sigma^{-1} \circ \pi^{-1}(\boldsymbol{z}^{-1})$ be the polynomial in $\mathcal{S}$ corresponding to $\boldsymbol{z}^{-1}$. The output ciphertext $(\boldsymbol{v}_r, \ell - r, 2p, \beta \cdot 2p)$ of the previous lemma can be also viewed as an encryption of $p^2 \cdot m^{-1}$. The error bound is increased by the convergence error $\|p^2 \cdot m^{-1} - m'\|_\infty^{\mathsf{can}} = \|p^2 \cdot m^{-1} \cdot (\hat{m}/p)^{2^r}\|_\infty^{\mathsf{can}} \leq 2^{-2^r} \cdot 2p$. Therefore, the ciphertext $(\boldsymbol{v}_r, \ell - r, 2p, (\beta + 2^{-2^r}) \cdot 2p)$ is a valid encryption of m' and its relative error is $\beta + 2^{-2^r} \leq \beta_0 + r\beta_* + 2^{-2^r}$, which is minimized when $r\beta_* \approx 2^{-2^r}$. Namely, $r = \lceil \log\log \beta_*^{-1} \rceil$ yields the inequality $\beta_0 + r\beta_* + 2^{-2^r} \leq \beta_0 + 2r\beta_* = \beta_0 + 2\lceil \log\log \beta_*^{-1} \rceil \cdot \beta_*$. Thus the precision loss during evaluation of multiplicative inverse is less than one bit if $2\lceil \log\log \beta_*^{-1} \rceil \cdot \beta_* \leq \beta_0$.

The optimal iterations number r can be changed upon more/less information about the magnitude of $\hat{m}$. Assume that we have an encryption of message $\hat{m}$ whose size is bounded by $\|\hat{m}\|_\infty^{\mathsf{can}} \leq \epsilon p$ for some $0 < \epsilon < 1$. By applying Lemma 8, we can compute an encryption of $p \cdot \prod_{i=0}^{r-1}(1 + (\hat{m}/p)^{2^i}) = (p^2 \cdot m^{-1}) \cdot (1 - (\hat{m}/p)^{2^r})$ with a relative error $\beta \leq \beta_0 + r\beta_*$, which is an approximate value of $p^2 \cdot m^{-1}$ with an error bounded by $\epsilon^{2^r} \cdot 2p$. Then the optimal iterations number is $r \approx \log\log \beta_*^{-1} - \log\log \epsilon^{-1}$ and the relative error becomes $\beta \leq \beta_0 + 2\lceil(\log\log \beta_*^{-1} - \log\log \epsilon^{-1})\rceil \cdot \beta_*$ when $r = \lceil(\log\log \beta_*^{-1} - \log\log \epsilon^{-1})\rceil$.

4.3 Fast Fourier Transform

Let d be a power of two integer and consider the complex primitive d-th root of unity $\zeta_d = \exp(2\pi i/d)$. For a complex vector $\boldsymbol{u} = (u_0, \ldots, u_{d-1})$, its discrete Fourier transform (DFT) is defined by the vector $\boldsymbol{v} = (v_0, \ldots, v_{d-1}) \leftarrow \mathrm{DFT}(\boldsymbol{u})$ where $v_k = \sum_{j=0}^{d-1} \zeta_d^{jk} \cdot u_j$ for $k = 0, \ldots, d-1$. The DFT has a numerous applications in mathematics and engineering such as signal processing technology. The basic idea is to send the data to Fourier space, carry out Hadamard operations and bring back the computation result to a original domain via the inverse DFT. We denote by $W_d(z) = (z^{j \cdot k})_{0 \leq j,k < d}$ the Vandermonde matrix generated by $\{z^k : 0 \leq k < d\}$. The DFT of $\boldsymbol{u}$ can be evaluated by the matrix multiplication $\mathrm{DFT}(\boldsymbol{u}) = W_d(\zeta_d) \cdot \boldsymbol{u}$, but the complexity of DFT can be reduced down to $O(d \log d)$ using FFT algorithm by representing the DFT matrix $W_d(\zeta_d)$ as a product of sparse matrices.

Recently, Costache et al. [15] suggested an encoding method which sends the complex d-th root of unity to the monomial $Y = X^{M/d}$ over cyclotomic ring $\mathcal{R} = \mathbb{Z}[X]/(\Phi_M(X))$ for cryptosystem. Then homomorphic evaluation of DFT is simply represented as a multiplication of the matrix $W_d(Y)$ to a vector of ciphertexts over polynomial ring.

On the other hand, our RLWE-based HE scheme can take advantage of batch technique as described in Sect. 3.2. In the slot of index $k \in T$, the monomial $Y = X^{M/d}$ and matrix $W_d(Y)$ are converted into ζ_d^k and the DFT matrix $W_d(\zeta_d^k)$, respectively, depending on primitive root of unity ζ_d^k. However, our batching scheme is still meaningful because the evaluation result of whole pipeline consisting of DFT, Hadamard operations, and inverse DFT is independent of index k, even though $W_d(Y)$ corresponds to the DFT matrices generated by different primitive d-th roots of unity.

It follows from the property of ordinary FFT algorithm that if $(\boldsymbol{c}_i, \ell, \nu, B)$ is an encryption of u_i for $i = 0, \ldots, d-1$ and $\boldsymbol{v} = (v_0, \ldots, v_{d-1}) \leftarrow W_d(Y) \cdot \boldsymbol{u}$, then the output of FFT algorithm using $X^{M/d}$ instead of ζ_d forms valid encryptions $(\boldsymbol{c}_i', \ell, \sqrt{d} \cdot \nu, \sqrt{d} \cdot B)$. Note that the precision of input ciphertexts is preserved as B/ν. Our FFT algorithm takes a similar time with [15] in the same parameter setting, but the amortized time is much smaller thanks to our own plaintext packing technique. In the evaluation of whole pipeline DFT-Hadamard multiplication-inverse DFT, one may scale down the transformed ciphertexts by $\sqrt{d}$ before Hadamard operations to maintain the magnitude of messages and reduce the required levels for whole pipeline.

The fast polynomial multiplication using the FFT algorithm is a typical example that computes the exact value using approximate arithmetic. In particular for the case of integral polynomials, the exact multiplication can be recovered from its approximate value since we know that their multiplication is also an integral polynomial. Likewise, when the output of a circuit has a specific format or property, it is possible to get the exact computation result from its sufficiently close approximation.

5 Implementation Results

In this section we describe how to select parameters for evaluating arithmetic circuits described in Sect. 4. We also provide implementation results with concrete parameters. Our implementation is based on the NTL C++ library running over GMP. Every experimentation was performed on a machine with an Intel Core i5 running at 2.9 GHz processor using a parameter set with 80-bit security level.

We need to set the ring dimension N that satisfies the security condition $N \geq \frac{\lambda+110}{7.2} \log(P \cdot q_L)$ to get λ-bit security level. [25,32] We note that $P \cdot q_L$ is the largest modulus to generate evaluation key and it suffices to assume that P is approximately equal to q_L. In our implementation, we used the Gaussian distribution of standard deviation $\sigma = 3.2$ to sample error polynomials, and set $h = 64$ as the number of nonzero coefficients in a secret key $s(X)$.

Evaluation of Typical Circuits. In Table 2, we present the parameter setting and performance results for computing a power of a ciphertext, the multiplicative inverse of a ciphertext and exponential function. The average running times are only for ciphertext operations, excluding encryption and decryption procedures. As described in Sect. 3.4, each ciphertext can hold $N/2$ plaintext slots and one can perform the computation in parallel in each slot. Here the amortized running time means a relative time per slot.

The homomorphic evaluation of the circuit x^{1024} with an input of 36-bit precision is hard to be implemented in practice over previous methods. Meanwhile, our scheme can compute this circuit simultaneously over 2^{14} slots in about 7.46 s, yielding an amortized rate of 0.43 ms per slot. Computation of the multiplicative inverse is done by evaluating the polynomial up to degree 8 as described in Algorithm 2. It gives an amortized time per slots of about 0.11 ms. In the case of exponential function, we used terms in its Taylor expansion up to degree 8 and it results in an amortized time per slots of 0.16 ms.

Table 2. Implementation results for homomorphic evaluation of typical circuits

Function	N	$\log q$	$\log p$	Consumed levels	Input precision	Total time	Amortized time
x^{16}	2^{13}	155	30	4	14 bits	0.31 s	0.07 ms
x^{-1}						0.45 s	0.11 ms
$\exp(x)$						0.65 s	0.16 ms
x^{1024}	2^{15}	620	56	10	36 bits	7.46 s	0.43 ms

Significance Loss. In Sect. 4, we analyzed the theoretical upper bounds on the growth of relative errors during evaluations. We can see from experimental result that initial precision is about 4 bits greater than theoretic bound of precision since we multiply 16 to the variance of encryption error to get a high probability bound. In Fig. 4, we depict bit precisions of output ciphertexts during the evaluation of homomorphic multiplications (e.g. x^{16} for the left figure and x^{1024} for the right figure). We can actually check that both theoretic bound and experimental result of precision loss during homomorphic multiplications is less than 4.1 (or resp. 10.1) when the depth of the circuit is 4 (or resp. 10).

Logistic Function. Let us consider the logistic function $f(x) = (1 + \exp(-x))^{-1}$, which is widely used in statistics, neural networks and machine

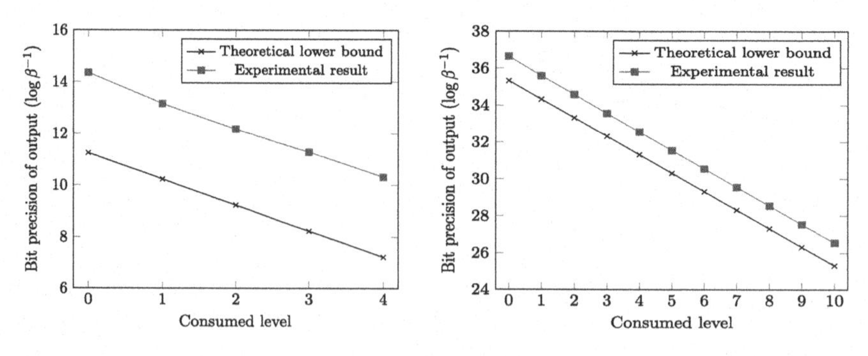

Fig. 4. The variation of bit precision of ciphertexts when $(f(x), N, \log p, \log q) = (x^{16}, 2^{13}, 30, 155)$ and $(x^{1024}, 2^{15}, 56, 620)$

learning as a probability function. For example, logistic regression is used for a prediction of the likelihood to have a heart attack in an unspecified period for men, as indicated in [3]. It was also used as a predictive equation to screen for diabetes, as described in [40]. This function can be approximated by its Taylor series

$$f(x) = \frac{1}{2} + \frac{1}{4}x - \frac{1}{48}x^3 + \frac{1}{480}x^5 - \frac{17}{80640}x^7 + \frac{31}{1451520}x^9 + O(x^{11}).$$

In [3,9], every real number is scaled by a predetermined factor to transform it as a binary polynomial before computation. The plaintext modulus t should be set large enough so that no reduction modulo t occurs in the plaintext space. The required bit size of plaintext modulus exponentially increases on the depth of the circuit, which strictly limits the performance of evaluation. On the other hand, the rescaling procedure in our scheme has the advantage that it significantly reduces the size of parameters (e.g. $(\log p, \log q) = (30, 155)$).

The parallelized computation for logistic function is especially important in real world applications such as statistic analysis using multiple data. In previous approaches, each slot of plaintext space should represent a larger degree than encoded polynomials so they could support only a few numbers of slots. On the other hand, we provide a parallelization method with an amortization amount independent from target circuit and get a better amortized time of evaluation (Table 3).

Table 3. Comparison of implementation results for homomorphic evaluation of logistic function

Method	N	$\log q$	Polynomial degree	Amortization amount	Total time	Amortized time
[3]	2^{14}	512	7	-	>30 s	-
[9]	17430	370	7	-	1.8 s	-
Ours	2^{13}	155	7	2^{12}	0.54 s	0.13 ms
	2^{14}	185	9	2^{13}	0.78 s	0.09 ms

Discrete Fourier Transform. With the parameters $(N, \log p) = (2^{13}, 50)$, we encrypt coefficients of polynomials and homomorphically evaluate the standard processing (FFT-Hadamard product of two vectors-inverse FFT) in 73 min (amortized 1.06 s per slot) when $d = 2^{13}$. We could reduce down the evaluation time to 22 min (amortized 0.34 s per slot) by adapting the multi-threading method on a machine with four cores, compared to 17 min of the previous work [15] on six cores. Since the rescaling procedure of transformed ciphertexts enables us to efficiently carry out higher degree Hadamard operations in Fourier space, the gap of parameter and running time between our scheme and previous methods grows very quickly as degree N and the depth of Hadamard operation increase. For instance, we also homomorphically evaluate the product of 8 polynomials, using pipeline consisting of FFT-Hadamard product of eight vectors-inverse FFT with parameters $(N, \log q) = (2^{14}, 250)$ in amortized time of 1.76 s (Table 4).

Table 4. Comparison of implementation results for homomorphic evaluation of a full image processing pipeline

Method	d	N	$\log q$	Degree of Hadamard operation	Amortization amount	Total time	Amortized time
[15][a]	2^4	2^{13}	150	2	-	0.46 s	-
	2^{13}	2^{14}	192	2	-	17 min	-
Ours[b]	2^4	2^{13}	120	2	2^{12}	0.88 s	0.21 ms
	2^{13}	2^{13}	130	2	2^{12}	22 min	0.34 s
	2^{13}	2^{14}	250	8	2^{13}	4 h	1.76 s

[a]This experiment was done on a machine with six Intel Xeon E5 2.7 GHz processors with 64 GB RAM.
[b]This experiment was done on a machine with four Intel Core i7 2.9 GHz processors with 16 GB RAM.

6 Conclusion

In this work, we presented a homomorphic encryption scheme which supports an approximate arithmetic over encryption. We also introduced a new batching technique for packing much information in a single ciphertext, so we could achieve practical performance advantage by parallelism. Another benefit of our scheme is the rescaling procedure, which enables us to preserve the precision of the message after approximate computation. Furthermore, it leads to reduce the size of ciphertext significantly so the scheme can be a reasonable solution for computation over large integers.

The primary open problem is finding way to convert our scheme to a fully homomorphic scheme using bootstrapping. The modulus of ciphertext decreases with homomorphic operations and our scheme can no longer support homomorphic computation at the lowest level. To overcome this problem, we aim to transform an input ciphertext into an encryption of almost the same plaintext with a much larger modulus.

Further improvement of our implementations are possible by other optimizations. We would like to enhance them to take advantage of Number Theoretical Transform (NTT) for fast polynomial multiplication.

Acknowledgments. This work was partially supported by IT R&D program of MSIP/KEIT (No. B0717-16-0098) and Samsung Electronics Co., Ltd. (No. 0421-20150074). The fourth author was supported by National Research Foundation of Korea (NRF) Grant funded by the Korean Government (No. NRF-2012H1A2A1049334). We would like to thank Kristin Lauter, Damien Stehlé and an anonymous ASIACRYPT referee for useful comments.

A LWE-based Construction

We start by adapting some notations from [5] to our context. Let n and q be positive integers. For a vector $\boldsymbol{x} \in \mathbb{Z}_q^N$, its bit decomposition and power of two are defined by $\mathsf{BD}(\boldsymbol{x}) = (\boldsymbol{u}_0, \ldots, \boldsymbol{u}_{\lceil \log q \rceil - 1}) \in \{0,1\}^{N\lceil \log q \rceil}$ with $\boldsymbol{x} = \sum_{i=0}^{\lceil \log q \rceil - 1} 2^i \boldsymbol{u}_i$, and $\mathcal{P}_2(\boldsymbol{x}) = (\boldsymbol{x}, \ldots, 2^{\lceil \log q \rceil - 1}\boldsymbol{x})$. Then we can see that $\langle \mathsf{BD}(\boldsymbol{x}), \mathcal{P}_2(\boldsymbol{y})\rangle = \langle \boldsymbol{x}, \boldsymbol{y}\rangle$. We also recall the definition of tensor product $\boldsymbol{u} \otimes \boldsymbol{v} = (u_1v_1, u_1v_2, \ldots, u_1v_m, \ldots, u_nv_1, \ldots, u_nv_m)$ on the vector space $\mathbb{R}^n \times \mathbb{R}^m$, and its relation with the inner product $\langle \boldsymbol{u} \otimes \boldsymbol{v}, \boldsymbol{u}' \otimes \boldsymbol{v}'\rangle = \langle \boldsymbol{u}, \boldsymbol{u}'\rangle \cdot \langle \boldsymbol{v}, \boldsymbol{v}'\rangle$.

- $\mathsf{KeyGen}(1^\lambda)$
  - Take an integer p and q_0. Let $q_\ell = p^\ell \cdot q_0$ for $\ell = 1, \ldots, L$. Choose the parameters $N = N(\lambda, q_L)$ and an error distribution $\chi = \chi(\lambda, q_L)$ appropriately for LWE problem of parameter (N, q_L, χ). Let $\tau = 2(N+1)\lceil \log q_L \rceil$. Output the parameters $params = (n, q_L, \chi, \tau)$.
  - Sample $\boldsymbol{s} \leftarrow \mathcal{HWT}(h)$ and set the secret key as $sk \leftarrow (1, \boldsymbol{s}) \in \mathbb{Z}_{q_L}^{N+1}$. For $1 \le i \le \tau$, sample $A \leftarrow \mathbb{Z}_{q_L}^{\tau \times N}$, $\boldsymbol{e} \leftarrow \chi^\tau$ and let $\boldsymbol{b} \leftarrow -A\boldsymbol{s} + \boldsymbol{e} \pmod{q_L}$. Set the public key as $pk \leftarrow (\boldsymbol{b}, A) \in \mathbb{Z}_{q_L}^{\tau \times (N+1)}$.
  - Let $\boldsymbol{s}' \leftarrow \mathcal{P}_2(\boldsymbol{s} \otimes \boldsymbol{s})$. Sample $A' \leftarrow \mathbb{Z}_{q_L}^{N^2\lceil \log q_L \rceil \times N}$ and $\boldsymbol{e}' \leftarrow \chi^{N^2\lceil \log q_L \rceil}$, and let $\boldsymbol{b}' \leftarrow -A'\boldsymbol{s}' + \boldsymbol{e}'$. Set the evaluation key as $evk \leftarrow (\boldsymbol{b}', A') \in \mathbb{Z}_{q_L}^{N^2\lceil \log q_L \rceil \times (N+1)}$.
- $\mathsf{Enc}(m)$. For an integer $m \in \mathbb{Z}$, sample a vector $\boldsymbol{r} \leftarrow \{0,1\}^\tau$. Output $\boldsymbol{c} \leftarrow (m, \boldsymbol{0}) + pk^T \cdot \boldsymbol{r} \in \mathbb{Z}_{q_L}^{N+1}$.
- $\mathsf{Add}(\boldsymbol{c}_1, \boldsymbol{c}_2)$. For $\boldsymbol{c}_1, \boldsymbol{c}_2 \in \mathbb{Z}_{q_\ell}^{N+1}$, output $\boldsymbol{c}_{\mathsf{add}} \leftarrow \boldsymbol{c}_1 + \boldsymbol{c}_2 \pmod{q_\ell}$.
- $\mathsf{Mult}(\boldsymbol{c}_1, \boldsymbol{c}_2)$. For $\boldsymbol{c}_1, \boldsymbol{c}_2 \in \mathbb{Z}_{q_\ell}^{N+1}$, let $\boldsymbol{c}' \leftarrow \mathsf{BD}(\boldsymbol{c}_1 \otimes \boldsymbol{c}_2)$. Output $\boldsymbol{c}_{\mathsf{mult}} \leftarrow evk^T \cdot \boldsymbol{c}' \pmod{q_\ell}$.
- $\mathsf{RS}_{\ell \to \ell'}(\boldsymbol{c})$. For a ciphertext $\boldsymbol{c} \in \mathbb{Z}_{q_\ell}^{N+1}$ at level ℓ, output the ciphertext $\boldsymbol{c}' \leftarrow \left\lfloor \frac{q_{\ell'}}{q_\ell} \boldsymbol{c} \right\rceil \in \mathbb{Z}_{q_{\ell'}}^{N+1}$.

B Noise Estimations

We follow the heuristic approach in [14,25]. Assume that a polynomial $a(X) \in \mathcal{R} = \mathbb{Z}[X]/(\Phi_M(X))$ is sampled from one of above distributions, so its nonzero entries are independently and identically distributed. Since $a(\zeta_M)$ is the inner product of coefficient vector of a and the fixed vector $(1, \zeta_M, \ldots, \zeta_M^{N-1})$ of Euclidean norm $\sqrt{N}$, the random variable $a(\zeta_M)$ has variance $V = \sigma^2 N$, where σ^2 is the variance of each coefficient of a. Hence $a(\zeta_M)$ has the variances $V_U = q^2N/12$, $V_G = \sigma^2 N$ and $V_Z = \rho N$, when a is sampled from $U(\mathcal{R}_q)$, $\mathcal{DG}(\sigma^2)$ and $\mathcal{ZO}(\rho)$, respectively. In particular, $a(\zeta_M)$ has the variance $V_H = h$ when $a(X)$ is chosen from $\mathcal{HWT}(h)$. Moreover, we can assume that $a(\zeta_M)$ is distributed similarly to a Gaussian random variable over complex plane since it is a sum of many independent and identically distributed random variables. Every evaluations at root of unity ζ_M^j share the same variance. Hence, we will use 6σ

as a high-probability bound on the canonical embedding norm of a when each coefficient has a variance σ^2. For a multiplication of two independent random variables close to Gaussian distributions with variances σ_1^2 and σ_2^2, we will use a high-probability bound $16\sigma_1\sigma_2$.

Proof of Lemma 1.

Proof. We choose $v \leftarrow \mathcal{ZO}(0.5)^2$ and $e_0, e_1 \leftarrow \mathcal{DG}(\sigma^2)$, then set $\boldsymbol{c} \leftarrow v \cdot pk + (m + e_0, e_1)$. The bound B_{clean} of encryption noise is computed by the following inequality:

$$\begin{aligned} \|\langle \boldsymbol{c}, sk\rangle - m \pmod{q_L}\|_\infty^{\mathsf{can}} &= \|v \cdot e + e_0 + e_1 \cdot s\|_\infty^{\mathsf{can}} \\ &\leq \|v \cdot e\|_\infty^{\mathsf{can}} + \|e_0\|_\infty^{\mathsf{can}} + \|e_1 \cdot s\|_\infty^{\mathsf{can}} \\ &\leq 8\sqrt{2} \cdot \sigma N + 6\sigma\sqrt{N} + 16\sigma\sqrt{hN}. \end{aligned}$$

For a vector $\boldsymbol{z} \in \mathbb{Z}[i]^{N/2}$, an encryption of $m = \mathsf{Ecd}(\boldsymbol{z}; \Delta)$ is also a valid encryption of $\Delta \cdot \sigma^{-1} \circ \pi^{-1}(\boldsymbol{z})$ with an increased error bound $B' = B_{\mathsf{clean}} + N/2$. If $\Delta^{-1} \cdot B' < 1/2$, then this error polynomial is removed by the rounding operation in decoding algorithm. □

Proof of Lemma 2.

Proof. It is satisfied that $\langle \boldsymbol{c}, sk\rangle = m + e \pmod{q_\ell}$ for some polynomial $e \in \mathcal{S}$ such that $\|e\|_\infty^{\mathsf{can}} \leq B$. The output ciphertext $\boldsymbol{c}' \leftarrow \left\lfloor \frac{q_{\ell'}}{q_\ell} \boldsymbol{c} \right\rceil$ satisfies $\langle \boldsymbol{c}', sk\rangle = \frac{q_{\ell'}}{q_\ell}(m+e) + e_{\mathsf{scale}} \pmod{q_{\ell'}}$ for the rounding error vector $\boldsymbol{\tau} = (\tau_0, \tau_1) = \boldsymbol{c}' - \frac{q_{\ell'}}{q_\ell}\boldsymbol{c}$ and the error polynomial $e_{\mathsf{scale}} = \langle \boldsymbol{\tau}, sk\rangle = \tau_0 + \tau_1 \cdot s$.

We may assume that each coefficient of τ_0 and τ_1 in the rounding error vector is computationally indistinguishable from the random variable in the interval $\frac{q_{\ell'}}{q_\ell}\mathbb{Z}_{q_\ell/q_{\ell'}}$ with variance $\approx 1/12$. Hence, the magnitude of scale error polynomial is bounded by $\|e_{\mathsf{scale}}\|_\infty^{\mathsf{can}} \leq \|\tau_0\|_\infty^{\mathsf{can}} + \|\tau_1 \cdot s\|_\infty^{\mathsf{can}} \leq 6\sqrt{N/12} + 16\sqrt{hN/12}$, as desired. □

Proof of Lemma 3.

Proof. Let $\boldsymbol{c}_i = (b_i, a_i)$ for $i = 1, 2$. Then $\langle \boldsymbol{c}_i, sk\rangle = m_i + e_i \pmod{q_\ell}$ for some polynomials $e_i \in \mathcal{S}$ such that $\|e_i\|_\infty^{\mathsf{can}} \leq B_i$. Let $(d_0, d_1, d_2) = (b_1 b_2, a_1 b_2 + a_2 b_1, a_1 a_2)$. This vector can be viewed as a level-ℓ encryption of $m_1 m_2$ with an error $m_1 e_2 + m_2 e_1 + e_1 e_2$ with respect to the secret vector $(1, s, s^2)$. It follows from Lemma 2 that the ciphertext $\boldsymbol{c}_{\mathsf{mult}} \leftarrow (d_0, d_1) + \lfloor P^{-1} \cdot (d_2 \cdot evk \pmod{Pq_\ell})\rceil$ contains an additional error $e'' = P^{-1} \cdot d_2 e'$ and a rounding error bounded by B_{scale}. We may assume that d_2 behaves as a uniform random variable on $\mathcal{R}_{q_\ell}$, so $P\|e''\|_\infty^{\mathsf{can}}$ is bounded by $16\sqrt{Nq_\ell^2/12}\sqrt{N\sigma^2} = 8N\sigma q_\ell/\sqrt{3} = B_{\mathsf{ks}} \cdot q_\ell$. Therefore, $\boldsymbol{c}_{\mathsf{mult}}$ is an encryption of $m_1 m_2$ with an error bounded by

$$\|m_1 e_2 + m_2 e_1 + e_1 e_2 + e''\|_\infty^{\mathsf{can}} + B_{\mathsf{scale}} \leq \nu_1 B_2 + \nu_2 B_1 + B_1 B_2 + P^{-1} \cdot q_\ell \cdot B_{\mathsf{ks}} + B_{\mathsf{scale}},$$

as desired. □

References

1. Arita, S., Nakasato, S.: Fully homomorphic encryption for point numbers. In: Chen, K., Lin, D., Yung, M. (eds.) Inscrypt 2016. LNCS, vol. 10143, pp. 253–270. Springer, Cham (2017). https://doi.org/10.1007/978-3-319-54705-3_16
2. Bos, J.W., Lauter, K., Loftus, J., Naehrig, M.: Improved security for a ring-based fully homomorphic encryption scheme. In: Stam, M. (ed.) IMACC 2013. LNCS, vol. 8308, pp. 45–64. Springer, Heidelberg (2013). https://doi.org/10.1007/978-3-642-45239-0_4
3. Bos, J.W., Lauter, K., Naehrig, M.: Private predictive analysis on encrypted medical data. J. Biomed. Inform. **50**, 234–243 (2014)
4. Brakerski, Z.: Fully homomorphic encryption without modulus switching from classical GapSVP. In: Safavi-Naini, R., Canetti, R. (eds.) CRYPTO 2012. LNCS, vol. 7417, pp. 868–886. Springer, Heidelberg (2012). https://doi.org/10.1007/978-3-642-32009-5_50
5. Brakerski, Z., Gentry, C., Vaikuntanathan, V.: (Leveled) fully homomorphic encryption without bootstrapping. In: Proceedings of ITCS, pp. 309–325. ACM (2012)
6. Brakerski, Z., Vaikuntanathan. V.: Efficient fully homomorphic encryption from (standard) LWE. In: Proceedings of the 2011 IEEE 52nd Annual Symposium on Foundations of Computer Science, FOCS 2011, pp. 97–106. IEEE Computer Society (2011)
7. Brakerski, Z., Vaikuntanathan, V.: Fully homomorphic encryption from ring-LWE and security for key dependent messages. In: Rogaway, P. (ed.) CRYPTO 2011. LNCS, vol. 6841, pp. 505–524. Springer, Heidelberg (2011). https://doi.org/10.1007/978-3-642-22792-9_29
8. Çetin, G.S., Doröz, Y., Sunar, B., Martin, W.J.: An investigation of complex operations with word-size homomorphic encryption. Cryptology ePrint Archive, Report 2015/1195 (2015). http://eprint.iacr.org/2015/1195
9. Cheon, J.H., Jung, J., Lee, J., Lee, K.: Privacy-preserving computations of predictive medical models with minimax approximation and non-adjacent form. In: WAHC 2017 (2017, to appear)
10. Cheon, J.H., Kim, A., Kim, M., Song, Y.: Implementation of HEA-AN (2016). https://github.com/kimandrik/HEAAN
11. Cheon, J.H., Kim, M., Lauter, K.: Homomorphic computation of edit distance. In: Brenner, M., Christin, N., Johnson, B., Rohloff, K. (eds.) FC 2015. LNCS, vol. 8976, pp. 194–212. Springer, Heidelberg (2015). https://doi.org/10.1007/978-3-662-48051-9_15
12. Cheon, J.H., Stehlé, D.: Fully homomophic encryption over the integers revisited. In: Oswald, E., Fischlin, M. (eds.) EUROCRYPT 2015. LNCS, vol. 9056, pp. 513–536. Springer, Heidelberg (2015). https://doi.org/10.1007/978-3-662-46800-5_20
13. Coron, J.-S., Lepoint, T., Tibouchi, M.: Scale-invariant fully homomorphic encryption over the integers. In: Krawczyk, H. (ed.) PKC 2014. LNCS, vol. 8383, pp. 311–328. Springer, Heidelberg (2014). https://doi.org/10.1007/978-3-642-54631-0_18
14. Costache, A., Smart, N.P.: Which ring based somewhat homomorphic encryption scheme is best? In: Sako, K. (ed.) CT-RSA 2016. LNCS, vol. 9610, pp. 325–340. Springer, Cham (2016). https://doi.org/10.1007/978-3-319-29485-8_19
15. Costache, A., Smart, N.P., Vivek, S.: Faster homomorphic evaluation of discrete fourier transforms. Cryptology ePrint Archive, Report 2016/1019 (2016). http://eprint.iacr.org/2016/1019

16. Costache, A., Smart, N.P., Vivek, S., Waller, A.: Fixed point arithmetic in SHE schemes. Cryptology ePrint Archive, Report 2016/250 (2016). http://eprint.iacr.org/2016/250
17. Damgård, I., Pastro, V., Smart, N., Zakarias, S.: Multiparty computation from somewhat homomorphic encryption. In: Safavi-Naini, R., Canetti, R. (eds.) CRYPTO 2012. LNCS, vol. 7417, pp. 643–662. Springer, Heidelberg (2012). https://doi.org/10.1007/978-3-642-32009-5_38
18. Dijk, M., Gentry, C., Halevi, S., Vaikuntanathan, V.: Fully homomorphic encryption over the integers. In: Gilbert, H. (ed.) EUROCRYPT 2010. LNCS, vol. 6110, pp. 24–43. Springer, Heidelberg (2010). https://doi.org/10.1007/978-3-642-13190-5_2
19. Doröz, Y., Hu, Y., Sunar, B.: Homomorphic AES evaluation using the modified LTV scheme. Des. Codes Crypt. **80**(2), 333–358 (2016)
20. Dowlin, N., Gilad-Bachrach, R., Laine, K., Lauter, K., Naehrig, M., Wernsing, J.: Manual for using homomorphic encryption for bioinformatics. Proc. IEEE **105**(3), 552–567 (2017)
21. Ducas, L., Micciancio, D.: FHEW: bootstrapping homomorphic encryption in less than a second. In: Oswald, E., Fischlin, M. (eds.) EUROCRYPT 2015. LNCS, vol. 9056, pp. 617–640. Springer, Heidelberg (2015). https://doi.org/10.1007/978-3-662-46800-5_24
22. Fan, J., Vercauteren, F.: Somewhat practical fully homomorphic encryption. IACR Cryptology ePrint Archive 2012/144 (2012)
23. Gentry, C.: A fully homomorphic encryption scheme. Ph.D. thesis, Stanford University (2009). http://crypto.stanford.edu/craig
24. Gentry, C., Halevi, S., Smart, N.P.: Fully homomorphic encryption with polylog overhead. In: Pointcheval, D., Johansson, T. (eds.) EUROCRYPT 2012. LNCS, vol. 7237, pp. 465–482. Springer, Heidelberg (2012). https://doi.org/10.1007/978-3-642-29011-4_28
25. Gentry, C., Halevi, S., Smart, N.P.: Homomorphic evaluation of the AES circuit. In: Safavi-Naini, R., Canetti, R. (eds.) CRYPTO 2012. LNCS, vol. 7417, pp. 850–867. Springer, Heidelberg (2012). https://doi.org/10.1007/978-3-642-32009-5_49
26. Gentry, C., Sahai, A., Waters, B.: Homomorphic encryption from learning with errors: conceptually-simpler, asymptotically-faster, attribute-based. In: Canetti, R., Garay, J.A. (eds.) CRYPTO 2013. LNCS, vol. 8042, pp. 75–92. Springer, Heidelberg (2013). https://doi.org/10.1007/978-3-642-40041-4_5
27. Jäschke, A., Armknecht, F.: Accelerating homomorphic computations on rational numbers. In: Manulis, M., Sadeghi, A.-R., Schneider, S. (eds.) ACNS 2016. LNCS, vol. 9696, pp. 405–423. Springer, Cham (2016). https://doi.org/10.1007/978-3-319-39555-5_22
28. Kim, J., Lee, C., Shim, H., Cheon, J.H., Kim, A., Kim, M., Song, Y.: Encrypting controller using fully homomorphic encryption for security of cyber-physical systems. IFAC-PapersOnLine **49**(22), 175–180 (2016)
29. Kim, M., Song, Y., Cheon, J.H.: Secure searching of biomarkers through hybrid homomorphic encryption scheme. BMC Med. Genomics **10**(2), 42 (2017)
30. Kim, M., Song, Y., Wang, S., Xia, Y., Jiang, X.: Privacy-preserving logistic regression based on homomorphic encryption. preprint
31. Lauter, K., López-Alt, A., Naehrig, M.: Private computation on encrypted genomic data. In: Aranha, D.F., Menezes, A. (eds.) LATINCRYPT 2014. LNCS, vol. 8895, pp. 3–27. Springer, Cham (2015). https://doi.org/10.1007/978-3-319-16295-9_1

32. Lindner, R., Peikert, C.: Better key sizes (and attacks) for LWE-based encryption. In: Kiayias, A. (ed.) CT-RSA 2011. LNCS, vol. 6558, pp. 319–339. Springer, Heidelberg (2011). https://doi.org/10.1007/978-3-642-19074-2_21
33. López-Alt, A., Tromer, E., Vaikuntanathan, V.: On-the-fly multiparty computation on the cloud via multikey fully homomorphic encryption. In: Proceedings of the 44th Symposium on Theory of Computing Conference, STOC 2012, pp. 1219–1234. ACM (2012)
34. Lyubashevsky, V., Peikert, C., Regev, O.: On ideal lattices and learning with errors over rings. In: Gilbert, H. (ed.) EUROCRYPT 2010. LNCS, vol. 6110, pp. 1–23. Springer, Heidelberg (2010). https://doi.org/10.1007/978-3-642-13190-5_1
35. Lyubashevsky, V., Peikert, C., Regev, O.: A toolkit for ring-LWE cryptography. In: Johansson, T., Nguyen, P.Q. (eds.) EUROCRYPT 2013. LNCS, vol. 7881, pp. 35–54. Springer, Heidelberg (2013). https://doi.org/10.1007/978-3-642-38348-9_3
36. Naehrig, M., Lauter, K., Vaikuntanathan, V.: Can homomorphic encryption be practical? In: Proceedings of the 3rd ACM Workshop on Cloud Computing Security Workshop, pp. 113–124. ACM (2011)
37. Paterson, M.S., Stockmeyer, L.J.: On the number of nonscalar multiplications necessary to evaluate polynomials. SIAM J. Comput. **2**(1), 60–66 (1973)
38. Smart, N.P., Vercauteren, F.: Fully homomorphic encryption with relatively small key and ciphertext sizes. In: Nguyen, P.Q., Pointcheval, D. (eds.) PKC 2010. LNCS, vol. 6056, pp. 420–443. Springer, Heidelberg (2010). https://doi.org/10.1007/978-3-642-13013-7_25
39. Smart, N.P., Vercauteren, F.: Fully homomorphic SIMD operations. Des. Codes Crypt. **71**(1), 57–81 (2014)
40. Tabaei, B.P., Herman, W.H.: A multivariate logistic regression equation to screen for diabetes development and validation. Diab. Care **25**(11), 1999–2003 (2002)
41. Wang, S., Zhang, Y., Dai, W., Lauter, K., Kim, M., Tang, Y., Xiong, H., Jiang, X.: Healer: homomorphic computation of exact logistic regression for secure rare disease variants analysis in GWAS. Bioinformatics **32**(2), 211–218 (2016)

Quantum Fully Homomorphic Encryption with Verification

Gorjan Alagic[1,2(✉)], Yfke Dulek[3], Christian Schaffner[3], and Florian Speelman[4]

[1] Joint Center for Quantum Information and Computer Science, University of Maryland, College Park, MD, USA
galagic@gmail.com

[2] National Institute of Standards and Technology, Gaithersburg, MD, USA

[3] CWI, QuSoft, and University of Amsterdam, Amsterdam, Netherlands

[4] QMATH, Department of Mathematical Sciences, University of Copenhagen, Copenhagen, Denmark

Abstract. Fully-homomorphic encryption (FHE) enables computation on encrypted data while maintaining secrecy. Recent research has shown that such schemes exist even for quantum computation. Given the numerous applications of classical FHE (zero-knowledge proofs, secure two-party computation, obfuscation, etc.) it is reasonable to hope that quantum FHE (or QFHE) will lead to many new results in the quantum setting. However, a crucial ingredient in almost all applications of FHE is *circuit verification.* Classically, verification is performed by checking a transcript of the homomorphic computation. Quantumly, this strategy is impossible due to no-cloning. This leads to an important open question: can quantum computations be delegated and verified in a non-interactive manner?

In this work, we answer this question in the affirmative, by constructing a scheme for QFHE with verification (vQFHE). Our scheme provides authenticated encryption, and enables arbitrary polynomial-time quantum computations without the need of interaction between client and server. Verification is almost entirely classical; for computations that start and end with classical states, it is completely classical. As a first application, we show how to construct quantum one-time programs from classical one-time programs and vQFHE.

1 Introduction

The 2009 discovery of fully-homomorphic encryption (FHE) in classical cryptography is widely considered to be one of the major breakthroughs of the field. Unlike standard encryption, FHE enables non-interactive computation on encrypted data even by parties that do not hold the decryption key. Crucially, the input, output, and all intermediate states of the computation remain encrypted, and thus hidden from the computing party. While FHE has some obvious applications (e.g., cloud computing), its importance in cryptography stems from its wide-ranging applications to other cryptographic scenarios. For instance, FHE can be used to construct

T. Takagi and T. Peyrin (Eds.): ASIACRYPT 2017, Part I, LNCS 10624, pp. 438–467, 2017.
https://doi.org/10.1007/978-3-319-70694-8_16

secure two-party computation, efficient zero-knowledge proofs for NP, and indistinguishability obfuscation [4, 14]. In fact, the breadth of its usefulness has led some to dub FHE "the swiss army knife of cryptography" [4].

Recent progress on constructing quantum computers has led to theoretical research on "cloud-based" quantum computing. In such a setting, it is natural to ask whether users can keep their data secret from the server that performs the quantum computation. A recently-constructed quantum fully-homomorphic encryption (QFHE) scheme shows that this can be done in a single round of interaction [12]. This discovery raises an important question: do the numerous classical applications of FHE have suitable quantum analogues? As it turns out, most of the classical applications require an additional property which is simple classically, but non-trivial quantumly. That property is *verification*: the ability of the user to check that the final ciphertext produced by the server is indeed the result of a particular computation, homomorphically applied to the initial user-generated ciphertext. In the classical case, this is a simple matter: the server makes a copy of each intermediate computation step, and provides the user with all these copies. In the quantum case, such a "transcript" would appear to violate no-cloning. The user simply checks a transcript generated by the server. In the quantum case, this would violate no-cloning. In fact, one might suspect that the no-cloning theorem prevents non-interactive quantum verification *in principle.*

In this work, we show that verification of homomorphic quantum computations is in fact possible. We construct a new QFHE scheme which allows the server to generate a "computation log" which can certify to the user that a particular homomorphic quantum computation was performed on the ciphertext. The computation log itself is purely classical, and most (in some cases, all) of the verification can be performed on a classical computer. Unlike in all previous quantum homomorphic schemes, the underlying encryption is now authenticated.

Verification immediately yields new applications of QFHE, e.g., allowing users of a "quantum cloud service" to certify the server's computations. Verified QFHE (or vQFHE) also leads to a simple construction of quantum one-time programs (qOTPs) [9]. In this construction, the qOTP for a functionality Φ consists of an evaluation key and a classical OTP which performs vQFHE verification for Φ only. Finding other applications of vQFHE (including appropriate analogues of all classical applications) is the subject of ongoing work.

Related Work. Classical FHE was first constructed by Gentry in 2009 [15]. For us, the scheme of Brakerski and Vaikuntanathan [5] is of note: it has decryption in $\mathbf{NC}^1$ and is believed to be quantum-secure. Quantumly, partially-homomorphic (or partially-compact) schemes were proposed by Broadbent and Jeffery [6]. The first fully-homomorphic (leveled) scheme was constructed by Dulek, Schaffner and Speelman [12]. Recently, Mahadev proposed a scheme, based on classical indistinguishability obfuscation, in which the user is completely classical [17]. A parallel line of work has attempted to produce QFHE with information-theoretic security [18, 19, 21, 23]. There has also been significant research on delegating quantum computation interactively (see, e.g., [1, 8, 11]). Another notable interactive approach is quantum computation on authenticated data (QCAD), which was used to construct quantum one-time programs from classical one-time programs [9] and zero-knowledge proofs for QMA [10].

Summary of Results. Our results concern a new primitive: verified QFHE. A standard QFHE scheme consists of four algorithms: KeyGen, Enc, Eval and Dec [6,12]. We define vQFHE similarly, with two changes: (i) Eval provides an extra classical "computation log" output; (ii) decryption is now called VerDec, and accepts a ciphertext, a circuit description C, and a computation log. Informally, correctness then demands that, for all keys k and circuits C acting on plaintexts,

$$\mathsf{VerDec}_k^C \circ \mathsf{Eval}_{\mathsf{evk}}^C \circ \mathsf{Enc}_k = \Phi_C. \tag{1}$$

A crucial parameter is the relative difficulty of performing C and VerDec_k^C. In a nontrivial scheme, the latter must be simpler. In our case, C is an arbitrary poly-size quantum circuit and VerDec_k^C is almost entirely classical.

Security of verified QFHE. Informally, security should require that, if a server deviates significantly from the map Eval_k^C in (1), then VerDec_k^C will reject.

1. **Semantic security (SEM-VER).** Consider a QPT adversary $\mathcal{A}$ which manipulates a ciphertext (and side info) and declares a circuit, as in Fig. 1 (top). This defines a channel $\Phi_{\mathcal{A}} := \mathsf{VerDec} \circ \mathcal{A} \circ \mathsf{Enc}$. A simulator $\mathcal{S}$ does not receive or output a ciphertext, but does declare a circuit; this defines a channel $\Phi_{\mathcal{S}}$ which first runs $\mathcal{S}$ and then runs a circuit on the plaintext based on the outputs of $\mathcal{S}$. We say that a vQFHE scheme is semantically secure (SEM-VER) if for all adversaries $\mathcal{A}$ there exists a simulator $\mathcal{S}$ such that the channels $\Phi_{\mathcal{A}}$ and $\Phi_{\mathcal{S}}$ are computationally indistinguishable.
2. **Indistinguishability (IND-VER).** Consider the following security game. Based on a hidden coin flip b, $\mathcal{A}$ participates in one of two protocols. For $b = 0$, this is normal vQFHE. For $b = 1$, this is a modified execution, where we secretly swap out the plaintext $\rho_{\mathcal{A}}$ to a private register (replacing it with a fixed state), apply the desired circuit to $\rho_{\mathcal{A}}$, and then swap $\rho_{\mathcal{A}}$ back in; we then discard this plaintext if VerDec rejects the outputs of $\mathcal{A}$. Upon receiving the final plaintext of the protocol, $\mathcal{A}$ must guess the bit b. A vQFHE scheme is IND-VER if, for all $\mathcal{A}$, the success probability is at most $1/2 + \mathrm{negl}(n)$.
3. **New relations between security definitions.** If we restrict SEM-VER to empty circuit case, we recover (the computational version of) the definition of quantum authentication [7,13]. SEM-VER (resp., IND-VER) generalizes computational semantic security SEM (resp., indistinguishability IND) for quantum encryption [2,6]. We generalize SEM $\Leftrightarrow$ IND [2] as follows.

Theorem 1. *A vQFHE scheme satisfies SEM-VER iff it satisfies IND-VER.*

A scheme for vQFHE for poly-size quantum circuits. Our main result is a vQFHE scheme which admits verification of arbitrary polynomial-size quantum circuits. The verification in our scheme is almost entirely classical. In fact, we can verify classical input/output computations using purely classical verification. The main technical ingredients are (i) classical FHE with $\mathbf{NC}^1$ decryption [5], (ii) the trap code for computing on authenticated quantum data [7,9,20], and (iii) the "garden-hose gadgets" from the first QFHE scheme [12]. The scheme is called TrapTP; a brief sketch is as follows.

1. **Key Generation** (KeyGen)**.** We generate keys for the classical FHE scheme, as well as some encrypted auxiliary states (see evaluation below). This procedure requires the generation of single-qubit and two-qubit states from a small fixed set, performing Bell measurements and Pauli gates, and executing the encoding procedure of a quantum error-correcting code on which the trap code is based.
2. **Encryption** (Enc)**.** We encrypt each qubit of the plaintext using the trap code, and encrypt the trap code keys using the FHE scheme. This again requires the ability to perform Paulis, execute an error-correcting encoding, and the generation of basic single-qubit states.
3. **Evaluation** (Eval)**.** Paulis and CNOT are evaluated as in the trap code; keys are updated via FHE evaluation. To measure a qubit, we measure all ciphertext qubits and place the outcomes in the log. To apply P or H, we use encrypted magic states (from the eval key) plus the aforementioned gates. Applying T requires a magic state and an encrypted "garden-hose gadget" (because the T-gate magic state circuit applies a P-gate conditioned on a measurement outcome). In addition to all of the measurement outcomes, the log also contains a transcript of all the classical FHE computations.
4. **Verified decryption** (VerDec)**.** We check the correctness and consistency of the classical FHE transcript, the measurement outcomes, and the claimed circuit. The result of this computation is a set of keys for the trap code, which are correct provided that Eval was performed honestly. We decrypt using these keys and output either a plaintext or reject. In terms of quantum capabilities, decryption requires executing the decoding procedure of the error-correcting code, computational-basis and Hadamard-basis measurements, and Paulis.

Our scheme is *compact*: the number of elementary quantum operations performed by VerDec scales only with the size of the plaintext, and *not* with the size of the circuit performed via Eval. We do require that VerDec performs a classical computation which can scale with the size of the circuit; this is reasonable since VerDec must receive the circuit as input. Like the other currently-known schemes for QFHE, our scheme is leveled, in the sense that pre-generated auxiliary magic states are needed to perform the evaluation procedure.

Theorem 2 (Main result, informal). *Let* TrapTP *be the scheme outlined above, and let* $\mathsf{VerDec}^{\equiv}$ *be* VerDec *for the case of verifying the empty circuit.*

1. *The vQFHE scheme* TrapTP *satisfies IND-VER security.*
2. *The scheme* $(\mathsf{KeyGen}, \mathsf{Enc}, \mathsf{VerDec}^{\equiv})$ *is authenticating* [13] *and IND-CPA* [6].

Application: quantum one-time programs. A one-time program (or OTP) is a device which implements a circuit, but self-destructs after the first use. OTPs are impossible without hardware assumptions, even with quantum states; OTPs that implement quantum circuits (qOTP) can be built from classical OTPs (cOTP) [9]. As a first application of vQFHE, we give another simple construction of qOTPs. Our construction is weaker, since it requires a computational assumption. On the other hand, it is conceptually very simple and serves to demonstrates the power of verification. In our construction, the qOTP for a quantum circuit C is simply a (vQFHE) encryption of C together with a cOTP for verifying the universal circuit. To use the resulting qOTP, the user attaches their desired input, homomorphically evaluates the universal circuit, and then plugs their computation log into the cOTP to retrieve the final decryption keys.

Preliminaries. Our exposition assumes a working knowledge of basic quantum information and the associated notation. As for the particular notation of quantum gates, the gates $(\mathsf{H}, \mathsf{P}, \mathsf{CNOT})$ generate the so-called Clifford group (which can also be defined as the normalizer of the Pauli group); it includes the Pauli gates X and Z. In order to implement arbitrary unitary operators, it is sufficient to add the T gate (also known as the $\pi/8$ gate). Finally, we can reach universal quantum computation by adding single-qubit measurements in the computational basis.

We will frequently make use of several standard cryptographic ingredients, as follows. The quantum one-time pad (QOTP) will be used for information-theoretically secret one-time encryption. In its encryption phase, two bits $a, b \in \{0, 1\}$ are selected at random, and the map $\mathsf{X}^a\mathsf{Z}^b$ is applied to the input, projecting it to the maximally-mixed state. We will also need the computational security notions for quantum secrecy, including indistinguishability (IND, IND-CPA) [6] and semantic security (SEM) [2]. For quantum authentication, we will refer to the security definition of Dupuis, Nielsen and Salvail [13]. We will also make frequent use of the trap code for quantum authentication, described below in Sect. 3. For a security proof and methods for interactive computation on this code, see [9]. Finally, we will also use classical fully-homomorphic encryption (FHE). In brief, an FHE scheme consists of classical algorithms (KeyGen, Enc, Eval, Dec) for (respectively) generating keys, encrypting plaintexts, homomorphically evaluating circuits on ciphertexts, and decrypting ciphertexts. We will use FHE schemes which are quantum-secure and whose Dec circuits are in $\mathbf{NC}^1$ (see, e.g., [5]).

2 A New Primitive: Verifiable QFHE

We now define verified quantum fully-homomorphic encryption (or vQFHE), in the symmetric-key setting. The public-key case is a straightforward modification.

Basic Definition. The definition has two parameters: the class $\mathcal{C}$ of circuits which the user can verify, and the class $\mathcal{V}$ of circuits which the user needs to perform in order to verify. We are interested in cases where $\mathcal{C}$ is stronger than $\mathcal{V}$.

Definition 1 (vQFHE). *Let $\mathcal{C}$ and $\mathcal{V}$ be (possibly infinite) collections of quantum circuits. A $(\mathcal{C}, \mathcal{V})$-vQFHE scheme is a set of four QPT algorithms:*

- $\mathsf{KeyGen} : \{1\}^\kappa \to \mathcal{K} \times \mathfrak{D}(\mathcal{H}_E)$ *(security parameter $\to$ private key, eval key);*
- $\mathsf{Enc} : \mathcal{K} \times \mathfrak{D}(\mathcal{H}_X) \to \mathfrak{D}(\mathcal{H}_C)$ *(key, ptext $\to$ ctext);*
- $\mathsf{Eval} : \mathcal{C} \times \mathfrak{D}(\mathcal{H}_{CE}) \to \mathcal{L} \times \mathfrak{D}(\mathcal{H}_C)$ *(circuit, eval key, ctext $\to$ log, ctext);*
- $\mathsf{VerDec} : \mathcal{K} \times \mathcal{C} \times \mathcal{L} \times \mathfrak{D}(\mathcal{H}_C) \to \mathfrak{D}(\mathcal{H}_X) \times \{\mathsf{acc}, \mathsf{rej}\}$

such that (i) the circuits of VerDec *belong to the class $\mathcal{V}$, and (ii) for all $(sk, \rho_{\mathsf{evk}}) \leftarrow \mathsf{KeyGen}$, all circuits $c \in \mathcal{C}$, and all $\rho \in \mathfrak{D}(\mathcal{H}_{XR})$,*

$$\left\| \mathsf{VerDec}_{sk}(c, \mathsf{Eval}(c, \mathsf{Enc}_k(\rho), \rho_{\mathsf{evk}})) - \Phi_c(\rho) \otimes |\mathsf{acc}\rangle\langle\mathsf{acc}|) \right\|_1 \leq \mathrm{negl}(\kappa),$$

where R is a reference and the maps implicitly act on appropriate spaces.

We will refer to condition (ii) as *correctness.* It is implicit in the definition that the classical registers $\mathcal{K}, \mathcal{L}$ and the quantum registers E, X, C are really infinite families of registers, each consisting of poly(κ)-many (qu)bits. In some later definitions, it will be convenient to use a version of VerDec which also outputs a copy of the (classical) description of the circuit c.

Compactness. We note that there are trivial vQFHE schemes for some choices of $(\mathcal{C}, \mathcal{V})$ (e.g., if $\mathcal{C} \subset \mathcal{V}$, then the user can simply authenticate the ciphertext and then perform the computation during decryption). Earlier work on quantum and classical homomorphic encryption required compactness, meaning that the size of the decrypt circuit should not scale with the size of the homomorphic circuit.

Definition 2 (Compactness of QFHE). *A QFHE scheme S is compact if there exists a polynomial $p(\kappa)$ such that for any circuit C with n_{out} output qubits, and for any input ρ_X, the complexity of applying $S.\mathsf{Dec}$ to $S.\mathsf{Eval}^C$ $(S.\mathsf{Enc}_{sk}(\rho_X), \rho_{evk})$ is at most $p(n_{\mathsf{out}}, \kappa)$.*

When considering QFHE *with* verification, however, some tension arises. On one hand, trivial schemes like the above still need to be excluded. On the other hand, verifying that a circuit C has been applied requires reading a description of C, which violates Definition 2. We thus require a more careful consideration of the relationship between the desired circuit $C \in \mathcal{C}$ and the verification circuit $V \in \mathcal{V}$. In our work, we will allow the number of classical gates in V to scale with the size of C. We propose a new definition of compactness in this context.

Definition 3 (Compactness of vQFHE (informal)). *A vQFHE scheme S is compact if $S.\mathsf{VerDec}$ is divisible into a classical verification procedure $S.\mathsf{Ver}$ (outputting only an accept/reject flag), followed by a quantum decryption procedure $S.\mathsf{Dec}$. The running time of $S.\mathsf{Ver}$ is allowed to depend on the circuit size, but the running time of $S.\mathsf{Dec}$ is not.*

The procedure S.Dec is not allowed to receive and use any other information from S.Ver than whether or not it accepts or rejects. This prevents the classical procedure S.Ver from de facto performing part of the decryption work (e.g., by computing classical decryption keys). In Sect. 3, we will see a scheme that does not fulfill compactness for this reason.

Definition 4 (Compactness of vQFHE (formal)). *A vQFHE scheme S is compact if there exists a polynomial p such that S.VerDec can be written as $S.\mathsf{Dec} \circ S.\mathsf{Ver}$, and the output ciphertext space $\mathfrak{D}(\mathcal{H}_C)$ can be written as a classical-quantum state space $\mathcal{A} \times \mathfrak{D}(\mathcal{H}_B)$, where (i.) $S.\mathsf{Ver} : \mathcal{K} \times \mathcal{C} \times \mathcal{L} \times \mathcal{A} \rightarrow \{\mathsf{acc}, \mathsf{rej}\}$ is a classical polynomial-time algorithm, and (ii.) $S.\mathsf{Dec} : \{\mathsf{acc}, \mathsf{rej}\} \times \mathcal{K} \times \mathfrak{D}(\mathcal{H}_C) \rightarrow \mathfrak{D}(\mathcal{H}_X) \times \{\mathsf{acc}, \mathsf{rej}\}$ is a quantum algorithm such that for any circuit C with n_{out} output qubits and for any input ρ_X, S.Dec runs in time $p(n_{\mathsf{out}}, \kappa)$ on the output of $S.\mathsf{Eval}^C(S.\mathsf{Enc}(\rho_X), \rho_{evk})$.*

Note that in the above definition, the classical registers $\mathcal{K}$ and $\mathcal{A}$ are copied and fed to both S.Dec and S.Ver.

For privacy, we say that a vQFHE scheme is private if its ciphertexts are indistinguishable under chosen plaintext attack (IND-CPA) [6,12].

Secure Verifiability. In this section, we formalize the concept of verifiability. Informally, one would like the scheme to be such that whenever VerDec accepts, the output can be trusted to be close to the desired output. We will consider two formalizations of this idea: a semantic one, and an indistinguishability-based one.

Our semantic definition will state that every adversary with access to the ciphertext can be simulated by a simulator that only has access to an ideal functionality that simply applies the claimed circuit. It is inspired by quantum authentication [7,13] and semantic secrecy [2].

The real-world scenario (Fig. 1, top) begins with a state $\rho_{XR_1R_2}$ prepared by a QPT ("message generator") $\mathcal{M}$. The register X (plaintext) is subsequently encrypted and sent to the adversary $\mathcal{A}$. The registers R_1 and R_2 contain side information. The adversary acts on the ciphertext and R_1, producing some output ciphertext $C_{X'}$, a circuit description c, and a computation log log. These outputs are then sent to the verified decryption function. The output, along with R_2, is sent to a distinguisher $\mathcal{D}$, who produces a bit 0 or 1.

In the ideal-world scenario (Fig. 1, bottom), the plaintext X is not encrypted or sent to the simulator $\mathcal{S}$. The simulator outputs a circuit c and chooses whether to accept or reject. The channel Φ_c implemented by c is applied to the input register X directly. If reject is chosen, the output register X' is traced out and replaced by the fixed state Ω; this controlled-channel is denoted $\overline{\mathsf{ctrl}}\text{-}\oslash$.

Definition 5 (κ-SEM-VER). *A vQFHE scheme* (KeyGen, Enc, Eval, VerDec) *is* semantically *κ-verifiable if for any QPT adversary $\mathcal{A}$, there exists a QPT $\mathcal{S}$ such that for all QPTs $\mathcal{M}$ and $\mathcal{D}$,*

$$\left| \Pr\left[\mathcal{D}\Big(\mathsf{Real}^{\mathcal{A}}_{sk}(\mathcal{M}(\rho_{evk}))\Big) = 1\right] - \Pr\left[\mathcal{D}\Big(\mathsf{Ideal}^{\mathcal{S}}_{sk}(\mathcal{M}(\rho_{evk}))\Big) = 1\right] \right| \leq \mathrm{negl}(\kappa),$$

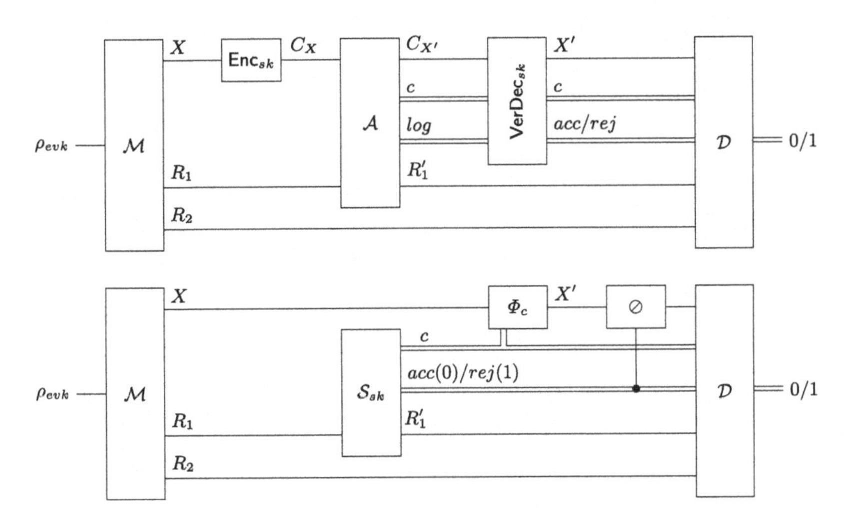

Fig. 1. The real-world (top) and ideal-world (bottom) for SEM-VER.

where $\mathsf{Real}^{\mathcal{A}}_{sk} = \mathsf{VerDec}_{sk} \circ \mathcal{A} \circ \mathsf{Enc}_{sk}$ *and* $\mathsf{Ideal}^{\mathcal{S}}_{sk} = \overline{\mathsf{ctrl}}\text{-}\oslash \circ \Phi_c \circ \mathcal{S}_{sk}$*, and the probability is taken over* $(\rho_{\mathsf{evk}}, sk) \leftarrow \mathsf{KeyGen}(1^\kappa)$ *and all QPTs above.*

Note that the simulator (in the ideal world) gets the secret key sk. We believe that this is necessary, because the actions of an adversary may depend on superficial properties of the ciphertext. In order to successfully simulate this, the simulator needs to be able to generate (authenticated) ciphertexts. He cannot do so with a fresh secret key, because the input plaintext may depend on the correlated evaluation key ρ_{evk}. Fortunately, the simulator does not become too powerful when in possession of the secret key, because he does not receive any relevant plaintexts or ciphertexts to encrypt or decrypt: the input register X is untouchable for the simulator.

Next, we present an alternative definition of verifiability, based on a security game motivated by indistinguishability.

Game 1. *For an adversary* $\mathcal{A} = (\mathcal{A}_1, \mathcal{A}_2, \mathcal{A}_3)$*, a scheme* S*, and a security parameter* κ*, the* $\mathsf{VerGame}_{\mathcal{A},S}(\kappa)$ *game proceeds as depicted in Fig. 2.*

The game is played in several rounds. Based on the evaluation key, the adversary first chooses an input (and some side information in R). Based on a random bit b this input is either encrypted and sent to $\mathcal{A}_2$ (if $b = 0$), or swapped out and replaced by a dummy input $|0^n\rangle\langle 0^n|$ (if $b = 1$). If $b = 1$, the ideal channel Φ_c is applied by the challenger, and the result is swapped back in right before the adversary (in the form of $\mathcal{A}_3$) has to decide on its output bit b'. If $\mathcal{A}_2$ causes a reject, the real result is also erased by the channel $\oslash$. We say that the adversary *wins* (expressed as $\mathsf{VerGame}_{\mathcal{A},S}(\kappa) = 1$) whenever $b' = b$.

Definition 6. (κ-IND-VER). *A vQFHE scheme* S *has* κ*-indistinguishable verification if for any QPT adversary* $\mathcal{A}$*,* $\Pr[\mathsf{VerGame}_{\mathcal{A},S}(\kappa) = 1] \leq \frac{1}{2} + \mathrm{negl}(\kappa)$.

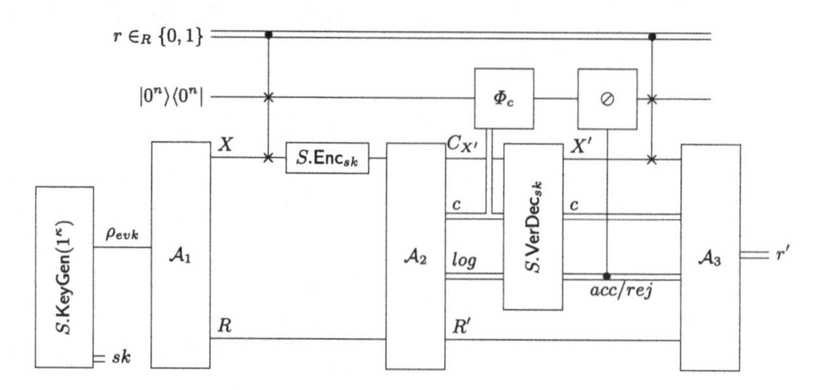

Fig. 2. The indistinguishability game $\mathsf{VerGame}_{\mathcal{A},S}(\kappa)$, as used in the definition of κ-IND-VER.

Theorem 3. *A vQFHE scheme is κ-IND-VER iff it is κ-SEM-VER.*

Proof (sketch). The forward direction is shown by contraposition. Given an adversary $\mathcal{A}$, define a simulator $\mathcal{S}$ that encrypts a dummy 0-state, then runs $\mathcal{A}$, and then VerDec. For this simulator, there exist $\mathcal{M}$ and $\mathcal{D}$ such that the difference in acceptance probability between the real and the ideal scenario is nonnegligible. The triple $(\mathcal{M}, \mathcal{A}, \mathcal{D})$ forms an adversary for the VER indistinguishability game.

For the reverse direction, we use the following approach. From an arbitrary adversary $\mathcal{A}$ for the IND-VER indistinguishability game, we define a semantic adversary, message generator, and distinguisher, that together simulate the game for $\mathcal{A}$. The fact that S is κ-SEM-VER allows us to limit the advantage of the semantic adversary over any simulator, and thereby the winning probability of $\mathcal{A}$.

For a detailed proof, see the full version [3]. □

3 TC: A partially-homomorphic scheme with verification

We now present a partially-homomorphic scheme with verification, which will serve as a building block for the fully-homomorphic scheme in Sect. 4. It is called TC (for "trap code"), and is homomorphic only for CNOT, (classically controlled) Paulis, and measurement in the computational and Hadamard basis. It does not satisfy compactness: as such, it performs worse than the trivial scheme where the client performs the circuit at decryption time. However, TC lays the groundwork for the vQFHE scheme we present in Sect. 4, and as such is important to understand in detail. It is a variant of the trap-code scheme presented in [9] (which requires classical interaction for T gates), adapted to our vQFHE framework. A variation also appears in [10], and implicitly in [20].

Setup and Encryption. Let CSS be a (public) self-dual $[[m, 1, d]]$ CSS code, so that H and CNOT are transversal. CSS can correct d_c errors, where $d = 2d_c + 1$.

We choose $m = \text{poly}(d)$ and large enough that $d_c = \kappa$ where κ is the security parameter. The concatenated Steane code satisfies all these requirements.

We generate the keys as follows. Choose a random permutation $\pi \in_R S_{3m}$ of $3m$ letters. Let n be the number of qubits that will be encrypted. For each $i \in \{1, \ldots, n\}$, pick bit strings $x[i] \in_R \{0,1\}^{3m}$ and $z[i] \in_R \{0,1\}^{3m}$. The secret key sk is the tuple $(\pi, x[1], z[1], \ldots, x[n], z[n])$, and ρ_{evk} is left empty.

Encryption is per qubit: (i) the state σ is encoded using CSS, (ii) m computational and m Hadamard 'traps' ($|0\rangle$ and $|+\rangle$ states, see [9]) are added, (iii) the resulting $3m$ qubits are permuted by π, and (iv) the overall state is encrypted with a quantum one-time pad (QOTP) as dictated by $x = x[i]$ and $z = z[i]$ for the ith qubit. We denote the ciphertext by $\widetilde{\sigma}$.

Evaluation. First, consider Pauli gates. By the properties of CSS, applying a logical Pauli is done by applying the same Pauli to all physical qubits. The application of Pauli gates (X and/or Z) to a state encrypted with a quantum one-time pad can be achieved without touching the actual state, by updating the keys to QOTP in the appropriate way. This is a classical task, so we can postpone the application of the Pauli to VerDec (recall it gets the circuit description) without giving up compactness for TC. So, formally, the evaluation procedure for Pauli gates is the identity map. Paulis conditioned on a classical bit b which will be known to VerDec at execution time (e.g., a measurement outcome) can be applied in the same manner.

Next, we consider CNOT. To apply a CNOT to encrypted qubits σ_i and σ_j, we apply CNOT transversally between the $3m$ qubits of $\widetilde{\sigma_i}$ and the $3m$ qubits of $\widetilde{\sigma_j}$. Ignoring the QOTP for the moment, the effect is a transversal application of CNOT on the pysical data qubits (which, by CSS properties, amounts to logical CNOT on $\sigma_i \otimes \sigma_j$), and an application of CNOT between the $2m$ pairs of trap qubits. Since $\mathsf{CNOT}|00\rangle = |00\rangle$ and $\mathsf{CNOT}|{++}\rangle = |{++}\rangle$, the traps are unchanged. Note that CNOT commutes with the Paulis that form the QOTP. In particular, for all $a, b, c, d \in \{0,1\}$, $\mathsf{CNOT}(\mathsf{X}_1^a \mathsf{Z}_1^b \otimes \mathsf{X}_2^c \mathsf{Z}_2^d) = (\mathsf{X}_1^a \mathsf{Z}_1^{b \oplus d} \otimes \mathsf{X}_2^{a \oplus c} \mathsf{Z}_2^d)\mathsf{CNOT}$. Thus, updating the secret-key bits (a, b, c, d) to $(a, b \oplus d, a \oplus c, d)$ finishes the job. The required key update happens in TC.VerDec (see below).

Next, consider computational-basis measurements. For CSS, logical measurement is performed by measurement of all physical qubits, followed by a classical decoding procedure [9]. In TC.Eval, we measure all $3m$ ciphertext qubits. During TC.VerDec, the contents of the measured qubits (now a classical string $a \in \{0,1\}^{3m}$) will be interpreted into a logical measurement outcome.

Finally, we handle Hadamard-basis measurements. A transversal application of H to all $3m$ relevant physical qubits precedes the evaluation procedure for the computational basis measurement. Since CSS is self-dual, this applies a logical H. Since $\mathsf{H}|0\rangle = |+\rangle$ and $\mathsf{H}|+\rangle = |0\rangle$, all computational traps are swapped with the Hadamard traps. This is reflected in the way TC.VerDec checks the traps (see the full version [3] for details). Note that this is a classical procedure (and thus its accept/reject output flag is classical).

Verification and Decryption. If a qubit is unmeasured after evaluation (as stated in the circuit), TC.VerDecQubit is applied. This removes the QOTP, undoes the permutation, checks all traps, and decodes the qubit. See the full version [3] for a specification of this algorithm.

If a qubit is measured during evaluation, TC.VerDec receives a list $\widetilde{w}$ of $3m$ physical measurement outcomes for that qubit. These outcomes are classically processed (removing the QOTP by flipping bits, undoing π, and decoding CSS) to produce the plaintext measurement outcome. Note that we only check the $|0\rangle$ traps in this case. Intuitively, this should not affect security, since any attack that affects only $|+\rangle$ but not $|0\rangle$ will be canceled by computational basis measurement.

The complete procedure TC.VerDec updates the QOTP keys according to the gates in the circuit description, and then decrypts all qubits and measurement results as described above (see the full version [3] for details).

Correctness, Compactness, and Privacy. For honest evaluation, TC.VerDec accepts with probability 1. Correctness is straightforward to check by following the description in Sect. 3. For privacy, note that the final step in the encryption procedure is the application of a (information-theoretically secure) QOTP with fresh, independent keys. If IND-CPA security is desired, one could easily extend TC by using a pseudorandom function for the QOTP, as in [2].

TC is not compact in the sense of Definition 4, however. In order to compute the final decryption keys, the whole gate-by-gate key update procedure needs to be executed, aided by the computation log and information about the circuit. Thus, we cannot break TC.VerDec up into two separate functionalities, Ver and Dec, where Dec can successfully retrieve the keys and decrypt the state, based on only the output ciphertext and the secret key.

Security of Verification. The trap code is proven secure in its application to one-time programs [9]. Broadbent and Wainewright proved authentication security (with an explicit, efficient simulator) [7]. One can use similar strategies to prove κ-IND-VER for TC. In fact, TC satisfies a stronger notion of verifiability, where the adversary is allowed to submit plaintexts in multiple rounds (letting the choice of the next plaintext depend on the previous ciphertext), which are either all encrypted or all swapped out. Two rounds (κ-IND-VER-2) are sufficient for us; the definitions and proof (see the full version [3]) extend straightforwardly to the general case κ-IND-VER-i for $i \in \mathbb{N}_+$.

Theorem 4. TC *is κ-IND-VER-2 for the above circuit class.*

4 TrapTP: Quantum FHE With Verification

In this section, we introduce our candidate scheme for verifiable quantum fully homomorphic encryption (vQFHE). In this section, we will define the scheme prove correctness, compactness, and privacy. We will show verifiability in Sect. 5.

Let $\kappa \in \mathbb{N}$ be a security parameter, and let $t, p, h \in \mathbb{N}$ be an upper bound on the number of T, P, and H gates (respectively) that will be in the circuit which is to be homomorphically evaluated. As in Sect. 3, we fix a self-dual $[[m, 1, d]]$ CSS code CSS which has $m = \text{poly}(d)$ and can correct $d_c := \kappa$ errors (e.g., the concatenated Steane code). We also fix a classical fully homomorphic public-key encryption scheme HE with decryption in LOGSPACE (see, e.g., [5]). Finally, fix a message authentication code MAC = (Tag, Ver) that is existentially unforgeable under adaptive chosen message attacks (EUF-CMA [16]) from a quantum adversary; for example, one may take the standard pseudorandom-function construction with a post-quantum PRF. This defines an authentication procedure $\mathsf{MAC.Sign}_k : m \mapsto (m, \mathsf{MAC.Tag}_k(m))$.

Key Generation and Encryption. The evaluation key will require a number of auxiliary states, which makes the key generation algorithm TrapTP.KeyGen somewhat involved (see Algorithms 1 and 2). Note that non-evaluation keys are generated first, and then used to encrypt auxiliary states which are included in the evaluation key (see TrapTP.Enc below). Most states are encrypted using the same 'global' permutation π, but all qubits in the error-correction gadget (except first and last) are encrypted using independent permutations π_i (see line 15). The T-gate gadgets are prepared by Algorithm 2, making use of garden-hose gadgets from [12].

Algorithm 1. $\mathsf{TrapTP.KeyGen}(1^\kappa, 1^t, 1^p, 1^h)$

1: $k \leftarrow \mathsf{MAC.KeyGen}(1^\kappa)$
2: $\pi \leftarrow_R S_{3m}$ ▷ S_{3m} is the permutation group on $3m$ elements
3: **for** i = 0, ..., t **do**
4: $\quad (sk_i, pk_i, evk_i) \leftarrow \mathsf{HE.KeyGen}(1^\kappa)$
5: $sk \leftarrow (\pi, k, sk_0, ..., sk_t, pk_0)$
6: **for** i = 1, ..., p **do** ▷ Magic-state generation for P
7: $\quad \mu_i^{\mathsf{P}} \leftarrow \mathsf{TrapTP.Enc}(sk, \mathsf{P}|+\rangle)$ ▷ See Algorithm 3 for TrapTP.Enc
8: **for** i = 1, ..., t **do** ▷ Magic-state generation for T
9: $\quad \mu_i^{\mathsf{T}} \leftarrow \mathsf{TrapTP.Enc}(sk, \mathsf{T}|+\rangle)$
10: **for** i = 1, ..., h **do** ▷ Magic-state generation for H
11: $\quad \mu_i^{\mathsf{H}} \leftarrow \mathsf{TrapTP.Enc}(sk, \frac{1}{\sqrt{2}}(\mathsf{H} \otimes \mathbb{I})(|00\rangle + |11\rangle))$
12: **for** i = 1, ..., t **do** ▷ Gadget generation for T
13: $\quad \pi_i \leftarrow_R S_{3m}$
14: $\quad (g_i, \gamma_i^{\mathsf{in}}, \gamma_i^{\mathsf{mid}}, \gamma_i^{\mathsf{out}}) \leftarrow \mathsf{TrapTP.GadgetGen}(sk_{i-1})$ ▷ See Algorithm 2
15: $\quad \Gamma_i \leftarrow \mathsf{MAC.Sign}(\mathsf{HE.Enc}_{pk_i}(g_i, \pi_i)) \otimes \mathsf{TrapTP.Enc}((\pi_i, k, sk_0, ..., sk_t, pk_i), \gamma_i^{\mathsf{mid}}) \otimes \mathsf{TrapTP.Enc}(sk, \gamma_i^{\mathsf{in}}, \gamma_i^{\mathsf{out}})$
16: $keys \leftarrow \mathsf{MAC.Sign}(evk_0, ..., evk_t, pk_0, ..., pk_t, \mathsf{HE.Enc}_{pk_0}(\pi))$
17: $\rho_{evk} \leftarrow (keys, \mu_0^{\mathsf{P}}, ..., \mu_p^{\mathsf{P}}, \mu_0^{\mathsf{T}}, ..., \mu_t^{\mathsf{T}}, \mu_0^{\mathsf{H}}, ..., \mu_h^{\mathsf{H}}, \Gamma_1, ..., \Gamma_t)$
18: **return** (sk, ρ_{evk})

Algorithm 2. $\mathsf{TrapTP.GadgetGen}(sk_i)$

1: $g_i \leftarrow g(sk_i)$ ▷ classical description of the garden-hose gadget, see [12], p. 13
2: $(\gamma^{\mathsf{in}}, \gamma^{\mathsf{mid}}, \gamma^{\mathsf{out}}) \leftarrow$ generate $|\Phi^+\rangle$ states and arrange them as described by g_i. Call the first qubit γ_i^{in} and the last qubit γ_i^{out}. The rest forms the state γ_i^{mid}.
3: **return** $(g_i, \gamma_i^{\mathsf{in}}, \gamma_i^{\mathsf{mid}}, \gamma_i^{\mathsf{out}})$

The encryption of a quantum state is similar to TC.Enc, only the keys to the QOTP are now chosen during encryption (rather than during key generation) and appended in encrypted and authenticated form to the ciphertext (see Algorithm 3). Note that the classical secret keys sk_0 through sk_t are not used.

Algorithm 3. $\mathsf{TrapTP.Enc}((\pi, k, sk_0, ..., sk_t, pk), \sigma)$

1: $\widetilde{\sigma} \leftarrow \sum_{x,z \in \{0,1\}^{3m}} \Big(\mathsf{TC.Enc}((\pi, x, z), \sigma) \otimes \mathsf{MAC.Sign}_k(\mathsf{HE.Enc}_{pk}(x, z))\Big)$ ▷ Algorithm 13
2: **return** $\widetilde{\sigma}$

Evaluation. Evaluation of gates is analogous to the interactive evaluation scheme using the trap code [9], except the interactions are replaced by classical homomorphic evaluation. Evaluation of a circuit c is done gate-by-gate, as follows.

In general, we will use the notation $\widetilde{\cdot}$ to denote encryptions of classical and quantum states. For example, in the algorithms below, $\widetilde{\sigma}$ is the encrypted input qubit for a gate and $\widetilde{x}$ and $\widetilde{z}$ are classical encryptions of the associated QOTP keys. We will assume throughout that HE.Enc and HE.Eval produce, apart from their actual output, a complete *computation log* describing all randomness used, all computation steps, and all intermediate results.

Measurements. Computational basis measurement is described in Algorithm 4. Recall that TC.VerDecMeasurement is a completely classical procedure that decodes the list of $3m$ measurement outcomes into the logical outcome and checks the relevant traps. Hadamard-basis measurement is performed similarly, except the qubits of $\widetilde{\sigma}$ are measured in the Hadamard basis and $\mathsf{HE.Enc}_{pk}(\times)$ is given as the last argument for the evaluation of TC.VerDecMeasurement.

Algorithm 4. $\mathsf{TrapTP.EvalMeasure}(\widetilde{\sigma}, \widetilde{x}, \widetilde{z}, \widetilde{\pi}, pk, evk)$

1: $a = (a_1, ..., a_{3m}) \leftarrow$ measure qubits of $\widetilde{\sigma}$ in the computational basis
2: $(\widetilde{a}, log_1) \leftarrow \mathsf{HE.Enc}_{pk}(a)$
3: $(\widetilde{b}, \widetilde{flag}, log_2) \leftarrow \mathsf{HE.Eval}_{evk}^{\mathsf{TC.VerDecMeasurement}}((\widetilde{\pi}, \widetilde{x}, \widetilde{z}), \widetilde{a}, \mathsf{HE.Enc}_{pk}(+))$
4: **return** $(\widetilde{b}, \widetilde{flag}, log_1, log_2)$ ▷ $b \in \{0, 1\}$ represents the output of the measurement

Pauli gates. A logical Pauli-X is performed by (homomorphically) flipping the X-key bits of the QOTP (see Algorithm 5). Since this is a classical operation, the functionality extends straightforwardly to a classically controlled Pauli-X (by specifying an additional bit b encrypted into $\widetilde{b}$ that indicates whether or not X should be applied; see Algorithm 6). The (classically controlled) evaluation of a Pauli-Z works the same way, only the relevant bits in $\widetilde{z}$ are flipped.

Algorithm 5. $\mathsf{TrapTP.EvalX}(\widetilde{\sigma}, \widetilde{x}, \widetilde{\pi}, pk, evk)$

1: $(\widetilde{x}, log_1) \leftarrow \mathsf{HE.Eval}_{evk}^{\mathsf{unpermute}}(\widetilde{\pi}, \widetilde{x})$
2: $(\widetilde{x}, log_2) \leftarrow \mathsf{HE.Eval}_{\mathsf{evk}}^{\oplus}(\widetilde{x}, \mathsf{HE.Enc}_{pk}(1^m 0^{2m}))$ ▷ this flips the first m bits
3: $(\widetilde{x}, log_3) \leftarrow \mathsf{HE.Eval}_{evk}^{\mathsf{permute}}(\widetilde{\pi}, \widetilde{x})$
4: **return** $(\widetilde{\sigma}, \widetilde{x}, log_1, log_2, log_3)$

Algorithm 6. $\mathsf{TrapTP.EvalCondX}(\widetilde{b}, \widetilde{\sigma}, \widetilde{x}, \widetilde{z}, \widetilde{\pi}, pk, evk)$

1: $(\widetilde{x}, log_1) \leftarrow \mathsf{HE.Eval}_{evk}^{\mathsf{unpermute}}(\widetilde{\pi}, \widetilde{x})$
2: $\widetilde{s} \leftarrow \mathsf{HE.Eval}_{evk}^{y \mapsto y^m 0^{2m}}(\widetilde{b})$
3: $(\widetilde{x}, log_2) \leftarrow \mathsf{HE.Eval}_{evk}^{\oplus}(\widetilde{x}, \widetilde{s})$ ▷ this conditionally flips the first m bits
4: $(\widetilde{x}, log_3) \leftarrow \mathsf{HE.Eval}_{evk}^{\mathsf{permute}}(\widetilde{\pi}, \widetilde{x})$
5: **return** $(\widetilde{\sigma}, \widetilde{x}, \widetilde{z}, log_1, log_2, log_3)$

CNOT *gates.* The evaluation of CNOT in TrapTP is analogous to TC, only the key updates are performed homomorphically during evaluation (see Algorithm 7).

Algorithm 7. $\mathsf{TrapTP.EvalCNOT}(\widetilde{\sigma_1}, \widetilde{\sigma_2}, \widetilde{x_1}, \widetilde{x_2}, \widetilde{z_1}, \widetilde{z_2}, \widetilde{\pi}, pk, evk)$

1: $(\widetilde{\sigma_1}, \widetilde{\sigma_2}) \leftarrow$ apply CNOT on all physical qubit pairs of $\widetilde{\sigma_1}, \widetilde{\sigma_2}$
2: $(\widetilde{x_1}, \widetilde{x_2}, \widetilde{z_1}, \widetilde{z_2}, log_1) \leftarrow \mathsf{HE.Eval}_{evk}^{\mathsf{CNOT-key-update}}(\widetilde{x_1}, \widetilde{x_2}, \widetilde{z_1}, \widetilde{z_2})$ ▷ for commutation rules, see Sect. 3
3: **return** $(\widetilde{\sigma_1}, \widetilde{\sigma_2}, \widetilde{x_1}, \widetilde{x_2}, \widetilde{z_1}, \widetilde{z_2}, log_1, log_2)$

Phase gates. Performing a P gate requires homomorphic evaluation of all the above gates: (classically controlled) Paulis, CNOTs, and measurements. We also consume the state μ_i^{P} (an encryption of the state $\mathsf{P}|+\rangle$) for the ith phase gate in the circuit. The circuit below applies P to the data qubit (see, e.g., [9]).

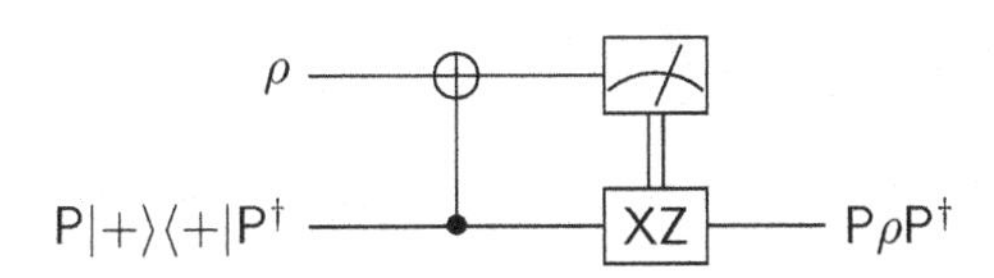

We define TrapTP.EvalP to be the concatenation of the corresponding gate evaluations. The overall computation log is just a concatenation of the logs.

Hadamard gate. The Hadamard gate can be performed using the same ingredients as the phase gate [9]. The ith gate consumes μ_i^{H}, an encryption of $(\mathsf{H}\otimes\mathbb{I})|\Phi^+\rangle$.

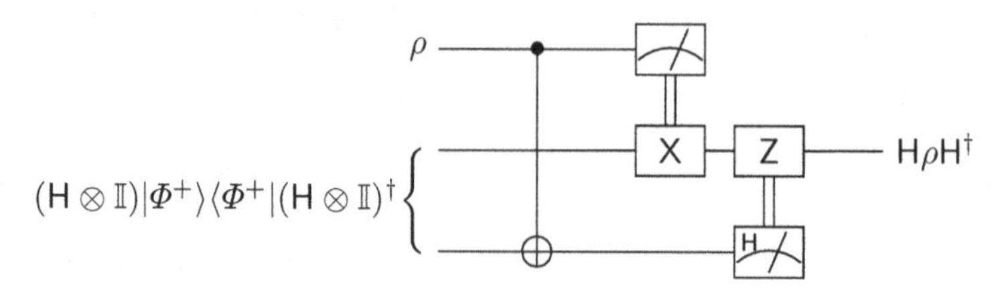

The T *gate.* A magic-state computation of T uses a similar circuit to that for P, using μ_i^{T} (an encryption of $\mathsf{T}|+\rangle$) as a resource for the ith T gate:

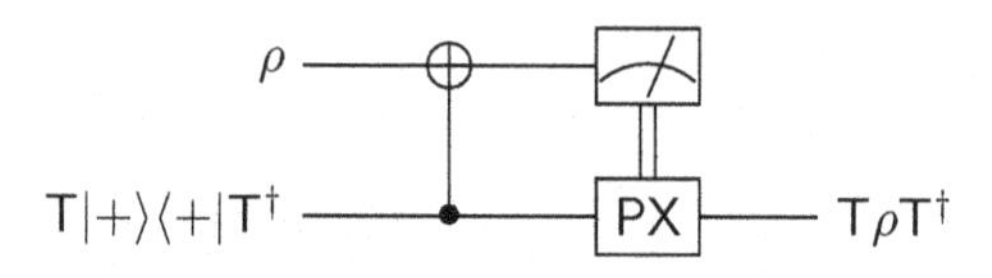

The evaluation of this circuit is much more complicated, since it requires the application of a classically-controlled phase correction P. We will accomplish this using the error-correction gadget Γ_i.

First, we remark on some subtleties regarding the encrypted classical information surrounding the gadget. Since the structure of Γ_i depends on the classical secret key sk_{i-1}, the classical information about Γ_i is encrypted under the (independent) public key pk_i (see Algorithm 1). This observation will play a crucial role in our proof that TrapTP satisfies IND-VER, in Sect. 5.

The usage of two different key sets also means that, at some point during the evaluation of a T gate, all classically encrypted information needs to be recrypted from the $(i-1)$st into the ith key set. This can be done because $\widetilde{sk_{i-1}}$ is included in the classical information g_i in Γ_i. The recryption is performed right before the classically-controlled phase gate is applied (see Algorithm 8).

Algorithm 8. $\mathsf{TrapTP.EvalT}(\widetilde{\sigma}, \widetilde{x}, \widetilde{z}, \widetilde{\pi}, \mu_i^{\mathsf{T}}, \Gamma_i, pk_{i-1}, evk_{i-1}, pk_i, evk_i)$

1: $(\widetilde{\sigma_1}, \widetilde{\sigma_2}, \widetilde{x_1}, \widetilde{z_1}, \widetilde{x_2}, \widetilde{z_2}, log_1) \leftarrow \mathsf{TrapTP.EvalCNOT}(\mu_i^{\mathsf{T}}, \widetilde{\sigma}, \widetilde{x}, \widetilde{z}, \widetilde{\pi}, pk_{i-1}, evk_{i-1})$
2: $(\widetilde{b}, log_2) \leftarrow \mathsf{TrapTP.EvalMeasure}(\widetilde{\sigma_2}, \widetilde{x_2}, \widetilde{z_2}, \widetilde{\pi}, pk_{i-1}, evk_{i-1})$
3: $log_3 \leftarrow$ recrypt *all* classically encrypted information (except $\widetilde{b}$) from key set $i-1$ into key set i.
4: $(\widetilde{\sigma}, log_4) \leftarrow \mathsf{TrapTP.EvalCondP}(\widetilde{b}, \widetilde{\sigma_1}, \widetilde{x_1}, \widetilde{z_1}, \Gamma_i, \widetilde{\pi}, pk_i, evk_i)$
5: **return** $(\widetilde{\sigma}, log_1, log_2, log_3, log_4)$

Algorithm 9 shows how to use Γ_i to apply logical P on an encrypted quantum state $\widetilde{\sigma}$, conditioned on a classical bit b for which only the encryption $\widetilde{b}$ is available. When TrapTP.EvalCondP is called, b is encrypted under the $(i-1)$st classical HE-key, while all other classical information (QOTP keys x and z, permutations π and π_i, classical gadget description g_i) is encrypted under the ith key. Note that we can evaluate Bell measurements using only evaluation of CNOT, computational-basis measurements, and H-basis measurements. In particular, no magic states are needed to perform a Bell measurement. After this procedure, the data is in qubit $\widetilde{\gamma_i^{\text{out}}}$. The outcomes a_1, a_2, a of the Bell measurements determine how the keys to the QOTP must be updated.

Algorithm 9. $\mathsf{TrapTP.EvalCondP}(\widetilde{b}, \widetilde{\sigma}, \widetilde{x}, \widetilde{z}, \Gamma_i = (\widetilde{g_i}, \widetilde{\pi_i}, \widetilde{\gamma_i^{\text{in}}}, \widetilde{\gamma_i^{\text{mid}}}, \widetilde{\gamma_i^{\text{out}}}), \widetilde{\pi}, pk_i, evk_i)$

1: $(\widetilde{a_1}, \widetilde{a_2}, log_1) \leftarrow$ evaluate Bell measurement between $\widetilde{\sigma}$ and $\widetilde{\gamma_i^{\text{in}}}$ ▷ $a_1, a_2 \in \{0, 1\}$
2: $(\widetilde{a}, log_2) \leftarrow$ evaluate Bell measurements in $\widetilde{\gamma_i^{\text{mid}}}$ as dictated by the ciphertext $\widetilde{b}$ and the garden-hose protocol for HE.Dec
3: $(\widetilde{x}, \widetilde{z}, log_3) \leftarrow \mathsf{HE.Eval}_{evk_i}^{\mathsf{T-key-update}}(\widetilde{x}, \widetilde{z}, \widetilde{a_1}, \widetilde{a_2}, \widetilde{a}, \widetilde{g_i})$
4: **return** $(\widetilde{\gamma_i^{\text{out}}}, \widetilde{x}, \widetilde{z}, log_1, log_2, log_3)$

Verified Decryption. The decryption procedure (Algorithm 10) consists of two parts. First, we perform several classical checks. This includes MAC-verification of all classically authenticated messages, and checking that the gates listed in the log match the circuit description. We also check the portions of the log which specify the (purely classical, FHE) steps taken during HE.Enc and HE.Eval; this is the standard transcript-checking procedure for FHE, which we call TrapTP.CheckLog. Secondly, we check all unmeasured traps and decode the remaining qubits. We reject if TrapTP.CheckLog rejects, or if the traps have been triggered.

Algorithm 10. $\mathsf{TrapTP.VerDec}(sk, \widetilde{\sigma}, (\widetilde{x[i]})_i, (\widetilde{z[i]})_i, log, c)$

1: Verify classically authenticated messages (in log) using k (contained in sk). If one of these verifications rejects, **reject**.
2: Check whether all claimed gates in log match the structure of c. If not, **return** $(\Omega, |\text{rej}\rangle)$. ▷ Recall that Ω is a dummy state.
3: $flag \leftarrow \mathsf{TrapTP.CheckLog}(log)$ If $flag$ = rej, **return** $(\Omega, |\text{rej}\rangle)$.
4: Check whether the claimed final QOTP keys in the log match $\widetilde{x}$ and $\widetilde{z}$. If not, **return** $(\Omega, |\text{rej}\rangle)$.
5: **for all** gates G of c **do**
6: **if** G is a measurement **then**
7: $\widetilde{x'}, \widetilde{z'} \leftarrow$ encrypted QOTP keys right before measurement (listed in log)
8: $\widetilde{w} \leftarrow$ encrypted measurement outcomes (listed in log)
9: $x', z', w \leftarrow \mathsf{HE.Dec}_{sk_t}(\widetilde{x'}, \widetilde{z'}, \widetilde{w})$
10: Execute $\mathsf{TC.VerDecMeasurement}((\pi, x', z'), w, basis)$, where $basis$ is the appropriate basis for the measurement, and store the (classical) outcome.
11: **if** a trap is triggered **then**
12: **return** $(\Omega, |\text{rej}\rangle)$.
13: **for all** unmeasured qubits $\widetilde{\sigma_i}$ in $\widetilde{\sigma}$ **do**
14: $x[i], z[i] \leftarrow \mathsf{HE.Dec}_{sk_t}(\widetilde{x[i]}, \widetilde{z[i]})$
15: $\sigma_i \leftarrow \mathsf{TC.VerDec}_{(\pi, x[i], z[i])}(\widetilde{\sigma_i})$. If TC.VerDec rejects, **return** $(\Omega, |\text{rej}\rangle)$.
16: $\sigma \leftarrow$ the list of decrypted qubits (and measurement outcomes) that are part of the output of c
17: **return** $(\sigma, |\text{acc}\rangle)$

4.1 Correctness, Compactness, and Privacy

If all classical computation was unencrypted, checking correctness of TrapTP can be done by inspecting the evaluation procedure for the different types of gates, and comparing them to the trap code construction in [9]. This suffices, since HE and the MAC authentication both satisfy correctness.

Compactness as defined in Definition 4 is also satisfied: verifying the computation log and checking all intermediate measurements (up until line 12 in Algorithm 10) is a completely classical procedure and runs in polynomial time in its input. The rest of TrapTP.VerDec (starting from line 13) only uses the secret key and the ciphertext $(\widetilde{\sigma}, \widetilde{x}, \widetilde{z})$ as input, not the log or the circuit description. Thus, we can separate TrapTP.VerDec into two algorithms Ver and Dec as described in Definition 4, by letting the second part (Dec, lines 13 to 17) reject whenever the first part (Ver, lines 1 to 12) does. It is worth noting that, because the key-update steps are performed homomorphically during the evaluation phase, skipping the classical verification step yields a QFHE scheme without verification that satisfies Definition 2 (and is authenticating). This is not the case for the scheme TC, where the classical computation is necessary for the correct decryption of the output state.

In terms of privacy, TrapTP satisfies IND-CPA (see Sect. 2). This is shown by reduction to IND-CPA of HE. This is non-trivial since the structure of the error-correction gadgets depends on the classical secret key. The reduction is done in steps, where first the security of the encryptions under pk_t is applied (no gadget depends on sk_t), after which the quantum part of the gadget Γ_t (which depends on sk_{t-1}) looks completely mixed from the point of view of the adversary. We then apply indistinguishability of the classical encryptions under pk_{t-1}, and repeat the process. After all classical encryptions of the quantum one-time pad keys are removed, the encryption of a state appears fully mixed. Full details of this proof can be found in Lemma 1 of [12], where IND-CPA security of an encryption function very similar to TrapTP.Enc is proven.

5 Proof of Verifiability for TrapTP

In this section, we will prove that TrapTP is κ-IND-VER. By Theorem 3, it then follows that TrapTP is also verifiable in the semantic sense. We will define a slight variation on the VER indistinguishability game, followed by several hybrid schemes (variations of the TrapTP scheme) that fit into this new game. We will argue that for any adversary, changing the game or scheme does not significantly affect the winning probability. After polynomially-many such steps, we will have reduced the adversary to an adversary for the somewhat homomorphic scheme TC, which we already know to be IND-VER. This will complete the argument that TrapTP is IND-VER. The IND-VER game is adjusted as follows.

Definition 7 (Hybrid game $\mathsf{Hyb}_{\mathcal{A},S}(\kappa)$). *For an adversary $\mathcal{A} = (\mathcal{A}_1, \mathcal{A}_2, \mathcal{A}_3)$, a scheme S, and security parameter κ, $\mathsf{Hyb}_{\mathcal{A},S}(\kappa)$ is the game in Fig. 3.*

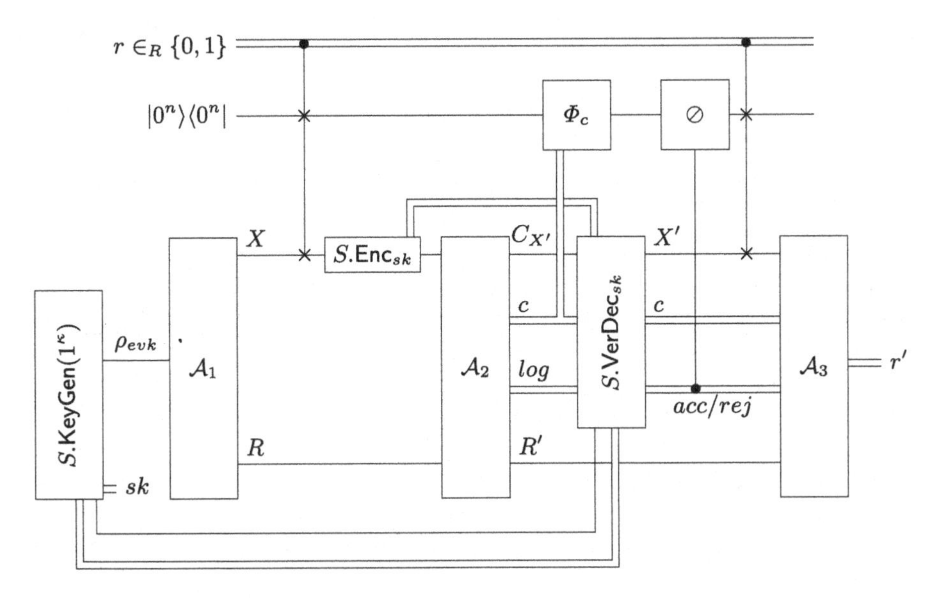

Fig. 3. The hybrid indistinguishability game $\mathsf{Hyb}_{\mathcal{A},S}(\kappa)$, which is a slight variation on $\mathsf{VerGame}_{\mathcal{A},S}(\kappa)$ from Fig. 2.

Comparing to Definition 1, we see that three new wires are added: a classical wire from $S.\mathsf{Enc}$ to $S.\mathsf{VerDec}$, and a classical and quantum wire from $S.\mathsf{KeyGen}$ to $S.\mathsf{VerDec}$. We will later adjust TrapTP to use these wires to bypass the adversary; TrapTP as defined in the previous section does not use them. Therefore, for any adversary, $\Pr[\mathsf{VerGame}_{\mathcal{A},\mathsf{TrapTP}}(\kappa) = 1] = \Pr[\mathsf{Hyb}_{\mathcal{A},\mathsf{TrapTP}}(\kappa) = 1]$.

Hybrid 1: Removing Classical MAC. In TrapTP, the initial keys to the QOTP can only become known to VerDec through the adversary. We thus use MAC to make sure these keys cannot be altered. Without this authentication, the adversary could, e.g., homomorphically use $\widetilde{\pi}$ to flip only those bits in $\widetilde{x}$ that correspond to non-trap qubits, thus applying X to the plaintext. In fact, all classical information in the evaluation key must be authenticated.

In the first hybrid, we argue that the winning probability of a QPT $\mathcal{A}$ in $\mathsf{Hyb}_{\mathcal{A},\mathsf{TrapTP}}(\kappa)$ is at most negligibly higher than in $\mathsf{Hyb}_{\mathcal{A},\mathsf{TrapTP}'}(\kappa)$, where TrapTP' is a modified version of TrapTP where the initial keys are sent directly from KeyGen and Enc to VerDec (via the extra wires above). More precisely, in $\mathsf{TrapTP}'.\mathsf{KeyGen}$ and $\mathsf{TrapTP}'.\mathsf{Enc}$, whenever $\mathsf{MAC.Sign}(\mathsf{HE.Enc}(x))$ or $\mathsf{MAC.Sign}(x)$ is called, the message x is also sent directly to $\mathsf{TrapTP}'.\mathsf{VerDec}$. Moreover, instead of decrypting the classically authenticated messages sent by the adversary, $\mathsf{TrapTP}'.\mathsf{VerDec}$ uses the information it received directly from $\mathsf{TrapTP}'.\mathsf{KeyGen}$ and $\mathsf{TrapTP}'.\mathsf{Enc}$. It still check whether the computation log provided by the adversary contains these values at the appropriate locations and whether the MAC signature is correct. The following fact is then a straightforward consequence of the EUF-CMA property of MAC.

Recall that all adversaries are QPTs, i.e., quantum polynomial-time uniform algorithms. Given two hybrid games H_1, H_2, and a QPT adversary $\mathcal{A}$, define

$$\mathsf{AdvHyb}_{H_1}^{H_2}(\mathcal{A},\kappa) := \left|\Pr[\mathsf{Hyb}_{\mathcal{A},\mathsf{H}_1}(\kappa)=1] - \Pr[\mathsf{Hyb}_{\mathcal{A},\mathsf{H}_2}(\kappa)=1]\right|.$$

Lemma 1. *For any QPT $\mathcal{A}$, $\mathsf{AdvHyb}_{\mathsf{TrapTP}}^{\mathsf{TrapTP}'}(\mathcal{A},\kappa) \leq \mathrm{negl}(\kappa)$.*

Hybrid 2: Removing Computation Log. In TrapTP and TrapTP′, the adversary (homomorphically) keeps track of the keys to the QOTP and stores encryptions of all intermediate values in the computation log. Whenever VerDec needs to know the value of a key (for example to check a trap or to decrypt the final output state), the relevant entry in the computation log is decrypted.

In TrapTP′, however, the plaintext initial values to the computation log are available to VerDec, as they are sent through the classical side channels. This means that whenever VerDec needs to know the value of a key, instead of decrypting an entry to the computation log, it can be computed by "shadowing" the computation log in the clear.

For example, suppose the log contains the encryptions $\widetilde{b_1}, \widetilde{b_2}$ of two initial bits, and specifies the homomorphic evaluation of XOR, resulting in $\widetilde{b}$ where $b = b_1 \oplus b_2$. If one knows the plaintext values b_1 and b_2, then one can compute $b_1 \oplus b_2$ directly, instead of decrypting the entry $\widetilde{b}$ from the computation log.

We now define a second hybrid, TrapTP″, which differs from TrapTP′ exactly like this: VerDec still verifies the authenticated parts of the log, checks whether the computation log matches the structure of c, and checks whether it is syntactically correct. However, instead of decrypting values from the log (as it does in TrapTP.VerDec, Algorithm 10, on lines 9 and 14), it computes those values from the plaintext initial values, by following the computation steps that are claimed in the log. By correctness of classical FHE, we then have the following.

Lemma 2. *For any QPT $\mathcal{A}$, $\mathsf{AdvHyb}_{\mathsf{TrapTP}'}^{\mathsf{TrapTP}''}(\mathcal{A},\kappa) \leq \mathrm{negl}(\kappa)$.*

Proof. Let s be the (plaintext) classical information that forms the input to the classical computations performed by the adversary: initial QOTP keys, secret keys and permutations, measurement results, etc. Let f be the function that the adversary computes on it in order to arrive at the final keys and logical measurement results. By correctness of HE, we have that

$$\Pr[\mathsf{HE.Dec}_{sk_t}(\mathsf{HE.Eval}^f_{evk_0,\ldots,evk_t}(\mathsf{HE.Enc}_{pk_0}(s))) \neq f(s)] \leq \mathrm{negl}(\kappa).$$

In the above expression, we slightly abuse notation and write $\mathsf{HE.Eval}_{evk_0,\ldots,evk_t}$ to include the t recryption steps that are performed during TrapTP.Eval. As long as the number of T gates, and thus the number of recryptions, is polynomial in κ, the expression holds.

Thus, the probability that TrapTP′.VerDec and TrapTP″.VerDec use different classical values (decrypting from the log vs. computing from the initial values) is negligible. Since this is the only place where the two schemes differ, the output of the two VerDec functions will be identical, except with negligible probability. Thus $\mathcal{A}$ will either win in both $\mathsf{Hyb}_{\mathcal{A},\mathsf{TrapTP}'}(\kappa)$ and $\mathsf{Hyb}_{\mathcal{A},\mathsf{TrapTP}''}(\kappa)$, or lose in both, again except with negligible probability. □

More Hybrids: Removing Gadgets. We continue by defining a sequence of hybrid schemes based on TrapTP''. In $4t$ steps, we will move all error-correction functionality from the gadgets to VerDec. This will imply that the adversary has no information about the classical secret keys (which are involved in constructing these gadgets). This will allow us to eventually reduce the security of TrapTP to that of TC.

We remove the gadgets back-to-front, starting with the final gadget. Every gadget is removed in four steps. For all $1 \leq \ell \leq t$, define the hybrids $\mathsf{TrapTP}_1^{(\ell)}$, $\mathsf{TrapTP}_2^{(\ell)}$, $\mathsf{TrapTP}_3^{(\ell)}$, and $\mathsf{TrapTP}_4^{(\ell)}$ (with $\mathsf{TrapTP}_4^{(t+1)} := \mathsf{TrapTP}''$) as follows:

1. $\mathsf{TrapTP}_1^{(\ell)}$ is the same as $\mathsf{TrapTP}_4^{(\ell+1)}$ (or, in the case that $\ell = t$, the same as TrapTP''), except for the generation of the state Γ_ℓ (see Algorithm 1, line 15). In $\mathsf{TrapTP}_1^{(\ell)}$, all classical information encrypted under pk_ℓ is replaced with encryptions of zeros. In particular, for $i \geq \ell$, line 15 is adapted to

$$\Gamma_i \leftarrow \mathsf{MAC.Sign}(\mathsf{HE.Enc}_{pk_i}(00\cdots 0)) \otimes \mathsf{TrapTP}''.\mathsf{Enc}'(sk', \gamma_i^{\mathsf{mid}}) \otimes \mathsf{TrapTP.Enc}(sk, \gamma_i^{\mathsf{in}} \otimes \gamma_i^{\mathsf{out}})$$

where $\mathsf{TrapTP}''.\mathsf{Enc}'$ also appends a signed encryption of zeros, effectively replacing line 1 in Algorithm 3 with

$$\widetilde{\sigma} \leftarrow \sum_{x,z \in \{0,1\}^{3m}} \Big(\mathsf{TC.Enc}((\pi, x, z), \sigma) \otimes \mathsf{MAC.Sign}_k(\mathsf{HE.Enc}_{pk}(00\cdots 0))\Big)$$

It is important to note that in both KeyGen and Enc', the information that is sent to VerDec through the classical side channel is *not* replaced with zeros. Hence, the structural and encryption information about Γ_ℓ is kept from the adversary, and instead is directly sent (only) to the verification procedure. Whenever VerDec needs this information, it is taken directly from this trusted source, and the all-zero string sent by the adversary will be ignored.

2. $\mathsf{TrapTP}_2^{(\ell)}$ is the same as $\mathsf{TrapTP}_1^{(\ell)}$, except that for the ℓth gadget, the procedure $\mathsf{TrapTP.PostGadgetGen}$ is called instead of $\mathsf{TrapTP.GadgetGen}$:

Algorithm 11. $\mathsf{TrapTP.PostGadgetGen}(sk_i)$

1: $g_i \leftarrow 0^{|g(sk_i)|}$
2: $(\gamma^{\mathsf{in}}, \gamma^{\mathsf{mid}}, \gamma^{\mathsf{out}}) \leftarrow$ halves of EPR pairs (send other halves to VerDec)
3: **return** $(g_i, \gamma_i^{\mathsf{in}}, \gamma_i^{\mathsf{mid}}, \gamma_i^{\mathsf{out}})$

This algorithm produces a 'gadget' in which all qubits are replaced with halves of EPR pairs. These still get encrypted in line 15 of Algorithm 1. All other halves of these EPR pairs are sent to VerDec through the provided quantum channel. $\mathsf{TrapTP}_2^{(\ell)}.\mathsf{VerDec}$ has access to the structural information g_ℓ (as this is sent via the classical side information channel from KeyGen to VerDec) and performs the necessary Bell measurements to recreate $\gamma_\ell^{\mathsf{in}}$, $\gamma_\ell^{\mathsf{mid}}$ and $\gamma_\ell^{\mathsf{out}}$ after the adversary has interacted with the EPR pair halves. Effectively, this postpones the generation of the gadget structure to decryption time. Of course, the measurement outcomes are taken into account by VerDec when calculating updates to the quantum one-time pad. As can be seen from the description of $\mathsf{TrapTP}_4^{(\ell)}$, all corrections that follow the ℓth one are unaffected by the fact that the server cannot hold the correct information about these postponed measurements, not even in encrypted form.

3. $\mathsf{TrapTP}_3^{(\ell)}$ is the same as $\mathsf{TrapTP}_2^{(\ell)}$, except that gadget generation for the ℓth gadget is handled by TrapTP.FakeGadgetGen instead of TrapTP.PostGadgetGen.

Algorithm 12. TrapTP.FakeGadgetGen(sk_i)

1: $g_i \leftarrow 0^{|g(sk_i)|}$
2: $(\gamma^{\mathsf{in}}, \gamma^{\mathsf{mid}}, \gamma^{\mathsf{out}}) \leftarrow$ halves of EPR pairs (send other halves to VerDec)
3: Send γ^{mid} to VerDec as well
4: **return** $(g_i, \gamma_i^{\mathsf{in}}, |00\cdots 0\rangle, \gamma_i^{\mathsf{out}})$

This algorithm prepares, instead of halves of EPR pairs, $|0\rangle$-states of the appropriate dimension for $\gamma_\ell^{\mathsf{mid}}$. (Note that this dimension does not depend on $sk_{\ell-1}$). For $\gamma_\ell^{\mathsf{in}}$ and $\gamma_\ell^{\mathsf{out}}$, halves of EPR pairs are still generated, as in $\mathsf{TrapTP}_2^{(\ell)}$. Via the side channel, the full EPR pairs for $\gamma_\ell^{\mathsf{mid}}$ are sent to VerDec. As in the previous hybrids, the returned gadget is encrypted in TrapTP.KeyGen.

$\mathsf{TrapTP}_3^{(\ell)}$.VerDec verifies that the adversary performed the correct Bell measurements on the fake ℓth gadget by calling TC.VerDec. If this procedure accepts, $\mathsf{TrapTP}_3^{(\ell)}$.VerDec performs the verified Bell measurements on the halves of the EPR pairs received from $\mathsf{TrapTP}_3^{(\ell)}$.KeyGen (and subsequently performs the Bell measurements that depend on g_ℓ on the other halves, as in $\mathsf{TrapTP}_2^{(\ell)}$). Effectively, $\mathsf{TrapTP}_3^{(\ell)}$.VerDec thereby performs a protocol for HE.Dec, removing the phase error in the process.

4. $\mathsf{TrapTP}_4^{(\ell)}$ is the same as $\mathsf{TrapTP}_3^{(\ell)}$, except that VerDec (instead of performing the Bell measurements of the gadget protocol) uses its knowledge of the initial QOTP keys and all intermediate measurement outcomes to compute whether or not a phase correction is necessary after the ℓth T gate. $\mathsf{TrapTP}_4^{(\ell)}$.VerDec then performs this phase correction on the EPR half entangled with $\gamma_\ell^{\mathsf{in}}$, followed by a Bell measurement with the EPR half entangled with $\gamma_\ell^{\mathsf{out}}$.

The first $\ell - 1$ gadgets in $\mathsf{TrapTP}_1^{(\ell)}$ through $\mathsf{TrapTP}_4^{(\ell)}$ are always functional gadgets, as in TrapTP. The last $t - \ell$ gadgets are all completely replaced by dummy states, and their functionality is completely outsourced to VerDec. In four steps described above, the functionality of the ℓth gadget is also transferred to VerDec. It is important to replace only one gadget at a time, because replacing a real gadget with a fake one breaks the functionality of the gadgets that occur later in the evaluation: the encrypted classical information held by the server does not correspond to the question of whether or not a phase correction is needed. By completely outsourcing the phase correction to VerDec, as is done for all gadgets after the ℓth one in all $\mathsf{TrapTP}_i^{(\ell)}$ schemes, we ensure that this incorrect classical information does not influence the outcome of the computation. Hence, correctness is maintained throughout the hybrid transformations. We now show that these transformations of the scheme do not significantly affect the adversary's winning probability in the hybrid indistinguishability game.

Lemma 3. *For any QPT $\mathcal{A}$, there exists a negligible function* negl *such that for all* $1 \leq \ell \leq t$, $\mathsf{AdvHyb}_{\mathsf{TrapTP}_1^{(\ell)}}^{\mathsf{TrapTP}_4^{(\ell+1)}}(\mathcal{A}, \kappa) \leq \text{negl}(\kappa)$.

Proof (sketch). In $\mathsf{TrapTP}_4^{(\ell+1)}$, no information about $sk_{(\ell)}$ is sent to the adversary. In the original TrapTP scheme, the structure of the quantum state $\Gamma_{\ell+1}$ depended on it, but this structure has been replaced with dummy states in several steps in $\mathsf{TrapTP}_2^{\ell+1}$ through $\mathsf{TrapTP}_4^{\ell+1}$.

This is fortunate, since if absolutely no secret-key information is present, we are able to bound the difference in winning probability between $\mathsf{Hyb}_{\mathcal{A},\mathsf{TrapTP}_4^{(\ell+1)}}(\kappa)$ and $\mathsf{Hyb}_{\mathcal{A},\mathsf{TrapTP}_1^{\ell}}(\kappa)$ by reducing it to the IND-CPA security against quantum adversaries [6] of the classical homomorphic encryption scheme HE.

The proof is closely analogous to the proof of Lemma 1 in [12], and on a high level it works as follows. Let $\mathcal{A} = (\mathcal{A}_1, \mathcal{A}_2, \mathcal{A}_3)$ be a QPT adversary for the game $\mathsf{Hyb}_{\mathcal{A},\mathsf{TrapTP}_1^{(\ell)}}(\kappa)$ or $\mathsf{Hyb}_{\mathcal{A},\mathsf{TrapTP}_4^{(\ell+1)}}(\kappa)$ (we do not need to specify for which one, since they both require the same input/output interface). A new quantum adversary $\mathcal{A}'$ for the classical IND-CPA indistinguishability game is defined by having the adversary taking the role of challenger in either the game $\mathsf{Hyb}_{\mathcal{A},\mathsf{TrapTP}_1^{(\ell)}}(\kappa)$ or the game $\mathsf{Hyb}_{\mathcal{A},\mathsf{TrapTP}_4^{(\ell+1)}}(\kappa)$. Which game is simulated depends on the coin flip of the challenger for the IND-CPA indistinguishability game, and is unknown to $\mathcal{A}'$. This situation is achieved by having $\mathcal{A}'$ send any classical plaintext that should be encrypted under pk_ℓ to the challenger, so that either that plaintext is encrypted or a string of zeros is.

Based on the guess of the simulated $\mathcal{A}$, which $\mathcal{A}'$ can verify to be correct or incorrect in his role of challenger, $\mathcal{A}'$ will guess which of the two games was just simulated. By IND-CPA security of the classical scheme against quantum adversaries, $\mathcal{A}'$ cannot succeed in this guessing game with nonnegligible advantage over random guessing. This means that the winning probability of $\mathcal{A}$ in both games cannot differ by a lot. For details, we refer the reader the proof of Lemma 5, in which a very similar approach is taken.

Technically, the success probability of $\mathcal{A}'$, and thus the function negl, may depend on ℓ. A standard randomizing argument, as found in e.g. the discussion of hybrid arguments in [16], allows us to get rid of this dependence by defining another adversary $\mathcal{A}''$ that selects a random value of j, and then bounding the advantage of $\mathcal{A}''$ by a negligible function that is independent of j. □

Lemma 4. *For $1 \leq \ell \leq t$ and any QPT $\mathcal{A}$,* $\mathsf{AdvHyb}^{\mathsf{TrapTP}_2^{(\ell)}}_{\mathsf{TrapTP}_1^{(\ell)}}(\mathcal{A}, \kappa) = 0$.

Proof. In $\mathsf{TrapTP}_1^{(\ell)}$, the ℓth error-correction gadget consists of a number of EPR pairs arranged in a certain order, as described by the garden-hose protocol for HE.Dec. For example, this protocol may dictate that the ith and jth qubit of the gadget must form an EPR pair $|\Phi^+\rangle$ together. This can alternatively be achieved by creating two EPR pairs, placing half of each pair in the ith and jth position of the gadget state, and performing a Bell measurement on the other two halves. This creates a Bell pair $\mathsf{X}^a\mathsf{Z}^b|\Phi^+\rangle$ in positions i and j, where $a, b \in \{0, 1\}$ describe the outcome of the Bell measurement.

From the point of view of the adversary, it does not matter whether these Bell measurements are performed during KeyGen, or whether the halves of EPR pairs are sent to VerDec for measurement – because the key to the quantum one-time pad of the ℓth gadget is not sent to the adversary at all, the same state is created with a completely random Pauli in either case. Of course, the teleportation correction Paulis of the form $\mathsf{X}^a\mathsf{Z}^b$ need to be taken into account when updating the keys to the quantum one-time pad on the data qubits after the gadget is used. VerDec has all the necessary information to do this, because it observes the measurement outcomes, and computes the key updates itself (instead of decrypting the final keys from the computation log).

Thus, with the extra key update steps in $\mathsf{TrapTP}_2^{(\ell)}$.VerDec, the inputs to the adversary are exactly the same in the games of $\mathsf{TrapTP}_1^{(\ell)}$ and $\mathsf{TrapTP}_2^{(\ell)}$. □

Lemma 5. *For any QPT $\mathcal{A}$, there exists a negligible function* negl *such that for all $1 \leq \ell \leq t$,* $\mathsf{AdvHyb}^{\mathsf{TrapTP}_3^{(\ell)}}_{\mathsf{TrapTP}_2^{(\ell)}}(\mathcal{A}, \kappa) \leq \text{negl}(\kappa)$.

Proof. We show this by reducing the difference in winning probabilities in the statement of the lemma to the IND-VER security of the somewhat homomorphic scheme TC. Intuitively, because TC is IND-VER, if $\mathsf{TrapTP}_2^{(\ell)}$ accepts the adversary's claimed circuit of Bell measurements on the EPR pair halves, the effective map on those EPR pairs is the claimed circuit. Therefore, we might just as well ask VerDec to apply this map, as we do in $\mathsf{TrapTP}_3^{(\ell)}$, to get the same output state. If $\mathsf{TrapTP}_2^{(\ell)}$ rejects the adversary's claimed circuit on those EPR pair halves, then $\mathsf{TrapTP}_3^{(\ell)}$ should reject too. This is why we let the adversary act on an encrypted dummy state of $|0\rangle$s.

Let $\mathcal{A} = (\mathcal{A}_1, \mathcal{A}_2, \mathcal{A}_3)$ be a set of QPT algorithms on the appropriate registers, so that we can consider it as an adversary for the hybrid indistinguishability game for either $\mathsf{TrapTP}_2^{(\ell)}$ or $\mathsf{TrapTP}_3^{(\ell)}$ (see Definition 7). Note the input/output wires to the adversary in both these games are identical, so we can evaluate $\Pr[\mathsf{Hyb}_{\mathcal{A},\mathsf{TrapTP}_2^{(\ell)}}(\kappa) = 1]$ and $\Pr[\mathsf{Hyb}_{\mathcal{A},\mathsf{TrapTP}_3^{(\ell)}}(\kappa) = 1]$ for the same $\mathcal{A}$.

Now define an adversary $\mathcal{A}' = (\mathcal{A}'_1, \mathcal{A}'_2, \mathcal{A}'_3)$ for the IND-VER game against TC, $\mathsf{VerGame}_{\mathcal{A}',\mathsf{TC}}(\kappa)$, as follows:

1. $\mathcal{A}'_1$: Run $\mathsf{TrapTP}_2^{(\ell)}$.KeyGen until the start of line 15 in the ℓth iteration of that loop. Up to this point, $\mathsf{TrapTP}_2^{(\ell)}$.KeyGen is identical to $\mathsf{TrapTP}_3^{(\ell)}$.KeyGen.

It has generated real gadgets Γ_1 through $\Gamma_{\ell-1}$, and halves of EPR pairs for $\gamma_\ell^{\mathsf{in}}$, $\gamma_\ell^{\mathsf{mid}}$ and $\gamma_\ell^{\mathsf{out}}$. Note furthermore that the permutation π_ℓ is used nowhere. Now send $\gamma_\ell^{\mathsf{mid}}$ to the challenger via the register X, and everything else (including sk) to $\mathcal{A}'_2$ via the side register R.

2. $\mathcal{A}'_2$: Continue $\mathsf{TrapTP}_2^{(\ell)}.\mathsf{KeyGen}$ using the response from the challenger instead of $\mathsf{TrapTP.Enc}'(sk', \gamma_\ell^{\mathsf{mid}})$ on line 15 in the ℓth iteration. Call the result ρ_{evk}. Again, this part of the key generation procedure is identical for $\mathsf{TrapTP}_2^{(\ell)}$ and $\mathsf{TrapTP}_3^{(\ell)}$. Start playing the hybrid indistinguishability game with $\mathcal{A}$:

- Flip a bit $r \in \{0, 1\}$.
- Send ρ_{evk} to $\mathcal{A}_1$. If $r = 0$, encrypt the response of $\mathcal{A}_1$ using the secret key sk generated by $\mathcal{A}'_1$. Note that for this, the permutation π_ℓ is also not needed. If $r = 1$, encrypt a $|0\rangle$ state of appropriate dimension instead.
- Send the resulting encryption, along with the side info from $\mathcal{A}_1$, to $\mathcal{A}_2$.
- On the output of $\mathcal{A}_2$, start running $\mathsf{TrapTP}_2^{(\ell)}.\mathsf{VerDec}$ until the actions on the ℓth gadget need to be verified. Since the permutation on the state $\gamma_\ell^{\mathsf{mid}}$ is unknown to $\mathcal{A}'_2$ (it was sent to the challenger for encryption), it cannot verify this part of the computation.
- Instead, send the relevant part of the computation log to the challenger for verification, along with the relevant part of the claimed circuit (the Bell measurements on the gadget state), and the relevant qubits, all received from $\mathcal{A}_2$, to the challenger for verification and decryption.
- In the meantime, send the rest of the working memory to $\mathcal{A}'_3$ via register R'.

3. $\mathcal{A}'_3$: Continue the simulation of the hybrid game with $\mathcal{A}$:

- If the challenger rejects, reject and replace the entire quantum state by the fixed dummy state Ω.
- If the challenger accepts, then we know that the challenger applies the claimed subcircuit to the quantum state it did not encrypt (either $|0\rangle$ or $\gamma_\ell^{\mathsf{mid}}$), depending on the bit the challenger flipped), and possibly swaps this state back in (again depending on which bit it flipped). Continue the $\mathsf{TrapTP}_2^{(\ell)}.\mathsf{VerDec}$ computation for the rest of the computation log.
- Send the result (the output quantum state, the claimed circuit, and the accept/reject flag) to $\mathcal{A}_3$, and call its output bit r'.

Output 0 if $r = r'$, and 1 otherwise. (i.e., output $NEQ(r, r')$)

Recall from Definition 7 that the challenger flips a coin (let us call the outcome $s \in \{0, 1\}$) to decide whether to encrypt the quantum state provided by $\mathcal{A}'$, or to swap in an all-zero dummy state before encrypting. Keeping this in mind while inspecting the definition of $\mathcal{A}'$, one can see that whenever $s = 0$, $\mathcal{A}'$ takes the role of challenger in the game $\mathsf{Hyb}_{\mathcal{A},\mathsf{TrapTP}_2^{(\ell)}}(\kappa)$ with $\mathcal{A}$, and whenever $s = 1$, they play $\mathsf{Hyb}_{\mathcal{A},\mathsf{TrapTP}_3^{(\ell)}}(\kappa)$. Now let us consider when the newly defined adversary $\mathcal{A}'$ wins the VER indistinguishability game for TC. If $s = 0$, $\mathcal{A}'$ needs to output a bit $s' = 0$ to win. This happens, by definition of $\mathcal{A}'$, if and only if $\mathcal{A}$ wins the game $\mathsf{Hyb}_{\mathcal{A},\mathsf{TrapTP}_2^{(\ell)}}(\kappa)$ (i.e. $r = r'$). On the other hand, if $s = 1$, $\mathcal{A}'$

needs to output a bit $s' = 1$ to win. This happens, by definition of $\mathcal{A}'$, if and only if $\mathcal{A}$ loses the game $\mathsf{Hyb}_{\mathcal{A},\mathsf{TrapTP}_3^{(\ell)}}(\kappa)$ (i.e. $r \neq r'$). Thus the winning probability of $\mathcal{A}'$ is:

$$\begin{aligned}
&\Pr[\mathsf{VerGame}_{\mathcal{A}',\mathsf{TC}}(\kappa) = 1] \\
&= \Pr[s = 0] \cdot \Pr[\mathsf{Hyb}_{\mathcal{A},\mathsf{TrapTP}_2^{(\ell)}}(\kappa) = 1] + \Pr[s = 1] \cdot \Pr[\mathsf{Hyb}_{\mathcal{A},\mathsf{TrapTP}_3^{(\ell)}}(\kappa) = 0] \\
&= \frac{1}{2} \Pr[\mathsf{Hyb}_{\mathcal{A},\mathsf{TrapTP}_2^{(\ell)}}(\kappa) = 1] + \frac{1}{2}\Big(1 - \Pr[\mathsf{Hyb}_{\mathcal{A},\mathsf{TrapTP}_3^{(\ell)}}(\kappa) = 1]\Big) \\
&= \frac{1}{2} + \frac{1}{2}\Big(\Pr[\mathsf{Hyb}_{\mathcal{A},\mathsf{TrapTP}_2^{(\ell)}}(\kappa) = 1] - \Pr[\mathsf{Hyb}_{\mathcal{A},\mathsf{TrapTP}_3^{(\ell)}}(\kappa) = 1]\Big)
\end{aligned}$$

From the IND-VER property of TC (see Theorem 4) we know that the above is at most $\frac{1}{2} + \text{negl}(\kappa)$. From this (and a randomizing argument similar to Lemma 3), the statement of the lemma follows directly. □

Lemma 6. *For any QPT $\mathcal{A}$, there exists a negligible function* negl *such that for all* $1 \leq \ell \leq t$, $\mathsf{AdvHyb}^{\mathsf{TrapTP}_4^{(\ell)}}_{\mathsf{TrapTP}_3^{(\ell)}}(\mathcal{A}, \kappa) \leq \text{negl}(\kappa)$.

Proof. Let $f(s)$ be the bit that, after the ℓth T gate, determines whether or not a phase correction is necessary. Here, s is all the relevant starting information (such as quantum one-time pad keys, gadget structure, permutations, and applied circuit), and f is some function that determines the X key on the relevant qubit right before application of the T gate.

In $\mathsf{TrapTP}_3^{(\ell)}$, a phase correction after the ℓth T gate is applied conditioned on the outcome of

$$\mathsf{HE.Dec}_{sk_{\ell-1}}(\mathsf{HE.Eval}^{f}_{evk_0,\ldots,evk_{\ell-1}}(\mathsf{HE.Enc}_{pk_0}(s))),$$

because the garden-hose computation in the gadget computes the classical decryption. In the above expression, we again slightly abuse notation, as in the proof of Lemma 2, and include recryption steps in $\mathsf{HE.Eval}_{evk_0,\ldots,\mathsf{evk}_{\ell-1}}$. As long as t is polynomial in κ, we have, by correctness of HE,

$$\Pr[\mathsf{HE.Dec}_{sk_{\ell-1}}(\mathsf{HE.Eval}^{f}_{evk_0,\ldots,evk_{\ell-1}}(\mathsf{HE.Enc}_{pk_0}(s))) \neq f(s)] \leq \text{negl}(\kappa).$$

In $\mathsf{TrapTP}_4^{(\ell)}$, the only difference from $\mathsf{TrapTP}_3^{(\ell)}$ is that, instead of performing the garden-hose computation on the result of the classical homomorphic evaluation procedure, the phase correction is applied directly by VerDec, conditioned on $f(s)$. The probability that in $\mathsf{TrapTP}_4^{(\ell)}$, a phase is applied (or not) when in $\mathsf{TrapTP}_3^{(\ell)}$ it is not (or is), is negligible. The claim follows directly. □

Final Hybrid: Removing All Classical FHE. In $\mathsf{TrapTP}_4^{(1)}$, all of the error-correction gadgets have been removed from the evaluation key, and the error-correction functionality has been redirected to VerDec completely. Effectively, $\mathsf{TrapTP}_4^{(1)}.\mathsf{KeyGen}$ samples a permutation π, generates a lot of magic states (for

P, H and T) and encrypts them using $\mathsf{TC.Enc}_\pi$, after which the keys to the quantum one-time pad used in that encryption are homomorphically encrypted under pk_0. The adversary is allowed to act on those encryptions, but while its homomorphic computations are syntactically checked in the log, VerDec does not decrypt and use the resulting values. This allows us to link $\mathsf{TrapTP}_4^{(1)}$ to a final hybrid, TrapTP^f, where all classical information is replaced with zeros before encrypting.

The proof of the following lemma is analogous to that of Lemma 3, and reduces to the IND-CPA security of the classical scheme HE:

Lemma 7. *For any QPT* $\mathcal{A}$, $\mathsf{AdvHyb}_{\mathsf{TrapTP}_4^{(1)}}^{\mathsf{TrapTP}^f}(\mathcal{A}, \kappa) \leq \mathrm{negl}(\kappa)$.

Proof of Main Theorem. Considering TrapTP^f in more detail, we can see that it is actually very similar to TC. This allows us to prove the following lemma, which is the last ingredient for the proof of verifiability of TrapTP.

Lemma 8. *For any QPT* $\mathcal{A}$, $\Pr[\mathsf{Hyb}_{\mathcal{A},\mathsf{TrapTP}^f}(\kappa) = 1] \leq \frac{1}{2} + \mathrm{negl}(\kappa)$.

Proof. To see the similarity with TC, consider the four algorithms of TrapTP^f.

In $\mathsf{TrapTP}^f.\mathsf{KeyGen}$, a permutation π is sampled, and magic states for P, H and T are generated, along with some EPR pair halves (to replace in_i and out_i). For all generated quantum states, random keys for QOTPs are sampled, and the states are encrypted using $\mathsf{TC.Enc}$ with these keys as secret keys. No classical FHE is present anymore. Thus, $\mathsf{TrapTP}^f.\mathsf{KeyGen}$ can be viewed as $\mathsf{TC.KeyGen}$, followed by $\mathsf{TC.Enc}$ on the magic states and EPR pair halves.

$\mathsf{TrapTP}^f.\mathsf{Enc}$ is identical to $\mathsf{TC.Enc}$, only the keys to the quantum one-time pad are sampled on the fly and sent to $\mathsf{TrapTP}^f.\mathsf{VerDec}$ via a classical side-channel, whereas $\mathsf{TC.VerDec}$ receives them as part of the secret key. Since the keys are used exactly once and not used anywhere else besides in Enc and VerDec, this difference does not affect the outcome of the game.

$\mathsf{TrapTP}^f.\mathsf{Eval}$ only requires CNOT, classically controlled Paulis, and computational/Hadamar basis measurements. For the execution of any other gate, it suffices to apply a circuit of those gates to the encrypted data, encrypted magic states and/or encrypted EPR halves.

$\mathsf{TrapTP}^f.\mathsf{VerDec}$ does two things: (i) it syntactically checks the provided computation log, and (ii) it runs $\mathsf{TC.VerDec}$ to verify that the evaluation procedure correctly applied the circuit of CNOTs and measurements.

An execution of $\mathsf{Hyb}_{\mathcal{A},\mathsf{TrapTP}^f}(\kappa)$ for any $\mathcal{A}$ corresponds to the two-round VER indistinguishability game for TC as follows. Let $\mathcal{A} = (\mathcal{A}_1, \mathcal{A}_2, \mathcal{A}_3)$ be a polynomial-time adversary for the game $\mathsf{Hyb}_{\mathcal{A},\mathsf{TrapTP}^f}(\kappa)$. Define an additional QPT $\mathcal{A}_0$ that produces magic states and EPR pair halves to the register X_1. The other halves of the EPR pairs are sent through R, and untouches by $\mathcal{A}_1$ and $\mathcal{A}_2$. The above analysis shows that the adversary $\mathcal{A}' = (\mathcal{A}_0, \mathcal{A}_1, \mathcal{A}_2, \mathcal{A}_3)$ can be viewed as an adversary for the VER-2 indistinguishability game $\mathsf{VerGame}^2_{\mathcal{A}',\mathsf{TC}}(\kappa)$ and wins whenever $\mathsf{Hyb}_{\mathcal{A},\mathsf{TrapTP}^f}(\kappa) = 1$. The other direction does not hold: $\mathcal{A}$

loses the hybrid indistinguishability game if $\mathsf{TrapTP}^f.\mathsf{VerDec}$ rejects check (i), but accepts check (ii) (see above). In this case, $\mathcal{A}'$ would still win the VER-2 indistinguishability game. Hence,

$$\Pr[\mathsf{Hyb}_{\mathcal{A},\mathsf{TrapTP}^f}(\kappa) = 1] \leq \Pr[\mathsf{VerGame}^2_{\mathcal{A}',\mathsf{TC}}(\kappa) = 1].$$

Theorem 4 yields $\Pr[\mathsf{VerGame}^2_{\mathcal{A}',\mathsf{TC}}(\kappa) = 1] \leq \frac{1}{2} + \mathrm{negl}(\kappa)$, and the result follows. □

Theorem 5. *The vQFHE scheme* TrapTP *satisfies* κ*-SEM-VER.*

Proof. From Lemmas 1, 2, 3, 4, 5, 6, and 7, we may conclude that if t (the number of T gates in the circuit) is polynomial in κ (the security parameter), then for any polynomial-time adversary $\mathcal{A}$,

$$\Pr[\mathsf{VerGame}_{\mathcal{A},\mathsf{TrapTP}}(\kappa) = 1] \quad - \quad \Pr[\mathsf{Hyb}_{\mathcal{A},\mathsf{TrapTP}^f}(\kappa) = 1] \quad \leq \quad \mathrm{negl}(\kappa),$$

since the sum poly-many negligible terms is negligible (it is important to note that there is only a constant number of *different* negligible terms involved). By Lemma 8, which reduces verifiability of TrapTP^f to verifiability of TC, $\Pr[\mathsf{Hyb}_{\mathcal{A},\mathsf{TrapTP}^f}(\kappa) = 1] \leq 1/2 + \mathrm{negl}(\kappa)$. It follows that $\Pr[\mathsf{VerGame}_{\mathcal{A},\mathsf{TrapTP}}(\kappa) = 1] \leq 1/2 + \mathrm{negl}(\kappa)$, i.e., that TrapTP is κ-IND-VER. By Theorem 3, TrapTP is also κ-SEM-VER. □

6 Application to Quantum One-Time Programs

One-Time Programs. We now briefly sketch an application of the vQFHE scheme to one-time programs. A classical one-time program (or cOTP) is an idealized object which can be used to execute a function once, but then self-destructs. In the case of a quantum OTP (or qOTP), the program executes a quantum channel Φ. In the usual formalization, Φ has two inputs and is public. One party (the sender) creates the qOTP by fixing one input, and the qOTP is executed by a receiver who selects the other input. To recover the intuitive notion of OTP, choose Φ to be a universal circuit. We will work in the universally-composable (UC) framework, following the approach of [9]. We thus first define the ideal functionality of a qOTP.

Definition 8 (Functionality 3 in [9]**).** *The ideal functionality* $\mathcal{F}^{\mathsf{OTP}}_{\Phi}$ *for a channel* $\Phi_{XY\to Z}$ *is the following:*

1. **Create:** *given register X from sender, store X and send* create *to receiver.*
2. **Execute:** *given register Y from receiver, send Φ applied to XY to receiver. Delete any trace of this instance.*

A qOTP is then a real functionality which "UC-emulates" the ideal functionality [22]. As in [9], we only allow corrupting receivers; unlike [9], we consider computational (rather than statistical) UC security. The achieved result is therefore slightly weaker. The construction within our vQFHE framework is however much simpler, and shows the relative ease with which applications of vQFHE can be constructed.

The Construction. Choose a vQFHE scheme $\Pi = (\mathsf{KeyGen}, \mathsf{Enc}, \mathsf{Eval}, \mathsf{VerDec})$ satisfying SEM-VER. For simplicity, we first describe the classical input/output case, i.e., the circuit begins and ends with full measurement of all qubits. Let C be such a circuit, for the map $\Phi_{XY \to Z}$. On **Create**, the sender generates keys $(k, \rho_{\mathsf{evk}}) \leftarrow \mathsf{KeyGen}$ and encrypts their input register X using k. The sender also generates a classical OTP for the public, classical function VerDec, choosing the circuit and key inputs to be C and k; the computation log is left open for the receiver to select. The qOTP is then the triple

$$\Xi_C^X := (\rho_{\mathsf{evk}}, \mathsf{Enc}_k(\rho_X), \mathsf{OTP}_{\mathsf{VerDec}}(C, k)) .$$

On **Execute**, the receiver computes as follows. The receiver's (classical) input Y together with the (public) circuit C defines a homomorphic computation on the ciphertext $\mathsf{Enc}_k(\rho_X)$, which the receiver can perform using Eval and ρ_{evk}. Since C has only classical outputs, the receiver measures the final state completely. At the end of that computation, the receiver holds the (completely classical) output of the computation log from Eval. The receiver plugs the log into $\mathsf{OTP}_{\mathsf{VerDec}}(C, k)$, which produces the decrypted output.

We handle the case of arbitrary circuits C (with quantum input and output) as follows. Following the ideas of [9], we augment the above quantum OTP with two auxiliary quantum states: an "encrypt-through-teleport" gadget σ_{in} and a "decrypt-through-teleport" gadget σ_{out}. These are maximally entangled states with the appropriate map (encrypt or decrypt) applied to one half. The receiver uses teleportation on $\sigma^{\mathsf{in}}_{Y_1 W_1}$ to encrypt their input register Y before evaluating, and places the teleportation measurements into the computation log. After evalution, the receiver uses $\sigma^{\mathsf{out}}_{W_2 Y_2}$ to teleport the plaintext out, combining the teleportation measurements with the output of $\mathsf{OTP}_{\mathsf{VerDec}}(C, k)$ to compute the final QOTP decryption keys.

Security Proof Sketch. Starting with a QPT adversary $\mathcal{A}$ which attacks the real functionality, we construct a QPT simulator $\mathcal{S}$ which attacks the ideal functionality (with similar success probability). We split $\mathcal{A}$ into $\mathcal{A}_1$ (receive input, output the OTP query and side information) and $\mathcal{A}_2$ (receive result of OTP query and side information, produce final output). The simulator $\mathcal{S}$ will generate its own keys, provide fake gadgets that will trick $\mathcal{A}$ into teleporting its input to $\mathcal{S}$, who will then use that input on the ideal functionality. Details follow.

The simulator first generates $(k, \rho_{\mathsf{evk}}) \leftarrow \mathsf{KeyGen}$ and encrypts the input X via Enc_k. Instead of the encrypt gadget $\sigma^{\mathsf{in}}_{Y_1 W_1}$, $\mathcal{S}$ provides half of a maximally entangled state in register Y and likewise in register W. The other halves Y_1' and W_1' of these entangled states are kept by $\mathcal{S}$. The same is done in place of the decrypt gadget $\sigma^{\mathsf{out}}_{W_2 Y_2}$, with $\mathcal{S}$ keeping Y_2' and W_2'. Then $\mathcal{S}$ runs $\mathcal{A}_1$ with input $\rho_{\mathsf{evk}}, \mathsf{Enc}_k(\rho_X)$ and registers Y and W. It then executes VerDec_k on the output (i.e., the query) of $\mathcal{A}_1$ to see if $\mathcal{A}_1$ correctly followed the Eval protocol. If it did not, then $\mathcal{S}$ aborts; otherwise, $\mathcal{S}$ plugs register Y_1' into the ideal functionality, and then teleports the output into register W_2'. Before responding to $\mathcal{A}_2$, it corrects the one-time pad keys appropriately using its teleportation measurements.

7 Conclusion

In this work, we devised a new quantum-cryptographic primitive: quantum fully-homomorphic encryption with verification (vQFHE). Using the trap code for quantum authentication [9] and the garden-hose gadgets of [12], we constructed a vQFHE scheme TrapTP which satisfies (i) correctness, (ii) compactness, (iii) security of verification, (iv) IND-CPA secrecy, and (v) authentication. We also outlined a first application of vQFHE, to quantum one-time programs.

We leave open several interesting directions for future research. Foremost is finding more applications of vQFHE. Another interesting question is whether vQFHE schemes exist where verification can be done publicly (i.e., without the decryption key), as is possible classically. Finally, it is unknown whether vQFHE (or even QFHE) schemes exist with evaluation key that does not scale with the size of the circuit at all.

Acknowledgements. This work was completed while GA was a member of the QMATH center at the Department of Mathematical Sciences at the University of Copenhagen. GA and FS acknowledge financial support from the European Research Council (ERC Grant Agreement no 337603), the Danish Council for Independent Research (Sapere Aude), Qubiz - Quantum Innovation Center, and VILLUM FONDEN via the QMATH Centre of Excellence (Grant No. 10059). CS is supported by an NWO VIDI grant.

References

1. Aharonov, D., Ben-Or, M., Eban, E.: Interactive proofs for quantum computations. arXiv preprint arXiv:0810.5375 (2008)
2. Alagic, G., Broadbent, A., Fefferman, B., Gagliardoni, T., Schaffner, C., St. Jules, M.: Computational security of quantum encryption. In: Nascimento, A.C.A., Barreto, P. (eds.) ICITS 2016. LNCS, vol. 10015, pp. 47–71. Springer, Cham (2016). https://doi.org/10.1007/978-3-319-49175-2_3
3. Alagic, G., Dulek, Y., Schaffner, C., Speelman, F.: Quantum fully homomorphic encryption with verification. arXiv preprint arXiv:1708.09156 (2017)
4. Barak, B., Brakerski, Z.: Windows on theory: the swiss army knife of cryptography (2012). URL https://windowsontheory.org/2012/05/01/the-swiss-army-knife-of-cryptography/
5. Brakerski, Z., Vaikuntanathan, V.: Efficient fully homomorphic encryption from (standard) LWE. In: 52nd Annual Symposium on Foundations of Computer Science (FOCS), pp. 97–106 (2011). https://doi.org/10.1109/FOCS.2011.12
6. Broadbent, A., Jeffery, S.: Quantum homomorphic encryption for circuits of low T-gate complexity. In: Gennaro, R., Robshaw, M. (eds.) CRYPTO 2015. LNCS, vol. 9216, pp. 609–629. Springer, Heidelberg (2015). https://doi.org/10.1007/978-3-662-48000-7_30
7. Broadbent, A., Wainewright, E.: Efficient simulation for quantum message authentication. arXiv preprint arXiv:1607.03075 (2016)
8. Broadbent, A., Fitzsimons, J., Kashefi, E.: Universal blind quantum computation. In: 50th Annual Symposium on Foundations of Computer Science (FOCS), pp. 517–526. IEEE (2009)

9. Broadbent, A., Gutoski, G., Stebila, D.: Quantum one-time programs. In: Canetti, R., Garay, J.A. (eds.) CRYPTO 2013. LNCS, vol. 8043, pp. 344–360. Springer, Heidelberg (2013). https://doi.org/10.1007/978-3-642-40084-1_20
10. Broadbent, A., Ji, Z., Song, F., Watrous, J.: Zero-knowledge proof systems for QMA. In: 57th Annual Symposium on Foundations of Computer Science (FOCS), pp. 31–40, October 2016. https://doi.org/10.1109/FOCS.2016.13
11. Coladangelo, A., Grilo, A., Jeffery, S., Vidick, T.: Verifier-on-a-leash: new schemes for verifiable delegated quantum computation, with quasilinear resources. arXiv preprint arXiv:1708.02130 (2017)
12. Dulek, Y., Schaffner, C., Speelman, F.: Quantum homomorphic encryption for polynomial-sized circuits. In: Robshaw, M., Katz, J. (eds.) CRYPTO 2016. LNCS, vol. 9816, pp. 3–32. Springer, Heidelberg (2016). https://doi.org/10.1007/978-3-662-53015-3_1
13. Dupuis, F., Nielsen, J.B., Salvail, L.: Actively secure two-party evaluation of any quantum operation. In: Safavi-Naini, R., Canetti, R. (eds.) CRYPTO 2012. LNCS, vol. 7417, pp. 794–811. Springer, Heidelberg (2012). https://doi.org/10.1007/978-3-642-32009-5_46
14. Garg, S., Gentry, C., Halevi, S., Raykova, M., Sahai, A., Waters, B.: Candidate indistinguishability obfuscation and functional encryption for all circuits. In: 54th Annual Symposium on Foundations of Computer Science (FOCS), pp. 40–49, October (2013). https://doi.org/10.1109/FOCS.2013.13
15. Gentry, C.: Fully homomorphic encryption using ideal lattices. In: 41st Annual ACM Symposium on Theory of Computing (STOC), pp. 169–178 (2009). https://doi.org/10.1145/1536414.1536440
16. Katz, J., Lindell, Y.: Introduction to Modern Cryptography. CRC Press, Boca Raton (2014)
17. Mahadev, U.: Classical homomorphic encryption for quantum circuits. arXiv preprint arXiv:1708.02130 (2017)
18. Newman, M., Shi, Y.: Limitations on transversal computation through quantum homomorphic encryption. arXiv e-prints, April 2017
19. Ouyang, Y., Tan, S.-H., Fitzsimons, J.: Quantum homomorphic encryption from quantum codes. arXiv preprint arXiv:1508.00938 (2015)
20. Shor, P.W., Preskill, J.: Simple proof of security of the BB84 quantum key distribution protocol. Phys. Rev. Lett. **85**, 441–444 (2000). https://doi.org/10.1103/PhysRevLett.85.441
21. Tan, S.-H., Kettlewell, J.A., Ouyang, Y., Chen, L., Fitzsimons, J.: A quantum approach to homomorphic encryption. Sci. Rep. **6**, 33467 (2016). https://doi.org/10.1038/srep33467
22. Unruh, D.: Universally composable quantum multi-party computation. In: Gilbert, H. (ed.) EUROCRYPT 2010. LNCS, vol. 6110, pp. 486–505. Springer, Heidelberg (2010). https://doi.org/10.1007/978-3-642-13190-5_25
23. Li, Y., Pérez-Delgado, C.A., Fitzsimons, J.F.: Limitations on information-theoretically-secure quantum homomorphic encryption. Phys. Rev. A **90**, 050303 (2014). https://doi.org/10.1103/PhysRevA.90.050303

Access Control

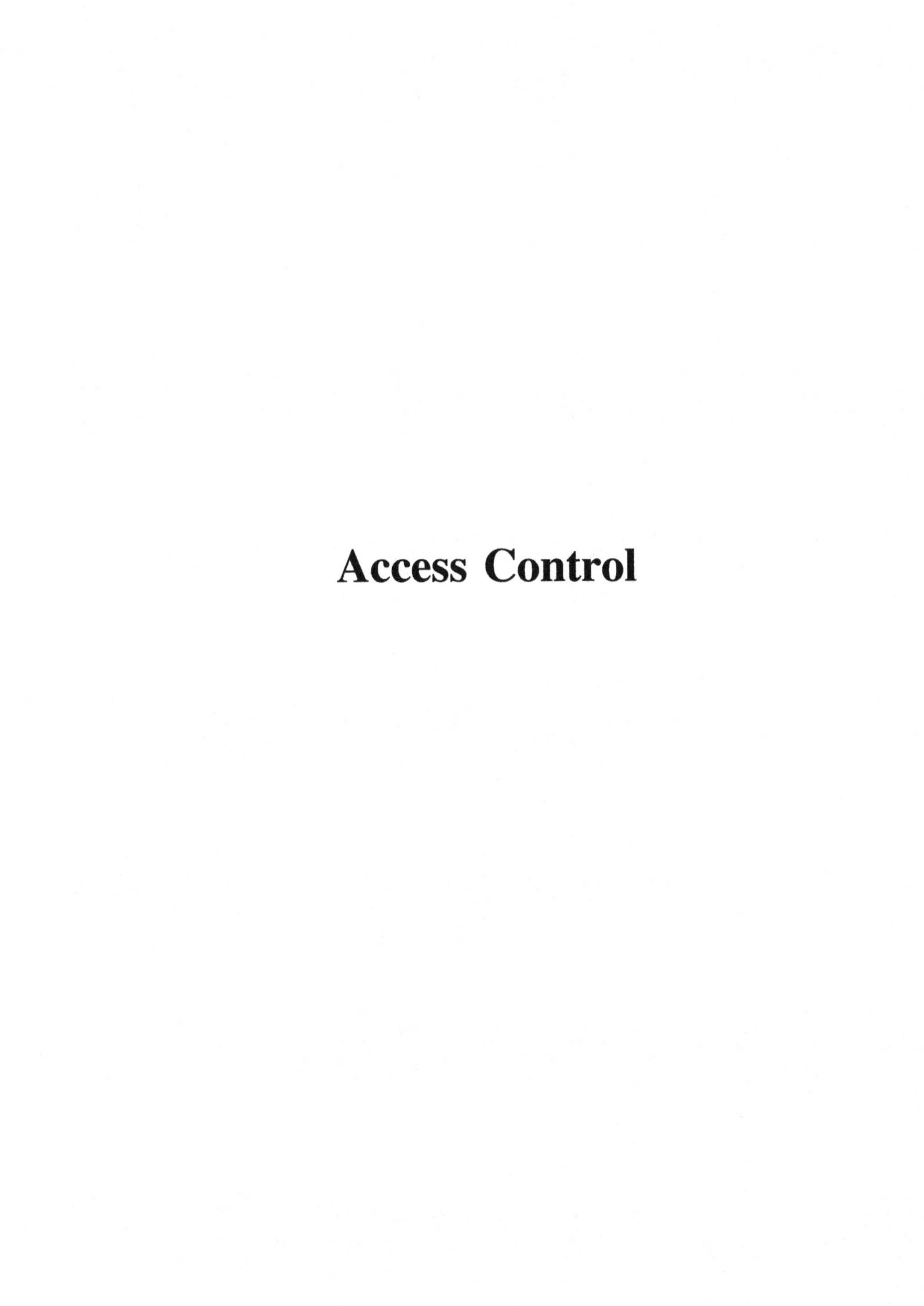

Access Control Encryption for General Policies from Standard Assumptions

Sam Kim(✉) and David J. Wu

Stanford University, Stanford, USA
skim13@cs.stanford.edu

Abstract. Functional encryption enables fine-grained access to encrypted data. In many scenarios, however, it is important to control not only what users are allowed to *read* (as provided by traditional functional encryption), but also what users are allowed to *send*. Recently, Damgård et al. (TCC 2016) introduced a new cryptographic framework called *access control encryption* (ACE) for restricting information flow within a system in terms of both what users can read as well as what users can write. While a number of access control encryption schemes exist, they either rely on strong assumptions such as indistinguishability obfuscation or are restricted to simple families of access control policies.

In this work, we give the first ACE scheme for *arbitrary* policies from standard assumptions. Our construction is generic and can be built from the combination of a digital signature scheme, a predicate encryption scheme, and a (single-key) functional encryption scheme that supports randomized functionalities. All of these primitives can be instantiated from standard assumptions in the plain model and therefore, we obtain the first ACE scheme capable of supporting general policies from standard assumptions. One possible instantiation of our construction relies upon standard number-theoretic assumptions (namely, the DDH and RSA assumptions) and standard lattice assumptions (namely, LWE). Finally, we conclude by introducing several extensions to the ACE framework to support dynamic and more fine-grained access control policies.

1 Introduction

In the last ten years, functional encryption [16,44] has emerged as a powerful tool for enforcing fine-grained access to encrypted data. But in many real-world scenarios, system administrators need to restrict not only what users are allowed to *read*, but also, what users are allowed to *send*—for example, users with top-secret security clearance in a system should not be able to make sensitive information publicly available. Recently, Damgård, Haagh, and Orlandi [23] introduced the notion of access control encryption (ACE) to enable cryptographic control of the information flow within a system.

Access control encryption. An access control encryption scheme [23] provides a cryptographic mechanism for restricting information flow in a system,

The full version of this paper is available at https://eprint.iacr.org/2017/467.pdf.

T. Takagi and T. Peyrin (Eds.): ASIACRYPT 2017, Part I, LNCS 10624, pp. 471–501, 2017.
https://doi.org/10.1007/978-3-319-70694-8_17

both in terms of what parties can read, as well as in terms of what parties can write. Of course, cryptography alone is insufficient here since a malicious sender can always broadcast sensitive messages in the clear. To address this, Damgård et al. [23] introduce an additional party called the *sanitizer*. All communication between senders and receivers is routed through the sanitizer, which performs some processing on the message before broadcasting it to the receivers. The goal in access control encryption is to simplify the operation of the sanitizer so that its function can be outsourced to a party that is only trusted to execute correctly (in particular, the sanitizer does not need to know either the identity of the sender or receiver of each message, nor the security policy being enforced).

Concretely, an ACE scheme is defined with respect to a set of senders $\mathcal{S}$, a set of receivers $\mathcal{R}$, and an access control policy $\pi \colon \mathcal{S} \times \mathcal{R} \to \{0,1\}$, where $\pi(S,R) = 1$ if a receiver $R \in \mathcal{R}$ is allowed to read messages from sender $S \in \mathcal{S}$ (and vice versa). Each sender S has an encryption key ek_S and each receiver R has a decryption key dk_R. To send a message m, the sender first encrypts $\mathsf{ct} \leftarrow \mathsf{ACE.Encrypt}(\mathsf{ek}_S, m)$ and sends ct to the sanitizer. The sanitizer performs some simple processing on ct to obtain a new ciphertext ct', which it broadcasts to all of the receivers. The correctness requirement of an ACE scheme is that if $\pi(S,R) = 1$, then $\mathsf{ACE.Decrypt}(\mathsf{dk}_R, \mathsf{ct}') = m$. Critically, the sanitizer does not know the identities of the sender or receiver, nor does it know the policy π.

The security requirements of an ACE scheme mirror those in the Bell-LaPadula [7] security model. In particular, the *no-read rule* requires that any set of unauthorized receivers $\{R_j\}$ (even in collusion with the sanitizer) cannot learn any information from a sender S if $\pi(S, R_j) = 0$ for all j. The *no-write rule* says that no set of (possibly malicious) senders $\{S_i\}$ can transfer any information to any set of (possibly malicious) receivers $\{R_j\}$ if $\pi(S_i, R_j) = 0$ for all i, j.

Existing constructions of ACE. Damgård et al. [23] gave two constructions of ACE capable of supporting arbitrary policies $\pi \colon \{0,1\}^n \times \{0,1\}^n \to \{0,1\}$ (here, the senders and receivers are represented as n-bit identities). Their first construction takes a brute-force approach and is based on standard number-theoretic assumptions such as the decisional Diffie-Hellman assumption (DDH) or the decisional composite residuosity assumption (DCR). The limitation, however, is that ciphertexts in their construction grow *exponentially* in n, thus rendering the scheme inefficient when the set of identities is large. Their second construction is more efficient (the ciphertext size is polylogarithmic in n), but relies on the full power of indistinguishability obfuscation ($i\mathcal{O}$) [6,29].

Subsequently, Fuchsbauer et al. [28] showed how to construct access control encryption for restricted classes of predicates (i.e., equality, comparisons, and interval membership) from standard assumptions on bilinear maps—namely, the symmetric external Diffie-Hellman assumption (SXDH). While their constructions are asymptotically efficient (their ciphertexts are linear in n), the functionalities they can handle are specialized to a restricted class of predicates.

Recently, Tan et al. [57] showed how to instantiate the Damgård et al. brute-force construction using the learning with errors (LWE) assumption. Since their construction follows the Damgård et al. approach, ciphertexts in their construction also grown exponentially in n.

A natural question is whether it is possible to construct an asymptotically-efficient ACE scheme for *arbitrary* functionalities without relying on powerful assumptions like indistinguishability obfuscation. In this work, we show that under standard assumptions (for instance, the DDH, RSA, and LWE assumptions suffice), we obtain an *asymptotically-efficient* ACE scheme for *general* policies.

1.1 Our Contributions

Our main contribution in this work is a new construction of access control encryption that is asymptotically efficient, supports arbitrary policies, and relies only on simple, well-studied assumptions. All previous constructions of ACE were either inefficient, restricted to simple policies, or relied on indistinguishability obfuscation. We refer to Table 1 for a comparison with the state-of-the-art.

Table 1. Concrete comparison of the ACE construction in this work with previous ACE constructions [23,28,57] for predicates $\pi\colon \{0,1\}^n \times \{0,1\}^n \to \{0,1\}$. For the predicate class, we write "arbitrary" if the scheme can support arbitrary access control policies and "restricted" if it can only handle a small set of access control policies (e.g., equality, comparisons, interval testing).

Construction	Predicate	Ciphertext size	Assumption
Damgård et al. [23, Sect. 3]	arbitrary	$O(2^n)$	DDH or DCR
Damgård et al. [23, Sect. 4]	arbitrary	$\mathsf{poly}(n)$	$i\mathcal{O}$
Fuchsbauer et al. [28]	restricted	$\mathsf{poly}(n)$	SXDH
Tan et al. [57]	arbitrary	$O(2^n)$	LWE
This work	arbitrary	$\mathsf{poly}(n)$	DDH, RSA, and LWE

In this work, we give a *generic construction* of access control encryption from three main ingredients: a digital signature scheme, a general-purpose predicate encryption scheme [32], and a (single-key) functional encryption scheme that supports randomized functionalities [1,33]. We give a high-level overview of our construction here and provide the formal description in Sect. 3. In Sect. 3.1, we show how to instantiate the underlying primitives to obtain an ACE scheme from standard assumptions. Our work thus resolves the main open question posed by Damgård et al. [23] on constructing asymptotically-efficient ACE schemes for arbitrary functionalities from standard assumptions.

Starting point: predicate encryption. First, we review the syntax of a predicate encryption scheme. In a predicate encryption scheme [17,36,56], ciphertexts are associated with a message m in addition to a set of attributes x, and secret keys are associated with functions f. Decrypting a ciphertext associated with an attribute-message pair (x, m) using a secret key for a function f outputs m if and only if $f(x) = 1$. Moreover, ciphertexts in a predicate encryption scheme hide *both* the attribute x as well as the message m from all parties that are not able to decrypt the ciphertext.[1] Not surprisingly, a predicate encryption scheme that supports general policies can be used to obtain a primitive that resembles an access control encryption scheme. Each sender's encryption key is just the public key for the predicate encryption scheme. To encrypt a message m, the sender encrypts m with its identity as the attribute (i.e., an n-bit string). The sanitizer would simply forward the ciphertext along. The decryption key for a receiver R is a predicate encryption key that implements the policy $\pi(\cdot, R)$. Of course, because the sanitizer simply broadcasts the sender's message to the receivers, this basic scheme does not satisfy the no-write rule. A malicious sender can simply broadcast the message in the clear.

Sanitizing the ciphertext. To provide security against malicious senders, the sanitizer must perform some kind of re-randomization of the sender's ciphertexts. Damgård et al. [23] achieve this by introducing the notion of "sanitizable functional encryption," which is a functional encryption scheme that supports re-randomization of ciphertexts. However, constructing sanitizable functional encryption seems to require indistinguishability obfuscation. In this work, we take a different strategy similar in spirit to proxy re-encryption [4]. Specifically, we view the sanitizer as implementing a "proxy" that takes as input a sender's ciphertext (under some encryption scheme) and then *re-encrypts* that ciphertext under the predicate encryption scheme (with the attribute set to the sender's identity). The guarantee we seek is that the output of the sanitizer is either $\perp$ (if the input ciphertext is invalid) or a *fresh* encryption of the sender's message under the predicate encryption scheme. With this guarantee, the no-read and no-write properties reduce to the security of the predicate encryption scheme.

The problem of building ACE thus reduces to constructing a suitable proxy re-encryption scheme. Here, we rely on a single-key functional encryption for randomized functionalities [1,33]. In a standard functional encryption [16,44] (FE) scheme, secret keys are associated with functions f and ciphertexts are associated with messages m. The combination of a decryption key for a function f and a ciphertext for a message m should together reveal $f(m)$ and nothing more. Recently, Alwen et al. [2] and Goyal et al. [33] extended the notion of functional encryption to also consider issuing keys for randomized functionalities.

A (general-purpose) FE scheme that supports randomized functionalities immediately gives a way of implementing the proxy re-encryption functionality for the sanitizer. First, to encrypt a message m, sender S encrypts the pair (S, m) under the FE scheme. The sanitizer is given a functional key for the re-encryption function that takes as input a pair (S, m) and outputs a predicate encryption of m with attribute S. The receivers' behavior is unchanged. By appealing to the correctness and security of the FE scheme, the sanitizer's output is distributed

[1] This is in contrast to the weaker notion of attribute-based encryption [12,34,52] where the attribute is public.

like a fresh predicate encryption ciphertext.[2] Importantly for our construction, the FE scheme only needs to support issuing a *single* decryption key (for the sanitizer). This means that it is possible to instantiate the FE scheme from standard assumptions (i.e., by applying the transformation in [1] to standard FE constructions such as [30,31,51]). Our construction is conceptually similar to the approach in [23] based on sanitizable FE. In Remark 3.1, we compare our approach to the one in [23] and highlight the key differences that allow us to avoid the need for indistinguishability obfuscation (as seemingly needed for sanitizable FE), and thus, base our construction on simple assumptions.

Signatures for policy enforcement. The remaining problem with the above construction is that the sender has the freedom to choose the identity S at encryption time. Thus, a malicious sender could choose an arbitrary identity and trivially break the no-write security property. We address this by requiring the sender "prove" its identity to the sanitizer when submitting its ciphertext (but without revealing its identity to the sanitizer in the process). This can be done using a standard technique described in [23] (and also applied in several other contexts [10,19]) by giving each sender S a signature σ_S on its identity (included as part of the sender's encryption key). Then, to encrypt a message m, the sender would construct an FE ciphertext for the tuple (S, σ_S, m) containing its identity, the signature on its identity, and the message. The sanitizer's FE key then implements a re-encryption function that first checks the validity of the signature on the identity before outputting a fresh predicate encryption of the message m (still with attribute S). Thus, a malicious sender is only able to produce valid ciphertexts associated with identities for which it possesses a valid signature. With this modification, we can show that the resulting construction is a secure ACE scheme (Theorems 3.2 and 3.3).

Instantiating our construction. Our construction above gives a generic construction of ACE from digital signatures, predicate encryption, and a single-key general-purpose functional encryption scheme for randomized functionalities. In Sect. 3.1, we show that all of the requisite building blocks of our generic construction can be instantiated from standard assumptions. In particular, security can be reduced to the decisional Diffie-Hellman (DDH) assumption [14], the RSA assumption [49], and the learning with errors (LWE) assumption [48]. This yields the first ACE scheme that supports general policies from standard assumptions.

Extending ACE. In Sect. 4, we describe several extensions to the notion of ACE that naturally follow from our generic construction. We primarily view these extensions as ways of augmenting the schema of access control encryption to provide

[2] In the actual construction, satisfying the no-write property requires the stronger property that decrypting a *maliciously-generated* ciphertext, say, from a corrupt sender, also yields a fresh ciphertext under the predicate encryption scheme. This is the notion of security against malicious encrypters first considered in [33] and subsequently extended in [1]. The work of [1] shows how to obtain functional encryption for randomized functionalities with security against malicious encrypters from any functional encryption scheme supporting deterministic functionalities in conjunction with standard number-theoretic assumptions.

increased flexibility or to support additional functionalities, and not as qualitatively new properties specific to our particular construction. Indeed, the $i\mathcal{O}$-based construction of Damgård et al. [23] can also be extended to achieve these properties. Our primary contribution is showing that we can achieve these stronger properties without relying on $i\mathcal{O}$. We briefly summarize our main extensions:

- **Dynamic policies:** In the standard notion of ACE [23], the access control policy is specified at the time of system setup. In realistic scenarios, senders and receivers may need to be added to the system, and moreover, access control policies can evolve over time. In Sect. 4.1, we show that our ACE construction allows us to associate an access control policy specific to each receiver's decryption key. Thus, each receiver's policy can be determined at the time of receiver key generation rather than system setup, which enables a dynamic specification of access control policies.
- **Fine-grained sender policies:** The standard notion of ACE only considers policies expressible as a function of the sender's and receiver's identities. In many scenarios, we may want to impose additional restrictions on the types of messages that a sender could send. For instance, a sender could be allowed to send messages to any receiver with top-secret security clearance, but we want to ensure that all of the messages they send contains a signature from both the sender as well as their supervisor (who would certify the contents of the message). In Sect. 4.2, we show that a straightforward extension of our construction allows us to additionally enforce policies on the types of messages a user is allowed to send. We also introduce a new security notion for ACE that captures the property that a sender should only be allowed to send messages that conform to their encryption policy.
- **Beyond all-or-nothing decryption:** In a standard ACE scheme, decryption is "all-or-nothing:" receivers who are authorized to decrypt a particular ciphertext are able to do so and learn the underlying message, while receivers who are not authorized to decrypt learn nothing about the message. Just as functional encryption extends beyond all-or-nothing encryption by enabling decrypters to learn partial information about an encrypted message, we can consider a functional encryption analog of access control encryption where receivers are allowed to learn only partial information about messages in accordance with the precise access control policies of the underlying scheme. As a concrete example, an analyst with secret security clearance might only be authorized to learn the metadata of a particular encrypted communication, while an analyst with top-secret security clearance might be authorized to recover the complete contents of the communication. In a "functional ACE" scheme, decryption keys are associated with functions and the decryption algorithm computes a function on the underlying message. In the full version [38], we show how our ACE scheme can be easily extended to obtain a functional ACE scheme.

Concurrent work. Concurrent to this work, Badertscher et al. [5] introduced several strengthened security notions for access control encryption such as security against chosen ciphertext attacks (CCA-security). They then show how to extend the ACE scheme (for restricted policies) in [28] to achieve their new security notions. In contrast, our focus in this work is constructing an ACE scheme (under the original security notions from [23]) for arbitrary policies from standard assumptions.

Open problems. We leave as an open problem the construction of an ACE scheme (for general policies) where the sanitizer key can be public. This is the case for the ACE construction for restricted policies in [28], but not the case for our construction or the $i\mathcal{O}$-based construction in [23]. Another open problem is constructing an ACE scheme that provides full sender anonymity (see Remark 2.6 for more details). Notably, this is possible from $i\mathcal{O}$ [23], but seems non-trivial from standard assumptions.

1.2 Additional Related Work

Information flow control is a widely studied topic in computer security (see, for instance [7,24,25,45,50,53,54] and the references therein). In particular, the "no read" and "no write" security notions for access control encryption are inspired by the "no read-up" and "no write-down" security policies first introduced in the seminal work of Bell and LaPadula [7]. In this work, we focus on designing cryptographic solutions for information flow control.

Numerous cryptographic primitives, starting with identity-based encryption [15,22,55], and progressing to attribute-based encryption [12,34,52], predicate encryption [17,36,40,43], and finally, culminating with functional encryption [16,44,51], have focused on ways of enabling fine-grained *access* to encrypted data (i.e., impose policies on the decryption capabilities of users in a system). Access control encryption seeks to simultaneously enforce policies on both the encryption capabilities of the sender as well as the decryption capabilities of the receiver.

A key challenge in access control encryption (and how it differs from traditional notions of functional encryption) is in preventing corrupt senders from communicating (covertly or otherwise) with unauthorized recipients. One way of viewing these goals is as a mechanism for protecting against steganography techniques [35]. Recent works on cryptographic reverse firewalls [26,42] have looked at preventing compromised or malicious software from leaking sensitive information. Raykova et al. [47] studied the problem of access control for outsourced data. Their goal was to hide access patterns from the cloud and preventing corrupt writers from updating files that they are not authorized to update. Their work considers a covert security model where malicious writers are caught; in contrast, with ACE, we require the stronger guarantee that communication between corrupt senders and unauthorized receivers are completely blocked.

Also related to access control encryption is the recent line of work on sanitizable signatures [3,20,27]. These works study the case where an intermediate party can sanitize messages and signatures that are sent over a channel while learning minimal information about the messages and signatures. The notion of sanitizable signatures is conceptually different from that of ACE since sanitizable signatures are not designed to prevent corrupt senders from leaking information to corrupt receivers.

2 Preliminaries

For $n \geq 1$, we write $[n]$ to denote the set of integers $\{1, \ldots, n\}$. For a distribution $\mathcal{D}$, we write $x \leftarrow \mathcal{D}$ to denote that x is a sampled from $\mathcal{D}$. For a finite set S, we write $x \xleftarrow{\text{R}} S$ to denote that x is sampled uniformly at random from S. For a randomized function f, we write $f(x; r)$ to denote an evaluation of f using randomness r. Unless otherwise noted, we always write λ for the security parameter. We say a function $f(\lambda)$ is negligible in the security parameter λ (denoted $f(\lambda) = \mathsf{negl}(\lambda)$) if $f(\lambda) = o(1/\lambda^c)$ for all $c \in \mathbb{N}$. We write $f(\lambda) = \mathsf{poly}(\lambda)$ to denote that f is a (fixed) polynomial in λ. An algorithm is efficient if it runs in polynomial time in the length of its input. For two ensembles of distributions $\mathcal{D}_1$ and $\mathcal{D}_2$, we write $\mathcal{D}_1 \stackrel{c}{\approx} \mathcal{D}_2$ if the two distributions are computationally indistinguishable (that is, no efficient algorithm can distinguish $\mathcal{D}_1$ from $\mathcal{D}_2$ except with negligible probability).

We now formally define the tools we need to build our ACE scheme. Due to space limitations, we defer the standard definitions of a digital signature scheme and predicate encryption scheme to the full version of this paper [38]. In Sect. 2.1, we review the notion of functional encryption for randomized functionalities, and in Sect. 2.2, we introduce the notion of an access control encryption scheme.

2.1 Functional Encryption for Randomized Functionalities

Functional encryption (FE) [16,44,51] is a generalization of predicate encryption. In an FE scheme, secret keys are associated with functions and ciphertexts are associated with messages. Given a secret key sk_f for a (deterministic) function f and a ciphertext ct_x encrypting a value x, the decryption function in an FE scheme outputs $f(x)$. The security guarantee roughly states that sk_f and ct_x together reveal $f(x)$, and nothing more. Alwen et al. [2] and Goyal et al. [33] extended the notion of functional encryption to include support for *randomized* functionalities (i.e., secret keys are associated with randomized functions). Subsequently, Komargodski et al. [39], as well as Agrawal and Wu [1] showed how to generically transform FE schemes that support deterministic functions into schemes that support randomized functions; the former transformation [39] applies in the secret-key setting while the latter [1] applies in the public-key setting.

Syntax. We now give the formal definition of a functional encryption for randomized functionalities in the public-key setting. Our definitions are adapted from those in [1,33]. A functional encryption for randomized functionalities for a function family $\mathcal{F}$ over a domain $\mathcal{X}$, range $\mathcal{Y}$, and randomness space $\mathcal{R}$ is a tuple of algorithms $\Pi_{\mathsf{rFE}} = (\mathsf{rFE.Setup}, \mathsf{rFE.KeyGen}, \mathsf{rFE.Encrypt}, \mathsf{rFE.Decrypt})$ with the following properties:

- $\mathsf{rFE.Setup}(1^\lambda) \to (\mathsf{pp}, \mathsf{msk})$: On input the security parameter λ, the setup algorithm outputs the public parameters pp and the master secret key msk.

- $\mathsf{rFE.KeyGen}(\mathsf{msk}, f) \rightarrow \mathsf{sk}_f$: On input the master secret key msk and the description of a (possibly randomized) function $f: \mathcal{X} \rightarrow \mathcal{Y}$, the key-generation algorithm outputs a secret key sk_f.
- $\mathsf{rFE.Encrypt}(\mathsf{pp}, x) \rightarrow \mathsf{ct}_x$: On input the public parameters pp and a message $x \in \mathcal{X}$, the encryption algorithm outputs a ciphertext ct_x.
- $\mathsf{rFE.Decrypt}(\mathsf{sk}, \mathsf{ct}) \rightarrow y$: On input a secret key sk, and a ciphertext ct, the decryption algorithm outputs a value $y \in \mathcal{Y} \cup \{\bot\}$.

Correctness. The correctness property for an FE scheme that supports randomized functionalities states that given a secret key sk_f for a randomized function f and a ciphertext ct_x encrypting a value x, the decryption function $\mathsf{rFE.Decrypt}(\mathsf{sk}_f, \mathsf{ct}_x)$ outputs a random draw from the output distribution of $f(x)$. Moreover, when multiple function keys are applied to multiple ciphertexts, decryption should output an *independent* draw from the output distribution for each ciphertext-key pair. This property should hold even given the public parameters as well as the function keys for the function encryption scheme. We give the formal definition below:

Definition 2.1 (Correctness *[1,33, adapted]***).** *A functional encryption scheme for randomized functionalities* $\Pi_{\mathsf{rFE}} = (\mathsf{rFE.Setup}, \mathsf{rFE.KeyGen}, \mathsf{rFE.Encrypt}, \mathsf{rFE.Decrypt})$ *over a message space* $\mathcal{X}$ *for a (randomized) function family* $\mathcal{F}$ *(operating over a randomness space* $\mathcal{R}$*) is correct if for every polynomial* $n = n(\lambda)$*, every collection of functions* $(f_1, \ldots, f_n) \in \mathcal{F}^n$*, and every collection of messages* $(x_1, \ldots, x_n) \in \mathcal{X}^n$*, setting* $(\mathsf{pp}, \mathsf{msk}) \leftarrow \mathsf{rFE.Setup}(1^\lambda)$*,* $\mathsf{sk}_i \leftarrow \mathsf{rFE.KeyGen}(\mathsf{msk}, f_i)$*,* $\mathsf{ct}_j \leftarrow \mathsf{rFE.Encrypt}(\mathsf{pp}, x_j)$*, and* $r_{i,j} \xleftarrow{\text{R}} \mathcal{R}$ *for* $i, j \in [n]$*, the following two distributions are computationally indistinguishable:*

$$\left(\mathsf{pp}, \{\mathsf{sk}_i\}_{i\in[n]}, \{\mathsf{rFE.Decrypt}(\mathsf{sk}_i, \mathsf{ct}_j)\}_{i,j\in[n]}\right)$$

$$\text{and} \quad \left(\mathsf{pp}, \{\mathsf{sk}_i\}_{i\in[n]}, \{f_i(x_j; r_{i,j})\}_{i,j\in[n]}\right).$$

Remark 2.2 (Weaker Correctness Notions). Existing constructions of functional encryption for randomized functionalities [1,33] consider a weaker correctness requirement that the joint distribution $\{\mathsf{rFE.Decrypt}(\mathsf{sk}_i, \mathsf{ct}_j)\}_{i,j\in[n]}$ be computationally indistinguishable from $\{f_i(x_j; r_{i,j})\}_{i,j\in[n]}$. In this work, we require the stronger property that these two distributions remain computationally indistinguishable even given the public parameters as well as the (honestly-generated) decryption keys. It is not difficult to see that existing constructions such as the Agrawal-Wu generic construction [1] satisfy this stronger correctness requirement.[3]

[3] Specifically, the generic construction of functional encryption for randomized functionalities from standard functional encryption in [1] uses a PRF key for derandomization. In their construction, they secret share the PRF key across the ciphertext and the decryption key. By appealing to related-key security of the underlying

Security. In this work, we use a simulation-based definition of security. Our access control encryption construction relies critically on our FE scheme providing robustness against *malicious* encrypters. This can be viewed as the analog of CCA-security in the context of public-key encryption [46], and is captured formally in the security game by giving the adversary access to a decryption oracle (much like in the CCA-security game). We give a simplified variant of the definition from [1,33] where the adversary is only allowed to issue key-queries before making challenge queries (i.e., the adversary is restricted to making *non-adaptive* key queries). In this non-adaptive setting, Gorbunov et al. [31] showed that security against an adversary who makes a single challenge query implies security against an adversary that makes a polynomial number of challenge queries. This is the definition we use in this work. Additionally, for the decryption queries, we also consider the simplified setting of [33] where the adversary can only submit a *single* ciphertext on each decryption query.[4] We now give the formal definition:

Definition 2.3 (q-NA-SIM **Security** *[1,33, adapted]*)**.** *Let* $\Pi_{\mathsf{rFE}} = (\mathsf{rFE.Setup}, \mathsf{rFE.KeyGen}, \mathsf{rFE.Encrypt}, \mathsf{rFE.Decrypt})$ *be a functional encryption scheme for randomized functionalities over a message space* $\mathcal{X}$ *for a (randomized) function family* $\mathcal{F}$ *(with randomness space* $\mathcal{R}$*). We say that* Π_{rFE} *is* q-NA-SIM*-secure against malicious encrypters if there exists an efficient (stateful) simulator* $\mathcal{S} = (\mathcal{S}_1, \mathcal{S}_2, \mathcal{S}_3, \mathcal{S}_4)$ *such that for all efficient adversaries* $\mathcal{A} = (\mathcal{A}_1, \mathcal{A}_2)$ *where* $\mathcal{A}_1$ *makes at most* q *key-generation queries, the outputs of the following two experiments are computationally indistinguishable:*

Experiment $\mathsf{Real}_{\Pi_{\mathsf{rFE}},\mathcal{A}}(1^\lambda)$:	**Experiment** $\mathsf{Ideal}_{\Pi_{\mathsf{rFE}},\mathcal{A},\mathcal{S}}(1^\lambda)$:
$(\mathsf{pp}, \mathsf{msk}) \leftarrow \mathsf{rFE.Setup}(1^\lambda)$	$(\mathsf{pp}, \mathsf{st}') \leftarrow \mathcal{S}_1(1^\lambda)$
$\mathsf{st} \leftarrow \mathcal{A}_1^{\mathcal{O}_1(\mathsf{msk},\cdot), \mathcal{O}_3(\mathsf{msk},\cdot,\cdot)}(1^\lambda, \mathsf{pp})$	$\mathsf{st} \leftarrow \mathcal{A}_1^{\mathcal{O}_1'(\mathsf{st}',\cdot), \mathcal{O}_3'(\mathsf{st}',\cdot,\cdot)}(1^\lambda, \mathsf{pp})$
$\alpha \leftarrow \mathcal{A}_2^{\mathcal{O}_2(\mathsf{pp},\cdot), \mathcal{O}_3(\mathsf{msk},\cdot,\cdot)}(\mathsf{st})$	$\alpha \leftarrow \mathcal{A}_2^{\mathcal{O}_2'(\mathsf{st}',\cdot), \mathcal{O}_3(\mathsf{st}',\cdot,\cdot)}(\mathsf{st})$
Output $(\{g\}, \{y\}, \alpha)$	**Output** $(\{g'\}, \{y'\}, \alpha)$

where the key-generation, encryption, and decryption oracles are defined as follows:

Real experiment $\mathsf{Real}_{\Pi_{\mathsf{rFE}},\mathcal{A}}(1^\lambda)$*:*

- ***Key-generation oracle:*** $\mathcal{O}_1(\mathsf{msk}, \cdot)$ *implements* $\mathsf{rFE.KeyGen}(\mathsf{msk}, \cdot)$.
- ***Encryption oracle:*** $\mathcal{O}_2(\mathsf{pp}, \cdot)$ *implements* $\mathsf{rFE.Encrypt}(\mathsf{pp}, \cdot)$.
- ***Decryption oracle:*** *On input* (g, ct) *where* $g \in \mathcal{F}$ *and* $\mathsf{ct} \in \{0,1\}^*$*, the decryption oracle* $\mathcal{O}_3(\mathsf{msk}, \cdot, \cdot)$ *computes* $\mathsf{sk}_g \leftarrow \mathsf{rFE.KeyGen}(\mathsf{msk}, g)$ *and outputs* $y = \mathsf{rFE.Decrypt}(\mathsf{sk}_g, \mathsf{ct})$. *The (ordered) set* $\{g\}$ *consists of the set of*

PRF [8,9,11,13], the randomness used for function evaluation during decryption is computationally indistinguishable from a random string. Moreover, this holds even if one of the key-shares is known (in our setting, this is the key-share embedded within the decryption key).

[4] Subsequent work [1] showed how to extend the security definition to also capture adversaries that can induce correlations across multiple ciphertexts, but this strengthened definition is not necessary in our setting.

functions that appear in the decryption queries of $\mathcal{A}$ *and the (ordered) set* $\{y\}$ *consist of the responses of* $\mathcal{O}_3$.

Ideal experiment $\mathsf{Ideal}_{\Pi_{\mathsf{rFE}},\mathcal{A},\mathcal{S}}(1^\lambda)$:

- ***Key-generation oracle:*** *On input a function* $f \in \mathcal{F}$*, the ideal key-generation oracle* $\mathcal{O}'_1$ *computes* $(\mathsf{sk}'_f, \mathsf{st}') \leftarrow \mathcal{S}_2(\mathsf{st}', f)$*, and returns* sk'_f*. The updated simulator state* st' *is carried over to future invocations of the simulator.*
- ***Encryption oracle:*** *On input a message* $x \in \mathcal{X}$*, the ideal encryption oracle* $\mathcal{O}'_2$ *samples* $r_1, \ldots, r_q \xleftarrow{\text{R}} \mathcal{R}$*, and sets* $y_i = f_i(x; r_i)$ *for* $i \in [q]$*, where* f_i *is the* i^{th} *key-generation query* $\mathcal{A}_1$ *made to the key-generation oracle. The oracle computes* $(\mathsf{ct}', \mathsf{st}') \leftarrow \mathcal{S}_3(\mathsf{st}', \{y_i\}_{i \in [q]})$ *and returns* ct'.
- ***Decryption oracle:*** *On input* (g', ct') *where* $g' \in \mathcal{F}$ *and* $\mathsf{ct}' \in \{0,1\}^*$*, the ideal decryption oracle* $\mathcal{O}'_3$ *invokes the simulator algorithm* $(x, \mathsf{st}') \leftarrow \mathcal{S}_4(\mathsf{st}', \mathsf{ct}')$*, where* $x \in \mathcal{X} \cup \{\perp\}$*. If* $x \neq \perp$*, the oracle samples* $r \xleftarrow{\text{R}} \mathcal{R}$ *and replies with* $g'(x; r)$*. Otherwise, if* $x = \perp$*, the oracle replies with* $\perp$*. The (ordered) set* $\{g'\}$ *denotes the functions in the decryption queries of* $\mathcal{A}$ *and* $\{y'\}$ *denotes the outputs of* $\mathcal{O}'_3$.

2.2 Access Control Encryption (ACE)

In this section, we review the definition of *access control encryption* (ACE) [23,28,57]. An access control encryption scheme over an identity space $\mathcal{I}$, a message space $\mathcal{M}$, and a ciphertext space $\mathcal{C}$ is defined by a tuple of algorithms $\Pi_{\mathsf{ACE}} =$ (ACE.Setup, ACE.EKGen, ACE.DKGen, ACE.Encrypt, ACE.Sanitize, ACE.Decrypt) with the following properties:

- $\mathsf{ACE.Setup}(1^\lambda, \pi) \to (\mathsf{sank}, \mathsf{msk})$: On input a security parameter λ and an access control policy $\pi\colon \mathcal{I} \times \mathcal{I} \to \{0,1\}$, the setup algorithm outputs the sanitizer key sank and the master secret key msk.
- $\mathsf{ACE.EKGen}(\mathsf{msk}, i) \to \mathsf{ek}_i$: On input the master secret key msk and a sender identity $i \in \mathcal{I}$, the encryption key-generation algorithm outputs an encryption key ek_i.
  $\mathsf{ACE.DKGen}(\mathsf{msk}, j) \to \mathsf{dk}_j$: On input the master secret key msk, and a receiver identity $j \in \mathcal{I}$, the decryption key-generation algorithm returns a decryption key dk_j.
- $\mathsf{ACE.Encrypt}(\mathsf{ek}, m) \to \mathsf{ct}$: On input an encryption key ek, and a message $m \in \mathcal{M}$, the encryption algorithm outputs a ciphertext ct.[5]
- $\mathsf{ACE.Sanitize}(\mathsf{sank}, \mathsf{ct}) \to \mathsf{ct}'$: On input the sanitizer key sank and a ciphertext ct, the sanitize algorithm outputs a ciphertext $\mathsf{ct}' \in \mathcal{C} \cup \{\perp\}$.
- $\mathsf{ACE.Decrypt}(\mathsf{dk}, \mathsf{ct}') \to m'$: On input a decryption key dk and a ciphertext $\mathsf{ct}' \in \mathcal{C}$, the decryption algorithm outputs a message $m' \in \mathcal{M} \cup \{\perp\}$.

[5] Note that we do not require that $\mathsf{ct} \in \mathcal{C}$. In particular, the ciphertexts output by the encryption algorithm can be syntactically different from those output by the sanitize algorithm. To simplify the notation, we only explicitly model the ciphertexts space $\mathcal{C}$ corresponding to those produced by the ACE.Sanitize algorithm.

Definition 2.4 (Correctness *[23]***).** *An ACE scheme* $\Pi_{\mathsf{ACE}} = (\mathsf{ACE.Setup}, \mathsf{ACE.EKGen}, \mathsf{ACE.DKGen}, \mathsf{ACE.Encrypt}, \mathsf{ACE.Sanitize}, \mathsf{ACE.Decrypt})$ *over an identity space* $\mathcal{I}$ *and a message space* $\mathcal{M}$ *is correct if for all messages* $m \in \mathcal{M}$*, all policies* $\pi \colon \mathcal{I} \times \mathcal{I} \to \{0,1\}$*, and all identities* $i, j \in \mathcal{I}$ *where* $\pi(i,j) = 1$*, setting* $(\mathsf{sank}, \mathsf{msk}) \leftarrow \mathsf{ACE.Setup}(1^\lambda, \pi)$*,* $\mathsf{ek}_i \leftarrow \mathsf{ACE.EKGen}(\mathsf{msk}, i)$*,* $\mathsf{dk}_j \leftarrow \mathsf{ACE.DKGen}(\mathsf{msk}, j)$*, we have that*

$$\Pr[\mathsf{ACE.Decrypt}(\mathsf{dk}_j, \mathsf{ACE.Sanitize}(\mathsf{sank}, \mathsf{ACE.Encrypt}(\mathsf{ek}_i, m))) = m] = 1 - \mathsf{negl}(\lambda).$$

Security definitions. Damgård et al. [23] introduced two security notions for an ACE scheme: the *no-read rule* and the *no-write rule*. The no-read rule captures the property that only the intended recipients of a message (namely, those authorized to decrypt it) should be able to learn anything about the message. In particular, a subset of unauthorized receivers should be unable to combine their respective decryption keys to learn something about a ciphertext they are not authorized to decrypt. Moreover, this property should hold even if the recipients collude with the sanitizer. The no-write rule captures the property that a sender can only encrypt messages to receivers that it is authorized to do so. Specifically, no sender with identity i should be able to form a ciphertext that can be decrypted by a receiver with identity j where $\pi(i,j) = 0$. Furthermore, this property should hold even when multiple senders and receivers collude. We now review the formal definitions introduced in [23].

Definition 2.5 (No-Read Rule *[23]***).** *Let* $\Pi_{\mathsf{ACE}} = (\mathsf{ACE.Setup}, \mathsf{ACE.EKGen}, \mathsf{ACE.DKGen}, \mathsf{ACE.Encrypt}, \mathsf{ACE.Sanitize}, \mathsf{ACE.Decrypt})$ *be an ACE scheme over an identity space* $\mathcal{I}$ *and a message space* $\mathcal{M}$*. Let* $\mathcal{A}$ *be an efficient adversary and* $\pi \colon \mathcal{I} \times \mathcal{I} \to \{0,1\}$ *be an access control policy. For a security parameter* λ *and a bit* $b \in \{0,1\}$*, we define the no-read rule experiment* $\mathsf{Expt}^{(\mathsf{Read})}_{\Pi_{\mathsf{ACE}}, \mathcal{A}, \pi}(\lambda, b)$ *as follows. The challenger first samples* $(\mathsf{sank}, \mathsf{msk}) \leftarrow \mathsf{ACE.Setup}(1^\lambda, \pi)$*, and gives the sanitizer key* sank *to* $\mathcal{A}$*. Then,* $\mathcal{A}$ *is given access to the following oracles:*

- ***Encryption oracle.*** *On input a message* $m \in \mathcal{M}$ *and a sender identity* $i \in \mathcal{I}$*, the challenger responds with a ciphertext* $\mathsf{ct} \leftarrow \mathsf{ACE.Encrypt}(\mathsf{ACE.EKGen}(\mathsf{msk}, i), m)$*.*
- ***Encryption key-generation oracle.*** *On input a sender identity* $i \in \mathcal{I}$*, the challenger responds with an encryption key* $\mathsf{ek}_i \leftarrow \mathsf{ACE.EKGen}(\mathsf{msk}, i)$*.*
- ***Decryption key-generation oracle.*** *On input a receiver identity* $j \in \mathcal{I}$*, the challenger responds with a decryption key* $\mathsf{dk}_j \leftarrow \mathsf{ACE.DKGen}(\mathsf{msk}, j)$*.*
- ***Challenge oracle.*** *On input a pair of messages* $(m_0, m_1) \in \mathcal{M} \times \mathcal{M}$ *and a pair of sender indices* $(i_0, i_1) \in \mathcal{I} \times \mathcal{I}$*, the challenger responds with* $\mathsf{ACE.Encrypt}(\mathsf{ACE.EKGen}(\mathsf{msk}, i_b), m_b)$*.*

At the end of the experiment, adversary $\mathcal{A}$ *outputs a bit* $b' \in \{0,1\}$*, which is the output of the experiment. An adversary* $\mathcal{A}$ *is admissible for the no-read rule security game if for all queries* $j \in \mathcal{I}$ *that* $\mathcal{A}$ *makes to the receiver key-generation* *oracle,* $\pi(i_0, j) = 0 = \pi(i_1, j)$. *We say that* Π_{ACE} *satisfies the no-read rule if for all policies* $\pi \colon \mathcal{I} \times \mathcal{I} \to \{0,1\}$, *and all efficient and admissible adversaries* $\mathcal{A}$,

$$\left|\Pr\left[\mathsf{Expt}^{(\mathsf{Read})}_{\Pi_{\mathsf{ACE}},\mathcal{A},\pi}(\lambda, 0) = 0\right] - \Pr\left[\mathsf{Expt}^{(\mathsf{Read})}_{\Pi_{\mathsf{ACE}},\mathcal{A},\pi}](\lambda, 1) = 1\right]\right| = \mathsf{negl}(\lambda).$$

Remark 2.6 (Sender Anonymity). The definition of the no-read rule given in [23] also imposes the stronger requirement of *sender anonymity,* which guarantees the anonymity of the sender even against adversaries that are able to decrypt the ciphertext. In contrast, our definition only ensures sender anonymity (in addition to message privacy) against a coalition of receivers that *cannot* decrypt the challenge ciphertext. This is akin to the notion of "weak attribute-hiding" in the context of predicate encryption [40,43], and was also the notion considered in [28] for building ACE for restricted classes of functionalities.

Definition 2.7 (No-Write Rule *[23]***).** *Let* $\Pi_{\mathsf{ACE}} = (\mathsf{ACE.Setup}, \mathsf{ACE.EKGen}, \mathsf{ACE.DKGen}, \mathsf{ACE.Encrypt}, \mathsf{ACE.Sanitize}, \mathsf{ACE.Decrypt})$ *be an ACE scheme over an identity space* $\mathcal{I}$ *and a message space* $\mathcal{M}$. *Let* $\mathcal{A}$ *be an efficient adversary, and let* $\pi \colon \mathcal{I} \times \mathcal{I} \to \{0,1\}$ *be an access control policy. For a security parameter* λ *and a bit* $b \in \{0,1\}$, *we define the no-write rule experiment* $\mathsf{Expt}^{(\mathsf{Write})}_{\Pi_{\mathsf{ACE}},\mathcal{A},\pi}(\lambda, b)$ *as follows. The challenger begins by sampling* $(\mathsf{sank}, \mathsf{msk}) \leftarrow \mathsf{ACE.Setup}(1^\lambda, \pi)$. *Then,* $\mathcal{A}$ *is given access to the following oracles:*

- ***Encryption oracle.*** *On input a message* $m \in \mathcal{M}$ *and a sender identity* $i \in \mathcal{I}$, *the challenger responds by first computing* $\mathsf{ek}_i \leftarrow \mathsf{ACE.EKGen}(\mathsf{msk}, i)$ *and returning* $\mathsf{ACE.Sanitize}(\mathsf{sank}, \mathsf{ACE.Encrypt}(\mathsf{ek}_i, m))$.
- ***Encryption key-generation oracle.*** *On input a sender index* $i \in \mathcal{I}$, *the challenger responds with an encryption key* $\mathsf{ek}_i \leftarrow \mathsf{ACE.EKGen}(\mathsf{msk}, i)$.
- ***Decryption key-generation oracle.*** *On input a receiver index* $j \in \mathcal{I}$, *the challenger responds with a decryption key* $\mathsf{dk}_j \leftarrow \mathsf{ACE.DKGen}(\mathsf{msk}, j)$.
- ***Challenge oracle.*** *On input a ciphertext* $\mathsf{ct}^* \in \{0,1\}^*$ *and a sender identity* $\mathsf{id}^* \in \mathcal{I}$, *the challenger sets* $\mathsf{ct}_0 = \mathsf{ct}^*$. *Then, the challenger samples* $m' \xleftarrow{\mathrm{R}} \mathcal{M}$, *computes* $\mathsf{ct}_1 \leftarrow \mathsf{ACE.Encrypt}(\mathsf{ACE.EKGen}(\mathsf{msk}, \mathsf{id}^*), m')$, *and responds with* $\mathsf{ACE.Sanitize}(\mathsf{sank}, \mathsf{ct}_b)$.

At the end of the experiment, adversary $\mathcal{A}$ *outputs a bit* $b' \in \{0,1\}$, *which is the output of the experiment. An adversary* $\mathcal{A}$ *is admissible for the no-write rule security game if the following conditions hold:*

- *The adversary* $\mathcal{A}$ *makes at most one query to the challenge oracle.*[6]
- *For all identities* $i \in \mathcal{I}$ *that* $\mathcal{A}$ *submits to the encryption key-generation oracle prior to its challenge and all identities* $j \in \mathcal{I}$ *that* $\mathcal{A}$ *submits to the decryption key-generation oracle,* $\pi(i, j) = 0$.
- *The adversary* $\mathcal{A}$ *makes an encryption key-generation query on the challenge identity* $\mathsf{id}^* \in \mathcal{I}$ *prior to making its challenge query.*

[6] We impose this restriction to simplify the security definition. A standard hybrid argument shows that security against an adversary that makes a single challenge query implies security against one that makes multiple challenge queries.

We say that Π_{ACE} satisfies the no-write rule if for all policies $\pi : \mathcal{I} \times \mathcal{I} \to \{0,1\}$, and all efficient and admissible adversaries $\mathcal{A}$,

$$\left| \Pr\left[\mathsf{Expt}^{(\mathsf{Write})}_{\Pi_{\mathsf{ACE}},\mathcal{A},\pi}(\lambda, 0) = 0\right] - \Pr\left[\mathsf{Expt}^{(\mathsf{Write})}_{\Pi_{\mathsf{ACE}},\mathcal{A},\pi}(\lambda, 1) = 1\right] \right| = \mathsf{negl}(\lambda).$$

3 Generic Construction of Access Control Encryption

In this section, we show how to generically construct access control encryption for general policies from a digital signature scheme, a predicate encryption scheme, and a general-purpose functional encryption scheme for randomized functionalities. Then, in Sect. 3.1, we describe our concrete instantiation of an ACE scheme that supports arbitrary policies from standard assumptions.

Construction 3.1 *Let $\mathcal{I}$ be the identity space and $\mathcal{M}$ be the message space. Our access control encryption for general access policies relies on the following primitives:*

- *Let $\Pi_{\mathsf{Sig}} = (\mathsf{Sig.Setup}, \mathsf{Sig.Sign}, \mathsf{Sig.Verify})$ be a signature scheme with message space $\mathcal{I}$. Let $\mathcal{T}$ denote the space of signatures output by the* Sig.Sign *algorithm.*
- *Let $\Pi_{\mathsf{PE}} = (\mathsf{PE.Setup}, \mathsf{PE.KeyGen}, \mathsf{PE.Encrypt}, \mathsf{PE.Decrypt})$ be a (public-key) predicate encryption scheme with attribute space $\mathcal{I}$ and message space $\mathcal{M}$. Let $\mathcal{C}$ denote the ciphertext space for Π_{PE}, and let $\mathcal{R}$ denote the space for the encryption randomness for* PE.Encrypt *(namely, the space of values from which the randomness used in* PE.Encrypt *is sampled).*
- *Let $\Pi_{\mathsf{rFE}} = (\mathsf{rFE.Setup}, \mathsf{rFE.KeyGen}, \mathsf{rFE.Encrypt}, \mathsf{rFE.Decrypt})$ be a general-purpose public-key functional encryption scheme for randomized functionalities (with security against malicious encrypters) with domain $\mathcal{I} \times \mathcal{T} \times \mathcal{M}$, range $\mathcal{C}$, and randomness space $\mathcal{R}$.*

We construct the ACE scheme $\Pi_{\mathsf{ACE}} = (\mathsf{ACE.Setup}, \mathsf{ACE.EKGen}, \mathsf{ACE.DKGen}, \mathsf{ACE.Encrypt}, \mathsf{ACE.Sanitize}, \mathsf{ACE.Decrypt})$ as follows:

- $\mathsf{ACE.Setup}(1^\lambda, \pi)$: *On input the security parameter λ and a policy $\pi : \mathcal{I} \times \mathcal{I} \to \{0,1\}$, the setup algorithm samples $(\mathsf{Sig.vk}, \mathsf{Sig.sk}) \leftarrow \mathsf{Sig.Setup}(1^\lambda)$, $(\mathsf{PE.pp}, \mathsf{PE.msk}) \leftarrow \mathsf{PE.Setup}(1^\lambda)$, and $(\mathsf{rFE.pp}, \mathsf{rFE.msk}) \leftarrow \mathsf{rFE.Setup}(1^\lambda)$. Next, it defines the function $F_{\mathsf{Sig.vk},\mathsf{PE.pp}} : \mathcal{I} \times \mathcal{T} \times \mathcal{M} \to \mathcal{C}$ as follows:*

$$F_{\mathsf{Sig.vk},\mathsf{PE.pp}}(i, \sigma, m; r) = \begin{cases} \mathsf{PE.Encrypt}(\mathsf{PE.pp}, i, m; r) & \text{if } \mathsf{Sig.Verify}(\mathsf{Sig.vk}, i, \sigma) = 1 \\ \bot & \text{otherwise.} \end{cases}$$

Then, it generates a decryption key $\mathsf{rFE.sk}_F \leftarrow \mathsf{rFE.KeyGen}(\mathsf{rFE.msk}, F_{\mathsf{Sig.vk},\mathsf{PE.pp}})$. Finally, it outputs the sanitizer key $\mathsf{sank} = \mathsf{rFE.sk}_F$ and the master secret key

$$\mathsf{msk} = (\pi, \mathsf{Sig.sk}, \mathsf{PE.msk}, \mathsf{rFE.pp}).$$

- $\mathsf{ACE.EKGen}(\mathsf{msk}, i)$: *On input the master secret key* $\mathsf{msk} = (\pi, \mathsf{Sig.sk}, \mathsf{PE.msk}, \mathsf{rFE.pp})$, *and an identity* $i \in \mathcal{I}$, *the encryption key-generation algorithm constructs a signature* $\sigma \leftarrow \mathsf{Sig.Sign}(\mathsf{Sig.sk}, i)$ *and outputs* $\mathsf{ek}_i = (\mathsf{rFE.pp}, i, \sigma)$.
- $\mathsf{ACE.DKGen}(\mathsf{msk}, j)$: *On input the master secret key* $\mathsf{msk} = (\pi, \mathsf{Sig.sk}, \mathsf{PE.msk}, \mathsf{rFE.pp})$, *and an identity* $j \in \mathcal{I}$, *the decryption key-generation algorithm generates a key* $\mathsf{PE.sk} \leftarrow \mathsf{PE.KeyGen}(\mathsf{PE.msk}, f_{\pi,j})$ *where* $f_{\pi,j}(i) \colon \mathcal{I} \to \{0,1\}$ *is defined as* $f_{\pi,j}(i) = \pi(i,j)$, *and outputs* $\mathsf{dk}_j = \mathsf{PE.sk}$.
- $\mathsf{ACE.Encrypt}(\mathsf{ek}_i, m)$: *On input the encryption key* $\mathsf{ek}_i = (\mathsf{rFE.pp}, i, \sigma)$ *and a message* $m \in \mathcal{M}$, *the encryption algorithm outputs* $\mathsf{rFE.Encrypt}(\mathsf{rFE.pp}, (i, \sigma, m))$.
- $\mathsf{ACE.Sanitize}(\mathsf{sank}, \mathsf{ct})$: *On input the sanitizer key* $\mathsf{sank} = \mathsf{rFE.sk}_F$ *and a ciphertext* ct, *the sanitize algorithm outputs* $\mathsf{rFE.Decrypt}(\mathsf{rFE.sk}_F, \mathsf{ct})$.
- $\mathsf{ACE.Decrypt}(\mathsf{dk}_j, \mathsf{ct}')$: *On input a decryption key* $\mathsf{dk}_j = \mathsf{PE.sk}$ *and a ciphertext* ct', *the decryption algorithm outputs* $\mathsf{PE.Decrypt}(\mathsf{PE.sk}, \mathsf{ct}')$.

We now state that our main correctness and security theorems. Specifically, we show that assuming correctness and security of the underlying primitive Π_{Sig}, Π_{PE}, and Π_{FE}, our access control encryption scheme satisfies correctness (Definition 2.4), no-read security (Definition 2.5), and no-write security (Definition 2.7). We give the proof of Theorem 3.1 in the full version [38] and the proofs of Theorems 3.2 and 3.3 in Sects. 3.2 and 3.3, respectively. We conclude this subsection with a remark comparing our construction to the Damgård et al. [23] construction of ACE from sanitizable FE.

Theorem 3.1 (Correctness). *Suppose* Π_{Sig} *is a correct signature scheme,* Π_{PE} *is a correct predicate encryption scheme, and* Π_{FE} *is a correct functional encryption scheme for randomized functionalities (Definition 2.1). Then, the access control encryption scheme from Construction 3.1 is correct (Definition 2.4).*

Theorem 3.2 (No-Read Rule). *Suppose* Π_{Sig} *is perfectly correct,* Π_{PE} *is a secure predicate encryption scheme and* Π_{rFE} *is an* 1-*NA-SIM-secure functional encryption scheme for randomized functionalities (Definition 2.3). Then, the access control encryption scheme from Construction 3.1 satisfies the no-read rule (Definition 2.5).*

Theorem 3.3 (No-Write Rule). *If* Π_{Sig} *is existentially unforgeable,* Π_{PE} *is a secure predicate encryption scheme, and* Π_{rFE} *is a* 1-*NA-SIM-secure functional encryption for randomized functionalities (Definition 2.3). Then, the access control encryption scheme from Construction 3.1 satisfies the no-write rule (Definition 2.7).*

Remark 3.1 (Comparison with Sanitizable FE). The high-level schema of our access control encryption scheme bears some similarities to the ACE construction from sanitizable functional encryption in [23]. Here, we highlight some of the key differences between our construction and that of [23]. In [23], the sanitizer key is used only to test whether a particular ciphertext is valid or not.

After validating the certificate, the sanitizer relies on the *algebraic* structure of the sanitizable FE scheme to *re-randomize* the ciphertext. In contrast, in our construction, the sanitizer actually performs a *re-encryption* of the incoming ciphertext under a *different* (predicate) encryption scheme, and moreover, the validation procedure (that the ciphertext originated from a valid sender) is embedded within the re-encryption key possessed by the sanitizer. As such, our construction only requires us to issue a single functional encryption key to the sanitizer. This means that we can base our construction on standard cryptographic assumptions. While it may be possible to build sanitizable FE from an FE scheme that supports randomized functionalities, it seems difficult to reduce security to standard assumptions (because the existing general-purpose FE schemes from standard assumptions [30,31,51] remain secure only if we give out an *a priori* bounded number of decryption keys). Thus, using re-encryption rather than re-randomization offers qualitatively better properties that enables a construction that does not rely on strong assumptions like indistinguishability obfuscation.

3.1 Concrete Instantiations

In this section, we describe one candidate instantiation of Construction 3.1 that yields an access control encryption scheme for arbitrary policies from standard assumptions. All of our primitives can be built from standard assumptions, namely the decisional Diffie-Hellman assumption (DDH) [14], the RSA assumption (RSA) [49], and the learning with errors assumption (LWE) [48]. The DDH and RSA assumptions are needed to leverage the generic construction of functional encryption for randomized functionalities from standard functional encryption (for deterministic functionalities) in [1]. The remaining primitives can be built from LWE. We now describe one possible instantiation of the primitives in Construction 3.1:

- The signature scheme Π_{Sig} can be instantiated using the standard-model construction of Cash et al. [21] based on LWE. Note that because our construction makes non-black-box use of the underlying signature scheme (in particular, we need to issue an FE key that performs signature verification), we are unable to instantiate our construction with a signature scheme that relies on a random oracle.
- The (general-purpose) predicate encryption scheme Π_{PE} can be instantiated using the construction of Gorbunov et al. [32] based on the LWE assumption.
- The (general-purpose) 1-NA-SIM-secure FE scheme Π_{rFE} for randomized functionalities that provides security against malicious encrypters can be instantiated by applying the Agrawal-Wu deterministic-to-randomized transformation [1] to a 1-NA-SIM-secure FE scheme for deterministic functionalities. The underlying 1-NA-SIM-secure FE scheme can in turn be based on any public-key encryption [31] or on the LWE assumption [30]. Applying the deterministic-to-randomized transformation to the former yields an FE scheme for randomized functionalities from the DDH and RSA assumptions

(cf. [1, Corollary 5.5]), while applying the transformation to the latter yields an FE scheme based on the DDH, RSA, and LWE assumptions.

Putting the pieces together, we obtain the following corollary to Theorems 3.2 and 3.3:

Corollary 3.1 *Under standard assumptions (namely, the DDH, RSA, and LWE assumptions), there exists an access control scheme for general policies over arbitrary identity spaces $\mathcal{I} = \{0,1\}^n$ where $n = \mathsf{poly}(\lambda)$ that satisfies the no-read and no-write security properties.*

3.2 Proof of Theorem 3.2

Our proof proceeds via a sequence of hybrid experiments between an adversary $\mathcal{A}$ and a challenger. First, fix an access control policy $\pi\colon \mathcal{I}\times\mathcal{I} \to \{0,1\}$. We now define our sequence of hybrid experiments:

- Hyb_0: This is the ACE security experiment $\mathsf{Expt}^{(\mathsf{Read})}_{\Pi_{\mathsf{ACE}},\mathcal{A},\pi}(\lambda,0)$ from Definition 2.5. Specifically, at the beginning of the game, the challenger samples keys $(\mathsf{Sig.vk},\mathsf{Sig.sk}) \leftarrow \mathsf{Sig.Setup}(1^\lambda)$, $(\mathsf{PE.pp},\mathsf{PE.msk}) \leftarrow \mathsf{PE.Setup}(1^\lambda)$, and $(\mathsf{rFE.pp},\mathsf{rFE.msk}) \leftarrow \mathsf{rFE.Setup}(1^\lambda)$. It then generates the sanitizer key $\mathsf{rFE.sk}_F \leftarrow \mathsf{rFE.KeyGen}(\mathsf{rFE.msk}, F_{\mathsf{Sig.vk},\mathsf{PE.pp}})$ and gives $\mathsf{sank} = \mathsf{rFE.sk}_F$ to the adversary. It sets $\mathsf{msk} = (\pi,\mathsf{Sig.sk},\mathsf{PE.msk},\mathsf{rFE.pp})$. During the query phase, the challenger answers the adversary's queries to the encryption and key-generation oracles by computing the encryption and key-generation algorithms exactly as in the real scheme. When the adversary makes a challenge oracle query with messages $(m_0,m_1) \in \mathcal{M}\times\mathcal{M}$ and identities $(i_0,i_1) \in \mathcal{I}\times\mathcal{I}$, the challenger responds with $\mathsf{ACE.Encrypt}(\mathsf{ACE.EKGen}(\mathsf{msk},i_0),m_0)$.
- Hyb_1: Same as Hyb_0, except that the challenger uses the simulator $\mathcal{S} = (\mathcal{S}_1,\mathcal{S}_2,\mathcal{S}_3,\mathcal{S}_4,)$ for Π_{rFE} to construct the public parameters, the sanitizer key sank, and in replying to the adversary's challenge queries. Specifically, we make the following changes to the challenger:
  - **Setup:** At the beginning of the game, instead of sampling $\mathsf{rFE.pp}$ using $\mathsf{rFE.Setup}$, the challenger instead runs the simulation algorithm $(\mathsf{rFE.pp},\mathsf{st}') \leftarrow \mathcal{S}_1(1^\lambda)$. For the sanitizer key, the challenger computes $\mathsf{rFE.sk}_F \leftarrow \mathcal{S}_2(\mathsf{st}', F_{\mathsf{Sig.vk},\mathsf{PE.pp}})$. It saves $\mathsf{rFE.pp}$ as part of the master secret key and gives $\mathsf{sank} = (\mathsf{rFE.sk}_F)$ to the adversary.
  - **Challenge queries:** When the adversary submits a challenge (m_0,m_1,i_0,i_1), the challenger first computes $\mathsf{ct}' \leftarrow \mathsf{PE.Encrypt}(\mathsf{PE.pp},i_0,m_0)$. Then it replies to the adversary with the simulated ciphertext $\mathsf{ct} \leftarrow \mathcal{S}_3(\mathsf{st}',\mathsf{ct}')$.

  The encryption and key-generation queries are handled exactly as in Hyb_0.
- Hyb_2: Same as Hyb_1, except when answering challenge queries (m_0,m_1,i_0,i_1), the challenger instead computes $\mathsf{ct}' \leftarrow \mathsf{PE.Encrypt}(\mathsf{PE.pp},i_1,m_1)$ and replies with the simulated ciphertext $\mathsf{ct} \leftarrow \mathcal{S}_3(\mathsf{st}',\mathsf{ct}')$.
- Hyb_3: Same as Hyb_2, except that the challenger constructs the public parameters $\mathsf{rFE.pp}$ and the sanitizer key sank as described in the real scheme. For challenge queries (m_0,m_1,i_0,i_1), the challenger replies with the ciphertext

$\mathsf{ACE.Encrypt}(\mathsf{ACE.EKGen}(\mathsf{msk}, i_1), m_1)$. This corresponds to the ACE security experiment $\mathsf{Expt}^{(\mathsf{Read})}_{\Pi_{\mathsf{ACE}}, \mathcal{A}, \pi}(\lambda, 1)$ from Definition 2.5.

We now argue that each pair of hybrid experiments are computationally indistinguishable. For an adversary $\mathcal{A}$, we write $\mathsf{Hyb}_i(\mathcal{A})$ to denote the output of Hyb_i. In the following, we implicitly assume that the adversary in each pair of hybrid arguments is admissible.

Lemma 3.1 *If Π_{Sig} is perfectly correct and Π_{rFE} is 1-NA-SIM-secure, then for all efficient adversaries $\mathcal{A}$,* $|\Pr[\mathsf{Hyb}_0(\mathcal{A}) = 1] - \Pr[\mathsf{Hyb}_1(\mathcal{A}) = 1]| = \mathsf{negl}(\lambda)$.

Proof Suppose there exists an adversary $\mathcal{A}$ that can distinguish between Hyb_0 and Hyb_1. We use $\mathcal{A}$ to construct an algorithm $\mathcal{B}$ that can distinguish between $\mathsf{Real}_{\Pi_{\mathsf{rFE}}, \mathcal{A}}(1^\lambda)$ and $\mathsf{Ideal}_{\Pi_{\mathsf{rFE}}, \mathcal{A}, \mathcal{S}}(1^\lambda)$. Algorithm $\mathcal{B}$ works as follows:

1. At the beginning of the game, algorithm $\mathcal{B}$ is given the public parameters $\mathsf{rFE.pp}$. It constructs the other components of the master secret key msk for the ACE scheme exactly as in Hyb_0 and Hyb_1.
2. Algorithm $\mathcal{B}$ makes a key-generation query for the function $F_{\mathsf{Sig.vk}, \mathsf{PE.pp}}$ and receives a key $\mathsf{rFE.sk}_F$. It sets $\mathsf{sank} = \mathsf{rFE.sk}_F$ and gives $\mathsf{rFE.sk}_F$ to $\mathcal{A}$.
3. Algorithm $\mathcal{B}$ answers the encryption and key-generation queries exactly as in Hyb_0 and Hyb_1 (this is possible because these queries only rely on $\mathsf{rFE.pp}$).
4. Whenever $\mathcal{A}$ makes a challenge query (m_0, m_1, i_0, i_1), algorithm $\mathcal{B}$ computes a signature $\sigma \leftarrow \mathsf{Sig.Sign}(\mathsf{Sig.sk}, i_0)$ and queries its encryption oracle on the value (i_0, σ, m_0) to obtain a challenge ciphertext ct. It gives ct to the adversary.
5. At the end of the game, algorithm $\mathcal{B}$ outputs whatever $\mathcal{A}$ outputs.

First, we note that $\mathcal{B}$ makes a single non-adaptive key query, so it is a valid adversary for the 1-NA-SIM security game. By construction, if the public parameters, the key-generation oracle and the encryption oracle are implemented according to $\mathsf{Real}_{\Pi_{\mathsf{rFE}}, \mathcal{A}}(1^\lambda)$, then $\mathcal{B}$ perfectly simulates Hyb_0 for $\mathcal{A}$. We claim that if the public parameters, the key-generation oracle, and the encryption oracle are implemented according to $\mathsf{Ideal}_{\Pi_{\mathsf{rFE}}, \mathcal{A}, \mathcal{S}}(1^\lambda)$, then $\mathcal{B}$ perfectly simulates Hyb_1. It suffices to check that the challenge queries are correctly simulated.

- In Hyb_1, on a challenge query (m_0, m_1, i_0, i_1), the challenger responds by computing $\mathcal{S}_3(\mathsf{st}', \mathsf{ct}')$ where $\mathsf{ct}' \leftarrow \mathsf{PE.Encrypt}(\mathsf{PE.pp}, i_0, m_0)$.
- In the reduction, if the encryption oracle is implemented according to $\mathsf{Ideal}_{\Pi_{\mathsf{rFE}}, \mathcal{A}, \mathcal{S}}(1^\lambda)$, then $\mathcal{B}$'s response ct to a challenge query (m_0, m_1, i_0, i_1) is the output of $\mathcal{S}_3(\mathsf{st}', \mathsf{ct}')$, where $\mathsf{ct}' \leftarrow F_{\mathsf{Sig.vk}, \mathsf{PE.pp}}(i_0, \sigma, m_0)$ and $\sigma \leftarrow \mathsf{Sig.Sign}(\mathsf{Sig.sk}, i_0)$. By perfect correctness of Π_{Sig} and definition of $F_{\mathsf{Sig.vk}, \mathsf{PE.pp}}$, the output distribution of $F_{\mathsf{Sig.vk}, \mathsf{PE.pp}}(i_0, \sigma, m_0)$ is exactly a fresh encryption $\mathsf{PE.Encrypt}(\mathsf{PE.pp}, i_0, m_0)$.

We conclude that if the oracles are implemented according to $\mathsf{Ideal}_{\Pi_{\mathsf{rFE}}, \mathcal{A}, \mathcal{S}}(1^\lambda)$, then $\mathcal{B}$ perfectly simulates Hyb_1 for $\mathcal{A}$. The claim then follows by 1-NA-SIM security of Π_{rFE}. □

Lemma 3.2 *If Π_{PE} is secure, then for all efficient adversaries $\mathcal{A}$,*

$$|\Pr[\mathsf{Hyb}_1(\mathcal{A}) = 1] - \Pr[\mathsf{Hyb}_2(\mathcal{A}) = 1]| = \mathsf{negl}(\lambda).$$

Proof Suppose there exists an adversary $\mathcal{A}$ that can distinguish between Hyb_1 and Hyb_2. We use $\mathcal{A}$ to construct an algorithm $\mathcal{B}$ that can break the security of the predicate encryption scheme Π_{PE}. Algorithm $\mathcal{B}$ works as follows:

1. At the beginning of the game, $\mathcal{B}$ receives $\mathsf{PE.pp}$ from the predicate encryption challenger. It samples the parameters for the signature scheme as well as the parameters for the functional encryption scheme as described in Hyb_1 and Hyb_2 (in particular, the simulator uses the honest key-generation algorithm to sample the parameters for Π_{Sig} and uses the simulator $\mathcal{S}$ for Π_{rFE} to construct the parameters $\mathsf{rFE.pp}$). Algorithm $\mathcal{B}$ constructs the sanitizer key sank as in Hyb_1 and Hyb_2 (using $\mathsf{PE.pp}$), and gives sank to the adversary. It also defines msk as in the real scheme, with the exception that it leaves $\mathsf{PE.msk}$ unspecified.
2. During the query phase, $\mathcal{B}$ answers the encryption and encryption key-generation queries exactly as in Hyb_1 and Hyb_2 (these queries only depend on quantities known to $\mathcal{B}$). The decryption key-generation and challenge queries are handled as follows:
   - **Decryption key-generation oracle**: When $\mathcal{A}$ queries for a decryption key for an identity $j \in \mathcal{I}$, algorithm $\mathcal{B}$ submits the function $f_{\pi,j} \colon \mathcal{I} \to \{0,1\}$ (where $f_{\pi,j}(i) = \pi(i,j)$) to the key-generation oracle for the predicate encryption game, and receives the key $\mathsf{PE.sk}_{f_{\pi,j}}$. It gives $\mathsf{PE.sk}_{f_{\pi,j}}$ to $\mathcal{A}$.
   - **Challenge oracle**: When $\mathcal{A}$ makes its challenge query (m_0, m_1, i_0, i_1), algorithm $\mathcal{B}$ submits the pairs (i_0, m_0), (i_1, m_1) as its challenge query to the predicate encryption challenger and receives a ciphertext ct'. It runs the simulator $\mathsf{ct} \leftarrow \mathcal{S}_3(\mathsf{st}', \mathsf{ct}')$ and returns ct to $\mathcal{A}$.

Since $\mathcal{A}$ is admissible for the no-read rule security game, $\pi(i_0, j) = 0 = \pi(i_1, j)$ for all identities j that the adversary submits to the decryption key-generation oracle. This means that each function $f_{\pi,j}$ that $\mathcal{B}$ submits to the predicate encryption challenger satisfies $f_{\pi,j}(i_0) = 0 = f_{\pi,j}(i_1)$. Thus, $\mathcal{B}$ is admissible for the predicate encryption security game. By construction, if $\mathcal{B}$ is interacting according to $\mathsf{Expt}^{\mathsf{PE}}_{\Pi_{\mathsf{PE}},\mathcal{B}}(\lambda, 0)$, then $\mathcal{B}$ perfectly simulates Hyb_1 for $\mathcal{A}$, and if $\mathcal{B}$ is interacting according to $\mathsf{Expt}^{\mathsf{PE}}_{\Pi_{\mathsf{PE}},\mathcal{B}}(\lambda, 1)$, then $\mathcal{B}$ perfectly simulates Hyb_2 for $\mathcal{A}$. Thus, if $\mathcal{A}$ is able to distinguish between Hyb_1 and Hyb_2 with non-negligible advantage, then $\mathcal{B}$ is able to break the security of Π_{PE} with the same advantage. □

Lemma 3.3 *If Π_{Sig} is perfectly correct, and Π_{rFE} is 1-NA-SIM-secure, then for all efficient adversaries $\mathcal{A}$, $|\Pr[\mathsf{Hyb}_2(\mathcal{A}) = 1] - \Pr[\mathsf{Hyb}_3(\mathcal{A}) = 1]| = \mathsf{negl}(\lambda)$.*

Proof Follows by a similar argument as that used in the proof of Lemma 3.1. □

Combining Lemmas 3.1 through 3.3, we conclude that the ACE scheme in Construction 3.1 satisfies the no-read rule. □

3.3 Proof of Theorem 3.3

Our proof proceeds via a sequence of hybrid experiments between an adversary $\mathcal{A}$ and a challenger.

- Hyb_0: This is the ACE security experiment $\mathsf{Expt}^{(\mathsf{Write})}_{\Pi_{\mathsf{ACE}},\mathcal{A},\pi}(\lambda, 0)$ from Definition 2.7. The challenger begins by sampling $(\mathsf{Sig.vk}, \mathsf{Sig.sk}) \leftarrow \mathsf{Sig.Setup}(1^\lambda)$, $(\mathsf{PE.pp}, \mathsf{PE.msk}) \leftarrow \mathsf{PE.Setup}(1^\lambda)$, and $(\mathsf{rFE.pp}, \mathsf{rFE.msk}) \leftarrow \mathsf{rFE.Setup}(1^\lambda)$. Then, it generates the decryption key $\mathsf{rFE.sk}_F \leftarrow \mathsf{rFE.KeyGen}(\mathsf{rFE.msk}, F_{\mathsf{Sig.vk},\mathsf{PE.pp}})$, and sets $\mathsf{sank} = \mathsf{rFE.sk}_F$ and $\mathsf{msk} = (\pi, \mathsf{Sig.sk}, \mathsf{PE.msk}, \mathsf{rFE.pp})$. During the query phase, the challenger answers the adversary's key-generation and encryption queries exactly as in the real scheme. When the adversary makes a challenge query on a ciphertext ct^* and an identity $\mathsf{id}^* \in \mathcal{I}$, the challenger responds with $\mathsf{ACE.Sanitize}(\mathsf{sank}, \mathsf{ct}^*)$.
- Hyb_1: Same as Hyb_0, except the challenger responds to the adversary's encryption queries with independently-generated predicate encryption ciphertexts. Specifically, for each encryption query on a message $m \in \mathcal{M}$ and identity $i \in \mathcal{I}$, the challenger responds with a fresh encryption $\mathsf{PE.Encrypt}(\mathsf{PE.pp}, i, m)$. The rest of the experiment remains unchanged.
- Hyb_2: Same as Hyb_1, except the challenger constructs the public parameters for the FE scheme, the sanitizer key, and its response to the challenge query using the simulator $\mathcal{S} = (\mathcal{S}_1, \mathcal{S}_2, \mathcal{S}_3, \mathcal{S}_4)$ for Π_{rFE} from Definition 2.3. Specifically, we make the following changes to the challenger:
  - **Setup:** At the beginning of the game, instead of sampling $\mathsf{rFE.pp}$ using $\mathsf{rFE.Setup}$, the challenger instead runs the simulation algorithm $(\mathsf{rFE.pp}, \mathsf{st}') \leftarrow \mathcal{S}_1(1^\lambda)$. For the sanitizer key, the challenger computes $\mathsf{rFE.sk}_F \leftarrow \mathcal{S}_2(\mathsf{st}', F_{\mathsf{Sig.vk},\mathsf{PE.pp}})$. The challenger samples $(\mathsf{Sig.vk}, \mathsf{Sig.sk})$ and $(\mathsf{PE.pp}, \mathsf{PE.msk})$ as in the real scheme.
  - **Challenge query:** For the challenge query $(\mathsf{ct}^*, \mathsf{id}^*)$, the challenger first invokes the simulator to obtain $y^* \leftarrow \mathcal{S}_4(\mathsf{st}', \mathsf{ct}^*)$. If $y^* \neq \perp$, it parses $y^* = (i^*, \sigma^*, m^*)$, and checks if $\mathsf{Sig.Verify}(\mathsf{Sig.vk}, i^*, \sigma^*) \stackrel{?}{=} 1$. If so, then the challenger returns $\mathsf{PE.Encrypt}(\mathsf{PE.pp}, i^*, m^*)$. In all other cases, the challenger outputs $\perp$.

  The rest of the experiment is identical to Hyb_1.
- Hyb_3: Same as Hyb_2, except the challenger aborts during the challenge phase if after computing $y^* \leftarrow \mathcal{S}_4(\mathsf{st}', \mathsf{ct}^*)$ and parsing $y^* = (i^*, \sigma^*, m^*)$, the following two conditions hold:
  - Adversary $\mathcal{A}$ did not previously make an encryption key-generation query for identity i^*.
  - $\mathsf{Sig.Verify}(\mathsf{Sig.vk}, i^*, \sigma^*) = 1$.

  Otherwise, the challenger proceeds as in Hyb_2.
- Hyb_4: Same as Hyb_3, except the challenger answers the challenge query with a sanitized encryption of a random message. Specifically, when the challenger receives a challenge query $(\mathsf{ct}^*, \mathsf{id}^*)$, it computes $y^* \leftarrow \mathcal{S}_4(\mathsf{st}', \mathsf{ct}^*)$ as usual and returns $\perp$ if $y^* = \perp$. Otherwise, it parses $y^* = (i^*, \sigma^*, m^*)$ and checks that $\mathsf{Sig.Verify}(\mathsf{Sig.vk}, i^*, \sigma^*) = 1$ (outputting $\perp$ if not). The challenger also checks

the abort condition in Hyb_3. If all the checks pass, the challenger samples a message $m' \xleftarrow{R} \mathcal{M}$ and returns $\mathsf{PE.Encrypt}(\mathsf{PE.pp}, \mathsf{id}^*, m')$ to the adversary. The rest of the experiment is unchanged.

- Hyb_5: Same as Hyb_4, except we remove the abort condition from the challenger.
- Hyb_6: Same as Hyb_5, except the challenger samples the public parameters for the FE scheme, the sanitizer key, and its response to the challenge query using the real algorithms Π_{rFE} rather than the simulator. In particular, when responding to the challenge query $(\mathsf{ct}^*, \mathsf{id}^*)$, the challenger responds with $\mathsf{rFE.Decrypt}(\mathsf{sank}, \mathsf{rFE.Encrypt}(\mathsf{rFE.pp}, (\mathsf{id}^*, \sigma, m')))$ where $m' \xleftarrow{R} \mathcal{M}$ and σ is a signature on id^* under $\mathsf{Sig.vk}$.
- Hyb_7: Same as Hyb_6, except the challenger responds to the adversary's encryption queries honestly as in the real scheme instead of responding with independently generated predicate encryption ciphertexts. This corresponds to the ACE security experiment $\mathsf{Expt}^{(\mathsf{Write})}_{\Pi_{\mathsf{ACE}}, \mathcal{A}, \pi}(\lambda, 1)$ from Definition 2.7.

Lemma 3.4 *If Π_{Sig} is perfectly correct and Π_{rFE} is correct, then for all efficient adversaries $\mathcal{A}$, we have that* $|\Pr[\mathsf{Hyb}_0(\mathcal{A}) = 1] - \Pr[\mathsf{Hyb}_1(\mathcal{A}) = 1]| = \mathsf{negl}(\lambda)$.

Proof The only difference between Hyb_0 and Hyb_1 is the way the challenger responds to the adversary's encryption queries. First, let $\mathsf{sank} = \mathsf{rFE.sk}_F \leftarrow \mathsf{rFE.KeyGen}(\mathsf{rFE.pp}, F_{\mathsf{Sig.vk},\mathsf{PE.pp}})$ be the sanitizer key generated by the challenger at setup. Suppose the adversary makes Q encryption queries on message-identity pairs $(m_1, i_1), \ldots, (m_Q, i_Q)$. In Hyb_0, the challenger responds to each query (m_k, i_k) by first computing the signature $\sigma_k \leftarrow \mathsf{Sig.Sign}(\mathsf{Sig.sk}, i_k)$ and the ciphertext $\mathsf{ct}_k \leftarrow \mathsf{rFE.Decrypt}(\mathsf{rFE.sk}_F, \mathsf{rFE.Encrypt}(\mathsf{rFE.pp}, (i_k, \sigma_k, m_k)))$. By correctness of Π_{rFE}, we have that

$$\left(\mathsf{rFE.pp}, \mathsf{rFE.sk}_F, \{\mathsf{ct}_k\}_{k \in [Q]}\right) \overset{c}{\approx} \left(\mathsf{rFE.pp}, \mathsf{rFE.sk}_F, \{F_{\mathsf{Sig.vk},\mathsf{PE.pp}}(i_k, \sigma_k, m_k; r_k)\}_{k \in [Q]}\right),$$

where $r_k \xleftarrow{R} \mathcal{R}$. Since σ_k is a signature on i_k, by perfect correctness of Π_{Sig} and definition of $F_{\mathsf{Sig.vk},\mathsf{PE.pp}}$, the output distribution of $F_{\mathsf{Sig.vk},\mathsf{PE.pp}}(i_k, \sigma_k, m_k; r_k)$ is precisely a fresh encryption $\mathsf{PE.Encrypt}(\mathsf{PE.pp}, i_k, m_k)$. This is the distribution in Hyb_1. Note that we include the sanitizer key $\mathsf{rFE.sk}_F$ in the joint distributions above because it is needed to simulate the response to the adversary's challenge query in Hyb_0 and Hyb_1. □

Lemma 3.5 *If Π_{rFE} is 1-NA-SIM-secure, then for all efficient adversaries $\mathcal{A}$, we have that* $|\Pr[\mathsf{Hyb}_1(\mathcal{A}) = 1] - \Pr[\mathsf{Hyb}_2(\mathcal{A}) = 1]| = \mathsf{negl}(\lambda)$.

Proof Suppose there exists an adversary $\mathcal{A}$ that can distinguish between Hyb_1 and Hyb_2. We use $\mathcal{A}$ to construct an algorithm $\mathcal{B}$ that can distinguish between $\mathsf{Real}_{\Pi_{\mathsf{rFE}}, \mathcal{A}}(1^\lambda)$ and $\mathsf{Ideal}_{\Pi_{\mathsf{rFE}}, \mathcal{A}, \mathcal{S}}(1^\lambda)$. Algorithm $\mathcal{B}$ works as follows:

1. At the beginning of the game, algorithm $\mathcal{B}$ is given the public parameters $\mathsf{rFE.pp}$. It constructs the other components of the master secret key msk for the ACE scheme exactly as in Hyb_1 and Hyb_2.

2. Algorithm $\mathcal{B}$ answers the encryption and key-generation queries exactly as in Hyb_1 and Hyb_2. These queries only depend on $\mathsf{rFE.pp}$ (and not $\mathsf{rFE.msk}$ and sank, both of which are unspecified).
3. When $\mathcal{A}$ makes a challenge query $(\mathsf{ct}^*, \mathsf{id}^*)$, algorithm $\mathcal{B}$ queries its decryption oracle on the pair $(F_{\mathsf{Sig.vk},\mathsf{PE.pp}}, \mathsf{ct}^*)$ to obtain a value z^*. It gives z^* to the adversary.
4. At the end of the game, algorithm $\mathcal{B}$ outputs whatever $\mathcal{A}$ outputs.

First, we note that $\mathcal{B}$ does not make any key queries or encryption queries, so it is trivially admissible for the 1-NA-SIM security game. By construction, if the public parameters, the key-generation oracle, the encryption oracle, and the decryption oracle are implemented according to $\mathsf{Real}_{\Pi_{\mathsf{rFE}},\mathcal{A}}(1^\lambda)$, then $\mathcal{B}$ perfectly simulates Hyb_1 for $\mathcal{A}$. In particular, we note that the sanitizer key sank is only needed when responding to the challenge query, and so, the key sampled by the decryption oracle in $\mathsf{Real}_{\Pi_{\mathsf{rFE}},\mathcal{A}}(1^\lambda)$ plays the role of sank. To conclude the proof, we show that if the public parameters, the key-generation oracle, the encryption oracle, and the decryption oracle are implemented according to $\mathsf{Ideal}_{\Pi_{\mathsf{rFE}},\mathcal{A},\mathcal{S}}(1^\lambda)$, then $\mathcal{B}$ perfectly simulates Hyb_2. It suffices to check that the challenge query is correctly simulated.

- In Hyb_2, on a challenge query $(\mathsf{ct}^*, \mathsf{id}^*)$, the challenger computes $y^* \leftarrow \mathcal{S}_4(\mathsf{st}', \mathsf{ct}^*)$. If $y^* = \perp$, then the challenger responds with $\perp$. Otherwise, it parses $y^* = (i^*, \sigma^*, m^*)$, and checks whether $\mathsf{Sig.Verify}(\mathsf{Sig.vk}, i^*, \sigma^*) \stackrel{?}{=} 1$ accepts. If so, it returns $\mathsf{PE.Encrypt}(\mathsf{PE.pp}, i^*, m^*; r)$ where $r \xleftarrow{\text{R}} \mathcal{R}$. Otherwise, it returns $\perp$. This logic precisely corresponds to evaluating $F_{\mathsf{Sig.vk},\mathsf{PE.pp}}(y^*; r)$.
- In the reduction, if the decryption oracle is implemented according to $\mathsf{Ideal}_{\Pi_{\mathsf{rFE}},\mathcal{A},\mathcal{S}}(1^\lambda)$, then the oracle first computes $y^* \leftarrow \mathcal{S}_4(\mathsf{st}', \mathsf{ct}^*)$. If $y^* = \perp$, the oracle returns $\perp$. Otherwise, it returns $F_{\mathsf{Sig.vk},\mathsf{PE.pp}}(y^*; r)$ where $r \xleftarrow{\text{R}} \mathcal{R}$. This is precisely the behavior in Hyb_2.

We conclude that if the oracles are implemented according to $\mathsf{Ideal}_{\Pi_{\mathsf{rFE}},\mathcal{A},\mathcal{S}}(1^\lambda)$, then $\mathcal{B}$ perfectly simulates Hyb_2 for $\mathcal{A}$. The claim then follows by 1-NA-SIM security of Π_{rFE}. □

Lemma 3.6 *If Π_{Sig} is existentially unforgeable, then for all efficient adversaries $\mathcal{A}$, we have that* $|\Pr[\mathsf{Hyb}_2(\mathcal{A}) = 1] - \Pr[\mathsf{Hyb}_3(\mathcal{A}) = 1]| = \mathsf{negl}(\lambda)$.

Proof Hybrids Hyb_2 and Hyb_3 are identical except for the extra abort condition in Hyb_3. Suppose there exists an adversary $\mathcal{A}$ that can distinguish between Hyb_2 and Hyb_3 with non-negligible advantage ε. Then, it must be the case that $\mathcal{A}$ can cause Hyb_3 to abort with probability at least ε (otherwise, the two experiments are identical). We use $\mathcal{A}$ to construct an algorithm $\mathcal{B}$ that breaks the security of Π_{Sig}. Algorithm $\mathcal{B}$ works as follows:

1. At the beginning of the existential unforgeability game, $\mathcal{B}$ is given the verification key $\mathsf{Sig.vk}$. Algorithm $\mathcal{B}$ chooses the parameters for the predicate encryption scheme and the functional encryption scheme as in Hyb_2 and Hyb_3. It constructs the sanitizer key sank and msk as in Hyb_2 and Hyb_3, except it leaves $\mathsf{Sig.sk}$ unspecified in msk.

2. During the query phase, $\mathcal{B}$ answers the encryption queries and the decryption key-generation queries exactly as in Hyb_2 and Hyb_3 (since none of these queries depend on knowledge of $\mathsf{Sig.sk}$). Algorithm $\mathcal{B}$ answers the encryption key-generation and challenge queries as follows:
   - **Encryption key-generation queries:** When $\mathcal{A}$ queries for an encryption key for an identity $i \in \mathcal{I}$, algorithm $\mathcal{B}$ submits i to its signing oracle and receives a signature σ. It gives $(\mathsf{rFE.pp}, i, \sigma)$ to $\mathcal{A}$.
   - **Challenge queries:** When $\mathcal{A}$ makes its challenge query (ct^*, i^*), algorithm $\mathcal{B}$ runs the simulator $y^* \leftarrow \mathcal{S}_4(\mathsf{st}', \mathsf{ct}^*)$. If $y^* = \perp$, then $\mathcal{B}$ replies with $\perp$. Otherwise, it parses $y^* = (i^*, \sigma^*, m^*)$, and submits (i^*, σ^*) as its forgery in the existential unforgeability game.

By construction, $\mathcal{B}$ perfectly simulates Hyb_2 and Hyb_3 for $\mathcal{A}$. Thus, with probability at least ε, algorithm $\mathcal{A}$ is able to produce a ciphertext ct^* that causes Hyb_3 to abort. This corresponds to the case where $\mathcal{A}$ never makes an encryption key-generation query for identity i^*, and yet, σ^* is a valid signature on i^*. Since $\mathcal{B}$ only queries the signing oracle when $\mathcal{A}$ makes an encryption key-generation query, by assumption, $\mathcal{B}$ never queries the signing oracle on the message i^*. In this case, σ^* is a valid forgery for the signature scheme, and $\mathcal{B}$ is able to break the security of the signature scheme with non-negligible advantage ε. □

Lemma 3.7 *If Π_{PE} is secure, then for all efficient adversaries $\mathcal{A}$, we have that* $|\Pr[\mathsf{Hyb}_3(\mathcal{A}) = 1] - \Pr[\mathsf{Hyb}_4(\mathcal{A}) = 1]| = \mathsf{negl}(\lambda)$.

Proof Suppose there exists an adversary $\mathcal{A}$ that can distinguish between Hyb_3 and Hyb_4. We use $\mathcal{A}$ to construct an algorithm $\mathcal{B}$ that can break the security of the predicate encryption scheme Π_{PE}. Algorithm $\mathcal{B}$ works as follows:

1. At the beginning of the game, $\mathcal{B}$ receives the public parameters $\mathsf{PE.pp}$ from the predicate encryption challenger. It samples $(\mathsf{Sig.vk}, \mathsf{Sig.sk})$, $\mathsf{rFE.pp}$, and sank exactly as in Hyb_3 and Hyb_4. It constructs msk exactly as in Hyb_3 and Hyb_4, except it leaves $\mathsf{PE.msk}$ unspecified.
2. During the query phase, $\mathcal{B}$ answers the encryption queries and the encryption key-generation queries exactly as in Hyb_3 and Hyb_4 (since they do not depend on $\mathsf{PE.msk}$). The decryption key-generation queries and the challenge queries are handled as follows:
   - **Decryption key-generation oracle**: When $\mathcal{A}$ queries for a decryption key for an identity $j \in \mathcal{I}$, algorithm $\mathcal{B}$ submits the function $f_{\pi,j} : \mathcal{I} \to \{0,1\}$ (where $f_{\pi,j}(i) = \pi(i,j)$) to the key-generation oracle for the predicate encryption game, and receives the key $\mathsf{PE.sk}_{f_{\pi,j}}$. It gives $\mathsf{PE.sk}_{f_{\pi,j}}$ to $\mathcal{A}$.
   - **Challenge oracle**: When $\mathcal{A}$ makes its challenge query $(\mathsf{ct}^*, \mathsf{id}^*)$, algorithm $\mathcal{B}$ first computes $y^* \leftarrow \mathcal{S}_4(\mathsf{st}', \mathsf{ct}^*)$. If $y^* = \perp$, algorithm $\mathcal{B}$ responds with $\perp$. Otherwise, it parses $y^* = (i^*, \sigma^*, m^*)$ and checks the abort condition. If $\mathcal{B}$ does not abort, then it samples a message $m' \xleftarrow{\text{R}} \mathcal{M}$, and submits the pairs (i^*, m^*), (id^*, m') as its challenge query for the predicate encryption security game. The predicate encryption challenger replies with a challenge ciphertext z which $\mathcal{B}$ sends to $\mathcal{A}$.

First, we argue that $\mathcal{B}$ is admissible for the predicate encryption security game. Since Hyb_3 and Hyb_4 behave identically if $\mathcal{B}$ aborts, it suffices to reason about the case where the experiment does not abort. We analyze each case individually:

- If $y^* = \perp$ or $\mathsf{Sig.Verify}(\mathsf{Sig.vk}, i^*, \sigma^*) \neq 1$, then the challenger responds with $\perp$ in both Hyb_3 and Hyb_4 (as does $\mathcal{B}$).
- If $\mathsf{Sig.Verify}(\mathsf{Sig.vk}, i^*, \sigma^*) = 1$, then $\mathcal{A}$ must have previously queried the encryption key-generation oracle on identity i^* (otherwise, the challenger in Hyb_3 and Hyb_4 would have aborted). Since $\mathcal{A}$ is admissible for the no-write security game, for all identities $j \in \mathcal{I}$ that $\mathcal{A}$ submits to the decryption key-generation oracle, it must be the case that $\pi(i^*, j) = 0$. Similarly, by admissibility of $\mathcal{A}$, it must have submitted its challenge identity id^* to the encryption key-generation oracle prior to making its challenge query. Thus, we also have that $\pi(\mathsf{id}^*, j) = 0$. This means that each function $f_{\pi,j}$, that $\mathcal{B}$ submits to the predicate encryption challenger satisfies $f_{\pi,j}(i^*) = 0 = f_{\pi,j}(\mathsf{id}^*)$.

We conclude that $\mathcal{B}$ is admissible. Moreover, if $\mathcal{B}$ is interacting according to $\mathsf{Expt}^{\mathsf{PE}}_{\Pi_{\mathsf{PE}},\mathcal{B}}(\lambda, 0)$, then $\mathcal{B}$ perfectly simulates Hyb_3 for $\mathcal{A}$ and if $\mathcal{B}$ is interacting according to $\mathsf{Expt}^{\mathsf{PE}}_{\Pi_{\mathsf{PE}},\mathcal{B}}(\lambda, 1)$, then $\mathcal{B}$ perfectly simulates Hyb_4 for $\mathcal{A}$. The lemma follows. □

Lemma 3.8 *If Π_{Sig} is existentially unforgeable, then for all efficient adversaries $\mathcal{A}$, we have that* $|\Pr[\mathsf{Hyb}_4(\mathcal{A}) = 1] - \Pr[\mathsf{Hyb}_5(\mathcal{A}) = 1]| = \mathsf{negl}(\lambda)$.

Proof Follows by a similar argument as that used in the proof of Lemma 3.6. □

Lemma 3.9 *If Π_{rFE} is 1-NA-SIM-secure, then for all efficient adversaries $\mathcal{A}$, we have that* $|\Pr[\mathsf{Hyb}_5(\mathcal{A}) = 1] - \Pr[\mathsf{Hyb}_6(\mathcal{A}) = 1]| = \mathsf{negl}(\lambda)$.

Proof Follows by a similar argument as that used in the proof of Lemma 3.5. □

Lemma 3.10 *If Π_{Sig} is perfectly correct and Π_{rFE} is correct, then for all efficient adversaries $\mathcal{A}$,* $|\Pr[\mathsf{Hyb}_6(\mathcal{A}) = 1] - \Pr[\mathsf{Hyb}_7(\mathcal{A}) = 1]| = \mathsf{negl}(\lambda)$.

Proof Follows by a similar argument as that used in the proof of Lemma 3.4. □

Combining Lemmas 3.4 through 3.10, we conclude that the ACE scheme in Construction 3.1 satisfies the no-write rule. □

4 Extensions

In this section, we describe several extensions to access control encryption that follow immediately from our generic ACE construction in Sect. 3. We present these extensions primarily as ways of extending the schema of access control encryption to provide increased flexibility, rather than as conceptually new properties achieved by our specific construction. Indeed, it is not too difficult to modify the $i\mathcal{O}$-based ACE construction from Damgård et al. [23] to also provide these properties.

4.1 Dynamic Policies

The access control encryption schema in Sect. 2.2 required that the access control policies be specified at setup time. In this section, we show how to modify Construction 3.1 so that policies can be associated with individual decryption keys rather than globally. This means that the access control policy no longer has to be fixed at the time of system setup, and moreover, different access control policies can be implemented for each receiver. Thus, the system can support new policies as new receivers are added to the system, and in addition, receivers can update their keys (i.e., obtain new keys from the key distributor) when the access control policies change. Notably, with this extension, changes to the access control policy do not require updating or re-issuing the sender keys. More formally, we would make the following two modifications to the schema of ACE scheme from Sect. 2.2:

- $\mathsf{ACE.Setup}(1^\lambda) \to (\mathsf{sank}, \mathsf{msk})$: On input the security parameter λ, the setup algorithm outputs the sanitizer key sank and the master secret key msk. Notably, the setup algorithm does *not* take the access control policy π as input.
- $\mathsf{ACE.DKGen}(\mathsf{msk}, j, \pi_j) \to \mathsf{dk}_{j,\pi_j}$: On input the master secret key msk, the receiver identity $j \in \mathcal{I}$, and an access control policy $\pi_j : \mathcal{I} \to \{0, 1\}$ (the access control policy takes in a sender identity $i \in \mathcal{I}$ and outputs a bit), the decryption key-generation algorithm outputs a decryption key $\mathsf{dk}_{j,\pi}$.

The usual notion of access control encryption from Sect. 2.2 just corresponds to the special case where the receiver-specific policy π_j is simply the global access control policy π (specialized to the particular receiver identity j). The correctness and security notions generalize accordingly.

Supporting dynamic policies. It is easy to modify Construction 3.1 to support dynamic policies according to the above schema. In fact, policy enforcement in Construction 3.1 is already handled by embedding the access control policy within the receiver's decryption keys. Thus, supporting receiver-specific policies $\pi_j : \mathcal{I} \to \{0, 1\}$ in $\mathsf{ACE.DKGen}$ can be implemented by simply generating the decryption key as $\mathsf{dk}_{j,\pi_j} \leftarrow \mathsf{PE.KeyGen}(\mathsf{PE.msk}, \pi_j)$. The correctness and security analysis remain unchanged.

4.2 Fine-Grained Sender Policies

As noted in Sect. 1.1, it is often desirable to support fine-grained sender policies that depend not only on the sender's identity, but also on the contents of the sender's message. In this section, we describe how to extend Construction 3.1 to support fine-grained sender policies. We also give a new security definition (Definition 4.1) to capture the property that a sender should only be able produce encryptions of messages that conform to its particular policy.

Schema changes. In the context of access control encryption, fine-grained sender policies can be captured by modifying the schema for the encryption key-generation algorithm to additionally take in a sender policy (which can be represented as a predicate on the message space of the encryption scheme). Formally, we write

- $\mathsf{ACE.EKGen}(\mathsf{msk}, i, \tau) \to \mathsf{ek}_{i,\tau}$: On input the master secret key msk, a sender identity $i \in \mathcal{I}$, and a sender policy $\tau \colon \mathcal{M} \to \{0,1\}$, the encryption key-generation algorithm outputs an encryption key $\mathsf{ek}_{i,\tau}$.

To support fine-grained sender policies, we first relax the correctness definition (Definition 2.4) by requiring that correctness only holds for messages $m \in \mathcal{M}$ that satisfy the sender's encryption policy. The no-read and no-write rules remain largely unchanged (they are defined with respect to the "always-accept" sender policy). To capture the property that a sender should only be able to encrypt messages for which it is authorized, we introduce a new "soundness" requirement that effectively states that a sender with encryption keys for some collection of policies $\tau_1, \ldots, \tau_Q$ cannot produce a new ciphertext ct that encrypts a message m (with respect to some decryption key dk) where $\tau_k(m) = 0$ for all $k \in [Q]$. More formally, we define the following soundness property:

Definition 4.1 (Soundness). *Let $\Pi_{\mathsf{ACE}} = (\mathsf{ACE.Setup}, \mathsf{ACE.EKGen}, \mathsf{ACE.DKGen}, \mathsf{ACE.Encrypt}, \mathsf{ACE.Sanitize}, \mathsf{ACE.Decrypt})$ be an ACE scheme over an identity space $\mathcal{I}$ and a message space $\mathcal{M}$. Let $\mathcal{A}$ be an efficient adversary and $\pi \colon \mathcal{I} \times \mathcal{I} \to \{0,1\}$ be an access control policy. For a security parameter λ, we define the soundness experiment $\mathsf{Expt}^{(\mathsf{Sound})}_{\Pi_{\mathsf{ACE}},\mathcal{A},\pi}(\lambda)$ as follows. The challenger begins by sampling $(\mathsf{sank}, \mathsf{msk}) \leftarrow \mathsf{ACE.Setup}(1^\lambda, \pi)$. The adversary $\mathcal{A}$ is then given access to the following oracles:*

- ***Encryption oracle.*** *On input a message $m \in \mathcal{M}$, and a sender identity $i \in \mathcal{I}$, the challenger first generates a sender key $\mathsf{ek}_i \leftarrow \mathsf{ACE.EKGen}(\mathsf{msk}, i, \tau)$, where $\tau(m) = 1$ for all $m \in \mathcal{M}$. The challenger responds with the ciphertext $\mathsf{ct} \leftarrow \mathsf{ACE.Sanitize}(\mathsf{sank}, \mathsf{ACE.Encrypt}(\mathsf{ek}_i, m))$.*
- ***Encryption key-generation oracle.*** *On input a sender identity $i \in \mathcal{I}$ and a sender policy $\tau \colon \mathcal{M} \to \{0,1\}$, the challenger responds with an encryption key $\mathsf{ek}_{i,\tau} \leftarrow \mathsf{ACE.EKGen}(\mathsf{msk}, i, \tau)$.*
- ***Decryption key-generation oracle.*** *On input a receiver identity $j \in \mathcal{I}$, the challenger responds with a decryption key $\mathsf{dk}_j \leftarrow \mathsf{ACE.DKGen}(\mathsf{msk}, j)$.*

At the end of the experiment, adversary $\mathcal{A}$ outputs a ciphertext $\mathsf{ct}^* \in \{0,1\}^*$, and a receiver identity $j^* \in \mathcal{I}$. The output of the experiment is 1 if and only if the following conditions hold:

- *$\mathsf{ACE.Decrypt}(\mathsf{ACE.DKGen}(\mathsf{msk}, j^*), \mathsf{ACE.Sanitize}(\mathsf{sank}, \mathsf{ct}^*)) = m^*$ for some $m^* \in \mathcal{M}$.*
- *Let $\{(i_k, \tau_k)\}_{k \in [Q]}$ be the queries $\mathcal{A}$ makes to the sender key-generation oracle. For all $k \in [Q]$ where $\pi(i_k, j^*) = 1$, $\tau_k(m^*) = 0$, where m^* is the decrypted message defined above.*

We say that Π_{ACE} is sound if for all policies $\pi\colon \mathcal{I} \times \mathcal{I} \to \{0,1\}$, and all efficient adversaries $\mathcal{A}$,

$$\Pr\left[\mathsf{Expt}^{(\mathsf{Sound})}_{\Pi_{\mathsf{ACE}},\mathcal{A},\pi}(\lambda) = 1\right] = \mathsf{negl}(\lambda).$$

Supporting sender policies. It is straightforward to extend Construction 3.1 to support *arbitrary* sender policies with little additional overhead. Concretely, we make the following changes to Construction 3.1:

- Instead of a signature on the identity i, the encryption key for an identity $i \in \mathcal{I}$ and sender policy $\tau\colon \mathcal{M} \to \{0,1\}$ contains a signature on the tuple (i, τ), as well as a description of the policy. Namely, $\mathsf{ek}_i = (\mathsf{rFE.pp}, i, \tau, \sigma)$ where $\sigma \leftarrow \mathsf{Sig.Sign}(\mathsf{Sig.sk}, (i, \tau))$.
- An encryption of a message $m \in \mathcal{M}$ under the encryption key $\mathsf{ek}_i = (\mathsf{rFE.pp}, i, \tau, \sigma)$ is an encryption of the tuple (i, τ, σ, m) using Π_{rFE}.
- The (randomized) sanitizer function $F_{\mathsf{Sig.vk},\mathsf{PE.pp}}$ now takes as input the tuple (i, τ, σ, m) and outputs $\mathsf{PE.Encrypt}(\mathsf{PE.pp}, i, m)$ if $\mathsf{Sig.Verify}(\mathsf{Sig.vk}, (i, \tau), \sigma) = 1$ and $\tau(m) = 1$. Otherwise, $F_{\mathsf{Sig.vk},\mathsf{PE.pp}}$ outputs $\perp$. The sanitizer key sank is then a decryption key $\mathsf{rFE.sk}_F$ for the modified sanitizer function: $\mathsf{rFE.sk}_F \leftarrow \mathsf{rFE.KeyGen}(\mathsf{msk}, F_{\mathsf{Sig.vk},\mathsf{PE.pp}})$.

At a high level, the sanitizer key implicitly checks that a sender's message is compliant with the associated policy, and outputs a ciphertext that can be decrypted only if this is the case. Here, the signature is essential in ensuring that the sender is only able to send messages that comply with one of its sending policies. In particular, we show the following theorem. We give the proof in the full version of this paper [38].

Theorem 4.1 *Suppose Π_{Sig} is existentially unforgeable and Π_{rFE} is a 1-NA-SIM-secure functional encryption scheme for randomized functionalities (Definition 2.3). Then the access control encryption scheme from Construction 3.1 with the above modifications satisfies soundness (Definition 4.1).*

Relation to constrained PRFs and constrained signatures. This notion of constraining the encryption key to only produce valid encryptions on messages that satisfy the predicate is very similar to the concept of constrained pseudorandom functions (PRF) [18,19,37] and constrained signatures [10,19,41]. Constrained PRFs (resp., constrained signatures) allow the holder of the secret key to issue a *constrained* key for a predicate that only allows PRF evaluation on inputs (resp., signing messages) that satisfy the predicate. When extending ACE to support fine-grained sender policies, the encryption key-generation algorithm can be viewed as giving out a constrained version of the corresponding sender key. Our technique for constraining the encryption key by including a signature of the predicate and having the encrypter "prove possession" of the signature is conceptually similar to the technique used in [19] to construct functional signatures and in [10] to construct policy-based signatures. In [10,19], this proof of possession is implemented by having the user provide a non-interactive zero-knowledge proof of knowledge of the signature, while in our setting, it is handled by having the user encrypt the signature under an FE scheme and giving out an FE key (to the sanitizer) that performs the signature verification.

Acknowledgments. We thank Shashank Agrawal and the anonymous reviewers for helpful comments. This work was funded by NSF, DARPA, a grant from ONR, and the Simons Foundation. Opinions, findings and conclusions or recommendations expressed in this material are those of the authors and do not necessarily reflect the views of DARPA.

References

1. Agrawal, S., Wu, D.J.: Functional encryption: deterministic to randomized functions from simple assumptions. In: Coron, J.-S., Nielsen, J.B. (eds.) EUROCRYPT 2017. LNCS, vol. 10211, pp. 30–61. Springer, Cham (2017). https://doi.org/10.1007/978-3-319-56614-6_2
2. Alwen, J., Barbosa, M., Farshim, P., Gennaro, R., Gordon, S.D., Tessaro, S., Wilson, D.A.: On the relationship between functional encryption, obfuscation, and fully homomorphic encryption. In: Cryptography and Coding (2013)
3. Ateniese, G., Chou, D.H., Medeiros, B., Tsudik, G.: Sanitizable signatures. In: Vimercati, S.C., Syverson, P., Gollmann, D. (eds.) ESORICS 2005. LNCS, vol. 3679, pp. 159–177. Springer, Heidelberg (2005). https://doi.org/10.1007/11555827_10
4. Ateniese, G., Fu, K., Green, M., Hohenberger, S.: Improved proxy re-encryption schemes with applications to secure distributed storage. In: NDSS (2005)
5. Badertscher, C., Matt, C., Maurer, U.: Strengthening access control encryption (2017)
6. Barak, B., Goldreich, O., Impagliazzo, R., Rudich, S., Sahai, A., Vadhan, S., Yang, K.: On the (im)possibility of obfuscating programs. In: Kilian, J. (ed.) CRYPTO 2001. LNCS, vol. 2139, pp. 1–18. Springer, Heidelberg (2001). https://doi.org/10.1007/3-540-44647-8_1
7. Bell, D.E., LaPadula, L.J.: Secure computer systems: mathematical foundations. Technical report, DTIC Document (1973)
8. Bellare, M., Cash, D.: Pseudorandom functions and permutations provably secure against related-key attacks. In: Rabin, T. (ed.) CRYPTO 2010. LNCS, vol. 6223, pp. 666–684. Springer, Heidelberg (2010). https://doi.org/10.1007/978-3-642-14623-7_36
9. Bellare, M., Cash, D., Miller, R.: Cryptography secure against related-key attacks and tampering. In: Lee, D.H., Wang, X. (eds.) ASIACRYPT 2011. LNCS, vol. 7073, pp. 486–503. Springer, Heidelberg (2011). https://doi.org/10.1007/978-3-642-25385-0_26
10. Bellare, M., Fuchsbauer, G.: Policy-based signatures. In: Krawczyk, H. (ed.) PKC 2014. LNCS, vol. 8383, pp. 520–537. Springer, Heidelberg (2014). https://doi.org/10.1007/978-3-642-54631-0_30
11. Bellare, M., Kohno, T.: A theoretical treatment of related-key attacks: RKA-PRPs, RKA-PRFs, and applications. In: Biham, E. (ed.) EUROCRYPT 2003. LNCS, vol. 2656, pp. 491–506. Springer, Heidelberg (2003). https://doi.org/10.1007/3-540-39200-9_31
12. Bethencourt, J., Sahai, A., Waters, B.: Ciphertext-policy attribute-based encryption. In: IEEE S&P (2007)

13. Biham, E.: New types of cryptanalytic attacks using related keys. In: Helleseth, T. (ed.) EUROCRYPT 1993. LNCS, vol. 765, pp. 398–409. Springer, Heidelberg (1994). https://doi.org/10.1007/3-540-48285-7_34
14. Boneh, D.: The decision Diffie-Hellman problem. In: Buhler, J.P. (ed.) ANTS 1998. LNCS, vol. 1423, pp. 48–63. Springer, Heidelberg (1998). https://doi.org/10.1007/BFb0054851
15. Boneh, D., Franklin, M.: Identity-based encryption from the weil pairing. In: Kilian, J. (ed.) CRYPTO 2001. LNCS, vol. 2139, pp. 213–229. Springer, Heidelberg (2001). https://doi.org/10.1007/3-540-44647-8_13
16. Boneh, D., Sahai, A., Waters, B.: Functional encryption: definitions and challenges. In: Ishai, Y. (ed.) TCC 2011. LNCS, vol. 6597, pp. 253–273. Springer, Heidelberg (2011). https://doi.org/10.1007/978-3-642-19571-6_16
17. Boneh, D., Waters, B.: Conjunctive, subset, and range queries on encrypted data. In: Vadhan, S.P. (ed.) TCC 2007. LNCS, vol. 4392, pp. 535–554. Springer, Heidelberg (2007). https://doi.org/10.1007/978-3-540-70936-7_29
18. Boneh, D., Waters, B.: Constrained pseudorandom functions and their applications. In: Sako, K., Sarkar, P. (eds.) ASIACRYPT 2013. LNCS, vol. 8270, pp. 280–300. Springer, Heidelberg (2013). https://doi.org/10.1007/978-3-642-42045-0_15
19. Boyle, E., Goldwasser, S., Ivan, I.: Functional signatures and pseudorandom functions. In: Krawczyk, H. (ed.) PKC 2014. LNCS, vol. 8383, pp. 501–519. Springer, Heidelberg (2014). https://doi.org/10.1007/978-3-642-54631-0_29
20. Brzuska, C., Fischlin, M., Freudenreich, T., Lehmann, A., Page, M., Schelbert, J., Schröder, D., Volk, F.: Security of sanitizable signatures revisited. In: Jarecki, S., Tsudik, G. (eds.) PKC 2009. LNCS, vol. 5443, pp. 317–336. Springer, Heidelberg (2009). https://doi.org/10.1007/978-3-642-00468-1_18
21. Cash, D., Hofheinz, D., Kiltz, E., Peikert, C.: Bonsai trees, or how to delegate a lattice basis. In: Gilbert, H. (ed.) EUROCRYPT 2010. LNCS, vol. 6110, pp. 523–552. Springer, Heidelberg (2010). https://doi.org/10.1007/978-3-642-13190-5_27
22. Cocks, C.: An identity based encryption scheme based on quadratic residues. In: Cryptography and Coding (2001)
23. Damgård, I., Haagh, H., Orlandi, C.: Access control encryption: enforcing information flow with cryptography. In: Hirt, M., Smith, A. (eds.) TCC 2016. LNCS, vol. 9986, pp. 547–576. Springer, Heidelberg (2016). https://doi.org/10.1007/978-3-662-53644-5_21
24. Denning, D.E.: A lattice model of secure information flow. Commun. ACM **19**(5), 236–243 (1976)
25. Denning, D.E., Denning, P.J.: Certification of programs for secure information flow. Commun. ACM **20**(7), 504–513 (1977)
26. Dodis, Y., Mironov, I., Stephens-Davidowitz, N.: Message transmission with reverse firewalls - secure communication on corrupted machines. In: CRYPTO (2016)
27. Fehr, V., Fischlin, M.: Sanitizable signcryption: sanitization over encrypted data (full version). IACR Cryptology ePrint Archive 2015 (2015)
28. Fuchsbauer, G., Gay, R., Kowalczyk, L., Orlandi, C.: Access control encryption for equality, comparison, and more. In: Fehr, S. (ed.) PKC 2017. LNCS, vol. 10175, pp. 88–118. Springer, Heidelberg (2017). https://doi.org/10.1007/978-3-662-54388-7_4
29. Garg, S., Gentry, C., Halevi, S., Raykova, M., Sahai, A., Waters, B.: Candidate indistinguishability obfuscation and functional encryption for all circuits. In: FOCS (2013)
30. Goldwasser, S., Kalai, Y.T., Popa, R.A., Vaikuntanathan, V., Zeldovich, N.: Reusable garbled circuits and succinct functional encryption. In: STOC (2013)

31. Gorbunov, S., Vaikuntanathan, V., Wee, H.: Functional encryption with bounded collusions via multi-party computation. In: Safavi-Naini, R., Canetti, R. (eds.) CRYPTO 2012. LNCS, vol. 7417, pp. 162–179. Springer, Heidelberg (2012). https://doi.org/10.1007/978-3-642-32009-5_11
32. Gorbunov, S., Vaikuntanathan, V., Wee, H.: Predicate encryption for circuits from LWE. In: Gennaro, R., Robshaw, M. (eds.) CRYPTO 2015. LNCS, vol. 9216, pp. 503–523. Springer, Heidelberg (2015). https://doi.org/10.1007/978-3-662-48000-7_25
33. Goyal, V., Jain, A., Koppula, V., Sahai, A.: Functional encryption for randomized functionalities. In: Dodis, Y., Nielsen, J.B. (eds.) TCC 2015. LNCS, vol. 9015, pp. 325–351. Springer, Heidelberg (2015). https://doi.org/10.1007/978-3-662-46497-7_13
34. Goyal, V., Pandey, O., Sahai, A., Waters, B.: Attribute-based encryption for fine-grained access control of encrypted data. In: ACM CCS (2006)
35. Hopper, N.J., Langford, J., Ahn, L.: Provably secure steganography. In: Yung, M. (ed.) CRYPTO 2002. LNCS, vol. 2442, pp. 77–92. Springer, Heidelberg (2002). https://doi.org/10.1007/3-540-45708-9_6
36. Katz, J., Sahai, A., Waters, B.: Predicate encryption supporting disjunctions, polynomial equations, and inner products. In: Smart, N. (ed.) EUROCRYPT 2008. LNCS, vol. 4965, pp. 146–162. Springer, Heidelberg (2008). https://doi.org/10.1007/978-3-540-78967-3_9
37. Kiayias, A., Papadopoulos, S., Triandopoulos, N., Zacharias, T.: Delegatable pseudorandom functions and applications. In: ACM CCS (2013)
38. Kim, S., Wu, D.J.: Access control encryption for general policies from standard assumptions. IACR Cryptology ePrint Archive 2017/467 (2017)
39. Komargodski, I., Segev, G., Yogev, E.: Functional encryption for randomized functionalities in the private-key setting from minimal assumptions. In: TCC (2015)
40. Lewko, A., Okamoto, T., Sahai, A., Takashima, K., Waters, B.: Fully secure functional encryption: attribute-based encryption and (hierarchical) inner product encryption. In: Gilbert, H. (ed.) EUROCRYPT 2010. LNCS, vol. 6110, pp. 62–91. Springer, Heidelberg (2010). https://doi.org/10.1007/978-3-642-13190-5_4
41. Maji, H.K., Prabhakaran, M., Rosulek, M.: Attribute-based signatures. In: Kiayias, A. (ed.) CT-RSA 2011. LNCS, vol. 6558, pp. 376–392. Springer, Heidelberg (2011). https://doi.org/10.1007/978-3-642-19074-2_24
42. Mironov, I., Stephens-Davidowitz, N.: Cryptographic reverse firewalls. In: Oswald, E., Fischlin, M. (eds.) EUROCRYPT 2015. LNCS, vol. 9057, pp. 657–686. Springer, Heidelberg (2015). https://doi.org/10.1007/978-3-662-46803-6_22
43. Okamoto, T., Takashima, K.: Hierarchical predicate encryption for inner-products. In: Matsui, M. (ed.) ASIACRYPT 2009. LNCS, vol. 5912, pp. 214–231. Springer, Heidelberg (2009). https://doi.org/10.1007/978-3-642-10366-7_13
44. O'Neill, A.: Definitional issues in functional encryption. IACR Cryptology ePrint Archive 2010 (2010)
45. Osborn, S.L., Sandhu, R.S., Munawer, Q.: Configuring role-based access control to enforce mandatory and discretionary access control policies. ACM Trans. Inf. Syst. Secur. **3**(2), 85–106 (2000)
46. Rackoff, C., Simon, D.R.: Non-interactive zero-knowledge proof of knowledge and chosen ciphertext attack. In: Feigenbaum, J. (ed.) CRYPTO 1991. LNCS, vol. 576, pp. 433–444. Springer, Heidelberg (1992). https://doi.org/10.1007/3-540-46766-1_35

47. Raykova, M., Zhao, H., Bellovin, S.M.: Privacy enhanced access control for outsourced data sharing. In: Keromytis, A.D. (ed.) FC 2012. LNCS, vol. 7397, pp. 223–238. Springer, Heidelberg (2012). https://doi.org/10.1007/978-3-642-32946-3_17
48. Regev, O.: On lattices, learning with errors, random linear codes, and cryptography. In: STOC (2005)
49. Rivest, R.L., Shamir, A., Adleman, L.M.: A method for obtaining digital signatures and public-key cryptosystems. Commun. ACM **21**(2), 120–126 (1978)
50. Sabelfeld, A., Myers, A.C.: Language-based information-flow security. IEEE J. Sel. Areas Commun. **21**(1), 5–19 (2003)
51. Sahai, A., Seyalioglu, H.: Worry-free encryption: functional encryption with public keys. In: ACM CCS (2010)
52. Sahai, A., Waters, B.: Fuzzy identity-based encryption. In: Cramer, R. (ed.) EUROCRYPT 2005. LNCS, vol. 3494, pp. 457–473. Springer, Heidelberg (2005). https://doi.org/10.1007/11426639_27
53. Sandhu, R.S.: Lattice-based access control models. IEEE Comput. **26**(11), 9–19 (1993)
54. Sandhu, R.S., Coyne, E.J., Feinstein, H.L., Youman, C.E.: Role-based access control models. IEEE Comput. **29**(2), 38–47 (1996)
55. Shamir, A.: Identity-based cryptosystems and signature schemes. In: Blakley, G.R., Chaum, D. (eds.) CRYPTO 1984. LNCS, vol. 196, pp. 47–53. Springer, Heidelberg (1985). https://doi.org/10.1007/3-540-39568-7_5
56. Shi, E., Bethencourt, J., Chan, H.T., Song, D.X., Perrig, A.: Multi-dimensional range query over encrypted data. In: IEEE S&P (2007)
57. Tan, G., Zhang, R., Ma, H., Tao, Y.: Access control encryption based on LWE. In: APKC@AsiaCCS (2017)

Strengthening Access Control Encryption

Christian Badertscher, Christian Matt(✉), and Ueli Maurer

Department of Computer Science, ETH Zurich, 8092 Zurich, Switzerland
{badi,mattc,maurer}@inf.ethz.ch

Abstract. Access control encryption (ACE) was proposed by Damgård et al. to enable the control of information flow between several parties according to a given policy specifying which parties are, or are not, allowed to communicate. By involving a special party, called the *sanitizer*, policy-compliant communication is enabled while policy-violating communication is prevented, even if sender and receiver are dishonest. To allow outsourcing of the sanitizer, the secrecy of the message contents and the anonymity of the involved communication partners is guaranteed.

This paper shows that in order to be resilient against realistic attacks, the security definition of ACE must be considerably strengthened in several ways. A new, substantially stronger security definition is proposed, and an ACE scheme is constructed which provably satisfies the strong definition under standard assumptions.

Three aspects in which the security of ACE is strengthened are as follows. First, CCA security (rather than only CPA security) is guaranteed, which is important since senders can be dishonest in the considered setting. Second, the revealing of an (unsanitized) ciphertext (e.g., by a faulty sanitizer) cannot be exploited to communicate more in a policy-violating manner than the information contained in the ciphertext. We illustrate that this is not only a definitional subtlety by showing how in known ACE schemes, a single leaked unsanitized ciphertext allows for an arbitrary amount of policy-violating communication. Third, it is enforced that parties specified to receive a message according to the policy cannot be excluded from receiving it, even by a dishonest sender.

Keywords: Access control encryption · Information flow control · Chosen-ciphertext attacks

1 Introduction

1.1 Access Control Encryption—Model and Security Requirements

The concept of *access control encryption (ACE)* was proposed by Damgård et al. [6] in order to enforce information flow using cryptographic tools rather than a standard access control mechanism (e.g., a reference monitor) within an

The full version of this paper is available at https://eprint.iacr.org/2017/429.

T. Takagi and T. Peyrin (Eds.): ASIACRYPT 2017, Part I, LNCS 10624, pp. 502–532, 2017.
https://doi.org/10.1007/978-3-319-70694-8_18

information system. If the encryption scheme provides certain operations (e.g., ciphertext sanitization) and satisfies an adequate security definition, then the reference monitor can be outsourced, as a component called the *sanitizer*, to an only partially trusted service provider. The goal of ACE is that the sanitizer learns nothing not intrinsically necessary. Security must also be guaranteed against dishonest users, whether senders or receivers of information, and against certain types of sanitizer misbehavior.

The information flow problem addressed by ACE is defined in a context with a set $\mathcal{R}$ of roles corresponding, for example, to different security clearances. Each user in a system can be assigned several roles. For example the users are employees of a company collaborating on a sensitive project, and they need to collaborate and exchange information by sending messages. Since the information is sensitive, which information a party can see must be restricted (hence the term *access control*), even if some parties are dishonest. In the most general form, the specification of which role may send to which other role corresponds to a relation (a subset of $\mathcal{R} \times \mathcal{R}$) or, equivalently, to a predicate $P \colon \mathcal{R} \times \mathcal{R} \to \{0, 1\}$, where $s \in \mathcal{R}$ is allowed to communicate to $r \in \mathcal{R}$ if and only if $P(s, r) = 1$. The predicate P is called the *(security) policy*. Typical examples of such policies arise from the Bell-LaPadula [2] model where roles are (partially) ordered, and the so-called "no-write-down" rule specifies that it is forbidden for a user to send information to another user with a lower role. Note that for this specific example, the relation is transitive, but ACE also allows to capture non-transitive security policies.

ACE was designed to work in the following setting. Users can communicate anonymously with a sanitizer. If a user wants to send a message, it is encrypted under a key corresponding to the sender's role. Then the ciphertext is sent (anonymously) to the sanitizer who applies a certain sanitization operation and writes the sanitized ciphertext on a publicly readable bulletin board providing anonymous read-access to the users (receivers). Users who are supposed to receive the message according to the policy (and only those users) can decrypt the sanitized ciphertext.

To ensure security in the described setting, the ACE scheme must at least provide the following guarantees:

1. The encryption must assure privacy and anonymity against dishonest receivers as well as the sanitizer, i.e., neither the sanitizer nor dishonest receivers without access allowed by the policy should be able to obtain information about messages or the sender's role.
2. A dishonest sender must be unable to communicate with a (potentially dishonest) receiver, unless this is allowed according to the policy. In other words, the system must not provide covert channels allowing for policy-violating communication.

As usual in a context with dishonest senders, the first goal requires security against chosen-ciphertext attacks (CCA) because dishonest users can send a ciphertext for which they do not know the contained message and by observing the effects the received message has on the environment, potentially obtain information about the message. This corresponds to the availability of a decryption oracle, as in the CCA-security definition.

Note that the second goal is only achievable if users cannot directly write to the repository or communicate by other means bypassing the sanitizer, and if the sanitizer is not actively dishonest because a dishonest sanitizer can directly write any information received from a dishonest sender to the repository. The assumption that a user cannot bypass the sanitizer and communicate to another party outside of the system can for example be justified by assuming that users, even if dishonest, want to avoid being caught communicating illegitimately, or if only a user's system (not the user) is corrupted, and the system can technically only send message to the sanitizer.

Since the sanitizer is not fully trusted in our setting, one should consider the possibility that an unsanitized ciphertext is leaked (intentionally or unintentionally) to a dishonest party. This scenario can be called *(unsanitized) ciphertext-revealing attack*. Obviously, all information contained in this ciphertext gets leaked to that party. While this cannot be avoided, such an attack should not enable dishonest parties to violate the security requirements beyond that.

We point out that previously proposed encryption techniques (before ACE), such as attribute-based encryption [11,17] and functional encryption [4], enable the design of schemes where a sender can encrypt messages such that only designated receivers (who possess the required key) can read the message. This captures the access control aspects of *read* permissions, but it does not allow to capture the control of *write/send* permissions. In other words, such schemes only achieve the first goal listed above, not the second one.

1.2 Contributions of this Paper

While the proposal of the ACE-concept and of efficient ACE-schemes were important first steps toward outsourcing access control, the existing security definition turns out to be insufficient for several realistic attack scenarios. The main contributions of this paper consist of uncovering issues with existing definitions and schemes, fixing these issues by proposing stronger security notions, and constructing a scheme satisfying our stronger notions.

Issues with existing definitions and schemes. As argued above, chosen-ciphertext attacks should be considered since the use case for ACE includes dishonest senders. Existing definitions, however, do not take this into account, i.e., the adversary does not have access to a decryption oracle in the security games.

Furthermore, existing notions do not consider ciphertext-revealing attacks. Technically speaking, the security game that is supposed to prevent dishonest senders from transmitting information to dishonest receivers (called no-write game), gives the adversary only access to an encryption oracle that sanitizes ciphertexts before returning them. This means that the adversary has no access to unsanitized ciphertexts. This is not only a definitional subtlety, but can completely break down any security guarantees. We demonstrate that existing ACE schemes allow the following attack: Assume there are three users A, M, and E in the system, where A is honest and by the policy allowed to send information to E, and M and E are dishonest and not allowed to communicate. If A sends an (innocent) message to E and the corresponding unsanitized ciphertext is leaked to M, malleability of the ciphertext can be exploited by M to subsequently communicate an arbitrary number of arbitrary messages chosen by M to E. Note that while this attack crucially exploits malleability of ciphertexts, it is not excluded by CCA security for two reasons: first, CCA security does not prevent an adversary from producing valid ciphertexts for *unrelated* messages, and second, the integrity should still hold if the adversary has the decryption key (but not the encryption key).

Finally, existing security definitions focus on preventing dishonest parties from communicating if disallowed by the policy, but they do not enforce information flow. For example, if user A only has a role such that according to the policy, users B and C can read what A sends, existing schemes do not prevent A from sending a message that can be read by B but not by C, or sending a message such that B and C receive different messages. This is not as problematic as the two issues above, and one can argue that A could anyway achieve something similar by additionally encrypting the message with another encryption scheme. Nevertheless, for some use cases, actually precisely enforcing the policy can be required (consider, e.g., a logging system), and one might intuitively expect that ACE schemes achieve this.

New security definitions. We propose new, stronger security definitions for ACE that exclude all issues mentioned above. First, we give the adversary access to a decryption oracle. More precisely, the oracle first sanitizes the given ciphertext and then decrypts it, since this is what happens in the application if a dishonest party sends a ciphertext to the sanitizer. Second, we incorporate ciphertext-revealing attacks by giving the adversary access to an encryption oracle that returns unsanitized ciphertexts for arbitrary roles. Finally, we introduce a new security game in which an adversary can obtain encryption keys and decryption keys from an oracle and has to output a ciphertext such that one of the following events occur: either the set of roles that can successfully decrypt the ciphertext (to an arbitrary message) is inconsistent with the policy for all sender roles for which the adversary has an encryption key (in this case, we say the adversary is not *role-respecting*); or the ciphertext can be successfully decrypted with two keys such that two different messages are obtained (in this case, we say the *uniform-decryption* property is violated).

Construction of an ACE scheme for our stronger notions. Our construction proceeds in three steps and follows the general structure of the generic construction by Fuchsbauer et al. [9]. Since we require much stronger security notions in all three steps, our constructions and proofs are consequently more involved than existing ones. First, we construct a scheme for a primitive we call *enhanced sanitizable public-key encryption (sPKE)*. Second, we use an sPKE scheme to construct an ACE scheme satisfying our strong security notion for the equality policy, i.e., for the policy that allows s to send to r if and only if $r = s$. Third, we show how to lift an ACE scheme for the equality policy to an ACE scheme for the disjunction of equalities policy. This policy encodes roles as vectors $\mathbf{x} = (x_1, \ldots, x_\ell)$ and allows role $\mathbf{x}$ to send to role $\mathbf{y}$ if and only if $x_1 = y_1 \vee \ldots \vee x_\ell = y_\ell$. As shown by Fuchsbauer et al. [9], useful policies including the inequality predicate corresponding to the Bell-LaPadula model can efficiently be implemented using this policy by encoding the roles appropriately.

Enhanced sanitizable PKE. An sPKE scheme resembles publicy-key encryption with an additional setup algorithm that outputs sanitizer parameters and a master secret key. The master secret key is needed to generate a public/private key pair and the sanitizer parameters can be used to sanitize a ciphertext. A sanitized ciphertext cannot be linked to the original ciphertext without the decryption key. We require the scheme to be CCA secure (with respect to a sanitize-then-decrypt oracle) and anonymous. Sanitization resembles rerandomization [12,15], also called universal re-encryption [10], but we allow sanitized ciphertexts to be syntactically different from unsanitized ciphertexts. This allows us to achieve full CCA security, which is needed for our ACE construction and unachievable for rerandomizable encryption.

Our scheme is based on ElGamal encryption [7], which can easily be rerandomized and is anonymous. We obtain CCA security using the technique of Naor and Yung [14], i.e., encrypting the message under two independent keys and proving in zero-knowledge that the ciphertexts are encryptions of the same message, which was shown by Sahai to achieve full CCA security if the zero-knowledge proof is simulation-sound [16]. A technical issue is that if the verification of the NIZK proof was done by the decrypt algorithm, the sanitization would also need to sanitize the proof. Instead, we let the sanitizer perform the verification. Since we want to preserve anonymity, this needs to be done without knowing under which public keys the message was encrypted. Therefore, the public keys are part of the witness in the NIZK proof. Now the adversary could encrypt the same message under two different public keys that were not produced together by the key-generation, which would break the reduction. To prevent this, the pair of public keys output by the key-generation is signed using a signature key that is contained in the master secret key and the corresponding verification key is contained in the sanitizer parameters.

ACE for equality. The basic idea of our ACE scheme for the equality policy is to use for each role, encryption and decryption keys of an sPKE scheme as the encryption and decryption keys of the ACE scheme, respectively. Since we need to prevent dishonest senders without an encryption key for some role from producing valid ciphertexts for that role even after seeing encryptions of other messages for this role and obtaining encryption keys for other roles, we add a signature key to the encryption key, sign this pair using a separate signing key, where the corresponding verification key is part of the sanitizer parameters, and let senders sign their ciphertexts. To preserve anonymity, this signature cannot be part of the ciphertext. Instead, senders prove in zero-knowledge that they know such a signature and that the encryption was performed properly.

ACE for disjunction of equalities. The first step of our lifting is identical to the lifting described by Fuchsbauer et al. [9]: for each component of the role-vector, the encryption and decryption keys contain corresponding keys of an ACE scheme for the equality policy. To encrypt a message, this message is encrypted under each of the key-components. In a second step, we enforce role-respecting security with the same trick we used in our ACE scheme for equality; that is, we sign encryption key-vectors together with a signing key for that role, and senders prove in zero-knowledge that they have used a valid key combination to encrypt and that they know a signature of the ciphertext vector.

1.3 Related Work

The concept of access control encryption has been introduced by Damgård et al. [6]. They provided the original security definitions and first schemes. Subsequent work by Fuchsbauer et al. [9], by Tan et al. [18], and by Kim and Wu [13] focused on new schemes that are more efficient, based on different assumptions, or support more fine grained access control policies. In contrast to our work, they did not attempt to strengthen the security guarantees provided by ACE.

2 Preliminaries

2.1 Notation

We write $x \leftarrow y$ for assigning the value y to the variable x. For a finite set X, $x \twoheadleftarrow X$ denotes assigning to x a uniformly random value in X. For $n \in \mathbb{N}$, we use the convention

$$[n] := \{1, \ldots, n\}.$$

By $\mathbb{Z}_n$ we denote the ring of integers modulo n, and by $\mathbb{Z}_n^*$ its multiplicative group of units. The probability of an event A in an experiment E is denoted by $\Pr^E[A]$, e.g., $\Pr^{x \twoheadleftarrow \{0,1\}}[x = 0] = \frac{1}{2}$. If the experiment is clear from the context, we omit the superscript. The conditional probability of A given B is denoted by $\Pr[A \mid B]$ and the complement of A is denoted by $\neg A$. For a probabilistic algorithm $\mathcal{A}$ and $r \in \{0,1\}^*$, we denote by $\mathcal{A}(x; r)$ the execution of $\mathcal{A}$ on input x with randomness r. For algorithms $\mathcal{A}$ and $\mathcal{O}$, $\mathcal{A}^{\mathcal{O}(\cdot)}(x)$ denotes the execution of $\mathcal{A}$ on input x, where $\mathcal{A}$ has oracle access to $\mathcal{O}$.

2.2 Security Definitions, Advantages, Efficiency, and Negligibility

We define the security of a scheme via a random experiment (or game) involving an adversary algorithm $\mathcal{A}$. For a given scheme $\mathcal{E}$ and adversary $\mathcal{A}$, we define the advantage of $\mathcal{A}$, which is a function of the security parameter κ. To simplify the notation, we omit the security parameter when writing the advantage, e.g., we write $\mathsf{Adv}^{\mathsf{Sig\text{-}EUF\text{-}CMA}}_{\mathcal{E},\mathcal{A}}$ instead of $\mathsf{Adv}^{\mathsf{Sig\text{-}EUF\text{-}CMA}}_{\mathcal{E},\mathcal{A}}(\kappa)$ for the advantage of $\mathcal{A}$ in the existential unforgeability game for the signature scheme $\mathcal{E}$. Such a scheme is considered *secure* if $\mathsf{Adv}^{\mathsf{Sig\text{-}EUF\text{-}CMA}}_{\mathcal{E},\mathcal{A}}$ is *negligible* for all *efficient* $\mathcal{A}$. An algorithm $\mathcal{A}$ is *efficient* if it runs in *probabilistic polynomial time (PPT)*, i.e., $\mathcal{A}$ has access to random bits and there is a polynomial p such that $\mathcal{A}(x)$ terminates after at most $p(|x|)$ steps (on some computational model, e.g., Turing machines) for all inputs x, where $|x|$ denotes the bit-length of x. A function f is *negligible* if for every polynomial p, there exists $n_0 \in \mathbb{N}$ such that $f(n) < 1/p(n)$ for all $n \geq n_0$. While these asymptotic definitions yield concise statements, we will in all proofs derive precise bounds on the advantages, following a concrete security approach.

2.3 Access Control Encryption

We recall the definition of access control encryption by Damgård et al. [6]. Following Fuchsbauer et al. [9], we do not have sanitizer keys and require Gen to be deterministic. The set of roles is assumed to be $\mathcal{R} = [n]$.

Definition 1. *An* access control encryption (ACE) scheme $\mathcal{E}$ *consists of the following five PPT algorithms:*

Setup: *The algorithm* Setup *on input a security parameter* 1^κ *and a* policy $P\colon [n] \times [n] \to \{0,1\}$, *outputs a* master secret key msk *and* sanitizer parameters sp. *We implicitly assume that all keys include the finite* message space $\mathcal{M}$ *and the* ciphertext spaces $\mathcal{C}, \mathcal{C}'$.

Key generation: *The algorithm* Gen *is deterministic and on input a master secret key* msk, *a role* $i \in [n]$, *and the type* $\mathtt{sen}$, *outputs an* encryption key ek_i; *on input* msk, $j \in [n]$, *and the type* $\mathtt{rec}$, *outputs a* decryption key dk_j.

Encryption: *The algorithm* Enc *on input an encryption key* ek_i *and a message* $m \in \mathcal{M}$, *outputs a ciphertext* $c \in \mathcal{C}$.

Sanitization: *The algorithm* San *on input sanitizer parameters* sp *and a ciphertext* $c \in \mathcal{C}$, *outputs a* sanitized ciphertext $c' \in \mathcal{C}' \cup \{\bot\}$.

Decryption: *The algorithm* Dec *on input a decryption key* dk_j *and a sanitized ciphertext* $c' \in \mathcal{C}'$, *outputs a message* $m \in \mathcal{M} \cup \{\bot\}$; *on input* dk_j *and* $\bot$, *it outputs* $\bot$.

For a probabilistic algorithm $\mathcal{A}$, *consider the experiment* $\mathsf{Exp}^{\mathsf{ACE\text{-}CORR}}_{\mathcal{E},\mathcal{A}}$ *that given a security parameter* 1^κ *and a policy* P, *executes* $(sp, msk) \leftarrow \mathsf{Setup}(1^\kappa, P)$, $(m, i, j) \leftarrow \mathcal{A}^{\mathsf{Gen}(msk,\cdot,\cdot)}(sp)$, $ek_i \leftarrow \mathsf{Gen}(msk, i, \mathtt{sen})$, *and* $dk_j \leftarrow \mathsf{Gen}(msk, j, \mathtt{rec})$. *We define the* correctness advantage *of* $\mathcal{A}$ *(for security parameter* κ *and policy* P*) as*

$$\mathsf{Adv}^{\mathsf{ACE\text{-}CORR}}_{\mathcal{E},\mathcal{A}} := \Pr\big[P(i,j) = 1 \ \wedge \ \mathsf{Dec}\big(dk_j, \mathsf{San}(sp, \mathsf{Enc}(ek_i, m))\big) \neq m\big],$$

where the probability is over the randomness in $\mathsf{Exp}_{\mathcal{E},\mathcal{A}}^{\mathsf{ACE\text{-}CORR}}$ *and the random coins of* Enc, San, *and* Dec. *The scheme* $\mathcal{E}$ *is called* correct *if* $\mathsf{Adv}_{\mathcal{E},\mathcal{A}}^{\mathsf{ACE\text{-}CORR}}$ *is negligible for all efficient* $\mathcal{A}$, *and* perfectly correct *if* $\mathsf{Adv}_{\mathcal{E},\mathcal{A}}^{\mathsf{ACE\text{-}CORR}} = 0$ *for all* $\mathcal{A}$.

Remark 1. Correctness of an encryption scheme is typically not defined via a game with an adversary, but by requiring that decryption of an encryption of m yields m with probability 1. This perfect correctness requirement is difficult to achieve for ACE schemes and not necessary for applications because it is sufficient if a decryption error only occurs with negligible probability in any execution of the scheme. Damgård et al. [6] define correctness by requiring that for all m, i, and j with $P(i,j) = 1$, the probability that a decryption fails is negligible, where the probability is over setup, key generation, encrypt, sanitize, and decrypt. While this definition is simpler than ours, it does not guarantee that decryption errors only occur with negligible probability in any execution of the scheme. For example, a scheme could on setup choose a random message m and embed it into all keys such that decryption always fails for encryptions of this particular message. This does not violate the definition by Damgård et al. since for any fixed message, the probability that this message is sampled during setup is negligible (if the message space is large). Nevertheless, an adversary can always provoke a decryption error by sending that particular message m, which is not desirable. The above example might at first sight seem somewhat artificial, and typically, schemes do not have such a structure. However, capturing correctness via an experiment is important when thinking of composition, since we expect that the correctness guarantee still holds when the ACE scheme is run as part of a larger system. In order to meet this expectation, and to exclude the above issue, we formalize correctness via an experiment.

Additionally, Fuchsbauer et al. have defined detectability, which guarantees that decrypting with a wrong key yields $\perp$ with high probability [9]. This allows receivers to detect whether a message was sent to them. As for correctness, we define it via an experiment. The notion is related to robustness for public-key encryption [1]. We additionally define strong detectability, in which the randomness for the encryption is adversarially chosen.

Definition 2. *Let* $\mathcal{E} = (\mathsf{Setup}, \mathsf{Gen}, \mathsf{Enc}, \mathsf{San}, \mathsf{Dec})$ *be an ACE scheme and let* $\mathcal{A}$ *be a probabilistic algorithm. Consider the experiment* $\mathsf{Exp}_{\mathcal{E},\mathcal{A}}^{\mathsf{ACE\text{-}DTCT}}$ *that given a security parameter* 1^κ *and a policy* P, *executes* $(sp) \leftarrow \mathsf{Setup}(1^\kappa, P)$, $(m, i, j) \leftarrow \mathcal{A}^{\mathsf{Gen}(msk,\cdot,\cdot)}(sp, msk)$, $ek_i \leftarrow \mathsf{Gen}(msk, i, \mathtt{sen})$, *and* $dk_j \leftarrow \mathsf{Gen}(msk, j, \mathtt{rec})$. *We define the* detectability advantage *of* $\mathcal{A}$ *as*

$$\mathsf{Adv}_{\mathcal{E},\mathcal{A}}^{\mathsf{ACE\text{-}DTCT}} := \Pr\big[P(i,j) = 0 \ \wedge \ \mathsf{Dec}\big(dk_j, \mathsf{San}(sp, \mathsf{Enc}(ek_i, m))\big) \neq \perp\big],$$

where the probability is over the randomness in $\mathsf{Exp}_{\mathcal{E},\mathcal{A}}^{\mathsf{ACE\text{-}DTCT}}$ *and the random coins of* Enc, San, *and* Dec. *The scheme* $\mathcal{E}$ *is called* detectable *if* $\mathsf{Adv}_{\mathcal{E},\mathcal{A}}^{\mathsf{ACE\text{-}DTCT}}$ *is negligible for all efficient* $\mathcal{A}$. *The experiment* $\mathsf{Exp}_{\mathcal{E},\mathcal{A}}^{\mathsf{ACE\text{-}sDTCT}}$ *is identical to*

$\mathsf{Exp}_{\mathcal{E},\mathcal{A}}^{\mathsf{ACE\text{-}DTCT}}$ *except that* $\mathcal{A}$ *returns* (m, r, i, j). *The* strong detectability advantage *of* $\mathcal{A}$ *is defined as*

$$\mathsf{Adv}_{\mathcal{E},\mathcal{A}}^{\mathsf{ACE\text{-}sDTCT}} := \Pr\big[P(i,j) = 0 \ \wedge \ \mathsf{Dec}\big(dk_j, \mathsf{San}(sp, \mathsf{Enc}(ek_i, m; r))\big) \neq \bot\big],$$

where the probability is over the randomness in $\mathsf{Exp}_{\mathcal{E},\mathcal{A}}^{\mathsf{ACE\text{-}sDTCT}}$ *and the random coins of* San *and* Dec. *The scheme* $\mathcal{E}$ *is called* strongly detectable *if* $\mathsf{Adv}_{\mathcal{E},\mathcal{A}}^{\mathsf{ACE\text{-}sDTCT}}$ *is negligible for all efficient* $\mathcal{A}$.

2.4 Existing Security Definitions

Existing notions for ACE specify two core properties: the so-called *no-read rule* and the *no-write rule*. The no-read rule formalizes privacy and anonymity: roughly, an honestly generated ciphertext should not leak anything about the message, except possibly its length, or about the role of the sender. The security game allows an adversary to interact with a key-generation oracle (to obtain encryption and decryption keys for selected roles), and an encryption oracle to obtain encryptions of chosen messages for roles for which the adversary does not possess the encryption key. This attack model reflects that an adversary cannot obtain useful information by observing the ciphertexts that are sent to the sanitizer. To exclude trivial attacks, it is not considered a privacy breach if the adversary knows a decryption key that allows to decrypt the challenge ciphertext according to the policy. Similarly, it is not considered an anonymity breach if the encrypted messages are different. We next state the definition of the no-read rule.[1]

Definition 3. *Let* $\mathcal{E} = (\mathsf{Setup}, \mathsf{Gen}, \mathsf{Enc}, \mathsf{San}, \mathsf{Dec})$ *be an ACE scheme and let* $\mathcal{A} = (\mathcal{A}_1, \mathcal{A}_2)$ *be a pair of probabilistic algorithms. Consider the experiment* $\mathsf{Exp}_{\mathcal{E},\mathcal{A}}^{\mathsf{ACE\text{-}no\text{-}read}}$ *in Fig. 1 and let* J *be the set of all* j *such that* $\mathcal{A}_1$ *or* $\mathcal{A}_2$ *issued the query* $(j, \texttt{rec})$ *to the oracle* $\mathcal{O}_G$. *The* payload-privacy advantage *and the* sender-anonymity advantage *of* $\mathcal{A}$ *are defined as*

$$\begin{aligned}\mathsf{Adv}_{\mathcal{E},\mathcal{A}}^{\mathsf{ACE\text{-}no\text{-}read,priv}} &:= 2 \cdot \Pr\big[b' = b \ \wedge \ |m_0| = |m_1| \\ &\qquad\qquad \wedge \ \forall j \in J \ P(i_0, j) = P(i_1, j) = 0\big] - 1, \\ \mathsf{Adv}_{\mathcal{E},\mathcal{A}}^{\mathsf{ACE\text{-}no\text{-}read,anon}} &:= 2 \cdot \Pr\big[b' = b \ \wedge \ m_0 = m_1 \\ &\qquad\qquad \wedge \ \forall j \in J \ P(i_0, j) = P(i_1, j)\big] - 1,\end{aligned}$$

respectively, where the probabilities are over the randomness of all algorithms in $\mathsf{Exp}_{\mathcal{E},\mathcal{A}}^{\mathsf{ACE\text{-}no\text{-}read}}$. *The scheme* $\mathcal{E}$ *satisfies the* payload-privacy no-read rule *and the* sender-anonymity no-read rule *if* $\mathsf{Adv}_{\mathcal{E},\mathcal{A}}^{\mathsf{ACE\text{-}no\text{-}read,priv}}$ *and* $\mathsf{Adv}_{\mathcal{E},\mathcal{A}}^{\mathsf{ACE\text{-}no\text{-}read,anon}}$ *are negligible for all efficient* $\mathcal{A}$, *respectively. If it satisfies both, it is said to satisfy the* no-read rule.

[1] For anonymity, we adopt here the definition of [6], which is stronger than the one used by Fuchsbauer et al. [9] since there, anonymity is not guaranteed against parties who can decrypt.

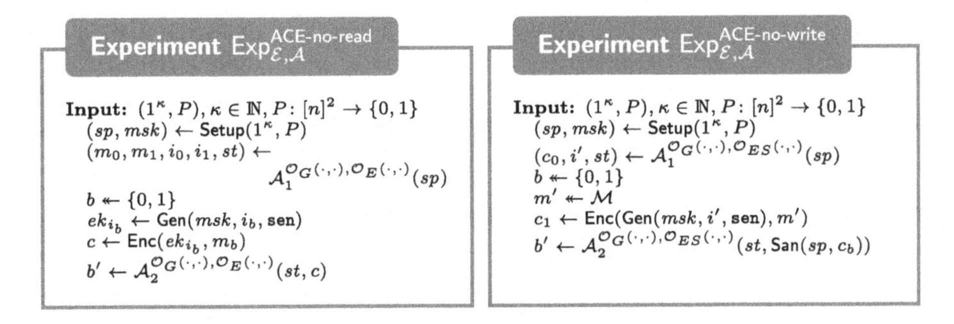

Fig. 1. The no-read and no-write experiments for an ACE scheme $\mathcal{E}$ and an adversary $\mathcal{A} = (\mathcal{A}_1, \mathcal{A}_2)$. The oracles are defined as $\mathcal{O}_G(\cdot,\cdot) := \mathsf{Gen}(msk,\cdot,\cdot)$, $\mathcal{O}_E(\cdot,\cdot) := \mathsf{Enc}(\mathsf{Gen}(msk,\cdot,\mathtt{sen}),\cdot)$, and $\mathcal{O}_{ES}(\cdot,\cdot) := \mathsf{San}(sp, \mathsf{Enc}(\mathsf{Gen}(msk,\cdot,\mathtt{sen}),\cdot))$.

The no-write rule of ACE is the core property to capture access control. In a nutshell, if the adversary only possesses encryption keys for roles i and decryption keys for roles j with $P(i,j) = 0$, then he should not be able to create a ciphertext from which, after being sanitized, he can retrieve any information. Technically, in the corresponding security game, the adversary is given a key-generation oracle as above, and in addition an oracle to obtain *sanitized* ciphertexts for selected messages and roles. This attack model corresponds to a setting where an adversary only sees the outputs of a sanitizer, but not its inputs, and in particular no unsanitized ciphertexts generated for roles for which he does not possess the encryption key. The adversary wins if he manages to distinguish the sanitized version of a ciphertext of his choice from a sanitized version of a freshly generated encryption of a random message, and if he does not obtain the encryption key for any role i and the decryption key of any role j for which $P(i,j) = 1$, as this would trivially allow him to distinguish.

Definition 4. *Let* $\mathcal{E} = (\mathsf{Setup}, \mathsf{Gen}, \mathsf{Enc}, \mathsf{San}, \mathsf{Dec})$ *be an ACE scheme and let* $\mathcal{A} = (\mathcal{A}_1, \mathcal{A}_2)$ *be a pair of probabilistic algorithms. Consider the experiment* $\mathsf{Exp}_{\mathcal{E},\mathcal{A}}^{\mathsf{ACE\text{-}no\text{-}write}}$ *in Fig. 1, let* I_1 *be the set of all* i *such that* $\mathcal{A}_1$ *issued the query* $(i, \mathtt{sen})$ *to* $\mathcal{O}_G$*, and let* J *be the set of all* j *such that* $\mathcal{A}_1$ *or* $\mathcal{A}_2$ *issued the query* $(j, \mathtt{rec})$ *to* $\mathcal{O}_G$*. We define the* no-write advantage *of* $\mathcal{A}$ *as*

$$\mathsf{Adv}_{\mathcal{E},\mathcal{A}}^{\mathsf{ACE\text{-}no\text{-}write}} := 2 \cdot \Pr\big[b' = b \ \wedge\ i' \in I_1 \ \wedge\ \forall i \in I_1\ \forall j \in J\ P(i,j) = 0 \\ \wedge\ \mathsf{San}(sp, c_0) \neq \bot\big] - 1,$$

where the probability is over the randomness of all algorithms in $\mathsf{Exp}_{\mathcal{E},\mathcal{A}}^{\mathsf{ACE\text{-}no\text{-}write}}$*. The scheme* $\mathcal{E}$ *satisfies the* no-write rule *if* $\mathsf{Adv}_{\mathcal{E},\mathcal{A}}^{\mathsf{ACE\text{-}no\text{-}write}}$ *is negligible for all efficient* $\mathcal{A}$*.*

Remark 2. Our definition follows the one by Fuchsbauer et al. [9] by requiring $\mathsf{San}(sp, c_0) \neq \bot$ in the winning condition for the no-write rule, which was not required in the original definition by Damgård et al. [6]. Schemes can be made secure with respect to the original definition by letting the algorithm San create a fresh ciphertext for a random message when given an invalid ciphertext.

The condition $i' \in I_1$ together with $\forall i \in I_1 \; \forall j \in J \; P(i,j) = 0$ ensures that $\mathcal{A}$ does not have a key to decrypt c_1, which would trivially allow to distinguish. Requiring that $\mathcal{A}$ obtains a key for i' however excludes adversaries that obtain no key at all. The original definitions [6] therefore include a special role 0 with $P(0,j) = 0$ for all j. One can then assume without loss of generality that anyone obtains a key for this role. Since assuming the existence of such a role appears to be a technicality that is only needed for the no-write rule, we do not make this assumption and present new security definitions in Sect. 4.2 that do not rely on such a role.

3 Ciphertext-Revealing Attacks Against Existing Schemes

3.1 Generic Description of Attack

We describe a fundamental practical issue of schemes which meet the above no-read and no-write definitions and show why the guarantees expected from an ACE scheme need to be strengthened. We show that schemes fulfilling the definition can suffer from what we call a malleability attack, which effectively bypasses the given policy and allows communication that is forbidden by the policy. The attack does not abuse any peculiarities of existing models and in fact only requires that the semi-honest sanitizer shares its inputs and outputs with colluding parties, which is arguably possible when the sanitizer is outsourced. In particular, security against such a sanitizer is desirable from a practical point of view.

We first give a high-level explanation of the attack, formalize it as a second step, and finally show that the "linear" scheme by Damgård et al. [6] based on ElGamal is vulnerable. In the full version, we also show this for the ElGamal-based scheme by Fuchsbauer et al. [9].

Assume there are three parties, Alice, Bob, and Charlie, each having a different role assigned. We denote by A, B, and C the associated roles. In our example, Alice and Charlie are always honest. Alice is allowed to communicate with Bob and Charlie. Bob is dishonest and forbidden to send messages to Charlie (and to Alice). The attack now proceeds as follows: When Alice sends her first message, Bob requests the corresponding ciphertext and the sanitized ciphertext from the semi-honest sanitizer. He then decrypts the sanitized ciphertext and thus receives the message Alice has sent. With the knowledge of this message, as we show below, he is able to create a valid ciphertext for a chosen message m', which will be correctly sanitized and later decrypted by Charlie, hence allowing unrestricted communication from Bob to Charlie. Details follow.

Consider the policy defined by

$$P(i,j) := \begin{cases} 1, & i = \mathsf{A}, \\ 0, & \text{otherwise.} \end{cases}$$

For the sake of presentation, we assume that the ACE scheme $\mathcal{E}$ under consideration enjoys perfect correctness. Also, we assume that the setup-phase has completed and the three parties thus possess the encryption and decryption keys, ek_i and dk_i, respectively. Now, imagine that the ACE scheme admits an efficient function $\mathsf{maul}_{\mathcal{E}}$ with the following property (later we show how to implement such a function for some existing schemes): For all messages m and m', any role i, and sanitizer parameters sp in the range of Setup, and for any fixed randomness r,

$$\mathsf{maul}_{\mathcal{E}}\big(\mathsf{Enc}(ek_i, m; r), sp, m, m')\big) = \mathsf{Enc}(ek_i, m'; r). \tag{1}$$

If such a malleability function exists, the communication policy can be bypassed as follows:

1. Alice encrypts a message $c \leftarrow \mathsf{Enc}(ek_\mathsf{A}, m)$ and the sanitizer computes $c' \leftarrow \mathsf{San}(sp, c)$ and gives c and c' to Bob.
2. Bob computes $m \leftarrow \mathsf{Dec}(dk_\mathsf{B}, c')$ and $\hat{c} \leftarrow \mathsf{maul}_{\mathcal{E}}(c, sp, m, m')$ and sends $\hat{c}$ to the sanitizer.
3. The ciphertext is sanitized $\hat{c}' \leftarrow \mathsf{San}(sp, \hat{c})$ and subsequently sent to Charlie. By the (perfect) correctness of the assumed ACE scheme and by our assumption on $\mathsf{maul}_{\mathcal{E}}$, $\hat{c}'$ is a valid ciphertext (under the encryption key of Alice) and Charlie is able to decrypt $m' \leftarrow \mathsf{Dec}(dk_\mathsf{C}, \hat{c}')$, effectively receiving Bob's message m'.

3.2 DHO Scheme Based on ElGamal

We briefly recall the ElGamal based ACE scheme for a single identity. The sanitizer parameters of the scheme contain the description of a finite cyclic group $G = \langle g \rangle$ and its group order q, and additionally an element $h = g^x$ for a uniform random $x \in \mathbb{Z}_q$. The encryption key for A is a random value $ek \in \mathbb{Z}_q$, and the decryption key is $-x$. The algorithm Enc on input an encryption key ek_i and a message $m \in \mathcal{M}$, samples $r_1, r_2 \in \mathbb{Z}_q$ uniformly at random and outputs the ciphertext

$$c = (c_0, c_1, c_2, c_3) := (g^{r_1}, h^{r_1} g^{ek_i}, g^{r_2}, m \cdot h^{r_2}).$$

We can define the function $\mathsf{maul}_{\mathsf{DHO}}$ as

$$\mathsf{maul}_{\mathsf{DHO}}\big((c_0, c_1, c_2, c_3), sp, m, m'\big) := \big(c_0, c_1, c_2, m' \cdot m^{-1} \cdot c_3\big).$$

Since the group order q is part of sp, this function is efficiently computable. For $c_3 = m \cdot h^{r_2}$, we thus get a new fourth component $c_3' = m' \cdot h^{r_2}$ and Eq. (1) is satisfied.

The malleability for more than one identity (and in particular in our scenario described above) follows since the scheme for several identities is composed of independent instances of the basic single-identity scheme.

4 A Stronger Notion of ACE

In this section, we introduce our new security definitions, which exclude the issues we have discovered. In the full version, we also show in which sense they imply the original notions.

4.1 ACE with Modification Detection

To be resilient against the ciphertext-revealing attacks described in Sect. 3, the sanitizer should ideally only sanitize fresh encryptions and block ciphertexts that are either replays or obtained by modifying previous ciphertexts. Therefore, we introduce an additional algorithm for detecting modified ciphertexts. If the sanitizer receives a ciphertext that is detected to be a modification of a previously received one, this ciphertext is blocked. Since such ciphertexts will not be stored in the repository and consequently not be decrypted, we define chosen-ciphertext security with respect to a decryption oracle that does not return a decryption if the received ciphertext is detected to be a modification of the challenge ciphertext. Our definitions can therefore be seen as a variant of publicly-detectable replayable-CCA security, which was introduced by Canetti et al. [5] for public key encryption. Before defining the security, we define the syntax of ACE schemes with this additional algorithm.

Definition 5. *An* access control encryption with modification detection scheme *is an ACE scheme $\mathcal{E}$ together with a PPT algorithm* DMod *that on input sanitizer parameters sp and two ciphertexts $c, \tilde{c} \in \mathcal{C}$, outputs a bit b (where $b = 1$ means that $\tilde{c}$ was obtained via modifying c).*

From now on, we will only consider ACE schemes with modification detection and thus often refer to them simply as ACE schemes.

The algorithm DMod should output 1 if $\tilde{c}$ is an adversarial modification of c, and 0 otherwise. We have the following intuitive requirements on DMod:

1. All ciphertexts $\tilde{c}$ an adversary can produce given ciphertexts $c_1, \ldots, c_l$ and no encryption key, are either invalid (i.e., sanitize to $\perp$) or we have $\mathsf{DMod}(sp, c_i, \tilde{c}) = 1$ for some $i \in \{1, \ldots, n\}$.
2. Given encryption and decryption keys, an adversary is unable to produce a ciphertext c such that a ciphertext produced by Enc for a message of the adversary's choice is detected to be a modification of c. In particular, independent encryptions of messages collide only with negligible probability.

The first requirement is captured by role-respecting security as defined in Definition 9, the second one by non-detection of fresh encryptions defined in Definition 8.

Remark 3. Canetti et al. (translated to our setting) additionally require that if $\mathsf{DMod}(sp, c, \tilde{c}) = 1$, then c and $\tilde{c}$ decrypt to the same message [5]. For our purpose, this is not needed. This means that we do not want to detect replays in the sense that the same message is replayed, but more generally, whether the given ciphertext was obtain via some modification of another ciphertext.

4.2 New Security Definitions

We formalize chosen-ciphertext attacks by giving the adversary access to an oracle $\mathcal{O}_{SD}$ that first sanitizes a given ciphertext and then decrypts the result. One could also consider *chosen-sanitized-ciphertext attacks* by providing the adversary access to an oracle $\mathcal{O}_D$ that only decrypts. This is potentially stronger since the adversary can emulate the oracle $\mathcal{O}_{SD}$ by first sanitizing the ciphertexts and then giving the result to $\mathcal{O}_D$, but given $\mathcal{O}_{SD}$, it is not necessarily possible to emulate $\mathcal{O}_D$. Since in the application, users can only send ciphertexts to the sanitizer but not directly write ciphertexts to the repository such that they are decrypted without being sanitized, the weaker notion is sufficient.

In principle, the adversary has in all definitions access to $\mathcal{O}_{SD}$, as well as to an encryption oracle and a key-generation oracle. To simplify the definitions, we omit the encryption or decryption oracles if the winning condition places no restriction on the encryption or decryption keys obtained from the key-generation oracle, respectively.

Privacy and anonymity. We first give definitions for (payload) privacy and sender-anonymity. The former guarantees that encryptions of different messages under the same encryption key cannot be distinguished as long as the adversary has no decryption key that allows to decrypt. We also require this for messages of different length, i.e., schemes satisfying our definition do not leak the length of the encrypted message, which means that the message space has to be bounded. Anonymity guarantees that encryptions of the same message under different keys cannot be distinguished. We distinguish a weak and a strong variant of anonymity, where the weak one provides no guarantees if the adversary can decrypt the ciphertext, and the strong one guarantees that even if the adversary has decryption keys, nothing is leaked about the sender role beyond which of the adversary's decryption keys can be used to decrypt.

Definition 6. *Let* $\mathcal{E} = (\mathsf{Setup}, \mathsf{Gen}, \mathsf{Enc}, \mathsf{San}, \mathsf{Dec}, \mathsf{DMod})$, *be an ACE with modification detection scheme and let* $\mathcal{A} = (\mathcal{A}_1, \mathcal{A}_2)$ *be a pair of probabilistic algorithms. Consider the experiment* $\mathsf{Exp}_{\mathcal{E},\mathcal{A}}^{\mathsf{ACE\text{-}PRV\text{-}ANON\text{-}CCA}}$ *in Fig. 2 and let* J *be the set of all* j *such that* $\mathcal{A}_1$ *or* $\mathcal{A}_2$ *issued the query* $(j, \mathtt{rec})$ *to the oracle* $\mathcal{O}_G$. *We define the* privacy under chosen-ciphertext attacks advantage *and the* sender-anonymity under chosen-ciphertext attacks advantages *of* $\mathcal{A}$ *as*

$$\mathsf{Adv}_{\mathcal{E},\mathcal{A}}^{\mathsf{ACE\text{-}PRV\text{-}CCA}} := 2 \cdot \Pr\big[b' = b \ \wedge\ i_0 = i_1 \ \wedge\ \forall j \in J \ P(i_0, j) = 0\big] - 1,$$

$$\mathsf{Adv}_{\mathcal{E},\mathcal{A}}^{\mathsf{ACE\text{-}wANON\text{-}CCA}} := 2 \cdot \Pr\big[b' = b \ \wedge\ m_0 = m_1 \\ \wedge\ \forall j \in J \ P(i_0, j) = P(i_1, j) = 0\big] - 1,$$

$$\mathsf{Adv}_{\mathcal{E},\mathcal{A}}^{\mathsf{ACE\text{-}sANON\text{-}CCA}} := 2 \cdot \Pr\big[b' = b \ \wedge\ m_0 = m_1 \\ \wedge\ \forall j \in J \ P(i_0, j) = P(i_1, j)\big] - 1,$$

respectively, where all probabilities are in $\mathsf{Exp}_{\mathcal{E},\mathcal{A}}^{\mathsf{ACE\text{-}PRV\text{-}ANON\text{-}CCA}}$. *The scheme* $\mathcal{E}$ *is called* private under chosen-ciphertext attacks (PRV-CCA secure), weakly sender-anonymous under chosen-ciphertext attacks (wANON-CCA secure), *and* strongly sender-anonymous under chosen-ciphertext attacks (sANON-CCA secure) *if* $\mathsf{Adv}^{\text{ACE-PRV-CCA}}_{\mathcal{E},\mathcal{A}}$, $\mathsf{Adv}^{\text{ACE-wANON-CCA}}_{\mathcal{E},\mathcal{A}}$, *and* $\mathsf{Adv}^{\text{ACE-sANON-CCA}}_{\mathcal{E},\mathcal{A}}$ *are negligible for all efficient* $\mathcal{A}$, *respectively.*

Experiment $\mathsf{Exp}^{\text{ACE-PRV-ANON-CCA}}_{\mathcal{E},\mathcal{A}}$

Input: $(1^\kappa, P), \kappa \in \mathbb{N}, P\colon [n]^2 \to \{0,1\}$
$(sp, msk) \leftarrow \mathsf{Setup}(1^\kappa, P)$
$(m_0, m_1, i_0, i_1, st) \leftarrow \mathcal{A}_1^{\mathcal{O}_G(\cdot,\cdot),\mathcal{O}_{SD}(\cdot,\cdot)}(sp)$
$b \twoheadleftarrow \{0,1\}$
$ek_{i_b} \leftarrow \mathsf{Gen}(msk, i_b, \mathbf{sen})$
$c^* \leftarrow \mathsf{Enc}(ek_{i_b}, m_b)$
$b' \leftarrow \mathcal{A}_2^{\mathcal{O}_G(\cdot,\cdot),\mathcal{O}_{SD^*}(\cdot,\cdot)}(st, c^*)$

Experiment $\mathsf{Exp}^{\text{ACE-SAN-CCA}}_{\mathcal{E},\mathcal{A}}$

Input: $(1^\kappa, P), \kappa \in \mathbb{N}, P\colon [n]^2 \to \{0,1\}$
$(sp, msk) \leftarrow \mathsf{Setup}(1^\kappa, P)$
$(c_0, c_1, st) \leftarrow \mathcal{A}_1^{\mathcal{O}_G(\cdot,\cdot),\mathcal{O}_{SD}(\cdot,\cdot)}(sp)$
$c_0' \leftarrow \mathsf{San}(sp, c_0)$; $c_1' \leftarrow \mathsf{San}(sp, c_1)$
$b \twoheadleftarrow \{0,1\}$
$b' \leftarrow \mathcal{A}_2^{\mathcal{O}_G(\cdot,\cdot),\mathcal{O}_{SD}(\cdot,\cdot)}(st, c_b')$
for $j \in [n]$ **do**
  $m_{0,j} \leftarrow \mathsf{Dec}(\mathsf{Gen}(msk, j, \mathbf{rec}), c_0')$
  $m_{1,j} \leftarrow \mathsf{Dec}(\mathsf{Gen}(msk, j, \mathbf{rec}), c_1')$

Experiment $\mathsf{Exp}^{\text{ACE-NDTCT-FENC}}_{\mathcal{E},\mathcal{A}}$

Input: $(1^\kappa, P), \kappa \in \mathbb{N}, P\colon [n]^2 \to \{0,1\}$
$(sp, msk) \leftarrow \mathsf{Setup}(1^\kappa, P)$
$(m, i, c) \leftarrow \mathcal{A}^{\mathcal{O}_G(\cdot,\cdot)}(sp)$
$ek_i \leftarrow \mathsf{Gen}(msk, i, \mathbf{sen})$
$c^* \leftarrow \mathsf{Enc}(ek_i, m)$
$b \leftarrow \mathsf{DMod}(sp, c, c^*)$

Experiment $\mathsf{Exp}^{\text{ACE-URR}}_{\mathcal{E},\mathcal{A}}$

Input: $(1^\kappa, P), \kappa \in \mathbb{N}, P\colon [n]^2 \to \{0,1\}$
$(sp, msk) \leftarrow \mathsf{Setup}(1^\kappa, P)$
$c \leftarrow \mathcal{A}^{\mathcal{O}_G(\cdot,\cdot),\mathcal{O}_E(\cdot,\cdot)}(sp)$
$\mathsf{dct} \leftarrow \mathbf{false}$
for $\tilde{c} \in \{\text{answers from } \mathcal{O}_E\}$ **do**
  $\mathsf{dct} \leftarrow \mathsf{dct} \vee \mathsf{DMod}(sp, \tilde{c}, c) = 1$
$c' \leftarrow \mathsf{San}(sp, c)$
for $j \in [n]$ **do**
  $m_j \leftarrow \mathsf{Dec}(\mathsf{Gen}(msk, j, \mathbf{rec}), c')$

Definitions of oracles

$$\mathcal{O}_G(i,t) := \mathsf{Gen}(msk, i, t)$$
$$\mathcal{O}_E(i,m) := \mathsf{Enc}(\mathsf{Gen}(msk, i, \mathbf{sen}), m)$$
$$\mathcal{O}_{SD}(j,c) := \mathsf{Dec}(\mathsf{Gen}(msk, j, \mathbf{rec}), \mathsf{San}(sp, c))$$
$$\mathcal{O}_{SD^*}(j,c) := \begin{cases} \mathsf{Dec}(\mathsf{Gen}(msk, j, \mathbf{rec}), \mathsf{San}(sp, c)), & \mathsf{DMod}(sp, c^*, c) = 0 \\ \mathbf{test}, & \text{else} \end{cases}$$

Fig. 2. Security experiments for an ACE with modification detection scheme $\mathcal{E}$ and an adversary $\mathcal{A}$, where $\mathcal{A} = (\mathcal{A}_1, \mathcal{A}_2)$ in the first two experiments.

Remark 4. Weak anonymity corresponds to the anonymity notion considered by Fuchsbauer et al. [9] and strong anonymity to the one considered by Damgård et al. [6]. We state both definitions because weak anonymity is easier to achieve but strong anonymity might be required by some applications. If anonymity is only required against the sanitizer or if all messages are anyway signed by the sender, weak anonymity is sufficient. Strong anonymity is required in settings where senders also want to retain as much anonymity as possible against legitimate receivers.

Sanitization security. We next define sanitization security, which excludes that dishonest parties can communicate via the ciphertexts. We formalize this by requiring that the output of the sanitizer for two different ciphertexts cannot be distinguished, as long as both sanitized ciphertexts are not $\perp$ and the adversary has no decryption key that decrypts one of the ciphertexts. This provides no security guarantees if the adversary can decrypt the ciphertexts, which does not seem to be an issue since in this case, the parties can anyway directly communicate via the messages. However, we additionally consider a stronger variant, where the adversary is allowed to possess a decryption key that decrypts the ciphertexts, as long as they both decrypt to the same message. This stronger variant excludes subliminal channels, i.e., even if the involved parties are allowed to communicated by the policy, they cannot exchange information via ciphertexts beyond the encrypted message.

Since the adversary provides the two ciphertexts that are sanitized, we do not know to which roles they correspond; they could even be particularly crafted without belonging to an existing role. Hence, we cannot state the requirement (in the weak variant) that the adversary should not be able to decrypt by only considering the policy and the obtained decryption keys, as in the no-write rule in Definition 4. Instead, we require that the decryption algorithm returns $\perp$ for all decryption keys the adversary possesses. For this to provide the intended security, we need that the decrypt algorithm returns $\perp$ whenever the receiver role corresponding to the used key is not supposed to read the message. This is guaranteed by role-respecting security which is defined later.

Definition 7. *Let* $\mathcal{E} = (\mathsf{Setup}, \mathsf{Gen}, \mathsf{Enc}, \mathsf{San}, \mathsf{Dec}, \mathsf{DMod})$ *be an ACE with modification detection scheme and let* $\mathcal{A} = (\mathcal{A}_1, \mathcal{A}_2)$ *be a pair of probabilistic algorithms. Consider the experiment* $\mathsf{Exp}^{\mathsf{ACE\text{-}SAN\text{-}CCA}}_{\mathcal{E},\mathcal{A}}$ *in Fig. 2 and let* J *be the set of all* j *such that* $\mathcal{A}_1$ *or* $\mathcal{A}_2$ *issued the query* $(j, \mathtt{rec})$ *to the oracle* $\mathcal{O}_G$*. We define the* sanitization under chosen-ciphertext attacks advantage *and the* strong sanitization under chosen-ciphertext attacks advantage *of* $\mathcal{A}$ *as*

$$\mathsf{Adv}^{\mathsf{ACE\text{-}SAN\text{-}CCA}}_{\mathcal{E},\mathcal{A}} := 2 \cdot \Pr\big[b' = b \ \wedge \ c_0' \neq \perp \neq c_1' \ \wedge \ \forall j \in J \ m_{0,j} = m_{1,j} = \perp\big] - 1,$$

$$\mathsf{Adv}^{\mathsf{ACE\text{-}sSAN\text{-}CCA}}_{\mathcal{E},\mathcal{A}} := 2 \cdot \Pr\big[b' = b \ \wedge \ c_0' \neq \perp \neq c_1' \ \wedge \ \forall j \in J \ m_{0,j} = m_{1,j}\big] - 1,$$

respectively, where the probability is over the randomness in $\mathsf{Exp}^{\mathsf{ACE\text{-}SAN\text{-}CCA}}_{\mathcal{E},\mathcal{A}}$*. The scheme* $\mathcal{E}$ *is called* sanitization under chosen-ciphertext attacks secure (SAN-CCA secure) *and* strongly sanitization under chosen-ciphertext attacks secure (sSAN-CCA secure) *if* $\mathsf{Adv}^{\mathsf{ACE\text{-}SAN\text{-}CCA}}_{\mathcal{E},\mathcal{A}}$ *and* $\mathsf{Adv}^{\mathsf{ACE\text{-}sSAN\text{-}CCA}}_{\mathcal{E},\mathcal{A}}$ *are negligible for all efficient* $\mathcal{A}$*, respectively.*

Non-detection of fresh encryptions. In the intended way of using a scheme satisfying our notions, the sanitizer only adds sanitized ciphertexts to the repository if the given ciphertext is not detected to be a modification of a previously received ciphertext. This means that if an adversary can find a ciphertext c such that another ciphertext c^* that is later honestly generated is detected as a modification of c, the delivery of the message at that later point can be prevented by sending the ciphertext c to the sanitizer earlier. We exclude this by the following definition, which can be seen as an extended correctness requirement.

Definition 8. *Let* $\mathcal{E} = (\mathsf{Setup}, \mathsf{Gen}, \mathsf{Enc}, \mathsf{San}, \mathsf{Dec}, \mathsf{DMod})$ *be an ACE with modification detection scheme and let* $\mathcal{A}$ *be a probabilistic algorithm. Consider the experiment* $\mathsf{Exp}^{\mathsf{ACE\text{-}NDTCT\text{-}FENC}}_{\mathcal{E},\mathcal{A}}$ *in Fig. 2. We define the* non-detection of fresh encryptions advantage *of* $\mathcal{A}$ *as*

$$\mathsf{Adv}^{\mathsf{ACE\text{-}NDTCT\text{-}FENC}}_{\mathcal{E},\mathcal{A}} := \Pr\big[b = 1\big],$$

where the probability is over the randomness in $\mathsf{Exp}^{\mathsf{ACE\text{-}NDTCT\text{-}FENC}}_{\mathcal{E},\mathcal{A}}$. *The scheme* $\mathcal{E}$ *has* non-detecting fresh encryptions (NDTCT-FENC) *if* $\mathsf{Adv}^{\mathsf{ACE\text{-}NDTCT\text{-}FENC}}_{\mathcal{E},\mathcal{A}}$ *is negligible for all efficient* $\mathcal{A}$.

Role-respecting and uniform-decryption security. We finally define role-respecting and uniform-decryption security. The former means that an adversary cannot produce a ciphertext for which the pattern of roles that can decrypt does not correspond to a role for which the adversary has an encryption key. For example, if the adversary has only an encryption key for the role i such that roles j_0 and j_1 are the only roles j with $P(i, j) = 1$, all ciphertexts produced by the adversary are either invalid (i.e., sanitized to $\perp$ or detected as a modification) or decrypt to a message different from $\perp$ precisely under the decryption keys for j_0 and j_1. On the one hand, this means that receivers who are not allowed to receive the message get $\perp$ and hence know that the message is not for them.[2] On the other hand, it also guarantees that the adversary cannot prevent receivers with role j_1 from receiving a message that is sent to receivers with role j_0. Furthermore, uniform decryption guarantees for all ciphertexts c output by an adversary that if c decrypts to a message different from $\perp$ for different decryption keys, it always decrypts to the same message. In the example above, this means that j_0 and j_1 not only both receive *some* message, but they both receive *the same* one.

Definition 9. *Let* $\mathcal{E} = (\mathsf{Setup}, \mathsf{Gen}, \mathsf{Enc}, \mathsf{San}, \mathsf{Dec}, \mathsf{DMod})$ *be an ACE with modification detection scheme and let* $\mathcal{A}$ *be a probabilistic algorithm. Consider the*

[2] Detectability (Definition 2) provides this guarantee for honest encryptions, role-respecting security extends this to maliciously generated ciphertexts. Note, however, that detectability is not implied by role-respecting security: If an adversary has encryption keys for two roles i and i', role-respecting security does not exclude that encrypting some message (depending on i') with the key for role i can be decrypted with keys for roles that are allowed to receive from i'.

experiment $\mathsf{Exp}_{\mathcal{E},\mathcal{A}}^{\mathsf{ACE\text{-}URR}}$ *in Fig. 2 and let* I *and* J *be the sets of all* i *and* j *such that* $\mathcal{A}$ *issued the query* $(i, \mathtt{sen})$ *and* $(j, \mathtt{rec})$ *to the oracle* $\mathcal{O}_G$*, respectively. We define the* role-respecting advantage *and the* uniform-decryption advantage *of* $\mathcal{A}$ *as*

$$\mathsf{Adv}_{\mathcal{E},\mathcal{A}}^{\mathsf{ACE\text{-}RR}} := \Pr\big[c' \neq \bot \ \wedge \ \mathtt{dct} = \mathtt{false} \\ \wedge \ \neg\big(\exists i \in I \ \forall j \in J \ (m_j \neq \bot \leftrightarrow P(i,j) = 1)\big)\big],$$

$$\mathsf{Adv}_{\mathcal{E},\mathcal{A}}^{\mathsf{ACE\text{-}UDEC}} := \Pr\big[\exists j, j' \in J \ m_j \neq \bot \neq m_{j'} \ \wedge \ m_j \neq \ m_{j'}\big],$$

respectively, where the probabilities are over the randomness in $\mathsf{Exp}_{\mathcal{E},\mathcal{A}}^{\mathsf{ACE\text{-}URR}}$*. The scheme* $\mathcal{E}$ *is* role-respecting (RR secure) *and* uniform-decryption (UDEC) secure *if* $\mathsf{Adv}_{\mathcal{E},\mathcal{A}}^{\mathsf{ACE\text{-}RR}}$ *and* $\mathsf{Adv}_{\mathcal{E},\mathcal{A}}^{\mathsf{ACE\text{-}UDEC}}$ *are negligible for all efficient* $\mathcal{A}$*, respectively.*

Remark 5. Note that in Definition 9, we only check the decryptions for receiver roles for which $\mathcal{A}$ has requested the corresponding decryption key. This means that an adversary in addition to producing a ciphertext that causes an inconsistency, also has to find a receiver role for which this inconsistency manifests. If the total number of roles n is small (say polynomial in the security parameter), $\mathcal{A}$ can simply query $\mathcal{O}_G$ on all receiver keys, but for large n this condition is non-trivial. For example, we consider a scheme secure if an adversary can efficiently produce a ciphertext such that there is a receiver role that can decrypt it even though the policy does not allow it, as long as this receiver role is hard to find. The rationale is that in this case, the inconsistency cannot be exploited and will only be observed with negligible probability in an execution of the protocol.

5 Enhanced Sanitizable Public-Key Encryption

5.1 Definitions

As a stepping stone toward ACE schemes satisfying our new security definitions, we introduce *enhanced sanitizable public-key encryption.* Sanitizable public-key encryption has been considered by Damgård et al. [6] and Fuchsbauer et al. [9] as a relaxation of universal re-encryption [10] and rerandomizable encryption [12,15]. It allows to *sanitize* a ciphertext to obtain a *sanitized ciphertext* that cannot be linked to the original ciphertext except that it decrypts to the correct message. In contrast to rerandomizable encryption, sanitized ciphertexts can have a different syntax than ciphertexts, i.e., it is not required that a sanitized ciphertext is indistinguishable from a fresh encryption. We introduce an enhanced variant with a different syntax and stronger security guarantees.

Definition 10. *An* enhanced sanitizable public-key encryption (sPKE) scheme *consists of the following five PPT algorithms:*

Setup: *The algorithm* Setup *on input a security parameter* 1^κ*, outputs* sanitizer parameters *sp, and a* master secret key *msk. We implicitly assume that all parameters and keys include the finite* message space $\mathcal{M}$ *and the* ciphertext spaces $\mathcal{C}, \mathcal{C}'$.

Key generation: *The algorithm* Gen *on input a master secret key msk, outputs an* encryption key *ek and a* decryption key *dk.*

Encryption: *The algorithm* Enc *on input an encryption key ek and a message* $m \in \mathcal{M}$*, outputs a ciphertext* $c \in \mathcal{C}$*.*

Sanitization: *The algorithm* San *on input sanitizer parameters sp and a ciphertext* $c \in \mathcal{C}$*, outputs a* sanitized ciphertext $c' \in \mathcal{C}' \cup \{\bot\}$.

Decryption: *The algorithm* Dec *on input a decryption key dk and a sanitized ciphertext* $c' \in \mathcal{C}'$*, outputs a message* $m \in \mathcal{M} \cup \{\bot\}$*; on input dk and* $\bot$*, it outputs* $\bot$*.*

For correctness, we require for all (sp, msk) *in the range of* Setup*, all* (ek, dk) *in the range of* $\mathsf{Gen}(msk)$*, and all* $m \in \mathcal{M}$ *that*

$$\mathsf{Dec}\big(dk, \mathsf{San}\big(sp, \mathsf{Enc}(ek, m)\big)\big) = m$$

with probability 1.

We require robustness in the sense that no ciphertext decrypts to a message different from $\bot$ for two different decryption keys (except with negligible probability). This is similar to detectability for ACE schemes, but we allow the adversary to directly output a ciphertext, instead of a message, which is then honestly encrypted. Our notion therefore closely resembles *unrestricted strong robustness (USROB)*, introduced by Farshim et al. [8] for public-key encryption, which also allows the adversary to choose a ciphertext and, in contrast to strong robustness by Abdalla et al. [1], gives the adversary access to decryption keys.

Definition 11. *Let* $\mathcal{E} = (\mathsf{Setup}, \mathsf{Gen}, \mathsf{Enc}, \mathsf{San}, \mathsf{Dec})$ *be an sPKE scheme. For a probabilistic algorithm* $\mathcal{A}$*, we define the experiment* $\mathsf{Exp}^{\mathsf{sPKE\text{-}USROB}}_{\mathcal{E},\mathcal{A}}$ *that executes* $(sp, msk) \leftarrow \mathsf{Setup}(1^\kappa)$ *and* $(c, i_0, i_1) \leftarrow \mathcal{A}^{\mathcal{O}_G(\cdot)}(sp)$*, where the oracle* $\mathcal{O}_G$ *on input* `getNew`*, outputs a fresh key pair* $(ek, dk) \leftarrow \mathsf{Gen}(msk)$*. Let q be the number of oracle queries and let for* $i \in \{1, \dots, q\}$*,* (ek_i, dk_i) *be the i-th answer from* $\mathcal{O}_G$*. We define the* (unrestricted strong) robustness advantage *of* $\mathcal{A}$ *as*

$$\begin{aligned}\mathsf{Adv}^{\mathsf{sPKE\text{-}USROB}}_{\mathcal{E},\mathcal{A}} := \Pr\big[&1 \le i_0, i_1 \le q \;\wedge\; i_0 \ne i_1 \\ &\wedge\; \mathsf{Dec}\big(dk_{i_0}, \mathsf{San}(sp, c)\big) \ne \bot \ne \mathsf{Dec}\big(dk_{i_1}, \mathsf{San}(sp, c)\big)\big],\end{aligned}$$

where the probability is over the randomness in $\mathsf{Exp}^{\mathsf{sPKE\text{-}USROB}}_{\mathcal{E},\mathcal{A}}$ *and the random coins of* San *and* Dec *(both executed independently twice). The scheme* $\mathcal{E}$ *is* (unrestricted strongly) robust (USROB secure) *if* $\mathsf{Adv}^{\mathsf{sPKE\text{-}USROB}}_{\mathcal{E},\mathcal{A}}$ *is negligible for all efficient* $\mathcal{A}$*.*

We next define IND-CCA security analogously to the definition for ordinary public-key encryption. In contrast to the usual definition, we do not require the adversary to output two messages of equal length, which implies that schemes satisfying our definition do not leak the length of the encrypted message.

Definition 12. *Let* $\mathcal{E} = (\mathsf{Setup}, \mathsf{Gen}, \mathsf{Enc}, \mathsf{San}, \mathsf{Dec})$ *be an sPKE scheme and let* $\mathcal{A} = (\mathcal{A}_1, \mathcal{A}_2)$ *be a pair of probabilistic algorithms. Consider the experiment*

$\mathsf{Exp}^{\mathsf{sPKE\text{-}IND\text{-}CCA}}_{\mathcal{E},\mathcal{A}}$ *in Fig. 3 and let* $C_{\mathcal{A}_2}$ *be the set of all ciphertexts that* $\mathcal{A}_2$ *queried to the oracle* $\mathcal{O}_{SD}$. *We define the* ciphertext indistinguishability under chosen-ciphertext attacks advantage *of* $\mathcal{A}$ *as*

$$\mathsf{Adv}^{\mathsf{sPKE\text{-}IND\text{-}CCA}}_{\mathcal{E},\mathcal{A}} := 2 \cdot \Pr\big[b' = b \ \wedge \ c^* \notin C_{\mathcal{A}_2}\big] - 1,$$

where the probability is over the randomness in $\mathsf{Exp}^{\mathsf{sPKE\text{-}IND\text{-}CCA}}_{\mathcal{E},\mathcal{A}}$. *The scheme* $\mathcal{E}$ *has* indistinguishable ciphertexts under chosen-ciphertext attacks (is IND-CCA secure) *if* $\mathsf{Adv}^{\mathsf{sPKE\text{-}IND\text{-}CCA}}_{\mathcal{E},\mathcal{A}}$ *is negligible for all efficient* $\mathcal{A}$.

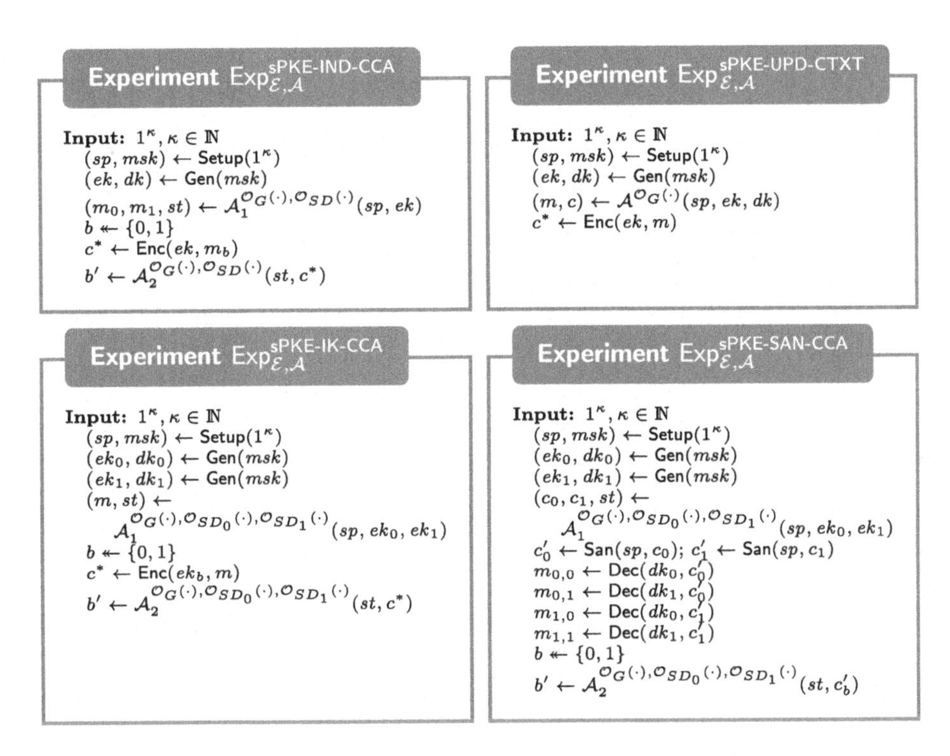

Fig. 3. Security experiments for an sPKE scheme $\mathcal{E}$ and an adversary $\mathcal{A}$, where $\mathcal{A} = (\mathcal{A}_1, \mathcal{A}_2)$ in the experiments $\mathsf{Exp}^{\mathsf{sPKE\text{-}IND\text{-}CCA}}_{\mathcal{E},\mathcal{A}}$, $\mathsf{Exp}^{\mathsf{sPKE\text{-}IK\text{-}CCA}}_{\mathcal{E},\mathcal{A}}$, and $\mathsf{Exp}^{\mathsf{sPKE\text{-}SAN\text{-}CCA}}_{\mathcal{E},\mathcal{A}}$. The oracle $\mathcal{O}_{SD}$ is defined as $\mathcal{O}_{SD}(c) = \mathsf{Dec}(dk, \mathsf{San}(sp, c))$ and the oracle $\mathcal{O}_{SD_j}$ as $\mathcal{O}_{SD_j}(c) = \mathsf{Dec}(dk_j, \mathsf{San}(sp, c))$. Moreover, the oracle $\mathcal{O}_G$ on input `getNew`, outputs a fresh key pair $(ek, dk) \leftarrow \mathsf{Gen}(msk)$.

We also need that it is hard to predict a ciphertext generated by Enc from a message of the adversary's choice given encryption and decryption keys. Note that this is not implied by IND-CCA security since the adversary here obtains the decryption key.

Definition 13. *Let* $\mathcal{E} = (\mathsf{Setup}, \mathsf{Gen}, \mathsf{Enc}, \mathsf{San}, \mathsf{Dec})$ *be an sPKE scheme and let* $\mathcal{A}$ *be a probabilistic algorithm. Consider the experiment* $\mathsf{Exp}^{\mathsf{sPKE\text{-}UPD\text{-}CTXT}}_{\mathcal{E},\mathcal{A}}$ *in Fig. 3. We define the* ciphertext unpredictability advantage *of* $\mathcal{A}$ *as*

$$\mathsf{Adv}^{\mathsf{sPKE\text{-}UPD\text{-}CTXT}}_{\mathcal{E},\mathcal{A}} := \Pr\big[c = c^*\big],$$

where the probability is over the randomness in $\mathsf{Exp}^{\mathsf{sPKE\text{-}UPD\text{-}CTXT}}_{\mathcal{E},\mathcal{A}}$. *The scheme* $\mathcal{E}$ *has* unpredictable ciphertexts (is UPD-CTXT secure) *if* $\mathsf{Adv}^{\mathsf{sPKE\text{-}UPD\text{-}CTXT}}_{\mathcal{E},\mathcal{A}}$ *is negligible for all efficient* $\mathcal{A}$.

We further define anonymity or indistinguishability of keys following Bellare et al. [3].

Definition 14. *Let* $\mathcal{E} = (\mathsf{Setup}, \mathsf{Gen}, \mathsf{Enc}, \mathsf{San}, \mathsf{Dec})$ *be an sPKE scheme and let* $\mathcal{A} = (\mathcal{A}_1, \mathcal{A}_2)$ *be a pair of probabilistic algorithms. Consider the experiment* $\mathsf{Exp}^{\mathsf{sPKE\text{-}IK\text{-}CCA}}_{\mathcal{E},\mathcal{A}}$ *in Fig. 3 and let* $C_{\mathcal{A}_2}$ *be the set of all ciphertexts that* $\mathcal{A}_2$ *queried to the oracle* $\mathcal{O}_{SD_0}$ *or* $\mathcal{O}_{SD_1}$. *We define the* indistinguishability of keys under chosen-ciphertext attacks advantage *of* $\mathcal{A}$ *as*

$$\mathsf{Adv}^{\mathsf{sPKE\text{-}IK\text{-}CCA}}_{\mathcal{E},\mathcal{A}} := 2 \cdot \Pr\big[b' = b \ \wedge \ c^* \notin C_{\mathcal{A}_2}\big] - 1,$$

where the probability is over the randomness in $\mathsf{Exp}^{\mathsf{sPKE\text{-}IK\text{-}CCA}}_{\mathcal{E},\mathcal{A}}$. *The scheme* $\mathcal{E}$ *has* indistinguishable keys under chosen-ciphertext attacks (is IK-CCA secure) *if* $\mathsf{Adv}^{\mathsf{sPKE\text{-}IK\text{-}CCA}}_{\mathcal{E},\mathcal{A}}$ *is negligible for all efficient* $\mathcal{A}$.

Sanitization security formalizes that given certain public keys and a sanitized ciphertext, it is hard to tell which of two adversarially chosen ciphertexts was actually sanitized. To exclude trivial attacks, we require that both ciphertexts are encryptions relative to the two challenge public keys ek_0 and ek_1. Otherwise, the adversary could use the oracle $\mathcal{O}_G$ to obtain a fresh key-pair (ek, dk) and encrypt two different messages under ek. It could then decrypt the challenge ciphertext using dk and win the game.

Definition 15. *Let* $\mathcal{E} = (\mathsf{Setup}, \mathsf{Gen}, \mathsf{Enc}, \mathsf{San}, \mathsf{Dec})$ *be an sPKE scheme and let* $\mathcal{A} = (\mathcal{A}_1, \mathcal{A}_2)$ *be a pair of probabilistic algorithms. Consider the experiment* $\mathsf{Exp}^{\mathsf{sPKE\text{-}SAN\text{-}CCA}}_{\mathcal{E},\mathcal{A}}$ *in Fig. 3. We define the* sanitization under chosen-ciphertext attacks advantage *of* $\mathcal{A}$ *as*

$$\mathsf{Adv}^{\mathsf{sPKE\text{-}SAN\text{-}CCA}}_{\mathcal{E},\mathcal{A}} := 2 \cdot \Pr\big[b' = b \ \wedge \ \exists j, j' \in \{0,1\} \ m_{0,j} \neq \bot \neq m_{1,j'}\big] - 1,$$

where the probability is over the randomness in $\mathsf{Exp}^{\mathsf{sPKE\text{-}IK\text{-}CCA}}_{\mathcal{E},\mathcal{A}}$. *We say the scheme* $\mathcal{E}$ *is* sanitization under chosen-ciphertext attacks (SAN-CCA) secure *if* $\mathsf{Adv}^{\mathsf{sPKE\text{-}SAN\text{-}CCA}}_{\mathcal{E},\mathcal{A}}$ *is negligible for all efficient* $\mathcal{A}$.

We finally define the probability that two independent executions of the key-generation algorithm produce the same encryption key. This probability has to be small for all IND-CCA-secure schemes because an attacker can otherwise obtain a new key pair from $\mathcal{O}_G$ and if the obtained encryption key matches the one with which the challenge ciphertext is generated, the attacker can decrypt and win the IND-CCA game. We anyway explicitly define this probability to simplify our reductions later.

Definition 16. *Let* $\mathcal{E} = (\mathsf{Setup}, \mathsf{Gen}, \mathsf{Enc}, \mathsf{San}, \mathsf{Dec})$ *be an sPKE scheme. We define the* encryption-key collision probability $\mathsf{Col}_{\mathcal{E}}^{\mathsf{ek}}$ *as*

$$\mathsf{Col}_{\mathcal{E}}^{\mathsf{ek}} := \Pr^{(sp,msk)\leftarrow\mathsf{Setup}(1^\kappa);\ (ek_0,dk_0)\leftarrow\mathsf{Gen}(msk);\ (ek_1,dk_1)\leftarrow\mathsf{Gen}(msk)}[ek_0 = ek_1].$$

5.2 Constructing an sPKE Scheme

We next construct an sPKE scheme satisfying our security definitions. Our construction resembles the weakly sanitizable PKE scheme by Fuchsbauer et al. [9]. We use a variant of ElGamal encryption and obtain security against chosen-ciphertext attacks using the technique of Naor and Yung [14], i.e., encrypting the message under two independent keys and proving in zero-knowledge that the ciphertexts are encryptions of the same message, which was shown to achieve full IND-CCA security if the zero-knowledge proof is one-time simulation sound by Sahai [16].

Let PKE be a (IND-CPA secure) public-key encryption scheme, let Sig be a (EUF-CMA-secure) signature scheme, and let NIZK be a (one-time simulation sound) NIZK proof system for the language $L := \{x \mid \exists w\ (x,w) \in R\}$, where the relation R is defined as follows: for $x = \big(g, ek^{\mathsf{PKE}}, vk^{\mathsf{Sig}}, c_1, c_2, c_\sigma\big)$ and $w = (m, g^a, g^b, r_1, s_1, r_2, s_2, \sigma, r)$, we have $(x,w) \in R$ if and only if

$$\begin{aligned} c_1 =& (g^{r_1}, g^{a\cdot r_1}, g^{s_1}, g^{a\cdot s_1}\cdot m) \ \wedge\ c_2 = (g^{r_2}, g^{b\cdot r_2}, g^{s_2}, g^{b\cdot s_2}\cdot m) \\ &\wedge\ \mathsf{Sig.Ver}\big(vk^{\mathsf{Sig}}, (g^a, g^b), \sigma\big) = 1 \ \wedge\ c_\sigma = \mathsf{PKE.Enc}\big(ek^{\mathsf{PKE}}, (g^a, g^b, \sigma); r\big). \end{aligned}$$

We define an sPKE scheme as follows:

Setup: The setup algorithm $\mathsf{sPKE.Setup}$ first generates

$$\begin{aligned} \big(ek^{\mathsf{PKE}}, dk^{\mathsf{PKE}}\big) &\leftarrow \mathsf{PKE.Gen}(1^\kappa), \\ \big(vk^{\mathsf{Sig}}, sk^{\mathsf{Sig}}\big) &\leftarrow \mathsf{Sig.Gen}(1^\kappa), \\ crs &\leftarrow \mathsf{NIZK.Gen}(1^\kappa). \end{aligned}$$

Let $G = \langle g \rangle$ be a cyclic group with prime order p generated by g, with $p \geq 2^\kappa$, and let $\mathcal{M} \subseteq G$ such that $|\mathcal{M}|/p \leq 2^{-\kappa}$. The sanitizer parameters sp^{sPKE} contain ek^{PKE}, vk^{Sig}, crs, and a description of G, including g and p. The master secret key msk^{sPKE} consists of ek^{PKE}, vk^{Sig}, sk^{Sig}, crs, and a description of G, including g and p.

Key generation: The algorithm $\mathsf{sPKE.Gen}$ on input msk^{sPKE}, samples two elements $dk_1, dk_2 \twoheadleftarrow \mathbb{Z}_p^*$ and computes $ek_1 \leftarrow g^{dk_1}$, $ek_2 \leftarrow g^{dk_2}$, as well as $\sigma \leftarrow \mathsf{Sig.Sign}\big(sk^{\mathsf{Sig}}, (ek_1, ek_2)\big)$. Finally, it outputs $ek^{\mathsf{sPKE}} := \big(g, p, crs, ek^{\mathsf{PKE}}, vk^{\mathsf{Sig}}, ek_1, ek_2, \sigma\big)$ and $dk^{\mathsf{sPKE}} := (dk_1, dk_2)$.

Encryption: The algorithm $\mathsf{sPKE.Enc}$ on input an encryption key $ek^{\mathsf{sPKE}} = \big(g, p, crs, ek^{\mathsf{PKE}}, vk^{\mathsf{Sig}}, ek_1, ek_2, \sigma\big)$ and a message $m \in \mathcal{M}$, samples randomness r, chooses $r_1, s_1, r_2, s_2 \twoheadleftarrow \mathbb{Z}_p^*$ uniformly at random, and computes

$$c_1 \leftarrow \left(g^{r_1}, ek_1^{r_1}, g^{s_1}, ek_1^{s_1} \cdot m\right),$$
$$c_2 \leftarrow \left(g^{r_2}, ek_2^{r_2}, g^{s_2}, ek_2^{s_2} \cdot m\right),$$
$$c_\sigma \leftarrow \mathsf{PKE.Enc}\left(ek^{\mathsf{PKE}}, (ek_1, ek_2, \sigma); r\right).$$

It then generates $\pi \leftarrow \mathsf{NIZK.Prove}\big(crs, x := (g, ek^{\mathsf{PKE}}, vk^{\mathsf{Sig}}, c_1, c_2, c_\sigma), w := (m, ek_1, ek_2, r_1, s_1, r_2, s_2, \sigma, r)\big)$, and outputs the ciphertext $c := (c_1, c_2, c_\sigma, \pi)$.

Sanitization: The algorithm $\mathsf{sPKE.San}$ on input sanitizer parameters sp^{sPKE} and a ciphertext $c = (c_1, c_2, c_\sigma, \pi)$, first verifies the NIZK proof by evaluating $\mathsf{NIZK.Ver}\big(crs, x := (g, ek^{\mathsf{PKE}}, vk^{\mathsf{Sig}}, c_1, c_2, c_\sigma), \pi\big)$. It then parses the first ciphertext component as $(c_{1,1}, c_{1,2}, c_{1,3}, c_{1,4}) \leftarrow c_1$. If the verification succeeds and $c_{1,1} \neq 1 \neq c_{1,2}$, then it chooses a random $t \twoheadleftarrow \mathbb{Z}_p^*$ and outputs the sanitized ciphertext

$$c' := \left((c_{1,1})^t \cdot c_{1,3},\ (c_{1,2})^t \cdot c_{1,4}\right).$$

If the verification fails or if $c_{1,1} = 1$ or $c_{1,2} = 1$, it outputs $\perp$.

Decryption: The algorithm $\mathsf{sPKE.Dec}$ on input a decryption key $dk^{\mathsf{sPKE}} = (dk_1, dk_2)$ and a sanitized ciphertext $c' = (c'_1, c'_2)$, computes the message $m \leftarrow c'_2 \cdot \left((c'_1)^{dk_1}\right)^{-1}$. It outputs m if $m \in \mathcal{M}$, and otherwise it outputs $\perp$. On input dk^{sPKE} and $\perp$, it outputs $\perp$.

The following proposition and theorem summarize the correctness and security properties of our scheme and are proven in the full version of this paper.

Proposition 1. *If* Sig *is correct and* NIZK *has perfect completeness, the scheme* sPKE *from above is correct, robust, has unpredictable ciphertexts, and negligible encryption-key collision probability.*

Theorem 1. *If the DDH assumption holds in the group G,* PKE *is IND-CPA secure,* Sig *is EUF-CMA secure, and if* NIZK *is zero-knowledge, computationally sound, and one-time simulation sound, then the scheme* sPKE *from above is IND-CCA secure, IK-CCA secure, and SAN-CCA secure.*

On a high level, our proof proceeds as follows. It is rather straightforward to show that our variant of ElGamal encryption satisfies the CPA versions of the three properties. The proof of CCA security follows the proof by Sahai for public-key encryption [16]: Since the NIZK ensures that both ciphertext components are encryptions of the same message, it does not matter which component is decrypted. In a reduction, where we assume an adversary $\mathcal{A}$ against the CCA variants of the desired properties, and we want to break the corresponding CPA variants, we only get one public key and no decryption oracle from the challenger. In order to emulate the view toward $\mathcal{A}$, the reduction chooses an additional public key and a CRS for the NIZK scheme. Since the reduction thus knows one of the secret keys, it can emulate a decryption oracle. To generate a challenge ciphertext, the reduction obtains one challenge ciphertexts from its CPA challenger, and encrypts another, arbitrary message to get a second ciphertext. The reduction uses the NIZK simulator to obtain an accepting proof that is indistinguishable from a "real proof", even if the underlying statement is not true. A crucial point here is that the NIZK scheme has to be *one-time simulation sound*. This ensures that even if the adversary sees one simulated (accepting) proof of a wrong statement, it is not capable of producing accepting proofs of wrong statements, except by reproducing the exact proof obtained within the challenge, but which $\mathcal{A}$ is not allowed to ask to the decryption oracle by the CCA definition. The fundamental result of Sahai [16] is that the above strategy successfully simulates a complete CCA attack toward $\mathcal{A}$.

An additional obstacle we have is that to preserve anonymity, the NIZK needs to be verified without knowing which encryption keys were used. On the other hand, the reduction only works if the two used keys "match", since otherwise, the emulated decryption oracle would use an incorrect key to decrypt. To prevent an adversary from mixing different key pairs for encryptions, the key-generation process signs valid key pairs, and the NIZK ensures that a signed pair was used. Due to anonymity, this signature cannot be directly contained in the ciphertexts. Instead, it is part of the witness. To prove that if a ciphertext is accepted, the used key pair was indeed signed by the key-generation process, we show that if $\mathcal{A}$ manages to produce a ciphertext that is accepted but the keys were not signed, we can break EUF-CMA security of the signature scheme. In this reduction, we have to provide a forgery. Hence, the reduction needs to extract the signature and the used encryption keys from the ciphertext. This could be achieved by assuming that the NIZK is extractable. Extractability and simulation-soundness at the same time is, however, a quite strong assumption. Instead, we add an encryption of the signature and the key pair under a separate PKE scheme to the ciphertexts. The reduction can then generate the keys for this PKE scheme itself and perform extraction by decrypting that ciphertext.

6 Construction of an ACE Scheme

6.1 Construction for Equality

Following Fuchsbauer et al. [9], we first construct an ACE scheme for the equality policy, i.e., $P(i,j) = 1 \Leftrightarrow i = j$, and then use such a scheme in another construction for richer policies. We base our construction on an sPKE scheme, which already has many important properties needed for a secure ACE scheme. A syntactical difference between sPKE and ACE schemes is that the key generation of the former on every invocation produces a fresh key pair, while the latter schemes allow the generation of keys for a given role. To bind key pairs to some role $i \in [n]$, we use the output of a pseudorandom function on input i as the randomness for the sPKE key generation. For role-respecting security, we have to ensure that an adversary can only produce ciphertexts for keys obtained from the key generation oracle. This is achieved by signing all keys with a signing key generated at setup. To prevent malleability attacks as the ones described in Sect. 3, the encryption algorithm additionally signs all ciphertexts with a separate signing key that is tied to the encryption key. To maintain anonymity, the signatures are not part of the ciphertext but the encrypters prove in zero-knowledge that they know such signatures. Finally, the modification detection simply checks whether the ciphertexts (without the NIZK proofs) are equal. Intuitively, this is sufficient since we assume the underlying sPKE scheme to be CCA secure, which implies that it is not possible to meaningfully modify a given ciphertext. Hence, a ciphertext is either equal to an existing one (and thus detected by the algorithm) or a fresh encryption.

Our construction. Let sPKE be a sanitizable public-key encryption scheme, let Sig be a signature scheme, and let F be a PRF. Further let NIZK be a NIZK proof of knowledge system for the language $L := \{x \mid \exists w\ (x, w) \in R\}$, where the relation R is defined as follows: for $x = (vk^{\mathsf{Sig}}, \tilde{c})$ and $w = (ek_i^{\mathsf{sPKE}}, m, r, vk_i^{\mathsf{Sig}}, \sigma_i^{\mathsf{Sig}}, \sigma_c^{\mathsf{Sig}})$, $(x, w) \in R$ if and only if

$$\tilde{c} = \mathsf{sPKE.Enc}(ek_i^{\mathsf{sPKE}}, m; r) \ \wedge\ \mathsf{Sig.Ver}(vk^{\mathsf{Sig}}, [ek_i^{\mathsf{sPKE}}, vk_i^{\mathsf{Sig}}], \sigma_i^{\mathsf{Sig}}) = 1 \\ \wedge\ \mathsf{Sig.Ver}(vk_i^{\mathsf{Sig}}, \tilde{c}, \sigma_c^{\mathsf{Sig}}) = 1.$$

We define an ACE with modification detection scheme ACE as follows:

Setup: On input a security parameter 1^κ and a policy $P\colon [n] \times [n] \to \{0,1\}$ with $P(i,j) = 1 \Leftrightarrow i = j$, the algorithm $\mathsf{ACE.Setup}$ picks a random PRF key K for a PRF F, and runs

$$\begin{aligned} (sp^{\mathsf{sPKE}}, msk^{\mathsf{sPKE}}) &\leftarrow \mathsf{sPKE.Setup}(1^\kappa), \\ (vk^{\mathsf{Sig}}, sk^{\mathsf{Sig}}) &\leftarrow \mathsf{Sig.Gen}(1^\kappa), \\ crs^{\mathsf{NIZK}} &\leftarrow \mathsf{NIZK.Gen}(1^\kappa). \end{aligned}$$

It outputs the master secret key $msk^{\mathsf{ACE}} := (K, msk^{\mathsf{sPKE}}, vk^{\mathsf{Sig}}, sk^{\mathsf{Sig}}, crs^{\mathsf{NIZK}})$ and the sanitizer parameters $sp^{\mathsf{ACE}} := (sp^{\mathsf{sPKE}}, vk^{\mathsf{Sig}}, crs^{\mathsf{NIZK}})$.

Key generation: On input a master secret key $msk^{\mathsf{ACE}} = (K, msk^{\mathsf{sPKE}}, vk^{\mathsf{Sig}}, sk^{\mathsf{Sig}}, crs^{\mathsf{NIZK}})$, a role $i \in [n]$, and a type $t \in \{\mathtt{sen}, \mathtt{rec}\}$, $\mathsf{ACE.Gen}$ computes

$$(ek_i^{\mathsf{sPKE}}, dk_i^{\mathsf{sPKE}}) \leftarrow \mathsf{sPKE.Gen}(msk^{\mathsf{sPKE}}; F_K([i, 0])).$$

If $t = \mathtt{sen}$, it further computes

$$\begin{aligned} (vk_i^{\mathsf{Sig}}, sk_i^{\mathsf{Sig}}) &\leftarrow \mathsf{Sig.Gen}(1^\kappa; F_K([i, 1])), \\ \sigma_i^{\mathsf{Sig}} &\leftarrow \mathsf{Sig.Sign}(sk^{\mathsf{Sig}}, [ek_i^{\mathsf{sPKE}}, vk_i^{\mathsf{Sig}}]; F_K([i, 2])). \end{aligned}$$

If $t = \mathtt{sen}$, it outputs the encryption key $ek_i^{\mathsf{ACE}} := (vk^{\mathsf{Sig}}, ek_i^{\mathsf{sPKE}}, vk_i^{\mathsf{Sig}}, sk_i^{\mathsf{Sig}}, \sigma_i^{\mathsf{Sig}}, crs^{\mathsf{NIZK}})$; if $t = \mathtt{rec}$, it outputs the decryption key $dk_i^{\mathsf{ACE}} := dk_i^{\mathsf{sPKE}}$.

Encryption: On input an encryption key $ek_i^{\mathsf{ACE}} = \big(vk^{\mathsf{Sig}}, ek_i^{\mathsf{sPKE}}, vk_i^{\mathsf{Sig}}, sk_i^{\mathsf{Sig}}, \sigma_i^{\mathsf{Sig}}, crs^{\mathsf{NIZK}}\big)$ and a message $m \in \mathcal{M}^{\mathsf{ACE}}$, the algorithm $\mathsf{ACE.Enc}$ samples randomness r and computes

$$\begin{aligned} \tilde{c} &\leftarrow \mathsf{sPKE.Enc}\big(ek_i^{\mathsf{sPKE}}, m; r\big), \\ \sigma_c^{\mathsf{Sig}} &\leftarrow \mathsf{Sig.Sign}\big(sk_i^{\mathsf{Sig}}, \tilde{c}\big), \\ \pi^{\mathsf{NIZK}} &\leftarrow \mathsf{NIZK.Prove}\big(crs^{\mathsf{NIZK}}, x := \big(vk^{\mathsf{Sig}}, \tilde{c}\big), \\ &\qquad\qquad w := \big(ek_i^{\mathsf{sPKE}}, m, r, vk_i^{\mathsf{Sig}}, \sigma_i^{\mathsf{Sig}}, \sigma_c^{\mathsf{Sig}}\big)\big). \end{aligned}$$

It outputs the ciphertext $c := \big(\tilde{c}, \pi^{\mathsf{NIZK}}\big)$.

Sanitization: On input sanitizer parameters $sp^{\mathsf{ACE}} = \big(sp^{\mathsf{sPKE}}, vk^{\mathsf{Sig}}, crs^{\mathsf{NIZK}}\big)$ and a ciphertext $c = \big(\tilde{c}, \pi^{\mathsf{NIZK}}\big)$, $\mathsf{ACE.San}$ outputs the sanitized ciphertext $c' \leftarrow \mathsf{sPKE.San}\big(sp^{\mathsf{sPKE}}, \tilde{c}\big)$ if $\mathsf{NIZK.Ver}\big(crs^{\mathsf{NIZK}}, x := \big(vk^{\mathsf{Sig}}, \tilde{c}, \big), \pi^{\mathsf{NIZK}}\big) = 1$; otherwise, it outputs $\perp$.

Decryption: The algorithm $\mathsf{ACE.Dec}$ on input a decryption key dk_j^{ACE} and a sanitized ciphertext c', outputs the message $m \leftarrow \mathsf{sPKE.Dec}(dk_j^{\mathsf{ACE}}, c')$.

Modification detection: The algorithm $\mathsf{ACE.DMod}$ on input sp^{ACE}, $c_1 = \big(\tilde{c}_1, \pi_1^{\mathsf{NIZK}}\big)$, and $c_2 = \big(\tilde{c}_2, \pi_2^{\mathsf{NIZK}}\big)$, outputs 1 if $\tilde{c}_1 = \tilde{c}_2$, and 0 otherwise.

The following proposition states that our scheme is correct and strongly detectable, and is proven in the full version of the paper.

Proposition 2. *Let* ACE *be the scheme from above. Then,* ACE *is perfectly correct, i.e.,* $\mathsf{Adv}_{\mathsf{ACE},\mathcal{A}}^{\mathsf{ACE\text{-}CORR}} = 0$ *for all* $\mathcal{A}$*. Moreover, if* F *is pseudorandom and* sPKE *is unrestricted strongly robust, then* ACE *is strongly detectable.*

The security of our scheme is summarized by the theorem below, which we prove in the full version.

Theorem 2. *If* F *is pseudorandom,* NIZK *is zero-knowledge and extractable,* Sig *is EUF-CMA secure, and* sPKE *is IND-CCA, IK-CCA, SAN-CCA, USROB, and UPD-CTXT secure and has negligible encryption-key collision probability, then the scheme* ACE *from above is PRV-CCA, sANON-CCA, SAN-CCA, UDEC, and RR secure, and has NDTCT-FENC.*

6.2 Lifting Equality to Disjunction of Equalities

We finally show how an ACE scheme for equality, as the one from Sect. 6.1, can be used to construct a scheme for the policy $P_{\mathsf{DEq}} : \mathcal{D}^\ell \times \mathcal{D}^\ell \to \{0,1\}$ with

$$P_{\mathsf{DEq}}\big(\mathbf{x} = (x_1, \ldots, x_\ell), \mathbf{y} = (y_1, \ldots, y_\ell)\big) = 1 \;:\Longleftrightarrow\; \bigvee_{i=1}^{\ell} x_i = y_i,$$

where $\mathcal{D}$ is some finite set and $\ell \in \mathbb{N}$.[3] This policy can for example be used to implement the no read-up and now write-down principle ($P(i,j) = 1 \Leftrightarrow i \leq j$) from the Bell-LaPadula model [2] via an appropriate encoding of the roles [9].

The intuition of our construction is as follows. A key for a role $\mathbf{x} = (x_1, \ldots, x_\ell)$ contains one key of the ACE scheme for equality for each component x_i of the role vector. To encrypt a message, this message is encrypted with each of these keys. To decrypt, one tries to decrypt each ciphertext component with the corresponding key. If at least one component of the sender and receiver roles match (i.e., if the policy is satisfied), one of the decryptions is successful. So far, the construction is identical to the one by Fuchsbauer et al. [9]. That construction is, however, not role-respecting, since a dishonest sender with keys for more than one role can arbitrarily mix the components of the keys for the encryption. Moreover, the construction does not guarantee uniform decryption, because different messages can be encrypted in different components. We fix these issues using the same techniques we used in our construction of the scheme for equality, i.e., we add a signature of the key vector to the encryption keys, sign the ciphertexts, and require a zero-knowledge proof of knowledge that a valid key combination was used to encrypt the same message for each component and that all signatures are valid.

Our construction. Let $\mathsf{ACE}_=$ be an ACE with modification detection scheme for the equality predicate on $\mathcal{D} \times [\ell]$, let Sig be a signature scheme, let F be a PRF, and let NIZK be a NIZK proof of knowledge system for the language $L := \{x \mid \exists w \ (x, w) \in R\}$, where the relation R is defined as follows: for $x = (vk^{\mathsf{Sig}}, c_1, \ldots, c_\ell)$ and $w = (ek^{\mathsf{ACE}_=}_{(x_1,1)}, \ldots, ek^{\mathsf{ACE}_=}_{(x_\ell,\ell)}, m, r_1, \ldots, r_\ell, vk^{\mathsf{Sig}}_{\mathbf{x}}, \sigma^{\mathsf{Sig}}_{\mathbf{x}}, \sigma^{\mathsf{Sig}}_{c})$, $(x, w) \in R$ if and only if

$$\bigwedge_{i=1}^{\ell} c_i = \mathsf{ACE}_=.\mathsf{Enc}(ek^{\mathsf{ACE}_=}_{(x_i,i)}, m; r_i) \ \wedge \ \mathsf{Sig}.\mathsf{Ver}(vk^{\mathsf{Sig}}_{\mathbf{x}}, [c_1, \ldots, c_\ell], \sigma^{\mathsf{Sig}}_{c}) = 1$$

$$\wedge \ \mathsf{Sig}.\mathsf{Ver}(vk^{\mathsf{Sig}}, [ek^{\mathsf{ACE}_=}_{(x_1,1)}, \ldots, ek^{\mathsf{ACE}_=}_{(x_\ell,\ell)}, vk^{\mathsf{Sig}}_{\mathbf{x}}], \sigma^{\mathsf{Sig}}_{\mathbf{x}}) = 1.$$

We define an ACE scheme $\mathsf{ACE}_{\mathsf{DEq}}$ as follows:

Setup: On input a security parameter 1^κ and the policy P_{DEq}, $\mathsf{ACE}_{\mathsf{DEq}}.\mathsf{Setup}$ picks a random key K for F and runs

$$\begin{aligned}(msk^{\mathsf{ACE}_=}, sp^{\mathsf{ACE}_=}) &\leftarrow \mathsf{ACE}_=.\mathsf{Setup}(1^\kappa),\\ (vk^{\mathsf{Sig}}, sk^{\mathsf{Sig}}) &\leftarrow \mathsf{Sig}.\mathsf{Gen}(1^\kappa),\\ crs^{\mathsf{NIZK}} &\leftarrow \mathsf{NIZK}.\mathsf{Gen}(1^\kappa).\end{aligned}$$

[3] In this section, we denote roles by $\mathbf{x}$ and $\mathbf{y}$ instead of i and j. To be compatible with our definitions that consider policies $[n] \times [n] \to \{0, 1\}$, one needs to identify elements of $\mathcal{D}^\ell$ with numbers in $[n]$. We will ignore this technicality to simplify the presentation.

It outputs the master secret key $msk^{\mathsf{ACE_{DEq}}} := \big(K, msk^{\mathsf{ACE}_=}, vk^{\mathsf{Sig}}, sk^{\mathsf{Sig}}, crs^{\mathsf{NIZK}}\big)$ and the sanitizer parameters $sp^{\mathsf{ACE_{DEq}}} := \big(sp^{\mathsf{ACE}_=}, vk^{\mathsf{Sig}}, crs^{\mathsf{NIZK}}\big)$.

Key generation: The algorithm $\mathsf{ACE_{DEq}.Gen}$ on input a master secret key $msk^{\mathsf{ACE_{DEq}}} = \big(K, msk^{\mathsf{ACE}_=}, vk^{\mathsf{Sig}}, sk^{\mathsf{Sig}}, crs^{\mathsf{NIZK}}\big)$, a role $\mathbf{x} \in \mathcal{D}^\ell$, and type $\mathtt{sen}$, generates

$$\begin{aligned} ek^{\mathsf{ACE}_=}_{(x_i,i)} &\leftarrow \mathsf{ACE}_=.\mathsf{Gen}\big(msk^{\mathsf{ACE}_=}, (x_i, i), \mathtt{sen}\big) \quad (\text{for } i \in [\ell]), \\ \big(vk^{\mathsf{Sig}}_{\mathbf{x}}, sk^{\mathsf{Sig}}_{\mathbf{x}}\big) &\leftarrow \mathsf{Sig.Gen}(1^\kappa; F_K([\mathbf{x}, 0])), \\ \sigma^{\mathsf{Sig}}_{\mathbf{x}} &\leftarrow \mathsf{Sig.Sign}\big(sk^{\mathsf{Sig}}, \big[ek^{\mathsf{ACE}_=}_{(x_1,1)}, \ldots, ek^{\mathsf{ACE}_=}_{(x_\ell,\ell)}, vk^{\mathsf{Sig}}_{\mathbf{x}}\big]; F_K([\mathbf{x}, 1])\big), \end{aligned}$$

and outputs the encryption key $ek^{\mathsf{ACE_{DEq}}}_{\mathbf{x}} := \big(vk^{\mathsf{Sig}}, ek^{\mathsf{ACE}_=}_{(x_1,1)}, \ldots, ek^{\mathsf{ACE}_=}_{(x_\ell,\ell)}, vk^{\mathsf{Sig}}_{\mathbf{x}}, sk^{\mathsf{Sig}}_{\mathbf{x}}, \sigma^{\mathsf{Sig}}_{\mathbf{x}}, crs^{\mathsf{NIZK}}\big)$; on input $msk^{\mathsf{ACE_{DEq}}}$, a role $\mathbf{y} \in \mathcal{D}^\ell$, and the type $\mathtt{rec}$, it generates for $i \in [\ell]$,

$$dk^{\mathsf{ACE}_=}_{(y_i,i)} \leftarrow \mathsf{ACE}_=.\mathsf{Gen}\big(msk^{\mathsf{ACE}_=}, (y_i, i), \mathtt{rec}\big),$$

and outputs the decryption key $dk^{\mathsf{ACE_{DEq}}}_{\mathbf{y}} := \big(dk^{\mathsf{ACE}_=}_{(y_1,1)}, \ldots, dk^{\mathsf{ACE}_=}_{(y_\ell,\ell)}\big)$.

Encryption: On input an encryption key $ek^{\mathsf{ACE_{DEq}}}_{\mathbf{x}} = \big(vk^{\mathsf{Sig}}, ek^{\mathsf{ACE}_=}_{(x_1,1)}, \ldots, ek^{\mathsf{ACE}_=}_{(x_\ell,\ell)}, vk^{\mathsf{Sig}}_{\mathbf{x}}, sk^{\mathsf{Sig}}_{\mathbf{x}}, \sigma^{\mathsf{Sig}}_{\mathbf{x}}, crs^{\mathsf{NIZK}}\big)$ and a message $m \in \mathcal{M}^{\mathsf{ACE_{DEq}}}$, $\mathsf{ACE_{DEq}.Enc}$ samples randomness $r_1, \ldots, r_\ell$ and computes

$$\begin{aligned} c_i &\leftarrow \mathsf{ACE}_=.\mathsf{Enc}\big(ek^{\mathsf{ACE}_=}_{(x_i,i)}, m; r_i\big) \quad (\text{for } i \in [\ell]), \\ \sigma^{\mathsf{Sig}}_c &\leftarrow \mathsf{Sig.Sign}\big(sk^{\mathsf{Sig}}_{\mathbf{x}}, [c_1, \ldots, c_\ell]\big), \\ \pi^{\mathsf{NIZK}} &\leftarrow \mathsf{NIZK.Prove}\big(crs^{\mathsf{NIZK}}, x := \big(vk^{\mathsf{Sig}}, c_1, \ldots, c_\ell\big), \\ &\qquad w := \big(ek^{\mathsf{ACE}_=}_{(x_1,1)}, \ldots, ek^{\mathsf{ACE}_=}_{(x_\ell,\ell)}, m, r_1, \ldots, r_\ell, vk^{\mathsf{Sig}}_{\mathbf{x}}, \sigma^{\mathsf{Sig}}_{\mathbf{x}}, \sigma^{\mathsf{Sig}}_c\big)\big). \end{aligned}$$

It outputs the ciphertext $c := \big(c_1, \ldots, c_\ell, \pi^{\mathsf{NIZK}}\big)$.

Sanitization: On input sanitizer parameters $sp^{\mathsf{ACE_{DEq}}} = \big(sp^{\mathsf{ACE}_=}, vk^{\mathsf{Sig}}, crs^{\mathsf{NIZK}}\big)$ and a ciphertext $c = \big(c_1, \ldots, c_\ell, \pi^{\mathsf{NIZK}}\big)$, the algorithm $\mathsf{ACE_{DEq}.San}$ checks whether $\mathsf{NIZK.Ver}\big(crs^{\mathsf{NIZK}}, x := \big(vk^{\mathsf{Sig}}, c_1, \ldots, c_\ell\big), \pi^{\mathsf{NIZK}}\big) = 1$. If this is the case, it runs $c'_i \leftarrow \mathsf{ACE}_=.\mathsf{San}(c_i)$ for $i \in [\ell]$. If $c'_i \neq \bot$ for all $i \in [\ell]$, it outputs the sanitized ciphertext $c' := \big(c'_1, \ldots, c'_\ell\big)$. If the verification fails or any of the sanitized ciphertexts is $\bot$, it outputs $\bot$.

Decryption: On input a decryption key $dk^{\mathsf{ACE_{DEq}}}_{\mathbf{y}} = \big(dk^{\mathsf{ACE}_=}_{(y_1,1)}, \ldots, dk^{\mathsf{ACE}_=}_{(y_\ell,\ell)}\big)$ and a sanitized ciphertext $c' := \big(c'_1, \ldots, c'_\ell\big)$, the algorithm $\mathsf{ACE_{DEq}.Dec}$ computes for $i \in [\ell]$ the message $m_i \leftarrow \mathsf{ACE}_=.\mathsf{Dec}\big(dk^{\mathsf{ACE}_=}_{(y_i,i)}, c'_i\big)$. If $m_i \neq \bot$ for some $i \in [\ell]$, $\mathsf{ACE_{DEq}.Dec}$ outputs the first such m_i; otherwise it outputs $\bot$.

Modification detection: On input sanitizer parameters $sp^{\mathsf{ACE_{DEq}}} := \big(sp^{\mathsf{ACE}_=}, vk^{\mathsf{Sig}}, crs^{\mathsf{NIZK}}\big)$ and two ciphertexts $c = \big(c_1, \ldots, c_\ell, \pi^{\mathsf{NIZK}}\big)$ and $\tilde{c} := \big(\tilde{c}_1, \ldots, \tilde{c}_\ell, \tilde{\pi}^{\mathsf{NIZK}}\big)$, $\mathsf{ACE_{DEq}.DMod}$ checks for $i \in [\ell]$ if $\mathsf{ACE}_=.\mathsf{DMod}\big(sp^{\mathsf{ACE}_=}, c_i, \tilde{c}_i\big) = 1$. If this is the case for some $i \in [\ell]$, it outputs 1; otherwise, it outputs 0.

Weak and strong anonymity. As we show below, our scheme enjoys weak anonymity. It is easy to see that it does not have strong anonymity: Given a decryption key for the role $(1, 2)$, one can decrypt ciphertexts encrypted under a key for the roles $(1, 1)$ and $(2, 2)$. One does, however, also learn which of the two components decrypted successfully. If it is the first one, the sender role must be $(1, 1)$, if it is the second one, the sender role must be $(2, 2)$. For similar reasons, we do not achieve strong sanitization security.

A similar construction can be used to achieve strong anonymity for less expressive policies: If a sender role still corresponds to a vector $(x_1, \ldots, x_\ell) \in \mathcal{D}^\ell$ but a receiver role only to one component $(j, y) \in [\ell] \times \mathcal{D}$, one can consider the policy that allows to receive if $x_j = y$. Now, we do not need several components for the decryption key and the problem sketched above disappears.

Proposition 3. *If* $\mathsf{ACE}_=$ *is correct and detectable, then the scheme* $\mathsf{ACE}_{\mathsf{DEq}}$ *from above is correct and detectable. If* $\mathsf{ACE}_=$ *is strongly detectable, then* $\mathsf{ACE}_{\mathsf{DEq}}$ *is also strongly detectable. More precisely, for all probabilistic algorithms* $\mathcal{A}$*, there exist probabilistic algorithms* $\mathcal{A}_{\mathsf{corr}}$*,* $\mathcal{A}_{\mathsf{dtct}}$*,* $\mathcal{A}'_{\mathsf{dtct}}$*, and* $\mathcal{A}_{\mathsf{sdtct}}$ *such that*

$$\begin{aligned}
\mathsf{Adv}^{\mathsf{ACE\text{-}CORR}}_{\mathsf{ACE}_{\mathsf{DEq}},\mathcal{A}} &\le \mathsf{Adv}^{\mathsf{ACE\text{-}CORR}}_{\mathsf{ACE}_=,\mathcal{A}_{\mathsf{corr}}} + (\ell - 1) \cdot \mathsf{Adv}^{\mathsf{ACE\text{-}DTCT}}_{\mathsf{ACE}_=,\mathcal{A}_{\mathsf{dtct}}}, \\
\mathsf{Adv}^{\mathsf{ACE\text{-}DTCT}}_{\mathsf{ACE}_{\mathsf{DEq}},\mathcal{A}} &\le \ell \cdot \mathsf{Adv}^{\mathsf{ACE\text{-}DTCT}}_{\mathsf{ACE}_=,\mathcal{A}'_{\mathsf{dtct}}}, \\
\mathsf{Adv}^{\mathsf{ACE\text{-}sDTCT}}_{\mathsf{ACE}_{\mathsf{DEq}},\mathcal{A}} &\le \ell \cdot \mathsf{Adv}^{\mathsf{ACE\text{-}sDTCT}}_{\mathsf{ACE}_=,\mathcal{A}_{\mathsf{sdtct}}}.
\end{aligned}$$

See the full version of this paper for proofs of this proposition and the following theorem, which summarizes the security properties of our scheme.

Theorem 3. *If* F *is pseudorandom,* NIZK *is zero-knowledge and extractable,* Sig *is EUF-CMA secure, and* $\mathsf{ACE}_=$ *is perfectly correct, strongly detectable, has NDTCT-FENC, and is PRV-CCA, wANON-CCA, SAN-CCA, RR, and UDEC secure, then the scheme* $\mathsf{ACE}_{\mathsf{DEq}}$ *from above has NDTCT-FENC and is PRV-CCA, wANON-CCA, SAN-CCA, RR, and UDEC secure.*

7 Conclusion and Directions for Future Work

In this paper, we have critically revisited existing notions for access control encryption, proposed stronger security definitions, and presented a new scheme that provably achieves our strong requirements. The need for stronger notions is not only a theoretical one as we have shown: In particular, we have described a practical attack based on the observation that a semi-honest sanitizer might leak an unsanitized ciphertext to a dishonest party.

An important question is whether all realistic attacks are excluded by our definitions. Furthermore, we would like to understand the fundamental limits of ACE. This includes investigating in which scenarios it can or cannot be used. To settle these questions, the authors are currently working on a theoretical model to capture the use case of ACE in a simulation-based framework. Another interesting research direction is to find more efficient schemes for useful policies.

References

1. Abdalla, M., Bellare, M., Neven, G.: Robust encryption. In: Micciancio, D. (ed.) TCC 2010. LNCS, vol. 5978, pp. 480–497. Springer, Heidelberg (2010). https://doi.org/10.1007/978-3-642-11799-2_28
2. Bell, D.E., LaPadula, L.J.: Secure computer systems: mathematical foundations. Technical report MTR-2547, MITRE (1973)
3. Bellare, M., Boldyreva, A., Desai, A., Pointcheval, D.: Key-privacy in public-key encryption. In: Boyd, C. (ed.) ASIACRYPT 2001. LNCS, vol. 2248, pp. 566–582. Springer, Heidelberg (2001). https://doi.org/10.1007/3-540-45682-1_33
4. Boneh, D., Sahai, A., Waters, B.: Functional encryption: definitions and challenges. In: Ishai, Y. (ed.) TCC 2011. LNCS, vol. 6597, pp. 253–273. Springer, Heidelberg (2011). https://doi.org/10.1007/978-3-642-19571-6_16
5. Canetti, R., Krawczyk, H., Nielsen, J.B.: Relaxing chosen-ciphertext security. In: Boneh, D. (ed.) CRYPTO 2003. LNCS, vol. 2729, pp. 565–582. Springer, Heidelberg (2003). https://doi.org/10.1007/978-3-540-45146-4_33
6. Damgård, I., Haagh, H., Orlandi, C.: Access control encryption: enforcing information flow with cryptography. In: Hirt, M., Smith, A. (eds.) TCC 2016. LNCS, vol. 9986, pp. 547–576. Springer, Heidelberg (2016). https://doi.org/10.1007/978-3-662-53644-5_21
7. Elgamal, T.: A public key cryptosystem and a signature scheme based on discrete logarithms. IEEE Trans. Inf. Theory **31**(4), 469–472 (1985)
8. Farshim, P., Libert, B., Paterson, K.G., Quaglia, E.A.: Robust encryption, revisited. In: Kurosawa, K., Hanaoka, G. (eds.) PKC 2013. LNCS, vol. 7778, pp. 352–368. Springer, Heidelberg (2013). https://doi.org/10.1007/978-3-642-36362-7_22
9. Fuchsbauer, G., Gay, R., Kowalczyk, L., Orlandi, C.: Access control encryption for equality, comparison, and more. In: Fehr, S. (ed.) PKC 2017. LNCS, vol. 10175, pp. 88–118. Springer, Heidelberg (2017). https://doi.org/10.1007/978-3-662-54388-7_4
10. Golle, P., Jakobsson, M., Juels, A., Syverson, P.: Universal re-encryption for mixnets. In: Okamoto, T. (ed.) CT-RSA 2004. LNCS, vol. 2964, pp. 163–178. Springer, Heidelberg (2004). https://doi.org/10.1007/978-3-540-24660-2_14
11. Goyal, V., Pandey, O., Sahai, A., Waters, B.: Attribute-based encryption for fine-grained access control of encrypted data. In: Proceedings of the 13th ACM Conference on Computer and Communications Security, CCS 2006, pp. 89–98. ACM (2006)
12. Groth, J.: Rerandomizable and replayable adaptive chosen ciphertext attack secure cryptosystems. In: Naor, M. (ed.) TCC 2004. LNCS, vol. 2951, pp. 152–170. Springer, Heidelberg (2004). https://doi.org/10.1007/978-3-540-24638-1_9
13. Kim, S., Wu, D.J.: Access control encryption for general policies from standard assumptions. In: Takagi, T., Peyrin, T. (eds.) ASIACRYPT 2017. LNCS, vol. 10624, pp. 471–501. Springer, Heidelberg (2017)
14. Naor, M., Yung, M.: Public-key cryptosystems provably secure against chosen ciphertext attacks. In: Proceedings of the Twenty-second Annual ACM Symposium on Theory of Computing, STOC 1990, pp. 427–437. ACM (1990)
15. Prabhakaran, M., Rosulek, M.: Rerandomizable RCCA encryption. In: Menezes, A. (ed.) CRYPTO 2007. LNCS, vol. 4622, pp. 517–534. Springer, Heidelberg (2007). https://doi.org/10.1007/978-3-540-74143-5_29
16. Sahai, A.: Non-malleable non-interactive zero knowledge and adaptive chosen-ciphertext security. In: 40th Annual Symposium on Foundations of Computer Science, pp. 543–553 (1999)

17. Sahai, A., Waters, B.: Fuzzy identity-based encryption. In: Cramer, R. (ed.) EUROCRYPT 2005. LNCS, vol. 3494, pp. 457–473. Springer, Heidelberg (2005). https://doi.org/10.1007/11426639_27
18. Tan, G., Zhang, R., Ma, H., Tao, Y.: Access control encryption based on LWE. In: Proceedings of the 4th ACM International Workshop on ASIA Public-Key Cryptography, APKC 2017, pp. 43–50. ACM (2017)

Adaptive Oblivious Transfer with Access Control from Lattice Assumptions

Benoît Libert[1,2(✉)], San Ling[3], Fabrice Mouhartem[2], Khoa Nguyen[3], and Huaxiong Wang[3]

[1] CNRS, Laboratoire LIP, Lyon, France
benoit.libert@ens-lyon.fr
[2] ENS de Lyon, Laboratoire LIP (U. Lyon, CNRS, ENSL, INRIA, UCBL), Lyon, France
[3] School of Physical and Mathematical Sciences, Nanyang Technological University, Singapore, Singapore

Abstract. Adaptive oblivious transfer (OT) is a protocol where a sender initially commits to a database $\{M_i\}_{i=1}^N$. Then, a receiver can query the sender up to k times with private indexes $\rho_1, \ldots, \rho_k$ so as to obtain $M_{\rho_1}, \ldots, M_{\rho_k}$ and nothing else. Moreover, for each $i \in [k]$, the receiver's choice ρ_i may depend on previously obtained messages $\{M_{\rho_j}\}_{j<i}$. Oblivious transfer with access control (OT-AC) is a flavor of adaptive OT where database records are protected by distinct access control policies that specify which credentials a receiver should obtain in order to access each M_i. So far, all known OT-AC protocols only support access policies made of conjunctions or rely on *ad hoc* assumptions in pairing-friendly groups (or both). In this paper, we provide an OT-AC protocol where access policies may consist of any branching program of polynomial length, which is sufficient to realize any access policy in $\mathsf{NC1}$. The security of our protocol is proved under the Learning-with-Errors (LWE) and Short-Integer-Solution (SIS) assumptions. As a result of independent interest, we provide protocols for proving the correct evaluation of a committed branching program on a committed input.

Keywords: Lattice assumptions · Standard assumptions · Zero-knowledge arguments · Adaptive oblivious transfer

1 Introduction

Oblivious transfer (OT) is a central cryptographic primitive coined by Rabin [45] and extended by Even *et al.* [20]. It involves a sender S with a database of messages $M_1, \ldots, M_N$ and a receiver R with an index $\rho \in \{1, \ldots, N\}$. The protocol allows R to retrieve the ρ-th entry M_ρ from S without letting S infer anything on R's choice ρ. Moreover, R only obtains M_ρ learns nothing about $\{M_i\}_{i \neq \rho}$.

In its adaptive flavor [40], OT allows the receiver to interact k times with S to retrieve $M_{\rho_1}, \ldots, M_{\rho_k}$ in such a way that, for each $i \in \{2, \ldots, k\}$, the i-th index ρ_i may depend on the messages $M_{\rho_1}, \ldots, M_{\rho_{i-1}}$ previously obtained by R.

T. Takagi and T. Peyrin (Eds.): ASIACRYPT 2017, Part I, LNCS 10624, pp. 533–563, 2017.
https://doi.org/10.1007/978-3-319-70694-8_19

OT is known to be a complete building block for cryptography (see, e.g., [24]) in that, if it can be realized, then any secure multiparty computation can be. In its adaptive variant, OT is motivated by applications in privacy-preserving access to sensitive databases (e.g., medical records or financial data) stored in encrypted form on remote servers, oblivious searches or location-based services.

As far as efficiency goes, adaptive OT protocols should be designed in such a way that, after an inevitable initialization phase with linear communication complexity in N and the security parameter λ, the complexity of each transfer is at most poly-logarithmic in N. At the same time, this asymptotic efficiency should not come at the expense of sacrificing ideal security properties. The most efficient adaptive OT protocols that satisfy the latter criterion stem from the work of Camenisch et al. [13] and its follow-ups [26–28].

In its basic form, (adaptive) OT does not restrict in any way the population of users who can obtain specific records. In many sensitive databases (e.g., DNA databases or patients' medical history), however, not all users should be able to download all records: it is vital access to certain entries be conditioned on the receiver holding suitable credentials delivered by authorities. At the same time, privacy protection mandates that authorized users be able to query database records while leaking as little as possible about their interests or activities. In medical datasets, for example, the specific entries retrieved by a given doctor could reveal which disease his patients are suffering from. In financial or patent datasets, the access pattern of a company could betray its investment strategy or the invention it is developing. In order to combine user-privacy and fine-grained database security, it is thus desirable to enrich adaptive OT protocols with refined access control mechanisms in many of their natural use cases.

This motivated Camenisch et al. [11] to introduce a variant named *oblivious transfer with access control* (OT-AC), where each database record is protected by a different access control policy $P : \{0,1\}^* \to \{0,1\}$. Based on their attributes, users can obtain credentials generated by pre-determined authorities, which entitle them to anonymously retrieve database records of which the access policy accepts their certified attributes: in other words, the user can only download the records for which he has a valid credential Cred_x for an attribute string $x \in \{0,1\}^*$ such that $P(x) = 1$. During the transfer phase, the user demonstrates possession of a pair (Cred_x, x) and simultaneously convinces the sender that he is querying some record M_ρ associated with a policy P such that $P(x) = 1$. The only information that the database holder eventually learns is that some user retrieved some record which he was authorized to obtain.

Camenisch *et al.* formalized the OT-AC primitive and provided a construction in groups with a bilinear map [11]. While efficient, their solution "only" supports access policies consisting of conjunctions: each policy P is specified by a list of attributes that a given user should obtain a credential for in order to complete the transfer. Several subsequent works [10,12,50] considered more expressive access policies while even hiding the access policies in some cases [10,12]. Unfortunately, all of them rely on non-standard assumptions (known as "q-type assumptions") in groups with a bilinear maps. For the sake of not putting all one's eggs in the same basket, a primitive as powerful as OT-AC ought to have alternative realizations based on firmer foundations.

In this paper, we propose a solution based on lattice assumptions where access policies consist of any branching program of width 5, which is known [6] to suffice for the realization of any access policy in $\mathsf{NC1}$. As a result of independent interest, we provide protocols for proving the correct evaluation of a committed branching program. More precisely, we give zero-knowledge arguments for demonstrating possession of a secret input $\mathbf{x} \in \{0,1\}^\kappa$ and a secret (and possibly certified) branching program BP such that $\mathsf{BP}(\mathbf{x}) = 1$.

Related Work. Oblivious transfer with adaptive queries dates back to the work of Naor and Pinkas [40], which requires $O(\log N)$ interaction rounds per transfer. Naor and Pinkas [42] also gave generic constructions of (adaptive) k-out-of-N OT from private information retrieval (PIR) [16]. The constructions of [40,42], however, are only secure in the half-simulation model, where simulation-based security is only considered for one of the two parties (receiver security being formalized in terms of a game-based definition). Moreover, the constructions of Adaptive OT from PIR [42] require a complexity $O(N^{1/2})$ at each transfer where Adaptive OT allows for $O(\log N)$ cost. Before 2007, many OT protocols (e.g., [3,41,49]) were analyzed in terms of half-simulation.

While several efficient fully simulatable protocols appeared the last 15 years (e.g., [37,44] and references therein), full simulatability remained elusive in the adaptive k-out-of-N setting [40] until the work [13] of Camenisch, Neven and shelat, who introduced the "assisted decryption" paradigm. The latter consists in having the sender obliviously decrypt a re-randomized version of one of the original ciphertexts contained in the database. This technique served as a blueprint for many subsequent protocols [26–28,31], including those with access control [1,10–12] and those presented in this paper. In the adaptive k-out-of-N setting (which we denote as $\mathcal{OT}_{k\times 1}^N$), the difficulty is to achieve full simulatability without having to transmit a $O(N)$ bits at each transfer. To our knowledge, except the oblivious-PRF-based approach of Jarecki and Liu [31], all known fully simulatable $\mathcal{OT}_{k\times 1}^N$ protocols rely on bilinear maps.[1]

A number of works introduced various forms of access control in OT. Priced OT [3] assigns variable prices to all database records. In conditional OT [18], access to a record is made contingent on the user's secret satisfying some predicate. Restricted OT [29] explicitly protects each record with an independent access policy. Still, none of these OT flavors aims at protecting the anonymity of users. The model of Coull et al. [17] does consider user anonymity via stateful credentials. For the applications of OT-AC, it would nevertheless require re-issuing user credentials at each transfer.

While efficient, the initial OT-AC protocol of Camenisch *et al.* [11] relies on non-standard assumptions in groups with a bilinear map and only realizes access policies made of conjunctions. Abe *et al.* [1] gave a different protocol which they

[1] Several pairing-free candidates were suggested in [33] but, as pointed out in [28], they cannot achieve full simulatability in the sense of [13]. In particular, the sender can detect if the receiver fetches the same record in two distinct transfers.

proved secure under more standard assumptions in the universal composability framework [14]. Their policies, however, remain limited to conjunctions. It was mentioned in [1,11] that disjunctions and DNF formulas can be handled by duplicating database entries. Unfortunately, this approach rapidly becomes prohibitively expensive in the case of (t, n)-threshold policies with $t \approx n/2$. Moreover, securing the protocol against malicious senders requires them to prove that all duplicates encrypt the same message. More expressive policies were considered by Zhang *et al.* [50] who gave a construction based on attribute-based encryption [47] that extends to access policies expressed by any Boolean formulas (and thus $\mathsf{NC1}$ circuits). Camenisch et al. [12] generalized the OT-AC functionality so as to hide the access policies. In [10], Camenisch *et al.* gave a more efficient construction with hidden policies based on the attribute-based encryption scheme of [43]. At the expense of a proof in the generic group model, [10] improves upon the expressiveness of [12] in that its policies extend into CNF formulas. While the solutions of [10,12] both hide the access policies to users (and the successful termination of transfers to the database), their policies can only live in a proper subset of $\mathsf{NC1}$. As of now, threshold policies can only be efficiently handled by the ABE-based construction of Zhang *et al.* [50], which requires *ad hoc* assumptions in groups with a bilinear map.

OUR RESULTS AND TECHNIQUES. We describe the first OT-AC protocol based on lattice assumptions. Our construction supports access policies consisting of any branching program of width 5 and polynomial length – which suffices to realize any $\mathsf{NC1}$ circuit – and prove it secure under the SIS and LWE assumptions. We thus achieve the same level of expressiveness as [50] with the benefit of relying on well-studied assumptions. In its initial version, our protocol requires the database holder to communicate $\Theta(N)$ bits to each receiver so as to prove that the database is properly encrypted. In the random oracle model, we can eliminate this burden via the Fiat-Shamir heuristic and make the initialization cost linear in the database size N, regardless of the number of users.

As a first step, we build an ordinary $\mathcal{OT}_{k\times 1}^{N}$ protocol (i.e., without access control) via the "assisted decryption" approach of [13]. In short, the sender encrypts all database entries using a semantically secure cryptosystem. At each transfer, the receiver gets the sender to obliviously decrypt one of the initial ciphertexts without learning which one. Security against malicious adversaries is achieved by adding zero-knowledge (ZK) or witness indistinguishable (WI) arguments that the protocol is being followed. The desired ZK or WI arguments are obtained using the techniques of [34] and we prove that this basic protocol satisfies the full imulatability definitions of [13] under the SIS and LWE assumptions. To our knowledge, this protocol is also the first $\mathcal{OT}_{k\times 1}^{N}$ realization to achieve the standard simulation-based security requirements under lattice assumptions.

So far, all known "beyond-conjunctions" OT-AC protocols [10,50] rely on ciphertext-policy attribute-based encryption (CP-ABE) and proceed by attaching each database record to a CP-ABE ciphertext. Our construction departs from the latter approach for two reasons. First, the only known LWE-based CP-ABE schemes are obtained by applying a universal circuit to a key-policy system, making them very inefficient. Second, the ABE-based approach requires a fully secure ABE (i.e., selective security and semi-adaptive security are insufficient) since the access policies are dictated by the environment *after* the generation of the issuer's public key, which contains the public parameters of the underlying ABE in [10,50]. Even with the best known LWE-based ABE candidates [25], a direct adaptation of the techniques in [10,50] would incur to rely on a complexity leveraging argument [8] and a universal circuit. Instead, we take a different approach and directly prove in a zero-knowledge manner the correct evaluation of a committed branching program for a hidden input.

At a high level, our OT-AC protocol works as follows. For each $i \in [N]$, the database entry $M_i \in \{0,1\}^t$ is associated with branching program BP_i. In the initialization step, the database holder generates a Regev ciphertext $(\mathbf{a}_i, \mathbf{b}_i)$ of M_i, and issues a certificate for the pair $((\mathbf{a}_i, \mathbf{b}_i), \mathsf{BP}_i)$, using the signature scheme from [34]. At each transfer, the user U who wishes to get a record $\rho \in [N]$ must obtain a credential $\mathsf{Cred}_{\mathbf{x}}$ for an attribute string $\mathbf{x} \in \{0,1\}^\kappa$ such that $\mathsf{BP}_\rho(\mathbf{x}) = 1$. Then, U modifies $(\mathbf{a}_\rho, \mathbf{b}_\rho)$ into an encryption of $M_\rho \oplus \mu \in \{0,1\}^t$, for some random string $\mu \in \{0,1\}^t$, and re-randomizes the resulting ciphertext into a fresh encryption $(\mathbf{c}_0, \mathbf{c}_1)$ of $M_\rho \oplus \mu$. At this point, U proves that $(\mathbf{c}_0, \mathbf{c}_1)$ was obtained by transforming one of the original ciphertexts $\{(\mathbf{a}_i, \mathbf{b}_i)\}_{i=1}^N$ by arguing possession of a valid certificate for $((\mathbf{a}_\rho, \mathbf{b}_\rho), \mathsf{BP}_\rho)$ and knowledge of all randomness used in the transformation that yields $(\mathbf{c}_0, \mathbf{c}_1)$. At the same time, U proves possession of $\mathsf{Cred}_{\mathbf{x}}$ for a string $\mathbf{x}$ which is accepted by the committed BP_ρ. To demonstrate these statements in zero-knowledge, we develop recent techniques [34,36,39] for lattice-based analogues [32,38] of Stern's protocol [48].

As a crucial component of our OT-AC protocol, we need to prove knowledge of an input $\mathbf{x} = (x_0, \ldots, x_{\kappa-1})^\top \in \{0,1\}^\kappa$ satisfying a hidden BP of length L, where L and κ are polynomials in the security parameter. For each $\theta \in [L]$, we need to prove that the computation of the θ-th state

$$\eta_\theta = \pi_{\theta,0}(\eta_{\theta-1}) \cdot (1 - x_{\mathrm{var}(\theta)}) + \pi_{\theta,1}(\eta_{\theta-1}) \cdot x_{\mathrm{var}(\theta)}, \tag{1}$$

is done correctly, for permutations $\pi_{\theta,0}, \pi_{\theta,1} : [0,4] \to [0,4]$ and for integer $\mathrm{var}(\theta) \in [0, \kappa - 1]$ specified by BP. To date, equations of the form (1) have not been addressed in the context of zero-knowledge proofs for lattice-based cryptography. In this work, we are not only able to handle L such equations, but also manage to do so with a reasonable asymptotic cost.

In order to compute η_θ as in (1), we have to fetch the value $x_{\mathrm{var}(\theta)}$ in the input $(x_0, \ldots, x_{\kappa-1})^\top$ and provide evidence that the searching process is conducted honestly. If we perform a naive left-to-right search in the array $x_0, \ldots, x_{\kappa-1}$, the expected complexity is $\mathcal{O}(\kappa)$ for each step, and the total complexity amounts to $\mathcal{O}(L \cdot \kappa)$. If we instead perform a dichotomic search over $x_0, \ldots, x_{\kappa-1}$, we can decrease the complexity at each step down to $\mathcal{O}(\log \kappa)$. However, in this case, we need to prove in zero-knowledge a statement "I obtained $x_{\mathrm{var}(\theta)}$ by conducting a correct dichotomic search in my secret array."

We solve this problem as follows. For each $i \in [0, \kappa-1]$, we employ a SIS-based commitment scheme [32] to commit to x_i as com_i, and prove that the committed bits are consistent with the ones involved in the credential $\mathsf{Cred}_{\mathbf{x}}$ mentioned above. Then we build a SIS-based Merkle hash tree [36] of depth $\delta_\kappa = \lceil \log \kappa \rceil$ on top of the commitments $\mathsf{com}_0, \ldots, \mathsf{com}_{\kappa-1}$. Now, for each $\theta \in [L]$, we consider the binary representation $d_{\theta,1}, \ldots, d_{\theta,\delta_\kappa}$ of $\mathrm{var}(\theta)$. We then prove knowledge of a bit y_θ such that these conditions hold: "If one starts at the root of the tree and follows the path determined by the bits $d_{\theta,1}, \ldots, d_{\theta,\delta_\kappa}$, then one will reach the leaf associated with the commitment opened to y_θ." The idea is that, if the Merkle tree and the commitment scheme are both secure, then it should be true that $y_\theta = x_{\mathrm{var}(\theta)}$. In other words, this enables us to provably perform a "binary search" for $x_{\mathrm{var}(\theta)} = y_\theta$. Furthermore, this process can be done in zero-knowledge, by adapting the recent techniques from [36]. As a result, we obtain a protocol with communication cost just $\mathcal{O}(L \cdot \log \kappa + \kappa)$.

2 Background and Definitions

Vectors are denoted in bold lower-case letters and bold upper-case letters will denote matrices. The Euclidean and infinity norm of any vector $\mathbf{b} \in \mathbb{R}^n$ will be denoted by $\|\mathbf{b}\|$ and $\|\mathbf{b}\|_\infty$, respectively. The Euclidean norm of matrix $\mathbf{B} \in \mathbb{R}^{m \times n}$ with columns $(\mathbf{b}_i)_{i \leq n}$ is $\|\mathbf{B}\| = \max_{i \leq n} \|\mathbf{b}_i\|$. When $\mathbf{B}$ has full column-rank, we let $\widetilde{\mathbf{B}}$ denote its Gram-Schmidt orthogonalization.

When S is a finite set, we denote by $U(S)$ the uniform distribution over S, and by $x \hookleftarrow U(S)$ the action of sampling x according to this distribution. Finally for any integers A, B, N, we let $[N]$ and $[A, B]$ denote the sets $\{1, \ldots, N\}$ and $\{A, A+1, \ldots, B\}$, respectively.

2.1 Lattices

A lattice L is the set of integer linear combinations of linearly independent basis vectors $(\mathbf{b}_i)_{i \leq n}$ living in $\mathbb{R}^m$. We work with q-ary lattices, for some prime q.

Definition 1. *Let $m \geq n \geq 1$, a prime $q \geq 2$ and $\mathbf{A} \in \mathbb{Z}_q^{n \times m}$ and $\mathbf{u} \in \mathbb{Z}_q^n$, define $\Lambda_q(\mathbf{A}) := \{\mathbf{e} \in \mathbb{Z}^m \mid \exists \mathbf{s} \in \mathbb{Z}_q^n \text{ s.t. } \mathbf{A}^T \cdot \mathbf{s} = \mathbf{e} \bmod q\}$ as well as*

$$\Lambda_q^\perp(\mathbf{A}) := \{\mathbf{e} \in \mathbb{Z}^m \mid \mathbf{A} \cdot \mathbf{e} = \mathbf{0}^n \bmod q\},\ \Lambda_q^{\mathbf{u}}(\mathbf{A}) := \{\mathbf{e} \in \mathbb{Z}^m \mid \mathbf{A} \cdot \mathbf{e} = \mathbf{u} \bmod q\}.$$

For a lattice L, let $\rho_{\sigma,\mathbf{c}}(\mathbf{x}) = \exp(-\pi \|\mathbf{x} - \mathbf{c}\|^2 / \sigma^2)$ for a vector $\mathbf{c} \in \mathbb{R}^m$ and a real $\sigma > 0$. The discrete Gaussian of support L, center $\mathbf{c}$ and parameter σ is $D_{L,\sigma,\mathbf{c}}(\mathbf{y}) = \rho_{\sigma,\mathbf{c}}(\mathbf{y}) / \rho_{\sigma,\mathbf{c}}(L)$ for any $\mathbf{y} \in L$, where $\rho_{\sigma,\mathbf{c}}(L) = \sum_{\mathbf{x} \in L} \rho_{\sigma,\mathbf{c}}(\mathbf{x})$. The distribution centered in $\mathbf{c} = \mathbf{0}$ is denoted by $D_{L,\sigma}(\mathbf{y})$.

It is well known that one can efficiently sample from a Gaussian distribution with lattice support given a sufficiently short basis of the lattice.

Lemma 1 [9, Lemma 2.3]. *There exists a PPT algorithm* GPVSample *that takes as inputs a basis $\mathbf{B}$ of a lattice $L \subseteq \mathbb{Z}^n$ and a rational $\sigma \geq \|\widetilde{\mathbf{B}}\| \cdot \Omega(\sqrt{\log n})$, and outputs vectors $\mathbf{b} \in L$ with distribution $D_{L,\sigma}$.*

We also rely on the trapdoor generation algorithm of Alwen and Peikert [4], which refines the technique of Gentry *et al.* [22].

Lemma 2 [4, Theorem 3.2]. *There is a PPT algorithm* TrapGen *that takes as inputs* 1^n, 1^m *and an integer* $q \geq 2$ *with* $m \geq \Omega(n \log q)$, *and outputs a matrix* $\mathbf{A} \in \mathbb{Z}_q^{n \times m}$ *and a basis* $\mathbf{T_A}$ *of* $\Lambda_q^{\perp}(\mathbf{A})$ *such that* $\mathbf{A}$ *is within statistical distance* $2^{-\Omega(n)}$ *to* $U(\mathbb{Z}_q^{n \times m})$, *and* $\|\widetilde{\mathbf{T_A}}\| \leq \mathcal{O}(\sqrt{n \log q})$.

We use the basis delegation algorithm [15] that inputs a trapdoor for $\mathbf{A} \in \mathbb{Z}_q^{n \times m}$ and produces a trapdoor for any $\mathbf{B} \in \mathbb{Z}_q^{n \times m'}$ containing $\mathbf{A}$ as a submatrix. A technique from Agrawal *et al.* [2] is sometimes used in our proofs.

Lemma 3 [15, Lemma 3.2]. *There is a PPT algorithm* ExtBasis *that inputs* $\mathbf{B} \in \mathbb{Z}_q^{n \times m'}$ *whose first* m *columns span* $\mathbb{Z}_q^n$, *and a basis* $\mathbf{T_A}$ *of* $\Lambda_q^{\perp}(\mathbf{A})$ *where* $\mathbf{A} \in \mathbb{Z}_q^{n \times m}$ *is a submatrix of* $\mathbf{B}$, *and outputs a basis* $\mathbf{T_B}$ *of* $\Lambda_q^{\perp}(\mathbf{B})$ *with* $\|\widetilde{\mathbf{T_B}}\| \leq \|\widetilde{\mathbf{T_A}}\|$.

Lemma 4 [2, Theorem 19]. *There is a PPT algorithm* SampleRight *that inputs* $\mathbf{A}, \mathbf{C} \in \mathbb{Z}_q^{n \times m}$, *a small-norm* $\mathbf{R} \in \mathbb{Z}^{m \times m}$, *a short basis* $\mathbf{T_C} \in \mathbb{Z}^{m \times m}$ *of* $\Lambda_q^{\perp}(\mathbf{C})$, *a vector* $\mathbf{u} \in \mathbb{Z}_q^n$ *and a rational* σ *such that* $\sigma \geq \|\widetilde{\mathbf{T_C}}\| \cdot \Omega(\sqrt{\log n})$, *and outputs* $\mathbf{b} \in \mathbb{Z}^{2m}$ *such that* $\left[\mathbf{A} \mid \mathbf{A} \cdot \mathbf{R} + \mathbf{C}\right] \cdot \mathbf{b} = \mathbf{u} \bmod q$ *and with distribution statistically close to* $D_{L,\sigma}$ *where* $L = \{\mathbf{x} \in \mathbb{Z}^{2m} : \left[\mathbf{A} \mid \mathbf{A} \cdot \mathbf{R} + \mathbf{C}\right] \cdot \mathbf{x} = \mathbf{u} \bmod q\}$.

2.2 Hardness Assumptions

Definition 2. *Let* n, m, q, β *be functions of* $\lambda \in \mathbb{N}$. *The Short Integer Solution problem* $\mathsf{SIS}_{n,m,q,\beta}$ *is, given* $\mathbf{A} \hookleftarrow U(\mathbb{Z}_q^{n \times m})$, *find* $\mathbf{x} \in \Lambda_q^{\perp}(\mathbf{A})$ *with* $0 < \|\mathbf{x}\| \leq \beta$.

If $q \geq \sqrt{n}\beta$ and $m, \beta \leq \mathsf{poly}(n)$, then standard worst-case lattice problems with approximation factors $\gamma = \widetilde{\mathcal{O}}(\beta\sqrt{n})$ reduce to $\mathsf{SIS}_{n,m,q,\beta}$ (see, e.g., [22, Sect. 9]).

Definition 3. *Let* $n, m \geq 1$, $q \geq 2$, *and let* χ *be a probability distribution on* $\mathbb{Z}$. *For* $\mathbf{s} \in \mathbb{Z}_q^n$, *let* $A_{\mathbf{s},\chi}$ *be the distribution obtained by sampling* $\mathbf{a} \hookleftarrow U(\mathbb{Z}_q^n)$ *and* $e \hookleftarrow \chi$, *and outputting* $(\mathbf{a}, \mathbf{a}^T \cdot \mathbf{s} + e) \in \mathbb{Z}_q^n \times \mathbb{Z}_q$. *The Learning With Errors problem* $\mathsf{LWE}_{n,q,\chi}$ *asks to distinguish* m *samples chosen according to* $\mathcal{A}_{\mathbf{s},\chi}$ *(for* $\mathbf{s} \hookleftarrow U(\mathbb{Z}_q^n)$*) and* m *samples chosen according to* $U(\mathbb{Z}_q^n \times \mathbb{Z}_q)$.

If q is a prime power, $B \geq \sqrt{n}\omega(\log n)$, $\gamma = \widetilde{\mathcal{O}}(nq/B)$, then there exists an efficient sampleable B-bounded distribution χ (i.e., χ outputs samples with norm at most B with overwhelming probability) such that $\mathsf{LWE}_{n,q,\chi}$ is as least as hard as SIVP_γ (see, e.g., [9,46]).

2.3 Adaptive Oblivious Transfer

In the syntax of [13], an adaptive k-out-of-N OT scheme $\mathcal{OT}_k^N$ is a tuple of stateful PPT algorithms $(\mathsf{S_I}, \mathsf{R_I}, \mathsf{S_T}, \mathsf{R_T})$. The sender $\mathsf{S} = (\mathsf{S_I}, \mathsf{S_T})$ consists of two interactive algorithms $\mathsf{S_I}$ and $\mathsf{S_T}$ and the receiver has a similar representation as algorithms $\mathsf{R_I}$ and $\mathsf{R_T}$. In the *initialization phase*, the sender and the receiver run interactive algorithms $\mathsf{S_I}$ and $\mathsf{R_I}$, respectively, where $\mathsf{S_I}$ takes as input messages $M_1, \ldots, M_N$ while $\mathsf{R_I}$ has no input. This phase ends with the two algorithms $\mathsf{S_I}$ and $\mathsf{R_I}$ outputting their state information S_0 and R_0 respectively.

During the i-th *transfer*, $1 \leq i \leq k$, both parties run an interactive protocol via the $\mathsf{R_T}$ and $\mathsf{S_T}$ algorithms. The sender starts runs $\mathsf{S_T}(S_{i-1})$ to obtain its updated state information S_i while the receiver runs $\mathsf{R_T}(R_{i-1}, \rho_i)$ on input of its previous state R_{i-1} and the index $\rho_i \in \{1, \ldots, N\}$ of the message it wishes to retrieve. At the end, $\mathsf{R_T}$ outputs an updated state R_i and a message M'_{ρ_i}.

Correctness mandates that, for all $M_1, \ldots, M_N$, for all $\rho_1, \ldots, \rho_k \in [N]$ and all coin tosses ϖ of the (honestly run) algorithms, we have $M'_{\rho_i} = M_{\rho_i}$ for all i.

We consider protocols that are secure (against static corruptions) in the sense of simulation-based definitions. The security properties against a cheating sender and a cheating receiver are formalized via the "real-world/ideal-world" paradigm. The security definitions of [13] are recalled in the full paper.

2.4 Adaptive Oblivious Transfer with Access Control

Camenisch *et al.* [11] define oblivious transfer with access control (OT-AC) as a tuple of PPT algorithms/protocols ($\mathsf{ISetup}, \mathsf{Issue}, \mathsf{DBSetup}, \mathsf{Transfer}$) such that:

ISetup**:** takes as inputs public parameters pp specifying a set $\mathcal{P}$ of access policies and generates a key pair (PK_I, SK_I) for the issuer.

Issue**:** is an interactive protocol between the issuer I and a stateful user U under common input (pp, x), where x is an attribute string. The issuer I takes as inputs its key pair (PK_I, SK_I) and a user pseudonym P_{U}. The user takes as inputs its state information st_{U}. The user U outputs either an error symbol $\perp$ or a credential $\mathsf{Cred_U}$, and an updated state st'_{U}.

$\mathsf{DBSetup}$**:** is an algorithm that takes as input the issuer's public key PK_I, a database $DB = (M_i, \mathsf{AP}_i)_{i=1}^N$ containing records M_i whose access is restricted by an access policy AP_i and outputs a database public key PK_{DB}, an encryption of the records $(ER_i)_{i=1}^N$ and a database secret key SK_{DB}.

$\mathsf{Transfer}$**:** is a protocol between the database DB and a user U with common inputs (PK_I, PK_{DB}). DB inputs SK_{DB} and U inputs $(\rho, st_{\mathsf{U}}, ER_\rho, \mathsf{AP}_\rho)$, where $\rho \in [N]$ is a record index to which U is requesting access. The interaction ends with U outputting $\perp$ or a string $M_{\rho'}$ and an updated state st'_{U}.

We assume private communication links, so that communications between a user and the issuer are authenticated, and those between a user and the database are anonymized: otherwise, anonymizing the $\mathsf{Transfer}$ protocol is impossible.

The security definitions formalize two properties called *user anonymity* and *database security*. The former captures that the database should be unable to tell which honest user is making a query and neither can tell which records are being accessed. This should remain true even if the database colludes with corrupted users and the issuer. As for database security, the intuition is that a cheating user cannot access a record for which it does not have the required credentials, even when colluding with other dishonest users. In case the issuer is colluding with these cheating users, they cannot obtain more records from the database than they retrieve. Precise security definitions [11] are recalled in the full version of the paper.

2.5 Vector Decompositions

We will employ the decomposition technique from [34,38], which allows transforming vectors with infinity norm larger than 1 into vectors with infinity norm 1.

For any $B \in \mathbb{Z}_+$, define the number $\delta_B := \lfloor \log_2 B \rfloor + 1 = \lceil \log_2(B+1) \rceil$ and the sequence $B_1, \ldots, B_{\delta_B}$, where $B_j = \lfloor \frac{B+2^{j-1}}{2^j} \rfloor$, $\forall j \in [1, \delta_B]$. This sequence satisfies $\sum_{j=1}^{\delta_B} B_j = B$ and any integer $v \in [0, B]$ can be decomposed to $\mathsf{idec}_B(v) = (v^{(1)}, \ldots, v^{(\delta_B)})^\top \in \{0,1\}^{\delta_B}$ such that $\sum_{j=1}^{\delta_B} B_j \cdot v_j = v$. We describe this decomposition procedure in a deterministic manner as follows:

1. Set $v' := v$; For $j = 1$ to δ_B do:
   If $v' \geq B_j$ then $v^{(j)} := 1$, else $v^{(j)} := 0$; $v' := v' - B_j \cdot v^{(j)}$.
2. Output $\mathsf{idec}_B(v) = (v^{(1)}, \ldots, v^{(\delta_B)})^\top$.

For any positive integers $\mathfrak{m}, B$, we define $\mathbf{H}_{\mathfrak{m},B} := \mathbf{I}_{\mathfrak{m}} \otimes [B_1 | \ldots | B_{\delta_B}] \in \mathbb{Z}^{\mathfrak{m} \times \mathfrak{m}\delta_B}$ and the following injective functions:

(i) $\mathsf{vdec}_{\mathfrak{m},B} : [0, B]^{\mathfrak{m}} \to \{0,1\}^{\mathfrak{m}\delta_B}$ that maps vector $\mathbf{v} = (v_1, \ldots, v_{\mathfrak{m}})$ to vector $\left(\mathsf{idec}_B(v_1)^\top \| \ldots \| \mathsf{idec}_B(v_{\mathfrak{m}})^\top\right)^\top$. Note that $\mathbf{H}_{\mathfrak{m},B} \cdot \mathsf{vdec}_{\mathfrak{m},B}(\mathbf{v}) = \mathbf{v}$.

(ii) $\mathsf{vdec}'_{\mathfrak{m},B} : [-B, B]^{\mathfrak{m}} \to \{-1,0,1\}^{\mathfrak{m}\delta_B}$ that maps vector $\mathbf{w} = (w_1, \ldots, w_{\mathfrak{m}})$ to vector $\left(\sigma(w_1) \cdot \mathsf{idec}_B(|w_1|)^\top \| \ldots \| \sigma(w_{\mathfrak{m}}) \cdot \mathsf{idec}_B(|w_{\mathfrak{m}}|)^\top\right)^\top$, where for each $i = 1, \ldots, \mathfrak{m}$: $\sigma(w_i) = 0$ if $w_i = 0$; $\sigma(w_i) = -1$ if $w_i < 0$; $\sigma(w_i) = 1$ if $w_i > 0$. Note that $\mathbf{H}_{\mathfrak{m},B} \cdot \mathsf{vdec}'_{\mathfrak{m},B}(\mathbf{w}) = \mathbf{w}$.

3 Building Blocks

We will use two distinct signature schemes because one of them only needs to be secure in the sense of a weaker security notion and can be more efficient. This weaker notion is sufficient to sign the database entries and allows a better efficiency in the scheme of Sect. 4.

3.1 Signatures Supporting Efficient Zero-Knowledge Proofs

We use a signature scheme proposed by Libert *et al.* [34] who extended the Böhl *et al.* signature [7] in order to sign messages comprised of multiple blocks while keeping the scheme compatible with zero-knowledge proofs.

Keygen$(1^\lambda, 1^{N_b})$: Given a security parameter $\lambda > 0$ and the number of blocks $N_b = \mathsf{poly}(\lambda)$, choose $n = \mathcal{O}(\lambda)$, a prime modulus $q = \widetilde{\mathcal{O}}(N \cdot n^4)$, a dimension $m = 2n\lceil \log q \rceil$; an integer $\ell = \mathsf{poly}(n)$ and Gaussian parameters $\sigma = \Omega(\sqrt{n \log q} \log n)$. Define the message space as $\mathcal{M} = (\{0,1\}^{m_I})^{N_b}$.

1. Run $\mathsf{TrapGen}(1^n, 1^m, q)$ to get $\mathbf{A} \in \mathbb{Z}_q^{n \times m}$ and a short basis $\mathbf{T_A}$ of $\Lambda_q^{\perp}(\mathbf{A})$. This basis allows computing short vectors in $\Lambda_q^{\perp}(\mathbf{A})$ with a Gaussian parameter σ. Next, choose $\ell + 1$ random $\mathbf{A}_0, \mathbf{A}_1, \ldots, \mathbf{A}_\ell \hookleftarrow U(\mathbb{Z}_q^{n \times m})$.
2. Choose random matrices $\mathbf{D} \hookleftarrow U(\mathbb{Z}_q^{n \times m/2})$, $\mathbf{D}_0 \hookleftarrow U(\mathbb{Z}_q^{n \times m}), \mathbf{D}_j \hookleftarrow U(\mathbb{Z}_q^{n \times m_I})$ for $j = 1, \ldots, N_b$, as well as a random vector $\mathbf{u} \hookleftarrow U(\mathbb{Z}_q^n)$.

The private signing key consists of $SK := \mathbf{T_A}$ while the public key is comprised of $PK := \big(\mathbf{A}, \{\mathbf{A}_j\}_{j=0}^{\ell}, \mathbf{D}, \{\mathbf{D}_k\}_{k=0}^{N_b}, \mathbf{u}\big)$.

Sign(SK, Msg)**:** To sign an N_b-block $\mathsf{Msg} = (\mathfrak{m}_1, \ldots, \mathfrak{m}_{N_b}) \in (\{0,1\}^{m_I})^{N_b}$,
1. Choose $\tau \hookleftarrow U(\{0,1\}^\ell)$. Using $SK := \mathbf{T_A}$, compute a short basis $\mathbf{T}_\tau \in \mathbb{Z}^{2m \times 2m}$ for $\Lambda_q^{\perp}(\mathbf{A}_\tau)$, where $\mathbf{A}_\tau = [\mathbf{A} \mid \mathbf{A}_0 + \sum_{j=1}^{\ell} \tau[j] \mathbf{A}_j] \in \mathbb{Z}_q^{n \times 2m}$.
2. Sample $\mathbf{r} \hookleftarrow D_{\mathbb{Z}^m, \sigma}$. Compute the vector $\mathbf{c}_M \in \mathbb{Z}_q^n$ as a chameleon hash of $(\mathfrak{m}_1, \ldots, \mathfrak{m}_{N_b})$. Namely, compute $\mathbf{c}_M = \mathbf{D}_0 \cdot \mathbf{r} + \sum_{k=1}^{N_b} \mathbf{D}_k \cdot \mathfrak{m}_k \in \mathbb{Z}_q^n$, which is used to define $\mathbf{u}_M = \mathbf{u} + \mathbf{D} \cdot \mathsf{vdec}_{n,q-1}(\mathbf{c}_M) \in \mathbb{Z}_q^n$. Using the delegated basis $\mathbf{T}_\tau \in \mathbb{Z}^{2m \times 2m}$, sample a vector $\mathbf{v} \in \mathbb{Z}^{2m}$ in $D_{\Lambda_q^{\mathbf{u}_M}(\mathbf{A}_\tau), \sigma}$.

Output the signature $sig = (\tau, \mathbf{v}, \mathbf{r}) \in \{0,1\}^\ell \times \mathbb{Z}^{2m} \times \mathbb{Z}^m$.

Verify(PK, Msg, sig)**:** Given $\mathsf{Msg} = (\mathfrak{m}_1, \ldots, \mathfrak{m}_{N_b}) \in (\{0,1\}^{m_I})^{N_b}$ and $sig = (\tau, \mathbf{v}, \mathbf{r}) \in \{0,1\}^\ell \times \mathbb{Z}^{2m} \times \mathbb{Z}^m$, return 1 if $\|\mathbf{v}\| < \sigma\sqrt{2m}$, $\|\mathbf{r}\| < \sigma\sqrt{m}$ and

$$\mathbf{A}_\tau \cdot \mathbf{v} = \mathbf{u} + \mathbf{D} \cdot \mathsf{vdec}_{n,q-1}(\mathbf{D}_0 \cdot \mathbf{r} + \sum_{k=1}^{N_b} \mathbf{D}_k \cdot \mathfrak{m}_k) \bmod q. \tag{2}$$

3.2 A Simpler Variant with Bounded-Message Security and Security Against Non-Adaptive Chosen-Message Attacks

We consider a stateful variant of the scheme in Sect. 3.1 where a bound $Q \in \mathsf{poly}(n)$ on the number of signed messages is fixed at key generation time. In the context of $\mathcal{OT}_{k \times 1}^N$, this is sufficient and leads to efficiency improvements. In the modified scheme hereunder, the string $\tau \in \{0,1\}^\ell$ is an ℓ-bit counter maintained by the signer to keep track of the number of previously signed messages. This simplified variant resembles the SIS-based signature scheme of Böhl *et al.* [7].

In this version, the message space is $\{0,1\}^{n\lceil \log q \rceil}$ so that vectors of $\mathbb{Z}_q^n$ can be signed by first decomposing them using $\mathsf{vdec}_{n,q-1}(.)$.

Keygen$(1^\lambda, 1^Q)$**:** Given $\lambda > 0$ and the maximal number $Q \in \mathsf{poly}(\lambda)$ of signatures, choose $n = \mathcal{O}(\lambda)$, a prime $q = \widetilde{\mathcal{O}}(Q \cdot n^4)$, $m = 2n\lceil \log q \rceil$, an integer $\ell = \lceil \log Q \rceil$ and Gaussian parameters $\sigma = \Omega(\sqrt{n \log q} \log n)$. The message space is $\{0,1\}^{m_d}$, for some $m_d \in \mathsf{poly}(\lambda)$ with $m_d \geq m$.
1. Run $\mathsf{TrapGen}(1^n, 1^m, q)$ to get $\mathbf{A} \in \mathbb{Z}_q^{n \times m}$ and a short basis $\mathbf{T_A}$ of $\Lambda_q^{\perp}(\mathbf{A})$, which allows sampling short vectors in $\Lambda_q^{\perp}(\mathbf{A})$ with a Gaussian parameter σ. Next, choose $\ell + 1$ random $\mathbf{A}_0, \mathbf{A}_1, \ldots, \mathbf{A}_\ell \hookleftarrow U(\mathbb{Z}_q^{n \times m})$.
2. Choose $\mathbf{D} \hookleftarrow U(\mathbb{Z}_q^{n \times m_d})$ as well as a random vector $\mathbf{u} \hookleftarrow U(\mathbb{Z}_q^n)$.

The counter τ is initialized to $\tau = 0$. The private key consists of $SK := \mathbf{T_A}$ and the public key is $PK := \big(\mathbf{A},\ \{\mathbf{A}_j\}_{j=0}^{\ell},\ \mathbf{D},\ \mathbf{u}\big)$.

Sign$(SK, \tau, \mathfrak{m})$**:** To sign a message $\mathfrak{m} \in \{0,1\}^{m_d}$,

1. Increment the counter by setting $\tau := \tau + 1$ and interpret it as a string $\tau \in \{0,1\}^{\ell}$. Then, using $SK := \mathbf{T_A}$, compute a short delegated basis $\mathbf{T}_\tau \in \mathbb{Z}^{2m \times 2m}$ for the matrix $\mathbf{A}_\tau = [\mathbf{A} \mid \mathbf{A}_0 + \sum_{j=1}^{\ell} \tau[j]\mathbf{A}_j] \in \mathbb{Z}_q^{n \times 2m}$.
2. Compute the vector $\mathbf{u}_M = \mathbf{u} + \mathbf{D} \cdot \mathfrak{m} \in \mathbb{Z}_q^n$. Then, using the delegated basis $\mathbf{T}_\tau \in \mathbb{Z}^{2m \times 2m}$, sample a short vector $\mathbf{v} \in \mathbb{Z}^{2m}$ in $D_{\Lambda_q^{\mathbf{u}_M}(\mathbf{A}_\tau), \sigma}$.

Output the signature $sig = (\tau, \mathbf{v}) \in \{0,1\}^{\ell} \times \mathbb{Z}^{2m}$.

Verify$(PK, \mathfrak{m}, sig)$**:** Given PK, $\mathfrak{m} \in \{0,1\}^{m_d}$ and a signature $sig = (\tau, \mathbf{v}) \in \{0,1\}^{\ell} \times \mathbb{Z}^{2m}$, return 1 if $\|\mathbf{v}\| < \sigma\sqrt{2m}$ and $\mathbf{A}_\tau \cdot \mathbf{v} = \mathbf{u} + \mathbf{D} \cdot \mathfrak{m} \bmod q$.

For our purposes, the scheme only needs to satisfy a notion of bounded-message security under non-adaptive chosen-message attack.

Theorem 1. *The scheme is bounded message secure under non-adaptive chosen-message attacks if the* SIS *assumption holds.* (The proof is given in the full version of the paper.)

4 A Fully Simulatable Adaptive OT Protocol

Our basic $\mathcal{OT}_{k \times 1}^{N}$ protocol builds on the "assisted decryption" technique [13]. The databases holder encrypts all entries using a multi-bit variant [44] of Regev's cryptosystem [46] and proves the well-formedness of its public key and all ciphertexts. In addition, all ciphertexts are signed using a signature scheme. At each transfer, the receiver statistically re-randomizes a blinded version of the desired ciphertext, where the blinding is done via the additive homomorphism of Regev. Then, the receiver provides a witness indistinguishable (WI) argument that the modified ciphertext (which is submitted for oblivious decryption) is a transformation of one of the original ciphertexts by arguing knowledge of a signature on this hidden ciphertext. In response, the sender obliviously decrypts the modified ciphertext and argues in zero-knowledge that the response is correct.

Adapting the technique of [13] to the lattice setting requires the following building blocks: (i) A signature scheme allowing to sign ciphertexts while remaining compatible with ZK proofs; (ii) A ZK protocol allowing to prove knowledge of a signature on some hidden ciphertext which belongs to a public set and was transformed into a given ciphertext; (iii) A protocol for proving the correct decryption of a ciphertext; (iv) A method of statistically re-randomizing an LWE-encrypted ciphertext in a way that enables oblivious decryption. The first three ingredients can be obtained from [34]. Since component (i) only needs to be secure against random-message attacks as long as the adversary obtains at most N signatures, we use the simplified SIS-based signature scheme of Sect. 3.2. The statistical re-randomization of Regev ciphertexts is handled via the noise flooding technique [5], which consists in drowning the initial noise with a super-polynomially larger noise. While recent results [19] provide potentially more efficient alternatives, we chose the flooding technique for simplicity because it does not require the use of FHE (and also because the known multi-bit version [30] of the GSW FHE [23] incurs an *ad hoc* circular security assumption).

Our scheme works with security parameter λ, modulus q, lattice dimensions $n = \mathcal{O}(\lambda)$ and $m = 2n\lceil \log q \rceil$. Let $B_\chi = \widetilde{\mathcal{O}}(\sqrt{n})$, and let χ be a B_χ-bounded distribution. We also define an integer B as a randomization parameter such that $B = n^{\omega(1)} \cdot (m+1)B_\chi$ and $B + (m+1)B_\chi \leq q/5$ (to ensure decryption correctness). Our basic $\mathcal{OT}_{k \times 1}^N$ protocol goes as follows.

Initialization$(\mathsf{S_I}(1^\lambda, \mathrm{DB}), \mathsf{R_I}(1^\lambda))$**:** In this protocol, the sender $\mathsf{S_I}$ has a database $\mathrm{DB} = (M_1, \ldots, M_N)$ of N messages, where $M_i \in \{0,1\}^t$ for each $i \in [N]$, for some $t \in \mathsf{poly}(\lambda)$. It interacts with the receiver $\mathsf{R_I}$ as follows.

1. Generate a key pair for the signature scheme of Sect. 3.2 in order to sign $Q = N$ messages of length $m_d = (n+t) \cdot \lceil \log q \rceil$ each. This key pair consists of $SK_{sig} = \mathbf{T_A} \in \mathbb{Z}^{m \times m}$ and $PK_{sig} := \big(\mathbf{A}, \{\mathbf{A}_j\}_{j=0}^{\ell}, \mathbf{D}, \mathbf{u}\big)$, where $\ell = \log N$ and $\mathbf{A}, \mathbf{A}_0, \ldots, \mathbf{A}_\ell \in U(\mathbb{Z}_q^{n \times m})$, $\mathbf{D} \in U(\mathbb{Z}_q^{n \times m_d})$. The counter is initialized to $\tau = 0$.
2. Choose $\mathbf{S} \hookleftarrow \chi^{n \times t}$ that will serve as a secret key for an LWE-based encryption scheme. Then, sample $\mathbf{F} \hookleftarrow U(\mathbb{Z}_q^{n \times m})$, $\mathbf{E} \hookleftarrow \chi^{m \times t}$ and compute
$$\mathbf{P} = [\mathbf{p}_1|\ldots|\mathbf{p}_t] = \mathbf{F}^\top \cdot \mathbf{S} + \mathbf{E} \ \in \mathbb{Z}_q^{m \times t}, \tag{3}$$
so that $(\mathbf{F}, \mathbf{P}) \in \mathbb{Z}_q^{n \times m} \times \mathbb{Z}_q^{m \times t}$ forms a public key for a t-bit variant of Regev's encryption scheme [46].
3. Sample vectors $\mathbf{a}_1, \ldots, \mathbf{a}_N \hookleftarrow U(\mathbb{Z}_q^n)$ and $\mathbf{x}_1, \ldots, \mathbf{x}_N \hookleftarrow \chi^t$ to compute
$$(\mathbf{a}_i, \mathbf{b}_i) = \big(\mathbf{a}_i,\ \mathbf{S}^\top \cdot \mathbf{a}_i + \mathbf{x}_i + M_i \cdot \lfloor q/2 \rfloor\big) \in \mathbb{Z}_q^n \times \mathbb{Z}_q^t \qquad \forall i \in [N]. \tag{4}$$
4. For each $i \in [N]$, generate a signature $(\tau_i, \mathbf{v}_i) \leftarrow \mathsf{Sign}(SK_{sig}, \tau, \mathfrak{m}_i)$ on the decomposition $\mathfrak{m}_i = \mathsf{vdec}_{n+t,q-1}(\mathbf{a}_i^\top|\mathbf{b}_i^\top)^\top \in \{0,1\}^{m_d}$.
5. $\mathsf{S_I}$ sends $R_0 = \big(PK_{sig},\ (\mathbf{F}, \mathbf{P}),\ \{(\mathbf{a}_i, \mathbf{b}_i), (\tau_i, \mathbf{v}_i)\}_{i=1}^N\big)$ to $\mathsf{R_I}$ and interactively proves knowledge of small-norm $\mathbf{S} \in \mathbb{Z}^{n \times t}$, $\mathbf{E} \in \mathbb{Z}^{m \times t}$, short vectors $\{\mathbf{x}_i\}_{i=1}^N$ and t-bit messages $\{M_i\}_{i=1}^N$, for which (3) and (4) hold. To this end, $\mathsf{S_I}$ plays the role of the prover in the ZK argument system described in Sect. 6.3. If the argument of knowledge does not verify or if there exists $i \in [N]$ such that $(\tau_i, \mathbf{v}_i)$ is an invalid signature on the message $\mathfrak{m}_i = \mathsf{vdec}_{n+t,q-1}(\mathbf{a}_i^\top|\mathbf{b}_i^\top)^\top$ w.r.t. PK_{sig}, then $\mathsf{R_I}$ aborts.
6. Finally $\mathsf{S_I}$ defines $S_0 = \big((\mathbf{S}, \mathbf{E}), (\mathbf{F}, \mathbf{P}), PK_{sig}\big)$, which it keeps to itself.

Transfer$(\mathsf{S_T}(S_{i-1}), \mathsf{R_T}(R_{i-1}, \rho_i))$**:** At the i-th transfer, the receiver $\mathsf{R_T}$ has state R_{i-1} and an index $\rho_i \in [1, N]$. It interacts as follows with the sender $\mathsf{S_T}$ that has state S_{i-1} in order to obtain M_{ρ_i} from DB.

1. $\mathsf{R_T}$ samples vectors $\mathbf{e} \hookleftarrow U(\{-1,0,1\}^m)$, $\mu \hookleftarrow U(\{0,1\}^t)$ and a random $\nu \hookleftarrow U([-B,B]^t)$ to compute
$$(\mathbf{c}_0, \mathbf{c}_1) = \big(\mathbf{a}_{\rho_i} + \mathbf{F} \cdot \mathbf{e},\ \mathbf{b}_{\rho_i} + \mathbf{P}^\top \cdot \mathbf{e} + \mu \cdot \lfloor q/2 \rfloor + \nu\big) \in \mathbb{Z}_q^n \times \mathbb{Z}_q^t, \tag{5}$$
which is a re-randomization of $(\mathbf{a}_{\rho_i}, \mathbf{b}_{\rho_i} + \mu \cdot \lfloor q/2 \rfloor)$. The ciphertext $(\mathbf{c}_0, \mathbf{c}_1)$ is sent to $\mathsf{S_T}$. In addition, $\mathsf{R_T}$ provides an interactive WI argument that $(\mathbf{c}_0, \mathbf{c}_1)$ is indeed a transformation of $(\mathbf{a}_{\rho_i}, \mathbf{b}_{\rho_i})$ for some $\rho_i \in [N]$, and $\mathsf{R_T}$ knows a signature on $\mathfrak{m} = \mathsf{vdec}_{n+1,q-1}(\mathbf{a}_{\rho_i}^\top|\mathbf{b}_{\rho_i}^\top)^\top \in \{0,1\}^{m_d}$. To this end, $\mathsf{R_T}$ runs the prover in the ZK argument system in Sect. 6.5.

2. If the argument of step 1 verifies, $\mathsf{S_T}$ uses $\mathbf{S}$ to decrypt $(\mathbf{c}_0, \mathbf{c}_1) \in \mathbb{Z}_q^n \times \mathbb{Z}_q^t$ and obtain $M' = \lfloor (\mathbf{c}_1 - \mathbf{S}^\top \cdot \mathbf{c}_0)/(q/2) \rceil \in \{0,1\}^t$, which is sent back to $\mathsf{R_T}$. In addition, $\mathsf{S_T}$ provides a zero-knowledge argument of knowledge of vector $\mathbf{y} = \mathbf{c}_1 - \mathbf{S}^\top \cdot \mathbf{c}_0 - M' \cdot \lfloor q/2 \rfloor \in \mathbb{Z}^t$ of norm $\|\mathbf{y}\|_\infty \leq q/5$ and small-norm matrices $\mathbf{E} \in \mathbb{Z}^{m \times t}$, $\mathbf{S} \in \mathbb{Z}^{n \times t}$ satisfying (modulo q)

$$\mathbf{P} = \mathbf{F}^\top \cdot \mathbf{S} + \mathbf{E}, \qquad \mathbf{c}_0^\top \cdot \mathbf{S} + \mathbf{y}^\top = \mathbf{c}_1^\top - M'^\top \cdot \lfloor q/2 \rfloor. \tag{6}$$

To this end, $\mathsf{S_T}$ runs the prover in the ZK argument system in Sect. 6.4.
3. If the ZK argument produced by $\mathsf{S_T}$ does not properly verify at step 2, $\mathsf{R_T}$ halts and outputs $\perp$. Otherwise, $\mathsf{R_T}$ recalls the random string $\mu \in \{0,1\}^t$ that was chosen at step 1 and computes $M_{\rho_i} = M' \oplus \mu$. The transfer ends with $\mathsf{S_T}$ and $\mathsf{R_T}$ outputting $S_i = S_{i-1}$ and $R_i = R_{i-1}$, respectively.

In the initialization phase, the sender has to repeat step 5 with each receiver to prove that $\{(\mathbf{a}_i, \mathbf{b}_i)\}_{i=1}^N$ are well-formed. Using the Fiat-Shamir heuristic [21], we can decrease this initialization cost from $O(N \cdot U)$ to $O(N)$ (regardless of the number of users U) by making the proof non-interactive. This modification also reduces each transfer to 5 communication rounds since, even in the transfer phase, the sender's ZK arguments can be non-interactive and the receiver's arguments only need to be WI, which is preserved when the basic ZK protocol (which has a ternary challenge space) is repeated $\omega(\log n)$ times in parallel. Details are given in the full version of the paper.

The security of the above $\mathcal{OT}_{k \times 1}^N$ protocol against static corruptions is proved in the full version of the paper under the SIS and LWE assumptions.

5 OT with Access Control for Branching Programs

In this section, we extend our protocol of Sect. 4 into a protocol where database entries can be protected by access control policies consisting of branching programs. In a nutshell, the construction goes as follows.

When the database is set up, the sender signs (a binary representation of) each database entry $(\mathbf{a}_i, \mathbf{b}_i)$ together with a hash value $\mathbf{h}_{\mathsf{BP},i} \in \mathbb{Z}_q^n$ of the corresponding branching program. For each possessed attribute $\mathbf{x} \in \{0,1\}^\kappa$, the user U obtains a credential $\mathsf{Cred}_{\mathsf{U},\mathbf{x}}$ from the issuer.

If U has a credential $\mathsf{Cred}_{\mathsf{U},\mathbf{x}}$ for an attribute $\mathbf{x}$ satisfying the ρ-th branching program, U can re-randomize $(\mathbf{a}_\rho, \mathbf{b}_\rho)$ into $(\mathbf{c}_0, \mathbf{c}_1)$, which is given to the sender, while proving that: (i) He knows a signature $(\tau, \mathbf{v})$ on some message $(\mathbf{a}_\rho, \mathbf{b}_\rho, \mathbf{h}_{\mathsf{BP},\rho})$ such that $(\mathbf{c}_0, \mathbf{c}_1)$ is a re-randomization of $(\mathbf{a}_\rho, \mathbf{b}_\rho)$; (ii) The corresponding $\mathbf{h}_{\mathsf{BP},\rho}$ is the hash value of (the binary representation of) a branching program BP_ρ that accepts an attribute $\mathbf{x} \in \{0,1\}^\kappa$ for which he has a valid credential $\mathsf{Cred}_{\mathsf{U},\mathbf{x}}$ (i.e., $\mathsf{BP}_\rho(\mathbf{x}) = 1$).

While statement (i) can be proved as in Sect. 4, handling (ii) requires a method of proving the possession of a (committed) branching program BP and a (committed) input $\mathbf{x} \in \{0,1\}^\kappa$ such that $\mathsf{BP}(\mathbf{x}) = 1$ while demonstrating possession of a credential for $\mathbf{x}$.

Recall that a branching program BP of length L, input space $\{0,1\}^\kappa$ and width 5 is specified by L tuples of the form $(\text{var}(\theta), \pi_{\theta,0}, \pi_{\theta,1})$ where

- var : $[L] \to [0, \kappa-1]$ is a function that associates the θ-th tuple with the coordinate $x_{\text{var}(\theta)} \in \{0,1\}$ of the input $\mathbf{x} = (x_0, \ldots, x_{\kappa-1})^\top$.
- $\pi_{\theta,0}, \pi_{\theta,1} : \{0,1,2,3,4\} \to \{0,1,2,3,4\}$ are permutations that determine the θ-th step of the evaluation.

On input $\mathbf{x} = (x_0, \ldots, x_{\kappa-1})^\top$, BP computes its output as follows. For each bit $b \in \{0,1\}$, let $\bar{b}$ denote the bit $1-b$. Let η_θ denote the state of computation at step θ. The initial state is $\eta_0 = 0$ and, for $\theta \in [1, L]$, the state η_θ is computed as

$$\eta_\theta = \pi_{\theta, x_{\text{var}(\theta)}}(\eta_{\theta-1}) = \pi_{\theta,0}(\eta_{\theta-1}) \cdot \bar{x}_{\text{var}(\theta)} + \pi_{\theta,1}(\eta_{\theta-1}) \cdot x_{\text{var}(\theta)}.$$

Finally, the output of evaluation is $\mathsf{BP}(\mathbf{x}) = 1$ if $\eta_L = 0$, otherwise $\mathsf{BP}(\mathbf{x}) = 0$.

We now let $\delta_\kappa = \lceil \log_2 \kappa \rceil$ and note that each integer in $[0, \kappa-1]$ can be determined by δ_κ bits. In particular, for each $\theta \in [L]$, let $d_{\theta,1}, \ldots, d_{\theta,\delta_\kappa}$ be the bits representing var(θ). Then, we consider the following representation of BP:

$$\begin{aligned}\mathbf{z}_{\mathsf{BP}} = \big(&d_{1,1}, \ldots, d_{1,\delta_\kappa}, \ldots, d_{L,1}, \ldots, d_{L,\delta_\kappa}, \pi_{1,0}(0), \ldots, \pi_{1,0}(4), \pi_{1,1}(0), \ldots, \\ &\pi_{1,1}(4), \ldots, \pi_{L,0}(0), \ldots, \pi_{L,0}(4), \pi_{L,1}(0), \ldots, \pi_{L,1}(4)\big)^\top \in [0,4]^\zeta, \end{aligned} \quad (7)$$

where $\zeta = L(\delta_\kappa + 10)$.

5.1 The OT-AC Protocol

We assume public parameters pp consisting of a modulus q, integers n, m such that $m = 2n\lceil \log q \rceil$, a public matrix $\bar{\mathbf{A}} \in \mathbb{Z}_q^{n \times m}$, the maximal length $L \in \mathsf{poly}(n)$ of branching programs and their desired input length $\kappa \in \mathsf{poly}(n)$.

ISetup(pp)**:** Given public parameters $\mathsf{pp} = \{q, n, m, \bar{\mathbf{A}}, L, \kappa\}$, generate a key pair $(PK_I, SK_I) \leftarrow \mathsf{Keygen}(\mathsf{pp}, 1)$ for the signature scheme in Sect. 3.1 in order to sign single-block messages (i.e., $N_b = 1$) of length $m_I = n \cdot \lceil \log q \rceil + \kappa$. Letting $\ell_I = \mathcal{O}(n)$, this key pair contains $SK_I = \mathbf{T}_{\mathbf{A}_I} \in \mathbb{Z}^{m \times m}$ and

$$PK_I := \big(\mathbf{A}_I,\ \{\mathbf{A}_{I,j}\}_{j=0}^{\ell_I},\ \mathbf{D}_I,\ \{\mathbf{D}_{I,0}, \mathbf{D}_{I,1}\},\ \mathbf{u}_I\big).$$

Issue$(\mathsf{I}(\mathsf{pp}, SK_I, PK_I, P_\mathsf{U}, \mathbf{x}) \leftrightarrow \mathsf{U}(\mathsf{pp}, \mathbf{x}, st_\mathsf{U}))$**:** On common input $\mathbf{x} \in \{0,1\}^\kappa$, the issuer I and the user U interact in the following way:

1. If $st_\mathsf{U} = \emptyset$, U creates a pseudonym $P_\mathsf{U} = \bar{\mathbf{A}} \cdot \mathbf{e}_\mathsf{U} \in \mathbb{Z}_q^n$, for a randomly chosen $\mathbf{e}_\mathsf{U} \hookleftarrow U(\{0,1\}^m)$, which is sent to I. It sets $st_\mathsf{U} = (\mathbf{e}_\mathsf{U}, P_\mathsf{U}, 0, \emptyset, \emptyset)$. Otherwise, U parses its state st_U as $(\mathbf{e}_\mathsf{U}, P_\mathsf{U}, f_{DB}, C_\mathsf{U}, \mathsf{Cred}_\mathsf{U})$.
2. The issuer I defines the message $\mathfrak{m}_{\mathsf{U},\mathbf{x}} = (\mathsf{vdec}_{n,q-1}(P_\mathsf{U})^\top | \mathbf{x}^\top)^\top \in \{0,1\}^{m_I}$. Then, it runs the signing algorithm of Sect. 3.1 to obtain and return $\mathsf{cert}_{\mathsf{U},\mathbf{x}} = (\tau_\mathsf{U}, \mathbf{v}_\mathsf{U}, \mathbf{r}_\mathsf{U}) \leftarrow \mathsf{Sign}(SK_I, \mathfrak{m}_{\mathsf{U},\mathbf{x}}) \in \{0,1\}^{\ell_I} \times \mathbb{Z}^{2m} \times \mathbb{Z}^m$, which binds U's pseudonym P_U to the attribute string $\mathbf{x} \in \{0,1\}^\kappa$.

3. U checks that $\mathsf{cert}_{\mathsf{U},\mathbf{x}}$ satisfies (2) and that $\|\mathbf{v}_\mathsf{U}\| \leq \sigma\sqrt{2m}, \mathbf{r}_\mathsf{U} \leq \sigma\sqrt{m}$. If so, U sets $C_\mathsf{U} := C_\mathsf{U} \cup \{\mathbf{x}\}$, $\mathsf{Cred}_\mathsf{U} := \mathsf{Cred}_\mathsf{U} \cup \{\mathsf{cert}_{\mathsf{U},\mathbf{x}}\}$ and updates its state $st_\mathsf{U} = (\mathbf{e}_\mathsf{U}, P_\mathsf{U}, f_{DB}, C_\mathsf{U}, \mathsf{Cred}_\mathsf{U})$. If $\mathsf{cert}_{\mathsf{U},\mathbf{x}}$ does not properly verify, U aborts the interaction and leaves st_U unchanged.

DBSetup$(PK_I, \mathrm{DB} = \{(M_i, \mathsf{BP}_i)\}_{i=1}^N)$**:** The sender DB has $\mathrm{DB} = \{(M_i, \mathsf{BP}_i)\}_{i=1}^N$ which is a database of N pairs made of a message $M_i \in \{0,1\}^t$ and a policy realized by a length-L branching program $\mathsf{BP}_i = \{\mathrm{var}_i(\theta), \pi_{i,\theta,0}, \pi_{i,\theta,1}\}_{\theta=1}^L$.

1. Choose a random matrix $\mathbf{A}_{\mathrm{HBP}} \hookleftarrow U(\mathbb{Z}_q^{n\times\varsigma})$ which will be used to hash the description of branching programs.
2. Generate a key pair for the signature scheme of Sect. 3.2 in order to sign $Q = N$ messages of length $m_d = (2n+t)\cdot\lceil\log q\rceil$ each. This key pair consists of $SK_{sig} = \mathbf{T_A} \in \mathbb{Z}^{m\times m}$ and $PK_{sig} := (\mathbf{A}, \{\mathbf{A}_j\}_{j=0}^{\ell}, \mathbf{D}, \mathbf{u})$, where $\ell = \lceil\log N\rceil$ and $\mathbf{A}, \mathbf{A}_0, \ldots, \mathbf{A}_\ell \in U(\mathbb{Z}_q^{n\times m})$, $\mathbf{D} \in U(\mathbb{Z}_q^{n\times m_d})$ with $m = 2n\lceil\log q\rceil$, $m_d = (2n+t)\lceil\log q\rceil$. The counter is initialized to $\tau = 0$.
3. Sample $\mathbf{S} \hookleftarrow \chi^{n\times t}$ which will serve as a secret key for an LWE-based encryption scheme. Then, sample $\mathbf{F} \hookleftarrow U(\mathbb{Z}_q^{n\times m})$, $\mathbf{E} \hookleftarrow \chi^{m\times t}$ to compute
$$\mathbf{P} = [\mathbf{p}_1|\ldots|\mathbf{p}_t] = \mathbf{F}^\top\cdot\mathbf{S} + \mathbf{E} \in \mathbb{Z}_q^{m\times t} \tag{8}$$
so that $(\mathbf{F}, \mathbf{P})$ forms a public key for a t-bit variant of Regev's system.
4. Sample vectors $\mathbf{a}_1, \ldots, \mathbf{a}_N \hookleftarrow U(\mathbb{Z}_q^n)$ and $\mathbf{x}_1, \ldots, \mathbf{x}_N \hookleftarrow \chi^t$ to compute
$$(\mathbf{a}_i, \mathbf{b}_i) = (\mathbf{a}_i,\ \mathbf{a}_i^\top\cdot\mathbf{S} + \mathbf{x}_i + M_i\cdot\lfloor q/2\rfloor) \in \mathbb{Z}_q^n \times \mathbb{Z}_q^t \qquad \forall i \in [N] \tag{9}$$
5. For each $i = 1$ to N, $(\mathbf{a}_i, \mathbf{b}_i)$ is bound to BP_i as follows.
   a. Let $\mathbf{z}_{\mathsf{BP},i} \in [0,4]^\varsigma$ be the binary representation of the branching program. Compute its digest $\mathbf{h}_{\mathsf{BP},i} = \mathbf{A}_{\mathrm{HBP}}\cdot\mathbf{z}_{\mathsf{BP},i} \in \mathbb{Z}_q^n$.
   b. Using SK_{sig}, generate a signature $(\tau_i, \mathbf{v}_i) \leftarrow \mathsf{Sign}(SK_{sig}, \tau, \mathfrak{m}_i)$ on the message $\mathfrak{m}_i = \mathsf{vdec}_{2n+t,q-1}(\mathbf{a}_i|\mathbf{b}_i|\mathbf{h}_{\mathsf{BP},i}) \in \{0,1\}^{m_d}$ obtained by decomposing $(\mathbf{a}_i^\top|\mathbf{b}_i^\top|\mathbf{h}_{\mathsf{BP},i}^\top)^\top \in \mathbb{Z}_q^{2n+t}$.
6. The database's public key is defined as $PK_{\mathrm{DB}} = (PK_{sig}, (\mathbf{F}, \mathbf{P}), \mathbf{A}_{\mathrm{HBP}})$ while the encrypted database is $\{ER_i = (\mathbf{a}_i, \mathbf{b}_i, (\tau_i, \mathbf{v}_i)), \mathsf{BP}_i\}_{i=1}^N$. The sender DB outputs $(PK_{\mathrm{DB}}, \{ER_i, \mathsf{BP}_i\}_{i=1}^N)$ and keeps $SK_{\mathrm{DB}} = (SK_{sig}, \mathbf{S})$.

Transfer$(\mathsf{DB}(SK_{\mathrm{DB}}, PK_{\mathrm{DB}}, PK_I), \mathsf{U}(\rho, st_\mathsf{U}, PK_I, PK_{\mathrm{DB}}, ER_\rho, \mathsf{BP}_\rho))$**:** Given an index $\rho \in [N]$, a record $ER_\rho = (\mathbf{a}_\rho, \mathbf{b}_\rho, (\tau_\rho, \mathbf{v}_\rho))$ and a policy BP_ρ, the user U parses st_U as $(\mathbf{e}_\mathsf{U}, P_\mathsf{U}, f_{DB}, C_\mathsf{U}, \mathsf{Cred}_\mathsf{U})$. If C_U does not contain any $\mathbf{x} \in \{0,1\}^\kappa$ s.t. $\mathsf{BP}_\rho(\mathbf{x}) = 1$ and Cred_U contains the corresponding $\mathsf{cert}_{\mathsf{U},\mathbf{x}}$, U outputs $\perp$. Otherwise, he selects such a pair $(\mathbf{x}, \mathsf{cert}_{\mathsf{U},\mathbf{x}})$ and interacts with DB:

1. If $f_{DB} = 0$, U interacts with DB for the first time and requires DB to prove knowledge of small-norm $\mathbf{S} \in \mathbb{Z}^{n\times t}$, $\mathbf{E} \in \mathbb{Z}^{m\times t}$, $\{\mathbf{x}_i\}_{i=1}^N$ and t-bit messages $\{M_i\}_{i=1}^N$ satisfying (8)–(9). To do this, DB uses the ZK argument in Sect. 6.3. If there exists $i \in [N]$ such that $(\tau_i, \mathbf{v}_i)$ is an invalid signature on $\mathsf{vdec}_{2n+t,q-1}(\mathbf{a}_i^\top|\mathbf{b}_i^\top|\mathbf{h}_{\mathsf{BP},i}^\top)^\top$ or if the ZK argument does not verify, U aborts. Otherwise, U updates st_U and sets $f_{DB} = 1$.

2. U re-randomizes the pair $(\mathbf{a}_\rho, \mathbf{b}_\rho)$ contained in ER_ρ. It samples vectors $\mathbf{e} \hookleftarrow U(\{-1,0,1\}^m)$, $\mu \hookleftarrow U(\{0,1\}^t)$ and $\nu \hookleftarrow U([-B,B]^t)$ to compute

$$(\mathbf{c}_0, \mathbf{c}_1) = \big(\mathbf{a}_\rho + \mathbf{F}\cdot\mathbf{e},\ \mathbf{b}_\rho + \mathbf{P}^\top\cdot\mathbf{e} + \mu\cdot\lfloor q/2\rfloor + \nu\big) \in \mathbb{Z}_q^n \times \mathbb{Z}_q^t, \qquad (10)$$

which is sent to DB as a re-randomization of $(\mathbf{a}_\rho, \mathbf{b}_\rho + \mu\cdot\lfloor q/2\rfloor)$. Then, U provides an interactive WI argument that $(\mathbf{c}_0, \mathbf{c}_1)$ is a re-randomization of some $(\mathbf{a}_\rho, \mathbf{b}_\rho)$ associated with a policy BP_ρ for which U has a credential $\mathsf{cert}_{\mathsf{U},x}$ for some $\mathbf{x}\in\{0,1\}^\kappa$ such that $\mathsf{BP}_\rho(\mathbf{x}) = 1$. In addition, U demonstrates possession of: (i) a preimage $\mathbf{z}_{\mathsf{BP},\rho} \in [0,4]^\varsigma$ of $\mathbf{h}_{\mathsf{BP},\rho} = \mathbf{A}_{\mathrm{HBP}}\cdot\mathbf{z}_{\mathsf{BP},\rho} \in \mathbb{Z}_q^n$; (ii) a credential $\mathsf{Cred}_{\mathsf{U},\mathbf{x}}$ for the corresponding $\mathbf{x}\in\{0,1\}^\kappa$ and the private key $\mathbf{e}_\mathsf{U}\in\{0,1\}^m$ for the pseudonym P_U to which $\mathbf{x}$ is bound; (iii) the coins leading to the randomization of some $(\mathbf{a}_\rho, \mathbf{b}_\rho)$. Then entire step is conducted by arguing knowledge of

$$\begin{cases} \mathbf{e}_\mathsf{U}\in\{0,1\}^m, \mathfrak{m}_{\mathsf{U},\mathbf{x}}\in\{0,1\}^{m_I},\ \mathbf{x}\in\{0,1\}^\kappa,\ \widehat{\mathfrak{m}}_{\mathsf{U},\mathbf{x}}\in\{0,1\}^{m/2} \\ \tau_\mathsf{U}\in\{0,1\}^{\ell_I},\ \mathbf{v}_\mathsf{U} = (\mathbf{v}_{\mathsf{U},1}^\top|\mathbf{v}_{\mathsf{U},2}^\top)^\top\in[-\beta,\beta]^{2m},\ \mathbf{r}_\mathsf{U}\in[-\beta,\beta]^m \\ \qquad\qquad\qquad // \text{ signature on } \mathfrak{m}_{\mathsf{U},\mathbf{x}} = (\mathsf{vdec}_{n,q-1}(P_\mathsf{U})^\top|\mathbf{x}^\top)^\top \\ \mathbf{z}_{\mathsf{BP},\rho}\in[0,4]^\varsigma \qquad // \text{ representation of } \mathsf{BP}_\rho \\ \mathfrak{m}\in\{0,1\}^{m_d},\ \tau\in\{0,1\}^\ell,\ \mathbf{v} = (\mathbf{v}_1^\top|\mathbf{v}_2^\top)^\top\in\mathbb{Z}^{2m} \\ \qquad\qquad\qquad // \text{ signature on } \mathfrak{m} = \mathsf{vdec}_{2n+t,q-1}(\mathbf{a}_i^\top|\mathbf{b}_i^\top|\mathbf{h}_{\mathsf{BP},\rho}^\top)^\top \\ \mathbf{e}\in\{-1,0,1\}^t,\ \mu\in\{0,1\}^t,\ \nu\in[-B,B]^t, \\ \qquad\qquad\qquad // \text{ coins allowing the re-randomization of } (\mathbf{a}_\rho,\mathbf{b}_\rho) \end{cases}$$

satisfying the relations (modulo q)

$$\begin{cases} \mathbf{H}_{2n+t,q-1}\cdot\mathfrak{m} + \left[\begin{array}{c|c|c|c} \mathbf{F} & & & \\ \hline \mathbf{P}^\top & \mathbf{I}_t\cdot\lfloor q/2\rfloor & \mathbf{I}_t & \\ \hline & & & -\mathbf{A}_{\mathrm{HBP}} \end{array}\right]\cdot\left[\begin{array}{c} \mathbf{e} \\ \hline \mu \\ \hline \nu \\ \hline \mathbf{z}_{\mathsf{BP},\rho} \end{array}\right] = \left[\begin{array}{c} \mathbf{c}_0 \\ \hline \mathbf{c}_1 \\ \hline \mathbf{0}^n \end{array}\right] \\ \qquad // \text{ (recall that } (\mathbf{a}_\rho^\top|\mathbf{b}_\rho^\top|\mathbf{h}_{\mathsf{BP},\rho}^\top)^\top = \mathbf{H}_{2n+t,q-1}\cdot\mathfrak{m}) \\ \mathbf{A}\cdot\mathbf{v}_1 + \mathbf{A}_0\cdot\mathbf{v}_2 + \sum_{j=1}^{\ell}\mathbf{A}_j\cdot(\tau[j]\cdot\mathbf{v}_2) - \mathbf{D}\cdot\mathfrak{m} = \mathbf{u} \\ \mathbf{A}_I\cdot\mathbf{v}_{\mathsf{U},1} + \mathbf{A}_{I,0}\cdot\mathbf{v}_{\mathsf{U},2} + \sum_{j=1}^{\ell_I}\mathbf{A}_{I,j}\cdot(\tau_\mathsf{U}[j]\cdot\mathbf{v}_{\mathsf{U},2}) - \mathbf{D}_I\cdot\widehat{\mathfrak{m}}_{\mathsf{U},\mathbf{x}} = \mathbf{u}_I \\ \mathbf{D}_{I,0}\cdot\mathbf{r}_\mathsf{U} + \mathbf{D}_{I,1}\cdot\mathfrak{m}_{\mathsf{U},\mathbf{x}} - \mathbf{H}_{n,q-1}\cdot\widehat{\mathfrak{m}}_{\mathsf{U},\mathbf{x}} = \mathbf{0} \\ \left[\begin{array}{c|c} \mathbf{H}_{n,q-1} & \mathbf{0} \\ \hline \mathbf{0} & \mathbf{I}_\kappa \end{array}\right]\cdot\mathfrak{m}_{\mathsf{U},\mathbf{x}} + \left[\begin{array}{c} -\bar{\mathbf{A}} \\ \hline \mathbf{0} \end{array}\right]\cdot\mathbf{e}_\mathsf{U} + \left[\begin{array}{c} \mathbf{0} \\ \hline -\mathbf{I}_\kappa \end{array}\right]\cdot\mathbf{x} = \mathbf{0} \end{cases} \qquad (11)$$

and such that $\mathbf{z}_{\mathsf{BP},\rho}\in[0,4]^\varsigma$ encodes BP_ρ such that $\mathsf{BP}_\rho(\mathbf{x}) = 1$. This is done by running the argument system described in Sect. 6.6.

3. If the ZK argument of step 2 verifies, DB decrypts $(\mathbf{c}_0,\mathbf{c}_1)\in\mathbb{Z}_q^n\times\mathbb{Z}_q^t$ to obtain $M' = \lfloor(\mathbf{c}_1 - \mathbf{S}^\top\cdot\mathbf{c}_0)/(q/2)\rceil\in\{0,1\}^t$, which is returned to U. Then, DB argues knowledge of $\mathbf{y} = \mathbf{c}_1 - \mathbf{S}^\top\cdot\mathbf{c}_0 - M'\cdot\lfloor q/2\rfloor\in\mathbb{Z}^t$ of norm $\|\mathbf{y}\|_\infty\le q/5$ and small-norm $\mathbf{E}\in\mathbb{Z}^{m\times t}$, $\mathbf{S}\in\mathbb{Z}^{n\times t}$ satisfying (modulo q)

$$\mathbf{P} = \mathbf{F}^\top\cdot\mathbf{S} + \mathbf{E}, \qquad \mathbf{c}_0^\top\cdot\mathbf{S} + \mathbf{y}^\top = \mathbf{c}_1^\top - M'^\top\cdot\lfloor q/2\rfloor.$$

To this end, DB uses the ZK argument system of Sect. 6.4.

4. If the ZK argument produced by DB does not verify, U outputs $\perp$. Otherwise, U recalls the string $\mu \in \{0,1\}^t$ and outputs $M_{\rho_i} = M' \oplus \mu$.

Like our construction of Sect. 4, the above protocol requires the DB to repeat a ZK proof of communication complexity $\Omega(N)$ with each user U during the initialization phase. By applying the Fiat-Shamir heuristic as shown in the full version of the paper, the cost of the initialization phase can be made independent of the number of users: the sender can publicize $\big(PK_{\mathrm{DB}}, \{ER_i, \mathsf{BP}_i\}_{i=1}^N\big)$ along with a with a universally verifiable non-interactive proof of well-formedness.

The security of the above protocol against static corruptions is proved in the full version of the paper, under the SIS and LWE assumptions.

6 Our Zero-Knowledge Arguments of Knowledge

This section provides all the zero-knowledge arguments of knowledge (ZKAoK) used as building blocks in our two adaptive OT schemes. Our argument systems operate in the framework of Stern's protocol [48], which was originally introduced in the context of code-based cryptography but has been developed [34–36,38,39].

In Sect. 6.1, we first recall Stern's protocol in a generalized, abstract manner suggested in [34]. Then, using various transformations, we will demonstrate that all the required ZKAoKs can be obtained from this abstract protocol. Our basic strategy and techniques are summarized in Sect. 6.2, while the details of the protocols are given in the next subsections. In particular, our treatment of hidden branching programs in Sect. 6.6 is rather sophisticated as it requires to handle a number of secret objects nested together via branching programs, commitments, encryptions, signatures and Merkle trees. This protocol introduces new techniques and insights of which we provide the intuition hereafter.

6.1 Abstracting Stern's Protocol

Let K, D, q be positive integers with $D \geq K$ and $q \geq 2$, and let VALID be a subset of $\mathbb{Z}^D$. Suppose that $\mathcal{S}$ is a finite set such that every $\phi \in \mathcal{S}$ can be associated with a permutation Γ_ϕ of D elements satisfying the following conditions:

$$\begin{cases} \mathbf{w} \in \mathsf{VALID} \iff \Gamma_\phi(\mathbf{w}) \in \mathsf{VALID}, \\ \text{If } \mathbf{w} \in \mathsf{VALID} \text{ and } \phi \text{ is uniform in } \mathcal{S}, \text{ then } \Gamma_\phi(\mathbf{w}) \text{ is uniform in } \mathsf{VALID}. \end{cases} \tag{12}$$

We aim to construct a statistical ZKAoK for the following abstract relation:

$$\mathrm{R}_{\mathrm{abstract}} = \big\{\big((\mathbf{M}, \mathbf{v}), \mathbf{w}\big) \in \mathbb{Z}_q^{K\times D} \times \mathbb{Z}_q^D \times \mathsf{VALID} : \mathbf{M}\cdot\mathbf{w} = \mathbf{v} \bmod q.\big\}$$

Stern's original protocol corresponds to the case $\mathsf{VALID} = \{\mathbf{w} \in \{0,1\}^D : \mathsf{wt}(\mathbf{w}) = k\}$, where $\mathsf{wt}(\cdot)$ denotes the Hamming weight and $k < D$ is a given integer, $\mathcal{S} = \mathcal{S}_D$ is the set of all permutations of D elements and $\Gamma_\phi(\mathbf{w}) = \phi(\mathbf{w})$.

The conditions in (12) play a crucial role in proving in ZK that $\mathbf{w} \in \mathsf{VALID}$. To this end, the prover samples a random $\phi \hookleftarrow U(\mathcal{S})$ and lets the verifier check that $\Gamma_\phi(\mathbf{w}) \in \mathsf{VALID}$ without learning any additional information about $\mathbf{w}$ due to the randomness of ϕ. Furthermore, to prove in a zero-knowledge manner that the linear equation is satisfied, the prover samples a masking vector $\mathbf{r}_w \hookleftarrow U(\mathbb{Z}_q^D)$, and convinces the verifier instead that $\mathbf{M} \cdot (\mathbf{w} + \mathbf{r}_w) = \mathbf{M} \cdot \mathbf{r}_w + \mathbf{v} \bmod q$.

The interaction between prover $\mathcal{P}$ and verifier $\mathcal{V}$ is described in Fig. 1. The protocol uses a statistically hiding and computationally binding string commitment scheme COM (e.g., the SIS-based scheme from [32]).

1. **Commitment:** Prover samples $\mathbf{r}_w \leftarrow U(\mathbb{Z}_q^D)$, $\phi \leftarrow U(\mathcal{S})$ and randomness ρ_1, ρ_2, ρ_3 for COM. Then he sends $\mathrm{CMT} = \big(C_1, C_2, C_3\big)$ to the verifier, where

$$C_1 = \mathsf{COM}(\phi, \mathbf{M} \cdot \mathbf{r}_w \bmod q; \rho_1), \;\; C_2 = \mathsf{COM}(\Gamma_\phi(\mathbf{r}_w); \rho_2),$$
$$C_3 = \mathsf{COM}(\Gamma_\phi(\mathbf{w} + \mathbf{r}_w \bmod q); \rho_3).$$

2. **Challenge:** The verifier sends a challenge $Ch \leftarrow U(\{1, 2, 3\})$ to the prover.
3. **Response:** Depending on Ch, the prover sends RSP computed as follows:
   - $Ch = 1$: Let $\mathbf{t}_w = \Gamma_\phi(\mathbf{w})$, $\mathbf{t}_r = \Gamma_\phi(\mathbf{r}_w)$, and $\mathrm{RSP} = (\mathbf{t}_w, \mathbf{t}_r, \rho_2, \rho_3)$.
   - $Ch = 2$: Let $\phi_2 = \phi$, $\mathbf{w}_2 = \mathbf{w} + \mathbf{r}_w \bmod q$, and $\mathrm{RSP} = (\phi_2, \mathbf{w}_2, \rho_1, \rho_3)$.
   - $Ch = 3$: Let $\phi_3 = \phi$, $\mathbf{w}_3 = \mathbf{r}_w$, and $\mathrm{RSP} = (\phi_3, \mathbf{w}_3, \rho_1, \rho_2)$.

Verification: Receiving RSP, the verifier proceeds as follows:

- $Ch = 1$: Check that $\mathbf{t}_w \in \mathsf{VALID}$, $C_2 = \mathsf{COM}(\mathbf{t}_r; \rho_2)$, $C_3 = \mathsf{COM}(\mathbf{t}_w + \mathbf{t}_r \bmod q; \rho_3)$.
- $Ch = 2$: Check that $C_1 = \mathsf{COM}(\phi_2, \mathbf{M} \cdot \mathbf{w}_2 - \mathbf{v} \bmod q; \rho_1)$, $C_3 = \mathsf{COM}(\Gamma_{\phi_2}(\mathbf{w}_2); \rho_3)$.
- $Ch = 3$: Check that $C_1 = \mathsf{COM}(\phi_3, \mathbf{M} \cdot \mathbf{w}_3; \rho_1)$, $C_2 = \mathsf{COM}(\Gamma_{\phi_3}(\mathbf{w}_3); \rho_2)$.

In each case, the verifier outputs 1 if and only if all the conditions hold.

Fig. 1. Stern-like ZKAoK for the relation $\mathrm{R}_{\mathrm{abstract}}$.

Theorem 2. *The protocol in Fig. 1 is a statistical* ZKAoK *with perfect completeness, soundness error* $2/3$, *and communication cost* $\mathcal{O}(D \log q)$. *Namely:*

- *There exists a polynomial-time simulator that, on input* $(\mathbf{M}, \mathbf{v})$, *outputs an accepted transcript statistically close to that produced by the real prover.*
- *There exists a polynomial-time knowledge extractor that, on input a commitment* CMT *and* 3 *valid responses* $(\mathrm{RSP}_1, \mathrm{RSP}_2, \mathrm{RSP}_3)$ *to all* 3 *possible values of the challenge* Ch, *outputs* $\mathbf{w}' \in \mathsf{VALID}$ *such that* $\mathbf{M} \cdot \mathbf{w}' = \mathbf{v} \bmod q$.

The proof of the theorem relies on standard simulation and extraction techniques for Stern-like protocols [32,34,38]. It is given in the full version of the paper.

6.2 Our Strategy and Basic Techniques, in a Nutshell

Before going into the details of our protocols, we first summarize our governing strategy and the techniques that will be used in the next subsections.

In each protocol, we prove knowledge of (possibly one-dimensional) integer vectors $\{\mathbf{w}_i\}_i$ that have various constraints (e.g., smallness, special arrangements of coordinates, or correlation with one another) and satisfy a system

$$\Big\{\sum_i \mathbf{M}_{i,j} \cdot \mathbf{w}_i = \mathbf{v}_j\Big\}_j, \tag{13}$$

where $\{\mathbf{M}_{i,j}\}_{i,j}$, $\{\mathbf{v}_j\}_j$ are public matrices (which are possibly zero or identity matrices) and vectors. Our strategy consists in transforming this entire system into one equivalent equation $\mathbf{M} \cdot \mathbf{w} = \mathbf{v}$, where matrix $\mathbf{M}$ and vector $\mathbf{v}$ are public, while the constraints of the secret vector $\mathbf{w}$ capture those of witnesses $\{\mathbf{w}_i\}_i$ and they are provable in zero-knowledge via random permutations. For this purpose, the Stern-like protocol from Sect. 6.1 comes in handy.

A typical transformation step is of the form $\mathbf{w}_i \to \bar{\mathbf{w}}_i$, where there exists public matrix $\mathbf{P}_{i,j}$ such that $\mathbf{P}_{i,j} \cdot \bar{\mathbf{w}}_i = \mathbf{w}_i$. This subsumes the decomposition and extension mechanisms which first appeared in [38].

- **Decomposition:** Used when $\mathbf{w}_i$ has infinity norm bound larger than 1 and we want to work more conveniently with $\bar{\mathbf{w}}_i$ whose norm bound is exactly 1. In this case, $\mathbf{P}_{i,j}$ is a decomposition matrix (see Sect. 2.5).
- **Extension:** Used when we insert "dummy" coordinates to $\mathbf{w}_i$ to obtain $\bar{\mathbf{w}}_i$ whose coordinates are somewhat balanced. In this case, $\mathbf{P}_{i,j}$ is a $\{0,1\}$-matrix with zero-columns corresponding to positions of insertions.

Such a step transforms the term $\mathbf{M}_{i,j} \cdot \mathbf{w}_i$ into $\overline{\mathbf{M}}_{i,j} \cdot \bar{\mathbf{w}}_i$, where $\overline{\mathbf{M}}_{i,j} = \mathbf{M}_{i,j} \cdot \mathbf{P}_{i,j}$ is a public matrix. Also, using the commutativity property of addition, we often group together secret vectors having the same constraints.

After a number of transformations, we will reach a system equivalent to (13):

$$\begin{cases} \mathbf{M}'_{1,1} \cdot \mathbf{w}'_1 + \mathbf{M}'_{1,2} \cdot \mathbf{w}'_2 + \cdots + \mathbf{M}'_{1,k} \cdot \mathbf{w}'_k = \mathbf{v}_1, \\ \qquad\qquad\qquad \vdots \\ \mathbf{M}'_{t,1} \cdot \mathbf{w}'_1 + \mathbf{M}'_{t,2} \cdot \mathbf{w}'_2 + \cdots + \mathbf{M}'_{t,k} \cdot \mathbf{w}'_k = \mathbf{v}_t, \end{cases} \tag{14}$$

where integers t, k and matrices $\mathbf{M}'_{i,j}$ are public. Defining

$$\mathbf{M} = \left(\begin{array}{c|c|c|c} \mathbf{M}'_{1,1} & \mathbf{M}'_{1,2} & \cdots & \mathbf{M}'_{1,k} \\ \hline \vdots & \vdots & \ddots & \vdots \\ \hline \mathbf{M}'_{t,1} & \mathbf{M}'_{t,2} & \cdots & \mathbf{M}'_{t,k} \end{array}\right); \quad \mathbf{w} = \begin{pmatrix} \mathbf{w}'_1 \\ \vdots \\ \mathbf{w}'_k \end{pmatrix}; \quad \mathbf{v} = \begin{pmatrix} \mathbf{v}_1 \\ \vdots \\ \mathbf{v}_t \end{pmatrix},$$

we obtain the unified equation $\mathbf{M} \cdot \mathbf{w} = \mathbf{v} \bmod q$. At this stage, we will use a properly defined composition of random permutations to prove the constraints of $\mathbf{w}$. We remark that the crucial aspect of the above process is in fact the manipulation of witness vectors, while the transformations of public matrices/vectors just follow accordingly. To ease the presentation of the next subsections, we will thus only focus on the secret vectors.

In the process, we will employ various extending and permuting techniques which require introducing some notations. The most frequently used ones are given in Table 1. Some of these techniques appeared (in slightly different forms) in previous works [34–36,38,39]. The last three parts of the table summarizes newly-introduced techniques that will enable the treatment of secret-and-correlated objects involved in the evaluation of hidden branching programs.

In particular, the technique of the last row will be used for proving knowledge of an integer $z = x \cdot y$ for some $(x, y) \in [0, 4] \times \{0, 1\}$ satisfying other relations.

6.3 Protocol 1

Let n, m, q, N, t, B_χ be the parameters defined in Sect. 4. The protocol allows the prover to prove knowledge of LWE secrets and the well-formedness of ciphertexts. It is summarized as follows.

Common input: $\mathbf{F} \in \mathbb{Z}_q^{n\times m}$, $\mathbf{P} \in \mathbb{Z}_q^{m\times t}$; $\{\mathbf{a}_i \in \mathbb{Z}_q^n,\ \mathbf{b}_i \in \mathbb{Z}_q^t\}_{i=1}^N$.
Prover's goal is to prove knowledge of $\mathbf{S} \in [-B_\chi, B_\chi]^{n\times t}$, $\mathbf{E} \in [-B_\chi, B_\chi]^{m\times t}$, $\{\mathbf{x}_i \in [-B_\chi, B_\chi]^t, M_i \in \{0,1\}^t\}_{i=1}^N$ such that the following equations hold:

$$\begin{cases} \mathbf{F}^\top \cdot \mathbf{S} + \mathbf{E} = \mathbf{P} \bmod q \\ \forall i \in [N] : \mathbf{S}^\top \cdot \mathbf{a}_i + \mathbf{x}_i + \lfloor q/2 \rfloor \cdot M_i = \mathbf{b}_i \bmod q. \end{cases} \tag{15}$$

For each $j \in [t]$, let $\mathbf{p}_j, \mathbf{s}_j, \mathbf{e}_j$ be the j-th column of matrices $\mathbf{P}, \mathbf{S}, \mathbf{E}$, respectively. For each $(i, j) \in [N] \times [t]$, let $\mathbf{b}_i[j], \mathbf{x}_i[j], M_i[j]$ denote the j-th coordinate of vectors $\mathbf{b}_i, \mathbf{x}_i, M_i$, respectively. Then, observe that (15) can be rewritten as:

$$\begin{cases} \forall j \in [t] : \mathbf{F}^\top \cdot \mathbf{s}_j + \mathbf{I}_m \cdot \mathbf{e}_j = \mathbf{p}_j \bmod q \\ \forall (i,j) \in [N] \times [t] : \mathbf{a}_i^\top \cdot \mathbf{s}_j + 1 \cdot \mathbf{x}_i[j] + \lfloor q/2 \rfloor \cdot M_i[j] = \mathbf{b}_i[j] \bmod q. \end{cases} \tag{16}$$

Then, we form the following vectors:

$$\mathbf{w}_1 = \left(\mathbf{s}_1^\top \mid \ldots \mid \mathbf{s}_t^\top \mid \mathbf{e}_1^\top \mid \ldots \mid \mathbf{e}_t^\top \mid (\mathbf{x}_1[1], \ldots, \mathbf{x}_N[t])\right)^\top \in [-B_\chi, B_\chi]^{(n+m+N)t};$$
$$\mathbf{w}_2 = (M_1[1], \ldots, M_N[t])^\top \in \{0,1\}^{Nt}.$$

Next, we run $\mathsf{vdec}'_{(n+m+N)t, B_\chi}$ to decompose $\mathbf{w}_1$ into $\bar{\mathbf{w}}_1$ and then extend $\bar{\mathbf{w}}_1$ to $\mathbf{w}_1^* \in \mathsf{B}^3_{(n+m+N)t\delta_{B_\chi}}$. We also extend $\mathbf{w}_2$ into $\mathbf{w}_2^* \in \mathsf{B}^2_{Nt}$ and we then form $\mathbf{w} = ((\mathbf{w}_1^*)^\top \mid (\mathbf{w}_2^*)^\top)^\top \in \{-1,0,1\}^D$, where $D = 3(n+m+N)t\delta_{B_\chi} + 2Nt$.

Observe that relations (16) can be transformed into *one* equivalent equation of the form $\mathbf{M} \cdot \mathbf{w} = \mathbf{v} \bmod q$, where $\mathbf{M}$ and $\mathbf{v}$ are built from the common input.

Having performed the above unification, we now define VALID as the set of all vectors $\mathbf{t} = (\mathbf{t}_1^\top \mid \mathbf{t}_2^\top)^\top \in \{-1,0,1\}^D$, where $\mathbf{t}_1 \in \mathsf{B}^3_{(n+m+N)t\delta_{B_\chi}}$ and $\mathbf{t}_2 \in \mathsf{B}^2_{Nt}$. Clearly, our vector $\mathbf{w}$ belongs to the set VALID.

Next, we specify the set $\mathcal{S}$ and permutations of D elements $\{\Gamma_\phi : \phi \in \mathcal{S}\}$, for which the conditions in (12) hold.

- $\mathcal{S} := \mathcal{S}_{3(n+m+N)t\delta_{B_\chi}} \times \mathcal{S}_{2Nt}$.
- For $\phi = (\phi_1, \phi_2) \in \mathcal{S}$ and for $\mathbf{t} = (\mathbf{t}_1^\top \mid \mathbf{t}_2^\top)^\top \in \mathbb{Z}^D$, where $\mathbf{t}_1 \in \mathbb{Z}^{3(n+m+N)t\delta_{B_\chi}}$ and $\mathbf{t}_2 \in \mathbb{Z}^{2Nt}$, we define $\Gamma_\phi(\mathbf{t}) = (\phi_1(\mathbf{t}_1)^\top \mid \phi_2(\mathbf{t}_2)^\top)^\top$.

By inspection, it can be seen that the desired properties in (12) are satisfied. As a result, we can obtain the required ZKAoK by running the protocol from Sect. 6.1 with common input $(\mathbf{M}, \mathbf{v})$ and prover's input $\mathbf{w}$. The protocol has communication cost $\mathcal{O}(D \log q) = \widetilde{\mathcal{O}}(\lambda) \cdot \mathcal{O}(Nt)$ bits.

While this protocol has linear complexity in N, it is only used in the initialization phase, where $\Omega(N)$ bits inevitably have to be transmitted anyway.

6.4 Protocol 2

Let n, m, q, N, t, B be system parameters. The protocol allows the prover to prove knowledge of LWE secrets and the correctness of decryption.

Common input: $\mathbf{F} \in \mathbb{Z}_q^{n\times m}$, $\mathbf{P} \in \mathbb{Z}_q^{m\times t}$; $\mathbf{c}_0 \in \mathbb{Z}_q^n, \mathbf{c}_1 \in \mathbb{Z}_q^t$, $M' \in \{0,1\}^t$.
Prover's goal is to prove knowledge of $\mathbf{S} \in [-B_\chi, B_\chi]^{n\times t}$, $\mathbf{E} \in [-B_\chi, B_\chi]^{m\times t}$ and $\mathbf{y} \in [-q/5, q/5]^t$ such that the following equations hold:

$$\mathbf{F}^\top \cdot \mathbf{S} + \mathbf{E} = \mathbf{P} \bmod q; \mathbf{c}_0^\top \cdot \mathbf{S} + \mathbf{y}^\top = \mathbf{c}_1^\top - M'^\top \cdot \lfloor q/2 \rfloor \bmod q. \quad (17)$$

For each $j \in [t]$, let $\mathbf{p}_j, \mathbf{s}_j, \mathbf{e}_j$ be the j-th column of matrices $\mathbf{P}, \mathbf{S}, \mathbf{E}$, respectively; and let $\mathbf{y}[j], \mathbf{c}_1[j], M'[j]$ be the j-th entry of vectors $\mathbf{y}, \mathbf{c}_1, M'$, respectively. Then, observe that (17) can be re-written as:

$$\forall j \in [t] : \begin{cases} \mathbf{F}^\top \cdot \mathbf{s}_j + \mathbf{I}_m \cdot \mathbf{e}_j = \mathbf{p}_j \bmod q \\ \mathbf{c}_0^\top \cdot \mathbf{s}_j + 1 \cdot \mathbf{y}[j] = \mathbf{c}_1[j] - M'[j] \cdot \lfloor q/2 \rfloor \bmod q. \end{cases} \quad (18)$$

Next, we form vector $\mathbf{w}_1 = (\mathbf{s}_1^\top \mid \ldots \mid \mathbf{s}_t^\top \mid \mathbf{e}_1^\top \mid \ldots \mid \mathbf{e}_t^\top)^\top \in [-B_\chi, B_\chi]^{(n+m)t}$, then decompose it to $\bar{\mathbf{w}}_1 \in \{-1,0,1\}^{(n+m)t\delta_{B_\chi}}$, and extend $\bar{\mathbf{w}}_1$ to $\mathbf{w}_1^* \in \mathsf{B}^3_{(n+m)t\delta_{B_\chi}}$.

At the same time, we decompose vector $\mathbf{y} = (\mathbf{y}[1], \ldots, \mathbf{y}[t])^\top \in [-q/5, q/5]^t$ to $\bar{\mathbf{y}} \subset \{-1,0,1\}^{t\delta_{q/5}}$, and then extend $\bar{\mathbf{y}}$ to $\mathbf{y}^* \in \mathsf{B}^3_{t\delta_{q/5}}$.

Defining the ternary vector $\mathbf{w} = ((\mathbf{w}_1^*)^\top \mid (\mathbf{y}^*)^\top)^\top \in \{-1,0,1\}^D$ of dimension $D = 3(n+m)t\delta_{B_\chi} + 3t\delta_{q/5}$, we finally obtain the equation $\mathbf{M} \cdot \mathbf{w} = \mathbf{v} \bmod q$, for public matrix $\mathbf{M}$ and public vector $\mathbf{v}$. Using similar arguments as in Sect. 6.3, we can obtain the desired zero-knowledge argument system. The protocol has communication cost $\mathcal{O}(D \log q) = \widetilde{\mathcal{O}}(\lambda) \cdot \mathcal{O}(t)$ bits.

6.5 Protocol 3

Let n, m, m_d, q, t, ℓ, B be the parameters defined in Sect. 4. The protocol allows the prover to argue that a given ciphertext is a correct randomization of some hidden ciphertext and that he knows a valid signature on that ciphertext. Let β be the infinity norm bound of these valid signatures.

Table 1. Basic notations and extending/permuting techniques used in our protocols.

Notation	Meaning/Property/Usage/Technique
$\mathsf{B}^2_{\mathfrak{m}}$	– The set of vectors in $\{0,1\}^{2\mathfrak{m}}$ with Hamming weight $\mathfrak{m}$
	– $\forall \phi \in \mathcal{S}_{2\mathfrak{m}}, \mathbf{x}' \in \mathbb{Z}^{2\mathfrak{m}} : \mathbf{x}' \in \mathsf{B}^2_{\mathfrak{m}} \Leftrightarrow \phi(\mathbf{x}') \in \mathsf{B}^2_{\mathfrak{m}}$
	– To prove $\mathbf{x} \in \{0,1\}^{\mathfrak{m}}$: Extend $\mathbf{x}$ to $\mathbf{x}' \in \mathsf{B}^2_{\mathfrak{m}}$, then permute $\mathbf{x}'$
$\mathsf{B}^3_{\mathfrak{m}}$	– The set of vectors in $\{-1,0,1\}^{3\mathfrak{m}}$ that have exactly $\mathfrak{m}$ coordinates equal to j, for every $j \in \{-1,0,1\}$
	– $\forall \phi \in \mathcal{S}_{3\mathfrak{m}}, \mathbf{x}' \in \mathbb{Z}^{3\mathfrak{m}} : \mathbf{x}' \in \mathsf{B}^3_{\mathfrak{m}} \Leftrightarrow \phi(\mathbf{x}') \in \mathsf{B}^3_{\mathfrak{m}}$
	– To prove $\mathbf{x} \in \{-1,0,1\}^{\mathfrak{m}}$: Extend $\mathbf{x}$ to $\mathbf{x}' \in \mathsf{B}^3_{\mathfrak{m}}$, then permute $\mathbf{x}'$
$\mathsf{ext}_2(\cdot)$ and $T_2[\cdot](\cdot)$	– For $c \in \{0,1\}$: $\mathsf{ext}_2(c) = (\bar{c}, c)^\top \in \{0,1\}^2$
	– For $b \in \{0,1\}$ and $\mathbf{x} = (x_0, x_1)^\top \in \mathbb{Z}^2$: $T_2[b](\mathbf{x}) = (x_b, x_{\bar{b}})^\top$
	– Property: $\mathbf{x} = \mathsf{ext}_2(c) \Leftrightarrow T_2[b](\mathbf{x}) = \mathsf{ext}_2(c \oplus b)$
	– To prove $c \in \{0,1\}$ simultaneously satisfies many relations: Extend it to $\mathbf{x} = \mathsf{ext}_2(c)$, then permute and use the *same* b at all appearances
$\mathsf{expand}\cdot,\cdot$ and $T_{\mathsf{exp}}[\cdot,\cdot](\cdot)$	– For $c \in \{0,1\}$ and $\mathbf{x} \in \mathbb{Z}^m$: $\mathsf{expand}(c, \mathbf{x}) = (\bar{c} \cdot \mathbf{x}^\top \mid c \cdot \mathbf{x}^\top)^\top \in \mathbb{Z}^{2m}$
	– For $b \in \{0,1\}, \phi \in \mathcal{S}_m$, $\mathbf{v} = \begin{pmatrix} \mathbf{v}_0 \\ \mathbf{v}_1 \end{pmatrix} \in \mathbb{Z}^{2m}$: $T_{\mathsf{exp}}[b, \phi](\mathbf{v}) = \begin{pmatrix} \phi(\mathbf{v}_b) \\ \phi(\mathbf{v}_{\bar{b}}) \end{pmatrix}$
	– Property: $\mathbf{v} = \mathsf{expand}(c, \mathbf{x}) \Leftrightarrow T_{\mathsf{exp}}[b, \phi](\mathbf{v}) = \mathsf{expand}(c \oplus b, \phi(\mathbf{x}))$
$[\cdot]_5$	For $k \in \mathbb{Z}$: $[k]_5$ denotes the integer $t \in \{0,1,2,3,4\}$, s.t. $t = k \bmod 5$
$\mathsf{ext}_5(\cdot)$ and $T_5[\cdot](\cdot)$	– For $x \in [0,4]$: $\mathsf{ext}_5(x) = ([x+4]_5, [x+3]_5, [x+2]_5, [x+1]_5, x)^\top \in [0,4]^5$
	– For $c \in [0,4]$ and $\mathbf{v} = (v_0, v_1, v_2, v_3, v_4)^\top \in \mathbb{Z}^5$: $T_5[c](\mathbf{v}) = (v_{[-c]_5}, v_{[-c+1]_5}, v_{[-c+2]_5}, v_{[-c+3]_5}, v_{[-c+4]_5})^\top$
	– Property: $\mathbf{v} = \mathsf{ext}_5(x) \Leftrightarrow T_5[c](\mathbf{v}) = \mathsf{ext}_5(x + c \bmod 5)$
	–To prove $x \in [0,4]$ simultaneously satisfies many relations: Extend it to $\mathbf{v} = \mathsf{ext}_5(x)$, then permute and use the *same* c at all appearances
unit_x	– $\forall x \in [0,4]$: unit_x is the 5-dim unit vector $(v_0, \ldots, v_4)^\top$ with $v_x = 1$
	– For $c \in [0,4], \mathbf{v} \in \mathbb{Z}^5$: $\mathbf{v} = \mathsf{unit}_x \Leftrightarrow T_5[c](\mathbf{v}) = \mathsf{unit}_{x+c \bmod 5}$
	$\rightarrow$ Allow proving $\mathbf{v} = \mathsf{unit}_x$ for some $x \in [0,4]$ satisfying other relations
$\mathsf{ext}_{5\times 2}(\cdot,\cdot)$ and $T_{5\times 2}[\cdot,\cdot](\cdot)$	– For $x \in [0,4]$ and $y \in \{0,1\}$: $\mathsf{ext}_{5\times 2}(x,y) = ([x+4]_5 \cdot \bar{y}, [x+4]_5 \cdot y, [x+3]_5 \cdot \bar{y}, [x+3]_5 \cdot y, [x+2]_5 \cdot \bar{y}, [x+2]_5 \cdot y, [x+1]_5 \cdot \bar{y}, [x+1]_5 \cdot y, x \cdot \bar{y}, x \cdot y)^\top \in [0,4]^{10}$
	For $(c,b) \in [0,4] \times \{0,1\}$ and $\mathbf{v} = (v_{0,0}, v_{0,1}, \ldots, v_{4,0}, v_{4,1})^\top \in \mathbb{Z}^{10}$: $T_{5\times 2}[c,b](\mathbf{v}) = (v_{[-c]_5,b}, v_{[-c]_5,\bar{b}}, v_{[-c+1]_5,b}, v_{[-c+1]_5,\bar{b}}, v_{[-c+2]_5,b}, v_{[-c+2]_5,\bar{b}}, v_{[-c+3]_5,b}, v_{[-c+3]_5,\bar{b}}, v_{[-c+4]_5,b}, v_{[-c+4]_5,\bar{b}})^\top$
	– Property: $\mathbf{v} = \mathsf{ext}_{5\times 2}(x,y) \Leftrightarrow T_{5\times 2}[c,b](\mathbf{v}) = \mathsf{ext}_{5\times 2}(x + c \bmod 5, y \oplus b)$ $\rightarrow$ Allow proving $z = x \cdot y$ for some $(x,y) \in [0,4] \times \{0,1\}$ satisfying other relations: Extend z to $\mathbf{v} = \mathsf{ext}_{5\times 2}(x,y)$, then permute and use the *same* c, b at all appearances of x, y, respectively

Common input: It consists of matrices $\mathbf{F} \in \mathbb{Z}_q^{n\times m}$, $\mathbf{P} \in \mathbb{Z}_q^{m\times t}$, $\mathbf{A}, \mathbf{A}_0, \mathbf{A}_1, \ldots, \mathbf{A}_\ell \in \mathbb{Z}_q^{n\times m}$, $\mathbf{D} \in \mathbb{Z}_q^{n\times m_d}$ and vectors $\mathbf{c}_0 \in \mathbb{Z}_q^n, \mathbf{c}_1 \in \mathbb{Z}_q^t, \mathbf{u} \in \mathbb{Z}_q^n$.

Prover's goal is to prove knowledge of $\mathfrak{m} \in \{0,1\}^{m_d}$, $\mu \in \{0,1\}^t$, $\mathbf{e} \in \{-1,0,1\}^t$, $\nu \in [-B,B]^t$, $\tau = (\tau[1], \ldots, \tau[\ell])^\top \in \{0,1\}^\ell$, $\mathbf{v}_1, \mathbf{v}_2 \in [-\beta,\beta]^m$ such that the following equations hold:

$$\begin{cases} \mathbf{A} \cdot \mathbf{v}_1 + \mathbf{A}_0 \cdot \mathbf{v}_2 + \sum_{j=1}^{\ell} \mathbf{A}_j \cdot (\tau[j] \cdot \mathbf{v}_2) - \mathbf{D} \cdot \mathfrak{m} = \mathbf{u} \bmod q; \\ \mathbf{H}_{n+t,q-1} \cdot \mathfrak{m} + \begin{pmatrix} \mathbf{F} \\ \mathbf{P}^\top \end{pmatrix} \cdot \mathbf{e} + \begin{pmatrix} \mathbf{0}^{n \times t} \\ \lfloor \frac{q}{2} \rfloor \cdot \mathbf{I}_t \end{pmatrix} \cdot \mu + \begin{pmatrix} \mathbf{0}^{n \times t} \\ \mathbf{I}_t \end{pmatrix} \cdot \nu = \begin{pmatrix} \mathbf{c}_0 \\ \mathbf{c}_1 \end{pmatrix} \bmod q. \end{cases} \tag{19}$$

For this purpose, we perform the following transformations on the witnesses.

Decompositions. Decompose vectors $\mathbf{v}_1, \mathbf{v}_2, \nu$ to vectors $\bar{\mathbf{v}}_1 \in \{-1,0,1\}^{m\delta_\beta}$, $\bar{\mathbf{v}}_2 \in \{-1,0,1\}^{m\delta_\beta}$, $\bar{\nu} \in \{-1,0,1\}^{t\delta_B}$, respectively.

Extensions/Combinations.

- Let $\mathbf{w}_1 = (\mathfrak{m}^\top \mid \mu^\top)^\top \in \{0,1\}^{m_d+t}$ and extend it into $\mathbf{w}_1^* \in \mathsf{B}^2_{m_d+t}$.
- Let $\mathbf{w}_2 = (\bar{\mathbf{v}}_1^\top \mid \bar{\nu}^\top \mid \mathbf{e}^\top)^\top \in \{-1,0,1\}^{m\delta_\beta + t\delta_B + t}$ and extend it into the vector $\mathbf{w}_2^* \in \mathsf{B}^3_{m\delta_\beta + t\delta_B + t}$.
- Extend $\bar{\mathbf{v}}_2$ into $\mathbf{s}_0 \in \mathsf{B}^3_{m\delta_\beta}$. Then, for each $j \in [\ell]$, define $\mathbf{s}_j = \mathsf{expand}(\tau[j], \mathbf{s}_0)$. (We refer to Table 1 for details about $\mathsf{expand}(\cdot,\cdot)$.)

Now, we form vector $\mathbf{w} = \left(\mathbf{w}_1^{*\top} \mid \mathbf{w}_2^{*\top} \mid \mathbf{s}_0^\top \mid \mathbf{s}_1^\top \mid \ldots \mid \mathbf{s}_\ell^\top\right)^\top \in \{-1,0,1\}^D$, where $D = (2\ell+2)3m\delta_\beta + 3t\delta_B + 3t + 2(m_d+t)$. At this point, we observe that the equations in (19) can be equivalently transformed into $\mathbf{M} \cdot \mathbf{w} = \mathbf{v} \bmod q$, where the matrix $\mathbf{M}$ and the vector $\mathbf{v}$ are built from the public input.

Having performed the above transformations, we now define VALID as the set of all vectors $\mathbf{t} = (\mathbf{t}_1^\top \mid \mathbf{t}_2^\top \mid \mathbf{t}_{3,0}^\top \mid \mathbf{t}_{3,1}^\top \mid \ldots \mid \mathbf{t}_{3,\ell}^\top)^\top \in \{-1,0,1\}^D$ for which there exists $\tau = (\tau[1], \ldots, \tau[\ell])^\top \in \{0,1\}^\ell$ such that:

$$\mathbf{t}_1 \in \mathsf{B}^2_{m_d+t};\ \mathbf{t}_2 \in \mathsf{B}^3_{m\delta_\beta + t\delta_B + t};\ \mathbf{t}_{3,0} \in \mathsf{B}^3_{m\delta_\beta};\ \forall j \in [\ell] : \mathbf{t}_{3,j} = \mathsf{expand}(\tau[j], \mathbf{t}_{3,0}).$$

It can be seen that $\mathbf{w}$ belongs to this tailored set. Now, let us specify the set $\mathcal{S}$ and permutations of D elements $\{\Gamma_\phi : \phi \in \mathcal{S}\}$ satisfying the conditions in (12).

- $\mathcal{S} := \mathcal{S}_{2(m_d+t)} \times \mathcal{S}_{3(m\delta_\beta + t\delta_B + t)} \times \mathcal{S}_{3m\delta_\beta} \times \{0,1\}^\ell$.
- For $\phi = \left(\phi_1, \phi_2, \phi_3, (b[1], \ldots, b[\ell])^\top\right) \in \mathcal{S}$, we define the permutation Γ_ϕ that transforms vector $\mathbf{t} = (\mathbf{t}_1^\top \mid \mathbf{t}_2^\top \mid \mathbf{t}_{3,0}^\top \mid \mathbf{t}_{3,1}^T \mid \ldots \mid \mathbf{t}_{3,\ell}^\top)^\top \in \mathbb{Z}^D$ as follows:

$$\Gamma_\phi(\mathbf{t}) = \left(\phi_1(\mathbf{t}_1)^\top \mid \phi_2(\mathbf{t}_2)^\top \mid \phi_3(\mathbf{t}_{3,0})^\top \mid \right. \\ \left. T_{\mathsf{exp}}[b[1], \phi_3](\mathbf{t}_{3,1})^\top \mid \ldots \mid T_{\mathsf{exp}}[b[\ell], \phi_3](\mathbf{t}_{3,\ell})^\top\right)^\top.$$

By inspection, it can be seen that the properties in (12) are indeed satisfied. As a result, we can obtain the required argument of knowledge by running the protocol from Sect. 6.1 with common input $(\mathbf{M}, \mathbf{v})$ and prover's input $\mathbf{w}$. The protocol has communication cost $\mathcal{O}(D \log q) = \widetilde{\mathcal{O}}(\lambda) \cdot \mathcal{O}(\log N + t)$ bits.

6.6 Protocol 4: A Treatment of Hidden Branching Programs

We now present the proof system run by the user in the OT-AC system of Sect. 5. It allows arguing knowledge of an input $\mathbf{x} = (x_0, \ldots, x_{\kappa-1})^\top \in \{0,1\}^\kappa$ satisfying a hidden branching program $\mathsf{BP} = \{(\mathrm{var}(\theta), \pi_{\theta,0}, \pi_{\theta,1})\}_{\theta=1}^{L}$ of length for $L \in \mathsf{poly}(\lambda)$. The prover should additionally demonstrate that: (i) He has a valid credential for $\mathbf{x}$; (ii) The hashed encoding of BP is associated with some hidden ciphertext of the database (and he knows a signature guaranteeing this link); (iii) A given ciphertext is a re-randomization of that hidden ciphertext.

Recall that, at each step $\theta \in [L]$ of the evaluation of $\mathsf{BP}(\mathbf{x})$, we have to look up the value $x_{\mathrm{var}(\theta)}$ in $\mathbf{x} = (x_0, \ldots, x_{\kappa-1})^\top$ to compute the θ-th state η_θ as per

$$\eta_\theta = \pi_{\theta, x_{\mathrm{var}(\theta)}}(\eta_{\theta-1}) = \pi_{\theta,0}(\eta_{\theta-1}) \cdot \bar{x}_{\mathrm{var}(\theta)} + \pi_{\theta,1}(\eta_{\theta-1}) \cdot x_{\mathrm{var}(\theta)}. \tag{20}$$

To prove that each step is done correctly, it is necessary to provide evidence that the corresponding search is honestly carried out without revealing $x_{\mathrm{var}(\theta)}$, $\mathrm{var}(\theta)$ nor $\{\pi_{\theta,b}\}_{b=0}^{1}$. To this end, a first idea is to perform a simple left-to-right search on $(x_0, \ldots, x_{\kappa-1})$: namely, (20) is expressed in terms of a matrix-vector relation where η_θ is encoded as a unit vector of dimension 5; $\{\pi_{\theta,b}\}_{b=0}^{1}$ are represented as permutation matrices; and $\mathbf{x}_{\mathrm{var}(\theta)} = \mathbf{M}_{\mathrm{var}(\theta)} \cdot \mathbf{x}$ is computed using a matrix $\mathbf{M}_{\mathrm{var}(\theta)} \in \{0,1\}^{\kappa \times \kappa}$ containing exactly one 1 per row. While this approach can be handled using proofs for matrix-vector relations using the techniques of [35], the expected complexity is $\mathcal{O}(\kappa)$ for each step, so that the total complexity becomes $\mathcal{O}(L\kappa)$. Fortunately, a better complexity can be achieved.

If we instead perform a dichotomic search on $\mathbf{x} = (x_0, \ldots, x_{\kappa-1})^\top$, we can reduce the complexity of each step to $\mathcal{O}(\log \kappa)$. To this end, we need to prove a statement "I performed a correct dichotomic search on my secret array $\mathbf{x}$".

In order to solve this problem, we will employ two existing lattice-based tools:

(i) A variant of the SIS-based computationally binding and statistically hiding commitment scheme from [32], which allows to commit to one-bit messages;
(ii) The SIS-based Merkle hash tree proposed in [36].

Let $\bar{\mathbf{A}} \hookleftarrow U(\mathbb{Z}_q^{n \times m})$ and $\mathbf{a}_{\mathsf{com}} \hookleftarrow U(\mathbb{Z}_q^n)$. For each $i \in [0, \kappa - 1]$, we let the receiver commit to $x_i \in \{0,1\}$ as $\mathsf{com}_i = \mathbf{a}_{\mathsf{com}} \cdot x_i + \bar{\mathbf{A}} \cdot \mathbf{r}_{\mathsf{com},i}$, with $\mathbf{r}_{\mathsf{com},i} \hookleftarrow U(\{0,1\}^m)$, and reveal $\mathsf{com}_1, \ldots, \mathsf{com}_{\kappa-1}$ to the sender. We build a Merkle tree of depth $\delta_\kappa = \lceil \log \kappa \rceil$ on top of the leaves $\mathsf{com}_0, \ldots, \mathsf{com}_{\kappa-1}$ using the SIS-based hash function $h_{\bar{\mathbf{A}}} : \{0,1\}^{n\lceil \log q \rceil} \times \{0,1\}^{n\lceil \log q \rceil} \rightarrow \{0,1\}^{n\lceil \log q \rceil}$ of [36]. Our use of Merkle trees is reminiscent of [36] in that the content of the leaves is public. The Merkle tree will actually serve as a "bridge" ensuring that: (i) The same string $\mathbf{x}$ is used in all steps while enabling dichotomic searches; (ii) At each step, the prover indeed uses some coordinate of $\mathbf{x}$ (without revealing which one), the choice of which is dictated by a path in the tree determined by $\mathrm{var}(\theta)$.

Since $\{\mathsf{com}_i\}_{i=0}^{\kappa-1}$ are public, both parties can deterministically compute the root $\mathbf{u}_{\mathsf{tree}}$ of the Merkle tree. For each $\theta \in [L]$, we consider the binary representation $d_{\theta,1}, \ldots, d_{\theta,\delta_\kappa}$ of $\mathrm{var}(\theta)$, which is part of the encoding of BP defined in (7). We then prove knowledge of a bit y_θ satisfying the statement "From the root $\mathbf{u}_{\mathsf{tree}} \in \{0,1\}^{n\lceil \log q \rceil}$ of the tree, the path determined by the bits $d_{\theta,1}, \ldots, d_{\theta,\delta_\kappa}$ leads to the leaf associated with the commitment opened to y_θ." If the Merkle tree and the commitment scheme are both secure, it should hold that $y_\theta = x_{\mathrm{var}(\theta)}$. Said otherwise, we can provably perform a "dichotomic search" for $x_{\mathrm{var}(\theta)} = y_\theta$. Moreover, the techniques from [36] can be adapted to do this in zero-knowledge manner, i.e., without revealing the path nor the reached leaf.

Now, our task can be divided into 3 steps: (i) Proving that the searches on Merkle tree yield $y_1, \ldots, y_L$; (ii) Proving that the branching program evaluates to $\mathsf{BP}(\mathbf{x}) = 1$ if $y_1, \ldots, y_L$ are used in the evaluation; (iii) Proving all the other relations mentioned above, as well as the consistency of $\{\mathsf{com}_i\}_{i=0}^{\kappa-1}$ and the fact that they open to a certified $\mathbf{x} \in \{0,1\}^\kappa$.

Thanks to dichotomic searches, the communication cost drops to $\mathcal{O}(L\delta_\kappa + \kappa)$. These steps can be treated as explained below.

THE MERKLE TREE STEP. At each step $\theta \in [L]$, the prover proves knowledge of a path made of δ_κ nodes $\mathbf{g}_{\theta,1}, \ldots, \mathbf{g}_{\theta,\delta_\kappa} \in \{0,1\}^{n\lceil \log q \rceil}$ determined by $d_{\theta,1}, \ldots, d_{\theta,\delta_\kappa}$, as well as their siblings $\mathbf{t}_{\theta,1}, \ldots, \mathbf{t}_{\theta,\delta_\kappa} \in \{0,1\}^{n\lceil \log q \rceil}$. Also, the prover argues knowledge of an opening $(y_\theta, \mathbf{r}_\theta) \in \{0,1\} \times \{0,1\}^m$ for the commitment of which $\mathbf{g}_{\theta,\delta_\kappa}$ is a binary decomposition. As shown in [36] (and recalled in the full paper), it suffices to prove the following relations (mod q):

$$\forall \theta \in [L] \begin{cases} \bar{\mathbf{A}} \cdot \mathsf{expand}(d_{\theta,1}, \mathbf{g}_{\theta,1}) + \bar{\mathbf{A}} \cdot \mathsf{expand}(\bar{d}_{\theta,1}, \mathbf{t}_{\theta,1}) = \mathbf{H}_{n,q-1} \cdot \mathbf{u}_{\mathsf{tree}}, \\ \bar{\mathbf{A}} \cdot \mathsf{expand}(d_{\theta,2}, \mathbf{g}_{\theta,2}) + \bar{\mathbf{A}} \cdot \mathsf{expand}(\bar{d}_{\theta,2}, \mathbf{t}_{\theta,2}) \\ \qquad\qquad\qquad\qquad\qquad\qquad - \mathbf{H}_{n,q-1} \cdot \mathbf{g}_{\theta,1} = \mathbf{0}, \\ \qquad\qquad\qquad\qquad \vdots \\ \bar{\mathbf{A}} \cdot \mathsf{expand}(d_{\theta,\delta_\kappa}, \mathbf{g}_{\theta,\delta_\kappa}) + \bar{\mathbf{A}} \cdot \mathsf{expand}(\bar{d}_{\theta,\kappa}, \mathbf{t}_{\theta,\kappa}) \\ \qquad\qquad\qquad\qquad\qquad\qquad - \mathbf{H}_{n,q-1} \cdot \mathbf{g}_{\theta,\delta_\kappa - 1} = \mathbf{0}, \\ \mathbf{a}_{\mathsf{com}} \cdot y_\theta + \bar{\mathbf{A}} \cdot \mathbf{r}_\theta - \mathbf{H}_{n,q-1} \cdot \mathbf{g}_{\theta,\delta_\kappa} = \mathbf{0}, \end{cases} \tag{21}$$

where $\mathsf{expand}(\cdot,\cdot)$ is defined in Table 1.

Extending.

- For each $(\theta, i) \in [L] \times [\delta_\kappa]$: Extend $\mathbf{g}_{\theta,i}, \mathbf{t}_{\theta,i} \in \{0,1\}^{m/2}$ to $\widetilde{\mathbf{g}}_{\theta,i}, \widetilde{\mathbf{t}}_{\theta,i} \in \mathsf{B}_{m/2}^2$, respectively. Then, let $\widehat{\mathbf{g}}_{\theta,i} = \mathsf{expand}(d_{\theta,i}, \widetilde{\mathbf{g}}_{\theta,i})$ and $\widehat{\mathbf{t}}_{\theta,i} = \mathsf{expand}(\bar{d}_{\theta,i}, \widetilde{\mathbf{t}}_{\theta,i})$.
- For each $\theta \in [L]$, extend the bit y_θ into the vector $\mathbf{y}_\theta = \mathsf{ext}_2(y_\theta) \in \{0,1\}^2$.
- Let $\widetilde{\mathbf{r}} = (\mathbf{r}_1^\top \mid \ldots \mid \mathbf{r}_L^\top)^\top \in \{0,1\}^{mL}$, then extend it into the vector $\widehat{\mathbf{r}} \in \mathsf{B}_{mL}^2$.

Combining. Next, we let $D_{\mathsf{tree}} = 5mL\delta_\kappa + 2L + 2mL$ and define

$$\mathbf{w}_{\mathsf{tree}} = \big(\widetilde{\mathbf{g}}_{1,1}^\top \mid \widehat{\mathbf{g}}_{1,1}^\top \mid \widehat{\mathbf{t}}_{1,1}^\top \mid \cdots \mid \widetilde{\mathbf{g}}_{1,\delta_\kappa}^\top \mid \widehat{\mathbf{g}}_{1,\delta_\kappa}^\top \mid \widehat{\mathbf{t}}_{1,\delta_\kappa}^\top \mid \cdots \mid \widetilde{\mathbf{g}}_{L,1}^\top \mid \widehat{\mathbf{g}}_{L,1}^\top \mid \widehat{\mathbf{t}}_{L,1}^\top \\ \mid \ldots \mid \widetilde{\mathbf{g}}_{L,\delta_\kappa}^\top \mid \widehat{\mathbf{g}}_{L,\delta_\kappa}^\top \mid \widehat{\mathbf{t}}_{L,\delta_\kappa}^\top \mid \mathbf{y}_1^\top \mid \ldots \mid \mathbf{y}_L^\top \mid \widehat{\mathbf{r}}^\top\big)^\top \in \{0,1\}^{D_{\mathsf{tree}}}. \tag{22}$$

Then, observe that, the above $L(\delta_\kappa + 1)$ equations can be combined into one:

$$\mathbf{M}_{\mathsf{tree}} \cdot \mathbf{w}_{\mathsf{tree}} = \mathbf{v}_{\mathsf{tree}} \bmod q, \tag{23}$$

where matrix $\mathbf{M}_{\mathsf{tree}}$ and vector $\mathbf{v}_{\mathsf{tree}}$ are built from the public input.

THE BRANCHING PROGRAM STEP. The last three parts of Table 1 describe the vector transformations that will be used to handle the secret vectors appearing in the evaluation of BP. The following equations emulate the evaluation process. In particular, for each $\theta \in [2, L]$, we introduce an extra vector $\mathbf{e}_\theta = (c_{\theta,0}, \ldots, c_{\theta,4}) \in \{0,1\}^5$ to enable the extraction of the values $\pi_{\theta,0}(\eta_{\theta-1})$, and $\pi_{\theta,1}(\eta_{\theta-1})$.

$$\begin{cases} \pi_{1,0}(0) \cdot \bar{y}_1 + \pi_{1,1}(0) \cdot y_1 - \eta_1 = 0, & \text{// computing } \eta_1 \text{ with } \eta_0 = 0 \\ \mathbf{e}_2 - \sum_{i=0}^{4} \mathsf{unit}_i \cdot c_{2,i} = (0,0,0,0,0)^\top, & \text{// we will also prove } \mathbf{e}_2 = \mathsf{unit}_{\eta_1} \\ f_{2,0} - \sum_{i=0}^{4} \pi_{2,0}(i) \cdot c_{2,i} = 0, & \text{// meaning: } f_{2,0} = \pi_{2,0}(\eta_1) \\ f_{2,1} - \sum_{i=0}^{4} \pi_{2,1}(i) \cdot c_{2,i} = 0, & \text{// meaning: } f_{2,1} = \pi_{2,1}(\eta_1) \\ f_{2,0} \cdot \bar{y}_2 + f_{2,1} \cdot y_2 - \eta_2 = 0, & \text{// computing } \eta_2 \\ \qquad \vdots & \\ \mathbf{e}_L - \sum_{i=0}^{4} \mathsf{unit}_i \cdot c_{L,i} = (0,0,0,0,0)^\top, & \text{// we will also prove } \mathbf{e}_L = \mathsf{unit}_{\eta_{L-1}} \\ f_{L,0} - \sum_{i=0}^{4} \pi_{L,0}(i) \cdot c_{L,i} = 0, & \text{// meaning: } f_{L,0} = \pi_{L,0}(\eta_{L-1}) \\ f_{L,1} - \sum_{i=0}^{4} \pi_{L,1}(i) \cdot c_{L,i} = 0, & \text{// meaning: } f_{L,1} = \pi_{L,1}(\eta_{L-1}) \\ f_{L,0} \cdot \bar{y}_L + f_{L,1} \cdot y_L = 0. & \text{// final state} \eta_L = 0 \end{cases} \tag{24}$$

Extending.

- For each $\theta \in [L-1]$, extend $\eta_\theta \in [0,4]$ to 5-dimensional vector $\mathbf{s}_\theta = \mathsf{ext}_5(\eta_\theta)$.
- For each $(\theta, j) \in [2, L] \times \{0,1\}$, extend $f_{\theta,j} \in [0,4]$ to $\mathbf{f}_{\theta,j} = \mathsf{ext}_5(f_{\theta,j})$.
- For each $(\theta, i) \in [2, L] \times [0,4]$, extend $c_{\theta,i} \in \{0,1\}$ to $\mathbf{c}_{\theta,i} = \mathsf{ext}_2(c_{\theta,i})$.
- Extend the products $\pi_{1,0}(0) \cdot \bar{y}_1$ and $\pi_{1,1}(0) \cdot y_1$ into 10-dimensional vectors $\mathbf{h}_{1,0} = \mathsf{ext}_{5\times 2}(\pi_{1,0}(0), \bar{y}_1)$ and $\mathbf{h}_{1,1} = \mathsf{ext}_{5\times 2}(\pi_{1,1}(0), y_1)$, respectively.
- For each $\theta \in [2, L]$, extend the products $f_{\theta,0} \cdot \bar{y}_\theta$ and $f_{\theta,1} \cdot y_\theta$ into 10-dimensional vectors $\mathbf{h}_{\theta,0} = \mathsf{ext}_{5\times 2}(f_{\theta,0}, \bar{y}_\theta)$ and $\mathbf{h}_{\theta,1} = \mathsf{ext}_{5\times 2}(f_{\theta,1}, y_\theta)$.
- For $(\theta, i) \in [2, L] \times [0,4]$, extend the products $\pi_{\theta,0}(i) \cdot c_{\theta,i}$ and $\pi_{\theta,1}(i) \cdot c_{\theta,i}$ into $\mathbf{z}_{\theta,0,i} = \mathsf{ext}_{5\times 2}(\pi_{\theta,0}(i), c_{\theta,i})$ and $\mathbf{z}_{\theta,1,i} = \mathsf{ext}_{5\times 2}(\pi_{\theta,1}(i), c_{\theta,i})$, respectively.

Combining. Let $D_{\mathsf{BP}} = 150L - 130$, and form $\mathbf{w}_{\mathsf{BP}} \in [0,4]^{D_{\mathsf{BP}}}$ of the form:

$$\begin{aligned} (\mathbf{s}_1^\top \mid \ldots \mid \mathbf{s}_{L-1}^\top \mid \mathbf{e}_2^\top \mid \ldots \mid \mathbf{e}_L^\top \mid \mathbf{c}_{2,0}^\top \mid \ldots \mid \mathbf{c}_{L,4}^\top \mid \mathbf{z}_{2,0,0}^\top \mid \ldots \mid \mathbf{z}_{L,1,4}^\top \mid \\ \mathbf{f}_{2,0}^\top \mid \ldots \mid \mathbf{f}_{L,1}^\top \mid \mathbf{h}_{1,0}^\top \mid \mathbf{h}_{1,1}^\top \mid \mathbf{h}_{2,0}^\top \mid \mathbf{h}_{2,1}^\top \mid \ldots \mid \mathbf{h}_{L,0}^\top \mid \mathbf{h}_{L,1}^\top)^\top. \end{aligned} \tag{25}$$

Then, observe that the vector $\mathbf{w}_{\mathsf{BP}}$ of (25) satisfies *one* equation of the form:

$$\mathbf{M}_{\mathsf{BP}} \cdot \mathbf{w}_{\mathsf{BP}} = \mathbf{v}_{\mathsf{BP}}, \tag{26}$$

where matrix $\mathbf{M}_{\mathsf{BP}}$ and vector $\mathbf{v}_{\mathsf{BP}}$ are obtained from the common input. Note that we work with integers in $[0,4]$, which are much smaller than q. As a result,

$$\mathbf{M}_{\mathsf{BP}} \cdot \mathbf{w}_{\mathsf{BP}} = \mathbf{v}_{\mathsf{BP}} \bmod q. \tag{27}$$

Conversely, if we can prove that (27) holds for a well-formed vector $\mathbf{w}_{\mathsf{BP}}$, then that vector should also satisfy (26).

THE THIRD STEP. In the third layer, we have to prove knowledge of:

$$\begin{cases} d_{1,1},\ldots,d_{L,\delta_\kappa} \in \{0,1\}, \pi_{1,0}(0),\ldots,\pi_{L,1}(4) \in [0,4],\ \mathsf{m} \in \{0,1\}^{m_d}, \\ \mathbf{x} = (x_0,\ldots,x_{\kappa-1})^\top \in \{0,1\}^\kappa,\ \mathsf{m}_{\mathsf{U},\mathbf{x}} \in \{0,1\}^{\frac{m}{2}+\kappa},\ \widehat{\mathsf{m}}_{\mathsf{U},\mathbf{x}} \in \{0,1\}^{\frac{m}{2}}, \\ \mathbf{e}_{\mathsf{U}} \in \{0,1\}^m,\ \mathbf{r}_{\mathsf{com},0},\ldots,\mathbf{r}_{\mathsf{com},\kappa-1} \in \{0,1\}^m,\ \mu \in \{0,1\}^t,\ \tau \in \{0,1\}^\ell, \\ \tau_{\mathsf{U}} \in \{0,1\}^{\ell_I}, \mathbf{v}_1,\mathbf{v}_2,\mathbf{v}_{\mathsf{U},1},\mathbf{v}_{\mathsf{U},2},\mathbf{r}_{\mathsf{U}} \in [-\beta,\beta]^m, \mathbf{e} \in \{-1,0,1\}^t, \nu \in [-B,B]^t, \end{cases} \tag{28}$$

which satisfy the equations of (11) for $\mathbf{z}_{\mathsf{BP},\rho} = (d_{1,1},\ldots,d_{L,\delta_\kappa},\pi_{1,0}(0),\ldots,\pi_{L,1}(4))^\top$ and, $\forall i \in [0,\kappa-1]$, the bit x_i is committed in com_i with randomness $\mathbf{r}_{\mathsf{com},i}$:

$$\begin{bmatrix} \mathbf{a}_{\mathsf{com}} & & \\ & \ddots & \\ & & \mathbf{a}_{\mathsf{com}} \end{bmatrix} \cdot \mathbf{x} + \begin{bmatrix} \bar{\mathbf{A}} & & \\ & \ddots & \\ & & \bar{\mathbf{A}} \end{bmatrix} \cdot \begin{pmatrix} \mathbf{r}_{\mathsf{com},0} \\ \vdots \\ \mathbf{r}_{\mathsf{com},\kappa-1} \end{pmatrix} = \begin{pmatrix} \mathsf{com}_0 \\ \vdots \\ \mathsf{com}_{\kappa-1} \end{pmatrix} \bmod q.$$

Decomposing. We use $\mathsf{vdec}'_{m,\beta}(\cdot)$ to decompose $\mathbf{v}_1,\mathbf{v}_2,\mathbf{v}_{\mathsf{U},1},\mathbf{v}_{\mathsf{U},2},\mathbf{r}_{\mathsf{U}} \in [-\beta,\beta]^m$ into $\bar{\mathbf{v}}_1,\bar{\mathbf{v}}_2,\bar{\mathbf{v}}_{\mathsf{U},1},\bar{\mathbf{v}}_{\mathsf{U},2},\bar{\mathbf{r}}_{\mathsf{U}} \in \{-1,0,1\}^{m\delta_\beta}$, respectively. Similarly, we decompose vector $\nu \in [-B,B]^t$ into vector $\bar{\nu} = \mathsf{vdec}'_{t,B}(\nu) \in \{-1,0,1\}^{t\delta_B}$.

Extending and Combining. Next, we perform the following steps:

- For each $(\theta,i) \in [L] \times [\delta_\kappa]$, extend $d_{\theta,i}$ to $\mathbf{d}_{\theta,i} = \mathsf{ext}_2(d_{\theta,i})$.
- For each $(\theta,j,i) \in [L] \times \{0,1\} \times [0,4]$, extend $\pi_{\theta,j}(i)$ to $\Pi_{\theta,j,i} = \mathsf{ext}_5(\pi_{\theta,j}(i))$.
- Let $\overline{\mathbf{w}}_{3,1} = (\mathbf{x}^\top|\mathbf{r}_{\mathsf{com},0}^\top|\cdots|\mathbf{r}_{\mathsf{com},\kappa-1}^\top|\mathsf{m}_{\mathsf{U},\mathbf{x}}^\top|\widehat{\mathsf{m}}_{\mathsf{U},\mathbf{x}}^\top|\mathsf{m}^\top \mid \mathbf{e}_{\mathsf{U}}^\top|\mu^\top)^\top \in \{0,1\}^{D_{3,1}}$, where $D_{3,1} = \kappa(m+2) + 2m + m_d + t$. Then extend $\overline{\mathbf{w}}_{3,1}$ to $\mathbf{w}_{3,1} \in \mathsf{B}^2_{D_{3,1}}$.
- Define the vector $\overline{\mathbf{w}}_{3,2} = (\bar{\mathbf{v}}_1^\top|\bar{\mathbf{v}}_{\mathsf{U},1}^\top|\bar{\mathbf{r}}_{\mathsf{U}}^\top|\bar{\nu}^\top|\mathbf{e}^\top)^\top \in \{-1,0,1\}^{D_{3,2}}$ of dimension $D_{3,2} = 3m\delta_\beta + t(\delta_B+1)$ and extend it into $\mathbf{w}_{3,2} \in \mathsf{B}^3_{D_{3,2}}$.
- Extend $\bar{\mathbf{v}}_2$ to $\mathbf{s}_0 \in \mathsf{B}^3_{m\delta_\beta}$. Then for $j \in [\ell]$, form vector $\mathsf{s}_j = \mathsf{expand}(\tau[j],\mathbf{s}_0)$.
- Extend $\bar{\mathbf{v}}_{\mathsf{U},2}$ to $\mathbf{s}_{\mathsf{U},0} \in \mathsf{B}^3_{m\delta_\beta}$. Then for $j \in [\ell_I]$, form $\mathsf{s}_{\mathsf{U},j} = \mathsf{expand}(\tau_{\mathsf{U}}[j],\mathsf{s}_{\mathsf{U},0})$.

Given the above transformations, let $D_3 = 2L(\delta_\kappa + 25) + 2D_{3,1} + 3D_{3,2} + 3m\delta_\beta(2\ell+1) + 3m\delta_\beta(2\ell_I+1)$ and construct vector $\mathbf{w}_3 \in [-1,4]^{D_3}$ of the form:

$$\begin{aligned} (\mathbf{d}_{1,1}^\top \mid \cdots \mid \mathbf{d}_{L,\delta_\kappa}^\top \mid \Pi_{1,0,0}^\top \mid \cdots \mid \Pi_{L,1,4}^\top \mid \mathbf{w}_{3,1}^\top \mid \mathbf{w}_{3,2}^\top \mid \\ \mathbf{s}_0^\top \mid \mathbf{s}_1^\top \mid \cdots \mid \mathbf{s}_\ell^\top \mid \mathbf{s}_{\mathsf{U},0}^\top \mid \mathbf{s}_{\mathsf{U},1}^\top \mid \cdots \mid \mathbf{s}_{\mathsf{U},\ell_I}^\top \mid)^\top. \end{aligned} \tag{29}$$

Observe that the given five equations can be combined into one of the form:

$$\mathbf{M}_3 \cdot \mathbf{w}_3 = \mathbf{v}_3 \bmod q, \tag{30}$$

where matrix $\mathbf{M}_3$ and vector $\mathbf{v}_3$ can be built from the public input.

PUTTING PIECES ALTOGETHER. At the final stage of the process, we connect the three aforementioned steps. Indeed, all the equations involved in our process are captured by (23), (27), and (30) - which in turn can be combined into:

$$\mathbf{M} \cdot \mathbf{w} = \mathbf{v} \bmod q, \tag{31}$$

where $\mathbf{w} = (\mathbf{w}_{\mathsf{tree}}^\top \mid \mathbf{w}_{\mathsf{BP}}^\top \mid \mathbf{w}_3^\top)^\top \in [-1, 4]^D$, for

$$D = D_{\mathsf{tree}} + D_{\mathsf{BP}} + D_3 = \widetilde{\mathcal{O}}(\lambda) \cdot (L \cdot \log \kappa + \kappa) + \widetilde{\mathcal{O}}(\lambda) \cdot (\log N + \lambda) + \widetilde{\mathcal{O}}(1) \cdot t.$$

The components of $\mathbf{w}$ all have constraints listed in Table 1. By construction, these blocks either belong to the special sets $\mathsf{B}_{\mathfrak{m}}^2$, $\mathsf{B}_{\mathfrak{m}}^3$ or they have the special forms $\mathsf{expand}(\cdot,\cdot)$, $\mathsf{ext}_2(\cdot)$, $\mathsf{ext}_5(\cdot)$, $\mathsf{ext}_{5\times 2}(\cdot,\cdot)$, which are invariant under the permutations defined in Table 1. As a result, we can specify suitable sets VALID, $\mathcal{S}$ and permutations of D elements $\{\Gamma_\phi : \phi \in \mathcal{S}\}$, for which the conditions of (12) are satisfied. The description of VALID, $\mathcal{S}$ and Γ_ϕ is detailed in the full paper.

Our desired argument system then works as follows. At the beginning of the interaction, the prover computes commitments $\mathsf{com}_0, \ldots, \mathsf{com}_{\kappa-1} \in \mathbb{Z}_q^n$ and send them once to the verifier. Both parties construct matrix $\mathbf{M}$ and vector $\mathbf{v}$ based on the public input as well as $\mathsf{com}_0, \ldots, \mathsf{com}_{\kappa-1}$, while the prover prepares vector $\mathbf{w}$, as described. Finally, they run the protocol of Sect. 6.1, which has communication cost $\mathcal{O}(D \log q) = \mathcal{O}(L \cdot \log \kappa + \kappa)$.

Acknowledgements. Part of this research was funded by Singapore Ministry of Education under Research Grant MOE2016-T2-2-014(S) and by the French ANR ALAMBIC project (ANR-16-CE39-0006).

References

1. Abe, M., Camenisch, J., Dubovitskaya, M., Nishimaki, R.: Universally composable adaptive oblivious transfer (with access control) from standard assumptions. In: ACM Workshop on Digital Identity Management (2013)
2. Agrawal, S., Boneh, D., Boyen, X.: Efficient lattice (H)IBE in the standard model. In: Gilbert, H. (ed.) EUROCRYPT 2010. LNCS, vol. 6110, pp. 553–572. Springer, Heidelberg (2010). https://doi.org/10.1007/978-3-642-13190-5_28
3. Aiello, B., Ishai, Y., Reingold, O.: Priced oblivious transfer: how to sell digital goods. In: Pfitzmann, B. (ed.) EUROCRYPT 2001. LNCS, vol. 2045, pp. 119–135. Springer, Heidelberg (2001). https://doi.org/10.1007/3-540-44987-6_8
4. Alwen, J., Peikert, C.: Generating shorter bases for hard random lattices. In: STACS 2009 (2009)
5. Asharov, G., Jain, A., López-Alt, A., Tromer, E., Vaikuntanathan, V., Wichs, D.: Multiparty computation with low communication, computation and interaction via threshold FHE. In: Pointcheval, D., Johansson, T. (eds.) EUROCRYPT 2012. LNCS, vol. 7237, pp. 483–501. Springer, Heidelberg (2012). https://doi.org/10.1007/978-3-642-29011-4_29
6. Barrington, D.: Bounded-width polynomial-size branching programs recognize exactly those languages in NC1. In: STOC 1986 (1986)

7. Böhl, F., Hofheinz, D., Jager, T., Koch, J., Striecks, C.: Confined guessing: new signatures from standard assumptions. J. Cryptol. **28**(1), 176–208 (2015)
8. Boneh, D., Boyen, X.: Efficient selective-ID secure identity-based encryption without random Oracles. In: Cachin, C., Camenisch, J.L. (eds.) EUROCRYPT 2004. LNCS, vol. 3027, pp. 223–238. Springer, Heidelberg (2004). https://doi.org/10.1007/978-3-540-24676-3_14
9. Brakerski, Z., Langlois, A., Peikert, C., Regev, O., Stehlé, D.: On the classical hardness of learning with errors. In: STOC (2013)
10. Camenisch, J., Dubovitskaya, M., Enderlein, R.R., Neven, G.: Oblivious transfer with hidden access control from attribute-based encryption. In: Visconti, I., Prisco, R. (eds.) SCN 2012. LNCS, vol. 7485, pp. 559–579. Springer, Heidelberg (2012). https://doi.org/10.1007/978-3-642-32928-9_31
11. Camenisch, J., Dubovitskaya, M., Neven, G.: Oblivious transfer with access control. In: ACM-CCS 2009 (2009)
12. Camenisch, J., Dubovitskaya, M., Neven, G., Zaverucha, G.M.: Oblivious transfer with hidden access control policies. In: Catalano, D., Fazio, N., Gennaro, R., Nicolosi, A. (eds.) PKC 2011. LNCS, vol. 6571, pp. 192–209. Springer, Heidelberg (2011). https://doi.org/10.1007/978-3-642-19379-8_12
13. Camenisch, J., Neven, G., Shelat, A.: Simulatable adaptive oblivious transfer. In: Naor, M. (ed.) EUROCRYPT 2007. LNCS, vol. 4515, pp. 573–590. Springer, Heidelberg (2007). https://doi.org/10.1007/978-3-540-72540-4_33
14. Canetti, R.: Universally composable security: a new paradigm for cryptographic protocols. In: FOCS 2001 (2001)
15. Cash, D., Hofheinz, D., Kiltz, E., Peikert, C.: Bonsai trees, or how to delegate a lattice basis. In: Gilbert, H. (ed.) EUROCRYPT 2010. LNCS, vol. 6110, pp. 523–552. Springer, Heidelberg (2010). https://doi.org/10.1007/978-3-642-13190-5_27
16. Chor, B., Goldreich, O., Kushilevitz, E., Sudan, M.: Private information retrieval. In: FOCS 1995 (1995)
17. Coull, S., Green, M., Hohenberger, S.: Controlling access to an oblivious database using stateful anonymous credentials. In: Jarecki, S., Tsudik, G. (eds.) PKC 2009. LNCS, vol. 5443, pp. 501–520. Springer, Heidelberg (2009). https://doi.org/10.1007/978-3-642-00468-1_28
18. Crescenzo, G., Ostrovsky, R., Rajagopalan, S.: Conditional oblivious transfer and timed-release encryption. In: Stern, J. (ed.) EUROCRYPT 1999. LNCS, vol. 1592, pp. 74–89. Springer, Heidelberg (1999). https://doi.org/10.1007/3-540-48910-X_6
19. Ducas, L., Stehlé, D.: Sanitization of FHE ciphertexts. In: Fischlin, M., Coron, J.-S. (eds.) EUROCRYPT 2016. LNCS, vol. 9665, pp. 294–310. Springer, Heidelberg (2016). https://doi.org/10.1007/978-3-662-49890-3_12
20. Even, S., Goldreich, O., Lempel, A.: A randomized protocol for signing contracts. Commun. ACM **28**(6), 637–647 (1985)
21. Fiat, A., Shamir, A.: How to prove yourself: practical solutions to identification and signature problems. In: Odlyzko, A.M. (ed.) CRYPTO 1986. LNCS, vol. 263, pp. 186–194. Springer, Heidelberg (1987). https://doi.org/10.1007/3-540-47721-7_12
22. Gentry, C., Peikert, C., Vaikuntanathan, V.: Trapdoors for hard lattices and new cryptographic constructions. In: STOC (2008)
23. Gentry, C., Sahai, A., Waters, B.: Homomorphic encryption from learning with errors: conceptually-simpler, asymptotically-faster, attribute-based. In: Canetti, R., Garay, J.A. (eds.) CRYPTO 2013. LNCS, vol. 8042, pp. 75–92. Springer, Heidelberg (2013). https://doi.org/10.1007/978-3-642-40041-4_5
24. Goldreich, O., Micali, S., Wigderson, A.: How to play any mental game or a completeness theorem for protocols with honest majority. In: STOC (1987)

25. Gorbunov, S., Vinayagamurthy, D.: Riding on asymmetry: efficient ABE for branching programs. In: Iwata, T., Cheon, J.H. (eds.) ASIACRYPT 2015. LNCS, vol. 9452, pp. 550–574. Springer, Heidelberg (2015). https://doi.org/10.1007/978-3-662-48797-6_23
26. Green, M., Hohenberger, S.: Blind identity-based encryption and simulatable oblivious transfer. In: Kurosawa, K. (ed.) ASIACRYPT 2007. LNCS, vol. 4833, pp. 265–282. Springer, Heidelberg (2007). https://doi.org/10.1007/978-3-540-76900-2_16
27. Green, M., Hohenberger, S.: Universally composable adaptive oblivious transfer. In: Pieprzyk, J. (ed.) ASIACRYPT 2008. LNCS, vol. 5350, pp. 179–197. Springer, Heidelberg (2008). https://doi.org/10.1007/978-3-540-89255-7_12
28. Green, M., Hohenberger, S.: Practical adaptive oblivious transfer from simple assumptions. In: Ishai, Y. (ed.) TCC 2011. LNCS, vol. 6597, pp. 347–363. Springer, Heidelberg (2011). https://doi.org/10.1007/978-3-642-19571-6_21
29. Herranz, J.: Restricted adaptive oblivious transfer. Theoret. Comput. Sci. **412**(46), 6498–6506 (2011)
30. Hiromasa, R., Abe, M., Okamoto, T.: Packing messages and optimizing bootstrapping in GSW-FHE. In: Katz, J. (ed.) PKC 2015. LNCS, vol. 9020, pp. 699–715. Springer, Heidelberg (2015). https://doi.org/10.1007/978-3-662-46447-2_31
31. Jarecki, S., Liu, X.: Efficient oblivious pseudorandom function with applications to adaptive OT and secure computation of set intersection. In: Reingold, O. (ed.) TCC 2009. LNCS, vol. 5444, pp. 577–594. Springer, Heidelberg (2009). https://doi.org/10.1007/978-3-642-00457-5_34
32. Kawachi, A., Tanaka, K., Xagawa, K.: Concurrently secure identification schemes based on the worst-case hardness of lattice problems. In: Pieprzyk, J. (ed.) ASIACRYPT 2008. LNCS, vol. 5350, pp. 372–389. Springer, Heidelberg (2008). https://doi.org/10.1007/978-3-540-89255-7_23
33. Kurosawa, K., Nojima, R., Phong, L.T.: Generic fully simulatable adaptive oblivious transfer. In: Lopez, J., Tsudik, G. (eds.) ACNS 2011. LNCS, vol. 6715, pp. 274–291. Springer, Heidelberg (2011). https://doi.org/10.1007/978-3-642-21554-4_16
34. Libert, B., Ling, S., Mouhartem, F., Nguyen, K., Wang, H.: Signature schemes with efficient protocols and dynamic group signatures from lattice assumptions. In: Cheon, J.H., Takagi, T. (eds.) ASIACRYPT 2016. LNCS, vol. 10032, pp. 373–403. Springer, Heidelberg (2016). https://doi.org/10.1007/978-3-662-53890-6_13
35. Libert, B., Ling, S., Mouhartem, F., Nguyen, K., Wang, H.: Zero-knowledge arguments for matrix-vector relations and lattice-based group encryption. In: Cheon, J.H., Takagi, T. (eds.) ASIACRYPT 2016. LNCS, vol. 10032, pp. 101–131. Springer, Heidelberg (2016). https://doi.org/10.1007/978-3-662-53890-6_4
36. Libert, B., Ling, S., Nguyen, K., Wang, H.: Zero-knowledge arguments for lattice-based accumulators: logarithmic-size ring signatures and group signatures without trapdoors. In: Fischlin, M., Coron, J.-S. (eds.) EUROCRYPT 2016. LNCS, vol. 9666, pp. 1–31. Springer, Heidelberg (2016). https://doi.org/10.1007/978-3-662-49896-5_1
37. Lindell, A.Y.: Efficient fully-simulatable oblivious transfer. In: Malkin, T. (ed.) CT-RSA 2008. LNCS, vol. 4964, pp. 52–70. Springer, Heidelberg (2008). https://doi.org/10.1007/978-3-540-79263-5_4
38. Ling, S., Nguyen, K., Stehlé, D., Wang, H.: Improved zero-knowledge proofs of knowledge for the ISIS problem, and applications. In: Kurosawa, K., Hanaoka, G. (eds.) PKC 2013. LNCS, vol. 7778, pp. 107–124. Springer, Heidelberg (2013). https://doi.org/10.1007/978-3-642-36362-7_8

39. Ling, S., Nguyen, K., Wang, H.: Group signatures from lattices: simpler, tighter, shorter, ring-based. In: Katz, J. (ed.) PKC 2015. LNCS, vol. 9020, pp. 427–449. Springer, Heidelberg (2015). https://doi.org/10.1007/978-3-662-46447-2_19
40. Naor, M., Pinkas, B.: Oblivious transfer with adaptive queries. In: Wiener, M. (ed.) CRYPTO 1999. LNCS, vol. 1666, pp. 573–590. Springer, Heidelberg (1999). https://doi.org/10.1007/3-540-48405-1_36
41. Naor, M., Pinkas, B.: Efficient oblivious transfer protocols. In: SODA (2001)
42. Naor, M., Pinkas, B.: Computationally secure oblivious transfer. J. Cryptol. **18**(1), 1–35 (2005)
43. Nishide, T., Yoneyama, K., Ohta, K.: Attribute-based encryption with partially hidden encryptor-specified access structures. In: Bellovin, S.M., Gennaro, R., Keromytis, A., Yung, M. (eds.) ACNS 2008. LNCS, vol. 5037, pp. 111–129. Springer, Heidelberg (2008). https://doi.org/10.1007/978-3-540-68914-0_7
44. Peikert, C., Vaikuntanathan, V., Waters, B.: A framework for efficient and composable oblivious transfer. In: Wagner, D. (ed.) CRYPTO 2008. LNCS, vol. 5157, pp. 554–571. Springer, Heidelberg (2008). https://doi.org/10.1007/978-3-540-85174-5_31
45. Rabin, M.: How to exchange secrets by oblivious transfer. Technical report TR-81, Aiken Computation Laboratory, Harvard University (1981)
46. Regev, O.: On lattices, learning with errors, random linear codes, and cryptography. In: STOC (2005)
47. Sahai, A., Waters, B.: Fuzzy identity-based encryption. In: Cramer, R. (ed.) EUROCRYPT 2005. LNCS, vol. 3494, pp. 457–473. Springer, Heidelberg (2005). https://doi.org/10.1007/11426639_27
48. Stern, J.: A new paradigm for public key identification. IEEE Trans. Inf. Theory **42**(6), 1757–1768 (1996)
49. Kalai, Y.T.: Smooth projective hashing and two-message oblivious transfer. In: Cramer, R. (ed.) EUROCRYPT 2005. LNCS, vol. 3494, pp. 78–95. Springer, Heidelberg (2005). https://doi.org/10.1007/11426639_5
50. Zhang, Y., Au, M.H., Wong, D.S., Huang, Q., Mamoulis, N., Cheung, D.W., Yiu, S.-M.: Oblivious transfer with access control: realizing disjunction without duplication. In: Joye, M., Miyaji, A., Otsuka, A. (eds.) Pairing 2010. LNCS, vol. 6487, pp. 96–115. Springer, Heidelberg (2010). https://doi.org/10.1007/978-3-642-17455-1_7

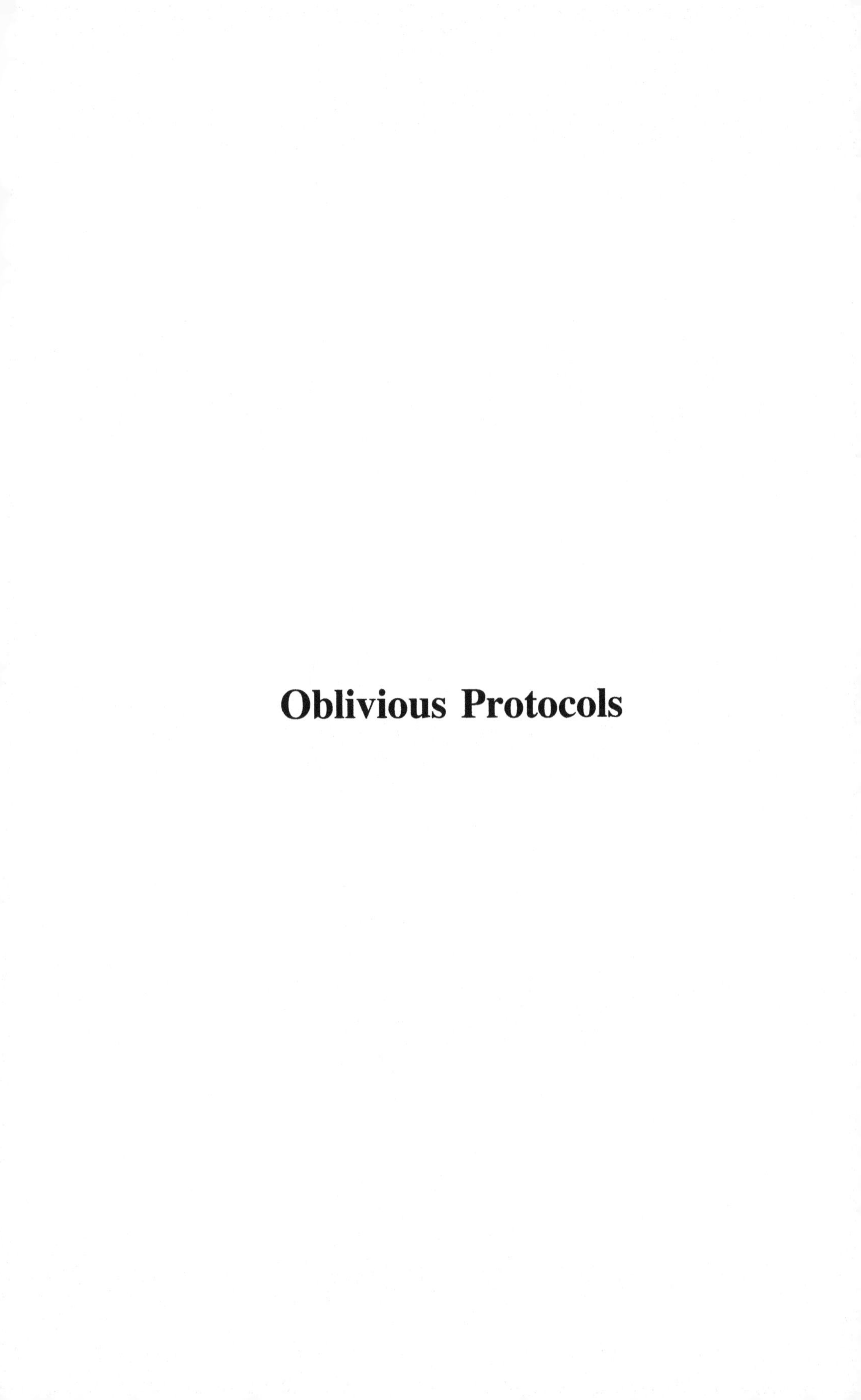

Oblivious Protocols

On the Depth of Oblivious Parallel RAM

T.-H. Hubert Chan[1(✉)], Kai-Min Chung[2], and Elaine Shi[3]

[1] The University of Hong Kong, Pokfulam, Hong Kong
hubert@cs.hku.hk
[2] Academia Sinica, Taipei, Taiwan
[3] Cornell University, Ithaca, USA

Abstract. Oblivious Parallel RAM (OPRAM), first proposed by Boyle, Chung, and Pass, is the natural parallel extension of Oblivious RAM (ORAM). OPRAM provides a powerful cryptographic building block for hiding the access patterns of programs to sensitive data, while preserving the paralellism inherent in the original program. All prior OPRAM schemes adopt a single metric of "simulation overhead" that characterizes the blowup in parallel runtime, assuming that oblivious simulation is constrained to using the *same* number of CPUs as the original PRAM.

In this paper, we ask whether oblivious simulation of PRAM programs can be further sped up if the OPRAM is allowed to have *more* CPUs than the original PRAM. We thus initiate a study to understand the true depth of OPRAM schemes (i.e., when the OPRAM may have access to unbounded number of CPUs). On the upper bound front, we construct a new OPRAM scheme that gains a logarithmic factor in depth and without incurring extra blowup in total work in comparison with the state-of-the-art OPRAM scheme. On the lower bound side, we demonstrate fundamental limits on the depth any OPRAM scheme—even when the OPRAM is allowed to have an unbounded number of CPUs and blow up total work arbitrarily. We further show that our upper bound result is optimal in depth for a reasonably large parameter regime that is of particular interest in practice.

Keywords: Oblivious parallel RAM · Oblivious RAM · Depth complexity

1 Introduction

Oblivious RAM (ORAM), originally proposed in the seminal works of Goldreich and Ostrovsky [8,9], is a powerful cryptographic building block that allows a program to hide access patterns to sensitive data. Since Goldreich and Ostrovsky's ground-breaking results, numerous subsequent works showed improved ORAM constructions [10,13,18,20,21] with better asymptotics and/or practical performance. ORAM has also been used in various practical and theoretical applications such as multi-party computation [11,22], secure processor design [14,17], and secure storage outsourcing [19,23].

The full version of this paper is available online [3].

T. Takagi and T. Peyrin (Eds.): ASIACRYPT 2017, Part I, LNCS 10624, pp. 567–597, 2017.
https://doi.org/10.1007/978-3-319-70694-8_20

Since most modern computing architectures inherently support parallelism (e.g., cloud compute clusters and modern CPU designs), a natural problem is how to hide sensitive access patterns in such a parallel computing environment. In a recent seminal work, Boyle et al. [1] first propose the notion of Oblivious Parallel RAM (OPRAM), which is a natural extension of ORAM to the parallel setting. Since then, several subsequent works have constructed efficient OPRAM schemes [5,6,15]. One central question in this line of research is whether there is an OPRAM scheme whose *simulation overhead* matches that of the best known ORAM scheme. Specifically, an OPRAM scheme with simulation overhead X means that if the original PRAM consumes m CPUs and runs in parallel time T, then we can obliviously simulate this PRAM also with m CPUs, and in parallel runtime $X \cdot T$. In a recent companion paper called Circuit OPRAM [5], we answered this question in the affirmative. In particular, if N is the number of distinct blocks that the CPUs can request, then Circuit OPRAM proposed a unifying framework where we can obtain statistically secure OPRAMs with $O(\log^2 N)$ simulation overhead, and computationally secure OPRAMs with $(\log^2 N/\log\log N)$ simulation overhead—thus matching the best known ORAM schemes in both settings [13,21].

All previous OPRAM schemes consider a single performance metric referred to as simulation overhead as mentioned above. It is immediate that an OPRAM scheme with X simulation overhead also immediately implies an ORAM construction with X simulation overhead. Thus, the recent Circuit OPRAM [5] also suggests that we have hit some road-block for constructing more efficient OPRAM schemes—unless we knew how to asymptotically improve the efficiency of sequential ORAM. Note also that in the regime of sufficiently large block sizes, Circuit OPRAM achieves $O(\alpha \log N)$ simulation overhead for any super-constant function α, and this is (almost) tight in light of Goldreich and Ostrovsky's logarithmic ORAM lower bound [8,9].

1.1 Our Results and Contributions

In this paper, we rethink the performance metrics for an OPRAM scheme. We argue that while adopting a single simulation overhead metric is intuitive, this single metric fails to capture the true "work-depth" of the oblivious simulation. In particular, we ask the questions:

1. *If the OPRAM is allowed to access more CPUs than the original PRAM, can we have oblivious simulations with smaller parallel runtime blowup than existing OPRAM schemes?*
2. *Are there any fundamental limits to an OPRAM's work-depth, assuming that the OPRAM can have access to an unbounded number of CPUs?*

To answer the above questions, we turn to the parallel algorithms literature, and adopt two classical metrics, that is, *total work* and *parallel runtime* in the study of OPRAMs. Like the parallel algorithms literature, we also refer to a(n) PRAM/OPRAM's parallel runtime as its *work-depth* (or *depth*). The depth metric represents the runtime of a PRAM when given ample CPUs—thus the depth is the inherently sequential part of a PRAM that cannot be further parallelized even with an arbitrarily large number of CPUs. The depth metric is commonly used in conjunction with total work—since we would like to design low-depth parallel algorithms that do not blow up total work by too much in comparison with the sequential setting (e.g., by repeating computations too many times). Using these classical metrics from the parallel algorithms literature, we can re-interpret the single "simulation overhead" metric adopted by previous OPRAM works as follows: an OPRAM with simulation overhead X has both total work blowup and parallel runtime blowup X in comparison with the original PRAM.

Note that when the OPRAM is constrained to using the same number of CPUs as the original PRAM, its parallel runtime blowup must be at least as large as the total work blowup. In this paper, however, we show that this need not be the case when the OPRAM can access more CPUs than the original PRAM. We design a new OPRAM scheme that gains a logarithmic factor in speed (i.e., depth) in comparison with the state-of-the-art [5] when given logarithmically many more CPUs than the original PRAM. In some sense, our new OPRAM scheme shows that the blowup in total work incurred due to obliviousness can be parallelized further (albeit through non-trivial techniques). Additionally, we prove new lower bounds that shed light on the inherent limits on any OPRAM scheme's depth. In light of our lower bounds, our new OPRAM scheme is optimal in depth for a wide range of parameters. We now present an informal overview of our results and contributions.

Upper Bounds. First, we show that for any PRAM running in time T and consuming W amount of total work, there exists a statistically secure oblivious simulation that consumes logarithmically many more CPUs than the original PRAM, and runs in parallel runtime $O(T \log N \log\log N)$ and total work $O(W \log^2 N)$.

In comparison, the best known (statistically secure) OPRAM scheme incurs both $O(\log^2 N)$ blowup in both total work and parallel runtime (i.e., $O(\log^2 N)$ simulation overhead). In this sense, while preserving the total work blowup, we improve existing OPRAMs' depth by a logarithmic factor.

We then extend our construction to the computationally secure setting by adapting an elegant trick originally proposed by Fletcher et al. [7], and show how to shave another $\log\log N$ factor off both the total work and parallel runtime, assuming that one-way functions exist. Our results are summarized in the following informal theorem.

Theorem 1 (Small-depth OPRAMs: Informal). *The following results are possible for small-depth OPRAMs where N denotes the original PRAM's total memory size, m denotes the original PRAM's number of CPUs, and the security failure must be negligible in N.*

- **Statistically secure, general block size.** *There exists a statistically secure OPRAM that achieves $O(\log^2 N)$ blowup in total work and $O(\log N \log\log N)$ blowup in parallel runtime for general block sizes of $\Omega(\log N)$ bits.*

- **Computationally secure, general block size.** *Assume the existence of one-way functions, then there exists a computationally secure OPRAM that achieves* $O(\frac{\log^2 N}{\log \log N})$ *total work blowup and* $O(\log N)$ *parallel runtime blowup for general block sizes of* $\Omega(\log N)$ *bits.*
- **Statistically secure, large block size.** *For any super-constant function* $\alpha(N) = \omega(1)$, *for any constant* $\epsilon > 0$, *there exists a statistically secure OPRAM that achieves* $O(\alpha \log N \log \log N)$ *total work blowup and* $O(\log m + \log \log N)$ *parallel runtime blowup for blocks of* N^ϵ *bits or larger.*

Lower Bounds. Next, we consider if there are any fundamental limits to an OPRAM scheme's work-depth. We prove a non-trivial lower bound showing that any *online* OPRAM scheme (i.e., with no a-priori knowledge of future requests) that does not perform encoding of data blocks and does not duplicate data blocks too extensively must suffer from at least $\Omega(\log m)$ depth blowup where m is the number of CPUs—and this lower bound holds even when the OPRAM scheme may access arbitrarily many CPUs and have arbitrarily large total work blowup. We stress that our lower bound employs techniques that are different in nature from those of Goldreich and Ostrovsky's classical ORAM lower bound [8,9]—in particular, theirs bounds total work rather than depth. Furthermore, our lower bound holds even for computational security.

Theorem 2 (Lower bound for an OPRAM's depth). *Any computationally or statistically secure online OPRAM scheme must incur at least* $\Omega(\log m)$ *blowup in parallel runtime, as long as the OPRAM (1) does not perform encoding of data blocks (i.e., in the "balls-and-bins" model); and (2) does not make more than* $m^{0.1}$ *copies of each data block.*

We note that the conditions our lower bound assumes (online, balls-and-bins, and bounded duplication) hold for all ORAM and OPRAM constructions.

On the Tightness of Our Upper and Lower Bounds. In light of our lower bound, our OPRAM constructions are optimal in depth in a reasonably large parameter regime. Specifically, our (computationally secure) OPRAM scheme is depth-optimal when $m = N^\epsilon$ for any constant $\epsilon > 0$ for general block sizes. For larger block sizes, our OPRAM scheme is depth-optimal for a larger range of m—in particular, when the block size is sufficiently large, our (statistically secure) OPRAM scheme is tight for m as small as $m = \mathsf{poly} \log N$.

Technical Highlights. Both our lower bounds and upper bounds introduce non-trivial new techniques. Since our lower bound studies the depth of parallel algorithms, it is of a very different nature than Goldreich and Ostrovsky's ORAM lower bounds for total work [8,9]. To prove the depth lower bound, we also depart significantly in technique from Goldreich and Ostrovsky [8,9]. In particular, our lower bound is of an *online* nature and considers the possible batches of requests that a low-depth access pattern can support in a single PRAM step; whereas in comparison, Goldreich and Ostrovksy's lower bound applies even to offline ORAM/OPRAM algorithms, and they perform a counting argument over many steps of the ORAM/OPRAM. The most difficult challenge in proving our lower bound is how to offset the large number of possibilities introduced by "preprocessing", i.e., the number of possible memory configurations before the PRAM step of concern starts. To deal with this challenge, our core idea is to devise a new method of counting that is *agnostic to preprocessing.*

For our new small-depth OPRAM, the main challenge we cope with is of a very different nature from known ORAM and OPRAM works. In particular, all previous ORAMs and OPRAMs that follow the tree-based paradigm [18] adopt a standard recursion technique such that the CPU need not store a large amount of metadata (referred to as the position map). Known schemes treat this recursion as a blackbox technique. Unfortunately, in our work, it turns out that this recursion becomes the main limiting factor to an OPRAM's depth. Thus, we open up the recursion, and our core technique for achieving small-depth OPRAM is to devise a novel offline/online paradigm, such that the online phase that is inherently sequential across recursion levels has small (i.e., $O(\log\log N)$) depth per recursion level; whereas all work that incurs logarithmic depth is performed in an offline phase in parallel across all recursion levels. Designing such an offline/online algorithm incurs several challenges which we explain in Sect. 5.2. We hope that these new techniques can also lend to the design of oblivious parallel algorithms in general.

Another way to view our small-depth OPRAM's contributions is the following. In our setting, we must address two challenges: (1) concurrency, i.e., how to coordinate a batch of m requests such that they can be served simultaneously without causing write conflicts; and (2) parallelism, i.e., how to make each request parallel by using more CPUs. Note that the concurrency aspect is applicable only to OPRAMs where multiple concurrent requests are involved, whereas the parallelism aspect is applicable even for parallelizing the operations of a sequential ORAM. Previous OPRAM constructions [1,6] are concerned only about the former concurrency aspect, but we need to take both into account—in this sense, we are in fact the *first* to investigate the "parallelism" aspect of ORAMs/OPRAMs.[1] In particular, in our fetch phase algorithm, the two aspects are intertwined for the case of general m, in the sense that we cannot separate our techniques into two phases involving one "concurrent compilation" and one "parallel compilation"—such intertwining allows us to construct more efficient algorithms. In the maintain phase, our divide-and-conquer strategy for eviction indeed can be used to parallelize a sequential ORAM.

Related work. Boyle, Chung, and Pass recently initiated the study of Oblivious Parallel RAM (OPRAM) [1]. They were also the first to phrase the simulation overhead metric for OPRAMs, i.e., the parallel runtime blowup of the OPRAM in comparison with the original PRAM, assuming that the OPRAM consumes the same number of CPUs as the original PRAM. Several subsequent

[1] We gratefully acknowledge the Asiacrypt reviewers for pointing out this aspect of our contribution.

works [1,5,6,15] have improved Boyle et al. [1]'s OPRAM construction. Most recently, Chan and Shi [5] show that we can construct statistically secure and computationally secure OPRAMs whose asymptotical performance match the best known sequential ORAM; and their approach is based on the tree-based paradigm [18]. A similar asymptotical result (but for the computationally secure setting only) was also shown by Chan et al. [4] using the hierarchical framework originally proposed by Goldreich and Ostrovsky [8,9]. In the OPRAM context, Goldreich and Ostrovsky's logarithmic lower bound [8,9] immediately implies that any OPRAM with constant blocks of CPU cache must suffer from at least logarithmic *total work* blowup. Thus far there is no other known OPRAM lower bound (and our depth lower bound departs significantly in techniques from Goldreich and Ostrovksy's lower bound).

In the interest of space, we refer the reader to our online full version [3] for additional discussions about the related work.

2 Definitions

2.1 Parallel Random-Access Machines

A *parallel random-access machine* (PRAM) consists of a set of CPUs and a shared memory denoted mem indexed by the address space $[N] := \{1, 2, \ldots, N\}$. In this paper, we refer to each memory word also as a *block*, and we use B to denote the bit-length of each block.

We use m to denote the number of CPUs. In each step t, each CPU executes a next instruction circuit denoted Π, updates its CPU state; and further, CPUs interact with memory through request instructions $\boldsymbol{I}^{(t)} := (I_i^{(t)} : i \in [m])$. Specifically, at time step t, CPU i's instruction is of the form $I_i^{(t)} := (\mathsf{op}, \mathsf{addr}, \mathsf{data})$, where the operation is $\mathsf{op} \in \{\mathtt{read}, \mathtt{write}\}$ performed on the virtual memory block with address addr and block value $\mathsf{data} \in \{0,1\}^B \cup \{\perp\}$. If $\mathsf{op} = \mathtt{read}$, then we have $\mathsf{data} = \perp$ and the CPU issuing the instruction should receive the content of block $\mathsf{mem}[\mathsf{addr}]$ at the initial state of step t. If $\mathsf{op} = \mathtt{write}$, then we have $\mathsf{data} \neq \perp$; in this case, the CPU still receives the initial state of $\mathsf{mem}[\mathsf{addr}]$ in this step, and at the end of step t, the content of virtual memory $\mathsf{mem}[\mathsf{addr}]$ should be updated to data.

Write conflict resolution. By definition, multiple `read` operations can be executed concurrently with other operations even if they visit the same address. However, if multiple concurrent `write` operations visit the same address, a conflict resolution rule will be necessary for our PRAM be well-defined. In this paper, we assume the following:

- The original PRAM supports concurrent reads and concurrent writes (CRCW) with an arbitary, parametrizable rule for write conflict resolution. In other words, there exists some priority rule to determine which `write` operation takes effect if there are multiple concurrent writes in some time step t.

– The compiled, oblivious PRAM (defined below) is a "concurrent read, exclusive write" PRAM (CREW). In other words, the design of our OPRAM construction must ensure that there are no concurrent writes at any time.

We note that a CRCW-PRAM with a parametrizable conflict resolution rule is among the most powerful CRCW-PRAM model, whereas CREW is a much weaker model. Our results are stronger if we allow the underlying PRAM to be more powerful but the compiled OPRAM uses a weaker PRAM model. For a detailed explanation on how stronger PRAM models can emulate weaker ones, we refer the reader to the work by Hagerup [12].

CPU-to-CPU communication. In the remainder of the paper, we sometimes describe our algorithms using CPU-to-CPU communication. For our OPRAM algorithm to be oblivious, the inter-CPU communication pattern must be oblivious too. We stress that such inter-CPU communication can be emulated using shared memory reads and writes. Therefore, when we express our performance metrics, we assume that all inter-CPU communication is implemented with shared memory reads and writes. In this sense, our performance metrics already account for any inter-CPU communication, and there is no need to have separate metrics that characterize inter-CPU communication. In contrast, Chen et al. [6] defines separate metrics for inter-CPU communication.

Additional assumptions and notations. Henceforth, we assume that each CPU can only store $O(1)$ memory blocks. Further, we assume for simplicity that the runtime of the PRAM is *fixed* a priori and *publicly known*. Therefore, we can consider a PRAM to be a tuple

$$\mathsf{PRAM} := (\Pi, N, m, T),$$

where Π denotes the next instruction circuit, N denotes the total memory size (in terms of number of blocks), m denotes the number of CPUs, and T denotes the PRAM's parallel time steps. *Without loss of generality, we assume that* $N \geq m$. We stress that henceforth in the paper, the notations N and m denote the number of memory blocks and the number of CPUs for the original PRAM—our OPRAM construction will consume $O(1)$ factor more memory and possibly more than m CPUs.

2.2 Oblivious Parallel Random-Access Machines

Randomized PRAM. A *randomized PRAM* is a special PRAM where the CPUs are allowed to generate private, random numbers. For simplicity, we assume that a randomized PRAM has a priori known, deterministic runtime.

Oblivious PRAM (OPRAM). A randomized PRAM parametrized with total memory size N is said to be *statistically oblivious*, iff there exists a negligible function $\epsilon(\cdot)$ such that for any inputs $x_0, x_1 \in \{0,1\}^*$,

$$\mathsf{Addresses}(\mathsf{PRAM}, x_0) \overset{\epsilon(N)}{\equiv} \mathsf{Addresses}(\mathsf{PRAM}, x_1),$$

where $\mathsf{Addresses}(\mathsf{PRAM}, x)$ denotes the joint distribution of memory accesses made by PRAM upon input x and the notation $\overset{\epsilon(N)}{\equiv}$ means the statistical distance is bounded by $\epsilon(N)$. More specifically, for each time step $t \in [T]$, $\mathsf{Addresses}(\mathsf{PRAM}, x)$ includes the memory addresses requested by the CPUs in time step t, as well as whether each memory request is a read or write operation. Henceforth we often use the notation OPRAM to denote a PRAM that satisfies statistical obliviousness.

Similarly, a randomized PRAM parametrized with total memory size N is said to be *computationally oblivious*, iff there exists a negligible function $\epsilon(\cdot)$ such that for any inputs $x_0, x_1 \in \{0,1\}^*$,

$$\mathsf{Addresses}(\mathsf{PRAM}, x_0) \overset{\epsilon(N)}{\equiv_{\mathsf{c}}} \mathsf{Addresses}(\mathsf{PRAM}, x_1)$$

Note the only difference from statistical security is that here the access patterns only need to be indistinguishable to computationally bounded adversaries, denoted by the notaiton $\overset{\epsilon(N)}{\equiv_{\mathsf{c}}}$.

Following the convention of most existing ORAM and OPRAM works [8,9, 13,20,21], we will require that the security failure probability to be negligible in the N, i.e., the PRAM's total memory size.

Oblivious simulation. We say that a given OPRAM *simulates* a PRAM if for every input $x \in \{0,1\}^*$, $\Pr[\mathsf{OPRAM}(x) = \mathsf{PRAM}(x)] = 1 - \mu(N)$ where the completeness error μ is a negligible function and the probability is taken over the randomness consumed by the OPRAM—in other words, we require that the OPRAM and PRAM output the same outcome on any input x.

Online OPRAM. In this paper we focus on *online* OPRAM that simulates a PRAM by processing memory request of each PRAM step in an online fashion. Namely, each PRAM memory request is processed by the OPRAM without knowing the future requests. Note that all known ORAM and OPRAM constructions satisfy the online property.

Performance measures. For an online OPRAM simulates a certain PRAM, we measure its performance by its *work-depth* and *total work* overhead. The work-depth overhead is defined to be the number of time steps d for OPRAM to simulate each PRAM step. Let W denote the total number of blocks accessed by OPRAM to simulate a PRAM step. The total work overhead is defined to be W/m, which captures the overhead to simulate a batch of memory request in a PRAM step. Note that both d and W are random variables.

3 Lower Bound on Work-Depth

We show a lower bound on the work-depth in terms of the number m of CPUs. We establish a $\Omega(\log m)$ depth lower bound for OPRAMs satisfying the following properties. We remark that our construction in Sect. 5 as well as all existing ORAM and OPRAM constructions satisfy these properties.

1. **Balls-and-bins storage.** As coined in the ORAM lower bound of Goldreich and Ostrovsky [2], data blocks are modeled as "balls," while shared memory locations and CPU registers are modeled as "bins". In particular, this means that every memory location stores at most one data block and the content of the data block can be retrieved from that location independent of other storage.
2. **Online OPRAM.** As defined in Sect. 2.2, we consider online OPRAM that only learns the logic memory request at the beginning of a PRAM step.
3. **s-bounded duplication.** We also need a technical condition on the bound of data duplication. Namely, there is a bound s such that every data block has at most s copies stored on the memory. All known ORAM and OPRAM constructions do not store duplications on the memory[2], i.e., $s = 1$.

It is worth comparing our depth lower bound for OPRAM with the ORAM lower bound of [2]. Both lower bounds assume the balls-and-bins model, but establish lower bound for different metrics and rely on very different arguments (in particular, as we discussed below, counting arguments do not work in our setting). We additionally require online and bounded duplication properties, which are not needed in [2]. On the other hand, our lower bound holds even for OPRAM with *computational security.* In contrast, the lower bound of [2] only holds for statistical security.

The setting for the lower bound. For simplicity, we consider the following setting for establishing the lower bound. First, we consider OPRAM with initialization, where n logical data blocks of the original PRAM are initialized with certain distinct content. This is not essential as we can view the initialization as the first n steps of the PRAM program. We also assume that the logical data size n is sufficiently larger than the total CPUs register size. Specifically, let α be a constant in $(0, 1/3)$ and r be the register size of a CPU. We assume $n \geq \Omega(r \cdot m^{1+(\alpha/4)})$. For any OPRAM satisfying the above three properties with $s \leq m^{(1/3)-\alpha}$, we show that the work-depth is at least $(\alpha/3) \cdot \log m$ with probability at least $1 - m^{-\alpha/4}$ for every PRAM step. In particular, the expected work-depth per step is at least $\Omega(\log m)$ as long as $s \leq m^{1/3-\Omega(1)}$.

Theorem 3 (Lower Bound on Work-Depth). *Let Π be a computationally-secure online OPRAM that satisfies the balls-and-bins model with s-bounded duplication for $s < m^{(1/3)-\alpha}$ for constant $\alpha \in (0, 1/3)$, where the number N of blocks is at least m. Let r be the register size of each CPU. Assume that $n \geq 4r \cdot m^{1+(\alpha/4)}$ and Π has correctness error $\mu \leq m^{-\alpha/4}/4$. Then for each PRAM step t, let $\mathsf{depth}(\Pi, t)$ denote the work-depth of Π for PRAM step t,*

$$\Pr[\mathsf{depth}(\Pi, t) \leq (\alpha/3) \cdot \log m] \leq m^{-\alpha/4},$$

where the probability is over the randomness of the OPRAM compiler Π.

[2] In some hierarchical ORAMs [10,13], there might be several copies of the same block on the server, but only one copy is regarded as *fresh,* while other copies are *stale* and may contain old contents.

Before proving Theorem 3, we first discuss the intuition behind the lower bound proof in Sect. 3.1, where under simplifying assumptions, we reduce the OPRAM lower bound to solving a "user-movie problem" that captures the main argument of our lower bound proof. We discuss how to remove the simplifying assumptions in the end of the section. We then present the formal proof of Theorem 3 in Sect. 3.2

3.1 Intuition: A User-Movie Problem

As a warmup, we first present an intuitive proof making a few simplifying assumptions: (1) the OPRAM compiler must be perfectly correct and perfectly secure; and (2) there is no data block duplication in memory. Later in our formal proofs in Sect. 3.2, these assumptions will be relaxed.

Let us consider how to prove the depth lower bound for a PRAM step t for an OPRAM. Recall that we consider online OPRAM that learn the logical memory requests at the beginning of the step. We can view what happened before the step t as a preprocessing phase that stores the logical memory blocks in different memory locations, and the step t corresponds to an online phase where the CPUs fetch the requested memory blocks with certain observed access pattern. Since the access pattern should hide the logical memory request, any fixed access pattern should allow the CPUs to complete any possible batch of m requests (assuming perfect correctness and perfect security). We say that an access pattern can support a batch of m requests, if there *exists a pre-processing* (i.e., packing of data blocks into memory), such that each CPU can "reach" its desired data block through this access pattern. Our goal is to show that if the access pattern is low depth, then it is impossible to satisfy every batch of m requests—even when one is allowed to enumerate all possible pre-processings to identify one that best matches the requests (given the fixed access pattern). To show this, our argument involves two main steps.

1. First, we show that for any access pattern of low depth, say, d, each CPU can reach at most 2^d memory locations.
2. Second, we show that if an access pattern can satisfy all possible batches of m requests (with possibly different pre-processing), then it must be that some CPU can reach many physical locations in memory.

The former is relatively easy to show. Informally speaking, consider the balls-and-bins model as mentioned earlier: in every PRAM step, each CPU can only access a single memory location (although each memory location can be accessed by many CPUs). This means that at the end of the PRAM step, the block held by each CPU can only be one of two choices: (1) the block previously held by the CPU; or (2) the block in the memory location the CPU just accessed. This means that the access pattern graph must have a small fan-in of 2 (although the fan-out may be unbounded). It is not difficult to formalize this intuition, and show that given any depth-d access pattern, only 2^d memory locations can "flow into" any given CPU. Henceforth, we focus on arguing why the latter is also true—and this requires a much more involved argument.

For ease of understanding, henceforth we shall refer to CPUs as *users*, and refer to data blocks in physical memory as *movies*. There are n distinct movies stored in a database of size N (without duplications) and m users. Each user wants to watch a movie and can access to certain 2^d locations in the database, but the locations the users access to cannot depend on the movies they want to watch. On the other hand, we can decide which location to store each movie to help the users to fetch their movies from the locations they access to. In other words, we first decide which 2^d locations each user access to, then learn which movie each user wants to watch. Then we decide the location to store each movie to help the users to fetch their movies. Is it possible to find a strategy to satisfy all possible movie requests?

We now discuss how to prove the impossibility for the user-movie problem. We first note that a simple counting argument does not work, since there are n^m possible movie requests but roughly $N^n \gg n^m$ possible ways to store the movies in physical memory. To prove the impossibility, we first observe that since we do not allow duplications, when two users request the same movie, they must have access to the same location that stores the movie. Thus, any pair of users must be able to reach a common movie location—henceforth we say that the two users "share" a movie location. This observation alone is not enough, since the users may all share some (dummy) location. If, however, two sets of users request two different movies, then not only must each set share a movie location, the two sets must share two *distinct* locations. More generally, the m users' movie requests induce a *partition* among users where all users requesting the same movie are in the same *part* (i.e., equivalence class), and users in two different parts request different movies. This observation together with carefully chosen partitions allow us to show the existence of a user that needs to access to a large number of locations, which implies an impossibility for the user-movie problem for sufficiently small depth d. We stress that this idea of "partitioning" captures the essence of what pre-processing *cannot* help with, and this explains why our proof works even when there are a large number of possible pre-processings.

Specifically, let $k = m/2$ and label the m users with the set $M := [2] \times [k]$. We consider the following k partitions that partition the users into k pairs. For each $i \in [k]$, we define partition $P_i = \{\{(1,a),(2,a+i)\} : a \in [k]\}$, where the addition is performed modulo k. Note that all k^2 pairs in the k partitions are distinct. By the above observation, for each partition P_i, there are k *distinct* locations $\ell_{i,1},\ldots,\ell_{i,k} \in [N]$ such that for each pair $\{(1,a),(2,a+i)\}$ for $a \in [k]$, both users $(1,a),(2,a+i)$ access to the location $\ell_{i,a}$. Now, for each location $\ell \in [N]$, let w_ℓ denote the number of $\ell_{i,a} = \ell$ and d_ℓ denote the number of users access to the location ℓ. Note that $w_\ell \leq k$ since user pairs in a partition access to distinct locations (i.e., $\ell_{i,a} \neq \ell_{i,a'}$ for every $i \in [k]$ and $a \neq a' \in [k]$). Also note that $d_\ell \geq \sqrt{2w_\ell}$ since there are only $\binom{d_\ell}{2}$ distinct pairs of users access to the location ℓ.

To summarize, we have (i) $\sum_\ell w_\ell = k^2$, (ii) $w_\ell \leq k$ for all $\ell \in [N]$, and (iii) $d_\ell \geq \sqrt{2w_\ell}$ for all $\ell \in [N]$, which implies $\sum_\ell d_\ell \geq k \cdot \sqrt{2k} = \sqrt{k/2} \cdot m$. Recall that d_ℓ denote the number of users access to the location ℓ and there are m users.

By averaging, there must exist a user who needs to access to at least $\sqrt{k/2}$ locations. Therefore, the user-movie problem is impossible for $d \leq 0.5 \cdot \log m - 2$. Note that the distinctness of the $\ell_{i,a}$'s induced by the partitions plays a crucial role to drive a non-trivial lower bound on the summation $\sum_{\ell} d_{\ell}$.

Removing the simplifying assumptions. In above intuitive proof we make several simplyfing assumptions such as perfect security and perfect correctness. We now briefly discuss how to remove these assumptions. The main non-trivial step is to handle computational security, which requires two additional observations. Following the above argument, let us say that an access pattern is compatible with a CPU/user partition if it can support a logic memory request with corresponding induces CPU/user partition.

- First, the above impossibility argument for the user-movie problem can be refined to show that if an access pattern has depth d, then it can be compatible with at most $2^{2(d+1)}$ partitions in $P_1, \ldots, P_k$ defined above.
- Second, whether an access pattern is compatible with a partition can be verified in polynomial time.

Based on these two observations, we show that if $d \leq 0.5 \cdot \log m - 4$ (with noticeable probability), then we can identify two efficiently distinguishable CPU partitions, which implies a depth lower bound for computationally-secure OPRAM. First, we consider the access pattern of partition P_1. Since $d \leq 0.5 \cdot \log m - 4$, it can only be compatible with at most $k/2$ partitions. By an averaging argument, there exists some partition P_i such that P_i is not compatible with the access pattern of P_1 with probability at least $1/2$. On the other hand, by perfect correctness, the access pattern of P_i is always compatible with P_i. Therefore, the access patterns of P_1 and P_i are efficiently distinguishable by an efficient distinguisher D that simply verifies if the access pattern is compatible with P_i.

We now briefly discuss how to remove the remaining assumptions. First, it is not hard to see that the above argument does not require perfect correctness and can tolerate a small correctness error. Second, we make an implicit assumption that the requested data blocks are not stored in the CPU registers so that the CPUs must fetch the requested data blocks from physical locations on the server. This can be handled by considering logic access requests with random logical address and assuming that the logic memory size n is sufficiently larger than the total CPU register size (as in the theorem statement).

We also implicitly assume that we can observe the beginning and end of the access pattern of a PRAM step t. For this, we note that by the online property, we can without loss of generality consider t as the last step so that we know the end of the access pattern for free. Furthermore, we observe that we do not need to know the beginning of the access pattern since the compatibility property is monotone in the following sense. If a partition P_i is compatible with the access pattern of the last d accesses, it is also compatible with the access pattern of the last $d + 1$ accesses. Thus, we can consider the access pattern of the last d accesses for certain appropriately chosen d.

Finally, to handle s-bounded duplication with $s > 1$, we consider CPU partitions where each part is a set of size $s+1$, instead of a pair. By the pigeonhole principle, each part can still certify a pair of CPUs with a shared memory location. However, some extra care is needed for defining the partitions to make sure that different partitions do not certify the same pair of CPUs, and the depth lower bound degrades when s increases. Nevertheless, the lower bound remains $\Omega(\log m)$ for $s \le m^{1/3-\Omega(1)}$.

3.2 Proof of Theorem 3

We now proceed with a formal proof. We first note that for proving lower bound of the PRAM step t, we can consider PRAM programs where t is the last step, since the behavior of an online OPRAM does not depend on the future PRAM steps. Thus, we can focus on proving lower bound of the last PRAM step. We prove the theorem by contradiction. Suppose that

$$\Pr[\mathsf{depth}(\Pi, t) \le (\alpha/3) \cdot \log m] > m^{-\alpha/4}, \tag{1}$$

we show two PRAM programs $\mathsf{P}_1, \mathsf{P}_2$ with identical first $t-1$ steps and different logic access request at step t such that the access pattern of $\Pi(\mathsf{P}_1)$ and $\Pi(\mathsf{P}_2)$, which denote the OPRAM simulation of $\mathsf{P}_1, \mathsf{P}_2$ respectively, are efficiently distinguishable. Towards this, we define the CPU partition of a memory request.

Definition 1 (CPU Partition). *Let* $\mathsf{addr} = (\mathsf{addr}_1, \dots, \mathsf{addr}_m) \in [n]^m$ *be a memory request.* addr *induces a partition* P *on the CPUs, where two CPUs* c_1, c_2 *are in the same part* iff *they request for the same logical address* $\mathsf{addr}_{c_1} = \mathsf{addr}_{c_2}$. *In other words,* P *partitions the CPUs according to the requested logical addresses.*

Recall that s is the bound on the number of duplication. We assume $m = (s+1) \cdot k$ for some prime k. This is without loss of generality, because any integer has a prime number that is within a multiplicative factor of 2. We label the m CPUs with the set $\mathrm{M} := [s+1] \times [k]$. We consider the following set of partitions $P_1, \dots, P_k$: For $i \in [k]$, the partition $P_i := \{S_i(a) : a \in [k]\}$ is defined such that each part has the form $S_i(a) := \{(b, a + bi) : b \in [s+1]\}$, where addition is performed modulo k. In other words, the parts in the partitions can be viewed as all possible distinct line segments in the $\mathbb{Z}_k^2$ plane.

We will show two programs where their last memory requests have induced partitions P_1 and P_i for some $i \in [k]$ such that their compiled access patterns are efficiently distinguishable. To show this, we model the view of the adversary with an *access pattern graph* and consider a *compatiability* property between an access pattern graph and a CPU partition, defined as follows.

Access pattern graphs and compatibility. Given the access pattern of $\Pi(\mathsf{P})$ for a PRAM program P and a depth parameter $d \in \mathbb{N}$, we define an access pattern graph G as follows.

(a) **Nodes.** The nodes are partitioned into $d+1$ *layers*. In layer 0, each node represents a physical location in the memory at the beginning of the last d-th time step of $\Pi(\mathsf{P})$.
For $1 \leq i \leq d$, each node in layer i represents a physical location in the memory or a CPU at the end of the last $(d-i+1)$-st time step.
Hence, we represent each node with (i, u), where i is the layer number and u is either a CPU or a memory location.

(b) **Edges.** Each edge is directed and points from a node in layer $i-1$ to one in layer i for some $i \geq 1$. For each CPU or a memory location, there is a directed edge from its copy in layer $i-1$ to one in layer i.
If a CPU c reads from some physical location ℓ in the last $(d-i)$-th time step, then there is a directed edge from $(i-1, \ell)$ to (i, c). Since we allow concurrent read, the out-degree of a node corresponding to a physical location can be unbounded.
If a CPU c writes to some physical location ℓ in the last $(d-i)$-th time step, then there is a directed edge from $(i-1, c)$ to (i, ℓ).
Observe that since we consider OPRAM with exclusive write, the in-degree of a node (either corresponding to a CPU or a memory location) is at most 2. In fact, the degree 2 bound holds even with concurrent write models as long as the write conflict resolution can be determined only by the access pattern.

The access pattern graph G captures the potential data flow of the last d time steps of the data access. Specifically, a path from $(0, \ell)$ to (d, c) means CPU c may learn the content of the memory location ℓ at the last d time step. If there is no such path, then CPU c cannot learn the content. This motivates the definition of compatible partitions.

Definition 2 (Compatible Partition). *Let G be an access pattern graph and $P_1, \ldots, P_k$ be the partitions defined above. We say $P_i = \{S_i(a) : a \in [k]\}$ is compatible with G if there exist k distinct physical locations $\ell_{i,1}, \ldots, \ell_{i,k}$ on the server such that for each $a \in [k]$, there are at least two CPUs c_1 and c_2 in $S_i(a)$ such that both nodes (d, c_1) and (d, c_2) are reachable from $(0, \ell_{i,a})$ in G.*

Intuitively, compatibility is a necessary condition for the last d time steps of data access to "serve" an access request with induced partition P_i, assuming the requested data blocks are not stored in the CPU registers at the last d-th time step. Recall that each data block has at most s copies in the server, and each part $S_i(a)$ has size $s+1$. By the Pigeonhole principle, for each part $S_i(a)$ in the induced partition, there must be at least two CPUs $c_1, c_2 \in S_i(a)$ obtaining the logical block from the same physical location ℓ_a on the server, which means the nodes (d, c_1) and (d, c_2) are reachable from $(0, \ell_a)$ in G. We note that verifying compatibility can be done in polynomial time.

Lemma 1 (Verifying Compatibility Takes Polynomial Time). *Given a CPU partition P and an access pattern graph G, it takes polynomial time to verity whether P is compatible with G.*

Proof. Given P and G as in the hypothesis of the lemma, we construct a bipartite graph H as follows. Each vertex in L is labeled with a memory location ℓ, and each vertex in R is labeled with a part S in P. There is an edge connecting a vertex ℓ in L to a vertex S in R *iff* there are at least two CPUs c_1 and c_2 in S such that both (d, c_1) and (d, c_2) are reachable from $(0, \ell)$ in G. This bipartite graph can be constructed in polynomial time.

Observe that P is compatible with G *iff* there is a matching in H such that all vertices's in R are matched. Hence, a maximum matching algorithm can be applied to H to decide if P is compatible with G.

Now, the following key lemma states that an access pattern graph G with small depth d cannot be compatible with too many partitions. We will use the lemma to show two programs with efficiently distinguishable access patterns.

Lemma 2. *Let G be an access pattern graph with the depth parameter d, and $P_1, \ldots, P_k$ be the partitions defined above. Among $P_1, \ldots, P_k$, there are at most $((s+1) \cdot 2^d)^2$ partitions that are compatible with G.*

Proof. Recall that the in-degree of each node is at most 2. Thus, for each node (d, c) in layer d, there are at most 2^d nodes $(0, \ell)$ in layer 0 that can reach the node (d, c). For the sake of contradiction, we show that if G is compatible with $u > ((s+1) \cdot 2^d)^2$ partitions, then there exists a node (d, c) that is reachable by more that 2^d nodes in layer 0.

For convenience, we define a bipartite graph $H = (L, R, E)$ from G as follows. Each vertex in L is labeled with a CPU c, and each vertex in R is labeled with a physical location ℓ of the memory. There is an edge (c, ℓ) in H iff $(0, \ell)$ reaches (d, c) in G. Our goal can be restated as showing that if G is compatible with $u > ((s+1) \cdot 2^d)^2$ partitions, then there exists $c \in L$ with degree $\deg(c) > 2^d$. We do so by lower bounding the number of edges $|E| > m \cdot 2^d$.

By definition, if P_i is compatible with G, then there exist k distinct physical locations $\ell_{i,1}, \ldots, \ell_{i,k}$ on the server such that for each $a \in [k]$, there are at least two CPUs $c_{i,a}, c'_{i,a} \in S_i(a)$ such that $(d, c_{i,a})$ and $(d, c'_{i,a})$ are reachable from $(0, \ell_{i,a})$ in G, which means there are edges $(c_{i,a}, \ell_{i,a})$ and $(c'_{i,a}, \ell_{i,a})$ in H. Thus, a compatible partition certifies $2k$ edges in H, although two different partitions may certify the same edges.

Let $P_{i_1}, \ldots, P_{i_u}$ be the set of compatible partitions. While they may certify the same edges, the set of CPU pairs $\{(c_{i_j,a}, c'_{i_j,a}) : j \in [u], a \in [k]\}$ are distinct for the following reason: Recall that the parts in partitions correspond to different line segments in $\mathbb{Z}_k^2$. Since two points define a line, the fact that the parts correspond to different lines implies that all CPU pairs are distinct.

For each memory location ℓ, let w_ℓ denote the number of $\ell_{i_j,a} = \ell$. It means that ℓ is connected to w_ℓ distinct pairs of CPUs in H, which implies that $\deg(\ell) \geq \sqrt{2w_\ell}$ since there must be at least $\sqrt{2w_\ell}$ distinct CPUs. Also, note that $\sum_\ell w_\ell = u \cdot k$ and $w_\ell \leq u$ for every ℓ since ℓ can appear in each partition at most once. It is not hard to see that the above conditions imply a lower bound on $|E| = \sum_\ell \deg(\ell) \geq k \cdot \sqrt{2u} > m \cdot 2^d$. This in turn implies the existence of $c \in L$ with degree $\deg(c) > 2^d$, a contradiction.

Let us now consider a PRAM program P_1 that performs dummy access in the first $t-1$ steps and a random access request at the t step with induced partition P_1. Specifically, in the first $t-1$ steps, all CPUs read the first logic data block. For the t-th step, let $(b_1, \ldots, b_k)$ be uniformly random k distinct logic data blocks. For $a \in [k]$, the CPUs in part $S_1(a)$ of P_1 read the block b_a at the t-th step. Let $d = (\alpha/3) \cdot \log m$ and $G(\Pi(\mathsf{P}_1), d)$ denote the access pattern graph of $\Pi(\mathsf{P}_1)$ with depth parameter d. The following lemma follows directly by Lemma 2 and an averaging argument.

Lemma 3. *There exists $i^* \in [k]$ such that*

$$\Pr[P_{i^*} \textit{ is compatible with } G(\Pi(\mathsf{P}_1), d)] \leq ((s+1) \cdot 2^d)^2/k \leq m^{-\alpha/3},$$

where the randomness is over Π and P_1.

Now, consider a PRAM program P_2 that is identical to P_1, except that the access request at the t-th step has induced partition P_{i^*} instead of P_1. Namely, for $a \in [k]$, the CPUs in part $S_{i^*}(a)$ of P_{i^*} read the block b_a at the t-th step, where $(b_1, \ldots, b_k)$ are uniformly random k distinct logic data blocks.

Lemma 4. *Suppose that Π satisfies Eq. (1), then*

$$\Pr[P_{i^*} \textit{ is compatible with } G(\Pi(\mathsf{P}_2), d)] > m^{-\alpha/4}/4,$$

where the randomness is over Π and P_2.

Proof. First note that since each CPU request a random data block at the t-th PRAM step, the probability that the requested data block is stored in the CPU register is at most r/n. By a union bound, with probability at least $1 - m \cdot (r/m) \geq 1 - m^{-\alpha/4}/4$, all data blocks requested at the t-th PRAM step are not in the corresponding CPU registers. In this case, the CPUs need to obtain the data blocks from the server. Furthermore, if the work-depth of the t-th PRAM step is $\leq d$, then the CPUs need to obtain the data blocks in the last d time steps of data access, which as argued above, implies compatibility. Therefore,

$$\Pr[P_{i^*} \text{ is compatible with } G(\Pi(\mathsf{P}_2), d)] > m^{-\alpha/4} - m^{-\alpha/4}/4 - \epsilon_c > m^{-\alpha/4}/4.$$

Recall by Lemma 1 that compatibility can be checked in polynomial time. The above two lemmas imply that assuming Eq. (1), $\Pi(\mathsf{P}_1)$ and $\Pi(\mathsf{P}_2)$ are efficiently distinguishable by a distinguisher D who checks the compatibility of P_{i^*} and the access pattern graph with depth parameter $d = (\alpha/3) \cdot \log m$. This is a contradiction and completes the proof of Theorem 3.

4 Background on Circuit OPRAM and Building Blocks

4.1 Preliminaries: Circuit OPRAM

As a warmup, we first briefly review the recent Circuit OPRAM algorithm [5] that we build on top of. For clarity, we make a few simplifying assumptions in this overview:

- We explain the non-recursive version of the algorithm where we assume that the CPU can store a position map for free that tracks the rough physical location of every block: this CPU-side position map is later removed using a standard recursion technique in Circuit OPRAM [5]—however, as we point out later, to obtain a small depth OPRAM in our paper, we must implement the recursion differently and thus in our paper we can no longer treat the recursion as blackbox technique.
- We assume that m is not too small and is at least polylogarithmic in N; and
- A standard conflict resolution procedure proposed by Boyle et al. [1] has been executed such that the incoming batch of m requests are for distinct real blocks (or dummy requests).

Core data structure: a pool and $2m$ *subtrees.* Circuit ORAM partitions the ORAM data structure in memory into $2m$ *disjoint subtrees.* Given a batch of m memory requests (from m CPUs), each request will be served from a random subtree. On average, each subtree must serve $O(1)$ requests in a batch; and due to a simple balls and bins argument, except with negligible probability, even the worst-case subtree serves only $O(\alpha \log N)$ incoming requests for any super-constant function α.

In addition to the $2m$ subtree, Circuit OPRAM also maintains an overflow pool that stores overflowing data blocks that fail to be evicted back into the $2m$ subtrees at the end of each batch of m requests.

It will help the reader to equivalently think of the $2m$ subtrees and the pool in the following manner: First, think of a single big Circuit ORAM [21] tree (similar to other tree-based ORAMs [18]). Next, identify a height with $2m$ buckets, which naturally gives us $2m$ *disjoint subtrees.* All buckets from smaller heights as well as the Circuit ORAM's stash form the *pool.* As proven in the earlier work [5], at any time, the pool contains at most $O(m + \alpha \log N)$ blocks.

Fetch. Given a batch of m memory requests, henceforth without loss of generality, we assume that the m requests are for distinct addresses. This is because we can adopt the conflict resolution algorithm by Boyle et al. [1] to suppress duplicates, and after data has been fetched, rely on oblivious routing to send fetched data to all request CPUs. Now, look up the requested blocks in two places, both the pool and the subtrees:

- *Subtree lookup:* For a batch of m requests, each request comes with a position label—and all m position labels define m random paths in the $2m$ subtrees. We can now fetch from the m path in parallel. Since each path is $O(\log N)$ in length, each fetch can be completed in $O(\log \log N)$ parallel steps with the help of $\log N$ CPUs.

  All fetched blocks are merged into the pool. Notice that at this moment, the pool size has grown by a constant factor, but later in a cleanup step, we will compress the pool back to its original size. Also, at this moment, we have not removed the requested blocks from the subtrees yet, and we will remove them later in the maintain phase.

- *Pool lookup:* At this moment, all requested blocks must be in the pool. Assuming that m is not too small, we can now rely on oblivious routing to route blocks back to each requesting CPU—and this can be completed in $O(\log m)$ parallel steps with m CPUs.

Maintain. In the maintain phase, perform the following: (1) remove all blocks fetched from the paths read; and (2) perform eviction on each subtree.

- *Efficient simultaneous removals.* After reading each subtree, we need to remove up to $\mu := O(\alpha \log N)$ blocks that are fetched. Such removal operations can lead to write contention when done in parallel: since the paths read by different CPUs overlap, up to $\mu := O(\alpha \log N)$ CPUs may try to write to the same location in the subtree. Circuit OPRAM employs a novel *simultaneous removal* algorithm to perform such removal in $O(\log N)$ parallel time with m CPUs. We refer the reader to the Circuit OPRAM paper for an exposition of the simultaneous removal algorithm. As noted in the Circuit OPRAM paper [5], simulatenous removal from m fetch paths can be accomplished in $O(\log m + \log \log N)$ parallel steps with $O(m \cdot \log N)$ total work.
- *Selection of eviction candidates and pool-to-subtree routing.* At this moment, we will select exactly one eviction candidate from the pool for each subtree. If there exists one or more blocks in the pool to be evicted to a certain subtree, then the *deepest* block (where deepest is precisely defined in Circuit ORAM [21]) with respect to the current eviction path will be chosen. Otherwise, a dummy block will be chosen for this subtree. Roughly speaking, using the above criterion as a preference rule, we can rely on oblivious routing to route the selected eviction candidate from the pool to each subtree. This can be accomplished in $O(\log m)$ parallel steps with m CPUs assuming that m is not too small.
- *Eviction.* Now, each subtree performs exactly 1 eviction. This can be accomplished in $O(\log N)$ runtime using the sequential procedure described in the original Circuit ORAM paper [21]. At the end of this step, each subtree will output an eviction leftover block: the leftover block is dummy if the chosen eviction candidate was successfully evicted into the subtree (or if the eviction candidate was dummy to start with); otherwise the leftover block is the orginal eviction candidate. All these eviction leftovers will be merged back into the central pool.
- *Pool cleanup.* Notice that in the process of serving a batch of requests, the pool size has grown—however, blocks that have entered the pool may be dummy. In particular, we shall prove that the pool's occupancy will never exceed $c \cdot m + \alpha \log N$ for an appropriate constant c except with $\mathsf{negl}(N)$ probability. Therefore, at the end of the maintain phase, we must compress the pool back to $c \cdot m + \alpha \log N$. Such compression can easily be achieved through oblivious sorting in $O(\log m)$ parallel steps with m CPUs, assuming that m is not too small.

Recursion. Thus far, we have assumed that the position map is stored on the CPU-side, such that the CPU knows where every block is in physical memory.

To get rid of the position map, Circuit OPRAM employs a standard recursion technique that comes with the tree-based ORAM/OPRAM framework [18]. At a high level, the idea of the recursion framework is very simple: instead of storing the position map on the CPU side, we recurse and store the position map in a smaller OPRAM in physical memory; and then we recurse again and store the position map of this smaller OPRAM in a yet smaller OPRAM in physical memory, and so on. If each block can store $\gamma > 1$ number of position labels, then every time we recurse, the OPRAM's size reduces by a factor of γ. Thus in at most $\log N$ recursion levels, the metadata size becomes at most $O(1)$ blocks—and at this moment, the CPU can store all the metadata locally in cache.

Although most prior tree-based ORAM/OPRAM papers typically treat this recursion as a standard, blackbox technique, in this paper we cannot—on the contrary, it turns out that the recursion becomes the most non-trivial part of our low-depth OPRAM algorithm. Thus, henceforth the reader will need to think of the recursion in an expanded form—we now explain what exactly happens in the recursion in an expanded form. Imagine that one of the memory requests among the batch of m requests asks for the logical address $(0101100)_2$ in binary format, and suppose that each block can store 2 position labels. Henceforth we focus on what happens for fetching this logical address $(0101100)_2$—but please keep in mind that there are m such addresses and thus the following process is repeated m times in parallel.

- First, the 0th recursion level (of constant size) will tell the 1st recursion level the position label for the address $(0*)_2$.
- Next, the 1st recursion level fetch the metadata block at level-1 address $(0*)_2$ and this fetched block contains the position labels for $(00*)_2$ and $(01*)_2$.
- Now, level-1 informs level-2 of the position label for $(01*)_2$; at this moment, level-2 fetches the metadata block for the level-2 address $(01*)_2$ and this fetched block contains the position labels for the addresses $(010*)_2$ and $(011*)_2$; and so on.
- This continues until the D-th recursion level (i.e., the final recursion level)—this final recursion level stores actual data blocks rather than metadata, and thus the desired data block will be fetched at the end.

As mentioned, the above steps are in fact replicated m times in parallel since there are m requests in a batch. This introduces a couple additional subtleties:

- First, notice that for obliviousness, conflict resolution must be performed upfront for each recursion level before the above procedure starts—this step can be parallelized across all recursion levels.
- Second, how do the m fetch CPUs at one recursion level *obliviously* route the fetched position labels to the m fetch CPUs waiting in the next recursion level? Circuit OPRAM relies on a standard oblivious routing procedure (initially described by Boyle et al. [1]) for this purpose, thus completely hiding which CPUs route to which.

Important observation. At this moment, we make an important observation. In the Circuit OPRAM algorithm, the fetch phase operations are inherently sequential across all recursion levels, and the maintain phase operations can be parallelized across all recursion levels. In particular, during the fetch phase, the m fetch CPUs at recursion level d must block waiting for recursion level $d-1$ to pass down the fetched position labels before its own operations can begin. Due to the sequential nature of the fetch phase, Circuit OPRAM incurs at least $(\log m + \log\log N)\log N$ depth, where the $\log m$ stems from level-to-level oblivious routing, $\log\log N$ stems the depth needed to parallel-fetch from a path of length $\log N$ (and other operations), and the $\log N$ factor is due to the number of recursion levels. In comparison, the depth of the maintain phase is not the limiting factor due to the ability to perform the operations in parallel across recursion levels.

4.2 Other Important Building Blocks

Permutation-related building blocks. We will rely on the following building blocks related to generating and applying permutations. In the interest of this space, we describe the abstractions of the building blocks but defer their full specification to our online full version [3].

1. **Apply a pre-determined permutation to an array.** It is not difficult to see that we can in parallel apply a pre-determined permutation to an array in a single parallel step (see our online full version [3] for the detailed algorithm).
2. **Permute an array by a secret random permutation.** One can generate a secret random permutation and apply it to an array obliviously, without revealing any information about the permutation—and this can be accomplished in $O(\log n)$ depth and $O(n \log n)$ work for an array of size n. The formal specification and proofs are deferred to the online full version [3].
3. **Obliviously construct a routing permutation that permutes a source to a destination array.** In our online full version [3] we show how to accomplish the following task: given a source array snd of length k containing distinct real elements and dummies (where each dummy element contains unique identifying information as well), and a destination array rcv also of length k containing distinct real elements and dummies, with the guarantee that the set of real elements in snd are the same as the set of real elements in rcv. Now, construct a routing permutation $\pi : [k] \to [k]$ (in an oblivious manner) such that for all $i \in [k]$, if $\mathsf{snd}[i]$ contains a real element, then $\mathsf{rcv}[\pi[i]] = \mathsf{snd}[i]$. This can be accomplished in $O(n \log n)$ work and $O(\log n)$ depth by calling oblivious sort $O(1)$ number of times.

Oblivious bin-packing. Oblivious bin-packing is the following primitive.

- *Inputs:* Let B denote the number of bins, and let Z denote the target bin capacity. We are given an input array denoted In, where each element is either a dummy denoted $\perp$ or a real element that is tagged with a bin number

$g \in [B]$. It is guaranteed that there are at most Z elements destined for each bin.
- *Outputs:* An array $\mathsf{Out}[1 : BZ]$ of length $B \cdot Z$ containing real and dummy elements, such that $\mathsf{Out}[(g-1)B+1 : gB]$ denotes contents of the g-th bin for $g \in [B]$. The output array Out must guarantee that the g-th bin contains all elements in the input array In tagged with the bin number g; and that all real elements in bin g must appear in the input array In and are tagged with g.

There is an oblivious parallel algorithm that accomplishes oblivious bin packing in total work $O(\widetilde{n} \log \widetilde{n})$ and parallel runtime $O(\log \widetilde{n})$ where $\widetilde{n} = \max(|\mathsf{In}|, B \cdot Z)$. The algorithm works as follows:

1. For each group $g \in [B]$, append Z filler elements of the form $(\texttt{filler}, g)$ to the resulting array—these filler elements ensure that every group will receive at least Z elements after the next step.
2. Obliviously sort the resulting array by the group number, placing all dummies at the end. When elements have the same group number, place filler elements after real elements.
3. By invoking an instance of the oblivious aggregation algorithm [1,16] (see full version [3] for the definition of oblivious aggretation), each element in the array finds the leftmost element in its own group. Now for each element in the array, if its offset within its own group is greater than Z, replace the element with a dummy $\perp$.
4. Oblivious sort the resulting array placing all dummies at the end. Truncate the resulting array and preserve only first $B \cdot Z$ blocks.
5. For every filler element in the resulting array, replace it with a dummy.

5 A Small-Depth OPRAM: Level-to-Level Routing Algorithm

5.1 Overview of Our OPRAM

We now show how we can improve the depth of OPRAM schemes [1] by a logarithmic factor, through employing the help of more CPUs; and importantly, we achieve this without incurring extra total work in comparison with the best known OPRAM scheme [5].

Challenges. As argued earlier in Sect. 4.1, for the case of general block sizes, the most sequential part of the Circuit OPRAM algorithm stems from the (up to) $\log N$ recursion levels. More specifically, (apart from the final data level), each recursion level's job is to fetch the metadata (referred to as position labels) necessary, and route this information to the next recursion level. In this way, the next recursion level will know where in physical memory to look for the metadata needed by its next recursion level, and so on (we refer the reader to Sect. 4.1 for a more detailed exposition of the recursion).

Thus, the fetch phase operations of Circuit OPRAM are inherently sequential among the D recursion levels, incurring $(D(\log m + \log\log N))$ in depth, where the $\log m$ term stems from the level-to-level oblivious routing of fetched metadata, and the $\log\log N$ term stems from fetching metadata blocks from a path of length $\log N$. Igoring the $\log\log N$ term, our goal therefore is to get rid of the $\log m$ depth that stems from level-to-level oblivious routing.

Our result. Our main contribution is to devise a low-depth algorithm to perform level-to-level routing of metadata. At first sight, this task seems unlikely to be successful—since each recursion level must *obliviously* route its metadata to the next level, it would seem like we are inherently subject to the depth necessary for an oblivious routing algorithm [1]. Since oblivious routing in some sense implies oblivious sorting, it would seem like we have to devise an oblivious sorting algorithm of less than logarthmic depth to succeed in our goal.

Perhaps somewhat surprisingly, we show that this need not be the case. In particular, we show that by (1) allowing a negligible statistical failure probability; (2) exploiting special structures of our routing problem; and (3) introducing an offline/online paradigm for designing parallel oblivious algorithms, we can devise a special-purpose level-to-level oblivious routing algorithm such that

1. all work that is inherently $\log m$ in depth is pushed to an offline phase that can be parallelized across all recursion levels; and
2. during the online phase that is inherently sequential among all $\log N$ recursion levels, we can limit the work-depth of each recursion level to only $\log\log N$ rather than $\log m$—note that for most interesting parameter regimes that we care about, $\log m \gg \log\log N$.

We defer the detailed introduction of this algorithm and its proofs to Sect. 5.2. As a result, we obtain a new, *statistically* secure OPRAM algorithm (for general block sizes) that achieves $O(\log N \log\log N)$ depth blowup and $O(\log^2 N)$ total work blowup. In comparison, under our new performance metrics, the best known OPRAM algorithm [5] achieves $O(\log^2 N)$ total work blowup and $O(\log^2 N)$ depth blowup. Thus we achieve a logarithmic factor improvement in terms of depth.

Extensions. We consider several extensions. First, using a standard technique described by Fletcher et al. [7] and extended to the OPRAM setting by Chan and Shi [5], we show how to obtain a *computationally* secure OPRAM scheme with $O(\log^2 N/\log\log N)$ total work blowup and $O(\log N)$ depth blowup, and supporting general block sizes. In light of our aforementioned OPRAM depth lower bound (which also applies to computationally secure OPRAMs), our OPRAM scheme is optimal for $m = N^\epsilon$ where $\epsilon > 0$ is an arbitrarily small constant.

Finally, we consider a setting with sufficiently large blocks, say, the block size is N^ϵ for any constant $\epsilon > 0$—in this case, the recursion depth becomes $O(1)$. In this case, the limiting factor to an OPRAM's work depth now is the eviction algorithm (rather than the level-to-level routing). We show how to leverage a non-trivial devise and conquer technique to devise a new, small-depth eviction algorithm, allowing us to perform eviction along a path of length $\log N$ in $\log\log N$ depth rather than $\log N$—however, this is achieved at the cost of a small $\log\log N$ blowup in total work. As a result, we show that for sufficiently large blocks, there is an OPRAM scheme with depth as small as $O(\log\log N + \log m)$ where the $\log\log N$ part arises from our low-depth eviction algorithm (and other operations), and the $\log m$ part arises from the conflict resolution and oblivious routing of fetched data back to requesting CPUs—thus tightly matching our depth lower bound as long as m is at least logarithmic in N.

5.2 Small-Depth Routing of Position Identifiers: Intuition

Problem statement. As we explained earlier, in each recursion level, m fetch CPUs fetch the metadata (i.e., position labels) required for the next recursion level. The next recursion level contains m fetch CPUs waiting to receive these position labels, before its own operations can begin. Circuit OPRAM performs such level-to-level routing using a standard oblivious routing building block, thus incurring at least $D \log m$ depth where D is the number of recursion levels which can be as large as $\log N$, and $\log m$ is the depth of standard oblivious routing. How can we reduce the depth necessary for level-to-level routing?

We will first clarify some details of the problem setup. Recall that in each PRAM step, we receive a batch of m memory requests, i.e., m logical addresses. Given these m logical addresses, we immediately know which level-d addresses to fetch for each recursion level d (see Sect. 4.1 for details). We assume that conflict resolution has been performed for each recursion level d on all of the m level-d addresses, and thus, every real (i.e., non-dummy) level-d address is distinct. Now, note that from all these level-d addresses (and even without fetching the actual metadata in each recursion level), we can already determine the routing topology from level to level: as an example, a level-2 CPU that needs to fetch the level-2 address $(010*)$ would like to receive position labels from the level-1 fetch CPU with the address $(01*)$.

Our goal here is to improve the OPRAM's depth to $O(\log N \log\log N)$ for general (worst-case) block sizes. We use the parameter Γ to denote the number of position labels that a block can store; we let $\gamma := \min\{\Gamma, m\}$ be an upper bound on the number of position labels in a block that is "useful" for the next recursion level. To achieve this, in the part of the algorithm that is sequential among all recursion levels (henceforth also referred to as the *online* part), we can only afford $O(\log\log N)$ depth rather than the $\log m$ necessary for oblivious routing. Indeed, for a general oblivious routing problem consisting of m senders and m receivers, it appears the best one can do is to rely on an oblivious routing network [1,6] that has $\log m$ depth—so how can we do better here? We rely on two crucial insights:

1. First, we observe that our routing problem has *small fan-in and fan-out*: each sender has at most γ recipients; and each recipient wants to receive from at most 1 sender. This is because that each fetched metadata block contains at

most γ position labels, and obviously each fetch CPU in the next level only needs one position label to proceed.
2. Second, we will rely on an *offline-online paradigm*—in the offline phase, we are allowed to perform preparation work that indeed costs $\log m$ depth; however, in the online phase, the depth is kept to be small. Later when we employ this offline/online oblivious routing building block in our full OPRAM algorithm, we will show that the offline phase does not depend on any fetched data, and thus can be paralellized across all recursion levels, whereas the online phase must still be sequential—but recall that the online phase has much smaller depth.

First insight: localized routing. Our first idea is to rely on this observation to restrict oblivious routing to happen only within small groups—as we shall explain later, for this idea to work, it is essential that our routing problem has small fan-in and fan-out. More specifically, we would like that each small group of senders talk to a corresponding small group of receivers, say, sender group S_i talks only to receiver group R_i, where both S_i and R_i are $\mu := \alpha\gamma^2 \log N$ in size, where the choice of μ is due to Lemma 5. If we do this, then oblivious routing within each small group costs only $\log \mu$ depth.

How can we arrange senders and receivers into such small groups? For correctness we must guarantee that for every i, each receiver in R_i will be able to obtain its desired item from some sender in S_i.

To achieve this, we rely on a randomized load balancing approach. The idea is very simple. First, we pad the sender array with dummy senders to a size of $2m$—recall that there are at most m real senders. Similarly, we pad the receiver array to a size of $2m$ as well. Henceforth if a receiver wants an item from a sender, we say that the sender and receiver are connected. Every dummy sender is obviously connected to 0 receivers.

Now, if we pick a random sender from the sender array, in expectation this sender will be connected to 0.5 receivers. Thus a random subset of μ senders will in expectation is connected to $0.5\,\mu$ receivers—using measure concentration techniques, it is not difficult to show that a random subset of μ senders is connected to μ receivers except with negligible probability—note that this measure concentration result holds only when our routing problem has small fan-in and fan-out (see Lemma 5 for details).

Our idea is to randomly permute the source array, and have the first μ sender be group 1, the second μ senders be group 2, and so on. By relying on $O(1)$ number of oblivious sorts, we can now arrange the receiver array to be "loosely aligned" with the sender array, i.e., all receivers connected to sender group 1 are in the first size-μ bucket of the receiver array, all receivers connected to sender group 2 are in the second size-μ bucket of the receiver array, and so on.

Using the above idea, the good news is that oblivious routing is now constrained to μ-sized groups (each containing γ addresses), thus costing only $\log \mu$ depth. However, our above algorithm still involves randomly permuting the sender array and oblivious routing to loosely align the receiver array with the sender array—these steps cost $\log m$ depth. Thus our idea is to perform these steps in an offline phase that can be parallelized across all recursion levels, and thus the depth does not blow up by the number of recursion levels. Nonetheless how to instantiate this offline/online idea is non-trivial as we explain below.

Second insight: online/offline paradigm. One challenge that arises is how to coordinate among all recursion levels. To help the reader understand the problem, let us first describe what would have happened if everything were performed online, sequentially level by level:

Imagine that each recursion has $2m$ fetch CPUs (among which at most m are real) first acting as receivers. Once these receivers have received the position labels, they will fetch data from the OPRAM's tree data structure. At this point, they hold the position labels desired by the next recursion level, and thus the receivers now switch roles and become senders with resepct to the next recursion level. Before the receivers become senders, it is important that they be randomly permuted for our earlier load balancing technique to work. Now, we can go ahead and prepare the next recursion level's receivers to be loosely aligned with the permuted senders, and proceed with the localized oblivious routing.

Now let us consider how to divide this algorithm into a parallel offline phase and a subsequent low-depth online phase. Clearly, the oblivious routing necessary for loosely aligning each recursion level's receivers with the last level's senders must be performed in the offline phase—and we must paralellize this step among all recursion levels. Thus, our idea is the following:

- First, for each recursion level d in parallel, we randomly permutate level d's fetch CPUs in an oblivious fashion (using a building block called oblivious random permutation), at the end of which we have specified the configuration of level d's sender array (that is, after level d's fetch CPUs switch roles and become senders).
- At this point, each recursion level d can prepare its receiver array based on the configuration of level $(d-1)$'s sender array. This can be done in parallel too.
- During the online phase, after fetching metadata from the OPRAM tree, the receivers must permute themselves to switch role to senders—since the offline stage has already dictated the sender array's configuration, this permutation step must respect the offline stage's decision.
  To achieve this in small online depth, our idea is that during the offline phase, each recursion level relies on an instance of oblivious routing to figure out exactly what permutation to apply (henceforth called the "routing permutation") to switch the receiver array to the sender array's configuration—and this can be done in parallel among all recursion levels once a level's receiver and sender arrays have both been determined. Once the offline stage has written down this routing permutation, in the online stage, the receivers can simply apply the permutation, i.e., each receiver writes itself to some array location as specified by the permutation that offline stage has written down. Applying the permutation online takes a single parallel step.

One observation is that during the online stage, the routing permutation is revealed in the clear. To see why this does not leak information, it suffices to see that the result of this routing permutation, i.e., the sender array, was obliviously randomly permuted to start with (using a building block called oblivious random permutation). Thus, even conditioned on having observed the oblivious random permutation's access patterns, each permutation is still equally likely—and thus the routing permutation that is revealed is indistinguishable from a random permutation (even when conditioned on having observed the oblivious random permutation's access patterns).

5.3 Core Subroutine: Localized Routing

Notations and informal explanation. In the OPRAM's execution, the instructions waiting to receive position labels at a recursion level d is denoted $\mathsf{Instr}^{\langle d \rangle}$. $\mathsf{Instr}^{\langle d \rangle}$ has been obliviously and randomly permuted in the offline phase. When these incomplete instructions have received position labels, they become complete and are now called $\mathsf{CInstr}^{\langle d \rangle}$ where $\mathsf{CInstr}^{\langle d \rangle}$ and $\mathsf{Instr}^{\langle d \rangle}$ are arranged in the same order. When data blocks are fetched in recursion level d, they are called $\mathsf{Fetched}^{\langle d \rangle}$, and $\mathsf{Fetched}^{\langle d \rangle}$ has the same order as $\mathsf{CInstr}^{\langle d \rangle}$. In the offline phase, $\mathsf{Instr}^{\langle d \rangle}$ is obliviously sorted to be loosely aligned with $\mathsf{Fetched}^{\langle d-1 \rangle}$ resulting in $\overline{\mathsf{Instr}}^{\langle d \rangle}$, such that $\overline{\mathsf{Instr}}^{\langle d \rangle}$ can receive position labels from $\mathsf{Fetched}^{\langle d-1 \rangle}$ through localized oblivious routing. The offline phase also prepares a routing permutation $\pi^{d \rightarrow d+1}$, that will permute $\overline{\mathsf{Instr}}^{\langle d \rangle}$ (after having received position labels) back to $\mathsf{CInstr}^{\langle d \rangle}$—and the online phase will apply this routing permutation $\pi^{d \rightarrow d+1}$ in a single parallel step. We now describe our algorithms more formally.

We consider the following problem where there is a source array and a destination array, and the destination array wants to receive position identifiers from the source. Specifically, the source array is a set of fetched blocks in randomly permuted order, where each block may contain up to γ position labels corresponding to γ addresses in the next recursion level. The destination array is an incomplete instruction array where each element contains the address of the block to be read at the next recursion level—and each address must receive its corresponding position label before the fetch operations at the next recursion level can be invoked.

- *Inputs:* The inputs contain a randomly permuted source array $\mathsf{Fetched}^{\langle d \rangle}$ that represent the fetched position identifier blocks at recursion level d, and a randomly permuted destination array $\mathsf{Instr}^{\langle d+1 \rangle}$ which represents the incomplete instruction array at recursion level $d + 1$.
  - The source array $\mathsf{Fetched}^{\langle d \rangle}$ contains $2m$ blocks, each of which contains up to γ (logical) pairs of the form $(\mathsf{addr}, \mathsf{pos})$ that are needed in the next recursion level. All the γ addresses in the same block comes from Γ contiguous addresses, and thus in reality the address storage is actually compressed—however, we think of each block in $\mathsf{Fetched}^{\langle d \rangle}$ as *logically* containing pairs of the form $(\mathsf{addr}, \mathsf{pos})$.

- The destination array $\mathsf{Instr}^{\langle d+1\rangle}$ contains m elements each of which is of the form $(\mathsf{addr}, \_)$, where "_" denotes a placeholder for receiving the position identifier for addr later. This array $\mathsf{Instr}^{\langle d+1\rangle}$ is also referred to as the incomplete instruction array.
- We assume that
  (1) all addresses in the destination array must occur in the source array;
  (2) the γ addresses contained in the same block come from Γ contiguous addresses; and
  (3) both the source array $\mathsf{Fetched}^{\langle d\rangle}$ and the destination array $\mathsf{Instr}^{\langle d+1\rangle}$ have been randomly permuted.

– *Outputs:* A complete instruction array denoted CInstr of length $2m$ where $\mathsf{CInstr}^{\langle d+1\rangle}[i]$ is of the form $(\mathsf{addr}_i, \mathsf{pos}_i)$ such that
  - $\mathsf{Instr}^{\langle d+1\rangle}[i] = (\mathsf{addr}_i, \_)$, i.e., the sequence of addresses contained in the output $\mathsf{CInstr}^{\langle d+1\rangle}$ agree with those contained in the input $\mathsf{Instr}^{\langle d+1\rangle}$; and
  - The tuple $(\mathsf{addr}_i, \mathsf{pos}_i)$ exists in some block in $\mathsf{Fetched}^{\langle d\rangle}$, i.e., the position identifier addr_i receives is correct (as defined by $\mathsf{Fetched}^{\langle d\rangle}$).

Offline phase. The inputs are the same as the above. In the offline phase, we aim to output the following arrays:

(a) A permuted destination array $\overline{\mathsf{Instr}}^{\langle d+1\rangle}$ that is a permutation of $\mathsf{Instr}^{\langle d+1\rangle}$ such that it is *somewhat aligned* with the source $\mathsf{Fetched}^{\langle d\rangle}$, where *somewhat aligned* means the following:
  [Somewhat aligned:] Fix $\alpha := \omega(1)$ to be any super-constant function. For each consecutive $\mu := \alpha\gamma^2 \log N$ contiguous source blocks denoted $\mathsf{Fetched}^{\langle d\rangle}[k\mu + 1 : (k + 1)\mu]$, there is a segment of μ contiguous destination blocks $\overline{\mathsf{Instr}}^{\langle d+1\rangle}[k\mu+1 : (k+1)\mu]$ such that all addresses in $\mathsf{Instr}^{\langle d+1\rangle}$ that are contained in $\mathsf{Fetched}^{\langle d\rangle}[k\mu + 1 : (k + 1)\mu]$ appear in the range $\overline{\mathsf{Instr}}^{\langle d+1\rangle}[k\mu + 1 : (k + 1)\mu]$.

(b) A routing permutation $\pi^{d\to d+1} : [2m] \to [2m]$.

In other words, the goal of the offline phase is to prepare the source and the destination arrays such that in the online phase, we only perform oblivious routing from every $\mu := \alpha\gamma^2 \log N$ blocks (each containing at most γ labels) in the source to every μ tuples in the destination where $\alpha = \omega(1)$ is any super-constant function. This way, the online phase has $O(\log \mu)$ parallel runtime.

Before explaining how to accomplish the above, we first prove that if the source array, i.e., $\mathsf{Fetched}^{\langle d\rangle}$ has been randomly permuted, then every μ contiguous blocks contain at most μ position identifiers needed by the destination.

Lemma 5. *Let* arr *denote an array of* $2m$ *randomly permuted blocks, each of which contains* γ *items such that out of the* $2m \cdot \gamma$ *items, at most* m *are real and the rest are dummy.*

Then, for any consecutive n *blocks in* arr, *with probability at least* $1 - \exp(-\frac{n}{2\gamma^2})$, *the number of real items contained in them is at most* n.

The proof of Lemma 5 follows by a standard concentration argument and is to the online full version [3].

We now explain the offline algorithm, i.e., permute the destination array to be somewhat aligned with the source array such that localized oblivious routing will be sufficient. We describe a parallel oblivious algorithm that completes in $O(m \log m)$ total work and $O(\log m)$ parallel runtime.

1. For each block in $\mathsf{Fetched}^{\langle d \rangle}$, write down a tuple $(\mathsf{minaddr}, \mathsf{maxaddr}, i)$ where $\mathsf{minaddr}$ is the minimum address contained in the block, $\mathsf{maxaddr}$ is the maximum address contained in the block, and i is the offset of the block within the $\mathsf{Fetched}^{\langle d \rangle}$ array.
   Henceforth we refer to the resulting array as $\mathsf{SrcMeta}$.
2. Imagine that the resulting array $\mathsf{SrcMeta}$ and the destination array $\mathsf{Instr}^{\langle d+1 \rangle}$ are concatenated. Now, oblivious sort this concatenated array such that each metadata tuple $(\mathsf{minaddr}, \mathsf{maxaddr}, i) \in \mathsf{SrcMeta}$ is immediately followed by all tuples from $\mathsf{Instr}^{\langle d+1 \rangle}$ whose addresses are contained within the range $[\mathsf{minaddr}, \mathsf{maxaddr}]$.
3. Relying on a parallel oblivious aggregate operation [1,16] (see full version [3] for the definition), each element in the array (resulting from the above step) learns the first metadata tuple $(\mathsf{minaddr}, \mathsf{maxaddr}, i)$ to its left. In this way, each address will learn which block (i.e., i) within $\mathsf{Fetched}^{\langle d \rangle}$ it will receive its position identifier from.
   The result of this step is an array such that each metadata tuple of the $(\mathsf{minaddr}, \mathsf{maxaddr}, i)$ is replaced with a dummy entry $\perp$, and each address addr is replaced with (addr, i), denoting that the address addr will receive its position identifier from the i-th block of $\mathsf{Fetched}^{\langle d \rangle}$.
4. For each non-dummy entry in the above array, tag the entry with a group number $\lfloor \frac{i}{\mu} \rfloor$. For each dummy entry, tag it with $\perp$.
5. Invoke an instance of the oblivious bin packing algorithm and pack the resulting array into $\lceil \frac{2m}{\mu} \rceil$ bins of capacity μ each. We refer to the resulting array as $\overline{\mathsf{Instr}}^{\langle d+1 \rangle}$.
6. Obliviously compute the routing permutation $\pi^{d \to d+1}$ that maps $\overline{\mathsf{Instr}}^{\langle d+1 \rangle}$ to $\mathsf{Instr}^{\langle d+1 \rangle}$.
7. Output $\overline{\mathsf{Instr}}^{\langle d+1 \rangle}$ and $\pi^{d \to d+1}$.

Online phase. The online phase consists of the following steps:

1. For every k, fork an instance of the oblivious routing algorithm such that $\overline{\mathsf{Instr}}^{\langle d+1 \rangle}[k\mu + 1 : (k+1)\mu]$ will receive its position identifiers from $\mathsf{Fetched}^{\langle d \rangle}[k\mu + 1 : (k+1)\mu]$.
   This completes in $O(m \log \mu)$ total work and $O(\log \mu)$ parallel runtime.
2. Apply the routing permutation $\pi^{d \to d+1}$ to $\overline{\mathsf{Instr}}^{\langle d+1 \rangle}$, and output the result as $\mathsf{CInstr}^{\langle d+1 \rangle}$.

5.4 Level-to-Level Routing

Given our core localized routing building block, the full level-to-level position identifier routing algorithm is straightforward to state.

Offline phase. Upon receiving a batch of m memory requests, for each recursion level d in parallel:

- Truncate the addresses to the first d bits and perform conflict resolution. The result is an array of length m containing distinct addresses and dummies to read from recursion level d.
- Randomly permute the resulting array, and obtain an incomplete instruction array $\mathsf{Instr}^{\langle d \rangle}$. It is important for security that the random permutation is performed obliviously such that no information is leaked to the adversary about the permutation.
  For $d = 0$, additionally fill in the position map identifiers and complete the instruction array to obtain $\mathsf{CInstr}^{\langle 0 \rangle}$.
- From the $\mathsf{Instr}^{\langle d \rangle}$ array, construct a corresponding incomplete $\mathsf{Fetched}^{\langle d \rangle}$ array where all position identifier fields are left blank as "_". The blocks in $\mathsf{Fetched}^{\langle d \rangle}$ are ordered in the same way as $\mathsf{Instr}^{\langle d \rangle}$.
- If d is not the data level, fork an instance of the localized routing algorithm with input arrays $\mathsf{Fetched}^{\langle d \rangle}$ and $\mathsf{Instr}^{\langle d+1 \rangle}$, and output a permuted version of $\mathsf{Instr}^{\langle d+1 \rangle}$ denoted $\overline{\mathsf{Instr}}^{\langle d+1 \rangle}$ a routing permutation $\pi^{d \to d+1}$.

Online phase. From each recursion level $d = 0, 1, \ldots D$ sequentially where $D = O(\frac{\log N}{\log \Gamma})$ is the total number of recursion levels:

- Based on the completed instruction $\mathsf{CInstr}^{\langle d \rangle}$, allocate an appropriate number of processors for each completed instruction and perform the fetch phase of the OPRAM algorithm. The result is a fetched array $\mathsf{Fetched}^{\langle d \rangle}$.
- Execute the online phase of the localized routing algorithm for recursion level d with the inputs $\mathsf{Fetched}^{\langle d \rangle}$, $\overline{\mathsf{Instr}}^{\langle d+1 \rangle}$, and $\pi^{d \to d+1}$. The result is a completed instruction array $\mathsf{CInstr}^{\langle d+1 \rangle}$ for the next recursion level.

5.5 Main Upper Bound Theorems

In the interest of space, we defer the full details of our OPRAM construction and proofs to the online full version [3]. Our main theorem is the following:

Theorem 4 (Statistically secure, small-depth OPRAM). *There exists a statistically secure OPRAM scheme (for general block sizes) with $O(\log^2 N)$ total work blowup, and $O(\log N \log\log N)$ parallel runtime blowup, where the OPRAM consumes only $O(1)$ blocks of CPU private cache.*

Using a standard PRF-and-counter compression trick first proposed by Fletcher et al. [7] and later improved and extended to the parallel setting by Chan and Shi [5], we obtain the following computationally secure variant.

Corollary 1 (Computationally secure, small-depth OPRAM). *Assume that one-way functions exist. Then, there exists a computationally secure OPRAM scheme (for general block sizes) with* $O(\log^2 N/\log\log N)$ *total work blowup and* $O(\log N)$ *parallel runtime blowup, where the OPRAM consumes only* $O(1)$ *blocks of CPU private cache.*

Finally, in our online full version [3], we include additional algorithmic results that specifically optimize our OPRAM's depth for sufficiently large block sizes.

Acknowledgments. We thank Rafael Pass for numerous helpful discussions and for being consistently supportive. We thank Feng-Hao Liu and Wei-Kai Lin for helpful conversations regarding the lower bound. This work is supported in part by NSF grants CNS-1314857, CNS-1514261, CNS-1544613, CNS-1561209, CNS-1601879, CNS-1617676, an Office of Naval Research Young Investigator Program Award, a DARPA Safeware grant (subcontract under IBM), a Packard Fellowship, a Sloan Fellowship, Google Faculty Research Awards, a Baidu Research Award, and a VMWare Research Award.

References

1. Boyle, E., Chung, K.-M., Pass, R.: Oblivious parallel RAM and applications. In: Kushilevitz, E., Malkin, T. (eds.) TCC 2016. LNCS, vol. 9563, pp. 175–204. Springer, Heidelberg (2016). https://doi.org/10.1007/978-3-662-49099-0_7
2. Boyle, E., Naor, M.: Is there an oblivious RAM lower bound? In: TCC (2016)
3. Chan, T.-H.H., Chung, K.-M., Shi, E.: On the depth of oblivious parallel RAM. Cryptology ePrint Archive, Report 2017/861 (2017). http://eprint.iacr.org/2017/861
4. Chan, T.-H.H., Guo, Y., Lin, W.-K., Shi, E.: Oblivious hashing revisited, and applications to asymptotically efficient ORAM and OPRAM. In: ASIACRYPT (2017)
5. Chan, T.-H.H., Shi, E.: Circuit OPRAM: unifying statistically and computationally secure ORAMs and OPRAMs. In: TCC (2017)
6. Chen, B., Lin, H., Tessaro, S.: Oblivious parallel RAM: improved efficiency and generic constructions. In: Kushilevitz, E., Malkin, T. (eds.) TCC 2016. LNCS, vol. 9563, pp. 205–234. Springer, Heidelberg (2016). https://doi.org/10.1007/978-3-662-49099-0_8
7. Fletcher, C.W., Ren, L., Kwon, A., van Dijk, M., Devadas, S.: Freecursive ORAM: [nearly] free recursion and integrity verification for position-based oblivious RAM. In: ASPLOS (2015)
8. Goldreich, O.: Towards a theory of software protection and simulation by oblivious RAMs. In: STOC (1987)
9. Goldreich, O., Ostrovsky, R.: Software protection and simulation on oblivious RAMs. J. ACM **43**, 431–473 (1996)
10. Goodrich, M.T., Mitzenmacher, M.: Privacy-preserving access of outsourced data via oblivious RAM simulation. In: Aceto, L., Henzinger, M., Sgall, J. (eds.) ICALP 2011. LNCS, vol. 6756, pp. 576–587. Springer, Heidelberg (2011). https://doi.org/10.1007/978-3-642-22012-8_46
11. Gordon, S.D., Katz, J., Kolesnikov, V., Krell, F., Malkin, T., Raykova, M., Vahlis, Y.: Secure two-party computation in sublinear (amortized) time. In: CCS (2012)

12. Hagerup, T.: Fast and optimal simulations between CRCW PRAMs. In: Finkel, A., Jantzen, M. (eds.) STACS 1992. LNCS, vol. 577, pp. 45–56. Springer, Heidelberg (1992). https://doi.org/10.1007/3-540-55210-3_172
13. Kushilevitz, E., Lu, S., Ostrovsky, R.: On the (in)security of hash-based oblivious RAM and a new balancing scheme. In: SODA (2012)
14. Liu, C., Hicks, M., Harris, A., Tiwari, M., Maas, M., Shi, E.: Ghostrider: a hardware-software system for memory trace oblivious computation. In: ASPLOS (2015)
15. Nayak, K., Katz, J.: An oblivious parallel ram with $O(\log^2 N)$ parallel runtime blowup. Cryptology ePrint Archive, Report 2016/1141 (2016)
16. Nayak, K., Wang, X.S., Ioannidis, S., Weinsberg, U., Taft, N., Shi, E.: GraphSC: parallel secure computation made easy. In: IEEE S&P (2015)
17. Ren, L., Yu, X., Fletcher, C.W., van Dijk, M., Devadas, S.: Design space exploration and optimization of path oblivious RAM in secure processors. In: ISCA, pp. 571–582 (2013)
18. Shi, E., Chan, T.-H.H., Stefanov, E., Li, M.: Oblivious RAM with $O((\log N)^3)$ worst-case cost. In: ASIACRYPT (2011)
19. Stefanov, E., Shi, E.: Oblivistore: high performance oblivious cloud storage. In: IEEE Symposium on Security and Privacy (S&P) (2013)
20. Stefanov, E., van Dijk, M., Shi, E., Fletcher, C., Ren, L., Yu, X., Devadas, S., Path ORAM – an extremely simple oblivious ram protocol. In: CCS (2013)
21. Wang, X.S., Chan, T.-H.H., Shi, E.: Circuit ORAM: on tightness of the Goldreich-Ostrovsky lower bound. In: ACM CCS (2015)
22. Wang, X.S., Huang, Y., Chan, T.-H.H., Shelat, A., Shi, E.: SCORAM: oblivious RAM for secure computation. In: CCS (2014)
23. Williams, P., Sion, R., Tomescu, A.: Privatefs: a parallel oblivious file system. In: CCS (2012)

Low Cost Constant Round MPC Combining BMR and Oblivious Transfer

Carmit Hazay[1(✉)], Peter Scholl[2], and Eduardo Soria-Vazquez[3]

[1] Bar-Ilan University, Ramat Gan, Israel
carmit.hazay@biu.ac.il
[2] Aarhus University, Aarhus, Denmark
peter.scholl@cs.au.dk
[3] University of Bristol, Bristol, UK
eduardo.soria-vazquez@bristol.ac.uk

Abstract. In this work, we present two new universally composable, actively secure, constant round multi-party protocols for generating BMR garbled circuits with free-XOR and reduced costs.

1. Our first protocol takes a generic approach using any secret-sharing based MPC protocol for binary circuits, and a correlated oblivious transfer functionality.
2. Our specialized protocol uses secret-sharing based MPC with information-theoretic MACs. This approach is less general, but requires no additional correlated OTs to compute the garbled circuit.

In both approaches, the underlying secret-sharing based protocol is only used for *one secure* $\mathbb{F}_2$ *multiplication per AND gate*. An interesting consequence of this is that, with current techniques, constant round MPC for binary circuits is not much more expensive than practical, non-constant round protocols.

We demonstrate the practicality of our second protocol with an implementation, and perform experiments with up to 9 parties securely computing the AES and SHA-256 circuits. Our running times improve upon the best possible performance with previous BMR-based protocols by 60 times.

1 Introduction

Secure multi-party computation (MPC) protocols allow a group of n parties to compute some function f on the parties' private inputs, while preserving a number of security properties such as *privacy* and *correctness*. The former property implies data confidentiality, namely, nothing leaks from the protocol execution but the computed output. The latter requirement implies that the protocol enforces the integrity of the computations made by the parties, namely, honest parties learn the correct output. Modern, practical MPC protocols typically fall into two main categories: those based on secret-sharing [5,13,15,18,22,35], and

P. Scholl—Work done whilst at University of Bristol, UK.

T. Takagi and T. Peyrin (Eds.): ASIACRYPT 2017, Part I, LNCS 10624, pp. 598–628, 2017.
https://doi.org/10.1007/978-3-319-70694-8_21

those based on garbled circuits [2,11,25–28,32,39]. When it comes to choosing a protocol, many different factors need to be taken into account, such as the function being evaluated, the latency and bandwidth of the network and the adversary model.

Secret-sharing based protocols such as [5,15,18] tend to have lower communication requirements in terms of bandwidth, but require a large number of rounds of communication, which increases with the complexity of the function. In this approach the parties first secret-share their inputs and then evaluate the circuit gate by gate while preserving privacy and correctness. In low-latency networks, they can have an extremely fast online evaluation stage, but the round complexity makes them much less suited to high-latency networks, when the parties may be far apart.

Garbled circuits, introduced in Yao's protocol [39], are the core behind all practical, constant round protocols for secure computation. In the two-party setting, one of the parties "encrypts" the circuit being evaluated, whereas the other party privately evaluates it. Garbled circuit-based protocols have recently become much more efficient, and currently give the most practical approach for actively secure computation of binary circuits [34,37]. With more than two parties, the situation is more complex, as the garbled circuit must be computed by all parties in a distributed manner using another (non-constant-round) MPC protocol, as in the BMR protocol from [2]. This still leads to a low depth circuit, hence a constant round protocol overall, because all gates can be garbled in parallel. We note that this paradigm has received very little attention, compared with two-party protocols. The original BMR construction uses generic zero-knowledge techniques for proving correct computation of PRG values, so is impractical. A different protocol, but only for three parties, was designed by Choi et al. [11] in the dishonest majority setting. More practical, actively secure protocols for any number of parties are the recent works of Lindell et al. [29,31], which use somewhat homomorphic encryption (SHE) or generic MPC to garble a circuit. Ben-Efraim et al. [4] recently presented and implemented an efficient multi-party garbling protocol based on oblivious transfer, but with only semi-honest security. Very recently, Katz et al. introduced in [23] protocols based on authenticated garbling, with a preprocessing phase that can be instantiated based on TinyOT [33].

1.1 Our Contributions

In this work, we present a practical, actively secure, constant round multi-party protocol for generating BMR garbled circuits with free-XOR in the presence of up to $n-1$ out of n corruptions. As in prior constructions, our approach has two phases: a preprocessing phase where the garbled circuit is mutually generated by all parties, and an online phase where the parties obtain the output of the computation. While the online phase is typically efficient and incurs no cost to achieve active security, the focus of recent works was on optimizing the preprocessing complexity, where the main bottleneck is with respect to garbling

AND gates. In that context, we present two new constant-round protocols for securely generating the garbled circuit:

1. A generic approach using any secret-sharing based MPC protocol for binary circuits, and a correlated oblivious transfer functionality.
2. A specialized protocol which uses secret-sharing based MPC with information-theoretic MACs, such as TinyOT [17,33]. This approach is less general, but requires no additional correlated OTs to compute the garbled circuit.

In both approaches, the underlying secret-sharing based protocol is only used for *one secure* $\mathbb{F}_2$ *multiplication per AND gate.*

In the first, more general method, every pair of parties needs to run one correlated OT per AND gate, which costs $O(\kappa)$ communication for security parameter κ. Combining this with the overhead induced by the correlated OTs in our protocol, we obtain total complexity $O(|C|\kappa n^2)$, assuming only symmetric primitives and $O(\kappa)$ seed OTs between every pair of parties. This gives an overall communication cost of $O(M + |C|\kappa n^2)$ to evaluate a circuit C, where M is the cost of evaluating $|C|$ AND gates in the secret-sharing based protocol, Π. To realize Π, we can define a functionality with multiplication depth 1 that computes all the AND gates in parallel (these multiplications can be computed in parallel as they are independent of the parties' inputs). Furthermore, the [21] compiler can be instantiated with semi-honest [18] as the inner protocol and [12] as the outer protocol. By Theorem 2, Sect. 5 from [21], for some constant number of parties $m \geq 2$, the functionality can be computed with communication complexity $O(|C|)$ plus low order terms that depend on a statistical parameter s, the circuit's depth and $\log|C|$. As in [21], this extends to the case of a non-constant number of parties n, in which case the communication complexity grows by an additional factor of $|C|\mathsf{poly}(n)$.

Another interesting candidate for instantiating Π would be to use an MPC protocol optimized for SIMD binary circuits such as MiniMAC [16]. This is because in our construction, all the AND gates can be computed in parallel. Currently, the only known preprocessing methods [17] for MiniMAC are not practical, but this seems to be an interesting future direction to explore.

TinyOT is currently the most practical approach to secret-sharing based MPC on binary circuits, so the second method leads to a highly practical protocol for constant-round secure computation. The complexity is essentially the same as TinyOT, as here we do not require any additional OTs. However, the protocol is less general and has worse asymptotic communication complexity, since TinyOT costs either $O(|C|B\kappa n^2)$ (with 2 parties or the recent protocol of [38]), or $O(|C|B^2\kappa n^2)$ (with [17]), where $B = O(1 + s/\log|C|)$ (and in practice is between 3–5), and s is the statistical security parameter.

Our constructions employ several very appealing features. For a start, we embed into the modeling of the preprocessing functionality, which computes the garbled circuit, an additive error introduced into the garbling by the adversary. Concretely, we extend the functionality from [29] so that it obtains a vector of additive errors from the adversary to be applied to each garbled gate, which captures the fact that the adversary may submit inconsistent keys and pseudorandom function (PRF) values. We further strengthen this by allowing the adversary to pick the error *adaptively* after seeing the garbled circuit (in prior constructions this error is independent of the garbling) and allowing corrupt parties to choose their own PRF keys, possibly not at random. This requires a new analysis and proof of the online phase.

Secondly, we devise a new consistency check to enforce correctness of inputs to correlated OT, which is based on very efficient linear operations similar to recent advances in homomorphic commitments [9]. This check, combined with our improved error analysis for the online phase, allows the garbled circuit to be created without authenticating any of the parties' keys or PRF values, which removes a significant cost from previous works (saving a factor of $\Omega(n)$).

Implementation. We demonstrate the practicality of our TinyOT-based protocol with an implementation, and perform experiments with up to 9 parties securely computing the AES and SHA-256 circuits. In a 1 Gbps LAN setting, we can securely compute the AES circuit with 9 parties in just 620 ms. This improves upon the best possible performance that would be attainable using [29] by around 60 times. The details of our implementation can be found in Sect. 6.

Comparison with Other Approaches. Table 1 shows how the communication complexity of our work compares with other actively secure, constant-round protocols. As mentioned earlier, most previous constructions express the garbling function as an arithmetic circuit over a large finite field. In these protocols, garbling even a single AND gate requires computing $O(n)$ multiplications over a large field with SHE or MPC. This means they scale at least cubically in the number of parties. In constrast, our protocol only requires one $\mathbb{F}_2$ multiplication per AND gate, so scales with $O(n^2)$. Previous SHE-based protocols also require zero-knowledge proofs of plaintext knowledge of SHE ciphertexts, which in practice are very costly. Note that the recent MASCOT protocol [24] for secure computation of arithmetic circuits could also be used in [29], instead of SHE, but this still has very high communication costs. We denote by MASCOT-BMR-FX an optimized variant of [29], modified to use free-XOR as in our protocol, with multiplications in $\mathbb{F}_{2^\kappa}$ done using MASCOT. Finally, the recent concurrent work by Katz et al. [23] is based on an optimized variant of TinyOT, with comparable performance to our approach.

None of these previous works have reported implementations at the time of writing, but our implementation of the TinyOT-based protocol improves upon the best times that would be achievable with SPDZ-BMR and MASCOT by up to 60x. This is because our protocol has lower communication costs than [29] (by at least 2 orders of magnitude) and the main computational costs are from standard symmetric primitives, so far cheaper than using SHE.

Overall, our protocols significantly narrow the gap between the cost of constant-round and many-round MPC protocols for binary circuits. More specifically, this implies that, with current techniques, constant round MPC for binary circuits is not much more expensive than practical, non-constant round protocols. Additionally, both of our protocols have potential for future improvement by optimizing existing non-constant round protocols: a practical implementation of MiniMAC [16] would lead to a very efficient approach with our generic protocol, whilst any future improvements to multi-party TinyOT would directly give a similar improvement to our second protocol.

Table 1. Comparison of actively secure, constant round MPC protocols. $B = O(1 + s/\log|C|)$ is a cut-and-choose parameter, which in practice is between 3–5. Our second protocol can also be based upon optimized TinyOT to obtain the same complexity as [23].

Protocol	Based on	Free XOR	Comms. per garbled gate
SPDZ-BMR [29]	SHE + ZKPoPK	✗	$O(n^4\kappa)$
SHE-BMR [31]	SHE (depth 4) + ZKPoPK	✗	$O(n^3\kappa)$
MASCOT-BMR-FX	OT	✓	$O(n^3\kappa^2)$
This work Sect. 3	OT + [21]	✓	$O(n^2\kappa + \mathsf{poly}(n))$
This work Sect. 4	TinyOT	✓	$O(n^2B^2\kappa)$
[23] (concurrent)	Optimized TinyOT	✓	$O(n^2B\kappa)$

1.2 Technical Overview

Our protocol is based on the recent free-XOR variant of BMR garbling used for semi-honest MPC in [4]. In that scheme, a garbling of the g-th AND gate with input wires u, v and output wire w, consists of the $4n$ values (where n is the number of parties):

$$\begin{aligned}\tilde{g}^j_{a,b} = &\left(\bigoplus_{i=1}^{n} F_{k^i_{u,a},k^i_{v,b}}(g\|j)\right) \oplus k^j_{w,0} \\ &\oplus \left(R^j((\lambda_u \oplus a)(\lambda_v \oplus b) \oplus \lambda_w)\right), \quad (a,b) \in \{0,1\}^2,\ j \in [n]\end{aligned} \tag{1}$$

Here, F is a double-key PRF, $R^j \in \{0,1\}^\kappa$ is a fixed correlation string for free-XOR known to party P_j, and the keys $k^j_{u,a}, k^j_{v,b} \in \{0,1\}^\kappa$ are also known to P_j. Furthermore, the wire masks $\lambda_u, \lambda_v, \lambda_w \in \{0,1\}$ are random, additively secret-shared bits known by no single party.

The main idea behind BMR is to compute the garbling, except for the PRF values, with a general MPC protocol. The analysis of [29] showed that it is not necessary to prove in zero-knowledge that every party inputs the correct PRF values to the MPC protocol that computes the garbling. This is because when evaluating the garbled circuit, each party P_j can check that the decryption of the j-th entry in every garbled gate gives one of the keys $k^j_{w,b}$, and this check would overwhelmingly fail if any PRF value was incorrect. It further implies that the adversary cannot flip the value transmitted through some wire as that would require from it to guess a key.

Our garbling protocol proceeds by computing a random, *unauthenticated*, additive secret sharing of the garbled circuit. This differs from previous works [29, 31], which obtain *authenticated* (with MACs, or SHE ciphertexts) sharings of the entire garbled circuit. Our protocol greatly reduces this complexity, since the PRF values and keys (on the first line of Eq. (1)) do not need to be authenticated. The main challenge, therefore, is to compute shares of the products on the second line of (1). Similarly to [4], a key observation that allows efficiency is the fact that these multiplications are either between two secret-shared bits, or a secret-shared bit and a fixed, secret string. So, we do not need the full power of an MPC protocol for arithmetic circuit evaluation over $\mathbb{F}_{2^\kappa}$ or $\mathbb{F}_p$ (for large p), as used in previous works.

To compute the bit product $\lambda_u \cdot \lambda_v$, we can use any actively secure GMW-style MPC protocol for binary circuits. This protocol is only needed for computing *one secure AND per garbled AND gate*, since all bit products in $\tilde{g}^j_{a,b}$ can be computed as linear combinations of $\lambda_u \cdot \lambda_v$, λ_u and λ_v. We then need to multiply the resulting secret-shared bits by the string R^j, known to P_j. We give two variants for computing this product, the first one being more general and the second more concretely efficient. In more details,

1. The first solution performs the multiplication by running actively secure correlated OT between P_j and every other party, where P_j inputs R^j as the fixed OT correlation. The parties then run a consistency check by applying a universal linear hash function to the outputs and sacrificing a few OTs, ensuring the correct inputs were provided to the OT. This protocol is presented in Sect. 3.
2. The second method requires using a 'TinyOT'-style protocol [6,17] based on information-theoretic MACs, and allows us to compute the bit/string products directly from the MACs, provided each party's MAC key is chosen to be the same string R^i used in the garbling. This saves interaction since we do not need any additional OTs. This protocol is presented in Sect. 4.

After creating shares of all these products, the parties can compute shares of the whole garbled circuit. These shares must then be rerandomized, before they can be broadcast. Opening the garbled circuit in this way allows a corrupt party to introduce further errors into the garbling by changing their share, even *after learning the correct garbled circuit*, since we may have a rushing adversary. Nevertheless, we prove that the BMR online phase remains secure when this type of error is allowed, as it would only lead to an abort. This significantly strengthens the result from [29], which only allowed corrupt parties to provide incorrect PRF values, and is an important factor that allows our preprocessing protocol to be so efficient.

Concurrent Work. Two recent works by Katz, Ranellucci and Wang introduced constant round, two-party [38] and multi-party [23] protocols based on *authenticated garbling*, with a preprocessing phase that can be instantiated based on TinyOT. At the time of writing, their two-party paper also reports on an implementation, but the multi-party version does not. Our work is conceptually quite similar, since both involve generating a garbled circuit in a distributed manner using TinyOT. The main difference seems to be that our protocol is symmetric, since all parties evaluate the same garbled circuit. With authenticated garbling, the garbled circuit is only evaluated by one party. This makes the garbled circuit slightly smaller, since there are $n-1$ sets of keys instead of n, but the online phase requires at least one more round of interaction (if all parties learn the output). The works of Katz et al. also contain concrete and asymptotic improvements to the two-party and multi-party TinyOT protocols, which improves upon the TinyOT protocol we give in the full version of this paper [20] by a factor of $O(s/\log|C|)$, where s is a statistical parameter. These improvements can be directly plugged into our second garbling protocol. We remark that the two-party protocol in [38] inspired our use of TinyOT MACs to perform the bit/string multiplications in our protocol from Sect. 4. The rest of our work is independent.

Another difference is that our protocol from Sect. 3 is more generic, since $\mathcal{F}_{\text{BitMPC}}$ can be implemented with *any* secret-sharing based bit-MPC protocol, rather than just TinyOT. This can be instantiated with [21] to obtain a constant-round protocol with complexity $O(|C|(\kappa n^2 + \mathsf{poly}(n)))$ in the OT-hybrid model. The multi-party paper [23] does not have an analogous generic result.

2 Preliminaries

We denote the security parameter by κ. We say that a function $\mu : \mathbb{N} \to \mathbb{N}$ is *negligible* if for every positive polynomial $p(\cdot)$ and all sufficiently large κ it holds that $\mu(\kappa) < \frac{1}{p(\kappa)}$. We use the abbreviation PPT to denote probabilistic polynomial-time. We further denote by $a \leftarrow A$ the uniform sampling of a from a set A, and by $[d]$ the set of elements $(1, \dots, d)$. We often view bit-strings in $\{0,1\}^k$ as vectors in $\mathbb{F}_2^k$, depending on the context, and denote exclusive-or by "$\oplus$" or "$+$". If $a, b \in \mathbb{F}_2$ then $a \cdot b$ denotes multiplication (or AND), and if $\boldsymbol{c} \in \mathbb{F}_2^\kappa$ then $a \cdot \boldsymbol{c} \in \mathbb{F}_2^\kappa$ denotes the product of a with every component of $\boldsymbol{c}$.

For vectors $\boldsymbol{x} = (x_1, \dots, x_n) \in \mathbb{F}_2^n$ and $\boldsymbol{y} \in \mathbb{F}_2^m$, the *tensor product* (or *outer product*) $\boldsymbol{x} \otimes \boldsymbol{y}$ is defined as the $n \times m$ matrix over $\mathbb{F}_2$ where the i-th row is $x_i \cdot \boldsymbol{y}$. We use the following property.

Fact 21. *If* $\boldsymbol{x} \in \mathbb{F}_2^n, \boldsymbol{y} \in \mathbb{F}_2^m$ *and* $\mathbf{M} \in \mathbb{F}_2^{m \times n}$ *then*

$$\mathbf{M} \cdot (\boldsymbol{x} \otimes \boldsymbol{y}) = (\mathbf{M} \cdot \boldsymbol{x}) \otimes \boldsymbol{y}.$$

Universal composability. We prove security of our protocols in the universal composability (UC) framework [7] (see also [8] for a simplified version of UC).

Communication model. We assume all parties are connected via authenticated communication channels, as well as secure point-to-point channels and a broadcast channel. The default method of communication in our protocols is authenticated channels, unless otherwise specified. Note that in practice, these can all be implemented with standard techniques (in particular, for broadcast a simple 2-round protocol suffices, since we allow abort [19]).

Adversary model. The adversary model we consider is a static, active adversary who corrupts up to $n-1$ out of n parties. This means that the identities of the corrupted parties are fixed at the beginning of the protocol, and they may deviate arbitrarily from the protocol.

2.1 Circular 2-Correlation Robust PRF

The BMR garbling technique from [29] is proven secure based on a pseudorandom function (PRF) with multiple keys. However, since our scheme supports free-XOR, we need to adapt the definition of correlation robustness with circularity from [10] given for hash functions to double-key PRFs. This definition captures the related key and circularity requirements induced by supporting the free-XOR technique. Formally, fix some function $F : \{0,1\}^n \times \{0,1\}^\kappa \times \{0,1\}^\kappa \mapsto \{0,1\}^\kappa$. We define an oracle Circ_R as follows:

- $\mathsf{Circ}_R(k_1, k_2, g, j, b_1, b_2, b_3)$ outputs $F_{k_1 \oplus b_1 R, k_2 \oplus b_2 R}(g \| j) \oplus b_3 R$.

The outcome of oracle Circ is compared with the a random string of the same length computed by an oracle Rand:

- $\mathsf{Rand}(k_1, k_2, g, j, b_1, b_2, b_3)$: if this input was queried before then return the answer given previously. Otherwise choose $u \leftarrow \{0,1\}^\kappa$ and return u.

Definition 21 (Circular 2-correlation robust PRF). *A PRF F is* circular 2-correlation robust *if for any non-uniform polynomial-time distinguisher $\mathcal{D}$ making legal queries to its oracle, there exists a negligible function* negl *such that:*

$$\left| \Pr[R \leftarrow \{0,1\}^\kappa; \mathcal{D}^{\mathsf{Circ}_R(\cdot)}(1^\kappa) = 1] - \Pr[\mathcal{D}^{\mathsf{Rand}(\cdot)}(1^\kappa) = 1] \right| < \mathsf{negl}(\kappa).$$

As in [10], some trivial queries must be ruled out. Specifically, the distinguisher is restricted as follows: (1) it is not allowed to make any query of the form $\mathcal{O}(k_1, k_2, g, j, 0, 0, b_3)$ (since it can compute $F_{k_1,k_2}(g\|j)$ on its own) and (2) it is not allowed to query both tuples $\mathcal{O}(k_1, k_2, g, j, b_1, b_2, 0)$ and $\mathcal{O}(k_1, k_2, g, j, b_1, b_2, 1)$ for any values k_1, k_2, g, j, b_1, b_2 (since that would allow it to trivially recover the global difference). We say that any distinguisher respecting these restrictions makes legal queries.

2.2 Almost-1-Universal Linear Hashing

We use a family of almost-1-universal linear hash functions over $\mathbb{F}_2$, defined by:

Definition 22 (Almost-1-Universal Linear Hashing). *We say that a family $\mathcal{H}$ of linear functions $\mathbb{F}_2^m \to \mathbb{F}_2^s$ is ε-almost 1-universal, if it holds that for every non-zero $\boldsymbol{x} \in \mathbb{F}_2^m$ and for every $\boldsymbol{y} \in \mathbb{F}_2^s$:*

$$\Pr_{\mathbf{H} \leftarrow \mathcal{H}}[\mathbf{H}(\boldsymbol{x}) = \boldsymbol{y}] \leq \varepsilon$$

where $\mathbf{H}$ is chosen uniformly at random from the family $\mathcal{H}$. We will identify functions $\mathbf{H} \in \mathcal{H}$ with their $s \times m$ transformation matrix, and write $\mathbf{H}(\boldsymbol{x}) = \mathbf{H} \cdot \boldsymbol{x}$.

This definition is slightly stronger than a family of almost-universal linear hash functions (where the above need only hold for $\boldsymbol{y} = 0$, as in [9]). However, this is still much weaker than *2-universality* (or *pairwise independence*), which a linear family of hash functions cannot achieve, because $\mathbf{H}(0) = 0$ always. The two main properties affecting the efficiency of a family of hash functions are the *seed size*, which refers to the length of the description of a random function $\mathbf{H} \leftarrow \mathcal{H}$, and the *computational complexity* of evaluating the function. The simplest family of almost-1-universal hash functions is the set of all $s \times m$ matrices; however, this is not efficient as the seed size and complexity are both $O(m \cdot s)$. Recently, in [9], it was shown how to construct a family with seed size $O(s)$ and complexity $O(m)$, which is asymptotically optimal. A more practical construction is a polynomial hash based on GMAC (used also in [34]), described as follows (here we assume that s divides m, for simplicity):

- Sample a random seed $\alpha \leftarrow \mathbb{F}_{2^s}$
- Define $\mathbf{H}_\alpha$ to be the function:

$$\mathbf{H}_\alpha : \mathbb{F}_{2^s}^{m/s} \to \mathbb{F}_{2^s}, \quad \mathbf{H}_\alpha(x_1, \ldots, x_{m/s}) = \alpha \cdot x_1 + \alpha^2 \cdot x_2 + \cdots + \alpha^{m/s} \cdot x_{m/s}$$

  Note that by viewing elements of $\mathbb{F}_{2^s}$ as vectors in $\mathbb{F}_2^s$, multiplication by a fixed field element $\alpha^i \in \mathbb{F}_{2^s}$ is linear over $\mathbb{F}_2$. Therefore, $\mathbf{H}_\alpha$ can be seen as an $\mathbb{F}_2$-linear map, represented by a unique matrix in $\mathbb{F}_2^{s \times m}$.

Here, the seed is short, but the computational complexity is $O(m \cdot s)$. However, in practice when $s = 128$ the finite field multiplications can be performed very efficiently in hardware on modern CPUs. Note that this gives a 1-universal family with $\varepsilon = \frac{m}{s} \cdot 2^{-s}$. This can be improved to 2^{-s} (i.e. perfect), at the cost of a larger seed, by using m/s distinct elements α_i, instead of powers of α.

2.3 Commitment Functionality

We require a UC commitment functionality $\mathcal{F}_{\text{Commit}}$ (Fig. 1). This can easily be implemented in the random oracle model by defining $\texttt{Commit}(x, P_i) = \mathsf{H}(x, i, r)$, where H is a random oracle and $r \leftarrow \{0, 1\}^\kappa$.

The Functionality $\mathcal{F}_{\text{Commit}}$

Commit: On input $(\texttt{Commit}, x, i, \tau_x)$ from P_i, store (x, i, τ_x) and output (i, τ_x) to all parties.

Open: On input $(\texttt{Open}, i, \tau_x)$ by P_i, output (x, i, τ_x) to all parties.
If instead $(\texttt{NoOpen}, i, \tau_x)$ is given by the adversary, and P_i is corrupt, the functionality outputs $(\perp, i, \tau_x)$ to all parties.

Fig. 1. Ideal commitments

Functionality $\mathcal{F}_{\text{Rand}}$

Upon receiving $(\texttt{rand}, S)$ from all parties, where S is any efficiently sampleable set, it samples $r \leftarrow S$ and outputs r to all parties.

Fig. 2. Coin-tossing functionality

2.4 Coin-Tossing Functionality

We use a standard coin-tossing functionality, $\mathcal{F}_{\text{Rand}}$ (Fig. 2), which can be implemented with UC commitments to random values.

2.5 Correlated Oblivious Transfer

In this work we use an actively secure protocol for oblivious transfer (OT) on correlated pairs of strings of the form $(a_i, a_i \oplus \Delta)$, where Δ is fixed for every OT. The TinyOT protocol [33] for secure two-party computation constructs such a protocol, and a significantly optimized version of this is given in [34]. The communication cost is roughly $\kappa + s$ bits per OT. The ideal functionality is shown in Fig. 3.

Fixed Correlation OT Functionality - $\mathcal{F}_{\text{COT}}$

Initialize: Upon receiving $(\texttt{init}, \Delta)$, where $\Delta \in \{0,1\}^\kappa$ from P_S and $(\texttt{init})$ from P_R, store Δ. Ignore any subsequent $(\texttt{init})$ commands.

Extend: Upon receiving $(\texttt{extend}, x_1, \ldots, x_m)$ from P_R, where $x_i \in \{0,1\}$, and $(\texttt{extend})$ from P_S, do the following:
- Sample $t_i \in \{0,1\}^\kappa$, for $i \in [m]$. If P_R is corrupted then wait for $\mathcal{A}$ to input t_i.
- Compute $q_i = t_i + x_i \cdot \Delta$, for $i \in [m]$.
- If P_S is corrupted then wait for $\mathcal{A}$ to input $q_i \in \{0,1\}^\kappa$ and recompute $t_i = q_i + x_i \cdot \Delta$.
- Output t_i to P_R and q_i to P_S, for $i \in [m]$.

Fig. 3. Fixed correlation oblivious transfer functionality

2.6 Functionality for Secret-Sharing-Based MPC

We make use of a general, actively secure protocol for secret-sharing-based MPC for binary circuits, which is modeled by the functionality $\mathcal{F}_{\mathrm{BitMPC}}$ in Fig. 4. This functionality allows parties to provide private inputs, which are then stored and can be added or multiplied internally by $\mathcal{F}_{\mathrm{BitMPC}}$, and revealed if desired. Note that we also need the **Multiply** command to output a random additive secret-sharing of the product to all parties; this essentially assumes that the underlying protocol is based on secret-sharing.

We use the notation $\langle x \rangle$ to represent a secret-shared value x that is stored internally by $\mathcal{F}_{\mathrm{BitMPC}}$, and define x^i to be party P_i's additive share of x (if it is known). We also define the $+$ and $\cdot$ operators on two shared values $\langle x \rangle, \langle y \rangle$ to call the **Add** and **Multiply** commands of $\mathcal{F}_{\mathrm{BitMPC}}$, respectively, and return the identifier associated with the result.

The Bit MPC Functionality - $\mathcal{F}_{\mathrm{BitMPC}}$

The functionality runs with parties $P_1, \ldots, P_n$ and an adversary $\mathcal{A}$. The functionality maintains a dictionary, $\mathtt{Val} \leftarrow \{\}$, to keep track of values in $\mathbb{F}_2$.

Input: On receiving $(\mathtt{Input}, \mathtt{id}_1, \ldots, \mathtt{id}_\ell, x_1, \ldots, x_\ell, P_j)$ from party P_j and $(\mathtt{Input}, \mathtt{id}_1, \ldots, \mathtt{id}_\ell, P_j)$ from all other parties, where $x_i \in \mathbb{F}_2$, store $\mathtt{Val}[\mathtt{id}_i] \leftarrow x_i$ for $i \in [\ell]$.

Add: On input $(\mathsf{Add}, \overline{\mathtt{id}}, \mathtt{id}_1, \ldots, \mathtt{id}_\ell)$ from all parties, where $(\mathtt{id}_1, \ldots, \mathtt{id}_\ell)$ are keys in $\mathtt{Val}$, set $\mathtt{Val}[\overline{\mathtt{id}}] \leftarrow \sum_{i=1}^{\ell} \mathtt{Val}[\mathtt{id}_i]$.

Multiply: On input $(\mathtt{multiply}, \overline{\mathtt{id}}, \mathtt{id}_1, \mathtt{id}_2)$ from all parties, where $(\mathtt{id}_1, \mathtt{id}_2)$ are keys in $\mathtt{Val}$, compute $y \leftarrow \mathtt{Val}[\mathtt{id}_1] \cdot \mathtt{Val}[\mathtt{id}_2]$. Receive shares $y^i \in \mathbb{F}_2$ from $\mathcal{A}$, for $i \in I$, then sample random honest parties' shares $y^j \in \mathbb{F}_2$, for $j \notin I$, such that $\sum_{i=1}^{n} y^i = y$. Send y^i to party P_i, for $i \in [n]$, and store the value $\mathtt{Val}[\overline{\mathtt{id}}] \leftarrow y$.

Open: On input $(\mathsf{Open}, \mathtt{id})$ from all parties, where $\mathtt{id}$ is a key in $\mathtt{Val}$, send $x \leftarrow \mathtt{Val}[\mathtt{id}]$ to $\mathcal{A}$. Wait for an input from $\mathcal{A}$. If it inputs $\mathtt{OK}$ then output x to all parties, otherwise output $\perp$ and terminate.

Fig. 4. Functionality for GMW-style MPC for binary circuits

2.7 BMR Garbling

The [2] garbling technique by Beaver, Micali and Rogaway involves garbling each gate separately using pseudorandom generators while ensuring consistency between the wires. This method was recently improved in a sequence of works [4,29,31], where the latter work further supports the free XOR property. The main task of generating the garbled circuit while supporting this property is to compute, for each AND gate g with input wires u, v and output wire w, the $4n$ values:

$$\tilde{g}^j_{a,b} = \left(\bigoplus_{i=1}^{n} F_{k^i_{u,a}, k^i_{v,b}}(g \| j) \right) \oplus k^j_{w,0}$$
$$\oplus \left(R^j \cdot ((\lambda_u \oplus a) \cdot (\lambda_v \oplus b) \oplus \lambda_w) \right), \quad (a,b) \in \{0,1\}^2,\ j \in [n] \tag{2}$$

where the wire masks $\lambda_u, \lambda_v, \lambda_w \in \{0,1\}$ are secret-shared between all parties, while the PRF keys $k^j_{u,a}, k^j_{v,b}$ and the global difference string R^j are known only to party P_j.

3 Generic Protocol for Multi-party Garbling

We now describe our generic method for creating the garbled circuit using any secret-sharing based MPC protocol (modeled by $\mathcal{F}_{\text{BitMPC}}$) and the correlated OT functionality $\mathcal{F}_{\text{COT}}$. We first describe the functionality in Sect. 3.1 and the protocol in Sect. 3.2, and then analyse its security in Sect. 3.4.

3.1 The Preprocessing Functionality

The preprocessing functionality, formalized in Fig. 5, captures the generation of the garbled circuit as well as an error introduced by the adversary. The adversary is allowed to submit an additive error, chosen adaptively after seeing the garbled circuit, that is added by the functionality to each entry when the garbled circuit is opened.

3.2 Protocol Overview

The garbling protocol, shown in Fig. 6, proceeds in three main stages. Firstly, the parties locally sample all of their keys and shares of wire masks for the garbled circuit. Secondly, the parties compute shares of the products of the wire masks and each party's global difference string; these are then used by each party to locally obtain a share of the entire garbled circuit. Finally, the bit masks for the output wires are opened to all parties. The opening of the garbled circuit is shown in Fig. 7.

Concretely, each party P_i starts by sampling a global difference string $R^i \leftarrow \{0,1\}^\kappa$, and for each wire w which is an output wire of an AND gate, or an input wire, P_i also samples the keys $k^i_{w,0}$, $k^i_{w,1} = k^i_{w,0} \oplus R^i$ and an additive share of the wire mask, $\lambda^i_w \leftarrow \mathbb{F}_2$. As in [4], we let P_i input the actual wire mask (instead of a share) for every input wire associated with P_i's input.

In step 3, the parties compute additive shares of the bit products $\lambda_{uv} = \lambda_u \cdot \lambda_v \in \mathbb{F}_2$, and then, for each $j \in [n]$, shares of:

$$\lambda_u \cdot R^j, \quad \lambda_v \cdot R^j, \quad \lambda_{uvw} \cdot R^j \in \mathbb{F}_2^\kappa \tag{3}$$

where $\lambda_{uvw} := \lambda_{uv} \oplus \lambda_w$, and u, v and w are the input and output wires of AND gate g. We note that (as observed in [4]) only one bit/bit product and $3n$ bit/string products are necessary, even though each gate has $4n$ entries, due to correlations between the entries, as can be seen below.

The Preprocessing Functionality

Let F be a circular 2-correlation robust PRF. The functionality runs with parties $P_1, \ldots, P_n$ and an adversary $\mathcal{A}$, who corrupts a subset $I \subset [n]$ of parties.

Garbling: On input $(\texttt{Garbling}, C_f)$ from all parties, where C_f is a boolean circuit, denote by W its set of wires and by G its set of AND gates. The functionality is defined as follows:

- Sample a global difference $R^j \leftarrow \{0,1\}^\kappa$, for each $j \notin I$, and receive corrupt parties' strings $R^i \in \{0,1\}^\kappa$ from $\mathcal{A}$, for $i \in I$.
- Passing topologically through all the wires $w \in W$ of the circuit:
  - If w is an input wire:
    1. Sample $\lambda_w \leftarrow \{0,1\}$. If P_j, the party who provides input on that wire in the online phase, is corrupt, instead receive λ_w from $\mathcal{A}$.
    2. Sample a key $k^j_{w,0} \leftarrow \{0,1\}^\kappa$, for each $j \notin I$, and receive corrupt parties' keys $k^i_{w,0}$ from $\mathcal{A}$, for $i \in I$. Define $k^i_{w,1} = k^i_{w,0} \oplus R^i$ for all $i \in [n]$.
  - If w is the output of an AND gate:
    1. Sample $\lambda_w \leftarrow \{0,1\}$.
    2. Sample a key $k^j_{w,0} \leftarrow \{0,1\}^\kappa$, for each $j \notin I$, and receive corrupt parties' keys $k^i_{w,0}$ from $\mathcal{A}$, for $i \in I$. Set $k^i_{w,1} = k^i_{w,0} \oplus R^i$, for $i \in [n]$.
  - If w is the output of a XOR gate, and u and v its input wires:
    1. Compute and store $\lambda_w = \lambda_u \oplus \lambda_v$.
    2. For $i \in [n]$, set $k^i_{w,0} = k^i_{u,0} \oplus k^i_{v,0}$ and $k^i_{w,1} = k^i_{w,0} \oplus R^i$.
- For every AND gate $g \in G$, the functionality computes the $4n$ entries of the garbled version of g as:

$$\tilde{g}^j_{a,b} = \left(\bigoplus_{i=1}^{n} F_{k^i_{u,a}, k^i_{v,b}}(g \| j) \right) \oplus k^j_{w,0}$$
$$\oplus \left(R^j \cdot ((\lambda_u \oplus a) \cdot (\lambda_v \oplus b) \oplus \lambda_w) \right), \quad (a,b) \in \{0,1\}^2,\ j \in [n].$$

  Set $\tilde{\mathbf{g}}_{a,b} = \tilde{g}^1_{a,b} \circ \ldots \circ \tilde{g}^n_{a,b} \quad (a,b) \in \{0,1\}^2$. The functionality stores the values $\tilde{\mathbf{g}}_{a,b}$.
- Wait for an input from $\mathcal{A}$. If it inputs $\texttt{OK}$ then output λ_w to all parties for each circuit-output wire w, and output to each P_i all the keys $\{k^i_{w,0}\}_{w \in W}$, and R^i. Otherwise, output $\perp$ and terminate.

Open Garbling: On receiving $(\texttt{OpenGarbling})$ from all parties, when the **Garbling** command has already run successfully, the functionality sends to $\mathcal{A}$ the values $\tilde{\mathbf{g}}_{a,b}$ for all $g \in G$ and waits for a reply.

- If $\mathcal{A}$ returns $\perp$ then the functionality aborts.
- Otherwise, the functionality receives $\texttt{OK}$ and an additive error $\mathbf{e} = \{e^{a,b}_g\}_{a,b \in \{0,1\}, g \in G}$ chosen by $\mathcal{A}$. Afterwards, it sends to all parties the garbled circuit $\tilde{\mathbf{g}}_{a,b} \oplus e^{a,b}_g$ for all $g \in G$ and $a,b \in \{0,1\}$.

Fig. 5. The preprocessing functionality $\mathcal{F}_{\text{Prepocessing}}$

The Preprocessing Protocol $\Pi_{\text{Preprocessing}}$ – Garbling Stage

Given a gate g, we denote by u (resp. v) its left (resp. right) input wire, and by w its output wire. $\langle \cdot \rangle^i$ denotes the i-th share of an authenticated bit and $(\cdot)^i$ the i-th share of a string. Let $F : \{0,1\}^{2\kappa} \times [|G|] \times [n] \to \{0,1\}^\kappa$ be a circular 2-correlation robust PRF, and $\mathcal{G} : \{0,1\}^\kappa \to \{0,1\}^{4n\kappa|G|}$ be a PRG.

Garbling:

1. Each party P_i samples a random key offset $R^i \leftarrow \mathbb{F}_2^\kappa$.
2. **Generate wire masks and keys:** For each wire w in topological order:
   - If w is a *circuit-input* wire from P_j:
     (a) P_j calls **Input** on $\mathcal{F}_{\text{BitMPC}}$ with a randomly sampled $\lambda_w \in \{0,1\}$ to obtain $\langle \lambda_w \rangle$. P_j defines the share $\lambda_w^j = \lambda_w$, every other P_i sets $\lambda_w^i = 0$.
     (b) Every P_i samples a key $k_{w,0}^i \leftarrow \{0,1\}^\kappa$ and sets $k_{w,1}^i = k_{w,0}^i \oplus R^i$.
   - If the wire w is the output of an AND gate:
     (a) Each P_i calls **Input** on $\mathcal{F}_{\text{BitMPC}}$ with a randomly sampled $\lambda_w^i \leftarrow \{0,1\}$. The parties then compute the secret-shared wire mask as $\langle \lambda_w \rangle = \sum_{i \in [n]} \langle \lambda_w^i \rangle$.
     (b) Every P_i samples a key $k_{w,0}^i \leftarrow \{0,1\}^\kappa$ and sets $k_{w,1}^i = k_{w,0}^i \oplus R^i$.
   - If the wire w is the output of a XOR gate:
     (a) The parties compute the mask on the output wire as $\langle \lambda_w \rangle = \langle \lambda_u \rangle + \langle \lambda_v \rangle$.
     (b) Every P_i sets $k_{w,0}^i = k_{u,0}^i \oplus k_{v,0}^i$ and $k_{w,1}^i = k_{w,0}^i \oplus R^i$.
3. **Secure product computations:**
   (a) For each AND gate $g \in G$, the parties compute $\langle \lambda_{uv} \rangle = \langle \lambda_u \rangle \cdot \langle \lambda_v \rangle$ by calling **Multiply** on $\mathcal{F}_{\text{BitMPC}}$.
   (b) Each P_i calls **Input** on $\mathcal{F}_{\text{BitMPC}}$ with randomly sampled bits $\hat{x}_1^i, \ldots, \hat{x}_s^i$. For $\ell \in [s]$, the parties compute secret-shared mask $\langle \hat{x}_\ell \rangle = \sum_{i \in [n]} \langle \hat{x}_\ell^i \rangle$.
   (c) For every $j \in [n]$, the parties run the subprotocol $\Pi_{\text{Bit}\times\text{String}}$, where P_j inputs R^j and everyone inputs the $3|G| + s$ shared bits:
   $$(\langle \lambda_u \rangle, \langle \lambda_v \rangle, \langle \lambda_{uv} \rangle + \langle \lambda_w \rangle)_{(u,v,w)} \quad \text{and} \quad (\langle \hat{x}_1 \rangle, \ldots, \langle \hat{x}_s \rangle).$$
   where the (u,v,w) indices are taken over the input/output wires of each AND gate $g \in G$.
   (d) For each AND gate g, party P_i obtains from $\Pi_{\text{Bit}\times\text{String}}$ an additive share of the $3n$ values (each defined as one row of the matrix $\mathbf{Z}_j$ in this subprotocol):
   $$\lambda_u \cdot R^j, \quad \lambda_v \cdot R^j, \quad \lambda_{uvw} \cdot R^j, \qquad \text{for } j \in [n]$$
   where $\lambda_{uvw} := \lambda_{uv} + \lambda_w$. Each P_i then uses these to compute a share of
   $$\rho_{j,a,b} = \lambda_{uvw} \cdot R^j \oplus a \cdot \lambda_v \cdot R^j \oplus b \cdot \lambda_u \cdot R^j \oplus a \cdot b \cdot R^j$$
4. **Garble gates:** For each AND gate $g \in G$, each $j \in [n]$, and the four combinations of $a, b \in \{0,1\}^2$, the parties compute shares of the j-th entry of the garbled gate $\tilde{g}_{a,b}$ as follows:
   - P_j sets $(\tilde{g}_{a,b}^j)^j = \rho_{j,a,b}^j \oplus F_{k_{u,a}^j, k_{v,b}^j}(g \| j) \oplus k_{w,0}^j$.
   - For every $i \neq j$, P_i sets $(\tilde{g}_{a,b}^j)^i = \rho_{j,a,b}^i \oplus F_{k_{u,a}^i, k_{v,b}^i}(g \| j)$.
5. **Reveal masks for output wires:** For every circuit-output-wire w, the parties call **Open** on $\mathcal{F}_{\text{BitMPC}}$ to reveal λ_w to all the parties.

Fig. 6. The preprocessing protocol that realizes $\mathcal{F}_{\text{Prepocessing}}$ in the $\{\mathcal{F}_{\text{COT}}, \mathcal{F}_{\text{BitMPC}}, \mathcal{F}_{\text{Rand}}\, \mathcal{F}_{\text{Commit}}\}$-hybrid model.

The Preprocessing Protocol $\Pi_{\text{Preprocessing}}$ – Open Garbling Stage

Open Garbling: Let $\tilde{C}^i = ((\tilde{g}^j_{a,b})^i)_{j,a,b,g} \in \{0,1\}^{4n\kappa|G|}$ be P_i's share of the whole garbled circuit.

1. Each party P_i samples random seeds $s^i_j \leftarrow \{0,1\}^\kappa, j \neq i$. P_i sends s^i_j to P_j over a private channel.
2. P_i computes the shares $S^i_i = \bigoplus_{i \neq j} \mathcal{G}(s^i_j)$, and $S^j_i = \mathcal{G}(s^j_i)$, for $j \neq i$.[a]
3. Each P_i, for $i = 2, \ldots, n$, sends $\tilde{C}^i \oplus \bigoplus_{j=1}^n S^j_i$ to P_1.
4. P_1 reconstructs the garbled circuit, $\tilde{C}$, and broadcasts this.

[a] Steps 1 to 2 are independent of $\tilde{C}^i$, so can be merged with previous rounds in the **Garbling** stage.

Fig. 7. Open garbling stage of the preprocessing protocol.

We compute the bit multiplications using the $\mathcal{F}_{\text{BitMPC}}$ functionality on the bits that are already stored by $\mathcal{F}_{\text{BitMPC}}$. To compute the bit/string multiplications in (3), we use correlated OT, followed by a consistency check to verify that the parties provided the correct shares of λ_w and correlation R^i to each $\mathcal{F}_{\text{COT}}$ instance; see Sect. 3.3 for details.

Using shares of the bit/string products, the parties can locally compute an unauthenticated additive share of the entire garbled circuit (steps 3d–4). First, for each of the four values $(a, b) \in \{0,1\}^2$, each party $P_i, i \neq j$ computes the share

$$\rho^i_{j,a,b} = \begin{cases} a \cdot (\lambda_v \cdot R^j)^i \oplus b \cdot (\lambda_u \cdot R^j)^i \oplus (\lambda_{uvw} \cdot R^j)^i & \text{if } i \neq j \\ a \cdot (\lambda_v \cdot R^j)^i \oplus b \cdot (\lambda_u \cdot R^j)^i \oplus (\lambda_{uvw} \cdot R^j)^i \oplus a \cdot b \cdot R^j & \text{if } i = j \end{cases}$$

These define additive shares of the values

$$\begin{aligned} \rho_{j,a,b} &= R^j \cdot (a \cdot \lambda_v \oplus b \cdot \lambda_u \oplus \lambda_{uvw} \oplus a \cdot b) \\ &= R^j \cdot ((\lambda_u \oplus a) \cdot (\lambda_v \oplus b) \oplus \lambda_w) \end{aligned}$$

Each party's share of the garbled circuit is then obtained by adding the appropriate PRF values and keys to the shares of each $\rho_{j,a,b}$. To conclude the **Garbling** stage, the parties reveal the masks for all output wires using $\mathcal{F}_{\text{BitMPC}}$, so that the outputs can be obtained in the online phase.

Before opening the garbled circuit, the parties must rerandomize their shares by distributing a fresh, random secret-sharing of each share to the other parties, via private channels. This is needed so that the shares do not leak any information on the PRF values, so we can prove security. This may seem unnecessary, since the inclusion of the PRF values in the shares should randomize them sufficiently. However, we cannot prove this intuition, as the same PRF values are used to compute the garbled circuit that is output by the protocol, so they cannot also be used as a one-time pad.[1] In steps 1 to 2 of Fig. 7, we show how to perform this extra rerandomization step with $O(n^2 \cdot \kappa)$ communication.

Finally, to reconstruct the garbled circuit, the parties sum up and broadcast the rerandomized shares and add them together to get $\tilde{g}^j_{a,b}$.

3.3 Bit/String Multiplications

Our method for this is in the subrotocol $\Pi_{\mathsf{Bit} \times \mathsf{String}}$ (Fig. 8). It proceeds in two stages: first the **Multiply** step creates the shared products, and then the **Consistency Check** verifies that the correct inputs were used to create the products.

Recall that the task is for the parties to obtain an additive sharing of the products, for each $j \in [n]$ and $(a, b) \in \{0, 1\}^2$:

$$R^j \cdot ((\lambda_u \oplus a) \cdot (\lambda_v \oplus b) \oplus \lambda_w) \tag{4}$$

where the string R^j is known only to P_j, and fixed for every gate. Denote by x one of the additively shared $\lambda_{(\cdot)}$ bits used in a single bit/string product and stored by $\mathcal{F}_{\mathsf{BitMPC}}$. We obtain shares of $x \cdot R^j$ using actively secure correlated OT (cf. Fig. 3), as follows:

1. For each $i \neq j$, parties P_i and P_j run a correlated OT, with choice bit x^i and correlation R^j. P_i obtains $T_{i,j}$ and P_j obtains $Q_{i,j}$ such that:

$$T_{i,j} = Q_{i,j} + x^i \cdot R^j.$$

2. Each P_i, for $i \neq j$, defines the share $Z^i = T_{i,j}$, and P_j defines $Z^j = \sum_{i \neq j} Q_{i,j} + x^j \cdot R^j$. Now we have:

$$\sum_{i=1}^{n} Z^i = \sum_{i \neq j} T_{i,j} + \sum_{i \neq j} Q_{i,j} + x^j \cdot R^j \quad = \sum_{i \neq j} (T_{i,j} + Q_{i,j}) + x^j \cdot R^j = x \cdot R^j$$

as required.

The above method is performed $3|G|$ times and for each P_j, to produce the shared bit/string products $x \cdot R^j$, for $x \in \{\lambda_u, \lambda_v, \lambda_{uv}\}$.

3.4 Consistency Check

We now show how the parties verify that the correct shares of x and correlations R^j were used in the correlated OTs, and analyse the security of this check. The parties first create $m + s$ bit/string products, where m is the number of products needed and s is a statistical security parameter, and then open random

[1] Furthermore, the environment sees all of the PRF keys of the honest parties, since these are outputs of the protocol, which seems to rule out any kind of computational reduction in the security proof.

Bit/string multiplication subprotocol – $\Pi_{\mathsf{Bit}\times\mathsf{String}}$

Inputs: Each P_j inputs the private global difference string $R^j \in \mathbb{F}_2^\kappa$, which was generated in the main protocol. All parties input $3|G|$ authenticated, additively shared bits, $\langle x_1 \rangle, \ldots, \langle x_{3|G|} \rangle$, and s additional, random shared bits, $\langle \hat{x}_1 \rangle, \ldots, \langle \hat{x}_s \rangle$, to be used as masking values and discarded.

I: Init: Every ordered pair of parties (P_i, P_j) calls **Initialize** on $\mathcal{F}_{\mathrm{COT}}$, where P_j, the sender, inputs the global difference string R^j.

II: Multiply: For each $j \in [n]$, the parties do as follows:

1. For every $i \neq j$, parties P_i and P_j call **Extend** on the $\mathcal{F}_{\mathrm{COT}}$ instance where P_j is sender, and P_i inputs the choice bits $\boldsymbol{x}^i = (x_1^i, \ldots, x_{3|G|}^i, \hat{x}_1^i, \ldots, \hat{x}_s^i)$.
   For each OT between (P_i, P_j), P_j receives $q \in \{0,1\}^\kappa$ and P_i receives $t \in \{0,1\}^\kappa$. P_i stores their $3|G| + s$ strings from this instance into the rows of a matrix $\mathbf{T}_{i,j}$, and P_j stores the corresponding outputs in $\mathbf{Q}_{i,j}$. These satisfy:
$$\mathbf{T}_{i,j} = \mathbf{Q}_{i,j} + \boldsymbol{x}^i \otimes R^j \in \mathbb{F}_2^{(3|G|+s)\times\kappa}.$$
2. Each P_i, for $i \neq j$, defines the matrix $\mathbf{Z}_j^i = \mathbf{T}_{i,j}$, and P_j defines $\mathbf{Z}_j^j = \sum_{i\neq j} \mathbf{Q}_{i,j} + \boldsymbol{x}^j \otimes R^j$.

Now, it should hold that $\sum_{i=1}^n \mathbf{Z}_j^i = \boldsymbol{x} \otimes R^j$, for each $j \in [n]$.

III: Consistency Check: The parties check correctness of the above as follows:

1. Each P_i removes the last s rows from $\mathbf{Z}_j^i$ (for $j \in [n]$) and places these 'dummy' masking values in a matrix $\hat{\mathbf{Z}}_j^i \in \mathbb{F}_2^{s\times\kappa}$. Similarly, redefine $\boldsymbol{x}^i = (x_1^i, \ldots, x_{3|G|}^i)$ and let $\hat{\boldsymbol{x}}^i = (\hat{x_1}^i, \ldots, \hat{x_s}^i)$.
2. The parties call $\mathcal{F}_{\mathrm{Rand}}$ (Figure 2) to sample a seed for a uniformly random, ε-almost 1-universal linear hash function, $\mathbf{H} \in \mathbb{F}_2^{s\times 3|G|}$.
3. All parties compute the vector:
$$\langle \mathbf{c}_x \rangle = \mathbf{H} \cdot \langle \boldsymbol{x} \rangle + \langle \hat{\boldsymbol{x}} \rangle \in \mathbb{F}_2^s$$
   and open $\mathbf{c}_x$ using the **Open** command of $\mathcal{F}_{\mathrm{BitMPC}}$. If $\mathcal{F}_{\mathrm{BitMPC}}$ aborts, the parties abort.
4. Each party P_i calls **Commit** on $\mathcal{F}_{\mathrm{Commit}}$ (Figure 1) with input the n matrices:
$$\mathbf{C}_j^i = \mathbf{H} \cdot \mathbf{Z}_j^i + \hat{\mathbf{Z}}_j^i, \quad \text{for } j \neq i, \text{ and} \quad \mathbf{C}_i^i = \mathbf{H} \cdot \mathbf{Z}_i^i + \hat{\mathbf{Z}}_i^i + \mathbf{c}_x \otimes R^i.$$
5. All parties open their commitments and check that, for each $j \in [n]$:
$$\sum_{i=1}^n \mathbf{C}_j^i = 0.$$
   If the check fails, the parties abort.
6. Each party P_i stores the matrices $\mathbf{Z}_1^i, \ldots, \mathbf{Z}_n^i$.

Fig. 8. Subprotocol for bit/string multiplication and checking consistency

linear combinations (over $\mathbb{F}_2$) of all the products and check correctness of the opened results. This is possible because the products are just a linear function of the fixed string R^j. In more detail, the parties first sample a random ε-almost 1-universal hash function $\mathbf{H} \leftarrow \mathbb{F}_2^{m \times s}$, and then open

$$\boldsymbol{c}_x = \mathbf{H} \cdot \boldsymbol{x} + \hat{\boldsymbol{x}}$$

using $\mathcal{F}_{\text{BitMPC}}$. Here, $\boldsymbol{x}$ is the vector of all m wire masks to be multiplied, whilst $\hat{\boldsymbol{x}} \in \mathbb{F}_2^s$ are the additional, random masking bits, used as a one-time pad to ensure that $\boldsymbol{c}_x$ does not leak information on $\boldsymbol{x}$.

To verify that a single shared matrix $\mathbf{Z}_j$ is equal to $\boldsymbol{x} \otimes R^j$ (as in Fig. 8), each party P_i, for $i \neq j$, then commits to $\mathbf{H} \cdot \mathbf{Z}_j^i$, whilst P_j commits to $\mathbf{H} \cdot \mathbf{Z}_j^i + \boldsymbol{c}_x \otimes R^j$. The parties then open all commitments and check that these sum to zero, which should happen if the products were correct.

The intuition behind the check is that any errors present in the original bit/string products will remain when multiplied by $\mathbf{H}$, except with probability ε, by the almost-1-universal property (Definition 22). Furthermore, it turns out that cancelling out any non-zero errors in the check requires either guessing an honest party's global difference R^j, or guessing the secret masking bits $\hat{\boldsymbol{x}}$.

We formalize this, by first considering the exact deviations that are possible by a corrupt P_j in $\Pi_{\text{Bit} \times \text{String}}$. These are:

1. Provide inconsistent inputs R^j when acting as sender in the **Initialize** command of the $\mathcal{F}_{\text{COT}}$ instances with two different honest parties.
2. Input an incorrect share x^j when acting as receiver in the **Extend** command of $\mathcal{F}_{\text{COT}}$.

Note that in both of these cases, we are only concerned when the other party in the $\mathcal{F}_{\text{COT}}$ execution is honest, as if both parties are corrupt then $\mathcal{F}_{\text{COT}}$ does not need to be simulated in the security proof.

We model these two attacks by defining $R^{j,i}$ and $x^{j,i}$ to be the *actual* inputs used by a corrupt P_j in the above two cases, and then define the errors (for $j \in I$ and $i \notin I$):

$$\Delta^{j,i} = R^{j,i} + R^j$$
$$\delta_\ell^{j,i} - x_\ell^{j,i} + x_\ell^j, \quad \ell \in [3|G|]$$

Note that $\Delta^{j,i}$ is fixed in the initialization of $\mathcal{F}_{\text{COT}}$, whilst $\delta_\ell^{j,i}$ may be different for every OT. Whenever P_i and P_j are both corrupt, or both honest, for convenience we define $\Delta^{j,i} = 0$ and $\delta^{j,i} = 0$.

This means that the outputs of $\mathcal{F}_{\text{COT}}$ with (P_i, P_j) then satisfy (omitting ℓ subscripts):

$$t_{i,j} = q_{i,j} + x^i \cdot R^j + \delta^{i,j} \cdot R^j + \Delta^{j,i} \cdot x^i$$

where $\delta^{i,j} \neq 0$ if P_i cheated, and $\Delta^{j,i} \neq 0$ if P_j cheated.

Now, as in step 1 of the first stage of $\Pi_{\text{Bit} \times \text{String}}$, we can put the $\mathcal{F}_{\text{COT}}$ outputs for each party into the rows of a matrix, and express the above as:

$$\mathbf{T}_{i,j} = \mathbf{Q}_{i,j} + \boldsymbol{x}^i \otimes R^j + \boldsymbol{\delta}^{i,j} \otimes R^j + \Delta^{j,i} \otimes x^i$$

where $\boldsymbol{\delta}^{j,i} = (\delta_1^{j,i}, \ldots, \delta_{3|G|}^{j,i})$, and the tensor product notation is defined in Sect. 2.

Accounting for these errors in the outputs of the **Multiply** step in $\Pi_{\mathsf{Bit} \times \mathsf{String}}$, we get:

$$\mathbf{Z}_j = \sum_{i=1}^{n} \mathbf{Z}_j^i = \boldsymbol{x} \otimes R^j + R^j \cdot \underbrace{\sum_{i \in I} \boldsymbol{\delta}^{i,j}}_{=\boldsymbol{\delta}^j} + \sum_{i \notin I} x_i \cdot \Delta^{j,i}. \quad (5)$$

The following lemma shows if a party cheated, then to pass the check they must either guess all of the shares $\hat{\boldsymbol{x}}^i \in \mathbb{F}_2^s$ for some honest P_i, or guess P_i's global difference R^i (except with negligible probability over the choice of the ε-almost 1-universal hash function, $\mathbf{H}$).

Lemma 31. *If the check in $\Pi_{\mathsf{Bit} \times \mathsf{String}}$ passes, then except with probability* $\max(2^{-s}, \varepsilon + 2^{-\kappa})$*, all of the errors $\boldsymbol{\delta}^j, \Delta^{i,j}$ are zero.*

The proof can be found in the full version of the paper [20].

We now give some intuition behind the security of the whole protocol. In the proof, the strategy of the simulator is to run an internal copy of the protocol, using dummy, random values for the honest parties' keys and wire mask shares. All communication with the adversary is simulated by computing the correct messages according to the protocol and the dummy honest shares, until the final output stage. In the output stage, we switch to fresh, random honest parties' shares, consistent with the garbled circuit received from $\mathcal{F}_{\mathsf{Prepocessing}}$ and the corrupt parties' shares.

Firstly, by Lemma 31, it holds that in the real execution, if the adversary introduced any non-zero errors then the consistency check fails with overwhelming probability. The same is true in the ideal execution; note that the errors are still well-defined in this case because the simulator can compute them by comparing all inputs received to $\mathcal{F}_{\mathsf{COT}}$ with the inputs the adversary should have used, based on its random tape. This implies that the probability of passing the check is the same in both worlds. Also, if the check fails then both executions abort, and it is straightforward to see that the two views are indistinguishable because no outputs are sent to honest parties (hence, also the environment).

It remains to show that the two views are indistinguishable when the consistency check passes, and the environment sees the outputs of all honest parties, as well as the view of the adversary during the protocol. The main point of interest here is the output stage. We observe that, without the final rerandomization step, the honest parties' shares of the garbled circuit would *not be uniformly random.* Specifically, consider an honest P_i's share, $(\tilde{g}_{a,b}^j)^i$, where P_j is corrupt. This is computed by adding some PRF value, v, to the $\mathcal{F}_{\mathsf{COT}}$ outputs where P_i was receiver and P_j was sender (step 2 of $\Pi_{\mathsf{Bit} \times \mathsf{String}}$). Since P_j knows both strings in each OT, there are only two possibilities for P_i's output (depending on the choice bit), so this is not uniformly random. It might be tempting to argue that v is a random PRF output, so serves as a one-time pad, but this proof attempt fails because v is also used to compute the final garbled circuit.

In fact, it seems difficult to rely on any reduction to the PRF, since all the PRF keys are included in the output to the environment. To avoid this issue, we need the rerandomization step using a PRG, and the additional assumption of secure point-to-point channels.

Theorem 31. *Protocol $\Pi_{\text{Preprocessing}}$ from Fig. 6 UC-securely computes $\mathcal{F}_{\text{Prepocessing}}$ from Fig. 5 in the presence of a static, active adversary corrupting up to $n-1$ parties in the $\{\mathcal{F}_{\text{COT}}, \mathcal{F}_{\text{BitMPC}}, \mathcal{F}_{\text{Rand}}, \mathcal{F}_{\text{Commit}}\}$-hybrid model.*

The proof can be found in the full version of the paper [20].

4 More Efficient Garbling with Multi-party TinyOT

We now describe a less general, but concretely more efficient, variant of the protocol in the previous section. We replace the generic $\mathcal{F}_{\text{BitMPC}}$ functionality with a more specialized one based on 'TinyOT'-style information-theoretic MACs. This is asymptotically worse, but more practical, than using [21] for $\mathcal{F}_{\text{BitMPC}}$. It also allows us to completely remove the bit/string multiplications and consistency checks in $\Pi_{\text{Bit}\times\text{String}}$, since we show that these can be obtained directly from the TinyOT MACs. This means the only cost in the protocol, apart from opening and evaluating the garbled circuit, is the single bit multiplication per AND gate in the underlying TinyOT-based protocol.

In the full version of this paper [20], we present a complete description of a suitable TinyOT-based protocol. This is done by combining the multiplication triple generation protocol (over $\mathbb{F}_2$) from [17] with a consistency check to enforce correct shared random bits, which is similar to the more general check from the previous section.

4.1 Secret-Shared MAC Representation

For $x \in \{0,1\}$ held by P_i, define the following two-party MAC representation, as used in 2-party TinyOT [33]:

$$[x]_{i,j} = (x, M_j^i, K_i^j), \qquad M_j^i = K_i^j + x \cdot R^j$$

where P_i holds x and a MAC M_j^i, and P_j holds a local MAC key K_i^j, as well as the fixed, global MAC key R^j.

Similarly, we define the n-party representation of an additively shared value $x = x^1 + \cdots + x^n$:

$$[x] = (x^i, \{M_j^i, K_j^i\}_{j\neq i})_{i\in[n]}, \quad M_j^i = K_i^j + x^i \cdot R^j$$

where each party P_i holds the $n-1$ MACs M_j^i on x^i, as well as the keys K_j^i on each x^j, for $j \neq i$, and a global key R^i. Note that this is equivalent to every pair (P_i, P_j) holding a representation $[x^i]_{i,j}$.

The key observation for this section, is that a sharing $[x]$ can be used to directly compute shares of all the products $x \cdot R^j$, as in the following claim.

Claim 41. *Given a representation $[x]$, the parties can locally compute additive shares of $x \cdot R^j$, for each $j \in [n]$.*

Proof. Write $[x] = (x^i, \{M_j^i, K_j^i\}_{j \neq i})_{i \in [n]}$. Each party P_i defines the n shares:

$$Z_i^i = x^i \cdot R^i + \sum_{j \neq i} K_j^i \quad \text{and} \quad Z_j^i = M_j^i, \quad \text{for each } j \neq i$$

We then have, for each $j \in [n]$:

$$\begin{aligned}\sum_{i=1}^{n} Z_j^i &= Z_j^j + \sum_{i \neq j} Z_j^i = (x^j \cdot R^j + \sum_{i \neq j} K_i^j) + \sum_{i \neq j} M_j^i \\ &= x^j \cdot R^j + \sum_{i \neq j} (M_j^i + K_i^j) = x^j \cdot R^j + \sum_{i \neq j} (x^i \cdot R^j) = x \cdot R^j.\end{aligned}$$

We define addition of two shared values $[x], [y]$, to be straightforward addition of the components. We define addition of $[x]$ with a public constant $c \in \mathbb{F}_2$ by:

- P_1 stores: $(x^1 + c, \{M_j^1, K_j^1\}_{j \neq 1})$
- P_i stores: $(x^i, (M_1^i, K_1^i + c \cdot R^i), \{M_j^i, K_j^i\}_{j \in [n] \setminus \{1,i\}})$, for $i \neq 1$

This results in a correct sharing of $[x + c]$.

We can create a sharing of the product of two shared values using a random multiplication triple $([x], [y], [z])$ such that $z = x \cdot y$ with Beaver's technique [1].

4.2 MAC-Based MPC Functionality

The functionality $\mathcal{F}_{\text{n-TinyOT}}$, which we use in place of $\mathcal{F}_{\text{BitMPC}}$ for the optimized preprocessing, is shown in the full version [20]. It produces authenticated sharings of random bits and multiplication triples. For both of these, $\mathcal{F}_{\text{n-TinyOT}}$ first receives corrupted parties' shares, MAC values and keys from the adversary, and then randomly samples consistent sharings and MACs for the honest parties.

Another important aspect of the functionality is the **Key Queries** command, which allows the adversary to try to guess the MAC key R^i of any party, and will be informed if the guess is correct. This is needed to allow the security proof to go through; we explain this in more detail in the full version. In that section we also present a complete description of a variant on the multi-party TinyOT protocol, which can be used to implement this functionality.

4.3 Garbling with $\mathcal{F}_{\text{n-TinyOT}}$

Following from the observation in Claim 41, if each party P_j chooses the global difference string in $\Pi_{\text{Preprocessing}}$ to be the same R^j as in the MAC representation, then given $[\lambda]$, additive shares of the products $\lambda \cdot R^j$ can be obtained at no extra cost. Moreover, the shares are guaranteed to be correct, and the honest party's shares will be random (subject to the constraint that they sum to the correct value), since they come directly from the $\mathcal{F}_{\text{n-TinyOT}}$ functionality. This means there is no need to perform the consistency check, which greatly simplifies the protocol.

The rest of the protocol is mostly the same as $\Pi_{\text{Preprocessing}}$ in Fig. 6, using $\mathcal{F}_{\text{n-TinyOT}}$ with $[\cdot]$-sharings instead of $\mathcal{F}_{\text{BitMPC}}$ with $\langle\cdot\rangle$-sharings. One other small difference is that because $\mathcal{F}_{\text{n-TinyOT}}$ does not have a private input command, we instead sample $[\lambda_w]$ shares for input wires using random bits, and later use a private output protocol to open the relevant input wire masks to P_i. This change is not strictly necessary, but simplifies the protocol for implementing $\mathcal{F}_{\text{n-TinyOT}}$—if $\mathcal{F}_{\text{n-TinyOT}}$ also had an **Input** command for sharing private inputs based on n-$\mathsf{Bracket}$, it would be much more complex to implement with the correct distribution of shares and MACs.

In more detail, the **Garbling** phase proceeds as follows.

1. Each party obtains a random key offset R^i by calling the **Initialize** command of $\mathcal{F}_{\text{n-TinyOT}}$.
2. For every wire w which is an input wire, or the output wire of an AND gate, the parties obtain a shared mask $[\lambda_w]$ using the **Bit** command of $\mathcal{F}_{\text{n-TinyOT}}$.
3. All the wire keys $k^i_{w,0}, k^i_{w,1} = k^i_{w,0} \oplus R^i$ are defined by P_i the same way as in $\Pi_{\text{Preprocessing}}$.
4. For XOR gates, the output wire mask is computed as $[\lambda_w] = [\lambda_u] + [\lambda_v]$.
5. For each AND gate, the parties compute $[\lambda_{uv}] = [\lambda_u \cdot \lambda_v]$.
6. The parties then obtain shares of the garbled circuit as follows:
   - For each AND gate $g \in G$ with wires (u, v, w), the parties use Claim 41 with the shared values $[\lambda_u], [\lambda_v], [\lambda_{uv} + \lambda_w]$, to define, for each $j \in [n]$, shares of the bit/string products:
$$\lambda_u \cdot R^j, \quad \lambda_v \cdot R^j, \quad (\lambda_{uv} + \lambda_w) \cdot R^j$$
   - These are then used to define shares of $\rho_{j,a,b}$ and the garbled circuit, as in the original protocol.
7. For every circuit-output-wire w, the parties run Π_{Open} to reveal λ_w to all the parties.
8. For every *circuit input wire* w corresponding to party P_i's input, the parties run Π^i_{Open} to open λ_w to P_i.

The only interaction introduced in the new protocol is in the multiply and opening protocols, which were abstracted away by $\mathcal{F}_{\text{BitMPC}}$ in the previous protocol. Simulating and proving security of these techniques is straightforward, due to the correctness and randomness of the multiplication triples and MACs produced by $\mathcal{F}_{\text{n-TinyOT}}$. One important detail is the **Key Queries** command of the $\mathcal{F}_{\text{n-TinyOT}}$ functionality, which allows the adversary to try to guess an honest party's global MAC key share, R^i, and learn if the guess is correct. To allow the proof to go through, we modify $\mathcal{F}_{\text{Prepocessing}}$ to also have a **Key Queries** command, so that the simulator can use this to respond to any key queries from the adversary. We denote this modified functionality by $\mathcal{F}^{\text{KQ}}_{\text{Prepocessing}}$.

The following theorem can be proven, similarly to the proof of Theorem 31 where we modify the preprocessing functionality to support key queries, and adjust the simulation as described above.

Theorem 41. *The modified protocol described above UC-securely computes* $\mathcal{F}^{\mathrm{KQ}}_{\mathrm{Prepocessing}}$ *from Fig. 5 in the presence of a static, active adversary corrupting up to* $n-1$ *parties in the* $\mathcal{F}_{\text{n-TinyOT}}$*-hybrid model.*

5 The Online Phase

Our final protocol, presented in Fig. 9, implements the online phase where the parties reveal the garbled circuit's shares and evaluate it. Our protocol is presented in the $\mathcal{F}_{\mathrm{Prepocessing}}$-hybrid model. Upon reconstructing the garbled circuit and obtaining all input keys, the process of evaluation is similar to that of [39], except here all parties run the evaluation algorithm, which involves each party computing n^2 PRF values per gate. During evaluation, the parties only see the randomly masked wire values and cannot determine the actual wire values. Upon completion, the parties compute the actual output using the output wire masks revealed from $\mathcal{F}_{\mathrm{Prepocessing}}$. We conclude with the following theorem.

Theorem 51. *Let* f *be an* n*-party functionality* $\{0,1\}^{n\kappa} \mapsto \{0,1\}^{\kappa}$ *and assume that* F *is a PRF. Then Protocol* Π_{MPC} *from Fig. 9, UC-securely computes* f *in the presence of a static, active adversary corrupting up to* $n-1$ *parties in the* $\mathcal{F}_{\mathrm{Prepocessing}}$*-hybrid.*

Proof overview. Our proof follows by first demonstrating that the adversary's view is computationally indistinguishable in both real and simulated executions. To be concrete, we consider an event for which the adversary successfully causes the bit transferred through some wire to be flipped and prove that this event can only occur with negligible probability (our proof is different to the proof in [29] as in our case the adversary may choose its additive error as a function of the garbled circuit). Then, conditioned on the event flip not occurring, we prove that the two executions are computationally indistinguishable via a reduction to the correlation robust PRF, inducing a garbled circuit that is indistinguishable. The complete proof can be found in the full version of the paper [20].

6 Performance

In this section we present implementation results for our protocol from Sect. 4 for up to 9 parties. We also analyse the concrete communication complexity of the protocol and compare this with previous, state-of-the-art protocols in a similar setting.

We have made a couple of tweaks to our protocol to simplify the implementation. We moved the **Open Garbling** stage to the preprocessing phase, instead of the online phase. This optimizes the online phase so that the amount of communication is independent of the size of the circuit. This change means that

The MPC Protocol - Π_{MPC}

On input a circuit C_f representing the function f and $\rho = (\rho_1, \ldots, \rho_n)$ where ρ_i is party's P_i input, the parties execute the following commands in sequence.

Preprocessing: This sub-task is performed as follows.
- Call **Garbling** on $\mathcal{F}_{\text{Prepocessing}}$ with input C_f.
- Each party P_i obtains the λ_w wire masks for every output wire and every wire associated with their input, and all the keys $\{k^i_{w,0}\}_{w \in W}$ and R^i.

Online Computation: This sub-task is performed as follows.
- For all input wires w with input from P_i, party P_i computes $\Lambda_w = \rho_w \oplus \lambda_w$, where ρ_w is P_i's input to C_f, and λ_w was obtained in the preprocessing stage. Then, P_i broadcasts the public value Λ_w to all parties.
- For all input wires w, each party P_i broadcasts the key k^i_w associated to Λ_w.
- The parties call **Open Garbling** on $\mathcal{F}_{\text{Prepocessing}}$ to reconstruct $\tilde{g}^j_{a,b}$ for every gate g and values a, b.
- Passing through the circuit topologically, the parties can now locally compute the following operations for each gate g. Let the gates input wires be labelled u and v, and the output wire be labelled w. Let a and b be the respective public values on the input wires.
  1. If g is a XOR gate, set the public value on the output wire to be $c = a + b$. In addition, for every $j \in [n]$, each party computes $k^j_{w,c} = k^j_{u,a} \oplus k^j_{v,b}$.
  2. If g is an AND gate , then each party computes, for all $j \in [n]$:
  $$k^j_{w,c} = \tilde{g}^j_{a,b} \oplus \left(\bigoplus_{i=1}^{n} F_{k^i_{u,a}, k^i_{v,b}}(g \| j) \right)$$
  3. If $k^i_{w,c} \notin \{k^i_{w,0}, k^i_{w,1} = k^i_{w,0} \oplus R^i\}$, then P_i outputs `abort`. Otherwise, it proceeds. If P_i aborts it notifies all other parties with that information. If P_i is notified that another party has aborted it aborts as well.
  4. If $k^i_{w,c} = k^i_{w,0}$ then P_i sets $c = 0$; if $k^i_{w,c} = k^i_{w,1}$ then P_i sets $c = 1$.
  5. The output of the gate is defined to be $(k^1_{w,c}, \ldots, k^n_{w,c})$ and the public value c.
- Assuming no party aborts, everyone will obtain a public value c_w for every circuit-output wire w. The party can then recover the actual output value from $\rho_w = c_w \oplus \lambda_w$, where λ_w was obtained in the preprocessing stage.

Fig. 9. The MPC protocol - Π_{MPC}

our standard model security proof would no longer apply, but we could prove it secure using a random oracle instead of the circular-correlation robust PRF, similarly to [3,30]. Secondly, when not working in a modular fashion with a separate preprocessing functionality, the share rerandomization step in the output stage is not necessary to prove security of the entire protocol, so we omit this.

6.1 Implementation

We implemented our variant of the multi-party TinyOT protocol (given in the full version) using the `libOTe` library [36] for the fixed-correlation OTs and tested it for between 3 and 9 parties. We benchmarked the protocol over a 1 Gbps LAN on 5 servers with 2.3 GHz Intel Xeon CPUs with 20 cores. For the experiments with more than 5 parties, we had to run more than one party per machine; this should not make much difference in a LAN, as the number of threads being used was still fewer than the number of cores. As benchmarks, we measured the time for securely computing the circuits for AES (6800 AND gates) and SHA-256 (90825 AND gates).

For the TinyOT bit and triple generation, every pair of parties needs two correlated OT instances running between them (one in each direction). We ran each OT instance in a separate thread with `libOTe`, so that each party uses $2(n-1)$ OT threads. This gave a small improvement ($\approx$6%) compared with running $n-1$ threads. We also considered a multiple execution setting, where many (possibly different) secure computations are evaluated. Provided the total number of AND gates in the circuits being evaluated is at least 2^{20}, this allows us to generate the TinyOT triples for all executions at once using a bucket size of $B = 3$, compared with $B = 5$ for one execution of AES or $B = 4$ for one execution of SHA-256. Since the protocol scales with B^2, this has a big impact on performance. The results for 9 parties, for the different choices of B, are shown in Table 2.

Table 2. Runtimes in ms for AES and SHA-256 evalution with 9 parties

	AES ($B = 5$)	AES ($B = 3$)	SHA-256 ($B = 5$)	SHA-256 ($B = 3$)
Prep.	1329	586.9	10443	6652
Online	35.34	33.30	260.58	252.8

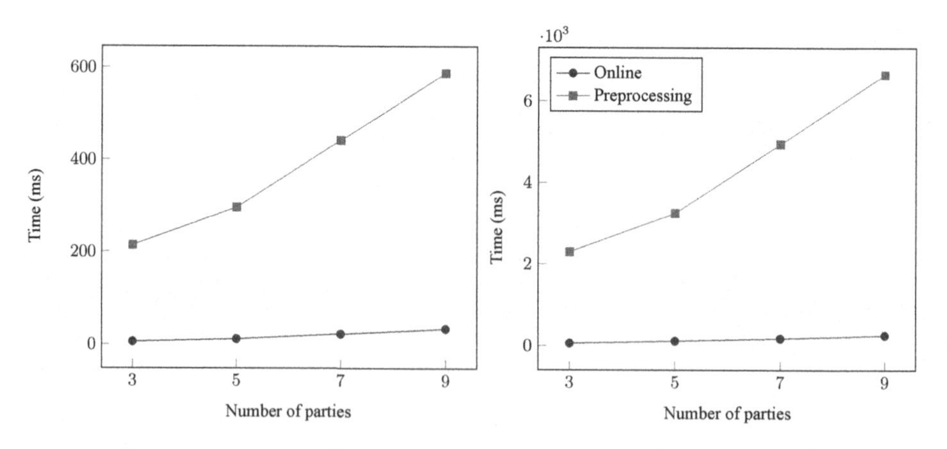

Fig. 10. AES performance (6800 AND gates).

Fig. 11. SHA-256 performance (90825 AND gates).

Figures 10 and 11 show how the performance of AES and SHA-256 scales with different numbers of parties, in the amortized setting. Although the asymptotic complexity is quadratic, the runtimes grow relatively slowly as the number of parties increases. This is because in the preprocessing phase, the amount of data sent *per party* is actually linear. However, the super-linear trend is probably due to the limitations of the total network capacity, and the computational costs.

Comparison with other works. We calculated the cost of computing the SPDZ-BMR protocol [29] using [24] to derive estimates for creating the SPDZ triples (the main cost). Using MASCOT over $\mathbb{F}_{2^\kappa}$ with free-XOR, SPDZ-BMR requires $3n+1$ multiplications per garbled AND gate. This gives an estimated cost of at least 14 s to evaluate AES, which is over 20x slower than our protocol.

The only other implementation of actively secure, constant-round, dishonest majority MPC is the concurrent work of [23], which presents implementation figures for up to 256 parties running on Amazon servers. Their runtimes with 9 parties in a LAN setting are around 200 ms for AES and 2200 ms for SHA-256, which is around 3 times faster than our results. However, their LAN setup has 10Gbps bandwidth, whereas we only tested on machines with 1Gbps bandwidth. Since the bottleneck in our implementation is mostly communication, it seems that our implementation could perform similar to or even faster than theirs in the same environment, despite our higher communication costs. However, it is not possible to make an accurate comparison without testing both implementations in the same environment.

Compared with protocols based solely on secret-sharing, such as SPDZ and TinyOT, the advantage of our protocol is the low round complexity. We have not yet managed to benchmark our protocol in a WAN setting, but since our total round complexity is less than 20, it should perform reasonably fast. With secret-sharing, using e.g. TinyOT, evaluating the AES circuit requires at least 40 rounds in just the online phase (it can be done with 10 rounds [14], but this uses a special representation of the AES function, rather than a general circuit), whilst computing the SHA-256 circuit requires *4000 rounds*. In a network with 100 ms delay between parties, the AES online time alone would be at least 4 s, whilst SHA-256 would take over 10 min to securely compute in that setting. If our protocol is run in this setting, we should be able to compute both AES and SHA-256 in just a few seconds (assuming that latency rather than bandwidth is the bottleneck).

6.2 Communication Complexity Analysis

We now focus on analysing the concrete communication complexity of the optimized variant of our protocol and compare it with the state of the art in constant-round two-party and multi-party computation protocols. We have not implemented our protocol, but since the underlying computational primitives are very simple, the communication cost will be the overall bottleneck. As a benchmark, we estimate the cost of securely computing the AES circuit (6800 AND gates, 25124 XOR gates), where we assume that one party provides a 128-bit plaintext or ciphertext and the rest of them have an XOR sharing of a 128-bit AES key. This implies we have $128 \cdot n$ input wires and an additional layer of XOR gates in the circuit to add the key shares together. We consider a single set of 128 output wires, containing the final encrypted or decrypted message.

Table 3. Communication estimates for secure AES evaluation with our protocol and previous works in the two-party setting. Cost is the maximum amount of data sent by any one party, per execution.

Protocol	# Executions	Function-indep. prep.	Function-dep. prep.	Online
[37]	32	–	3.75 MB	25.76 kB
	128	–	2.5 MB	21.31 kB
	1024	–	1.56 MB	16.95 kB
[34]	1	14.94 MB	227 kB	16.13 kB
	32	8.74 MB	227 kB	16.13 kB
	128	7.22 MB	227 kB	16.13 kB
	1024	6.42 MB	227 kB	16.13 kB
[38]	1	2.86 MB	570 kB	4.86 kB
	32	2.64 MB	570 kB	4.86 kB
	128	2.0 MB	570 kB	4.86 kB
	1024	2.0 MB	570 kB	4.86 kB
Ours + [38]	1	2.86 MB	872 kB	4.22 kB
	32	2.64 MB	872 kB	4.22 kB
	128	2.0 MB	872 kB	4.22 kB
	1024	2.0 MB	872 kB	4.22 kB

Two Parties. In Table 3 we compare the cost of our protocol in the two-party case, with state-of-the-art secure two-party computation protocols. We instantiate our TinyOT-based preprocessing method with the optimized, two-party TinyOT protocol from [38], lowering the previous costs further. For consistency with the other two-party protocols, we divide the protocol costs into three phases: function-independent preprocessing, which only depends on the size of the circuit; function-dependent preprocessing, which depends on the exact structure of the circuit; and the online phase, which depends on the parties' inputs. As with the implementation, we move the garbled circuit opening to the function-dependent preprocessing, to simplify the online phase.

The online phase of the modified protocol is just two rounds of interaction, and has the lowest online cost of *any* actively secure two-party protocol.[2] The

[2] If counting the *total* amount of data sent, in both directions, our online cost would be larger than [38], which is highly asymmetric. In practice, however, the latency depends on the largest amount of communication from any one party, which is why we measure in this way.

main cost of the function-dependent preprocessing is opening the garbled circuit, which requires each party to send 8κ bits per AND gate. This is slightly larger than the best Yao-based protocols, due to the need for a set of keys for every party in BMR.

In the batch setting, where many executions of the *same circuit* are needed, protocols such as [37] clearly still perform the best. However, if many circuits are required, but they may be different, or not known in advance, then our multi-party protocol is highly competitive with two-party protocols.

Comparison with Multi-party Protocols. In Table 4 we compare our work with previous constant-round protocols suitable for any number of parties, again for evaluating the AES circuit. We do not present the communication complexity of the online phase as we expect it to be very similar in all of the protocols. We denote by MASCOT-BMR-FX an optimized variant of [29], modified to use free-XOR as in our protocol, with multiplications done using the OT-based MASCOT protocol [24].

Table 4. Comparison of the cost of our protocol with previous constant-round MPC protocols in a range of security models, for secure AES evaluation. Costs are the amount of data sent over the network per party.

Protocol	Security	Function-indep. prep.		Function-dep. prep.	
		$n = 3$	$n = 10$	$n = 3$	$n = 10$
SPDZ-BMR	Active	25.77 GB	328.94 GB	61.57 MB	846.73 MB
SPDZ-BMR	Covert, pr. $\frac{1}{5}$	7.91 GB	100.98 GB	61.57 MB	846.73 MB
MASCOT-BMR-FX	Active	3.83 GB	54.37 GB	12.19 MB	178.25 MB
[23]	Active	4.8 MB	20.4 MB	1.3 MB	4.4 MB
Ours	Active	14.01 MB	63.22 MB	1.31 MB	4.37 MB

As in the previous section, we move the cost of opening the garbled circuit to the preprocessing phase for all of the presented protocols (again relying on random oracles). By applying this technique the online phase of our work is just two rounds, and has exactly the same complexity as the current most efficient *semi-honest* constant-round MPC protocol for any number of parties [4], except we achieve active security. We see that with respect to other actively secure protocols, we improve the communication cost of the preprocessing by around 2–4 orders of magnitude. Moreover, our protocol scales much better with n, since the complexity is $O(n^2)$ instead of $O(n^3)$. The concurrent work of Katz et al. [23] requires around 3 times less communication than our protocol, which is due to their optimized version of the multi-party TinyOT protocol.

Acknowledgements. We are grateful to Moriya Farbstein and Lior Koskas for their valuable assistance with implementation and experiments. We also thank Yehuda Lindell for helpful feedback.

The first author was supported by the European Research Council under the ERC consolidators grant agreement No. 615172 (HIPS), and by the BIU Center for Research in Applied Cryptography and Cyber Security in conjunction with the Israel National Cyber Bureau in the Prime Minister's Office. The second author was supported by the Defense Advanced Research Projects Agency (DARPA) and Space and Naval Warfare Systems Center, Pacific (SSC Pacific) under contract No. N66001-15-C-4070, and by the Danish Independent Research Council, Grant-ID DFF-6108-00169. The third author was supported by the European Union's Horizon 2020 research and innovation programme under the Marie Skłodowska-Curie grant agreement No. 643161.

References

1. Beaver, D.: Efficient multiparty protocols using circuit randomization. In: Feigenbaum, J. (ed.) CRYPTO 1991. LNCS, vol. 576, pp. 420–432. Springer, Heidelberg (1992). https://doi.org/10.1007/3-540-46766-1_34
2. Beaver, D., Micali, S., Rogaway, P.: The round complexity of secure protocols (extended abstract). In: 22nd ACM STOC, pp. 503–513. ACM Press, May 1990
3. Bellare, M., Hoang, V.T., Rogaway, P.: Foundations of garbled circuits. In: Yu, T., Danezis, G., Gligor, V.D. (eds.) ACM CCS 12, pp. 784–796. ACM Press, October 2012
4. Ben-Efraim, A., Lindell, Y., Omri, E.: Optimizing semi-honest secure multiparty computation for the internet. In: Weippl, E.R., Katzenbeisser, S., Kruegel, C., Myers, A.C., Halevi, S. (eds.) ACM CCS 16, pp. 578–590. ACM Press, October 2016
5. Ben-Or, M., Goldwasser, S., Wigderson, A.: Completeness theorems for non-cryptographic fault-tolerant distributed computation (extended abstract). In: 20th ACM STOC, pp. 1–10. ACM Press, May 1988
6. Burra, S.S., Larraia, E., Nielsen, J.B., Nordholt, P.S., Orlandi, C., Orsini, E., Scholl, P., Smart, N.P.: High performance multi-party computation for binary circuits based on oblivious transfer. Cryptology ePrint Archive, Report 2015/472 (2015). http://eprint.iacr.org/2015/472
7. Canetti, R.: Universally composable security: a new paradigm for cryptographic protocols. In: 42nd FOCS, pp. 136–145. IEEE Computer Society Press, October 2001
8. Canetti, R., Cohen, A., Lindell, Y.: A simpler variant of universally composable security for standard multiparty computation. In: Gennaro, R., Robshaw, M. (eds.) CRYPTO 2015. LNCS, vol. 9216, pp. 3–22. Springer, Heidelberg (2015). https://doi.org/10.1007/978-3-662-48000-7_1
9. Cascudo, I., Damgård, I., David, B., Döttling, N., Nielsen, J.B.: Rate-1, linear time and additively homomorphic UC commitments. In: Robshaw, M., Katz, J. (eds.) CRYPTO 2016. LNCS, vol. 9816, pp. 179–207. Springer, Heidelberg (2016). https://doi.org/10.1007/978-3-662-53015-3_7
10. Choi, S.G., Katz, J., Kumaresan, R., Zhou, H.-S.: On the security of the "Free-XOR" technique. In: Cramer, R. (ed.) TCC 2012. LNCS, vol. 7194, pp. 39–53. Springer, Heidelberg (2012). https://doi.org/10.1007/978-3-642-28914-9_3
11. Choi, S.G., Katz, J., Malozemoff, A.J., Zikas, V.: Efficient three-party computation from cut-and-choose. In: Garay, J.A., Gennaro, R. (eds.) CRYPTO 2014. LNCS, vol. 8617, pp. 513–530. Springer, Heidelberg (2014). https://doi.org/10.1007/978-3-662-44381-1_29

12. Damgård, I., Ishai, Y.: Scalable secure multiparty computation. In: Dwork, C. (ed.) CRYPTO 2006. LNCS, vol. 4117, pp. 501–520. Springer, Heidelberg (2006). https://doi.org/10.1007/11818175_30
13. Damgård, I., Nielsen, J.B.: Scalable and unconditionally secure multiparty computation. In: Menezes, A. (ed.) CRYPTO 2007. LNCS, vol. 4622, pp. 572–590. Springer, Heidelberg (2007). https://doi.org/10.1007/978-3-540-74143-5_32
14. Damgård, I., Nielsen, J.B., Nielsen, M., Ranellucci, S.: Gate-scrambling revisited - or: the TinyTable protocol for 2-party secure computation. Cryptology ePrint Archive, Report 2016/695 (2016). http://eprint.iacr.org/2016/695
15. Damgård, I., Pastro, V., Smart, N.P., Zakarias, S.: Multiparty computation from somewhat homomorphic encryption. In: Safavi-Naini, R., Canetti, R. (eds.) CRYPTO 2012. LNCS, vol. 7417, pp. 643–662. Springer, Heidelberg (2012). https://doi.org/10.1007/978-3-642-32009-5_38
16. Damgård, I., Zakarias, S.: Constant-overhead secure computation of boolean circuits using preprocessing. In: Sahai, A. (ed.) TCC 2013. LNCS, vol. 7785, pp. 621–641. Springer, Heidelberg (2013). https://doi.org/10.1007/978-3-642-36594-2_35
17. Frederiksen, T.K., Keller, M., Orsini, E., Scholl, P.: A unified approach to MPC with preprocessing using OT. In: Iwata, T., Cheon, J.H. (eds.) ASIACRYPT 2015. LNCS, vol. 9452, pp. 711–735. Springer, Heidelberg (2015). https://doi.org/10.1007/978-3-662-48797-6_29
18. Goldreich, O., Micali, S., Wigderson, A.: How to play any mental game or a completeness theorem for protocols with honest majority. In: Aho, A. (ed.) 19th ACM STOC, pp. 218–229. ACM Press, May 1987
19. Goldwasser, S., Lindell, Y.: Secure multi-party computation without agreement. J. Cryptol. **18**(3), 247–287 (2005)
20. Hazay, C., Scholl, P., Soria-Vazquez, E.: Low cost constant round MPC combining BMR and oblivious transfer. IACR Cryptology ePrint Archive, 2017:214 (2017)
21. Ishai, Y., Prabhakaran, M., Sahai, A.: Founding cryptography on oblivious transfer – efficiently. In: Wagner, D. (ed.) CRYPTO 2008. LNCS, vol. 5157, pp. 572–591. Springer, Heidelberg (2008). https://doi.org/10.1007/978-3-540-85174-5_32
22. Ishai, Y., Prabhakaran, M., Sahai, A.: Secure arithmetic computation with no honest majority. In: Reingold, O. (ed.) TCC 2009. LNCS, vol. 5444, pp. 294–314. Springer, Heidelberg (2009). https://doi.org/10.1007/978-3-642-00457-5_18
23. Katz, J., Ranellucci, S., Wang, X.: Authenticated garbling and efficient maliciously secure multi-party computation. IACR Cryptology ePrint Archive, 2017:189 (2017)
24. Keller, M., Orsini, E., Scholl, P.: MASCOT: faster malicious arithmetic secure computation with oblivious transfer. In: Weippl, E.R., Katzenbeisser, S., Kruegel, C., Myers, A.C., Halevi, S. (eds.) ACM CCS 16, pp. 830–842. ACM Press, October 2016
25. Kolesnikov, V., Schneider, T.: Improved garbled circuit: free XOR gates and applications. In: Aceto, L., Damgård, I., Goldberg, L.A., Halldórsson, M.M., Ingólfsdóttir, A., Walukiewicz, I. (eds.) ICALP 2008. LNCS, vol. 5126, pp. 486–498. Springer, Heidelberg (2008). https://doi.org/10.1007/978-3-540-70583-3_40
26. Lindell, Y., Pinkas, B.: An efficient protocol for secure two-party computation in the presence of malicious adversaries. In: Naor, M. (ed.) EUROCRYPT 2007. LNCS, vol. 4515, pp. 52–78. Springer, Heidelberg (2007). https://doi.org/10.1007/978-3-540-72540-4_4
27. Lindell, Y., Pinkas, B.: A proof of security of Yao's protocol for two-party computation. J. Cryptol. **22**(2), 161–188 (2009)

28. Lindell, Y., Pinkas, B.: Secure two-party computation via cut-and-choose oblivious transfer. In: Ishai, Y. (ed.) TCC 2011. LNCS, vol. 6597, pp. 329–346. Springer, Heidelberg (2011). https://doi.org/10.1007/978-3-642-19571-6_20
29. Lindell, Y., Pinkas, B., Smart, N.P., Yanai, A.: Efficient constant round multi-party computation combining BMR and SPDZ. In: Gennaro, R., Robshaw, M. (eds.) CRYPTO 2015. LNCS, vol. 9216, pp. 319–338. Springer, Heidelberg (2015). https://doi.org/10.1007/978-3-662-48000-7_16
30. Lindell, Y., Riva, B.: Blazing fast 2PC in the offline/online setting with security for malicious adversaries. In: Ray, I., Li, N., Kruegel, C. (eds.) ACM CCS 15, pp. 579–590. ACM Press, October 2015
31. Lindell, Y., Smart, N.P., Soria-Vazquez, E.: More efficient constant-round multi-party computation from BMR and SHE. In: Hirt, M., Smith, A. (eds.) TCC 2016. LNCS, vol. 9985, pp. 554–581. Springer, Heidelberg (2016). https://doi.org/10.1007/978-3-662-53641-4_21
32. Mohassel, P., Rosulek, M., Zhang, Y.: Fast and secure three-party computation: the garbled circuit approach. In: Ray, I., Li, N., Kruegel, C. (eds.) ACM CCS 15, pp. 591–602. ACM Press, October 2015
33. Nielsen, J.B., Nordholt, P.S., Orlandi, C., Burra, S.S.: A new approach to practical active-secure two-party computation. In: Safavi-Naini, R., Canetti, R. (eds.) CRYPTO 2012. LNCS, vol. 7417, pp. 681–700. Springer, Heidelberg (2012). https://doi.org/10.1007/978-3-642-32009-5_40
34. Nielsen, J.B., Schneider, T., Trifiletti, R.: Constant round maliciously secure 2PC with function-independent preprocessing using lego. In: 24th NDSS Symposium. The Internet Society (2017). http://eprint.iacr.org/2016/1069
35. Rabin, T., Ben-Or, M.: Verifiable secret sharing and multiparty protocols with honest majority (extended abstract). In: 21st ACM STOC, pp. 73–85. ACM Press, May 1989
36. Rindal, P.: libOTe: an efficient, portable, and easy to use Oblivious Transfer Library. https://github.com/osu-crypto/libOTe
37. Rindal, P., Rosulek, M.: Faster malicious 2-party secure computation with online/offline dual execution. In: 25th USENIX Security Symposium (USENIX Security 16), pp. 297–314. USENIX Association, Austin (2016)
38. Wang, X., Ranellucci, S., Katz, J.: Authenticated garbling and communication-efficient, constant-round, secure two-party computation. IACR Cryptology ePrint Archive, 2017:30 (2017)
39. Yao, A.C.-C.: How to generate and exchange secrets (extended abstract). In: 27th FOCS, pp. 162–167. IEEE Computer Society Press, October 1986

Maliciously Secure Oblivious Linear Function Evaluation with Constant Overhead

Satrajit Ghosh(✉), Jesper Buus Nielsen, and Tobias Nilges

Aarhus University, Aarhus, Denmark
{satrajit,jbn,tobias.nilges}@cs.au.dk

Abstract. In this work we consider the problem of oblivious linear function evaluation (OLE). OLE is a special case of oblivious polynomial evaluation (OPE) and deals with the oblivious evaluation of a linear function $f(x) = ax + b$. This problem is non-trivial in the sense that the sender chooses a, b and the receiver x, but the receiver may only learn $f(x)$. We present a highly efficient and UC-secure construction of OLE in the OT-hybrid model that requires only $O(1)$ OTs per OLE. The construction is based on noisy encodings introduced by Naor and Pinkas (STOC'99) and used for passive secure OLEs by Ishai, Prabhakaran and Sahai (TCC'09). A result asymptotically similar to ours is known by applying the IPS compiler to the mentioned passive secure OLE protocol, but our protocol provides better constants and would be considerably simpler to implement. Concretely we use only 16 OTs to generate one active secure OLE, and our protocol achieves active security by adding fairly simple checks to the passive secure protocol. We therefore believe our protocol takes an important step towards basing practical active-secure arithmetic computations on OLEs. Our result requires novel techniques that might be of independent interest. As an application we present the currently most efficient OPE construction.

1 Introduction

The oblivious evaluation of functions is an essential building block in cryptographic protocols. The first and arguably most famous result in the area is oblivious transfer (OT), which was introduced in the seminal work of Rabin [31]. Here, a sender can specify two bits s_0, s_1, and a receiver can learn one of the bits s_c depending on his choice bit c. It is guaranteed that the sender does not learn c, while the receiver learns nothing about s_{1-c}. Kilian [27] subsequently showed that OT allows the (oblivious) evaluation of *any* function.

While there has been tremendous progress in the area of generic MPC over the last three decades, there are certain classes of functions that can be evaluated more efficiently by direct constructions instead of taking the detour via

S. Ghosh and T. Nilges—Supported by the European Union's Horizon 2020 research and innovation programme under grant agreement #669255 (MPCPRO).
J.B. Nielsen—Supported by the European Union's Horizon 2020 research and innovation programme under grant agreement #731583 (SODA).

T. Takagi and T. Peyrin (Eds.): ASIACRYPT 2017, Part I, LNCS 10624, pp. 629–659, 2017.
https://doi.org/10.1007/978-3-319-70694-8_22

MPC. In this context, Naor and Pinkas [29] introduced oblivious polynomial evaluation (OPE) as an useful primitive. OPE deals with the problem of evaluating a polynomial P on an input α obliviously, i.e., in such a way that the sender specifies the polynomial P but does not learn α, while the receiver learns $P(\alpha)$ but nothing else about P. OPE has many applications, ranging from secure set intersection [18,30] over RSA key generation [16] to oblivious keyword search [13]. Due to its versatility, OPE has received considerable attention in recent years [7,14,18,19,26,28,33,34].

A special case of OPE, called oblivious linear function evaluation (OLE, sometimes also referred to as OLFE, or OAFE) has more recently been introduced as an essential building block in (the preprocessing of) MPC protocols for arithmetic circuits [8] or garbled arithmetic circuits [1]. Instead of evaluating an arbitrary polynomial P, the receiver wants to evaluate a linear or affine function $f(x) = ax + b$. Ishai et al. [22] propose a passively secure protocol for oblivious multiplication which uses a similar approach as [29], and can be easily modified to give a passively secure OLE. Their approach to achieve active security is to apply a compiler like [21] to the passive protocol. Another approach was taken by [10], who built an unconditionally UC-secure protocol for OLE based on tamper-proof hardware [23] as a setup assumption.

Currently, all of the above mentioned actively secure realizations of OPE or OLE require rather expensive computations or strong setup assumptions. In contrast, the most efficient passively secure constructions built from noisy encodings and OT require only simple field operations. However, to date a direct construction of a maliciously secure protocol in this setting has been elusive. While passive-to-active compilers such as [21] yield actively secure protocols with a constant overhead, such a transformation typically incurs a prohibitively large constant, resulting in efficiency only in an asymptotic sense.[1] Thus, the most efficient realizations currently follow from applying the techniques used for the precomputation of multiplied values in arithmetic MPC protocols such as SPDZ [8] or more recently MASCOT [25].

1.1 Our Contribution

We present a UC-secure protocol for oblivious linear function evaluation in the OT-hybrid model based on noisy encodings. The protocol is based on the semi-honest secure implementation of OLE by Ishai et al. [22], which is the most efficient protocol for passively secure OLE that we are aware of. Our actively secure protocol only has a constant overhead of 2 compared to the passively secure original construction. In numbers, this means:

- We need 16 OTs per OLE, compared to 8 for the semi-honest protocol [22].
- We communicate $16 \cdot (2 + c_{\mathsf{OT}}) \cdot n + 8$ field elements for n multiplications, compared to $8 \cdot (2 + c_{\mathsf{OT}}) \cdot n$ in [22], where c_{OT} is the cost for one OT.
- The computational overhead is twice that of the semi-honest case.

[1] We will compare in more detail to protocols obtained via [21] after comparing to known direct constructions.

One nice property of [22] and the main reason for its efficiency is that it directly allows to multiply a batch of values. This property is preserved in our construction, i.e., we can simultaneously evaluate several linear functions.

In order to achieve our result we solve the long standing open problem of finding an actively secure OLE/OPE protocol which can directly be reduced to the security of noisy encodings (and OT). This problem was not solved in [29] and has only been touched upon in follow-up work [22,26]. The key technical contribution of the paper is a reduction which shows that noisy encodings are robust against leakage in a strong sense, which allows their application in a malicious setting. As a matter of fact, our robustness results are more general and extend to *all* noisy encodings.

An immediate application of our UC-secure batch-OLE construction is a UC-secure OPE construction. The construction is very simple and has basically no overhead over the OLE construction. We follow the approach taken in [30], i.e., we use the fact that a polynomial of degree d can be decomposed into d linear functions. Such a decomposed polynomial is evaluated with the batch-OLE and then reconstructed. UC-security against the sender directly follows from the UC-security of the batch-OLE. In order to make the protocol secure against a cheating receiver, we only have to add one additional check that ensures that the receiver chooses the same input for each linear function. Table 1 compares the efficiency of our result with existing solutions in the literature.

Table 1. Overview of OPE realizations, where d is the degree of the polynomial, k is a computational security parameter and s a statistial security parameter ([19] propose $s \approx 160$). We compare the number of OTs and exponentiations in the respective protocols.

	Assumption	OTs	Expon.	Security
[7]	OT	$O(d\kappa)$	0	Passive
[30]	OT and Noisy encodings	$O(d\kappa \log \kappa)$	0	Passive
[22]	OT and Noisy encodings	$O(d)$	0	Passive
[19]	CRS and DCRP	0	$O(ds)$	UC
[18]	DDH	0	$O(d)^*$	Active
This work	OT and Noisy encodings	$O(d)$	0	UC

We point out that [18] only realizes OPE in the exponent, which is still sufficient for many applications, but requires additional work to yield a full fledged OPE. In particular, this might entail additional expensive operations. Another important factor regarding the efficiency of OT-based protocols is the cheap extendability of OT [20,24] which shows that the asymptotic price of OT is only small constant number of applications of a symmetric primitive like for instance AES. Therefore, the concrete cost of the OTs is much less than the price of exponentiations if d is sufficiently large, or if several OPEs have to be carried out. In such a scenario, we get significant improvements over previous (actively secure) solutions, which always require expensive operations in the degree of the polynomial.

Comparing to IPS. In [22], which provides the passive secure OLE protocol from which we depart, the authors propose to get active security using the IPS compiler [21]. Our protocol is inspired by this approach but we have spared a lot of the generic mechanism of the IPS compiler and obtained a protocol which both in terms of number of OTs and implementation is considerable simpler.

In the IPS-compiler one needs to specify an outer protocol for n servers which should be active secure against a constant fraction of corruptions. Since our goal here is to generate active secure OLEs, we will need an outer protocol computing OLEs and therefore an arithmetic protocol is natural. One should also specify an inner protocol, which is a passive secure two-party computation protocol allowing to emulate the individual parties in the outer protocol. This protocol should be based on passive secure OLEs, as this is our underlying primitive.

For the outer protocol it seems as a good choice to use a variant of [2]. The protocol has perfect active security and therefore fits the [22] framework and avoids complications with coinflipping into the well. Since the IPS compiler does not need error recovery, but only error detection, the entire dispute-control mechanism of [2] can be skipped, yielding a very simple protocol. By setting the number n of servers very high, the number of corrupted servers that the outer protocol should tolerate can become an arbitrarily small constant ϵ. In that case one could probably get a packed version of [2], where one computes the same circuit on $(n - 3\epsilon)/2$ different values in parallel. This means that each secure multiplication done by the outer protocol will involve just more than 2 local multiplications of the servers. For simplicity we ignore all other costs than these two multiplications.

Using the natural inner protocol based on the passive secure OLE from [22], each of the two emulated multiplications will consume 2 passive secure OLEs. Each passive secure OLE consumes 8 OTs, for a total cost of 32 OTs per active secure OLE generated. Our protocol uses 16 OTs per OLE.

Putting the inner and outer protocols together in the IPS framework involves considerably additional complications like setting up watch-lists and encrypting and sending all internal messages of emulated servers on these watch-list channels. In comparison our protocol has almost no additional mechanisms extra to the underlying passive secure protocol. Our overhead on local computation of the parties compared to the passive secure OLE protocol is 2. It seems unlikely that the IPS compiler would have an overhead even close to just 2.

In summary, our protocol saves a factor of 2 in consumptions of OTs and is much simpler and therefore easier to implement and has considerable lower computational cost.

1.2 Technical Overview

At the heart of our constructions are noisy encodings. These were introduced by Naor and Pinkas [29] in their paper on OPE and provide a very efficient means to obliviously compute multiplications. A noisy encoding is basically an encoding of a message via a linear code that is mixed with random values in such a way that the resulting vector hides which elements belong to the codeword and which elements are random, thereby hiding the initial message. In a little more detail, the input $\mathbf{x} \in \mathbb{F}^t$ is used as t sampling points on locations α_i of an otherwise random polynomial P of some degree $d > t$. Then the polynomial is evaluated at e.g., $4d$ positions β_i, and half of these positions are replaced by uniformly random values, resulting in the encoding $\mathbf{v}$. It is assumed that this encoding is indistinguishable from a uniformly random vector.[2]

Robustness of noisy encodings. The main problem of using noisy encodings in maliciously secure protocols is that the encoding is typically used in a non-black-box way. On one hand this allows for very efficient protocols, but on the other hand a malicious party obtains knowledge that renders the assumption that is made on the indistinguishability of noisy encodings useless. In a little more detail, consider a situation where the adversary obtains the encoding and manipulates it in a way that is not specified by the protocol. The honest party only obtains part of the encoding (this is usually necessary even in the passively secure case). In order to achieve active security, a check is performed which is supposed to catch a deviating adversary. But since the check is dependent on which part of the encoding the honest party learned, this check actually leaks some non-trivial information to the adversary, typically noisy positions of the codeword.

We show that noisy encodings as defined by [22,30] are very robust with respect to leakage. In particular, we show the following theorem that is basically a stronger version of a result previously obtained by Kiayias and Yung [26].

Theorem *(informal). For appropriate choices of parameters, noisy encodings are resilient against non-adaptive leakage of $O(\log \kappa)$ noisy positions.*

In a little more detail, we show that a noisy encoding generated as described above remains indistinguishable from a random vector of field elements, even for an adversary that is allowed to *fix* the position of f noisy positions. Fixing f positions is of course stronger than being able to leak f positions. The security loss incurred by the fixing of f positions is 3^f.

We then show that an adversary which is given a noisy encoding cannot identify a super-logarithmic sized set consisting of only noisy positions.

Theorem *(informal). For appropriate choices of parameters, an adversary cannot identify more than $O(\log \kappa)$ noisy positions in a noisy encoding.*

---

[2] The problem is related to efficient polynomial reconstruction, i.e., decoding Reed-Solomon codes, and as such well researched. The parameters have to be chosen in such a way that all known decoding algorithms fail.

These theorems together show that we can tolerate the leakage of any number of noisy positions that might be guessed. This is the basis for the security of our protocol. Note that tolerance to leakage of a set of noisy positions that might be guessed is not trivial, as we are working with an indistinguishability notion. Hence leakage of a single bit might *a priori* break the assumption.

We describe the main idea behind our reduction proving the first theorem. Assume that there are a total of ρ noisy positions. Consider an adversary that is allowed to submit a set F. Then we generate a noisy encoding as above, except that all positions $i \in F$ are fixed to be noisy. The remaining $\rho - |F|$ noisy positions are picked at random. Denote the distribution by $\boldsymbol{v}^{\rho,F}$. Let $\boldsymbol{v}^{\rho} = \boldsymbol{v}^{\rho,\emptyset}$. Let $\boldsymbol{v}^{\$}$ denote a vector with all positions being uniformly random. We start from the assumption that $\boldsymbol{v}^{\rho} \approx \boldsymbol{v}^{\$}$. Let n be the total number of positions. Then clearly $\boldsymbol{v}^{n,F} = \boldsymbol{v}^{\$}$ for all F.

We want to prove that $\boldsymbol{v}^{\rho,F} \approx \boldsymbol{v}^{\$}$ for small F. Let F be a set of size f. Assume that we managed to prove that $\boldsymbol{v}^{\rho,F'} \approx \boldsymbol{v}^{\$}$ for all sets F' of size $f-1$. Assume also that we have managed to prove that $\boldsymbol{v}^{\rho+1,F} \approx \boldsymbol{v}^{\$}$.

For a set F let i be the smallest index in F and let $F = F' \cup \{i\}$. Consider the reduction which is given $\boldsymbol{v}$ from $\boldsymbol{v}^{\rho,F'}$ or $\boldsymbol{v}^{\$}$ and which adds noise to position i in $\boldsymbol{v}$ and outputs the result $\boldsymbol{v}'$. If $\boldsymbol{v} \sim \boldsymbol{v}^{\$}$, then $\boldsymbol{v}' \sim \boldsymbol{v}^{\$}$. If $\boldsymbol{v} \sim \boldsymbol{v}^{\rho,F'}$, then $\boldsymbol{v}' \sim \alpha \boldsymbol{v}^{\rho+1,F} + (1-\alpha)\boldsymbol{v}^{\rho,F}$, where α is the probability that i is not already a noisy position. Putting these together we get that $\boldsymbol{v}^{\rho,F'} \approx \boldsymbol{v}^{\$}$ implies that $\alpha \boldsymbol{v}^{\rho+1,F} + (1-\alpha)\boldsymbol{v}^{\rho,F} \approx \boldsymbol{v}^{\$}$. We then use that $\boldsymbol{v}^{\rho+1,F} \approx \boldsymbol{v}^{\$}$ to get that $\alpha \boldsymbol{v}^{\$} + (1-\alpha)\boldsymbol{v}^{\rho,F} \approx \boldsymbol{v}^{\$}$, which implies that $\boldsymbol{v}^{\rho,F} \approx \boldsymbol{v}^{\$}$, when α is not too large.

We are then left with proving that $\boldsymbol{v}^{\rho,F'} \approx \boldsymbol{v}^{\$}$ and $\boldsymbol{v}^{\rho+1,F} \approx \boldsymbol{v}^{\$}$. These are proven by induction. The basis for $\boldsymbol{v}^{\rho,F'} \approx \boldsymbol{v}^{\$}$ is $\boldsymbol{v}^{\rho,\emptyset} \approx \boldsymbol{v}^{\$}$. The basis for $\boldsymbol{v}^{\rho+1,F} \approx \boldsymbol{v}^{\$}$ is $\boldsymbol{v}^{n,F} = \boldsymbol{v}^{\$}$. Controlling the security loss in these polynomially deep and nested inductions is tricky. We give the full details later.

We now give the intuition behind the proof of the second theorem. Assume that some adversary can guess a set S of s noisy positions with polynomial probability p_1 given an encoding $\boldsymbol{v}^{\rho} \approx \boldsymbol{v}^{\$}$ and assume that s is super-logarithmic and that ρ/n is a non-zero constant. We prove the theorem for noise level ρ but have to start with the assumption that $\boldsymbol{v}^{\rho - c\kappa} \approx \boldsymbol{v}^{\$}$ for an appropriate constant $c \in (0,1)$ and where κ is the security parameter.

Consider the reduction which is given a sample $\boldsymbol{v}$ from $\boldsymbol{v}^{\rho - c\kappa}$ or $\boldsymbol{v}^{\$}$. It starts by adding κ random positions R to $\boldsymbol{v}$ to get $\boldsymbol{v}'$. Then it feeds $\boldsymbol{v}'$ to $\mathcal{A}$ to get a set S. Then it uses its knowledge of R to sample the size of the intersection between S and the noisy positions in $\boldsymbol{v}'$. If it is "large" we guess that $\boldsymbol{v} \sim \boldsymbol{v}^{\rho - c\kappa}$. Otherwise we guess that $\boldsymbol{v} \sim \boldsymbol{v}^{\$}$. We pick c such that the total number of random positions in $\boldsymbol{v}'$ is ρ with polynomial probability when $\boldsymbol{v} \sim \boldsymbol{v}^{\rho - c\kappa}$, in which case S is a subset of the noisy positions with probability p_1, which will give a large intersection. If $\boldsymbol{v} \sim \boldsymbol{v}^{\$}$, then R is uniformly hidden to the adversary, and the expected size of the intersection will be smaller by a constant factor depending on ρ and c. The calibration of c and "small" is done as to allow a formal proof using a Chernoff bound. The details are given in the following.

Efficient OLE from noisy encodings. We build a UC-secure OLE protocol inspired by the passively secure multiplication protocol of Ishai et al. [22]. Let us briefly recall their construction on an intuitive level. One party, let us call it the sender, has as input t values $a_1, \dots, a_t \in \mathbb{F}$, while the receiver has an input $b_1, \dots, b_t \in \mathbb{F}$. A set of distinct points $\alpha_1, \dots, \alpha_{n/4}$ is fixed. The high-level idea is as follows: both sender and receiver interpolate a degree $n/4 - 1$ polynomial through the points (α_i, a_i) and (α_i, b_i) (picking a_i, b_i for $i > t$ randomly), to obtain $A(x)$ and $B(x)$, respectively. They also agree on n points $\beta_1, \dots, \beta_n$. Now the receiver replaces half of the points $B(\beta_i)$ with uniformly random values (he creates a noisy encoding) and sends these n values $\bar{B}(\beta_i)$ to the sender. He keeps track of the noiseless positions using an index set L. The sender draws an additional random polynomial R of degree $2(n/4 - 1)$ to mask the output. He then computes $Y(\beta_i) = A(\beta_i) \cdot \bar{B}(\beta_i) + R(\beta_i)$ and uses these points as input into a $n/2 - 1$-out-of-n OT, from which the receiver chooses the $n/2 - 1$ values in L. He can then interpolate the obtained points of $Y(\beta_i)$ to reconstruct Y and learn $a_i \cdot b_i + r_i$ in the positions α_i. This also directly yields an OLE: the polynomial R is simply used as another input of the sender, since it can be generated identically to A.

The passive security for the sender follows from the fact that the receiver obtains only $n/2 - 1$ values and thus R completely masks the inputs $a_1, \dots, a_t$. Passive security of the receiver follows from the noisy encoding, i.e., the sender cannot learn B from the noisy encoding.

In order to achieve actively secure OLE from the above protocol, we have to ensure several things: first of all, we need to use an actively secure k-out-of-n OT. But instead of using a black-box realization, which incurs an overhead of $n \log n$ on the number of OTs, we use n 1-out-of-2 OTs and ensure that the right number of messages was picked via a secret sharing, which the receiver has to reconstruct. This protocol first appeared in [32]. It does not have active security against the sender, who can guess some choice bits. A less efficient but active secure version was later given in [9], using verifiable secret sharing. We can, however, use the more efficient but less secure original variant as we can tolerate leakage of a few choice bits in the overall protocol.

Secondly, we also need to make sure that the parties used the right inputs in the computation, i.e., valid polynomials A, B and R. In order to catch deviations, we add two checks—one in each direction. The check is fairly simple: one party selects a random point z and the other party sends a pair $A(z), R(z)$, or $B(z), Y(z)$ respectively. Each party can now locally verify that the values satisfy the equation $A(z) \cdot B(z) + R(z) = Y(z)$.

As it turns out, both of these additions to the protocol, while ensuring protocol compliance w.r.t. the inputs, are dependent on the encoding. But this also means that a malicious sender can do selective failure attacks, e.g., it inputs incorrect shares for the secret sharing, and gets some leakage on the "secret key" of the encoding. This problem does not occur when considering semi-honest security.

2 Preliminaries

We use the standard notions of probabilistic polynomial time (PPT) algorithms, negligible and overwhelming functions. Further, we denote by $\mathbf{x} \in \mathbb{F}^n$ a vector of length n, x_i as the ith element of $\mathbf{x}$ and $\mathbf{x}_{|I}$ all x_i for $i \in I$. Unless noted otherwise, $P(x)$ denotes a polynomial in $\mathbb{F}[X]$, and X denotes a distribution.

We will typically denote a value $\hat{x}$ chosen or extracted by the simulator, while x^* is chosen by the adversary $\mathcal{A}$.

2.1 Universal Composability Framework

We state and prove our results in the Universal Composability (UC) framework of Canetti [5]. Security is defined via the comparison of an *ideal model* and a *real model*. In the real model, a protocol Π between the protocol participants is executed, while in the ideal model the parties only communicate with an ideal functionality $\mathcal{F}$ that is supposed to model the ideal security guarantees of the protocol. For an adversary $\mathcal{A}$ in the real protocol who coordinates the behavior of all malicious parties, there has to exist a simulator $\mathcal{S}$ for $\mathcal{A}$ in the ideal protocol. An environment $\mathcal{Z}$, which is plugged to both the real and the ideal protocol, provides the inputs to the parties and can read the outputs. The simulator $\mathcal{S}$ has to ensure that $\mathcal{Z}$ is not able to distinguish these models. Thus, even with concurrently executed protocols (running in the environment) the security holds. Usually, we assume that $\mathcal{A}$ is a dummy adversary controlled by $\mathcal{Z}$, which means that $\mathcal{Z}$ can adaptively choose its inputs depending on protocol messages it received and send messages on behalf of a (corrupted) protocol party.

More formally, let $\mathsf{Real}_{\Pi}^{\mathcal{A}}(\mathcal{Z})$ denote the random variable describing the output of $\mathcal{Z}$ when interacting with the real model, and let $\mathsf{Ideal}_{\mathcal{F}}^{\mathcal{S}}(\mathcal{Z})$ denote the random variable describing the output of $\mathcal{Z}$ when interacting with the ideal model. A protocol Π is said to *UC-realize* a functionality $\mathcal{F}$ if for any (PPT) adversary $\mathcal{A}$, there exists a PPT simulator $\mathcal{S}$ such that for any (PPT) environment $\mathcal{Z}$, $\mathsf{Real}_{\Pi}^{\mathcal{A}}(\mathcal{Z}) \approx \mathsf{Ideal}_{\mathcal{F}}^{\mathcal{S}}(\mathcal{Z})$.

For our constructions we assume active adversaries and static corruption. We prove security in the hybrid model access to oblivious transfer (OT).

2.2 Commitment Scheme

A commitment scheme COM consists of two algorithms $(\mathsf{COM.Commit}, \mathsf{COM.Open})$. It is a two party protocol between a sender and a receiver. In the commit phase of the protocol, when the sender wants to commit to some secret value m, it runs $\mathsf{COM.Commit}(m)$ and gets back two values $(\mathsf{com}, \mathsf{unv})$. It sends com to the receiver. Later on in the unveil phase, the sender sends the unveil information unv to the receiver, who can use $\mathsf{COM.Open}$ to verify that the commitment com contains the actual secret m.

A commitment scheme must satisfy two security properties; (1) *Hiding*: The receiver cannot learn any information about the committed secret before the unveil phase, and (2) *Binding*: The sender must not be able to change the committed secret after the commit phase. For our purpose we need efficient UC-secure commitment schemes that can be realized in $\mathcal{F}_{\mathrm{OT}}$-hybrid model and in $\mathcal{F}_{\mathrm{OLE}}$-hybrid model.

In [6] the authors proposed an UC-secure commitment scheme in the $\mathcal{F}_{\mathrm{OT}}$-hybrid model. Their protocol gives the first UC commitment scheme with "optimal properties" of rate approaching 1 and linear time complexity, in a "amortized sense". In our UC-secure OLE protocol also we need to commit to many values at a time, so we can use their UC commitment scheme in our protocol.

3 Noisy Encodings

The security of our protocols is based on a noisy encoding assumption. Very briefly, a noisy encoding is an encoding of a message, e.g., via a linear code, that is mixed with random field elements. It is assumed that such a codeword, and in particular the encoded message, cannot be recovered. This assumption seems reasonable due to its close relationship to decoding random linear codes or the efficient decoding of Reed-Solomon codes with a large fraction of random noise.

Noisy encodings were first introduced by Naor and Pinkas [29], specifically for the purpose of realizing OPE. Their encoding algorithm basically generates a random polynomial P of degree $k-1$ with $P(0) = x$. The polynomial is evaluated at $n > k$ locations, and then $n - k$ positions are randomized. Generalizing the approach of [30], Ishai et al. [22] proposed a more efficient encoding procedure that allows to encode several field elements at once instead of a single element, using techniques of [12]. Basically, they use Reed-Solomon codes and then artificially blind the codeword with random errors in order to mask the location of the codeword elements in the resulting string.

The encoding procedure depicted in Fig. 1 is nearly identical to the procedure given in [22], apart from the fact that we do not fix the signal-to-noise ratio (because this will be dependent on the protocol). We also allow to pass a set of points $\mathcal{P}$ as an argument to Encode to simplify the description of our protocol later on. This change has no impact on the assumption, since these points are made public anyway via G.

[29] propose two different encodings and related assumptions, tailored to their protocols. One of these assumptions was later broken by [3,4], and a fixed version was presented in [30]. We are only interested in the unbroken assumption. The same assumption was used by [22] and we will adopt their notation in the following.

Assumption 1. *Let κ be a security parameter and $n, \rho \in \mathsf{poly}(\kappa)$. Further let $x, y \in \mathbb{F}^{t(\kappa)}$. Then the ensembles $\{\mathsf{Encode}_{n,\rho}(x)\}_\kappa$ and $\{\mathsf{Encode}_{n,\rho}(y)\}_\kappa$ are computationally indistinguishable, for $t \leq \frac{\ell}{4}$.*

As it is, our security reductions do not hold with respect to Assumption 1, but rather a variant of Assumption 1 which was already discussed in [29]. Instead of requiring indistinguishability of two encodings, we require that an encoding is

$\mathsf{Encode}_{n,\rho}(\mathbf{x}, \mathcal{P})$

Generator $\mathcal{G}(n, \rho)$ for Reed-Solomon code: Let the output be (G, H, L).
- Let $\ell = n - \rho$ and $k = \frac{\ell-1}{2} + 1$.
- Let $\alpha_1, \ldots, \alpha_k, \beta_1, \ldots, \beta_n$ be the points in $\mathcal{P}$. If $\mathcal{P} = \emptyset$, pick these points randomly from $\mathbb{F}$.
- Define the $n \times k$ matrix G such that for any $\mathbf{u} \in \mathbb{F}^k$, $(G\mathbf{u})_i = P(\beta_i)$ for $i = 1, \ldots, n$, where P is the unique degree $k-1$ polynomial such that $P(\alpha_i) = u_i$ for $i = 1, \ldots, k$.
- Pick $L \subset [n]$ with $|L| = \ell$ uniformly at random.
- Let H be the $k \times 2k - 1$ matrix such that $(H\mathbf{v}_{|L})_i = Q(\alpha_i)$, where Q is the unique degree $2(k-1)$ polynomial such that $Q(\beta_j) = \mathbf{v}_j$ for all $j \in L$.

Encoding: Let the private output be $(G, H, L, \mathbf{v})$ and the public output be $(G, \mathbf{v})$.
- Let $(G, H, L) \leftarrow \mathcal{G}(n, \rho)$.
- Pick a random $\mathbf{u} \in \mathbb{F}^k$ conditioned on $u_i = x_i$ for $i = 1, \ldots, t$. Compute $G\mathbf{u} \in \mathbb{F}^n$.
- Pick a random vector $v \in \mathbb{F}^n$, conditioned on $v_i = (G\mathbf{u})_i$ for $i \in L$. Output $\mathbf{v}$.

Fig. 1. Encoding procedure for noisy encodings.

pseudorandom. In order for these assumption to hold, [22] propose $n = 4\kappa, \rho = 2\kappa + 1$ as parameters, or $n = 8\kappa, \rho = 6\kappa + 1$ on the conservative side.

Assumption 2. *Let κ be a security parameter and $n, \rho \in \mathsf{poly}(\kappa)$. Further let $x \in \mathbb{F}^{t(\kappa)}$. Then the ensembles $\{\mathsf{Encode}_{n,\rho}(x)\}_\kappa$ and $\{G \leftarrow \mathcal{G}(n, \rho), v \leftarrow \mathbb{F}^n\}_\kappa$ are computationally indistinguishable, for $t \leq \frac{\ell}{4}$.*

Clearly, Assumption 2 implies Assumption 1, while the other direction is unclear. Apart from being a very natural assumption, Kiayias and Yung [26] provide additional evidence that Assumption 2 is valid. They show that if an adversary cannot decide for a random position i of the encoding whether it is part of the codeword or not, then the noisy codeword is indeed pseudorandom.

4 Noisy Encodings Are Robust Against Leakage

In this section we show that a large class of computational assumptions can be extended to allow some leakage without loss of asymptotic security. This is one of the main technical contributions of the paper and we deem the reductions to be of independent interest. We first define the class of assumptions we consider.

Definition 1. *Let a finite field $\mathbb{F}$ be given. For a positive integer n we use U^n to denote the uniform distribution on $\mathbb{F}^n$. Let n be the length of a codeword, ρ the number of randomised positions, G a generator for the encoding and F some fixed random positions with $|F| \leq \rho$. The distribution $\mathsf{Y}^{n,\rho,G,F}$ is sampled as follows.*

1. *Sample* $(x_1, \ldots, x_n) \leftarrow G(1^\kappa)$, *where* κ *is the security parameter.*
2. *Sample uniform* $R \subseteq [n]$ *under the restriction* $|R| = \rho$ *and* $F \subseteq R$.
3. *Sample uniform* $(e_1, \ldots, e_n) \leftarrow \mathsf{U}^n$ *under the restriction* $e_i = 0$ *for* $i \notin R$.
4. *Output* $\mathbf{y} = \mathbf{x} + \mathbf{e}$.

Clearly it holds that the above defined Encode generalises Definition 1, i.e., all of the following results hold for noisy encodings as well.

We will mostly consider the case that $n = \Theta(\kappa)$ and $\rho = \Theta(\kappa)$. Typically n and G are fixed, in which case we denote the distribution of $\mathbf{y}$ by $\mathsf{Y}^{\rho,F}$. Note that $\mathsf{Y}^{n,F} = \mathsf{U}^n$.

We are going to assume that for sufficiently large ρ it holds that $\mathsf{Y}^\rho \approx \mathsf{Y}^n$, where $\approx$ means the distributions are computationally indistinguishable. For example, this is given by Assumption 2 for noisy encodings with appropriate parameters. We will use this to prove that the same result holds given limited leakage on R and that it is hard to compute a lot of elements of R given only $\mathbf{y}$.

When we prove the first result, we are going to do an argument with two nested recursive reductions. To make it easier to track the security loss, we are going to shift to a concrete security notation for a while and then switch back to asymptotic security when we have control over the security loss.

Given two distributions A_0 and A_1 and a distinguisher D let $\mathsf{Adv}_D(A_0, A_1) = \Pr[A_1 = 1] - \Pr[A_0 = 0]$. We use $A_0 \approx^t_\epsilon A_1$ to denote that it holds for all distinguishers computable in time t that $\mathsf{Adv}_D(A_0, A_1) \leq \epsilon$.

Given two distributions A_0 and A_1 and $0 \leq \alpha_0 \leq 1$ and $\alpha_1 = 1 - \alpha_0$ we use $B = \alpha_0 A_0 + \alpha_1 A_1$ to denote the distribution generated by the following procedure. Sample $c \in \{0, 1\}$ with $\Pr[c = i] = \alpha_i$. Then sample $b \leftarrow A_c$ and output b.

We will use the following simple facts in our proofs.

Lemma 1. *Let* A_0, A_1 *and* Z *be distributions.*

$$A_0 \approx^t_\epsilon A_1 \;\Rightarrow\; \alpha_0 A_0 + \alpha_1 Z \approx^t_{\alpha_0 \epsilon} \alpha_0 A_1 + \alpha_1 Z.$$

$$\alpha_0 A_0 + \alpha_1 Z \approx^t_\epsilon \alpha_0 A_1 + \alpha_1 Z \;\Rightarrow\; A_0 \approx^t_{\alpha_0^{-1}\epsilon} A_1.$$

Proof. Let $B_i = \alpha_0 A_0 + \alpha_1 Z$. We first prove the first implication. Consider a distinguisher D for B_0 and B_1. Then

$$\begin{aligned}
\mathsf{Adv}_D(B_0, B_1) &= \Pr[D(B_1) = 1] - \Pr[D(B_0) = 1] \\
&= \sum_i \alpha_i \Pr[D(B_1) = 1 \mid c = i] - \sum_i \alpha_i \Pr[D(B_0) = 1 \mid c = i] \\
&= \alpha_0 \Pr[D(B_1) = 1 \mid c = 0] - \alpha_0 \Pr[D(B_0) = 1 \mid c = 0] \\
&= \alpha_0 \Pr[D(A_1) = 1] - \alpha_0 \Pr[D(A_0) = 1] \\
&= \alpha_0 \Pr[D(A_1) = 1] - \alpha_0 \Pr[D(A_0) = 1],
\end{aligned}$$

from which it follows that

$$\mathsf{Adv}_D(B_0, B_1) = \alpha_0 \mathsf{Adv}_D(A_0, A_1). \tag{1}$$

From (1) it follows that $\mathsf{Adv}_D(B_0, B_1) \leq \alpha_0 \epsilon$ for all D, which proves the claim in the lemma. Consider a distinguisher D for A_0 and A_1. It can also act as distinguisher for B_0 and B_1, so from (1) we have that

$$\mathsf{Adv}_D(A_0, A_1) = \alpha_0^{-1}\mathsf{Adv}_D(B_0, B_1).$$

From this the second claim follows. □

In the following we will show that if $\mathsf{Y}^\rho \approx_\epsilon^{t+t'} \mathsf{Y}^n$ and F is not too large, then $\mathsf{Y}^{\rho,F} \approx_{\epsilon'}^t \mathsf{Y}^{n,F}$ for ϵ' polynomially related to ϵ and t' a small fixed polynomial. Since the reduction will be recursive and will modify ϵ multiplicatively, we will keep explicit track of ϵ to ensure the security loss is not too large. As for t, each reduction will only *add* a small t' to t, namely the time to sample a distribution. The time will therefore clearly grow by at most a polynomial term. We therefore do not keep track of t, for notational convenience.

Lemma 2. *If $\rho - |F| \geq 2n/3$ and $\mathsf{Y}^j \approx_\epsilon \mathsf{Y}^n$ for all $j \geq \rho$, then $\mathsf{Y}^{\rho,F} \approx_{\sigma_\rho} \mathsf{Y}^{n,F}$ for $\sigma_\rho = 3^{|F|}\epsilon$.*

Proof. We prove the claim by induction in the size of F. If $|F| = 0$ it follows by assumption. Consider then the following randomised function f with inputs $(y_1, \ldots, y_n)$ and F. Let i be the largest element in F and let $F' = F \setminus \{i\}$. Sample uniformly random $y_i' \in \mathbb{F}$. For $j \neq i$, let $y_j' = y_j$. Output $\mathbf{y}'$. Consider the distribution $\mathsf{Y}^{\rho,F'}$. Let R denote the randomised positions. If $i \in R$, then $f(\mathsf{Y}^{\rho,F'}) = \mathsf{Y}^{\rho,F}$. If $i \notin R$, then $f(\mathsf{Y}^{\rho,F'}) = \mathsf{Y}^{\rho+1,F}$, as we added one more noisy point. The point i is a fixed point not in F'. There are $n - |F| + 1$ points not in F'. There are ρ randomised points, i.e., $|R| = \rho$. Exactly $|F| - 1$ of these points are the points of F'. The other points are uniform outside F'. So there are $\rho - |F| + 1$ such points. Therefore the probability that $i \in R$ is $p = \frac{\rho - |F| + 1}{n - |F| + 1}$. It follows that

$$f(\mathsf{Y}^{n,F'}) = \mathsf{Y}^{n,F}, \quad f(\mathsf{Y}^{\rho,F'}) = p\mathsf{Y}^{\rho,F} + (1-p)\mathsf{Y}^{\rho+1,F}.$$

It then follows from $\mathsf{Y}^{n,F'} \approx_{\epsilon'} \mathsf{Y}^{\rho,F'}$ (where $\epsilon' = 3^{|F'|}\epsilon$) that

$$\mathsf{Y}^{n,F} \approx_{\epsilon'} p\mathsf{Y}^{\rho,F} + (1-p)\mathsf{Y}^{\rho+1,F}.$$

We now claim that $\mathsf{Y}^{\rho,F} \approx_{3\epsilon'} \mathsf{Y}^{n,F}$. The claim is trivially true for $\rho = n$, so we can assume that $\rho < n$ and assume the claim is true for all $\rho' > \rho$. Using Lemma 1 and the induction hypothesis we get that

$$p\mathsf{Y}^{\rho,F} + (1-p)\mathsf{Y}^{\rho+1,F} \approx_{(1-p)3\epsilon'} p\mathsf{Y}^{\rho,F} + (1-p)\mathsf{Y}^{n,F}.$$

Clearly

$$\mathsf{Y}^{n,F} \approx_0 p\mathsf{Y}^{n,F} + (1-p)\mathsf{Y}^{n,F}.$$

Putting these together we get that

$$p\mathsf{Y}^{n,F} + (1-p)\mathsf{Y}^{n,F} \approx_{\epsilon' + (1-p)3\epsilon'} p\mathsf{Y}^{\rho,F} + (1-p)\mathsf{Y}^{n,F}.$$

Using Lemma 1 we get that $\mathsf{Y}^{n,F} \approx_{p^{-1}(\epsilon'+(1-p)3\epsilon')} \mathsf{Y}^{\rho,F}$. We have that $p \geq 2/3$ so

$$p^{-1}(\epsilon' + (1-p)3\epsilon') \leq \frac{3}{2}(\epsilon' + \frac{1}{3}3\epsilon') = 3\epsilon' = 3^{|F|}\epsilon.$$

□

In the rest of the section, assume that n and ρ are functions of a security parameter κ and that $n, \rho = \Theta(\kappa)$. Also assume that $\rho \geq 2n/3$ and that $n - \rho = \Theta(\kappa)$. We say that $\mathsf{Y}^{\rho} \approx \mathsf{Y}^{n}$ if $\mathsf{Y}^{\rho'} \approx^{q(\kappa)}_{1/p(\kappa)} \mathsf{Y}^{n(\kappa)}$ for all polynomials p and q and all sufficiently large κ and all $\rho' \geq \rho$.

From $3^{O(\log \kappa)}$ being a polynomial in κ we get that

Corollary 1. *If* $\mathsf{Y}^{\rho} \approx \mathsf{Y}^{n}$ *and* $F \subseteq [n]$ *has size* $O(\log \kappa)$, *then* $\mathsf{Y}^{\rho,F} \approx \mathsf{Y}^{n}$.

Now assume that $\mathsf{Y}^{\rho} \approx \mathsf{Y}^{n}$. Let $\mathsf{Y}^{\rho,F,\neg}$ be defined as $\mathsf{Y}^{\rho,F}$ except that R is sampled according to the restriction that $F \not\subseteq R$, i.e., F has at least one element outside R. Let $p(F)$ be the probability that $F \subseteq R$ for a uniform R. Then by definition and the law of total probability $\mathsf{Y}^{\rho} = p\mathsf{Y}^{\rho,F} + (1-p)\mathsf{Y}^{\rho,F,\neg}$. We have that $\mathsf{Y}^{\rho} \approx \mathsf{Y}^{n}$ and that $\mathsf{Y}^{\rho,F} \approx_{3^{|F|}} \mathsf{Y}^{n,F} = \mathsf{Y}^{n}$. Putting these together we have that $\mathsf{Y}^{n} \approx_{p3^{|F|}} p\mathsf{Y}^{n} + (1-p)\mathsf{Y}^{\rho,F,\neg}$. We then get that $\mathsf{Y}^{n} \approx_{3^{|F|}} \mathsf{Y}^{\rho,F,\neg}$.

Corollary 2. *If* $\mathsf{Y}^{\rho} \approx \mathsf{Y}^{n}$ *and* $F \subseteq [n]$ *has size* $O(\log \kappa)$, *then* $\mathsf{Y}^{\rho,F,\neg} \approx \mathsf{Y}^{n}$.

We now prove that given one small query on R does not break the security.

Definition 2. *Let* $\mathcal{A}$ *be a PPT algorithm and* Y *as defined in Definition 1. The game* $\mathcal{G}_{\mathsf{leak}}$ *is defined as follows.*

1. *Run* $\mathcal{A}$ *to get a subset* $Q \subseteq [n]$.
2. *Sample a uniformly random bit* c.
   - *If* $c = 0$, *then sample* $\mathbf{y} \leftarrow \mathsf{Y}^{\rho}$ *and let* R *be the subset used in the sampling*
   - *If* $c = 1$, *then sample* $\mathbf{y} \leftarrow \mathsf{Y}^{n}$ *and let* $R \subseteq [n]$ *be a uniformly random subset of size* ρ.
3. *Let* $r \in \{0, 1\}$ *be* 1 *iff* $Q \subseteq R$ *and input* $(r, \mathbf{y})$ *to* $\mathcal{A}$.
4. *Run* $\mathcal{A}$ *to get a guess* $g \in \{0, 1\}$.

The advantage of $\mathcal{A}$ *is* $\mathsf{Adv}_{\mathcal{A}} = \Pr[g = 1 \mid c = 1] - \Pr[g = 1 \mid c = 0]$.

Theorem 1. *Assume that* $n, \rho = \Theta(\kappa)$ *and* $\rho \geq \frac{2}{3}n$. *If* $\mathsf{Y}^{\rho} \approx \mathsf{Y}^{n}$, *then* $\mathsf{Adv}_{\mathcal{A}} \approx 0$.

Proof. Let p be the probability that $Q \subseteq R$. If p is negligible, then in item 3 of the game we could send the constant $r = 0$ to $\mathcal{A}$ and it would only change the advantage by a negligible amount. But in the thus modified game $\mathsf{Adv}_{\mathcal{A}} \approx 0$ because $\mathsf{Y}^{n} \approx \mathsf{Y}^{\rho}$. So assume that p is a polynomial.[3] Let Y_0 be the distribution

[3] Formally we should consider the case where it is a polynomial for infinitely many κ, but the following argument generalises easily to this case.

of (r, y) when $c = 0$. Let Y_1 be the distribution of $(r, \mathbf{y})$ when $c = 1$. If $c = 0$, then $(r, \mathbf{y})$ is distributed as follows

$$Y_0 = p \cdot (1, \mathsf{Y}^{\rho,Q}) + (1-p) \cdot (0, \mathsf{Y}^{\rho,Q,\neg}).$$

When p is polynomial, then $|F| = O(\log \kappa)$ as $n - \rho = \Theta(\kappa)$. From this we get

$$Y_0 \approx p \cdot (1, \mathsf{Y}^n) + (1-p) \cdot (0, \mathsf{Y}^n) = Y_1,$$

using the above asymptotic corollaries. □

We will then prove that it is hard to compute a lot of elements of R.

Definition 3. *Let $\mathcal{A}$ be a PPT algorithm and Y as defined in Definition 1. The game $\mathcal{G}_{\mathsf{ident}}$ is defined as follows.*

1. *Sample $\mathbf{y} \leftarrow \mathsf{Y}^\rho$ and let R denote the randomized positions.*
2. *Input $\mathbf{y}$ to $\mathcal{A}$.*
3. *Run $\mathcal{A}$ and denote the output by $Q \subseteq [n]$. We require that $|Q| = s$.*
4. *Let $r \in \{0, 1\}$ be 1 iff $Q \subseteq R$.*
5. *Output r.*

The advantage of $\mathcal{A}$ is $\mathsf{Adv}_{\mathcal{A}}^{\rho,s} = \Pr[r = 1]$.

Theorem 2. *Let $n = \Theta(\kappa)$ and assume that $\mathsf{Y}^\sigma \approx \mathsf{Y}^n$. Then $\mathsf{Adv}_{\mathcal{A}}^{\rho,s} \approx 0$ is true in both of these cases:*

1. *Let $\sigma = n\frac{\rho-\kappa}{n-\kappa}$ and $s = \kappa$.*
2. *Let $\sigma = \frac{n\kappa}{n-\rho-\kappa}$ and $s \in \omega(\log \kappa)$.*

Proof. Let $\mathcal{A}$ be an adversary such that when $Q \leftarrow \mathcal{A}(\mathsf{Y}^\rho)$, then $Q \subseteq R$ with non-negligible probability p. The argumentation is similar for both cases. For the first part of the theorem, consider the following adversary $\mathcal{B}(\mathbf{y})$ receiving $\mathbf{y} \in \mathbb{F}^n$. It samples a uniform $X \subset [n]$ of size κ. For $i \in X$ let y'_i be uniformly random. For $i \notin X$ let $y'_i = y_i$. Compute $Q \leftarrow \mathcal{A}(\mathbf{y})$. If $|Q \cap X| \geq \frac{\kappa^2}{\rho}$, then output 1. Otherwise output 0.

We now prove that $\Pr[\mathcal{B}(\mathsf{Y}^n) = 1] \approx 0$ and that $\Pr[\mathcal{B}(\mathsf{Y}^\sigma) = 1]$ is non-negligible, which proves the first statement of the theorem.

Let R be the positions that were randomised in $\mathbf{y}$. Let $R' = R \cup X$. Note that if $\mathbf{y} \leftarrow \mathsf{Y}^\sigma$, then

$$\mathsf{E}[|R'|] = \kappa + \sigma - \mathsf{E}[|X \cap R|] = \kappa + \sigma - \kappa\frac{\sigma}{n} = \rho.$$

It is straight forward to verify that $\Pr[|R'| = \rho] = 1/O(\kappa)$, which implies that $\Pr[Q \subseteq R] = p/O(\kappa)$, which is non-negligible when p is non-negligible. Let E denote the event that $Q \subseteq R$. By a simple application of linearity of expectation we have that

$$\mathsf{E}[|Q \cap X| \,|\, E] = \frac{\kappa^2}{\rho},$$

as X is a uniformly random subset $X \subseteq R$ given the view of $\mathcal{A}$. From this it follows that $\Pr[\mathcal{B}(\mathsf{Y}^\sigma) = 1]$ is non-negligible.

Then consider $\mathcal{B}(\mathsf{Y}^n)$. Note that now $R = [n]$ and again X is a uniformly random subset of R independent of the view of $\mathcal{A}$. Therefore

$$\mathsf{E}[|Q \cap X|] = \frac{\kappa^2}{n} =: \mu.$$

Then

$$\Pr[|Q \cap X| \geq \frac{s(\rho - \kappa)}{\rho}] = \Pr[|Q \cap X| \geq \frac{n}{\rho}\mu] = \Pr[|Q \cap X| \geq (1+\delta)\mu]$$

for $\delta \in (0,1)$. It follows that

$$\Pr[|Q \cap X| \geq \frac{\kappa^2}{\rho}] \leq e^{-\frac{\mu\delta^2}{3}} = e^{-\Theta(\mu)} = e^{-\Theta(\kappa)} = \mathsf{negl}(\kappa).$$

To see this let $X = \{x_1, \ldots, x_\kappa\}$ and let X_i be the indicator variable for the event that the i'the element of X ends up in Q. Then $\Pr[X_i = 1] = \frac{\kappa}{n}$ and $|X \cap Q| = \sum_i X_i$. Consider then the modified experiment called *Independent Sampling*, where we sample the κ elements for X uniformly at random from $[n]$ and independently, i.e., it may happen that two of them are identical. In that case the inequality is a simple Chernoff bound. It is easy to see that when we go back to *Dependent Sampling*, where we sample x_i uniformly at random except that they must be different from $x_1, \ldots, x_{i-1}$, then we only lower the variance of the sum $\sum_i X_i$ compared to Independent Sampling, so $\Pr[|Q \cap X| \geq (1+\delta)\mu]$ will drop. Too see this, consider the sequence $x, x_1 + x_2, \ldots, \sum_i x_i$ as a random walk. In the Dependent Sampling case, when $\sum_i x_i$ is larger than the expectation, then x_{i+1} is less likely to be in Q compared to the Independent Sampling case, as an above expectation number of slots in Q is already taken. Similarly, when $\sum_i x_i$ is smaller than the expectation, then x_{i+1} is more likely to be in Q compared to the Independent Sampling case, as a below expectation number of slots in Q is already taken. Therefore the random walk in the Dependent Sampling case will always tend closer to average compared to the Independent Sampling random walk.

The second statement of Theorem 2 follows by setting X to be a uniform subset of size $\rho - \kappa$. As above, if $\mathcal{A}$ outputs Q such that $|Q \cap X| \geq \frac{s(\rho-\kappa)}{\rho}$, then $\mathcal{B}(\mathbf{y})$ outputs 1. Otherwise it outputs 0. Let again R be the positions that were randomised in $\mathbf{y}$. Let $R' = R \cup X$. If $\mathbf{y} \leftarrow \mathsf{Y}^\sigma$, then

$$\mathsf{E}[|R'|] = \rho - \kappa + \sigma - \mathsf{E}[|X \cap R|] = \rho - \kappa + \sigma - (\rho - \kappa)\frac{\sigma}{n} = \rho.$$

Let E denote the event that $Q \subseteq R$. Following the above argumentation,

$$\mathsf{E}[|Q \cap X| \,|\, E] = \frac{s(\rho - \kappa)}{\rho}.$$

From this it follows that $\Pr[\mathcal{B}(\mathsf{Y}^\sigma) = 1]$ is non-negligible. Then consider $\mathcal{B}(\mathsf{Y}^n)$. Note that now $R = [n]$ and again X is a uniformly random subset of R independent of the view of $\mathcal{A}$. Therefore

$$\mathsf{E}[|Q \cap X|] = \frac{s(\rho - \kappa)}{n} =: \mu.$$

It follows that

$$\Pr[|Q \cap X| \geq \frac{s(\rho - \kappa)}{\rho}] \leq e^{-\frac{\mu\delta^2}{3}} = e^{-\Theta(\mu)} = e^{-\omega(\log \kappa)} = \mathsf{negl}(\kappa).$$

□

5 Constant Overhead Oblivious Linear Function Evaluation

Oblivious linear function evaluation (OLE) is the task of computing a linear function $f(x) = ax + b$ in the following setting. One party, lets call it the sender S, provides the function, namely the values a and b. The other party, the receiver R, wants to evaluate this function on his input x. This task becomes non-trivial if the parties want to evaluate the function in such a way that the sender learns nothing about x, while the receiver learns only $f(x)$, but not a and b. OLE can be seen as a special case of oblivious polynomial evaluation (OPE) as proposed by Naor and Pinkas [29], where instead of a linear function f, the sender provides a polynomial p.

5.1 Ideal Functionality

The efficiency of our protocol follows in part from the fact that we can directly perform a batch of multiplications. This is reflected in the ideal UC-functionality for $\mathcal{F}_{\mathrm{OLE}}^{\mathrm{t}}$ (cf. Fig. 2), which allows both sender and receiver to input vectors of size t.

Functionality $\mathcal{F}_{\mathrm{OLE}}^{\mathrm{t}}$

1. Upon receiving a message $(\texttt{inputS}, \mathbf{a}, \mathbf{b})$ from S with $\mathbf{a}, \mathbf{b} \in \mathbb{F}^t$, verify that there is no stored tuple, else ignore that message. Store $\mathbf{a}$ and $\mathbf{b}$ and send a message $(\texttt{input})$ to $\mathcal{A}$.
2. Upon receiving a message $(\texttt{inputR}, \mathbf{x})$ from R with $\mathbf{x} \in \mathbb{F}^t$, verify that there is no stored tuple, else ignore that message. Store $\mathbf{x}$ and send a message $(\texttt{input})$ to $\mathcal{A}$.
3. Upon receiving a message $(\texttt{deliver}, \mathsf{S})$ from $\mathcal{A}$, check if both $\mathbf{a}, \mathbf{b}$ and $\mathbf{x}$ are stored, else ignore that message. Send $(\texttt{delivered})$ to S.
4. Upon receiving a message $(\texttt{deliver}, \mathsf{R})$ from $\mathcal{A}$, check if both $\mathbf{a}, \mathbf{b}$ and $\mathbf{x}$ are stored, else ignore that message. Set $y_i = a_i \cdot x_i + b_i$ for $i \in [t]$ and send $(\texttt{output}, \mathbf{y})$ to R.

Fig. 2. Ideal functionality for an oblivious linear function evaluation.

5.2 Our Protocol

Our starting point is the protocol of Ishai et al. [22] for *passively* secure batch multiplication. Their protocol is based on noisy encodings, similar to our construction. We will now briefly sketch their construction (with minor modifications) and then present the high-level ideas that are necessary to make the construction actively secure.

In their protocol, the receiver first creates a noisy encoding $(G, H, L, \mathbf{v}) \leftarrow \mathsf{Encode}(\mathbf{x})$ (as described in Sect. 3, Fig. 1) and sends $(G, \mathbf{v})$ to the sender. At this point, the locations $i \in L$ of $\mathbf{v}$ hide a degree $\frac{\ell-1}{2}$ polynomial over the points $\beta_1, \ldots, \beta_n$ which evaluates to the input $\mathbf{x} = x_1, \ldots, x_t$ in the positions $\alpha_1, \ldots, \alpha_t$. The sender picks two random polynomials A and B with the restriction that $A(\alpha_i) = a_i$ and $B(\alpha_i) = b_i$ for $i \in [t]$. The degree of A is $\frac{\ell-1}{2}$, and the degree of B is $\ell - 1$.[4] This means that B completely hides A and therefore the inputs of the sender. Now the sender simply computes $w_i = A(\beta_i) \cdot v_i + B(\beta_i)$. Sender and receiver engage in an ℓ-out-of-n OTs, and the receiver picks the ℓ positions in L. He applies H to the obtained values and interpolates a polynomial Y which evaluates in position α_i to $a_i \cdot x_i + b_i$.

We keep the generic structure of the protocol of [22] in our protocol. In order to ensure correct and consistent inputs, we have to add additional checks. The complete description is given in Fig. 3, and we give a high-level description of the ideas in the following paragraph.

First, we need to ensure that the receiver can only learn ℓ values, otherwise he could potentially reconstruct part of the input. Instead of using an expensive ℓ-out-of-n OT, we let the sender create a (ρ, n)-secret sharing (remember that $\rho + \ell = n$) of a random value e and the share s_i in the i'th OT, letting tge other message offered be a random value t_i. Depending on his set L, the receiver chooses t_i or the share s_i. Then he uses the shares to reconstruct e and sends it to the sender. This in turn might leak some information on L to the sender, if he can provide an inconsistent secret sharing. We thus force the sender to commit to e and later provide an unveil. Here the sender can learn some information on L, if he cheats but is not caught, but we can use our results from the previous section to show that this leakage is tolerable. The receiver can proceed and provide the encoding $\mathbf{v}$, which allows the sender to compute $\mathbf{w}$.

Second, we have to make sure that the sender computes the correct output. In order to catch a cheating sender, we add a check to the end of the protocol. Recall that the receiver knows the output Y. He can compute another polynomial X of his input and then pick a uniformly random challenge z_{R}. He sends it to the sender, who has to answer with $A(z_{\mathsf{R}}), B(z_{\mathsf{R}})$. Now the receiver can verify that $Y(z_{\mathsf{R}}) = A(z_{\mathsf{R}})X(z_{\mathsf{R}}) + B(z_{\mathsf{R}})$, i.e., the sender did not cheat in the noiseless positions. Again this leaks some information to the sender, but with the correct choice of parameters this leakage is inconsequential.

---

[4] The value ℓ is fixed by the encoding, but we require that ℓ is uneven due to the fact that we have to reconstruct a polynomial of even degree $\frac{\ell-1}{2} + \frac{\ell-1}{2} = \ell - 1$, which requires ℓ values.

Security against a malicious receiver basically follows from the passively secure protocol. We only have to make sure that the extraction of his input is correct and that no information about the sender's inputs is leaked if e is incorrect. We thus mask the w_i by a one-time-pad and add the following check. This time the sender chooses z_S and the receiver has to answer with $X(z_\mathsf{S}), Y(z_\mathsf{S})$, which enforces correct extraction.

Protocol Π_{OLE}

Let $\mathcal{P} = \{\alpha_1, \ldots, \alpha_{\frac{\ell+1}{2}}, \beta_1, \ldots, \beta_n\}$ be a set of publicly known distinct points in $\mathbb{F}$. Further, let SS be a (ρ, n) secret sharing and COM be an OT-based commitment scheme. Set $\rho = \frac{3}{4}n$, $n = 8k$ and $\ell = n - \rho$.

1. Sender (Input $\mathbf{a}, \mathbf{b} \in \mathbb{F}^t$):
   - Draw a random polynomial A of degree $\frac{\ell-1}{2}$ with $A(\alpha_i) = a_i$ and a random polynomial B of degree $\ell - 1$ with $B(\alpha_i) = b_i \ \forall i \in \{1, \ldots, t\}$.
   - Draw a uniformly random vector $\mathbf{t} \in \mathbb{F}^n$.
   - Draw a random value $e \in \mathbb{F}$ and compute $\mathbf{s} \leftarrow \mathsf{SS.Share}(e)$. Further compute $(\mathsf{com}, \mathsf{unv}) \leftarrow \mathsf{COM.Commit}(e)$.
   - Send com to the receiver and engage in n OT instances with input (t_i, s_i) for instance i.
2. Receiver (Input $\mathbf{x} \in \mathbb{F}^t$):
   - Start the encode procedure $\mathsf{Encode}_{n,\rho}(\mathbf{x}, \mathcal{P})$ and obtain $(G, H, L, \mathbf{v})$. Interpolate a polynomial X through the points (β_i, v_i) for $i \in L$.
   - For each OT instance i, if $i \in L$, set $\mathtt{choice}_i = 0$, otherwise set $\mathtt{choice}_i = 1$.
   - Obtain ℓ values t_i and ρ values $\tilde{s}_i$. Compute $\tilde{e} = \mathsf{SS.Reconstruct}(\tilde{\mathbf{s}}_{|\neg L})$.
   - Send $\tilde{e}$ to the sender.
3. Sender: Check if $\tilde{e} = e$, if not abort. Send unv to the receiver.
4. Receiver: Check if $\mathsf{COM.Open}(\mathsf{com}, \mathsf{unv}, \tilde{e}) = 1$, abort if not. Send $(G, \mathbf{v})$ to the sender.
5. Sender: Compute $\tilde{w}_i = A(\beta_i) \cdot v_i + B(\beta_i) + t_i$ for $i \in \{1, \ldots, n\}$. Send $\tilde{\mathbf{w}} = (\tilde{w}_1, \ldots, \tilde{w}_n)$ to the receiver.
6. Receiver: Set $w_i = \tilde{w}_i - t_i$ for $i \in L$ and interpolate the degree $\ell - 1$ polynomial Y through the points (β_i, w_i) for $i \in L$. Draw $z_\mathsf{R} \in \mathbb{F} \setminus \mathcal{P}$ uniformly at random and send z_R to the sender.
7. Sender: Draw $z_\mathsf{S} \in \mathbb{F} \setminus \mathcal{P}$ uniformly at random and send $\big(A(z_\mathsf{R}), B(z_\mathsf{R}), z_\mathsf{S}\big)$ to the receiver.
8. Receiver:
   - Check if $A(z_\mathsf{R}) \cdot X(z_\mathsf{R}) + B(z_\mathsf{R}) = Y(z_\mathsf{R})$ and abort if not.
   - Send $(X(z_\mathsf{S}), Y(z_\mathsf{S}))$ to the sender and output $\mathbf{y} = H\mathbf{w}_{|L}$.
9. Sender: Check if $A(z_\mathsf{S}) \cdot X(z_\mathsf{S}) + B(z_\mathsf{S}) = Y(z_\mathsf{S})$ and abort if not.

Fig. 3. Actively secure realization of $\mathcal{F}_{\mathrm{OLE}}$ in the OT-hybrid model.

Theorem 3. *The protocol Π_{OLE} UC-realizes $\mathcal{F}_{\mathrm{OLE}}$ in the* OT*-hybrid model with computational security.*

Proof. **Corrupted sender:** In the following we present a simulator $\mathcal{S}_\mathsf{S}$ which provides a computationally indistinguishable simulation of Π_{OLE} to a malicious sender $\mathcal{A}_\mathsf{S}$ (cf. Fig. 4).

Simulator $\mathcal{S}_\mathsf{S}$

1. Let com^* be the message from $\mathcal{A}_\mathsf{S}$. Upon receiving input from $\mathcal{F}_{\text{OLE}}^{\text{t}}$, select a random value $\hat{\mathbf{x}} \in \mathbb{F}^t$ and compute $(\hat{G}, \hat{H}, \hat{L}, \hat{\mathbf{v}}) \leftarrow \mathsf{Encode}(\hat{\mathbf{x}})$. Further interpolate a polynomial $\hat{X}$ such that $\hat{X}(\beta_i) = \hat{v}_i$ for $i \in \hat{L}$.
2. Learn all of $\mathcal{A}_\mathsf{S}$'s inputs $(t_1^*, \ldots, t_n^*)$ and $\mathbf{s}^* = (s_1^*, \ldots, s_n^*)$ sent to the n OT instances.
   - Compute $\hat{e} \leftarrow \mathsf{SS.Reconstruct}(\mathbf{s}^*_{|\neg\hat{L}})$.
   - Send $\hat{e}$ to $\mathcal{A}_\mathsf{S}$.
3. Upon receiving unv^*, check if $\mathsf{COM.Open}(\mathsf{com}^*, \mathsf{unv}^*, \hat{e}) = 1$, if not abort. Send $(\hat{G}, \hat{\mathbf{v}})$ to $\mathcal{A}_\mathsf{S}$.
4. Upon receiving $\tilde{\mathbf{w}}^*$, do the following:
   - Compute $\hat{w}_i = \tilde{w}_i^* - t_i$ for all $i \in [n]$.
   - Interpolate the degree $\ell - 1$ polynomial $\hat{Y}$ such that $\hat{Y}(\beta_i) = \hat{w}_i$ for $i \in \hat{L}$.
   - Draw a random $\hat{z}_\mathsf{R} \in \mathbb{F} \setminus \mathcal{P}$ and send it to $\mathcal{A}_\mathsf{S}$.
5. Upon receiving $(A(\hat{z}_\mathsf{R})^*, B(\hat{z}_\mathsf{R})^*, z_\mathsf{S}^*)$, check if $A(\hat{z}_\mathsf{R})^* \cdot \hat{X}(\hat{z}_\mathsf{R}) + B(\hat{z}_\mathsf{R})^* = \hat{Y}(\hat{z}_\mathsf{R})$ and abort if not. Proceed as follows:
   - For all $i \notin \hat{L}$, set $\bar{a}_i = \frac{\hat{Y}(\beta_i) - \hat{w}_i}{\hat{X}(\beta_i) - \hat{v}_i}$.
   - Interpret the ρ points $\bar{a}_i$ as a Reed-Solomon encoded codeword. Decode $(\bar{a}_1, \ldots, \bar{a}_\rho)$ into $(\tilde{a}_1, \ldots, \tilde{a}_\rho)$ and abort if the codeword $(\bar{a}_1, \ldots, \bar{a}_\rho)$ contains more than κ errors. Interpolate a polynomial $\hat{A}$ such that $\hat{A}(\beta_i) = \tilde{a}_i$. Obtain $\hat{a}_1, \ldots, \hat{a}_t$ by evaluating $\hat{A}$ in $\alpha_1, \ldots, \alpha_t$.
   - Compute $\hat{b}_i = \hat{Y}(\beta_i) - \hat{X}(\beta_i)\hat{A}(\beta_i)$ for $i \in \hat{L}$. Interpolate a polynomial $\hat{B}$ such that $\hat{B}(\beta_i) = \hat{b}_i$. Obtain $\hat{b}_1, \ldots, \hat{b}_t$ by evaluating $\hat{B}$ in $\alpha_1, \ldots, \alpha_t$.
6. Set $\hat{\mathbf{a}} = (\hat{a}_1, \ldots, \hat{a}_t)$ and $\hat{\mathbf{b}} = (\hat{b}_1, \ldots, \hat{b}_t)$. Send $(\mathtt{inptS}, \hat{\mathbf{a}}, \hat{\mathbf{b}})$ to $\mathcal{F}_{\text{OLE}}^{\text{t}}$. Proceed with the simulation according to protocol.

Fig. 4. Simulator against a corrupted sender in Π_{OLE}.

The main idea behind the extraction is the following. Since $\mathcal{S}_\mathsf{S}$ learns all inputs into the OTs, it can use the now available noisy elements $\hat{v}_i$ with $i \notin L$ to learn the input $\mathbf{a}$. The noiseless $\hat{v}_i, i \in L$ can be extrapolated to the noisy positions via a polynomial $\hat{Y}$ ($\hat{w}_i$ values imply a degree $\ell - 1$ polynomial for $i \in L$, and the receiver always learns ℓ values).

Ignoring for the moment that $\mathcal{A}_\mathsf{S}$ might provide inconsistent inputs, the simulator now knows two values for each position $\beta_i, i \notin L$: $\hat{w}_i = a_i \cdot \hat{v}_i + b_i$ and $\hat{Y}(\beta_i)$. Therefore, assuming that $\mathcal{A}_\mathsf{S}$ is honest, by computing $\hat{w}_i - \hat{Y}(\beta_i)$ the simulator gets $a_i \cdot \hat{v}_i + b_i - a_i \cdot \hat{x}_i + b_i = a_i(\hat{v}_i - \hat{x}_i)$, where $\hat{x}_i$ is the value that his input $\hat{\mathbf{x}} \in \mathbb{F}^t$ would imply according to the encoding $\hat{\mathbf{v}}_{|L}$ on position β_i. Since the simulator knows $\hat{v}_i$ and $\hat{x}_i$, it can simply compute a_i. From $\frac{\ell-1}{2} + 1$ of these points it can then compute the degree-$\frac{\ell-1}{2}$ polynomial A. From $Y = AZ + B$, it can then compute B and therefore the b_is. For this to work we only need $\frac{\ell-1}{2} + 1$

points. Therefore, if the corrupted sender sends incorrect values in at most κ positions $i \notin L$ and $\frac{\ell-1}{2} + 2\kappa < n$ there are still enough points to at least define a *correct* A and therefore also a *correct* $B = Y - AX$.

We now show that for every PPT environment $\mathcal{Z}$, the two distributions $\mathsf{Real}_{\Pi_{\mathrm{OLE}}}^{\mathcal{A}_\mathsf{S}}(\mathcal{Z})$ and $\mathsf{Ideal}_{\mathcal{F}_{\mathrm{OLE}}}^{\mathcal{S}_\mathsf{S}}(\mathcal{Z})$ are indistinguishable. Consider the following series of hybrid experiments.

Hybrid 0: This is the real protocol.
Hybrid 1: Identical to Hybrid 0, except that $\mathcal{S}_1$ extracts all inputs (s_i, t_i) input into the OT's by $\mathcal{A}_\mathsf{S}$.
Hybrid 2: Identical to Hybrid 1, except that $\mathcal{S}_2$ extracts the values $\bar{a}$ as shown in Fig. 4 and aborts if the check in Step 8 is passed, but $\bar{a}_1, \ldots, \bar{a}_\rho$ has more than κ errors.
Hybrid 3: Identical to Hybrid 2, except that $\mathcal{S}_3$ encodes a random value $\hat{\mathbf{x}}$ as its input.

The indistinguishability of Hybrids 0 and 1 is immediate. We show that Hybrid 1 and Hybrid 2 are computationally indistinguishable in Lemma 1, and then we prove indistinguishability of Hybrid 2 and Hybrid 3 in Lemma 2.

Lemma 1. *Hybrids 1 and 2 are computationally indistinguishable from $\mathcal{Z}$'s view given that Assumption 2 holds.*

Proof. In order to prove the lemma, we have to show the following two statements.

1. $\mathcal{S}_2$ correctly extracts the input $\hat{\mathbf{a}}, \hat{\mathbf{b}}$, if there are less than κ errors in noisy positions.
2. The probability that $\mathcal{S}_2$ aborts due to more than κ errors in noisy positions is negligible in κ.

There are two ways in which $\mathcal{A}_\mathsf{S}$ can cheat and prevent the correct extraction: (1) it uses an inconsistent input for a noiseless value $\hat{v}_i, i \in L$ which leads to a wrong polynomial $\hat{Y}$ (and also an incorrect $\bar{a}_i$); (2) it uses an inconsistent input for a noisy value $\hat{v}_i, i \notin L$, which leads to incorrectly extracted values $\bar{a}_i$.

In case (1), the honest party will abort due to the check in Step 8 with overwhelming probability. It has to hold that $A(z)^* \cdot \hat{X}(z) + B(z)^* = \hat{Y}(z)$ for a uniformly chosen z. From Assumption 2 it follows that $\hat{X}$ (and thus $\hat{Y}$) are unknown to $\mathcal{Z}$, as they would be unconditionally hidden by a completely random vector. By the fundamental theorem of algebra there are at most $\deg(\hat{Y}) = \ell - 1$ possible values z for which $A(z)^* \cdot \hat{X}(z) + B(z)^* = \hat{Y}(z)$ for incorrect $A(z)^*, B(z)^*$. Since z_R is chosen uniformly at random from $\mathbb{F}$, the probability that the check succeeds with incorrect $A(z)^*, B(z)^*$ is thus upper bounded by $\frac{\ell-1}{|\mathbb{F}|}$, which is negligible in the security parameter. This means that the check in Step 8 ensures that *all* the values $\hat{w}_i$ for $i \in L$ are correct.

For case (2), we first argue that the extraction also succeeds if $\mathcal{A}_\mathsf{S}$ adds less than κ errors in noisy positions (the simulator will abort if more than κ

errors occur). By the choice of parameters it holds that $\rho > 3\kappa = 6\ell$, and the simulator learns ρ values a_i that are supposed to represent a degree $\frac{\ell-1}{2}$ polynomial. Applying a standard Reed-Solomon decoder then yields the correct values a_i, i.e., if less than κ errors occur, $\mathcal{S}_\mathsf{S}$ extracts the correct $\mathbf{a} \in \mathbb{F}^t$ (and thus also the correct $\mathbf{b} \in \mathbb{F}^t$).

This shows that as long as there are less than κ errors in noisy positions, the extracted values are correct.

We claim that a $\mathcal{Z}$ that places more than κ errors in noisy positions breaks Assumption 2. The scenario of $\mathcal{Z}$ in the simulation is identical to the game $\mathcal{G}_{\mathsf{ident}}$: $\mathcal{Z}$ gets an encoding $\mathbf{v} \leftarrow \mathsf{Encode}_{n,\rho}(\mathbf{x})$ with ρ noisy positions and has to output a set of positions $Q \subseteq [n]$ such that $Q \subseteq R$ and $|Q| \geq \kappa$.

As discussed in Sect. 3, we can assume that $\mathsf{Encode}_{n,n/2}$ yields encodings that are indistinguishable from $\mathsf{Encode}_{n,n}$, i.e., truly random strings. In order to meet the requirements of Theorem 2, it therefore has to hold that $\sigma = n\frac{\rho-\kappa}{n-\kappa} \geq \frac{n}{2}$. Thus, we get that ρ has to be larger than $\frac{n+\kappa}{2}$, which by our choice of parameters is the case. Thus the claim directly follows from Theorem 2. □

Lemma 2. *Hybrids 2 and 3 are computationally indistinguishable from $\mathcal{Z}$'s view given that Assumption 2 holds and* COM *is a UC commitment scheme.*

Proof. Assume that there exists a PPT $\mathcal{Z}$ that distinguishes Hybrids 2 and 3 with non-negligible probability ϵ. We will show that $\mathcal{Z}$ breaks Assumption 2 with non-negligible probability.

We have to consider all the messages that $\mathcal{A}_\mathsf{S}$ receives during a protocol run. First note that $\mathcal{S}_\mathsf{S}$ (resp. R) outputs either e or aborts in Step 4. Assume for the sake of contradiction that $\mathcal{A}_\mathsf{S}$ manages to create two secret sharings $s_{1,1}, \ldots, s_{1,n}$ and $s_{2,1}, \ldots, s_{2,n}$ for values e, e' such that R outputs both of e or e' with non-negligible probability ϵ without aborting depending on the set L and L', respectively. Then we create an adversary $\mathcal{B}$ from $\mathcal{Z}$ that breaks the binding property of COM. $\mathcal{B}$ simulates the protocol and learns all values s_i^*, then draws two uniformly random sets L, L'. $\mathcal{B}$ samples via L and L' two subsets of secret sharings that reconstruct to e and e', respectively, with non-negligible probability. It must hold for both values that $\mathsf{COM.Open}(\mathsf{com}, \mathsf{unv}, e) = \mathsf{COM.Open}(\mathsf{com}, \mathsf{unv}, e') = 1$, otherwise $\mathcal{B}$ aborts as the real R would. Since $\mathcal{A}_\mathsf{S}$ achieves that R outputs e or e' with non-negligible probability, $\mathcal{B}$ outputs com, c, c' with non-negligible probability to the binding experiment and thereby breaks the binding property of COM.

The next message he receives is the encoding. Recall that the choice bits into the OTs are derived from the set L of the encoding, i.e., a cheating $\mathcal{A}_\mathsf{S}$ might try to use inconsistent inputs (e.g., incorrect s_i values in positions that are supposedly not in L) in the OT such that R aborts depending on the set L. However, $\mathcal{A}_\mathsf{S}$ has to cheat before knowing the encoding $\mathbf{v}$ and as shown above always learns the same e, thus he can obtain at most 1 bit of leakage, namely whether the cheating was detected or not. We will now show that the leakage in Step 4 does not help $\mathcal{Z}$ to distinguish. The situation for a malicious $\mathcal{Z}$ is identical to game $\mathcal{G}_{\mathsf{leak}}$. First, $\mathcal{A}_\mathsf{S}$ has to decide on a set of values which he believes are not in L. Then he is informed (via a successful check) that his guess was correct, and given the encoding. Now he has to decide whether he is given a random encoding or not. We can directly apply Theorem 1, given that $\rho \geq \frac{2}{3}n$, and get that $\mathcal{Z}$'s distinguishing advantage is negligible.

After learning $\mathbf{v}$, $\mathcal{A}_\mathsf{S}$ has to compute the values $\mathbf{w}$, which are checked in Step 8. By cheating in noisy positions, Step 8 will succeed, but $\mathcal{A}_\mathsf{S}$ learns some noisy positions by learning the bit whether the check succeeded. This case is more involved than the above step, since now $\mathcal{A}_\mathsf{S}$ can decide on the set Q after seeing the encoding $\mathbf{v}$. We argue that the distinguishing advantage of $\mathcal{Z}$ remains negligible. It is obvious that $\mathcal{A}_\mathsf{S}$ can always find $O(\log \kappa)$ noisy positions with polynomial probability simply by guessing. However, Theorem 2 guarantees that in this scenario $\mathcal{A}_\mathsf{S}$ cannot find more than $O(\log \kappa)$ noisy positions, if $\mathsf{Y}^\sigma \approx \mathsf{Y}^n$ for $\sigma = \frac{n\kappa}{n-\rho-\kappa}$. From Theorem 1 we know that if $Q = O(\log \kappa)$ and $\sigma > \frac{2}{3}n$, then $\mathsf{Y}^\sigma \approx \mathsf{Y}^n$. Combined, we have that for $\rho = n - \frac{\kappa}{2}$, $\mathcal{A}_\mathsf{S}$ cannot find more than $O(\log n)$ noisy positions and the distinguishing advantage of $\mathcal{Z}$ is negligible. This concludes the proof. $\square$

Corrupted receiver. In the following we present a simulator $\mathcal{S}_\mathsf{R}$ which provides a statistically indistinguishable simulation of Π_{OLE} to a malicious receiver $\mathcal{A}_\mathsf{R}$ (cf. Fig. 5). Conceptually the simulation is straight forward. The simulator learns all choice bits and thus can reconstruct the set L, which is sufficient to decode the codeword $\mathbf{v}$. Knowing X, $\mathcal{S}_\mathsf{R}$ can easily derive consistent inputs A, B. Care has to be taken since $\mathcal{A}_\mathsf{R}$ obtains one additional pair of values related to the polynomials A and B, thus he can tamper with the extraction. In a little more detail, he obtains one more value than necessary to reconstruct Y and can therefore play both with the degree of his input as well as with the correctness of L and $\mathbf{v}$. We describe and analyze a subtle attack in Lemma 4, which makes the analysis a bit more complex.

We now show the indistinguishability of the simulation in a series of hybrid experiments. For every PPT environment $\mathcal{Z}$, the two distributions $\mathsf{Real}_{\Pi_{\mathrm{OLE}}}^{\mathcal{A}_\mathsf{S}}(\mathcal{Z})$ and $\mathsf{Ideal}_{\mathcal{F}_{\mathrm{OLE}}}^{\mathcal{S}_\mathsf{S}}(\mathcal{Z})$ are indistinguishable.

Hybrid 0: This is the real protocol.

Hybrid 1: Identical to Hybrid 0, except that $\mathcal{S}_1$ extracts all inputs $\mathtt{choice}_i$ input into OT by $\mathcal{A}_\mathsf{R}$.

Hybrid 2: Identical to Hybrid 1, except that $\mathcal{S}_2$ aborts if $\mathcal{A}_\mathsf{R}$ passes the check in Step 3, although he selects less than ρ values s_i.

Hybrid 3: Identical to Hybrid 2, except that $\mathcal{S}_3$ reconstructs $\hat{X}$ as shown in Fig. 5 and aborts if $\hat{Y}(\hat{z}_\mathsf{S}) \neq Y^*(\hat{z}_\mathsf{S})$, $\hat{X}(\hat{z}_\mathsf{S}) \neq X^*(\hat{z}_\mathsf{S})$ or $\hat{Y} = R$.

Indistinguishability of Hybrids 0 and 1 is trivial. We show the indistinguishability of Hybrids 1 and 2 in Lemma 3, based on the privacy of the secret sharing and the hiding property of the commitment. In Lemma 4 we show that we can always extract the correct input of $\mathcal{A}_\mathsf{R}$ and thus Hybrid 2 and Hybrid 3 are statistically indistinguishable.

Simulator $\mathcal{S}_{\mathsf{R}}$

1. Upon receiving a message `input` from $\mathcal{F}^{\mathrm{t}}_{\mathrm{OLE}}$, simulate the first part of Π_{OLE} with random inputs.
   - Draw a uniformly random vector $\hat{\mathbf{t}} \in \mathbb{F}^n$ and a random value $\hat{e} \in \mathbb{F}$.
   - Compute $\hat{\mathbf{s}} \leftarrow \mathsf{SS.Share}(\hat{e})$ and $(\widehat{\mathsf{com}}, \widehat{\mathsf{unv}}) \leftarrow \mathsf{COM.Commit}(\hat{e})$.
   - Send $\widehat{\mathsf{com}}$ to $\mathcal{A}_{\mathsf{R}}$ and engage in n OT instances with input $(\hat{t_i}, \hat{s_i})$ in OT i.
2. Learn all choice bits $(\texttt{choice}^*_1, \ldots, \texttt{choice}^*_n)$ of $\mathcal{A}_{\mathsf{R}}$ from the n OT instances. Reconstruct $\hat{L}$ as follows: for each $i \in [n]$, if $\texttt{choice}^*_i = 0$ then $i \in \hat{L}$.
3. Upon receiving e^*, check if $\hat{e} = e^*$, otherwise abort. Send $\widehat{\mathsf{unv}}$ to $\mathcal{A}_{\mathsf{R}}$.
4. Upon receiving $(G^*, \mathbf{v}^*)$ from $\mathcal{A}_{\mathsf{R}}$, proceed as follows.
   (a) Let $\deg(P_{\hat{L}})$ denote the degree of the polynomial defined by $\mathbf{v}_{|\hat{L}}$.
      - If $|\hat{L}| = \ell - 1$, interpolate the polynomial $P_{\hat{L}}$ defined over $\mathbf{v}_{|\hat{L}}$. If $\deg(P_{\hat{L}}) \leq \frac{\ell-1}{2}$, set $\hat{X} = P_{\hat{L}}$.
      - If $|\hat{L}| = \ell$, interpolate the polynomial $P_{\hat{L}}$ defined over $\mathbf{v}_{|\hat{L}}$. If $\deg(P_{\hat{L}}) \leq \frac{\ell-1}{2}$, set $\hat{X} = P_{\hat{L}}$. If $\deg(P_{\hat{L}}) > \frac{\ell+1}{2}$, try for all $\hat{i} \in \hat{L}$ if it holds that for $\hat{L}' = \hat{L} \setminus \hat{i}$, $\deg(P_{\hat{L}'}) \leq \frac{\ell-1}{2}$. If such an $\hat{i}$ exists, set $\hat{X} = P_{\hat{L}'}$ and $\hat{L} = \hat{L}'$.

   (b) Compute $\hat{x}_i = \hat{X}(\alpha_i), i \in [t]$ and send $(\texttt{inputR}, \hat{\mathbf{x}})$ to $\mathcal{F}^{\mathrm{t}}_{\mathrm{OLE}}$. Let $(\texttt{output}, \hat{\mathbf{y}})$ be the result. Pick a random polynomial $\hat{Y}$ such that $\deg(\hat{Y}) = \deg(\hat{X}) + \frac{\ell-1}{2}$ and $\hat{Y}(\alpha_i) = \hat{y}_i, i \in [t]$. If no $\hat{X}$ was extracted in Step 4a, set $\hat{Y}$ to be a random degree $\ell - 1$ polynomial R.

   (c) For $i \in \hat{L}$, set $\hat{w}_i = \hat{Y}(\beta_i) + t_i$, otherwise pick a uniform $\hat{w}_i$ and send $\hat{\mathbf{w}}$ to $\mathcal{A}_{\mathsf{R}}$.
5. Upon receiving z^*_{R}, draw $\hat{z}_{\mathsf{S}} \in \mathbb{F}$ and proceed as follows:
   - If $\hat{Y} \neq R$, compute $\hat{X}(z^*_{\mathsf{R}}), \hat{Y}(z^*_{\mathsf{R}})$ and sample a random $\hat{b} \in \mathbb{F}$. Set $\hat{a} = \frac{\hat{Y}(z^*_{\mathsf{R}}) - \hat{b}}{\hat{X}(z^*_{\mathsf{R}})}$ and send $(\hat{a}, \hat{b}, \hat{z}_{\mathsf{S}})$ to $\mathcal{A}_{\mathsf{R}}$.
   - If $\hat{Y} = R$, pick random $\hat{a}, \hat{b} \in \mathbb{F}$ and send $(\hat{a}, \hat{b}, \hat{z}_{\mathsf{S}})$ to $\mathcal{A}_{\mathsf{R}}$.
6. Upon receiving $(X^*(\hat{z}_{\mathsf{S}}), Y^*(\hat{z}_{\mathsf{S}}))$, proceed as follows:
   - If $\hat{Y} \neq R$, check if $Y^*(z_{\mathsf{S}}) = \hat{Y}(z_{\mathsf{S}})$ and $X^*(z_{\mathsf{S}}) = \hat{X}(z_{\mathsf{S}})$ and abort if not.
   - If $\hat{Y} = R$, abort.

Fig. 5. Simulator against a corrupted receiver in Π_{OLE}.

Lemma 3. *Hybrids 1 and 2 are statistically indistinguishable from $\mathcal{Z}$'s view given that* SS *is a perfectly private secret sharing and* COM *is a statistically hiding commitment scheme.*

Proof. Assume for the sake of contradiction that there exists an environment $\mathcal{Z}$ that distinguishes the hybrids, i.e., $\mathcal{Z}$ has to make $\mathcal{S}_2$ abort with non-negligible probability ε. We will construct from $\mathcal{Z}$ an adversary $\mathcal{B}$ that breaks the hiding property of COM with non-negligible probability. $\mathcal{B}$ simulates the protocol exactly like $\mathcal{S}_2$, but creates a secret sharing of a random r and picks two random e, e', which he sends to the hiding experiment. The hiding experiment returns a commitment com on one of these values. Then $\mathcal{B}$ integrates the commitment and secret sharing into the simulation and checks whether $\mathcal{Z}$ inputs less than ρ values $\texttt{choice}_i = 1$ into OT, otherwise $\mathcal{B}$ aborts. Since SS is a perfectly private secret sharing and $\mathcal{Z}$ obtains less than ρ values s_i, these values leak nothing about r and the simulation of $\mathcal{B}$ is indistinguishable from $\mathcal{S}_2$'s simulation. Let now e^* be $\mathcal{Z}$'s answer in the simulated protocol. $\mathcal{B}$ simply forwards e^* to the hiding experiment. Since it has to hold that $e^* = e$ or $e^* = e'$ with non-negligible probability ε (otherwise the check in Step 3 fails), $\mathcal{B}$ breaks the hiding property of COM with the same probability. From this contradiction it follows that $\mathcal{A}_\mathsf{R}$ learns at most ℓ values t_i through OT. □

Lemma 4. *Hybrids 2 and 3 are statistically indistinguishable from $\mathcal{Z}$'s view.*

Proof. In order to distinguish Hybrids 2 and 3, $\mathcal{Z}$ must pass the check in Step 9, even though it holds that $\mathcal{S}_3$ picked a random polynomial R (allowing to distinguish the simulation from the real protocol). First note that the result $\mathbf{w}$ always defines a polynomial of degree $\ell - 1$ if $\mathcal{A}_\mathsf{R}$'s input polynomial has degree less than $\frac{\ell-1}{2}$. As we know from Lemma 3, $\mathcal{A}_\mathsf{R}$ learns at most ℓ values through the OTs and then one additional pair (a, b) via the check in Step 9.

Before we look at the details of the extraction, let us first describe an generic adversarial strategy that we have to cover. The adversary gets 1 free query and might try to use this query to prevent extraction. Say he picks a polynomial of degree $\frac{\ell-1}{2}$, but only uses $\ell - 1$ values of L. In the choice phase, he selects a random index $i^* \notin L$ and sets $\mathtt{choice}_{i^*} = 0$, i.e., $\mathcal{S}_3$ will assume this index is also in L. Towards the same goal, $\mathcal{A}_\mathsf{R}$ can simply set the value v_i for a random index i to a random value. $\mathcal{S}$ will then extract a wrong polynomial (with degree greater than $\frac{\ell+1}{2}$), while $\mathcal{A}_\mathsf{R}$ can still reconstruct Y via the additional values. However, since $\mathcal{A}_\mathsf{R}$ can only add exactly 1 random element, $\mathcal{S}_3$ can identify the index by trying for each $i \in L$ whether the set $L' = L \setminus i$ defines a polynomial of degree $\frac{\ell-1}{2}$ over the v_i. Here it is essential that there are no two sets L_1, L_2 with $|L_1| = \ell - 1, |L_2| = \ell$ such that $L_1 \subset L_2$ and $\deg(P_{L_1}) = \frac{\ell-1}{2}, \deg(P_{L_2}) = \frac{\ell+1}{2}$, i.e., there is only one possible index i that can be removed. This follows from the fact that the polynomial $P = P_{L_2} - P_{L_1}$ has only $\frac{\ell+1}{2}$ roots, but L_1 and L_2 have to agree on $\ell - 1$ positions. If that scenario were possible, $\mathcal{S}_3$ would not be able to distinguish these cases.

Let in the following $\deg(P_{\hat{L}})$ denote the degree of the polynomial that is defined by the points v_i for $i \in \hat{L}$.

- $|\hat{L}| \leq \ell - 2$: $\mathcal{A}_\mathsf{R}$ obtains at most $\ell - 2 + 1$ points, but Y is of degree $\ell - 1$ and thus underspecified. Clearly $\mathcal{A}_\mathsf{R}$'s probability of answering the check in Step 9 with a correct $X^*(z_\mathsf{S}), Y^*(z_\mathsf{S})$ is negligible in $\mathbb{F}$. Since $\mathcal{S}_3$ aborts as well, Hybrids 2 and 3 are indistinguishable in this case.
- $|\hat{L}| = \ell - 1$: In this case it holds that $\hat{Y} = R$ only if $\deg(P_{\hat{L}}) \geq \frac{\ell+1}{2}$.
  - $\deg(P_{\hat{L}}) = \frac{\ell-1}{2}$: In this case $\mathcal{A}_\mathsf{R}$ can reconstruct Y and pass the check in Step 9, but $\mathcal{S}_3$ extracts the correct $\hat{X}$. From the argument above, there cannot exist another polynomial X' that fits with the set $\hat{L}$ and thus Hybrids 2 and 3 are indistinguishable.
  - $\deg(P_{\hat{L}}) = \frac{\ell+1}{2}$: In this case $\mathcal{A}_\mathsf{R}$ obtains $\ell - 1 + 1$ points, but the resulting Y is of degree $\frac{\ell-1}{2} + \frac{\ell+1}{2} = \ell$, i.e., $\mathcal{A}_\mathsf{R}$ needs $\ell + 1$ points to reconstruct Y. By the same argument as above, Hybrids 2 and 3 are indistinguishable.

- $\deg(P_{\hat{L}}) > \frac{\ell+1}{2}$: In this case $\mathcal{A}_{\mathsf{R}}$ can behave as described above, i.e., add a random i to the set $\hat{L}$ and thereby artificially increase $\deg(P_{\hat{L}})$. But since $|\hat{L}| = \ell - 1$, removing an additional value from $\hat{L}$ leads to the case $|\hat{L}| \leq \ell - 2$ and thus indistinguishability of Hybrids 2 and 3.

– $|\hat{L}| = \ell$: In this case it holds that $\hat{Y} = R$ only if $\deg(P_{\hat{L}}) > \frac{\ell+1}{2}$ and no index i can be identified to reduce $\deg(P_{\hat{L}})$ to $\frac{\ell-1}{2}$.
  - $\deg(P_{\hat{L}}) = \frac{\ell-1}{2}$: In this case $\mathcal{A}_{\mathsf{R}}$ can reconstruct Y and pass the check in Step 9, but $\mathcal{S}_3$ extracts the correct $\hat{X}$.
  - $\deg(P_{\hat{L}}) = \frac{\ell+1}{2}$: In this case $\mathcal{A}_{\mathsf{R}}$ obtains $\ell + 1$ points, and the resulting Y is of degree $\frac{\ell-1}{2} + \frac{\ell+1}{2} = \ell$. Thus $\mathcal{A}_{\mathsf{R}}$ can reconstruct Y and pass the check, but $\mathcal{S}_3$ extracts the correct $\hat{X}$.
  - $\deg(P_{\hat{L}}) > \frac{\ell+1}{2}$: In this case $\mathcal{A}_{\mathsf{R}}$ can behave as described above, i.e., add a random i to the set $\hat{L}$ and thereby artificially increase $\deg(P_{\hat{L}})$. Removing an additional value from $\hat{L}$ leads to the case $|\hat{L}| = \ell - 1$, i.e., $\mathcal{S}_3$ will simulate correctly. Otherwise, $\mathcal{S}_3$ will abort, but $\mathcal{A}_{\mathsf{R}}$ cannot reconstruct Y and thus fails the check in Step 9.

– $|\hat{L}| > \ell$: $\mathcal{S}_3$ aborts, and from Lemma 3 it follows that Hybrids 2 and 3 are indistinguishable.

The correctness of the simulation follows from the fact that either $\mathcal{S}_3$ extracts the correct input $\hat{X}$, or the check in Step 9 fails with overwhelming probability, in which case $\hat{X} = R$. Thus, the event that $\mathcal{Z}$ can provoke an abort is negligible, i.e., Hybrids 2 and 3 are indistinguishable. □

This concludes the proof. □

6 Efficient Oblivious Polynomial Evaluation

The ideal functionality $\mathcal{F}_{\mathrm{OPE}}$ for OPE is depicted in Fig. 6. It allows the sender S to input a polynomial P and the receiver R to input $\alpha \in \mathbb{F}$. In the remainder of this section we will establish the following theorem.

Functionality $\mathcal{F}_{\mathrm{OPE}}$

1. Upon receiving a message $(\texttt{inputS}, P)$ from S where $P \in \mathbb{F}[X]$, verify that there is no stored tuple, else ignore that message. Store P and send a message $(\texttt{input})$ to $\mathcal{A}$.
2. Upon receiving a message $(\texttt{inputR}, \alpha)$ from R with $\alpha \in \mathbb{F}$, verify that there is no stored tuple, else ignore that message. Store α and send a message $(\texttt{input})$ to $\mathcal{A}$.
3. Upon receiving a message $(\texttt{deliver}, \mathsf{S})$ from $\mathcal{A}$, check if both P and α are stored, else ignore that message. Send $(\texttt{delivered})$ to S.
4. Upon receiving a message $(\texttt{deliver}, \mathsf{R})$ from $\mathcal{A}$, check if both P and α are stored, else ignore that message. Send $(\texttt{output}, P(\alpha))$ to R.

Fig. 6. Ideal functionality for an oblivious polynomial evaluation.

Theorem 4. *There exists a (constant-round) protocol Π_{OPE} that UC-realizes $\mathcal{F}_{\mathrm{OPE}}$ with unconditional security in the $\mathcal{F}_{\mathrm{OLE}}^{\mathrm{t}}$-hybrid model. In particular, for a polynomial P of degree d, $t = d + 2$.*

Our roadmap is as follows. We first show how to reduce $\mathcal{F}_{\mathrm{OPE}}$ to an intermediate OLE-based functionality $\mathcal{F}_{\mathrm{OLE}}^{\mathrm{t},1}$. After establishing this we present an efficient reduction of $\mathcal{F}_{\mathrm{OLE}}^{\mathrm{t},1}$ to $\mathcal{F}_{\mathrm{OLE}}^{\mathrm{t}}$ (or $\mathcal{F}_{\mathrm{OLE}}$).

We follow the generic idea of Naor and Pinkas [30] of using the linearization technique from [17] to construct an oblivious polynomial evaluation protocol. They decompose a polynomial P of degree d into d *linear* functions. These functions can then be evaluated using our OLE with input α for each of the functions, and the receiver can reconstruct the value $P(\alpha)$. We state the lemma here, a proof can be found in [30] and the full version of this paper [15].

Lemma 3 [17]. *For every polynomial P of degree d, there exist d linear polynomials $P_1, \ldots, P_d$, such that an OPE of P can be reduced to a parallel execution of an OLE of each of $P_1, \ldots, P_d$, where all the linear polynomials are evaluated at the same point.*

In the semi-honest case, this approach directly works with the $\mathcal{F}_{\mathrm{OLE}}^{\mathrm{t}}$ for $t = d$. But unlike the construction of [30], our batch-OLE does not enforce the receiver to use the same input α in all of the OLEs. Therefore we cannot use the reduction of [30] that shows malicious security against a receiver. In particular, a malicious receiver might learn some non-trivial linear combinations of the coefficients of P.

Functionality $\mathcal{F}_{\mathrm{OLE}}^{\mathrm{t},1}$

1. Upon receiving a message $(\texttt{inputS}, \mathbf{a}, \mathbf{b})$ from S with $\mathbf{a}, \mathbf{b} \in \mathbb{F}^t$, verify that there is no stored tuple, else ignore that message. Store $\mathbf{a}$ and $\mathbf{b}$ and send a message $(\texttt{input})$ to $\mathcal{A}$.
2. Upon receiving a message $(\texttt{inputR}, x)$ from R with $x \in \mathbb{F}$, verify that there is no stored tuple, else ignore that message. Store x and send a message $(\texttt{input})$ to $\mathcal{A}$.
3. Upon receiving a message $(\texttt{deliver}, \mathsf{S})$ from $\mathcal{A}$, check if both $\mathbf{a}, \mathbf{b}$ and x are stored, else ignore that message. Send $(\texttt{delivered})$ to S.
4. Upon receiving a message $(\texttt{deliver}, \mathsf{R})$ from $\mathcal{A}$, check if both $\mathbf{a}, \mathbf{b}$ and x are stored, else ignore that message. Set $y_i = a_i \cdot x + b_i$ for $i \in [t]$ and send $(\texttt{output}, \mathbf{y})$ to R.

Fig. 7. Ideal functionality for a $(t, 1)$-oblivious linear function evaluation.

Reducing $\mathcal{F}_{\mathrm{OPE}}$ to $\mathcal{F}_{\mathrm{OLE}}^{\mathrm{t},1}$. As a first step we reduce OPE to a variant of OLE where the receiver has only one input $x \in \mathbb{F}$, while the sender inputs two vectors $\mathbf{a}, \mathbf{b}$. This is depicted in Fig. 7.

The reduction of $\mathcal{F}_{\mathrm{OPE}}$ to $\mathcal{F}_{\mathrm{OLE}}^{\mathrm{t},1}$ is straightforward, given Lemma 3. The sender decomposes his polynomial P into d linear functions $f_1, \ldots, f_d$ with coefficients (a_i, b_i) and inputs these into $\mathcal{F}_{\mathrm{OLE}}^{\mathrm{d},1}$. The receiver chooses his input α

and obtains d linear evaluations, from which he can reconstruct $P(\alpha)$. The number of OLEs required is only dependent on the realization of $\mathcal{F}_{\mathrm{OLE}}^{\mathrm{d},1}$. A formal description of the protocol is given in Fig. 8.

Lemma 4. *The protocol Π_{OPE} UC-realizes $\mathcal{F}_{\mathrm{OPE}}$ in the $\mathcal{F}_{\mathrm{OLE}}^{\mathrm{d},1}$-hybrid model with unconditional security.*

Proof. The security of Π_{OPE} is immediate: the simulator simulates $\mathcal{F}_{\mathrm{OLE}}^{\mathrm{d},1}$ and learns all inputs, which it simply forwards to $\mathcal{F}_{\mathrm{OPE}}$ (and reconstructs if necessary). The correctness of the decomposition of P follows from Lemma 3. □

Note that by taking our approach, we also remove the need for the stronger assumption of [30], while having a comparable efficiency in the resulting protocol.

Protocol Π_{OPE}

1. Sender (Input $P \in \mathbb{F}[X]$ of degree d):
   - Generate d linear polynomials of the form $f_i(x) = a_i x + b_i$, $\forall i \in [d]$, where $a_i, b_i \in \mathbb{F}$ according to Lemma 3.
   - Construct $\mathbf{a}, \mathbf{b} \in \mathbb{F}^d$, such that $\mathbf{a} = \{a_1, \ldots, a_d\}$ and $\mathbf{b} = \{b_1, \ldots, b_d\}$.
   - Send $(\mathtt{inputS}, (\mathbf{a}, \mathbf{b}))$ to $\mathcal{F}_{\mathrm{OLE}}^{\mathrm{d},1}$.
2. Receiver (Input $\alpha \in \mathbb{F}$):
   - Send $(\mathtt{inputR}, \alpha)$ into $\mathcal{F}_{\mathrm{OLE}}^{\mathrm{d},1}$.
   - Obtain $(\mathtt{output}, \mathbf{y})$ from $\mathcal{F}_{\mathrm{OLE}}^{\mathrm{d},1}$.
   - Compute $P(\alpha)$ from $\mathbf{y} = f_1(\alpha), \ldots, f_d(\alpha)$. Output $P(\alpha)$.

Fig. 8. Reduction of $\mathcal{F}_{\mathrm{OPE}}$ to $\mathcal{F}_{\mathrm{OLE}}^{\mathrm{d},1}$.

Reducing $\mathcal{F}_{\mathrm{OLE}}^{\mathrm{t},1}$ to $\mathcal{F}_{\mathrm{OLE}}^{\mathrm{t+2}}$. As a second step, we need to realize $\mathcal{F}_{\mathrm{OLE}}^{\mathrm{t},1}$ from $\mathcal{F}_{\mathrm{OLE}}^{\mathrm{t}}$. Döttling et al. [11] describe a black-box protocol that realizes $\mathcal{F}_{\mathrm{OLE}}^{\mathrm{t},1}$ from $\mathcal{F}_{\mathrm{OLE}}$ (or our batch variant) with unconditional UC-security. The protocol has a constant multiplicative overhead of $2 + \varepsilon$ in the number of OLEs, and works for any field $\mathbb{F}$. While this protocol basically solves our problem, we propose a more efficient variant that makes essential use of the fact that we only consider a large field $\mathbb{F}$. Our new approach requires only two additional OLEs and thus has overhead $1 + \epsilon$.

Our solution for $\mathcal{F}_{\mathrm{OLE}}^{\mathrm{t},1}$ is as follows. Let $\mathbf{a}, \mathbf{b} \in \mathbb{F}^t$ be given as input to the sender. It now needs to choose one additional pair of inputs (a_{t+1}, b_{t+1}) such that $\sum_{i=1}^{t+1} a_i = 0$ and b_{t+1} is uniformly random in $\mathbb{F}$. The sender inputs $\mathbf{a}', \mathbf{b}' \in \mathbb{F}^{t+1}$ into $\mathcal{F}_{\mathrm{OLE}}^{\mathrm{t+1}}$, while the receiver inputs $\mathbf{x}' = (x, \ldots, x) \in \mathbb{F}^{t+1}$. Now the receiver locally computes $c = \sum_{i=1}^{t+1} y_i = \sum_{i=1}^{t+1} a_i x + \sum_{i=1}^{t+1} b_i = \sum_{i=1}^{t+1} b_i$ and sends a commitment to c to the sender. This commitment can also be based on OLE, even in such a way that we can use $\mathcal{F}_{\mathrm{OLE}}^{\mathrm{t+2}}$ by precomputing the commitment (a detailed description is given in the full version [15]). The sender answers with $c' = \sum_{i=1}^{t+1} b_i$, which the receiver can verify. This makes sure that the sender chose $\mathbf{a}'$ correctly, while c' itself does not give the receiver any new information. Now the receiver unveils, which shows the sender whether the receiver used the same x in each invocation. There is one small problem left: if the receiver cheated, he will be caught, be he might still learn some information about the sender's inputs that cannot be simulated. In order to solve this issue, we let $\mathbf{a}'$ and $\mathbf{b}'$ be uniformly random and then replace these with the inputs after the check succeeded. A detailed description of the protocol is given in Fig. 9.

Protocol $\Pi_{\mathrm{OLE}}^{\mathrm{t},1}$

Let COM be an OLE-based commitment.

1. Sender (Input $\mathbf{a}, \mathbf{b} \in \mathbb{F}^t$): Choose $\mathbf{u}, \mathbf{v} \in \mathbb{F}^{t+2}$ uniformly random such that $\sum_{i=1}^{t+1} u_i = 0$ and sends $(\mathtt{inputS}, (\mathbf{u}, \mathbf{v}))$ to $\mathcal{F}_{\mathrm{OLE}}^{\mathrm{t}+2}$. Store (u_{t+2}, v_{t+2}) as the auxiliary receiver inputs for COM.
2. Receiver (Input $x \in \mathbb{F}$):
   - Set $\mathbf{x} = (x, \ldots, x, w) \in \mathbb{F}^{t+2}$ with w random and send $(\mathtt{inputR}, \mathbf{x})$ into $\mathcal{F}_{\mathrm{OLE}}^{\mathrm{t}+2}$.
   - Obtain $(\mathtt{output}, \mathbf{z})$ from $\mathcal{F}_{\mathrm{OLE}}^{\mathrm{t}+2}$. Let $\bar{\mathbf{z}} = (z_1, \ldots, z_t)$.
   - Let (w, z_{t+2}) be the auxiliary sender input for COM. Compute $c = \sum_{i=1}^{t+1} z_i$, $(\mathsf{com}, \mathsf{unv}) \leftarrow \mathsf{COM}.\mathsf{Commit}(c)$ and send com to the sender.
3. Sender: Send $c' = \sum_{i=1}^{t+1} v_i$ to the receiver.
4. Receiver: Check if $c' = c$ and abort if not. Send unv to the sender.
5. Sender: Check if $\mathsf{COM}.\mathsf{Open}(\mathsf{com}, \mathsf{unv}, c') = 1$ and abort if not. Send $\mathbf{u}' = \mathbf{a} - \bar{\mathbf{u}}$ and $\mathbf{v}' = \mathbf{b} - \bar{\mathbf{v}}$ to the receiver, where $\bar{\mathbf{v}}$, $\bar{\mathbf{u}}$ contain the first t values of $\mathbf{u}$,$\mathbf{v}$.
6. Receiver: Compute $\mathbf{y} = \mathbf{u}'x + \mathbf{v}' + \bar{\mathbf{z}} = \mathbf{a}x + \mathbf{b}$ and output $\mathbf{y}$.

Fig. 9. Reduction of $\mathcal{F}_{\mathrm{OLE}}^{\mathrm{t},1}$ to $\mathcal{F}_{\mathrm{OLE}}^{\mathrm{t}+2}$.

Lemma 5. *The protocol $\Pi_{\mathrm{OLE}}^{\mathrm{t},1}$ UC-realizes $\mathcal{F}_{\mathrm{OLE}}^{\mathrm{t},1}$ in the $\mathcal{F}_{\mathrm{OLE}}^{\mathrm{t}+2}$-hybrid model with unconditional security.*

Proof. **Corrupted sender:** The simulator $\mathcal{S}_\mathsf{S}$ simulates $\mathcal{F}_{\mathrm{OLE}}^{\mathrm{t}+2}$ for the corrupted sender $\mathcal{A}_\mathsf{S}$. It extracts all the inputs, namely $\hat{\mathbf{u}}$ and $\hat{\mathbf{v}}$. We do not need to extract the commitment, which also uses $\mathcal{F}_{\mathrm{OLE}}^{\mathrm{t}+2}$. $\mathcal{S}_\mathsf{S}$ sends a commitment to $\hat{c} = \sum_{i=1}^{t+1} \hat{v}_i$ to the receiver. If it holds that $\sum_{i=1}^{t+1} u_i \neq 0$, but the check in Step 4 succeeds, $\mathcal{S}_\mathsf{S}$ aborts. Otherwise, it computes $\hat{\mathbf{a}} = \hat{\mathbf{u}}'^* + \hat{\mathbf{u}}$ and $\hat{\mathbf{b}} = \hat{\mathbf{v}}'^* + \hat{\mathbf{v}}$ and inputs the first t elements of each into $\mathcal{F}_{\mathrm{OLE}}^{\mathrm{t},1}$.

First note that if $\sum_{i=1}^{t+1} u_i \neq 0$, the commitment $\widehat{\mathsf{com}}$ contains an incorrect value. As long as the receiver always aborts in this case, the hiding property of COM guarantees indistinguishability of the simulation. So the only way that a malicious environment $\mathcal{Z}$ can distinguish the simulation from the real protocol is by forcing an abort. Note that if $\sum_{i=1}^{t+1} u_i = e \neq 0$, then c depends on x and is thus uniformly distributed, since

$$c = \sum_{i=1}^{t+1} z_i = \sum_{i=1}^{t+1} u_i x + \sum_{i=1}^{t+1} v_i = ex + c'.$$

Thus, the probability that $c' = c$ is negligible.

Corrupted receiver. The simulator $\mathcal{S}_\mathsf{R}$ against the corrupted receiver $\mathcal{A}_\mathsf{R}$ simulates $\mathcal{F}_{\mathrm{OLE}}^{t+2}$ and learns $\hat{\mathbf{x}}$. It chooses $\hat{\mathbf{u}}, \hat{\mathbf{v}} \in \mathbb{F}^{t+2}$ according to $\Pi_{\mathrm{OLE}}^{\mathrm{t,1}}$, and computes $\hat{\mathbf{z}} \in \mathbb{F}^{t+2}$, where $\hat{z}_i = \hat{u}_i \hat{x}_i + \hat{v}_i \; \forall i \in [1, t+2]$. $\mathcal{S}_\mathsf{R}$ sends $\hat{\mathbf{z}}$ to $\mathcal{A}_\mathsf{R}$. After receiving the commitment, $\mathcal{S}_\mathsf{R}$ sends $\hat{c}' = \sum_{i=1}^{t+1} \hat{z}_i$. It aborts if the commitment unveils correctly, even though $x_i \neq x_j$ for some $i, j \in [t+1]$. If that is not the case, it inputs $\hat{x}$ into $\mathcal{F}_{\mathrm{OLE}}^{\mathrm{t,1}}$ and obtains $\mathbf{y}$. $\mathcal{S}_\mathsf{R}$ picks $\hat{\mathbf{v}}' \in \mathbb{F}^t$ uniformly at random, sets $\hat{u}_i' = \frac{y_i - \hat{z}_i - \hat{v}_i'}{x} \forall i \in [t]$. It sends $\hat{\mathbf{u}}', \hat{\mathbf{v}}'$ to $\mathcal{A}_\mathsf{R}$.

For an honest receiver, the check in Step 5 always succeeds. A malicious $\mathcal{Z}$ can only distinguish between the simulation and the real protocol by producing a correct commitment on c, even though $x_i \neq x_j$ for some $i, j \in [t+1]$. Since the commitment is binding, $\mathcal{A}_\mathsf{R}$ must commit to some value c before seeing c'. Let w.l.o.g. $x_j = (x + e) \neq x$ for some j. Then we have

$$c = \sum_{i=1}^{t+1} z_i = \sum_{\substack{i=1 \\ i \neq j}}^{t+1} u_i x + u_j(x+e) + \sum_{i=1}^{t+1} v_i = \sum_{i=1}^{t+1} u_i x + u_j e + \sum_{i=1}^{t+1} v_i = c' + u_j e.$$

But this means that c' is uniformly distributed from $\mathcal{A}_\mathsf{R}$'s point of view, because u_j is chosen uniformly and unknown to $\mathcal{A}_\mathsf{R}$. As a consequence, the probability that $\mathcal{Z}$ can distinguish the simulation from the real protocol is negligible. □

Combining the results from this section we get that $\mathcal{F}_{\mathrm{OPE}}$ for a polynomial P of degree d requires $\mathcal{F}_{\mathrm{OLE}}^{\mathrm{d,1}}$, which in turn can be based on $\mathcal{F}_{\mathrm{OLE}}^{\mathrm{d+2}}$. This establishes Theorem 4.

Remark 1. It is possible to evaluate several polynomials in parallel with the batch-OLE functionality, given that t is chosen of appropriate size. Then, for each polynomial the above described protocol is carried out (including making sure that the receiver uses the same α in all OLEs relevant to the respective polynomial).

References

1. Applebaum, B., Ishai, Y., Kushilevitz, E.: How to garble arithmetic circuits. In: Ostrovsky, R. (ed.) 52nd FOCS, pp. 120–129. IEEE Computer Society Press, October 2011
2. Beerliová-Trubíniová, Z., Hirt, M.: Perfectly-secure MPC with linear communication complexity. In: Canetti, R. (ed.) TCC 2008. LNCS, vol. 4948, pp. 213–230. Springer, Heidelberg (2008). https://doi.org/10.1007/978-3-540-78524-8_13
3. Bleichenbacher, D., Nguyen, P.Q.: Noisy polynomial interpolation and noisy chinese remaindering. In: Preneel, B. (ed.) EUROCRYPT 2000. LNCS, vol. 1807, pp. 53–69. Springer, Heidelberg (2000). https://doi.org/10.1007/3-540-45539-6_4
4. Boneh, D.: Finding smooth integers in short intervals using CRT decoding. In: 32nd ACM STOC, pp. 265–272. ACM Press, May 2000
5. Canetti, R.: Universally composable security: a new paradigm for cryptographic protocols. In: 42nd FOCS, pp. 136–145. IEEE Computer Society Press, October 2001

6. Cascudo, I., Damgård, I., David, B., Döttling, N., Nielsen, J.B.: Rate-1, linear time and additively homomorphic UC commitments. In: Robshaw, M., Katz, J. (eds.) CRYPTO 2016. LNCS, vol. 9816, pp. 179–207. Springer, Heidelberg (2016). https://doi.org/10.1007/978-3-662-53015-3_7
7. Chang, Y.-C., Lu, C.-J.: Oblivious polynomial evaluation and oblivious neural learning. In: Boyd, C. (ed.) ASIACRYPT 2001. LNCS, vol. 2248, pp. 369–384. Springer, Heidelberg (2001). https://doi.org/10.1007/3-540-45682-1_22
8. Damgård, I., Pastro, V., Smart, N., Zakarias, S.: Multiparty computation from somewhat homomorphic encryption. In: Safavi-Naini, R., Canetti, R. (eds.) CRYPTO 2012. LNCS, vol. 7417, pp. 643–662. Springer, Heidelberg (2012). https://doi.org/10.1007/978-3-642-32009-5_38
9. David, B.M., Nishimaki, R., Ranellucci, S., Tapp, A.: Generalizing efficient multiparty computation. In: Lehmann, A., Wolf, S. (eds.) ICITS 2015. LNCS, vol. 9063, pp. 15–32. Springer, Cham (2015). https://doi.org/10.1007/978-3-319-17470-9_2
10. Döttling, N., Kraschewski, D., Müller-Quade, J.: David and Goliath oblivious affine function evaluation - asymptotically optimal building blocks for universally composable two-party computation from a single untrusted stateful tamper-proof hardware token. Cryptology ePrint Archive, Report 2012/135 (2012). http://eprint.iacr.org/2012/135
11. Döttling, N., Kraschewski, D., Müller-Quade, J.: Statistically secure linear-rate dimension extension for oblivious affine function evaluation. In: Smith, A. (ed.) ICITS 2012. LNCS, vol. 7412, pp. 111–128. Springer, Heidelberg (2012). https://doi.org/10.1007/978-3-642-32284-6_7
12. Franklin, M.K., Yung, M.: Communication complexity of secure computation (extended abstract). In: 24th ACM STOC, pp. 699–710. ACM Press, May 1992
13. Freedman, M.J., Ishai, Y., Pinkas, B., Reingold, O.: Keyword search and oblivious pseudorandom functions. In: Kilian, J. (ed.) TCC 2005. LNCS, vol. 3378, pp. 303–324. Springer, Heidelberg (2005). https://doi.org/10.1007/978-3-540-30576-7_17
14. Freedman, M.J., Nissim, K., Pinkas, B.: Efficient private matching and set intersection. In: Cachin, C., Camenisch, J.L. (eds.) EUROCRYPT 2004. LNCS, vol. 3027, pp. 1–19. Springer, Heidelberg (2004). https://doi.org/10.1007/978-3-540-24676-3_1
15. Ghosh, S., Nielsen, J.B., Nilges, T.: Maliciously secure oblivious linear function evaluation with constant overhead. IACR Cryptology ePrint Archive 2017, 409 (2017). http://eprint.iacr.org/2017/409
16. Gilboa, N.: Two party RSA key generation. In: Wiener, M.J. (ed.) CRYPTO 1999. LNCS, vol. 1666, pp. 116–129. Springer, Heidelberg (1999). https://doi.org/10.1007/3-540-48405-1_8
17. Gilboa, N.: Topics in private information retrieval. Ph.D. thesis, Thesis (Doctoral)-Technion - Israel Institute of Technology, Faculty of Computer Science, Haifa (2001)
18. Hazay, C.: Oblivious polynomial evaluation and secure set-intersection from algebraic PRFs. In: Dodis, Y., Nielsen, J.B. (eds.) TCC 2015. LNCS, vol. 9015, pp. 90–120. Springer, Heidelberg (2015). https://doi.org/10.1007/978-3-662-46497-7_4
19. Hazay, C., Lindell, Y.: Efficient oblivious polynomial evaluation with simulation-based security. Cryptology ePrint Archive, Report 2009/459 (2009). http://eprint.iacr.org/2009/459
20. Ishai, Y., Kilian, J., Nissim, K., Petrank, E.: Extending oblivious transfers efficiently. In: Boneh, D. (ed.) CRYPTO 2003. LNCS, vol. 2729, pp. 145–161. Springer, Heidelberg (2003). https://doi.org/10.1007/978-3-540-45146-4_9

21. Ishai, Y., Prabhakaran, M., Sahai, A.: Founding cryptography on oblivious transfer – efficiently. In: Wagner, D. (ed.) CRYPTO 2008. LNCS, vol. 5157, pp. 572–591. Springer, Heidelberg (2008). https://doi.org/10.1007/978-3-540-85174-5_32
22. Ishai, Y., Prabhakaran, M., Sahai, A.: Secure arithmetic computation with no honest majority. In: Reingold, O. (ed.) TCC 2009. LNCS, vol. 5444, pp. 294–314. Springer, Heidelberg (2009). https://doi.org/10.1007/978-3-642-00457-5_18
23. Katz, J.: Universally composable multi-party computation using tamper-proof hardware. In: Naor, M. (ed.) EUROCRYPT 2007. LNCS, vol. 4515, pp. 115–128. Springer, Heidelberg (2007). https://doi.org/10.1007/978-3-540-72540-4_7
24. Keller, M., Orsini, E., Scholl, P.: Actively secure OT extension with optimal overhead. In: Gennaro, R., Robshaw, M. (eds.) CRYPTO 2015. LNCS, vol. 9215, pp. 724–741. Springer, Heidelberg (2015). https://doi.org/10.1007/978-3-662-47989-6_35
25. Keller, M., Orsini, E., Scholl, P.: MASCOT: faster malicious arithmetic secure computation with oblivious transfer. In: Weippl, E.R., Katzenbeisser, S., Kruegel, C., Myers, A.C., Halevi, S. (eds.) ACM CCS 16, pp. 830–842. ACM Press, October 2016
26. Kiayias, A., Yung, M.: Cryptographic hardness based on the decoding of reed-solomon codes. IEEE Trans. Inf. Theory **54**(6), 2752–2769 (2008)
27. Kilian, J.: Founding cryptography on oblivious transfer. In: 20th ACM STOC, pp. 20–31. ACM Press, May 1988
28. Lindell, Y., Pinkas, B.: Privacy preserving data mining. In: Bellare, M. (ed.) CRYPTO 2000. LNCS, vol. 1880, pp. 36–54. Springer, Heidelberg (2000). https://doi.org/10.1007/3-540-44598-6_3
29. Naor, M., Pinkas, B.: Oblivious transfer and polynomial evaluation. In: 31st ACM STOC, pp. 245–254. ACM Press, May 1999
30. Naor, M., Pinkas, B.: Oblivious polynomial evaluation. SIAM J. Comput. **35**(5), 1254–1281 (2006)
31. Rabin, M.O.: How to exchange secrets with oblivious transfer. Technical report TR-81, Aiken Computation Lab, Harvard University (1981)
32. Shankar, B., Srinathan, K., Rangan, C.P.: Alternative protocols for generalized oblivious transfer. In: Rao, S., Chatterjee, M., Jayanti, P., Murthy, C.S.R., Saha, S.K. (eds.) ICDCN 2008. LNCS, vol. 4904, pp. 304–309. Springer, Heidelberg (2007). https://doi.org/10.1007/978-3-540-77444-0_31
33. Tonicelli, R., Nascimento, A.C.A., Dowsley, R., Müller-Quade, J., Imai, H., Hanaoka, G., Otsuka, A.: Information-theoretically secure oblivious polynomial evaluation in the commodity-based model. Int. J. Inf. Secur. **14**(1), 73–84 (2015). http://dx.doi.org/10.1007/s10207-014-0247-8
34. Zhu, H., Bao, F.: Augmented oblivious polynomial evaluation protocol and its applications. In: Vimercati, S.C., Syverson, P., Gollmann, D. (eds.) ESORICS 2005. LNCS, vol. 3679, pp. 222–230. Springer, Heidelberg (2005). https://doi.org/10.1007/11555827_13

Oblivious Hashing Revisited, and Applications to Asymptotically Efficient ORAM and OPRAM

T.-H. Hubert Chan[1(✉)], Yue Guo[2], Wei-Kai Lin[2], and Elaine Shi[2]

[1] The University of Hong Kong, Pokfulam, Hong Kong
hubert@cs.hku.hk
[2] Cornell University, Ithaca, USA
yg393@cornell.edu, wklin@cs.cornell.edu, elaine@cs.cornell.edu

Abstract. Oblivious RAM (ORAM) is a powerful cryptographic building block that allows a program to provably hide its access patterns to sensitive data. Since the original proposal of ORAM by Goldreich and Ostrovsky, numerous improvements have been made. To date, the best asymptotic overhead achievable for general block sizes is $O(\log^2 N/\log\log N)$, due to an elegant scheme by Kushilevitz et al., which in turn relies on the oblivious Cuckoo hashing scheme by Goodrich and Mitzenmacher.

In this paper, we make the following contributions: we first revisit the prior $O(\log^2 N/\log\log N)$-overhead ORAM result. We demonstrate the somewhat incompleteness of this prior result, due to the subtle incompleteness of a core building block, namely, Goodrich and Mitzenmacher's oblivious Cuckoo hashing scheme.

Even though we do show how to patch the prior result such that we can fully realize Goodrich and Mitzenmacher's elegant blueprint for oblivious Cuckoo hashing, it is clear that the extreme complexity of oblivious Cuckoo hashing has made understanding, implementation, and proofs difficult. We show that there is a conceptually simple $O(\log^2 N/\log\log N)$-overhead ORAM that dispenses with oblivious Cuckoo hashing entirely.

We show that such a conceptually simple scheme lends to further extensions. Specifically, we obtain the first $O(\log^2 N/\log\log N)$ Oblivious Parallel RAM (OPRAM) scheme, thus not only matching the performance of the best known sequential ORAM, but also achieving super-logarithmic improvements in comparison with known OPRAM schemes.

Keywords: Oblivious RAM · Oblivious PRAM

1 Introduction

Oblivious RAM [19,20,37], originally proposed in the seminal work by Goldreich and Ostrovsky [19,20], is a powerful cryptographic primitive that provably obfuscates a program's access patterns to sensitive data. Since Goldreich

The full version of this paper is available on Cryptology ePrint Archive [7].

T. Takagi and T. Peyrin (Eds.): ASIACRYPT 2017, Part I, LNCS 10624, pp. 660–690, 2017.
https://doi.org/10.1007/978-3-319-70694-8_23

and Ostrovsky's original work [19,20], numerous subsequent works have proposed improved constructions, and demonstrated a variety of ORAM applications in both theoretical contexts (e.g., multiparty computation [23,27], Garbled RAMs [18,28]) as well as in secure hardware and software systems (e.g., secure processors [15,16,29,36], and cloud outsourcing [22,35,38,39,43]). To hide access patterns, an ORAM scheme typically involves reading, writing, or shuffling multiple blocks for every data request. Suppose that on average, for each data request, an ORAM scheme must read/write X blocks. In this paper, we refer to X as the overhead (or the total work blowup) of the ORAM scheme.

Goldreich and Ostrovsky [19,20] showed that, roughly speaking, any "natural" ORAM scheme that treats each block as an "opaque ball" must necessarily suffer from at least logarithmic overhead. The recent Circuit ORAM [41] work demonstrated an almost matching upper bound for large enough blocks. Let N denote the total memory size. Circuit ORAM showed the existence of a *statistically* secure ORAM scheme that achieves $O(\alpha \log N)$ overhead for N^{ϵ}-bit blocks for any constant $\epsilon > 0$ and any super-constant function $\alpha = \omega(1)$. To date, the existence of an almost logarithmic ORAM scheme is only known for large blocks. For general block sizes, the state of affairs is different: the best known construction (asymptotically speaking) is a *computationally* secure scheme by Kushilevitz et al. [26], which achieves $O(\frac{\log^2 N}{\log \log N})$ overhead assuming block sizes of $\Omega(\log N)$[1]. We note that all known ORAM schemes assume that a memory block is at least large enough to store its own address, i.e., at least $\Omega(\log N)$ bits long. Therefore, henceforth in this paper, we use the term "general block size" to refer to a block size of $\Omega(\log N)$.

Although most practical ORAM implementations (in the contexts of secure multi-party computation, secure processors, and storage outsourcing) opted for tree-based ORAM constructions [37,40,41] due to tighter practical constants, we note that hierarchical ORAMs are nonetheless of much theoretical interest: for example, when the CPU has $O(\sqrt{N})$ private cache, hierarchical ORAMs can achieve $O(\log N)$ simulation overhead while a comparable result is not known in the tree-based framework. Recent works [3,8] have also shown how hierarchical ORAMs can achieve asymptotically better locality and IO performance than known tree-based approaches.

Our contributions. In this paper, we make the following contributions:

- **Revisit $O(\log^2 N/\log\log N)$ ORAMs.** We revisit how to construct a *computationally* secure ORAM with $O(\frac{\log^2 N}{\log \log N})$ overhead for general block sizes. First, we show why earlier results along this front [22,26] are somewhat incomplete due to the incompleteness of a core building block, oblivious Cuckoo hashing, that is proposed and described by Goodrich and Mitzenmacher [22]. Next, besides fixing and restating the earlier results regarding the existence

[1] This $O(\frac{\log^2 N}{\log \log N})$ result for computational security was later matched in the tree-based ORAM framework [9,14] although tree-based ORAMs were initially investigated for the case of statistical security.

of an $O(\log^2 N/\log\log N)$ ORAM, perhaps more compellingly, we show how to obtain an ORAM with the same asymptotic overhead, but in a conceptually much simpler manner, completely obviating the need to perform oblivious Cuckoo hashing [22] which is the center of complexity in the earlier result [26].
- **New results on efficient OPRAMs.** Building on our new ORAM scheme, we next present the first Oblivious Parallel RAM (OPRAM) construction that achieves $O(\frac{\log^2 N}{\log\log N})$ simulation overhead. To the best of our knowledge, our OPRAM scheme is the first one to asymptotically match the best known sequential ORAM scheme for general block sizes. Moreover, we achieve a super-logarithmic factor improvement over earlier works [5,10] and over the concurrent work by Nayak and Katz [31] (see further clarifications in Sect. 1.3).

We stress that our conceptual simplicity and modular approach can open the door for possible improvements. For example, our OPRAM results clearly demonstrate the benefits of having a conceptually clean hierarchical ORAM framework: had we tried to make (a corrected variant of) Kushilevitz et al. [26] into an OPRAM, it is not clear whether we could have obtained the same performance. In particular, achieving $O(\log^2 N/\log\log N)$ worst-case simulation overhead requires deamortizing a parallel version of their oblivious cuckoo hash rebuilding algorithm, and moreover, work and depth have to be deamortized at the same time—and we are not aware of a way to do this especially due to the complexity of their algorithm.

1.1 Background on Oblivious Hashing and Hierarchical ORAMs

In this paper, we consider the hierarchical framework, originally proposed by Goldreich and Ostrovsky [19,20], for constructing ORAM schemes. At a high level, this framework constructs an ORAM scheme by having exponentially growing levels of capacity $1, 2, 4, \ldots, N$ respectively, where each smaller level can be regarded as a "stash" for larger levels. Each level in the hierarchy is realized through a core abstraction henceforth called *oblivious hashing* in the remainder of this paper. Since oblivious hashing is the core abstraction we care about, we begin by explicitly formulating oblivious hashing as the following problem:

- *Functional abstraction.* Given an array containing n possibly dummy elements where each non-dummy element is a (key, value) pair, design an efficient algorithm that builds a hash table data structure, such that after the building phase, each element can be looked up by its key consuming a small amount of time and work. In this paper, we will assume that all non-dummy elements in the input array have distinct keys.
- *Obliviousness.* The memory access patterns of both the building and lookup phases do not leak any information (to a computationally bounded adversary) about the initial array or the sequence of lookup queries Q—as long as *all non-dummy queries in Q are distinct.* In particular, obliviousness must hold even when Q may contain queries for elements *not contained* in the array in which case the query should return the result $\perp$. The correct answer to a dummy query is also $\perp$ by convention.

Not surprisingly, the performance of a hierarchical ORAM crucially depends on the core building block, oblivious hashing. Here is the extent of our knowledge about oblivious hashing so far:

- Goldreich and Ostrovsky [19,20] show an oblivious variant of normal balls-and-bins hashing that randomly throws n elements into n bins. They show that obliviously building a hash table containing n elements costs $O(\alpha n \log n \log \lambda)$ work, and each query costs $O(\alpha \log \lambda)$ work. If α is any super-constant function, we can attain a failure probability $\mathsf{negl}(\lambda)$. This leads to an $O(\alpha \log^3 N)$-overhead ORAM scheme, where N is the total memory size[2].
- Subsequently, Goodrich and Mitzenmacher [22] show that the Cuckoo hashing algorithm can be made oblivious, incurring $O(n \log n)$ total work for building a hash table containing n elements, and only $O(1)$ query cost (later we will argue why their oblivious hashing scheme is somewhat incomplete). This leads to an ORAM scheme with $O(\log^2 N)$-overhead.
- Kushilevitz et al. [26] in turn showed an elegant reparametrization trick atop the Goodrich and Mitzenmacher ORAM, thus improving the overhead to $O(\frac{\log^2 N}{\log \log N})$. Since Kushilevitz et al. [26] crucially rely on Goodrich and Mitzenmacher's oblivious Cuckoo hashing scheme, incompleteness of the hashing result in some sense carries over to their $O(\frac{\log^2 N}{\log \log N})$ overhead ORAM construction.

1.2 Technical Roadmap

Revisit oblivious Cuckoo hashing. Goodrich and Mitzenmacher [22]'s blueprint for obliviously building a Cuckoo hash table is insightful and elegant. They express the task of Cuckoo hash table rebuilding as a MapReduce task (with certain nice properties), and they show that any such MapReduce algorithm has an efficient oblivious instantiation.

Fundamentally, their construction boils down using a sequence of oblivious sorts over arrays of (roughly) exponentially decreasing lengths. To achieve full privacy, it is necessary to hide the true lengths of these arrays during the course of the algorithm. Here, Goodrich and Mitzenmacher's scheme description and their proof appear inconsistent: their scheme seems to suggest padding each array to the maximum possible length for security—however, this would make their scheme $O(\log^3 N)$ overhead rather than the claimed $O(\log^2 N)$. On the other hand, their proof appears only to be applicable, if the algorithm reveals the true lengths of the arrays—however, as we argue in detail in the online full version [7], the array lengths in the cuckoo hash rebuilding algorithm contain information about the size of each connected component in the cuckoo graph.

[2] Henceforth in this paper, we use n to denote the size of a hash table and λ to denote its security parameter. For our ORAM construction, we use N to denote both the logical memory size as well as the ORAM's security parameter. This distinction is necessary since the ORAM will employ hash tables of varying n.

Thus leaking array lengths can lead to an explicit attack that succeeds with non-negligible probability: at a high level, this attack tries to distinguish two request sequences, one repeatedly requesting the same block whereas the other requests disctinct blocks. The latter request sequence will cause the cuckoo graph in the access phase to resemble the cuckoo graph in the rebuild phase, whereas the former request sequence results in a fresh random cuckoo hash graph for the access phase (whose connected component sizes are different than the rebuild phase with relatively high probability).

As metioned earlier, the incompleteness of oblivious Cuckoo hashing also makes the existence proof of an $O(\log^2 N/\log\log N)$-overhead ORAM somewhat incomplete.

Is oblivious Cuckoo hashing necessary for efficient hierarchical ORAM? Goodrich and Mitzenmacher's oblivious Cuckoo hashing scheme is extremely complicated. Although we do show in our online full version [7] that the incompleteness of Goodrich and Mitzemacher's construction and proofs can be patched, thus correctly and fully realizing the elegant blueprint they had in mind—the resulting scheme nonetheless suffers from large constant factors, and is unsuitable for practical implementation. Therefore, a natural question is, *can we build efficient hierarchical ORAMs without oblivious Cuckoo hashing?*

Our first insight is that perhaps oblivious Cuckoo hashing scheme is an overkill for constructing efficient hierarchical ORAMs after all. As initial evidence, we now present an almost trivial modification of the original Goldreich and Ostrovsky oblivious balls-and-bins hashing scheme such that we can achieve an $O(\alpha \log^2 N)$-overhead ORAM for any super-constant function α.

Recall that Goldreich and Ostrovsky [19,20] perform hashing by hashing n elements into n bins, each of $O(\alpha \log \lambda)$ capacity, where λ is the security parameter. A simple observation is the following: instead of having n bins, we can have $\frac{n}{\alpha \log \lambda}$ bins—it is not hard to show that each bin's occupancy will still be upper bounded by $O(\alpha \log \lambda)$ except with $\mathsf{negl}(\lambda)$ probability. In this way, we reduce the size of the hash table by a $\log \lambda$ factor, and thus the hash table can be obliviously rebuilt in logarithmically less time. Plugging in this new hash table into Goldreich and Ostrovsky's ORAM construction [19,20], we immediately obtain an ORAM scheme with $O(\alpha \log^2 N)$ overhead.

This shows that through a very simple construction we can almost match Goodrich and Mitzenmacher's ORAM result [22]. This simple scheme does not quite get us to where we aimed to be, but we will next show that oblivious Cuckoo hashing is likewise an overkill for constructing $(\frac{\log^2 N}{\log\log N})$-overhead ORAMs.

Conceptually simple $(\frac{\log^2 N}{\log\log N})$-overhead ORAM. Recall that a hierarchical ORAM's overhead is impacted by two cost metrics of the underlying oblivious hashing scheme, i.e., the cost of building the hash-table, and the cost of each lookup query. Goodrich and Mitzenmacher's oblivious Cuckoo hashing scheme [22] minimizes the lookup cost to $O(1)$, but this complicates the building of the hash-table.

Our key insight is that in all known hashing-based hierarchical ORAM constructions [19,20,22,26], the resulting ORAM's cost is dominated by the hash-table rebuilding phase, and thus it may be okay if the underlying hashing scheme is more expensive in lookup. More specifically, to obtain an $O(\frac{\log^2 N}{\log\log N})$ ORAM, we would like to apply Kushilevitz et al. [26]'s reparametrized version of the hierarchical ORAM. Kushilevitz et al. [26] showed that their reparametrization technique works when applied over an oblivious Cuckoo hashing scheme. We observe that in fact, Kushilevitz et al. [26]'s reparametrization technique is applicable for a much broader parameter range, and concretely for any oblivious hashing scheme with the following characteristics:

- It takes $O(n \log n)$ total work to build a hash table of n elements—in other words, the per-element building cost is $O(\log n)$.
- The lookup cost is asymptotically smaller than the per-element building cost—specifically, $O(\log^{\epsilon} \lambda)$ lookup cost suffices where $\epsilon \in (0.5, 1)$ is a suitable constant.

This key observation allows us to relax the lookup time on the underlying oblivious hashing scheme. We thus propose a suitable oblivious hashing scheme that is conceptually simple. More specifically, our starting point is a (variant of a) two-tier hashing scheme first described in the elegant work by Adler et al. [1]. In a two-tier hashing scheme, there are two hash tables denoted H_1 and H_2 respectively, each with $\frac{n}{\log^{\epsilon} \lambda}$ bins of $O(\log^{\epsilon} \lambda)$ capacity, where $\epsilon \in (0.5, 1)$ is a suitable constant. To hash n elements (non-obliviously), we first throw each element into a random bin in H_1. For all the elements that overflow its bin capacity, we throw them again into the second hash table H_2. Stochastic bounds show that the second hash table H_2 does not overflow except with $\mathsf{negl}(\lambda)$ probability. Clearly, the lookup cost is $O(\log^{\epsilon} \lambda)$; and we will show that the hash table building algorithm can be made oblivious through $O(1)$ number of oblivious sorts.

New results on oblivious parallel RAM. The conceptual simplicity of our ORAM scheme not only makes it easier to understand and implement, but also lends to further extensions. In particular, we construct a computationally secure OPRAM scheme that has $O(\log^2 N/ \log\log N)$ overhead—to the best of our knowledge, this is the first OPRAM scheme that matches the best known sequential ORAM in performance for general block sizes. Concretely, the hierarchical lookup phase can be parallelized using the standard conflict resolution (proposed by Boyle et al. [5]) as this phase is read-only. In the rebuild phase, our two-tier oblivious hashing takes only $O(1)$ number of oblivious sort and linear scan that marks excess elements, which can be parallelized with known algorithms, i.e. range prefix sum.

As mentioned earlier, our modular approach and conceptual simplicity turned out to be a crucial reason why we could turn our ORAM scheme into an OPRAM—it is not clear whether (a corrected version of) Kushilevitz et al. [26] is amenable to the same kind of transformation achieving the same overhead due to complications in deamortizing their cuckoo hash rebuilding algorithm. Thus we argue that our conceptually simple framework can potentially lend to other possible applications and improvements.

1.3 Related Work

ORAMs. ORAM was first proposed in a seminal work by Goldreich and Ostrovsky [19,20] who showed a computationally secure scheme with $O(\alpha \log^3 N)$ overhead for general block sizes and for any super-constant function $\alpha = \omega(1)$. Subsequent works improve the hierarchical ORAM [22,26] and show that $O(\frac{\log^2 N}{\log\log N})$ overhead can be attained under computational security—our paper points out several subtleties and the incompleteness of the prior results; additionally, we show that it is possible to obtain such an $O(\frac{\log^2 N}{\log\log N})$ overhead in a conceptually much simpler manner.

Besides the hierarchical framework, Shi et al. [37] propose a tree-based paradigm for constructing ORAMs. Numerous subsequent works [11,40,41] improved tree-based constructions. With the exception of a few works [14], the tree-based framework was primarily considered for the construction of *statistically secure* ORAMs. The performance of tree-based ORAMs depend on the block size, since with a larger block size we can reduce the number of recursion levels in these constructions. The recent Circuit ORAM work [41] shows that under block sizes as large as N^ϵ for any arbitrarily small constant ϵ, we can achieve $\alpha \log N$ bandwidth overhead for an arbitrary super-constant function $\alpha = \omega(1)$—this also shows the (near) tightness of the Goldreich-Ostrovsky lower bound [19,20] showing that any ORAM scheme must necessarily incur logarithmic overhead. Note that under block sizes of at least $\log^{1+\epsilon} N$ for an arbitrarily small constant ϵ, Circuit ORAM [41] can also attain $O(\frac{\log^2 N}{\log\log N})$ overhead and it additionally achieves statistical security rather than computational.

OPRAMs. Since modern computing architectures such as cloud platforms and multi-core architectures exhibit a high degree of parallelism, it makes sense to consider the parallel counterpart of ORAM. Oblivious Parallel ORAM (OPRAM) was first proposed by Boyle et al. [5], who showed a construction with $O(\alpha \log^4 N)$ overhead for any super-constant function α. Boyle et al.'s result was later improved by Chen et al. [10], who showed how to achieve $O(\alpha \log^3 N)$ overhead with poly-logarithmic CPU private cache—their result also easily implies an $O(\alpha \log^3 N \log\log N)$ overhead OPRAM with $O(1)$ CPU private cache, the setting that we focus on in this paper for generality.

A concurrent and independent manuscript by Nayak et al. [31] further improves the CPU-memory communication by extending Chen et al.'s OPRAM [10]. However, their scheme still requires $O(\alpha \log^3 N \log\log N)$ CPU-CPU communication which was the dominant part of the overhead in Chen et al. [10]. Therefore, under a general notion of overhead that includes both CPU-CPU communication and CPU-memory communication, Nayak et al.'s scheme still has the same asymptotic overhead[3] as Chen et al. [10] which is more than a logarithmic factor more expensive in comparison with our new OPRAM construction.

[3] The title of their paper [31] suggests $O(\log^2 N)$ overhead, since they did not account for the cost of CPU-CPU communication when describing the overhead.

In a companion paper, Chan et al. [9] showed how to obtain statistically secure and computationally secure OPRAMs in the tree-based ORAM framework. Specifically, they showed that for general block sizes, we can achieve statistically secure OPRAM with $O(\log^2 N)$ simulation overhead and computationally secure OPRAM with $O(\log^2 N/\log\log N)$ simulation overhead. For the computationally secure setting, Chan et al. [9] achieves the same asymptotical overhead as this paper, but the two constructions follow different paradigms so we believe that they are both of value. In another recent work, Chan et al. [6] proposed a new notion of depth for OPRAMs where the OPRAM is allowed to have more CPUs than the original PRAM to further parallelize the computation. In this paper, an OPRAM's simulation overhead is defined as its runtime blowup assuming that the OPRAM consumes the same number of CPUs as the PRAM.

Non-oblivious techniques for hashing. Many hashing schemes [4,12,17,25,30] were considered in the (parallel) algorithms literature. Unfortunately, most of them are *not* good candidates for constructing efficient ORAM and OPRAM schemes since there is no known efficient and oblivious counterpart for the algorithm. We defer detailed discussions of these related works to our online full version [7].

2 Definitions and Building Blocks

2.1 Parallel Random Access Machines

We define a Parallel Random Access Machine (PRAM) and an Oblivious Parallel Random Access Machine (OPRAM) in a similar fashion as Boyle et al. [5] as well as Chan and Shi [9]. Some of the definitions in this section are borrowed verbatim from Boyle et al. [5] or Chan and Shi [9].

Although we give definitions only for the parallel case, we point out that this is without loss of generality, since a sequential RAM can be thought of as a special-case PRAM.

Parallel Random Access Machine (PRAM). A *parallel random-access machine* (PRAM) consists of a set of CPUs and a shared memory denoted mem indexed by the address space $[N] := \{1, 2, \ldots, N\}$. In this paper, we refer to each memory word also as a *block*, and we use D to denote the bit-length of each block.

We support a more general PRAM model where the number of CPUs in each time step may vary. Specifically, in each step $t \in [T]$, we use m_t to denote the number of CPUs. In each step, each CPU executes a next instruction circuit denoted Π, updates its CPU state; and further, CPUs interact with memory through request instructions $\boldsymbol{I}^{(t)} := (I_i^{(t)} : i \in [m_t])$. Specifically, at time step t, CPU i's instruction is of the form $I_i^{(t)} := (\mathtt{read}, \mathsf{addr})$, or $I_i^{(t)} := (\mathtt{write}, \mathsf{addr}, \mathsf{data})$ where the operation is performed on the memory block with address addr and the block content $\mathsf{data} \in \{0,1\}^D \cup \{\perp\}$.

If $I_i^{(t)} = (\texttt{read}, \mathsf{addr})$ then the CPU i should receive the contents of $\mathsf{mem}[\mathsf{addr}]$ at the beginning of time step t. Else if $I_i^{(t)} = (\texttt{write}, \mathsf{addr}, \mathsf{data})$, CPU i should still receive the contents of $\mathsf{mem}[\mathsf{addr}]$ at the beginning of time step t; further, at the end of step t, the contents of $\mathsf{mem}[\mathsf{addr}]$ should be updated to data.

Write conflict resolution. By definition, multiple `read` operations can be executed concurrently with other operations even if they visit the same address. However, if multiple concurrent `write` operations visit the same address, a conflict resolution rule will be necessary for our PRAM be well-defined. In this paper, we assume the following:

- The original PRAM supports concurrent reads and concurrent writes (CRCW) with an arbitary, parametrizable rule for write conflict resolution. In other words, there exists some priority rule to determine which `write` operation takes effect if there are multiple concurrent writes in some time step t.
- Our compiled, oblivious PRAM (defined below) is a "concurrent read, exclusive write" PRAM (CREW). In other words, our OPRAM algorithm must ensure that there are no concurrent writes at any time.

We note that a CRCW-PRAM with a parametrizable conflict resolution rule is among the most powerful CRCW-PRAM model, whereas CREW is a much weaker model. Our results are stronger if we allow the underlying PRAM to be more powerful but the our compiled OPRAM uses a weaker PRAM model. For a detailed explanation on how stronger PRAM models can emulate weaker ones, we refer the reader to the work by Hagerup [24].

CPU-to-CPU communication. In the remainder of the paper, we sometimes describe our algorithms using CPU-to-CPU communication. For our OPRAM algorithm to be oblivious, the inter-CPU communication pattern must be oblivious too. We stress that such inter-CPU communication can be emulated using shared memory reads and writes. Therefore, when we express our performance metrics, we assume that all inter-CPU communication is implemented with shared memory reads and writes. In this sense, our performance metrics already account for any inter-CPU communication, and there is no need to have separate metrics that characterize inter-CPU communication. In contrast, some earlier works [10] adopt separate metrics for inter-CPU communication.

Additional assumptions and notations. Henceforth, we assume that each CPU can only store $O(1)$ memory blocks. Further, we assume for simplicity that the runtime of the PRAM, the number of CPUs activited in each time step and which CPUs are activited in each time step are *fixed* a priori and *publicly known* parameters. Therefore, we can consider a PRAM to be a tuple

$$\mathsf{PRAM} := (\Pi, N, T, \ (P_t : t \in [T])),$$

where Π denotes the next instruction circuit, N denotes the total memory size (in terms of number of blocks), T denotes the PRAM's total runtime, and P_t denotes the set of CPUs to be activated in each time step $t \in [T]$, where $m_t := |P_t|$.

Finally, in this paper, we consider PRAMs that are *stateful* and can evaluate a sequence of inputs, carrying state across in between. Without loss of generality, we assume each input can be stored in a single memory block.

2.2 Oblivious Parallel Random-Access Machines

Randomized PRAM. A *randomized PRAM* is a special PRAM where the CPUs are allowed to generate private random numbers. For simplicity, we assume that a randomized PRAM has a priori known, deterministic runtime, and that the CPU activation pattern in each time step is also fixed a priori and publicly known.

Memory access patterns. Given a PRAM program denoted PRAM and a sequence of inputs $(\mathsf{inp}_1, \ldots, \mathsf{inp}_d)$, we define the notation $\mathsf{Addresses}[\mathsf{PRAM}](\mathsf{inp}_1, \ldots, \mathsf{inp}_d)$ as follows:

- Let T be the total number of parallel steps that PRAM takes to evaluate inputs $(\mathsf{inp}_1, \ldots, \mathsf{inp}_d)$.
- Let $A_t := \{(\mathsf{cpu}_1^t, \mathsf{addr}_1^t), (\mathsf{cpu}_2^t, \mathsf{addr}_2^t) \ldots, (\mathsf{cpu}_{m_t}^t, \mathsf{addr}_{m_t}^t)\}$ be the list of (CPU id, address) pairs such that cpu_i^t accessed memory address addr_i^t in time step t.
- We define $\mathsf{Addresses}[\mathsf{PRAM}](\mathsf{inp}_1, \ldots, \mathsf{inp}_d)$ to be the random variable $[A_t]_{t\in[T]}$.

Oblivious PRAM (OPRAM). A randomized PRAM is said to be computationally *oblivious*, iff there exists a p.p.t. simulator Sim, and a negligible function $\epsilon(\cdot)$ such that for any input sequence $(\mathsf{inp}_1, \ldots, \mathsf{inp}_d)$ where $\mathsf{inp}_i \in \{0,1\}^D$ for $i \in [d]$,

$$\mathsf{Addresses}[\mathsf{PRAM}](\mathsf{inp}_1, \ldots, \mathsf{inp}_d) \overset{\epsilon(N)}{\approx} \mathsf{Sim}(1^N, d, T, (P_t : t \in [T]))$$

where $\overset{\epsilon(N)}{\approx}$ means that no p.p.t. adversary can distinguish the two probability ensembles except with $\epsilon(N)$ probability.

In other words, obliviousness requires that there is a polynomial-time simulator Sim that can simulate the memory access patterns knowing only the memory size N, the number of inputs d, the parallel runtime T for evaluating the inputs, as well as the a-priori fixed CPU activation pattern $(P_t : t \in [T])$. In particular, the simulator Sim does not know anything about the sequence of inputs.

Oblivious simulation and simulation overhead. We say that a oblivious PRAM, denoted as OPRAM, *simulates* a PRAM if for every input sequence $(\mathsf{inp}_1, \ldots, \mathsf{inp}_d)$, $\mathsf{OPRAM}(\mathsf{inp}_1, \ldots, \mathsf{inp}_d) = \mathsf{PRAM}(\mathsf{inp}_1, \ldots, \mathsf{inp}_d)$, i.e., OPRAM and PRAM output the same outcomes on any input sequence. In addition, an *OPRAM scheme* is a randomized PRAM algorithm such that, given any PRAM, the scheme compiles PRAM into an oblivious PRAM, OPRAM, that simulates PRAM.

For convenience, we often adopt two intermediate metrics in our descriptions, namely, *total work blowup* and *parallel runtime blowup*. We say that an OPRAM scheme has a total work blowup of x and a parallel runtime blowup of y, iff for every PRAM step t in which the PRAM consumes m_t CPUs, the OPRAM can complete this step with $x \cdot m_t$ total work and in y parallel steps—if the OPRAM is allowed to consume any number of CPUs (possibly greater than m_t).

Fact 1. *If there exists an OPRAM with x total work blowup and y parallel runtime blowup such that $x \geq y$, then there exists an OPRAM that has $O(x)$ simulation overhead when consuming the same number of CPUs as the orginal PRAM for simulating at PRAM step.*

In the interest of space, we defer the proof of this simple fact to the online full version [7].

2.3 Oblivious Hashing Scheme

Without loss of generality, we define only the parallel version, since the sequential version can be thought of the parallel version subject to executing on a single CPU.

A parallel oblivious hashing scheme contains the following two parallel, possibly randomized algorithms to be executed on a *Concurrent Read, Exclusive Write* PRAM:

- $\mathsf{T} \leftarrow \mathsf{Build}(1^\lambda, \{(k_i, v_i) \mid \mathsf{dummy}\}_{i\in[n]})$: given a security parameter 1^λ, and a set of n elements, where each element is either a dummy denoted dummy or a (key, value) pair denoted (k_i, v_i), the Build algorithm outputs a memory data structure denoted T that will later facilitate query. For an input array $S := \{(k_i, v_i) \mid \mathsf{dummy}\}_{i\in[n]}$ to be *valid*, we require that any two non-dummy elements in S must have distinct keys.
- $v \leftarrow \mathsf{Lookup}(\mathsf{T}, k)$: takes in the data structure T and a (possibly dummy) query k, outputs a value v.

Correctness. Correctness is defined in a natural manner: given a valid initial set $S := \{(k_i, v_i) \mid \mathsf{dummy}\}_{i\in[n]}$ and a query k, we say that v is the correct answer for k with respect to S, iff

- If $k = \mathsf{dummy}$ (i.e., if k is a dummy query) or if $k \notin S$, then $v = \bot$.
- Else, it must hold that $(k, v) \in S$.

More informally, the answer to any dummy query must be $\bot$; if a query searches for an element non-existent in S, then the answer must be $\bot$. Otherwise, the answer returned must be consistent with the initial set S.

We say that a parallel oblivious hashing scheme is correct, if for any valid initial set S, for any query k, and for all λ, it holds that

$$\Pr\left[\mathsf{T} \leftarrow \mathsf{Build}(1^\lambda, S), v \leftarrow \mathsf{Lookup}(\mathsf{T}, k) : \ v \text{ is correct for } k \text{ w.r.t. } S\right] = 1$$

where the probability space is taken over the random coins chosen by the Build and Lookup algorithms.

Obliviousness. A query sequence $\mathbf{k} = (k_1, \ldots, k_j)$ is said to be *non-recurrent*, if all non-dummy queries in $\mathbf{k}$ are distinct.

A parallel hashing scheme denoted (Build, Lookup) is said to be *oblivious*, if there exists a polynomial-time simulator Sim, such that for any security parameter λ, for any valid initial set S, for any *non-recurrent* query sequence $\mathbf{k} := (k_1, \ldots, k_j)$ of polynomial length, it holds that

$$\mathsf{Addresses}[\mathsf{Build}, \mathsf{Lookup}](1^\lambda, S, \mathbf{k}) \stackrel{c}{\equiv} \mathsf{Sim}(1^\lambda, |S|, |\mathbf{k}|)$$

where $\stackrel{c}{\equiv}$ denotes computationally indistinguishability, i.e., a computationally bounded adversary can distinguish between the two distributions with an advantage at most $\mathsf{negl}(\lambda)$. Intuitively, this security definition says that a simulator, knowing only the length of the input set and the number of queries, can simulate the memory access patterns.

Definition 1 **($(W_{\text{build}}, T_{\text{build}}, W_{\text{lookup}}, T_{\text{lookup}})$-parallel oblivious hashing scheme).** *Let $W_{\text{build}}(\cdot,\cdot)$, $W_{\text{lookup}}(\cdot,\cdot)$, $T_{\text{build}}(\cdot,\cdot)$, and $T_{\text{lookup}}(\cdot,\cdot)$ be functions in n and λ. We say that* (Build, Lookup) *is a $(W_{\text{build}}, T_{\text{build}}, W_{\text{lookup}}, T_{\text{lookup}})$-parallel oblivious hashing scheme, iff* (Build, Lookup) *satisfies correctness and obliviousness as defined above; and moreover, the scheme achieves the following performance:*

- *Building a hash table with n elements takes $n \cdot W_{\text{build}}(n, \lambda)$ total work and $T_{\text{build}}(n, \lambda)$ time with all but $\mathsf{negl}(\lambda)$ probability. Note that $W_{\text{build}}(n, \lambda)$ is the per-element amount of work required for preprocessing.*
- *A lookup query takes $W_{\text{lookup}}(n, \lambda)$ total work and $T_{\text{lookup}}(n, \lambda)$ time.*

As a special case, we say that (Build, Lookup) *is a $(W_{\text{build}}, W_{\text{lookup}})$-oblivious hashing scheme, if it is a $(W_{\text{build}}, \_, W_{\text{lookup}}, \_)$-parallel oblivious hashing scheme for any choice of the wildcard field "_"—in other words, in the sequential case, we do not care about the scheme's parallel runtime, and the scheme's total work is equivalent to the runtime when running on a single CPU.*

[Read-only lookup assumption.] When used in ORAM, observe that elements are inserted in a hash table in a batch only in the Build algorithm. Moreover, we will assume that the Lookup algorithm is read-only, i.e., it does not update the hash table data structure T, and no state is carried across between multiple invocations of Lookup.

A note on the security parameter. Since later in our application, we will need to apply oblivious hashing to different choices of n (including possibly small choices of n), throughout the description of the oblivious hashing scheme, we distinguish the security parameter denoted λ and the size of the set to be hashed denoted n.

2.4 Building Blocks

Duplicate suppression. Informally, duplicate suppression is the following building block: given an input array X of length n consisting of (key, value) pairs and possibly dummy elements where each key can have multiple occurrences, and additionally, given an upper bound n' on the number of distinct keys in X, the algorithm outputs a duplicate-suppressed array of length n' where only one occurrence of each key is preserved, and a preference function priority is used to choose which one.

Earlier works have [5,19,20] proposed an algorithm that relies on oblivious sorting to achieve duplicate suppression in $O(n \log n)$ work and $O(\log n)$ parallel runtime where $n := |X|$.

Oblivious select. $\mathsf{Select}(X, k, \mathsf{priority})$ takes in an array X where each element is either of the form (k, v) or a dummy denoted $\perp$, a query k, and a priority function priority which defines a total ordering on all elements with the same key; and outputs a value v such that $(k, v) \in X$ and moreover there exists no $(k, v') \in X$ such that v' is preferred over v for the key k by the priority function priority.

Oblivious select can be accomplished using a simple tree-based algorithm [9] in $O(\log n)$ parallel runtime and $O(n)$ total work where $n = |X|$.

Oblivious multicast. Oblivious multicast is the following building block. Given the following inputs:

- a source array $X := \{(k_i, v_i) \mid \mathsf{dummy}\}_{i \in [n]}$ where each element is either of the form (k, v) or a dummy denoted dummy, and further all real elements must have a distinct k; and
- a destination array $Y := \{k'_i\}_{i \in [n]}$ where each element is a query k' (possibly having duplicates).

the oblivious multicast algorithm outputs an array $\mathsf{ans} := \{v_i\}_{i \in [n]}$ such that if $k'_i \notin X$ then $v_i := \perp$; else it must hold that $(k'_i, v_i) \in X$.

Boyle et al. [5] propose an algorithm based on $O(1)$ oblivious sorts that achieves oblivious multicast in $O(\log n)$ parallel runtime and $O(n \log n)$ total work.

Range prefix sum. We will rely on a parallel range prefix sum algorithm which offers the following abstraction: given an input array $X = (x_1, \dots, x_n)$ of length n where each element of X is of the form $x_i := (k_i, v_i)$, output an array $Y = (y_1, \dots, y_n)$ where each y_i is defined as follows:

- Let $i' \leq i$ be the smallest index such that $k_{i'} = k_{i'+1} = \dots = k_i$;
- $y_i := \sum_{j=i'}^{i} v_j$.

In the GraphSC work, Nayak et al. [32] provide an oblivious algorithm that computes the range prefix sum in $O(\log n)$ parallel runtime and $O(n \log n)$ total work—in particular [32] defines a building block called "longest prefix sum" which is a slight variation of the range prefix sum abstraction we need. It is easy to see that Nayak et al.'s algorithm for longest prefix sum can be modified in a straightforward manner to compute our notion of range prefix sum.

3 Oblivious Two-Tier Hashing Scheme

In this section, we present a simple oblivious two-tier hashing scheme. Before we describe our scheme, we make a couple important remarks that the reader should keep in mind:

- Note that our security definition implies that the adversary can only observe the memory access patterns, and we require simulatability of the memory access patterns. Therefore our scheme description does not explicitly encrypt data. When actually deploying an ORAM scheme, all data must be encrypted if the adversary can also observe the contents of memory.
- In our oblivious hashing scheme, we use λ to denote the security parameter, and use n to denote the hash table's size. Our ORAM application will employ hash tables of varying sizes, so n can be small. Observe that an instance of hash table building can fail with $\mathsf{negl}(\lambda)$ probability; when this happens in the context of ORAM, the hash table building is restarted. This ensures that the ORAM is always correct, and the security parameter is related to the running time of the ORAM.
- For small values of n, we need special treatment to obtain $\mathsf{negl}(\lambda)$ security failure probability—specifically, we simply employ normal balls-and-bins hashing for small values of n. Instead of having the ORAM algorithm deal with this issue, we wrap this part inside the oblivious hashing scheme, i.e., the oblivious hashing scheme will automatically decide whether to employ normal hashing or two-tier hashing depending on n and λ.
  This modular approach makes our ORAM and OPRAM algorithms conceptually simple and crystallizes the security argument as well.

The goal of this section is to give an oblivious hashing scheme with the following guarantee.

Theorem 1 (Parallel oblivious hashing). *For any constant $\epsilon > 0.5$, for any $\alpha(\lambda) := \omega(1)$, there exists a $(W_{\text{build}}, T_{\text{build}}, W_{\text{lookup}}, T_{\text{lookup}})$-parallel oblivious hashing scheme where*

$$W_{\text{build}} = O(\log n), \qquad T_{\text{build}} = O(\log n),$$
$$W_{\text{lookup}} = \begin{cases} O(\alpha \log \lambda) & \text{if } n < e^{3\log^{\epsilon}\lambda} \\ O(\log^{\epsilon}\lambda) & \text{if } n \geq e^{3\log^{\epsilon}\lambda} \end{cases}, \quad T_{\text{lookup}} = O(\log\log\lambda)$$

3.1 Construction: Non-oblivious and Sequential Version

For simplicity, we first present a non-oblivious and sequential version of the hashing algorithm, and we can use this version of the algorithm for the purpose of our stochastic analysis. Later in Sect. 3.2, we will show how to make the algorithm both oblivious and parallel. Henceforth, we fix some $\epsilon \in (0.5, 1)$.

Case 1: $n < e^{3\log^{\epsilon}\lambda}$. When n is sufficiently small relative to the security parameter λ, we simply apply normal hashing (i.e., balls and bins) in the following manner. Let each bin's capacity $Z(\lambda) = \alpha \log \lambda$, for any $\alpha = \omega(1)$ superconstant function in λ.

For building a hash table, first, generate a secret PRF key denoted $\mathsf{sk} \xleftarrow{\$} \{0,1\}^\lambda$. Then, store the n elements in $B := \lceil 5n/Z \rceil$ bins each of capacity Z, where each element $(k, \_)$ is assigned to a pseudorandom bin computed as follows:

$$\text{bin number} := \mathsf{PRF}_{\mathsf{sk}}(k).$$

Due to a simple application of the Chernoff bound, the probability that any bin overflows is negligible in λ as long as Z is superlogarithmic in λ.

To look up an element with the key k, compute the bin number as above and read the entire bin.

Case 2: $n \geq e^{3 \log^\epsilon \lambda}$. This is the more interesting case, and we describe our two-tier hashing algorithm below.

- **Parameters and data structure.** Suppose that our memory is organized into two hash tables named H_1 and H_2 respectively, where each hash table has $B := \lceil \frac{n}{\log^\epsilon \lambda} \rceil$ bins, and each bin can store at most $Z := 5 \log^\epsilon \lambda$ blocks.
- $\mathsf{Build}(1^\lambda, \{(k_i, v_i) \mid \mathsf{dummy}\}_{i \in [n]})$:
  a) Generate a PRF key $\mathsf{sk} \xleftarrow{\$} \{0,1\}^\lambda$.
  b) For each element $(k_i, v_i) \in S$, try to place the element into the bin numbered $\mathsf{PRF}_{\mathsf{sk}}(1||k_i)$ in the first-tier hash table H_1. In case the bin is full, instead place the element in the overflow pile henceforth denoted Buf.
  c) For each element (k, v) in the overflow pile Buf, place the element into the bin numbered $\mathsf{PRF}_{\mathsf{sk}}(2||k)$ in the second-tier hash table H_2.
  d) Output $\mathsf{T} := (\mathsf{H}_1, \mathsf{H}_2, \mathsf{sk})$.
- $\mathsf{Lookup}(\mathsf{T}, k)$: Parse $\mathsf{T} := (\mathsf{H}_1, \mathsf{H}_2, \mathsf{sk})$ and perform the following.
  a) If $k = \perp$, i.e., this is a dummy query, return $\perp$.
  b) Let $i_1 := \mathsf{PRF}_{\mathsf{sk}}(1||k)$. If an element of the form (k, v) is found in $\mathsf{H}_1[i_1]$, return v. Else, let $i_2 := \mathsf{PRF}_{\mathsf{sk}}(2||k)$, look for an element of the form (k, v) in $\mathsf{H}_2[i_2]$ and return v if found.
  c) If still not found, return $\perp$.

Overflow event. If in the above algorithm, an element happens to choose a bin in the second-tier hash table H_2 that is full, we say that a bad event called overflow has happened. When a hash building is called in the execution of an ORAM, recall that if an overflow occurs, we simply discard all work thus far and restart the build algorithm from the beginning.

In Sect. 3.4, we will prove that indeed, overflow events occur with negligible probability. Therefore, henceforth in our ORAM presentation, *we will simply pretend that* overflow *events never happen during hash table building.*

Remark 1. Since the oblivious hashing scheme is assumed to retry from scratch upon overflows, we guarantee perfect correctness and computational security failure (due to the use of a PRF). Similarly, our resulting ORAM and OPRAM schemes will also have perfect correctness and computational security. Obviously, the algorithms may execute longer if overflows and retries take place—henceforth in the paper, whenever we say that an algorithm's total work or runtime is bounded by x, we mean that it is bounded by x except with negligible probability over the randomized execution.

3.2 Construction: Making It Oblivious

Oblivious Building. To make the building phase oblivious, it suffices to have the following Placement building block.

Let B denote the number of bins, let Z denote each bin's capacity, and let R denote the maximum capacity of the overflow pile. Placement is the following building block. Given an array $\mathsf{Arr} = \{(\mathsf{elem}_i, \mathsf{pos}_i) \mid \mathsf{dummy}\}_{i\in[n]}$ containing n possibly dummy elements, where each non-dummy element elem_i is tagged with a pseudo-random bin number $\mathsf{pos}_i \in [B]$, output B arrays $\{\mathsf{Bin}_i\}_{i\in[B]}$ each of size exactly Z and an overflow pile denoted Buf of size exactly R. The placement algorithm must output a valid assignment if one exists. Otherwise if no valid assignment exists, the algorithm should abort outputting hash-failure.

We say that an assignment is valid if the following constraints are respected:

(i) Every non-dummy $(\mathsf{elem}_i, \mathsf{pos}_i) \in \mathsf{Arr}$ exists either in some bin or in the overflow pile Buf.
(ii) For every Bin_i, every non-dummy element in Bin_i is of the form $(\_, i)$. In other words, non-dummy elements can only reside in their targeted bin or the overflow pile Buf.
(iii) For every Bin_i, if there exists a dummy element in Bin_i, then no element of the form $(\_, i)$ appears in Buf. In other words, no elements from each bin should overflow to Buf unless the bin is full.

[*Special case*]. A special case of the placement algorithm is when the overflow pile's targeted capacity $R = 0$. This special case will be used when we create the second-tier hash table.

Below, we show that using standard oblivious sorting techniques [2], Placement can be achieved in $O(n \log n)$ total work:

1. For each $i \in [B]$, add Z copies of filler elements $(\diamond, i)$ where $\diamond$ denotes that this is a filler element. These filler elements are there to make sure that each bin is assigned at least Z elements. Note that filler elements and dummy elements are treated differently.
2. Oblivious sort all elements by their bin number. For elements with the same bin number, break ties by placing real elements to the left of filler elements.
3. In a single linear scan, for each element that is not among the first Z elements of its bin, tag the element with the label "`excess`".
4. Oblivious sort all elements by the following ordering function:
   - All dummy elements must appear at the very end;
   - All non-excess elements appear before excess elements;

- For two non-excess elements, the one with the smaller bin number appears first (breaking ties arbitrarily).
- For excess elements, place real elements to the left of filler elements.

Oblivious lookups. It remains to show how to make lookup queries oblivious. To achieve this, we can adopt the following simple algorithm:

- If the query $k \neq \perp$: compute the first-tier bin number as $i_1 := \mathsf{PRF}_{\mathsf{sk}}(1||k)$. Read the entire bin numbered i_1 in the first-tier hash table H_1. If found, read an entire random bin in H_2; else compute $i_2 := \mathsf{PRF}_{\mathsf{sk}}(2||k)$ and read the entire bin numbered i_2 in the second-tier hash table H_2. Finally, return the element found or $\perp$ if not found.
- If the query $k = \perp$, read an entire random bin in H_1, and an entire random bin in H_2. Both bin numbers are selected freshly and independently at random. Finally, return $\perp$.

3.3 Construction: Making It Parallel

To make the aforementioned algorithm parallel, it suffices to make the following observations:

(i) Oblivious sorting of n elements can be accomplished using a sorting circuit [2] that involves $O(n \log n)$ total work and $O(\log n)$ parallel runtime.
(ii) Step 3 of the oblivious building algorithm involves a linear scan of the array marking each excessive element that exceeds its bin's capacity.
This linear scan can be implemented in parallel using the oblivious "range prefix sum" algorithm in $O(n \log n)$ total work and $O(\log n)$ parallel runtime. We refer the reader to Sect. 2.4 for a definition of the range prefix sum algorithm.
(iii) Finally, observe that the oblivious lookup algorithm involves searching in entire bin for the desired block. This can be accomplished obliviously and in parallel through our "oblivious select" building block defined in Sect. 2.4. Since each bin's capacity is $O(\log^\epsilon n)$, the oblivious select algorithm can be completed in $O(\log \log n)$ parallel runtime and tight total work.

Remark 2 (The case of small n). So far, we have focused our attention on the (more interesting) case when $n \geq e^{3 \log^\epsilon \lambda}$. When $n < e^{3 \log^\epsilon \lambda}$, we rely on normal hashing, i.e., balls and bins. In this case, hash table building can be achieved through a similar parallel oblivious algorithm that completes in $O(n \log n)$ total work and $O(\log n)$ parallel runtime; further, each lookup query completes obliviously in $O(\alpha \log \lambda)$ total work and $O(\log \log \lambda)$ parallel runtime.

Performance of our oblivious hashing scheme. In summary, the resulting algorithm achieves the following performance:

- Building a hash table with n elements takes $O(n \log n)$ total work and $O(\log n)$ parallel runtime with all but $\mathsf{negl}(\lambda)$ probability, regardless of how large n is.

– Each lookup query takes $O(\log^{\epsilon} \lambda)$ total work when $n \geq e^{3\log^{\epsilon}\lambda}$ and $O(\alpha \log \lambda)$ total work when $n < e^{3\log^{\epsilon}\lambda}$ where $\alpha(\lambda) = \omega(1)$ can be any super-constant function. Further, regardless of how large n is, each lookup query can be accomplished in $O(\log \log \lambda)$ parallel runtime.

3.4 Overflow Analysis

We give the overflow analysis of the two-tier construction in Sect. 3.1. We use the following variant of Chernoff Bound.

Fact 2 (Chernoff Bound for Binomial Distribution). *Let X be a random variable sampled from a binomial distribution (with any parameters). Then, for any $k \geq 2E[X]$, $Pr[X \geq k] \leq e^{-\frac{k}{6}}$.*

Utilization of first-tier hash. Recall that the number of bins is $B := \left\lceil \frac{n}{\log^{\epsilon}\lambda} \right\rceil$. For $i \in [B]$, let X_i denote the number of items that are sent to bin i in the first-tier hash. Observe that the expectation $E[X_i] = \frac{n}{B} \geq \log^{\epsilon} \lambda$.

Overflow from first-tier hash. For $i \in [B]$, let $\widehat{X}_i$ be the number of items that are sent to bin i in the first-tier but have to be sent to the overflow pile because bin i is full. Recall that the capacity of a bin is $Z := 5\log^{\epsilon}\lambda$. Then, it follows that $\widehat{X}_i$ equals $X_i - Z$ if $X_i > Z$, and 0 otherwise.

Tail bound for overflow pile. We next use the standard technique of moment generating function to give a tail inequality for the number $\sum_i \widehat{X}_i$ of items in the overflow pile. For sufficiently small $t > 0$, we have $E[e^{t\widehat{X}_i}] \leq 1 + \sum_{k\geq 1} \Pr[X_i = Z + k] \cdot e^{tk} \leq 1 + \sum_{k\geq 1} \Pr[X_i \geq Z + k] \cdot e^{tk} \leq 1 + \frac{\exp(-\frac{Z}{6})}{e^{\frac{1}{6}-t}-1}$, where the last inequality follows from Fact 2 and a standard computation of a geometric series. For the special case $t = \frac{1}{12}$, we have $E[e^{\frac{\widehat{X}_i}{12}}] \leq 1 + 12\exp(-\frac{Z}{6})$.

Lemma 1 (Tail Inequality for Overflow Pile). *For $k \geq 288Be^{-\frac{Z}{6}}$, $\Pr[\sum_{i\in[B]} \widehat{X}_i \geq k] \leq e^{-\frac{k}{24}}$.*

Proof. Fix $t := \frac{1}{12}$. Then, we have $\Pr[\sum_{i\in[B]} \widehat{X}_i \geq k] = \Pr[t\sum_{i\in[B]} \widehat{X}_i \geq tk] \leq e^{-tk} \cdot E[e^{t\sum_{i\in[B]}\widehat{X}_i}]$, where the last inequality follows from the Markov's inequality.

As argued in [13], when n balls are thrown independently into n bins uniformly at random, then the numbers X_i's of balls received in the bins are negatively associated. Since $\widehat{X}_i$ is a monotone function of X_i, it follows that the $\widehat{X}_i$'s are also negatively associated. Hence, it follows that $E[e^{t\sum_{i\in[B]}\widehat{X}_i}] \leq \prod_{i\in[B]} E[e^{t\widehat{X}_i}] \leq \exp(12Be^{-\frac{Z}{6}})$.

Finally, observing that $k \geq 288Be^{-\frac{Z}{6}}$, we have $\Pr[\sum_{i\in[B]} \widehat{X}_i \geq k] \leq \exp(12Be^{-\frac{Z}{6}} - \frac{k}{12}) \leq e^{-\frac{k}{24}}$, as required.

In view of Lemma 1, we consider $N := 288Be^{-\frac{Z}{6}}$ as an upper bound on the number of items in the overflow pile. The following lemma gives an upper bound on the probability that a particular bin overflows in the second-tier hash.

Lemma 2 (Overflow Probability in the Second-Tier Hash). *Suppose the number of items in the overflow pile is at most $N := 288Be^{-\frac{Z}{6}}$, and we fix some bin in the second-tier hash. Then, the probability that this bin receives more than Z items in the second tier hash is at most $e^{-\frac{Z^2}{6}}$.*

Proof. Observe that the number of items that a particular bin receives is stochastically dominated by a binomial distribution with N items and probability $\frac{1}{B}$. Hence, the probability that it is at least Z is at most $\binom{N}{Z} \cdot (\frac{1}{B})^Z \leq (\frac{Ne}{Z})^Z \cdot (\frac{1}{B})^Z \leq e^{-\frac{Z^2}{6}}$, as required.

Corollary 1 (Negligible Overflow Probability). *Suppose the number n of items is chosen such that both $Be^{-\frac{Z}{6}}$ and Z^2 are $\omega(\log \lambda)$, where $B := \left\lceil \frac{n}{\log^{\epsilon} \lambda} \right\rceil$ and $Z := \lceil 5 \log^{\epsilon} \lambda \rceil$. Then, the probability that the overflow event happens in the second-tier hash is negligible in λ.*

Proof. Recall that $B = \lceil \frac{n}{\log^{\epsilon} \lambda} \rceil$, where $n \geq e^{3 \log^{\epsilon} \lambda}$ in Theorem 1. By choosing $N = 288Be^{-\frac{Z}{6}}$, from Lemma 1, the probability that there are more than N items in the overflow pile is $\exp(-\Theta(N))$, which is negligible in λ.

Given that the number of items in the overflow pile is at most N, according to Lemma 2, the probability that there exists some bin that overflows in the second-tier hash is at most $Be^{-\frac{Z^2}{6}}$ by union bound, which is also negligible in λ, because we assume $B \leq \text{poly}(\lambda)$.

3.5 Obliviousness

If there is no overflow, for any valid input, Build accesses fixed addresses. Also, Lookup fetches a fresh pseudorandom bin for each dummy or non-dummy request. Hence, the simulator is just running Build and Lookup with all dummy requests. See the online full version [7] for the formal proof.

4 Modular Framework for Hierarchical ORAM

4.1 Preliminary: Hierarchical ORAM from Oblivious Hashing

Goldreich and Ostrovsky [19,20] were the first to define Oblivious RAM (ORAM) and they provide an elegant solution to the problem which was since referred to as the "hierarchical ORAM". Goldreich and Ostrovsky [19,20] describe a special-case instantiation of a hierarchical ORAM where they adopt an oblivious variant of naïve hashing. Their scheme was later extended and improved by several subsequent works [22,26,42].

In this section, we will present a generalized version of Goldreich and Ostrovsky's hierarchical ORAM framework. Specifically, we will show that Goldreich and Ostrovsky's core idea can be interpreted as the following: take any oblivious hashing scheme satisfying the abstraction defined in Sect. 2.3, we can construct a corresponding ORAM scheme that makes blackbox usage of the oblivious hashing scheme.

From our exposition, it will be clear why such a modular approach is compelling: it makes both the construction and the security proof simple. In comparison, earlier hierarchical ORAM works do not adopt this modular approach, and their conceptual complexity could sometimes confound the security proof [34].

Data structure. There are $\log N + 1$ *levels* numbered $0, 1, \ldots, L$ respectively, where $L := \lceil \log_2 N \rceil$ is the maximum level. Each level is a hash table denoted $\mathsf{T}_0, \mathsf{T}_1, \ldots, \mathsf{T}_L$ where T_i has capacity 2^i. At any time, each table T_i can be in two possible states, *available* or *full*. Available means that this level is currently empty and does not contain any blocks, and thus one can rebuild into this level. Full means that this level currently contains blocks, and therefore an attempt to rebuild into this level will effectively cause a cascading merge.

ORAM operations. Upon any memory access request $(\mathtt{read}, \mathsf{addr})$ or $(\mathtt{write}, \mathsf{addr}, \mathsf{data})$, perform the following procedure. For simplicity, we omit writing the security parameter of the algorithms, i.e., let $\mathsf{Build}(\cdot) := \mathsf{Build}(1^N, \cdot)$, and let $\mathsf{Lookup}(\cdot) := \mathsf{Lookup}(1^N, \cdot)$.

1. $\mathsf{found} := \mathtt{false}$.
2. For each $\ell = 0, 1, \ldots L$ in increasing order,
   - If not found, $\mathsf{fetched} := \mathsf{Lookup}(\mathsf{T}_\ell, \mathsf{addr})$: if $\mathsf{fetched} \neq \perp$, let $\mathsf{found} := \mathtt{true}$, $\mathsf{data}^* := \mathsf{fetched}$.
   - Else $\mathsf{Lookup}(\mathsf{T}_\ell, \perp)$.
3. Let $\mathsf{T}^\emptyset := \{(\mathsf{addr}, \mathsf{data}^*)\}$ if this is a $\mathtt{read}$ operation; else let $\mathsf{T}^\emptyset := \{(\mathsf{addr}, \mathsf{data})\}$. Now perform the following hash table rebuilding:
   - Let ℓ be the smallest level index such that T_ℓ is marked available. If all levels are marked full, then $\ell := L$. In other words, ℓ is the target level to be rebuilt.
   - Let $S := \mathsf{T}^\emptyset \cup \mathsf{T}_0 \cup \mathsf{T}_1 \cup \ldots \cup \mathsf{T}_{\ell-1}$; if all levels are marked full, then additionally let $S := S \cup \mathsf{T}_L$. Further, tag each non-dummy element in S with its level number, i.e., if a non-dummy element in S comes from T_i, tag it with the level number i.
   - $\mathsf{T}_\ell := \mathsf{Build}(\mathsf{SuppressDuplicate}(S, 2^\ell, \mathsf{pref}))$, and mark T_ℓ as full. Further, let $\mathsf{T}_0 = \mathsf{T}_1 = \ldots = \mathsf{T}_{\ell-1} := \emptyset$ and their status bits set to available. Here we adopt the following priority function pref:
     - When two or more real blocks with the same address (i.e., key) exist, the one with the smaller level number is preferred (and the algorithm maintains the invariant that no two blocks with the same address and the same level number should exist).
4. Return data^*.

Deamortization. In the context of hierarchical ORAM, a hash table of capacity n is rebuilt every n memory requests, and we typically describe the ORAM's overhead in terms of the amortized cost per memory request. As one may observe, every now and then, the algorithm needs to rebuild a hash table of size N, and thus a small number of memory requests may incur super-linear cost to complete.

A standard deamortization technique was described by Ostrovsky and Shoup [33] to evenly spread the cost of hash table rebuilding over time, and this deamortization framework only blows up the total work of the ORAM scheme by a small constant factor; the details are in the online full version [7]. In the rest of the paper, we assume that every instance of hash table used in an ORAM scheme is rebuilt in the background using this deamortization technique without explicitly mentioning so. Further, the stated costs in the theorems are applicable to worst-case performance (not just amortized).

Obliviousness. To show obliviousness of the above construction, we make the following observations.

Fact 3 (Non-recurrent queries imply obliviousness). *In the aforementioned ORAM construction, as long as lookup queries to every instance of hash table satisfies the non-recurrent condition specified in Sect. 2.3, the resulting ORAM scheme satisfies obliviousness.*

The proof of this fact is deferred to our online full version [7].

Fact 4 (Non-recurrence condition is preserved). *In the above ORAM construction, it holds that for every hash table instance, all lookup queries it receives satisfy the non-recurrence condition.*

Proof. Due to our ORAM algorithm, every 2^ℓ operations, the old instance of hash table T_ℓ is destroyed and a new hash table instance is created for T_ℓ. It suffices to prove the non-recurrence condition in between every two rebuilds for T_ℓ. Suppose that after T_ℓ is rebuilt in some step, now we focus on the time steps going forward until the next rebuild. Consider when a block block^* is first found in T_ℓ where $\ell \in [L]$, block^* is entered into $\mathsf{T}^\emptyset$. Due to the definition of the ORAM algorithm, until the next time T_ℓ is rebuilt, block^* exists in some $\mathsf{T}_{\ell'}$ where $\ell' < \ell$. Due to the way the ORAM performs lookups—in particular, we would look up a dummy element in T_ℓ if block^* is found in a smaller level—we conclude that until T_ℓ is rebuilt, no lookup query will ever be issued again for block^* to T_ℓ.

Lemma 3 (Obliviousness). *Suppose that the underlying hashing scheme satisfies correctness and obliviousness as defined in Sect. 2.3, then it holds that the above ORAM scheme satisfies obliviousness as defined in Sect. 2.2.*

Proof. Straightforward from Facts 3 and 4.

Theorem 2 (Hierarchical ORAM from oblivious hashing). *Assume the existence of one-way functions and a* $(W_{\text{build}}, W_{\text{lookup}})$*-oblivious hashing scheme.*

Then, there exists an ORAM scheme that achieves the following blowup for block sizes of $\Omega(\log N)$ bits:

$$\textit{ORAM's blowup} := \max\left(\sum_{\ell=0}^{\log N} W_{\text{build}}(2^\ell, N),\ \sum_{\ell=0}^{\log N} W_{\text{lookup}}(2^\ell, N)\right) + O(\log^2 N)$$

This theorem is essentially proved by Goldreich and Ostrovsky [19,20]—however, they proved it only for a special case. We generalize their hierarchical ORAM construction and express it modularly to work with any oblivious hashing scheme as defined in Sect. 2.3.

Remark 3. We point out that due to the way we define our oblivious hashing abstraction, each instance of oblivious hash table will independently generate a fresh PRF key during Build, and this PRF key is stored alongside the resulting hash table data structure in memory. Throughout this paper, we assume that each PRF operation can be evaluated in $O(1)$ runtime on top of our RAM. *We stress that this implicit assumption (or equivalent) was made by all earlier ORAM works* [19,20,22,26] *that rely on a PRF for security.*

4.2 Preliminary: Improving Hierarchical ORAM by Balancing Reads and Writes

Subsequent to Goldreich and Ostrovsky's ground-breaking result [19,20], Kushilevitz et al. [26] propose an elegant optimization for the hierarchical ORAM framework such that under some special conditions to be specified later, they can shave a (multiplicative) $\log \log N$ factor off the total work for a hierarchical ORAM scheme. Similarly, Kushilevitz et al. [26] describe a special-case instantiation of an ORAM scheme based on oblivious Cuckoo hashing which was proposed by Goodrich and Mitzenmacher [22].

In this section, we observe that the Kushilevitz et al.'s idea can be generalized. For the sake of exposition, we will first ignore the smaller ORAM levels that employ normal hashing in the following discussion, i.e., we assume that the smaller levels that employ normal hashing will not be a dominating factor in the cost. Now, imagine that there is an oblivious hashing scheme such that for sufficiently large n, the per-element cost for preprocessing is more expensive than the cost of a lookup by a $\log^\delta n$ factor for some constant $\delta > 0$. In other words, imagine that there exists a constant $\delta > 0$ such that the following condition is met for sufficiently large n:

$$\frac{W_{\text{build}}(n, \lambda)}{W_{\text{lookup}}(n, \lambda)} \geq \log^\delta n.$$

If the underlying oblivious hashing scheme satisfies the above condition, then Kushilevitz et al. [26] observes that Goldreich and Ostrovsky's hierarchical ORAM construction is suboptimal in the sense that the cost of fetch phase is asymptotically smaller than the cost of the rebuild phase. Hence, the resulting

ORAM's total work will be dominated by the rebuild phase, which is then determined by the building cost of the underlying hashing scheme, i.e., $W_{\text{build}}(n, \lambda)$.

Having observed this, Kushilevitz et al. [26] propose the following modification to Goldreich and Ostrovsky's hierarchical ORAM [19,20]. In Goldreich and Ostrovsky's ORAM, each level is a factor of 2 larger than the previous level—henceforth the parameter 2 is referred to the *branching factor*. Kushilevitz et al. [26] proposes to adopt a branching factor of $\mu := \log N$ instead of 2, and this would reduce the number of levels to $O(\log N/\log\log N)$—in this paper, we will adopt a more general choice of $\mu := \log^{\phi} N$ for a suitable positive constant ϕ. To make this idea work, they allow up to $\mu - 1$ simultaneous hash table instances for any ORAM level. If for all levels below ℓ, all instances of hash tables are full, then all levels below ℓ will be merged into a new hash table residing at level $\ell + 1$. The core idea here is to *balance the cost of the fetch phase and the rebuild phase by having a larger branching factor*; and as an end result, we could shave a $\log\log N$ factor from the ORAM's total work.

We now elaborate on this idea more formally.

Data structure. Let $\mu := \log^{\phi} N$ for a suitable positive constant ϕ to be determined later. There are $O(\log N/\log\log N)$ *levels* numbered $0, 1, \ldots, L$ respectively, where $L = \lceil \log_{\mu} N \rceil$ denotes the maximum level. Except for level L, for every other $\ell \in \{0, 1, \ldots, L-1\}$: the ℓ-th level contains up to $\mu - 1$ hash tables each of capacity μ^{ℓ}. Henceforth we use the notation T_{ℓ} to denote level ℓ, and T_{ℓ}^{i} to denote the i-th hash table within level ℓ. The largest level L contains a single hash table of capacity N denoted T_{L}^{0}. Finally, every level $\ell \in \{0, 1, \ldots, L\}$ has a counter c_{ℓ} initialized to 0. Effectively, for every level $\ell \neq L$, if $c_{\ell} = \mu - 1$, then the level is considered full; else the level is considered available.

ORAM operations. Upon any memory access query $(\mathtt{read}, \mathsf{addr})$ or $(\mathtt{write}, \mathsf{addr}, \mathsf{data})$, perform the following procedure.

1. $\mathsf{found} := \mathtt{false}$.
2. For each $\ell = 0, 1, \ldots L$ in increasing order, for $\tau = c_{\ell} - 1, c_{\ell} - 2 \ldots 0$ in decreasing order:
   If not found: $\mathsf{fetched} := \mathsf{Lookup}(\mathsf{T}_{\ell}^{\tau}, \mathsf{addr})$; if $\mathsf{fetched} \neq \bot$, let $\mathsf{found} := \mathtt{true}$, $\mathsf{data}^* := \mathsf{fetched}$. Else $\mathsf{Lookup}(\mathsf{T}_{\ell}^{\tau}, \bot)$.
3. Let $\mathsf{T}^{\emptyset} := \{(\mathsf{addr}, \mathsf{data}^*)\}$ if this is a $\mathtt{read}$ operation; else let $\mathsf{T}^{\emptyset} := \{(\mathsf{addr}, \mathsf{data})\}$. Now, perform the following hash table rebuilding.
   - Let ℓ be the smallest level index such that its counter $c_{\ell} < \mu - 1$. If no such level index exists, then let $\ell := L$. In other words, we plan to rebuild a hash table in level ℓ.
   - Let $S := \mathsf{T}^{\emptyset} \cup \mathsf{T}_0 \cup \mathsf{T}_1 \cup \ldots, \cup \mathsf{T}_{\ell-1}$; and if $\ell = L$, additionally, let $S := S \cup \mathsf{T}_{L}^{0}$ and let $c_L = 0$. Further, in the process, tag each non-dummy element in S with its level number and its hash table number within the level. For example, if a non-dummy element in S comes from T_{i}^{τ}, i.e., the τ-th table in the i-th level, tag it with (i, τ).
   - Let $\mathsf{T}_{\ell}^{c_{\ell}} := \mathsf{Build}(\mathsf{SuppressDuplicate}(S, \mu^{\ell}, \mathsf{pref}))$, and let $c_{\ell} := c_{\ell} + 1$.

Here we adopt the following priority function pref: when two or more blocks with the same address (i.e., key) exist, the one with the *smaller* level number is preferred; if there is a tie in level number, the one with the *larger* hash table number is preferred.

- Let $\mathsf{T}_0 = \mathsf{T}_1 = \ldots = \mathsf{T}_{\ell-1} := \emptyset$ and set $c_0 = c_1 = \ldots = c_{\ell-1} := 0$.

4. Return data^*.

Goldreich and Ostrovsky's ORAM scheme [19,20] is a special case of the above for $\mu = 2$.

Deamortization. The deamortization technique of Ostrovsky and Shoup [33] (described in the online full version [7]) applies in general to hierarchical ORAM schemes for which each level is some data structure that is rebuilt regularly. Therefore, it can be applied to our scheme as well, and thus the work of rebuilding hash tables is spread evenly across memory requests.

Obliviousness. The obliviousness proof is basically identical to that presented in Sect. 4.1, since the only change here from Sect. 4.1 is that the parameters are chosen differently due to Kushilevitz et al.'s elegant idea [26].

Theorem 3 (Hierarchical ORAM variant.). *Assume the existence of one-way functions and a $(W_{\text{build}}, W_{\text{lookup}})$-oblivious hashing scheme. Then, there exists an ORAM scheme that achieves the following blowup for block sizes of $\Omega(\log N)$ bits where $L = O(\log N / \log\log N)$:*

$$\textit{ORAM's blowup} := \max\left(\sum_{\ell=0}^{L} W_{\text{build}}(\mu^\ell, N),\ \log^\phi N \cdot \sum_{\ell=0}^{L} W_{\text{lookup}}(\mu^\ell, N)\right) + O(L \log N)$$

We note that Kushilevitz et al. [26] proved a special case of the above theorem, we now generalize their technique and describe it in the most general form.

4.3 Conceptually Simpler ORAM for Small Blocks

In the previous section, we presented a hierarchical ORAM scheme, reparametrized using Kushilevitz et al. [26]'s technique, consuming any oblivious hashing scheme with suitable performance characteristics as a blackbox.

To obtain a conceptually simple ORAM scheme with $O(\log^2 N / \log\log N)$ overhead, it suffices to plug in the oblivious two-tier hashing scheme described earlier in Sect. 3.

Corollary 2 (Conceptually simpler ORAM for small blocks). *There exists an ORAM scheme with $O(\log^2 N / \log\log N)$ runtime blowup for block sizes of $\Omega(\log N)$ bits.*

Proof. Using the simple oblivious two-tier hashing scheme in Sect. 3 with $\epsilon = \frac{3}{4}$, we can set $\phi = \frac{1}{4}$ in Theorem 3 to obtain the result.

4.4 IO Efficiency and the Case of Large CPU Cache

Besides the ORAM's runtime, we often care about its IO performance as well, where IO-cost is defined as the number of cache misses as in the standard external-memory algorithms literature. When the CPU has a large amount of private cache, e.g., N^ϵ blocks where $\epsilon > 0$ is an arbitrarily small constant, several works have shown that oblivious sorting $n \leq N$ elements can be accomplished with $O(n)$ IO operations [8,21,22]. Thus, a direct corollary is that for the case of N^ϵ CPU cache, we can construct a computationally secure ORAM scheme with $O(\log N)$ IO-cost (by using the basic hierarchical ORAM construction with $O(\log N)$ levels with an IO-efficient oblivious sort).

5 Asymptotically Efficient OPRAM

In this section, we show how to construct an $O(\frac{\log^2 N}{\log\log N})$ OPRAM scheme. To do this, we will show how to parallelize our new $O(\frac{\log^2 N}{\log\log N})$-overhead ORAM scheme. Here we benefit tremendously from the conceptual simplicity of our new ORAM scheme. In particular, as mentioned earlier, our oblivious two-tier hashing (Build, Lookup) algorithms have efficient parallel realizations. We will now present our OPRAM scheme. For simplicity, we first present a scheme assuming that the number of CPUs in each step of the computation is fixed and does not change over time. In this case, we show that parallelizing our earlier ORAM construction boils down to parallelizing the (Build and Lookup) algorithms of the oblivious hashing scheme. We then extend our construction to support the case when the number of CPUs varies over time.

5.1 Intuition

Warmup: uniform number of CPUs. We first describe the easier case of uniform m, i.e., the number of CPUs in the PRAM does not vary over time. Further, we will consider the simpler case when the branching factor $\mu := 2$.

- *Data structure.* Recall that our earlier ORAM scheme builds an exponentially growing hierarchy of oblivious hash tables, of capacities $1, 2, 4, \ldots, N$ each. Here, we can do the same, but we can start the level of hierarchy at capacity $m = 2^i$ (i.e., skip the smaller levels).
- *OPRAM operations.* Given a batch of m simultaneous memory requests, suppose that all addresses requested are distinct—if not, we can run a standard conflict resolution procedure as described by Boyle et al. [5] incurring only $O(\log m)$ parallel steps consuming m CPUs. We now need to serve these requests in parallel. In our earlier ORAM scheme, each request has two stages: (1) reading one block from each level of the exponentially growing hierarchy; and (2) perform necessary rebuilding of the levels. It is not hard to see that the fetch phase can be parallelized easily—particularly, observe that the fetch

phase is read-only, and thus having m CPUs performing the reads in parallel will not lead to any write conflicts.

It remains to show how to parallelize the rebuild phase. Recall that in our earlier ORAM scheme, each level has a status bit whose value is either *available* or *full*. Whenever we access a single block, we find the available (i.e., empty) level ℓ and merge all smaller levels as well as the updated block into level ℓ. If no such level ℓ exists, we simply merge all levels as well as the updated block into the largest level.

Here in our OPRAM construction, since the smallest level is of size m, we can do something similar. We find the smallest available (i.e., empty) level ℓ, and merge all smaller levels as well as the possibly updated values of the m fetched blocks into level ℓ. If no such level ℓ exists, we simply merge all levels as well as possibly updated values of the m fetched blocks into the largest level. Rebuilding a level in parallel effectively boils down to rebuilding a hash table in parallel (which boils down to performing $O(1)$ number of oblivious sorts in parallel)—which we have shown to be possible earlier in Sect. 3.

Varying number of CPUs. Our definitions of PRAM and OPRAMs allow the number of CPUs to vary over time. In this case, oblivious simulation of a PRAM is more sophisticated. First, instead of truncating the smaller levels whose size are less than m, here we have to preserve all levels—henceforth we assume that we have an exponentially growing hierarchy with capacities $1, 2, 4, \ldots, N$ respectively. The fetch phase is simple to parallelize as before, since the fetch phase does not make modifications to the data structure. We now describe a modified rebuild phase when serving a batch of $m = 2^\gamma$ requests: note that in the following, γ is a level that matches the current batch size, i.e., the number of CPUs in the present PRAM step of interest:

(a) Suppose level γ is marked available. Then, find the first available (i.e., empty) level ℓ greater than γ. Merge all levels below γ and the updated values of the newly fetched m blocks into level ℓ.
If no such level ℓ exists, then merge all blocks and the updated values of the newly fetched m blocks into the largest level L.

(b) Suppose level γ is marked as full. Then, find the first available (i.e., empty) level ℓ greater than γ. Merge all levels below or equal to γ (but not the updated values of the m fetched blocks) into level ℓ; rebuild level γ to contain the updated values of the m fetched blocks.
Similarly, if no such level ℓ exists, then merge all blocks and the updated values of the newly fetched m blocks into the largest level L.

One way to view the above algorithm is as follows: let us view the concatenation of all levels' status bits as a binary counter (where full denotes 1 and available denotes 0). If a single block is accessed like in the ORAM case, the counter is incremented, and if a level flips from 0 to 1, this level will be rebuilt. Further, if there would be a carry-over to the $(L + 1)$-st level, then the largest level L is rebuilt. However, now m blocks may be requested in a single batch—in this case, the above procedure for rebuilding effectively can be regarded as incrementing the counter by some value v where $v \leq 2m$—in particular, the value v is chosen such that only $O(1)$ levels must be rebuilt by the above rule.

We now embark on describing the full algorithm—specifically, we will describe for a general choice of the branching factor μ that is not necessarily 2. Further, our description supports the case of varying number of CPUs.

5.2 Detailed Algorithm

Data structure. Same as in Sect. 4.2. Specifically, there are $O(\log N / \log\log N)$ *levels* numbered $0, 1, \ldots, L$ respectively, where $L = \lceil \log_\mu N \rceil$ denotes the maximum level. Except for level L, for every other $\ell \in \{0, 1, \ldots, L-1\}$: the ℓ-th level contains up to $\mu - 1$ hash tables each of capacity μ^ℓ. Henceforth, we use the notation T_ℓ to denote level ℓ. Moreover, for $0 \leq i < \mu - 1$, we use T_ℓ^i to denote the i-th hash table within level ℓ. The largest level L contains a single hash table of capacity N denoted T_L^0. Finally, every level $\ell \in \{0, 1, \ldots, L\}$ has a counter c_ℓ initialized to 0.

We say that a level $\ell < L$ is *available* if its counter $c_\ell < \mu - 1$; otherwise, $c_\ell = \mu - 1$, and we say that the level $\ell < L$ is *full*. For the largest level L, we say that it is available if $c_L = 0$; else we say that it is full. Note that for the case of general $\mu > 2$, available does not necessarily mean that the level's empty.

OPRAM operations. Upon a batch of m memory access requests $Q := \{\mathsf{op}_p\}_{p \in [m]}$ where each op_p is of the form $(\mathtt{read}, \mathsf{addr}_p)$ or $(\mathtt{write}, \mathsf{addr}_p, \mathsf{data}_p)$, perform the following procedure. Henceforth we assume that $m = 2^\gamma$ where γ denotes the level whose capacity matches the present batch size.

1. **Conflict resolution.** $Q' := \mathsf{SuppressDuplicate}(Q, m, \mathsf{PRAM.priority})$, i.e., perform conflict resolution on the batch of memory requests Q, and obtain a batch Q' of the same size but where each distinct address appears only once—suppressing duplicates using the PRAM's priority function priority, and padding the resulting set with dummies to length m.
2. **Fetch phase.** For each $\mathsf{op}_i \in Q'$ in parallel where $i \in [m]$, parse $\mathsf{op}_i = \perp$ or $\mathsf{op}_i = (\mathtt{read}, \mathsf{addr}_i)$ or $\mathsf{op}_i = (\mathtt{write}, \mathsf{addr}_i, \mathsf{data}_i)$:
   (a) If $\mathsf{op}_i = \perp$, let found $:= \mathtt{true}$; else let found $:= \mathtt{false}$.
   (b) For each $\ell = 0, 1, \ldots L$ in increasing order, for $\tau = c_\ell - 1, c_\ell - 2 \ldots 0$ in decreasing order:
      - If not found: fetched $:= \mathsf{Lookup}(\mathsf{T}_\ell^\tau, \mathsf{addr}_i)$; if fetched $\neq \perp$, let found $:= \mathtt{true}$, $\mathsf{data}_i^* :=$ fetched.
      - Else, $\mathsf{Lookup}(\mathsf{T}_\ell^\tau, \perp)$.
3. **Rebuild phase.** For each $\mathsf{op}_i \in Q'$ in parallel where $i \in [m]$: if op_i is a $\mathtt{read}$ operation add $(\mathsf{addr}_i, \mathsf{data}_i^*)$ to $\mathsf{T}^\emptyset$; else if op_i is a $\mathtt{write}$ operation, add $(\mathsf{addr}_i, \mathsf{data}_i)$ to $\mathsf{T}^\emptyset$; else add $\perp$ to $\mathsf{T}^\emptyset$.
   Perform the following hash table rebuilding—recall that γ is the level whose capacity matches the present batch size:

(a) If level γ is full, then skip this step; else, perform the following: Let $S := \mathsf{T}_0 \cup \mathsf{T}_1 \cup \ldots \cup \mathsf{T}_{\gamma-1}$, and $\mathsf{T}_\gamma^{c_\gamma} := \mathsf{Build}(\mathsf{SuppressDuplicate}(S, \mu^\gamma, \mathsf{pref}))$ where pref prefers a block from a smaller level (i.e., the fresher copy) if multiple blocks of the same address exists. Let $c_\gamma := c_\gamma + 1$, and for every $j < \gamma$, let $c_j := 0$.

(b) At this moment, if level γ is still available, then let $\mathsf{T}_\gamma^{c_\gamma} := \mathsf{Build}(\mathsf{T}^\emptyset)$, and $c_\gamma := c_\gamma + 1$.
Else, if level γ is full, perform the following:
Find the first available level $\ell > \gamma$ greater than γ that is available; if no such level ℓ exists, let $\ell := L$ and let $c_L := 0$.
Let $S := \mathsf{T}^\emptyset \cup \mathsf{T}_0 \cup \ldots \cup \mathsf{T}_{\ell-1}$; if $\ell = L$, additionally include $S := S \cup \mathsf{T}_L$.
Let $\mathsf{T}_\ell^{c_\ell} := \mathsf{Build}(\mathsf{SuppressDuplicate}(S, \mu^\ell, \mathsf{pref}))$, and let $c_\ell := c_\ell + 1$. For every $j < \ell$, reset $c_j := 0$.

Deamortization. The deamortization technique (described in the online full version [7]) of Ostrovsky and Shoup [33] applies here as well, and thus the work of rebuilding hash tables are spread evenly across memory requests.

Obliviousness. The obliviousness proof is basically identical to that presented in Sect. 4.1. Since we explicitly resolve conflict before serving a batch of m requests, we preserve the non-recurrence condition. The only remaining differences here in comparison with Sect. 4.1 is that (1) here we use a general branching factor of μ rather than 2 (as in Sect. 4.1); and (2) here we consider the parallel setting. It is clear that neither of these matter to the obliviousness proof.

Theorem 4 (OPRAM from oblivious parallel hashing). *Assume the existence of one-way functions and a $(W_{\text{build}}, T_{\text{build}}, W_{\text{lookup}}, T_{\text{lookup}})$-oblivious hashing scheme. Then, there exists an ORAM scheme that achieves the following performance for block sizes of $\Omega(\log N)$ bits where $L = O(\frac{\log N}{\log \log N})$:*

$$\textit{total work blowup} := \max\left(\sum_{\ell=0}^{L} W_{\text{build}}(\mu^\ell, N),\ \log^\phi N \cdot \sum_{\ell=0}^{L} W_{\text{lookup}}(\mu^\ell, N)\right) + O(L \log N),$$

and para. runtime blowup :=

$$\max\left(\{T_{\text{build}}(\mu^\ell, N)\}_{\ell \in [L]},\ \log^\phi N \cdot \sum_{\ell=0}^{L} T_{\text{lookup}}(\mu^\ell, N)\right) + O(L).$$

Proof. Basically, the proof is our explicit OPRAM construction from any parallel oblivious hashing scheme described earlier in this section. For total work and parallel runtime blowup, we basically take the maximum of the ORAM's fetch phase and rebuild phase. The additive term $O(L \log N)$ in the total work stems from additional building blocks such as parallel duplicate suppression and other steps in our OPRAM scheme; and same for the additive term $O(L)$ in the parallel runtime blowup.

Using the simple oblivious hashing scheme in Sect. 3 with $\epsilon = \frac{3}{4}$, we can set $\phi = \frac{1}{4}$ to obtain the following corollary.

Corollary 3 (Asympototically efficient OPRAM for small blocks). *Assume that one-way functions exist. Then, there exists a computationally secure OPRAM scheme that achieves* $O(\log^2 N/\log\log N)$ *simulation overhead when the block size is at least* $\Omega(\log N)$ *bits.*

Acknowledgments. This work is supported in part by NSF grants CNS-1314857, CNS-1514261, CNS-1544613, CNS-1561209, CNS-1601879, CNS-1617676, an Office of Naval Research Young Investigator Program Award, a Packard Fellowship, a DARPA Safeware grant (subcontractor under IBM), a Sloan Fellowship, Google Faculty Research Awards, a Baidu Research Award, and a VMWare Research Award.

References

1. Adler, M., Chakrabarti, S., Mitzenmacher, M., Rasmussen, L.E.: Parallel randomized load balancing. Random Struct. Algorithms **13**(2), 159–188 (1998)
2. Ajtai, M., Komlós, J., Szemerédi, E.: An O(N Log N) sorting network. In: STOC (1983)
3. Asharov, G., Chan, H., Nayak, K., Pass, R., Ren, L., Shi, E.: Oblivious computation with data locality. IACR Cryptology ePrint Archive 2017/772 (2017)
4. Bast, H., Hagerup, T.: Fast and reliable parallel hashing. In: SPAA (1991)
5. Boyle, E., Chung, K.-M., Pass, R.: Oblivious parallel RAM and applications. In: Kushilevitz, E., Malkin, T. (eds.) TCC 2016. LNCS, vol. 9563, pp. 175–204. Springer, Heidelberg (2016). https://doi.org/10.1007/978-3-662-49099-0_7
6. Chan, T.-H.H., Chung, K.-M., Shi, E.: On the depth of oblivious parallel RAM. In: Asiacrypt (2017)
7. Chan, T.-H.H., Guo, Y., Lin, W.-K., Shi, E.: Oblivious hashing revisited, and applications to asymptotically efficient ORAM and OPRAM. Online full version of this paper, IACR Cryptology ePrint Archive 2017/924 (2017)
8. Chan, T.-H.H., Guo, Y., Lin, W.-K., Shi, E.: Cache-oblivious and data-oblivious sorting and applications. In: SODA (2018)
9. Chan, T.-H.H., Shi, E.: Circuit OPRAM: unifying statistically and computationally secure ORAMs and OPRAMs. In: TCC (2017)
10. Chen, B., Lin, H., Tessaro, S.: In: TCC, pp. 205–234 (2016)
11. Chung, K.-M., Liu, Z., Pass, R.: Statistically-secure ORAM with $\tilde{O}(\log^2 n)$ overhead. In: Asiacrypt (2014)
12. Dietzfelbinger, M., Karlin, A., Mehlhorn, K., Meyer auF der Heide, F., Rohnert, H., Tarjan, R.E.: Dynamic perfect hashing: upper and lower bounds. SIAM J. Comput. **23**(4), 738–761 (1994)
13. Dubhashi, D.P., Ranjan, D.: Balls and bins: a study in negative dependence. Random Struct. Algorithms **13**(2), 99–124 (1998)
14. Fletcher, C.W., Ren, L., Kwon, A., van Dijk, M., Devadas, S.: Freecursive ORAM: [nearly] free recursion and integrity verification for position-based oblivious RAM. In: ASPLOS (2015)
15. Fletcher, C.W., Ren, L., Kwon, A., van Dijk, M., Stefanov, E., Devadas, S.: RAW Path ORAM: a low-latency, low-area hardware ORAM controller with integrity verification. IACR Cryptology ePrint Archive 2014/431 (2014)

16. Fletcher, C.W., Ren, L., Yu, X., van Dijk, M., Khan, O., Devadas, S.: Suppressing the oblivious RAM timing channel while making information leakage and program efficiency trade-offs. In: HPCA, pp. 213–224 (2014)
17. Fredman, M.L., Komlós, J., Szemerédi, E.: Storing a sparse table with O(1) worst case access time. J. ACM **31**(3), 538–544 (1984)
18. Gentry, C., Halevi, S., Lu, S., Ostrovsky, R., Raykova, M., Wichs, D.: Garbled RAM revisited. In: Nguyen, P.Q., Oswald, E. (eds.) EUROCRYPT 2014. LNCS, vol. 8441, pp. 405–422. Springer, Heidelberg (2014). https://doi.org/10.1007/978-3-642-55220-5_23
19. Goldreich, O.: Towards a theory of software protection and simulation by oblivious RAMs. In: STOC (1987)
20. Goldreich, O., Ostrovsky, R.: Software protection and simulation on oblivious RAMs. J. ACM **43**, 431–473 (1996)
21. Goodrich, M.T.: Data-oblivious external-memory algorithms for the compaction, selection, and sorting of outsourced data. In: SPAA (2011)
22. Goodrich, M.T., Mitzenmacher, M.: Privacy-preserving access of outsourced data via oblivious RAM simulation. In: Aceto, L., Henzinger, M., Sgall, J. (eds.) ICALP 2011. LNCS, vol. 6756, pp. 576–587. Springer, Heidelberg (2011). https://doi.org/10.1007/978-3-642-22012-8_46
23. Gordon, S.D., Katz, J., Kolesnikov, V., Krell, F., Malkin, T., Raykova, M., Vahlis, Y.: Secure two-party computation in sublinear (amortized) time. In: CCS (2012)
24. Hagerup, T.: Fast and optimal simulations between CRCW PRAMs. In: Finkel, A., Jantzen, M. (eds.) STACS 1992. LNCS, vol. 577, pp. 45–56. Springer, Heidelberg (1992). https://doi.org/10.1007/3-540-55210-3_172
25. Hagerup, T.: The log-star revolution. In: Finkel, A., Jantzen, M. (eds.) STACS 1992. LNCS, vol. 577, pp. 257–278. Springer, Heidelberg (1992). https://doi.org/10.1007/3-540-55210-3_189
26. Kushilevitz, E., Lu, S., Ostrovsky, R.: On the (in)security of hash-based oblivious RAM and a new balancing scheme. In: SODA (2012)
27. Liu, C., Wang, X.S., Nayak, K., Huang, Y., Shi, E.: ObliVM: a programming framework for secure computation. In S & P (2015)
28. Lu, S., Ostrovsky, R.: How to garble ram programs. In: Eurocrypt (2013)
29. Maas, M., Love, E., Stefanov, E., Tiwari, M., Shi, E., Asanovic, K., Kubiatowicz, J., Song, D.: Phantom: practical oblivious computation in a secure processor. In: CCS (2013)
30. Matias, Y., Vishkin, U.: Converting high probability into nearly-constant time—with applications to parallel hashing. In: STOC, pp. 307–316 (1991)
31. Nayak , K., Katz, J.: An oblivious parallel ram with $o(\log^2 n)$ parallel runtime blowup. https://eprint.iacr.org/2016/1141
32. Nayak, K., Wang, X.S., Ioannidis, S., Weinsberg, U., Taft, N., Shi, E.: GraphSC: parallel secure computation made easy. In: IEEE S & P (2015)
33. Ostrovsky, R., Shoup, V.: Private information storage (extended abstract). In: STOC, pp. 294–303 (1997)
34. Pinkas, B., Reinman, T.: Oblivious RAM revisited. In: Rabin, T. (ed.) CRYPTO 2010. LNCS, vol. 6223, pp. 502–519. Springer, Heidelberg (2010). https://doi.org/10.1007/978-3-642-14623-7_27
35. Ren, L., Fletcher, C.W., Kwon, A., Stefanov, E., Shi, E., van Dijk, M., Devadas, S.: Constants count: practical improvements to oblivious RAM. In: USENIX Security Symposium, pp. 415–430 (2015)
36. Ren, L., Yu, X., Fletcher, C.W., van Dijk, M., Devadas, S.: Design space exploration and optimization of path oblivious RAM in secure processors. In: ISCA (2013)

37. Shi, E., Chan, T.-H.H., Stefanov, E., Li, M.: Oblivious RAM with $O((\log N)^3)$ worst-case cost. In: Asiacrypt (2011)
38. Stefanov, E., Shi, E.: Multi-cloud oblivious storage. In: CCS (2013)
39. Stefanov, E., Shi, E.: ObliviStore: high performance oblivious cloud storage. In: IEEE Symposium on Security and Privacy (S & P) (2013)
40. Stefanov, E., van Dijk, M., Shi, E., Fletcher, C., Ren, L., Yu, X., Devadas, S.: Path ORAM - an extremely simple oblivious ram protocol. In: CCS (2013)
41. Wang, X., Chan, H., Shi, E.: Circuit ORAM: on tightness of the Goldreich-Ostrovsky lower bound. In: CCS (2015)
42. Williams, P., Sion, R., Carbunar, B.: Building castles out of mud: practical access pattern privacy and correctness on untrusted storage. In: CCS, pp. 139–148 (2008)
43. Williams, P., Sion, R., Tomescu, A.: PrivateFS: a parallel oblivious file system. In: CCS (2012)

Side Channel Analysis

Authenticated Encryption in the Face of Protocol and Side Channel Leakage

Guy Barwell[1], Daniel P. Martin[2,3](✉), Elisabeth Oswald[1], and Martijn Stam[1]

[1] Department of Computer Science, University of Bristol, Merchant Venturers Building, Woodland Road, Bristol BS8 1UB, UK
rgb.crypto@gmail.com, {elisabeth.oswald,martijn.stam}@bris.ac.uk
[2] School of Mathematics, University of Bristol, Bristol BS8 1TW, UK
[3] The Heilbronn Institute for Mathematical Research, Bristol, UK
dan.martin@bris.ac.uk

Abstract. Authenticated encryption schemes in practice have to be robust against adversaries that have access to various types of leakage, for instance decryption leakage on invalid ciphertexts (protocol leakage), or leakage on the underlying primitives (side channel leakage). This work includes several novel contributions: we augment the notion of nonce-base authenticated encryption with the notion of continuous leakage and we prove composition results in the face of protocol and side channel leakage. Moreover, we show how to achieve authenticated encryption that is simultaneously both misuse resistant and leakage resilient, based on a sufficiently leakage resilient PRF, and finally we propose a concrete, pairing-based instantiation of the latter.

Keywords: Provable security · Authenticated encryption · Generic composition · Leakage resilience · Robustness

1 Introduction

Authenticated Encryption (AE) has arisen out of (practical) necessity: historic modes-of-operation for symmetric encryption [33] implicitly target confidentiality against passive adversaries, but most realistic threat models also demand security against active adversaries. Thwarting adversaries trying to modify ciphertexts is best captured by requiring ciphertext integrity; encryption schemes that offer both this and a suitable passive indistinguishability notion are said to provide authenticated encryption. Today, authenticated encryption has become the primitive of choice to enable secure communication. AE schemes can be constructed from components that individually provide either confidentiality or authenticity, both in a traditional probabilistic setting [6] and a more modern nonce-based one [32]. As a result, there exist several black-box constructions of authenticated encryption schemes based on simpler, keyed primitives such as pseudorandom functions or permutations, including MACs and blockciphers.

T. Takagi and T. Peyrin (Eds.): ASIACRYPT 2017, Part I, LNCS 10624, pp. 693–723, 2017.
https://doi.org/10.1007/978-3-319-70694-8_24

Unfortunately, in practice neither the composition nor the underlying components behave as black-boxes: side-channel attacks often leak additional information to an adversary, leading to real-life breaks (e.g. [47]). Invariably, these attacks are possible by exploiting a discrepancy between the capabilities of a theoretical adversary and an actual, real-life one. Thus, these attacks neither violate the security assumptions on the primitive nor do they invalidate the security claims: rather, they render these claims insufficient and the existing security models as inadequate.

In response, a number of works have tried to capture more closely how protocols behave when implemented [10,16,19]. We are particularly interested in *subtle* authenticated encryption [4] which augments the authenticated encryption security game with an implementation-dependent leakage oracle that provides an adversary deterministic decryption leakage on *invalid* ciphertexts only. Subtle authenticated encryption encompasses earlier notions such as multiple decryption errors [9] and the release of unverified plaintexts [2]; it can be regarded as *protocol* leakage.

Orthogonally, *primitives* can leak. Kocher (et al.) [24,25] showed how both timing and power measurements lead to a side-channel, enabling the extraction of secret data out of cryptographic devices. Primitives believed to be secure, such as AES, were broken without actually violating the assumption that AES is a secure pseudorandom permutation. Such attacks are captured in the framework of leakage resilient cryptography. Here an adversary can adaptively choose a leakage function that is restricted in scope as only computation is assumed to leak information [31], and in size. The latter is captured by leaking only a certain number of bits per call. If the overall leakage remains unbounded the model is referred to as continuous leakage. For a variety of schemes and security notions, resilience against certain classes of leakage can be proven [12,23,46], but dealing with adaptivity that allows leakage after an adversary has received a challenge is often problematic.

The current theory of authenticated encryption is not suited to take this additional leakage resource into account. In this work we provide a framework for dealing with AE in the presence of leakage, which then allows us to determine the constraints on primitives and constructions alike to yield AE secure against classes of leakage functions. Moreover, we propose a concrete instantiation of a leakage-resilient pseudorandom function suitable to be used to form the first leakage-resilient, nonce-based authenticated encryption scheme.

1.1 Our Contributions

Augmenting nonce-base authenticated encryption with leakage. We start by augmenting the nonce-based authenticated encryption security notion (Sect. 2.1) with leakage (Sect. 3). This new notion, which we will refer to as LAE, can be regarded as a generalization of the SAE framework by Barwell et al. [4], yet it also captures leakage-resilience as introduced by Dziembowski and Pietrzak [14]. We provide corresponding leakage notions for the primitives used by the composition results by Namprempre et al. [32] (henceforth NRS), namely nonce- or iv-based encryption, pseudorandom functions, and message authentication codes.

For the traditional AE notion by Rogaway and Shrimpton [42], an adversary has to distinguish between a world with a real encryption and decryption oracle on the one hand, and a world with a random ciphertext generator and a rejection oracle on the other. In the LAE game the number of oracles available to the adversary increased from two to four: both worlds are augmented with true encryption and decryption oracles and we will allow (only) these additional oracles to leak.

For the leakage mechanism, we adopt the approach originally suggested by Micali and Reyzin [31] and later adapted for leakage resilience [14] where an adversary can provide a leakage function to be evaluated on the internal variables of the oracle, with the leakage output to be returned to the adversary alongside the normal output. The model is very powerful, allowing the adversary to adaptively choose which leakage function they would like evaluated on a query by query basis.

To avoid trivial wins, the leakage functions that are allowed need to be restricted to prevent, for instance, leaking the entire key in one go. We model this by explicitly defining security relative to a class of leakage functions (as is common for instance in the contexts for related-key or key-dependent message attacks). By appropriately setting the class of leakage functions, we show that our notion generalises previous strengthened AE security notions, including SAE, RUP and distinguishable decryption errors [2,4,9], and previous leakage notions, including the simulatable leakage, auxiliary input and probing models [12,20,46].

Generic composition with leakage. Our second contribution (Sect. 5) is an investigation on how to perform generic composition in the presence of leakage by extending the results of NRS [32]. We establish that schemes susceptible to release of unverified plaintext are unsuitable even for much more modest types of leakage and we confirm modern folklore that this affects all schemes that are roughly of the type Encrypt-and-MAC or MAC-then-Encrypt (cf. [2]). Conversely, we show that Encrypt-then-MAC style schemes *are* secure against a large class of leakage functions, where we express this class in terms of the leakage classes against which the underlying primitives are secure. For this composition of leakage from different primitives, we effectively just concatenate the leakage of the constituent parts, which implicitly assumes that only computation leaks (cf. [31]).

In particular, we show security of the N2 and A5 constructions of NRS against nonce-respecting adversaries (Theorem 1 and Corollary 1), and of A6 against adversaries who never repeat a nonce and associated-data pair (Corollary 2).

The above result imply that *none* of the NRS schemes achieve misuse resistant LAE security (mrLAE), hence we propose a novel generic construction that *does* meet this strongest definition of security, albeit at the cost of further ciphertext expansion (Theorem 3). Our result gives ciphertexts that are two blocks longer than the messages (rather than the single block expansion of an NRS scheme): we leave open whether mrLAE security can be achieved with less ciphertext expansion.

Moreover, we show that instantiating CFB mode with a pseudorandom function yields a secure iv-based encryption scheme even under leakage (Theorem 4). This allows us to apply our generic composition results to construct the first AE scheme secure against continuous leakage based on a pseudorandom function actively secure against continuous leakage and a MAC scheme secure against continuous leakage of both tagging and verification.

Instantiation using a new leakage resilient PRF. Our final contribution (in the full version [3]) is the construction of these latter two primitives. To this end, we extend the MAC of Martin et al. [30] in two directions. First, we show how it can be adapted such that it may leak under verification answering an open question from their work. Then, we show how to implement the tagging function such that it is a PRF in the face of leakage. While the previous implementation of the MAC is a pseudorandom function when no leakage is present, already small amounts of leakage are disastrous for the pseudorandomness property. It turns out that the underlying key update mechanism due to Kiltz and Pietrzak [23] is intrinsically unsuitable to create an actively secure pseudorandom function: the mechanism shares a key out in two which allows a form of leak-in-the-middle attack. The solution we propose is to use three shares instead and we prove that the resulting construction is indeed a pseudorandom function that is leakage-resilient even against adaptive adversaries.

1.2 Related Work

Authenticated encryption. One of the earliest symmetric works on concrete security of AE was by Bellare and Namprempre [6]. Working within the probabilistic model, they formalised what it meant to be both confidential and authentic, and investigated how one could achieve this through generic composition, combining two schemes (one with each security property) such that their composition achieved both. Yet, modern authenticated encryption is a stateless and deterministic notion, taking in any randomness or state as an extra parameter termed the nonce. It was formalised across a number of papers, culminating in Rogaway and Shrimpton's 2006 work on DAE [42] and only recently a comprehensive study of all the ways one could combine a PRF with an encryption scheme was completed in the nonce-based setting [32].

The CAESAR competition [7] has driven further research into AE, and particularly into the concept of robustness, namely the idea that a scheme should be more resistant to common problems faced in the real-world. One branch of this research has been into designing schemes that are resistant to certain forms of leakage. Prior to the competition, Boldyreva et al. [9] had investigated how to model a scheme from which decryption failures are not identical, such as under a timing attack. Andreeva et al. [2] (RUP) considered the release of unverified plaintexts, where the decryption oracle releases candidate plaintexts even if they fail verification. The robust authenticated encryption notion of Hoang et al. [19] also implies security against the leakage of these candidate plaintexts, among other goals. Barwell et al. [4] defined the SAE framework as a generalisation of these notions, and used it to compare the three previous works. However, in each of these cases the adversary only receives leakage from decryption, and this leakage is modelled as a fixed, deterministic function, rather than a more general set of functions available to an adaptive side-channel attacker.

Leakage resilient constructions. Within the leakage resilient literature, there are several works towards providing leakage resilient encryption, but most of them have been in the bounded leakage model [18,37]. In the bounded retrieval model, Bellare et al. [5] proved the security of a symmetric encryption scheme that provides authenticated encryption in the leak free case, and indistinguishability when leakage is involved. Pereira et al. [34] proposed what is, to our knowledge, the first and only leakage resilient encryption scheme in the simulatable leakage model. However, the construction requires a leak free component and in practice relies on the existence of efficient simulators of the leakage from (e.g.) AES, simulators that Longo et al. [27] demonstrate are unlikely to exist.

Following on from Pereira et al. [34], the recent work by Berti et al. [8] also attempts to construct leakage resilient misuse-resistant authenticated encryption, albeit from a very different direction. In some respects, our work is "top-down", setting a clear objective and evaluating what this demands of the underlying primitives, while theirs is "bottom-up", beginning with well understood primitives and asking what can be constructed. Motivated by this, the two papers adopt very different leakage models: we work in full generality, whereas different sections of Berti follow different leakage models. More generally, their work assumes a single (completely) leak free component, whereas ours allows any of the components to leak as long as the overall leakage is not too great. They hypothesis that (without many leak-free components) leakage resilient misuse resistant authenticated encryption is impossible, while we show that this can be achieved. Furthermore, their work does not consider associated data.

Another manner to ensure that the adversary cannot progressively leak the key material is to update the keys themselves (instead of their representation). Previous leakage resilient works in this direction include the MAC of Schipper [44], or the DH-ratcheting concept [11,35]. However, these tend to require that all parties to the communication hold modifiable state and remain perfectly in sync, a demand we are able to avoid.

Each of the models above severely restricts the information or computations that an adversary may be able to perform, thereby limiting their utility for modelling active side-channel attacks. The continuous leakage model mitigates these problems, which is why we focus on that when instantiating our AE scheme. To the best of our knowledge, ours is the first leakage resilient encryption scheme in the continuous leakage model.

Our generic composition results allow us to combine leakage resilient components, for which we provide candidates built around a PRF secure against leakage. Currently there are two leakage resilient PRGs, due to Pietrzak (and

Dziembowski) [14,36], from which it may be possible to build a leakage resilient stream cipher, although issues arise with restarting using the same key. Works of Dodis and Pietrzak [13], and Faust et al. [15] describe two PRFs secure under non-adaptive leakage: each requires that the leakage (functions) are fixed at the start of the game, while the latter also requires the inputs to be fixed. For a PRF to be used within a composition theorem, adaptive security is required. Finally, Martin et al. [30] provide a MAC which is secure against leakage on the tagging function only. We will use this as the basis of our instantiations, and extend it to achieve security against leakage on verification queries, resolving an open question from their work.

2 Preliminaries

General notation. For assignment of a value U to the variable T we will write $T \leftarrow U$, where U may also be the outcome of some computation. If the variable is a set, we use the shorthand $S \leftarrow_{\cup} U$ for $S \leftarrow S \cup \{U\}$. To assign a value drawn uniformly at random from some finite set B to variable A, we write $A \leftarrow_{\$} B$. By convention, arrays and lists are initialised empty. We use $=$ for equality testing. We write $\mathbb{A} \rightarrow b$, to denote that adversary $\mathbb{A}$ outputs some value b. To define notions etc. we will write $X := Y$ to say that X is defined as some expression Y. The distinguished symbol $\lightning$ denotes an invalid query. The symbol $||$ denotes an *unambiguous* encoding, meaning if $Z \leftarrow X||Y$ it must be possible given Z to uniquely recover X and Y, notated $X||Y \leftarrow Z$, no matter what types X, Y may take. The length $|A|$ is the length of A when expressed as a string of elements of some underlying alphabet Σ (usually $\Sigma = \{0, 1\}$).

Whenever a function is described with a subscript, this will define the first parameter, meaning $f_k(\cdot, \cdot) = f(k, \cdot, \cdot)$. For consistency and clarity of notation, we refer to security definitions in capitals (e.g. IND–CPA) and typeset functions in calligraphic ($\mathcal{E}$), spaces in sans serif (K), "secret" elements in lower case (k), known elements in upper case (M), and adversaries in blackboard bold ($\mathbb{A}$). When we introduce implementations, these will be denoted in bold ($\boldsymbol{\mathcal{E}}$).

Adversarial advantages. We will define our security notions through indistinguishability games where an adversary is given access to one of two collections of oracles. The adversary $\mathbb{A}$ may make queries to these oracles, and eventually outputs a bit. Instead of writing the games in code, we adopt shorthand notation [2] so that the *distinguishing advantage* of $\mathbb{A}$ between two collections of n oracles $(\mathcal{O}_1, \ldots, \mathcal{O}_n)$ and $(\mathcal{P}_1, \ldots, \mathcal{P}_n)$ is defined as

$$\underset{\mathbb{A}}{\Delta} \begin{pmatrix} \mathcal{O}_1, \ldots, \mathcal{O}_n \\ \mathcal{P}_1, \ldots, \mathcal{P}_n \end{pmatrix} := \left| \Pr\left[\mathbb{A}^{\mathcal{O}_1, \ldots, \mathcal{O}_n} \rightarrow 1\right] - \Pr\left[\mathbb{A}^{\mathcal{P}_1, \ldots, \mathcal{P}_n} \rightarrow 1\right] \right|,$$

where the probabilities are taken over the randomness of the oracles, and key $k \leftarrow_{\$} \mathsf{K}$ (note that multiple oracles will often use the same key). We may refer to the oracles by their numerical position: the ith oracle implements either $\mathcal{O}_i$ or $\mathcal{P}_i$ depending which collection the adversary is interacting with.

A scheme is considered secure with respect to a particular security goal if the relevant adversarial advantage is small for all adversaries running within reasonable resources. We do not draw judgement as to what "small" may mean, nor what constitutes "reasonable resources", since these depend heavily on context.

2.1 Authenticated Encryption

Core definitions. Early works to formalize symmetric encryption (cf. [21]) closely followed the precedent for public key encryption. Over the years understanding of what should be expected of symmetric encryption evolved considerably, both in terms of syntax and security. The basis for our work will be the widely accepted nonce-based model using indistinguishability from random bits for confidentiality [39–41]. After introducing this model, we will briefly refer back to an older, non-authenticated version of encryption as it is one of the building blocks later on.

Syntax. An authenticated encryption scheme consists of a pair of deterministic functions Enc and Dec, called encryption and decryption, respectively. Encryption Enc takes four inputs, resulting in a single ciphertext $C \in \mathsf{C}$. Besides the key $k \in \mathsf{K}$ and the message $M \in \mathsf{M}$, the inputs are some associated data $A \in \mathsf{A}$ that will be authenticated but not encrypted, and finally a nonce $N \in \mathsf{N}$ used to ensure that repeat encryptions will not result in repeat ciphertexts. Decryption Dec takes as input again the key, the nonce, and the associated data, in addition to the ciphertext. It outputs a purported message or an error message $\perp \notin \mathsf{M}$.

This syntax can be summarized as

$$\begin{aligned}\mathsf{Enc}\colon \mathsf{K} \times \mathsf{N} \times \mathsf{A} \times \mathsf{M} &\to \mathsf{C}\\ \mathsf{Dec}\colon \mathsf{K} \times \mathsf{N} \times \mathsf{A} \times \mathsf{C} &\to \mathsf{M} \cup \{\perp\}.\end{aligned}$$

In practice, the key space K, nonce space N, associated data A, message space M, and ciphertext space C are generally bitstrings of various lengths. It is common to have $\mathsf{A} = \mathsf{M} = \mathsf{C} = \{0,1\}^*$, and $\mathsf{K} = \mathsf{N} = \{0,1\}^n$ for some security parameter n. That said, our implementation, given in the full version [3] instantiates the various spaces with more general groups (linked to pairings).

We require that an authenticated encryption scheme is both *correct* and *tidy*. These two properties are satisfied iff, for all k, N, A, M, C in the appropriate spaces:

Correctness: $\mathsf{Dec}_k(N, A, \mathsf{Enc}_k, (N, A, M)) = M$
Tidiness: if $\mathsf{Dec}_k(N, A, C) \neq \perp$ then $\mathsf{Enc}_k(N, A, \mathsf{Dec}_k(N, A, C)) = C$

Together, tidiness and correctness imply that decryption is wholly specified by the encryption routine.

Additionally, we require encryption to be *length regular*, which is satisfied if there exists some stretch function $\tau\colon \mathbb{N} \to \mathbb{N}$ such that for all inputs the ciphertext length $|\mathsf{Enc}_k(N, A, M)| = |M| + \tau(|M|)$.

Security notions. Ever since Rogaway and Shrimpton's treatment of deterministic authenticated encryption, it is customary to capture both confidentiality and integrity requirements in a single game. Here the adversary gets oracle access either to the "real" world or to the "ideal" world and needs to distinguish between these two worlds. In the real world, oracle access consists of the encryption and decryption functionalities Enc_k and Dec_k, using a randomly drawn and secret key k. In the ideal world, the encryption oracle is replaced with an oracle \$ that generates randomly drawn ciphertexts and the decryption oracle with an oracle $\perp$ that rejects all ciphertexts. Irrespective of which world the adversary is in, we will refer to the Enc_k vs. \$ oracle as the challenge encryption oracle or as the first oracle (based on the oracle ordering) and to the Dec_k vs. $\perp$ oracle as the challenge decryption (or second) oracle.

We will use a slightly different, but equivalent, formulation where an adversary additionally has access to the true encryption and decryption oracles in both worlds. Thus the adversary will have access to *four* oracles in each world: the challenge encryption oracle, the challenge decryption oracle, the true encryption oracle, and finally the true decryption oracle. Having these extra oracles will help us later on to add leakage, which will only ever be on the true oracles and never on one of the challenge oracles. One could even argue that the additional oracles provide a more representative and expressive framework: the honest oracles describe how an adversary may "learn" about a system, while the challenge ones allow them to "prove" they have done so (cf. a similar, more detailed argument for subtle authenticated encryption [4]).

As our reference point we will use the oracles defined in Fig. 1, with all probabilities taken over randomness of the key and sampling within the oracle.

function $\$^F(X)$	**function** $\perp^G(X)$
$\quad C_0 \leftarrow F(X)$	$\quad$ **return** $\perp$
$\quad C_1 \leftarrow_{\$} \Sigma^{\lvert C_0 \rvert}$	
$\quad$ **return** C_1	

Fig. 1. The generic oracles $\$^F$ and $\perp^G$ idealise the output of F as random elements of Σ, and of G as always rejecting. They are used to define the reference world in our security definitions, for various choices of (F, G), which will be omitted whenever clear. Usually $\Sigma = \{0, 1\}$, with $|C_0|$ the length of C_0 as a bitstring.

Queries. Already in the leak-free setting, certain combinations of queries will easily distinguish the two worlds. To avoid these trivial wins, we will therefore prohibit certain queries—or in some cases simply assume adversaries refrain from making prohibited queries. For example, if an adversary can send a challenge encryption to decryption they can trivially win. As a general rule, we prohibit the same query being made to oracles which take the same inputs (such as the honest and challenge encryption oracles), and also prohibit performing the inverse of previous queries. For example, the ciphertext output from the challenge encryption oracle cannot be passed into the decryption oracle.

If an adversary has made a query (N, A, M) to an encryption oracles (either challenge or true) receiving output C, then making the same query again to one of the encryption oracles or making the query (N, A, C) to one of the decryption oracles (either challenge or true) the original and the new queries are deemed *equivalent*. For any query, we refer to the process of later making an equivalent query as *forwarding* the query, i.e. to make a second query whose inputs were inputs or outputs from the first query. A special case of forwarding a query is *repeating* the query, namely making the same query again to the same oracle. Forwarding queries from challenge to true oracles (or vice versa) or from challenge encryption to challenge decryption oracles (or vice versa) will lead to trivial wins unless oracle behaviour is adapted. Without loss of generality, we will restrict the adversary from making problematic queries instead.

Nonce selection requirements. Our security games will be agnostic over how the nonce is selected, with this property enforced by restricting the adversary. An adversary against an (authenticated) encryption scheme is called *nonce respecting* if whenever making a new query they do not use a nonce more than once to any oracle matching the syntax of Enc_k or $\mathcal{E}_k$. They are *random-iv respecting*, or simply *iv respecting*, if for any new query with these oracles their nonce N (which we term an IV and will generally write as I instead) is sampled uniformly from N immediately prior to querying the oracle (and thus not involved in the logic used to select other elements of the query). These requirements do not apply when interacting with oracles matching the syntax of Dec_k or $\mathcal{D}_k$. A scheme is called *(nonce) misuse resistant* if the adversary does not have to be nonce respecting, providing that the adversary does not make multiple queries using the same (N, A, M) triple.

Definition 1. *Let* Enc *be an authenticated encryption scheme,* $\mathbb{A}$ *an adversary who does forward queries to or from his first or second oracle (and thus does not repeat first oracle queries). Then, the nAE advantage of an adversary* $\mathbb{A}$ *against* Enc *is*

$$\mathbf{Adv}^{\mathrm{AE}}_{\mathsf{Enc}}(\mathbb{A}) := \underset{\mathbb{A}}{\Delta}\begin{pmatrix}\mathsf{Enc}_k, \mathsf{Dec}_k, \mathsf{Enc}_k, \mathsf{Dec}_k \\ \$ \;\;,\;\; \perp \;\;, \mathsf{Enc}_k, \mathsf{Dec}_k\end{pmatrix}.$$

Following our earlier convention, we will refer to a secure *nAE* scheme (or simply nAE) if this nAE advantage is small for all nonce-respecting adversaries running within reasonable resources, and mrAE if it is small for all adversaries running within reasonable resources that might repeat nonces.

Building blocks: Encryption, MACs and PRFs. An authenticated encryption scheme is often constructed out of simpler components, with authenticated encryption security derived from that of its constituent parts. The most common of these are "simple" symmetric encryption (ivE), MACs and PRFs. Here we omit the relevant syntax and security notions of these notions, though in the full version [3] we provide a treatment analogous to that for authenticated encryption above.

Generic composition for nAE. NRS [32] investigated how to construct an nAE scheme by composing two PRFs with an ivE scheme. The IV of the ivE scheme is derived from the nAE's inputs using the first PRF call; the optional second PRF call may be used to create an authentication tag. Different schemes emerge by changing which variables are provided to each of the components. NRS identify eight schemes, dubbed A1–A8, with strong security bounds. For a further four schemes (A9–A12) neither strong security bounds nor insecurity was established. Additionally, NRS investigated mechanisms for combining a PRF with an nE scheme. Three schemes (N1–N3) were found secure, with that of a fourth (N4) remaining unresolved.

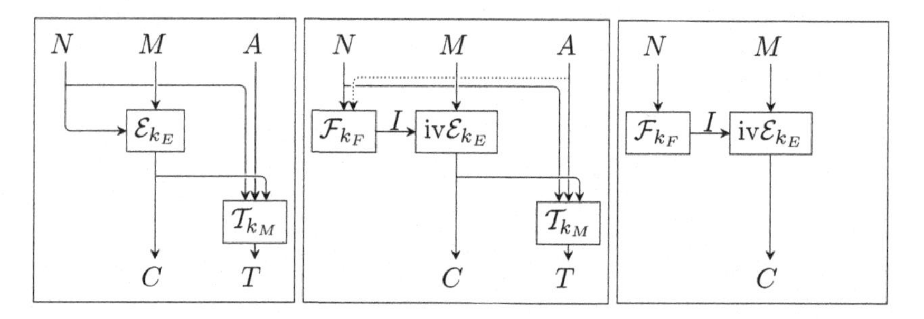

Fig. 2. Graphical representations of the encryption directions of generic composition mechanisms. On the left, N2 converts a nonce-based encryption algorithm $\mathcal{E}$ and MAC scheme $(\mathcal{T}, \mathcal{V})$ into an nAE scheme. On the right, iv2n converts an iv-based encryption scheme iv$\mathcal{E}$ and a PRF into a nonce-based encryption algorithm. Composing these yields A5, shown in the middle ignoring the dotted input, while A6 includes the dotted input. Overall decryption of A5, A6, and N2 will recompute and verify the tag first, only proceeding with further decryption of C if this verification is successful.

Figure 2's middle panel shows the schemes A5 and A6. For these two schemes, as well as for N2 (on the left), the ciphertext is input to the second PRF, which means they classify as Encrypt-then-MAC (EtM). The schemes A4, A7–A12, as well as N3 and N4 only use a single PRF and release the IV as tag; for that reason we refer to them as MAC-then-Encrypt (MtE). Finally, the schemes A1–A3 and N1 use two PRFs that can be called in parallel, leading to their classification as Encrypt-and-MAC (E&M). We refer to NRS for full descriptions and graphical illustrations of all schemes mentioned above.

3 Security Notions Involving Leakage

Authenticated encryption, as defined above, is deterministic. In a leakage-free setting, this provides a stronger notion than the older probabilistic notion of encryption (as implicitly still used for ivE). When introducing leakage, deterministic schemes are problematic both from a practical and a theoretical perspective.

On the one hand, a practical side-channel attack such as differential power analysis can effectively recover keys from unprotected blockciphers and their AE modes with near certainty. Randomized masking based on secret sharing is one of the main countermeasures against these attacks.

On the other hand, theoretical leakage is often modelled as a function on the inputs of the computation, which will include the key. If with each invocation of the scheme an adversary can let the scheme leak a different key bit of its choice, the full key is easily recovered. To prevent such devastating yet simple leakage, a typical design strategy is to split the key in two shares and update the shares on-the-fly using fresh randomness, mimicking the practical approach.

3.1 Implementations versus Functions

In both the practical and the theoretical approaches mentioned above, a deterministic scheme is implemented in a randomized fashion in order to provide resistance against leakage. Therefore, when arguing about leakage, we will need to make a distinction between the *scheme* (a collection of deterministic functions) and its probabilistic *implementation.*

For our definition of the implementations of a function we take our cue from the secret-sharing approach, where a redundant representation of the key is used and this representation is rerandomized as part of the implementation. To enable this rerandomization, we provide the implementation of a function with explicit randomness in Definition 2 below, where we use a bold font to denote either the implementation of a function or the representation of a key used by the implementation.

Definition 2. *An implementation of a function* $f : \mathsf{K} \times \mathsf{X} \to \mathsf{Y}$ *is a deterministic function* $\boldsymbol{f} \colon \mathbf{K} \times \mathsf{X} \times \mathsf{R} \to \mathbf{K} \times \mathsf{Y}$ *along with a probabilistic key initialisation function* $\iota \colon \mathsf{K} \to \mathbf{K}$ *such that* $\iota(k) = \iota(l) \Rightarrow k = l$. *We define the inverse of* ι *as the function* $\iota^{-1} \colon \mathbf{K} \to \mathsf{K} \cup \{\bot\}$ *such that* $\iota^{-1}(\boldsymbol{k}) = k$ *if* $\iota(k)$ *could have resulted in* $\boldsymbol{k}$*, and* $\bot$ *if no such* k *exists.*

The implementation is correct *iff for all* $k \in \mathsf{K}, X \in \mathsf{X}$*, and* $r \in \mathsf{R}$*, setting* $\boldsymbol{k} \leftarrow \iota(k)$ *and* $(\boldsymbol{k}', Y) \leftarrow \boldsymbol{f}(\boldsymbol{k}, X; r)$ *guarantees both* $Y = f(k, X)$ *and* $\iota^{-1}(\boldsymbol{k}') = k$.

The initial representation of the key is generated using the function ι, which maps a key $k \in \mathsf{K}$ to a suitable representation $\boldsymbol{k} \in \mathbf{K}$ for the implementation. We assume that ι is performed only once, and in a leak-free manner, during setup (straight after key generation). Moreover, its inverse ι^{-1} induces an equivalence relation on the space $\mathbf{K}$; in other words, the implementation keys $\boldsymbol{k}$ can be thought of as alternative representations of the key. During evaluation of $\boldsymbol{f}$ the auxiliary input $r \in \mathsf{R}$ is used to refresh the representation; typically this requires a good randomness source to draw r from.

Discussion. Correctness implies that an implementation is identical to the original function when restricted to the second output and that the new key representation $\boldsymbol{k}'$ is equivalent to the initial one $\boldsymbol{k}$. We make no demands of

$\boldsymbol{k}$ or $\boldsymbol{k}'$ beyond these, so it is permissible to set $\boldsymbol{k} = \boldsymbol{k}' = k$ and thus recover the traditional syntax. Our security definitions will be such that for correct schemes and assuming "trivial" leakage, the corresponding leak-free security notions from the preceding section will emerge.

Definition 2 can be linked to practice in a straightforward manner. Recall that practical implementations of blockciphers often use masking based on secret sharing schemes. In this case, the implementation of the blockcipher describes how to evaluate the blockcipher based on the shares of the key as well as how the sharing is refreshed using external randomness r (which need not be leak-free). Furthermore, ι is exactly the function that creates the initial secret sharing of the key.

Syntactically the implementation $\boldsymbol{f}$ may appear stateful: after all they take in some $\boldsymbol{k}$ and output an updated $\boldsymbol{k}'$ for the next invocation. However, since the implementation is of a stateless function f, there is no need to synchronize state between communication parties. Instead, each party can use its own, independent representation of the key.

Implementation of an nAE scheme. For concreteness, we now explicitly define the implementation of an nAE scheme. We assume that **Enc** and **Dec** syntactically use the same representations $\mathbf{K}$ (and key initialisation function ι), which we later use for expressing our security notions.

By correctness of the implementation, one can see that the ciphertext output by **Enc** (resp. message by **Dec**) will always be independent of the randomness r, since they are equal to the corresponding output of Enc (resp. Dec). Definitions for the implementations of other security primitives are written accordingly.

Definition 3. *Let* (Enc, Dec) *be an authenticated encryption scheme. An* AE implementation *is a pair of deterministic functions*

$$\mathbf{Enc}\colon \mathbf{K} \times \mathsf{N} \times \mathsf{A} \times \mathsf{M} \times \mathsf{R} \rightarrow \mathbf{K} \times \mathsf{C}$$
$$\mathbf{Dec}\colon \mathbf{K} \times \mathsf{N} \times \mathsf{A} \times \mathsf{C} \times \mathsf{R} \rightarrow \mathbf{K} \times (\mathsf{M} \cup \{\bot\})$$

along with $\iota\colon \mathsf{K} \rightarrow \mathbf{K}$ *satisfying* $\iota(k) = \iota(l) \Rightarrow k = l$ *and* $\iota^{-1}\colon \mathbf{K} \rightarrow \mathsf{K} \cup \{\bot\}$ *such that* $\iota^{-1}(\boldsymbol{k}) = k$ *if* $\iota(k)$ *could have resulted in* $\boldsymbol{k}$*, and* $\bot$ *if no such* k *exists. The implementation is* correct *iff for any* k, N, A, M, C, r *from the appropriate spaces and* $\boldsymbol{k} \leftarrow_{\$} \iota(k)$*, setting*

$$(\boldsymbol{k}', C') \leftarrow \mathbf{Enc}(\boldsymbol{k}, N, A, M; r) \text{ and } (\boldsymbol{k}'', M') \leftarrow \mathbf{Dec}(\boldsymbol{k}, N, A, C; r),$$

$(\boldsymbol{k}', C') \leftarrow \mathbf{Enc}(\boldsymbol{k}, N, A, M; r)$ *and* $(\boldsymbol{k}'', M') \leftarrow \mathbf{Dec}(\boldsymbol{k}, N, A, C; r)$*, the following properties hold:*

$$k = \iota^{-1}(\boldsymbol{k}) = \iota^{-1}(\boldsymbol{k}') = \iota^{-1}(\boldsymbol{k}'')$$
$$C' = \mathsf{Enc}_k(N, A, M) \text{ and } M' = \mathsf{Dec}_k(N, A, C).$$

3.2 What Constitutes Leakage

Following Micali and Reyzin's approach, we will model leakage by allowing an adversary to specify a leakage function in conjunction with an oracle query. The input signature of the leakage function matches that of the implementation $\boldsymbol{f}$ it relates to, allowing it to wholly simulate the implementation. A leakage set is a collection of leakage functions for an implementation.

Definition 4. *A* leakage function *of an implementation* $\boldsymbol{f}\colon \mathsf{K} \times \mathsf{X} \times \mathsf{R} \rightarrow \mathsf{K} \times \mathsf{Y}$ *is a function* $L\colon \mathsf{K} \times \mathsf{X} \times \mathsf{R} \rightarrow \mathsf{L}$ *for some output leakage space* L*. A* leakage set *of an implementation* $\boldsymbol{f}$ *is a set of leakage functions.*

The choice of leakage set should contain all plausible (functions of) inputs to the implementation that an adversary can compute, and may be probabilistic. This might include functions of any intermediate variables, since these are computable from the inputs simply by simulating the construction. Broadly speaking, the larger the leakage set the more powerful the adversary is likely to be. The leakage set $\emptyset$ allows us to model the leak-free case. Technically we define it to be the set containing just the null function, meaning the adversary can always select a leakage function, thus maintaining the correct syntax for our security games.

3.3 Security Notions Incorporating Leakage

We are now in a position to define the security of an implementation in the presence of leakage. We do so by reframing the classical notions given to work on the implementation of a function, and by extending the notions such that the honest oracles are allowed to leak. The adversary wins the game if they can distinguish whether their leak-free challenge oracles implement the scheme honestly or are idealised. We differentiate our notions from the classic variant by prefixing an "L", for *leakage*.

In the classical setting, each oracle simply evaluates the appropriate function with the game's secret key. For an implementation, a similar, but slightly more complicated, approach is required. The oracle must draw randomness, and provide this to the implementation to update the key representation. This same randomness, along with all other inputs, must be provided to the leakage function. The new representation must then be stored, and the two outputs returned to the adversary. For any implementation $\boldsymbol{f}$, the corresponding leakage oracle is denoted $\ell[\boldsymbol{f}]_{\boldsymbol{k}}$, when initialised with representation $\boldsymbol{k} = \iota(k)$. Code-based descriptions for certain leaky implementations related to authenticated encryption are given in Fig. 3. If an adversary has access to multiple oracles based on the same key, say Enc_k and Dec_k, then we will assume that their respective implementation oracles (so $\ell[\mathbf{Enc}]_{\boldsymbol{k}}$ and $\ell[\mathbf{Dec}]_{\boldsymbol{k}}$) will operate on the same representation $\boldsymbol{k}$, which hence will be initialized only once. Such a shared representation corresponds to a setting where both $\mathbf{Enc}$ and $\mathbf{Dec}$ are implemented on the same device. Needless to say, our security definitions below can be strengthened by allowing an adversary to interact with multiple implementations each using their own representation of the same key.

function $\ell[\boldsymbol{\mathcal{E}}]_k(M;L)$
  $r \leftarrow_{\$} \mathsf{R}$
  $\Lambda \leftarrow L(\boldsymbol{k}, M; r)$
  $C, \boldsymbol{k} \leftarrow \boldsymbol{\mathcal{E}}(\boldsymbol{k}, M; r)$
  return (C, Λ)

function $\ell[\boldsymbol{\mathcal{D}}]_k(C;L)$
  $r \leftarrow_{\$} \mathsf{R}$
  $\Lambda \leftarrow L(\boldsymbol{k}, C; r)$
  $M, \boldsymbol{k} \leftarrow \boldsymbol{\mathcal{D}}(\boldsymbol{k}, C; r)$
  return $(\perp, \Lambda)$

function $\ell[\mathbf{Enc}]_k(N, A, M; L)$
  $r \leftarrow_{\$} \mathsf{R}$
  $\Lambda \leftarrow L(\boldsymbol{k}, N, A, M; r)$
  $C, \boldsymbol{k} \leftarrow \mathbf{Enc}(\boldsymbol{k}, N, A, M; r)$
  return (C, Λ)

function $\ell[\mathbf{Dec}]_k(N, A, C; L)$
  $r \leftarrow_{\$} \mathsf{R}$
  $\Lambda \leftarrow L(\boldsymbol{k}, N, A, C; r)$
  $M, \boldsymbol{k} \leftarrow \mathbf{Dec}(\boldsymbol{k}, N, A, C; r)$
  return (M, Λ)

Fig. 3. Honest leakage oracles an adversary may use to help them distinguish. All inputs are taken from the appropriate spaces, with leakage functions chosen from the relevant leakage set. For $\mathcal{L}_{\mathsf{E}}$-IND–CPLA, the adversary has access to $\ell[\boldsymbol{\mathcal{E}}]_k$, and for the augmented notion $(\mathcal{L}_{\mathsf{E}}, \mathcal{L}_{\mathsf{D}})$-IND-aCPLA they are also given very limited access to $\ell[\boldsymbol{\mathcal{D}}]_k$. LAE security, $(\mathcal{L}_{\mathsf{Enc}}, \mathcal{L}_{\mathsf{Dec}})$-LAE provides access to $(\ell[\mathbf{Enc}]_k, \ell[\mathbf{Dec}]_k)$.

As in the leakage free definitions, security is taken over the randomness of the initial keys, and of the oracles. Notice that this choice includes the sampling from R. We assume the adversary only ever makes queries for which his inputs are selected from the appropriate spaces. For leakage, this means some leakage set that will be specified in the security notion.

For the purposes of defining forwarding of queries, we will ignore the additional input associated to the leakage. For instance, after a query (N, A, M) to the challenge encryption oracle, the query (N, A, M, L) to the true encryption oracle will be considered equivalent—and would constitute forwarding—irrespective of L.

Definition 5. *Let* $(\mathbf{Enc}, \mathbf{Dec})$ *be an implementation of an authenticated encryption scheme* $\mathsf{Enc}, \mathsf{Dec}$*, and* $\mathbb{A}$ *an adversary who does not forward queries to or from his first or second oracles (and thus does not repeat such queries). Then, the* $(\mathcal{L}_{\mathsf{Enc}}, \mathcal{L}_{\mathsf{Dec}})$-LAE advantage *of an adversary* $\mathbb{A}$ *against* $(\mathbf{Enc}, \mathbf{Dec})$ *under leakage* $(\mathcal{L}_{\mathsf{Enc}}, \mathcal{L}_{\mathsf{Dec}})$ *is*

$$\mathbf{Adv}^{\mathrm{LAE}}_{\mathbf{Enc},\mathbf{Dec};\mathcal{L}_{\mathsf{Enc}},\mathcal{L}_{\mathsf{Dec}}}(\mathbb{A}) := \underset{\mathbb{A}}{\Delta}\begin{pmatrix} \mathsf{Enc}_k, \mathsf{Dec}_k, \ell[\mathbf{Enc}]_k, \ell[\mathbf{Dec}]_k \\ \$ \ \ , \ \ \perp \ \ , \ell[\mathbf{Enc}]_k, \ell[\mathbf{Dec}]_k \end{pmatrix}.$$

Definition 6. *Let* $\boldsymbol{\mathcal{E}}$ *be an implementation of an encryption scheme* $\mathcal{E}$*, and* $\mathbb{A}$ *an adversary who never forwards queries to or from his first oracle (and thus does not repeat first oracle queries). The* $\mathcal{L}_{\mathsf{E}}$*-IND–CPLA advantage (named for chosen-plaintext-with-leakage-attack) of* $\mathbb{A}$ *against* $\boldsymbol{\mathcal{E}}$ *is*

$$\mathbf{Adv}^{\mathrm{IND-CPLA}}_{\boldsymbol{\mathcal{E}};\mathcal{L}_{\mathsf{E}}}(\mathbb{A}) := \underset{\mathbb{A}}{\Delta}\begin{pmatrix} \mathcal{E}_k, \ell[\boldsymbol{\mathcal{E}}]_k \\ \$ \ , \ell[\boldsymbol{\mathcal{E}}]_k \end{pmatrix}.$$

We next provide an additional encryption notion, IND–aCPLA, that will be required for our composition results later. It describes a modified version of the IND–CPLA game in which the adversary is also allowed leakage from the decryption implementation $\ell[\boldsymbol{\mathcal{D}}]_k$ (see Fig. 3), but *only* on ciphertexts they have previously received from $\ell[\boldsymbol{\mathcal{E}}]_k$. At first glance, this appears to be more similar to an IND–CCA style notion, but we emphasise this is not the case since the possible decryption queries are heavily restricted. Thus it should be thought of as IND–CPA under the most general form of leakage. Indeed, when the leakage sets are empty, the resulting security notion is equivalent to IND–CPA.

Definition 7. *Let* $(\boldsymbol{\mathcal{E}}, \boldsymbol{\mathcal{D}})$ *be an implementation of an encryption scheme,* $\mathbb{A}$ *an adversary who does not forward queries to or from his first oracle, and only makes queries to their third oracle that were forwarded from the second. Then the* $(\mathcal{L}_{\mathsf{E}}, \mathcal{L}_{\mathsf{D}})$*-IND–aCPLA advantage of* $\mathbb{A}$ *against* $\boldsymbol{\mathcal{E}}$ *is*

$$\mathbf{Adv}^{\mathrm{IND-aCPLA}}_{\boldsymbol{\mathcal{E}},\boldsymbol{\mathcal{D}};\mathcal{L}_{\mathsf{E}},\mathcal{L}_{\mathsf{D}}}(\mathbb{A}) := \underset{\mathbb{A}}{\Delta}\begin{pmatrix}\mathcal{E}_k, \ell[\boldsymbol{\mathcal{E}}]_k, \ell[\boldsymbol{\mathcal{D}}]_k \\ \$, \ell[\boldsymbol{\mathcal{E}}]_k, \ell[\boldsymbol{\mathcal{D}}]_k\end{pmatrix}.$$

The IND–aCPLA notion is required for the general composition, where the goal is to construct an LAE scheme from an ivLE scheme (and other components). However, for decryption of the LAE scheme to leak (as we want the leakage to be as powerful as possible), the decryption of ivLE scheme would have to leak. The IND–CPLA security notion does not capture this. Consider an IND–CPA scheme where encryption does not leak, but the leakage from decrypting the zero string returns the key. Clearly the scheme is also IND–CPLA but will trivially break when the adversary is given decryption leakage. The IND–aCPLA notion is trying to capture that decryption "does not leak too much information", so that limited decryption queries made by the LAE scheme will be able to leak.

Against many natural choices of leakage sets, $(\mathcal{L}_{\mathsf{E}}, \mathcal{L}_{\mathsf{D}})$-IND–aCPLA and $\mathcal{L}_{\mathsf{E}}$-IND–CPLA are equivalent, since the encryption oracle often suffices to simulate any leakage from decryption. In the nonce-abusing setting (where the adversary is free to select nonces however they wish) there is an obvious mechanism for proving the equivalence, using repeat encryption queries to simulate leaking decryption queries, but even this requires rather strong assumptions on the leakage sets.

In the nonce respecting or iv respecting scenarios such a general reduction is not possible, because there is no way to allow the adversary to use the same nonce multiple times, something a decryption oracle would allow. If the leakage is independent of the nonce (for example) similar results can be recovered, but these are much more restrictive scenarios. It is an interesting open problem to describe sets $\mathcal{L}_{\mathsf{ED}}$ that are in some sense "minimal" for various pairs of leakage sets $(\mathcal{L}_{\mathsf{E}}, \mathcal{L}_{\mathsf{D}})$ taken from some general function classes.

LMAC and LPRF. Here we give the PRF and MAC notions a similar treatment to the encryption definitions by enhancing the standard definitions to incorporate leakage.

The LPRF definition below strengthens earlier definitions by Dodis and Pietrzak [13], and by Faust et al. [15]. In our definition (Fig. 4) both the leakage functions and the inputs can be chosen adaptively based on outputs already seen by the adversary.

function $\ell[\boldsymbol{\mathcal{T}}]_k(M; L)$	**function** $\ell[\boldsymbol{\mathcal{V}}]_k(M, T; L)$
$r \leftarrow_\$ \mathsf{R}$	$r \leftarrow_\$ \mathsf{R}$
$\Lambda \leftarrow L(\boldsymbol{k}, M; r)$	$\Lambda \leftarrow L(\boldsymbol{k}, M, T; r)$
$T, \boldsymbol{k} \leftarrow \boldsymbol{\mathcal{T}}(\boldsymbol{k}, M; r)$	$V, \boldsymbol{k} \leftarrow \boldsymbol{\mathcal{V}}(\boldsymbol{k}, M, T; r)$
return (T, Λ)	**return** (V, Λ)

Fig. 4. Honest leakage oracles an adversary may use to help them distinguish. All inputs are taken from the appropriate spaces, with leakage functions chosen from the relevant leakage set. $(\mathcal{L}_\mathsf{T}, \mathcal{L}_\mathsf{V})$-LMAC security gives access to $(\ell[\boldsymbol{\mathcal{T}}]_k, \ell[\boldsymbol{\mathcal{V}}]_k)$. Since PRFs and the tagging function of a MAC have the same syntax, the LPRF game provides access to $\ell[\boldsymbol{\mathcal{F}}]_k$, which is identical to $\ell[\boldsymbol{\mathcal{T}}]_k$.

Definition 8. *Let $\boldsymbol{\mathcal{F}}$ be an implementation of a function $\mathcal{F}$, and $\mathbb{A}$ an adversary who never forwards or repeats queries. Then the $\mathcal{L}_\mathsf{F}$-PRLF advantage of $\mathbb{A}$ against $\boldsymbol{\mathcal{F}}$ under leakage $\mathcal{L}_\mathsf{F}$ is*

$$\mathbf{Adv}^{\mathrm{PRLF}}_{\boldsymbol{\mathcal{F}};\mathcal{L}_\mathsf{F}}(\mathbb{A}) := \underset{\mathbb{A}}{\Delta}\begin{pmatrix} \mathcal{F}_k, \ell[\boldsymbol{\mathcal{F}}]_k \\ \$\ , \ell[\boldsymbol{\mathcal{F}}]_k \end{pmatrix}.$$

Our notion of strong existential unforgeability under chosen message with leakage (below) strengthens both the classical definition, and the leakage definition of Martin et al. [30] (they only allow tagging to leak; setting $\mathcal{L}_\mathsf{V} = \emptyset$ recovers their definition). Allowing meaningful leakage on $\boldsymbol{\mathcal{T}}$ hampers direct use of a secure LPRF as a MAC as typically during verification the "correct" tag would be recomputed as output of the PRF and could consequently be leaked upon (effectively yielding a surreptitious tagging algorithm).

Definition 9. *Let $(\boldsymbol{\mathcal{T}}, \boldsymbol{\mathcal{V}})$ be an implementation of a MAC $(\mathcal{T}, \mathcal{V})$, and $\mathbb{A}$ an adversary who does not forward queries from his second oracle to the first. Then the $(\mathcal{L}_\mathsf{T}, \mathcal{L}_\mathsf{V})$-sEUF-CMLA advantage of $\mathbb{A}$ against $(\boldsymbol{\mathcal{T}}, \boldsymbol{\mathcal{V}})$ under leakage $(\mathcal{L}_\mathsf{T}, \mathcal{L}_\mathsf{V})$ is*

$$\mathbf{Adv}^{\mathrm{sEUF-CMLA}}_{\boldsymbol{\mathcal{T}},\boldsymbol{\mathcal{V}};\mathcal{L}_\mathsf{T},\mathcal{L}_\mathsf{V}}(\mathbb{A}) := \underset{\mathbb{A}}{\Delta}\begin{pmatrix} \mathcal{V}_k, \ell[\boldsymbol{\mathcal{T}}]_k, \ell[\boldsymbol{\mathcal{V}}]_k \\ \bot, \ell[\boldsymbol{\mathcal{T}}]_k, \ell[\boldsymbol{\mathcal{V}}]_k \end{pmatrix}.$$

Note that we cast unforgeability as a distinguishing game, rather than as a more usual computational game ("adversary must forge a tag"), but it is straightforward to show equivalence (even in the face of leakage).

4 Applying LAE to Attacks in Theory and Practice

A security framework is not much use if it does not highlight the difference between schemes for which strong security results are known, and those against which efficient attacks exist. In this section we discuss the types of leakage normally considered within the literature. We show how previous leakage models can be captured by our leakage set style notion. In the literature there is focus on two types of leakage; protocol leakage (by the AE literature) and side channel leakage (by the leakage resilient literature). We believe that these two notions are highly related and thus we discuss how to capture both. For example, termination of an algorithm at different points (distinguishable decryption failures) is normally detected by a side-channel; timing can be used to capture this if the failures terminate the algorithm at different points in time and power can be used to detect if conditional branches were taken.

Below we recast existing leakage resilience work within our general framework. For completeness, in the full version [3] we describe an existing attack (against GCM) within our setting.

4.1 Theoretical Leakage Models

We observe that our model is in many ways the most general possible, and that many previous leakage notions can be captured as version of the $(\mathcal{L}_{\mathsf{E}}, \mathcal{L}_{\mathsf{D}})$-LAE security game for suitable choice of leakage sets $(\mathcal{L}_{\mathsf{E}}, \mathcal{L}_{\mathsf{D}})$. Reassuringly, by setting $(\mathcal{L}_{\mathsf{E}}, \mathcal{L}_{\mathsf{D}}) = (\emptyset, \emptyset)$ we recover the traditional leakage-free security notions, with $(\emptyset, \emptyset)$-nLAE equivalent to nAE, and both $\emptyset$-IND–CPLA and $(\emptyset, \emptyset)$-IND–aCPLA equivalent to IND–CPA, meaning a secure nE scheme is $\emptyset$-nLE secure.

The deterministic decryption leakage notions from the AE literature can be recovered by choosing the appropriate leakage set. The SAE framework generalises both the RUP model, and (nonce-based analogues of) the Distinguishable Decryption Failure notions of Boldyreva et al. [2,4,9]. The security notions are parametrised by a deterministic decryption leakage function Λ, corresponding to security under the leakage sets $(\mathcal{L}_{\mathsf{E}}, \mathcal{L}_{\mathsf{D}}) = (\emptyset, \{\Lambda\})$. Thus the strongest notions available in these settings are equivalent to $(\emptyset, \{\Lambda\})$–LAE. Several of their weaker notions translate to the corresponding weakening of this, including authenticity under deterministic leakage, (known variously as CTI–sCPA, INT–RUP or an extended form of INT–CTXT), which translates to a variant of $(\emptyset, \{\Lambda\})$–LAE in which the adversary cannot query the encryption challenge oracle (and thus does not interact with either $\mathcal{E}_k$ or \$).

In the simulatable leakage model (e.g. [46]), the adversary receives leakage in addition to their query, but is restricted to leakage functions that can be simulated without the key. The simulatable model considered by Standaert et al. (for example) can be captured by our model by having set of leakage functions contain the single function which provides the power trace to the adversary. The auxiliary input model [12] gives the adversary the output of a hard to invert function applied to the key, alongside the normal security notion interactions. The only computation leaks model [31] (discussed in more detail in Sect. 5.1) restricts the adversary to leakage functions that can be locally computed: any step of the algorithm can only leak on variables being used at that point. In the following sections we show how this leakage set can be defined for our given constructions.

In the probing model [20] the adversary can gain access to the values of t of the internal wires from the computation. A scheme is secure if an adversary with the knowledge of t internal wires can do no better than if they had access to the function in a black box manner. If there are n internal wires, this leakage can be captured by our set notation by constructing a set with n choose t leakage functions, each giving the complete value of the relevant wires.

Our leakage sets incorporate the bounded leakage model (e.g. [18,22,26]) by restricting the set of allowable adversaries to those who only make sufficiently few queries to the leakage oracles.

One mechanism that need not rely on randomness is to instead use a leak-free component [48]. Although instantiating such components in practice is between hard and impossible [29], our framework nonetheless supports it (by suitable choice of leakage set).

Another idea to provide security is frequent rekeying. However, such a solution relies on synchronized states between encryption and decryption which can be difficult to maintain, thereby restricting applicability of this approach. However, in specific contexts such as secure channels, synchronization might not be too onerous.

5 Generic Composition for LAE

5.1 Modelling Composed Leakage

Our challenge is to establish to what extent the NRS schemes remain secure when taking leakage into account. Ideally, we would like to claim that if both the ivE and the PRFs are secure in the presence of leakage, then so will the composed nAE be. To make such a statement precise, the leakage classes involved need to be specified. We opt for an approach where the leakage classes for the components are given (and can be arbitrary) and then derive a leakage class for the resulting nAE for which we can prove security.

Encryption leakage. In a nutshell, we define the leakage of the composition as the composition of the leakage. As an example, consider an implementation of A5 (Fig. 2). When encrypting, the leakage may come from any of the components: the PRF $\boldsymbol{\mathcal{F}}$ may leak some information $L_F(\boldsymbol{k}_F, N; r_F)$; the IV-encryption routine $\mathbf{iv}\boldsymbol{\mathcal{E}}$ might leak some information $L_E(\boldsymbol{k}_E, I, M; r_E)$; the Tag function $\boldsymbol{\mathcal{T}}$ may leak some information $L_T(\boldsymbol{k}_M, N, A, C_e; r_M)$. To ease notation, we will use the shorthand $L_F(\star), L_E(\star)$, and $L_T(\star)$ respectively for these leakages. In that case, we say that the leakage on the authenticated encryption operation as a whole consists of the triple $(L_F(\star), L_E(\star), L_T(\star))$. Under the hood, this implies some parsing and forwarding of the AE's key $(\boldsymbol{k}_F, \boldsymbol{k}_E, \boldsymbol{k}_M)$, randomness (r_F, r_E, r_M)

and inputs N, A, M, including the calculated values I and C_e, to the component leakage functions L_F, L_E, and L_T.

Expanding the above to classes of functions is as follows. Let $\mathcal{L}_\mathsf{F}, \mathcal{L}_\mathsf{E}$, and $\mathcal{L}_\mathsf{T}$ be the respective leakage classes for $\boldsymbol{\mathcal{F}}, \mathbf{iv}\boldsymbol{\mathcal{E}}$, and $\boldsymbol{\mathcal{T}}$. Then the leakage class $\mathcal{L}_\mathsf{Enc}$ for the resulting authenticated encryption scheme is defined as

$$\{(L_F, L_E, L_T)|L_F \in \mathcal{L}_\mathsf{F}, L_E \in \mathcal{L}_\mathsf{E}, L_T \in \mathcal{L}_\mathsf{T}\}.$$

Since an adversary has to select a leakage function in $\mathcal{L}_\mathsf{Enc}$ the moment it queries the encryption oracle, it will not be possible to adaptively select for instance the leakage function L_T based on the leakage received from L_E *of that encryption query*.

Decryption leakage. In order to describe leakage from decryption, we take a closer look at the role of the two PRFs in the generic constructions. The first one, $\boldsymbol{\mathcal{F}}$, computes the initial vector which is needed both for encryption and decryption. This makes it inevitable that during decryption $\mathcal{F}$ is again computed as a PRF, presumably using the same implementation $\boldsymbol{\mathcal{F}}$. On the other hand, the second PRF, $\boldsymbol{\mathcal{T}}$, is used to create a tag T during encryption. Normally during decryption one would recompute the tag (again using $\boldsymbol{\mathcal{T}}$) and check whether the recomputed tag T' equals the received tag T. Yet, in the leakage setting this approach is problematic: T' is the correct tag and its recomputation might well leak it, even when used (repeatedly) to check an incorrect and completely unrelated T. Hence, during decryption we will not use a recompute-and-check model, but rather refer directly to a tag-verification implementation $\boldsymbol{\mathcal{V}}$ (that hopefully leaks less).

When considering the decryption leakage of A5, we will assume that, on invalid ciphertexts, the computation terminates as soon as the verification algorithm returns $\perp$. This implies that for invalid ciphertexts only leakage on $\boldsymbol{\mathcal{V}}$ will be available, whereas for valid ciphertexts all three components ($\boldsymbol{\mathcal{V}}, \boldsymbol{\mathcal{F}}$, and $\mathbf{iv}\boldsymbol{\mathcal{E}}$) might leak.

Overview and interpretation. Recall that we divided the NRS schemes in three categories: MtE, M&E, and EtM. Figure 5 shows how the composed leakage will leak for each of these schemes. For completeness, we also listed the leakage for the EtM scheme (such as A5) in case full decryption will always take place, even for invalid ciphertexts (where one could have aborted early).

Our choice to model the leakage from the authenticated encryption scheme as completely separate components from the three underlying primitives is rooted in the assumption that only computation leaks. This assumption was first formalized by Micali and Reyzin [31] and, although there are counterexamples to the assumption at for instance the gate level [38], we believe that implementations of the three primitives result in large enough physical components, which can be suitably segregated to avoid cross-leakage.

Leakage on the wire (for instance of the initial vector I) can be captured as leakage of the PRF computing the I or alternatively as that of the ivE. In

Structure	Leakage	Inverse	Inverse Leakage
MtE	$L_T(\star)$, $L_F(\star)$, $L_E(\star)$	DtV	$L_F(\star)$, $L_D(\star)$, $L_V(\star)$
M&E	$L_T(\star)$, $L_F(\star)$, $L_E(\star)$		
EtM	$L_F(\star)$, $L_E(\star)$, $L_T(\star)$	D&V	$L_F(\star)$, $L_D(\star)$, $L_V(\star)$
		VtD	$\begin{cases} L_V(\star) & \text{if } \mathcal{V}(\star) = \bot \\ L_V(\star), L_F(\star), L_D(\star) & \text{if } \mathcal{V}(\star) = \top \end{cases}$

Fig. 5. The structure of a leakage function from a composition scheme based on the order of its primitives. The exact input parameters to the leakage function vary per scheme, so have been replaced with $\star$: the different $\star$ variables are not the same. On the left are the encryption structures MtE, M&E and EtM, along with the associated leakage function. The right gives the associated inverse: DtV (Decrypt then Verify) is the only way of inverting MtE or M&E schemes. EtM schemes can be inverted in any order, as DtV, D&V (Decrypt and Verify) or VtD (Verify then Decrypt). All constructions have the same encryption leakage, and most have the same decryption leakage. The only one that is different is an EtM–VtD scheme, where the decryption leakage format depends on the validity of the ciphertext.

particular, by letting the decryption of the $\mathbf{iv}\boldsymbol{\mathcal{E}}$ component leak its full output (while not allowing any further leakage), we capture the release of unverified plaintext. Furthermore, distinguishable decryption failures on MtE and M&E schemes invariably arise from verification, which might incorporate a padding check as well. This is modelled by allowing $\mathcal{V}$ to leak, but not any of the other components.

5.2 MAC-and/then-Encrypt are Brittle Under Leakage

For schemes where the plaintext is input to the MAC (i.e. MtE and M&E schemes), decryption is inevitably of the form DtV. Consequently, during decryption a purported message M is computed before the tag can be verified. Leaking this message M corresponds to the release of unverified plaintext [2], but even more modest leakage, such as the first bit of the candidate message, can be insecure as we show by the following example.

Let us assume for a moment that the encryption routine $\mathbf{iv}\boldsymbol{\mathcal{E}}$ is online, so that reencrypting a slightly modified plaintext using the same I will only affect a change in the ciphertext after the modification in the plaintext. CBC and CFB modes are well-known examples of online $\mathbf{iv}\boldsymbol{\mathcal{E}}$ schemes. Additionally, assume that $\mathbf{iv}\boldsymbol{\mathcal{E}}$'s decryption routine indeed leaks the first bit of the message. Then the authenticated encryption scheme is not secure in the presence of leakage (for the leakage class derived according to the principles outlined previously), which an attack demonstrates.

The adversary first submits a message M to its challenge encryption oracle, receiving a ciphertext C which either is an encryption $\mathcal{E}_k(M||T)$ or, in the ideal world, a uniformly random string. The adversary subsequently queries its decryption-with-leakage oracle on C with its final bit flipped. In the real world, where $C = \mathcal{E}_k(M||T)$, the leakage will then equal the first bit of M with probability 1. Yet in the ideal world, C is independent of M, so the leakage will equal the first bit of M with probability half. Thus, testing whether the decryption leakage equals the first bit of M leads to a distinguisher with a significant advantage. However, this does not invalidate IND–aCPLA security of $\mathbf{iv}\boldsymbol{\mathcal{E}}$ as in that game decryption only leaks on valid ciphertexts with known plaintexts.

The above observation implies that for schemes where decryption follows a DtV or D&V structure proving generic composition secure in the presence of leakage is impossible. This affects the NRS compositions A1–A4, A7–A12, N1, N3 and N4; none of which can be regarded as generically secure under leakage and *all* are insecure when using online iv$\mathcal{E}$ and releasing unverified plaintext.

Less general composition results might still be possible, for instance by restricting the leakage classes of the primitives. After all, in the trivial case that the leakage classes are all $\emptyset$, the original NRS results hold directly. We leave open whether significantly larger realistic leakage classes exist leading to secure MtE constructions.

Alternatively, stronger assumptions on $\mathcal{E}$ could help. For instance, if $\mathcal{E}$'s security matches that of a tweakable (variable input length) cipher, the MAC-then-Encrypt constructions become a sort of encode-then-encipher. The latter is secure against release of unverified plaintext [19]. We leave open the identification of sufficient conditions on $\mathcal{E}$ for a generic composition result in the presence of leakage to pull through for EtM or E&M; relatedly, we leave open the extension of our work to the encode-then-encipher setting.

5.3 Encrypt-then-MAC is Secure Under Leakage

The iv-based schemes A5 and A6, as well as the nonce-based N2, all fall under the EtM design. The inverse of an EtM scheme can be D&V or VtD, but as just discussed for the D&V variant no meaningful generic security is possible; henceforth we restrict attention to the VtD variant only. These schemes, along with the iv2n mechanism for building a nonce-based encryption scheme out of an iv-based one, are all represented in Fig. 2. Before proving their security, we begin with some observations about EtM–VtD designs in the face of leakage.

Initial observations. Since the final ciphertext will be formed from an encryption ciphertext and a tag, if the overall output is to be indistinguishable from random bits, then so must the tag. Thus we require both that $(\boldsymbol{\mathcal{T}}, \boldsymbol{\mathcal{V}})$ is a secure $(\mathcal{L}_{\mathsf{T}}, \mathcal{L}_{\mathsf{V}})$-LMAC, and that $\boldsymbol{\mathcal{T}}$ is a secure $\mathcal{L}_{\mathsf{T}}$-LPRF. Shrimpton and Terashima [45] defined a (weaker) authenticated encryption notion where the "recovery information" does not need to be random—only the ciphertext—in which case one may drop the second requirement.

In the traditional case, it is possible to build secure EtM schemes from an encryption scheme that is IND–CPA secure. After all, by assumption on the security of the MAC, the only output the adversary can ever receive from the internal decryption function $\mathcal{D}$ is a plaintext corresponding to a previous $\mathcal{E}$ query.

However, when leakage is involved, this previously harmless decryption query suddenly allows the adversary to evaluate a leakage function $L \in \mathcal{L}_{\mathsf{D}}$, albeit on a (N, C) pair for which they already know the corresponding plaintext. If $\mathcal{L}_{\mathsf{D}}$ contained functions revealing sufficient information about the key, this would render the composed scheme completely broken, notwithstanding any IND–CPLA security. Luckily, the augmented IND–aCPLA game in which the adversary is allowed to leak on select decryption queries, is sufficiently nuanced to capture relevant weaknesses in the decryption's implementation.

Security of EtM composition schemes. We now describe the security of the composition schemes A5, A6 and N2, and the iv2n construction. Working under the assumption of OCLI-style leakage, as described in Sect. 5.1, we will reduce the security of the composition to the security of its components. Technically the bound includes a term quantifying any additional weaknesses due to the composition scheme, but in all cases this term is zero. The proofs can be found in the full version [3]. We begin with N2, and show it is essentially as secure as the weakest of its components, by constructing explicit adversaries against each.

Theorem 1. *Let* $(\mathcal{L}_{\mathsf{E}}, \mathcal{L}_{\mathsf{D}}, \mathcal{L}_{\mathsf{T}}, \mathcal{L}_{\mathsf{V}})$ *be leakage sets for the appropriate primitives, and define* $(\mathcal{L}_{\mathsf{Enc}}, \mathcal{L}_{\mathsf{Dec}})$ *as in Sect. 5.1. Let* $\mathbb{A}$ *be an adversary against the* $(\mathcal{L}_{\mathsf{Enc}}, \mathcal{L}_{\mathsf{Dec}})$*-nLAE security of* $\mathsf{N2}[\boldsymbol{\mathcal{E}}, \boldsymbol{\mathcal{D}}; \boldsymbol{\mathcal{T}}, \boldsymbol{\mathcal{V}}]$*. Then, there exist adversaries* $\mathbb{A}_{\mathrm{CPA}}$*,* $\mathbb{A}_{\mathrm{PRF}}$ *and* $\mathbb{A}_{\mathrm{MAC}}$ *against the* $(\mathcal{L}_{\mathsf{E}}, \mathcal{L}_{\mathsf{D}})$*-nLE security of* $(\boldsymbol{\mathcal{E}}, \boldsymbol{\mathcal{D}})$*, the* $\mathcal{L}_{\mathsf{T}}$*-LPRF security of* $\boldsymbol{\mathcal{T}}$ *and the* $(\mathcal{L}_{\mathsf{T}}, \mathcal{L}_{\mathsf{V}})$*-LMAC security of* $(\boldsymbol{\mathcal{T}}, \boldsymbol{\mathcal{V}})$ *such that:*

$$\mathbf{Adv}^{\mathrm{nLAE}}_{\mathsf{N2};\mathcal{L}_{\mathsf{Enc}},\mathcal{L}_{\mathsf{Dec}}}(\mathbb{A}) \leq$$
$$\mathbf{Adv}^{\mathrm{IND-aCPLA}}_{\boldsymbol{\mathcal{E}},\boldsymbol{\mathcal{D}};\mathcal{L}_{\mathsf{E}},\mathcal{L}_{\mathsf{D}}}(\mathbb{A}_{\mathrm{CPA}}) + \mathbf{Adv}^{\mathrm{LPRF}}_{\boldsymbol{\mathcal{T}};\mathcal{L}_{\mathsf{T}}}(\mathbb{A}_{\mathrm{PRF}}) + 2 \cdot \mathbf{Adv}^{\mathrm{sEUF-CMLA}}_{\boldsymbol{\mathcal{T}},\boldsymbol{\mathcal{V}};\mathcal{L}_{\mathsf{T}},\mathcal{L}_{\mathsf{V}}}(\mathbb{A}_{\mathrm{MAC}}).$$

As the following result shows, the intuitive mechanism for building a nLE scheme from a secure ivLE scheme and a secure LPRF is itself secure. While unsurprising, this will allow us to instantiate the N2 construction with the more common object of an ivLE scheme.

Theorem 2. *Let* $(\mathcal{L}_{\mathsf{ivE}}, \mathcal{L}_{\mathsf{ivD}}, \mathcal{L}_{\mathsf{F}})$ *be leakage sets for the appropriate primitives, and define* $(\mathcal{L}_{\mathsf{E}}, \mathcal{L}_{\mathsf{D}})$ *as in Sect. 5.1. Let* $\mathbb{A}$ *be an adversary against the* $(\mathcal{L}_{\mathsf{E}}, \mathcal{L}_{\mathsf{D}})$*-nLE security of* $\mathsf{iv2n}[\mathbf{iv}\boldsymbol{\mathcal{E}}, \mathbf{iv}\boldsymbol{\mathcal{D}}; \boldsymbol{\mathcal{F}}]$*. Then, there exist* $\mathbb{A}_{\mathrm{CPA}}$*,* $\mathbb{A}_{\mathrm{PRF}}$ *against the* $(\mathcal{L}_{\mathsf{ivE}}, \mathcal{L}_{\mathsf{ivD}})$*-ivLE security of* $(\mathbf{iv}\boldsymbol{\mathcal{E}}, \mathbf{iv}\boldsymbol{\mathcal{D}})$*, and the* $\mathcal{L}_{\mathsf{F}}$*-LPRF security of* $\boldsymbol{\mathcal{F}}$ *respectively, such that:*

$$\mathbf{Adv}^{\mathrm{IND-aCPLA}}_{\mathsf{iv2n};\mathcal{L}_{\mathsf{E}},\mathcal{L}_{\mathsf{D}}}(\mathbb{A}) \leq \mathbf{Adv}^{\mathrm{LPRF}}_{\boldsymbol{\mathcal{F}};\mathcal{L}_{\mathsf{F}}}(\mathbb{A}_{\mathrm{PRF}}) + \mathbf{Adv}^{\mathrm{IND-aCPLA}}_{\mathbf{iv}\boldsymbol{\mathcal{E}},\mathbf{iv}\boldsymbol{\mathcal{D}};\mathcal{L}_{\mathsf{ivE}},\mathcal{L}_{\mathsf{ivD}}}(\mathbb{A}_{\mathrm{CPA}}).$$

Pulling these two results together and taking the maximum over the similar adversaries, we are able to prove the security of the A5 construction. The security of A6 against adversaries who never repeat the pair (N, A) can be easily recovered from this by considering it as an equivalent representation of the A5 scheme acting on nonce space $\mathsf{N}' = \mathsf{N} \times \mathsf{A}$ but with no associated data.

Corollary 1 (nLAE from ivLE and LPRF via A5 composition). *Let* $(\mathcal{L}_{\mathsf{ivE}}, \mathcal{L}_{\mathsf{ivD}}, \mathcal{L}_{\mathsf{T}}, \mathcal{L}_{\mathsf{V}}, \mathcal{L}_{\mathsf{F}})$ *be leakage sets for the appropriate primitives, and define* $(\mathcal{L}_{\mathsf{Enc}}, \mathcal{L}_{\mathsf{Dec}})$ *as in Sect. 5.1. Let* $\mathbb{A}$ *be an adversary against the* $(\mathcal{L}_{\mathsf{Enc}}, \mathcal{L}_{\mathsf{Dec}})$*-nLAE security of* $\mathsf{A5}[\mathbf{iv}\boldsymbol{\mathcal{E}}, \mathbf{iv}\boldsymbol{\mathcal{D}}; \boldsymbol{\mathcal{F}}; \boldsymbol{\mathcal{T}}, \boldsymbol{\mathcal{V}}]$*. Then, there exist adversaries* $\mathbb{A}_{\mathrm{CPA}}$, $\mathbb{A}_{\mathrm{PRF}}$, $\mathbb{A}'_{\mathrm{PRF}}$, *and* $\mathbb{A}_{\mathrm{MAC}}$ *against the* $(\mathcal{L}_{\mathsf{ivE}}, \mathcal{L}_{\mathsf{ivD}})$*-ivLE security of* $(\mathbf{iv}\boldsymbol{\mathcal{E}}, \mathbf{iv}\boldsymbol{\mathcal{D}})$*, the* $\mathcal{L}_{\mathsf{F}}$*-LPRF security of* $\boldsymbol{\mathcal{F}}$*, the* $\mathcal{L}_{\mathsf{T}}$*-LPRF security of* $\boldsymbol{\mathcal{T}}$ *and the* $(\mathcal{L}_{\mathsf{T}}, \mathcal{L}_{\mathsf{V}})$*-LMAC security of* $(\boldsymbol{\mathcal{T}}, \boldsymbol{\mathcal{V}})$ *such that*

$$\begin{aligned}\mathbf{Adv}^{\mathrm{nLAE}}_{\mathsf{A5};\mathcal{L}_{\mathsf{Enc}},\mathcal{L}_{\mathsf{Dec}}}(\mathbb{A}) \le \mathbf{Adv}^{\mathrm{IND-aCPLA}}_{\boldsymbol{\mathcal{E}},\boldsymbol{\mathcal{D}};\mathcal{L}_{\mathsf{ivE}},\mathcal{L}_{\mathsf{ivD}}}(\mathbb{A}_{\mathrm{CPA}}) + \mathbf{Adv}^{\mathrm{LPRF}}_{\boldsymbol{\mathcal{F}};\mathcal{L}_{\mathsf{F}}}(\mathbb{A}_{\mathrm{PRF}})\\ + \mathbf{Adv}^{\mathrm{LPRF}}_{\boldsymbol{\mathcal{T}};\mathcal{L}_{\mathsf{T}}}(\mathbb{A}'_{\mathrm{PRF}}) + 2\cdot \mathbf{Adv}^{\mathrm{EUF-CMLA}}_{\boldsymbol{\mathcal{T}},\boldsymbol{\mathcal{V}};\mathcal{L}_{\mathsf{T}},\mathcal{L}_{\mathsf{V}}}(\mathbb{A}_{\mathrm{MAC}}).\end{aligned}$$

Corollary 2 (nLAE from ivLE and LPRF via A6 composition). *Let* $(\mathcal{L}_{\mathsf{ivE}}, \mathcal{L}_{\mathsf{ivD}}, \mathcal{L}_{\mathsf{T}}, \mathcal{L}_{\mathsf{V}}, \mathcal{L}_{\mathsf{F}})$ *be leakage sets for the appropriate primitives, and define* $(\mathcal{L}_{\mathsf{Enc}}, \mathcal{L}_{\mathsf{Dec}})$ *as in Sect. 5.1. Let* $\mathbb{A}$ *be an adversary against the* $(\mathcal{L}_{\mathsf{Enc}}, \mathcal{L}_{\mathsf{Dec}})$*-LAE security of* $\mathsf{A6}[\mathbf{iv}\boldsymbol{\mathcal{E}}, \mathbf{iv}\boldsymbol{\mathcal{D}}; \boldsymbol{\mathcal{F}}; \boldsymbol{\mathcal{T}}, \boldsymbol{\mathcal{V}}]$ *who does not make two encryption queries with the same* (N, A) *pair. Then, there exist explicit adversaries* $\mathbb{A}_{\mathrm{CPA}}$, $\mathbb{A}_{\mathrm{PRF}}$, $\mathbb{A}'_{\mathrm{PRF}}$, *and* $\mathbb{A}_{\mathrm{MAC}}$ *against the* $(\mathcal{L}_{\mathsf{ivE}}, \mathcal{L}_{\mathsf{ivD}})$*-ivLE security of* $(\mathbf{iv}\boldsymbol{\mathcal{E}}, \mathbf{iv}\boldsymbol{\mathcal{D}})$*, the* $\mathcal{L}_{\mathsf{F}}$*-LPRF security of* $\boldsymbol{\mathcal{F}}$*, the* $\mathcal{L}_{\mathsf{T}}$*-LPRF security of* $\boldsymbol{\mathcal{T}}$ *and the* $(\mathcal{L}_{\mathsf{T}}, \mathcal{L}_{\mathsf{V}})$*-LMAC security of* $(\boldsymbol{\mathcal{T}}, \boldsymbol{\mathcal{V}})$ *such that*

$$\begin{aligned}\mathbf{Adv}^{\mathrm{nLAE}}_{\mathsf{A6};\mathcal{L}_{\mathsf{Enc}},\mathcal{L}_{\mathsf{Dec}}}(\mathbb{A}) \le \mathbf{Adv}^{\mathrm{IND-aCPLA}}_{\boldsymbol{\mathcal{E}},\boldsymbol{\mathcal{D}};\mathcal{L}_{\mathsf{ivE}},\mathcal{L}_{\mathsf{ivD}}}(\mathbb{A}_{\mathrm{CPA}}) + \mathbf{Adv}^{\mathrm{LPRF}}_{\boldsymbol{\mathcal{F}};\mathcal{L}_{\mathsf{F}}}(\mathbb{A}_{\mathrm{PRF}})\\ + \mathbf{Adv}^{\mathrm{LPRF}}_{\boldsymbol{\mathcal{T}};\mathcal{L}_{\mathsf{T}}}(\mathbb{A}'_{\mathrm{PRF}}) + 2\cdot \mathbf{Adv}^{\mathrm{sEUF-CMLA}}_{\boldsymbol{\mathcal{T}},\boldsymbol{\mathcal{V}};\mathcal{L}_{\mathsf{T}},\mathcal{L}_{\mathsf{V}}}(\mathbb{A}_{\mathrm{MAC}}).\end{aligned}$$

5.4 Achieving Misuse Resistant LAE Security

In Sect. 5.2 we discussed why no composition scheme can be (generically) secure against leakage if its decryption begins by calculating a candidate plaintext. This meant ruling out every NRS construction secure in the nonce misuse model, an important feature for a modern robust AE schemes [7,19,42]. Roughly speaking, for MRAE security a scheme must be MtE (to ensure maximum diffusion) yet for leakage resilience it must be EtM (to ensure minimal leakage).

The Synthetic IV and Tag (SIVAT) scheme, defined in Fig. 6, addresses the combined mrLAE goal, by essentially using an MtEtM approach. It can be seen as composing the SIV construction [42] (referred to as A4 in NRS) with a secure MAC, or alternatively as the natural strengthening of A6 towards nonce misuse security, by adding the message to the IV calculation and making the appropriate modifications to enable decryption.

Our additional feature does come at a cost. While schemes in the traditional setting achieve misuse resistance for the same ciphertext expansion as non-resistant schemes, the SIVAT scheme requires essentially twice the expansion. It also has a large number of internal wires, with each function taking in a large number of inputs, although removing any one leads to incorrectness or insecurity. For encryption calls, all inputs must go into the LPRF (for misuse resistance) and for decryption they must go into verification (to prevent RUP attacks).

The proof (in the full version) is very similar to that for A5 or A6 (Corollaries 1 and 2), since the additional element of a SIVAT ciphertext (I) is present in those settings, and might already be available to the adversary through leakage.

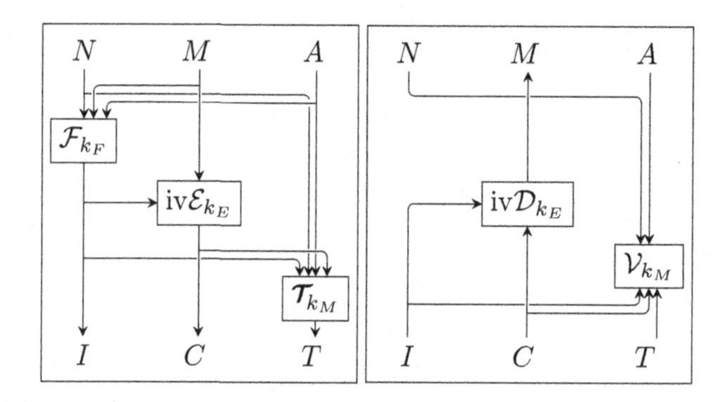

Fig. 6. The Synthetic-IV-and-Tag (SIVAT) scheme. On the left, the encryption routine runs from top to bottom, outputting a ciphertext $I||C||T$. Decryption (on the right) runs from bottom to top. If during decryption verification fails, and $\mathcal{V}_{k_m}$ returns $\perp$, no further computations are performed. In the decryption direction, the PRF $\mathcal{F}$ is not required.

Theorem 3. *Let* $(\mathcal{L}_{\mathsf{ivE}}, \mathcal{L}_{\mathsf{ivD}}, \mathcal{L}_{\mathsf{T}}, \mathcal{L}_{\mathsf{V}}, \mathcal{L}_{\mathsf{F}})$ *be leakage sets for the appropriate primitives, and define* $(\mathcal{L}_{\mathsf{Enc}}, \mathcal{L}_{\mathsf{Dec}})$ *as in Sect. 5.1. Let* $\mathbb{A}$ *be an adversary against the* $(\mathcal{L}_{\mathsf{Enc}}, \mathcal{L}_{\mathsf{Dec}})$*-mrLAE security of* $\mathsf{SIVAT}[\mathbf{iv\mathcal{E}}, \mathbf{iv\mathcal{D}}; \boldsymbol{\mathcal{F}}; \boldsymbol{\mathcal{T}}, \boldsymbol{\mathcal{V}}]$. *Then, there exist explicit adversaries* $\mathbb{A}_{\mathrm{CPA}}$, $\mathbb{A}_{\mathrm{PRF}}$, $\mathbb{A}'_{\mathrm{PRF}}$, *and* $\mathbb{A}_{\mathrm{MAC}}$ *against the* $(\mathcal{L}_{\mathsf{ivE}}, \mathcal{L}_{\mathsf{ivD}})$*-ivLE security of* $(\mathbf{iv\mathcal{E}}, \mathbf{iv\mathcal{D}})$*, the* $\mathcal{L}_{\mathsf{F}}$*-LPRF security of* $\boldsymbol{\mathcal{F}}$*, the* $\mathcal{L}_{\mathsf{T}}$*-LPRF security of* $\boldsymbol{\mathcal{T}}$ *and the* $(\mathcal{L}_{\mathsf{T}}, \mathcal{L}_{\mathsf{V}})$*-LMAC security of* $(\boldsymbol{\mathcal{T}}, \boldsymbol{\mathcal{V}})$ *such that*

$$\begin{aligned}\mathbf{Adv}^{\mathrm{nLAE}}_{\mathsf{SIVAT};\mathcal{L}_{\mathsf{Enc}},\mathcal{L}_{\mathsf{Dec}}}(\mathbb{A}) \leq \mathbf{Adv}^{\mathrm{IND-aCPLA}}_{\boldsymbol{\mathcal{E}},\boldsymbol{\mathcal{D}};\mathcal{L}_{\mathsf{ivE}},\mathcal{L}_{\mathsf{ivD}}}(\mathbb{A}_{\mathrm{CPA}}) + \mathbf{Adv}^{\mathrm{LPRF}}_{\boldsymbol{\mathcal{F}};\mathcal{L}_{\mathsf{F}}}(\mathbb{A}_{\mathrm{PRF}}) \\ + \mathbf{Adv}^{\mathrm{LPRF}}_{\boldsymbol{\mathcal{T}};\mathcal{L}_{\mathsf{T}}}(\mathbb{A}'_{\mathrm{PRF}}) + 2 \cdot \mathbf{Adv}^{\mathrm{sEUF-CMLA}}_{\boldsymbol{\mathcal{T}},\boldsymbol{\mathcal{V}};\mathcal{L}_{\mathsf{T}},\mathcal{L}_{\mathsf{V}}}(\mathbb{A}_{\mathrm{MAC}}).\end{aligned}$$

5.5 A Leakage Resilient IV-Based Encryption Scheme

A crucial component required for our composition is an encryption scheme iv$\mathcal{E}$, whose implementation $(\mathbf{iv\mathcal{E}}, \mathbf{iv\mathcal{D}})$ is IND-aCPLA secure against a rich class $(\mathcal{L}_{\mathsf{ivE}}, \mathcal{L}_{\mathsf{ivD}})$ of leakage functions. As generic composition relies on a secure PRLF implementation $\boldsymbol{\mathcal{F}}$ anyway, we will investigate to what extent this PRLF can be used to bootstrap some iv$\mathcal{E}$ implementation as well. Here we turn to the classical mode of operation CFB (Fig. 7), which has the advantage that only the forward direction of the underlying primitive $\mathcal{F}$ is required, even for decryption (relevant if one would instantiate with a blockcipher). When we move from the standard CFB[$\mathcal{F}$] to its implementation CFB[$\boldsymbol{\mathcal{F}}$] (by replacing $\mathcal{F}$ with its

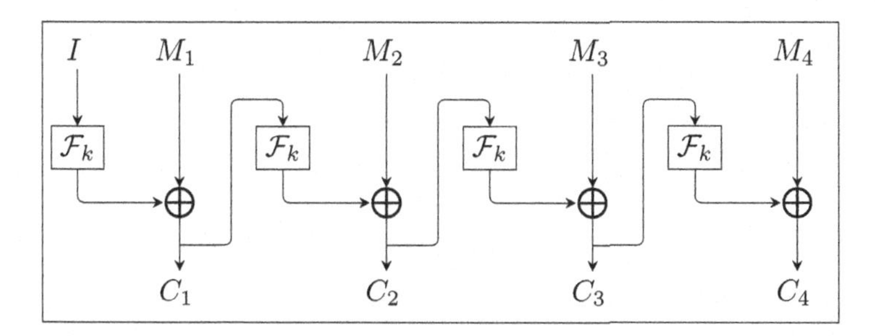

Fig. 7. CFB Mode of Operation based on $\mathcal{F} : \mathsf{K} \times \mathsf{X} \rightarrow \mathsf{X}$. The message M is parsed into blocks or elements $M_1 || \ldots || M_m$, and fed through to output ciphertext $C_1 || \ldots || C_m$. The operation $\oplus$ can be any group operation on X.

implementation $\boldsymbol{\mathcal{F}}$), processing multi-block plaintexts (or ciphertexts) will result in multiple refreshes of the key's representation $\boldsymbol{k}$ (one for each call to $\boldsymbol{\mathcal{F}}$). We will show that CFB is secure against leakage when instantiated with a PRLF, using an adaptation of the classical proof for CFB security [1].

Our first task is to express the leakage sets $(\mathcal{L}_{\mathsf{ivE}}, \mathcal{L}_{\mathsf{ivD}})$ for scheme $\mathbf{iv}\boldsymbol{\mathcal{E}}$ in terms of that of the PRF $\boldsymbol{\mathcal{F}}$, namely $\mathcal{L}_{\mathsf{F}}$. When tracing through the operation of CFB-encryption, we will make two assumptions. Firstly, that leakage for each of the $\boldsymbol{\mathcal{F}}$ calls is local (cf. OCLI), which in particular means leakage will be restricted to the representation of $\boldsymbol{k}$ specific for the $\boldsymbol{\mathcal{F}}$ call at hand (and $\boldsymbol{k}$ is expected to be refreshed during a single $\mathbf{iv}\boldsymbol{\mathcal{E}}$ call). Secondly, that all visible wires in Fig. 7, corresponding to the $\mathbf{iv}\boldsymbol{\mathcal{E}}$'s public inputs and outputs, will leak. Note that longer messages will lead to more leakage for an adversary, which matches practice (where the size of the power trace might be linear in the size of the message).

Decryption closely matches encryption and, under the same assumptions as above, leakage on decryption of a ciphertext can be expressed instead as leakage on the encryption of the corresponding plaintext. Hence we refer to decryption leakage as $\mathcal{L}_{\mathsf{ivE}}{}'$ (where the prime connotes the syntactical malarkey to deal with the different input spaces for encryption and decryption).

Concluding, we define the leakage set $\mathcal{L}_{\mathsf{ivE}}$ to be the collection of all functions $L_{\mathrm{CFB}} : \mathsf{K} \times \mathsf{N} \times \mathsf{M} \times \mathsf{R} \rightarrow \{0,1\}^*$ that are of the form

$$L_{\mathrm{CFB}}(\boldsymbol{k}, I, M; r) = (M, C, L_i(\boldsymbol{k}_i, C_i; r_i)_{i \in \{0..n-1\}})$$

with $L_i \in \mathcal{L}_{\mathsf{T}}$ (for $i \in \{0 \ldots n-1\}$) and where M is an n-block message, $C = \mathrm{iv}\mathcal{E}_k(I, M)$ is an $(n+1)$-block ciphertext constituted of blocks C_i ($i \in \{0, \ldots n\}$), r is the concatenation of the random values r_i passed to the ith $\boldsymbol{\mathcal{F}}$-call ($i \in \{1 \ldots n\}$), and $\boldsymbol{k}_{i-1}$ is the key representation for the ith $\boldsymbol{\mathcal{F}}$-call ($i \in \{1 \ldots n\}$).

Theorem 4. *Let* $\boldsymbol{\mathcal{F}} : \mathsf{T}^* \rightarrow \mathsf{T}$ *be a PRF with leakage class* $\mathcal{L}_{\mathsf{F}}$ *and let* $(\mathbf{iv}\boldsymbol{\mathcal{E}}, \mathbf{iv}\boldsymbol{\mathcal{D}})$ *be the symmetric encryption scheme* $\mathrm{CFB}[\boldsymbol{\mathcal{F}}]$ *with derived leakage* $(\mathcal{L}_{\mathsf{ivE}}, \mathcal{L}_{\mathsf{ivE}}{}')$. *Let* $\mathbb{A}$ *be an iv-respecting adversary against the* $(\mathcal{L}_{\mathsf{ivE}}, \mathcal{L}_{\mathsf{ivE}}{}')$*-IND–aCPLA security* *of* $(\mathbf{iv}\boldsymbol{\mathcal{E}}, \mathbf{iv}\boldsymbol{\mathcal{D}})$. *Then there exists an adversary* $\mathbb{A}_{\mathrm{PRF}}$ *of similar complexity to* $\mathbb{A}$ *against the* $\mathcal{L}_{\mathsf{F}}$*-PRLF security of* $\boldsymbol{\mathcal{F}}$ *such that*

$$\mathbf{Adv}^{\mathrm{IND-aCPLA}}_{\mathbf{iv}\boldsymbol{\mathcal{E}},\mathbf{iv}\boldsymbol{\mathcal{D}};\mathcal{L}_{\mathsf{ivE}},\mathcal{L}_{\mathsf{ivE}}'}(\mathbb{A}) \leq 2 \cdot \mathbf{Adv}^{\mathrm{PRLF}}_{\boldsymbol{\mathcal{F}};\mathcal{L}_{\mathsf{F}}}(\mathbb{A}_{\mathrm{PRF}}) + \frac{3}{4} \cdot \frac{\sigma^2}{|\mathsf{T}|},$$

where σ *is the total number of blocks encrypted, and the blocksize is* $|\mathsf{T}|$.

The proof can be found in the full version [3].

6 mrLAE Security by Instantiating the PRF and MAC

The A5 and SIVAT composition mechanisms can be instantiated with any suitably secure primitives to yield secure nLAE or mrLAE schemes. Together with using CFB[$\boldsymbol{\mathcal{F}}$] as underlying $\mathbf{iv}\boldsymbol{\mathcal{E}}$, these allow us to construct a secure mrLAE scheme through any PRF $\mathcal{F}$ with a secure implementation $\boldsymbol{\mathcal{F}}$ and a secure MAC implementation $(\boldsymbol{\mathcal{T}}, \boldsymbol{\mathcal{V}})$. The remaining questions therefore are what can be said about securely implementing these primitives and what conclusions for the overall scheme can subsequently be drawn. We will answer these questions from two perspectives: a practical side-channel one (for those favouring masked AES) and a more theoretical, yet eminently implementable one in the continuous leakage model.

A side-channel perspective. Our result provides a roadmap for obtaining a side-channel misuse-resistant AE scheme by selecting reasonable practical primitives (and implementations) for the PRF and the MAC (say a suitably masked AES, respectively KMAC) and subsequently gauging to what extent actual leakage on the *primitive implementations* can be used to break the relevant PRLF or EUF–CLMA notions as well as whether leakage on the *full* implementation is cleanly segregated or whether undesired correlation indicates bleeding of leakage from the values or variables from one component into say part of the power trace associated with another component.

The result above no longer explicitly takes into account leakage classes; these have effectively become implicit artefacts of the attack. We assume that a successful attack on the full scheme will be recognized as such: our result essentially says that if such an attack is found then either the leakage is not cleanly separated or one of the primitive implementations is already insecure (or both).

A leakage resilience perspective. A complementary approach to the practical one above is to design the primitives and their implementations with a provable level of resistance against leakage functions from a specific class. As already explained in the introduction, a multitude of models exist depending on the class of functions under consideration. One of the stronger models is that of continuous leakage: here the leakage functions can be arbitrary, subject to the constraint that their range is bounded. A usual refinement is to use a split-state model, where the key's representation $\boldsymbol{k}$ is operated upon in two (or more) tranches and each tranche can only leak on that part of $\boldsymbol{k}$ in scope for the operation at hand (assuming only computation leaks, as usual).

While there are PRFs that have been proven secure in the continuous leakage model, as far as we can tell this has always come at the price of adaptivity. In order for our constructions to be implemented a new PRF is called for, with an implementation secure in the stronger, adaptive continuous leakage model. In the full version [3] we provide such a function and implementation, and prove the latter secure in the generic group model (against adaptive continous leakage in a split-state setting). Additionally, we show how to create a related MAC such that leaking on the verification's implementation is ok.

Our construction is an evolution of the MAC of Martin et al. [30], itself inspired by a scheme by Kiltz and Pietrzak [23]. The key enabling novelty is the use of *three* shares instead of the customary two. A thorough discussion of the design choices, specifications, and security justification can be found in the full version [3] but for completeness we provide the final theorem statement below.

Theorem 5. *Let* SIVAT *be the SIVAT mechanism instantiated with the implementations described in the full version [3] over a generic group of* p *elements, and assume that each share of the internal PRF leaks at most* λ *per call following the associated leakage functions, as described by leakage sets* $(\mathcal{L}_{\mathsf{Enc}}, \mathcal{L}_{\mathsf{Dec}})$. *Let* $\mathbb{A}$ *be an adversary making at most* g *direct queries to the generic group oracle (including the complexity of all chosen leakage queries) and making* q *construction queries totalling* σ *blocks. Then,*

$$\mathbf{Adv}^{\mathrm{LAE}}_{\mathsf{SIVAT};\mathcal{L}_{\mathsf{Enc}},\mathcal{L}_{\mathsf{Dec}}}(\mathbb{A}) \leq \frac{7}{p}\left(2^{4\lambda}\cdot\sigma^2\cdot(g+9q+5\sigma)^2 + 8(g+9q+5\sigma)^2\right).$$

To get a feel for the practical security level, let's look at parameters if the schemes are instantiated over a 512 bit elliptic curves, and we want the keep the attack success probability below 2^{-60} (a common limit in the real world, e.g. [28]). Let's assume that each internal leakage function leaks at most $\lambda = 85$ bits, which is approximately a sixth of a group element. Then the scheme would remain secure until the adversary has encrypted or decrypted around 2^{25} blocks, and made a similar number of queries to the generic group.

This result comes with a few caveats, covered in more detail by the full version [3]. For instance, to ensure security against the leakage of arbitrary functions of the key, to process q queries of total σ blocks the construction must sample $4q + \sigma$ random group elements in a leakage-resilient manner, which can be complicated [30]. Nonetheless, our construction is proof positive of the existence of leakage resilient authenticated encryption in a very strong sense.

7 Conclusions and Open Problems

We introduced notions for strengthened AE when considering leakage, discussed generic composition under leakage, and showed the EtM type constructions can be proven secure in this context. We give a new scheme, SIVAT, that achieves misuse resistance and leakage resilience simultaneously, and show how this can be bootstrapped from a PRF secure against leakage. Finally, we give a concrete instantiation for the SIVAT mechanism. Our research unveils several interesting open problems, which we summarise subsequently.

IND–aCPLA. If one allows nonce-reuse, then for any leakage set $\mathcal{L}_{\mathsf{E}}$ security against $\mathcal{L}_{\mathsf{E}}$-IND–CPLA adversary implies $(\mathcal{L}_{\mathsf{E}}, \mathcal{L}_{\mathsf{E}}')$-IND–aCPLA security, where $\mathcal{L}_{\mathsf{E}}'$ is the essentially the same set as $\mathcal{L}_{\mathsf{E}}$ with some minor bookkeeping to ensure correct syntax. The implication is trivial as the leakage on any valid $\mathcal{D}$-query can be perfectly simulated by repeating the corresponding $\mathcal{E}$-query instead. In the the nonce or iv respecting cases the implication remains open (as repeating encryption queries including nonce is no longer allowed). Nonetheless, we conjecture that even in these two settings for many reasonable leakage sets $\mathcal{L}_{\mathsf{E}}$, $\mathcal{L}_{\mathsf{E}}$-IND–CPLA does imply $(\mathcal{L}_{\mathsf{E}}, \mathcal{L}_{\mathsf{E}}')$-IND–aCPLA. We leave it as an interesting question to formalise this or find a counter-example. More generally, is there some way of defining $\mathcal{L}_{\mathsf{ED}}$ as a function of some general sets $\mathcal{L}_{\mathsf{E}}, \mathcal{L}_{\mathsf{D}}$ such that $\mathcal{L}_{\mathsf{ED}}$-IND–CPLA $\implies$ $(\mathcal{L}_{\mathsf{E}}, \mathcal{L}_{\mathsf{D}})$-IND–aCPLA?

MtE with restricted leakage sets. The insecurity of the majority of the MtE schemes when considering leakage comes from a generic attack against any schemes whose inverse follows the decrypt-then-verify or decrypt-and-verify structure. We leave it as an interesting open question to investigate the leakage security under other more restricted leakage sets.

Misuse resistance with minimal message expansion. We demonstrate that misuse resistance can be achieved through generic composition, at the cost of additional message expansion, using a MAC-then-Encrypt-then-MAC structure (leading to SIVAT). We believe that dedicated constructions are likely to exist that can achieve mrLAE security with minimal expansion, or more generally LAE without requiring independent keys.

Acknowledgements. Initial work was conducted while Dan Martin was employed by the Department of Computer Science, University of Bristol. Guy Barwell was supported by an EPSRC grant; Elisabeth Oswald and Dan Martin were in part supported by EPSRC via grants EP/I005226/1 (SILENT) and EP/N011635/1 (LADA).

References

1. Alkassar, A., Geraldy, A., Pfitzmann, B., Sadeghi, A.-R.: Optimized self-synchronizing mode of operation. In: Matsui, M. (ed.) FSE 2001. LNCS, vol. 2355, pp. 78–91. Springer, Heidelberg (2002). https://doi.org/10.1007/3-540-45473-X_7
2. Andreeva, E., Bogdanov, A., Luykx, A., Mennink, B., Mouha, N., Yasuda, K.: How to securely release unverified plaintext in authenticated encryption. In: Sarkar and Iwata [43], pp. 105–125

3. Barwell, G., Martin, D.P., Oswald, E., Stam, M.: Authenticated encryption in the face of protocol and side channel leakage. IACR Cryptology ePrint Archive 2017/068 (2017). http://eprint.iacr.org/2017/068
4. Barwell, G., Page, D., Stam, M.: Rogue decryption failures: reconciling AE robustness notions. In: Groth [17], pp. 94–111
5. Bellare, M., Kane, D., Rogaway, P.: Big-Key symmetric encryption: resisting key exfiltration. In: Robshaw, M., Katz, J. (eds.) CRYPTO 2016. LNCS, vol. 9814, pp. 373–402. Springer, Heidelberg (2016). https://doi.org/10.1007/978-3-662-53018-4_14
6. Bellare, M., Namprempre, C.: Authenticated encryption: relations among notions and analysis of the generic composition paradigm. In: Okamoto, T. (ed.) ASIACRYPT 2000. LNCS, vol. 1976, pp. 531–545. Springer, Heidelberg (2000). https://doi.org/10.1007/3-540-44448-3_41
7. Bernstein, D.J.: CAESAR competition call (2013). http://competitions.cr.yp.to/caesar-call-3.html
8. Berti, F., Koeune, F., Pereira, O., Peters, T., Standaert, F.X.: Leakage-resilient and misuse-resistant authenticated encryption. Cryptology ePrint Archive, Report 2016/996 (2016). http://eprint.iacr.org/2016/996
9. Boldyreva, A., Degabriele, J.P., Paterson, K.G., Stam, M.: On symmetric encryption with distinguishable decryption failures. In: Moriai, S. (ed.) FSE 2013. LNCS, vol. 8424, pp. 367–390. Springer, Heidelberg (2014). https://doi.org/10.1007/978-3-662-43933-3_19
10. Boldyreva, A., Degabriele, J.P., Paterson, K.G., Stam, M.: Security of symmetric encryption in the presence of ciphertext fragmentation. Cryptology ePrint Archive, Report 2015/059 (2015). http://eprint.iacr.org/2015/059
11. Cohn-Gordon, K., Cremers, C., Dowling, B., Garratt, L., Stebila, D.: A formal security analysis of the signal messaging protocol. Cryptology ePrint Archive, Report 2016/1013 (2016). http://eprint.iacr.org/2016/1013
12. Dodis, Y., Kalai, Y.T., Lovett, S.: On cryptography with auxiliary input. In: Mitzenmacher, M. (ed.) 41st ACM STOC, pp. 621–630. ACM Press, May/June 2009
13. Dodis, Y., Pietrzak, K.: Leakage-resilient pseudorandom functions and side-channel attacks on feistel networks. In: Rabin, T. (ed.) CRYPTO 2010. LNCS, vol. 6223, pp. 21–40. Springer, Heidelberg (2010). https://doi.org/10.1007/978-3-642-14623-7_2
14. Dziembowski, S., Pietrzak, K.: Leakage-resilient cryptography. In: 49th FOCS, pp. 293–302. IEEE Computer Society Press, October 2008
15. Faust, S., Pietrzak, K., Schipper, J.: Practical leakage-resilient symmetric cryptography. In: Prouff, E., Schaumont, P. (eds.) CHES 2012. LNCS, vol. 7428, pp. 213–232. Springer, Heidelberg (2012). https://doi.org/10.1007/978-3-642-33027-8_13
16. Fischlin, M., Günther, F., Marson, G.A., Paterson, K.G.: Data is a stream: security of stream-based channels. In: Gennaro, R., Robshaw, M. (eds.) CRYPTO 2015. LNCS, vol. 9216, pp. 545–564. Springer, Heidelberg (2015). https://doi.org/10.1007/978-3-662-48000-7_27
17. Groth, J. (ed.): IMACC 2015. LNCS, vol. 9496. Springer, Cham (2015). https://doi.org/10.1007/978-3-319-27239-9
18. Hazay, C., López-Alt, A., Wee, H., Wichs, D.: Leakage-resilient cryptography from minimal assumptions. In: Johansson, T., Nguyen, P.Q. (eds.) EUROCRYPT 2013. LNCS, vol. 7881, pp. 160–176. Springer, Heidelberg (2013). https://doi.org/10.1007/978-3-642-38348-9_10

19. Hoang, V.T., Krovetz, T., Rogaway, P.: Robust authenticated-encryption AEZ and the problem that it solves. In: Oswald, E., Fischlin, M. (eds.) EUROCRYPT 2015. LNCS, vol. 9056, pp. 15–44. Springer, Heidelberg (2015). https://doi.org/10.1007/978-3-662-46800-5_2
20. Ishai, Y., Sahai, A., Wagner, D.: Private circuits: securing hardware against probing attacks. In: Boneh, D. (ed.) CRYPTO 2003. LNCS, vol. 2729, pp. 463–481. Springer, Heidelberg (2003). https://doi.org/10.1007/978-3-540-45146-4_27
21. Katz, J., Lindell, Y.: Introduction to Modern Cryptography. Chapman & Hall/CRC, Boca Raton (2008)
22. Katz, J., Vaikuntanathan, V.: Signature schemes with bounded leakage resilience. In: Matsui, M. (ed.) ASIACRYPT 2009. LNCS, vol. 5912, pp. 703–720. Springer, Heidelberg (2009). https://doi.org/10.1007/978-3-642-10366-7_41
23. Kiltz, E., Pietrzak, K.: Leakage resilient ElGamal encryption. In: Abe, M. (ed.) ASIACRYPT 2010. LNCS, vol. 6477, pp. 595–612. Springer, Heidelberg (2010). https://doi.org/10.1007/978-3-642-17373-8_34
24. Kocher, P.C.: Timing attacks on implementations of Diffie-Hellman, RSA, DSS, and other systems. In: Koblitz, N. (ed.) CRYPTO 1996. LNCS, vol. 1109, pp. 104–113. Springer, Heidelberg (1996). https://doi.org/10.1007/3-540-68697-5_9
25. Kocher, P., Jaffe, J., Jun, B.: Differential power analysis. In: Wiener, M. (ed.) CRYPTO 1999. LNCS, vol. 1666, pp. 388–397. Springer, Heidelberg (1999). https://doi.org/10.1007/3-540-48405-1_25
26. Kurosawa, K., Trieu Phong, L.: Leakage resilient IBE and IPE under the DLIN assumption. In: Jacobson, M., Locasto, M., Mohassel, P., Safavi-Naini, R. (eds.) ACNS 2013. LNCS, vol. 7954, pp. 487–501. Springer, Heidelberg (2013). https://doi.org/10.1007/978-3-642-38980-1_31
27. Longo, J., Martin, D.P., Oswald, E., Page, D., Stam, M., Tunstall, M.: Simulatable leakage: analysis, pitfalls, and new constructions. In: Sarkar and Iwata [43], pp. 223–242
28. Luykx, A., Paterson, K.: Limits on authenticated encryption use in TLS (2016). http://www.isg.rhul.ac.uk/kp/TLS-AEbounds.pdf
29. Mangard, S., Oswald, E., Popp, T.: Power Analysis Attacks: Revealing the Secrets of Smart Cards. Springer, Heidelberg (2008). https://doi.org/10.1007/978-0-387-38162-6
30. Martin, D.P., Oswald, E., Stam, M., Wójcik, M.: A leakage resilient MAC. In: Groth [17], pp. 295–310
31. Micali, S., Reyzin, L.: Physically observable cryptography. In: Naor, M. (ed.) TCC 2004. LNCS, vol. 2951, pp. 278–296. Springer, Heidelberg (2004). https://doi.org/10.1007/978-3-540-24638-1_16
32. Namprempre, C., Rogaway, P., Shrimpton, T.: Reconsidering generic composition. In: Nguyen, P.Q., Oswald, E. (eds.) EUROCRYPT 2014. LNCS, vol. 8441, pp. 257–274. Springer, Heidelberg (2014). https://doi.org/10.1007/978-3-642-55220-5_15
33. NIST: FIPS 81: DES Modes of Operation. Issued December **2**, 63 (1980)
34. Pereira, O., Standaert, F.X., Vivek, S.: Leakage-resilient authentication and encryption from symmetric cryptographic primitives. In: Ray, I., Li, N., Kruegel, C. (eds.) ACM CCS 15, pp. 96–108. ACM Press, October 2015
35. Perrin, T.: Double Ratchet algorithm (2014). https://github.com/trevp/double_ratchet/wiki. Accessed 10 Sept 2016
36. Pietrzak, K.: A leakage-resilient mode of operation. In: Joux, A. (ed.) EUROCRYPT 2009. LNCS, vol. 5479, pp. 462–482. Springer, Heidelberg (2009). https://doi.org/10.1007/978-3-642-01001-9_27

37. Qin, B., Liu, S.: Leakage-flexible CCA-secure public-key encryption: simple construction and free of pairing. In: Krawczyk, H. (ed.) PKC 2014. LNCS, vol. 8383, pp. 19–36. Springer, Heidelberg (2014). https://doi.org/10.1007/978-3-642-54631-0_2
38. Renauld, M., Standaert, F.-X., Veyrat-Charvillon, N., Kamel, D., Flandre, D.: A formal study of power variability issues and side-channel attacks for nanoscale devices. In: Paterson, K.G. (ed.) EUROCRYPT 2011. LNCS, vol. 6632, pp. 109–128. Springer, Heidelberg (2011). https://doi.org/10.1007/978-3-642-20465-4_8
39. Rogaway, P.: Authenticated-encryption with associated-data. In: Atluri, V. (ed.) ACM CCS 02, pp. 98–107. ACM Press, November 2002
40. Rogaway, P.: Nonce-based symmetric encryption. In: Roy, B., Meier, W. (eds.) FSE 2004. LNCS, vol. 3017, pp. 348–358. Springer, Heidelberg (2004). https://doi.org/10.1007/978-3-540-25937-4_22
41. Rogaway, P., Bellare, M., Black, J., Krovetz, T.: OCB: a block-cipher mode of operation for efficient authenticated encryption. In: ACM CCS 01, pp. 196–205. ACM Press, November 2001
42. Rogaway, P., Shrimpton, T.: A provable-security treatment of the key-wrap problem. In: Vaudenay, S. (ed.) EUROCRYPT 2006. LNCS, vol. 4004, pp. 373–390. Springer, Heidelberg (2006). https://doi.org/10.1007/11761679_23
43. Sarkar, P., Iwata, T. (eds.): ASIACRYPT 2014. LNCS, vol. 8873. Springer, Heidelberg (2014). https://doi.org/10.1007/978-3-662-45611-8
44. Schipper, J.: Leakage-resilient authentication. Ph.D. thesis, Utrecht University (2010)
45. Shrimpton, T., Terashima, R.S.: A modular framework for building variable-input-length tweakable ciphers. In: Sako, K., Sarkar, P. (eds.) ASIACRYPT 2013. LNCS, vol. 8269, pp. 405–423. Springer, Heidelberg (2013). https://doi.org/10.1007/978-3-642-42033-7_21
46. Standaert, F.-X., Pereira, O., Yu, Y.: Leakage-resilient symmetric cryptography under empirically verifiable assumptions. In: Canetti, R., Garay, J.A. (eds.) CRYPTO 2013. LNCS, vol. 8042, pp. 335–352. Springer, Heidelberg (2013). https://doi.org/10.1007/978-3-642-40041-4_19
47. Yau, A.K.L., Paterson, K.G., Mitchell, C.J.: Padding oracle attacks on CBC-mode encryption with secret and random IVs. In: Gilbert, H., Handschuh, H. (eds.) FSE 2005. LNCS, vol. 3557, pp. 299–319. Springer, Heidelberg (2005). https://doi.org/10.1007/11502760_20
48. Yu, Y., Standaert, F.X., Pereira, O., Yung, M.: Practical leakage-resilient pseudorandom generators. In: Al-Shaer, E., Keromytis, A.D., Shmatikov, V. (eds.) ACM CCS 10, pp. 141–151. ACM Press, October 2010

Consolidating Inner Product Masking

Josep Balasch[1(✉)], Sebastian Faust[2,3], Benedikt Gierlichs[1], Clara Paglialonga[2,3], and François-Xavier Standaert[4]

[1] imec-COSIC KU Leuven, Leuven, Belgium
{josep.balasch,benedikt.gierlichs}@esat.kuleuven.be
[2] Ruhr-Universität Bochum, Bochum, Germany
{sebastian.faust,clara.paglialonga}@rub.de
[3] Technische Universität Darmstadt, Darmstadt, Germany
[4] Université catholique de Louvain, ICTEAM/ELEN/Crypto Group, Louvain-la-Neuve, Belgium
fstandae@uclouvain.be

Abstract. Masking schemes are a prominent countermeasure to defeat power analysis attacks. One of their core ingredients is the encoding function. Due to its simplicity and comparably low complexity overheads, many masking schemes are based on a Boolean encoding. Yet, several recent works have proposed masking schemes that are based on alternative encoding functions. One such example is the inner product masking scheme that has been brought towards practice by recent research. In this work, we improve the practicality of the inner product masking scheme on multiple frontiers. On the conceptual level, we propose new algorithms that are significantly more efficient and have reduced randomness requirements, but remain secure in the t-probing model of Ishai, Sahai and Wagner (CRYPTO 2003). On the practical level, we provide new implementation results. By exploiting several engineering tricks and combining them with our more efficient algorithms, we are able to reduce execution time by nearly 60% compared to earlier works. We complete our study by providing novel insights into the strength of the inner product masking using both the information theoretic evaluation framework of Standaert, Malkin and Yung (EUROCRYPT 2009) and experimental analyses with an ARM microcontroller.

1 Introduction

Physical side-channel attacks where the adversary exploits, e.g., the power consumption [34] or the running time [33] of a cryptographic device are one of the most powerful cyberattacks. Researchers have shown that they can extract secret keys from small embedded devices such as smart cards [22,34], and recent reports illustrate that also larger devices such as smart phones and computers can be attacked [4,24]. Given the great threat potential of side-channel attacks there has naturally been a large body of work proposing countermeasures to defeat them [35]. One of the most well-studied countermeasures against side-channel attacks – and in particular, against power analysis – are masking schemes [12,29].

T. Takagi and T. Peyrin (Eds.): ASIACRYPT 2017, Part I, LNCS 10624, pp. 724–754, 2017.
https://doi.org/10.1007/978-3-319-70694-8_25

The basic idea of a masking scheme is simple. Since side-channel attacks attempt to learn information about the intermediate values that are produced by a cryptographic algorithm during its evaluation, a masking scheme conceals these values by hiding them with randomness.

Masking schemes have two important ingredients: a randomized encoding function and a method to securely compute with these encodings without revealing sensitive information. The most common masking scheme is Boolean masking [12,32], which uses a very simple additive n-out-of-n secret sharing as its encoding function. More concretely, to encode a bit b we sample uniformly at random bits $(b_1, \ldots, b_n)$ such that $\sum_i b_i = b$ (where the sum is in the binary field). The basic security property that is guaranteed by the encoding function is that if the adversary only learns up to $n-1$ of the shares then nothing is revealed about the secret b. The main challenge in developing secure masking schemes has been in lifting the security properties guaranteed by the encoding function to the level of the entire masked computation. To this end, we usually define masked operations for field addition and field multiplication, and show ways to compose them securely.

The most standard security property that we want from a masking scheme is to resist t-probing attacks. To analyze whether a masking scheme is secure against t-probing attacks we can carry out a security analysis in the so-called t-probing model – introduced in the seminal work of Ishai et al. [32]. In the t-probing model the adversary is allowed to learn up to t intermediate values of the computation, which shall not reveal anything about the sensitive information, and in particular nothing about the secret key used by a masked cryptographic algorithm. In the last years, there has been a flourishing literature surrounding the topic of designing better masking schemes, including many exciting works on efficiency improvements [11,14,16,37,42], stronger security guarantees [18, 21,39,41] and even fully automated verification of masking schemes [5,6] – to just name a few.

As mentioned previously, the core ingredient of any masking scheme is its encoding function. We can only hope to design secure masking schemes if we start with a strong encoding function at first place. Hence, it is natural to ask what security guarantees can be offered by our encoding functions and to what extent these security properties can be lifted to the level of the masked computation. Besides the Boolean masking which is secure in the t-probing model, several other encoding functions which can be used for masking schemes have been introduced in the past. This includes the affine masking [23], the polynomial masking [28,40] and the inner product masking [1,2,20]. Each of these masking functions offers different trade-offs in terms of efficiency and what security guarantees it can offer.

The goal of this work, is to provide novel insights into the inner product masking scheme originally introduced by Dziembowski and Faust [20] and Goldwasser and Rothblum [25], and later studied in practice by Balasch et al. [1,2]. Our main contribution is to consolidate the work on inner product masking thereby improving the existing works of Balasch et al. [1,2] on multiple frontiers and providing several novel insights. Our contributions can be summarized as follows.

New algorithms with t–SNI security property. On a conceptual level we propose simplified algorithms for the multiplication operation protected with inner product masking. In contrast to the schemes from [1,2] they are resembling the schemes originally proposed by Ishai et al. [32] (and hence more efficient and easier to implement than the schemes in [1]), but work with the inner product encoding function. We prove that our new algorithms satisfy the property of t-strong non-interference (t–SNI) introduced by Barthe et al. [5,6], and hence can safely be used for larger composed computation. An additional contribution is that we provide a new secure multiplication algorithm – we call it $\texttt{IPMult}_{L}^{(2)}$ shown in Algorithm 7 – that can result in better efficiency when composed with certain other masked operations. Concretely, when we want to compose a linear function $g()$ with a multiplication, then either we can use $\texttt{IPMult}_{L}^{(1)}$ and require an additional refreshing operation at the output of $g()$, or we use our new algorithm $\texttt{IPMult}_{L}^{(2)}$ that eliminates the need for the additional refreshing. This can save at least $\mathcal{O}(n^2)$ in randomness.

New implementation results. We leverage on the proposed algorithms for the multiplication operation to build new software implementations of AES-128 for embedded AVR architectures. Compared to earlier works [1], we are able to reduce the execution times by nearly a factor 60% (for 2 shares) and 55% (for 3 shares). The improvements stem not only from a decrease in complexity of the new algorithms, but also from an observation that enables the tabulation of the AES affine transformation. We additionally provide various flavors of AES-128 implementations protected with Boolean masking, using different addition chains that have been proposed to compute the field inversion. Our performance evaluation allow us to quantify the current gap between Boolean and IP masking schemes in terms of execution time as well as non-volatile storage.

Information theoretic evaluation. We continue our investigations with a comprehensive information theoretic evaluation of the inner product encoding. Compared to the previous works of Balasch et al., we consider the mutual information between a sensitive variable and the leakage of its inner product shares for an extended range of noises, for linear and non-linear leakage functions and for different values of the public vector of the encoding. Thanks to these evaluations, we refine the understanding of the theoretical pros and cons of such masking schemes compared to the mainstream Boolean masking. In particular, we put forward interesting properties of inner product masking regarding "security order amplification" in the spirit of [9,10,31] and security against transition-based leakages [3,15]. We also highlight that these interesting properties are quite highly implementation-dependent.

Experimental evaluation. Eventually, we confront our new algorithms and their theoretical analyses with practice. In particular, we apply leakage detection techniques on measurements collected from protected AES-128 routines running on an ARM Cortex-M4 processor. Our results reveal the unequivocal presence of leakage (univariate, first-order) in the first-order Boolean masked implementation. In contrast the first-order inner product masked implementation shows significantly less evidence of leakage (with the same number of measurements). Combined with the previous proofs and performance evaluations, these results therefore establish inner product masking as an interesting alternative to Boolean masking, with good properties for composability, slight performance overheads and significantly less evidence of leakage.

2 Notation

In the following we denote by $\mathcal{K}$ a field of characteristic 2. We denote with uppercase letters the elements of the field $\mathcal{K}$ and with bold notation that one in the $\mathcal{K}$-vector spaces. The field multiplication is represented by the dot $\cdot$ while the standard inner product over $\mathcal{K}$ is denoted as $\langle \boldsymbol{X}, \boldsymbol{Y} \rangle = \sum_i X_i \cdot Y_i$, where X_i and Y_i are the components of the vectors $\boldsymbol{X}$ and $\boldsymbol{Y}$.

The symbol δ_{ij} corresponds to the element 0 when $i = j$ and 1 otherwise.

3 New Algorithm

Our new multiplication scheme is based on the inner product construction of Dziembowski and Faust [20] and constitutes an improvement to the works [1, 2]. The encoding of a variable $S \in \mathcal{K}$ consists of a vector $\boldsymbol{S} \in \mathcal{K}^n$ such that $S = \langle \boldsymbol{L}, \boldsymbol{S} \rangle$, where $\boldsymbol{L}$ is a freely chosen, public non-zero parameter with first component $L_1 = 1$.

The algorithms for initialization and masking are depicted in the `IPSetup` and `IPMask` procedures. The subroutine `rand`($\mathcal{K}$) samples an element uniformly at random from the field $\mathcal{K}$. The algorithms for addition and refreshing are kept

Algorithm 1. Setup the masking scheme: $\boldsymbol{L} \leftarrow \texttt{IPSetup}_n(\mathcal{K})$

Input: field description $\mathcal{K}$
Output: random vector $\boldsymbol{L}$

```
L_1 = 1;
for i = 2 to n do
    L_i ← rand(K \ {0});
end for
```

the same as in [1], while a new multiplication scheme $\texttt{IPMult}^{(1)}$ is proposed in Algorithm 3. The schemes achieves security order $t = n - 1$ in the t-probing model.

Our starting point for the Algorithm 3 is the multiplication scheme from [32]. We reuse the idea of summing the matrix of the inner products of the inputs with a symmetric matrix of random elements, in order to compute the shares of

Algorithm 2. Masking a variable: $\boldsymbol{S} \leftarrow \texttt{IPMask}_{\boldsymbol{L}}(S)$

Input: variable $S \in \mathcal{K}$
Output: vector $\boldsymbol{S}$ such that $S = \langle \boldsymbol{L}, \boldsymbol{S} \rangle$

```
for i = 2 to n do
    S_i ← rand(K);
end for
S_1 = S + Σ_{i=2}^{n} L_i · S_i;
```

the output in a secure way. In particular we design these two matrices ($\boldsymbol{T}$ and $\boldsymbol{U}'$ in the algorithm) to be consistent with our different masking model.

Algorithm 3. Multiply masked values: $\boldsymbol{C} \leftarrow \texttt{IPMult}_{\boldsymbol{L}}^{(1)}(\boldsymbol{A}, \boldsymbol{B})$

Input: vectors $\boldsymbol{A}$ and $\boldsymbol{B}$ of length n
Output: vector $\boldsymbol{C}$ such that $\langle \boldsymbol{L}, \boldsymbol{C} \rangle = \langle \boldsymbol{L}, \boldsymbol{A} \rangle \cdot \langle \boldsymbol{L}, \boldsymbol{B} \rangle$

```
▷ Computation of the matrix T
for i = 1 to n do
    for j = 1 to n do
        T_{i,j} = A_i · B_j · L_j;
    end for
end for
▷ Computation of the matrices U and U'
for i = 1 to n do
    for j = 1 to n do
        if i < j then
            U'_{ij} ← rand(K);
        end if
        if i > j then
            U'_{i,j} = −U'_{j,i};
        end if
        U_{i,j} = U'_{i,j} · δ_{ij} L_i^{−1};
    end for
end for
▷ Computation of the matrix V
V = T + U;
▷ Computation of the output vector C
for i = 1 to n do
    C_i = Σ_j V_{i,j};
end for
```

The correctness of the scheme is proved in the following lemma.

Lemma 1. *For any* $\boldsymbol{L}, \boldsymbol{A}, \boldsymbol{B} \in \mathcal{K}^n$ *and* $\boldsymbol{C} = \texttt{IPMult}_{\boldsymbol{L}}^{(1)}(\boldsymbol{A}, \boldsymbol{B})$, *we have*

$$\langle \boldsymbol{L}, \boldsymbol{C} \rangle = \langle \boldsymbol{L}, \boldsymbol{A} \rangle \cdot \langle \boldsymbol{L}, \boldsymbol{B} \rangle.$$

Proof. For all $i \neq j$ it holds:

$$\begin{aligned}\langle \boldsymbol{L}, \boldsymbol{C}\rangle &= \sum_i L_i \cdot C_i = \sum_i L_i \sum_j V_{i,j} = \sum_i L_i \sum_j (T_{ij} + U_{ij}) \\ &= \sum_i L_i \sum_j (A_i B_j L_j + U'_{ij} L_i^{-1}) = \sum_i L_i \sum_j A_i B_j L_j + \sum_{ij} U'_{ij} \\ &= \sum_i L_i A_i \sum_j B_j L_j = \langle \boldsymbol{L}, \boldsymbol{A}\rangle \langle \boldsymbol{L}, \boldsymbol{B}\rangle\end{aligned}$$

□

3.1 Security Proof

We analyze the security of our new multiplication scheme in the t-probing model, introduced in the seminal work of Ishai et al. [32], in which the adversary is allowed to learn up to t intermediate values that are produced during the computation. In particular we prove our algorithm to be secure also when composed with other gadgets in more complex circuits, by proving the stronger property of $t-$ *Strong Non-Interference* (t–SNI) defined by Barthe et al. in [5] and recalled in the following.

Definition 1 ($t-$ Strong Non-Interferent). *An algorithm $\mathcal{A}$ is $t-$* Strong Non-Interferent *(t–SNI) if and only if for any set of t_1 probes on intermediate variables and every set of t_2 probes on output shares such that $t_1 + t_2 \leq t$, the totality of the probes can be simulated by only t_1 shares of each input.*

In a few words the property requires not only that an adversary can simulate $d < t$ probes with d inputs, like in the classical t-probing model, but also that the number of input shares needed in the simulation are independent from the number of probes on the output shares.

The following lemma shows the t–SNI security of $\texttt{IPMult}_{\boldsymbol{L}}^{(1)}$.

Lemma 2. *The algorithm* $\texttt{IPMult}_{\boldsymbol{L}}^{(1)}$ *is t–SNI with $t = n - 1$.*

Proof. Let $\Omega = (\mathcal{I}, \mathcal{O})$ be a set of t observations respectively on the internal and on the output wires, where $|\mathcal{I}| = t_1$ and in particular $t_1 + |\mathcal{O}| \leq t$. We construct a perfect simulator of the adversary's probes, which makes use of at most t_1 shares of the secrets $\boldsymbol{A}$ and $\boldsymbol{B}$.

Let $w_1, \ldots, w_t$ be the probed wires. We classify the internal wires in the following groups:

(1) A_i, B_i,
(2) $U_{i,j}, U'_{i,j}$,
(3) $A_i \cdot B_j, T_{i,j}, V_{i,j}$,
(4) $C_{i,j}$, which represents the value of C_i at iteration i, j of the last **for** loop.

We define two sets of indices I and J such that $|I| \leq t_1$, $|J| \leq t_1$ and the values of the wires w_h with $h = 1, \ldots, t$ can be perfectly simulated given only the knowledge of $(A_i)_{i \in I}$ and $(B_i)_{i \in J}$. The sets are constructed as follows.

- Initially I and J are empty.
- For every wire as in the groups (1), (2) and (4), add i to I and to J.
- For every wire as in the group (3) if $i \notin I$ add i to I and if $j \notin J$ add j to J.

Since the adversary is allowed to make at most t_1 internal probes, we have $|I| \leq t_1$ and $|J| \leq t_1$.

We now show how the simulator behaves, by starting to consider the internal observed wires.

1. For each observation as in the group (1), by definition of I and J the simulator has access to A_i, B_i and then the values are perfectly simulated.
2. For each observation as in the group (2), we distinguish two possible cases:
   - If $i \in I, J$ and $j \notin J$, the simulator assigns a random and independent value to $U'_{i,j}$: if $i < j$ this is what would happen in the real algorithm, otherwise since $j \notin J$ the element U'_{ij} will never enter into the computation of any w_h (otherwise j would be in J).
   - If $i \in I, J$ and $j \in J$, the values $U'_{i,j}$ and $U'_{j,i}$ can be computed as in the actual circuit: one of them (say $U'_{j,i}$) is assigned to a random and independent value and the other $U'_{i,j}$ to $-U'_{i,j}$.

   The value $U_{i,j}$ is computed using the simulated $U'_{i,j}$ and the public value L_i.
3. For each observation as in the group (3), by definition of the sets I and J and for the previous points, the simulator has access to A_i, A_j, B_i, B_j, to the public value L_j and $U_{i,j}, U'_{i,j}$ can be simulated. Therefore $A_i \cdot B_j, T_{i,j}$ and $V_{i,j}$ can be computed as in the real algorithm.
4. For each observation as in the group (4), by definition $i \in I, J$. At first we assign a random value to every summand V_{ik}, with $k \leq j$ and $k \notin J$, entering in the computation of any observed C_{ij}. Then if one of the addends V_{ik} with $k \leq j$ composing C_{ij} has been probed, since by definition $k \in J$, we can simulate it as in Step 3. Otherwise V_{ik} has been previously assigned at the beginning of the current Step 4.

We now simulate the output wires C_i. We have to take into account the following cases.

1. If the attacker has already observed some intermediate values of the output share C_i, we note that each C_i depends on the random values in the ith row of the matrix $\boldsymbol{U}'$, i.e. U'_{il} for $l < i$ and U'_{li} for $l > i$. In particular each of the U'_{il} appears a second time in one of the remaining $C_1, \cdots, C_{i-1}, C_{i+1}, \cdots, C_n$, as shown in the following matrix.

$$\begin{pmatrix} 0 & U'_{1,2} & U'_{1,3} & \dots & U'_{1,n} \\ -U'_{1,2} & 0 & U'_{2,3} & \dots & U'_{2,n} \\ -U'_{1,3} & -U'_{2,3} & 0 & \dots & U'_{3,n} \\ \vdots & \vdots & \vdots & \ddots & \vdots \\ -U'_{1,n} & -U'_{2,n} & -U'_{3,n} & \dots & 0 \end{pmatrix} \begin{matrix} \leftarrow C_1 \\ \leftarrow C_2 \\ \\ \\ \leftarrow C_n \end{matrix}$$

Since each C_i depends on $n-1$ random values and the adversary may have probed at most $n-2$ of that, then independently of the intermediate elements probed, at least one of the U'_{il} doesn't enter into the computation of $C_{i,j}$ and so C_i can be simulated as a random value.

2. If all the partial sums have been observed, we can use the values previously simulated and add them according to the algorithm. Finally it remains to simulate a C_i when no partial sum C_{ij} has been observed. By definition, at least one of the U'_{il} involved in the computation of C_i is not used in any other observed wire. Therefore we can assign a random value to C_i.

□

4 Application to AES Sbox

Since $\texttt{IPMult}^{(1)}_{\boldsymbol{L}}$ is proved to be t–SNI, it can be securely composed with other t–SNI or affine gadgets. In the following we analyze more in detail the algorithm for the exponentiation to the power 254 in GF(2^8), which constitutes the non-linear part of the AES Sbox. We consider Rivain and Prouff's algorithm from [17,42]. We recall the squaring routine $\texttt{IPSquare}_{\boldsymbol{L}}$ and the refreshing scheme from [1]. We give in particular a t–SNI refreshing $\texttt{SecIPRefresh}_{\boldsymbol{L}}$, which essentially consists in the execution of $\texttt{IPRefresh}_{\boldsymbol{L}}$ n times. In [1] the authors already remarked that such a scheme ensures security even if composed with other gadgets, but no formal proof was provided. In the following we formally analyze the security of the algorithm, by giving the proof of t–SNI.

Algorithm 4. Square masked variable: $\boldsymbol{Y} \leftarrow \texttt{IPSquare}_{\boldsymbol{L}}(\boldsymbol{X})$

Input: vector $\boldsymbol{X}$
Output: vector $\boldsymbol{Y}$ such that $\langle \boldsymbol{L}, \boldsymbol{Y} \rangle = \langle \boldsymbol{L}, \boldsymbol{X} \rangle \cdot \langle \boldsymbol{L}, \boldsymbol{X} \rangle$
  for $i = 1$ to n **do**
    $Y_i \leftarrow (X_i)^2 \cdot L_i$;
  end for

Algorithm 5. Refresh vector: $\boldsymbol{X}' \leftarrow \texttt{IPRefresh}_{\boldsymbol{L}}(\boldsymbol{X})$

Input: vector $\boldsymbol{X}$
Output: vector $\boldsymbol{X}'$ such that $\langle \boldsymbol{L}, \boldsymbol{X} \rangle = \langle \boldsymbol{L}, \boldsymbol{X}' \rangle$

$(A_2, \cdots A_n) \leftarrow \texttt{rand}(\mathcal{K}^{n-1})$
$A_1 \leftarrow \sum_{i=2}^{n} A_i \cdot L_i$;
$\boldsymbol{X}' = \boldsymbol{X} + \boldsymbol{A}$;

Algorithm 6. Refresh vector: $\boldsymbol{Y} \leftarrow \texttt{SecIPRefresh}_{\boldsymbol{L}}(\boldsymbol{X})$

Input: vector $\boldsymbol{X}$
Output: vector $\boldsymbol{Y}$ such that $\langle \boldsymbol{L}, \boldsymbol{X} \rangle = \langle \boldsymbol{L}, \boldsymbol{Y} \rangle$

$\boldsymbol{Y}_0 = \boldsymbol{X}$;
for $i = 1$ to n **do**
  $\boldsymbol{Y}_i = \texttt{IPRefresh}_{\boldsymbol{L}}(\boldsymbol{Y}_{i-1})$;
end for
$\boldsymbol{Y} = \boldsymbol{Y}_n$;

Lemma 3. *The algorithm* $\texttt{SecIPRefresh}_{\boldsymbol{L}}$ *is* t–SNI *with* $t = n - 1$.

Proof. Let $\Omega = (\mathcal{I}, \mathcal{O})$ be a set of t observations respectively on the internal and on the output wires, where $|\mathcal{I}| = t_1$ and in particular $t_1 + |\mathcal{O}| \leq t$. We point out the existence of a perfect simulator of the adversary's probes, which makes use of at most t_1 shares of the secret $\boldsymbol{X}$.

The internal wires w_h are classified as follows:

(1) X_i
(2) $A_{i,j}$, which is the component i of the vector $\boldsymbol{A}$ in the jth $\texttt{IPRefresh}_{\boldsymbol{L}}$
(3) $Y_{i,j} = X_i + \sum_{k=1}^{j} A_{i,k}$, which is the component i of $\boldsymbol{Y}$ in the jth $\texttt{IPRefresh}_{\boldsymbol{L}}$

We define a set of indices I such that $|I| \leq t_1$ as follows: for every observation as in the group (1), (2) or (3) add i to I.

Now we construct a simulator that makes use only of $(X_i)_{i \in I}$.

- For each observation as in the group (1), $i \in I$ and then by definition of I the simulator has access to the value of X_i.
- For each observation as in the group (2), $A_{i,j}$ can be sample uniformly at random. Indeed, this is what happens in the real execution of the algorithm for the shares $A_{i,j}$ with $i = 2, \ldots, n$. Otherwise, since we have at most $n - 1$ probes, the adversary's view of $A_{1,j}$ is also uniformly random.
- For each observation as in the group (3), X_i can be perfectly simulated, $A_{i,j}$ can be sampled as in the real execution of the algorithm, and then all the partial sums $Y_{i,j}$ can be computed.

As for the output wires, we distinguish two cases. If some partial sum has already been observed, we remark that each output share $Y_{i,n}$ involves the computation of $n - 1$ random bits $A_{i,1}, \ldots, A_{i,n-1}$. The situation can be better understood from the following matrix, which shows the use of the random bits for each output share.

$$\begin{pmatrix} A_{1,1} & A_{1,2} & \dots & A_{1,n-1} & \left(\sum_{k=1}^{n-1} A_{1,k} L_i\right) L_n^{-1} \leftarrow Y_{1,n} \\ A_{2,1} & A_{2,2} & \dots & A_{2,n-1} & \left(\sum_{k=1}^{n-1} A_{1,k} L_i\right) L_n^{-1} \leftarrow Y_{2,n} \\ \vdots & \vdots & & \vdots & \vdots \\ A_{n,1} & A_{n,2} & \dots & A_{n,n-1} & \left(\sum_{k=1}^{n-1} A_{1,k} L_i\right) L_n^{-1} \leftarrow Y_{n,n} \end{pmatrix}$$

Now, since the adversary can have just other $n-2$ observations, there exists at least one non-observed random bit and we can simulate $Y_{i,n}$ as a uniform and independent random value. Moreover, if all the partial sums have been observed, we can use the values previously simulated and add them according to the algorithm. Otherwise, if no partial sum has been probed, since the random values involved in the computation of $Y_{1,n}, \dots, Y_{i-1,n}, Y_{i+1,n}, \dots, Y_{n,n}$ are picked at random independently from that one of $Y_{i,n}$, we can again simulate $Y_{i,n}$ as a uniform and independent random value, completing the proof. □

Now, considering that the multiplication gadget $\texttt{IPMult}_{\boldsymbol{L}}^{(1)}$ and the refreshing $\texttt{SecIPRefresh}_{\boldsymbol{L}}$ are both t–SNI and since the exponentiations $.^2, .^4$ and $.^{16}$ are linear functions in GF(2^8), we can claim that the entire algorithm for the computation of $.^{254}$ is t–SNI, according to the arguments in [5].

4.1 A More Efficient Scheme

We underline that for achieving $(n-1)^{\text{th}}$-order security the masked inputs $\boldsymbol{A}$ and $\boldsymbol{B}$ of $\texttt{IPMult}_{\boldsymbol{L}}^{(1)}$ must be mutually independent. If this is not the case, a refreshing of one of the factors is needed before processing the multiplication.

In this section we present an extended multiplication scheme $\texttt{IPMult}_{\boldsymbol{L}}^{(2)}$, illustrated in Algorithm 7, which can securely receive in input two values of the form $\boldsymbol{A}$ and $g(\boldsymbol{A})$, where g is a linear function. Thanks to this property, in case of mutual dependence of the inputs the refreshing is no longer needed and we can save on the number of random bits. The main idea of the new algorithm is to introduce a vector $\boldsymbol{u}$ sampled at random at the beginning of the execution and used to internally refresh the shares of the secrets.

The correctness of $\texttt{IPMult}_{\boldsymbol{L}}^{(2)}$ is again quite simple and we leave it to the reader.

Lemma 4. *For any $\boldsymbol{L}, \boldsymbol{A} \in \mathcal{K}^n$ and $\boldsymbol{C} = \texttt{IPMult}_{\boldsymbol{L}}^{(2)}(\boldsymbol{A}, g(\boldsymbol{A}))$, we have*

$$\langle \boldsymbol{L}, \boldsymbol{C} \rangle = \langle \boldsymbol{L}, \boldsymbol{A} \rangle \cdot \langle \boldsymbol{L}, g(\boldsymbol{A}) \rangle.$$

Lemma 5 provides the security analysis of $\texttt{IPMult}_{\boldsymbol{L}}^{(2)}$.

Lemma 5. *Let g be a linear function over $\mathcal{K}$. The algorithm $\texttt{IPMult}_{\boldsymbol{L}}^{(2)}(\boldsymbol{A}, g(\boldsymbol{A}))$ is t–SNI, with $t = n-1$.*

Algorithm 7. Multiply dependent masked values: $\boldsymbol{C} \leftarrow \texttt{IPMult}_{\boldsymbol{L}}^{(2)}(\boldsymbol{A}, g(\boldsymbol{A}))$

Input: vector $\boldsymbol{A}$ of length n

Output: vector $\boldsymbol{C}$ satisfying $\langle \boldsymbol{L}, \boldsymbol{C} \rangle = \langle \boldsymbol{L}, \boldsymbol{A} \rangle \cdot \langle \boldsymbol{L}, g(\boldsymbol{A}) \rangle$, for g linear function

▷ Sampling at random of the vector $\boldsymbol{u}$
for $i = 1$ to n **do**
  $u_i \leftarrow \texttt{rand}(\mathcal{K})$;
end for
▷ Computation of the matrix $\boldsymbol{A}'$
for $i = 1$ to n **do**
  for $j = 1$ to n **do**
    $A'_{i,j} = A_i + \delta_{ij} u_j$;
  end for
end for
▷ Computation of the vector $\boldsymbol{B}'$
for $i = 1$ to n **do**
  $B'_i = g(A_i) \cdot u_i \cdot L_i$;
end for
▷ Computation of the matrix $\boldsymbol{T}$
for $i = 1$ to n **do**
  for $j = 1$ to n **do**
    $T_{i,j} = A'_{i,j} \cdot g(A_j) \cdot L_j$;
  end for
end for
▷ Computation of the matrices $\boldsymbol{U}$ and $\boldsymbol{U}'$
for $i = 1$ to n **do**
  for $j = 1$ to n **do**
    if $i < j$ **then**
      $U'_{ij} \leftarrow \texttt{rand}(\mathcal{K})$;
    end if
    if $i > j$ **then**
      $U'_{ij} = -U'_{ji}$;
    end if
    $U_{i,j} = U'_{i,j} \cdot \delta_{ij} L_i^{-1}$;
  end for
end for
▷ Computation of the matrix $\boldsymbol{V}$
for $i = 1$ to n **do**
  for $j = 1$ to n **do**
    $V_{i,j} = (T_{i,j} + U_{i,j}) - \delta_{ij} B'_j$;
  end for
end for
▷ Computation of the output vector $\boldsymbol{C}$
for $i = 1$ to n **do**
  $C_i = \sum_j V_{i,j}$;
end for

Proof. Let $\Omega = (\mathcal{I}, \mathcal{O})$ be a set of t observations respectively on the internal and on the output wires, where $|\mathcal{I}| = t_1$ and in particular $t_1 + |\mathcal{O}| \leq t$. We point out the existence of a perfect simulator of the adversary's probes, which makes use of at most t_1 shares of the secret $\boldsymbol{A}$.

Let $w_1, \ldots, w_t$ be the probed wires. We classify the internal wires in the following groups:

(1) $A_i, g(A_i), g(A_i) \cdot u_i, B'_i, u_i$
(2) $U_{i,j}, U'_{i,j}$
(3) $A'_{i,j}, A'_{i,j} \cdot g(A_j), T_{i,j}, T_{i,j} + U_{i,j}, V_{i,j}$
(4) $C_{i,j}$, which represents the value of C_i at iteration i, j of the last **for**

We now define the set of indices I with $|I| \leq t_1$ such that the wires w_h can be perfectly simulated given only the knowledge of $(A_i)_{i \in I}$. The procedure for constructing the set is the following:

- Initially I is empty.
- For every wire as in the groups (1), (2) and (4), add i to I.
- For every wire as in the group (3), if $i \notin I$ add i to I and if $i \in I$ add j to I.

Since the adversary is allowed to make at most t_1 internal probes, we have that $|I| \leq t_1$.

In the simulation phase, at first we assign a random value to every u_i entering in the computation of any observed w_h. Then the simulation for any internal wires w_h proceeds as follows.

1. For each observation in category (1), then $i \in I$ and by definition we can directly compute from A_i, u_i and the public value L_i.
2. For each observation in category (2), then $i \in I$ and we distinguish two possible cases:
   - If $j \notin I$, then we can assign a random and independent value to $U'_{i,j}$. Indeed if $i < j$ this is what would happen in the real execution of the algorithm and if $i > j$, since $j \notin I$, $U'_{i,j}$ will never be used in the computation of other observed values. We compute $U_{i,j}$ using $U'_{i,j}$ and the public value L_i.
   - If $j \in I$, the values $U'_{i,j}$ and $U'_{j,i}$ can be computed as in the actual circuit: we assign one of them (say $U'_{j,i}$) to a random and independent value and the other $U'_{i,j}$ to $-U'_{i,j}$. We compute $U_{i,j}$ using $U'_{i,j}$ and the public value L_i.
3. For each observation in category (3), then $i \in I$ and we distinguish two possible cases:
   - If $j \notin I$, then we can assign a random and independent value to w_h. Indeed, since $j \notin I$, one of the values composing w_h has not been observed (otherwise by construction j would be in I) and for the same reason also any of the w_h does not enter in the expression of any other observed wire.
   - If $j \in I$, the value w_h can be perfectly simulated by using the accessible values $A_i, g(A_j), u_i, u_j, L_i, L_j$ and the values $U_{i,j}, U'_{i,j}$ assigned in Step 2.

4. For each observation as in the group (4), by definition $i \in I$. At first we assign a random value to every summand V_{ik}, with $k \leq j$ and $k \notin I$, entering in the computation of any observed C_{ij}. Then if one of the addends V_{ik} with $k \leq j$ composing C_{ij} has been probed, since by definition $k \in I$, we can simulate it as in Step 3. Otherwise V_{ik} has been previously assigned at the beginning of the current Step 4.

As for the probed output wires, we distinguish the following cases.

1. If the attacker has already observed some intermediate values of C_i, using a similar argument to the one in the proof of Lemma 2, we point out that C_i can be simulated as a random value.
2. If all the partial sums have been observed, we can use the values previously simulated and add them according to the algorithm. Finally, when no partial sum C_{ij} has been observed, again as before, by definition at least one of the U'_{il} involved in the computation of C_i is not used in any other observed wire and then we can assign to C_i a random value.

□

We can now exploit this new scheme in the $.^{254}$ algorithm, by eliminating the first two refreshing and substituting the first two multiplications with our $\mathtt{IPMult}_L^{(2)}(\cdot, \cdot^2)$ and $\mathtt{IPMult}_L^{(2)}(\cdot, \cdot^4)$, while using in the rest the $\mathtt{IPMult}_L^{(1)}$. In particular, according to the squaring routine in Algorithm 4, we point out that in $\mathtt{IPMult}_L^{(2)}(\cdot, \cdot^2)$ the shares $g(A_i)$ correspond to the products $A_i^2 \cdot L_i$ and in $\mathtt{IPMult}_L^{(2)}(\cdot, \cdot^4)$ the shares $g(A_i)$ correspond to the products $A_i^4 \cdot L_i \cdot L_i \cdot L_i$. The implementation of the gadget $.^{254}$ is depicted in Fig. 1 and in Lemma 6 we prove that it is t–SNI, using the techniques presented in [5].

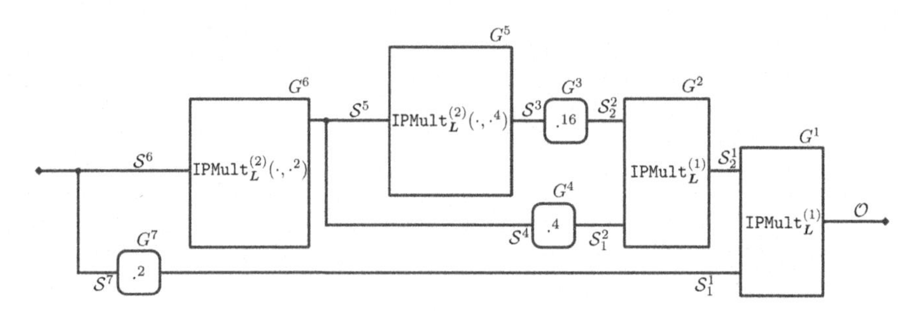

Fig. 1. Gadget $.^{254}$ which makes use of $\mathtt{IPMult}_L^{(1)}$ and $\mathtt{IPMult}_L^{(2)}$

Lemma 6. *Gadget* $.^{254}$, *shown in Fig. 1, is* t–SNI.

Proof. Let $\Omega = (\bigcup_{i=1}^{7} \mathcal{I}^i, \mathcal{O})$ a set of t observations respectively on the internal and output wires. In particular $\mathcal{I}^i$ are the observations on the gadget G^i and

$\sum_{i=1}^{7} |\mathcal{I}^i| + |\mathcal{O}| \leq t$. In the following we construct a simulator which makes use of at most $\sum_{i=1}^{7} |\mathcal{I}^i|$ shares of the secret, by simulating each gadget in turn.

Gadget G^1 Since $\mathtt{IPMult}_{\boldsymbol{L}}^{(1)}$ is t–SNI and $|\mathcal{I}^1 \cup \mathcal{O}| \leq t$, then there exist two sets of indices $\mathcal{S}_1^1, \mathcal{S}_2^1$ such that $|\mathcal{S}_1^1| \leq |\mathcal{I}^1|$, $|\mathcal{S}_2^1| \leq |\mathcal{I}^1|$ and the gadget can be perfectly simulated from its input shares corresponding to the indices in $\mathcal{S}_1^1$ and $\mathcal{S}_2^1$.

Gadget G^2 Since $\mathtt{IPMult}_{\boldsymbol{L}}^{(1)}$ is t–SNI and $|\mathcal{I}^2 \cup \mathcal{S}_2^1| \leq |\mathcal{I}^1| + |\mathcal{I}^2| \leq t$, then there exist two sets of indices $\mathcal{S}_1^2, \mathcal{S}_2^2$ such that $|\mathcal{S}_1^2| \leq |\mathcal{I}^2|$, $|\mathcal{S}_2^2| \leq |\mathcal{I}^2|$ and the gadget can be perfectly simulated from its input shares corresponding to the indices in $\mathcal{S}_1^2$ and $\mathcal{S}_2^2$.

Gadget G^3 Since $.^{16}$ is affine, there exists a set of indices $\mathcal{S}^3$ such that $|\mathcal{S}^3| \leq |\mathcal{I}^3| + |\mathcal{S}_2^2|$ and the gadget can be perfectly simulated from its input shares corresponding to the indices in $\mathcal{S}^3$.

Gadget G^4 Since $.^4$ is affine, there exists a set of indices $\mathcal{S}^4$ such that $|\mathcal{S}^4| \leq |\mathcal{I}^4| + |\mathcal{S}_1^2|$ and the gadget can be perfectly simulated from its input shares corresponding to the indices in $\mathcal{S}^4$.

Gadget G^5 Since $\mathtt{IPMult}_{\boldsymbol{L}}^{(2)}$ is t–SNI and $|\mathcal{I}^5 \cup \mathcal{S}^3| \leq |\mathcal{I}^5| + |\mathcal{I}^3| + |\mathcal{I}^2| \leq t$, then there exists a set of indices $\mathcal{S}^5$ such that $|\mathcal{S}^5| \leq |\mathcal{I}^5|$ and the gadget can be perfectly simulated from its input shares corresponding to the indices in $\mathcal{S}^5$.

Gadget G^6 Since $\mathtt{IPMult}_{\boldsymbol{L}}^{(2)}$ is t–SNI and $|\mathcal{I}^6 \cup \mathcal{S}^5| \leq |\mathcal{I}^6| + |\mathcal{I}^5| \leq t$, then there exists a set of indices $\mathcal{S}^6$ such that $|\mathcal{S}^6| \leq |\mathcal{I}^6|$ and the gadget can be perfectly simulated from its input shares corresponding to the indices in $\mathcal{S}^6$.

Gadget G^7 Since $.^2$ is affine, there exists a set of indices $\mathcal{S}^7$ such that $|\mathcal{S}^7| \leq |\mathcal{I}^7| + |\mathcal{S}_1^1| \leq |\mathcal{I}^7| + |\mathcal{I}^1|$ and the gadget can be perfectly simulated from its input shares corresponding to the indices in $\mathcal{S}^7$.

Each of the previous steps guarantee the existence of a simulator for the respective gadgets. The composition of them allows us to construct a simulator of the entire circuit which uses $\mathcal{S}^6 \cup \mathcal{S}^7$ shares of the input. Since $|\mathcal{S}^6 \cup \mathcal{S}^7| \leq |\mathcal{I}^7| + |\mathcal{I}^1| + |\mathcal{I}^6| \leq \sum_{i=1}^{7} |\mathcal{I}^i|$ we can conclude that the gadget $.^{254}$ is t–SNI. □

The advantage of the use of $\mathtt{IPMult}_{\boldsymbol{L}}^{(2)}$ mostly consists in amortizing the randomness complexity. Indeed the new scheme requires only n (for the vector $\boldsymbol{u}$) plus $\frac{n(n-1)}{2}$ (for the matrix $\boldsymbol{U}$) random bits, while the previous one uses a larger amount of randomness, corresponding to n^2 (for the $\mathtt{SecIPRefresh}_{\boldsymbol{L}}$) plus $\frac{n(n-1)}{2}$ (for the $\mathtt{IPMult}_{\boldsymbol{L}}^{(1)}$) bits. We summarize in Table 1 the complexities of the two schemes. The issue of providing a secure multiplication of two dependent operands was first addressed by Coron et al. in [17]. In their work the authors proposed a new algorithm which requires $n(n-1)$ random bits and that has later been proved to be t–SNI in [5]. By analyzing the amount of random generations and comparing with $\mathtt{IPMult}_{\boldsymbol{L}}^{(2)}$, we can see that our scheme is more efficient whenever $n > 3$, while it requires the same amount of randomness for $n = 3$ and more random bits for $n < 3$. On the other hand, from a complexity point of view the scheme in [17] is better optimized in terms of field multiplications since it makes use of look-up tables.

A more detailed performance analysis is provided in the next section.

Table 1. Complexity of $\mathtt{IPMult}_L^{(1)}$ and $\mathtt{IPMult}_L^{(2)}$ and comparison with the multiplication algorithms of [8,17]

	#Additions	# Multiplications	#Random bits
$\mathtt{IPMult}_L^{(1)}$	$2n^2$	$3n^2$	$\frac{n(n-1)}{2}$
$\mathtt{IPMult}_L^{(2)}$	$4n^2$	$3n^2 + 2n$	$\frac{n(n+1)}{2}$
$\mathtt{SecIPRefresh}_L$	$2n^2 - n$	$2n$	n^2
$\mathtt{IPMult}_L^{(1)}$ and $\mathtt{SecIPRefresh}_L$	$4n^2 - n$	$3n^2 + 2n$	$\frac{n(3n-1)}{2}$
Algorithm 5 in [17]	$4n(n-1)$	–	$n(n-1)$
Algorithm 3 in [8]	$4n(n-1)$	$\frac{1}{4}(n-1)(7n+3)$ (n odd) $\frac{1}{4}n(7n-6)$ (n even)	$n(n-1)$

5 Performance Evaluations

In this section we analyze the performance of our improved IP masking construction. Following the lines in [1,2], we opt to protect a software implementation of AES-128 encryption for AVR architectures. We develop protected implementations using either our new multiplication algorithm $\mathtt{IPMult}_L^{(1)}$ alone, or in combination with $\mathtt{IPMult}_L^{(2)}$. In order to compare performances, we also develop protected instances of AES-128 with Boolean masking. All our implementations have a constant-flow of operations and share the same underlying blocks. In particular, we use `log-alog` tables for field multiplication and look-up tables to implement raisings to a power. The most challenging operation to protect is the nonlinear `SubBytes` transformation, which is also the bottleneck of our implementations. Similar to earlier work, we take advantage of the algebraic structure of the AES and compute `SubBytes` as the composition of a power function x^{254} and an affine transformation. The remaining operations are straightforward to protect and are thus omitted in what follows. Our codes can be downloaded from http://homes.esat.kuleuven.be/~jbalasch.

Implementation of the power function. Rivain and Prouff proposed in [42] an algorithm to compute the inversion in $\mathbb{F}_2^8$ as x^{254} using an addition chain with only 4 multiplications. We select this algorithm for our implementations protected with IP masking. Recall that to ensure t–SNI it is necessary to execute the $\mathtt{SecIPRefresh}_L$ algorithm when using only $\mathtt{IPMult}_L^{(1)}$, but this can be omitted when using also $\mathtt{IPMult}_L^{(2)}$ as depicted in Fig. 1.

The same technique is used in our Boolean masking implementations, only in this case we employ the mask refreshing algorithm proposed by Duc et al. [18]. Additionally, we provide a faster implementation using the addition chain proposed by Grosso et al. [30], which leverages on the algorithm introduced by Coron et al. [17] to securely evaluate functions of the form $x \cdot g(x)$, where g is a linear function. This approach demands only 1 multiplication and 3 secure evaluations, and thus achieves significant performance gains. Note that further optimizations are possible by combining [30] with recent techniques, e.g. the common shares approach proposed by Coron et al. [16] or the multiplication gadget put forward by Belaïd et al. [8]. We expect however the gains to be relatively small (see results in [16]), and therefore have a limited impact for the purposes of comparison.

Implementation of the affine transformation. Securing the affine transformation using Boolean masking can be done in a highly efficient way by applying it to every input share separately, that is, by computing $\boldsymbol{A}x_1+\ldots+\boldsymbol{A}x_n+b$. Hence, each share x_i of x is only involved in one matrix-vector multiplication, which in practice can be tabulated. Unfortunately, such an approach is not directly applicable to IP masking, since the sharing of x consists of two vectors $\boldsymbol{L}$ and $\boldsymbol{R}$ with each n elements in $\mathbb{F}_2^8$. The affine transformation can be computed through a polynomial evaluation over $\mathbb{F}_2^8$, which is known to perform rather poorly when compared to Boolean masking (see [1,2]).

In this work we note that since $\boldsymbol{L}$ is fixed it is possible to change the representation of $\boldsymbol{A}$ depending on the values L_i. More precisely, we define n matrices $\boldsymbol{A_i}$ and compute the affine transformation as $\boldsymbol{A_1}x_1 + \ldots + \boldsymbol{A_n}x_n + b$. The matrices $\boldsymbol{A_i}$ need only to be pre-computed once, at initialization time. Given $L_i \in \mathbb{F}_2^8$ we first construct an 8×8 matrix $\boldsymbol{M_i}$ over $\mathbb{Z}_2$. Notice that L_i is represented by a polynomial $a_0+a_1 \cdot x+\ldots+a_7 \cdot x^7$, where $\{1, x, \ldots, x^7\}$ form the basis of $\mathbb{F}_2^8$. The j-th column of $\boldsymbol{M_i}$ corresponds to the coefficients of the polynomial $L_i \times x^{j-1}$. Given the matrix $\boldsymbol{M_i}$ as described above, we can then compute $\boldsymbol{A_i} = \boldsymbol{A} \times \boldsymbol{M_i}$ by simple matrix multiplication, and take advantage of tabulation in the implementation. In contrast to Boolean masking, the memory requirements of this tabulation increase linearly with the number of shares. However, the overheads remain reasonable for practical values of n.

Implementation results. We have developed assembly implementations for $n = 2, 3$ shares tailored to the target AVR architecture and optimized for speed. Results are summarized in Table 2. The implementation protected by IP masking using only $\texttt{IPMult}_{\boldsymbol{L}}^{(1)}$ requires roughly 157 k cycles and 372 k cycles for security levels $n = 2$ and $n = 3$, respectively. This represents a significant improvement over earlier work [1] which demanded 375 k and 815 k cycles to protect instances of AES-128 for the same security levels. The implementation protected by IP masking using $\texttt{IPMult}_{\boldsymbol{L}}^{(2)}$ in conjunction with $\texttt{IPMult}_{\boldsymbol{L}}^{(1)}$ performs slightly poorer in terms of cycles but, as mentioned earlier, has the advantage of demanding less randomness. The results for Boolean masking with the same number of secret shares are 110 k and 230 k, respectively. The timing gap with respect to IP masking stems exclusively from the computation of x^{254}, as the rest of AES operations execute in a similar number of cycles. The reason why IP masking is slower is mainly due to the extra operations in the multiplication gadgets. Note that since $\boldsymbol{L}$ is fixed, it is possible to tabulate the field multiplications with elements L_i and L_i^{-1}, given that the number of shares n is small. We have performed this optimization which allows to reduce the cycle count at the cost of more non-volatile storage. Thanks to this, we are able to decrease the gap between Boolean and IP masking implementations to slightly more than a factor

2 when compared to the implementation using the addition chain from [30]. We leave as open work whether a similar algorithm as in [17] to efficiently evaluate functions of the form $x \cdot g(x)$ can be devised for IP masking.

Table 2. Performance evaluation of protected AES-128 implementations on AVR architectures (optimized in assembly code). Timings in clock cycles, memory and randomness requirements in bytes.

Masking		Timings		Memory	Randomness
		x^{254}	AES-128		
IP masking	n = 2	709	157 196	2 816	1 632
(only $\mathtt{IPMult}_L^{(1)}$*)*	n = 3	1 752	372 225	3 328	4 864
IP masking	n = 2	763	167 996	2 816	1 632
($\mathtt{IPMult}_L^{(1)}$ & $\mathtt{IPMult}_L^{(2)}$*)*	n = 3	1 766	375 025	3 328	3 664
Boolean masking	n = 2	459	110 569	2 048	1 232
(addition chain [42])	n = 3	1 043	230 221	2 048	3 664
Boolean masking	n = 2	275	73 769	1 792	1 432
(addition chain [30])	n = 3	676	160 357	1 792	4 264

Lastly, we illustrate in Fig. 2 the performance trend of our implementations for larger values of n. Cycle counts correspond in this case to a single SBox operation. Note that the results for $n = 2, 3$ are significantly higher than those provided in Table 2, the reason being that the implementations are now written in C language (and are thus less optimized than their assembly counterparts). Note also that the gap between Boolean and IP masking protected versions increases almost to a factor 4. This is because we do not take advantage of

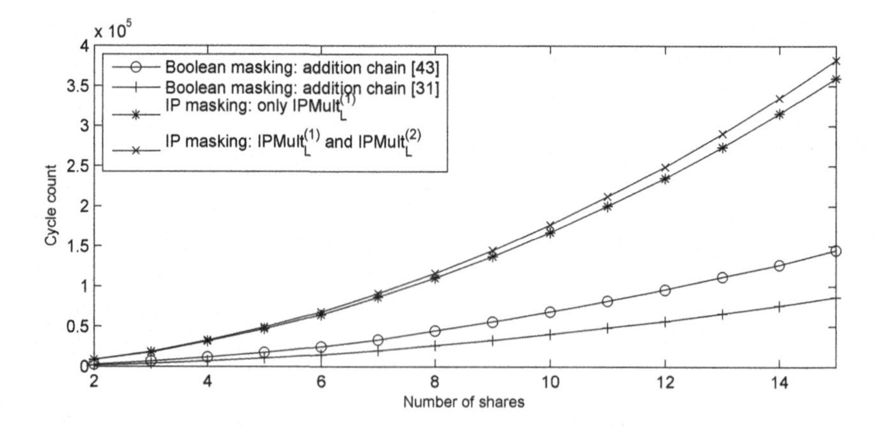

Fig. 2. Performance evaluation of protected AES Sbox implementations on AVR architectures (in C code) for increasing number of shares.

the tabulation of the field multiplications with elements L_i and L_i^{-1}, since the memory requirements would grow considerably for non-small values of n. In spite of this, we observe that the performance ratio between Boolean and IP masking protected implementations remains constant as the number of shares increases.

6 Information Theoretic Evaluation

As a complement to the previous proofs and performance evaluations, we now provide results regarding the information theoretic analysis of inner product masking. As first motivated in [44], the mutual information between a secret variable and its corresponding leakages can serve as a figure of merit for side-channel security, since it is proportional to the success rate of a (worst-case) Bayesian adversary exploiting these leakages (see [19] for a recent discussion). Such a metric has been used already for the evaluation of Boolean masking [45], affine masking [23], polynomial masking [28,40] and inner product masking [1, 2]. In this respect, and despite the encoding of our consolidated inner product masking schemes has not changed compared to the latter two references, we aim to improve their results in three important directions:

- *Extended noise range.* In [1,2], the mutual information of the inner product encoding was evaluated for a Hamming weight leakage function and noise variances up to 4. While this is sufficient to discuss the positive impact of the increased algebraic complexity of inner product masking for low noise levels, it is not sufficient to exhibit the security order (which corresponds to the lowest key-dependent statistical moment of the leakage distribution minus one [7], and is reflected by the slope of the information theoretic curves for high noise levels). Therefore, we generalize the improved numerical integration techniques from [19] to inner product encodings and compute the mutual information metric for noise variances up to 1000 (which allows us to exhibit and discuss security orders).
- *Other (public) L values.* In [1,2], the inner product encoding was evaluated based on a single value of the public $\boldsymbol{L}$. However, it was recently shown in [46] that for linear leakage functions (such as the Hamming weight leakage function), an appropriate choice of $\boldsymbol{L}$ may improve the security order of an implementation. In other words, it was shown that security in the bounded moment model (as recently formalized in [7]) can be higher than the probing security order in this case. Therefore, we evaluate the mutual information for different $\boldsymbol{L}$ vectors for our 8-bit targets (rather than 4-bit S-boxes in [46], which is again made possible by our exploitation of improved numerical integration techniques).
- *Non-linear leakage functions.* Since the previous security order amplification is highly dependent on the fact that the leakage function is linear, we finally complement our results by evaluating the information leakage of the inner product encoding for non-linear leakage functions.

Building on our experimental observations, we also highlight other interesting implementation properties of the inner product encoding (regarding the risk of transition-based leakages [3,15]) in Sect. 6.3. And we conclude the section by discussing general (theoretical) limitations of both the security order amplification and these implementation properties.

6.1 Linear (e.g., Hamming Weight) Leakages

We first analyze the information leakage of the inner product encoding of Algorithm 2 for $n = 2$ shares and a Hamming weight leakage function. More precisely, we consider a target intermediate secret variable $A \in \mathrm{GF}(2^8)$ that is encoded as $A = A_1 + L_2 \cdot A_2$ such that $\boldsymbol{A} = [A_1, A_2]$. The adversary is given the leakage (next denoted with the variable $\boldsymbol{O}$ for observation, to avoid confusion with the L values) corresponding to these two shares. That is, $\boldsymbol{O} = [O_1, O_2]$ with $O_1 = \mathsf{HW}(A_1) \boxplus N_1$, $O_2 = \mathsf{HW}(A_2) \boxplus N_2$, HW the Hamming weight function, N_1, N_2 two normally distributed (independent) noise random variables and $\boxplus$ the addition in the reals (in contrast with the group addition $+$). The mutual information between A and the observation $\boldsymbol{O}$ is expressed as:

$$\mathrm{MI}(A;\boldsymbol{O}) = \mathrm{H}[A] \boxplus \sum_{a \in \mathcal{A}} \Pr[a] \times \sum_{a_2 \in \mathcal{A}} \Pr[a_2] \times \sum_{\boldsymbol{o} \in \mathcal{O}^2} \mathsf{f}[\boldsymbol{o}|a] \times \log_2 \Pr[a|\boldsymbol{o}], \quad (1)$$

where $\mathsf{f}[\boldsymbol{o}|a]$ is the conditional Probability Density Function (PDF) of the observation $\boldsymbol{o}$ given the secret a, which is computed as a sum of normal PDFs (denoted as N) evaluated for all the (unknown) random shares: $\mathsf{f}[\boldsymbol{o}|a] = \sum_{a_2 \in \mathcal{A}} \mathsf{N}[\boldsymbol{o}|a, a_2] \cdot \Pr[a_2]$. The conditional probability $\Pr[a|\boldsymbol{o}]$ is obtained via Bayes' law: $\Pr[a|\boldsymbol{o}] = \frac{\mathsf{f}[\boldsymbol{o}|a]}{\sum_{a^* \in \mathcal{A}} \mathsf{f}[\boldsymbol{o}|a^*]}$ where the a^* notation is used for the secret a candidates.[1]

The result of our information theoretic analysis for Hamming weight leakages, for vectors $L_2 = 17, 5, 7$ and noise variances between 10^{-2} and 10^3 is given in Fig. 3, where we additionally report the leakage of an unprotected A (i.e., for which the adversary can observe $O = \mathsf{HW}(A) \boxplus N$) and of a Boolan encoding (which is a special case of inner product encoding such that $L_1 = L_2 = 1$) for illustration. For low noise levels, we reach the same conclusions as previous works [1,2]. Namely, the increased algebraic complexity of inner product masking allows significantly lower leakages than Boolean masking. Intuitively, this is simply explained by the fact that knowing one bit of each share directly leads to one bit of secret in Boolean masking, while it only leads to a (smaller) bias on the secret variable distribution in inner product masking.

For large noise levels, and as expected, we now clearly observe the security order (in the bounded moment model) of the masking schemes based on the slope of the information theoretic curves, which moves from -1 for an unprotected implementation to -2 for Boolean masking (the latter therefore corresponds to a security order 1 in the bounded moment model). Interestingly, our results also

---

[1] Note that despite our simulated leakages are coming from a continuous distribution, we estimate the mutual information by sampling (following the open source code of [19]), which explains why Eq. (1) uses sums rather than integrals.

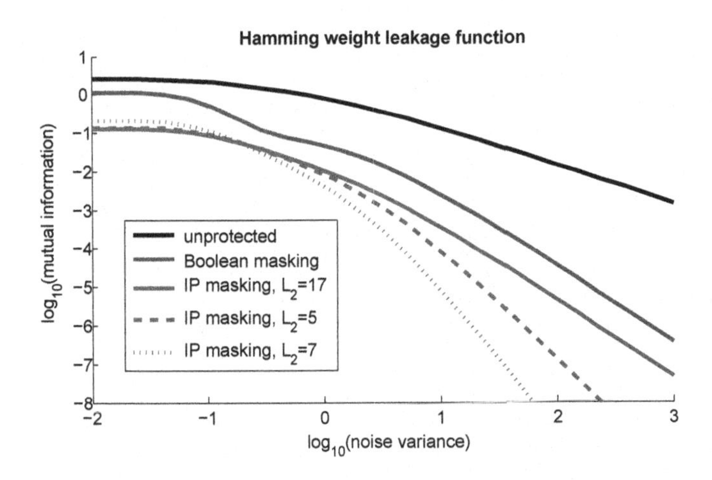

Fig. 3. Information theoretic evaluation of an inner product encoding.

show that by tuning the public L_2 value of the inner product encoding, we can reach much better results. Namely, the slope of the information theoretic curves can be reduced to -3 (which corresponds to a security order 2 in the bounded moment model) and even -4 (which corresponds to a security order 3 in the bounded moment model), despite the first-order security of this encoding in the probing model (proved in Sect. 3.1) has not changed.

The reason for this phenomenon has been given in a recent CARDIS 2016 paper [46] and is simply summarized by observing that the multiplication in GF(2^8) that is performed by the inner product encoding can be represented as a multiplication with an 8×8 matrix in $\mathsf{GF}(2)$. Roughly, depending on the number of linearly independent lines in this matrix, and assuming that the leakage function will only mix the bits of the encoding linearly (which is the case for the Hamming weight leakage function), the multiplication will XOR more shares together, implying a higher security order in the bounded moment model. And this "security order amplification" is limited to a slope of -4 (which corresponds to the attack exploiting the multiplication of the squares of all the shares).

6.2 Non-linear (e.g., Random) Leakages

In view of the previous positive observations obtained for the inner product encoding in the context of linear (e.g., Hamming weight) leakages, a natural next step is to investigate the consequences of a deviation from this assumption. For this purpose, we study an alternative scenario where the Hamming weight leakages are replaced by a random leakage function G with similar output range $\{0, 1, \dots, 8\}$, such that the adversary now observes $O_1 = \mathsf{G}(A_1) \boxplus N_1$ and $O_2 = \mathsf{G}(A_2) \boxplus N_2$. Note that the choice of an output range similar to the Hamming weight function allows the two types of leakages to provide signals of similar amplitude to the adversary (which makes them directly comparable).

The result of our information theoretic analysis for random leakages, vectors $L_2 = 17, 5, 7$ and noise variances between 10^{-2} and 10^3 is given in Fig. 4, where we again report the leakage of an unprotected A and a Boolean encoding. Our observations are twofold. First, for large noise levels the security order amplification vanishes and all the information theoretic curves corresponding to $d = 2$ shares have slope -2, as predicted by the proofs in the probing model. This is expected in view of the explanation based on the 8×8 matrix in $\mathsf{GF}(2)$ given in the previous section and actually corresponds to conclusions made in [31] for low entropy masking schemes. That is, because of the non-linear leakage function, the GF(2) shares that are mixed thanks to the inner product encoding are actually recombined which reduces the security order. So as in this previous work, the reduction of the security order actually depends on the degree of the leakage function. But in contrast with low entropy masking schemes, the security cannot collapse below what is guaranteed by the security order in the probing model.

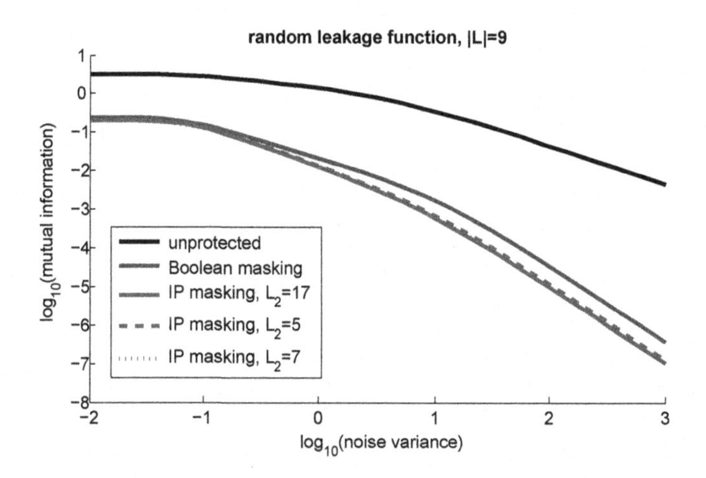

Fig. 4. Information theoretic evaluation of an inner product encoding.

Second and more surprisingly, we see that a non-linear leakage function also has a negative impact for the interest of the inner product encoding in the low noise region. This is explained by the fact that by making the leakage function non-linear, we compensate the low algebraic complexity of the Boolean encoding (so the distance between Boolean and inner product encodings vanishes).

From these experiments, we conclude that the security order amplification of inner product masking is highly implementation-dependent. We will further discuss the impact of this observation and the general limitations of the security order amplification in Sects. 6.4 and 7, but start by exhibiting another interesting property of the inner product encoding.

6.3 Transition-Based Leakages

In general, any masking scheme provides security guarantees under the condition that the leakages of each share are sufficiently noisy and independent. Yet, ensuring the independence condition usually turns out to be very challenging both in hardware and software implementations. In particular in the latter case, so-called transition-based leakages can be devastating. For illustration, let us consider a Boolean encoding such that $A = A_1 + A_2$. In case of transition-based leakages, the adversary will not only receive noisy versions of A_1 and A_2 but also of their distance. For example, in the quite standard case of Hamming distance leakages, the adversary will receive $\mathsf{HW}([A_1] + [A_2]) \boxplus N = \mathsf{HW}(A) \boxplus N$, which annihilates the impact of the secret sharing. Such transition-based leakages frequently happen in microcontrollers when the same register is used to consecutively store two shares of the same sensitive variable (which typically causes a reduction of the security order by a factor 2 [3]).

Interestingly, we can easily show that inner product masking provides improved tolerance against transition-based leakages. Taking again the example of an encoding $A = A_1 + L_2 \cdot A_2$, the Hamming distance between the two shares A_1 and A_2 (where $A_1 = A + L_2 \cdot A_2$) equals $\mathsf{HW}([A + L_2 \cdot A_2] + [A_2])$. Since for uniformly distributed A_2 and any $L_2 \neq 1$, we have that $A_2 + A_2 \cdot L_2$ is also uniformly distributed, this distance does not leak any information on A. Of course, and as in the previous section, this nice property only holds for certain combinations of shares (such as the group operation + in our example).

6.4 Limitations: A Negative Result

The previous sections showed that the inner product encodings offer interesting features for security order amplification and security against transition-based leakages in case the physical leakages are "kind" (e.g., linear functions, transitions based on a group operation). Independent of whether this condition holds in practice, which we discuss in the next section, one may first wonder whether these properties are maintained beyond the inner product encoding. Unfortunately, we answer to this question negatively. More precisely, we show that whenever non-linear operations are performed (such as multiplications), the security order of the inner product encoding gets back to the one of Boolean masking (and therefore is also divided by two in case transitions are observed).

Concretely, and assuming we want to multiply two shared secrets $A = A_1 + L_2 \cdot A_2$ and $B = B_1 + L_2 \cdot B_2$, a minimum requirement is to compute the cross products $A_i \cdot B_j$. So for example, an adversary can observe the pair of leakages $(A_1, A_2 \cdot B_2)$ which depends on A. Defining a function $\mathsf{F}_{B_2}(A_2) = A_2 \cdot B_2$, and assuming a (linear) Hamming weight leakage function HW, we see that the adversary obtains two leakage samples $O_1 = \mathsf{HW}(A_1) \boxplus N_1$ and $O_2 = \mathsf{HW}(\mathsf{F}_{B_2}(A_2)) \boxplus N_2$. In other words, it is in fact the composition of the functions F_{B_2} and HW that is subject to noise, the latter being non-linear and informative (because of the standard "zero issue" in multiplicative masking [26]). So whenever the implementation has to perform secure multiplications with inner product masking, we are in fact in a situation similar to the non-linear leakages of Sect. 6.2. A similar observation holds for the result of Sect. 6.3 regarding transition-based leakages. Taking exactly the previous example, observing the Hamming distance between A_1 and $\mathsf{F}_{B_2}(A_2)$ directly halves the security order, just as for the Boolean encodings in [3].

One natural scope for further research is to look for new solutions in order to maintain the security order guarantees even for non-linear operations (e.g., thanks to a different sequence of operations or additional refreshings). Nevertheless, even with the current algorithm and non perfectly linear leakages, inner product masking should reduce the number and informativeness of the key-dependent tuples of leakage samples in a protected implementation, which is not captured by the notion of (probing or bounded moment) security order. So overall, the improved theoretical understanding allowed by our investigations calls for a the concrete evaluation of an inner product masked AES. The next section makes a step in this direction, and discusses how these potential advantages translate into practice for a 32-bit ARM microcontroller.

7 Empirical Side-Channel Leakage Evaluation

In order to further complement the analysis we provide concrete results of empirical side-channel leakage evaluations for both Boolean masking with two shares and IP masking with $n = 2$ ($L_1 = 1, L_2 = 7$). Security proofs are valid only for the assumed and possibly idealized (e.g. simplified) leakage model, but real device leakage behaviour can be complex and hard to model. For instance, transition leakages are known to be difficult to deal with when moving from theory to practice. Similarly, the information theoretic analysis based on simulations is of course valid only for the simulated leakage behavior, and its results strongly vary for different leakage behaviours as we have shown, and it is limited to the encoding function.

We therefore assess and compare the leakage behavior of our implementations in practice with real measurements of our code running on a physical platform to round off our analysis. This evaluation allows us to reason about the leakage under typical conditions and without making modeling assumptions. Note also that this practical evaluation covers both the encoding as well as computation in the masked domain.

We use generic code that follows the guidelines of the masking algorithms provided in this paper but leave freedom to the compiler to perform register/memory allocations, optimizations, etc. The implementations are hence neither hand-optimized for the target platform nor adapted to its specific leakage behavior. The security of the implementations therefore depends in part on the compiler tool-chain.

Our target platform is an STM32 Nucleo board equipped with an ARM Cortex-M4 processor core. The processor runs at 168 MHz and features a built-in RNG capable of generating 32-bit random numbers every 40 clock cycles. The presence of the RNG is the main motivation for using this platform rather than an AVR. We have ported our generic (coded in C language) protected implementations of AES-128 using the addition chain from [42] to this platform using arm-none-eabi-gcc (v4.8.4) and verified that they run in constant time independent of the input values. Power measurements are obtained in a contactless fashion by placing a Langer RF-B 3-2 h-field (magnetic field) probe over a decoupling capacitor near the chip package, similar to [4]. The antenna output signal is amplified with a Langer PA-303 30 dB amplifier before we sample it with a Tektronix DPO 7254c oscilloscope and transfer it to a computer for analysis. We use a trigger signal generated from within the Nucleo board prior to each encryption routine to synchronize the power measurements.

Each power measurement comprises 500 000 samples that cover a time window of 4 ms. During this time the Boolean masked implementation executes slightly more than eight rounds of AES while the IP masking protected implementation executes about 2.5 rounds of AES. The timing difference of roughly a factor of four is in line with the data shown in Fig. 2. The time period covered by the measurements is a tradeoff between the amount of measurement data we need to handle on the one hand (shorter measurements give less data) and the complexity of the executed code on the other hand (we do not want to use a too simple toy example; two rounds of AES give full diffusion).

We use state-of-the-art leakage assessment techniques [13,27,36] to evaluate the leakage behavior of our masked implementations. Note that such an evaluation is independent of any adversarial strategy and hence it is not a security evaluation, i.e. it is not about testing resistance to certain attacks. Leakage assessment is a convenient tool to assess leakage regardless whether it is exploitable by a certain adversary.

In practice the most widely used methodology in the literature is Test Vector Leakage Assessment, first introduced in [27], and in particular the non-specific fixed versus random test. See for instance [43] for details. In brief, this particular test checks if the distribution of the measured side-channel leakage depends on the data processed by the device. If not, we can strongly ascertain that no adversary will be able to exploit the measurements to recover secret data.

To perform the test we collect two sets of measurements. For the first set we used a fixed input plaintext and we denote this set $\mathcal{S}_{fixed}$. For the second set the input plaintexts are drawn at random from uniform. We denote this set $\mathcal{S}_{random}$. Note that we obtain the measurements for both sets randomly interleaved (by flipping a coin before each measurement) to avoid time-dependent external and internal influences on the test result. The AES encryption key is fixed for all measurements.

We then compute Welch's (two-tailed) t-test:

$$t = \frac{\mu(\mathcal{S}_{fixed}) - \mu(\mathcal{S}_{random})}{\sqrt{\frac{\sigma^2(\mathcal{S}_{fixed})}{\#\mathcal{S}_{fixed}} + \frac{\sigma^2(\mathcal{S}_{random})}{\#\mathcal{S}_{random}}}}, \tag{2}$$

(where μ is the sample mean, σ^2 is the sample variance and # denotes the sample size) to determine if the samples in both sets were drawn from the same population (or from populations with the same mean). The *null hypothesis* is that the samples in both sets were drawn from populations with the same mean. In our context, this means that the masking is effective. The alternative hypothesis is that the samples in both sets were drawn from populations with different means. In our context, this means that the masking is not effective. A threshold for the t-score of ±4.5 is typically applied in the literature (corresponding roughly to a 99.999% confidence) to determine if the null hypothesis is rejected and the implementation is considered to leak. However, our primary intention is a relative comparison of the leakage of the different masked implementations.

7.1 RNG Deactivated

We first evaluate both implementations with the RNG deactivated (all random numbers are zero).

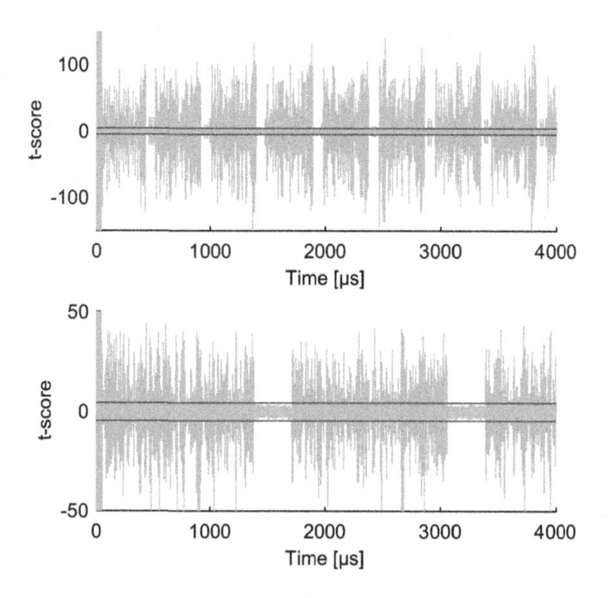

Fig. 5. t-test results for Boolean masking (top) and IP masking (bottom) with RNG deactivated; each based on 10 000 measurements. The red lines mark the ±4.5 threshold. (Color figure online)

In this scenario we expect both implementations to leak and we can use it to verify our measurement setup, analysis scripts, etc. We take 10 000 measurements from each implementation. Figure 5 shows plots of the t-scores for Boolean masking (top) and IP masking (bottom) in gray. The red lines mark the ±4.5 threshold.

As expected both implementations leak significantly. The repetitive patterns in the plots of the t-scores allow to recognize the rounds of AES as areas with high t-scores, interleaved by the key scheduling which does not leak in this test because the key is fixed. However, already in this scenario with deactivated RNG we can observe that the implementation protected with IP masking shows less evidence of leakage (lower t-scores).

7.2 RNG Activated

Next we repeat the evaluation with activated RNG.

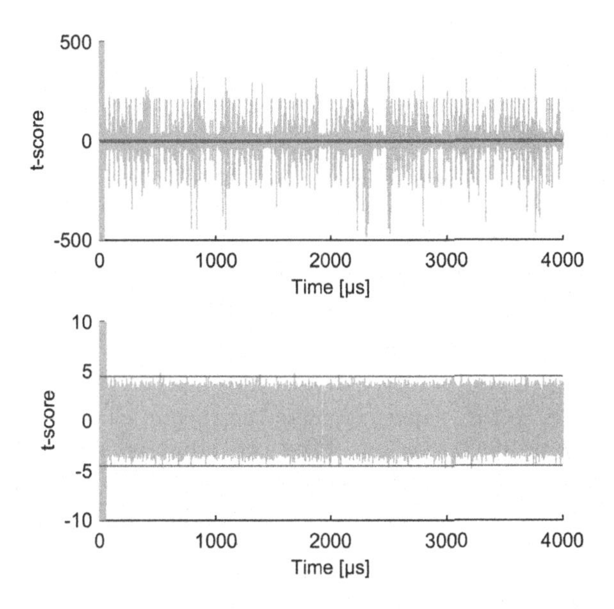

Fig. 6. t-test results for Boolean masking (top) and IP masking (bottom) with RNG activated; each based on 1 million measurements. The red lines mark the ±4.5 threshold. (Color figure online)

In this scenario we expect both implementations to leak less and we take more measurements (1 million from each implementation). Figure 6 shows the results for Boolean masking (top) and IP masking (bottom).

In this scenario we can observe a striking difference between the test results. The implementation protected with Boolean masking leaks. The t-scores are even higher than in the scenario with deactivated RNG, but this is due to the much larger number of measurements, which appears as sample size in the denominator of Eq. 2. The IP masking protected implementation on the other hand shows significantly less evidence of leakage than the implementation protected with Boolean masking, and is not deemed to leak for this number of measurements (a few t-scores slightly exceed the threshold but this is expected given that we have 500 000 t-scores and 99.999% confidence).

So based on these experiments and results, we can conclude that as expected from our theoretical investigations, IP masking allows reducing both the number of leaking samples in the implementation (which is assumably due to the better resistance to transition-based leakages) and the informativeness of these leaking samples (which is assumably due to the quite linear nature of our target leakage function). We insist that we make no claims on the fact that our IP masking implementation is first-order secure. We only conclude that it shows significantly less evidence of leakage than our Boolean masking implementation. Admittedly, first-order information could theoretically appear with larger number of measurements. For example, transition-based leakages implying a non-linear operation could lead to a flaw, which we did not observe. This could be because our specific code does not contain such a combination, or because it will only appear with more measurements. But our results anyway show that the more complex algebraic structure of the inner product encoding brings an interesting alternative (tradeoff) to Boolean masking with slight performance overheads compensated by less evidence of leakage in practice. We leave the careful investigation of the concrete leakages of the IP masking with advanced statistical tools (e.g., higher-order and multivariate attacks) as an interesting scope for further research.

8 Conclusions

Overall, the results in this paper complete the theoretical and practical understanding of inner product masking. First, we proposed new (simplified) multiplication algorithms that are conceptually close to the standard proposal of Ishai et al. [32], and have good properties for composability. Second we showed that these simplified algorithms allow better performance than reported in the previous works on inner product masking of the AES [1,2]. Third, we extended previous information theoretic evaluations in order to discuss the pros and cons of inner product masking in idealized implementations, and confronted these evaluations with first empirical experiments.

Acknowledgments. Benedikt Gierlichs is a Postdoctoral Fellow of the Fund for Scientific Research - Flanders (FWO). Sebastian Faust and Clara Paglialonga are partially funded by the Emmy Noether Program FA 1320/1-1 of the German Research Foundation (DFG). Franois-Xavier Standaert is a senior research associate of the Belgian Fund for Scientific Research (FNRS-F.R.S.). This work has been funded in parts by the European Commission through the CHIST-ERA project SECODE and the ERC project 724725 (acronym SWORD) and by the Research Council KU Leuven: C16/15/058 and Cathedral ERC Advanced Grant 695305.

References

1. Balasch, J., Faust, S., Gierlichs, B.: Inner product masking revisited. In: Oswald, E., Fischlin, M. (eds.) EUROCRYPT 2015. LNCS, vol. 9056, pp. 486–510. Springer, Heidelberg (2015). https://doi.org/10.1007/978-3-662-46800-5_19

2. Balasch, J., Faust, S., Gierlichs, B., Verbauwhede, I.: Theory and practice of a leakage resilient masking scheme. In: Wang, X., Sako, K. (eds.) ASIACRYPT 2012. LNCS, vol. 7658, pp. 758–775. Springer, Heidelberg (2012). https://doi.org/10.1007/978-3-642-34961-4_45
3. Balasch, J., Gierlichs, B., Grosso, V., Reparaz, O., Standaert, F.-X.: On the cost of lazy engineering for masked software implementations. In: Joye, M., Moradi, A. (eds.) CARDIS 2014. LNCS, vol. 8968, pp. 64–81. Springer, Cham (2015). https://doi.org/10.1007/978-3-319-16763-3_5
4. Balasch, J., Gierlichs, B., Reparaz, O., Verbauwhede, I.: DPA, bitslicing and masking at 1 GHz. In: Güneysu, T., Handschuh, H. (eds.) CHES 2015. LNCS, vol. 9293, pp. 599–619. Springer, Heidelberg (2015). https://doi.org/10.1007/978-3-662-48324-4_30
5. Barthe, G., Belaïd, S., Dupressoir, F., Fouque, P.-A., Grégoire, B.: Compositional verification of higher-order masking: application to a verifying masking compiler. IACR Cryptology ePrint Archive, 506 (2015)
6. Barthe, G., Belaïd, S., Dupressoir, F., Fouque, P.-A., Grégoire, B., Strub, P.-Y.: Verified proofs of higher-order masking. In: Oswald, E., Fischlin, M. (eds.) EUROCRYPT 2015. LNCS, vol. 9056, pp. 457–485. Springer, Heidelberg (2015). https://doi.org/10.1007/978-3-662-46800-5_18
7. Barthe, G., Dupressoir, F., Faust, S., Grégoire, B., Standaert, F.-X., Strub,P.-Y.: Parallel implementations of masking schemes and the bounded moment leakage model. Cryptology ePrint Archive, report 2016/912 (2016). http://eprint.iacr.org/2016/912
8. Belaïd, S., Benhamouda, F., Passelègue, A., Prouff, E., Thillard, A., Vergnaud, D.: Randomness complexity of private circuits for multiplication. In: Fischlin, M., Coron, J.-S. (eds.) EUROCRYPT 2016. LNCS, vol. 9666, pp. 616–648. Springer, Heidelberg (2016). https://doi.org/10.1007/978-3-662-49896-5_22
9. Carlet, C., Danger, J.-L., Guilley, S., Maghrebi, H.: Leakage squeezing of order two. In: Galbraith, S., Nandi, M. (eds.) INDOCRYPT 2012. LNCS, vol. 7668, pp. 120–139. Springer, Heidelberg (2012). https://doi.org/10.1007/978-3-642-34931-7_8
10. Carlet, C., Danger, J.-L., Guilley, S., Maghrebi, H.: Leakage squeezing: optimal implementation and security evaluation. J. Math. Cryptol. **8**(3), 249–295 (2014)
11. Carlet, C., Prouff, E., Rivain, M., Roche, T.: Algebraic decomposition for probing security. In: Gennaro, R., Robshaw, M. (eds.) CRYPTO 2015. LNCS, vol. 9215, pp. 742–763. Springer, Heidelberg (2015). https://doi.org/10.1007/978-3-662-47989-6_36
12. Chari, S., Jutla, C.S., Rao, J.R., Rohatgi, P.: Towards sound approaches to counteract power-analysis attacks. In: Wiener, M. (ed.) CRYPTO 1999. LNCS, vol. 1666, pp. 398–412. Springer, Heidelberg (1999). https://doi.org/10.1007/3-540-48405-1_26
13. Cooper, J., DeMulder, E., Goodwill, G., Jaffe, J., Kenworthy, G., Rohatgi, P.: Test vector leakage assessment (TVLA) methodology in practice. In: International Cryptographic Module Conference (2013). http://icmc-2013.org/wp/wp-content/uploads/2013/09/goodwillkenworthtestvector.pdf
14. Coron, J.-S.: Higher order masking of look-up tables. In: Nguyen, P.Q., Oswald, E. (eds.) EUROCRYPT 2014. LNCS, vol. 8441, pp. 441–458. Springer, Heidelberg (2014). https://doi.org/10.1007/978-3-642-55220-5_25
15. Coron, J.-S., Giraud, C., Prouff, E., Renner, S., Rivain, M., Vadnala, P.K.: Conversion of security proofs from one leakage model to another: a new issue. In: Schindler, W., Huss, S.A. (eds.) COSADE 2012. LNCS, vol. 7275, pp. 69–81. Springer, Heidelberg (2012). https://doi.org/10.1007/978-3-642-29912-4_6

16. Coron, J.-S., Greuet, A., Prouff, E., Zeitoun, R.: Faster evaluation of SBoxes via common shares. In: Gierlichs, B., Poschmann, A.Y. (eds.) CHES 2016. LNCS, vol. 9813, pp. 498–514. Springer, Heidelberg (2016). https://doi.org/10.1007/978-3-662-53140-2_24
17. Coron, J.-S., Prouff, E., Rivain, M., Roche, T.: Higher-order side channel security and mask refreshing. In: Moriai, S. (ed.) FSE 2013. LNCS, vol. 8424, pp. 410–424. Springer, Heidelberg (2014). https://doi.org/10.1007/978-3-662-43933-3_21
18. Duc, A., Dziembowski, S., Faust, S.: Unifying leakage models: from probing attacks to noisy leakage. In: Nguyen, P.Q., Oswald, E. (eds.) EUROCRYPT 2014. LNCS, vol. 8441, pp. 423–440. Springer, Heidelberg (2014). https://doi.org/10.1007/978-3-642-55220-5_24
19. Duc, A., Faust, S., Standaert, F.-X.: Making masking security proofs concrete. In: Oswald, E., Fischlin, M. (eds.) EUROCRYPT 2015. LNCS, vol. 9056, pp. 401–429. Springer, Heidelberg (2015). https://doi.org/10.1007/978-3-662-46800-5_16
20. Dziembowski, S., Faust, S.: Leakage-resilient circuits without computational assumptions. In: Cramer, R. (ed.) TCC 2012. LNCS, vol. 7194, pp. 230–247. Springer, Heidelberg (2012). https://doi.org/10.1007/978-3-642-28914-9_13
21. Dziembowski, S., Faust, S., Skorski, M.: Noisy leakage revisited. In: Oswald, E., Fischlin, M. (eds.) EUROCRYPT 2015. LNCS, vol. 9057, pp. 159–188. Springer, Heidelberg (2015). https://doi.org/10.1007/978-3-662-46803-6_6
22. Eisenbarth, T., Kasper, T., Moradi, A., Paar, C., Salmasizadeh, M., Shalmani, M.T.M.: On the power of power analysis in the real world: a complete break of the KeeLoq code hopping scheme. In: Wagner, D. (ed.) CRYPTO 2008. LNCS, vol. 5157, pp. 203–220. Springer, Heidelberg (2008). https://doi.org/10.1007/978-3-540-85174-5_12
23. Fumaroli, G., Martinelli, A., Prouff, E., Rivain, M.: Affine masking against higher-order side channel analysis. In: Biryukov, A., Gong, G., Stinson, D.R. (eds.) SAC 2010. LNCS, vol. 6544, pp. 262–280. Springer, Heidelberg (2011). https://doi.org/10.1007/978-3-642-19574-7_18
24. Genkin, D., Pipman, I., Tromer, E.: Get your hands off my laptop: physical side-channel key-extraction attacks on PCs - extended version. J. Cryptograph. Eng. **5**(2), 95–112 (2015)
25. Goldwasser, S., Rothblum, G.N.: How to compute in the presence of leakage. In: FOCS 2012, pp. 31–40 (2012)
26. Golić, J.D., Tymen, C.: Multiplicative masking and power analysis of AES. In: Kaliski, B.S., Koç, K., Paar, C. (eds.) CHES 2002. LNCS, vol. 2523, pp. 198–212. Springer, Heidelberg (2003). https://doi.org/10.1007/3-540-36400-5_16
27. Goodwill, G., Jun, B., Jaffe, J., Rohatgi, P.: A testing methodology for side channel resistance validation. In: NIST non-invasive attack testing workshop (2011). http://csrc.nist.gov/news_events/non-invasive-attack-testing-workshop/papers/08_Goodwill.pdf
28. Goubin,L., Martinelli, A.: Protecting AES with Shamir's secret sharing scheme. In: Preneel and Takagi [38], pp. 79–94 (2011)
29. Goubin, L., Patarin, J.: DES and differential power analysis the "Duplication" method. In: Koç, Ç.K., Paar, C. (eds.) CHES 1999. LNCS, vol. 1717, pp. 158–172. Springer, Heidelberg (1999). https://doi.org/10.1007/3-540-48059-5_15
30. Grosso, V., Prouff, E., Standaert, F.-X.: Efficient masked S-Boxes processing – a step forward –. In: Pointcheval, D., Vergnaud, D. (eds.) AFRICACRYPT 2014. LNCS, vol. 8469, pp. 251–266. Springer, Cham (2014). https://doi.org/10.1007/978-3-319-06734-6_16

31. Grosso, V., Standaert, F.-X., Prouff, E.: Low entropy masking schemes, revisited. In: Francillon, A., Rohatgi, P. (eds.) CARDIS 2013. LNCS, vol. 8419, pp. 33–43. Springer, Cham (2014). https://doi.org/10.1007/978-3-319-08302-5_3
32. Ishai, Y., Sahai, A., Wagner, D.: Private circuits: securing hardware against probing attacks. In: Boneh, D. (ed.) CRYPTO 2003. LNCS, vol. 2729, pp. 463–481. Springer, Heidelberg (2003). https://doi.org/10.1007/978-3-540-45146-4_27
33. Kocher, P.C.: Timing attacks on implementations of Diffie-Hellman, RSA, DSS, and other systems. In: Koblitz, N. (ed.) CRYPTO 1996. LNCS, vol. 1109, pp. 104–113. Springer, Heidelberg (1996). https://doi.org/10.1007/3-540-68697-5_9
34. Kocher, P., Jaffe, J., Jun, B.: Differential power analysis. In: Wiener, M. (ed.) CRYPTO 1999. LNCS, vol. 1666, pp. 388–397. Springer, Heidelberg (1999). https://doi.org/10.1007/3-540-48405-1_25
35. Mangard, S., Oswald, E., Popp, T.: Power Analysis Attacks: Revealing the Secrets of Smart Cards (Advances in Information Security). Springer-Verlag, New York Inc (2007). https://doi.org/10.1007/978-0-387-38162-6
36. Mather, L., Oswald, E., Bandenburg, J., Wójcik, M.: Does my device leak information? an *a priori* statistical power analysis of leakage detection tests. In: Sako, K., Sarkar, P. (eds.) ASIACRYPT 2013. LNCS, vol. 8269, pp. 486–505. Springer, Heidelberg (2013). https://doi.org/10.1007/978-3-642-42033-7_25
37. Nikova, S., Rechberger, C., Rijmen, V.: Threshold implementations against side-channel attacks and glitches. In: Ning, P., Qing, S., Li, N. (eds.) ICICS 2006. LNCS, vol. 4307, pp. 529–545. Springer, Heidelberg (2006). https://doi.org/10.1007/11935308_38
38. Preneel, B., Takagi, T. (eds.): CHES 2011. LNCS, vol. 6917. Springer, Heidelberg (2011). https://doi.org/10.1007/978-3-642-23951-9
39. Prouff, E., Rivain, M.: Masking against side-channel attacks: a formal security proof. In: Johansson, T., Nguyen, P.Q. (eds.) EUROCRYPT 2013. LNCS, vol. 7881, pp. 142–159. Springer, Heidelberg (2013). https://doi.org/10.1007/978-3-642-38348-9_9
40. Prouff, E., Roche, T.: Higher-order glitches free implementation of the AES using secure multi-party computation protocols. In: Preneel and Takagi [38], pp. 63–78 (2011)
41. Reparaz, O., Bilgin, B., Nikova, S., Gierlichs, B., Verbauwhede, I.: Consolidating masking schemes. In: Gennaro, R., Robshaw, M. (eds.) CRYPTO 2015. LNCS, vol. 9215, pp. 764–783. Springer, Heidelberg (2015). https://doi.org/10.1007/978-3-662-47989-6_37
42. Rivain, M., Prouff, E.: Provably secure higher-order masking of AES. In: Mangard, S., Standaert, F.-X. (eds.) CHES 2010. LNCS, vol. 6225, pp. 413–427. Springer, Heidelberg (2010). https://doi.org/10.1007/978-3-642-15031-9_28
43. Schneider, T., Moradi, A.: Leakage assessment methodology. In: Güneysu, T., Handschuh, H. (eds.) CHES 2015. LNCS, vol. 9293, pp. 495–513. Springer, Heidelberg (2015). https://doi.org/10.1007/978-3-662-48324-4_25
44. Standaert, F.-X., Malkin, T.G., Yung, M.: A unified framework for the analysis of side-channel key recovery attacks. In: Joux, A. (ed.) EUROCRYPT 2009. LNCS, vol. 5479, pp. 443–461. Springer, Heidelberg (2009). https://doi.org/10.1007/978-3-642-01001-9_26
45. Standaert, F.-X., Veyrat-Charvillon, N., Oswald, E., Gierlichs, B., Medwed, M., Kasper, M., Mangard, S.: The world is not enough: another look on second-order DPA. In: Abe, M. (ed.) ASIACRYPT 2010. LNCS, vol. 6477, pp. 112–129. Springer, Heidelberg (2010). https://doi.org/10.1007/978-3-642-17373-8_7

46. Wang, W., Standaert, F.-X., Yu, Y., Pu, S., Liu, J., Guo, Z., Gu, D.: Inner product masking for bitslice ciphers and security order amplification for linear leakages. In: Lemke-Rust, K., Tunstall, M. (eds.) CARDIS 2016. LNCS, vol. 10146, pp. 174–191. Springer, Cham (2017). https://doi.org/10.1007/978-3-319-54669-8_11

The First Thorough Side-Channel Hardware Trojan

Maik Ender[1], Samaneh Ghandali[2], Amir Moradi[1(✉)], and Christof Paar[1,2]

[1] Horst Görtz Institute for IT Security, Ruhr-Universität Bochum, Bochum, Germany
{maik.ender,amir.moradi,christof.paar}@rub.de
[2] University of Massachusetts Amherst, Amherst, USA
samaneh@umass.edu

Abstract. Hardware Trojans have gained high attention in academia, industry and by government agencies. The effective detection mechanisms and countermeasures against such malicious designs are only possible when there is a deep understanding of how hardware Trojans can be built in practice. In this work, we present a mechanism which shows how easily a stealthy hardware Trojan can be inserted in a provably-secure side-channel analysis protected implementation. Once the Trojan is triggered, the malicious design exhibits exploitable side-channel leakage leading to successful key recovery attacks. Such a Trojan does not add or remove any logic (even a single gate) to the design which makes it very hard to detect. In ASIC platforms, it is indeed inserted by subtle manipulations at the sub-transistor level to modify the parameters of a few transistors. The same is applicable on FPGA applications by changing the routing of particular signals, leading to **null** resource utilization overhead. The underlying concept is based on a secure masked hardware implementation which does not exhibit any detectable leakage. However, by running the device at a particular clock frequency one of the requirements of the underlying masking scheme is not fulfilled anymore, i.e., the Trojan is triggered, and the device's side-channel leakage can be exploited.

Although as a case study we show an application of our designed Trojan on an FPGA-based threshold implementation of the PRESENT cipher, our methodology is a general approach and can be applied on any similar circuit.

1 Introduction

Cryptographic devices are those pieces of (usually) hardware that implement cryptographic algorithm(s) providing different aspects of security. Since such devices often deal with secret information and/or privacy of the users, hardware Trojans have gained high attention in academia and industry as well as government agencies, and can leak the secrets in a particular fashion without the notice of the end users. Indeed, both bodies of research concerning the Trojan design and Trojan detection are large and active. Nevertheless, these two topics

T. Takagi and T. Peyrin (Eds.): ASIACRYPT 2017, Part I, LNCS 10624, pp. 755–780, 2017.
https://doi.org/10.1007/978-3-319-70694-8_26

are closely related. The effective detection mechanisms and countermeasures are only possible when there is an understanding of how hardware Trojans can be built.

Amongst several different ways to insert a Trojan into an IC, we can refer to those conducted *(i)* by an untrusted semiconductor foundry during manufacturing, *(ii)* by the original hardware designer who is pressured by the government bodies, and *(iii)* in the third-party IP cores. Most of the hardware Trojans are inserted by modifying a few gates (can be done at different abstraction levels). In short, one of the main goals of the Trojans is to be designed/implemented in such a way that the chance of detection becomes very low. Our focus in this article is those Trojans which leak out the secrets through a side channel. The first such a Trojan has been introduced in [35,36] which stealthily leaks out the cryptographic key using leakages through power consumption side channel. The underlying scheme is independent of the cryptographic algorithm and deals only with the secret key. This Trojan, made by a moderately large circuit including an LFSR and leaking circuit, is inserted at the netlist or HDL level. Therefore, it is likely detected by a Trojan inspector. Further, the designs in these works [35,36] are not parametric Trojans, i.e., they always leak through a side channel, which might be exploited by anybody not only the Trojan attacker.

On the other hand, the cryptographic devices – if pervasive and/or ubiquitous – are in danger of side-channel analysis (SCA) attacks. After around two decades since introduction of such physical attacks [32,33], integration of dedicated SCA countermeasures is a must for devices which deal with security. Therefore, if the design is not protected against SCA threats, any SCA adversary would be able to reveal the secrets independent of the existence of such a Trojan [36].

In a follow-up work [30], the authors expressed a relatively similar concept on an SCA-protected implementation. Their technique is based on inserting a logical circuit forming an LFRS-based Trojan leaking the internal state of the PRNG. As a side note, random number generators are necessary modules for those SCA-protected implementations which are based on masking [38]. Hence, the Trojan adversary would detect the internal state of the PRNG by means of SCA leakages and can conduct DPA attacks knowing the masks. It should be noted that those products which need to be protected against physical attacks are usually evaluated by a third-party certification body, e.g., through a common criteria evaluation lab. Therefore, due to its relatively large circuit, such a Trojan is very likely detected by an inspector.

As another work in this domain, we should refer to [2], where the Trojan is inserted by changing the dopant polarity of a few transistors in a circuit realized by the DPA-resistant logic style iMDPL [47]. However, none of such logic styles can perfectly provide security, and the leakage of an iMDPL circuit can still be exploited by ordinary SCA adversaries [40].

Our Contribution. In short, integrating an SCA Trojan into an SCA-protected design is challenging, if the device is supposed to be evaluated by a third-party certification body. It is because the device should provide the desired SCA protection under a white-box scenario, i.e., all design details including the netlist are known to the evaluation lab. In this work, we present a mechanism to design a provably- and practically-secure SCA-protected implementation which can be turned into an unprotected implementation by a Trojan adversary. Our Trojan does not add any logic (even a single gate) to the design, making it very hard to detect. In case of ASIC platforms, it is done by slightly changing the characteristic of a few transistors, and for FPGA platforms by changing the routing of particular signals. Most notably, our technique is **not** based on the leakage of the PRNG, and it does not affect the provable-security feature of the underlying design unless the Trojan is triggered. More precisely, our technique leads to inserting a *parametric* Trojan, i.e., under normal condition the device does not exhibit any SCA leakage to be detected by an evaluation lab. By increasing the clock frequency of the malicious device (or by decreasing its supply voltage) the Trojan is triggered and exhibits exploitable leakage. Note that such a high clock frequency is beyond the maximum frequency that the device can correctly operate. Hence, the device is not expected to be evaluated under such a condition by evaluation labs. As we show in the following sections, there is a gap between the maximum clock frequency of the device and the clock frequency where the Trojan is triggered. In other words, by increasing the clock frequency (violating its critical path delay) the device starts to operate faulty; by even more increasing the clock frequency the device operates again correctly while exhibiting SCA leakage (i.e., our inserted Trojan becomes active).

Outline. Section 2 deals with necessary background and definitions in the areas of hardware Trojan and threshold implementation as an SCA countermeasure. Afterwards, in Sect. 3 we express our core idea how to insert our Trojan into a secure threshold implementation. In Sect. 4 we give details on how to apply such a technique on a threshold implementation of the PRESENT cipher, and in Sect. 5 the corresponding result of FPGA-based SCA evaluations are exhibited.

2 Background

2.1 Hardware Trojan

Malicious and intentional modification of integrated circuit (IC) during manufacturing in untrusted foundry is an emerging security concern. This problem exists because the majority of ICs are fabricated abroad, and a government agency can force a foundry to manipulate the design maliciously. Also, an IC designer can be pressured by her own country government to modify the ICs maliciously, e.g., those ICs that are used in overseas products. Another possible insertion point are 3rd party IP cores. In general, a hardware Trojan is a back-door that can be inserted into an integrated circuit as an undesired and malicious modification, which makes the behavior of the IC incorrect.

There are many ways to categorize Trojans such as categorizing based on physical characteristics, design phase, abstraction level, location, triggering mechanism, and functionality. But a common Trojan categorization is based on the activation mechanism (Trojan trigger) and the effect on the circuit functionality (Trojan payload). A set of conditions that cause a Trojan to be activated is called trigger. Trojans can combinationally or sequentially be triggered. An attacker chooses a rare trigger condition so that the Trojan would not be triggered during conventional design-time verification and manufacturing test. Sequentially-triggered Trojans (time bombs) are activated by the occurrence of a sequence of rare events, or after a period of continuous operation [18].

The goal of the Trojan can be achieved by payload which can change the circuit functionally or leak its secret information. In [27] a categorization method according to how the payload of a Trojan works has been defined; some Trojans after triggering, propagate internal signals to output ports which can reveal secret information to the attackers (explicit payload). Other Trojans may make the circuit malfunction or destroy the whole chip (implicit payload). Another categorization for actions of hardware Trojans has been presented in [59], in which the actions can be categorized into classes of modify functionality, modify specification, leak information, and denial of service.

The work in [29] introduced a manipulation which makes an error detection module to work incorrectly and accept inputs that should be rejected. They showed how a Trojan could be used to change the instruction order in which CPU executes them, leak data by side-channel analysis, and change the content of programmable read only memory. The work in [31] presented how a hardware Trojan, which is inserted into a CPU by adding extra logic into its HDL code, can give an attacker unlimited access to the CPU.

Threats posed by hardware Trojans and the methods of deterring them have been analyzed in [18]. For example a bridging fault by insertion of a resistor and by increasing a net delay by enlarging its capacitance load has been introduced in this work. The works in [19,53] discussed about efficient generation of test patterns for hardware Trojans triggered by rare input signals. Hardware Trojans in wireless cryptographic ICs have been discussed in [28]. The goal is to design Trojans to leak secret information through the wireless channel. Detection challenges of such Trojans were discussed in this work and some improvements were proposed based on side-channel signals analysis. The work in [36] proposed a hardware Trojan that leaks the cryptographic key through side channel analysis attack. Similar to the hardware Trojans that were designed as part of a student hardware Trojan challenge at [51], this hardware Trojan was inserted at the netlist or HDL level. The work in [2] presented building stealthy Trojans at the layout-level. A hardware Trojan was inserted into a cryptographically-secure PRNG and into a side-channel resistant Sbox by manipulating the dopant polarity of a few registers. Building hardware Trojans that are triggered by aging was presented in [57]. These Trojans only become active after the IC has been working for a long time.

A class of hardware Trojans – Malicious Off-chip Leakage Enabled by Side-channels (MOLES) – has been presented in [35], which can retrieve secret information through side channels. They formulated the mechanism and detection methods of MOLES in theory and provided a verification process for multi-bit key extractions. A parametric Trojan has been introduced in [34] which triggers with a probability increasing under reduced supply voltage. In [21] a design methodology for building stealthy parametric hardware Trojans and its application to Bug Attacks [4,5] has been proposed. The Trojans are based on increasing delay of gates of a very rare-sensitized path in a combinatorial circuit, such as an arithmetic multiplier circuit. The Trojans are stealthy and have rare trigger conditions, so that the faulty behavior of the circuit under attack only occurs for very few combinations of the input vectors. Also an attack on the ECDH key agreement protocol by this Trojan has been presented in this work.

2.2 Threshold Implementation

It can be definitely said that *masking* is the most-studied countermeasure against SCA attacks. It is based on the concept of *secret sharing*, where a secret $\boldsymbol{x}$ (e.g., intermediate values of a cipher execution) is represented by a couple of shares $(\boldsymbol{x}^1, \ldots, \boldsymbol{x}^n)$. In case of an (n, n)-threshold secret sharing scheme, having access to $t < n$ does not reveal any information about $\boldsymbol{x}$. Amongst those is Boolean secret sharing, known as Boolean masking in the context of SCA, where $\boldsymbol{x} = \bigoplus_{i=1}^{n} \boldsymbol{x}^i$. Hence, if the entire computation of a cipher is conduced on such a shared representation, its SCA leakage will be (in average) independent of the secrets as long as no function (e.g., combinatorial circuit) operates on all n shares.

Due to the underlying Boolean construction, application of a linear function $\mathcal{L}(.)$ over the shares is straightforward since $\mathcal{L}(\boldsymbol{x}) = \bigoplus_{i=1}^{n} \mathcal{L}(\boldsymbol{x}^i)$. All the difficulties belong to implementing non-linear functions over such a shared representation. This concept has been applied in hardware implementation of AES (mainly with $n = 2$) with no success [16,39,41,46] until the Threshold Implementation (TI) – based on sound mathematical foundations – has been introduced in [45], which defines minimum number shares $n \geq t + 1$ with t the algebraic degree of the underlying non-linear function. For simplicity (and as our case study is based on) we focus on quadratic Boolean functions, i.e., $t = 2$, and minimum number of shares $n = 3$. Suppose that the TI of the non-linear function $y = \mathcal{F}(\boldsymbol{x})$ is desired, i.e., $(\boldsymbol{y}^1, \boldsymbol{y}^2, \boldsymbol{y}^3) = \mathcal{F}^{*}(\boldsymbol{x}^1, \boldsymbol{x}^2, \boldsymbol{x}^3)$, where

$$\boldsymbol{y}^1 \oplus \boldsymbol{y}^2 \oplus \boldsymbol{y}^3 = \mathcal{F}(\boldsymbol{x}^1 \oplus \boldsymbol{x}^2 \oplus \boldsymbol{x}^3).$$

Indeed, each output share $y^{i \in \{1,2,3\}}$ is provided by a component function $\mathcal{F}^i(.,.)$ which receives only two input shares. In other words, one input share is definitely missing in every component function. This, which is a requirement defined by TI as *non-completeness*, supports the aforementioned concept that "no function (e.g., combinatorial circuit) operates on all n shares", and implies the given formula $n \geq t + 1$. Therefore, three component functions $\left(\mathcal{F}^1\left(\boldsymbol{x}^2, \boldsymbol{x}^3\right), \mathcal{F}^2\left(\boldsymbol{x}^3, \boldsymbol{x}^1\right), \mathcal{F}^3\left(\boldsymbol{x}^1, \boldsymbol{x}^2\right)\right)$ form the shared output $(\boldsymbol{y}^1, \boldsymbol{y}^2, \boldsymbol{y}^3)$.

Uniformity. In order to fulfill the above-given statement that "having access to $t < n$ does not reveal any information about $\boldsymbol{x}$", the shares need to follow a uniform distribution. For simplicity suppose that $n = 2$, and the shares $(\boldsymbol{x}^1, \boldsymbol{x}^2)$ represent secret $\boldsymbol{x}$. If the distribution of x^1 has a bias (i.e., not uniform) which is known to the adversary, he can observe the distribution of $\boldsymbol{x}^2 = \boldsymbol{x} \oplus \boldsymbol{x}^1$ and guess $\boldsymbol{x}$. Hence, the security of the entire masking schemes[1] relies on the uniformity of the masks. More precisely, when $\boldsymbol{x}^1 = \boldsymbol{m}$, $\boldsymbol{x}^2 = \boldsymbol{x} \oplus \boldsymbol{m}$, and $\boldsymbol{m}$ is taken from a randomness source (e.g., a PRNG), the distribution of $\boldsymbol{m}$ should be uniform (or let say with full entropy).

The same holds for higher-order masking, i.e., $n > 2$. However, not only the distribution of every share but also the joint distribution of every $t < n$ shares is important. In case of $\mathcal{F}^*(.,.,.)$ as a TI of a bijective function $\mathcal{F}(.)$, the *uniformity* property of TI is fulfilled if $\mathcal{F}^*(.,.,.)$ forms a bijection. Otherwise, the security of such an implementation cannot be guaranteed. Note that fulfilling the uniformity property of TI constructions is amongst its most difficult challenges, and it has been the core topic of several articles like [3,9,12,45,48]. Alternatively, the shares can be remasked at the end of every non-uniform shared non-linear function (see [8,42]), which requires a source to provide fresh randomness at every clock cycle. Along the same line, another type of masking in hardware (which reduces the number of shares) has been developed in [23,52], which (almost always) needs fresh randomness to fulfill the uniformity.

We should emphasize that the above given expressions illustrate only the first-order TI of bijective quadratic functions. For the other cases including higher-order TI we refer the interested reader to the original articles [9,12,45].

3 Technique

As explained in former section – by means of TI – it is possible to realize hardware cryptographic devices secure against certain SCA attacks. Our goal is to provide a certain situation that an SCA-secure device becomes insecure while it still operates correctly. Such a dynamic transition from secure to insecure should be available and known only to the Trojan attacker. To this end, we target the uniformity property of a secure TI construction. More precisely, we plan to construct a secure and uniform TI design which becomes non-uniform (and hence insecure) at particular environmental conditions. In order to trigger the Trojan (or let say to provide such a particular environmental conditions) for example we select higher clock frequency than the device maximum operation frequency, or lower power supply than the device nominal supply voltage. It should not be forgotten that under such conditions the underlying device should still maintain its correct functionality.

To realize such a scenario – inspired from the stealthy parametric Trojan introduced in [21] – we intentionally lengthen certain paths of a combinatorial circuit. This is done in such a way that – by increasing the device clock frequency or lowering its supply voltage – such paths become faulty earlier than the other paths.

[1] Except those which are based on low-entropy masking [17,37].

We would achieve our goal if (*i*) the faults cancel each others' effect, i.e., the functionality of the design is not altered, and (*ii*) the design does not fulfill the uniformity property anymore.

In order to explain our technique – for simplicity without loss of generality – we focus on a 3-share TI construction. As explained in Sect. 2.2 – ignoring the uniformity – achieving a non-complete shared function $\mathcal{F}^*(.,.,.)$ of a given quadratic function $\mathcal{F}(.)$ is straightforward. Focusing on one output bit of $\mathcal{F}(\boldsymbol{x})$, and representing $\boldsymbol{x}$ by s input bits $\langle x_s, \ldots, x_1\rangle$, we can write

$$\begin{aligned}\mathcal{F}_i(\langle x_s, \ldots, x_1\rangle) =& k_0 \oplus k_1x_1 \oplus k_2x_2 \oplus \ldots \oplus k_sx_s \oplus \\ & k_{1,2}x_1x_2 \oplus k_{1,3}x_1x_3 \oplus \ldots \oplus k_{s-1,s}x_{s-1}x_s.\end{aligned}$$

The coefficients $k_0, \ldots, k_{s-1,s} \in \{0,1\}$ form the Algebraic Normal Form (ANF) of the quadratic function $\mathcal{F}_i : \{0,1\}^s \rightarrow \{0,1\}$. By replacing every input bit x_i by the sum of three corresponding shares $x_i^1 \oplus x_i^2 \oplus x_i^3$, the remaining task is just to split the terms in the ANF to three categories in such a way that each category is independent of one share. This can be done by a method denoted by *direct sharing* [12] as

- $\mathcal{F}_i^1(.,.)$ contains the linear terms x_i^2 and the quadratic terms $x_i^2x_j^2$ and $x_i^2x_j^3$.
- $\mathcal{F}_i^2(.,.)$ contains the linear terms x_i^3 and the quadratic terms $x_i^3x_j^3$ and $x_i^3x_j^1$.
- $\mathcal{F}_i^3(.,.)$ contains the linear terms x_i^1 and the quadratic terms $x_i^1x_j^1$ and $x_i^1x_j^2$.

The same is independently applied on each output bit of $\mathcal{F}(.)$ and all three component functions $\mathcal{F}^1\left(\boldsymbol{x}^2,\boldsymbol{x}^3\right)$, $\mathcal{F}^2\left(\boldsymbol{x}^3,\boldsymbol{x}^1\right)$, $\mathcal{F}^3\left(\boldsymbol{x}^1,\boldsymbol{x}^2\right)$ are constructed that fulfill the non-completeness, but nothing about its uniformity can be said.

There are indeed two different ways to obtain a uniform TI construction:

- If s (the underlying function size) is small, i.e., $s \leq 5$, it can be found that $\mathcal{F}(.)$ is affine equivalent to which s-bit class. More precisely, there is a quadratic class $\mathcal{Q}$ which can represent $\mathcal{F}$ as $\mathcal{A}' \circ \mathcal{Q} \circ \mathcal{A}$ (see [13] for an algorithm to find $\mathcal{A}$ and $\mathcal{A}'$ given $\mathcal{F}$ and $\mathcal{Q}$). A classification of such classes for $s = 3$ and $s = 4$ are shown in [12] and for $s = 5$ in [15]. Since the number of existing quadratic classes are restricted, it can exhaustively be searched to find their uniform TI. Note that while for many quadratic classes the direct sharing (explained above) can reach to a uniform TI, for some quadratic classes no uniform TI exists unless the class is represented by a composition of two other quadratic classes [12]. Supposing that $\mathcal{Q}^*(.,.,.)$ is a uniform TI of $\mathcal{Q}(.)$, applying the affine functions $\mathcal{A}'$ and $\mathcal{A}$ accordingly on each input and output of the component function $\mathcal{Q}^*$ would give a uniform TI of $\mathcal{F}(.)$:

$$\begin{aligned}\mathcal{F}^1(\boldsymbol{x}^2,\boldsymbol{x}^3) &= \mathcal{A}' \circ \mathcal{Q}^1\left(\mathcal{A}\left(\boldsymbol{x}^2\right), \mathcal{A}\left(\boldsymbol{x}^3\right)\right),\\ \mathcal{F}^2(\boldsymbol{x}^3,\boldsymbol{x}^1) &= \mathcal{A}' \circ \mathcal{Q}^2\left(\mathcal{A}\left(\boldsymbol{x}^3\right), \mathcal{A}\left(\boldsymbol{x}^1\right)\right),\\ \mathcal{F}^3(\boldsymbol{x}^1,\boldsymbol{x}^2) &= \mathcal{A}' \circ \mathcal{Q}^3\left(\mathcal{A}\left(\boldsymbol{x}^1\right), \mathcal{A}\left(\boldsymbol{x}^2\right)\right).\end{aligned}$$

This scenario has been followed in several works, e.g., [6,11,43,44,54].

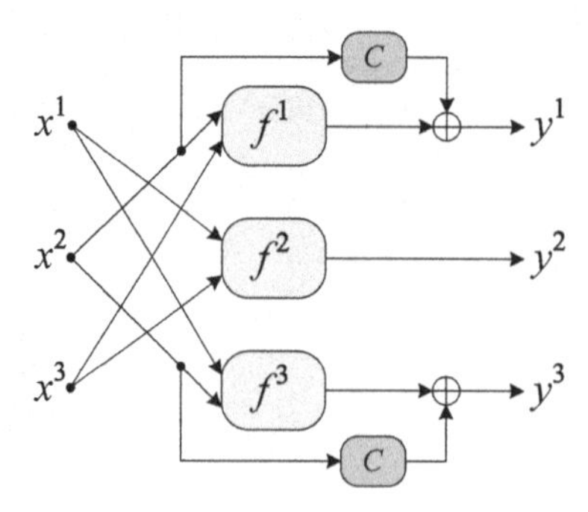

Fig. 1. Exemplary TI construction with a correction term C.

- Having a non-uniform TI construction, e.g., obtained by direct sharing, we can add *correction terms* to the component functions in such a way that the correctness and non-completeness properties are not altered, but the uniformity may be achieved. For example, the linear terms x_i^2 and/or the quadratic terms $x_i^2 x_j^2$ as correction terms can be added to the same output bit of **both** component functions $\mathcal{F}^1\left(\boldsymbol{x}^2, \boldsymbol{x}^3\right)$ and $\mathcal{F}^3\left(\boldsymbol{x}^1, \boldsymbol{x}^2\right)$. Addition of any correction term changes the uniformity of the design. Hence, by repeating this process – up to examining all possible correction terms and their combination, which is not feasible for large functions – a uniform construction might be obtained. Such a process has been conducted in [7,48] to construct uniform TI of PRESENT and Keccak non-linear functions.
  We should here refer to a similar approach called remasking [12,42] where – instead of correction terms – fresh randomness is added to the output of the component functions to make the outputs uniform. In this case, obviously a certain number of fresh mask bits are required at every clock cycle (see [10,42]).

Our technique is based on the second scheme explained above. If we make the paths related to the correction terms the longest path, by increasing the clock frequency such paths are the first whose delay are violated. As illustrated, each correction term must be added to two component functions (see Fig. 1). The paths must be very carefully altered in such a way that the path delay of both instances of the targeted correction term are the longest in the entire design and relatively the same. Hence, at a particular clock frequency both instances of the correction terms are not correctly calculated while all other parts of the design are fault free. This enables the design to still work properly, i.e., it generates correct ciphertext assuming that the underlying design realizes an encryption function. It means that the design operates like an alternative design where no correction terms exists. Hence, the uniformity of the TI construction is not fulfilled and SCA leakage can be exploited. To this end, we should keep a margin between (i) the path delay of the correction terms and (ii) the critical path delay of the rest of the circuit, i.e., that of the circuit without correction terms. This margin guarantees that at a certain high clock frequency the correction terms are canceled out but the critical path delay of the remaining circuit is not violated.

We would like to emphasize that in an implementation of a cipher once one of the TI functions generates non-uniform output (by violating the delay of correction terms), the uniformity is not maintained in the next TI functions and it leads to first-order leakage in all further rounds. If the uniformity is achieved by remasking (e.g., in [24]), the above-expressed technique can have the same effect by making the XOR with fresh mask the longest path. Hence, violating its delay in one TI function would make its output non-uniform, but the fresh randomness may make the further rounds of the cipher again uniform.

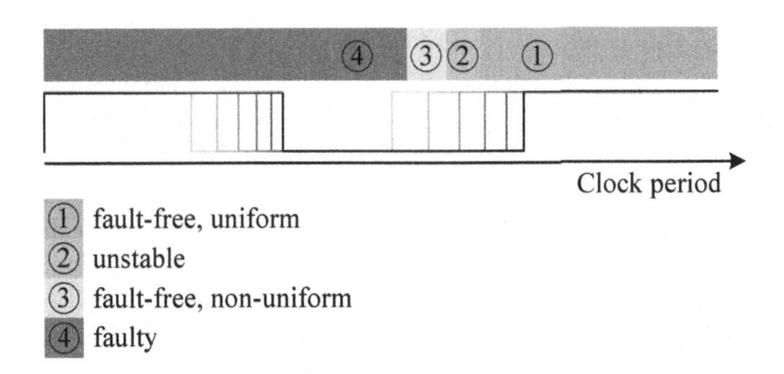

Fig. 2. Status of the design with Trojan at different clock frequencies.

Based on Fig. 2, which shows a corresponding timing diagram, the device status can be categorized into four states:

- at a low clock frequency (denoted by ①) the device operates fault free and maintains the uniformity,
- by increasing the clock frequency (in the ② period), the circuit first starts to become unstable, when indeed the correction terms do not fully cancel each others' effect, and the hold time and/or setup time of the registers are violated,
- by more increasing the clock frequency (in the ③ period), the delay of both instances of the correction term are violated and the circuit operates fault free, but does not maintain the uniformity, and
- by even more increasing the clock frequency (marked by ④), the clock period becomes smaller than the critical path delay of the rest of the circuit, and the device does not operate correctly.

The aforementioned margin defines the length of the ② period, which is of crucial importance. If it is very wide, the maximum operation frequency of the resulting circuit is obviously reduced, and the likelihood of the inserted Trojan to be detected by an evaluator is increased.

Correct functionality of the circuit is requited to enable the device being operated in the field. Otherwise, the faulty outputs might be detected (e.g., in a communication protocol) and the device may stop operating and prevent collecting SCA traces.

4 Application

In order to show an application of our technique, we focus on a first-order TI design of PRESENT cipher [14] as a case study. The PRESENT Sbox is 4-bit cubic bijection $\mathcal{S}$: C56B90AD3EF84712. Hence, its first-order TI needs at least $n = 4$ shares. Alternatively, it can be decomposed to two quadratic bijections $\mathcal{S}: \mathcal{F} \circ \mathcal{G}$ enabling the minimum number of shares $n = 3$ at the cost of having extra register between $\mathcal{F}^*$ and $\mathcal{G}^*$ (i.e., TI of $\mathcal{F}$ and $\mathcal{G}$). As shown in [12], $\mathcal{S}$ is affine equivalent to class $\mathcal{C}_{266}$: 0123468A5BCFED97, which can be decomposed to quadratic bijections with uniform TI. The works reported in [44,54,55] have followed this scenario and represented the PRESENT Sbox as $\mathcal{S}: \mathcal{A}'' \circ \mathcal{Q}' \circ \mathcal{A}' \circ \mathcal{Q} \circ \mathcal{A}$, with many possibilities for the affine functions $\mathcal{A}''$, $\mathcal{A}'$, $\mathcal{A}$ and the quadratic classes $\mathcal{Q}'$ and $\mathcal{Q}$ whose uniform TI can be obtained by direct sharing (see Sect. 3).

However, the first TI of PRESENT has been introduced in [48], where the authors have decomposed the Sbox by $\mathcal{G}$: 7E92B04D5CA1836F and $\mathcal{F}$: 08B7A31C46F9ED52. They have accordingly provided uniform TI of each of such 4-bit quadratic bijections. We focus on this decomposition, and select $\mathcal{G}$ as the target where our Trojan is implemented. Compared to all other related works, we first try to find a **non-uniform** TI of $\mathcal{G}(.)$, and we later make it uniform by means of correction terms. We start with the ANF of $\mathcal{G}(\langle d,c,b,a\rangle) = \langle g_3, g_2, g_1, g_0\rangle$:

$$g_0 = 1 \oplus a \oplus dc \oplus db \oplus cb, \qquad g_2 = 1 \oplus c \oplus b,$$
$$g_1 = 1 \oplus d \oplus b \oplus ca \oplus ba, \qquad g_3 = c \oplus b \oplus a.$$

One possible sharing of $\boldsymbol{y} = \mathcal{G}(\boldsymbol{x})$ can be represented by $\left(\boldsymbol{y}^1, \boldsymbol{y}^2, \boldsymbol{y}^3\right) = \left(\mathcal{G}^1\left(\boldsymbol{x}^2, \boldsymbol{x}^3\right), \mathcal{G}^2\left(\boldsymbol{x}^3, \boldsymbol{x}^1\right), \mathcal{G}^3\left(\boldsymbol{x}^1, \boldsymbol{x}^2\right)\right)$ as

$$y_0^1 = 1 \oplus a^2 \oplus d^2c^3 \oplus d^3c^2 \oplus d^2b^3 \oplus d^3b^2 \oplus c^2b^3 \oplus c^3b^2 \oplus d^2c^2 \oplus d^2b^2 \oplus c^2b^2,$$
$$y_1^1 = 1 \oplus b^2 \oplus d^3 \oplus c^2a^3 \oplus c^3a^2 \oplus b^2a^3 \oplus b^3a^2 \oplus c^2a^2 \oplus b^2a^2,$$
$$y_2^1 = 1 \oplus c^2 \oplus b^2, \qquad y_3^1 = c^2 \oplus b^2 \oplus a^2,$$

$$y_0^2 = a^3 \oplus d^3c^3 \oplus d^1c^3 \oplus d^3c^1 \oplus d^3b^3 \oplus d^1b^3 \oplus d^3b^1 \oplus c^3b^3 \oplus c^1b^3 \oplus c^3b^1,$$
$$y_1^2 = b^3 \oplus d^1 \oplus c^1a^3 \oplus c^3a^1 \oplus b^1a^3 \oplus b^3a^1 \oplus c^3a^3 \oplus b^3a^3,$$
$$y_2^2 = c^3 \oplus b^3, \qquad y_3^2 = c^3 \oplus b^3 \oplus a^3,$$

$$y_0^3 = a^1 \oplus d^1c^1 \oplus d^1c^2 \oplus d^2c^1 \oplus d^1b^1 \oplus d^1b^2 \oplus d^2b^1 \oplus c^1b^1 \oplus c^1b^2 \oplus c^2b^1,$$
$$y_1^3 = b^1 \oplus d^2 \oplus c^1a^2 \oplus c^2a^1 \oplus b^1a^2 \oplus b^2a^1 \oplus c^1a^1 \oplus b^1a^1,$$
$$y_2^3 = c^1 \oplus b^1, \qquad y_3^3 = c^1 \oplus b^1 \oplus a^1,$$

with $\boldsymbol{x}^{i\in\{1,2,3\}} = \langle d^i, c^i, b^i, a^i\rangle$. This is not a uniform sharing of $\mathcal{G}(.)$, and by searching through possible correction terms we found three correction terms c^1b^1,

c^2b^2, and c^3b^3 to be added to the second bit of the above-expressed component functions, that lead us to a uniform TI construction. More precisely, by defining

$$\mathcal{C}^1(\boldsymbol{x}^2, \boldsymbol{x}^3) = c^2b^2 \oplus c^3b^3,$$
$$\mathcal{C}^2(\boldsymbol{x}^3, \boldsymbol{x}^1) = c^1b^1 \oplus c^3b^3,$$
$$\mathcal{C}^3(\boldsymbol{x}^1, \boldsymbol{x}^2) = c^1b^1 \oplus c^2b^2,$$

and adding them respectively to y_1^1, y_1^2, and y_1^3, the resulting TI construction becomes uniform. If any of such correction terms is omitted, the uniformity is not maintained. In the following we focus on a single correction term c^2b^2 which should be added to $\mathcal{G}^1(.,.)$ and $\mathcal{G}^3(.,.)$. Note that for the sake of completeness a uniform sharing of $\mathcal{F}$ is given in Appendix A.

4.1 Inserting the Trojan

We realize the Trojan functionality by path delay fault model [58], without modifying the logic circuit. The Trojan is triggered by violating the delay of the combinatorial logic paths that pass through the targeted correction terms c^2b^2. It is indeed a parametric Trojan, which does not require any additional logic. The Trojan is inserted by modifying a few gates during manufacturing, so that their delay increase and add up to the path delay faults.

Given in [21], the underlying method to create a triggerable and stealthy delay-based Trojan consists of two phases: path selection and delay distribution. In the first phase, a set of uniquely-sensitized paths are found that passes through a combinatorial circuit from primary inputs to the primary outputs. Controllability and observability metrics are used to guide the selection of which gates to include in the path to make sure that the path(s) are uniquely sensitized[2]. Furthermore, a SAT-based check is performed to make sure that the path remains sensitizable each time a gate is selected to be added to the path. After a set of uniquely-sensitized paths is selected, the overall delay of the path(s) must be increased so that a delay fault occurs when the path is sensitized. However, any delay added to the gates of the selected path may also cause delay faults on intersecting paths, which would cause undesirable errors and affect the functionality of the circuit. The delay distribution phase addresses this problem by smartly choosing delays for each gate of the selected path to minimize the number of faults caused by intersecting paths. At the same time, the approach ensures that the overall path delay is sufficient for the selected paths to make it faulty.

ASIC Platforms. In an ASIC platform, such Trojans are introduce by slightly modifications on the sub-transistor level so that the parameters of a few transistors of the design are changed. To increase the delays of transistors in stealthy

[2] It means that the selected paths are the only ones in the circuit whose critical delay can be violated.

ways, there are many possible ways in practice. However, such Trojan is very difficult to be detected by e.g., functional testing, visual inspection, and side-channel profiling, because not a single transistor is removed or added to the design and the changes to the individual gates are minor. Also, full reverse-engineering of the IC would unlikely reveal the presence of the malicious manipulation in the design. Furthermore, this Trojan would not present at higher abstraction levels and hence cannot be detected at those levels, because the actual Trojan is inserted at the sub-transistor level.

A path delay fault in a design is sensitized by a sequence of (at least two) consecutive input vectors on consecutive clock cycles. Its reason is charging/discharging of output capacitances of gates of the path. The delay of each gate is determined by its speed in charging or discharging of its output capacitance. Therefore, if the state of the capacitances of gates (belonging to the targeted path) is not changed (i.e., the capacitances do not charge or discharge), the effect of the path delay fault cannot be propagated along the path. Therefore, to trigger the path delay fault, the consecutive input vectors should change the state of the capacitances of the targeted path.

There are several stealthy ways to change slightly the parameters of transistors of a gate and make it slower in charging/discharging its output capacitance (load capacitance). Exemplary, we list three methods below.

Decrease the Width. Usually a standard cell library has different drive strengths for each logic gate type, which correspond to various transistor widths. Current of a transistor is linearly proportional to the transistor width, therefore a transistor with smaller width is slower to charge its load capacitance. One way to increase the delay of a gate is to substitute it with its weaker version in the library which has smaller width, or to create a custom version of the gate with a narrow width, if the lower level information of the gate is available in the library (e.g., SPICE model). The problem here is that an inspector who test the IC optically, may detect the gate downsizing depending on how much the geometry has been changed.

Raise the Threshold. A common way of increasing delay of a gate is to increase the threshold voltage of its transistors by body biasing or doping manipulation. Using high and low threshold voltages at the same time in a design (i.e., Dual-Vt design) is very common approach and provides for designer to have more options to satisfy the speed goals of the design. Devices with low threshold voltage are fast and used where delay is critical; devices with high threshold voltage are slow and used where power consumption is important. Body biasing can change the threshold voltage and hence the delay of a gate through changing the voltage between body and source of the transistor [29]. A reverse body bias in which body is at lower voltage than the source, increases the threshold voltage and makes the device slow. In general, transistors with high threshold voltage will response later when an input switches, and conduct less current. Therefore, the load capacitances of the transistors will be charged or discharged more slowly. Dopant manipulation and body biasing, are both very difficult to detect.

Increase the Gate Length. Gate length biasing can increase delay of a gate by reducing the current of its transistors [26]. The likelihood of detection of this kind of manipulation depends on the degree of the modification.

FPGA Platforms. In case of the FPGAs, the combinatorial circuits are realized by Look-Up Tables (LUT), in currently-available Xilinx FPGAs, by 6-to-1 or 5-to-2 LUTs and in former generations by 4-to-1 LUTs. The delay of the LUTs cannot be changed by the end users; alternatively we offer the following techniques to make certain paths longer.

Through Switch Boxes. The routings in FPGA devices are made by configuring the switch boxes. Since the switch boxes are made by active components realizing logical switches, a signal which passes through many switch boxes has a longer delay compared to a short signal. Therefore, given a fully placed-and-routed design we can modify the routings by lengthening the selected signals. This is for example feasible by means of `Vivado Design Suite` as a standard tool provided by Xilinx for recent FPGA families and `FPGA Editor` for the older generations. It is in fact needs a high level of expertise, and cannot be done at HDL level. Interestingly, the resulting circuit would not have any additional resource consumption, i.e., the number of utilized LUTs, FFs and Slices, hence hard to detect particularly if the utilization reports are compared.

Through Route-Thrus LUTs. Alternatively, the LUTs can be configured as logical buffer. This, which is called *route-thrus*, is a usual technique applied by Xilinx tools to enable routing of non-straightforward routes. Inserting a route-thrus LUT into any path, makes its delay longer. Hence, another feasible way to insert Trojans by delay path fault is to introduce as many as required route-thrus LUTs into the targeted path. It should be noted that the malicious design would have more LUT utilization compared to the original design, and it may increase the chance of being detected by a Trojan inspector. However, none of such extra LUTs realizes a logic, and all of them are seen as *route-thrus* LUTs which are very often (almost in any design) inserted by the FPGA vendor's place-and-route tools. Compared to the previous method, this can be done at HDL level (by hard instantiating route-thrus LUTs).

Focusing on our target, i.e., correction term c^2b^2 in $\mathcal{G}^1(.,.)$ and $\mathcal{G}^3(.,.)$, by applying the above-explained procedure, we found the situation which enables introducing delay path fault into such routes:

- Considering Fig. 1, the XOR gate which receives the $\mathcal{F}^1$ and $\mathcal{C}$ output should be the last gate in the combinatorial circuit generating y_1^1, i.e., the second bit of $\mathcal{G}^1(.,.)$. The same holds for y_1^3, i.e., the second bit of $\mathcal{G}^3(.,.)$.
- The only paths which should be lengthened are both instances of c^2b^2. Therefore, in case of the FPGA platform we followed both above-explained methods to lengthen such paths, i.e., between (i) the output of the LUT generating c^2b^2 and (ii) the input of the aforementioned final XOR gate.

We have easily applied the second method (through route-thrus LUTs) at the HDL level by instantiating a couple of LUTs as buffer between the selected path. More detailed results with respect to the number of required route-thrus LUTs and the achieved frequencies to trigger the Trojan are shown in next Sect. 5. For the first method (through switch boxes) – since our target platform is a Spartan-6 FPGA – we made use of `FPGA Editor` to manually modify the selected routes (see Appendix C for two routes of a signal with different length). We should emphasize that this approach is possible if the correction term c^2b^2 is realized by a unique LUT (can be forced at HDL level by hard instantiating or placing such a module in a deeper hierarchy). Otherwise, the logic generating c^2b^2 might be merged with other logic into a LUT, which avoids having a separate path between c^2b^2 and a LUT that realizes the final XOR gate.

5 Practical Results

5.1 Design Architecture

We made use of the above-explained malicious PRESENT TI Sbox in a design with full encryption functionality. The underlying design is similar to the *Profile 2* of [48], where only one instance of the Sbox is implemented. The nibbles are serially shifted through the state register as well as through the Sbox module while the `PLayer` is performed in parallel in one clock cycle. Following its underlying first-order TI, the 64-bit plaintext is provided by three shares, i.e., second-order Boolean masking, while the 80-bit key is not shared (similar to that of [10,48]). Figure 3 shows an overview of the design architecture, which needs 527 clock cycles for a full encryption after the plaintext and key are serially shifted into the state (resp. key) registers.

We should here emphasize that the underlying TI construction is a first-order masking, which can provably provide security against first-order SCA attacks. However, higher-order attacks are expected to exploit the leakage, but they are sensitive to noise [50] since accurately estimating higher-order statistical

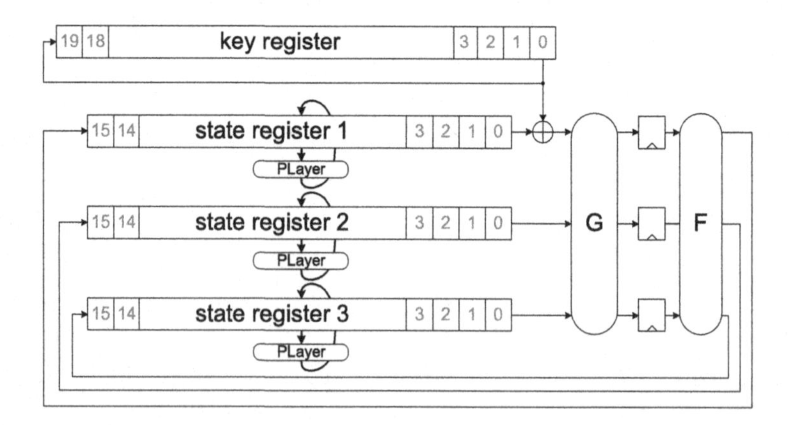

Fig. 3. Design architecture of the PRESENT TI as the case study.

moments needs huge amount of samples compared to lower-order moments. It is indeed widely known that such masking schemes should be combined with hiding techniques (to reduce the SNR) to practically harden (hopefully disable) the higher-order attacks. As an example we can refer to [44], where a TI construction is implemented by a power-equalization technique. We instead integrated a noise generator module into our target FPGA to increase the noise and hence decrease the SNR. The details of the integrated noise generator module is given in Appendix B. Note that without such a noise generator module, our design would be vulnerable to higher-order attacks and no Trojan would be required to reveal the secret. Therefore, the existence of such a hiding countermeasure to make higher-order attacks practically hard is essential.

The design is implemented on a Spartan-6 FPGA board SAKURA-G, as a platform for SCA evaluations [1]. In order to supply the PRESENT core with a high clock frequency, a Digital Clock Manager (DCM) has been instantiated in the target FPGA to multiply the incoming clock by a factor of 8. The external clock was provided by a laboratory adjustable signal generator to enable evaluating the design under different high clock frequencies.

Table 1 shows the resource utilization (excluding the noise generator) as well as the achieved margins for the clock frequency considering (*i*) the original design, (*ii*) malicious design made by *through switch boxes* method and (*iii*) malicious design made by *through route-thrus LUTs* technique. It is noticeable that the first malicious design does not change the utilization figures at all since lengthening the routes are done only through the switch boxes (see Appendix C). Using the second method – in order to achieve the same frequency margins – we added 4 route-thru LUTs (at the HDL level) to each path of the targeted correction term. This led to 8 extra LUT utilization and 4 more Slices; we would like to mention that the combinatorial circuit of the entire TI Sbox (both $\mathcal{G}^*$ $\mathcal{F}^*$) would fit into 29 LUTs (excluding the route-thru ones).

Regarding the frequency ranges, shown in Table 1, it can be seen that the maximum clock frequency of the malicious design is decreased from 219.2 MHz

Table 1. Performance figure of our PRESENT-80 encryption designs.

Design	Method	FF	Utilization: LUT *logic*	Utilization: LUT *route-thrus*	Utilization: Slice	Frequency [MHz]
Original	-	299	291	35	226	④ ① ; 219.2
Malicious	*switch box*	299	291	35	226	④ ③ ② ① ; 212.8, 219.2, 196
Malicious	*route-thru LUT*	299	291	43	230	

to 196 MHz, i.e., around 10% reduction. However, both ② and ③ periods are very narrow, that makes it hard to be detected either by a Trojan inspector or by an SCA evaluator.

5.2 SCA Evaluations

Measurement Setup. For SCA evaluations we collected power consumption traces (at the Vdd path) of the target FPGA by means of a digital oscilloscope at sampling rate of 1 GS/s. It might be thought that when the target design

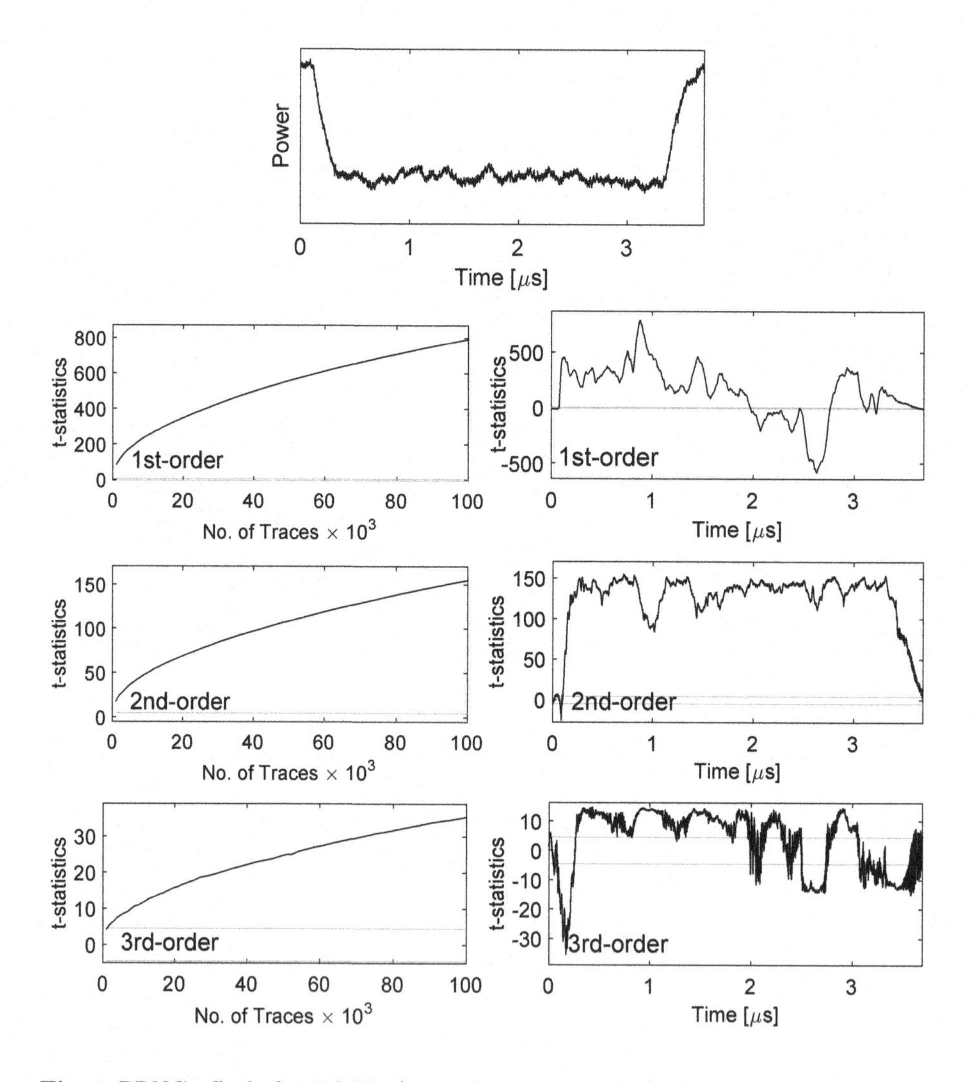

Fig. 4. PRNG off, clock 168 MHz (Trojan not triggered), (top) a sample power trace, t-test results (right) with 100,000 traces, (left) absolute maximum over the number of traces.

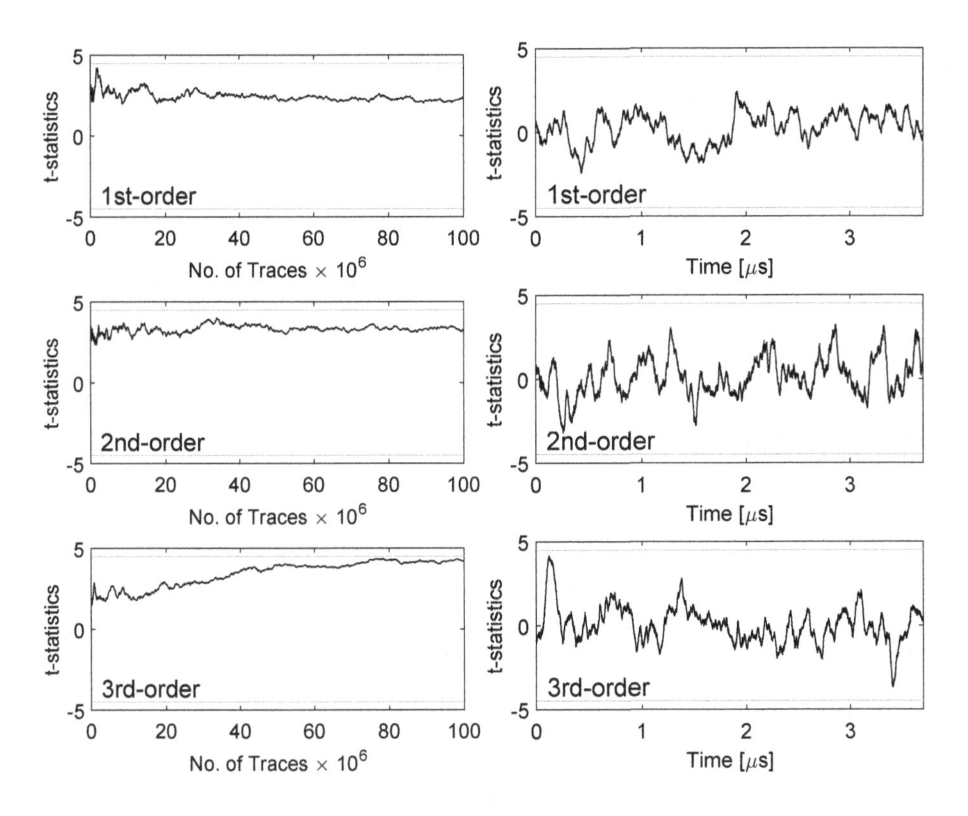

Fig. 5. PRNG on, clock 168 MHz (Trojan not triggered), t-test results (right) with 100,000,000 traces, (left) absolute maximum over the number of traces.

runs at a high frequency >150 MHz, such a sampling rate does not suffice to capture all leaking information. However, power consumption traces are already filtered due to the PCB, shunt resistor, measurement setup, etc. Hence, higher sampling rate for such a setting does not improve the attack efficiency[3], and often the bandwidth of the oscilloscope is even manually limited for noise reduction purposes (see [49]).

Methodology. In order to examine the SCA resistance of our design(s) in both settings, i.e., whether the inserted Trojan is triggered or not, we conducted two evaluation schemes. We first performed non-specific t-test (fixed versus random) [22,56] to examine the existence of detectable leakage. Later in case where the Trojan is triggered, we also conduct key-recovery attacks.

It should be mentioned that both of our malicious designs (see Table 1) operate similarly. It means that when the Trojan is triggered, the evaluation of both designs led to the same results. Therefore, below we exemplary show the result of the one formed by *through route-thrus LUTs*.

[3] It is not the case for EM-based analyses.

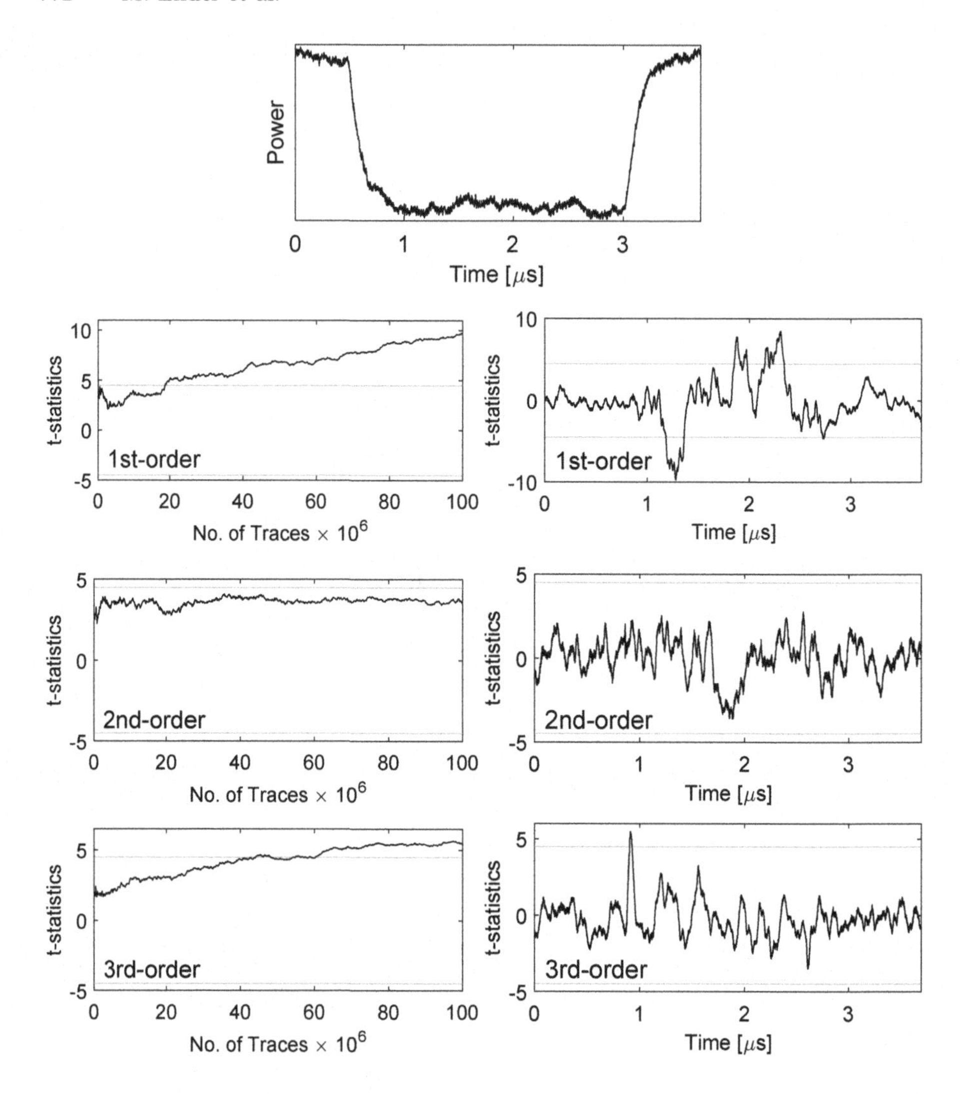

Fig. 6. PRNG on, clock 216 MHz (Trojan triggered), (top) a sample power trace, t-test results (right) with 100,000,000 traces, (left) absolute maximum over the number of traces.

To validate the setup, we start with a non-specific t-test when the PRNG of the target design (used to share the plaintext for the TI PRESENT encryption) is turned off, i.e., generating always zero instead of random numbers. To this end, we collected 100,000 traces when the design is operated at 168 MHz, i.e., the Trojan is not triggered. We followed the concept given in [56] for the collection of traces belonging to fixed and random inputs. The result of the t-test (up to third-order) is shown in Fig. 4, confirming the validity of the setup and the developed evaluation tools.

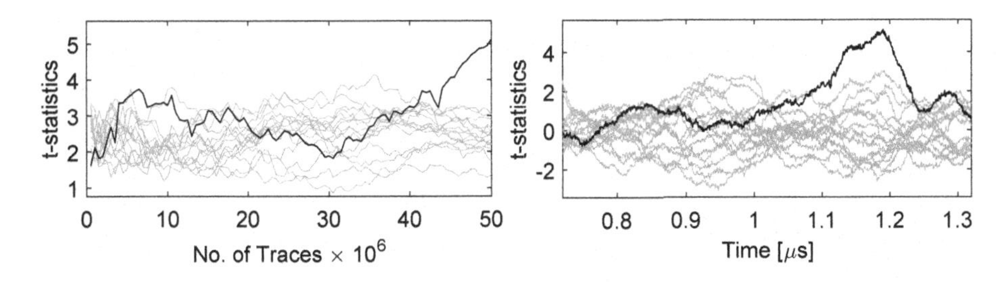

Fig. 7. PRNG on, clock 216 MHz (Trojan triggered), 50,000,000 traces, DPA attack result targeting a key nibble based on an Sbox output bit at the first round.

To repeat the same process when the PRNG is turned on, i.e., the masks for initial sharing of the plaintext are uniformly distributed, we collected 100,000,000 traces for non-specific t-test evaluations. In this case, the device still operates at 168 MHz, i.e., the Trojan is not triggered. The corresponding results are shown in Fig. 5. Although the underlying design is a realization of a first-order secure TI, it can be seen from the presented results that second- and third-order leakages are also not detectable. As stated before, this is due to the integration of the noise generator module which affects the detectability of higher-order leakages (see Appendix B).

As the last step, the same scenario is repeated when the clock frequency is increased to 216 MHz, where the design is in the ③ period, i.e., with correct functionality and without uniformity. Similar to the previous experiment, we collected 100,000,000 traces for a non-specific t-test, whose results are shown in Fig. 6. As shown by the graphics, there is detectable leakage through all statistical moments but with lower t-statistics compared to the case with PRNG off. Therefore, we have also examine the feasibility of key recovery attacks. To this end, we made use of those collected traces which are associated with random inputs, i.e., around 50,000,000 traces of the last non-specific t-test. We conducted several different CPA and DPA attacks considering intermediate values of the underlying PRESENT encryption function. The most successful attack was recognized as classical DPA attacks [33] targeting a key nibble by predicting an Sbox output bit at the first round of the encryption. As an example, Fig. 7 presents an exemplary corresponding result.

6 Conclusions

In this work it is shown how to insert a parametric hardware Trojan with very low overhead into SCA-resistance designs. The presented Trojan is capable of being integrated into both ASIC and FPGA platforms. Since it does not add any logic into the design (particularly its resource utilization in FPGAs can be null), the chance of being detected is expected to be very low. Compared to the original design, its only footprint is around 10% decrease in the maximum clock frequency.

We have shown that by increasing the clock frequency, the malicious threshold implementation design starts leaking exploitable information through side channels. Hence, the Trojan adversary can trigger the Trojan and make use of the exploitable leakage, while the design can pass SCA evaluations when the Trojan is not triggered. More precisely, suppose that the maximum clock frequency of the malicious device is 196 MHz. Hence, in an evaluation lab its SCA leakage will not be examined at 200 MHz because the device does not operate correctly. However, the Trojan adversary runs the device at 216 MHz and the SCA leakage becomes exploitable. To the best of our knowledge, compared to the previous works in the areas of side-channel hardware Trojans, our construction is the only one which is applied on a provably-secure SCA countermeasure, and is parametric with very low overhead.

A raising question is whether a control over the clock frequency by the Trojan adversary is a practical assumption. Such a control is usually available in FPGA designs since they are mainly externally clocked, and internally multiplied by PLL or DCM. In ASIC or embedded designs, the clock is more often generated internally, hence no control. Nevertheless, by decreasing the supply voltage the same effect can be seen. It can also be criticized that when the attacker has control over the clock, fault-injection attacks by clock glitch can also be a threat. As a message of this paper, overclocking and – at the same time – power supply reduction should be internally monitored to avoid such an SCA-based Trojan being activated. Related to this topic we should refer to [20], where the difficulties of embedding a "clock frequency monitor" in presence of supply voltage changes are shown.

Acknowledgments. The work was partially funded through grants ERC Advanced 695022 and NSF CNS-1421352.

A Uniform TI of $\mathcal{F}$

Considering $\boldsymbol{y} = \mathcal{F}(\boldsymbol{x})$ and $\boldsymbol{x}^{i\in\{1,2,3\}} = \langle d^i, c^i, b^i, a^i \rangle$ – derived by direct sharing – we present one of its uniform sharing $\left(\boldsymbol{y}^1, \boldsymbol{y}^2, \boldsymbol{y}^3\right) = \left(\mathcal{F}^1\left(\boldsymbol{x}^2, \boldsymbol{x}^3\right), \mathcal{F}^2\left(\boldsymbol{x}^3, \boldsymbol{x}^1\right), \mathcal{F}^3\left(\boldsymbol{x}^1, \boldsymbol{x}^2\right)\right)$ as

$$\begin{aligned}
y_0^1 &= b^2 \oplus c^2a^2 \oplus c^2a^3 \oplus c^3a^2, \\
y_1^1 &= c^2 \oplus b^2 \oplus d^2a^2 \oplus d^2a^3 \oplus d^3a^2, \\
y_2^1 &= d^2 \oplus b^2a^2 \oplus b^2a^3 \oplus b^3a^2, \\
y_3^1 &= c^2 \oplus b^2 \oplus a^2 \oplus d^2a^2 \oplus d^2a^3 \oplus d^3a^2,
\end{aligned}$$

$$\begin{aligned}
y_0^2 &= b^3 \oplus c^3a^3 \oplus c^1a^3 \oplus c^3a^1, \\
y_1^2 &= c^3 \oplus b^3 \oplus d^3a^3 \oplus d^1a^3 \oplus d^3a^1, \\
y_2^2 &= d^3 \oplus b^3a^3 \oplus b^1a^3 \oplus b^3a^1, \\
y_3^2 &= c^3 \oplus b^3 \oplus a^3 \oplus d^3a^3 \oplus d^1a^3 \oplus d^3a^1,
\end{aligned}$$

$$y_0^3 = b^1 \oplus c^1 a^1 \oplus c^1 a^2 \oplus c^2 a^1,$$
$$y_1^3 = c^1 \oplus b^1 \oplus d^1 a^1 \oplus d^1 a^2 \oplus d^2 a^1,$$
$$y_2^3 = d^1 \oplus b^1 a^1 \oplus b^1 a^2 \oplus b^2 a^1,$$
$$y_3^3 = c^1 \oplus b^1 \oplus a^1 \oplus d^1 a^1 \oplus d^1 a^2 \oplus d^2 a^1.$$

B Noise Generator

We have built a noise generator as an independent module, i.e., it does not have any connection to the target PRESENT design and operates independently. We followed one the concepts introduced in [25]. As shown by Fig. 8, it is made as a combination of a ring oscillator, an LFSR, and several shift registers. The actual power is consumed by the shift registers. Every shift register instantiates a `SRLC32E` primitive, which is a 32-bit shift register within a single LUT inside a `SLICEM`. The shift registers are initialized with the consecutive values of 01. Every shift register's output is feedback to its input and shifted by one at every clock cycle when enabled. Thus, every shift operation toggles the entire bits inside the registers, which maximizes the power consumption of the shift register.

The ring oscillator, made of 31 inverter LUTs, acts as the clock source inside the noise module for both the LFSR and the shift registers. The LFSR realizes the irreducible polynomial $x^{19} + x^{18} + x^{17} + x^{14} + 1$ to generate a pseudo-random clock enable signal for the shift registers.

We instantiated 4×8 instances of the shift register LUTs, fitting into 8 Slices. The ring oscillator required 17 Slices (as stated, made of 31 inverters), and the LFSR fits into 2 Slices, made by 1 LUT for the feedback function, 2 FFs and 2 shift register LUTs. Overall, the entire independent noise generator module required 27 Slices.

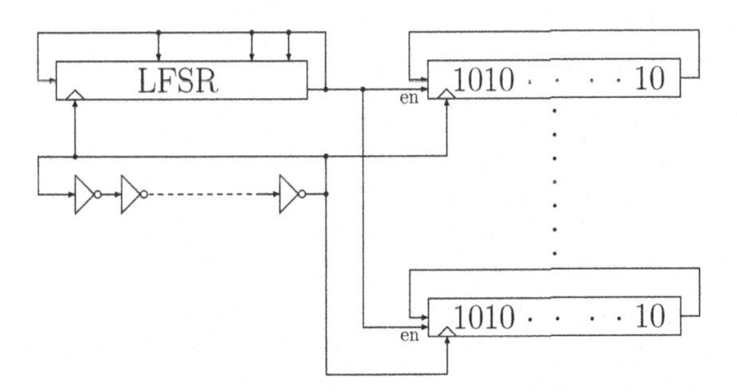

Fig. 8. Block diagram of the noise generator.

C Different Routings in FPGA

See Fig. 9.

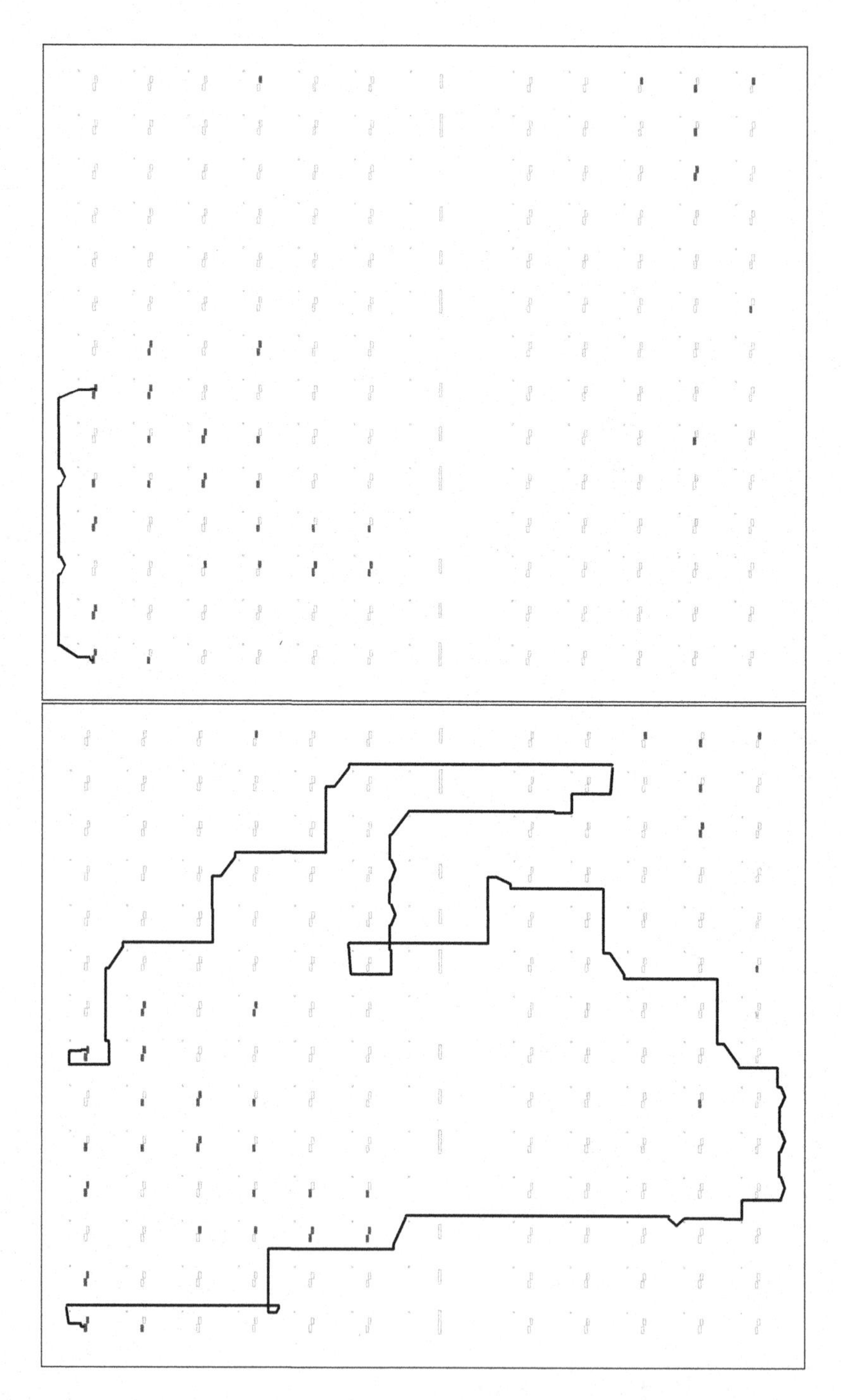

Fig. 9. Two routes of the same signal in a Spartan-6 FPGA, manually performed by `FPGA Editor`.

References

1. Side-channel AttacK User Reference Architecture. http://satoh.cs.uec.ac.jp/SAKURA/index.html
2. Becker, G.T., Regazzoni, F., Paar, C., Burleson, W.P.: Stealthy dopant-level hardware trojans. In: Bertoni, G., Coron, J.-S. (eds.) CHES 2013. LNCS, vol. 8086, pp. 197–214. Springer, Heidelberg (2013). https://doi.org/10.1007/978-3-642-40349-1_12
3. Beyne, T., Bilgin, B.: Uniform first-order threshold implementations. In: Avanzi, R., Heys, H. (eds.) SAC 2016. LNCS, vol. 10532, pp. 79–98. Springer, Cham (2017). https://doi.org/10.1007/978-3-319-69453-5_5
4. Biham, E., Carmeli, Y., Shamir, A.: Bug attacks. In: Wagner, D. (ed.) CRYPTO 2008. LNCS, vol. 5157, pp. 221–240. Springer, Heidelberg (2008). https://doi.org/10.1007/978-3-540-85174-5_13
5. Biham, E., Carmeli, Y., Shamir, A.: Bug attacks. J. Cryptol. **29**(4), 775–805 (2016)
6. Bilgin, B., Bogdanov, A., Knežević, M., Mendel, F., Wang, Q.: FIDES: lightweight authenticated cipher with side-channel resistance for constrained hardware. In: Bertoni, G., Coron, J.-S. (eds.) CHES 2013. LNCS, vol. 8086, pp. 142–158. Springer, Heidelberg (2013). https://doi.org/10.1007/978-3-642-40349-1_9
7. Bilgin, B., Daemen, J., Nikov, V., Nikova, S., Rijmen, V., Van Assche, G.: Efficient and first-order DPA resistant implementations of KECCAK. In: Francillon, A., Rohatgi, P. (eds.) CARDIS 2013. LNCS, vol. 8419, pp. 187–199. Springer, Cham (2014). https://doi.org/10.1007/978-3-319-08302-5_13
8. Bilgin, B., Gierlichs, B., Nikova, S., Nikov, V., Rijmen, V.: A more efficient AES threshold implementation. In: Pointcheval, D., Vergnaud, D. (eds.) AFRICACRYPT 2014. LNCS, vol. 8469, pp. 267–284. Springer, Cham (2014). https://doi.org/10.1007/978-3-319-06734-6_17
9. Bilgin, B., Gierlichs, B., Nikova, S., Nikov, V., Rijmen, V.: Higher-order threshold implementations. In: Sarkar, P., Iwata, T. (eds.) ASIACRYPT 2014. LNCS, vol. 8874, pp. 326–343. Springer, Heidelberg (2014). https://doi.org/10.1007/978-3-662-45608-8_18
10. Bilgin, B., Gierlichs, B., Nikova, S., Nikov, V., Rijmen, V.: Trade-offs for threshold implementations illustrated on AES. IEEE Trans. CAD Integr. Circuits Syst. **34**(7), 1188–1200 (2015)
11. Bilgin, B., Nikova, S., Nikov, V., Rijmen, V., Stütz, G.: Threshold implementations of all 3 × 3 and 4 × 4 S-boxes. In: Prouff, E., Schaumont, P. (eds.) CHES 2012. LNCS, vol. 7428, pp. 76–91. Springer, Heidelberg (2012). https://doi.org/10.1007/978-3-642-33027-8_5
12. Bilgin, B., Nikova, S., Nikov, V., Rijmen, V., Tokareva, N., Vitkup, V.: Threshold implementations of small S-boxes. Cryptogr. Commun. **7**(1), 3–33 (2015)
13. Biryukov, A., De Cannière, C., Braeken, A., Preneel, B.: A toolbox for cryptanalysis: linear and affine equivalence algorithms. In: Biham, E. (ed.) EUROCRYPT 2003. LNCS, vol. 2656, pp. 33–50. Springer, Heidelberg (2003). https://doi.org/10.1007/3-540-39200-9_3
14. Bogdanov, A., Knudsen, L.R., Leander, G., Paar, C., Poschmann, A., Robshaw, M.J.B., Seurin, Y., Vikkelsoe, C.: PRESENT: an ultra-lightweight block cipher. In: Paillier, P., Verbauwhede, I. (eds.) CHES 2007. LNCS, vol. 4727, pp. 450–466. Springer, Heidelberg (2007). https://doi.org/10.1007/978-3-540-74735-2_31
15. Bozilov, D., Bilgin, B., Sahin, H.A.: A note on 5-bit quadratic permutations' classification. IACR Trans. Symmetric Cryptol. **2017**(1), 398–404 (2017)

16. Canright, D., Batina, L.: A very compact "perfectly masked" S-box for AES. In: Bellovin, S.M., Gennaro, R., Keromytis, A., Yung, M. (eds.) ACNS 2008. LNCS, vol. 5037, pp. 446–459. Springer, Heidelberg (2008). https://doi.org/10.1007/978-3-540-68914-0_27
17. Carlet, C., Danger, J.-L., Guilley, S., Maghrebi, H.: Leakage squeezing of order two. In: Galbraith, S., Nandi, M. (eds.) INDOCRYPT 2012. LNCS, vol. 7668, pp. 120–139. Springer, Heidelberg (2012). https://doi.org/10.1007/978-3-642-34931-7_8
18. Chakraborty, R.S., Narasimhan, S., Bhunia, S.: Hardware Trojan: threats and emerging solutions. In: HLDVT 2009, pp. 166–171. IEEE Computer Society (2009)
19. Chakraborty, R.S., Wolff, F., Paul, S., Papachristou, C., Bhunia, S.: *MERO*: a statistical approach for hardware Trojan detection. In: Clavier, C., Gaj, K. (eds.) CHES 2009. LNCS, vol. 5747, pp. 396–410. Springer, Heidelberg (2009). https://doi.org/10.1007/978-3-642-04138-9_28
20. Endo, S., Li, Y., Homma, N., Sakiyama, K., Ohta, K., Fujimoto, D., Nagata, M., Katashita, T., Danger, J., Aoki, T.: A silicon-level countermeasure against fault sensitivity analysis and its evaluation. IEEE Trans. VLSI Syst. **23**(8), 1429–1438 (2015)
21. Ghandali, S., Becker, G.T., Holcomb, D., Paar, C.: A Design methodology for stealthy parametric Trojans and its application to bug attacks. In: Gierlichs, B., Poschmann, A.Y. (eds.) CHES 2016. LNCS, vol. 9813, pp. 625–647. Springer, Heidelberg (2016). https://doi.org/10.1007/978-3-662-53140-2_30
22. Goodwill, G., Jun, B., Jaffe, J., Rohatgi, P.: A testing methodology for side channel resistance validation. In: NIST Non-invasive Attack Testing Workshop (2011). http://csrc.nist.gov/news_events/non-invasive-attack-testing-workshop/papers/08_Goodwill.pdf
23. Gross, H., Mangard, S., Korak, T.: An efficient side-channel protected AES implementation with arbitrary protection order. In: Handschuh, H. (ed.) CT-RSA 2017. LNCS, vol. 10159, pp. 95–112. Springer, Cham (2017). https://doi.org/10.1007/978-3-319-52153-4_6
24. Groß, H., Wenger, E., Dobraunig, C., Ehrenhöfer, C.: Suit up! - made-to-measure hardware implementations of ASCON. In: DSD 2015, pp. 645–652. IEEE Computer Society (2015)
25. Güneysu, T., Moradi, A.: Generic side-channel countermeasures for reconfigurable devices. In: Preneel, B., Takagi, T. (eds.) CHES 2011. LNCS, vol. 6917, pp. 33–48. Springer, Heidelberg (2011). https://doi.org/10.1007/978-3-642-23951-9_3
26. Gupta, P., Kahng, A.B., Sharma, P., Sylvester, D.: Gate-length biasing for runtime-leakage control. IEEE Trans. CAD Integr. Circuits Syst. **25**(8), 1475–1485 (2006)
27. Jin, Y., Makris, Y.: Hardware Trojan detection using path delay fingerprint. In: HOST 2008, pp. 51–57. IEEE Computer Society (2008)
28. Jin, Y., Makris, Y.: Hardware Trojans in wireless cryptographic ICs. IEEE Des. Test Comput. **27**(1), 26–35 (2010)
29. Karri, R., Rajendran, J., Rosenfeld, K., Tehranipoor, M.: Trustworthy hardware: identifying and classifying hardware trojans. IEEE Comput. **43**(10), 39–46 (2010)
30. Kasper, M., Moradi, A., Becker, G.T., Mischke, O., Güneysu, T., Paar, C., Burleson, W.: Side channels as building blocks. J. Cryptogr. Eng. **2**(3), 143–159 (2012)
31. King, S.T., Tucek, J., Cozzie, A., Grier, C., Jiang, W., Zhou, Y.: Designing and implementing malicious hardware. In: USENIX Workshop on Large-Scale Exploits and Emergent Threats, LEET 2008. USENIX Association (2008)

32. Kocher, P.C.: Timing attacks on implementations of Diffie-Hellman, RSA, DSS, and other systems. In: Koblitz, N. (ed.) CRYPTO 1996. LNCS, vol. 1109, pp. 104–113. Springer, Heidelberg (1996). https://doi.org/10.1007/3-540-68697-5_9
33. Kocher, P., Jaffe, J., Jun, B.: Differential power analysis. In: Wiener, M. (ed.) CRYPTO 1999. LNCS, vol. 1666, pp. 388–397. Springer, Heidelberg (1999). https://doi.org/10.1007/3-540-48405-1_25
34. Kumar, R., Jovanovic, P., Burleson, W.P., Polian, I.: Parametric Trojans for fault-injection attacks on cryptographic hardware. In: FDTC 2014, pp. 18–28. IEEE Computer Society (2014)
35. Lin, L., Burleson, W., Paar, C.: MOLES: malicious off-chip leakage enabled by side-channels. In: ICCAD 2009, pp. 117–122. ACM (2009)
36. Lin, L., Kasper, M., Güneysu, T., Paar, C., Burleson, W.: Trojan side-channels: lightweight hardware Trojans through side-channel engineering. In: Clavier, C., Gaj, K. (eds.) CHES 2009. LNCS, vol. 5747, pp. 382–395. Springer, Heidelberg (2009). https://doi.org/10.1007/978-3-642-04138-9_27
37. Maghrebi, H., Guilley, S., Danger, J.-L.: Leakage Squeezing countermeasure against high-order attacks. In: Ardagna, C.A., Zhou, J. (eds.) WISTP 2011. LNCS, vol. 6633, pp. 208–223. Springer, Heidelberg (2011). https://doi.org/10.1007/978-3-642-21040-2_14
38. Mangard, S., Oswald, E., Popp, T.: Power Analysis Attacks: Revealing the Secrets of Smart Cards. Springer, Heidelberg (2007). https://doi.org/10.1007/978-0-387-38162-6
39. Mangard, S., Pramstaller, N., Oswald, E.: Successfully attacking masked AES hardware implementations. In: Rao, J.R., Sunar, B. (eds.) CHES 2005. LNCS, vol. 3659, pp. 157–171. Springer, Heidelberg (2005). https://doi.org/10.1007/11545262_12
40. Moradi, A., Kirschbaum, M., Eisenbarth, T., Paar, C.: Masked dual-rail precharge logic encounters state-of-the-art power analysis methods. IEEE Trans. VLSI Syst. **20**(9), 1578–1589 (2012). https://doi.org/10.1109/TVLSI.2011.2160375
41. Moradi, A., Mischke, O., Eisenbarth, T.: Correlation-enhanced power analysis collision attack. In: Mangard, S., Standaert, F.-X. (eds.) CHES 2010. LNCS, vol. 6225, pp. 125–139. Springer, Heidelberg (2010). https://doi.org/10.1007/978-3-642-15031-9_9
42. Moradi, A., Poschmann, A., Ling, S., Paar, C., Wang, H.: Pushing the limits: a very compact and a threshold implementation of AES. In: Paterson, K.G. (ed.) EUROCRYPT 2011. LNCS, vol. 6632, pp. 69–88. Springer, Heidelberg (2011). https://doi.org/10.1007/978-3-642-20465-4_6
43. Moradi, A., Schneider, T.: Side-channel analysis protection and low-latency in action – case study of PRINCE and Midori. In: Cheon, J.H., Takagi, T. (eds.) ASIACRYPT 2016. LNCS, vol. 10031, pp. 517–547. Springer, Heidelberg (2016). https://doi.org/10.1007/978-3-662-53887-6_19
44. Moradi, A., Wild, A.: Assessment of hiding the higher-order leakages in hardware – what are the achievements versus overheads? In: Güneysu, T., Handschuh, H. (eds.) CHES 2015. LNCS, vol. 9293, pp. 453–474. Springer, Heidelberg (2015). https://doi.org/10.1007/978-3-662-48324-4_23
45. Nikova, S., Rijmen, V., Schläffer, M.: Secure hardware implementation of nonlinear functions in the presence of glitches. J. Cryptol. **24**(2), 292–321 (2011)
46. Oswald, E., Mangard, S., Pramstaller, N., Rijmen, V.: A side-channel analysis resistant description of the AES S-box. In: Gilbert, H., Handschuh, H. (eds.) FSE 2005. LNCS, vol. 3557, pp. 413–423. Springer, Heidelberg (2005). https://doi.org/10.1007/11502760_28

47. Popp, T., Kirschbaum, M., Zefferer, T., Mangard, S.: Evaluation of the masked logic style MDPL on a prototype chip. In: Paillier, P., Verbauwhede, I. (eds.) CHES 2007. LNCS, vol. 4727, pp. 81–94. Springer, Heidelberg (2007). https://doi.org/10.1007/978-3-540-74735-2_6
48. Poschmann, A., Moradi, A., Khoo, K., Lim, C., Wang, H., Ling, S.: Side-channel resistant crypto for less than 2, 300 GE. J. Cryptol. **24**(2), 322–345 (2011)
49. Merino del Pozo, S., Standaert, F.-X.: Getting the most out of leakage detection. In: Guilley, S. (ed.) COSADE 2017. LNCS, vol. 10348, pp. 264–281. Springer, Cham (2017). https://doi.org/10.1007/978-3-319-64647-3_16
50. Prouff, E., Rivain, M., Bevan, R.: Statistical analysis of second order differential power analysis. IEEE Trans. Comput. **58**(6), 799–811 (2009)
51. Rajendran, J., Jyothi, V., Karri, R.: Blue team red team approach to hardware trust assessment. In: ICCD 2011, pp. 285–288. IEEE Computer Society (2011)
52. Reparaz, O., Bilgin, B., Nikova, S., Gierlichs, B., Verbauwhede, I.: Consolidating masking schemes. In: Gennaro, R., Robshaw, M. (eds.) CRYPTO 2015. LNCS, vol. 9215, pp. 764–783. Springer, Heidelberg (2015). https://doi.org/10.1007/978-3-662-47989-6_37
53. Saha, S., Chakraborty, R.S., Nuthakki, S.S., Anshul, Mukhopadhyay, D.: Improved test pattern generation for hardware Trojan detection using genetic algorithm and boolean satisfiability. In: Güneysu, T., Handschuh, H. (eds.) CHES 2015. LNCS, vol. 9293, pp. 577–596. Springer, Heidelberg (2015). https://doi.org/10.1007/978-3-662-48324-4_29
54. Sasdrich, P., Moradi, A., Güneysu, T.: Affine equivalence and its application to tightening threshold implementations. In: Dunkelman, O., Keliher, L. (eds.) SAC 2015. LNCS, vol. 9566, pp. 263–276. Springer, Cham (2016). https://doi.org/10.1007/978-3-319-31301-6_16
55. Sasdrich, P., Moradi, A., Güneysu, T.: Hiding higher-order side-channel leakage – randomizing cryptographic implementations in reconfigurable hardware. In: Handschuh, H. (ed.) CT-RSA 2017. LNCS, vol. 10159, pp. 131–146. Springer, Cham (2017). https://doi.org/10.1007/978-3-319-52153-4_8
56. Schneider, T., Moradi, A.: Leakage assessment methodology – a clear roadmap for side-channel evaluations. In: Güneysu, T., Handschuh, H. (eds.) CHES 2015. LNCS, vol. 9293, pp. 495–513. Springer, Heidelberg (2015). https://doi.org/10.1007/978-3-662-48324-4_25
57. Shiyanovskii, Y., Wolff, F.G., Rajendran, A., Papachristou, C.A., Weyer, D.J., Clay, W.: Process reliability based Trojans through NBTI and HCI effects. In: Adaptive Hardware and Systems AHS 2010, pp. 215–222. IEEE (2010)
58. Smith, G.L.: Model for delay faults based upon paths. In: International Test Conference 1985, pp. 342–351. IEEE Computer Society (1985)
59. Wang, X., Salmani, H., Tehranipoor, M., Plusquellic, J.F.: Hardware Trojan detection and isolation using current integration and localized current analysis. In: DFT 2008, pp. 87–95. IEEE Computer Society (2008)

Amortizing Randomness Complexity in Private Circuits

Sebastian Faust[1,2(✉)], Clara Paglialonga[1,2], and Tobias Schneider[1,3]

[1] Ruhr-Universität Bochum, Bochum, Germany
{sebastian.faust,clara.paglialonga,tobias.schneider-a7a}@rub.de
[2] Technische Universität Darmstadt, Darmstadt, Germany
[3] Université Catholique de Louvain, Louvain-la-Neuve, Belgium

Abstract. Cryptographic implementations are vulnerable to Side Channel Analysis (SCA), where an adversary exploits physical phenomena such as the power consumption to reveal sensitive information. One of the most widely studied countermeasures against SCA are masking schemes. A masking scheme randomizes intermediate values thereby making physical leakage from the device harder to exploit. Central to any masking scheme is the use of randomness, on which the security of any masked algorithm heavily relies. But since randomness is very costly to produce in practice, it is an important question whether we can reduce the amount of randomness needed while still guaranteeing standard security properties such as t-probing security introduced by Ishai, Sahai and Wagner (CRYPTO 2003). In this work we study the question whether internal randomness can be re-used by several gadgets, thereby reducing the total amount of randomness needed. We provide new techniques for masking algorithms that significantly reduce the amount of randomness and achieve better overall efficiency than known constructions for values of t that are most relevant for practical settings.

1 Introduction

Masking schemes are one of the most common countermeasures against physical side-channel attacks, and have been studied intensively in the last years by the cryptographic community (see, e.g., [7,9,10,12,15,17,18] and many more). Masking schemes prevent harmful physical side-channel leakage by concealing all sensitive information by encoding the computation carried out on the device. The most widely studied masking scheme is the Boolean masking [7,15], which encodes each intermediate value produced by the computation using an n-out-of-n secret sharing. That is, a bit b is mapped to a bit string $(b_1, \ldots, b_n)$ such that b_i is random subject to the constraint that $\sum_i b_i = b$ (where the sum is taken in the binary field). To mask computation, the designer of a masking scheme then has to develop masked operations (so-called gadgets) that enable to compute with encodings in a secure way. The security of masking schemes is typically analyzed by carrying out a security proof in the t-probing model [15], where an adversary that learns up to t intermediate values gains no information about the underlying encoded secret values.

T. Takagi and T. Peyrin (Eds.): ASIACRYPT 2017, Part I, LNCS 10624, pp. 781–810, 2017.
https://doi.org/10.1007/978-3-319-70694-8_27

While due to the linearity of the encoding function protecting linear operations is easy, the main challenge is to develop secure masked non-linear operations, and in particular a masked version of the multiplication operation. To this end, the masked multiplication algorithm internally requires additional randomness to securely carry out the non-linear operation in the masked domain. Indeed, it was shown by Belaid et al. [4] that any t-probing secure masked multiplication requires internally $O(t)$ fresh randomness. Notice that complex cryptographic algorithms typically consists of many non-linear operations that need to be masked, and hence the amount of randomness needed to protect the entire computation grows not only with the probing parameter t, but also with the number of operations that are used by the algorithm. Concretely, the most common schemes for masking the non-linear operation require $O(t^2)$ randoms, and algorithms such as a masked AES typically require hundreds of masked multiplication operations.

Unfortunately, the generation and usage of randomness is very costly in practice, and typically requires to run a TRNG or PRNG. In fact, generating the randomness and shipping it to the place where it is needed is one of the main challenge when masking schemes are implemented in practice. There are two possibilities in which we can save randomness when masking algorithms. The first method is in spirit of the work of Belaid et al. [4] who design masked non-linear operations that require less randomness. However, as discussed above there are natural lower bounds on the amount of randomness needed to securely mask the non-linear operation (in fact, the best known efficient masked multiplication still requires $O(t^2)$ randomness). Moreover, such an approach does not scale well, when the number of non-linear operations increases. Indeed, in most practical cases the security parameter t is relatively small (typically less than 10), while most relevant cryptographic algorithms require many non-linear operations. An alternative approach is to amortize randomness by re-using it over several masked operations. This is the approach that we explore in this work, which despite being a promising approach has gained only very little attention in the literature so far.

On amortizing randomness. At first sight, it may seem simple to let masked operations share the same randomness. However, there are two technical challenges that need to be addressed to make this idea work. First, we need to ensure that when randomness is re-used between multiple operations it does not cancel out accidentally during the masked computation. As an illustrative example suppose two secret bits a and b are masked using the same randomness r. That is, a is encoded as $(a + r, r)$ and b is encoded as $(b + r, r)$ (these may, for instance, be outputs of a masked multiplication). Now, if at some point during the computation the algorithm computes the sum of these two encodings, then the randomness cancels out, and the sensitive information $a + b$ can be attacked (i.e., it is not protect by any random mask). While this issue already occurs when $t = 1$, i.e., the adversary only learns one intermediate value, the situation gets much more complex when t grows and we want to reduce randomness between multiple masked operations. In this case, we must guarantee that the computation happening in the algorithm does not cancel out the randomness, but also we need to ensure that any set of t intermediate values produced by the masked algorithm does not allow the adversary to cancel out the (potentially shared) randomness. Our main contribution is to initiate the study of masking schemes where multiple gadgets share randomness, and show that despite the above challenges amortizing randomness over multiple operations is possible and can lead in certain cases to significantly more efficient masked schemes. We provide a more detailed description of our main contributions in the next section.

1.1 Our Contributions

Re-using randomness for $t > 1$. We start by considering the more challenging case when $t > 1$, i.e., when the adversary is allowed to learn multiple intermediate values. As a first contribution we propose a new security notion of gadgets that we call t–SCR which allows multiple gadgets (or blocks of gadgets) to securely re-use randomness. We provide a composition result for our new notion and show sufficient requirements for constructing gadgets that satisfy our new notion. To this end, we rely on ideas that have been introduced in the context of threshold implementations [6].

Finding blocks of gadgets for re-using randomness. Our technique for sharing randomness between multiple gadgets requires to structure a potentially complex algorithm into so-called blocks, where the individual gadgets in these blocks share their randomness. We devise a simple tool that depending on the structure of the algorithm identifies blocks which can securely share randomness. Our tool follows a naive brute-force approach, and we leave it as an important question for future work to develop more efficient tools for identifying blocks of gadgets that are suitable for re-using randomness.

Re-using randomness for $t = 1$. We design a new scheme that achieves security against one adversarial probe and requires only 2 randoms for arbitrary complex masked algorithms. Notice that since randomness can cancel out when it is re-used such a scheme needs to be designed with care, and the security analysis cannot rely on a compositional approach such as the 1-SNI property [2].[1] Additionally, we provide a counterexample that securing arbitrary computation with only one random is not possible if one aims for a general countermeasure.

Implementation results. We finally complete our analysis with a case study by applying our new countermeasures to masking the AES algorithm. Our analysis shows that for orders up to $t = 5$ (resp. $t = 7$ for a less efficient TRNG) we can not only significantly reduce the amount of randomness needed, but also improve on efficiency. We also argue that if we could not use a dedicated TRNG (which would be the case for most inexpensive embedded devices), then

[1] The compositional approach of Barthe et al. [2] requires that all gadgets use independent randomness.

our new countermeasure would outperform state-of-the-art solutions even up to $t = 21$. We leave it as an important question for future research to design efficient masking schemes with shared randomness when $t > 21$.

1.2 Related Work

Despite being a major practical bottleneck, there has been surprisingly little work on minimizing the amount of randomness in masking schemes. We already mentioned the work of Belaid et al. [4], which aim on reducing the amount of randoms needed in a masked multiplication. Besides giving lower bounds on the minimal amount needed to protect a masked multiplication, the authors also give new constructions that reduce the concrete amount of randomness needed for a masked multiplication. However, the best known construction still requires randomness that is quadratic in the security parameter. Another approach for reducing the randomness complexity of first-order threshold implementations of Keccak was also investigated in [5].

From a practical point of view, the concept of "recycled" randomness was briefly explored in [1]. The authors practically evaluated the influence of reusing some of the masks on their case studies and concluded that in some cases the security was reduced. However, these results do not negatively reflect on our methodology as their reuse of randomness lacked a formal proof of security.

From a theoretical point of view it is known that any circuit can be masked using polynomial in t randoms (and hence the amount of randoms needed is independent from the size of the algorithm that we want to protect). This question was studied by Ishai et al. [14]. The constructions proposed in these works rely on bipartite expander graphs and are mainly of interests as feasibility results (i.e., they become meaningful when t is very large), while in our work we focus on the practically more relevant case when t takes small values.

Finally, we want to conclude by mentioning that while re-using randoms is not a problem for showing security in the t-probing model, and hence for security with respect to standard side-channel attacks, it may result in schemes that are easier to attack by so-called horizontal attacks [3]. Our work opens up new research directions for exploring such new attack vectors.

2 Preliminaries

In this section we recall basic security notions and models that we consider in this work. In the following we will use bold and lower case to indicate vectors and bold and upper case for matrices.

2.1 Private Circuits

The concept of *private circuits* was introduced in the seminal work of Ishai et al. [15]. We start by giving the definition of *deterministic* and *randomized circuit*, as provided by Ishai et al. A *deterministic circuit* C is a direct acyclic graph whose vertices are Boolean gates and whose edges are wires. A *randomized circuit* is a circuit augmented with random-bit gates. A random-bit gate is a gate with fan-in 0 that produces a random bit and sends it along its output wire; the bit is selected uniformly and independently. As pointed out in [14], a *t-private circuit* is a randomized circuit which transforms a randomly encoded input into a randomly encoded output while providing the guarantee that the joint values of any t wires reveal nothing about the input. More formally a *private circuit* is defined as follows.

Definition 1 (Private circuit [14]). *A* private circuit *for* $f : \mathbb{F}_2^{m_i} \rightarrow \mathbb{F}_2^{m_o}$ *is defined by a triple* (I, C, O)*, where*

- $I : \mathbb{F}_2^{m_i} \rightarrow \mathbb{F}_2^{\hat{m}_i}$ *is a randomized input encoder;*
- C *is a randomized Boolean circuit with input in* $\mathbb{F}_2^{\hat{m}_i}$*, output in* $\mathbb{F}_2^{\hat{m}_o}$ *and uniform randomness* $r \in \mathbb{F}_2^n$*;*
- $O : \mathbb{F}_2^{\hat{m}_o} \rightarrow \mathbb{F}_2^{m_o}$ *is an output decoder.*

C is said to be a t-private implementation of f with encoder I and decoder O if the following requirements hold:

- *Correctness*: For any input $w \in \mathbb{F}_2^{m_i}$ we have $\Pr[O(C(I(w), \rho)) = f(w)] = 1$, where the probability is over the randomness of I and ρ;
- *Privacy*: For any $w, w' \in \mathbb{F}_{m_i}$ and any set $\mathcal{P}$ of t wires (also called *probes*) in C, the distributions $C_{\mathcal{P}}(I(w), \rho)$ and $C_{\mathcal{P}}(I(w'), \rho)$ are identical, where $C_{\mathcal{P}}$ denotes the set of t values on the wires from $\mathcal{P}$ (also called *intermediate values*).

The goal of a *t-limited attacker*, i.e. an attacker who can probe at most t wires, is then to find a set of probes $\mathcal{P}$ and two values $w, w' \in \mathbb{F}_2^{m_i}$ such that the distributions $C_{\mathcal{P}}(I(w), \rho)$ and $C_{\mathcal{P}}(I(w'), \rho)$ are not the same.

Privacy of a circuit is defined by showing the existence of a *simulator*, which can simulate the adversary's observations without having access to any internal values of the circuit.

According to the description in [15], the input encoder I maps every input value x into n binary values $(r_1, \ldots, r_n)$ called *shares* or *mask*, where the first $n-1$ values are chosen at random and $r_n = x \oplus r_1 \oplus \cdots \oplus r_{n-1}$. On the other hand, the output decoder O takes the n bits $y_1, \ldots, y_n$ produced by the circuit and decodes the values in $y = y_1 \oplus \cdots \oplus y_n$. In its internal working a private circuit is composed by *gadgets*, namely transformed gates which perform functions which take as input a set of masked inputs and output a set of masked outputs. In particular, we distinguish between linear operations (e.g. XOR), which can be performed by applying the operation to each share separately, and non-linear functions (e.g. AND), which process all the shares together and make use of additional random bits. A particular case of randomized gadget is the *refreshing* gadget, which takes as input the sharing of a value x and outputs randomized sharing of the same x. Another interesting gadget is the multiplicative one, which takes as input two values, say a and b shared in $(a_1, \ldots, a_n)$ and $(b_1, \ldots, b_n)$, and outputs a value c shared in $(c_1, \ldots, c_n)$ such that $\bigoplus_{i=1}^{n} c_i = a \cdot b$. We indicate in particular with $\mathsf{g}(x, \boldsymbol{r})$ a gadget which takes as input a value x and internally uses a vector $\boldsymbol{r}$ of random bits, where $\boldsymbol{r}$ is of the form $(\boldsymbol{r}_1, \ldots, \boldsymbol{r}_n)$ and each $\boldsymbol{r}_i$ is the vector of the random bits involved in the computation of the i-th output share. For example, referring to Algorithm 3, $\boldsymbol{r}_1$ is the vector (r_{13}, r_1, r_8, r_7). In the rest of the paper, if not needed otherwise, we will mainly specify a gadget with only its random component $\boldsymbol{r}$, so it will be indicated as $\mathsf{g}(\boldsymbol{r})$. Moreover we suppose that all the gadgets $\mathsf{g}(\boldsymbol{r})$ are such that every intermediate value used in the computation of the i-th output share contains only random bits in $\boldsymbol{r}_i$.

The following definitions and lemma from [2] formalize t-probing security with the notion of *t-Non Interference* and show that this is also equivalent to the concept of *simulatability*.

Definition 2 (($\mathcal{S}, \Omega$)-Simulatability, ($\mathcal{S}, \Omega$)-Non Interference). *Let* g *be a gadget with m inputs $(a^{(1)}, \ldots, a^{(m)})$ each composed by n shares and Ω be a set of t adversary's observations. Let $\mathcal{S} = (\mathcal{S}_1, \ldots, \mathcal{S}_m)$ be such that $\mathcal{S}_i \subseteq \{1, \ldots, n\}$ and $|\mathcal{S}_i| \leq t$ for all i.*

1. *The gadget* g *is called* $(\mathcal{S}, \Omega)$-simulatable *(or $(\mathcal{S}, \Omega)$–SIM) if there exists a simulator which, by using only $(a^{(1)}, \ldots, a^{(m)})_{|\mathcal{S}} = (a^{(1)}_{|\mathcal{S}_1}, \ldots, a^{(m)}_{|\mathcal{S}_m})$ can simulate the adversary's view, where $a^{(k)}_{|\mathcal{S}_j} := (a^{(k)}_i)_{i \in \mathcal{S}_j}$.*
2. *The gadget* g *is called* $(\mathcal{S}, \Omega)$-Non Interfering *(or $(\mathcal{S}, \Omega)$–NI) if for any $\boldsymbol{s}_0, \boldsymbol{s}_1 \in (\mathbb{F}_2^m)^n$ such that ${\boldsymbol{s}_0}_{|\mathcal{S}} = {\boldsymbol{s}_1}_{|\mathcal{S}}$ the adversary's views of* g *respectively on input $\boldsymbol{s}_0$ and $\boldsymbol{s}_1$ are identical, i.e.* $\mathsf{g}(\boldsymbol{s}_0)_{|\Omega} = \mathsf{g}(\boldsymbol{s}_1)_{|\Omega}$.

In the rest of the paper, when we will talk about *simulatability* of a gadget we will implicitly mean that for every observation set Ω with $|\Omega| \leq t$, where t is the security order, there exists a set $\mathcal{S}$ as in Definition 2 such that the gadget is $(\mathcal{S}, \Omega)$–SIM.

Lemma 1. *For every gadget* g *with m inputs, set $\mathcal{S} = (\mathcal{S}_1, \ldots, \mathcal{S}_m)$, with $\mathcal{S}_i \subseteq \{1, \ldots, n\}$ and $|\mathcal{S}_i| \leq t$, and observation set Ω, with $|\Omega| \leq t$,* g *is* $(\mathcal{S}, \Omega)$–SIM *if and only if* g *is* $(\mathcal{S}, \Omega)$–NI, *with respect to the same sets $(\mathcal{S}, \Omega)$.*

Definition 3 (t–NI). *A gadget* g *is* t-non-interfering *(t–NI) if and only if for every observation set Ω, with $|\Omega| \leq t$, there exists a set $\mathcal{S}$, with $|\mathcal{S}| \leq t$, such that* g *is* $(\mathcal{S}, \Omega)$–NI.

When applied to composed circuits, the definition of t–NI is not enough to guarantee the privacy of the entire circuit. Indeed, the notion of t–NI is not sufficient to argue about secure composition of gadgets. In [2], Barthe et al. introduced the notion of *t–Strong Non-Interference* (t–SNI), which allows for guaranteeing a secure composition of gadgets.

Definition 4 (t–Strong Non-Interference). *An algorithm $\mathcal{A}$ is t–*Strong Non-Interferent *(t–SNI) if and only if for any set of t_1 probes on intermediate values and every set of t_2 probes on output shares with $t_1 + t_2 \leq t$, the totality of the probes can be simulated by only t_1 shares of each input.*

Informally, it means that the simulator can simulate the adversary's view, using a number of shares of the inputs that is independent from the number of probed output wires. An example of t–SNI multiplication algorithm is the famous ISW scheme in Algorithm 1, introduced in [15] and proven to be t–SNI in [2], and a t–SNI refreshing scheme is Algorithm 2, introduced in [10] by Duc et al. and proven to be t–SNI by Barthe et al. in [2].

Algorithm 1. ISW multiplication algorithm with $n \geq 2$ shares.

Input: shares $(a_i)_{1 \leq i \leq n}$ and $(b_i)_{1 \leq i \leq n}$, such that $\bigoplus_i a_i = a$ and $\bigoplus_i b_i = b$.
Output: shares $(c_i)_{1 \leq i \leq n}$, such that $\bigoplus_i c_i = a \cdot b$.

```
for i = 1 to n do
  for j = i + 1 to n do
    r_{i,j} <-$- F_{2^n};
    r_{j,i} <- (r_{i,j} + a_i ⊙ b_j) + a_j · b_i;
  end for
end for
for i = 1 to n do
  c_i <- a_i · b_i + Σ_{j=1, j≠i}^{n} r_{i,j};
end for
```

Algorithm 2. Refreshing $\mathcal{R}$

Input: shares $(a_i)_{1 \leq i \leq n}$, such that $\bigoplus_i a_i = a$; random shares $(r_{ij})_{1 \leq i \leq n, i+1 \leq j \leq n}$.
Output: shares $(c_i)_{1 \leq i \leq n}$, such that $\bigoplus_i c_i = a$.

```
for i = 1 to n do
  c_i = a_i;
end for
for i = 1 to n do
  for j = i + 1 to n do
    c_i = c_i + r_{i,j}
    c_j = c_j - r_{i,j}
  end for
end for
```

As pointed out in [9,18], secure multiplication schemes, like ISW, require that the two masks in input are mutually *independent*. This condition is satisfied in two cases: when at least one of the two inputs is taken uniformly at random or when at least one of the two inputs is refreshed by means of a secure refreshing using completely fresh and independent randomness, as shown in Algorithm 2. In this paper, whenever we talk about independence of two inputs, we refer to the mutual independence of the masks, as specified above.

2.2 Threshold Implementation

As shown in [11,18], the probing model presented in the last section covers attacks such as the High Order Differential Power Analysis (HO-DPA) attack. The latter, introduced by Kocher et al. in [16], uses power consumption measurements of a device to extract sensitive information of processed operations. The following result from [6] specifies the relation between the order of a DPA attack and the one of a probing attack.

Lemma 2 [6]. *The attack order in a Higher-order DPA corresponds to the number of wires that are probed in the circuit (per unmasked bit).*

Threshold Implementation (TI) schemes are a t−order countermeasure against DPA attacks. It is based on secret sharing and multi party computation, and in addition takes into account physical effects such as glitches.

In order to implement a Boolean function $f : \mathbb{F}_2^{m_i} \to \mathbb{F}_2^{m_o}$, every input value x has to be split into n shares $(x_1, \dots, x_n)$ such that $x = x_1 \oplus \dots \oplus x_n$, using the same procedure seen in private circuits. We denote with $\boldsymbol{C}$ is the output distribution $f(\boldsymbol{X})$, where $\boldsymbol{X}$ is the distribution of the encoding of an input x. The function f is then shared in a vector of functions $f_1, \dots, f_n$, called component functions, which must satisfy the following properties:

1. **Correctness:** $f(x) = \bigoplus_{i=1}^{n} f_i(x_1, \dots, x_n)$.
2. $t-$ **Non-Completeness:** any combination of up to t component functions f_i of f must be independent of at least one input share x_i.
3. **Uniformity:** the probability $\Pr(\boldsymbol{C} = \boldsymbol{c} | c = \bigoplus_{i=1}^{n} c_i)$ is a fixed constant for every $\boldsymbol{c}$, where $\boldsymbol{c}$ denotes the vector of the output shares.

The last property requires that the distribution of the output is always a random sharing of the output, and can be easily satisfied by refreshing the output shares.

TI schemes are strongly related to private circuits. First, they solve a similar problem of formalizing privacy against a t-limited attacker and moreover, as shown in [17], the TI algorithm for multiplication is equivalent to the scheme proposed by ISW.

We additionally point out that the TI aforementioned properties imply simulatability of the circuit. Indeed, if a function f satisfies properties 1 and 2, then an adversary who probes t or fewer wires will get information from all but at least one input share. Therefore, the gadget g implementing such a function is t-NI and due to Lemma 1 is simulatable.

3 Probing Security with Common Randomness

In this section we analyze privacy of a particular set of gadgets $\mathsf{g}_1, \dots, \mathsf{g}_d$ having independent inputs, in which the random component is substituted by a set of bits $\boldsymbol{r} = (r_1, \dots, r_l)$ taken at random, but reused by each of the gadgets $\mathsf{g}_1, \dots, \mathsf{g}_d$. In particular, we introduce a new security definition, which formalizes the conditions needed in order to guarantee t-probing security in a situation where randomness is shared among the gadgets.

Definition 5 (t–SCR). *Let $\boldsymbol{r}$ be a set of random bits. We say that the gadgets $\mathsf{g}_1(\boldsymbol{r}),\ldots,\mathsf{g}_d(\boldsymbol{r})$ receiving each m inputs split into n shares are $t-$*secure with common randomness *(t–SCR) if*

1. *their inputs are mutually independent;*
2. *for each set $\mathcal{P}_i$ of t_i probes on g_i such that $\sum_i t_i \leq t$, the probes in $\mathcal{P}_i$ can be simulated by at most $n-1$ shares of the input of g_i and the simulation is consistent with the shared random component.*

Let us introduce some notation that we will use in the rest of the paper. With the term *block of gadgets* we define a sub-circuit composed by gadgets, with input an encoding of a certain x and output an encoding of y. Since our analysis focuses on the randomness, when we refer to such a block we only consider the randomized gadgets. In particular, we indicate a block of gadgets as $\mathcal{G}(\boldsymbol{R}) = \{\mathsf{g}_1(\boldsymbol{r_1}),\ldots,\mathsf{g}_d(\boldsymbol{r_d})\}$, where the g_i represent the randomized gadgets in the block and $\boldsymbol{R} = (\boldsymbol{r_1},\ldots,\boldsymbol{r_d})$ constitutes the total amount of randomness used by $\mathcal{G}$. We assume without loss of generality that the input of such a $\mathcal{G}$ is the input of the first randomized gadget g_1. Indeed, even if actually the first gadget of the block was a non-randomized one (i.e. a linear gadget), then this would change the actual value of the input, but not its properties related to the independence. We call *dimension* of a block $\mathcal{G}$ the number of randomized gadgets g_i composing the block. In Fig. 1 are represented N blocks of gadgets of dimension 4 each.

The following lemma gives a simple compositional result for multiple blocks of gadgets, where each such block uses the same random component $\boldsymbol{R}$. Slightly informally speaking, let $\mathcal{G}_j$ be multiple sets of gadgets, where all gadgets in $\mathcal{G}_j$ share the same randomness. Then, the lemma below shows that if the gadgets in $\mathcal{G}_j$ are t–SCR, then also the composition of the gadgets in all sets $\mathcal{G}_j$ are t–SCR. We underline that such a block constitutes itself a gadget. For simplicity, we assume that the blocks of gadgets that we consider in the lemma below all have the same dimension d. But our analysis can easily be generalized to a setting where each block has a different dimension.

Lemma 3 (composition of t–SCR gadgets). *For every $d \in \mathbb{N}$, consider $\mathcal{G}_1(\boldsymbol{R}) = \{\mathsf{g}_{1,1}(\boldsymbol{r_1}),\ldots,\mathsf{g}_{1,d}(\boldsymbol{r_d})\},\ldots,\mathcal{G}_N(\boldsymbol{R}) = \{\mathsf{g}_{N,1}(\boldsymbol{r_1}),\ldots,\mathsf{g}_{N,d}(\boldsymbol{r_d})\}$ N blocks of gadgets sharing the same random component $\boldsymbol{R} = (\boldsymbol{r_1},\ldots,\boldsymbol{r_d})$ and masking their input into n shares. Suppose $\mathcal{G}_i$ be t–NI for each $i = 1,\ldots,N$. If for all $j = 1\ldots,d$ the gadgets $\mathsf{g}_{1,j}(\boldsymbol{r_j}),\ldots,g_{N,j}(\boldsymbol{r_j})$ are t–SCR, then the blocks of gadgets $\{\mathcal{G}_1,\ldots,\mathcal{G}_N\}$ are t–SCR.*

Proof. First it is easy to see that, since $\mathsf{g}_{1,1},\ldots,g_{N,1}$ are t–SCR then their inputs have independent masks and so the same holds for the inputs of blocks $\mathcal{G}_1,\ldots,\mathcal{G}_N$. Let us next discuss the second property given in the t–SCR definition. We can prove the statement with an inductive argument on the dimension of the blocks.

- If $d = 1$, then by hypothesis $\{\mathsf{g}_{1,1},\ldots,g_{N,1}\}$ are t–SCR and then $\{\mathcal{G}_1,\ldots,\mathcal{G}_N\}$ are t–SCR.

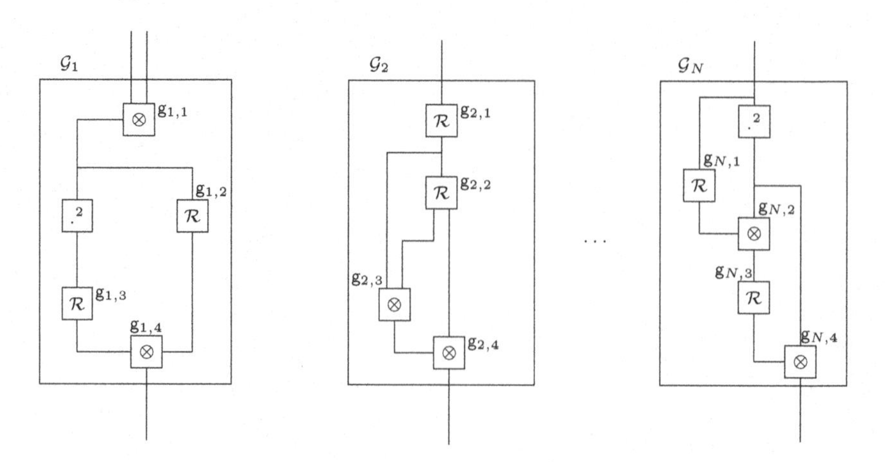

Fig. 1. A set of N blocks of gadgets with dimension $d = 4$ each.

– If $d > 1$ and $\{\{g_{1,1}, \dots, g_{1,d-1}\}, \dots, \{g_{N,1}, \dots, g_{N,d-1}\}\}$ are t–SCR, then by hypothesis $\{g_{1,d}, \dots, g_{N,d}\}$ are t–SCR. Now the following cases hold.
  - The probes are placed on the $\{\{g_{1,1}, \dots, g_{1,d-1}\}, \dots, \{g_{N,1}, \dots, g_{N,d-1}\}\}$: in this case, by the inductive hypothesis, the adversary's view is simulatable in the sense of Definition 5 of t–SCR.
  - The probes are placed on $\{g_{1,d}, \dots, g_{N,d}\}$: in this case, since by hypothesis $\{g_{1,d}, \dots, g_{N,d}\}$ are t–SCR, the adversary's view is simulatable in the sense of Definition 5.
  - A set of the probes $\mathcal{P}$ is placed on $\{g_{1,d}, \dots, g_{N,d}\}$ and a set of probes $\mathcal{Q}$ is placed on $\{\{g_{1,1}, \dots, g_{1,d-1}\}, \dots, \{g_{N,1}, \dots, g_{N,d-1}\}\}$: in this case, since the probes in $\mathcal{P}$ and in $\mathcal{Q}$ use different random bits, they can be simulated independently each other. The simulatability of the probes in $\mathcal{P}$ according to Definition 5 is guaranteed by the t–SCR of $\{g_{1,d}, \dots, g_{N,d}\}$ and the simulatability of the probes in $\mathcal{Q}$ is guaranteed by the t–SCR of $\{\{g_{1,1}, \dots, g_{1,d-1}\}, \dots, \{g_{N,1}, \dots, g_{N,d-1}\}\}$.

Therefore for the inductive step we conclude that for every dimension d of the blocks $\mathcal{G}_i$, with $i = 1, \dots, N$, the set $\{\mathcal{G}_1, \dots, \mathcal{G}_N\}$ is t–SCR. □

We point out that the t–SCR property itself is not sufficient for guaranteeing also a sound composition. The reason for this is that t–SCR essentially is only t–NI. Therefore, when used in combination with other gadgets, a t–SCR scheme needs additionally to satisfy the t–SNI property. We summarize this observation in the following theorem which gives a global result for circuits designed in blocks of gadgets sharing the same randomness.

Theorem 1. *Let C be a circuit composed by N blocks of gadgets $\mathcal{G}_1(\boldsymbol{R}), \dots, \mathcal{G}_N(\boldsymbol{R})$ where $\mathcal{G}_i(\boldsymbol{R}) = \{g_{i,1}(\boldsymbol{r_1}), \dots, g_{i,d}(\boldsymbol{r_d})\}$ for each $i = 1, \dots, N$ and with inputs masked with n shares and such that the gadgets outside such blocks are either linear or t–SNI ones. If*

– *the outputs of* $\mathcal{G}_1, \ldots, \mathcal{G}_N$ *are independent*
– $\forall j = 1, \ldots, N \quad \mathcal{G}_j$ *is* t–SNI *and*
– $\forall j = 1, \ldots, d \quad \mathsf{g}_{1j}, \ldots, g_{Nj}$ *are* t–SCR

then the circuit $\mathcal{C}$ *is* $t-$*probing secure.*

Proof. The proof of the theorem is straightforward. Indeed, Lemma 3 implies that $\mathcal{G}_1, \ldots, \mathcal{G}_N$ are t–SCR. Moreover, we point out that the t–SNI of the $\mathcal{G}_i$, for every $i = 1, \ldots, N$, and the independence of the outputs guarantees a secure composition

– among the blocks $\mathcal{G}_i$
– of the $\mathcal{G}_i$ with other randomized and t–SNI gadgets using fresh randomness
– of the $\mathcal{G}_i$ with linear gadgets.

This is sufficient to prove that the circuit $\mathcal{C}$ is t probing secure. □

To sum up, we showed in this section that, under certain conditions, it is possible to design a circuit which internally reuses the random bits involved and remains probing secure. Therefore, if used in an appropriate way, this result allows us to decrease the amount of randomness necessary in order to have a private circuit (because all the blocks share the same randomness). Nevertheless, we remark that, when designing such circuits, even if on the one hand the randomness involved in the gadgets can be completely reused, we require on the other hand additional refreshing schemes to guarantee the independence of the inputs and outputs of each block. Notice that independence is needed for ensuring t–SCR and, as recalled in Sect. 2.1, it is satisfied by refreshing via Algorithm 2.

For these reasons, in order to have an actual reduction in the amount of randomness, it is needed to take a couple of precautions when structuring a circuit into blocks of gadgets. First of all, it is necessary to construct these blocks such that the number of the outputs which are inputs of other blocks do not exceed the number of gadgets in the block; otherwise we would require more randomness for refreshing than what was saved by the reusing of randomness within the block. In addition, it is important to find a good trade-off between the dimension of the blocks and the number of them in the circuit.

More formally speaking if N is the number of randomized gadgets of the original circuit, N_{C} is the number of gadgets which use the same random bits in the restructured circuit and N_{R} is the number of new refreshing schemes that we need to add to it for guaranteeing the independence of the inputs of the blocks, then the total saving in the randomness of the circuit is given by the difference $N-(N_{\mathrm{C}}+N_{\mathrm{R}})$. To illustrate how this quantity changes according to the different dimension of the blocks let us take a look at some concrete cases. Suppose for simplicity that each block of gadget has only one input and one output. If we divide the circuit into many small blocks, then on the one hand we reuse a small amount of randomness, and so N_{C} is smaller, on the other hand, since at every block corresponds one output which needs to be refreshed before being input of another block, the number of new randomness involved increases, and then

N_R is bigger. Otherwise, if the circuit is designed in few large blocks of gadgets, then since we have fewer blocks, there are also fewer outputs to be refreshed, therefore the amount of fresh randomness N_R is reduced. On the other hand, more random bits are needed for refreshing for the common randomness in the blocks, and so the amount N_C increases. A more concrete example can be found in Figs. 2, 3 and 4, where the same circuit is structured in blocks of gadgets in two different ways.

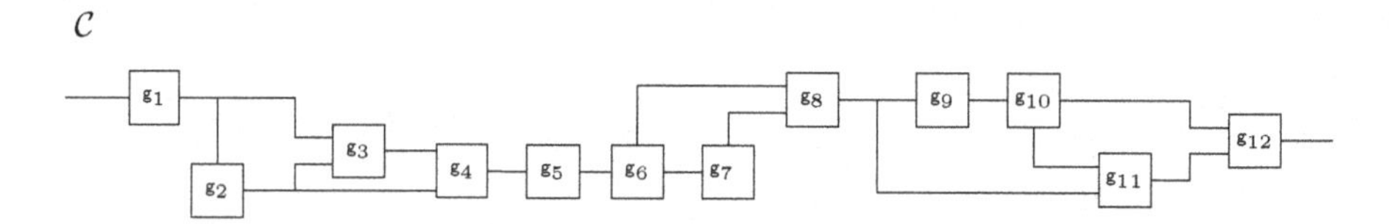

Fig. 2. The original circuit $\mathcal{C}$ composed by $N = 12$ randomized gadgets.

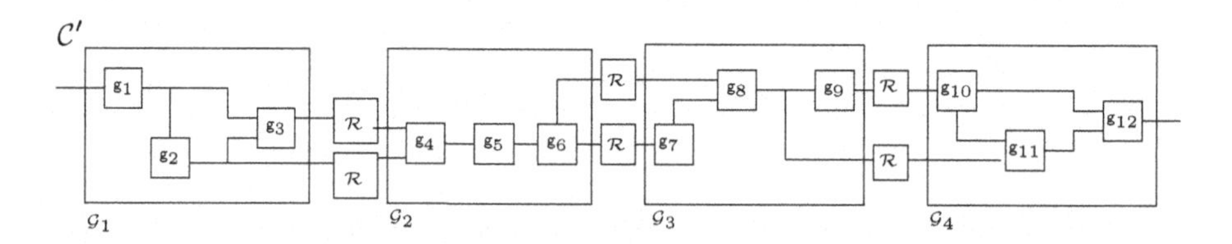

Fig. 3. The circuit $\mathcal{C}'$ representing $\mathcal{C}$ structured into 4 blocks of gadgets, where $N = 12, N_C = 3, N_R = 6$ and the saving consists of 3 randomized gadgets.

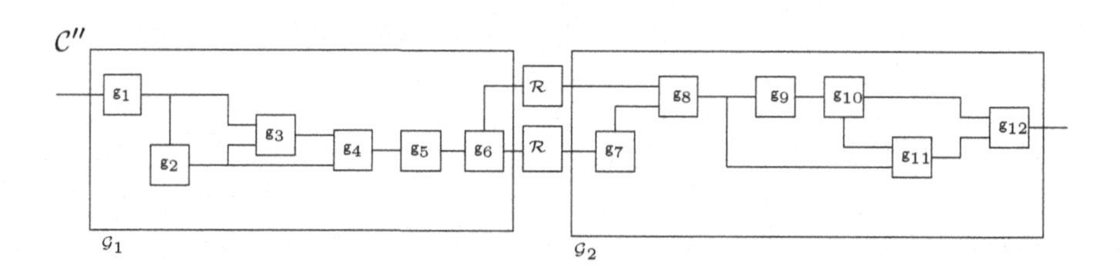

Fig. 4. The circuit $\mathcal{C}''$ representing $\mathcal{C}$ structured into 2 blocks of gadgets, where $N = 12, N_C = 6, N_R = 2$ and the saving consists of 4 randomized gadgets.

In Sect. 4, we will present a naive method to restructure a circuit in such a way that these conditions are satisfied and in order to find an efficient grouping in blocks of gadgets.

3.1 A t-SCR Multiplication Scheme

In this subsection, we introduce a multiplication scheme, which can be combined with other gadgets sharing the same randomness and remains t–SCR. In particular, our multiplication schemes are based on two basic properties (i.e., $\lfloor \frac{t}{2} \rfloor$-non-completeness and t–SNI) and we discuss how to construct instantiations of our multiplication according to these properties.

First, we construct a multiplication scheme in accordance with $\lfloor \frac{t}{2} \rfloor$-non-completeness. This process is similar to finding a $\lfloor \frac{t}{2} \rfloor$-order TI of the AND-gate [17] or multiplication [8]. However, for our application we additionally require that the number of output shares is equal to the number of input shares. Most higher-order TI avoid this restriction with additional refreshing- and compression-layers. Since the $\lfloor \frac{t}{2} \rfloor$-non-completeness should be fulfilled without fresh randomness, we have to construct a $\lfloor \frac{t}{2} \rfloor$-non-complete $\mathsf{Mult} : \mathbb{F}_2^n \rightarrow \mathbb{F}_2^n$ and cannot rely on compression of the output shares. Unfortunately, this is only possible for very specific values of n. Due to this minor difference, we cannot directly use the bounds from the original publications related to higher-order TI. In the following, we derive an equation for n given an arbitrary t for which there exist a $\lfloor \frac{t}{2} \rfloor$-non-complete Mult.

Initially, due to the $\lfloor \frac{t}{2} \rfloor$-non-completeness the number of shares for which we can construct a scheme with the above properties is given by

$$\left\lfloor \frac{t}{2} \right\rfloor \cdot l + 1 = n \tag{1}$$

where l denotes the number of input shares which are leaked by each of the output shares, i.e., even the combination of $\lfloor \frac{t}{2} \rfloor$ output shares is still independent of one input share. To construct a $\lfloor \frac{t}{2} \rfloor$-non-complete multiplication, we need to distribute $\binom{n}{2}$ terms of the form $a_i b_j + a_j b_i$ over n output shares, i.e., each output share is made up of the sum of $\frac{n-1}{2}$ terms. Each of these terms leaks information about the tuples (a_i, a_j) and (b_i, b_j), and we assume the encodings a and b are independent and randomly chosen. For a given l, the maximum number of possible terms, which can be combined without leaking about more than l shares of a or b, is $\binom{l}{2}$. The remaining $a_i b_i$ are equally distributed over the output shares without increasing l. By combining these two observations, we derive the relation

$$\frac{n-1}{2} = \frac{l^2 - l}{2}. \tag{2}$$

Based on Eq. (1), the minimum number of shares for $\lfloor \frac{t}{2} \rfloor$-non-completeness is $n = \lfloor \frac{t}{2} \rfloor \cdot l + 1$. We combine this with Eq. (2) and derive

$$n = \left\lfloor \frac{t}{2} \right\rfloor^2 + \left\lfloor \frac{t}{2} \right\rfloor + 1. \tag{3}$$

We use Eq. (3) to compute the number of shares for our t–SCR multiplication scheme with $t > 3$. For $t \leq 3$, the number of shares is bounded by the requirement for the multiplication to be t–SNI, i.e., $n > t$.

To achieve t–SCR, it is necessary to include randomness in the multiplications. Initially, $\frac{tn}{2}$ random components r_i need to be added for the multiplication to be t–SNI. A subset of t random components is added to each output share equally distributed over the sum, and each of these random bits is involved a second time in the computation of a single different output share. This ensures the simulatability of the gadget by using a limited number of input shares as required by the definition of t–SNI. In particular, the clever distribution of the random bits allows to simulate the output probes with a random and independent value. Furthermore, we include additional random elements $rx_{j=1,\ldots,n}$ which only occur in one output share each and enable a simple simulation of the gadget even in the presence of shared randomness.

The construction of a t–SCR multiplication scheme following the aforementioned guidelines is easy for small t. However, finding a distribution of terms that fulfils $\lfloor \frac{t}{2} \rfloor$-non-completeness becomes a complex task due to the large number of possible combinations for increasing t. For $t = 4$, one possible t–SCR multiplication is defined in Algorithm 3 and it requires $n = 7$ shares. A complete description of a multiplication algorithm for higher orders fulfilling the properties aforementioned can be found in the full version of the paper.

Algorithm 3. Mult for order $t = 4$ with $n = 7$ shares.

Input: shares $a_1, \ldots, a_7$ such that $\bigoplus a_i = a$, shares $b_1, \ldots, b_7$ such that $\bigoplus b_i = b$
Output: shares $c_1, \ldots, c_7$ such that $\bigoplus c_i = a \cdot b$

$c_1 = ((((((((((((rx_1 + a_1b_1) + r_{13}) + a_1b_2) + a_2b_1) + r_1) + a_1b_3) + a_3b_1) + r_8) + a_2b_3) + a_3b_2) + r_7) + rx_1);$

$c_2 = ((((((((((((rx_2 + a_4b_4) + r_{14}) + a_1b_4) + a_4b_1) + r_2) + a_1b_5) + a_5b_1) + r_9) + a_4b_5) + a_5b_4) + r_1) + rx_2);$

$c_3 = ((((((((((((rx_3 + a_7b_7) + r_8) + a_1b_6) + a_6b_1) + r_3) + a_1b_7) + a_7b_1) + r_{10}) + a_6b_7) + a_7b_6) + r_2) + rx_3);$

$c_4 = ((((((((((((rx_4 + a_2b_2) + r_9) + a_2b_4) + a_4b_2) + r_4) + a_2b_6) + a_6b_2) + r_{11}) + a_4b_6) + a_6b_4) + r_3) + rx_4);$

$c_5 = ((((((((((((rx_5 + a_5b_5) + r_{10}) + a_2b_5) + a_5b_2) + r_5) + a_2b_7) + a_7b_2) + r_{12}) + a_5b_7) + a_7b_5) + r_4) + rx_5);$

$c_6 = ((((((((((((rx_6 + a_3b_3) + r_{11}) + a_3b_4) + a_4b_3) + r_6) + a_3b_7) + a_7b_3) + r_{13}) + a_4b_7) + a_7b_4) + r_5) + rx_6);$

$c_7 = ((((((((((((rx_7 + a_6b_6) + r_{12}) + a_3b_5) + a_5b_3) + r_7) + a_3b_6) + a_6b_3) + r_{14}) + a_5b_6) + a_6b_5) + r_6) + rx_7);$

Now we present the security analysis of this multiplication scheme and we show that it can be securely composed with the refreshing scheme in Algorithm 2 in blocks of gadgets sharing the same random component. Due to size constraints, we only give a sketch of the proof and refer to the full version of the paper for the complete proof.

Lemma 4. *Let* $\mathsf{Mult}_1, \ldots, \mathsf{Mult}_N$ *be a set of* N *multiplication schemes as in Algorithm 3, with outputs* $\mathbf{c}^{(1)}, \ldots, \mathbf{c}^{(N)}$*. Suppose that the maskings of the inputs* *are independent and uniformly chosen and that for* $k = 1, \dots, N$ *each* Mult_k *uses the same random bits* $(r_i)_{i=1,\dots,tn/2}$. *Then* $\mathsf{Mult}_1, \dots, \mathsf{Mult}_N$ *are* t–SCR *and in particular* Mult *is* t–SNI.

Proof. In the first case, all probes are placed in the same Mult and it is sufficient to show t–SNI of Mult. We indicate with $p_{l,m}$ the m-th sum of the output c_l. We can classify the probes in the following groups.

(1) $a_i b_j + r_k =: p_{l,1}$
(2) $a_i, b_j, a_i b_j$
(3) r_k
(4) $p_{l,m} + a_i b_j =: q$
(5) $p_{l,m} + r_k =: s$
(6) output shares c_i

Suppose an adversary corrupts at most t wires $w_1, \dots, w_t$. We define two sets I, J with $|I| < n$ $|J| < n$ such that the values of the wires w_h can be perfectly simulated given the values $(a_i)_{i \in I}$, $(b_i)_{i \in J}$.

The procedure to construct the sets is the following:

1. We first define a set K such that for all the probes containing a random bit r_k, we add k to K.
2. Initially I, J are empty and the w_i unassigned.
3. For every wire in the group (1), (2), (4) and (5) add i to I and j to J.

Now we simulate the wires w_h using only the values $(a_i)_{i \in I}$ and $(b_i)_{i \in J}$.

- For every probe in group (2), then $i \in I$ and $i \in J$ and the values are perfectly simulated.
- For every probe in group (3), r_k can be simulated as a random and independent value.
- For every probe in group (1), if $k \notin K$, we can assign a random independent value to the probe, otherwise, if r_k has already been simulated we can simulate the probe by taking the r_k previously simulated, simulating the shares of a and b by using the needed indices in I and J and performing the inner products and additions as in the real execution of the algorithm.
  For every probe in group (4) if $p_{l,m}$ was already probed, we can compute q by using $p_{l,m}$ and the needed indices of a and b in I and J. Otherwise, we can pick q as a uniform and random value.
- For every probe in group (5), if $p_{l,m}$ was already probed and $k \in K$, we can compute s by using $p_{l,m}$ and the already simulated r_k. Otherwise, we can pick s as a uniform and random value.

Finally, we simulate the output wires c_i in group (6) using only a number of input shares smaller or equal to the number of internal probes. We have to take into account two cases.

- If the attacker has already observed a partial value of the output shares, we note that by construction, independently of the intermediate elements probed, at least one of the r_k does not enter into the computation of the probed internal values and so $\boldsymbol{c}_i$ can be simulated as a random value.
- If the adversary has observed all the partial sums of c_i, then, since these probes have been previously simulated, the simulator now add these simulated values for reconstructing the c_i.
- If no partial value fo c_i has been probed. By definition, at least one of the r_k involved in the computation of $\boldsymbol{c}_i$ is not used in any other observed wire. Therefore, $\boldsymbol{c}_i$ can be assigned to a random and independent value.

In the second case, the probes are placed into different Mult_i. However, the number of probes in one particular gadget does not exceed $\lfloor \frac{t}{2} \rfloor$. In this case, security is given by the $\lfloor \frac{t}{2} \rfloor$-non-completeness property of our multiplication schemes.

In the third case, the number of probes for one Mult_i does exceed $\lfloor \frac{t}{2} \rfloor$. For this, we base our proof strategy on the two observations. First, since all Mult_i share the same randomness, it is possible to probe the same final output share c_i in two gadgets to remove all random elements and get information about all the input shares used in the computation of c_i. Secondly, a probe in any intermediate sum of c_i is randomized by rx_i. Therefore, this probe can always be simulated as uniform random if not another probe is placed on rx_i or on a different intermediate sum of c_i (including in a different Mult_j). Therefore, any probe of an intermediate sum of c_i can be reduced to a probe of the final output share c_i, since in the latter case one receives information about more or an equal number of input shares with the same number of probes (i.e., two). Therefore, to get information about the maximum number of input shares the probes need to be placed in the same $\lfloor \frac{t}{2} \rfloor$ output shares in two multiplications. Based on the $\lfloor \frac{t}{2} \rfloor$-non-completeness, this can be easily simulated. The remaining probe, given that t is odd, can be simulated as uniform random, since it is either

- an intermediate sum of an unprobed output share c_k. This can be simulated as uniform random due to the unprobed rx_k.
- an unprobed output share c_k. This can be also simulated as uniform random, as by construction there is always at least one random element r_i which is not present in one of the $\lfloor \frac{t}{2} \rfloor$ probed output shares.

For the special case of $t < 4$, it is possible to avoid the extra rx_i per output share. This is based on the limited number of probes. For $t = 2$, 1-non-completeness (for the case of one probe in two multiplications) and t–SNI (for the case of two probes in one multiplication) are sufficient to enable t–SCR. The same applies to $t = 3$ as for the last probe there is always one unknown random r_i masking any required intermediate sum. □

In the following lemma we show that the t–SNI refreshing scheme in Algorithm 2 is also t–SCR. Due to size constraints, we again only provide a proof sketch and refer to the full version of the paper for the complete proof.

Lemma 5. *Let $\mathcal{R}_1, \ldots, \mathcal{R}_N$ be a set of N refreshing schemes, as in Algorithm 2, with inputs $\boldsymbol{a}^{(1)} \ldots, \boldsymbol{a}^{(N)}$ and outputs $\mathbf{c}^{(1)} \ldots, \mathbf{c}^{(N)}$. Suppose that $\left(a_i^{(1)}\right)_{i=1,\ldots,n}, \ldots, \left(a_i^{(N)}\right)_{i=1,\ldots,n}$ are independent and randomly chosen maskings of the input values and for $k = 1, \ldots, N$ each $\mathcal{R}_k$ uses the same random bits $\left(r_{i,j}\right)_{i,j=1,\ldots,n}$. Then $\mathcal{R}_1, \ldots, \mathcal{R}_N$ are t–*SCR.

Proof. Since according to Algorithm 2 every output share contains only one single share of the input and since the inputs are encoded in $n > t$ shares, it is not possible to probe all of the input shares of one $\mathcal{R}_i$ with t probes. Therefore, the simulation can be done easily. □

We remark that, due to the use of $n > t + 1$ shares in the multiplication algorithm for order $t > 3$, the refreshing scheme in Algorithm 2 makes use of a not optimal amount of randomness, since it requires $\frac{n^2}{2}$ random bits. We depict in Algorithm 4 a more efficient refreshing scheme which uses only $\frac{t \cdot n}{2}$ random bits. It essentially consist in multiplying the input value times 1, by means of Algorithm 3 as subroutine. It is easy to see that the security of the scheme relies on the security of the multiplication algorithm Mult, and therefore Algorithm 4 is t–SNI and it can securely share randomness with other multiplication gadgets.

Algorithm 4. Refreshing scheme with optimal amount of randomness

Input: shares $a_1, \ldots, a_n$ such that $\bigoplus a_i = a$
Output: shares $c_1, \ldots, c_n$ such that $\bigoplus c_i = a$

```
for i = 1 to n do
    u_i = 1;
end for
if n is even then
    u_n = 0;
end if
(c_1, ..., c_n) = Mult(a, (u_1, ..., u_n));
```

An example of blocks of gadgets using multiplication and refreshing schemes is given in Fig. 5, where are depicted two blocks of gadgets of dimension 6 involving the multiplication scheme Mult and the refreshing $\mathcal{R}$ of Algorithm 2 secure even if sharing the same randomness.

4 A Tool for General Circuits

The results from the previous sections essentially show that it is possible to transform a circuit $\mathcal{C}$ in another circuit $\mathcal{C}'$ performing the same operation, but using a reduced amount of randomness. To this end, according to Theorem 1, it is sufficient to group the gadgets composing the circuit $\mathcal{C}$ in blocks $\mathcal{G}_i$ sharing the same component of random bits and having independent inputs, i.e. values refreshed by Algorithm 2. As pointed out in Sect. 3, the actual efficiency of this

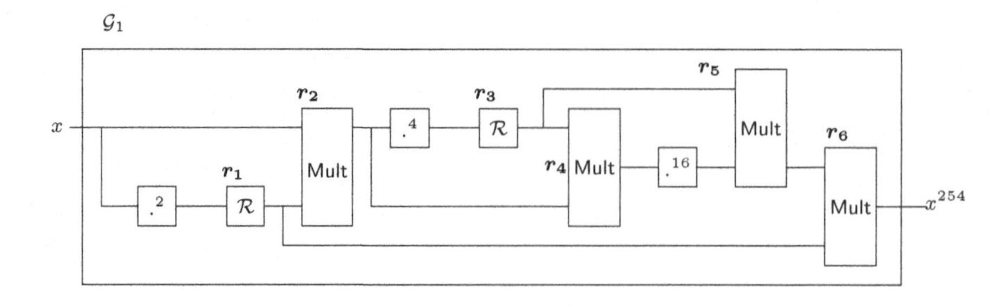

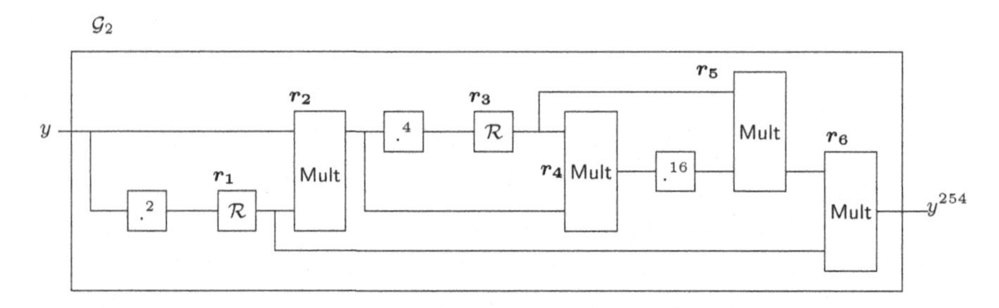

Fig. 5. Two blocks of gadgets $\mathcal{G}_1, \mathcal{G}_2$ composed by the same gadgets using the random components r_i with independent inputs x and y

procedure is not straightforward, but it is given by the right trade off between the dimension of the blocks and the number of extra refreshing schemes needed in order to guarantee the independence of their inputs.

In the following we give a tool, depicted in Algorithm 7, which allows to perform this partitioning and amortize the randomness complexity of a given circuit.

A circuit $\mathcal{C}$ is represented as a directed graph where the *nodes* constitute the randomized gadgets and the *edges* are input or output wires of the related gadget, according to the respective direction. In particular, if the same output wire is used as input several time in different gates, it is represented with a number of edges equivalent to the number of times it is used. The linear gates are not represented. The last node is assigned to the label "End" and every intersection node with parallel branches is marked as "Stop".

The idea at the basis of our algorithm is quite primitive. We empirically noticed that for a circuit composed by N randomized gadgets a balanced choice for the dimension of the blocks of gadgets can be the *central divisors* (d_1 and d_2 in the algorithms) of N, where if for instance $N = 12$ and then the vector of its divisors is $(1, 2, 3, 4, 6, 12)$, with *central divisors* we identify the values 3 and 4. Therefore, we aim at dividing the circuit in d_1 blocks of gadgets of dimension d_2 (and vice versa). We start taking the first d_1 nodes and we verify that the number of outputs do not exceed the one of randomized gadgets in the block.

Indeed, if it would be so, since each output needs to be refreshed before being input of another block, then the number of reused random bits is inferior to the one of new random bits which need to be refreshed. In case the condition is not verified the algorithm adds a new node, i.e. a new randomized gadget, to the block and check again the property, until it is verified. Then it takes again the next d_1 nodes and repeats the procedure. At last, we compare the saved randomness respectively when the algorithm tries to divide the circuit in d_1 blocks and in d_2 blocks and we output the transformed circuit with the best amortizing rate.

More technically, at first we give the subroutine in Algorithm 5, which chooses two divisors of a given integer. With $\boldsymbol{V}$ we indicate the vector composed by all the divisors of a given number N (which in the partitioning algorithm will be the number of the randomized gadgets of a circuit) and with $|\boldsymbol{V}|$ the length of $\boldsymbol{V}$, i.e. the number of its elements.

Algorithm 5. Divisors

Input: positive integer N
Output: divisors d_1 and d_2

$\boldsymbol{V} \leftarrow$ divisors of N (by look up table);
$n \leftarrow |\boldsymbol{V}|$;
if n is even **then**
  $i \leftarrow \frac{n}{2}$;
else
  $i \leftarrow \frac{n-1}{2}$;
end if
$d_1 \leftarrow \boldsymbol{V}[i]$;
$d_2 \leftarrow \boldsymbol{V}[i+1]$;
return d_1, d_2

Algorithm 6 constructs a block of gadgets $\mathcal{G}$ of dimension at least d, such that the number of extra refreshing needed does not exceed the number of randomized gadgets in the block. In the algorithm, the integers $m_o^{(j)}$ and $m_g^{(j)}$ represent respectively the number of output edges and the amount of nodes contained in the block of gates $\mathcal{G}_j$.

The procedure Partition in Algorithm 7 partitions a circuit $\mathcal{C}$ in sub-circuits $\mathcal{G}_i$ followed by a refreshing gate $\mathcal{R}$ per each output edge. In the algorithm, $\boldsymbol{O}$ and $\boldsymbol{M}$ are two vectors such that the j-th position represents respectively the number of output wires and the amount of nodes of the block $\mathcal{G}_j$. With $\mathcal{R}$ it is indicated the refreshing scheme of Algorithm 2. The integers n_R and n'_R count the total number of refreshing gadgets needed in the first and second partition of $\mathcal{C}$ respectively. The integers n_G and n'_G count the total number of gadgets (multiplications and refreshing) which need to refresh the random bits once in the circuit. The integers n_{TOT} and n'_{TOT} represent the total amount of randomness needed, computed as the number of gadgets which need fresh

Algorithm 6. FindBlock

Input: circuit $\mathcal{C}$ performing a function $f(x)$, first node v_j of the block, d
Output: block of gates $\mathcal{G}_i$, node v, integers $m_o^{(j)}, i$

$\mathcal{G}_j \leftarrow \{v_{j+1}, v_{j+2}, \ldots, v_{j+d}\}$;
$i \leftarrow 0$;
while $m_o^{(j)} \geq m_g^{(j)}$ **do**
  $i \leftarrow i + 1$;
  if $v_{j+d+i} \neq$ "Stop" **then**
    $\mathcal{G}_j \leftarrow \mathcal{G}_j \cup \{v_{j+d+i}\}$;
  else
    end while
    return $\mathcal{G}_j$, $v_{j+d+i}, m_o^{(j)}, i - 1$;
  end if
end while
return $\mathcal{G}_j$, $v_{j+d+i+1}, m_o^{(j)}, i$;

random bits once. By comparing these two values, the algorithm decides which is the best partition in terms of amortized randomness. In particular the notation $\boldsymbol{O}[i] \cdot \mathcal{R}$ means that the block $\mathcal{G}_i$ is followed by $\boldsymbol{O}[i]$ refreshing schemes (one per output edge).

We conclude this section by emphasizing that our algorithm is not designed to provide the optimal solution (as in finding the grouping which requires the least amount of randomness). Nevertheless, it can help to decompose an arbitrary circuit without a regular structure and serve as a starting point for further optimizations. However, for circuits with an obvious structure (e.g., layers for symmetric ciphers) which contain easily-exploitable regularities to group the gadgets, the optimal solutions can be usually found by hand.

5 1-Probing Security with Constant Amount of Randomness

The first order ISW scheme is not particularly expensive in terms of randomness, because it uses only one random bit. Unfortunately, when composed in more complicated circuits, the randomness involved increases with the size of the circuit, because we need fresh randomness for each gadget. Our idea is to avoid injecting new randomness in each multiplication and instead alternatively use the same random bits in all gadgets. In particular, we aim at providing a lower bound to the minimum number of bits needed in total to protect any circuit, and moreover show a matching upper bound, i.e., that it is possible to obtain a 1-probing secure private circuit, which uses only a constant amount of randomness. We emphasize that this means that the construction uses randomness that is *independent* of the circuit size, and in particular uses only 2 random bits in total per execution.

We will present a modified version of the usual gadgets for refreshing, multiplication and the linear ones, which, in place of injecting new randomness, use a

Algorithm 7. Partition

Input: circuit $\mathcal{C}$ performing a function $f(x)$, N total number of randomized gadgets
Output: circuit $\mathcal{C}'$ performing a function $f(x)$ with a reduced amount of randomness

$d_1, d_2 \leftarrow \mathsf{Divisors}(N)$;
$i \leftarrow 1$;
$\mathcal{G}_1, v, \boldsymbol{O}[1], \boldsymbol{M}[1] \leftarrow \mathsf{FindBlock}(\mathcal{C}, v_1, d_1)$;
while $v \neq$ "End" **do**
  $i \leftarrow i + 1$;
  $\mathcal{G}_i, v, \boldsymbol{O}[i], \boldsymbol{M}[i] \leftarrow \mathsf{FindBlock}(\mathcal{C}, v, d_1)$
end while
$n_R = \boldsymbol{O}[1] + \cdots + \boldsymbol{O}[i]$;
$n_G = \max(\boldsymbol{M}[1], \ldots, \boldsymbol{M}[i])$;
$n_{TOT} = n_R + n_G$;
$k \leftarrow 1$;
$\mathcal{G}'_1, v', \boldsymbol{O}'[1], \boldsymbol{M}'[1] \leftarrow \mathsf{FindBlock}(\mathcal{C}, v_1, d_2)$;
while $v \neq$ "End" **do**
  $k \leftarrow k + 1$;
  $\mathcal{G}'_i, v', \boldsymbol{O}'[k], \boldsymbol{M}'[k] \leftarrow \mathsf{FindBlock}(\mathcal{C}, v', d_2)$;
end while
$n'_R = \boldsymbol{O}'[1] + \cdots + \boldsymbol{O}'[k]$;
$n'_G = \max(\boldsymbol{M}'[1], \ldots, \boldsymbol{M}'[k])$;
$n'_{TOT} = n'_R + n'_G$;
if $n_{TOT} \leq n'_{TOT}$ **then**
  $\mathcal{C}' \leftarrow (\mathcal{G}_1, \boldsymbol{O}[1] \cdot \mathcal{R}, \ldots, \mathcal{G}_i, \boldsymbol{O}[i] \cdot \mathcal{R})$;
else
  $\mathcal{C}' \leftarrow (\mathcal{G}'_1, \boldsymbol{O}'[1] \cdot \mathcal{R}, \ldots, \mathcal{G}'_i, \boldsymbol{O}'[k] \cdot \mathcal{R})$;
end if
return $\mathcal{C}'$

value taken from a set of two bits chosen at the beginning of each evaluation of the masked algorithm. In particular, we will design these schemes such that they will produce outputs depending on at most one random bit and such that every value in the circuit will assume a fixed form. The most crucial change will be the one at the multiplication and refreshing schemes, which are the randomized gadgets, and so responsible for the accumulation of randomness. On the other hand, even tough the gadget for the addition does not use random bits, it will be subjected at some modifications as well, in order to avoid malicious situations that the reusing of the same random bits in the circuit can cause. As for the other linear gadgets, such as the powers $.^2$, $.^4$, etc., they will be not affected by any change, but will perform as usual share-wise computation.

We proceed by showing step by step the strategy to construct such circuits. First, we fix a set of bits $R = \{r_0, r_1\}$ where r_0 and r_1 are taken uniformly at random. The first randomized gadget of the circuit does not need to be substantially modified, because there is no accumulation of randomness to be avoided yet. The only difference with the usual multiplication and refreshing gadgets is that, in place of the random component, we need to use one of the random bits in R, as shown in Algorithms 8 and 9. Notice that when parts of the operations are written in parenthesizes, then this means that these operations are executed first.

Algorithm 8. 1-SecMult case (i)

Input: shares a_1, a_2 such that $a_1 \oplus a_2 = a$, shares b_1, b_2 such that $b_1 \oplus b_2 = b$
Output: shares c_i depending on a random number $r_k \in R$ such that $c_1 \oplus c_2 = a \cdot b$, the value r_k

$r_k \xleftarrow{\$} R$;
$c_1 \leftarrow a_1 b_1 + (a_1 b_2 + r_k)$;
$c_2 \leftarrow a_2 b_1 + (a_2 b_2 - r_k)$;

Algorithm 9. Refreshing case (i)

Input: shares a_1, a_2 such that $a_1 \oplus a_2 = a$
Output: shares c_i depending on the random number $r_k \in R$ such that $c_1 \oplus c_2 = a$, the value r_k

$r_k \xleftarrow{\$} R$;
$c_1 \leftarrow a_1 + r_k$;
$c_2 \leftarrow a_2 - r_k$;

Secondly, we analyze the different configurations that an element can take when not more than one randomized gadget has been executed, i.e. when only one random bit has been used in the circuit. The categories listed below are then the different forms that such an element takes if it is respectively the first input of the circuit, the output of the first refreshing scheme as in Algorithm 2 and the one of the first ISW multiplication scheme as in Algorithm 1 between two values x and y:

(1) $a = (a_1, a_2)$;
(2) $a = (a_1 + r, a_2 - r)$, where r is a random bit in R;
(3) $a = (x_1 y_1 + x_1 y_2 + r, x_2 y_1 + x_2 y_2 - r)$, where r is a random bit in R.

This categorization is important because according to the different form of the values that the second randomized gadget takes in input, the scheme will accumulate randomness in different ways. Therefore, we need to modify the gadgets by taking into account the various possibilities for the inputs, i.e. distinguish if:

(i) both the inputs are in category (1);
(ii) the first input is as in category (1), i.e. $a = (a_1, a_2)$, and the second one in category (2), i.e. $b = (b_1 + r_1, b_2 - r_1)$;
(iii) the first input is as in category (1), i.e. $a = (a_1, a_2)$, and the second one in category (3), i.e. $b = (c_1 d_1 + c_1 d_2 + r_1, c_2 d_1 + c_2 d_2 - r_1)$;
(iv) the first input is in category (3), i.e. $a = (c_1 d_1 + c_1 d_2 + r_0, c_2 d_1 + c_2 d_1 - r_0)$, and second one in category (2), i.e. $b = (b_1 + r_1, b_2 - r_1)$;

(v) both inputs are in category (2), i.e. $a = (a_1 + r_1, a_2 - r_1)$ and $b = (b_1 + r_0, b_2 - r_0)$;
(vi) both inputs values are in category (3), i.e. $a = (c_1d_1 + c_1d_2 + r_1, c_2d_1 + c_2d_2 - r_1)$ and $b = (c'_1d'_1 + c'_1d'_2 + r_0, c'_2d'_1 + c'_2d'_2 - r_0)$.

where for the moment we suppose that the two inputs depend on two different random bits each, but a more general scenario will be analyzed later. The goal of the modified gadgets that we will present soon will be not only to reuse the same random bits, avoiding an accumulation at every execution, but also to produce outputs in the groups (1), (2) or (3), in order to keep such a configuration of the wires unchanged throughout the circuit. In this way we guarantee that every wire depends only on one random bit and that we can use the same multiplication schemes in the entire circuit. According to this remark we modify the ISW as depicted in Algorithms 10 and 11.

Algorithm 10. 1-SecMult case (ii) and (iii)

Input: shares a_1, a_2 such that $a_1 \oplus a_2 = a$, shares b_1, b_2 depending on a random number $r_i \in R$ such that $b_1 \oplus b_2 = b$, the set $R = \{r_0, r_1\}$, r_i
Output: shares c_i depending on the random number r_{1-i} such that $c_1 \oplus c_2 = a \cdot b$, the value r_{1-i}

$c_1 \leftarrow a_1b_1 + (a_1b_2 + r_{1-i})$;
$c_2 \leftarrow a_2b_1 + (a_2b_2 - r_{1-i})$;

Algorithm 11. 1-SecMult case (iv), (v) and (vi)

Input: shares a_1, a_2 depending on the random number r_i such that $a_1 \oplus a_2 = a$, shares b_1, b_2 depending on the random number r_{1-i} satisfying $b_1 \oplus b_2 = b$, the set $R = \{r_0, r_1\}$
Output: shares c_i depending on the random number $r_{1-i} \in R$ satisfying $c_1 \oplus c_2 = a \cdot b$, the value r_{1-i}

$\delta \leftarrow -r_{1-i}$;
$\delta \leftarrow \delta + r_ib_1$;
$\delta \leftarrow \delta + r_ib_2$;
$c_1 \leftarrow a_1b_1 + (a_1b_2 - \delta)$;
$c_2 \leftarrow a_2b_1 + (a_2b_2 + \delta)$;

It is easy to prove that the new multiplication algorithms are such that their outputs always belong to group (3).

Lemma 6. *Let a and b be two input values of Algorithm 10 or of Algorithm 11. Then the output value $e = a \cdot b$ is of the form* (3).

As specified before, in the previous analysis we supposed to have as input of the multiplication schemes values depending on different random bits. Since this is not always the case in practice, we need to introduce a modified refreshing scheme, which replaces the random bit on which the input depends with the other random bit of the set R. The scheme is presented in Algorithm 12 and it has to be applied to one of the input values of a multiplication scheme every time that they depend on the same randomness. Algorithm 12 is also useful before a XOR gadget with inputs depending on the same random bit, because it avoids that the randomness is canceled out. The proof of correctness is quite

---

Algorithm 12 Modified refreshing $\mathcal{R}'$

Input: shares a_1, a_2 such that $a_1 \oplus a_2 = a$ depending on a random bit r_i, the value r_i
Output: shares c_i depending on the random number r_{1-i} such that $c_1 \oplus c_2 = a$, the value r_{1-i}

$c_1 \leftarrow (a_1 + r_{1-i}) - r_i$;
$c_2 \leftarrow (a_2 - r_{1-i}) + r_i$;

---

straightforward, therefore we provide only an exemplary proof for a value in category (3).

Lemma 7. *Let a be an input value of the form* (3) *depending on a random bit $r_i \in R$ for Algorithm 12. Then the output value is of the form* (3) *and depends on the random bit r_{1-i}.*

Proof. Suppose without loss of generality that the input a depends on the random bit r_1, so that $a = (c_1d_1 + c_1d_2 + r_0, c_2d_1 + c_2d_1 - r_0)$. Then the output $e = \mathcal{R}'(a)$ is:

$$e_1 = (c_1d_1 + c_1d_2 + r_0 + r_1) - r_0 = c_1d_1 + c_1d_2 + r_1$$
$$e_2 = (c_2d_1 + c_2d_1 - r_0 - r_1) + r_0 = c_2d_1 + c_2d_1 - r_1$$

completing the proof. □

Lastly, in Algorithm 13 we define a new scheme for addition, which allows to have outputs in one of the three categories (1), (2) or (3). Note that thanks to the use of the refreshing $\mathcal{R}'$, we can avoid having a dependence on the same random bit in the input of an addition gadget. The proof of correctness is again quite simple.

Algorithm 13 Modified addition XOR′

Input: shares a_1, a_2 such that $a_1 \oplus a_2 = a$ depending on a random bit r_i, shares b_1, b_2 such that $b_1 \oplus b_2 = b$ depending on a random bit r_{1-i}
Output: shares c_i depending on a random number $r_k \in R$ such that $c_1 \oplus c_2 = a + b$, the value r_k

$r_k \xleftarrow{\$} R$;
$c_1 \leftarrow a_1 + b_1 - r_k$;
$c_2 \leftarrow a_2 + b_2 + r_k$;

In conclusion, we notice that by using the schemes above and composing them according to the instructions just given, we obtain a circuit where each wire carries a value of a fixed form (i.e. in one of the categories (1), (2) or (3)) and therefore we can always use one of the multiplication schemes given in the Algorithms 8, 10 and 11 without accumulating randomness and without the risk of canceling the random bits. Moreover, it is easy to see that all the schemes just presented are secure against a 1-probing attack.

5.1 Impossibility of the 1-Bit Randomness Case

In the following we show that is impossible in general to have a 1st-order probing secure circuit, which uses only 1 bit of randomness in total. In particular, we present a counterexample which breaks the security of a circuit using only one random bit.

Let us consider c and c' two outputs of two multiplication schemes between the values a, b and a', b' respectively, and let r be the only random bit which is used in the entire circuit. Then c and c' are of the form

$$\begin{cases} c_1 = a_1b_1 + a_1b_2 + r \\ c_2 = a_2b_1 + a_2b_2 + r \end{cases} \quad \text{and} \quad \begin{cases} c'_1 = a'_1b'_1 + a'_1b'_2 + r \\ c'_2 = a'_2b'_1 + a'_2b'_2 + r \end{cases}.$$

Suppose now that these two values are inputs of an additive gadget, as in Fig. 6. Such a gadget could either use no randomness at all and just add the components each other, or involve in the computation the bit r maintaining the correctness. In the first case we obtain

$$\begin{cases} c'_1 + c_1 = a_1b_1 + a_1b_2 + a'_1b'_1 + a'_1b'_2 = a_1b + a'_1b' \\ c'_2 + c_2 = a_2b_1 + a_2b_2 + a'_2b'_1 + a'_2b'_2 = a_2b + a'_2b' \end{cases}$$

and then the randomness r will be completely canceled out, revealing the secret. In the second case, if we inject in the computation another r, then, in whatever point of the computation we put it, it will cancel out again one of the two r revealing one of the secrets during the computation of the output. For example, we can have

$$\begin{cases} c'_1 + c_1 = r + a_1b_1 + a_1b_2 + r + a'_1b'_1 + a'_1b'_2 + r = a_1b + a'_1b'_1 + a'_1b'_2 + r \\ c'_2 + c_2 = r + a_2b_1 + a_2b_2 + r + a'_2b'_1 + a'_2b'_2 + r = a_2b + a'_2b'_1 + a'_2b'_2 + r \end{cases}.$$

In view of this counterexample, we can conclude that the minimum number of random bits needed in order to have a 1st-order private circuit is 2.

6 Case Study: AES

To evaluate the impact of our methodology on the performance of protected implementations, we implemented AES-128 without and with common randomness. In particular, we consider the inversion of each Sbox call (cf. Fig. 5) as a block of gadgets $\mathcal{G}_{i=1,\ldots,200}$ using the same random components and each of

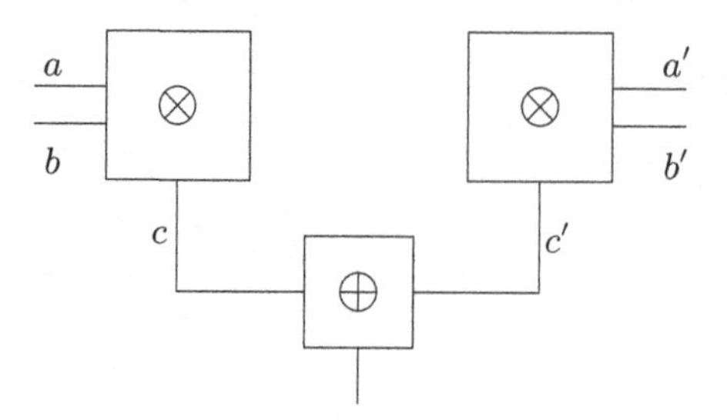

Fig. 6. The sum $(a \otimes b) \oplus (a' \otimes b')$

these inversions is followed by a refresh $\mathcal{R}_{i=1,\ldots,200}$. For the implementation without common randomness, we use the multiplication algorithm from [18] and the refresh from [10] (cf. Algorithm 2). To enable the use of common randomness, we replace the multiplication with our t–SCR multiplication, the refresh with Algorithm 4 for $t > 3$, and increase the number of shares accordingly. Table 1 summarizes the randomness requirements of both types of refresh and multiplication algorithms for increasing orders.

Table 1. Number of random elements required for the multiplication and refresh algorithms with and without common randomness from $t = 1$ to $t = 11$.

t	Without common randomness			With common randomness		
	n	Multiplication	Refresh	n	Multiplication	Refresh
1	2	1	1	2	1	1
2	3	3	3	3	3	3
3	4	6	6	4	6	6
4	5	10	10	7	21	21
5	6	15	15	7	25	25
6	7	21	21	13	52	52
7	8	28	28	13	59	59
8	9	36	36	21	105	105
9	10	45	45	21	116	116
10	11	55	55	31	186	186
11	12	66	66	31	202	202

Both types of protected AES were implemented on an ARM Cortex-M4F running at 168 MHz using C. The random components were generated using the TRNG of the evaluation board (STM32F4 DISCOVERY) which generates 32 bits of randomness every 40 clock cycles running in parallel at 48 MHz. To assess the influence of the TRNG performance on the result, we considered two modes of operation for the randomness generation. For $\texttt{TRNG}_{32}$, we use all 32 bits provided by the TRNG by storing them in a buffer and reading them in 8-bit parts when necessary. To simulate a slower TRNG, we also evaluated the performance of our implementations using $\texttt{TRNG}_8$ which only uses the least significant 8 of the 32 bits resulting in more idle states waiting for the TRNG to generate a fresh value. We applied the same degree of optimization on both implementations to allow a fair comparison. While it is possible to achieve better performances using Assembly (as recently shown by Goudarzi and Rivain in [13]) our implementations still suffice as a proof of concept. The problem of randomness generation affects a majority of implementations independent of the degree of optimization and can pose a bottleneck, especially if no dedicated TRNG is available. Therefore, we argue that our performance results can be transferred to other types of implementations and platforms, and we expect a similar performance improvement if the run time is not completely independent of the randomness generation (e.g., pre-computed randomness).

As shown in Table 2, the implementations with common randomness requires fewer calls to the TRNG for all considered t. Only after $t \geq 22$, the randomness complexity of the additional refreshes $\mathcal{R}_{i=1,\dots,200}$ becomes too high. The runtime benefit of common randomness strongly depends on the performance of the random number generator. While for the efficient $\texttt{TRNG}_{32}$ our approach leads to faster implementations only until $t = 5$, it is superior until $t = 7$ for the slower $\texttt{TRNG}_8$[2]. The dependency on the performance of the randomness generation is visualized in Fig. 7. For $\texttt{TRNG}_8$, the curve is shifted downwards compared to the faster generator. In theory, an even slower randomness generator could move the

Table 2. Cycle counts of our AES implementations on an ARM Cortex-M4F with $\texttt{TRNG}_{32}$. In addition, we provide the required number of calls to the TRNG for each t.

t	Without common randomness				With common randomness			
	n	TRNG calls	Cycle count		n	TRNG calls	Cycle count	
			$\texttt{TRNG}_{32}$	$\texttt{TRNG}_8$			$\texttt{TRNG}_{32}$	$\texttt{TRNG}_8$
1	2	1,200	112,919	187,519	2	206	70,262	70,196
2	3	3,600	308,600	548,477	3	618	173,490	199,063
3	4	7,200	496,698	1,089,092	4	1,236	309,844	412,887
4	5	12,000	751,670	1,812,213	7	4,326	737,260	1,206,558
5	6	18,000	1,051,323	2,729,052	7	5,150	808,412	1,358,560
6	7	25,200	1,403,243	3,836,006	13	10,712	1,973,885	3,134,628
7	8	33,600	1,779,403	5,125,072	13	12,154	2,147,190	3,467,553
8	9	43,200	2,286,003	6,603,199	21	21,630	4,647,611	7,017,148
9	10	54,000	2,814,435	8,257,996	21	23,896	4,877,985	7,498,022
10	11	66,000	3,459,684	10,096,735	31	38,316	8,282,630	12,467,274
11	12	79,200	4,046,836	12,112,375	31	41,612	8,640,018	13,211,240

[2] For $t = 1$, our implementation with common randomness is faster for $\texttt{TRNG}_8$ than for $\texttt{TRNG}_{32}$. This is due to the small number of TRNG calls and the extra logic required to access the randomness buffer of $\texttt{TRNG}_{32}$.

break-even point to after $t = 23$ for our scenario, i.e., until the implementation with common randomness requires more TRNG calls.

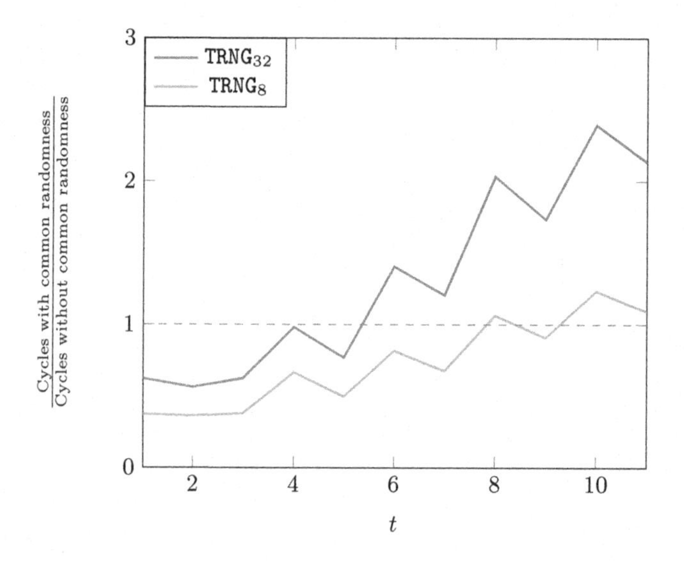

Fig. 7. Ratio between the cycle counts of the AES implementations from Table 2 with and without common randomness for each t.

For the special case of $t = 1$, we presented a solution (cf. Sect. 5) with constant randomness independent of the circuit size. Following the aforementioned procedure, we realized an 1-probing secure AES implementation with only two TRNG calls. Overall, the implementation using the constant randomness scheme requires more cycles than the one with common randomness, mostly due to additional operations in the multiplication, addition, and refresh algorithms. This is especially apparent for the key addition layer which is 40% slower. In general, however, the approach with constant randomness could lead to better performances for implementations with many TRNG calls and a slower source of randomness.

7 Conclusion

Since the number of shares n for our t–SCR multiplication grows in $\mathcal{O}(t^2)$ and $\mathcal{R}$ requires $\mathcal{O}(nt)$ random elements, the practicability our proposed methodology becomes limited for increasing t. Nevertheless, our case study showed that for small t our approach results in significant performance improvement for the masked implementations. The improvement factor could potentially be even larger, if we replace our efficient TRNG with a common PRNG. Additionally, an improved $\mathcal{R}$ with a smaller randomness complexity, e.g., $\mathcal{O}(t^2)$, could lead to better performances even for $t \geq 22$ and is an interesting starting point for future work. This would be of interest as with time larger security orders might be required to achieve long-term security.

Another interesting aspect for future work is the automatic application of our methodology to an arbitrary circuit. While we provide a basic heuristic approach in Sect. 4, further research might be able to derive an algorithm which finds the optimal grouping for any given design. This would help to create a compiler which automatically applies masking to an unprotected architecture in the most efficient way removing the requirement for a security-literate implementer and reducing the chance for human error.

Acknowledgments. Sebastian Faust and Clara Paglialonga are partially funded by the Emmy Noether Program FA 1320/1-1 of the German Research Foundation (DFG). Tobias Schneider is partially funded by European Unions Horizon 2020 program under project number 645622 PQCRYPTO. This work is also partially supported by the VeriSec project 16KIS0634 - 16KIS0602 from the Federal Ministry of Education and Research (BMBF).

References

1. Balasch, J., Gierlichs, B., Grosso, V., Reparaz, O., Standaert, F.-X.: On the cost of lazy engineering for masked software implementations. In: Joye, M., Moradi, A. (eds.) CARDIS 2014. LNCS, vol. 8968, pp. 64–81. Springer, Cham (2015). https://doi.org/10.1007/978-3-319-16763-3_5
2. Barthe, G., Belaïd, S., Dupressoir, F., Fouque, P.-A., Grégoire, B.: Compositional verification of higher-order masking: application to a verifying masking compiler. Technical report, Cryptology ePrint Archive, Report 2015/506 (2015)
3. Battistello, A., Coron, J.-S., Prouff, E., Zeitoun, R.: Horizontal side-channel attacks and countermeasures on the ISW masking scheme. In: Gierlichs, B., Poschmann, A.Y. (eds.) CHES 2016. LNCS, vol. 9813, pp. 23–39. Springer, Heidelberg (2016). https://doi.org/10.1007/978-3-662-53140-2_2
4. Belaïd, S., Benhamouda, F., Passelgue, A., Prouff, E., Thillard, A., Vergnaud, D.: Randomness complexity of private circuits for multiplication. Cryptology ePrint Archive, Report 2016/211 (2016). http://eprint.iacr.org/2016/211
5. Bilgin, B., Daemen, J., Nikov, V., Nikova, S., Rijmen, V., Van Assche, G.: Efficient and first-order DPA resistant implementations of Keccak. In: Francillon, A., Rohatgi, P. (eds.) CARDIS 2013. LNCS, vol. 8419, pp. 187–199. Springer, Cham (2014). https://doi.org/10.1007/978-3-319-08302-5_13
6. Bilgin, B., Gierlichs, B., Nikova, S., Nikov, V., Rijmen, V.: Higher-order threshold implementations. In: Sarkar, P., Iwata, T. (eds.) ASIACRYPT 2014. LNCS, vol. 8874, pp. 326–343. Springer, Heidelberg (2014). https://doi.org/10.1007/978-3-662-45608-8_18
7. Chari, S., Jutla, C.S., Rao, J.R., Rohatgi, P.: Towards sound approaches to counteract power-analysis attacks. In: Wiener, M. (ed.) CRYPTO 1999. LNCS, vol. 1666, pp. 398–412. Springer, Heidelberg (1999). https://doi.org/10.1007/3-540-48405-1_26
8. De Cnudde, T., Bilgin, B., Reparaz, O., Nikov, V., Nikova, S.: Higher-order threshold implementation of the AES S-Box. In: Homma, N., Medwed, M. (eds.) CARDIS 2015. LNCS, vol. 9514, pp. 259–272. Springer, Cham (2016). https://doi.org/10.1007/978-3-319-31271-2_16

9. Coron, J.-S., Prouff, E., Rivain, M., Roche, T.: Higher-order side channel security and mask refreshing. In: Moriai, S. (ed.) FSE 2013. LNCS, vol. 8424, pp. 410–424. Springer, Heidelberg (2014). https://doi.org/10.1007/978-3-662-43933-3_21
10. Duc, A., Dziembowski, S., Faust, S.: Unifying leakage models: from probing attacks to noisy leakage. In: Nguyen, P.Q., Oswald, E. (eds.) EUROCRYPT 2014. LNCS, vol. 8441, pp. 423–440. Springer, Heidelberg (2014). https://doi.org/10.1007/978-3-642-55220-5_24
11. Faust, S., Rabin, T., Reyzin, L., Tromer, E., Vaikuntanathan, V.: Protecting circuits from leakage: the computationally-bounded and noisy cases. In: Gilbert, H. (ed.) EUROCRYPT 2010. LNCS, vol. 6110, pp. 135–156. Springer, Heidelberg (2010). https://doi.org/10.1007/978-3-642-13190-5_7
12. Goubin, L., Patarin, J.: DES and differential power analysis the "Duplication" method. In: Koç, Ç.K., Paar, C. (eds.) CHES 1999. LNCS, vol. 1717, pp. 158–172. Springer, Heidelberg (1999). https://doi.org/10.1007/3-540-48059-5_15
13. Goudarzi, D., Rivain, M.: How fast can higher-order masking be in software? In: Coron, J.-S., Nielsen, J.B. (eds.) EUROCRYPT 2017. LNCS, vol. 10210, pp. 567–597. Springer, Cham (2017). https://doi.org/10.1007/978-3-319-56620-7_20
14. Ishai, Y., Kushilevitz, E., Li, X., Ostrovsky, R., Prabhakaran, M., Sahai, A., Zuckerman, D.: Robust pseudorandom generators. In: Fomin, F.V., Freivalds, R., Kwiatkowska, M., Peleg, D. (eds.) ICALP 2013. LNCS, vol. 7965, pp. 576–588. Springer, Heidelberg (2013). https://doi.org/10.1007/978-3-642-39206-1_49
15. Ishai, Y., Sahai, A., Wagner, D.: Private circuits: securing hardware against probing attacks. In: Boneh, D. (ed.) CRYPTO 2003. LNCS, vol. 2729, pp. 463–481. Springer, Heidelberg (2003). https://doi.org/10.1007/978-3-540-45146-4_27
16. Kocher, P.C., Jaffe, J., Jun, B.: Differential power analysis. In: Wiener, M. (ed.) CRYPTO 1999. LNCS, vol. 1666, pp. 388–397. Springer, Heidelberg (1999). https://doi.org/10.1007/3-540-48405-1_25
17. Reparaz, O., Bilgin, B., Nikova, S., Gierlichs, B., Verbauwhede, I.: Consolidating masking schemes. In: Gennaro, R., Robshaw, M. (eds.) CRYPTO 2015. LNCS, vol. 9215, pp. 764–783. Springer, Heidelberg (2015). https://doi.org/10.1007/978-3-662-47989-6_37
18. Rivain, M., Prouff, E.: Provably secure higher-order masking of AES. In: Mangard, S., Standaert, F.-X. (eds.) CHES 2010. LNCS, vol. 6225, pp. 413–427. Springer, Heidelberg (2010). https://doi.org/10.1007/978-3-642-15031-9_28

Author Index

Printed in the United States
By Bookmasters